Mineralogy
and Optical Mineralogy

Second Edition

Mineralogy and Optical Mineralogy

Second Edition

Written by Melinda Darby Dyar
Mount Holyoke College, South Hadley, Massachusetts, U.S.A.
and Mickey E. Gunter
University of Idaho, Moscow, Idaho, U.S.A.
Illustrated by Dennis Tasa
DK Tasa, Inc., Taos, New Mexico, U.S.A.

Mineralogical Society of America
Chantilly, VA

Mineralogy and Optical Mineralogy
Second Edition
ISBN 978-1-946850-02-7

3635 Concorde Pkwy., Suite 500
Chantilly, VA 20151 U.S.A.
www.minsocam.org

This material is based upon work supported by the National Science Foundation under Grants DUE-9952377 and DUE-0127191. Any opinions, findings, and conclusions or recommendations expressed in this material are those of the authors and do not necessarily reflect the views of the National Science Foundation.

Cover photo by Dennis Tasa. Fluorite from Tincroft Mine, Cornwall, England, 7.5 cm across. Harvard Mineralogical Museum #93206.

Production by DK Tasa, Inc.: Dennis Tasa, Karen Tasa, Cindy Robison, and Dan Pilkenton
Publication coordination by J. Alexander Speer, Executive Director, Mineralogical Society of America
Printing coordination by J. Alexander Speer, Executive Director, Mineralogical Society of America
Printed by Cenveo, Stamford, CT

Printed in the United States of America

10 9 8 7 6 5 4 3 2 1

We dedicate this book to our
professional mentors:

Roger, Charlie, and George
and
Don, Jerry, and Paul

To the people who supported
us personally throughout this project:

Peter, Duncan, Lindy, John, and Diane
and
Marion, Maycel (in spirit), and Suzanne

And to Karen Tasa whose perfection
with the book and patience with us
helped bring about this colorful
Second Edition

Brief Contents

Contents

Preface

Introduction

Welcome to the world of mineralogy! Mineralogy, while usually associated with geology, is really a stand-alone discipline that weaves itself into such diverse fields as chemistry, art, forensic science, wine production, and health-related issues, to name only a few. While this book is geared toward mineralogy as it applies to geology, it will also address mineralogy as a discipline in itself, and show you how it is related to the other sciences, arts, and everyday life. We developed the second edition of this textbook based on our now over 60 years of research and teaching in mineralogy, and many more years spent *thinking* about the best way to teach mineralogy in the 20th century.

This book is an outgrowth (possibly an *overgrowth* would be a better term!) of two CD-ROMs produced by Tasa Graphic Arts, Inc. The first, *The Study of Minerals*, was written by M.D.D. in conjunction with two of her former colleagues (Gil Wiswall and Rich Busch) at West Chester University (Dyar et al., 1997, 1998). Its purpose was to illustrate the major concepts needed to teach mineralogy in a graphical, interactive fashion. *Hands-On Mineral Identification*, a collaboration between M.D.D. and Dennis Tasa, soon followed, because we wanted to share the fun of mineralogy with the general public (Dyar, 1997). However, in writing these CD-ROMs, we became motivated to expand into a medium that would allow more background information and detailed explanations of our graphics, so the idea of this textbook was born.

From that point on, the book gradually became a reality through a combination of serendipitous circumstances: the genius and willingness of Dennis Tasa, who came up with the idea of animating every illustration in this book on an accompanying web tool; the chance meeting of Darby and Mickey in a van on a field trip at the NSF-sponsored Teaching Mineralogy Workshop in 1996, where they discovered their similar teaching philosophies; and our mutual desire to incorporate newer pedagogy into the teaching and learning of mineralogy.

As teachers, we were both excited by the prospect of seeing the material we teach in mineralogy come alive through colorful, 3-D animations. We can think of no other topic in the geosciences that is so inherently dependent on 3-D concepts. For years, we have both struggled with perspective drawings on a chalkboard that just don't do this material justice.

Further motivation was supplied by funding from the National Science Foundation, whose support of this project made it possible for M.D.D. to spend a semester at the University of Idaho and share M.E.G.'s Mineralogy class. We were fortunate to have this time together to dissect this material and have endless discussions about the best way to teach this course. This text represents a merging of our approaches, and it has benefited greatly from the insights made possible by spending time together in the same classroom.

How This Book Differs from the Others

This book differs in several ways from a traditional mineralogy textbook: (1) it is supported by a set of animations available on the web; (2) a searchable mineral database has been created to avoid cumbersome tables; (3) we use modern pedagogy; (4) it is written so that the more advanced chapters build on information learned in the earlier chapters.

It is fairly commonplace now for textbooks to include animations. While these can often be useful as stand-alone items, integrating them with ones in the textbook will help you better learn the material that requires 3-D animations to understand.

The comprehensive mineral database, which is often included in beginning mineralogy textbooks, is contained in our mineral database app. This has several other advantages. First, the database is more easily searched than one in a standard paper text. Second, you can easily have it with you always, whether you are in the lab or the field, and no paper database has rotatable structures.

Very few science textbooks are taking advantage of advances in pedagogy that have occurred over the past 10 to 20 years. Instead, these textbooks tend to present the material in the same way it has been taught for the past 100 years. These methods worked in the past, and may still work well for some students. However, research has shown that newer pedagogies can result in better learning, which is why we have incorporated them into our text.

Who Cares About Mineralogy?

We hope that this book will help you appreciate the role of mineralogy as it applies to geology, the other sciences, and more broadly speaking, our everyday lives. As a geology student, your thoughts about minerals are probably like ours when we took mineralogy. First, we heard that the course was very hard, and we thought the only goal of the course would be to learn how to identify minerals. Admittedly the course and the course material can be difficult because of the need to visualize things in three dimensions for the first time in your academic career. At the same time, you have to recall (or learn for the first time) principles from other courses such as mathematics, physics, and chemistry. The many animations on the MSA website should help you deal with the first of these problems. To help overcome the need for background knowledge in chemistry, physics, and mathematics, we have utilized new pedagogical methods that use spiral learning, concept maps, and inquiry-based learning to present this material.

Probably the biggest problem with mineralogy is your expectation that the main thing you will learn is how to identify minerals. You probably think that mineralogy is really only useful to identify minerals and, in turn, rocks. This could not be further from the truth. One of the goals of this book will be to broaden not only your views, but the views of others on the importance of mineralogy outside the field of geology. We'll do that by providing you with many relevant, everyday, uses and applications of minerals.

Our Reasons for Writing this Book

Mineralogy is of fundamental importance to the geosciences (solid earth, planetary, soil, hydrological, environmental, and ocean sciences) because the composition, structure, and physical properties of minerals ultimately control natural chemical and mechanical processes. Mineralogy has traditionally been one of the cornerstones of the geoscience curriculum. We cannot hope to understand how the Earth or other planets work if we do not know what they are made of! Whether representing melting reactions near the Moho on a pressure-temperature diagram, or examining Eh-pH relations that control acid mine drainage while using minerals to immobilize hazardous waste, the relationships between minerals and their local environments have important implications for all geosciences and for society. Ignorance of mineralogy has cost our society dearly, as witnessed by the asbestos problem of the 1980s (Gunter, 1994) and, more recently, the ruling that quartz is a human carcinogen (Gunter, 1999). Mineralogy is also particularly needed by K-12 educators: the first questions asked by students about geology are usually based on pebbles picked up on the playground. Elementary school children are always asking their teachers "What is this?"

The challenge we face is to take advantage of this natural curiosity about minerals. However, the subject of mineralogy, when taught at the college level, has historically been less than inspiring to the majority of students. A large part of the problem, we feel, has been the lack of an adequate textbook. The issue is not that mineralogy is uninteresting, but rather that new methods and materials for teaching mineralogy are needed to demonstrate fundamental principles and to present this information in meaningful contexts to create a better learning environment. In this book, we attempt to face these challenges.

When we started teaching mineralogy, each of us taught the course the way we had learned it as students. The class is usually taught as a series of subjects (crystallography, crystal chemistry, classification, etc.) in a very linear fashion, starting with a set of supposedly-simple material and progressing in a straight line to complex material. For example, a mineralogy course traditionally starts with crystallography. While it is incredibly important for many untold reasons, crystallography is very difficult to learn without being placed in some context. There is a huge amount of vocabulary that goes with learning crystallography, and it is often the first time that students face college-level studies of abstract visualization in three dimensions. After several weeks of crystallography (often without even the mention of a mineral name), the class moves on to crystal chemistry. Again, a large amount of material is introduced, completely distinct from what has come before.

About half-way through the semester, the first mention of real minerals occurs, and the remaining studies of minerals and mineral classes typically march through the progression of systematic mineralogy beginning with elements, sulfides, oxides, and ending with silicates. Depending on the experiences of the instructor, these subjects may not be interrelated, but merely presented as a set of facts that together constitute the science of mineralogy. All too often, with time running short at the end of the semester, coverage of the silicates is limited.

After several years of teaching in exactly this sequence, *it became very apparent to us that this was not a good approach.* Clearly portions of the course were improperly balanced. The majority of the minerals we asked our students to learn do not occur commonly as rock-forming minerals; e.g., a small number of silicates make up over 90% of the Earth's crust. Why not start with the really important silicates (i.e., quartz, feldspars, micas, amphiboles, and pyroxenes) the first day? Why are we devoting so much time to crystallography and crystal chemistry without relating them to each other (or perhaps more importantly, to geologic processes, and to the types of minerals that will occur in various parageneses)? So we began to rethink our curriculum and teach this course differently. Eventually this led to the creation of this book.

Using this Book: Professors Take Note!

This book is designed to have great flexibility in its use, so that it can serve the needs of introductory or advanced level mineralogy courses. It is written at three "levels" with increasingly amounts of complexity (Table 0.1). *No one* will cover all the material in this book in a single semester course, but we hope that *everyone* will find specific chapters, which are written to be somewhat modular, that fit their curricula.

The text begins with an introduction to mineralogical concepts, which we call "Round One." This chapter contains everything a student really needs to know about mineralogy to succeed in life. It is our hope that ten years after taking this course, a student will still remember and understand the concepts presented there. This chapter usually takes us a week or two to cover in class.

The second section of the book, which we will refer to as "Round Two," includes all the basic material on minerals that a geologist needs to be exposed to. This section is divided into chapters on elements, important minerals, crystal systems, symmetry, and optics. This material usually requires about 6–8 weeks to cover in lecture form, and thus constitutes what might be needed to cover mineralogy in a course (like Rocks and Minerals) that also includes discussion of rocks.

The third section of the book represents the most advanced material coverage ("Round Three"), where we meet the course material for the third time! Here we use a more conventional approach to present this material. Where possible, concepts are derived from basic principles, and the interconnectedness of ideas is stressed. This section can be used as a standalone textbook all to itself for a junior- or senior-level mineralogy course. However, we use it as the basis for the last

Table 0.1. Organizational Structure of the Text, and Suggested Use in a One-semester Course

Round one (1 week) Chapter 1	*Round two (3-4 weeks) Chapters 2-6*	*Round three (8-10 weeks) Chapters 7-24*
The Essence of Mineralogy	Ch.2. Hand Sample ID	
Big Ten Minerals	Ch.3. Crystal Chemistry	Ch.7. Chemistry of the Elements Ch.8. Bonding and Packing in Minerals Ch.9. Chemical Analysis of Minerals Ch.10. Mineral Formulas
Elements	Ch.4. Crystallography	Ch.11. Introduction to Symmetry Ch.12. Symmetry Ch.13. Mathematical Crystallography Ch.14. Representation of Crystal Structures Ch.15. Diffraction
Crystal Systems	Ch.5. Optical Mineralogy	Ch.16. Introduction to Optics Ch.17. Optical Crystallography Ch.18. Optical Crystal Chemistry Ch.19. Mineral Identification
Optical Systems	Ch.6. Systematic Mineralogy	Ch.20. Environments of Mineral Formation Ch.21. Nomenclature and Classification Ch.22. Silicate Minerals Ch.23. Non-Silicate Minerals Ch.24. Mineralogy Outside of Geology

half of our mineralogy classes, giving students a third, in-depth exposure to the material. We find that having the previous exposure (from Rounds One and Two) to the material at gradually increasing levels of complexity makes it easy for students to engage the most complicated concepts in mineralogy with relative ease.

So, you can use this book in one of three ways:

1. Use only the first two sections to cover mineralogy, as part of a Rocks and Minerals type course.
2. Use only the last section as part of an advanced course.
3. Mix and match chapters to suit your needs.

We encourage instructors to try the combined approach of using all three levels of the book to teach mineralogy. Modern pedagogy suggests that students learn best when curricula continually build upon previous learning experiences. A growing body of pedagogical research supports the idea that knowledge acquisition occurs in a non-linear, spiral fashion involving repeated exposure to concepts (e.g., Wals and van der Leij, 1997). This is called "spiral learning" (Figure 0.1) as described above.

We encourage you to refer to the MSA website dedicated to this book. A conventional textbook with 2-D illustrations cannot present the critical material needed in a modern mineralogy course. We also encourage you to use the programs (i.e., CrystalViewer™, CrystalDiffract®, Single Crystal™—the latter two running in demo mode) and input files on the MSA website to "interact" with minerals. Of course, we also encourage you to supplement this textbook with a rich assortment of hands-on encounters with minerals using hand samples, thin sections, grain mounts and whatever other techniques you have around.

Course Goals

Before we leave this introduction, we would like to present a suggested pedagogical framework for a mineralogy course. These thoughts are the result of stepping back and questioning what we really want students to learn from our one semester mineralogy course. Together we have arrived at the course goals that are discussed below. We realize that learning comes from repetition of material moving from simple to complex, and from establishing connections between material as learning occurs. Thus we have incorporated several different teaching strategies into our courses, which are summarized here. We hope that this methodology will stimulate others to re-

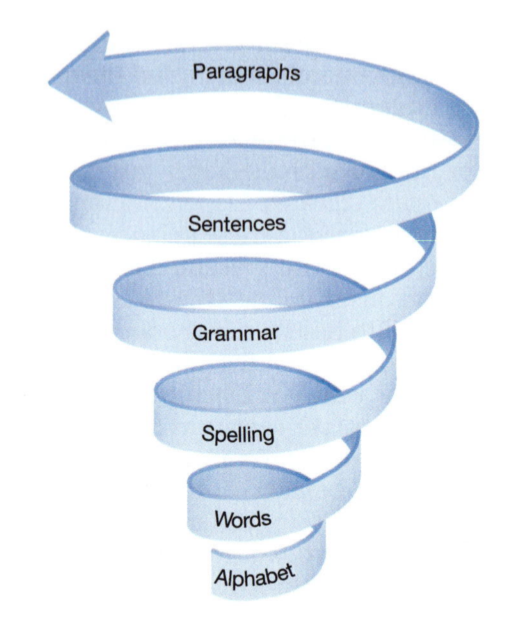

Figure 0.1. Spiral learning curve for a language. The curve builds from simple at the bottom to more complex at the top. The lower material must be mastered before the upper material can be learned.

evaluate their own goals and methods for teaching this course.

The first step in developing a course is to set forth its goals. Most professors instinctively model their courses on the ones they took as undergraduates. For a number of reasons, this is probably not the best approach: we are not clones of our professors, each class has its own level of student intellectual ability, and the material we are teaching is constantly changing. Accordingly, the following course objectives were developed. We believe this set should serve many, if not most, mineralogy courses currently being taught.

Introduce crystallography, crystal chemistry, and systematic mineralogy. Our goal is to help students attain a working knowledge of these basic concepts in mineralogy. What do these words mean? In this context, crystallography is simply the study of atomic arrangement. It is the science of how atoms arrange themselves to make crystals, and it has profound applications not just in mineralogy, but in chemistry, biology, physics, material science, and even mathematics. Crystal chemistry involves understanding the chemical make-up of those atomic arrangements. In other words, we want to know which kinds of atoms are where in the mineral, and why. Systematic mineralogy involves mineral classification and descriptions of minerals' physical properties.

Relate the physical properties of minerals to their crystal structures. Before students take mineralogy, they have probably thought about minerals from the perspective of hand samples; color, hardness, streak, and other properties are taught in middle, junior, and high school as well as in most introductory geology courses. Thus, most students enter the course focused on physical properties, which in this context include any observable or measurable characteristics. We can use these physical properties by relating them to characteristics of mineral structure or chemical composition. To understand minerals is to use these interrelationships to our advantage, to help not only in mineral identification, but also in relating mineralogy to geology.

Introduce analytical methods used in modern mineralogy, especially the polarizing light microscope. Ultimately in this course we must move away from hand sample diagnostic properties, which often yield incorrect mineral identifications, and into more unequivocal types of mineral characterizations. Analytical methods are an important part of mineralogy because they help identify and characterize minerals! If students have access to polarizing-light microscopes, they can learn one of the oldest and most useful tools in our field. It is also a useful skill to have when applying for jobs: for example, there are countless jobs outside the field of geology that require the use of light microscopy! Depending on the accessibility of other analytical equipment, students can also have the opportunity to learn how modern mineralogists work.

Learn how minerals are classified and named. As in biology, paleontology, and other fields where hierarchies are important, there is a formal classification system for minerals. It is based on the kind of anion or anionic complex in a mineral's structure, and, to a lesser extent, on crystal structure itself. Sadly, a large percentage of geologists do not understand the difference between a mineral species and a mineral group, leading to constant (and unnecessary!) confusion in petrologic studies and even raising important legal issues (e.g., Gunter et al., 2001). Mineral species names come from localities (where they are first found), appearance, chemical composition, and people's names. There are more than 4,300 officially-recognized mineral species names, and about 14,000 other mineral variety names that are in common usage. Through this course, we hope to give students familiarity with mineral nomenclature so they can use it correctly in their future lives as geoscientists, lawyers, medical professionals, etc.

Identify minerals in hand specimen and thin section, and with the aid of various analytical techniques. Mineral identification is a skill that is fundamental to many kinds of geology. By the end of this course, students should have the ability to identify many minerals with a hand lens on an outcrop, or in the lab with a thin section. The students' abilities to recognize minerals will progress as they gain experience and have access to different analytical facilities!

Appreciate the influence of crystal chemistry on mineral assemblages and mineral weathering. This is a course about minerals, so we try to avoid talking about rocks whenever possible (that's the province of petrology!). However, sometimes the process of mineral identification is aided by knowledge of where the sample comes from. So we'll look into the ways in which mineral chemistry affects rock-forming and rock-breakdown processes.

Develop the ability to research and learn mineralogical topics individually and in groups. When we teach this class, we encourage our students to explore this field through a series of individual and group learning projects. We hope that they will be inspired to look beyond classical mineralogy, and to do some research of their own on some aspect of mineralogy that they find interesting. We also explicitly recognize and do assessments based upon group and cooperative learning (e.g., Srogi and Baloche, 1997), and in doing so we remind our students that group learning is a necessary skill for employment in the 21st century. This approach incorporates the essential features of inquiry. Students are engaged by scientifically oriented questions. They give priority to evidence allowing them to develop and evaluate explanations that address these questions. Students must also evaluate their explanations in light of alternative explanations, and they must communicate and justify their explanations (National Research Council, 2000).

How to Accomplish the Goals

To accomplish the goals of learning mineralogy, we use four basic forms of pedagogy (i.e., learning methods and teaching styles) that should help students learn the material. These forms are: (1) spiral learning, (2) inquiry-based learning, (3) concept maps, and (4) interactive models and visualization. Each of these will be discussed in detail below, with pertinent mineralogical examples. We have selected these methods because research shows that they are some of the best strategies for learning.

These are the methods that most of us actually use to learn in other contexts, and they are well documented as learning strategies. For example, spiral learning involves beginning with a new concept and continuously reinforcing it as new, more advanced concepts are introduced. Consider a non-mineralogical example: when we first learned to cook, we learned the ingredients before we learned the recipes, but in making the recipes we relearn the characteristics of the ingredients. The same is true for mineralogy, where we first learn the ingredients (i.e., elements) and next learn how those ingredients are "mixed" together to make things (i.e., minerals). One step further in this analogy would take the different dishes (i.e., minerals) and combine them to make a meal (i.e., rocks), all the while considering the elements and minerals that make up the rocks. Continuing with the cooking analogy, inquiry-based learning involves questioning what might happen when we do something (substituting baking soda for baking powder in a recipe, for example), and then doing it to see the outcome (flat cookies!). For concept maps, the cooking analogy would show linkages between different types of foods, and interactive multimedia might be used to show the interrelationships in three dimensions. Incorporating these new methods into a mineralogy course becomes intuitive (and quite liberating!) once we dismiss our preconceptions (and previous experiences) based on our own backgrounds, and think about how we all learn!

Spiral learning. The idea of spiral curriculum was first developed by Bruner (who was in turn inspired by Piaget) as part of what is called "constructivist theory" starting in the 1960s (e.g., Bruner, 1960, 1966, 1973, 1990), and it has been widely adapted in K-12 curricula around the world (Texley and Wild, 1996). It is based on the idea that learning is an active process in which students always construct new ideas and concepts based upon their current and previous knowledge. Its underlying tenet is that basic scientific concepts can be introduced to children in a form that is easily comprehensible to them at early stages of their education. The same concepts can then be revisited repeatedly at successively higher levels, enhancing and deepening students' understanding of the concepts, so that students are continually building upon what they have already learned. A spiral curriculum is a very powerful educational tool, as it enables educators to carefully stage their teaching of often quite complex concepts in a way that makes it intelligible and interesting to their students (see Tobin, 1993 for examples). However, constructive approaches are rarely used in college-level science courses, which are typically linear in the way they present material.

Figure 0.1 shows an example of spiral learning in language. There are a series of words along the spiral. These words start with the simplest components of speech at the bottom, and end with the more complex at the top. The first thing we must learn before we can learn any language is its alphabet. Once this is mastered, we start to make words from these letters, and then form sentences from these words. There are rules we use to make words (spelling), and there are also rules we use to build sentences (grammar). When we begin to compose sentences, we are using words, and in using words we must know the alphabet. Thus each concept reintroduces and builds on the one below it. Finally, we put sentences together to form paragraphs. At each level, we are using the material from the level below, so we are reinforcing our knowledge of it.

We learn mathematics in a similar way (in fact, did you ever notice how your algebra skills improved during calculus?) (Figure 0.2). The first things we learn in mathematics are numbers; this is of course the "alphabet" of mathematics. The next whirl of the spiral is counting, which reinforces acquaintance with numbers, and the next logical step is addition. Once we can add, then we can subtract, but we cannot learn subtraction until we know addition. This process continues up the spiral. In fact, we often joke that when learning mathematics you never actually learn the last mathematics course you took because you only learn the previous course while taking the current one!

Figure 0.3 shows a spiral learning curve for crystal chemistry and Figure 0.4 shows a spiral learning curve for crystallography. A similar theme can be seen in these two figures when compared to the previous figures on language and mathematics. In all cases we start with something very simple and build to more complex knowledge. However, in crystal chemistry and crystallography these two spirals are also linked to each other; we must understand both of them and their integration to understand either of them. This linkage of material is formalized in education theory based on a concept maps (see following section).

In our own courses, we also envision the interrelationships among concepts with a learning spiral (Figure 0.5). Students continually reinforce the concepts learned previously, and they master simple subsets of material (e.g., ten minerals and six crystal systems) before moving on to more involved material (e.g., >150 minerals and 32 crystal classes). In fact this is the method we use when we need to learn new material (i.e., obtain several publications on the subject and read each one in increasing detail).

Concept maps. Concept maps show how material can be linked in a nonlinear fashion. Concept maps, like spiral learning, have become very popular in modern learning theory, and have been applied to everything from foreign languages to law, and even the sciences. Novak (1991, 1995) provides a recent overview of concept maps.

Figure 0.6 shows a concept map of geology and mineralogy. Within the boxes on a concept map there are usually given some objects, and a relationship is shown between these objects by using lines to connect them to other boxes. Geology is the central theme of this concept map. Geology is broken into six separate disciplines in this concept map, though of course many more could be included. One of these disciplines is mineralogy. Mineralogy is in turn broken into two separate subdisciplines: descriptive mineralogy and crystallography. Then the concept map shows how crystallography uses various techniques such as X-rays and microscopes, and in turn how other disciplines such as chemistry, physics, and art use the same instruments. Thus, links are established between art and geology. This map is of course not all-inclusive, because it could also show links between other subdisciplines in the field of geology to other disciplines outside the field of geology.

Another concept map (Figure 0.7), shows how we link the goals of our mineralogy course. In this concept map, minerals are the central theme. The upper four boxes show how minerals relate

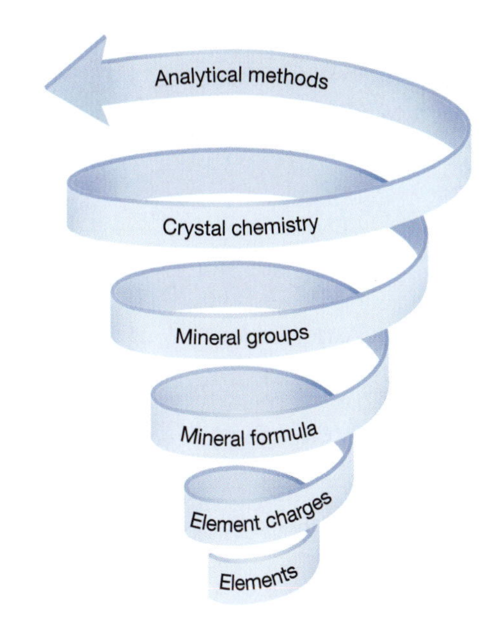

Figure 0.3. Spiral learning curve for crystal chemistry.

to other things we see in the world. First, minerals make up rocks; this is important for geology and is one of the main reasons students take mineralogy! Minerals also weather to form soils, though this is more the domain of agriculture rather than geology. The other two boxes show that minerals make up bones and teeth, and minerals are also used in many important industrial applications. The lower portion of the concept map shown in Figure 0.7 hopefully clearly defines the goals for our course. It shows how minerals are classified, named, and identified.

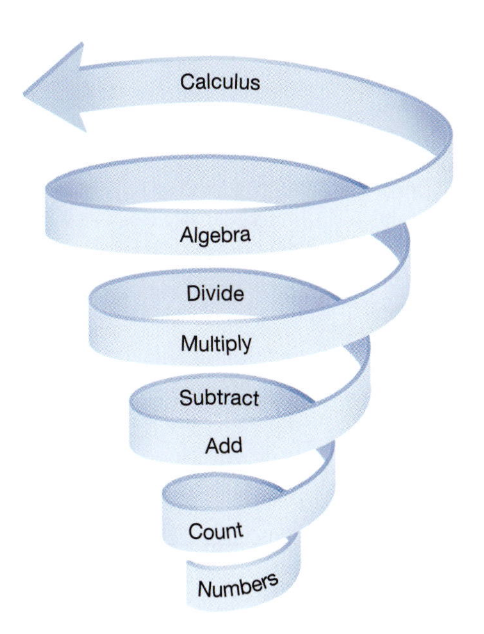

Figure 0.2. Spiral learning curve for mathematics. The simple mathematical concepts are on the bottom, the more complex on the top. Each spiral will reinforce the material below.

Inquiry-based learning. The inquiry-based approach (Fuller, 1980; Renner et al., 1985; Lawson et al., 1989; Wheeler, 2000; Bybee, 2000) is the centerpiece of the National Science Education Standards (National Research Council, 1996): "Inquiry into authentic questions generated from student experiences is the central strategy for teaching science" (p. 31). This approach is being increasingly used in the social science and education literature (Cangelosi, 1982 is an early example). The need for this approach is nicely summarized by Alberts (2000) as follows, in a paragraph where the word "Mineralogy" could easily be substituted for "Biology":

> "Where in a typical Biology 1 college course is the "science as inquiry" that is recommended for K–12 science classes in the National Science Education Standards (National Research Council, 1996)? These courses generally attempt to cover all of biology in a single year, a task that becomes evermore impossible with every passing year, as the amount of new knowledge explodes. Yet old habits die hard, and most Biology 1 courses are

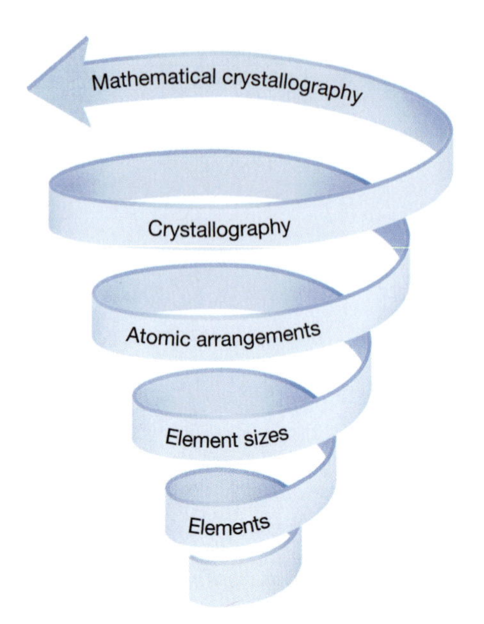

Figure 0.4. Spiral learning curve for crystallography.

still given as a fact-laden rush of lectures. These lectures leave no time for inquiry: they fail to provide students with any sense of what science is, or why science as a way of knowledge has been so successful in improving our understanding of the natural world…" (Alberts, 2000, p. 9–10).

Alberts (2000) goes on to say that inquiry-based learning is by far more efficient for long-term knowledge than the more traditional ways of simply being told how something works. In inquiry-based learning, students must be asking questions why and, in turn, trying to figure out the answers for themselves. Eventually students may be told the answers in class or read them in books, but if they have already figured them out, or at least thought about them, they will remember them much longer, and it will make more sense to them than if someone simply told them.

To be honest, we ourselves have not yet become comfortable enough with this concept to structure the entire course around it, but we use a combination of presentations of material that can be learned by experience, and material that can be learned through traditional methods. This accommodates students with different learning styles, as clearly stated by Welch et al. (1981, p. 46):

> "Our stance is that all students should not be expected to attain competence in all inquiry-related outcomes, which science educators (including ourselves) have advocated in the past. For some students and in some school environments it may not be appropriate to expect any inquiry-related outcomes at all."

As an example, the lecture on hand sample identification may begin by passing around a set of hand samples of different minerals. We then ask the students to think about a classification scheme that would allow them to organize the minerals

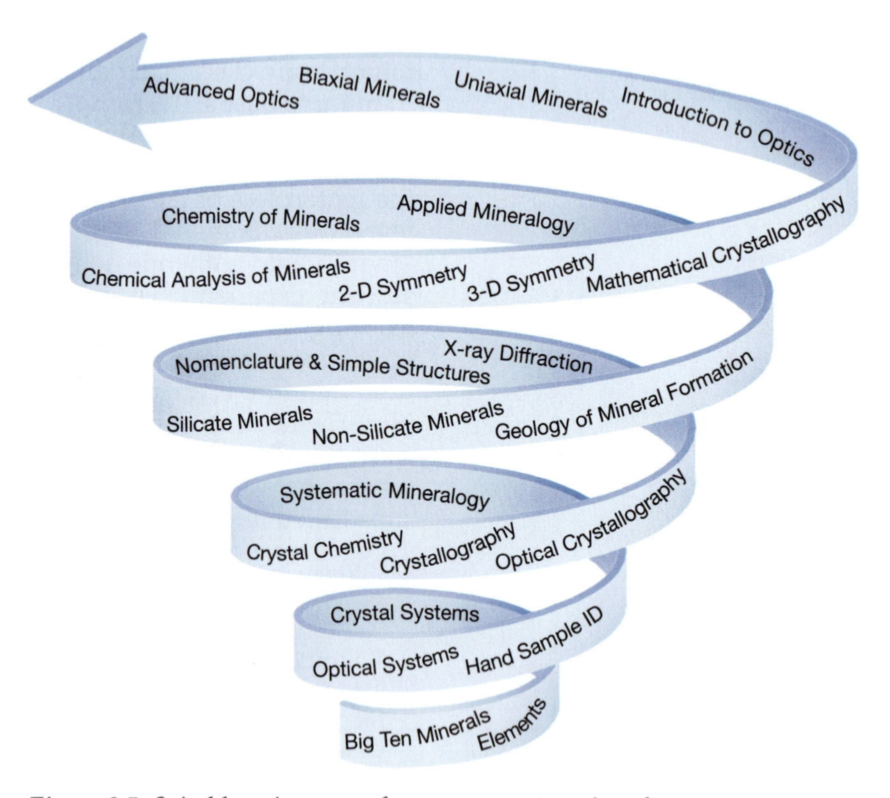

Figure 0.5. Spiral learning curve for a one semester mineralogy course.

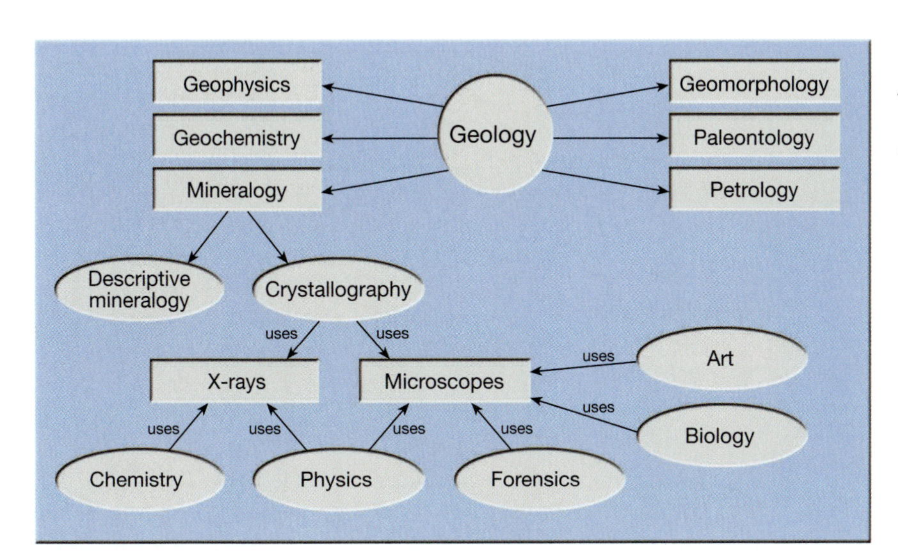

Figure 0.6. Example of a concept map for geology and mineralogy. The map shows the different subdisciplines of geology and how mineralogy relates to other disciplines.

into logical groups. Most students will instinctively group minerals by color and, to a lesser degree, morphology. A list of descriptive criteria for organizing the minerals is developed, and possible distinctions are discussed. Next, the chemical formulas for the minerals are written on the board, and students are asked to search for clues in the formulas to explain the differences in color (almost always caused by the presence of iron in silicate minerals). They are then given the opportunity to reconsider their putative classification scheme in light of the new chemical information. Ultimately they usually come up with a rational scheme that is not too far from the conventional one. Eventually, we present the "recognized" ways of identifying and classifying minerals, and end by summarizing the appropriate terminology in this area (which is often not far from the descriptive terminology used by students). So, we start off by asking students to conceive of their own schemes, and conclude with the "official" scheme, making the transition from inductive to traditional learning. This is a very different approach from what we normally use in formal lecturing. It takes a lot of time and patience, and it is very different from the way the material was taught to us. It also requires that the professor have a thorough understanding of the material being covered, because it is sometimes necessary to defend the rationale behind whatever it is that you are teaching. We love rising to this challenge!

Interactive models and visualization. One of the biggest problems in teaching mineralogy is the need to work with inherently three-dimensional course material, especially in a classroom with a two-dimensional blackboard. Most of us already rely extensively on visual aids including ball and stick models, coordination polyhedra, and optical indicatrices. However, unless the students interact with these materials, they are essentially static, and their teaching value is greatly diminished. The ideal is to pursue engagement with these materials, involving "student thought and interaction that goes beyond simple manipulation or movement via computer prompts" (Libarkin and Brick, 2002).

In many existing mineralogy classes, interactive tools are already used for many activities. Several of these are described in the *Journal of Geoscience Education* (e.g., Beaudoin, 1999) and in the MSA workbook on this topic (Brady et al., 1997). The infamous wooden blocks used to teach symmetry provide a good example of an exercise that helps students understand complex processes through direct manipulation. Many instructors employ styrofoam balls and wooden sticks to teach lessons about coordination polyhedra. Short of having students assemble their own crystal structure models (Gunter and Downs, 1991), visualizing and interacting with mineral structure models is more difficult. For these and many other abstract or inherently three-dimensional concepts in mineralogy, computer animations may provide a means for directly interacting with course material.

Use of interactive video- and computer-based learning programs has clear advantages for students. Research shows that the use of 3-D models will allow students to discern patterns more quickly and to detect relationships between patterns or structures that are not obvious in 2-D (e.g., Brodie et al., 1992; Kaufmann and Smarr, 1993). As noted in Bransford et al. (1999), "the ability of the human mind to quickly process and remember visual information suggests that concrete graphics and other visual representations of information can help people learn (Gordin and

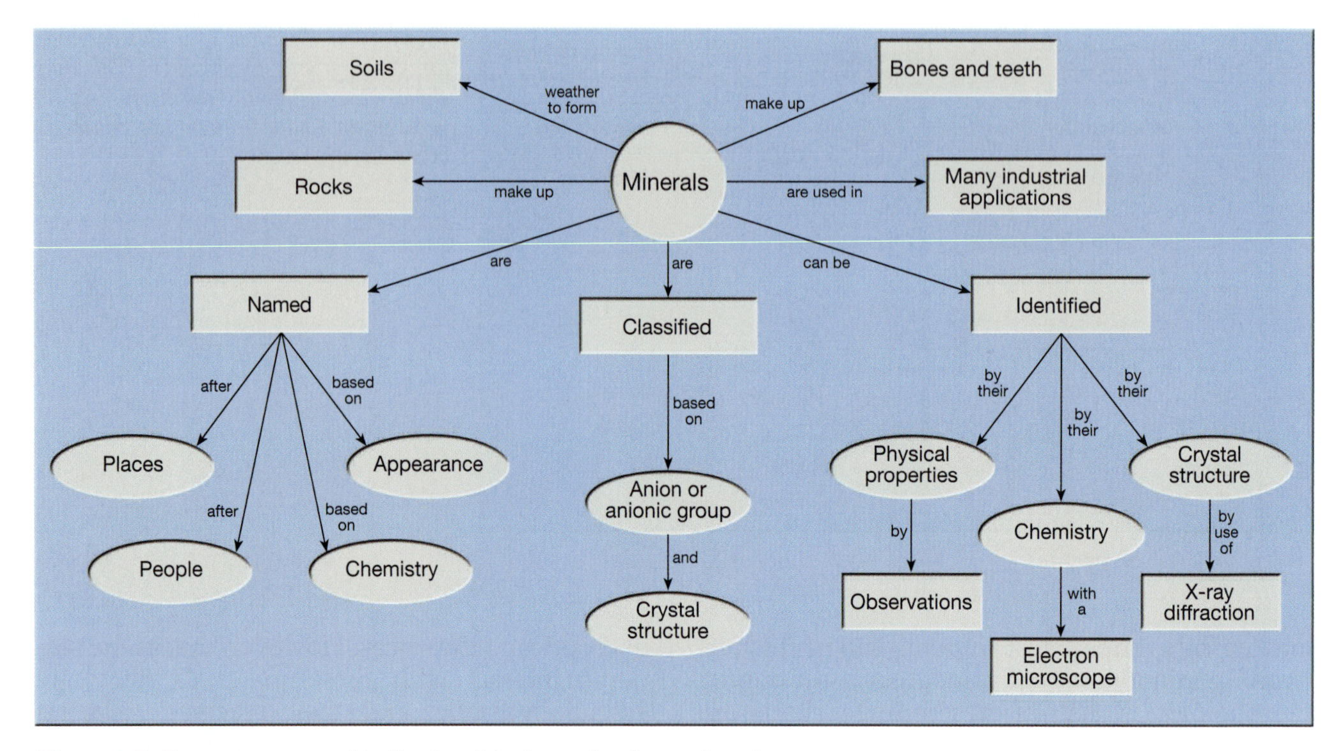

Figure 0.7. Concept map graphically showing the goals of our mineralogy course.

Pea, 1995), as well as help scientists in their work (Miller, 1986)." Interactive exercises have the potential to act as "tutors" to give students feedback on their understanding of the material. For example, the XTALDRAW program (Bartelmehs, 2002) or CrystalMaker (Palmer, 2002) begins with a simple listing of symmetry and atomic coordinates and from these, displays animated drawings of mineral structures. If used in lectures, these animations are basically an extension of static images with the advantage of better illustration (Libarkin and Brick, 2002). A more engaging student activity is to let the students build or manipulate the input files themselves, so they can understand the effects of varying crystal class or atomic locations. Of course, the MSA website should serve as a vehicle for in-depth interactions with the course content.

Finally...

Both Darby and Mickey (Figure 0.8) were fortunate to learn mineralogy the first time around from some wonderful instructors, despite the fact that traditional approaches were employed. However, the mineralogy courses of the 1960s and 1970s are still being regurgitated nearly verbatim in geology departments around the world, and this is, to us, a sad state of affairs. Writing this book is our response to the situation! As both research mineralogists and teaching faculty, we feel obligated to do what we can to teach our mineralogy courses in the most effective way possible. We believe that awareness of ongoing research into the effectiveness of new pedagogies, as well as use of this textbook, will convince instructors who teach mineralogy to consider making some changes for the betterment of the discipline.

We hope you enjoy this textbook, and that it helps make the study of minerals the most enjoyable class you've ever taken. We both feel that minerals are the most interesting, intriguing, and *important* aspect of geology: as Prof. Howie once said "If you take away the minerals, the rocks will fall down!" (Howie, 1999). We believe that rocks without minerals are just air. We firmly believe that *mineralogy can be fun*, and we hope you agree with us by the end of this book!

Acknowledgments

Production of this textbook was supported by NSF grants DUE-9952377 and DUE-0127191. We would like to thank all those mineralogists, as well as all the students and colleagues, from whom we have had the pleasure to learn.

This project would never have happened without the dedication and hard work of the staff at Tasa Graphic Arts, Inc.: Dennis Tasa and Karen

Tasa, Dan Pilkenton, Cindy Robison, and Holly Sievers. Work by David Palmer and the staff at CrystalMaker® Software (www.crystalmaker.com) made it possible to include viewer versions of their CrystalMaker®, CrystalDiffract®, and SingleCrystal™ software packages. We highly recommend that you purchase the full implementations from them.

We also acknowledge the support of our program directors at NSF, including Jill Singer, Jeffrey Ryan, and Keith Sverdrup. Special thanks go to Kylie Hanify and Don Halterman for proofreading the mineral database, and to Omar Davalos and Olivia Thomson for assistance with the second printing. We also thank our friends and colleagues at the Mineralogical Society of America, including Alex Speer and Rachel Russell, as well as the MSA subcommittee on this textbook, which was chaired by Peter Heaney. Paul Ribbe deserves special thanks for his early (and ongoing!) encouragement of this project.

In addition to those mentioned earlier, we also thank everyone on this list who contributed advice, photomicrographs, and reviews of this book.

Richard Abbott
Ross Angel
Bryan Bandli
David Barnett
Rachel Beane
Monte Boisen
John Brady
Kramer Campen
Ma Chi
Brian Cooper
Chuck Douthitt
Bob Downs
Drummond Earley
Eric Essene
Carl Francis
Greg Gerbi
Michael Glascock
Tim Glotch
Edward Grew
Bernard Grobéty
Steve Guggenheim
Bill Hames
Kylie Hanify
Rick Hervig
Kurt Hollocher
John Hughes
John Jaszczak
David Jenkins
Manfred Kampf
Dan Kile
Andrew Knudsen
Matt Kohn

__Figure 0.8.__ Darby Dyar and Mickey Gunter.

André Lalonde
Pierre Le Roch
Donald Lindsley
Heather Lowers
Jerry Marchand
Hap McSween
David Mogk
Olaf Medenbach
Bill Metropolis
Stephen Nelson

Jill Pasteris
Ron Peterson
Bob Reynolds
Michael Rhodes
Russell Rizzo
David Robertson
Matt Sanchez
Martha Schaefer
Sheila Seaman
Jane Selverstone
Eli Sklute
Michail Taran
Jenny Thomson
Bill Turner
David Von Bargen
Chris Voci
Kenneth Windom
Tom Williams
James Wittke
Thomas Witzke
Brigitte Wopenka
Bernhardt Wuensch
Andy Wulff

Finally, the photomicrographs of thin sections on the Mineral Database app (see the MSA website) are predominately the work of Peter Crowley of Amherst College. Peter also contributed ideas and acted as a sounding board for discussions of this text over many years. This project would not have been the same without his input.

In particular, M.D.D. also thanks those who inspired and encouraged her to pursue this very interesting field. Jim Besancon (Wellesley College) first tortured me with undergraduate mineralogy and then convinced me that I actually was good at it. Bernhardt Wuensch (M.I.T.) will always serve as my role model for brilliant teaching; thanks to him, I understand crystallography. My graduate advisor Roger Burns (M.I.T.) convinced me to switch from structural geology to mineralogy; he became a collaborator and the closest of friends. George Rossman (Caltech) took me in as a post-doc and has supported me in many ways ever since. I also had the good fortune to work closely with Charles Guidotti (University of Maine); from him I learned to understand the importance of thinking about minerals in petrologic contexts. Most importantly, I thank my children, Duncan and Lindy Crowley, and my ever-patient husband, Peter Crowley, for supporting me during this project.

M.E.G. would like to thank all those mineralogists who have encouraged him to complete this project and all those from who he has learned, especially his graduate professors at Virginia Tech—F.D. Bloss, G.V. Gibbs, and P.H. Ribbe: Bloss for writing several books that I learned from and for being my graduate advisor, mentor, and friend; Gibbs for all the lectures in mathematical crystallography; and Ribbe for all of those lectures in crystal chemistry. Most importantly, I thank my wife Suzanne Aaron, for all the support and proof-reading she's done over the years.

Finally, we would like to thank each other, because no co-authors could ever have been more patient, understanding, and helpful. We treasure the friendship that writing this book together has built.

References

Alberts, B. (2000) Some thoughts of a scientists on inquiry: Inquiring into Inquiry Learning and Teaching Science. J. Minstrell and E.H. van Zee, Eds., AAAS, Washington, D.C., 3–13.

Bartelmehs, K. (2002) XTALDRAW. (http://www.infotech.ns.utexas.edu/crystal/).

Beaudoin, G. (1999) EXPLORE; simulation of a mineral exploration campaign. Journal of Geoscience Education, 47, 469–472.

Brady, J.B., Mogk, D.W., and Perkins, D. III. (1997) Teaching Mineralogy. Mineralogical Society of America, Washington, D.C., 406 pp.

Bransford, J.D., Brown, A.L., and Cocking, R.R. (1999) How people learn: brain, mind, experience, and school. National Academy Press, Washington, D.C., 374 pp.

Brodie, K.W., Carpenter, L.A., Earnshaw, R.A., Gallop, J.R., Hubbold, R.J., Mumford, A.M., Osland, C.D., and Quarendon, P. (1992) Scientific Visualization, Springer-Verlag, Berlin.

Bruner, J. (1960) The Process of Education. Harvard University Press, Cambridge, MA, 97 pp.

Bruner, J. (1966) Toward a Theory of Instruction. Harvard University Press, 192 pp.

Bruner, J. (1973) Beyond the Information Given: Studies in the Psychology of Knowing. New York, Norton Press, 502 pp.

Bruner, J. (1990) Acts of Meaning. Harvard University Press, 208 pp.

Bybee, R.W. (2000) Teaching science as inquiry, in Inquiring into Inquiry Learning and Teaching in Science. J. Minstrell and E.H. van Zee, Eds., AAAS, Washington, D.C., p. 20–45.

Cangelosi, J.S. (1982) Measurement and Evaluation: An Inductive Approach for Teachers. W.C. Brown, Dubuque, IA, 421 pp.

Dyar, M.D., Busch, R.M., and Wiswall, G. (1997, 1998) The Study of Minerals, CD-ROM. Tasa Graphic Arts, Inc., Albuquerque, N.M.

Dyar, M.D. (1997) Hands-On Mineral Identification, CD-ROM. Tasa Graphic Arts, Inc., Albuquerque, N.M.

Fuller, R.G. (1980) Piagetian problems in higher education. ADAPT, University of Nebraska, Lincoln, 183 pp.

Gordin, D.N. and Pea, R.D. (1995) Prospects for scientific visualization as an educational technology. Journal of the Learning Sciences, 4(3), 249–258.

Gunter, M.E. (1994) Asbestos as a metaphor for teaching risk perception. Journal of Geological Education, 42, 17–24.

Gunter, M.E. (1999) Quartz - the most abundant mineral species in the earth's crust and a human carcinogen? Journal of Geoscience Education, 47, 341–349.

Gunter, M.E. and Downs, R.T. (1991) DRILL: A computer program to aid in the construction of ball and spoke crystal models. American Mineralogist, 76, 293–294.

Gunter, M.E., Brown, B.M., Bandli, B.R., and Dyar, M.D. (2001) Amphibole asbestos, vermiculite mining, and Libby, Montana: What's in a name?. Eleventh Annual Goldschmidt Conference, #3435.

Howie, R.A. (1999) Remarks made in Roebling Medal acceptance speech, Mineralogical Society of America annual awards luncheon, Salt Lake City, UT.

Kaufmann, W.J., II. and Smarr, L.L. (1993) Supercomputing and transformation of science. Scientific American Library, NY, 238 pp.

Lawson, A.E., Abraham, M.R., and Renner, J.W. (1989) A theory of instruction: Using the learning cycle to teach science concepts and thinking skills. NARST Monograph, Number One, National Association for Research in Science teaching, 136 pp.

Libarkin, J.C. and Brick, C. (2002) Research methodologies in science education: Visualization and the geosciences. Journal of Geoscience Education, 50, 449–455.

Miller, A.I. (1986) Imagery in Scientific Thought. MIT Press, Cambridge, MA, 355 pp.

National Research Council (1996) National Science Education Standards. Washington, D.C., National Academy Press, 260 pp.

National Research Council (2000) Inquiry on the National Science Education Standards, A guide for teaching and learning. Washington, D.C., National Academy Press, 202 pp.

Novak, J. (1991) Clarify with concept maps. The Science Teacher, 58, 45–49.

Novak, J. (1995) Concept mapping to facilitate teaching and learning. Prospects, 25, 79–86.

Palmer, D. (2002) CrystalMaker: Interactive crystallography for the Macintosh. CrystalMaker Software, Oxfordshire, UK.

Renner, J.W., Cater, J.M., Grybowski, E.B., Atkinson, L.J., Surber, C., and Marek, E.A. (1985) Investigation in natural science: Biology Teachers' Guide, Norman. Science Education Center, College of Education, University of Oklahoma.

Srogi, L.A. and Baloche, L. (1997) Using cooperative learning to teach mineralogy (and other courses, too!) In J.B. Brady, Mogk, D.W., and Perkins, D., eds., Teaching Mineralogy, Mineralogical Society of America, p. 1–25.

Texley, J., and Wild, A. (1996) NSTA Pathways to the Science Standards. NSTA, Arlington, VA, 208 pp.

Tobin, K. (1993) The Practice of Constructivism in Science Education, Washington, D.C., American Association for the Advancement of Science, 360 pp.

Wals, A.E.J. and van der Leij, T. (1997) Alternatives to national standards in environmental education: Process-based quality assessment. Canadian Journal of Environmental Education, 2, 7–27.

Welch, W.W., Klopfer, L.E., Aikenhead, G.S., and Robinson, J.T. (1981) The role of inquiry into science education: Analysis and recommendations. Science Education, 65, 33–50.

Wheeler, G.F. (2000) The three faces of inquiry: In Inquiring into Inquiry Learning and Teaching in Science. J. Minstrell and E.H. van Zee, Eds., AAAS, Washington, D.C., p. 14–19.

Chapter 1

The Essence of Mineralogy

Even though this chapter is labeled as Chapter 1, you really should start reading this book at the Preface. So if you have not read the preface, please do so now. Even if you've read the Preface once already, it might be worthwhile to go back and read it again. The reason why we say this here and now is that this chapter will seem very strange and out of place for a textbook, unless you have some idea of the context in which to place it.

So, we assume you've gone back and read the Preface. With that behind us, we can now explain the logic in the way we teach mineralogy. The central theme is repetition of material moving from simple to complex, as discussed in the Preface. What we don't say there is that both Darby and Mickey have realized that this is the way we ourselves learn new material. In fact, each of us actually learned a lot of new material in writing this book!

Thus comes the content of this first chapter in the book. This material is basically what we hope our students will remember 10 years from now. As we like to say in class, this is the guts of mineralogy. The point of this chapter is to provide some fairly basic content that reappears on our learning spiral time and time again. We hope you will agree that what we've picked to include here is very important.

There will be very little theory in this chapter for why many of the mineralogical concepts are the way they are. You may become frustrated by this, but bear with us! The best way to understand mineralogical concepts is first to have a basic grasp of the vocabulary. Thus this chapter contains the bottom elements on the spirals in the Preface. We present here the fundamental building blocks of mineralogy. For that reason, much of the material in this chapter must simply be memorized (sorry about that). When we move on to the following chapters, we will explain ways that you can remember this material. We promise to provide you with the knowledge and techniques to actually derive this information for yourselves, without the need for memorization.

You are already familiar with this way of learning. For example, in a foreign language class you start out with basic vocabulary words that you have to memorize. As you build your vocabulary, you begin to see patterns and interrelationships among the words you have memorized. Eventually, you learn to connect those words in complete sentences, and before you know it, the vocabulary becomes second nature to you.

That kind of assimilation of the fundamentals of mineralogy is our goal for you in this chapter. We even urge you to come back and re-read this chapter in three months time (or even more frequently), so you will realize how far you have come! By the end of the semester, the material in this chapter will become second nature to you. You will understand how you can derive much of this material without having to simply memorize it. Once you've accomplished that, this material will be in your long-term memory, and you will be able to recall it in your future professional career as a mineralogist, geologist, lawyer, or other profession. So, let's get started!

M.E.G. and M.D.D.

What's a Mineral?

Before we start discussing minerals, it would seem appropriate to define what one is. However, we will find that it is very difficult to give a precise definition for a mineral. This is not a limitation of minerals or mineralogy, but is common in many aspects of science, especially the natural sciences. This happens because we, as humans, often lack the ability to define natural phenomena precisely. Once you've mastered the material in this textbook, you'll understand that it is a natural shortcoming of our abilities to describe nature!

So for the time being, we'll have the following list of the five major characteristics of a mineral, and proceed to describe each in some detail. As we move through the rest of the textbook, your knowledge of minerals and mineralogy will increase. Then you can expand on the characteristics listed below, and better understand what this book is all about!

So here are five basic characteristics of a mineral:

1. **Naturally occurring.** A mineral must form by natural processes; this is one of the least debated aspects of the definition, or description, of a mineral. What this means is we cannot go into the lab and make a mineral. A mineral must form naturally somewhere on Earth, or in the rest of the universe (e.g., we have found several new mineral species on the Earth's Moon). This does not mean that we cannot go into the lab and make materials that are for all purposes identical to a mineral. You might even own a gemstone that was created in a laboratory. However, in order for a material to be called a mineral, it must also occur naturally. It is for this reason that mineralogy is usually included in the field of geology, because minerals make up rocks and rocks are the domain of a geologist. However, as we will learn in this book, minerals are important in many, many other disciplines than geology. In fact, societally speaking, the significance of minerals to human beings is much greater outside the field of geology, than it is within the field.
2. **Stable at room temperature.** In order for a crystalline material to be considered as a mineral, it must be stable or metastable at 25°C. However, there are exceptions to this rule: liquid mercury and water ice. The first exception was made because mercury is often found in its liquid form in nature. Water ice is considered to be a mineral species for historical reasons. Ice was approved as a mineral species a long time ago, and the keepers of mineral nomenclature (an international commission—read all about them in the next chapter) have not chosen to change its designation for what are probably sentimental reasons.
3. **Typically inorganic.** Inorganic basically means the opposite of organic. However, the term "organic" has its roots in chemistry, and is much like the word mineral: it can be very difficult to describe. The original derivation of the word "organic" as related to science was that only living things could produce organic compounds. However, as was discovered many years ago, organic compounds can be synthesized in the lab, and do not require a living creature to produce them. For some, the simplest way to define "inorganic" is to note that the minerals do not have C-C double bonds, or that materials that only contain carbon and hydrogen are usually organic. A last complication is that many living organisms (including humans and clams) produce minerals. Bones, for example, are made of intimate mixtures of collagen and apatite (Figure 1.1). So the requirement that a mineral species must be inorganic may not be a good one!
4. **Can be represented by a chemical formula.** Minerals can be described using a chemical formula that is either fixed or variable. This is one of the reasons that knowledge of chemistry is very important for understanding of mineralogy. In fact, you may find that studying mineral crystal chemistry will help you better understand your general chemistry courses. The chemical formula for mineral is simply a single

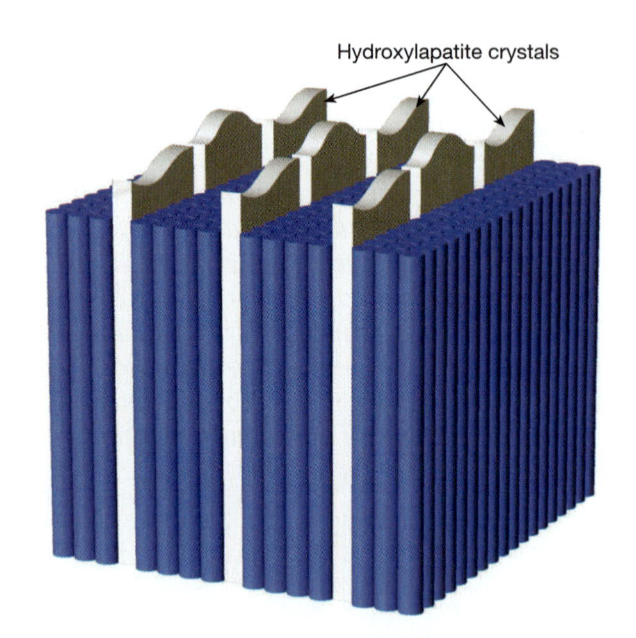

Figure 1.1. *A theoretical model for the location of platy hydroxylapatite crystals in human bones. After Young and Brown (1982).*

element or a collection of elements. We can view the elements as the alphabet for minerals. Just as there are simple words, some with only one letter or two letters, there are simple minerals that can be described with a chemical formula of only one or two elements. Likewise, there are words composed of many, many letters and there are minerals composed of many, many elements.

There is yet one more complexity with chemical formulas of minerals that has no good analogy with the alphabet and words. Instead, a food analogy may be more appropriate. Imagine that you walk into a restaurant and order vegetable soup. Your waiter was late for work and just started her shift, and the cook is out back taking a break, so she has no idea what's in that pot. So all she can tell you is that it's vegetable soup, and that may be all you want to know in order to make your decision. However, you're a reviewer for the local newspaper's food column, and you want to know *what kind* of vegetable soup you're getting. You would like to know whether it's gazpacho (tomato soup), borscht (beet soup), or even leek soup. Knowing if the soup is a mixture of many ingredients or a small number, is important to you because it affects the complexity of the dish!

Now consider yourself to be a NASA geologist. You're the first person to actually look at the first samples to be returned from the surface of Venus. It's difficult because the samples are just little pebbles, you have to view them through the window in an isolation chamber, and you can't tell much from the first glance. But the reporters are watching by camera from another room, waiting to hear what minerals are there! So, you identify a gemmy, green mineral as olivine! Everyone applauds!

But you're a geologist, and you want to know what kind of olivine it is! You are aware that the chemical composition of the olivine can give you important information about the types of rocks from which it came. You know that olivine is a mineral group, with several specific mineral species (Figure 1.2). It is possible that your sample is the mineral species fayalite, which has the specific formula Fe_2SiO_4, or maybe forsterite (Mg_2SiO_4). But in natural occurrences, we very rarely find pure fayalite or forsterite, because nature usually finds other elements to throw in. Unfortunately, with only a magnifying glass you can't tell anything about the composition of the olivine, so when a reporter asks you what that mineral's formula is, you tell her it is $(Mg,Fe)_2SiO_4$. Written this way, the formula implies that some combination of Mg and Fe atoms adds up to 2 in the formula. Thus, the formula for the olivine group minerals allows for some variability in the composition. Later, when you have had a chance to do chemical analyses or detailed optics on the sample, you'll know more about what it is!

Fayalite (Fe_2SiO_4)

Forsterite (Mg_2SiO_4)

Figure 1.2. *Several different mineral species make up the mineral group olivine, including fayalite (Fe_2SiO_4) and forsterite (Mg_2SiO_4) as shown here. Olivine group minerals make up a significant proportion of the interiors of terrestrial planets like Earth and (probably) Mars.*

5. **Has an ordered atomic arrangement.** "Ordered" means that the atoms are arranged in a repeating pattern, like the blocks in a quilt, only in three dimensions. Again, there is a catch to this definition—you have to specify at what scale the arrangement is "orderly." Most materials that are solid and otherwise meet the definition of a mineral are organized on a local scale. For example, volcanic glasses (and window glasses as well) contain repeatable units of SiO_4 tetrahedra linked together, but their

arrangement is only systematic on the scale of a few tetrahedra. So a glass is ordered only on a local scale. In order to be a mineral, a material must be made up of an orderly arrangement of atoms on scales of tens of units put together.

Before we continue, there are two important points to be made about number 4 and 5 above. In many ways these two descriptions of a mineral form the basis for understanding minerals. The chemical formula portion of the mineral description forms the field of crystal chemistry, while the atomic arrangement characteristic forms the field of crystallography. In fact, an earlier book on the subject of mineralogy was titled *Crystallography and Crystal Chemistry: An Introduction* (Bloss, 1994). The importance of the concepts of crystallography and crystal chemistry and their interrelationships cannot be over stressed. We can view crystal chemistry as the materials that nature has provided (i.e., the elements) and crystallography as the way nature has put these elements together to form minerals. Our goal will be to describe the basic rules that are followed in selecting which elements to use to build which minerals, and in turn how those elements are arranged in these minerals. The criteria that nature has chosen to make these selections will be based upon certain physical characteristics (e.g., pressure and temperature) present at the time of formation of the mineral. We can in turn determine the physical conditions at the time when the mineral was formed, perhaps tens of millions of years ago. This is really the foundation of the field of petrology, which is based on the study of formation conditions of rocks using characteristics we can learn from their minerals and mineral assemblages.

Given these descriptions for defining a mineral, we can now proceed to thinking about what's really important to learn about minerals. In this chapter, our goal is to present the concepts and vocabulary that we feel every geologist should know about mineralogy. These terms are part of the daily language of geology, and thus are worthy of memorization. As noted in the Preface, they form four groups that are the basic building blocks for this textbook: **chemical elements, crystal systems, optical systems**, and the **Big Ten minerals**. Each of these topics will be revisited with its own chapter in the next section of this book, and with several chapters in the advanced latter portion of the text. Our goal here is to let you know, right at the beginning, what's really important about mineralogy and to get you used to some of its terminology. That's why we have called this chapter "The Essence of Mineralogy."

Chemical Elements

As stated above, the elements are the basic building blocks for minerals, and for that matter, all materials. There are about 90 known naturally occurring elements. The good news is that only eight of these—oxygen, silicon, aluminum, iron, calcium, sodium, potassium, and magnesium—compose over 98% of the Earth's crust by weight (Figure 1.3). If you haven't yet had a chemistry course, or you've forgotten these, you'll want to become familiar with the chemical symbols for these elements, which are shown in Table 1.1 and will be used throughout this book.

Because this group of eight elements forms the majority of the crust, these elements, in turn, make up the building blocks for the vast majority of rock-forming minerals. Let's consider the information in Table 1.1 and see what conclusions can be drawn from it.

Notice that each of the elements in the list has a number in the column to the right of it to indicate the amount and type of charge on that ion. Atoms like oxygen, which have a negative charge, are called **anions** (think of it as "**a n**egative **ion**"). These atoms have an overall negative charge because they have more electrons (which are negatively charged) around the nucleus than they have protons (with positive charges) in the nucleus. On the other hand, some atoms have lost electrons relative to the normal stable element, and they are called **cations**. Such atoms have a positive charge because the number of protons in the nucleus is greater than the number of electrons around the nucleus (Figure 1.4).

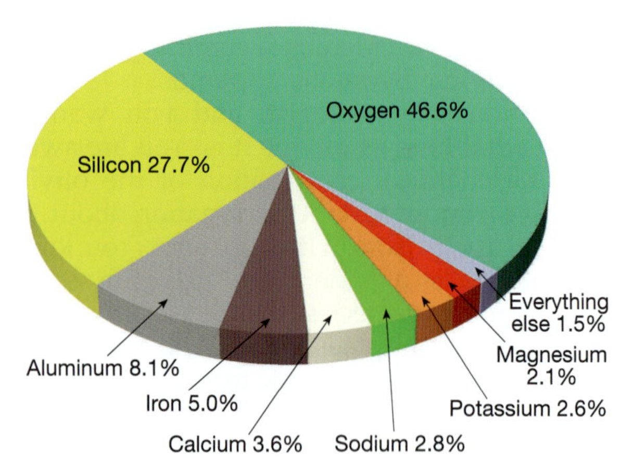

Figure 1.3. *A pie chart showing the weight percents of the eight most abundant elements in the Earth's crust. Silicon and oxygen are by far the most abundant elements in the Earth's crust, so it is no surprise that combinations of Si and O are the fundamental building blocks of most common minerals.*

Table 1.1. Eight Most Abundant Elements In the Earth's Crust and Some of Their Characteristics

Name	Chemical symbol	Charge (usually superscripted)	Common coordination number with oxygen
oxygen	O	2–	–
silicon	Si	4+	4
aluminum	Al	3+	4–6
iron	Fe	3+ or 2+	4–6
magnesium	Mg	2+	6
calcium	Ca	2+	6–8
sodium	Na	1+	6–8
potassium	K	1+	8–12

Because oxygen is so abundant in the Earth, it is the most important anion, or possibly element, in mineralogy. Most of this class will be concerned with minerals in which oxygen bonds with an assortment of cations. For the time being, memorize these eight elements and their charges. Later, we'll show you systematic methods for recalling them without having to resort to memorization.

By the way, these abundances are probably very similar on other terrestrial planets in our own solar system and on extrasolar planets as well. The crustal abundances of elements on a body depend on two factors: (1) the distance away from the sun of the body when the planets are condensing from the solar nebula, and (2) the amount of radioactive elements that are present to provide heat to drive the post-formation evolution of the planetary interior. The Earth's Moon, for example, has a crust composed primarily of feldspar that may have floated to the top of an immense magma ocean when the Moon melted. Thus, the most common elements in its crust are those in feldspar: O, Si, Al, and Ca, along with heavier elements found in basalts, such as Fe and Mg. See Dreibus and Wänke (1987) and Haskin and Warren (1991) for more information on planetary compositions. Mars is more like the Earth.

Returning to Table 1.1, the next column is titled "Common Coordination Number with Oxygen." This number tells you how many oxygen atoms (or other anions or ions) will surround the cation. In other words, negatively-charged anions will bond to positively-charged cations. For example, Si^{4+} will almost always be located inside a tetrahedron of four oxygen atoms (Figure 1.5); this is called 4-fold coordination. Only under situations of great pressure will Si^{4+} bond with six oxygens in what is called an octahedral coordination polyhedron. Al^{3+}, on the other hand, will surround itself with four, five, or six oxygens depending on pressure, temperature, and the other cations that are its neighbors in the structure.

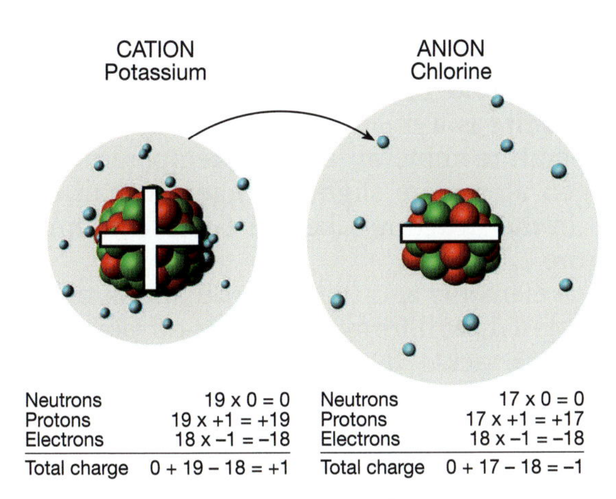

Figure 1.4. Naturally-occurring potassium chloride is the mineral species sylvite, which is used as a salt substitute (though it has a rather bitter taste!). The potassium atom has lost one of its electrons, so it has an overall positive (+) charge. The chlorine atom has gained an electron, which makes it a negative ion, or anion. The two ions bond together in the mineral structure because their opposite charges attract.

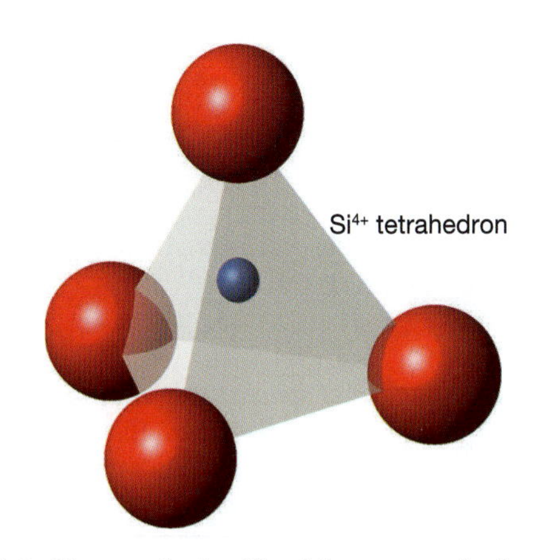

Figure 1.5. Si^{4+} tetrahedra like this one are the fundamental building blocks of silicate minerals.

Iron is a special case because it commonly occurs in two different valence states, Fe^{2+} and Fe^{3+}. The Fe^{3+} cation, which is smaller and more highly charged, prefers 4-coordinated sites (Fe^{3+} surrounded by four oxygen atoms) when they are available. The larger Fe^{2+} cation favors 6-coordinated sites because they have more room (Figure 1.6).

Mg^{2+} is almost always found surrounded by six oxygens. Na^{1+} and Ca^{2+} are happiest with six–eight oxygens around them, and K^{1+} prefers large sites with eight–twelve oxygens. For the time being, you should memorize the coordination numbers.

Why do these differences exist? For now, we can answer this question by using some simple logic. Note that as we go from Si^{4+} to Al^{3+} to Mg^{2+} to K^{1+}, the coordination number gets bigger. This implies that cations themselves are getting bigger, because they need larger sites (i.e., more oxygens around them) to fit into. So, as the charge *decreases*, the size *increases*. Size is controlled to some extent by the relationship between the number of electrons and the number of protons. In neutral atoms (e.g., Si^0), the number of negatively-charged electrons equals the number of positively charged protons. Cations have an overall positive charge because they have lost electrons. The more that the number of protons exceeds the number of electrons, the smaller the size of the cation overall, because the charge imbalance allows the protons to "pull in" the electron cloud closer to the nucleus. Thanks to angular momentum, the electrons do not go crashing into the nucleus, but remain circulating around it.

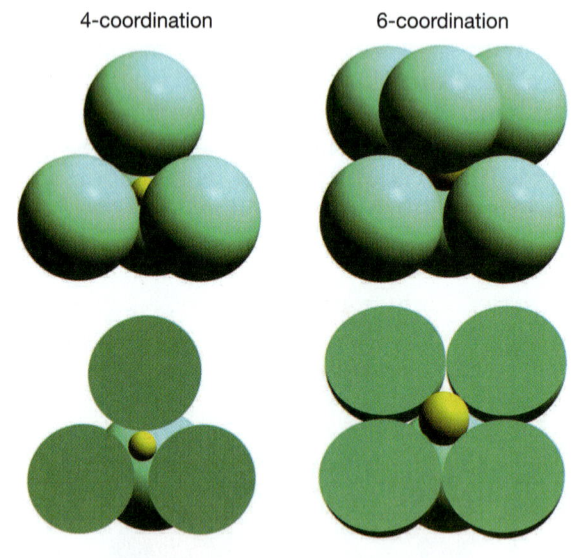

Figure 1.6. *The amount of space inside a tetrahedron of four oxygens is significantly smaller than the space inside an octahedron of six oxygens. This difference governs the size of the cation that can fit in the sites. Small, highly-charged cations like Si^{4+} and Al^{3+} like the tetrahedral site, while larger cations such as Fe^{2+} and Mg^{2+} prefer the octahedral, 6-coordinated site.*

From these examples, you might think that charge is the primary factor in determining the number of anions necessary to satisfy the charge on any cation. Although this is more-or-less true, charge is *not* the only factor. Cation coordination numbers vary for reasons we will discuss in more detail throughout the book. Many mineralogical peculiarities are explained by these variations in coordination number. In the Earth's mantle, for example, at depths of >670 km, the increased pressure is high enough to force many elements into larger coordination environments because the difference in size between the cations and the surrounding oxygens gets smaller as everything compresses (i.e., coordination number increases as pressure increases). In fact, olivine does not exist in the lower mantle, but is replaced by a mineral called silicate perovskite in which the Si^{4+} cations are in 6-fold coordination. We can even observe this behavior in the laboratory by putting olivine and garnet together in a huge press and watching them change into perovskite.

Another good example of this dependence on formation conditions is found in the aluminosilicate minerals, which have the basic formula Al_2SiO_5. Three different minerals have this same composition, but the coordination of oxygens around the Al^{3+} in the structure is dependent upon the pressure and temperature conditions during formation. In fact, this change in crystal structure makes a very useful indicator of geologic conditions! **Kyanite** is generally found in rocks from high pressure, low temperature environments, so all the Al^{3+} is most comfortable in 6-fold coordination. **Andalusite** forms at very low temperatures and pressures, so half its Al^{3+} cations assume 5-fold coordination geometries. Finally, **sillimanite** is a relatively low pressure but high temperature mineral; at high temperatures, the oxygen atoms are slightly bigger, so half of the Al^{3+} in the structure takes on 4-fold coordination (Figure 1.7).

The elements and their coordinations, as listed in Table 1.1, will form a "leaping-off point" from which to tackle the chemistry of all minerals. Once you have a feeling for which elements occupy which types of sites in which types of geologic conditions, then you're really empowered to start looking at a mineral formula and imagine what type of structure it might have, and in turn what its physical properties might be! Furthermore, knowing the charges assumed by various common cations will help immensely when it comes time to memorize mineral formulas for this class. In fact, you are not *memorizing*

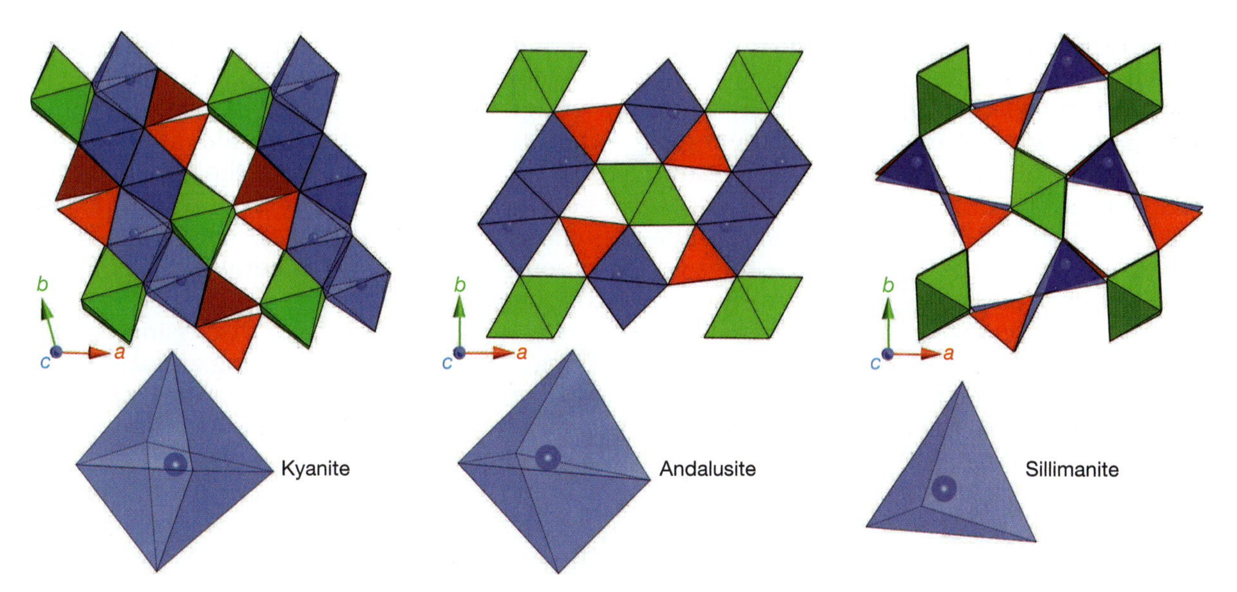

Figure 1.7. The minerals kyanite, andalusite, and sillimanite all have the same formula, Al_2SiO_5. The difference between them is the number of oxygen atoms that surround one of the Al sites, shown here as the polyhedron that is somewhat transparent. There are oxygen atoms at the corners off all the polyhedra shown; see the MSA website for a clearer depiction of these structures. In kyanite, there are six oxygens around each Al^{3+}. In andalusite, the Al^{3+} atoms are 5-coordinated, and in sillimanite, each Al^{3+} is surrounded by only four oxygens. These differences reflect the pressure and temperature conditions under which each mineral forms.

the chemical formulas when you understand coordination numbers and charges, you are *figuring them out* based on the first principles of crystal chemistry. So, we think you will find that the charges and coordination numbers given in Table 1.1 are well worth memorization. In future chapters, we will show a method whereby you can figure out the charges and coordination numbers rather than relying on simple memorization. Eventually, you will work with them enough that they become part of your background knowledge in this field.

Crystal Systems

How do atoms fit together to make crystals with different shapes? The answers to this question lie in the field of crystallography, which is the study of crystal systems. Crystal systems describe the organization of atoms in a mineral, which often can be observed in the external morphology of the mineral. An understanding of crystallography will allow us to explain the nearly infinite long-range order that is characteristic of minerals. Crystal systems are the underlying basis for crystallography. In this book we will use the non-British system that includes exactly six distinctive crystal classes that are given in Table 1.2 (in other countries, there are seven because systems based on 3-fold and 6-fold axes are treated separately).

The rationale for these classifications is fairly simple. From the preceding section, we know that one important aspect of mineralogy is going to be describing where the atoms are located within a crystal structure. It is not enough to say that all the Si^{4+} cations are in tetrahedral coordination (i.e., have coordination number of 4) with oxygen; we also need to say *where* that Si^{4+} tetrahedron is located in the structure: in the center? On the corners? To do this, we need to develop a terminology that permits us to say where atoms are located.

You are probably already familiar with one way to do this: using Cartesian coordinates, which are sets of *x*, *y*, and *z* axes in 3-D. We could specify the location of each atom in a mineral in terms of coordinates relative to a set of *x*, *y*, and *z* coordinates, and then plot the results to see how the structure looks. The Cartesian system is a way of defining atomic positions in three dimensions

Table 1.2. The Six Crystal Systems, the Three Optical Classes, and Their Relationships

Crystal systems	*Optical classes*
isometric ($a=b=c$)	isotropic
tetragonal ($a=b\neq c$)	uniaxial
hexagonal ($a=b\neq c$)	
orthorhombic ($a\neq b\neq c$)	biaxial
monoclinic ($a\neq b\neq c$)	
triclinic ($a\neq b\neq c$)	

using three different axes for reference; we call this a **basis vector set**. The Cartesian system is a specific kind of basis vector set in which all the axes are mutually perpendicular, and the repeat distance, or **unit length**, on each axis is the same. In other words, one unit on one axis is the same length as one unit on all the other axes. It probably seems logical for the repeats to be the same along the three different axes. Unfortunately, nature has not organized itself in this logical manner.

Instead, nature has chosen six different basis vector sets in which to organize the structures of minerals. To explain nature, we must understand these different basis vector sets. As a first step in doing this, we will have to create a new terminology that differs from the standard Cartesian basis vector set terminology. Because the x, y, z terminology assumes Cartesian constraints, we use a, b, and c instead to allow other permutations to occur. For example, what if the axes are *not* at right angles to each other? Or if a unit of length=1 is not the *same* length on all the axes? You can begin to see that there are many possible permutations on basis vector sets.

Luckily, only six produce the infinitely repeatable arrays necessary to build crystal structures in three dimensions: the six **crystal systems**. Our interest here is only to note the effect that difference basis vector sets will have on physical properties.

Begin with the simplest situation: one in which we *do* have a Cartesian system, so each axis is at right angles to both the other axes, and one unit length on each axis is the same as on all the others. This is the **isometric** system (sometimes called cubic). Minerals that are isometric have the same physical properties in all directions: color, hardness, and cleavage are the same in all orientations. Materials with this characteristic are called **isotropic** (from the Greek *iso*=same, *tropos*=way, manner, or turn); so their appearance is the same in all directions. Table salt (its mineral species name is halite) and garnet (the red mineral often used in jewelry) are both in the isometric class (Figure 1.8). The fact that all properties are the same in any direction is a very useful diagnostic tool in recognizing and characterizing isometric minerals.

Two other crystal systems have at least one pair of orthogonal (at right angles to each other) axes, but one of the three axes has a different length than the other two (i.e., $a = b \neq c$). This means that the unit of repeat is different on one of the axes. These systems are called **tetragonal** and **hexagonal**. These minerals are termed **anisotropic** because they do *not* have the same physical properties in all directions. The c axis direction is always distinct. The difference between the tetragonal and hexagonal systems is the angle between the a and b axes, which is 90° for **tetrag**onal crystal systems (from the Greek, *tetra*=four, because 90° is 360°/4) and 60° or 120° for **hex**agonal (*hex*=six, because 60°=360°/6). The most common hexagonal mineral species is quartz (Figure 1.9).

Let's take a break from discussing crystal systems and point out something about the way we learn new terminology. We are explaining here the derivation of some of the specific terminology used in mineralogy. An understanding of the root meaning of the words will help you remember a word. For example, research shows that people often find it difficult to remember other people's names. This happens because a person's name usually contains nothing specific to relate it to that person (unless the person happens to be Goldilocks). For instance, Mickey Gunter does not look (much) like that Disney Mouse—he doesn't have big black mouse ears. The same research also shows that if you can relate some physical characteristic of a person or some derivation of their name with that person, you will recall their name. So now, from this point forward, you'll remember Mickey's name by relating it to one of those Walt Disney characters with

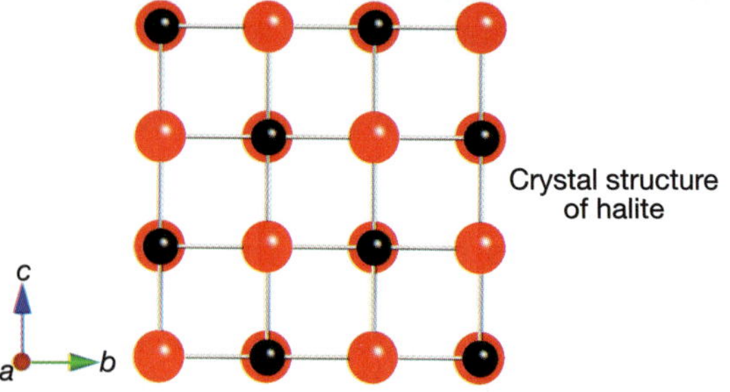

Figure 1.8. The cubic morphology of the mineral halite, NaCl, occurs because of the arrangement of the atoms that form the mineral. This is a good example of how the crystal structure of minerals control their physical properties.

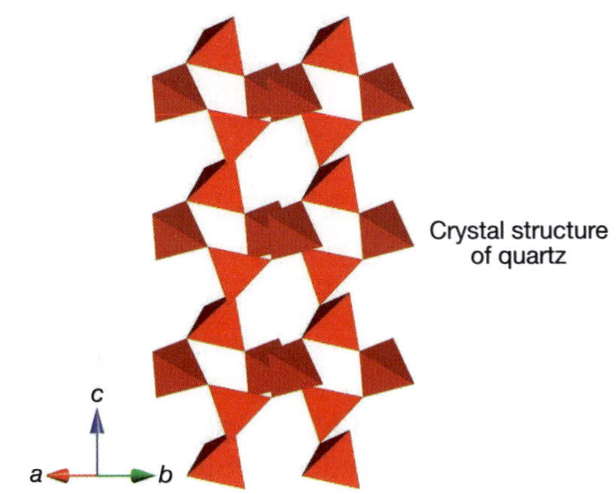

Figure 1.9. The crystal structure and shape of quartz, the most commonly occurring mineral species in the Earth's crust. In this representation we have used Si^{4+} tetrahedra (which is a Si^{4+} cation surrounded by four oxygens) to show the structure instead of individual atoms (as shown in Figure 1.8 for halite). Check the MSA website to get a better 3-D view of these mineral structures. Regardless of how we view the structure of quartz, it is more complicated than that of halite.

big ears. As we introduce mineral names, we will often give the derivation of these names, especially when the derivation of the name helps to remember some characteristic of that mineral. For instance, the feldspar mineral albite derives its name from the Latin *albus*, which means white; the mineral is usually white. There are other things that you may know that have that same prefix, for instance, the word "albino." From this point on you will remember that albite is white.

Now we return to the crystal systems. There are three more crystal systems in which none of the axes is the same length: $a \neq b \neq c$. There are three possible permutations here based on how many of the (three) angles between the axes are the same (i.e., are 90°). In the **orthorhombic** system, all three angles are 90°. The word orthorhombic comes from the Greek root, where *ortho*=90° angles and *rhombic*=solid. In the **monoclinic** system, two axes are 90°, and the third is not. The word monoclinic comes from the Greek root, where *mono*=one and *clinic*=angle. In other words, minerals in the monoclinic crystal system have one non-90° angle. Crystals in the **triclinic** system (Figure 1.10) have three non-90° angles between their axes (from the Greek root *tri*=three and *clinic*=angle again). Incidentally, there is no such thing as a biclinic crystal—you can't make blocks that fit together and fully fill space if two of the angles are non-90°. So the criterion for basic building blocks of minerals (which we call the **unit cell**) is that they must fill space without gaps in between the blocks. These tiny, basic building blocks are added one next to the other for an almost infinite collection, in order to build a mineral. The axial system relationships for all six crystal systems are shown in Figure 1.11.

All minerals (and all crystalline materials) must crystallize in one of these six crystal systems. Minerals in the isometric class are said to have high symmetry. As you remove equivalence in axis length and angles, the symmetry gets "lower," so minerals in the triclinic class are said to have the lowest symmetry of all. As a general rule, simple structures (like most industrial mate-

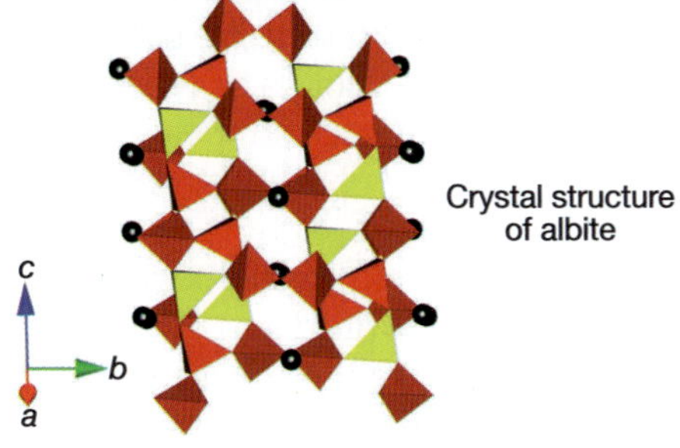

Figure 1.10. The crystal structure and shape of albite. To view this structure we have used a combination of polyhedra (as used for quartz in Figure 1.9) and individual atoms (as used for halite in Figure 1.8). This structure is more complicated than either halite or quartz and, in turn, the crystal system is now triclinic.

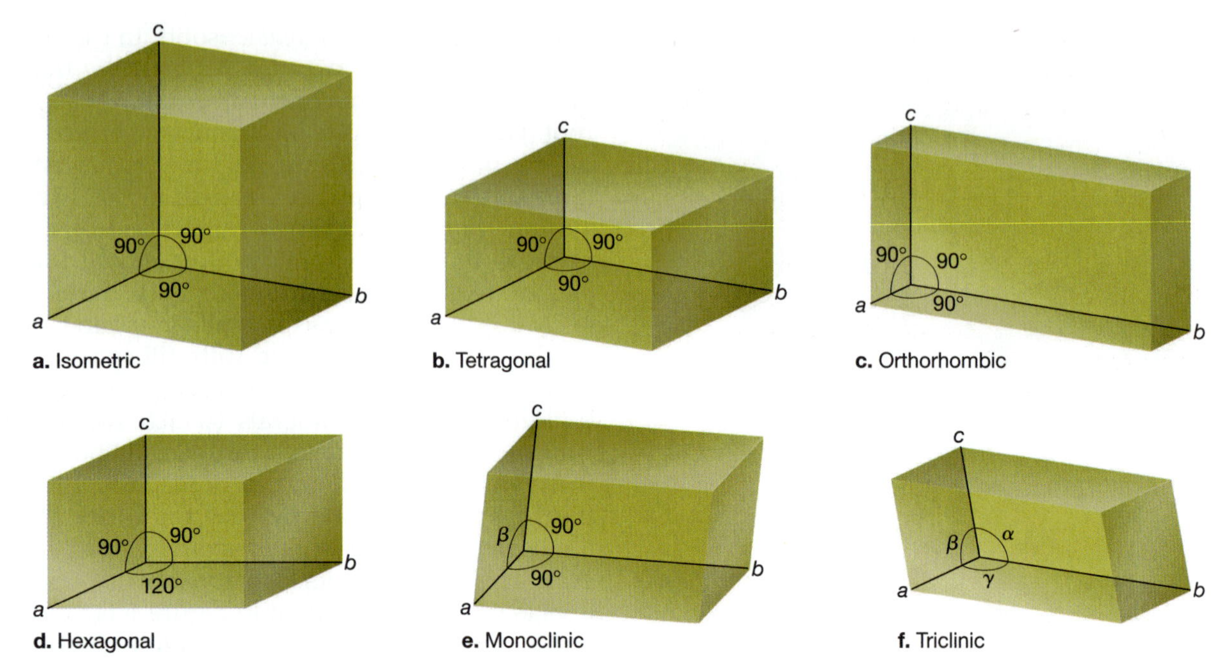

Figure 1.11. The six different crystal systems found in minerals. Notice the relationship between the axis lengths, the angles between them, and the resultant shapes.

rials such as metals and oxides) have high symmetry. Minerals with complex chemistries have lower symmetries as they distort themselves to try to fit all their constituents into the structure.

To summarize the six crystal systems and their associated basis vector sets, we could draw an analogy between them and 3-D solids such as blocks of wood. The blocks of wood would be three-dimensional objects similar to minerals. If we cut the length of the wood to different lengths in three directions, then these lengths would correspond to the *a*, *b*, and *c* repeat lengths in the mineral. We could also vary the angles between these repeats starting with the usual 90° angles and then relaxing the constraints for the angles to become non-90°. In so doing, we would create blocks of wood that represent the six different crystal systems, and in turn, the six different basis vector sets. These blocks of wood could represent the basic geometric shapes of the unit cells formed for one of these six crystal systems (Figure 1.12).

Let's start by taking a block of wood and making a cube. We would cut it the same length in three directions where these three directions all make 90° angles with one another; this block of wood would represent the unit cell of a mineral belonging to the isometric crystal system (Figure 1.12a).

Now we could take this block of wood and cut it in half; this would create the unit cell for a tetragonal mineral (Figure 1.12b). In this case, the lengths of the *a* and *b* axes are still the same, but the *c* axis is now reduced by half. Also, in this tetragonal block of wood the angles between *a*, *b*, and *c* remain 90°, (Figure 1.12b).

Next, we could take the tetragonal block of wood and cut its *a* axis to one-quarter of its original length. This then would form an orthorhombic block of wood, in which the *a*, *b*, and *c* axes are all different lengths. The angles between the axes are still 90°. Thus, we have these three separate wood blocks representing crystal systems, which could be represented by three mutually perpendicular vectors, differing only in the relationships between the length of their sides.

Returning to our blocks of wood and unit cell crystal system analogy, we now have to cut the blocks of wood at non-90° angles to make the remaining three crystal systems (we'll need a miter box for this!). Using our isometric block (Figure 1.12a), the crystal system with a highest symmetry, we can first cut the block in half, reducing the isometric symmetry to tetragonal symmetry (Figure 1.12b) by cutting the *c* axis to different lengths than *a* and *b*. Next we have to do some fancy cutting on the wood block. We need cut it to create an angle between the *a* and *b* axes that is 120°, all the while keeping *a* and *b* of equal lengths. Figure 1.12d shows the resulting wood block. Check out the animations on the MSA website (http://www.minsocam.org/msa/Figures/) to actually watch the saw make the necessary cuts to obtain this block from the isometric wood block.

To make a monoclinic crystal, we need to cut a non-90° angle between *a* and *c* while maintaining 90° angles between *a* and *b* and *b* and *c*. During our cutting operation, we will also need to cut different lengths of *a*, *b*, and *c* axes. The resultant block is shown in Figure 1.12e and animated on the MSA website. Finally, for the triclinic block, we could take a piece of wood and make three cuts through it at non-90° angles and at differing lengths in the three different directions, so as to arrive at a representation of unit cell for a mineral in the triclinic system (Figure 1.12f).

Optical Classes

If we only want to identify minerals that are big ("hand sample identification"), we would not have to worry about optical classes. However, many of the students who use this book will include optical mineralogy as part of their coursework, and so we feel that it is important to tie this material into the context of the other "must-know" concepts of mineralogy.

The optical classes can be simplified by virtue of the fact that they are inextricably related to the crystal systems, and to the physical properties of the minerals. Recall from the Preface of the book that as you learn new material based on previously-learned material, it helps to reinforce the previously learned material. This is the spiral learning concept, which we are using in this book. Learning the optical classes based on the crystal systems is a prime example of this. As you learn the optical classes, you will be using the terminology and relationships you just learned in dealing with the crystal systems.

Optical classes are based on the way in which light interacts with the crystal structure of a material as it passes through it. The speed of light and the color of the mineral are two such physical properties of the mineral: they are related to the interaction of light with the mineral's crystal structure.

Physical properties of materials must obey the rules of symmetry set forth by the previously learned crystal systems. In the **isotropic** optical class (*iso*=same, *tropic*=way), the optical properties are the same in all directions. No matter how you turn the crystal, the speed of light or color will be the same in all directions. Not surprisingly, all isometric minerals are isotropic. Unfortunately, not all members of the isotropic optical class are isometric (sorry!); the rationale for this will be discussed in future chapters.

In the uniaxial and biaxial optical classes, the speed of light is different depending on which direction you are looking (Figure 1.13). **Uniaxial** crystals have one direction that is unique (*uni*=one, *axial*=axis); this is always the *c* axis. Because the *c* axis is different in these minerals, materials in the tetragonal and hexagonal crystal systems are uniaxial. With a similar analogy, because the *a* and *b* crystallographic axis are equivalent in tetragonal and hexagonal minerals, the physical properties associated with light are the same in the *a-b* plane.

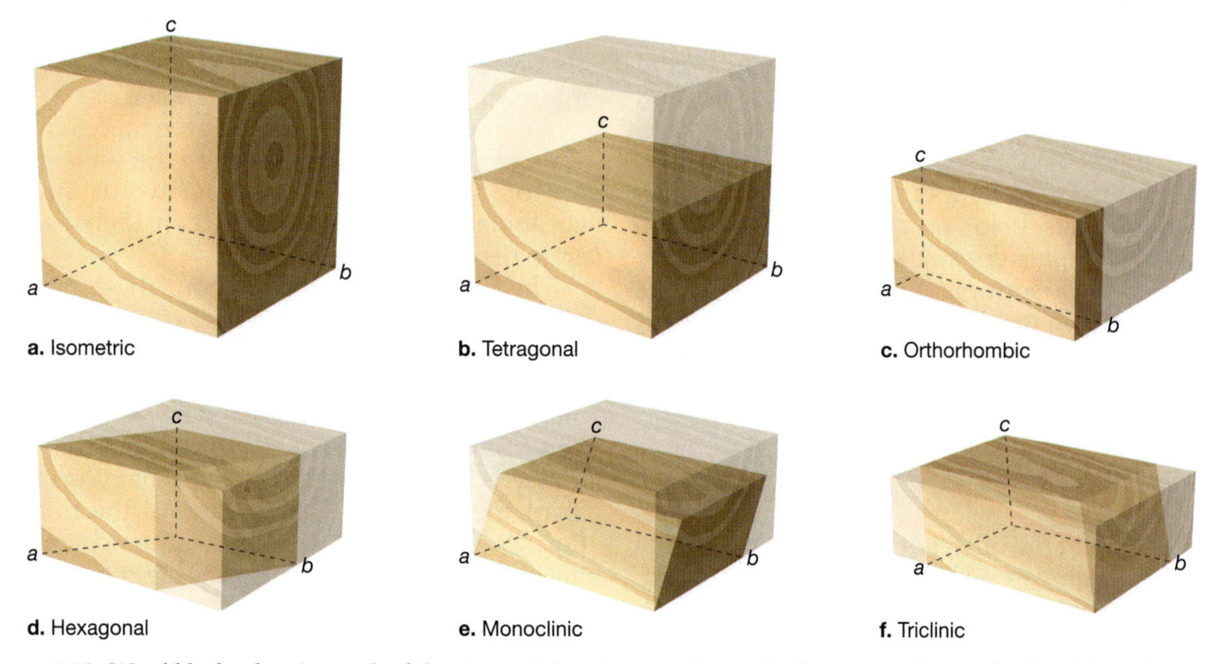

***Figure 1.12.** Wood blocks showing each of the six crystal systems: **a.** isometric; **b.** tetragonal; **c.** orthorhombic; **d.** hexagonal; **e.** monoclinic; **f.** triclinic. Consult the MSA website for animations of these blocks.*

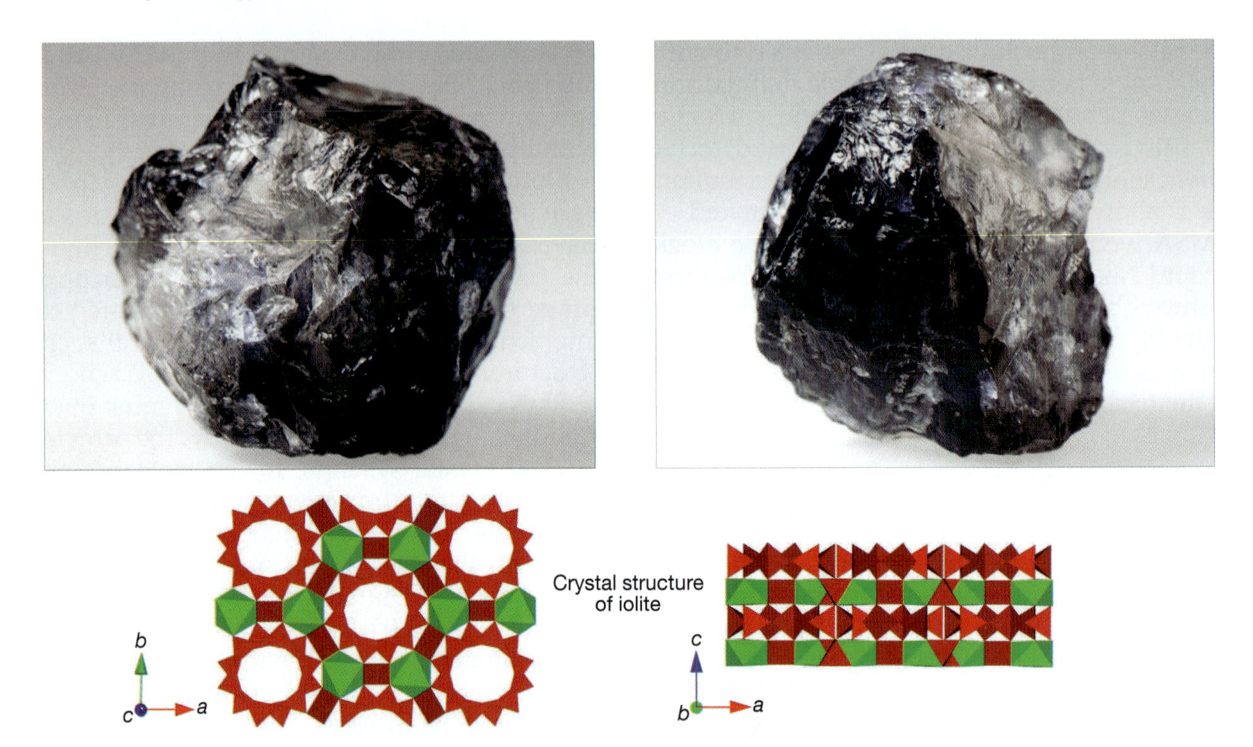

Figure 1.13. The biaxial mineral iolite (which is a variety of the mineral species cordierite) is either clear, blue, or yellow depending on which direction you look. This occurs because light interacts with the atoms in the crystal structure differently at different orientations. The structure drawing looking down the c *axis looks very different from the structure projected down the* b *axis. Because the structure is very different in these different directions, the physical properties can also differ, in this case producing different colors when viewed in different orientations.*

If you are keeping track of crystal systems, at this point you know we have three left. For these, the crystallographic axes have different lengths. Because they have different lengths, the physical properties associated with light can be different along all three directions. Thus there are no symmetry constraints on the physical properties of the interaction of light in these remaining crystal systems. The **biaxial** optical system is composed of materials from the orthorhombic, monoclinic, and triclinic crystal systems. The physical properties associated with light can vary in all directions within minerals belonging to a biaxial optical class. All minerals crystallize in one of the three optical classes: isotropic, uniaxial, or biaxial. As we'll see later in this book, the way in which light interacts with minerals at various angles is extremely useful for mineral identification.

The Big Ten Minerals

As we stated at the outset of this chapter, our goal here is to empower you with the most basic fundamental knowledge of mineralogy. Thus, we have selected a set of ten minerals that are important to this field (Table 1.3 and Figure 1.14). There is no doubt that it would be difficult to get all geologists, or mineralogists, to agree on what they thought were the ten most important minerals. Even your own instructor might not pick the exact same Big Ten, depending on his/her background and geological perspective, but we bet they will be close to ours! Be sure to ask your instructor to explain why his/her Big Ten minerals are different from ours—that might be very interesting!

We think everyone would agree these ten minerals are very, very important to the understanding of geology. Without these words in your vocabulary, you would not be considered a mineralogist or a geologist.

The first thing to notice about these minerals and their compositions is that they are composed primarily of the most abundant elements in the Earth's crust—this is no coincidence! Some of these mineral names might even be familiar to you from earlier classes. We think that this set of minerals is so critically important that you should learn the formulas. Using some of the information we've already learned about the charge and the coordination number for the eight most abundant elements in the Earth's crust, will help immensely in learning these chemical formulas.

Begin with the mineral quartz, which is the most abundant mineral *species* in the Earth's crust

(Figure 1.9). Not surprisingly, it is composed of the two most abundant elements in the Earth's crust, Si and O. We already know that Si^{4+} likes to occupy tetrahedra in which the four corners are oxygen atoms. Such tetrahedra would have an electronic charge of –4, because:

$$\begin{aligned} Si^{4+} \times 1 &= +4 \\ O^{2-} \times 4 &= -8 \\ \text{sum} &= -4. \end{aligned}$$

Minerals must be charge balanced (i.e., have a net zero charge), so sometimes the Si^{4+} tetrahedra link up with other Si^{4+} tetrahedra to share corners. If all the corners are shared, then each Si^{4+} is bonded to four oxygens, each of which is shared between two tetrahedra (Figure 1.15e). The net result is:

$$\begin{aligned} Si^{4+} \times 1 &= +4 \\ O^{2-} \times 2 &= -4 \\ \text{sum} &= 0. \end{aligned}$$

So the formula for quartz is SiO_2. As we will see later, all quartz has the chemical formula SiO_2. However, quartz is not the only mineral with a formula of SiO_2. We have already seen this with aluminosilicate minerals discussed above, and we will return to the subject of minerals with different crystal structures but the same chemistry later in the book.

Now, suppose you are crystallizing a mineral in an environment where there is abundant Al available (it's the next most abundant element, so this is likely!). Al has a 3+ charge, so it doesn't perfectly substitute for Si^{4+}. The charges are not

Figure 1.14. Darby and Mickey's Big Ten minerals.

the same, but the sizes are close. So let's see how Al^{3+} might substitute for Si^{4+} in quartz.

First we'll multiply the formula for quartz by 4 (we'll see why we picked this number later):

$$4 \times SiO_2 = Si_4O_8,$$

and then we could replace one of the Si^{4+} atoms by:

$$AlSi_3O_8^{1-}.$$

Because this has a charge of –1, we need to add something else to the formula to make it electronically neutral—a prerequisite for a mineral. From our list of common elements, we find two elements with charges of +1 that will make our formulas complete:

$NaAlSi_3O_8$, also known as **albite**, and
$KAlSi_3O_8$, which is the mineral **orthoclase**.

These minerals are end-members of a group of minerals called feldspars.

Actually, as in the case of the aluminosilicates just mentioned, there are three different minerals with the formula $KAlSi_3O_8$ (also a feldspar):

1. **sanidine**, which is found in volcanic rocks;
2. **microcline**, found in deep-seated igneous rocks, and
3. **orthoclase**, found in intermediate-depth plutons.

So here's another case where being able to identify a mineral gives you very important geologic insight (and vice versa as well—if someone hands you a rock from Mt. Fuji with a white mineral in it, it's probably sanidine and not microcline!)

What happens if we decide to substitute another Al into the structure structure? $AlSi_3O_8^{1-}$ would become $Al_2Si_2O_8^{2-}$, and then we need an element with a +2 charge that would fit into the feldspar structure (i.e., that would have a coordination number similar to K^{1+} or Na^{1+}). From our list of cations with preferred coordination numbers of "8," we see that Ca^{2+} would fit (pun intended) this bill. So we would arrive at:

$CaAl_2Si_2O_8$, which is the mineral **anorthite**.

So, by substitution of either one or two Al^{3+} into the modified quartz formula, Si_4O_8, we've now created the most abundant mineral *group* in the Earth's crust: the feldspars. Note that the word "feldspars" ends with the letter "s." This brings up some important points about our list of

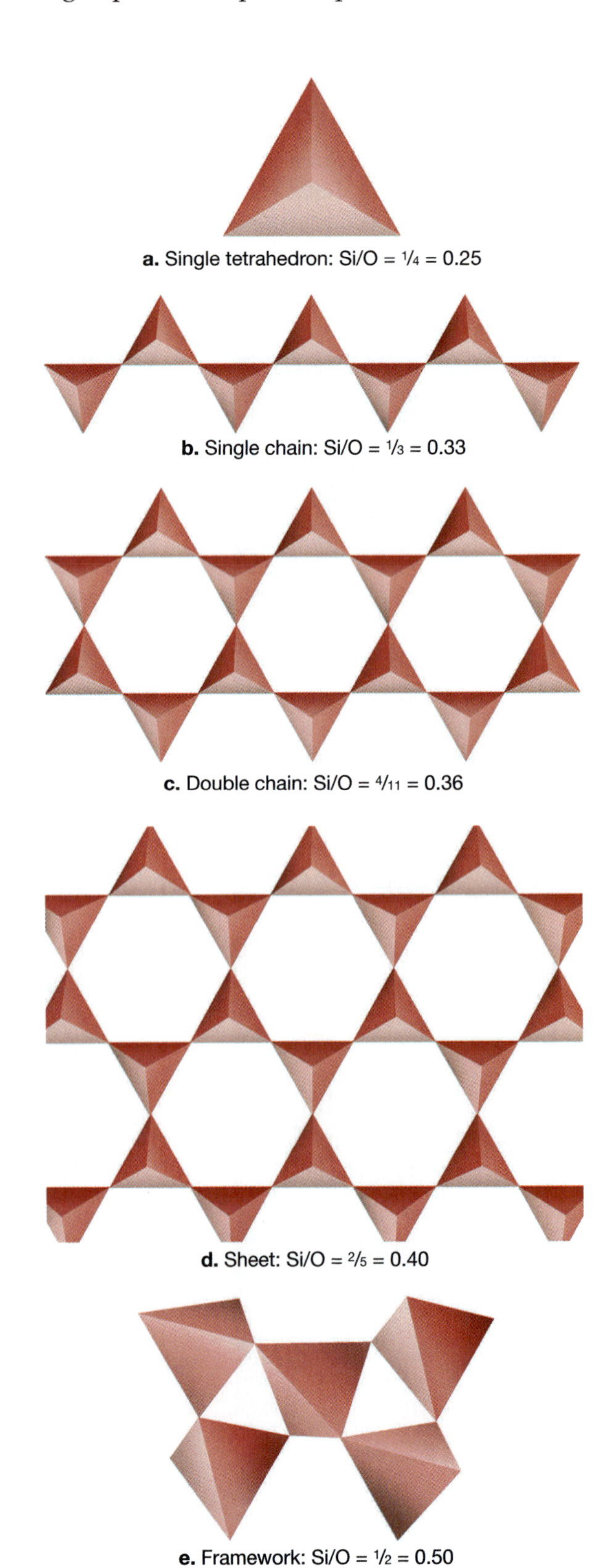

Figure 1.15. *Illustration (and animation on the MSA website) of the polymerization (i.e., joining) of Si^{4+} tetrahedra to form the different groups of silicate minerals discussed in the text. Note that the Si/O ratio listed for each group changes as a function of polymerization; the ratio becomes larger as polymerization increases. This implies that the number of oxygens decreases as polymerization increases because of sharing of oxygens between tetrahedra. Assuming the Si^{4+} tetrahedra from the main structure units for the silicate minerals, you should be able to predict the external morphology (i.e., shape) of these minerals based on underlying shape of the Si^{4+} tetrahedra groups.*

minerals. First of all, mineral names are not proper nouns, and *so they are never capitalized* unless they appear in a title or at the beginning of a sentence. Second, there are several kinds of mineral names: here we have **species** names, which represent specific chemical compositions, and **group** names, which represent groups of mineral species with related structures and compositions. Generally speaking, mineral names that end in the letter "s" like "micas," represent mineral groups; species names are usually given as singular nouns, like quartz or orthoclase. Thus we say that quartz is the most abundant mineral *species* in the Earth's crust, while feldspars are the most abundant mineral *group* in the Earth's crust.

The next two minerals on our Big Ten list represent members of the mica group, or the micas. Because of the "s" at the end of the word, we know this is a group name and there are going to be subdivisions such as series and species in the group. The two we have selected are common micas: biotite and muscovite. Micas form in environments where the element hydrogen (H) is present. They can be derived by "adding" $(OH)^{1-}$, to the Al^{3+}-substituted, quartz-modified formula above:

$$AlSi_3O_8 + (OH)_2.$$

We can further modify this formula by adding two more oxygen atoms, so that the basic formula for micas is (we'll explain the reasoning for this in detail in later chapters):

$$AlSi_3O_{10}(OH)_2^{7-}.$$

This formula needs help, because it is so far from neutrality (i.e., charge balance). So, we can add two trivalent (3+) cations (i.e., Al^{3+}) and a K^{1+}, to obtain a balanced formula for the mineral **muscovite**:

$$KAl_2(AlSi_3O_{10})(OH)_2.$$

Table 1.3. The "Big Ten" Minerals

	Mineral group	*Mineral species or series*	*Formula*
1		quartz	SiO_2
2	feldspars	orthoclase	$KAlSi_3O_8$
3		albite	$NaAlSi_3O_8$
4		anorthite	$CaAl_2Si_2O_8$
5	micas	muscovite	$KAl_2(AlSi_3O_{10})(OH)_2$
6		biotite	$K(Mg,Fe)_3(AlSi_3O_{10})(OH)_2$
7	amphiboles		$(Ca,Mg,Fe)_7Si_8O_{22})(OH)_2$
8	pyroxenes		$(Ca,Mg,Fe)_2Si_2O_6$
9	olivines		$(Mg,Fe)_2SiO_4$
10		calcite	$CaCO_3$

Why do you think we wrote the chemical formula for muscovite with two separate aluminum atoms instead of combining the two as follows:

$$KAl_3Si_3O_{10}(OH)_2?$$

Hopefully, that is the sort of question you will ask yourself often, because there are very good reasons for most things that we do. We will try to explain those reasons as we go along. We'll find as we progress through discussions about minerals and their chemical formulas that there is a "grammar" for the way that we write chemical formulas. By learning some basic mineral-grammar rules, you will understand the way we write chemical formulas as we do. You can then use this information to help you not only identify minerals but to predict their physical properties.

So, ask yourself the question: what might be the difference between the two ways of writing the formula for muscovite? The answer is that the two Al^{3+}'s have different coordination numbers. The first Al^{3+} has a coordination number of six, while the second has coordination number four. We will see in later chapters of this book that this is a big deal. Making a distinction based upon the number of oxygens around a cation is very important for understanding the bases of crystal chemistry.

Returning to the other mica, **biotite**, it can be made from the muscovite formula by replacing the six units of positive charge from the $Al^{3+}{}_2$ with six units of positive charge from $(Mg^{2+},Fe^{2+})_3$, to arrive at:

$$K(Mg,Fe)_3(AlSi_3O_{10})(OH)_2.$$

Thus if you know the chemical formula for muscovite, it is easy to determine the chemical formula for biotite.

Look more closely at the chemical formula of biotite. We have been talking about how to predict the physical properties of minerals based on a chemical formulas. Biotite is a great example of this capability. Muscovite and biotite have similar physical properties (i.e., cleavage, hardness); a similarity can even be seen in their chemical formulas. Because the chemical formulas are almost identical, one would assume that their physical properties would be almost identical.

Take a look at the photographs of muscovite and biotite in Figure 1.14. Muscovite is basically clear, while biotite is black. Why?

Look at the chemical formulas again. What is the difference? Biotite contains Mg and Fe, while muscovite does not. The conclusion you might reach is that because of the differences in chemistry, these two minerals have different colors. If you reach that conclusion, then you now have the ability to predict an important physical property of minerals based upon their chemical formulas. One more question, which will remain unanswered at this time, is this: which element do you think causes the dark color in biotite, Mg or Fe?

Another comment is needed to explain the expression $(Mg^{2+},Fe^{2+})_3$. Notice that these two cations are enclosed in parentheses and separated by a comma; we use this type of notation to indicate a couple of things. First, the sum of these two elements must equal three. For example, there could be one Mg^{2+} cation and two Fe^{2+} cations, or 2.5 Mg^{2+} cations and 0.5 Fe^{2+} cations. This notation indicates that these atoms can exist in a range in the mineral formula, and that they can substitute for one another.

The second important point here is that these atoms are approximately the same size (i.e., they would have the same, or overlapping, coordination numbers). This characteristic makes it easier for the atoms to occupy the same site!

Hopefully you are beginning to get the idea that mineral formulas are nothing but a game of mix and match with commonly-occurring elements, with two basic rules: (1) charge balance must be maintained (i.e., the sum of charges on the cations and anions must be zero), and (2) the atoms must "fit" into the structure. The charge balance in the chemical formulas is the domain of crystal chemistry, while the fitting of the atoms is more the domain of crystallography.

By changing the basic ratio of Si^{4+} (and/or Al^{3+}) to O^{2-} anions, nature creates a whole spectrum of mineral species with slightly different compositions and Si/O ratios. We began by considering minerals with a Si to O ratio of 1:2, or 0.50 (quartz and the feldspars), where both Si^{4+} and Al^{3+} occupy tetrahedral with all four corners shared. Next, we looked at the micas, where three of the four tetrahedral corners are shared. The ratio of tetrahedral (T) cations such as Si^{4+} and Al^{3+} to oxygens was 4:10 or 2:5, or 0.40. As we unlink or break apart the tetrahedra (i.e., depolymerize them), the T:O ratio decreases (Figure 1.15d).

A further depolymerization of the tetrahedra would break the micas into chains. What started out as a 3-D framework of Si^{4+} tetrahedra gets broken into sheets of tetrahedra (micas) and then into chains (Figure 1.15b, c). The amphiboles and pyroxenes are examples of these types of chain silicates. The structural difference between the two is that the pyroxenes consist of a single chain of tetrahedra, while the amphiboles consist of two cross-linked chains. This may seem like a subtle difference, but it will have a drastic effect on their physical properties as we will see in future chapters. Also, the difference will allow the amphiboles to accommodate hydrogen in their structures (as $(OH)^{1-}$), which the pyroxenes cannot.

Figure 1.15b is a schematic representation of a single chain of polymerized tetrahedra. The basic building block (i.e., the repeat unit) is two tetrahedra. If we block out these two tetrahedra and count the number of Si's and O's, we arrive at two Si's and six O's. Two of the O's are counted as 1/2's because they are shared with tetrahedra outside the "basic building block." This makes a basic formula:

$$(Si_2O_6)^{4-}.$$

This is the building block for the single-chain silicates. Notice a couple of things in the formula. First, the Si:O ratio is 2:6 or 1:3 or 0.33, the lowest so far. In fact, as depolymerization (unlinking of tetrahedra) continues to occur, this value will become lower. This is an important systematic trend in the silicates. Notice that the building block is not charge balanced; it needs +4 charges. It turns out that the size of the cations used in the structure for charge balance will tend to have a coordination number of six for the simplest of the pyroxenes. Thus, we would expect Mg^{2+} and Fe^{2+} to be major components of this mineral group. Also, for reasons we'll discuss later, some Ca^{2+} can enter the structure.

How many of +2 cations are needed to charge balance the basic Si_2O_6 building block?

The Si_2O_6 has a total charge of $(2 \times +4) + (6 \times -2)$, which is –4. So, 4 additional units of positive charge are needed. Thus, a general formula for the pyroxenes would be:

$$(Ca,Mg,Fe)_2Si_2O_6.$$

By the same line of reasoning, a simplified, general formula for the amphiboles (double chain silicates) is:

$$(Ca,Mg,Fe)_7Si_8O_{22}(OH)_2.$$

Note immediately the similarity between this formula and the one for pyroxenes. Both contain the same elements, but in different amounts. The ratio of Si to O controls how much additional substitution is needed. Amphiboles, like micas, contain $(OH)^{1-}$. There is also a structural similarity between the amphiboles and the micas that allows $(OH)^{1-}$ to occur in them but not the pyroxenes.

The difference in the proportions of atoms between the pyroxenes and amphiboles is based upon the difference in polymerization (linking) of the tetrahedra. Observation of Figure 1.15c shows that the basic building block for the amphiboles would yield a Si:O ratio of 4:11 or 0.36, a value between that of pyroxenes and the layer silicates. For reasons that we will discuss later, the basic building of amphiboles is extended to be:

$$[Si_8O_{22}(OH)_2]^{14-}.$$

Basically there is a doubling of the Si:O values and the addition of the appropriate amount of $(OH)^{1-}$. Note, as usual, the negative charge imbalance of –14. Given the same charge-balancing divalent cations as in the pyroxenes, it will take seven of them for charge balance.

If we continue with the depolymerization (breaking Si-O bonds) until we are left with a single, isolated tetrahedron (i.e., a tetrahedron whose O^{2-} anions bond only with single Si^{4+} cations). We have as the basic building block (Figure 1.15a):

$$(SiO_4)^{4-}.$$

This yields a Si:O ratio of 1:4 or 0.25, the lowest yet, and the lowest possible. Thus the range of Si:O ratios falls between 0.50, when the tetrahedra are completely linked (or fully polymerized), and 0.25, when they are completely depolymerized, or isolated. Again, we must add +4 charges for charge balance. Sticking with Mg^{2+} and Fe^{2+} we would need two atoms and would have as an example the mineral formula:

$$(Mg,Fe)_2SiO_4.$$

This is the general formula for the olivine mineral group. There are other examples of minerals that are formed based on isolated tetrahedra (e.g., the garnets), and we will discuss them in future chapters. But the olivines are the most geologically common minerals with isolated tetrahedra, which is why we list them as Big Ten mineral number 9, and the final silicate mineral in our list.

This leaves us with only one more important mineral on our Big Ten list, which is in some ways an oddball because it does not contain Si. This is the mineral **calcite**, which is $CaCO_3$. It makes the list because it is so critical to maintaining the Earth's atmospheric CO_2 balance by storing CO_2 in the form of $CaCO_3$ on the ocean floors and because it is so common. Without the help of calcite on the Earth's surface and in its oceans, CO_2 would build up in the atmosphere, and the Earth might look like Venus instead!

Concluding Remarks

The goal of this chapter was to introduce you to the major areas in mineralogy and the future contents of this book and its coverage of this discipline. The topics we have just discussed represent the first revolution or two on our learning spiral. We have introduced many important concepts in this chapter, and we will need to return to all of these at least once or twice more in the class and the book. As you progress through the course, it would be very helpful to return to this chapter often to review these very basic concepts and vocabulary that are presented here for the first time. By the end of the course, we hope that everything in this chapter has indeed become second nature to you, at which point you will be well on your way to understanding the important concepts of mineralogy!

P.S. Ten years from now, promise us you'll reread this chapter, and send us e-mail to let us know if you still remember what we've written here!

References

Bloss, F.D. (1994) Crystallography and Crystal Chemistry. Mineralogical Society of America, Washington, D.C. 239 pp.

Chapman, C.R. (1976) Asteroids as meteorite parent bodies: The astronomical perspective. Geochimica et Cosmochimica Acta, 40, 701–719.

Dreibus, G., and Wänke, H. (1987) Volatiles on Earth and Mars: A comparison. Icarus, 71, 225–240.

Haskin, L., and Warren, P. (1991) Lunar chemistry, in Lunar Sourcebook, G.H. Heiken, D.T. Vaniman, and B.M. French, eds. Cambridge University Press.

Young, R.A., and Brown, W.E. (1982) Structures of biological minerals. In Biological mineralization and demineralization, G.H. Nancollas, ed. Life Sciences Research Reports, 23, 101–141.

Chapter 2

Hand Sample Identification

Nearly every young child has the experience of picking up a rock and wondering over its smoothness, or its angularity, or simply the way it feels. To remember this tactile sensation is to understand the fundamental appeal of the science of mineralogy. When I was three years old, I spent an Easter morning bypassing eggs in favor of the far more interesting rocks in the park, so I remember this feeling very well. Although subsequent experiences with ugly brown, nondescript, and (to me) shapeless mineral specimens in junior high school took the joy out of minerals for me, I was pleased to rediscover them on (mostly) happier terms in college. I firmly believe that if we could all reconnect with our childhood curiosity about rocks and minerals, the world would be a better place (and not just because it would have more mineralogists!)

But really, who cares about mineral identification? Is this some obscure hobby practiced by enthusiastic retirees? Why is this so important? It all comes down to this: hand sample mineral identification is the first *tool used by geologists in any kind of field setting; if you can't identify minerals, you'll never understand the rocks they constitute.*

I personally learned this lesson the hard way. Because I went to college in New England, every field trip I took as a student was to igneous and metamorphic localities—I could spot a garnet at 50 yards with ease. When I went to field camp in Montana with a bunch of students from the Midwest, I was hopelessly lost. I'll never forget the assignment they gave us on the very first day: a simple fold, well-defined by layers of sandstone and limestone. I had never seen a sedimentary rock in the field, and to me, the whole road-cut looked like quartz. Needless to say, I didn't see the contact that defined the fold, and I failed the assignment... Eventually, I got out my hand lens and acid bottle, and figured out that I was looking at quartz and calcite. I ended up loving field geology. It just never occurred to me that mineralogy was something you would use in field camp!

Now, of course, I know that identification of minerals in outcrop is a basic and necessary skill for studying rocks in outcrop. Ancient marine environments are mapped on the basis of limestone (calcite), dolomite, and sandstone (quartz) distributions. Metamorphic isograds, which represent lines of equal temperature in regionally-metamorphosed rocks, are mapped on the basis of groups (assemblages) of minerals that are stable under similar conditions. In igneous rocks, you'll never distinguish a peridotite from a dunite if you can't tell green pyroxene and olivine apart.

So before we plunge into the nitty gritty of what makes minerals tick, and the advanced methods that may be required to identify an unknown, it seems appropriate to begin with what almost everyone wants to know right away: what can I learn from a hands-on encounter with a mineral?

M.D.D.

Introduction

So what is a mineral? Many natural phenomena occur in regular polyhedral shapes, such as dried-up mud beds, snow flakes, and columnar basalt columns. However, the concept of a crystal and mineral is restricted to a *homogeneous* body. Therefore, mud beds and basalt columns don't count because, for example, the basalt column is composed of feldspar, pyroxene, and olivine. But the snowflake *does* count.

It is the *regularity* of internal structure that determines the unique characteristics of crystalline matter. We are accustomed to thinking of the states of matter as gas, liquid, and solid; a more rational way of thinking of it would be gas, liquid, and crystal. Such definitions are somewhat arbitrary. Scientists who study crystallization processes from melts do observe a gradual or continuous process of short-to-long-range ordering which culminates in crystallization. They define a *glass transition* as the turning point between melt and crystal. However, the first appearance of crystals is a difficult thing to quantify and is, of course, really a function of the detection limit of the technique used to look for the crystals, as is obvious from the following definition.

Crystal = regions of matter composed of the same kind of molecule systematically repeated.

Note that this connotes a definite composition or range of compositions, as would be possessed by any homogeneous, solid body of a chemical element or compound. The ordered internal arrangement of atoms might be outwardly manifested by planar faces on a crystal. The scale of the systematic repetition is somewhat arbitrary but is generally considered to be on the order of at least tens of microns.

Mineral = a naturally-occurring, typically inorganic solid (at room temperature) that has an ordered atomic arrangement and can be represented by a chemical formula.

The definition includes any element or chemical compound that is crystalline and that has formed as the result of geologic processes. Mercury, which is a liquid at room temperature, is an approved exception to this definition, and is considered a real mineral because it is a native element. Water is not a mineral (although ice and snowflakes are), and crystalline biologic and artificial materials are generally not considered minerals.

Materials formed as the result of geologic processes from artificial substances are no longer accepted as true minerals. Similarly, anthropogenic substances (those formed "by human intervention") are not minerals either. For example, crystalline substances that form via interactions between existing minerals and human-made materials like blasting powder or industrial waste products, while interesting, do not qualify as mineral species (Nickel and Grice, 1998).

The definition of what is a mineral and what is not is in the hands of an organization called the Commission on New Minerals, Nomenclature, and Classification (CNMNC) of the International Mineralogical Association. People who think they have discovered a new mineral must apply to this commission with a detailed description of their finding; the discoverers are allowed to choose a name for their new mineral (Table 2.1). Minerals species have been named after people (45% are in honor of a person), type localities (usually the first locality at which the mineral was discovered), appearance (usually from some Greek or Latin root), and chemical composition (Blackburn and Dennen, 1997). Approved new mineral species are usually published in *The Mineralogical Record,* with new mineral *descriptions* in other journals such as the *American Mineralogist.* The number of valid mineral species is now well over 4,300, and constant research is being done to revise the list. There are also approximately 30,000 old, discredited, or synonym names for minerals; each has been shown to be a repetition or variety of an existing mineral name. A comprehensive collection of old and new mineral names and synonyms is tabulated in de Fourestier (1999).

Mineralogists also make a distinction between the types of mineral names (Figure 2.1):

Mineral species = a mineral distinguished from others by its specific, unique chemical and physical properties.

Species include most of the familiar mineral names, such as quartz, muscovite, and magnetite. Each species has a designated composition based on its original approval as a mineral name; those are used in this text and are listed in the *Glossary of Mineral Species* published by *The Mineralogical Record.* All minerals specimens are unique ("impure") to some extent because nature rarely makes things perfectly; therefore a mineral species may have small (or in some cases, significant) substitutions of cations that are not in its "formal" formula. Compositional and structural defects are very common in all minerals. In fact, the presence of such imperfections is often used to distinguish between naturally-occurring and synthetic gemstones!

Mineral variety = specific type of a mineral species with some distinguishing characteristic such as color, habit, or other external physical characteristics.

Table 2.1. You Too Can Name a New Mineral!
The Commission on New Minerals, Nomenclature, and Classification requires the following information in order to process a request for naming a new mineral: 1. Proposed name and the reason for its selection. 2. Description of the occurrence, including geographic and geologic setting and a list of coexisting minerals. 3. Chemical composition of the mineral, and the method of its analysis. 4. Chemical formula. 5. Crystallography, including crystal system, crystal class, space group, and unit cell parameters. 6. Crystal structure, and descriptions of which atoms are which sites in the structure. 7. General appearance and physical properties. 8. Optical properties. 9. Type material, i.e., the museum where the new mineral is located. 10. Relationship of the new species to other mineral species. 11. Publications on the new species, or those related to it.
Many/most new minerals are found by amateurs, who then work with scientists who have access to analytical facilities to obtain the necessary characterization. So it's very possible for any determined individual to find a new mineral! For more information, consult Nickel and Grice (1998), and check out the examples published in the *American Mineralogist.*

For example, sapphire and ruby are varieties of the mineral species corundum. Mineral variety names are not regulated by the CNMNC, and this has led to an unfortunate proliferation of them (see Martin, 1998) and a great deal of confusion over mineral names in general. The term variety may also be used to indicate the presence of substantial substitutions by ions that are not essential constituents of the mineral species. These are common for mineral groups like amphibole, where both adjectival modifiers (used for minor amounts of chemical substitution) and prefixes (used when the amount of substitution is so great that the original cation in the composition is sometimes completely replaced) are employed. Prefixes are considered an inseparable part of the name and are attached to the species names with a hyphen; adjectival modifiers are simply adjectives.

Examples from the nomenclature include:

Prefix	**Adjective**
chromium-	chromian
ferri-	ferrian
ferro-	ferroan
magnesio-	magnesian
manganese-	manganoan
potassium-	potassian
Cation	**Meaning**
chromium	Cr-rich
ferric iron	Fe^{3+}-rich
ferrous iron	Fe^{2+}-rich
magnesium	Mg-rich
manganese	Mn-rich
potassium	K-rich.

Some minerals with a constant composition have different structures under different conditions, and these differences may be noted by the addition of a Greek letter prefix to their species name. For example, you may encounter references to α-quartz (alpha-quartz) and β-quartz (beta-quartz), which are low and high-tempera-

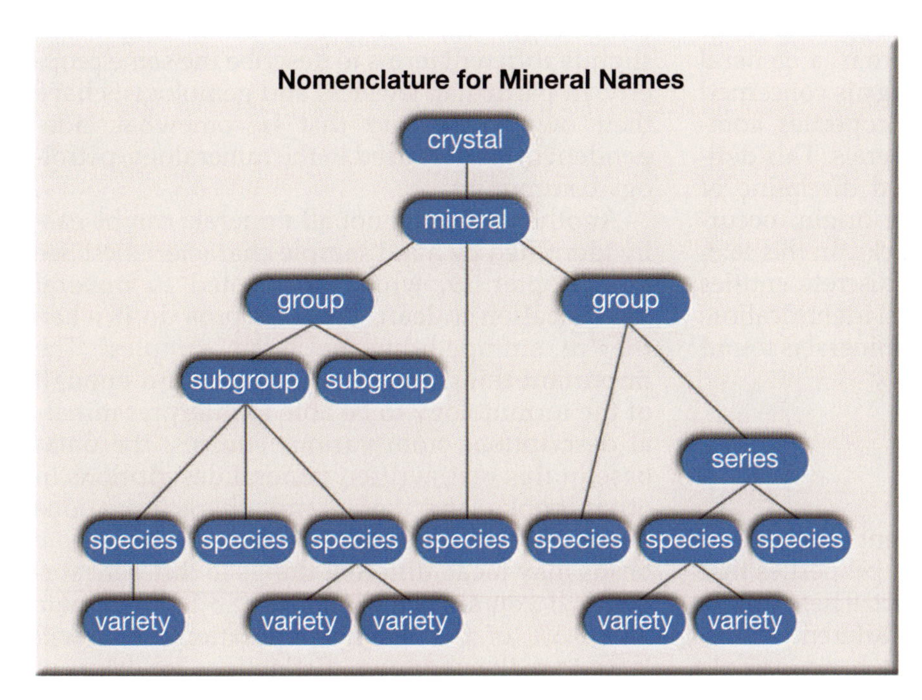

Figure 2.1. Nomenclature for mineral names is hierarchical, and only mineral species names are officially recognized. This vocabulary differs from animal or plant taxonomy because all species do not belong to a group or sub-group and only some species have varieties.

ture forms of SiO_2 with different structures and symmetry. Specific information on nomenclature for different mineral groups is given in the latter chapters (22–23) of this book.

Mineral series = two minerals species between which there is a complete range of naturally-occurring composition.

The minerals albite ($NaAlSi_3O_8$) and anorthite ($CaAl_2Si_2O_8$) form a series called plagioclase, which is commonly given the formula $(Na,Ca)Al(Si,Al)_3O_8$. Similarly, the biotite series represents mixtures of phlogopite, siderophyllite, annite, and eastonite. Series names are very useful in situations where the individual end-members can't be easily distinguished.

Mineral group = group of mineral species with the same type of crystal structure and different compositions.

Amphibole, garnet, olivine, and mica are all mineral groups; although they are often called "minerals" in common usage, this is technically incorrect. For example, the barite group is composed of the following mineral species:

anglesite	$PbSO_4$
celestine	$SrSO_4$
hashemite	$Ba(Cr,S)O_4$
barite	$BaSO_4$.

So barite is both a species name and a group name, and it's necessary to determine from context which term is being used.

Individual minerals are, of course, the primary component of rocks:

Rock = a cemented aggregate of one or more mineral grains, or a body of undifferentiated mineral matter, or of solid inorganic matter.

Thus, a mineral can be a rock, but most rocks are not considered minerals. From a general standpoint, the study of **mineralogy** is concerned with the formation, occurrence, properties, composition, and classification of minerals. This definition is peripheral to the related discipline of **petrology**, which deals with the origin, occurrence, structure, and history of rocks. In this text, we will consider minerals to be discrete entities until we reach the point of mineral identification, wherein the rock type in which a mineral is found can provide insight into its identity.

Mineral Properties

Mineral identification is dependent on observations of the physical and chemical properties that make each mineral species distinct. These observations can be made at many different scales, ranging from submicroscopic to hand sample size measurements. The rest of this book is dedicated to explaining *what* these characteristic properties of minerals are, *how* they can be used to identify and understand minerals, and *how* they relate to the crystal structure and chemical composition of the mineral. In the remainder of this chapter, the common terms used to describe minerals in hand sample are listed, described, and illustrated (note that there are almost as many descriptive terms as there are minerals, so only the most common ones are included!). These terms are used in the mineral database that accompanies this book. In later chapters, we will describe the physics and chemistry behind each of these characteristics. One of the most important skills that can be learned in any mineralogy course is the ability to pick up a mineral sample and identify it in hand.

The physical properties of minerals can be loosely divided into categories as follows; this is the structure used on the accompanying form (Table 2.2):

1. Color and streak
2. Luster
3. Hardness
4. Fracture and tenacity
5. Crystal form and system
6. Crystal shape and habit
7. Cleavage and parting
8. Twinning
9. Specific gravity (density)
10. Other properties, such as taste, radioactivity, and magnetism.

Keep in mind that the usage of many of these terms is *very subjective*. Even within the field of mineralogy, different professionals might use slightly different terms to describe the same property. In particular, jewelers and gemologists have their own vocabulary that is somewhat independent of what is used in the mineralogy/petrology community.

Another warning: not all minerals can be easily identified by hand sample characteristics! See the Chapter 19, which is devoted to mineral identification to learn how the pros do it when they're stumped by the hand samples! The important thing is that you understand enough of the terminology to be able to interpret mineral descriptions from various sources: the database in this text, written mineral descriptions in other books and texts, and characterizations given in the technical literature. Also, some terms may mean different things to different students: if you are color blind (like 8% of the male and 0.5% of the female population), you will learn to make your own distinctions for colors of

Table 2.2. Hands On Mineral Identification

Fill in this chart based on your own observations about the hand sample you are examining. Then use the database that comes with this textbook, and any other references, to identify your unknown mineral.

Color: ______________

Streak color: ______________________________

Other optical characteristics: ____

Luster:

metallic (submetallic)	nonmetallic	adamantine (subadamantine or splendent)	
dull	earthy	felty	glassy or vitreous
greasy	pearly	pitchy	resinous
satiny	silky	velvety	subvitreous
waxy			

Mohs Hardness (1–10): _______

Fracture and Tenacity:

brittle	conchoidal (subconchoidal)		ductile
elastic	even	fibrous	flexible
fragile	hackly	inelastic	malleable
plastic	sectile	splintery	tough
uneven	__________		

Crystal system:

isometric	orthorhombic	monoclinic	tetragonal
hexagonal	triclinic		

Shape or habit:

acicular	arborescent	asbestiform	banded
bladed	blocky	botryoidal	capillary
colloform	columnar	concentric	compact
concretion	coralloid	cryptocrystalline	dendritic
divergent	drusy	equant	featherlike
fibrous	filliform	foliated	geniculated
geode	globular	granular	gwindle
helicitic	lamellar	laminated	lath-like
massive	micaceous	nodular	pisolitic
platy	plumose	powdery	prismatic
radiated	radiating spherulitic	ramified	reniform
reticulated	rod-like	rosette	scaly
selliform	spherulitic	spiral	stalactitic
stalacmitic	stellate	striated	stubby
sugary	tabular	tufted	wiry

Cleavage and parting:

none	90° (isometric, cubic)	120° (rhombohedral)	basal
prismatic	octahedral	dodecahedral	pinacoidal
distinct	indistinct	micaceous	

Twinning:

contact (swallowtail)	geniculated (elbow)	interpenetrating	multiple (polysynthetic)
reticulated	cyclic	______________	

Specific Gravity: _____________

Other (magnetic, electrical, and radioactive properties?):

different minerals. Many successful petrologists and mineralogists (including at least one curator of a prominent U.S. mineral museum) have varying degrees of color-deficiency; in most cases, they have compensated for this defect to the extent that they can discern details, gradations in hue, and subtleties in samples that are not observed by normal-sighted persons!

Note that inspection of older mineralogy texts will reveal many archaic terms no longer in common usage; in this book we have tabulated only those descriptive terms that seem to us to be worth perpetuating!

Color and Streak

Color is generally the most conspicuous property of any mineral, although it can be irritatingly non-diagnostic for some minerals. (One of the low points of M.D.D.'s academic career was a mineralogy lab exam involving nine black minerals, all apparently different species.) Tourmaline group minerals, for example, can occur in just about any color ranging from black and brown to yellow, pink, purple, and blue. So although color is an easy characteristic to assess, its usefulness as a diagnostic property is often less than ideal! In fact, color is considered to be a diagnostic property of only a few minerals, such as rhodochrosite and epidote, that are always the same color. If the element (usually a transition metal) that is causing the color is a major constituent of the mineral's composition, it is said to be **idiochromatic** as in the case of uvarovite garnet, malachite, and lazulite. If only trace amounts of impurities are the source of color, the mineral is **allochromatic** as in sapphire or ruby.

Streak is the color of a mineral in the powdered form, usually obtained by running the mineral across an unglazed porcelain plate and observing the mark it leaves (Figure 2.2). A streak test shows the color of freshly powdered sample, which may not be the same as the color of the bulk hand sample. It is especially useful in identifying minerals with surfaces that are vulnerable to alteration. The mineral hematite, for example, may have a black, silver, or red appearance in hand sample, but its streak is always a reddish brown. Note that streak is not a useful property of minerals with hardnesses greater than about 6–7 on the Mohs scale (see below) because that is the hardness of the ceramic in the streak plate! Minerals harder than 7 will simply powder the streak plate, and not the sample! If you want to know the "streak color" of a mineral harder than 7, you have to physically remove and grind a piece into its powdered form (don't try this with the family gemstones…). Fortunately, there are not that many hard-to-identify minerals with hardnesses greater than a streak plate!

A number of other properties are related to interactions with light. These include the following:

Luminescence is the emission of visible light from non-incandescent material. The energy that is the stimulus for the emission may come from many different types of sources: nuclear, electrical, mechanical, chemical, biochemical etc.—at least fifteen different kinds of luminescence are known! The most common type of luminescence is **photoluminescence**, which is stimulated by light with ultraviolet wavelengths. It can be further subdivided into various types:

Fluorescence is caused by the *instantaneous* emission of light in response to absorption of energy from an external source, usually X-rays or ultraviolet light. Haunted houses at amusement parks commonly use ultraviolet lighting to make white objects glow in the dark! Scheelite, willemite, autunite, calcite, scapolite, diamond, and fluorite (after which this property was named) have this characteristic, as do many other minerals.

Phosphorescence is the *continuing* emission of light in response to absorption of energy from an external source, again usually X-rays or ultraviolet light. Objects with this property will continue to give off light long after the source of energy is turned off. "Glow in the dark" t-shirts and the minerals calcite, willemite, and pectolite may exhibit phosphorescence.

Thermoluminescent materials will become luminous when heated between the temperatures of about 50–475°C. Tourmaline, fluorite, calcite, apatite, scapolite, and some feldspars can have this property. **Triboluminescence** is a related property

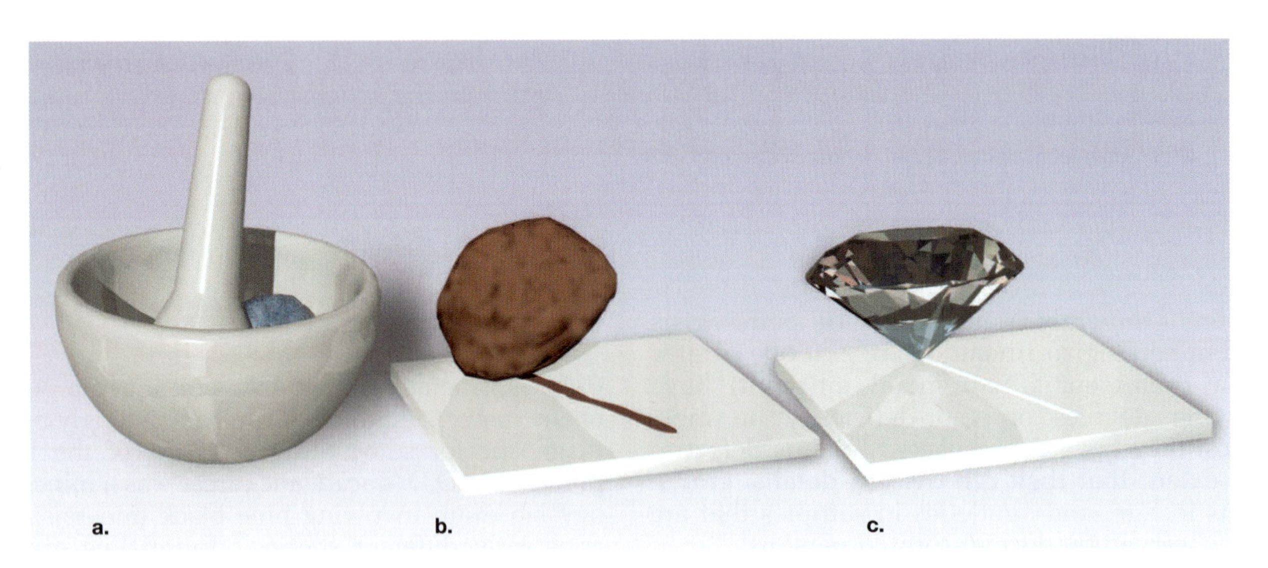

Figure 2.2. ***a.*** *Streak is the color of a mineral in the powdered form. Often, the color of a mineral is lighter when it is powdered because the powder is thinner than the bulk mineral sample.* ***b.*** *To test the streak of a mineral, we usually use a streak plate, which is a piece of unglazed porcelain. The mineral is scratched across the streak plate to show the color of its powder.* ***c.*** *Because the porcelain in the streak plate has a hardness of about 6–7, minerals that are harder than it will not be powdered. Instead, the powder you see may be ground up porcelain!*

possessed by any material that luminesces or sparks when scratched, rubbed, crushed, or hit with a hammer. Try this in a dark room with milky quartz, fluorite, lepidolite, sphalerite, or some breath mints! **Chemiluminescence** is observed when a mineral is reacted with a chemical, resulting in the release of energy in the form of light.

The **diaphaneity** of a mineral is its ability to transmit light (Figure 2.3). **Transparent** minerals allow light to pass through them without significantly diminishing in intensity; this is true for thin sheets of muscovite mica as well as for most gemstones and many quartz crystals. **Translucent** and **subtranslucent** minerals allow successively less light to be transmitted, such that only the outline of an object may be seen through it (like a shadow on a shower curtain!). Gypsum is usually translucent, as are jadeite and nephrite (the mineral forms of jade). Minerals that do not transmit any visible light at all are termed **opaque**; this group includes most minerals. Note that diaphaneity depends to some extent on the thickness of the sample being studied; nearly all minerals can be thinned to the point of at least some translucency. Notable exceptions to this rule are iron oxide minerals such as magnetite; the phenomenon that causes their black color (electrons hopping back and forth between sites surrounded by oxygen atoms) is so intense that it's nearly impossible to make magnetite thin enough to transmit light! To learn more about the fascinating subject of the causes of color in minerals, see Chapter 7.

Play of colors is the property by which different colors are reflected as a sample is turned. The classic example of a mineral having this property is opal (Figure 2.4).

Asterism results in the appearance of a star that is visible within the mineral grain (Figure 2.5). It is common in gemmy crystals of the corundum group of minerals; the Star of Asia Sapphire and the Rosser Reeves Star Ruby in the collections at the Smithsonian are beautiful examples of these. More commonly, a thin plate of muscovite or phlogopite mica, when held in front of a point source of light, will also exhibit asterism.

Chatoyancy (commonly called "cat's eye") results from the presence of tiny parallel bundles of fibers within the structure. Optical bands of wavy color appear to move as the sample is turned. Tiger's eye, which is a variety of the mineral species quartz, contains fibers of crocidolite or an iron oxide (Figure 2.6).

Iridescence is the play of colors caused by the scattering of light off (1) a thin coating on the surface of a mineral, as in goethite or hematite (Figure 2.7), (2) partially-open cleavages within the mineral, or (3) closely-spaced layers of slightly variable compositions. **Schiller** (formerly called **labradorescence** or **adularescence**) is a special type of iridescence found exclusively in feldspars such as moonstone and labradorite (Figure 2.8).

Opalescence is the appearance of milky or pearly reflections from the interior of a sample, as seen in the gemstones opal (Figure 1.4) and moonstone (a variety of the feldspar group minerals).

Tarnish (Figure 2.9) is a characteristic property of the mineral forms of silver and copper, and results from a chemical reaction between the minerals and the surrounding air. It can produce brightly colored thin coatings (such as those that are obvious on the bottom of a copper pan after cooking) or dull coatings.

Aventurine minerals look like they contain glitter, because they have fine-scale inclusions of reflective mineral grains such as mica or hematite (Figure 2.10). Orthoclase and quartz are the most common minerals displaying an aventurine appearance.

Pleochroism is the term applied to minerals that possess different colors in the same sample when light passes through them from different directions (Figure 2.11). Tourmaline, kyanite, epidote, spodumene (especially the variety kunzite), and cordierite often show obvious pleochroism. Many other minerals exhibit pleochroism when viewed through a filter that allows only plane-

Figure 2.3. *The diaphaneity of a mineral is its ability to transmit light. A mineral may be transparent like aquamarine (left), translucent like nephrite (center), or opaque like magnetite (right).*

Figure 2.4. Precious opal shows a play of colors when turned, making it a very attractive gemstone. Minerals with this appearance are said to be opalescent.

polarized light to be transmitted. Excessively pleochroic minerals are called **dichroic** if they exhibit two colors (e.g., the gemstone alexandrite, that changes from red to green) and **trichroic** if they exhibit three colors. This property will be used to great advantage in mineral identification using a polarizing microscope. See Chapter 5 for more information.

Luster

Luster refers to the appearance of light reflected from the surface of a mineral, and is described in terms of quality and intensity. This can be quantified by measuring the degree of light absorption of the mineral, and the amount of reflectivity it possess (see Chapter 5 for more information on the related term, refractive index). Note that luster is independent of color. Qualitatively, there are numerous hand sample terms for this property, as follows:

Metallic luster is just what it sounds like: the surface of the mineral reflects light brightly (Figure 2.12), like the shiny surface of the chrome bumper on your grandparents' 1960s car. It is found on materials such as steel (which is not a mineral), lead, tin, gold, silver, galena, and copper. Metallic luster is generally diagnostic, although there are a few minerals (notably hematite) that occur in both metallic and nonmetallic lusters. The shiny appearance occurs because electrons in the crystal structure are shared among atoms through metallic or covalent bonding, or when free or loosely attached electrons (as are found in conductors or semiconductors) exist in the material (see Chapter 8). Note that light reflected off of metallic surfaces is nearly pure reflectance (i.e., all the light that hits the surfaces is absorbed and then instantly re-emitted back in the form of visible light). Thus, minerals with metallic luster appear opaque. **Submetallic** luster is observed on minerals such as columbite, in which the metallic surface looks slightly less shiny than a old car bumper (Figure 2.13).

Nonmetallic luster is a duller light reflected from the surface of most minerals. The lack of bright reflectance indicates that most of the light is being absorbed or transmitted by the mineral, and only a little is being reflected back. Most min-

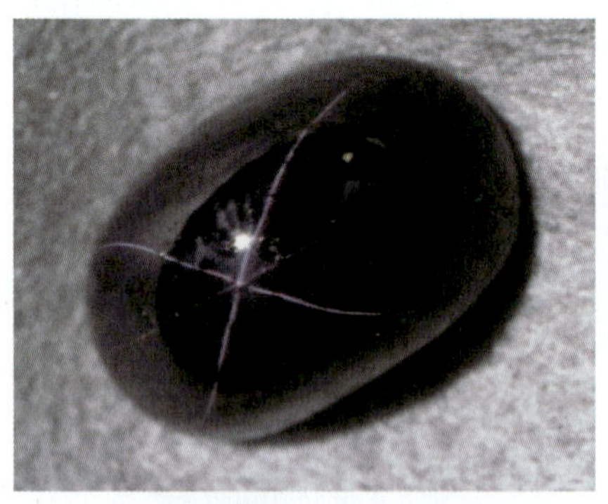

Figure 2.5. Asterism results in the appearance of a star, visible within the mineral grain, as in this star garnet.

Figure 2.6. Tiger's eye quartz contains fibers of crocidolite and exhibits the property of chatoyancy.

Figure 2.7. Iridescence is the play of colors caused by the scattering of light off a thin coating on the surface of a mineral, as in this hematite.

Figure 2.8. *Schiller (formerly called labradorescence or adularescence) is a special type of iridescence found in feldspars such this labradorite.*

Figure 2.9. *Tarnish is a characteristic property of the mineral forms of copper, silver, and some other minerals. It results from a chemical reaction between the minerals and the surrounding air.*

Figure 2.10. *Aventurine minerals like these feldspars look like they contain glitter, but actually they have fine-scale inclusions of reflective mineral grains such as of mica or hematite.*

Figure 2.11. *Pleochroism is the term applied to minerals like this cordierite crystal that appear to possess different colors in the same sample when viewed from different directions. This sample changes from colorless to blue to yellow.*

Figure 2.12. *The surface of a metallic mineral reflects light brightly, like this antimony.*

Figure 2.13. *Submetallic luster is observed on minerals such as this specimen of molybdenite, in which the metallic surface looks like slightly tarnished metal.*

erals have nonmetallic luster, so there are numerous specific terms to cover the range of nonmetallic appearances.

Adamantine minerals are extremely shiny (brilliant) and they reflect light very strongly (Figure 2.14). This luster is typical of very hard minerals such as diamond and corundum, and of lead-bearing minerals such as cerussite and anglesite. Lead is often added to the raw material used to create crystal goblets and decanters because it adds to the brilliance of the finished product! **Subadamantine** minerals are slightly less "brilliant" than adamantine ones; the mineral species zincite is frequently in this category. The extreme end of the "shininess" scale is sometimes described by the term **splendent** or **specular**, which means that a mineral has the brightest possible luster. A freshly-broken piece of the mineral galena will exhibit splendent luster (Figure 2.15), as will a piece of the specular variety of hematite.

The range of highly lustrous minerals can be directly quantified on a scale using refractive index, *n*, which is the ratio of the velocity of light in a vacuum to the velocity of light in a material. Povarennykh (1964) distinguished the following categories for transparent minerals: splendent minerals have n = 2.7–3.4, adamantine minerals range from n = 2.2–2.7, subadamantine phases have n = 1.8–2.2, and vitreous minerals range from 1.5–1.8. This calibration has the advantage of attaching a discrete, quantifiable value to the otherwise qualitative terms used for luster. See Chapter 5 for a more detailed discussion of the relationship between reflectance and refractive index.

Dull samples do not reflect much light (i.e., they scatter light in many different directions) because they are very fine-grained and porous (Figure 2.16). Minerals may sometimes be described as **earthy** if they resemble dirt or clay

(Figure 2.17). The clay minerals illite and montmorillonite are usually characterized as earthy.

Greasy luster resembles a thin coating of oil on the surface of a mineral (Figure 2.18); milky quartz and nepheline often have this appearance on fracture surfaces.

Pearly minerals have the appearance of the surface of a pearl (thus the name). The iridescent surfaces of micas (Figure 2.19) and the base of apophyllite, often have this type of luster.

Resinous, pitchy, or waxy minerals resemble the surface of a wax candle (Figure 2.20) or a piece of tree resin or pitch (such as pine sap). This type of luster is typical of most forms of sphalerite. **Subresinous** minerals (e.g., vanadinite) are more opaque.

Silky luster resembles the shine of a piece of silk (Figure 2.21), **satiny** luster mimics satin, **felty** looks like felt, and **velvety** luster looks like velvet. Any distinctions among these terms are purely subjective! They are all characteristic of minerals that tend to form in aggregates of fibers, such as gypsum (especially the variety called satin spar) and asbestos.

Vitreous or **glassy** minerals look like the surface of a piece of broken glass (Figure 2.22). Quartz and the majority of other nonmetallic minerals (about 70% of all minerals, in fact) have glassy luster. **Subvitreous** minerals (e.g., tephroite) look glassy but somewhat coarser or cloudier.

Hardness

A mineral's hardness is defined as its resistance to scratching. The most common quantifier of this property is **Mohs Hardness scale (H)** of ten minerals, which was conceived in 1812 by Friedrich Mohs; this is the preferred scale for hardness (Table 2.3). For completeness, however, it is necessary to mention some other scales devised for this measurement. The **technical scale** of 15 minerals, as given below, is often used in the materials industry, where there is a need to relate hardness to commonly manufactured materials. The **Mohs-Woodel scale**, introduced by Charles Woodel in 1935, is the same as the Mohs scale except that dia-

Figure 2.14. *Adamantine minerals are extremely shiny (brilliant) like this cassiterite; they reflect light very strongly.*

Figure 2.15. *The term splendent, which means that a mineral has the brightest possible luster, is well represented by this specimen of the mineral galena.*

Figure 2.16. *Dull samples do not reflect much light (i.e., they scatter light in different directions) like this sample of kaolinite.*

Figure 2.17. *Minerals may sometimes be described as earthy if they resemble dirt or clay like this piece of carnotite.*

Figure 2.18. *Greasy luster resembles a thin coating of oil on the surface of a mineral, as seen on this sample of talc.*

Figure 2.19. *Pearly minerals have the appearance of the surface of a pearl, as seen on this cleavage fragment of mica.*

Figure 2.20. Resinous, pitchy, or waxy minerals like this sample of smithsonite resemble the surface of a wax candle or a piece of tree resin or pitch.

Figure 2.21. This sample of gypsum has silky luster; it resembles the shine of a piece of silk.

Figure 2.22. Vitreous or glassy minerals like this amblygonite resemble the surface of a piece of broken glass.

mond is assigned a hardness of 42.5. This recognizes the fact that the difference in hardness between, say, topaz and corundum is much smaller than the difference between corundum and diamond. The **Knoop scale** was devised by the National Institute for Standards and Technology (then called the National Bureau of Standards). It uses a measure of the actual force required to make a permanent indentation in a polished surface. Obviously this is a very sensitive measurement, and it can even be used to distinguish between the hardnesses of different faces on a gemstone. Finally, another modification was suggested in 1963 by Povarennykh in recognition of the fact that some minerals on the Mohs scale have different hardnesses in different directions; this system indicates which crystallographic directions are to be used in assessing hardness (this terminology is covered in Chapter 12).

Several qualitative relationships between hardness, mineral composition, and structure are known. All of them reflect the strength and type of bonds in the mineral structure (Chapter 8). In general (and there are numerous exceptions to these generalizations), native metals such as gold, copper, silver, and lead, are relatively soft (Mohs hardnesses <3), other non-silicates and hydrous minerals tend to have hardnesses <5, and oxides and anhydrous silicates tend to be the hardest (H>5.5). Minerals made up of small atoms (especially highly-charged cations) and/or densely packed atoms tend to be harder because they have short, strong bonds. Hardness in general tends to correlate with bond density. Accordingly, some minerals (notably calcite and kyanite) have different hardnesses along different crystallographic directions.

The hardness scales range from soft to hard on a logarithmic scale, as shown in Table 2.3. The property of hardness is easily measured using a "test kit" composed of small pieces of each of the minerals on the scale (diamond is usually omitted, for obvious reasons!). Irregularly-shaped pieces of the

Table 2.3. Scales of Hardness

	Mohs scale	*Technical scale*	*Mohs-Woodel scale*	*Knoop scale*	*Povarennykh scale*
softest	1. talc	1. talc	1. talc	—	1. talc [001]
	2. gypsum	2. gypsum	2. gypsum	32	2. halite [100]
	3. calcite	3. calcite	3. calcite	135	3. galena [100]
	4. fluorite	4. fluorospar	4. fluorite	163	4. fluorite [111]
	5. apatite	5. apatite	5. apatite	430	5. scheelite 111]
	6. orthoclase	6. orthoclase	6. orthoclase	560	6. magnetite [111]
	7. quartz	7. pure silica glass	7. quartz	820	7. quartz [$10\bar{1}0$]
	8. topaz	8. quartz	8. topaz	1340	8. topaz [001]
	9. corundum	9. topaz	9. corundum	2100	9. corundum [$1\bar{1}20$]
	10. diamond	10. garnet	42.5 diamond	7000	10. TiC
		11. fused zircon			11. boron nitride
		12. corundum			12. B_4C
		13. silicon carbide			13. $B_{12}C_2$
		14. boron carbide			14. carbon diamond
hardest		15. diamond			15. bort diamond[111]

standard minerals are best, because it is desirable to use a corner or sharp edge of the standard for the test. Each standard mineral is scratched across the surface of an unknown, and the resultant powder is then examined to determined if the standard has scratched the unknown (implying $hardness_{standard} > hardness_{unknown}$) or vice versa (Figure 2.23). A hand lens or magnifying glass may prove useful in assessing the results of this test!

As a "field test", other commonly-occurring materials can be used to test hardness. These include your fingernail, coins (copper pennies work the best), pieces of glass, keys, a knife blade, and safety pins. If you have access to a Mohs kit, you can calibrate the things you commonly carry with you to determine their hardness. Note that similar materials will sometimes have different hardnesses (for example, your fingernail's hardness depends on what you eat!), so customization of your personal hardness kit is necessary! Simple tests for mineral hardness are given in Table 2.4.

One other term is sometimes used to describe the texture of a mineral and is related to its hardness. **Unctuous** or **soapy** minerals have the slippery feeling of a bar of soap. This texture is typical of minerals like talc.

Figure 2.23. *Hardness is determined by scratching an unknown mineral with a series of materials of known hardnesses. The hardnesses of two minerals can be compared by scratching an edge or corner of one mineral against another. The mineral that makes a groove in the other mineral is harder than the one being scratched.*

Fracture and Tenacity

The terms tenacity and fracture both refer to different aspects of what happens when you try to break a mineral into pieces: tenacity describes how difficult it is to make the material break, and fracture describes the surface that results when you do break it.

Tenacity (also called **tensile strength**) is the property of a substance to resist separation—its so-called "cohesiveness." It is related to hardness because it describes the way a mineral breaks when it is stressed or deformed (as when, for example, you scratch it with something from your test kit). It is also influenced by the crystallinity of the material: polycrystalline masses with interlocking crystals have generally higher tenacity than single crystals because they do not have extended surfaces on which cleavage can occur. The term toughness is also often used interchangeably with tenacity, though it implies a slightly different property. Specifically, **toughness** refers to the amount of energy that can be absorbed by a material before it fractures. Some minerals are so tough that they are almost impossible to break (for people who do field work, these types of minerals can prove to be very frustrating, because they are difficult to break off and bring home for further study!). Chalcedony (a variety of quartz) and rhodonite are extremely tough minerals. Several terms are used to describe tenacity, including:

Brittle or **fragile** minerals will break into pieces (Figure 2.24) or form powders under stress (i.e., when you hit them with a hammer). Most commonly-occurring rock-forming minerals are

Table 2.4. Simple Hardness Tests (Apply with Caution!)

Mohs Hardness Scale	*Criteria*
1	rubs off onto skin in tiny flakes, easily scratched by a fingernail
2	easily scratched by fingernail
3	scratched by nail, knife, or copper coin may be scratched by fingernail of some people
4	easily scratched by a nail or knife (never by a fingernail)
5	scratched by a nail or knife with pressure applied
6	NOT scratched by a knife but will scratch typical window glass
7	scratches window glass (but not most kitchen ceramics made from glass) can be scratched by topaz, corundum, or diamond (but don't try the latter)
8–10	difficult to distinguish except with diamond or corundum for scratch testing (try diamond sandpaper!)

brittle, because they are held together by mostly ionic bonding (see Chapter 8 to learn more about bonding). The term **friable** is also used for fragile minerals that disintegrate into fine grains under slight pressure.

Ductile minerals can be shaped or drawn out into wires (Figure 2.25). This property only occurs in native metals such as gold, silver, and copper because they are held together by metallic bonds.

Malleable materials can be pounded into a sheet with a hammer (Figure 2.26). Generally this a property associated only with native metals and metallic bonding (Chapters 3, 8). A few silver-containing minerals such as acanthite may also be slightly malleable. Because gold is malleable, it can be made into thin sheets that are used to cover objects in gold leaf (such as picture frames, religious objects, and paintings from the Middle Ages).

Sectile materials can be cut with a knife, or cut into thin shavings like chocolate curls (Figure 2.27). They are not completely malleable, however, and may break easily if hit with a hammer. These materials are gradational between the malleable metals and most brittle minerals. Minerals possessing this property include chlorargyrite, gypsum and chalcocite.

Flexible minerals can be bent without breaking but will not return to their original shapes after bending (Figure 2.28). Examples include stibnite, talc, and chlorite.

Elastic minerals can be bent but will return to their original shapes afterwards (Figure 2.29). This property is characteristic of both muscovite and biotite, and allowed them to be used in the manufacture of lampshades (and other household objects!) in previous centuries. **Inelastic** minerals will bend slightly and then break.

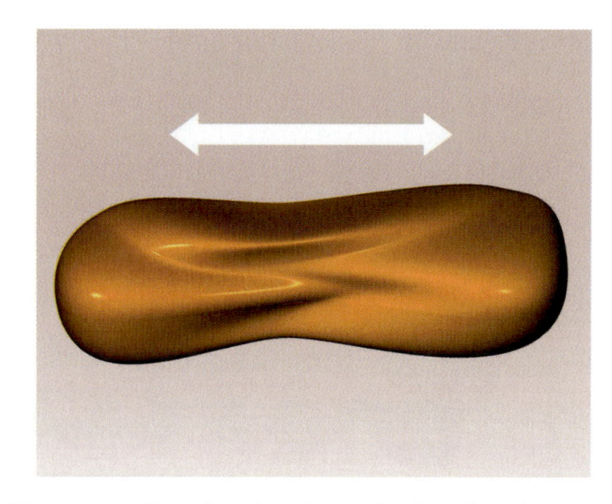

Figure 2.25. Ductile minerals can be shaped or drawn out into wires.

A comparison of brittle vs. ductile behavior can be observed during a scratch test: if the material excavated from the unknown mineral flies out into your face, then the mineral is brittle. If the scratched material simply flows or moves out of the way (as when you scratch a wax candle), then the material is behaving in a ductile fashion.

Fracture is formally defined as the texture that results when a mineral is broken in any direction *other* than along planes of cleavage. Terms used to describe it include:

Even fracture refers to a fracture surface that is relatively smooth and, in most cases, planar. Even fracture planes are usually cleavage planes.

Uneven fracture describes a breakage surface that is rough and uneven. The features of the breakage surface can be further described as:

Conchoidal fracture is fracturing along rounded surfaces marked by smooth, curving lines like

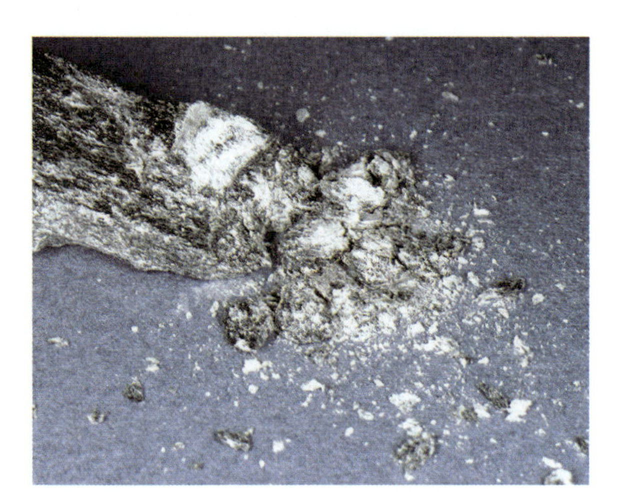

Figure 2.24. Brittle or fragile minerals will break into pieces or form powders under stress.

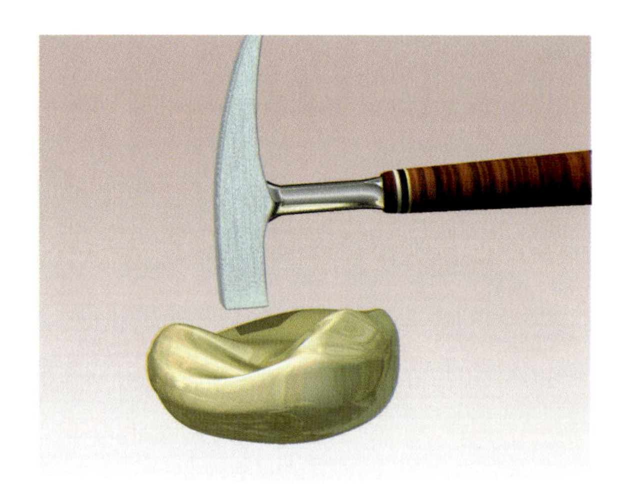

Figure 2.26. Malleable materials can be pounded into a sheet with a hammer.

Figure 2.27. Sectile materials can be cut with a knife or cut into thin shavings like chocolate curls.

Figure 2.29. Elastic minerals like many of the micas can be bent but will return to their original shape.

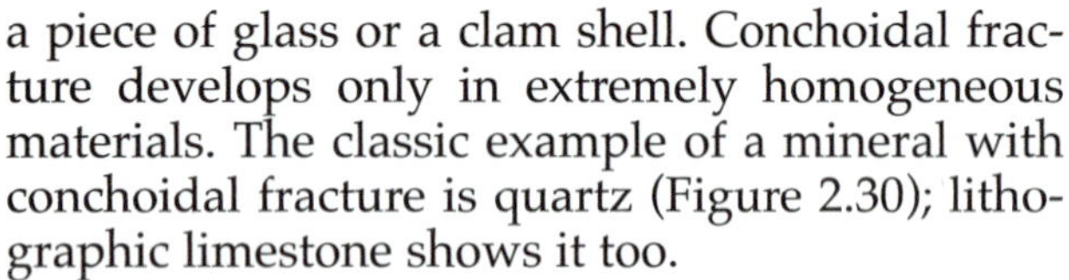

a piece of glass or a clam shell. Conchoidal fracture develops only in extremely homogeneous materials. The classic example of a mineral with conchoidal fracture is quartz (Figure 2.30); lithographic limestone shows it too.

Subconchoidal fracture is nearly conchoidal, with a rougher surface to the fracture plane. The minerals nepheline, petalite, and topaz exhibit subconchoidal fracture.

Fibrous fracture looks like the broken end of a frayed piece of rope (Figure 2.31). Minerals that crystallize in long, slender crystals such as brucite and asbestos (crocidolite and chrysotile) usually exhibit fibrous fracture.

Splintery fracture closely resembles fibrous fracture, except that the breakage surface more closely resembles the broken ends of a tree limb or piece of lumber. The minerals kyanite, boulangerite, pectolite, and bismuthinite may show splintery fracture.

Hackly fracture, commonly found on the breakage surfaces of copper and other metals, is breakage along a rough, jagged surface (Figure 2.32). The edge of your muffler pipe when it rusts and falls off your car will show hackly fracture.

Crystal Form

A **crystal form** is a group of similar faces on a crystal related by some type of symmetry, all of which have identical physical and chemical properties. In mineralogy, this is a very specific term, and it should not be used to describe the general shape of a crystal despite its general usage in the vernacular (instead, "habit" or "shape" are more appropriate). Many people confuse the terms "form," "fracture," "cleavage," and "habit." The terms are distinct because (1) form is the specific

Figure 2.28. Flexible minerals like stibnite can be bent by gliding twinning.

Figure 2.30. Conchoidal fracture of quartz shows smooth, curving lines like a piece of glass or a clam shell.

Figure 2.31. Fibrous fracture looks like the broken end of a frayed piece of rope.

Figure 2.32. Hackly fracture is breakage along a rough, jagged surface as seen on this piece of copper.

term used to describe the external crystal shape that reflects the internal arrangement of atoms, (2) fracture and cleavage describe surfaces along which a crystal has been *broken*, and (3) habit refers to the shape of a crystal without reference to atomic or internal structure. Try to remember that all minerals have a crystal form (though they may not always show it in hand sample, except in museum quality specimens), but not all minerals exhibit cleavage or even fracture.

Crystal forms are described using a specific set of vocabulary that can be found in older reference books; the terminology is rarely-used, so we won't provide details here. However, a few related terms are useful in the context of hand sample identification.

If a specimen exhibits crystal faces that show off its form, we say it is **euhedral**, or possessing a distinctive shape. In cases where opposite ends of the crystal have different forms, the

*Figure 2.33. **a.** Euhedral crystals like this quartz possess a distinctive shape, which may display a shape or form that is made up of faces related by symmetry. **b.** If opposite ends of the crystal have different forms, the crystal is hemimorphic like this wurtzite (photo by John A. Jaszczak). **c.** Samples that lack recognizable faces are anhedral.*

crystal is termed **hemimorphic**. If a sample has a few recognizable faces, we define it as **euhedral** (Figure 2.33); if there are no crystal faces, it is **anhedral**.

Crystal System

The **unit cell**, or fundamental atomic form, of any mineral may have a diagnostic shape that corresponds to one of the six crystal systems; the geometry of the macroscopic crystal then mimics that of the submicroscopic unit cell. As noted in the previous chapter, these classes are defined on the basis of a set of imaginary crystallographic axes. The relative axis lengths of the shapes, as well as the angles between those axes, describe each crystal class uniquely (Table 2.5; Figure 2.34).

Crystal Shape and Habit

The overall shape of a crystal is called its **habit**, and is described by terms that refer to crystal form and regular development such as tabular, prismatic, equant, etc. Crystal aggregates may also be distinctive, with names like granular and plumose, etc. Strictly speaking, the latter terms are not "habits," but they are grouped together here for convenience. There are a plethora of different useful terms and not all of them are included here by any means (Figure 2.35 to 2.81). Table 2.6 contains some of the common terms.

Table 2.5. Crystal Systems

	Relative lengths of axes	*Angles between axes*
isometric	$a = b = c$	all 90°
tetragonal	$a = b \neq c$	all 90°
hexagonal	$a = b \neq c$	$\alpha = \beta = 90°, \gamma = 120°$
orthorhombic	$a \neq b \neq c$	all 90°
monoclinic	$a \neq b \neq c$	$\alpha = \gamma = 90°, \beta \neq 90°$
triclinic	$a \neq b \neq c$	$\alpha \neq \beta \neq \gamma$

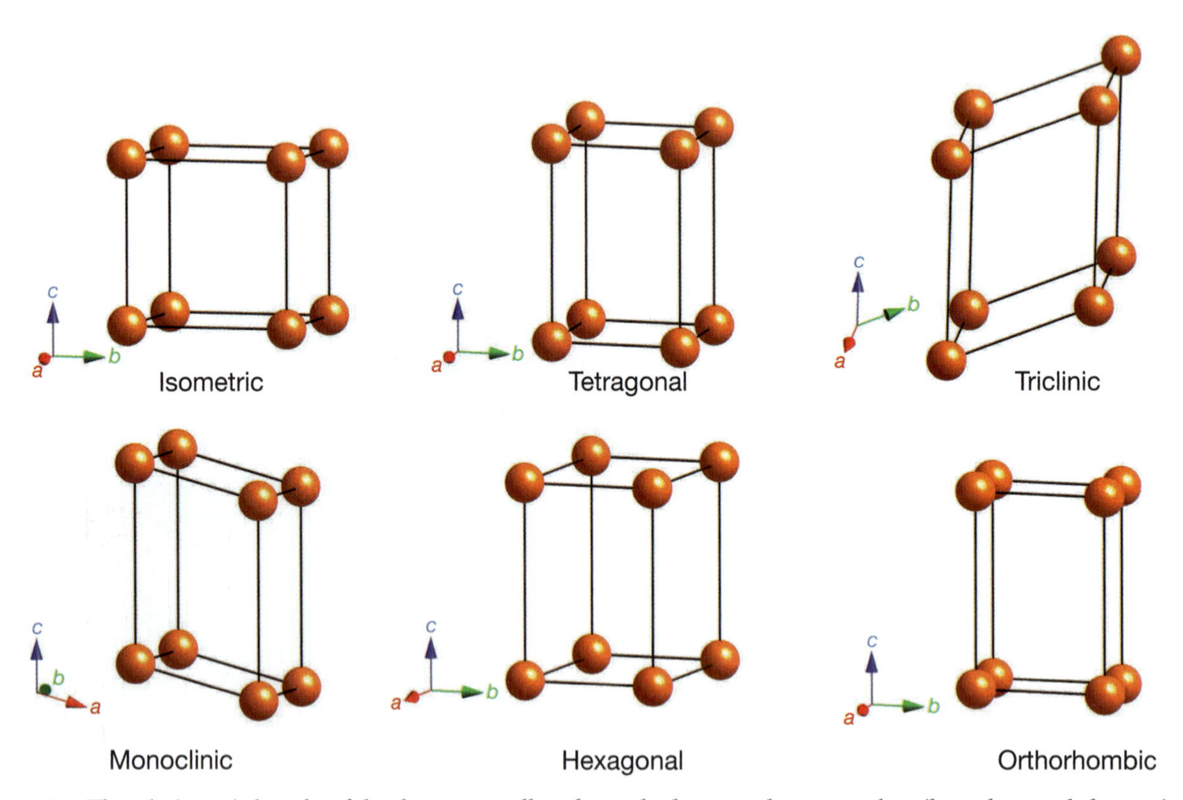

Figure 2.34. The relative axis lengths of the shapes, as well as the angles between those axes, describe each crystal class uniquely.

Figure 2.35. *Acicular boulangerite.*

Figure 2.36. *Arborescent copper.*

Figure 2.37. *Asbestiform chrysotile.*

Figure 2.38. *Banded agate.*

Figure 2.39. *Colorless blades of scolecite.*

Figure 2.40. *Blocky orthoclase.*

Figure 2.41. *Botryoidal hematite.*

Figure 2.42. *Capillary millerite.*

Figure 2.43. *Colloform graphite spherules.*

Figure 2.44. *Columnar elbaite.*

Figure 2.45. *Ironstone concretion permeated by opal. Photo by Russell G. Rizzo.*

Figure 2.46. *Coralloid or helicitic aragonite. Photo by Jorge M. Alves.*

Figure 2.47. *Cryptocrystalline chalcedony.*

Figure 2.48. *Dendritic manganese oxide (formerly called pyrolusite).*

Figure 2.49. *Divergent or radiated manganite.*

Figure 2.50. *Felted crystals of drusy aurichalcite with larger calcite crystals on the outer edges.*

Figure 2.51. *Equant cobaltite pyritohedron.*

Figure 2.52. *Featherlike or plumose silver.*

Figure 2.53. *Fibrous brucite.*

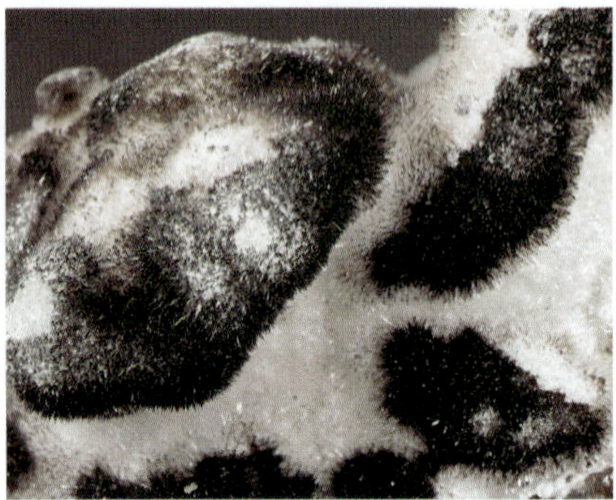

Figure 2.54. *Filiform hydrozincite.*

Figure 2.55. *Foliated lepidolite.*

Figure 2.56. *Geniculated rutile.*

Figure 2.57. *Amethyst geode.*

Figure 2.58. *Globular smithsonite.*

Figure 2.59. *Granular or sugary cryolite.*

Figure 2.60. *Gwindle or spiral, quartz.*

Figure 2.61. *Lamellar brucite.*

Figure 2.62. *Laminated clinochlore.*

Figure 2.63. *Massive graphite.*

Figure 2.64. *Micaceous biotite.*

Figure 2.65. *Pisolitic siderite.*

Figure 2.66. *Platy magnesite (mesitite variety).*

Figure 2.67. *Powdery vivianite.*

Figure 2.68. *Prismatic epidote.*

Figure 2.69. *Radiating spherulitic or spherulitic wavellite.*

Figure 2.70. *Reniform (kidney-shaped) hematite.*

Figure 2.71. *Reticulated taenite and kamacite.*

Figure 2.72. *Barite rose.*

Figure 2.73. *Scaly celadonite.*

Figure 2.74. *Stalactitic calcite.*

Figure 2.75. *Stalagmitic calcite.*

Figure 2.76. *Stellate muscovite twins.*

Figure 2.77. *Striated arsenopyrite.*

Figure 2.78. *Stubby manganotantalite.*

Figure 2.79. *Tabular albite.*

Figure 2.80. *Tufted rosasite.*

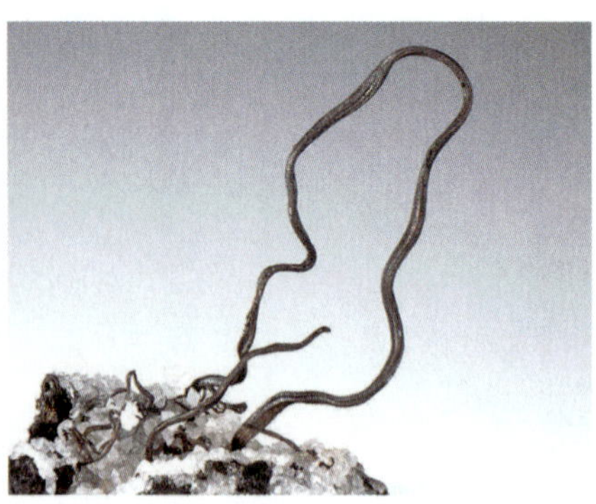

Figure 2.81. *Wiry silver with acanthite.*

Table 2.6. Mineral Habits

Term	*Description*	*Example*	*Figure Number*
acicular	thin, needle-shaped crystals (capillary)	epidote, natrolite, boulangerite	2.35
arborescent	long, thin branching crystals that form tree-like, three-dimensional growth pattern	ice, silver, copper	2.36
asbestiform	long, thin fibers like asbestos that separate easily	chrysotile, grunerite	2.37
banded	stripes or bands of different color or texture	agate	2.38
bladed	long, flat crystals shaped like the blade of a knife	stibnite, kyanite, scolecite	2.39
blocky	massive, block or brick-shaped	orthoclase	2.40
botryoidal	resembling bunches of grapes	smithsonite, conichalcite, hematite	2.41
capillary	very thin, delicate, hair-like crystals (acicular)	millerite	2.42
colloform	general term for crystals with spherical groups of any size; more specific terms include botryoidal, globular, and reniform habits	hemimorphite, graphite	2.43
columnar	crystal faces with linear intersections between them, resulting in a column or prism shape	beryl, elbaite, dravite, schorl	2.44
concretion	rounded layers around a small center, resulting in an onion-like growth pattern	ironstone	2.45
coralloid	twisted, curved branch-like shapes resembling coral	aragonite	2.46
cryptocrystalline	crystals smaller than can be seen with the human eye	variscite, spodumene, chalcedony	2.47
dendritic	plant-like or moss-like growth patterns, more two-dimensional than arborescent habits	copper, acanthite	2.48
divergent	radiating groups of crystals	mesolite, manganite	2.49
drusy	a coating of small crystals	calcite, stilpnomelane, aurichalcite, quartz	2.50
equant	crystals with equal size in all dimensions	garnet, zircon, anhydrite, cobaltite	2.51
featherlike	overlapping fine scales resembling a feather (plumose)	galena, descloizite, silver	2.52
fibrous	bundles of thin fibers in either parallel or radiating groups	chrysotile, brucite, strontianite	2.53
filiform	thin, thread-like crystals	rutile, hydrozincite	2.54
foliated	general term for crystals forming thin, easily separated sheets or plates; including leaflike, laminar, and micaceous	biotite, lepidolite, micas	2.55
geniculated	knee-like crystals	chalcocite, rutile	2.56
geode	spherical hollow structure lined with small crystals	amethyst	2.57
globular	radiating individuals form small globes or spheres	pyromorphite, smithsonite	2.58
granular	all crystals roughly equal in size, resembling granulated sugar	celestine, monazite, cryolite	2.59
gwindle	growing in a spiral shape	quartz	2.60
helicitic	same as coralloid	aragonite	2.46
lamellar	tabular, flat, platelike crystals stacked on each other and resembling a book shape	gypsum, barite, tilleyite, brucite	2.61
laminated	forming thin cleavable sheets	biotite, clinochlore	2.62

Cleavage and Parting

Cleavage is the property of breaking along crystallographically-controlled flat planes. It is described by the number of resultant planes and by specifying the angles between them (Figure 2.82):

Basal or **pinacoidal** cleavage is perfect in exactly one direction, as found in micas (Figure 2.83). This is also sometimes called **platy** cleavage.

Prismatic cleavage describes two intersecting cleavage planes, which may be at exactly 90° (wernerite) or 120° (amphibole; Figure 2.84).

Table 2.6. (continued)			
Term	Description	Example	Figure Number
massive	no distinctive shape or other characteristics	purpurite, iron, graphite	2.63
micaceous	very thin sheets that are easily separated	muscovite, biotite	2.64
pisolitic	pea-size spherical aggregates	siderite, bauxite	2.65
platy	forming thin rounded plates	diaspore, lepidocrocite, magnesite	2.66
plumose	same as featherlike	galena, descloizite, silver	2.52
powdery	tiny, powder-like crystals	vivianite	2.67
prismatic	long, slender crystals with parallel faces forming in a column or prism shape	proustite, natrolite, crocoite	2.68
radiated	same as divergent	mesolite, manganite	2.49
radiating spherulitic	slender crystals in rounded masses	wavellite, gibbsite	2.69
reniform	kidney-shaped crystals or groups of crystals	arsenic, sulfur, hematite	2.70
reticulated	net-like lattice pattern of crisscrossing crystals	silver, bismuth, taenite, kamacite	2.71
rosette	shaped like a rose	hematite, barite, gypsum	2.72
scaly	forming thin overlapping plates or scales	aurichalcite, celadonite	2.73
spherulitic	same as radiating spherulitic	wavellite, gibbsite	2.69
spiral	same as gwindle	quartz	2.60
stalactitic	cylinders or cones that grow with the pointed end down, as when forming on a cave ceiling	calcite, arsenic	2.74
stalagmitic	cylinders or cones that grow with the pointed end up, as when growing from a cave floor	calcite	2.75
stellate	radiating star or circle shapes	muscovite, astrophyllite	2.76
striated	crystal faces with fine, parallel lines	pyrite, arsenopyrite, gahnite	2.77
stubby	short, fat prisms	manganotantalite, borax, benitoite	2.78
sugary	same as granular	celestine, monazite	2.59
tabular	shaped like tablets, or flat, rectangles	cerussite, barite, gypsum, albite	2.79
tufted	forming small hair-like tufts	hydromagnesite, rosasite	2.80
wiry	forming thin wires	silver	2.81

Blocky cleavage occurs when a crystal is completely bounded by planar surfaces. **Isometric** or **cubic** cleavage is breakage along three mutually perpendicular planes (i.e., at perfect 90° angles; this is found in the species halite and galena). **Rhombohedral** cleavage, which occurs in calcite and rhodochrosite, is also characterized by fracture along three planes, but two of the three are at 120° angles to each other. **Octahedral** cleavage, observed in the minerals fluorite and diamond, occurs along four planes in a crystal structure. **Dodecahedral** cleavage occurs along six planes, and is found in sphalerite.

A special type of protractor called a contact goniometer can be used to measure the angles between crystal faces with some degree of accuracy (Figure 2.85).

Parting is a related term that refers to planes of weakness in a crystal structure (Figure 2.86). It is sometimes called "false cleavage." However, parting is caused by some kind of stress to the crystal, such as incipient planes of alteration, externally-imposed deformation, twinning (see below), or exsolution (re-equilibration of a single stable phase into two co-existing phases as a result of changing external conditions such as pressure or temperature). Parting is often observed in corundum, hematite, pyroxene, and magnetite. It is distinguished from cleavage by the fact that only some minerals exhibit partings, while all minerals can show cleavage.

Twinning

Twinning is a symmetrical intergrowth of two or more single crystals of the same mineral. The geometry of the pattern of intergrowth is restricted by symmetry. Its occurrence is controlled by the twin law, which is a symmetry operator that occurs in addition to point group symmetry (to be described in Chapters 11 and 12). The different parts can be related by (1) a *mirror* (reflection) *plane,* such that each and every atom in the crystal also exists an equal and opposite distance across and through an imaginary mirror plane; (2) an *axis of rotation* (see Chapters 4 and 11–12), or (3) a common center that is not related to the symmetry of the rest of that crystal. In a twinned crystal, some layers of atoms may be parallel to each other, and others are in a reversed position. Twinning may be caused by processes that occur

Number of Cleavage Directions	Description	Sketch	Illustration of Cleavage Directions
0 No cleavage, only fracture	No cleavage Irregular masses with no flat surfaces		None
1	Basal cleavage "Books" that split apart along flat sheets		
2 at 90°	Prismatic cleavage Elongated form with rectangular cross sections (prisms) and parts of such forms		
2 not at 90°	Prismatic cleavage Elongated form with parallelogram cross sections (prisms) and parts of such forms		
3 at 90°	Cubic cleavage Shapes made of cubes and parts of cubes		
3 not at 90°	Rhombohedral cleavage Shapes made of rhombohedra and parts of rhombohedra	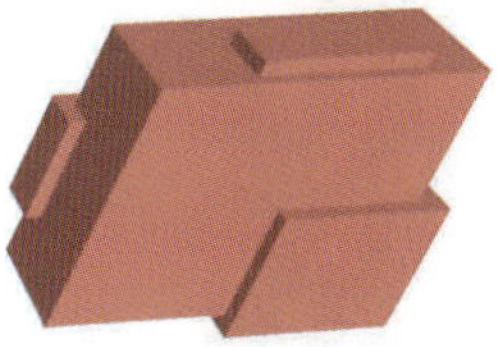	
4	Octahedral cleavage Shapes made of octahedra and parts of octahedra		
6	Dodecahedral cleavage Shapes made of dodecahedra and parts of dodecahedra		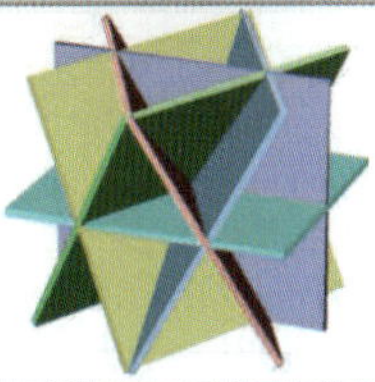

Figure 2.82. *Mineral cleavage can occur in many different directions and different angles.*

during growth or after growth. The latter case may involve either transformation (which is a change in mineral symmetry) or crystal gliding (a type of rock deformation).

Twinning in specific shapes or geometries is characteristic of some mineral species. Different types of twins include:

Contact twins, also called **juxtaposition twins** or **swallowtail twins**, are formed by two crystals growing symmetrically across a twin plane (Figure 2.87). The resultant pair of crystals shares a common base and resemble a bird's tail (thus the name). Contact twins are exhibited by the mineral species such as spinel, calcite, manganotantalite and gypsum.

Reticulated twins form a network of interlocking crystals like netting; this pattern is commonly found in the minerals hambergite and rutile.

Geniculated ("elbow") twins are formed when the initiation of twinning causes a crystal to bend abruptly in the middle; these "knee-like" twins may be observed in crystals of the mineral rutile (Figure 2.56).

Interpenetrating or **penetration** twins form crosses that look as though one crystal has grown right through the other one. Penetration twins are common in fluorite and microcline. When the twins form a cross shape with crystals at right angles to each other (as in the mineral staurolite), the terms "Greek cross" and "St. Andrew's cross" are often used (Figure 2.88). Pyrite and arsenopyrite may also form crosses.

Cyclic or **repeated twins** result from repeated twinning around a three, four, five, six, or eight-fold rotation axis (Figure 2.89), resulting in a pat-

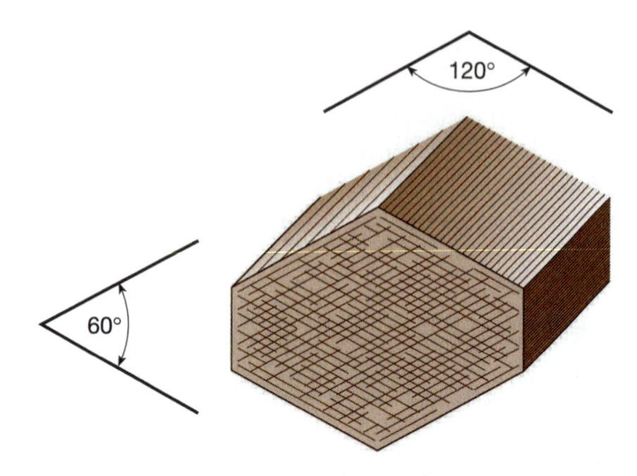

Figure 2.84. Another diagnostic cleavage occurs in the minerals of the amphibole group, which show cleavages at 60° and 120° angles.

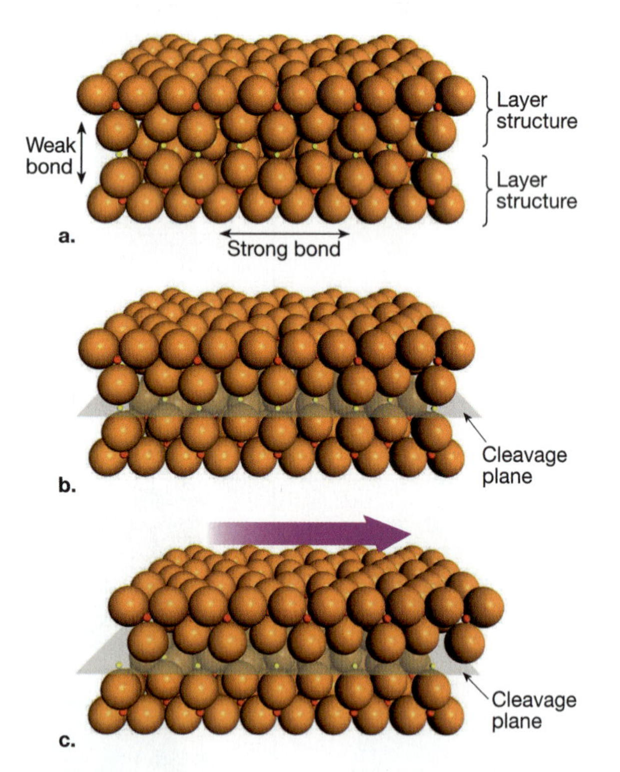

*Figure 2.83. **a.** Basal, or one-directional, cleavage occurs in mica group minerals... **b.** ...because there is only weak bonding between the structural layers of the mica structure. **c.** Movement, or cleavage, can occur easily along those planes of weak bonds.*

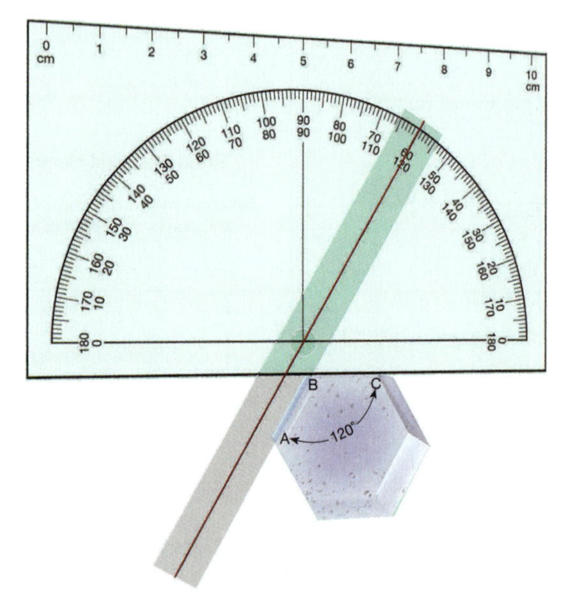

Figure 2.85. A simple instrument called a contact goniometer is used to measure the angles between crystal faces. It is a combination of a 180° protractor with an extended arm to be placed against a crystal face.

Figure 2.86. Parting is caused by mechanical deformation, or stress applied to the crystal, and results in planes of breakage like those shown here in ruby. Photo by Manfred Kampf.

tern called a threeling, fourling, fiveling, sixling (common in aragonite), or eightling, respectively. **Trilling** is the special case of cyclic twins that results in intergrowth of three individual crystals (sometimes observed in witherite, chrysoberyl, and aragonite); these twins are also called trills.

Multiple and **polysynthetic twins** are similar to cyclic twinning, and are characterized by repeated twinning along parallel planes, usually on a microscopic scale. The macroscopic manifestation is often the presence of striations on the surface of crystal faces (as in albite); is also found in arsenopyrite (Figure 2.77).

Tartan twinning is a microscopic intergrowth of albite and pericline that is found in the mineral microcline (Figure 2.90).

Specific Gravity

Specific gravity is the term most often used to describe the density of a mineral. Although it is numerically the same as density (or *d*, which is mass per unit of volume, usually g/cm^3), specific gravity (*s.g.*, *S.G.*, or sometimes just *G*) is defined on the basis of the amount of water it displaces. It is a more convenient unit of measure for minerals because naturally-occurring samples rarely have convenient dimensions that allow easy measurement of their volume!

An instrument called a **pycnometer**, which is a glass ball that can be filled with water and then weighed before and after the mineral is added to it, is often used to measure specific gravity. Specific gravity is defined as:

$$\frac{M}{P + M - P'}$$

where M is the mass, or weight of the material (mineral) of interest, P is the weight of the pycnometer filled exactly to a reference line with distilled water, and P' is the weight of the pycnometer after the mineral sample has been added and enough water has been removed so that it is again filled precisely to the reference line. Essentially this is the weight of the sample in air divided by the weight in air minus the weight in water! Note that this definition formally assumes that distilled water is used, and that the experiment is being performed at a temperature of 4°C (39.2°F), which is the temperature at which water has maximum density. However, most workers make these measurements at room temperature, noting the additional small error that results.

Other instruments used for this measurement include a specific gravity bottle and a hydrostatic, or **Jolly balance** (Figure 2.91). This balance uses the weights of the sample in air and in water to determine specific gravity:

$$\frac{\text{weight of a sample in air}}{\text{weight of an equal volume of water}}$$

Intuitively, specific gravity is easy to understand. Pick up your unknown mineral sample and heft it around a little. Try to envision how much heavier the rock is than an equivalent volume of water! If it seems "light" in weight, it has a low specific gravity. Good common examples of low *s.g.* minerals are quartz (*s.g.* = 2.65) and halite (2.1–2.2). Both of these are more that twice as heavy as the equivalent volume of water they might displace. Silver (10–12), gold (15.3–19.3), and mercury (13.6) have large (i.e., heavy) specific gravities. They are a *lot* heavier that an equivalent volume of water.

Figure 2.87. Contact or swallowtail twins, are common in the mineral phase chrysoberyl, shown here.

Minerals with metallic or adamantine lusters tend to have high specific gravities, while dull or non-metallic minerals are usually much lower. Minerals with high specific gravities tend to be composed of atoms with high mass (those that occur high up on the periodic chart) and strong bonds (see Chapters 7 and 8).

Other Properties

Occasionally some other properties come into play for mineral identification. Perhaps the most obvious of these is the sense of **taste**; some minerals taste salty (halite), salty/bitter (like sylvite, which is used as a salt substitute), or almost sweet (alum). **Smell** can also be an important tool in mineral identification. Limestone (a rock type composed of calcite) may often host small amounts of sulfur compounds, which give off a characteristic rotten egg smell when samples are fractured. Some types of coal minerals have a petroleum smell to them, and arsenic-bearing minerals like arsenopyrite release a garlic smell when hit with a hammer.

The **acid test** is used to determine whether or not a mineral contains carbonate, or $CaCO_3$, in its composition. In the field, this test is usually done by applying a few drops of very dilute hydrochloric acid, or HCl, to the surface of a mineral (note from experience: a leaky bottle of even very dilute HCl in your pocket will quickly eat a hole through your blue jeans; use with caution!). The HCl will react by fizzing wildly if $CaCO_3$ is present.

Magnetic properties are significant for only a few mineral species, but the terminology bears mention here because these terms are often found in reference volumes on hand sample identification. All of them are defined on the basis of the elements they contain (transition metals, normally iron, must be present for magnetism to occur; see Chapter 23), but simple explanations can be used to differentiate them.

Figure 2.89. Cyclic twinning is shown here in chrysoberyl.

Diamagnetic minerals have no attraction to a magnet, and may in fact be repelled very slightly by a magnet if they contain small amounts of transition metals.

Paramagnetic minerals such as diopside and gedrite contain transition metals, but spins of their electrons are randomly oriented in the crystal structures, and there is no tendency for the electrons in those atoms to become aligned. Paramagnetic minerals may sometimes show a faint attraction to a magnet.

Figure 2.88. Penetration twins are commonly formed by the mineral staurolite; the two crystals look as though they have grown right through each other.

Figure 2.90. Microcline displays a plaid or "tartan" twinning pattern formed by intersecting crystals of albite and pericline, as shown here at microscopic scale.

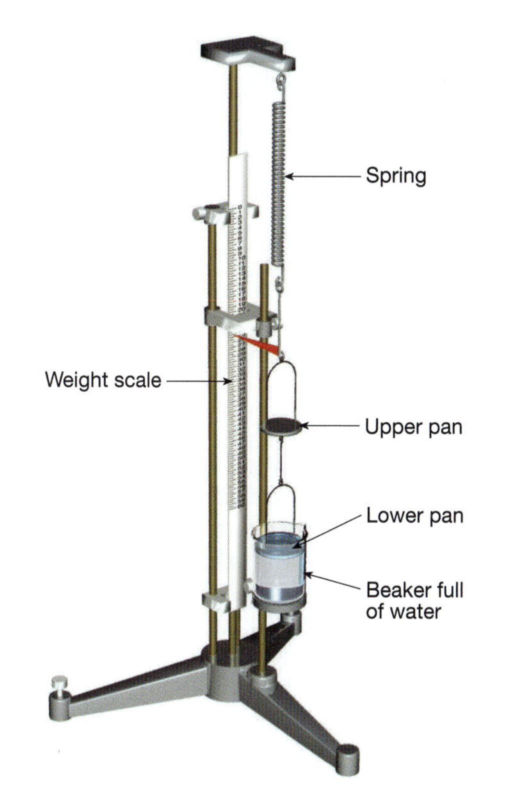

Figure 2.91. *A Jolly balance is a special type of scale that is used to measure specific gravity.*

Antiferromagnetism occurs in minerals such as hematite and goethite that contain transition metals, in which the electrons are all spinning together in pairs in each electron orbital: one electron spins in a clock-wise direction, and the other electron spins in a counterclockwise direction. Because the electrons are paired in this way, these materials are not attracted to a magnet.

Ferrimagnetism occurs in those rare situations when two sets of magnetic fields are created within the same crystal structure. In magnetite, for example, one-third of the unpaired Fe electrons spin in one direction, and two-thirds of the unpaired Fe electrons spin in the opposite direction, creating a net magnetic moment. Ferrimagnetism may also occur in impure forms of the species pyrrhotite, and maghemite, as well as in hematite. **Ferromagnetism**, in which all the unpaired electrons spin in the same direction, occurs rarely in kamacite and in iron-rich platinum.

Electrical properties are rarely employed in hand sample mineral identification, but are mentioned here in the interest of thoroughness. Their most important use is in the discrimination between cubic zirconia and diamond, which have identical appearances but radically different electrical conductivities. **Conductivity** is a measure of how easily electrical current passes through a material (also defined as the reciprocal of resistance in a circuit). Jewelers commonly employ a small device to measure this property and identify valuable diamonds on the basis of their high conductivity. Most of the native elements and some oxides and sulfides are also good conductors of electricity.

Most minerals are poor conductors (**dielectrics** or **insulators**) because they are held together by ionic or covalent bonds. However, there are two special cases of silicates in which either pressure or temperature can cause stress-induced conductivity:

Piezoelectric minerals develop an electric charge when pressure (compressional, traction, or torsion) is applied along certain crystallographic axes in their structures. One end become positively charged while the other end becomes negatively charged. Because quartz possesses strong piezoelectricity, it is used as a radio oscillator in many types of devices including watches, electrical components, and lighters.

Pyroelectricity occurs when a crystal acquires electric charge as a result of the thermal stress of being heated. Tourmaline is the best example of a mineral species with this property.

Radioactive minerals contain unstable elements such as uranium, thorium, rubidium, and potassium that are in the process of decaying, or spontaneously emitting energetic particles and changing into different nuclei. These minerals can be positively identified by use of a Geiger counter (usually called a survey meter), and must be handled and stored with care. Uraninite, autinite and carnotite all contain large amounts of the radioactive element uranium, and are therefore extremely radioactive. Over a long period of time, halos can develop around radioactive minerals due to the damage they induce on neighboring minerals. More extensive long-term radiation can cause the complete breakdown, or metamictization, of a mineral's structure.

A Concluding Analogy

As stated throughout this chapter, the physical properties of minerals are an important (and easy) first step in mineral identification because the physical properties of minerals are directly related to their crystal structures. Different minerals usually have different crystal structures. However, different minerals can have very similar physical properties and a positive identification may not always be possible based solely on physical properties.

An analogy to human recognition may help make this point more clear. We distinguish

Table 2.7. Examples of Physical Properties Used in Human Identification

Name	Height	Weight	Hair color	Sex
Bob	5'10"	250	blond	M
Fred	5'10"	160	red	M
Jack	5'10"	150	blond	M
Jill	5'5"	150	blond	F
Sue	5'5"	150	red	F

among our friends based on their physical attributes (e.g., height, weight, hair color, sex), to name but a few. These human attributes are analogous to the physical properties exhibited by minerals.

Five of my friends are listed in Table 2.7 with their associated physical properties. I could easily pick Bob out of the group because he weighs the most. It would be hard to tell Fred from Jack based on weight, at least by looking them, because their weights are so similar. If I had a scale I could weigh them to tell them apart, but this would probably prove embarrassing to me, and be a waste of an "analytical measurement," because Fred and Jack have different hair color that I could use to tell them apart. Likewise, if I weighed my five friends when they came in my house I could not tell Jack from Jill from Sue. However, if I just looked at the three of them I could probably tell one was a male and the two females could be differentiated based on hair color. There are several other combinations of properties in Table 2.7 that would prove useful and useless in telling these five individuals apart. See if you can find them.

Of course, if these people were truly my friends, I could just look at their faces and recognize them. In fact, as you acquire skills in hand sample and thin section identification, many minerals will be like old friends, and you will recognize them just on their appearance.

But what if after they leave my house I find red hair in my food and want to know who it came from? This is when an analytical method, like DNA testing, might be needed to identify the individual.

There are several points in this seemingly silly analogy that relate to mineral identification. Certain physical properties can be very useful and others useless (e.g., the black color of the nine mineral samples on my mineralogy exam). There are occasions (e.g., when given only a very small, nondescript sample) that analytical methods must be used. There are other times when a very simple diagnostic test (e.g., the acid test to tell calcite from quartz) makes much more sense than using a million-dollar instrument! One of the goals of this book is to teach you the skills required to identify minerals in the most efficient manner (i.e., to know what set of physical properties or analytical methods to use for the sample at hand).

References

Blackburn, W.H., and Dennen, W.H. (1997) Encyclopedia of Mineral Names. Special Publication 1, The Canadian Mineralogist, Mineralogical Association of Canada, Ottawa, Ontario, Canada, 360 pp.

de Fourestier, J. (1999) Glossary of Mineral Synonyms. Special Publication 2, The Canadian Mineralogist, Mineralogical Association of Canada, Ottawa, Ontario, Canada, 360 pp.

Nickel, E.H., and Grice, J.D. (1998) The IMA Commission on New Minerals and Mineral Names: Procedures and guidelines on mineral nomenclature, 1998. The Canadian Mineralogist, vol. 36, pp. 913–926.

Povarennykh, A.S. (1964) On the scale of lustre of minerals and the chemical bond. In Battey, M.H., and Tomkeiff, S.I., Aspects of Theoretical Mineralogy in the U.S.S.R. New York, Macmillan, pp. 488–495.

Chapter 3

Crystal Chemistry

The lowest grade on my college transcript was from second semester Introductory Chemistry, where we talked about elements, the periodic table, and structures of atoms. I passed, but barely, and only because my sympathetic mineralogy professor tutored me through every problem set. I just didn't get it. In hindsight I know exactly what my problem was: the information was presented as isolated facts without interconnections or context, and the instructor, who had doubtlessly taught this material a million times before, was as bored with it as I was. I could not imagine anything less important than the study of particles that I could not see, and random elements (with symbols I could never remember) with mysterious desires to gain or lose electrons.

What eventually saved me was, surprisingly, my interest in the Moon rocks from the Apollo missions. It was (and still is) a tremendous thrill to handle samples from the Earth's Moon. When I hold them in my hands, I remember sitting glued to the television with my family, watching grainy black and white images of astronauts bouncing around on the surface of the moon. I can still picture the aircraft carriers and helicopters hovering, waiting for the capsules to splash down in the deep ocean, and I will never forget standing in line to touch a Moon rock. So for me, my desire to study and understand the Moon rocks was the carrot that finally made me realize that maybe chemistry might be worth pursuing after all.

My first real student research project studied the occurrence of Ti^{3+} (titanium) in orange glass from the lunar surface (by comparison, on Earth, Ti is almost always oxidized, as Ti^{4+}). I learned that the color, as well as the structure (yes glass has structure!) of the glass and coexisting minerals, was a function of its chemical composition, and the environment in which they all formed. I discovered that chemistry was the basis for the technology they used (and still use) to map the geology of the Moon, the planets, and many other bodies in our solar system from satellites. In that field, if you don't understand how electronic transitions as explained by the structure of the atom (i.e., the motions of electrons) affect the appearance of minerals to remote sensing instruments, you could never map the geology of planetary surfaces.

So I finally found my reason to learn chemistry in the context of something I really wanted to understand. That's my wish for you, as well. Mineral chemistry touches our lives as Earth citizens in many ways. Most things around you are made from minerals, including the computer, and all of the non-wood material in your house. (As one saying goes "If you can't mine it, you have to grow it".) The same phosphate minerals that are mined and used as fertilizers also make up our bones and teeth! Did you ever wonder what role fluoride plays in reducing tooth decay? Such questions can only be answered with an understanding of mineral chemistry.

Thus, the goal of this chapter is to introduce you to the aspects of chemistry that are important to minerals. For some, this will be a refresher. For others, this will be all new material. Remember that at this point in the book, we are only half-way up the learning spiral for mineralogy. So all of these points will be expanded upon in Round Three chapters in great detail, where you will find derivations for all this information. My goal for now is to present an overview of how and why atoms make up minerals—and why you should care!

M.D.D.

Introduction

In Chapter 1, we introduced the idea that all the minerals found in nature can be assembled by putting together elements with different charges to produce neutral, charge-balanced mineral structures that can be repeated in three dimensions. We initially focused on the Big Eight elements. Hopefully by now, you have memorized their usual valence states and coordination numbers. Now we will figure out where those numbers come from, and why they happen that way!

To understand mineral chemistry, we need to step back and review some fundamentals of chemistry. You probably studied some of this material as early as eighth grade, but don't be tempted to skip this chapter. You will find that mineralogists have a different perspective on this topic than chemists. For example, chemists care a lot about the size and weight of individual atoms, and how they behave in gases. Mineralogists by definition usually don't care about atoms in gases (though this topic does come back to haunt us in petrologic systems), and we are not usually interested in the sizes or weights of atoms in the neutral state. Furthermore, we are mostly interested in atoms that occur commonly on the surface of the Earth (Figure 1.3). So you won't have to worry, here, about knowing the chemical symbol for technetium. Instead, we will focus again on the atomic structures of the Big Eight elements, and how those structures affect mineral formation.

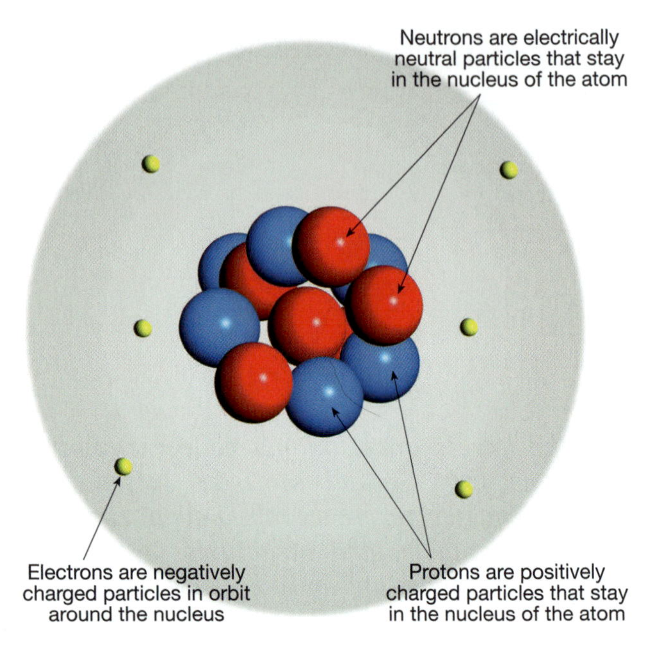
Neutrons are electrically neutral particles that stay in the nucleus of the atom
Electrons are negatively charged particles in orbit around the nucleus
Protons are positively charged particles that stay in the nucleus of the atom

***Figure 3.1**. Fundamental particles called atoms are composed of neutrons and protons in the nucleus (except for hydrogen), and electrons around the nucleus.*

Structures of Atoms

The structure of an atom was first suggested by a New Zealand physicist named Ernest Rutherford, who proposed that atoms consist of a nucleus with positive charge surrounded by negatively-charged electrons, with the whole array held together by electrostatic attractions. Soon thereafter, a Danish theoretical physicist named Niels Bohr formalized a model for the structure of the hydrogen atom, in which a single electron moved in an elliptical orbit around a nucleus. We now know that atoms are composed of the three fundamentally different particles shown in Figure 3.1:

1. Neutrons are electrically neutral particles in the nucleus.
2. Protons are positively-charged particles in the nucleus.
3. Electrons are negatively-charged particles that surround the nucleus.

Electrons move around the nucleus, and the geometrical shape describing the probable location of these electrons is called an **orbital**. This term is rather unfortunate, because it *incorrectly* implies that electrons move around the nucleus in "orbits" like the motion of the Earth around the Sun. A better way to think about this is to refer to an orbital as *the space in an atom occupied by an electron*. For any atom, that space has a characteristic set of shapes. As we'll see in Chapter 7, the cause of color in minerals is explained by how the orbitals change shape and size when they are encroached upon by surrounding atoms. There we will examine those shapes in detail and look at the mathematical derivation of orbital configurations using the Schrödinger equation, but for now, we need only consider some of their fundamental characteristics.

To do this, a simple analogy may help you understand the way electrons occupy the space around the nucleus. Imagine living in the Atomic Apartment Complex, which is portrayed in Figure 3.2. This is one of many buildings in a low budget dormitory complex for undergraduate students. It's a strange place to live because:

1. Each type of room has a characteristic shape that gets more elaborate as you higher up.
2. The rooms get bigger as you go up to the higher floors.
3. Each floor has an odd number of rooms: either 1, 3, 5, 7, 9.
4. The rooms are labeled first by the floor number, second by the letters: *s*, *p*, *d*, *f* (which means the

rooms have been subdivided pretty darn fine), and third (in some cases) by some strange combination of x, y, and z.
5. All the rooms are doubles.
6. The beds in each room face in opposite directions.
7. This dorm has no elevator, so the students are forced to climb to their rooms using ropes on the outside of the building.

Because the students are so tired after a hard day of classes, they prefer to live in rooms closest to the ground. Thus, the first apartments to fill are always on the ground floor, and the whole structure fills from the bottom up. In other words, Room 1*s* fills first, then 2*s*, then $2p_x$, $2p_y$, $2p_z$, then 3*s*, and so on (see Figure 3.2). On really windy days, the whole complex tends to be a little unstable, especially if the top floor is only partially occupied. Students living in the top-most, partially-occupied floor are easily convinced to move out to nearby apartment buildings (or they're constantly trying to find more friends to move in so that their floor will be completely occupied, making the building more stable).

Now let's consider electrons and orbitals instead of students and rooms (respectively). This

Figure 3.2. *In the Atomic Apartment Complex, all rooms are doubles, there is no elevator, and the rooms get bigger as you get higher up. It takes a whole lot less energy to live on the bottom floors than higher up! Rooms can be rented filling from the bottom up only (and who wants to live on the top floor anyway?) This resembles the orbital structure of an atom in that all orbitals are "double occupancy"—that is, each can accommodate two electrons. Also, the lowest energy orbitals are those closest to the nucleus, like the "low energy" students occupying the lowest floors of the apartment complex. As with electron orbitals, the rooms do get bigger as you go higher up.*

gives the real set of rules that describes how electrons occupy the space around the nucleus:

1. Different types of orbitals exist, and each has a characteristic shape (Figure 3.3). These shapes are predicted by the Schrödinger equation (mathematically, these are non-trivial). These pictures represent probability distributions: 90–99% of the time, the electron(s) in a given orbital will be found within the boundaries of the given shape. In actuality, the electron clouds are fuzzy and have no true boundaries. But these pictures give a good idea of where an electron is likely to be found at any time. The simplest of the orbital configurations is the *s* orbital, whose boundary surface is a sphere. The shapes become increasingly complex as the number of electrons increases.
2. The sizes of the orbitals increase as they get farther away from the nucleus, but the shapes remain roughly the same (Figure 3.4). Just as it takes more energy for students to climb to the higher floors of the apartment, electrons have different energies that increase as the distance from the nucleus increases. Orbitals are numbered as they get farther away from the nucleus. For example, 1*s* is the closest to the nucleus (or ground) and is therefore the smallest orbital and the lowest in energy. The 2*s* is larger, 3*s* is even larger, etc., in just the same way that the floors increase in size in the Atomic Apartment.
3. Each group of orbital shapes (Figure 3.3), like the different floors of our apartment building, is designated *s*, *p*, *d*, and *f* and has a different number of orbitals. Thus, a different total number of electrons can be accommodated (Table 3.1) on each level, just as each floor in the apartment can accommodate a different number of students.
4. Each orbital, no matter how big it is, can only be occupied by two electrons spinning in opposite directions. This is why each room in the apartment could only have two students, with beds facing in opposite directions.
5. The orbitals surround the nucleus in a specific order, with increasingly high energy needed to get farther away from the nucleus (like the energy needed to climb a rope to the higher floors). The order for filling orbitals in neutral atoms is the same as the sequence of floors in our apartment complex. It is known to be the following for most neutral atoms: 1*s*, 2*s*, 2*p*, 3*s*, 3*p*, 4*s*, 3*d*, 4*p*, 5*s*, 4*d*, 5*p*, 6*s*, $4f \approx 5d$, 6*p*, 7*s*, $5f \approx 6d$, and so on (we'll explain this in more detail in Chapter 7). Thus, if you only have eight electrons (that's our friend, the oxygen atom,

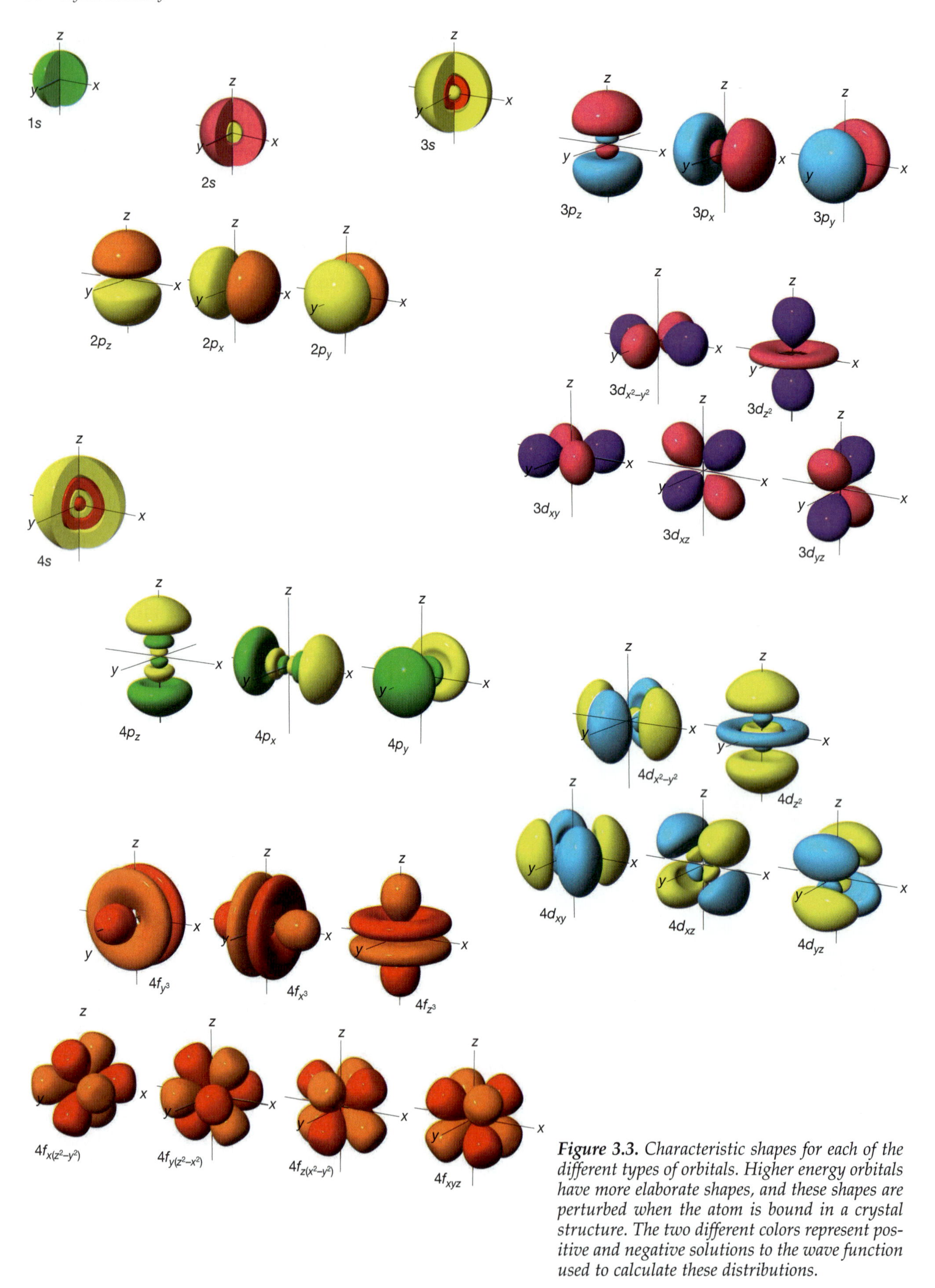

Figure 3.3. *Characteristic shapes for each of the different types of orbitals. Higher energy orbitals have more elaborate shapes, and these shapes are perturbed when the atom is bound in a crystal structure. The two different colors represent positive and negative solutions to the wave function used to calculate these distributions.*

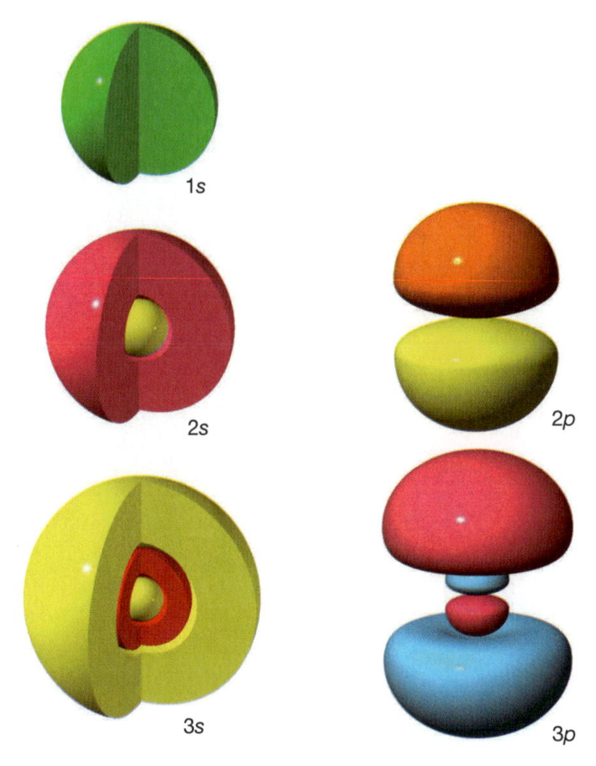

Figure 3.4. *As you increase the energy of the orbitals, the distributions get bigger and the shapes get more elaborate. Note the increasing size of the* s *orbitals, as well as the extra lobes in the* p *orbitals shown here. Shading as in Figure 3.3.*

which is by far the most abundant element in the Earth's crust; Figure 1.3), you will have two electrons in 1*s* orbitals, two electrons in 2*s* orbitals, and the remaining four electrons in 2*p* orbitals. The 2*p* orbitals won't be full (they can accommodate up to six electrons) but that's OK. As we'll learn in a few pages, oxygen atoms in most minerals like to steal two electrons (comparable to having a few more students move in to fill an apartment floor) in order to fill that 2*p* orbital up completely.

Armed with all this knowledge, we can now build the periodic chart of the elements (Figure 3.5). The first row of the periodic chart contains the elements that only have 1*s* orbitals: hydrogen and helium (this is the first floor of the apartment). The second row has the elements that fill the 1*s*, 2*s*, and 2*p* orbitals. The third row is the elements with 1*s*, 2*s*, 2*p*, 3*s*, and 3*p* orbitals filled, and so on. The idea is that as you add electrons (and there are always an equal number of electrons and protons in this example, to make the atoms be neutral in charge), you make increasingly heavier atoms, and you move higher up on the periodic chart. For this reason, we've chosen to depict the periodic chart upside down from the way it is usually displayed—to make the point that elements low down on the periodic chart have fewer electrons, neutrons, and protons (and they are lower in energy, like the bottom floors of the apartment).

While looking at the symbols on the periodic chart, you may be wondering how and why the elements were named. [I personally still have trouble remembering that the chemical symbol for the element tin is Sn.] The etymology of the chemical elements is a fascinating historical topic all to itself, because many of the element names are really old, and their roots are in languages other than English. Some of these elements are also minerals that were named in the days when folks thought that the only three elements were earth, wind, and fire. Of course this was also the time period that we thought the Sun revolved around the Earth. For example, "gold" (known from at least 4000 BC) derives from the German word *gelt*, which probably derives from the Anglo-Saxon term *gold*, both of which may ultimately come from the Sanskrit word *jyal* (Sinkakas, 1966). Its symbol, Au, like those of many elements, derives from Latin roots. Table 3.2 gives a list of the roots for some of the ones that give me trouble. For more information on this fascinating topic, take a look at Appendix A in the *Encyclopedia of Mineral Names* (Blackburn and Dennen, 1997) on the origins of the chemical element names and symbols [and it's also a great book for researching origins of mineral names!].

From Atoms to Ions

So far in this chapter, we've been talking only about atoms in the neutral states (i.e., how they behave when they are gases and when the number of protons equals the number of electrons). As noted earlier, gases are rarely relevant to mineralogists, so we need to consider what happens when atoms actually bond with each other. An atom that has an unequal number of electrons and protons is no longer electrically neutral, but becomes a charged particle. We call this type of particle an ion. Here, our analogy to the Atomic

Table 3.1. Electron orbitals made easy

Type of orbital	*Number of orbitals*	*Total number of electrons that can be accommodated*
s	1	2
p	3	6
d	5	10
f	7	14

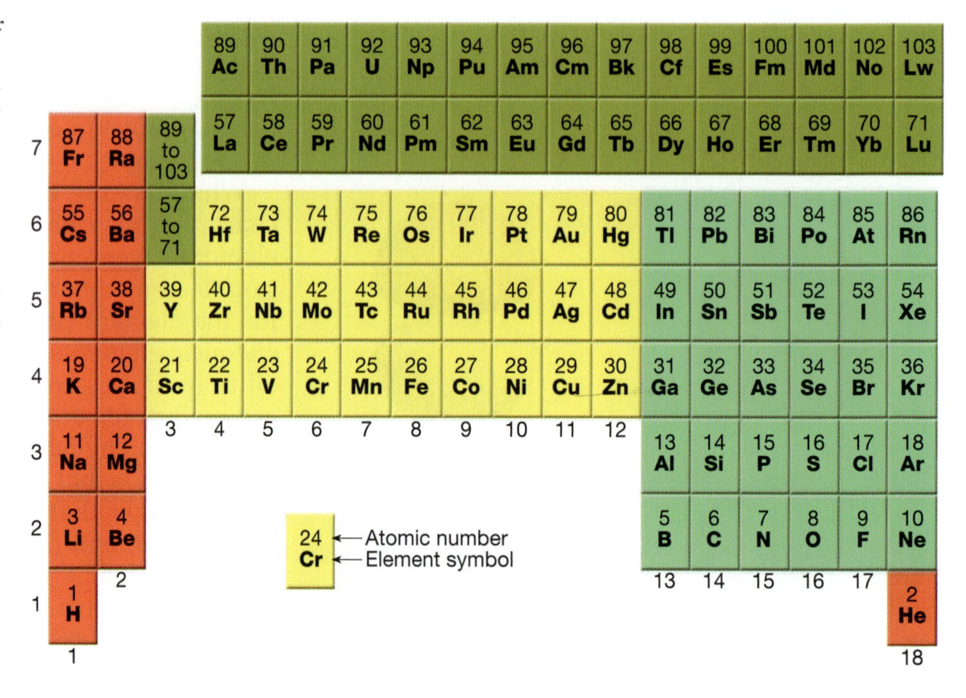

Figure 3.5. *The periodic chart of the elements, simplified. This is shown inverted, with H and He on the bottom row, to remind you that these represent atoms with the lowest number of neutrons, protons, and electrons. As you move up this periodic chart, the number of all these particles increases, and electrons occupy successively higher energy orbitals. The number on each box is the atomic number, which is the number of protons each atom contains.*

Apartment Complex serves us well. Remember that the whole complex tends to be a little unstable *when the top floor is only partially occupied* (Figure 3.6). Students are always trying to move out of the rooms on the top floor, or get their friends to move in, so that the top floor will be either completely full or completely empty. This is how atoms behave. The world of atoms is similar to the world of students, in that they always seek to expend the least amount of energy to accomplish a task.

Imagine that our apartment complex has only eight students left at the end of the academic year (Figure 3.6). Two live in the ground floor room (1*s*), two on the second floor (2*s*), and the remaining four students occupy the three rooms on the

Table 3.2. Some Hard to Remember Element Symbols and their Derivation

Element name	*Element symbol*	*Derivation*
antimony	Sb	From the Latin word *antimonium* and the Greek word *antimonos*, which means "against solitude," which refers to the fact that antimony is usually found with other elements. The symbol comes from the Latin *stibium*, or "mark" because it was used as eye shadow by the Romans.
gold	Au	From the Latin word *aurum*. Known from 4000 BC.
iron	Fe	From the Anglo-Saxon term *iren*. The symbol refers to the Latin word for iron, *ferrum*. Known from 3000 BC.
lead	Pb	Directly from the Anglo-Saxon word *lead*. The symbol refers to the Latin word for lead, which was *plumbum*. Known from at least 3500 BC.
mercury	Hg	Dating from the 6th century, the Latin term *mercurius* was used to refer to both the element and the planet! The symbol comes from the Latin for liquid silver, or *hydrargyrum*. Known from at least 1500 BC.
potassium	K	The Latin word for pot ash was *kalium*, which was probably based on the Arabic term *qaliy* for burnt ashes. Potassium was not isolated until 1809 (by Sir Humphrey Davy).
silver	Ag	Based on the Anglo-Saxon term for this element, *siolfor*. The symbol comes from the Latin term for silver, which was *argentium*. Known from antiquity.
sodium	Na	From the Arabic term for headache, or *suda*. The symbol is from the Latin term *natrium*, or the Greek word *natros*, referring to the mineral form of sodium carbonate. First isolated (and named) by Sir Humphrey Davy in 1807.
tin	Sn	From the German word for tin, *zinn*. The symbol is from the Latin word *stannum*. Known from at least 3500 BC as a constituent of bronze.
tungsten	W	The word comes from the Swedish terms *tung* (heavy) and *sten* (stone), referring to the high density of this element. Named in 1755 by A.F. Cronstedt; isolated by J.J. De Elhuyar and F. de Elhuyar in 1783. The symbol comes from the German words for wolf (*wolf*) and soot (*rahm*).

After Blackburn and Dennen (1997)

Figure 3.6. Neutral atoms sometimes have an "unstable" number of electrons; in other words, the highest energy orbitals are not completely filled. In this example, we see an oxygen atom with eight "electrons." To create a situation where the outmost orbitals are full, either four electrons must be given away (creating an O^{4+}) or two electrons are gained from some nearby atom, creating an O^{2-}. In minerals, oxygen is usually found as O^{2-}.

p-level, which is above 2*s*. Because only two students can live in each room, the floor is lopsided. To remedy the situation, two solutions are possible: either all four students move out, *or* two new students move in, so that all the *p*-level rooms will be full, and the floor balanced. The latter solution is probably easiest, because it disrupts only two students instead of four. Less energy overall is expended to move *in* two students than for four to move *out*.

Anions. In the oxygen atom (which has eight electrons), the preceding example is exactly what happens. The oxygen atom prefers to gain two electrons in order to fill its outermost 2*p* shell. Because it takes on two additional electrons, it now has a negative charge. It has two more electrons (which count as negative charges) than protons (which count as positive charges). Thus, O^{2-} is an example of an **anion**, which is an atom that has gained one or more electrons to become negatively-charged. It has also increased in size.

Figure 3.7 shows a group of dads and kids walking in the park. In this example, there are the same number of dads (protons) as kids (electrons)—eight of each. When protons = electrons, the kids move around the dads in a somewhat random pattern. If more kids join in (electrons are added), with the same number of dads, the dads can't keep the kids reigned in as well and the whole mess expands. Similarly, a neutral atom gets bigger when it becomes an anion. The electron cloud enlarges because there is more negative charge than positive charge.

Cations. Now consider a different example: the sodium (Na) atom, which has 11 electrons. The 11th electron is all by itself in the 3*s* orbital. This is a relatively unstable situation, so Na likes to give away this electron whenever it can. Na^{1+} is now missing an electron, so it has more protons than electrons. The resultant ion has an extra unit of positive charge. Na^{1+} is an example of a **cation**, an atom that has lost one or more electrons and is a positively-charged ion. Na^{1+} is smaller than Na^0 in just the same way that the apartment grows lighter when an upper floor is emptied.

Figure 3.8 uses the same analogy as above, but this time we are looking at a cation. In this example, the dads are so busy talking that they don't even notice that one of the kids has left. But if the number of dads exceeds the number of kids, the dads can keep the kids better reigned in, so that the whole mess contracts. In this case the neutral atom gets smaller when it becomes a cation. This time, the electron cloud contracts because there is now more positive charge than negative.

Atomic Bonds

In order to put together a mineral structure, we need a way to "glue" together various atoms: atomic bonds! Bonding is important because it is very directly associated with the physical properties of a mineral. For example, one of the simplest physical properties to determine is hardness, and it relates directly to bond types. While both diamond and graphite are made of carbon, diamond is the hardest material known and graphite is one of the softest. Did you ever wonder why this occurs? The answer is: Different bond types!

More complex properties can also be related to bonding. For instance, properties such as the speed of light through a mineral and thermal and electric conductivity are affected by bonding. Here we'll provide a brief discussion of the different types of bonding: ionic, covalent, metallic, van der Waals, and hydrogen bonds (there will be more on this topic later in Chapter 8). But we will also learn that there is continuum between some of these bond types. If "black" is ionic, and "white" is covalent, then there are many shades of "gray" between the two. The important thing to remember here is that all of these bonds are, in some way, based on attractions between atoms that occur because of the electrons. So the electrons are the "glue" that bonds different atoms together. The names help us describe some of the different ways that electrons behave in atomic bonding. But the names are

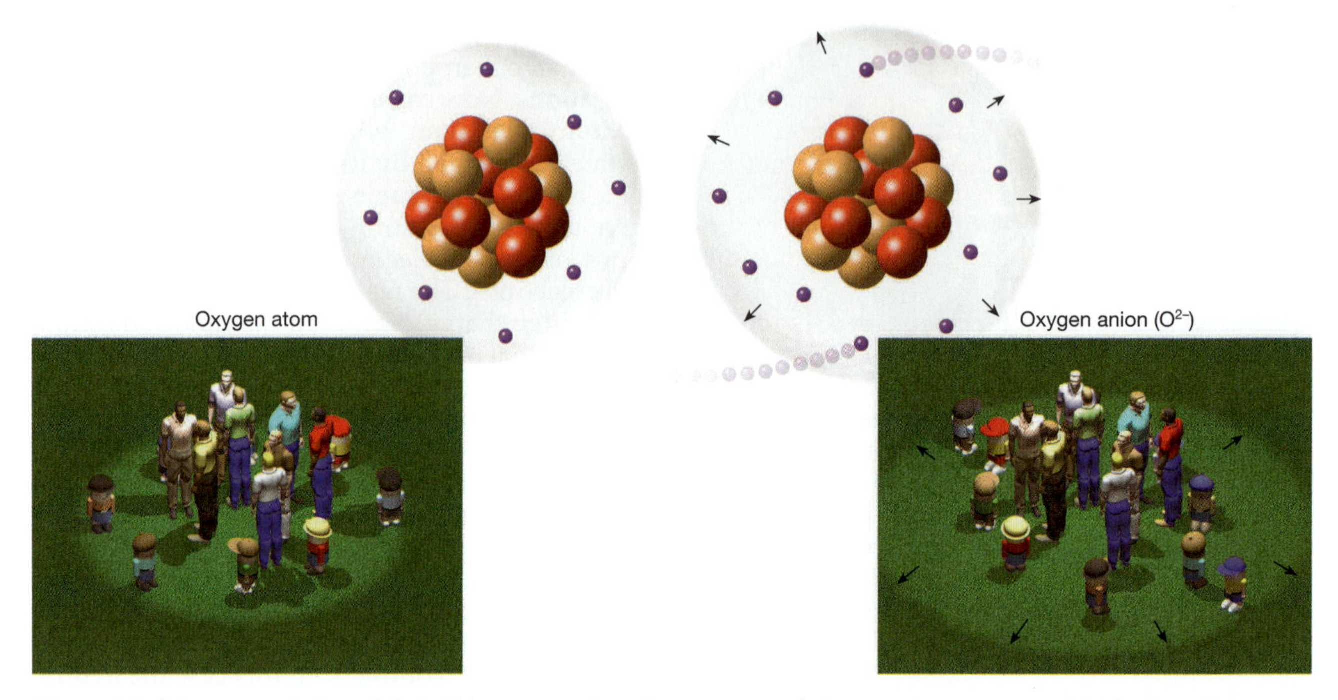

Figure 3.7. *We can use dads and their kids as an analogy for protons and electrons in an atom and ionic size. Assume we have eight dads (what atom would this be?) and each of them has one child. The dads' goal is to keep the kids bunched together as they move through the park. With an equal number of dads and kids they can keep them in the circle on the left side. However, if the dads pick up a few more kids (i.e., electrons) the sphere would get bigger, just as ion size increases when the number of electrons exceeds the number of protons.*

much less important than understanding the bonding processes and how these processes affect the physical properties of the minerals.

Probably the most important type of bonding in minerals is ionic. The old saying that opposites attract is true of ions as well: positively-charged

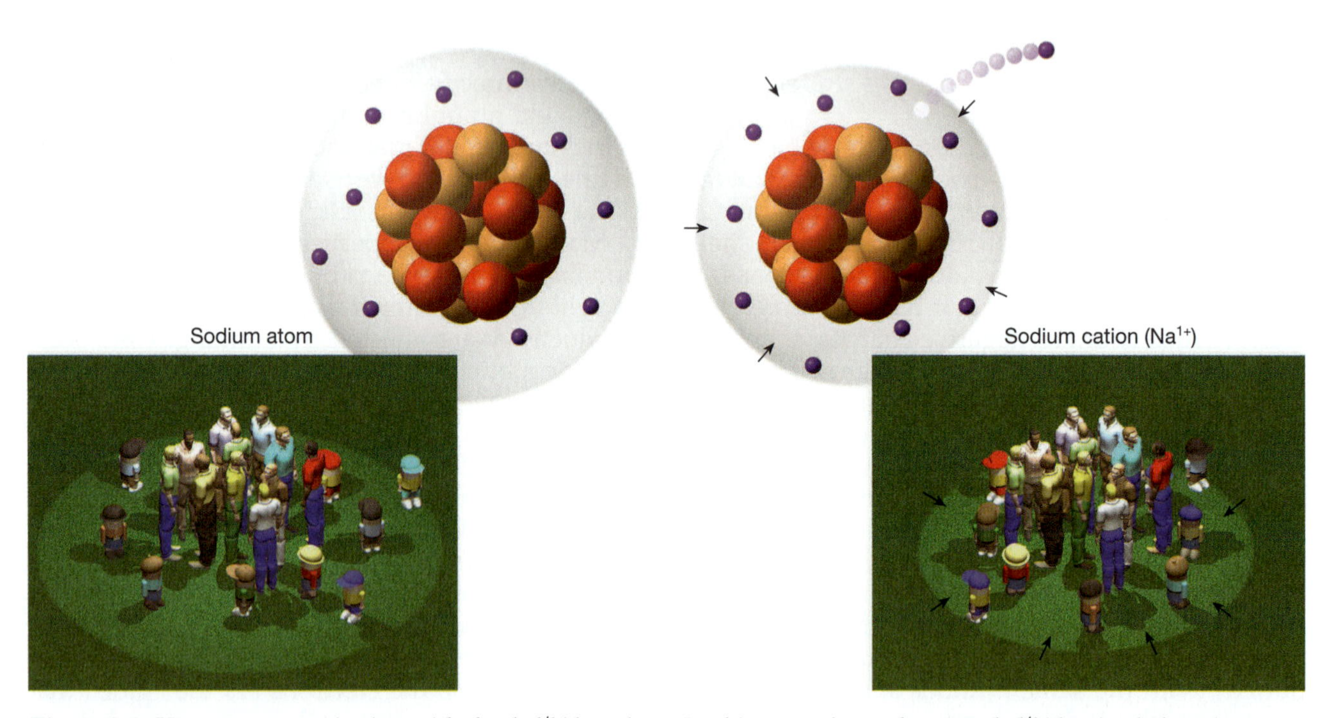

Figure 3.8. *Here we are continuing with the dad/kid analogy. In this example, we have 11 dad/kid pairs (what element is this?). Now if one of those kids wanders off, there are more dads than kids so the whole bunch shrinks somewhat because there is an extra dad to keep the kids bunched up.*

cations are attracted to negatively-charged anions. This electrostatic attraction between ions involves the loss and gain of electrons between adjacent ions (Figure 3.9), and is referred to as **ionic bonding**. Note that in this type of bonding, there is transfer of electrons from one atom to an adjacent atom, with no sharing. This is a common type of bond for ions that are far apart in the periodic chart.

Sometimes, electrons behave in just the opposite manner from what is described above, and are shared equally between two atoms. We refer to this as **covalent bonding** (Figure 3.10). These are bonds that form when adjacent atoms have similar, or equal electronegativities (see Chapter 8), and both atoms share the outermost electrons equally. Diamond has this type of bonding. It is this sharing of electrons that produces very strong bonds in diamond in three dimensions, and, in turn, make it the hardest material known.

If you need help in remembering the distinction of these two bonds types, maybe this will help: Covalent really means to share (the meaning of "co" as in cohabitate) valence electrons (that's where "valiant" comes from).

But what about the most common bond type found in minerals—the bond between the most common anion, O^{2-}, and the most common cation, Si^{4+}, in the Earth's crust? Is it ionic or covalent? If you guessed either, you would be 50% correct! It turns out the Si^{4+}-O^{2-} bond is about 50% ionic and 50% covalent. So what do we call it?

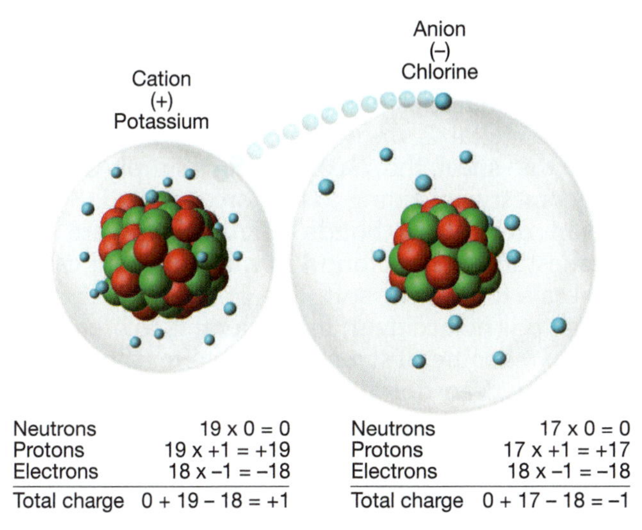

Figure 3.9. Ionic bonds occur when atoms with opposite charges are electrostatically attracted to each other. In this example, the potassium atom (K^{1+}) has lost an electron, so it is a cation. The chlorine atom (Cl^{1-}) is an anion because it has gained an electron. Notice the difference in size between the two atoms: K^{1+}, which has lost an electron, is smaller than Cl^{1-}, which has gained one. K^{1+} and Cl^{1-} bond together to form the mineral sylvite, which is sometimes used as a bitter but slightly palatable salt substitute. If someone in your family is on a low salt diet, check their "salt"—it's probably potassium chloride, or sylvite!

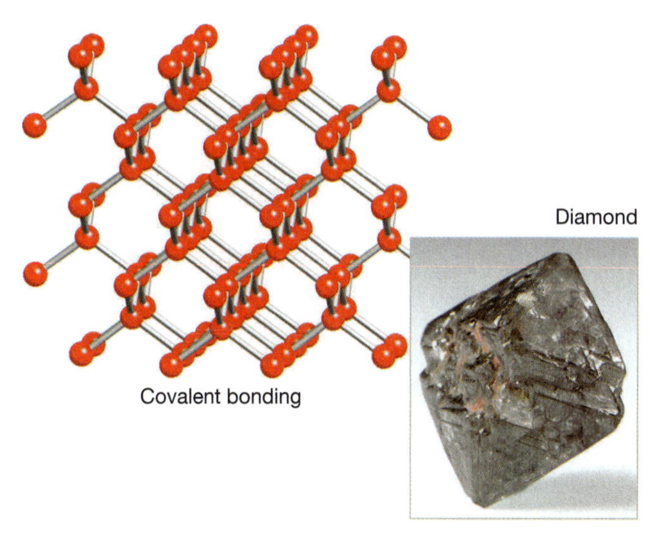

Figure 3.10. The carbon atoms in diamond are bound together by covalent bonding in three dimensions. In graphite, two-dimensional sheets are also covalently bonded within the sheets (see Figure 3.13).

This is a major problem in teaching and learning about the natural world. We don't have the ability to place precise words on all aspects found in nature, because many processes in nature are a continuum between two (or more!) end points. We apply names to the end points, but have no real name for all the intermediate conditions. We'll find that this "continuum" occurs over and over again in mineralogy, and in nature in general. We'll also find that when we place a name on a process and define that name, we are often in danger of placing more importance on knowing the name than on understanding the process. Ionic and covalent bonding are prime examples of this. We assign names to the end-members, but in reality there is a continuum in electron behavior between a pure ionic bond and a pure covalent bond.

An analogy might help to explain this continuum from pure ionic to pure covalent bonding more clearly. Figure 3.11 shows a room in a house with two heat sources and several kids. The situation is this. It's very cold outside and the kids don't have jackets. The kids (atoms) will want to stay inside, and they will distribute themselves to have a desirable level of warmth (energy). If both the stoves are set at the same temperature, the kids will tend to spend most of their time shared between them—this would be the example of covalent bonding. The kids are the electrons and the stoves the atoms. On the other hand, if one of the stoves is turned off, then the kids will head to the one left on—this is the analogy for ionic bonding. The cold stove is the cation and the warm stove the anion. To simulate mixed ionic/covalent bonding, one stove may just be warm, while the other is hot—the warm stove in this case is Si and the hot stove is O. This would result in the 50/50 character of the very important Si^{4+}-O^{2-} bond. This analogy also can be used to think of the probability of finding kids next to stoves (i.e., electrons near atoms). The probability of finding a child is the highest near the stove, but the kids can move to the edge of the room, even though the probability of finding one there is less.

Before leaving ionic and covalent bonding, we want to introduce one last concept. Light travels very slowly through covalently-bonded diamond because of the light's interaction with the electrons between the carbon atoms. Light travels faster though materials with ionic bonding (such as fluorite) because the electrons are bunched up close to the anions. So light is a useful tool for probing the type of bonding that occurs in minerals. An analogy for this is to imagine you are running through a woods. If the trees are tall without low branches you can go fast (the case of ionic bonding). If the trees have lower branches that interfere with your running, you would slow down (the case of covalent bonding).

Covalent bonding

Ionic bonding

Figure 3.11. The behavior of electrons around nuclei can be modeled by using kids to represent the electrons and a couple of stoves to represent the nuclei of the two bonded atoms. It's winter time and cold outside. The stoves and kids are placed in a room. The kids start doing what kids do, moving around! As they move around, they spend the majority of time between the two stoves if the stove temperatures are the same. This would be analogous to covalent bonding. On the other hand, if one of stoves goes out, the kids would all huddle near the other one (ionic bonding). There could also be a continuum of temperature differences between the two stoves. This difference would then be similar to intermediate cases between pure covalent bonding and pure ionic bonding (Modified from Gunter 1993).

Less common in rock-forming minerals, but *very important* in minerals like copper and gold, are **metallic bonds** (Figure 3.12). Valence electrons are stripped away from their nuclei, so the electrons no longer "belong" to a particular atom. The analogy for this type of bonding is that the atoms float in "a sea of electrons." The electrons behave like water and can move freely. So when an electrical charge is placed on one end of a copper wire, the electrons flow to conduct electricity. This is a relatively strong type of bonding, as indicated by the high melting points of most metals (in other words, it takes high temperatures to break such strong bonds).

The strength of a metallic bond is also reflected in two common properties of pure metals such as copper, gold, and silver: malleability and ductility. Both of these properties require that the atoms in the mineral structure be easily moved relative to one other but still remain strongly bonded (sort of like boats attached to anchors in a pond; the boats can move somewhat in the wind, but return to their original position when the wind stops). When a metal is drawn into a wire or pounded into a thin sheet, the sea of electrons simply adjusts its geometry according to the force applied.

Van der Waals bonds (Figure 3.13) are the weakest attractions found in minerals, and cause minerals to be very soft, like graphite. This relationship is no accident: the easier it is to break the bonds in a mineral structure, the softer the mineral will be. **Van der Waals** attractions result from the interplay between two types of forces: the repulsions between electrons in the filled orbitals of neighboring molecules, and the attractions that occurs when electrons in the occupied orbitals of adjacent atoms synchronize their motions to avoid each other as much as possible. A less descriptive way to view these bonds is first to recall that all bond types deal with charges, and charge differences are set up by the electrons. In van der Waals bonds, these charges are very weak, especially when compared to covalent bonds. For example, covalent bonding between two H atoms is 5000 times stronger than the van der Waals bonding between two He atoms (Gray,

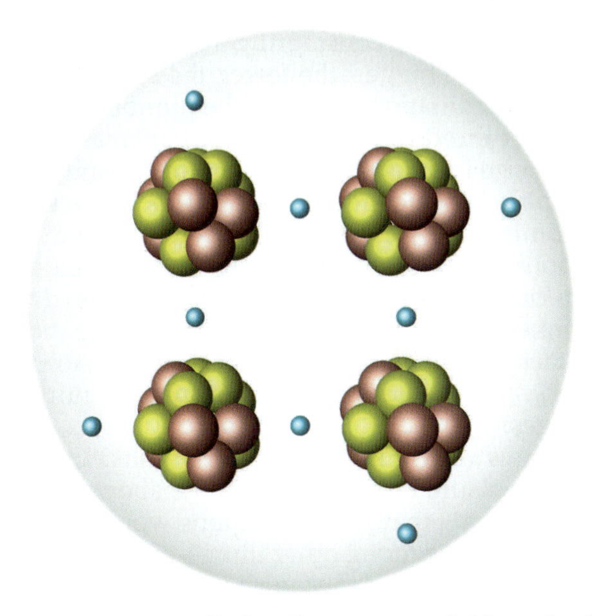

Figure 3.12. In metallic bonding, atoms are held together by a sea of electrons shared by a group of cations. Metallic copper has this type of bonding.

1994). Minerals with van der Waals bonding also melt at low temperatures because relatively little energy of thermal motion is required to break the van der Waals bonds. Crystalline sulfur, for example, melts at about 119°C.

Another weak but mineralogically-important bond type is the **hydrogen bond**, which results from interactions between a positively-charged hydrogen atom and a large, negatively-charged anion like oxygen (Figure 3.14). In this scenario, the lone electron on the hydrogen atom is strongly attracted toward the anion. On the opposite side of the hydrogen atom, the electron is rarely found, so the positive charge from the nucleus is exposed, and may weakly attract nearby anions. Hydrogen bonds are relatively weak, about 5% of the bond energy of a covalent bond though they are stronger than van der Waals bonds. The structure of ice is held together by hydrogen bonds (with a little help from covalent bonds), which explains why its hardness is only 1.5 on the Mohs scale. H bonds occur in many other minerals, such as the chlorites, which are important rock-forming minerals.

Generally speaking, minerals with low hardness tend to have a significant amount of hydrogen, or van der Waals bonding somewhere in their crystal structures, while harder minerals are held together by dominantly ionic or covalent bonds. Just as a chain is only as strong as its weakest link, a mineral is only as hard as its weakest bond! This is one of the most obvious demonstrations of how the physical properties of minerals are controlled by their crystal chemistry.

Ionic Sizes

Armed with our knowledge of atomic structure and the bonding of atoms, we need only one more piece of the puzzle to assemble mineral structures. In essence, we know what bricks to use (these are the atoms), and we have the mortar mixed and ready to go (this is atomic bonding), but we need to know the size of the bricks (i.e.,

Figure 3.13. Van der Waals bonds occur in situations where adjacent atoms are unlikely to gain or lose electrons. For example, frozen inert gases that have completely filled outer electron shells are held together (loosely) by Van der Waals attractions, as are the layers in the arsenic, antimony, and bismuth structures. In the graphite structure shown here, covalent bonds within the (horizontal) sheets are very strong (in fact, stronger than the C-C bonds in diamond), with each carbon bonded to three others. Van der Waals forces occur (vertically) between the sheets, which can be displaced very easily. This is why graphite has a hardness of only 1–2 on the Mohs scale.

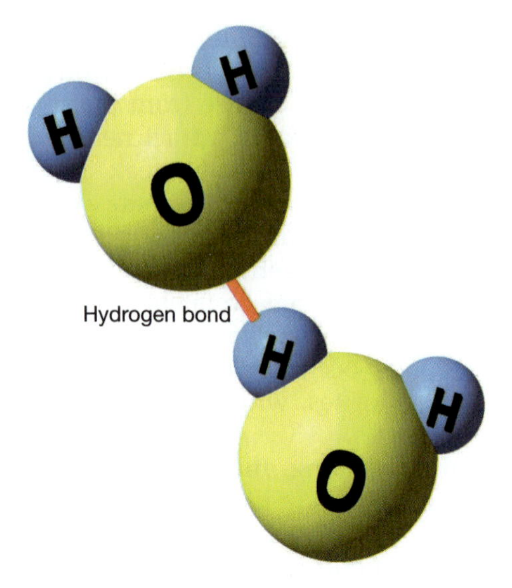

Figure 3.14. *In a water molecule, the electron distribution around the nucleus is rather thin, because there is only one lonely electron. When hydrogen (H) bonds with oxygen (O), that electron is strongly attracted to the oxygen, and its probability distribution is skewed from the side of the H atom towards the oxygen. This leaves the positively-charged hydrogen nucleus poorly shielded on the opposite side, so it gives off a weak positive charge. This charge in turn can be attracted to nearby anions such as the oxygen atoms from a neighboring water molecule. The geometry of these linkages gives H_2O a very rare property: it has higher volume as a solid than as a liquid.*

the ionic sizes) before we can build our house (a mineral), because the bricks need to fit together! In fact, there are really two major requirements that must be satisfied before elements can organize themselves into minerals: (1) the charges on the anions and cations must sum to zero and (2) the anions and cations must fit together.

Assigning a size, or even a shape, to ions is complicated. It may have seemed simple in introductory chemistry class, but in reality there are many factors that influence the size of an ion: the number of protons, neutrons, and electrons; the relative position of the atom on the periodic chart; the amount of covalency in the bonds; pressure; temperature; the coordination number of the element; and its oxidation state, or the charge on the ion. We will leave the discussion of most of these factors to Chapter 8, and focus here on the last two because they are so important in understanding minerals.

It stands to reason that electrons will play a major role in determining the size of an ion, because it is the electrons that are involved in bonding. The radius of the ion must be somehow related to the distance of its electrons from its nucleus. So the charge on the ion will be very important when considering its size: the more electrons, the bigger; the fewer, the smaller. Also, as the coordination number (the number of other ions surrounding a central ion) increases, the cation size increases. The cation can expand to fill the space in which it finds itself. This implies that an ion is not a rigid sphere! A better analogy for an ion might be a ball made of a sponge-like material. When wet it gets bigger and when dry it shrinks. Like a cation, this ball can also expand or contact to fill the space in which it is placed as when, for instance, the coordination number changes.

So how do we know the size of ions? Can we really measure them? Based on the above discussion, it seems that individual ions would change size based on, among other things, oxidation state (= ionic charge) and coordination number. So what good are these sizes if they vary? The last question is the most important to answer. Even though atomic sizes do vary, we can use them to help us predict which ions will fit into a structure and what elements can substitute for each other. As we discussed in Chapter 1, elements like Si^{4+} and Al^{3+} can substitute for each other in the feldspars, despite having slightly different charges. Mg^{2+} and Fe^{2+}, which are approximately the same size, easily substitute for each other in the pyroxenes, amphiboles, and micas. But the difference in size between elements like K^{1+} and Si^{4+} is so big that they never share the same site in a mineral.

Let's say you wear a dress size 8 (Al^{3+}), and your roommate is a size 6 (Si^{4+}). In cases of emergency, you can probably squeeze into her size 6's. But if you wear a size 12 (K^{1+}) you'll probably never fit into her size 6 cocktail dress (unless you lose a few electrons!).

So how do we measure the sizes of ions? Empirically. Indirectly. This is done by using X-ray diffraction, which will be discussed in Chapter 15. We can find the distance between two atoms very precisely by determining the crystal structure of the material, and then calculating the distance between the two atoms of interest. For instance, by using powder X-ray diffraction, we can find that the distance between Na^{1+} and Cl^{1-} in halite is about 2.8 Å (that's the abbreviation for an **ångstrom**, which is 1×10^{-10} meters; most atoms are between 1 and 5 Å in diameter.). So we know that the radius of Na^{1+} plus the radius of Cl^{1-} equals 2.8 Å. (This is really a simple experiment. If you have an X-ray powder diffractometer, you can easily demonstrate this based on the discussion in Brady et al., 1995). After NaCl, you might use KCl and find that the distance between K^{1+} and Cl^{1-} is 3.2 Å. This would lead you to conclude that the

K^{1+} ion is 0.3 Å bigger than Na^{1+}. This of course assumes the size of Cl^{1-} is the same in both NaCl and KCl (which is a logical assumption made in chemistry class, but probably incorrect).

In general, we are more interested in the rock-forming minerals. They contain predominantly O^{2-}, the most abundant anion in the Earth's crust, bonded to Si^{4+} the most abundant cation in the Earth's crust. Based on many measurements, the O^{2-} radius was assigned a value of 1.4 Å. Given this, it is then straightforward to determine the radius of other elements. For instance, in Figure 3.15 an Si^{4+}-O^{2-} bond is shown with a Si^{4+}-O^{2-} distance of 1.66 Å, so Si^{4+} has a radius of 0.26 Å. In the Al^{3+}-O^{2-} example, Al^{3+} bonded to O^{2-} would have a radius of 0.39 Å. This again assumes that O^{2-} does not change in size (though Gibbs et al., 1992 believe that it does).

We can precisely measure the *distance* between two ions, but it becomes difficult to *assign* that distance to each of the two ions, because the electrons from each ion overlap each other. If we solve this problem by assigning a fixed size to O^{2-}, then we must realize the limitations of the actual numbers we calculate for ionic size. So what good are these numbers? They allow us to predict what elements fit into different sizes of sites in a relative way. When you buy a new pair of shoes you have to try them on first, and your (assumed) shoe size is just a good starting point!

Table 3.3 lists the sizes, or ionic radii, of the Big Eight elements, while Table 3.4 gives a more comprehensive list of measured ionic radii. These are average numbers based on 100's of crystal structure analyses with the ions in different oxidation states and with different coordination numbers. On the basis of first inspection of Table 3.3, you will notice that the cation with the highest charge (Si^{4+}) has the smallest size, and the cations with the lowest charges (such as Na^{1+} and K^{1+}) are the largest on the list, as expected from the above analogies. To a first-order approximation, this occurs because Si^{4+} has lost four electrons, while Na^{1+} and K^{1+} have lost only one each. Of course, they don't all start out with the same number of electrons, and their sizes are also affected by the number of protons they contain. However, it is generally true that the commonly-occurring, highly-charged cations are smaller than those with lower charges.

There is a positive correlation between the size of the cation and its coordination number (Table 3.3). In other words, the larger the cation, the larger the coordination number. The bigger the cation, the higher the number of oxygen atoms that will be able to crowd around it (recall Figure 1.6).

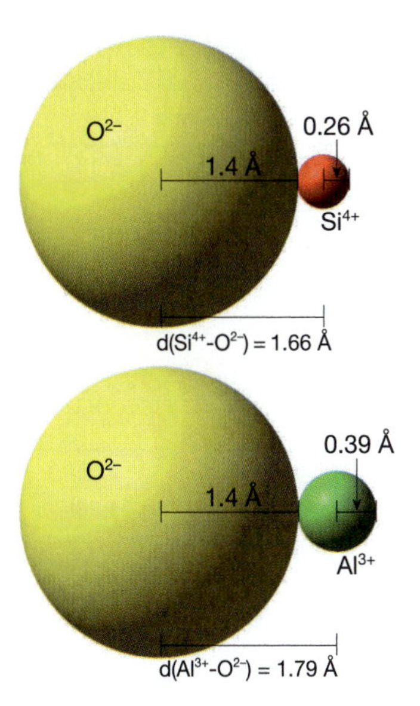

Figure 3.15. *The above sketches represent the geometric method that is used to determine ionic radii based on determining the bond distance from crystal structure data and assuming a fixed radii for O^{2-} of 1.40 Å. The top example is for Si^{4+} and the bottom for Al^{3+}.*

Coordination number and size. A simple food analogy may help you visualize this situation (try this in the dining hall or grocery store, but don't tell them who gave you the idea). Begin by collecting a plate of twelve oranges and one apple about the same size, which will represent twelve oxygen anions and a large cation; while you are at it, grab a pile of toothpicks. *[In reality, any similarly-sized spherical objects would work (i.e., soccer balls, pool balls, 00 buck, jawbreakers, marbles, or anything round).]* Back at your table, put six oranges in a ring (at the corners of a hexagon, see Figure 3.16), and then add a central apple. If all the fruits are the same size (and fairly spherical), then all six oranges should touch each other, and the apple should be exactly the right size to fill the gap (Figure 3.16).

Table 3.3. The Big Eight Elements, and their Ionic Radii

Preferred Ionic Form	*Common coordination number with oxygen*	*Ionic radius(Å)*
O^{2-}	-	1.40
Si^{4+}	4	0.26
Al^{3+}	4–6	0.39–0.54
Fe^{3+}	4–6	0.49–0.65
Mg^{2+}	6	0.72
Fe^{2+}	6	0.78
Ca^{2+}	6–8	1.00–1.12
Na^{1+}	6–8	1.02–1.18
K^{1+}	8–12	1.51–1.64

From Shannon and Prewitt (1969) and Bloss (1994); see Table 3.4 for complete table.

Table 3.4. Oxidation states, coordination number, and ionic radii (in Å) of the elements

Z	element	symbol	valance	2	3	4 tet	4 sp	5	6	7	8	9	10	11	12
1	hydrogen	H	+1	-0.18											
3	lithium	Li	+1			0.590			0.76		0.92				
4	beryllium	Be	+2		0.16	0.27			0.45						
5	boron	B	+3		0.01	0.11			0.27						
6	carbon	C	+4		-0.08	0.15			0.16						
7	nitrogen	N	-3			1.46									
7		N	+3						0.16						
7		N	+5		-0.104				0.13						
8	oxygen	O	-2	1.35	1.36	1.38			1.40		1.42				
9	fluorine	F	-1	1.285	1.30	1.31			1.33						
11	sodium	Na	+1			0.99	1.00		1.02	1.12	1.18	1.24			1.39
12	magnesium	Mg	+2			0.57	0.66		0.720		0.89				
13	aluminum	Al	+3			0.39		0.48	0.535						
14	silicon	Si	+4			0.26			0.400						
15	phosphorus	P	+3						0.44						
15		P	+5			0.17		0.29	0.38						
16	sulfur	S	-2						1.84						
16		S	+4						0.37						
16		S	+6			0.12			0.29						
17	chlorine	Cl	-1						1.81						
17		Cl	+5		0.12[PY]										
17		Cl	+7			0.08			0.27						
19	potassium	K	+1			1.37			1.38	1.46	1.51	1.55	1.59		1.64
20	calcium	Ca	+2						1.00	1.06	1.12	1.18	1.23		1.34
21	scandium	Sc	+3						0.745		0.870				
22	titanium	Ti	+2						0.86						
22		Ti	+3						0.670						
22		Ti	+4			0.42		0.51	0.605		0.74				
23	vanadium	V	+2						0.79						
23		V	+3						0.640						
23		V	+4					0.53	0.58		0.72				
23		V	+5			0.355		0.46	0.54						
24	chromium	Cr HS	+2						0.80						
24		Cr LS	+2						0.73						
24		Cr	+3						0.615						
24		Cr	+4			0.41			0.55						
24		Cr	+5			0.345			0.49		0.57				
24		Cr	+6			0.26			0.44						
25	manganese	Mn	+2								0.98*				
25		Mn HS	+2			0.66	0.75		0.830	0.90					
25		Mn LS	+2						0.67						
25		Mn	+3					0.58							
25		Mn HS	+3						0.645						
25		Mn LS	+3						0.58						
25		Mn	+4			0.39			0.530						
25		Mn	+5			0.33									
25		Mn	+6			0.255									
25		Mn	+7			0.25			0.46						
26	iron	Fe HS	+2			0.63	0.64		0.780		0.92				
26		Fe LS	+2						0.61						
26		Fe	+3					0.58							

Adapted from Bloss (1994)
HS indicates a high spin state; see Chapter 7.
LS indicates a low spin state; see Chapter 7.
PY indicates a cation located above the center of four ions arranged in a square (like the top of a four-sided pyramid).
SQ indicates a central ion with four neighbor ions arranged around it at the corners of a square.

Table 3.4. (continued)

Z	*element*	*symbol*	*valance*	*2*	*3*	*4 tet*	*4 sp*	*5*	*6*	*7*	*8*	*9*	*10*	*11*	*12*
26		Fe HS	+3			0.49			0.645		0.78				
26		Fe LS	+3						0.55						
26		Fe	+4						0.585						
26		Fe	+6			0.25									
27	cobalt	Co	+2								0.90				
27		Co HS	+2			0.58		0.67	0.745						
27		Co LS	+2						0.65						
27		Co HS	+3						0.61						
27		Co LS	+3						0.545						
27		Co	+4			0.40									
27		Co HS	+4						0.53						
28	nickel	Ni	+2			0.55	0.49	0.63	0.690						
28		Ni HS	+3						0.600						
28		Ni LS	+3						0.56						
28		Ni LS	+4						0.48						
29	copper	Cu	+1	0.46		0.60			0.77						
29		Cu	+2			0.57	0.57	0.65	0.73						
29		Cu LS	+3						0.54						
30	zinc	Zn	+2			0.60		0.68	0.740		0.90				
31	gallium	Ga	+3			0.47		0.55	0.620						
32	germanium	Ge	+2						0.73						
32		Ge	+4			0.390			0.530						
33	arsenic	As	+3						0.58						
33		As	+5			0.335			0.46						
34	selenium	Se	-2						1.98						
34		Se	+4						0.50						
34		Se	+6			0.28			0.42						
35	bromine	Br	-1						1.96						
35		Br	+3				0.59								
35		Br	+5		0.31[PY]										
35		Br	+7			0.25			0.39						
37	rubidium	Rb	+1						1.52	1.56	1.61	1.63	1.66	1.69	1.72
38	strontium	Sr	+2						1.18	1.21	1.26	1.31	1.36		1.44
39	yttrium	Y	+3						0.900	0.96	1.019	1.075			
40	zirconium	Zr	+4			0.59		0.66	0.72	0.78	0.84	0.89			
41	niobium	Nb	+3						0.72						
41		Nb	+4						0.68		0.79				
41		Nb	+5			0.48			0.64	0.59	0.74				
42	molybdenum	Mo	+3						0.69						
42		Mo	+4						0.650						
42		Mo	+5			0.46			0.61						
42		Mo	+6			0.41	0.50		0.59	0.73					
43	technetium	Tc	+4						0.645						
43		Tc	+5						0.60						
43		Tc	+6			0.37			0.56						
44	ruthenium	Ru	+3						0.68						
44		Ru	+4						0.620						
44		Ru	+5						0.565						
44		Ru	+7			0.38									
44		Ru	+8			0.36									
45	rhodium	Rh	+3						0.665						

HS indicates a high spin state; see Chapter 7.
LS indicates a low spin state; see Chapter 7.
PY indicates a cation located above the center of four ions arranged in a square (like the top of a four-sided pyramid).
SQ indicates a central ion with four neighbor ions arranged around it at the corners of a square.

Table 3.4. (continued)

Z	element	symbol	valance	2	3	4 tet	4 sp	5	6	7	8	9	10	11	12
45		Rh	+4						0.60						
45		Rh	+5						0.55						
46	palladium	Pd	+1	0.59											
46		Pd	+2			0.64^{SQ}			0.86						
46		Pd	+3						0.76						
46		Pd	+4						0.615						
47	silver	Ag	+1	0.67		1.00	1.02	1.09	1.15	1.22	1.28				
47		Ag	+2				0.79		0.94						
47		Ag	+3				0.67		0.75						
48	cadmium	Cd	+2			0.78		0.87	0.95	1.03	1.10				1.31
49	indium	In	+3			0.62			0.800		0.92				
50	tin	Sn	+4			0.55		0.62	0.690	0.75	0.81				
51	antimony	Sb	+3			0.76^{PY}		0.80	0.76						
51		Sb	+5						0.60						
52	tellurium	Te	-2						2.21						
52		Te	+4		0.52	0.66			0.97						
52		Te	+6			0.43			0.56						
53	iodine	I	-1						2.20						
53		I	+5		0.44^{PY}				0.95						
53		I	+7			0.42			0.53						
54	xenon	Xe	+8			0.40			0.48						
55	cesium	Cs	+1						1.67		1.74	1.78	1.81	1.85	1.88
56	barium	Ba	+2						1.35	1.38	1.42	1.47	1.52	1.57	1.61
57	lanthanum	La	+3						1.032	1.10	1.160	1.216	1.27	1.36	
58	cerium	Ce	+3						1.01	1.07	1.143	1.196	1.25		1.34
58		Ce	+4						0.87		0.97		1.07		1.14
59	praseodymium	Pr	+3						0.99		1.126	1.179			
59		Pr	+4						0.85		0.96				
60	neodymium	Nd	+2								1.29	1.35			
60		Nd	+3						0.983		1.109	1.163			1.27
61	promethium	Pm	+3						0.97		1.093	1.144			
62	samarium	Sm	+2							1.22	1.27	1.32			
62		Sm	+3						0.958	1.02	1.079	1.132			1.24
63	europium	Eu	+2						1.17	1.20	1.25	1.30	1.35		
63		Eu	+3						0.947	1.01	1.065	1.120			
64	gadolinium	Gd	+3						0.938	1.00	1.053	1.107			
65	terbium	Tb	+3						0.923	0.98	1.040	1.095			
65		Tb	+4						0.76		0.88				
66	dysprosium	Dy	+2						1.07	1.13	1.19				
66		Dy	+3						0.912	0.97	1.027	1.083			
67	holmium	Ho	+3						0.901		1.015	1.072	1.12		
68	erbium	Er	+3						0.890	0.945	1.004	1.062			
69	thulium	Tm	+2						1.03	1.09					
69		Tm	+3						0.880		0.994				1.052
70	ytterbium	Yb	+2						1.02	1.08	1.14				
70		Yb	+3						0.868	0.925	0.985	1.042			
71	lutetium	Lu	+3						0.861		0.977	1.032			
72	hafnium	Hf	+4			0.58			0.71	0.76	0.83				
73	tantalum	Ta	+3						0.72						
73		Ta	+4						0.68						
73		Ta	+5						0.64	0.69	0.74				

HS indicates a high spin state; see Chapter 7.
LS indicates a low spin state; see Chapter 7.
PY indicates a cation located above the center of four ions arranged in a square (like the top of a four-sided pyramid).
SQ indicates a central ion with four neighbor ions arranged around it at the corners of a square.

Table 3.4. (continued)

Z	*element*	*symbol*	*valance*	*2*	*3*	*4 tet*	*4 sp*	*5*	*6*	*7*	*8*	*9*	*10*	*11*	*12*
74	tungsten	W	+4						0.66						
74		W	+5						0.62						
74		W	+6			0.42		0.51	0.60						
75	rhenium	Re	+4						0.63						
75		Re	+5						0.58						
75		Re	+6						0.55						
75		Re	+7			0.38			0.53						
76	osmium	Os	+4						0.630						
76		Os	+5						0.575						
76		Os	+6					0.49	0.545						
76		Os	+7						0.525						
76		Os	+8			0.39									
77	iridium	Ir	+3						0.68						
77		Ir	+4						0.625						
77		Ir	+5						0.57						
78	platinum***	Pt	+2				0.60		0.80						
78		Pt	+4						0.625						
78		Pt	+5						0.57						
79	gold	Au	+1			1.37									
79		Au	+3				0.68		0.85						
79		Au	+5						0.57						
80	mercury	Hg	+1		0.97	1.19									
80		Hg	+2	0.69		0.96			1.02		1.14				
81	thallium	Tl	+1						1.50		1.59				1.70
81		Tl	+3			0.75			0.885		0.98				
82	lead	Pb	+2			0.98^{PY}			1.19	1.23	1.29	1.35	1.40	1.45	1.49
82		Pb	+4			0.65		0.73	0.775		0.94				
83	bismuth	Bi	+3					0.96	1.03		1.17				
83		Bi	+5						0.76						
84	polonium	Po	+4						0.94		1.08				
84		Po	+6						0.67						
85	astatine	At	+7						0.62						
87	francium	Fr	+1						1.80						
88	radium	Ra	+2								1.48				1.70
89	actinium	Ac	+3						1.12						
90	thorium	Th	+4						0.94		1.05	1.09	1.13	1.18	1.21
91	protactinium	Pa	+3						1.04						
91		Pa	+4						0.90		1.01				
91		Pa	+5						0.78		0.91	0.95			
92	uranium	U	+3						1.025						
92		U	+4						0.89	0.95	1.00	1.05			1.17
92		U	+5						0.76	0.84					
92		U	+6	0.45		0.52			0.73	0.81	0.86				
93	neptunium	Np	+2						1.10						
93		Np	+3						1.01						
93		Np	+4						0.87		0.98				
93		Np	+5						0.75						
93		Np	+6						0.72						
93		Np	+7						0.71						
94	plutonium	Pu	+3						1.00						
94		Pu	+4						0.86		0.96				

HS indicates a high spin state; see Chapter 7.
LS indicates a low spin state; see Chapter 7.
PY indicates a cation located above the center of four ions arranged in a square (like the top of a four-sided pyramid).
SQ indicates a central ion with four neighbor ions arranged around it at the corners of a square.

Table 3.4. (continued)															
Z	element	symbol	valance	2	3	4 tet	4 sp	5	6	7	8	9	10	11	12
94		Pu	+5						0.74						
94		Pu	+6						0.71						
95	americium	Am	+2							1.21	1.26	1.31			
95		Am	+3						0.975		1.09				
95		Am	+4						0.85		0.95				
96	curium	Cm	+3						0.97						
96		Cm	+4						0.85		0.95				
97	berkelium	Bk	+3						0.96						
97		Bk	+4						0.83		0.93				
98	californium	Cf	+3						0.95						
98		Cf	+4						0.821		0.92				
102	nobelium	No	+2						1.10						

HS indicates a high spin state; see Chapter 7.
LS indicates a low spin state; see Chapter 7.
PY indicates a cation located above the center of four ions arranged in a square (like the top of a four-sided pyramid).
SQ indicates a central ion with four neighbor ions arranged around it at the corners of a square.

Go ahead and attach the oranges to the apple with the toothpicks or something similar so that they don't roll away. Notice that we still haven't packed the maximum possible number of oranges around the central apple; at this point we are only working with one layer and not above or below. Next, see how many oranges you can put on in a second layer, still touching the central apple (i.e., place the oranges on top with each new orange touching the central apple). The answer should be exactly three. You could imagine that you could duplicate your top layer of three oranges, and put it on the bottom. If the whole thing is stuck together well enough, you might try turning it over. You will find that exactly twelve oranges will fit around a central apple, assuming they are all identical in size. In other words, if you pack twelve oxygen atoms together around a central object, the only size cation that will *perfectly* fit between them is something that is the same size as an oxygen atom (say, a potassium!). You could repeat this entire process with Styrofoam balls and glue them together.

Now let's explore how smaller balls (i.e., cations) might fit into some of the voids made by the oranges (i.e., oxygens). Return to the single layer of six oranges with a central apple. Replace the central apple with an orange. This would now simulate a layer of oxygen atoms packed as closely together as possible. Place another orange on top of the layer of oranges as shown in Figure 3.17. Look at the size of the hole beneath it. What would be the coordination number of a cation that would fit in the void below this orange? Next, try to find some piece of fruit, or any sphere, that will just fit into this void. A small grape is probably about the right size. This much smaller sphere (relative to the oranges) would represent the size difference between a large anion and a small cation.

There is another type of void created in the orange layer, but it is harder to envision. You'll need three or four more oranges to build this one. In Figure 3.18, the orange layer has been extended to the upper left by adding one orange. Next, three oranges are placed on top of the first layer as shown in the figure. Notice that a new void is created in the middle of three oranges on the bottom layer and the three on the top. What would be the coordination number for the cation that would fit in this void? As you did above, try to find a sphere that fits into this hole. Is it bigger or smaller than the sphere from Figure 3.17? You could assign the radius of the oranges to be 1.4 Å by measuring one and calculating a scale conversion. Then, given that scale, you could calculate

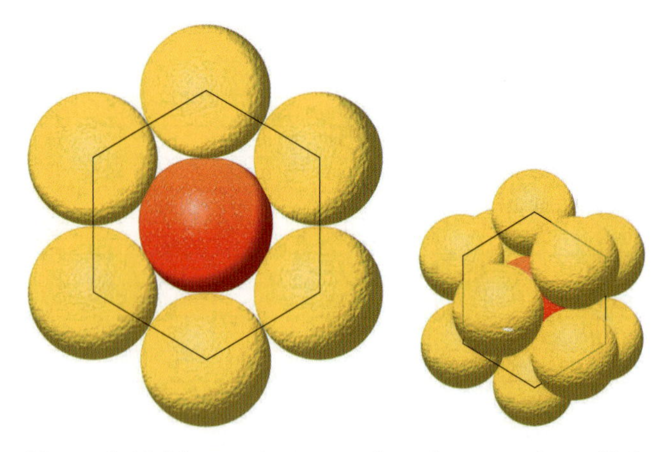

Figure 3.16. The maximum number of oranges that will fit exactly around a similarly-sized apple, with everything touching, is exactly 12.

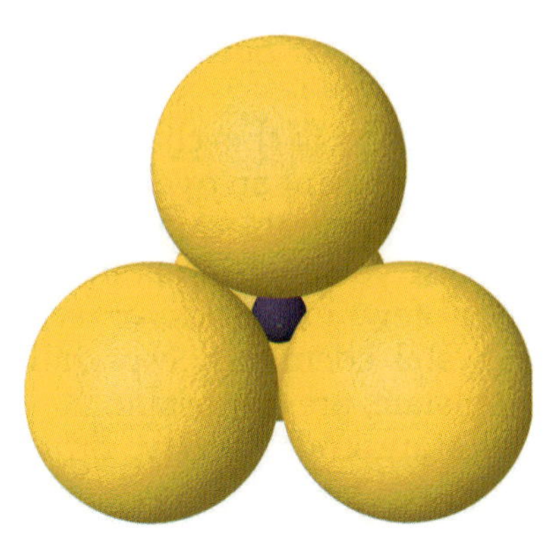

Figure 3.17. Notice that other voids are created when we place the oranges together. What size spheres will fit in these holes? Experiment and find a sphere (piece of fruit) that will fit in the void between four stacked oranges.

the size of the cations in the two examples. Check Table 3.3 to see how close you came to Si^{4+} in 4-coordination and Al^{3+} in 6-coordination.

As we just hinted, this entire geometric puzzle can be quantified. We can calculate the size of spheres that would fit in voids formed by packing spheres of equal size together. Something called the "radius-ratio rules" are the result, and will be discussed in more detail in Chapter 8. By using them, we can predict the size of cations that will fit into each of these voids, which are called coordination polyhedra. This concept works well for simple compounds that are basically anions packed together with cations fitting in the voids. As we move to the more complicated minerals, the analogy of close-packed spheres no longer works as well. However, the concept of the radius-ratio is still important in predicting what size range of cations will fit between differing numbers of oxygens.

Other trends are apparent from closer inspection of Table 3.4, which shows valences and coordination numbers for elements relevant to mineralogy. Consider the ionic radii for the aluminum cation, Al^{3+}. Al^{3+} can be found in minerals in three coordination states:

Coordination number	Ionic radius (Å)
4	0.39
5	0.48
6	0.54

Al^{3+} is the third most abundant element in the Earth's crust and the second most abundant cation behind Si^{4+}. So we are going be seeing a lot of it in minerals! One thing that makes Al^{3+} interesting is that it occurs both with a coordination number of four, as in Figure 3.17, and six, as in Figure 3.18. (1) It can substitute for Si^{4+}, which it does in the feldspars, where it would have a coordination number of four. (2) Or it can be in two different coordination sites in the same mineral, as in micas, where it has both a four and a six coordination number. (3) It could be entirely in 6-coordination, as in kyanite.

You may recall another example from Chapter 1. The three polymorphs of Al_2SiO_5 are also called the aluminosilicates: kyanite, andalusite, and sillimanite. More precisely, their formulas are $^{VI}Al^{VI}AlSiO_5$, $^{VI}Al^{V}AlSiO_5$, and $^{VI}Al^{IV}AlSiO_5$, respectively, where the superscripts represent the coordination numbers. Careful observation of these formulas shows that the only difference is in the coordination number of the second Al^{3+}. Thus, for this group Al^{3+} is 6-, 5-, and 4-coordinated. But why is this so? We already mentioned that pressure and temperature can have an influence on coordination number. The aluminosilicates are the best example of this. In fact, geologists use these minerals to predict the ancient conditions that were present when the minerals formed. Figure 3.19 is a **phase diagram** showing these relationships. You'll see more of these later in mineralogy and, if you continue in geology, in igneous and metamorphic petrology, and geochemistry. The phase diagram shows which phase (in this case, which one of the three aluminosilicates polymorphs) is stable under a given set of pressures and temperatures. You can see that kyanite would be stable at high pressure and low temperature. Can you think where this type of geologic setting might occur? (Hint: Think about different types of plate boundaries). In general,

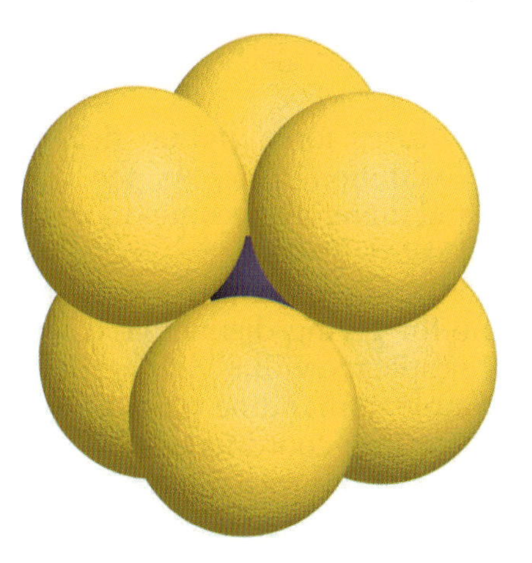

Figure 3.18. There is another way to place the second layer of oranges on top of the first. In Figure 3.17 we put the next layer (which was just a single orange) directly on top of the void in the first layer. What if we placed three oranges "around" the void in the first layer? Given this situation we now have six oranges (i.e., oxygens) surrounding the void. What is the size of the sphere that will fit in this hole as compared to the one in Figure 3.17 when only four oranges surrounded it?

higher pressure favors increased coordination numbers. Thus we can use an atomic level phenomenon to predict global scale geologic settings!

Oxidation state and size. A last important influence on ionic size is the oxidation state of the cation (also referred to as the valence state). **Oxidation state** is formally defined as the difference between the numbers of electrons associated with an ion relative to the atom in its elemental form. For example, native silver (Ag^0) is in the zero oxidation state, because it has the same number of electrons as atomic Ag. Earlier in this chapter, we discussed the fact that some elements are more stable when they have lost or gained electrons, in order to *completely fill* or *completely empty* their outermost orbitals. Another possibility is that enough electrons will be gained or lost to *half fill* all the outermost orbitals (that would be exactly one electron in each orbital) or even partially fill them. For some elements, multiple valence states are possible depending on the circumstances under which they form.

In minerals, some common elements that occur in multiple oxidation states are the metals titanium (Ti), vanadium (V), chromium (Cr), manganese (Mn), iron (Fe), and cobalt (Co), though many others exist. When these elements occupy 6-coordinated sites, their ionic radii are as follows:

element	2+	3+	4+	5+	6+
Ti	0.86	0.67	0.61		
V	0.79	0.64	0.58	0.54	
Cr	0.80	0.62	0.55	0.49	0.44
Mn	0.83	0.65	0.53		
Fe	0.78	0.65			
Co	0.75	0.61	0.53		

In every case, ionic radius decreases with increasing oxidation state. As the cation loses electrons (and therefore becomes more positively-charged), it gets smaller.

Incidentally, these multivalent cations are responsible for giving distinctive color to many minerals, a topic we will revisit again in Chapter 7. For example, almandine garnet owes distinctive red color to 8-coordinated Fe^{2+}, while pure andradite garnet has a characteristic green color from 6-coordinated Fe^{3+}. So the same element can give rise to very different colors! Thus, important information about the oxidation state and coordination of Fe in these garnets can be discerned from their color. This is just another case where a physical property of a mineral can give you information about its crystal chemistry. The physical properties of a mineral are directly related to its crystal structure (yes, this phrase is repeated several times because it is so important!).

In the introduction to this chapter, we told you our goal was to "present an overview of how and why atoms make minerals—and why you should care!" Just in case you still don't *care enough*, what follows are a few examples of why various aspects of crystal chemistry, especially ion size and oxidation state, are important.

Oxidation state and toxicity. *Erin Brockovich* (2000) was a very popular movie starring Julia Roberts, dealing with environmental contamination. The scientific basis of the story dealt with the oxidation state of Cr. Cr^{6+} is harmful if inhaled. In cooling towers used for heating systems, Cr is added to water so that unwanted deposits do not form. These are often placed on top of buildings. When the water evaporates, Cr^{6+} can be released as dust into the air. Cooling towers with fans also produce considerable water aerosol. If Cr^{6+} is inhaled, it is a health risk (Cr^{3+} presents less of a problem). If you saw the movie, you may be thinking that drinking water was a concern, but actually, the major health problem appears to come from inhaling Cr^{6+} in air, not from consuming Cr^{6+} in water. Other elements also have different toxic effects depending on their oxidation state.

Goiter, table salt, and iodine. You probably use table salt on your food every day. If you've ever looked at a salt container, you may notice that most salt is "iodized," which means that it contains the element iodine. It is less likely that

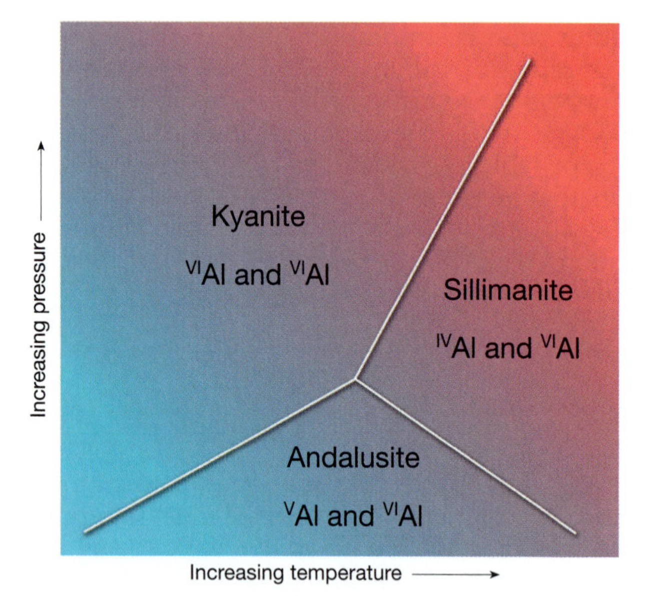

Figure 3.19. There are three minerals that have the chemical formula Al_2SiO_5: kyanite, andalusite, and sillimanite. The pressure and temperature controls which will form (e.g., lower temperature and higher pressures favor kyanite).

you've heard of goiter. Goiter is a disease caused by a lack of iodine in the diet. Outwardly it results in an enlargement of the thyroid gland (Figure 3.20); it also leads to lower IQ's, stunted growth, and stillbirths, to mention only a few of the health problems. The disease was eradicated in the United States in the mid-1900s. There is enough iodine in the soil near the world's oceans that food grown in those areas contains enough of this necessary nutrient. But food grown inland lacks iodine. Thus, approximately 1/3 of the world's population is at risk of this easily preventable disease.

So how did the U.S. eradicate this disease? By simply adding iodine to table salt, a change that cost only pennies. A small amount of KI is added to the NaCl. Check out the label on the container of Morton's salt (Figure 3.21). If the developed nations really wanted to help the developing nations, distributing iodized salt would be one of the most important things we do at a very low cost—but politics get in the way. Another crystal chemical wonder of table salt, although of much less significance than preventing goiter in millions of people, comes from adding "calcium silicate and dextrose," which serve as an anticaking agent in areas of high humidity. From this advance came Morton's trademark, "when it rains, it pours."

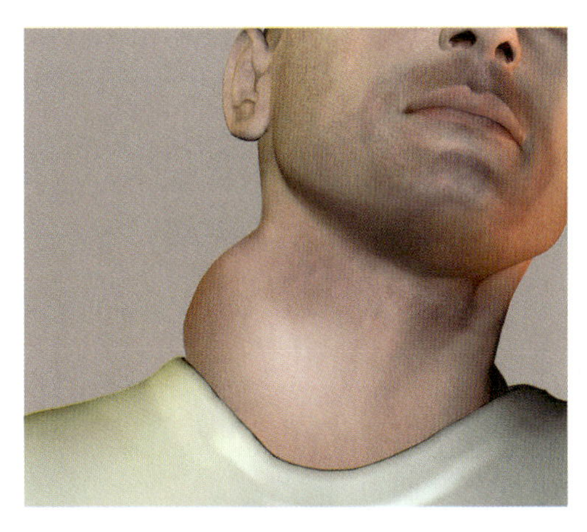

Figure 3.20. Image showing an enlarged thyroid gland, which results from iodine deficiency.

Fluoride and drinking water. Since the 1950s, most municipal water supplies in the U.S. have added 1 ppm F to their water. Although there remains some debate on how it works to prevent tooth decay, we know that it does. In fact, many of you will never have a cavity because of this. Ask your parents and grandparents how many cavities they have. Your teeth and bones contain the mineral apatite, which has the formula $Ca_5(PO_4)_3(OH,F,Cl)$. There are three different anions that can substitute for one another: Cl^{1-}, $(OH)^{1-}$, and F^{1-}. Your teeth and bones are predominately made of hydroxylapatite, which is a mineral that has mostly $(OH)^{1-}$ in that site. But F^{1-} will substitute for $(OH)^{1-}$ in the enamel of your teeth, and make it slightly less soluble (especially in acids), in such a way as to reduce cavities. Sadly, some (few) cities in the U.S. are still fighting to prevent their water systems from being fluoridated. Again, politics and junk science (claiming fluoride poses a health risk) get in the way.

Aging and oxidation. Although this may be a little off the subject of mineralogy, it deals with a

Figure 3.21. Photos of a Morton salt container. a. The front of the container stating the salt is "iodized" and the company's trademark "girl with umbrella in the rain with the salt pouring out." As explained in the text, the salt pours because of a mineral additive, and the "iodizing" has eliminated goiter in the US.

process we all do—age! For centuries, humans have tried to understand what causes us to age and how to slow, or prevent, aging. One of the processes deals with oxidation of cell walls. Once oxidation proceeds too far, our immune system thinks the cell walls are invaders and begins to attack them. No doubt you've also heard that taking antioxidants or eating food high in antioxidants might help reduce the risk of cancer. So oxidation is not only all around us, it's also in us!

References

Blackburn, W.H. and Dennen, W.H. (1997) Encyclopedia of Mineral Names. The Canadian Mineralogist, Special Publication 1. Mineralogical Association of Canada, Ontario. 360 pp.

Bloss, F.D. (1994) Crystallography and Crystal Chemistry. Mineralogical Society of America, Washington D.C. 293 pp.

Brady, J.B., Newton, R.M., and Boardman, S.J. (1995) New uses for powder X-ray diffraction experiments in the undergraduate curriculum. Journal of Geological Education, 43, 466–470.

Gibbs, G.V., Spackman, and Boisen, M.B., Jr. (1992) Bonded and promolecule radii for molecules and crystals. American Mineralogist, 77, 741–750.

Gray, H.B. (1964) Electrons and chemical bonding. Addison-Wesley Pub. Co.

Gunter, M.E. (1993) Some thoughts on teaching beginning geology. Journal of Geological Education, 41, 133–139.

Shannon, R.D. and Prewitt, C.T. (1969) Effective ionic radii and in oxides and fluorides. Acta Crystallographie, B25, 925–946.

Sinkakas, J. (1966) Mineralogy: A First Course. Van Nostrand, NY, NY. 587 pp.

Chapter 4

Crystallography

By the time you finish with this chapter and Chapter 11 of this book, it will become apparent to you that one of us is a quiltmaker. So I will confess right up front that I am the guilty party. My interest in quiltmaking really began before birth, because my mother was working on a king-size quilt when I was in utero. Sometime just before I left for college, the quiltmaking bug bit me really hard, and I could be found in all my spare time with a pile of fabric squares on my lap and a needle in my hand. I have never been a person who could sit still easily!

While I was in graduate school, I satisfied my artistic urges by studying painting three mornings a week at nearby Wellesley College, which I had attended as an undergraduate. It was a relief to escape the relentless pace of M.I.T. for a few hours, and to use the other half of my brain (while enjoying the company of women, a rare treat for me in those days). Looking back on it, I should not be surprised that the things I painted ended up being rather geometrical; my artistic expressions involved shape, color, and pattern. When I moved to Pasadena and then on to Oregon, I returned to quilt making, applying the lessons I learned from oil and acrylics to the medium of fabric. Now I make art quilts in my free time (what little free time I have!); hopefully someday I will find the time to exhibit them in shows.

Many years ago I took a quilt-making class from a wonderful artist named Margaret Miller. As an exercise, she assigned the task of arranging our triangular pieces on a wall in various patterns: all facing the same way, rotating in alternate directions, pointing up and pointing down, etc. Somewhere in this exercise I realized that we were simply running the gamut of possible symmetry operations. It was then that I had one of those personal epiphanies about myself. For me, science and art are interconnected. What I love about quilt-making is the same thing I love about crystallography: the variety of combinations that can be made from some very simple, geometrically-constrained actions like rotation and inversion. Soon thereafter I discovered a book by Ruth B. McDowell called Symmetry: A Design System for Quiltmakers *(1994), as well as an old library book of Escher patterns converted into quilt patterns, and my happiness was complete. Since then, I have found artistic expression in endless arrangements and rearrangements of bits of fabric—and I am always conscious of the symmetry of the pattern I'm working on. If a quilt design isn't working, I perform a symmetry operation or two, and something good always happens. The results have even been known to appear as exercises on mineralogy exams. Yes, science can be warm and fuzzy!*

The moral of the story is that the rules governing crystallography are timeless and broadly applicable. Don't ever assume, as I once did, that the things you learn in mineralogy (or any) class are bits of knowledge to be compartmentalized and sealed up for use exclusively in final exams and, perhaps, stored away for subsequent classes within some narrow discipline. The study of crystallography should enhance your awareness of the world of patterns that exists all around you: in brick walls, wallpaper, floor tiles, bedspreads, all kinds of art, and even plant life. There are intellectual connections in the offering everywhere, if only you take the time to look for them.

M.D.D.

Introduction

So, you want to build a mineral structure? In the previous chapter, we learned all about the structure and bonding capabilities of atoms, but now we need some guidance on how best to arrange them in the three-dimensional, long-range ordered structures that make up minerals. This chapter is an introduction to the science of crystallography, which is a symbolic language for describing repeating patterns in 2-D, 3-D, or any dimension of space one wants. It just so happens that the crystal structures of minerals are repeating patterns in 3-D, and we can take advantage of all the work that has been done to study patterns to help us understand minerals. Much later on in this book, you will have a chance to learn the mathematical basis for everything we talk about, so don't be fooled: what seems descriptive here is actually a very precise, mathematically-elegant way of describing space. In fact, it is really much easier to describe and understand symmetry based on mathematics than any other way.

The basis for the science of crystallography really began in 1669 with Nicolaus Steno, who noticed that quartz crystals, no matter where they came from or what size they were, always had the same set of characteristic angles between their faces. This observation, which is called Steno's Law, laid the foundation for subsequent studies of crystals in the 18th, 19th, and 20th centuries (Table 4.1). By the 1920s, the fundamental rules of crystallography, including its mathematical basis, had been worked out fairly well. Subsequent improvements in our ability to analyze crystals down to the atomic and subatomic levels have only confirmed what the early workers theorized—all on the basis of crystal morphology! However, they could only "theorize" structures based on morphology for fairly simple minerals like halite. This is just another example of how understanding a physical characteristic of a mineral (in this case, its shape), tells you something profound about the fine-scale details of its interior structure, because the internal structure really controls the shape.

All types of crystalline matter are based upon repetitions in space of identical structural units. These can be a single atom or a group of atoms, and are represented by the chemical formula of the material. This structural unit can be referred to as a **motif** or **basis**. In simple materials like the metals copper, silver, and gold, there is only one element in the crystal structure. In ice, the motif is the H_2O molecule, and in halite, Na^{1+} and Cl^{1-} atoms. In silicates such as muscovite ($KAl_2(AlSi_3O_{10})(OH)_2$), a large number of atoms or groups of atoms makes up the structural unit. In proteins there may be 10^4 atoms making up the motif!

We can describe the structure of crystals in terms of single periodic lattices that are repeated in three dimensions with our structural or chemical unit (motif or basis) repeated at or surrounding each lattice point (see Chapter 13). Crystal structures are therefore built up from various combinations of lattices (i.e., the repeating patterns) and structural units (i.e., the elements present).

Two-Dimensional Space

To make this manageable, let's begin with the notion of *two*-dimensional space. Quilt-makers have for years struggled with the basic concepts of arranging patterns in two dimensions. In fact, patterns can be conveniently explained in terms of simple geometry using the idea of **basis vector sets**, which is a fancy way of saying **coordinate systems**. Think of a sheet of graph paper, which gives you a large array of squares on a 2-D sheet of paper. We can describe those squares by saying that there are two basis vectors corresponding to the x and y directions, where x and y have the same units. The vectors are the same length, or the same number of units long. We could describe this by using length a for the number of units in the x direction, and length b for the number of units in the y direction. Thus, for the case of our graph paper, length $a = b$. The angles between those vectors will be 90°. This combination of descriptors gives you an array of squares.

What happens if we change the relationships between those vector lengths and the 90° angles between then—you would have a hard time finding graph paper with rectangles or triangles (though they do exist!). There are several ways of doing this. What happens if we change the lengths of the vectors? So now:

x and y have the same units,
$a \neq b$, and
angles = 90°.

This combination of basic vectors gives you a periodic pattern of rectangles.

We could also leave the lengths the same, but vary the angle between the basis vectors. What kind of pattern will we get if we use the following? Let

x and y have the same units,
$a = b$, and
angles = 60°.

If you sketch this out on a sheet of paper, you will find that it results in an array of equilateral diamonds, which can form a repeating pattern.

Table 4.1. A Brief History of Crystallography

1597	Libavius finds that the geometry of crystals is a fundamental characteristic of the salts he is studying.
1611	Johannes Kepler writes that the six-sided shape of snowflakes is due to "regular packing of the constituent particles."
1665	Hooke proposes that crystals are made up of "spheroids."
1669	Steno finds that quartz crystals always have the same angles between faces.
1780	Carangeot invents the contact goniometer, which can measure the angles between crystal faces.
1783	On the basis of studies of crystal cleavage, Bergman proposes that crystals are made up of packed rhombohedral units.
1783	de'Lisle confirms Steno's Law with the "Law of Constancy of Interfacial Angles."
1801	Haüy proposes that mineral shapes are due to stacking of basic structural units.
1809	Wollaston develops the reflection goniometer for the measurement of reflecting surfaces, allowing interfacial angles to be measured with great accuracy.
1819–1822	Mitscherlich discovers isomorphism and polymorphism.
1839	Miller devises Miller Indices to describe crystal faces.
1848	Pasteur describes enantiomorphous crystals.
1880–1900	Sohncke, Federov, Schönflies, and Barlow proposed theories of internal symmetry of crystals.
1906–1919	Groth publishes "Chemische Krystallographie," a summary of physical properties of >7,000 crystalline substances.
1907	Barlow and Pope suggest that ions are hard spheres that touch each other.
1912	Discovery of X-ray diffraction by Friedrich, Knipping, and von Laue.
1913	W.H. and W.L. Bragg use X-ray diffraction to solve the first crystal structure of halite.
1913	Ewald invents the concept of the reciprocal lattice.
1914	Debye proposes the theory of thermal motion of atoms in solids.
1916	Debye and Scherrer begin powder diffraction studies.

Our thanks to Stephen Heyes, University of Oxford, for his permission to borrow this compilation.

The last manipulation would be to let the angles and the repeat distances vary at the same time. We can express this as:

x and y have the same units,
$a \neq b$, and
angles $\neq 90^\circ$.

Again, if you try drawing patterns that satisfy these characteristics, you'll find that there are many ways you can make a periodic array. I have never seen graph paper with this arrangement of lines, though. In the end, it's important to realize that what we are working toward here is the fact that minerals make patterns with their atoms as they crystallize. There are going to be several different types of patterns.

Symmetry Operations

In the previous section, we discussed the creation of 2-D patterns by varying axis lengths and angles. Embedded in that discussion is the idea that you can vary the repetition by changing the angles between the axes. Another way of thinking about this is to say that you are taking the axes and rotating them. That **rotation** is part of a group of what we call **symmetry operations**. Symmetry operations involve a geometric movement that brings one point in a pattern into "coincidence" with another point. That's a fancy way of saying that when you perform a symmetry operation, the points all look the same as they did *before* you did the operation. Later in Chapters 11–13, we'll say that one axis, or point, is "mapped" to another by a particular operation.

To clarify this, let's work through the two symmetry operations we will need in this chapter (the others can wait until Chapter 11). The first operation is called **rotation**, which means just what you might think: turning, like a revolving door! We'll start with some two-dimensional examples. First, consider the wallpaper pattern shown in Figure 4.1. Imagine that you put a thumbtack in the very center of the pattern, and then rotated the piece of wallpaper around it. How much (how many degrees) would you have to rotate in order for the pattern to once again look the same? If you answered "120°," then you are correct. After three rotations, you would be back in your original starting place. So we would say that this pattern possesses a 3-fold rotation axis.

The quilt shown in Figure 4.2 is a traditional pattern called London Square. If only two solid fabrics were used in this quilt (red and white), how many degrees could you rotate this to have it look the same again? In reality, all the fabrics in this quilt are different; there are >100 different red-patterned fabrics and 100 white-patterned ones here. Given this reality, what type of rotation axis is present in the actual quilt? Take a look at the MSA website to find out the answer.

The other important symmetry operation for now is **reflection**. You deal with this type of symmetry every day when you look into the mirror. Every part of your face, zits and all, is reflected back at you. Every feature is reflected across the plane that is your mirror and into the space behind. Objects that are closest in the mirror (like your nose) have reflections that are close, and objects that seem far away in the mirror (your ears) look far away. So the symmetry operation of reflection occurs when a point is repeated across a **mirror plane**. Every part of the pattern is projected an equal and opposite distance onto the other side of the mirror plane (Figure 4.3). As part of this process, the **handedness** of the object changes. Handedness is a term that refers to any two objects that are alike except that they are related by a mirror reflection. Hold up your right hand in front of a mirror and its reflection looks like a left hand. The change in handedness explains why you can't read writing when it's reflected in a mirror.

Now stand in front of a mirror where you can see your entire body. When you look directly at the mirror, your body is reflected, and the mirror acts as a mirror plane. The next question is this: do you see another mirror plane, running down the center of your body, perpendicular to the mirror on your wall? Some (very) rare people are symmetrical in this way: their left side is a mirror image of their right side. But for most of us, there are subtle differences in body structure (including asymmetrical internal organs, right/left dominance, and the muscle changes these entail) that keep us from being perfectly symmetrical. To test this, take a digital photo of yourself and cut it in half with a photo editing program. Now flip the remaining side over and paste it back to make a body. See how different this looks? This exercise is particularly interesting to try with a photo of your face!

Figure 4.2. London Square, a traditional quilt pattern. How many different rotation axes can you find?

Figure 4.1. Wallpaper pattern with a 3-fold rotation axis.

The same principles apply to three-dimensional objects. In Figure 4.4, three perspectives of the same cube are shown, each representing a different axis. Let's look first at the 4-fold rotation axis. A line that is perpendicular to and in the very center of any face on a cube is a 4-fold axis, because when you rotate the cube 90° around that line, it will look the same (i.e., 360°/4 = 90°). Similar logic can be applied to the 3-fold and 2-fold axes on a cube, as shown in Figure 4.4.

Three-Dimensional Space

Now let's move on to explore in detail the variations that are possible in *three* dimensions. The simple rules we established for basis vector sets in 2-D will serve us well here also. To describe 3-D space, we need three vectors, which we will designate as *a, b,* and *c*. Note that Europeans use *x, y,* and *z* for same purpose.

Recall that we've already discussed some of this before, back in Chapter 1, where we first introduced the six crystal systems. In fact, it

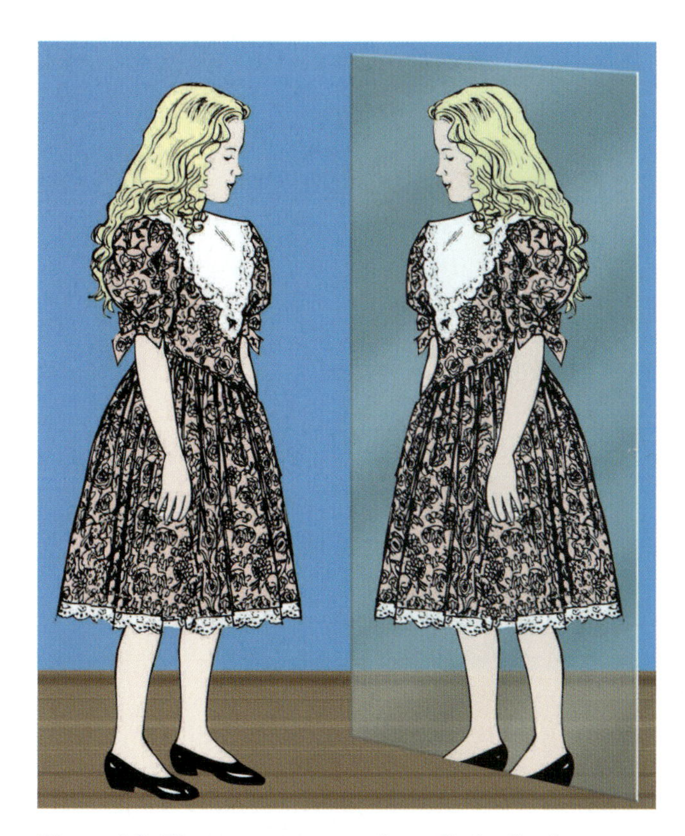

Figure 4.3. The symmetry operation called reflection occurs when every part of a pattern is projected an equal and opposite direction across a mirror plane.

might be a good idea to go back and review that section, or at least take a look at Figure 1.11 which shows the six different shapes that result from the six different crystal systems.

One of the problems of Figure 1.11 is that we are trying to project 3-D space on a 2-D plane. To do this we make perspective drawings, as in Figure 1.11, but these can be hard to interpret and visualize. Because much of mineralogy involves 3-D angular data, we need to devise some graphical methods to help us. You may remember the technique of spherical projections from an earlier social studies class. Imagine that you want to show the locations of various cities on a two-dimensional circular map of the Earth. Each city could be represented by a *vector* made by drawing a line from the center of the Earth to the location of the city. To make your map, you look down on the Earth from the North Pole, and project each city's location vector onto the equatorial plane. The equatorial plane cuts through the center of the Earth perpendicular to the North (and South) Poles. If you projected the endpoint of each vector directly down onto the circle, you'd get a weird sort of map, because all the cities at low latitudes (near the equator) would be compressed onto the outside of the plot (Figure 4.5).

There is a clever way of getting around this problem. If you project each city's location (i.e., the end of the vector) along a line that runs to the Pole of the *opposite* hemisphere (either the South or North Pole), you will get a more even distribution of cities on your round map. This is called a **stereographic projection** and it is a commonly used tool in geology, especially structural geology (Figure 4.5).

So let's plot the following locations on our equatorial section of the Earth: the North Pole, Greenwich, England (home of 0° longitude), South Hadley, Massachusetts (home of Mount Holyoke College), Moscow, Idaho (home of the University of Idaho), Bern, Switzerland, and Kyoto, Japan (Figure 4.6). The result is a two-dimensional drawing that specifies the locations of these cities very precisely by their angular coordinates. For each location, the angle relative

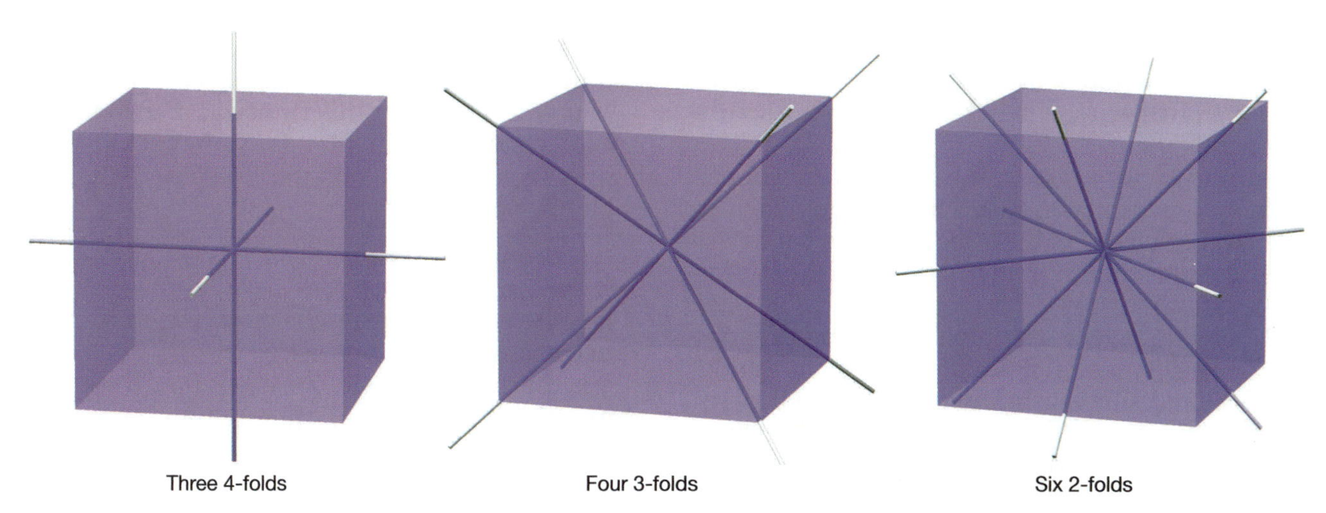

Figure 4.4. Rotation axes contained in a cube: three 4-folds running normal to the faces, four 3-folds running from corner to corner, and six 2-folds running from edge to edge.

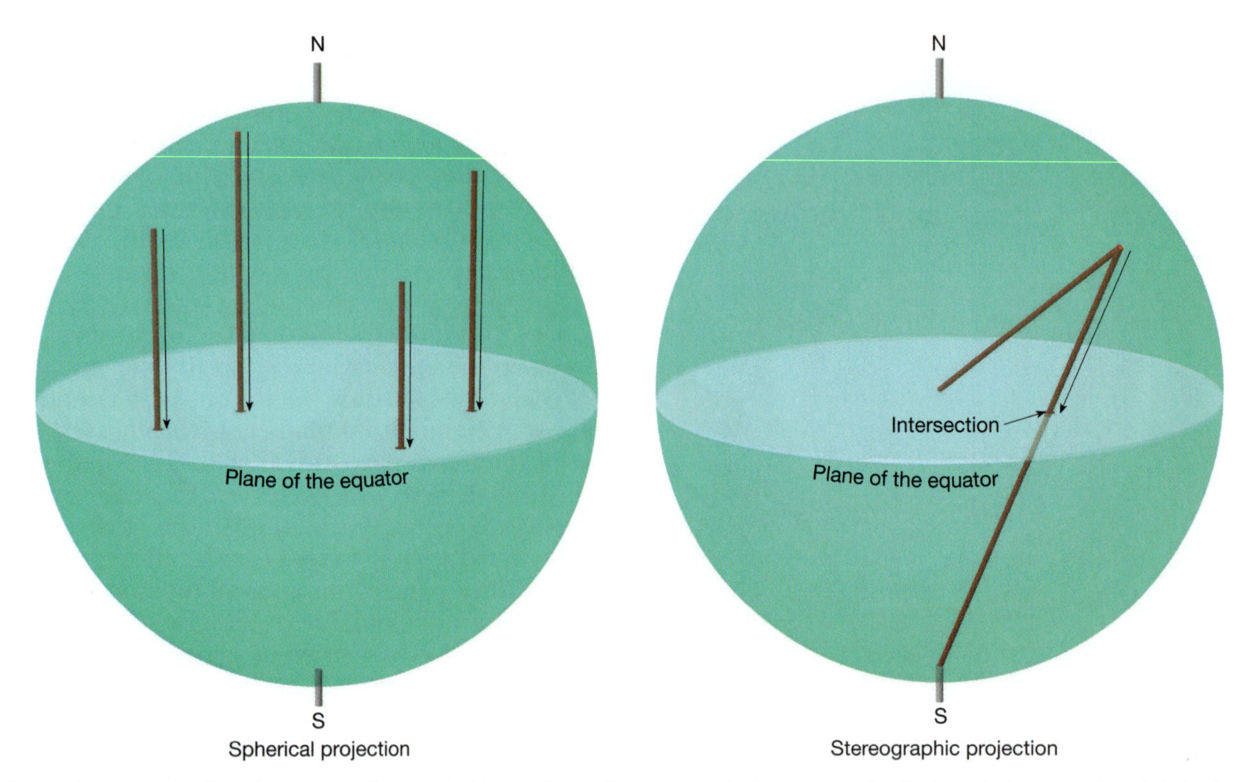

Figure 4.5. Spherical and stereographic projections of a vector onto a circle. For a spherical projection, each point is simply projected straight down onto the plane of the equator. In a stereographic projection, the end point of each vector is projected onto the opposite pole, and the location where it intersects the circle is labeled.

to Greenwich, England gives the longitude, and the distance of each point in from the edge of the circle tells you its latitude. Now, let's add the Galapagos Islands (on the equator) and Buenos Aries, Argentina, Jakarta, Indonesia, and Sydney, Australia, which all plot on the southern (or lower) hemisphere. To make the distinction between north and south (or upper and lower) hemisphere, we use a solid dot for the north, or upper, hemisphere vs. a hollow circle for the south, or lower, hemisphere, on our equatorial section. To help distinguish the two, we can think of the hollow circle as an open man-hole cover and then realize we'd fall through it to the southern hemisphere.

As stated above, we are using a polar coordinate system that you may have already seen in some other class. As we'll see later, we could use this to determine the angular distance between these cities.

Now, we have a new tool to help us describe the possible angular relationships in three-dimensional space using two-dimensional plots. Later in this book (Chapter 12), we'll find that it is handy to use stereographic projections to show the axis vectors for three-dimensional crystals, and to illustrate their relationships to other geometrical features of minerals.

Isometric System

Let's begin with the situation where all axes are the same length, and all angles are 90°. That's the highest amount of symmetry that a crystal can possess! This is the isometric crystal system, where lengths $a = b = c$, and vectors a, b, and c are at right angles to each another (Figure 4.7). In fact the derivation of the name "isometric" gives you a clue to its axes properties. "Iso" means "same" and "metric" means "measure," so isometric literally means "same measures" referring to the fact that the axes are the same length.

You will remember this from our discussion of crystal systems in Chapter 1. We have added to Figure 4.8 the proper notation for describing the angles between the axes:

α is the angle opposite the a axis (i.e., the angle between c and b).

β is the angle opposite the b axis (i.e., the angle between a and c).

γ is the angle opposite the c axis (i.e., the angle between a and b).

Another way to help you remember the relationships between a, b, c (the axis lengths, which are the cell parameters of a mineral) and α, β, γ (the interaxial angles) is sort of an odd-man-out rela-

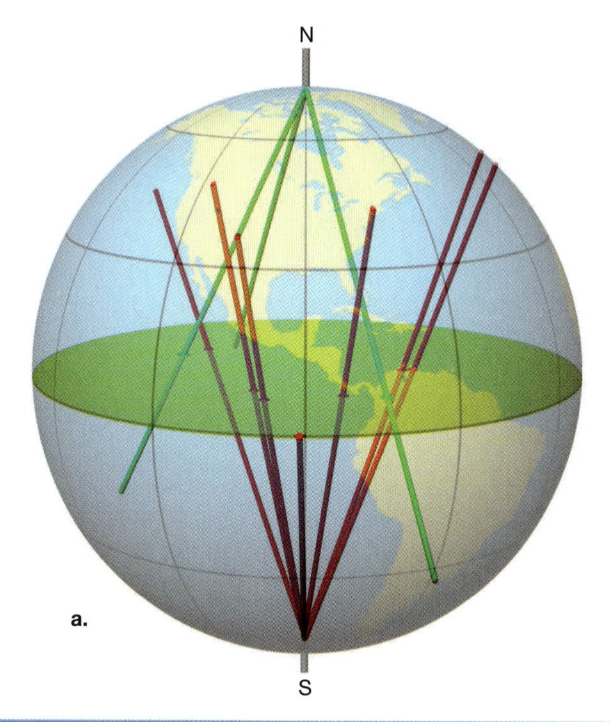

	Latitude	Longitude
Greenwich, England	51°N	0°
Taos, New Mexico	36°N	106°W
South Hadley, Massachusetts	42°N	73°W
Moscow, Idaho	47°N	117°W
Bern, Switzerland	47°N	7°E
Kyoto, Japan	35°N	136°E
Galapagos Islands	0°	91°W
Buenos Aires, Argentina	34°S	58°W
Jakarta, Indonesia	6°S	107°E
Sydney, Australia	34°S	151°E

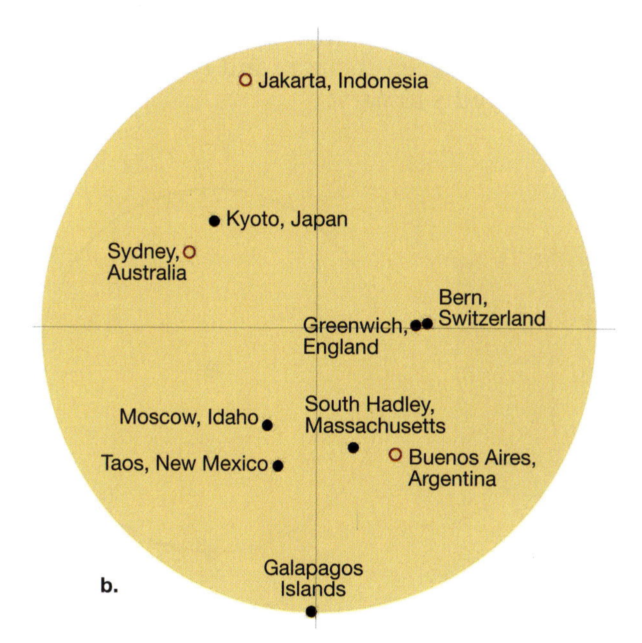

tionship. First realize that α, β, γ are the first three letters (in lower case) of the Greek alphabet. [Aside: we'll be using lots of Greek characters in mineralogy!] The α is pronounced alpha, β is beta, and γ is gamma. Thus, for the vectors *a* and *b*, γ would be the odd-man-out; for *a* and *c*, β is odd-man-out; and lastly for *b* and *c*, α is odd-man-out.

If you visualize these vectors inside a sphere, you can then appreciate the resultant stereographic projection, which will have *c* in the middle, *b* on the right-hand side, and *a* at the bottom (Figure 4.7). Each of the vectors is related to the other two by a 90° rotation, so we would label those rotation axes with the symbol for 4-fold rotation, which is a square. Notice that there are three unique 4-fold rotation axes that coincide with the three crystallographic axes. Some shapes with isometric symmetry include not only the cube (hexahedron), but a tetrahedron, an octahedron, and a dodecahedron. Minerals that crystallize with isometric symmetry include members of the garnet group, galena, pyrite, and fluorite.

We know from our discussion that a cube has additional rotation axes: 2-fold and 3-fold (Figure 4.4). How would we designate these on a stereo projection? Each of those rotation axes is a line in 3-D space, and so those lines can be represented on the stereo projection. Similarly, a cube has a bunch of mirror planes (up to nine of them!) in its shape, and these can be indicated as lines on the diagram. Thus, the stereographic projection for a cube is shown in Figure 4.7. There are other, different diagrams for other isometric shapes like a dodecahedron, but we'll meet them again later in Chapter 11. The lesson for now is that stereographic projections are a useful way to represent collections of symmetry operations and crystallographic axes.

Tetragonal System

In the tetragonal system, we let all the angles remain at 90°. Again, thinking about the name

Figure 4.6. a. *A 3-D view of the process used in stereographic projection to project points on the surface of a sphere onto its equatorial plane (i.e., the circle that would be formed by bisecting the globe with a plane parallel to the equator).* ***b.*** *The stereographic projection of the cities shown in the table. The cities that occur in the northern hemisphere are indicated by solid dots, while those in the lower hemisphere are indicated with hollow circles.*

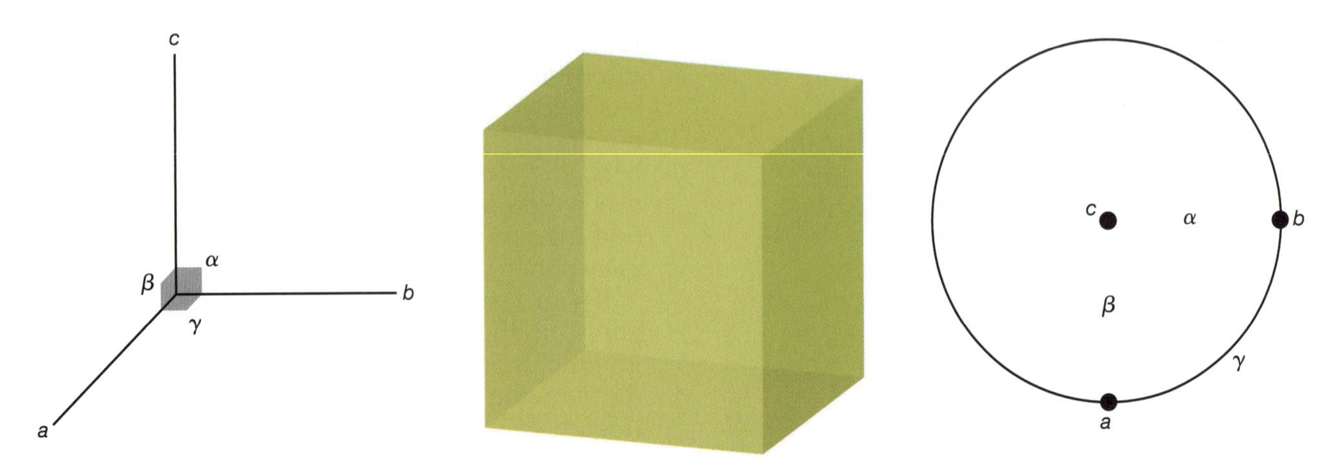

Figure 4.7. Representations of the isometric crystal system. Note how the lengths of a, b, *and* c *are all the same, and the angles between them, α, β, and γ, are all 90° (left): A 3-D perspective sketch showing the relationship between the three crystallographic axes. (center): A solid composed by these three axes showing one possible morphology in this crystal system. (right): A stereographic projection showing the angular relationships between the three crystallographic axes.*

will give us a clue to the axis system. "Tetra" means four, so the materials in this crystal system have 4-fold axes. Also, we can release the constraint that all the axes have to be the same length—this time, we'll let one of them be different. So mathematically, we would write this as:

lengths $a = b \neq c$, and
angles $\alpha = \beta = \gamma = 90°$.

Recall that the stereographic projection is used to show the angular relationships, but not the lengths of the axes, so in Figure 4.8 we show both the stereographic projection and a perspective sketch of the axis set. Crystals that form in the tetragonal system have a 3-D shape that looks like spaghetti boxes with square ends, hat boxes, and pieces of 4x4" fence posts. Their unifying characteristic is that they have 90° angles, and a 4-fold axes in only the *c* direction. In Chapter 13 we will show why this occurs mathematically. Some shapes with tetragonal symmetry include dipyramids (two pyramids stacked with their bases touching) and disphenoids (like tetrahedra, where two upper faces alternate with two lower faces). Minerals that crystallize with tetragonal symmetry include members of the scapolite group, chalcopyrite, and zircon.

Orthorhombic System

The prefix "ortho" means perpendicular, or with 90° angles, so the orthorhombic system keeps the 90° angles between the *a*, *b*, and *c*, but allows all the axis lengths to vary:

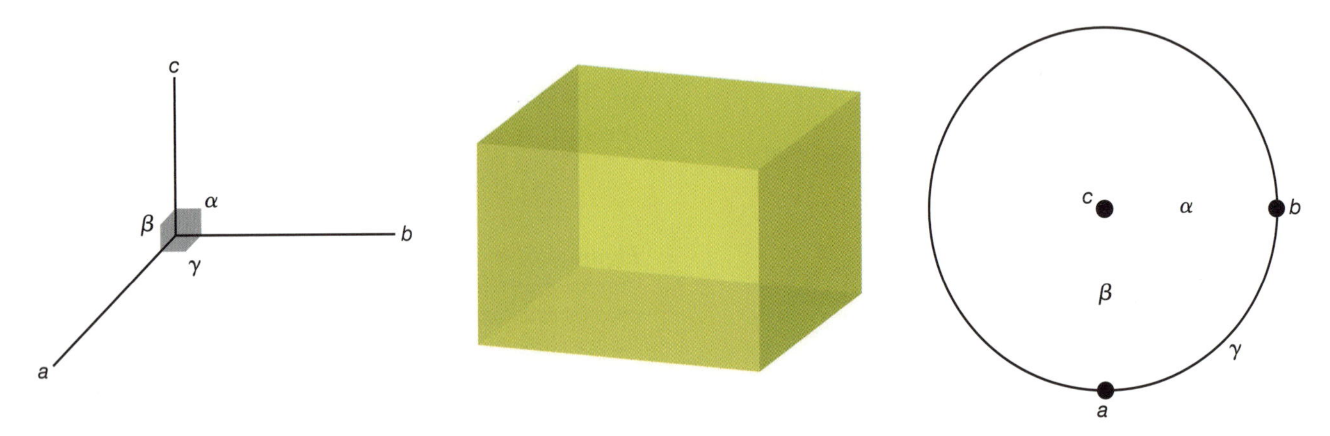

Figure 4.8. Representations of the tetragonal crystal system. Note how the lengths of a *and* b *are the same and* c *is different, but the interaxial angles are all still equal to 90°(left): A 3-D perspective sketch showing the relationship between the three crystallographic axes. (center): A solid composed by these three axes showing one possible morphology in this crystal system. (right): A stereographic projection showing the angular relationships between the three crystallographic axes.*

lengths $a \neq b \neq c$, and
angles $\alpha = \beta = \gamma = 90°$.

The resultant shape is relatively rare in minerals, because objects with orthorhombic symmetry can have nothing higher than the 2-fold rotation axes along *a*, *b*, and *c* (Figure 4.9), but many minerals crystallize in this system. Typical orthorhombic shapes include a cereal box and a brick, assuming they have no writing on them. The formal geometrical shapes are called rhombic prisms or rhombic dipyramids. Examples of minerals that crystallize in this system are olivine and some of the amphiboles and pyroxenes.

Hexagonal System

Now we can try another permutation of varying axes lengths and angles. Like the "tetra" in tetragonal, the "hexa" should give us a clue to the symmetry in this crystal system. Indeed, we'll find that we have 60° angles, or their multiples, involved in this system. In the hexagonal system, the axial relationships are the same as in the tetragonal system, but now one of the angles has different constraints:

lengths $a = b \neq c$, and
angles $\alpha = \beta = 90°$, $\gamma = 120°$.

Shapes in the hexagonal system are made up of prisms where the top and the bottom consist of rhombuses with 120° angles (Figure 4.10). The geometrical term for this is "right prism with a rhombus base." Hexagonal objects always have either a 3-fold or a 6-fold axis parallel to the c axis. The most abundant mineral species in the Earth's crust, quartz, crystallizes in this system. Calcite and corundum can crystallize in hexagonal scalenohedra (forms with twelve faces occurring in pairs).

Monoclinic System

In the name of this crystal system, "mono" means one and "clinic" refers to an angle (in this case a non-90°, inclined angle). Thus this crystal system has one-non 90° angle as follows:

lengths $a \neq b \neq c$, and
angles $\alpha = \gamma = 90°$, $\beta \neq 90°$.

An example of this relationship is shown in Figure 4.11. Note that shapes with monoclinic symmetry can have only a single mirror plane or 2-fold axes. The shapes of these crystals can only be described by saying that they are combinations of prisms with **pinacoids** (pairs of parallel faces). Many, if not most, silicate minerals crystallize in the monoclinic system, including many of the feldspars, layer silicates, amphiboles, and pyroxenes.

Triclinic System

In the name of this crystal system, "tri" means three. Thus as in monoclinic above, there are three non-90° angles in this crystal system. This is the least-*constrained* crystal system, with all the

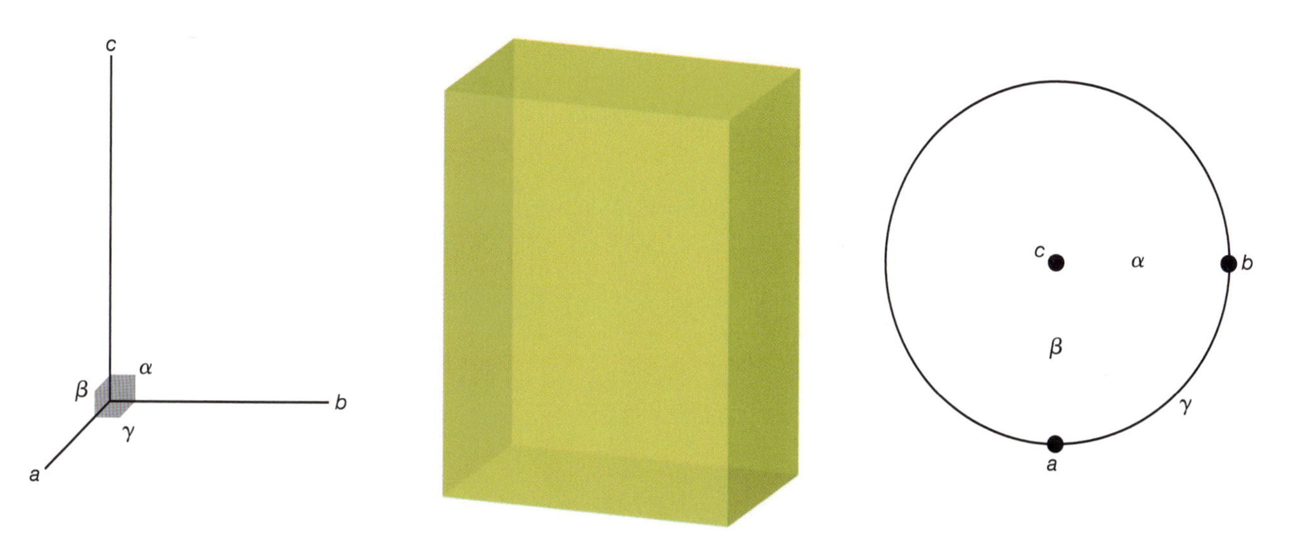

Figure 4.9. *Representations of the orthorhombic crystal system. For this crystal system* a, b, *and* c *can differ in length, while the interaxial angles are still constrained to be 90° (left): A 3-D perspective sketch showing the relationship between the three crystallographic axes. (center): A solid composed by these three axes showing one possible morphology in this crystal system. (right): A stereographic projection showing the angular relationships between the three crystallographic axes.*

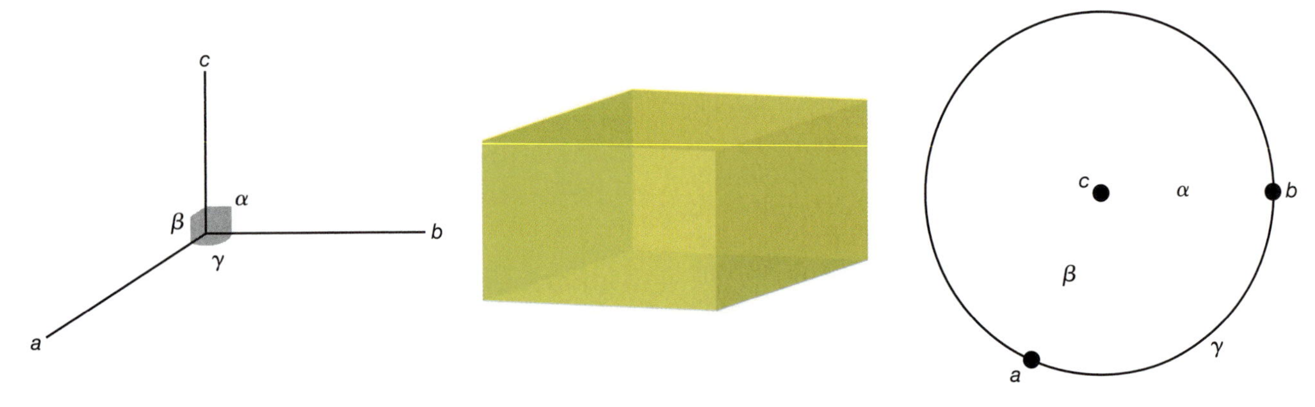

Figure 4.10. Representations of the hexagonal crystal system. This system is very similar to the tetragonal system, except the angle between a *and* b *(i.e.,* γ*) is 120° to reflect the 3- or 6-fold rotation axes parallel to* c *(left): A 3-D perspective sketch showing the relationship between the three crystallographic axes. (center): A solid composed by these three axes showing one possible morphology in this crystal system. (right): A stereographic projection showing the angular relationships between the three crystallographic axes.*

axes of different lengths, and none of the angles being 90°. So:

lengths $a \neq b \neq c$, and
angles $\alpha \neq \beta \neq \gamma \neq 90°$.

In this system, all the shapes have only 1-fold symmetry (i.e., a rotation of 360°) or a center of symmetry, with no mirror planes (Figure 4.12). Some of the feldspars (e.g., anorthite) crystallize in this system. Their crystal shape can be thought of as a set of three pinacoids or parallelohedra (any 3-D repeatable space-filling shape).

So the main point of this chapter is this: there are six possible basis vector sets that can be used to describe the ways in which materials crystallize. We first studied and derived these systems based on the external morphology of minerals and solids, but now one of the major uses of symmetry is to explain the internal arrangement of atoms in a crystalline material. Each system is unique, based upon its combination of symmetry elements such as rotation axes and mirror planes. There are even more interesting symmetry elements to be found in some of these systems, so stay tuned to Chapter 11 and 12. Then in Chapter 13 we will put all of our knowledge of symmetry to use to represent the crystal structures of minerals.

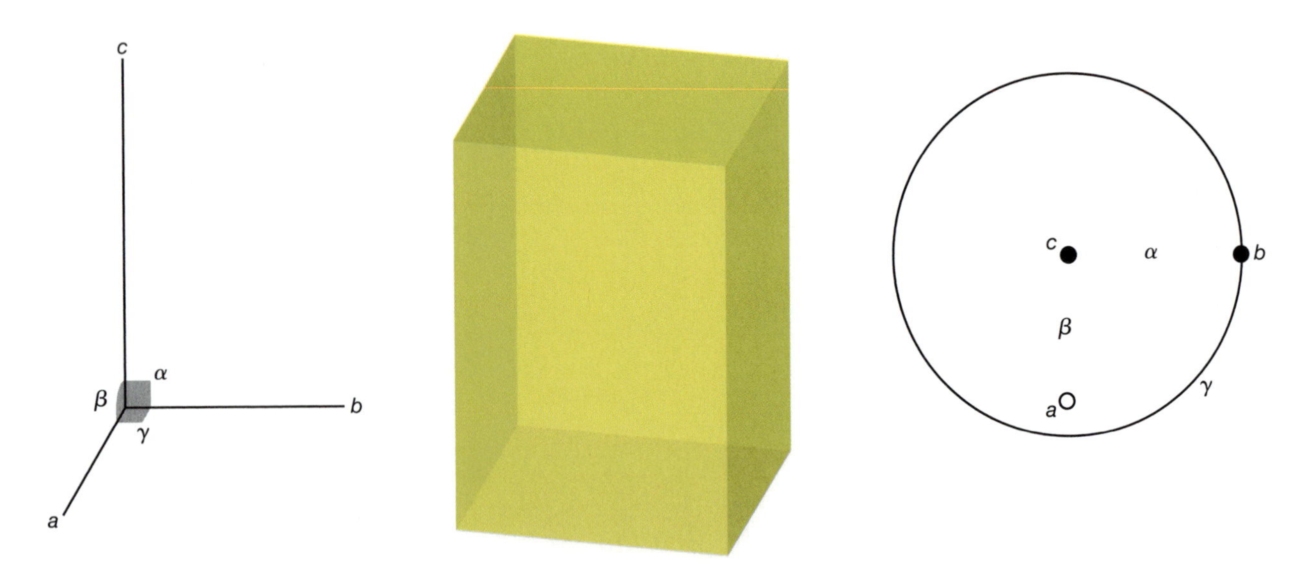

Figure 4.11. Representations of the monoclinic crystal system. In this crystal system, no axes are the same length, and β *is non-90° (left): A 3-D perspective sketch showing the relationship between the three crystallographic axes. (center): A solid composed by these three axes showing one possible morphology in this crystal system. (right): A stereographic projection showing the angular relationships between the three crystallographic axes.*

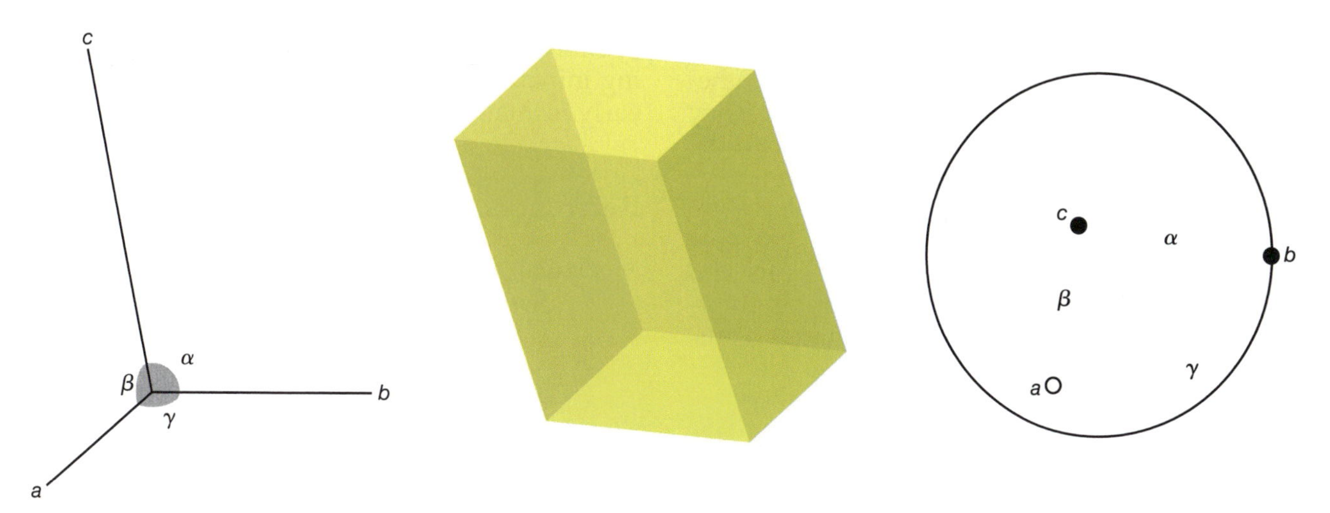

***Figure 4.12.** Representations of the triclinic crystal system. For this least-symmetric class of all, none of the axes lengths is equal and none of the interaxial angles is constrained to certain values (left): A 3-D perspective sketch showing the relationship between the three crystallographic axes. (center): A solid composed by these three axes showing one possible morphology in this crystal system. (right): A stereographic projection showing the angular relationships between the three crystallographic axes.*

In conclusion, recall that we ended the last chapter with some "real-life" reasons for why you should care about crystal chemistry. We'll end this chapter with some examples of why you should care about symmetry. Take a look at Figure 4.13. This was a poster for a guest lecturer in Japan who had won a Nobel prize in chemistry for figuring out how to grow compounds that were mirror images of each other; these are often termed right- and left-handed in reference to the fact that our right and left hands are mirror images of each other. Some forms (say, the one of the right side of the Figure 4.13) are cures for certain human diseases, while the other forms (the mirror image) are inert or worse yet, harmful. A current example of this is taxol, a treatment for breast cancer in women. Taxol comes in both right- and left-handed forms; and only one is useful for treatment.

Another sadder example comes from the 1950s and deals with the organic compound thalidomide. Thalidomide was used to control morning sickness in pregnant women. The compound was synthesized and used by many women. Unfortunately, birth defects started occurring in some children. It was found that the thalidomide occurred in both right- and left-handed forms. While the right-handed form treated morning sickness effectively, the left-handed form caused

***Figure 4.13.** Even in a different language, this poster should make some sense. It was posted on a door at Kyoto University in Japan announcing a guest lecturer who had recently won the Nobel Prize in Chemistry. Take a look at the molecular drawings in the drawing. Do you notice a symmetry relation between the two molecules (i.e., the right and left sides in the lower portion of the image)? If you said that they are mirror images of each other, then you are correct. The gentlemen in the photograph figured out a way to control which of these molecules could be grown. This is critical because many organic compounds used as medicine are inert or harmful in one form (or "handedness"), and a successful cure for a disease in the other form.*

birth defects. Currently, pharmaceutical industries spend billions of dollars to make sure they synthesize the correct handedness of their drugs. They often do this by using mineral surfaces (with the correct handedness) as growth templates.

Finally, there is currently some interesting research that links the first forms of life, possibly types of bacteria, with growth on minerals and the fact that these mineral surfaces exhibit favorable handedness for growth of bacteria. Thus, symmetry, while important to our of understanding minerals, may have also played a role in the early formation of life on the planet Earth.

References

McDowell, R.B. (1994) Symmetry: A Design System for Quiltmakers. C&T Publishing, Concord, California, 144 pp.

Chapter 5

Optical Mineralogy

"I confess; I used to fake it. I never saw the Becke Line for years. It wasn't until I had to teach it myself that I really saw what all these lecturers were on about. Having two or more people looking down a microscope can be a problem and, combined with equipment that might not have been optimally adjusted, it is perhaps not surprising that I was never properly convinced in the early part of my petrographic career.

Becke Lines, interference figures, dispersion, to name but a few of the wonders of optical mineralogy; tricky to see and trickier to explain without good teaching and good teaching aids—optical mineralogy was probably invented as a test bed for multimedia courseware and virtual environments." (Browning, 1996)

With these words, Paul Browning describes what is an all too common experience for frustrated mineralogy students in their first encounter with an optical microscope: the perception of the Becke line, which is used to measure refractive index of material and understand the concept of "relief" (and we don't mean the feeling you get when you walk out of mineralogy lab!).

For many students, mineralogy class represents the first opportunity to work extensively with a microscope (or an actual analytical instrument). This chair-bound, indoor experience may be exactly the opposite of what you envisioned when you signed up to be a geology major! But microscopy is a vital part of many disciplines of geology, not to mention all the other disciplines that deal with the identification and characterization of solids. One of the best things about geology is the fact that it explains natural phenomena at many different scales, from subatomic to macroscopic. How do we identify minerals with a polarizing light microscope? What causes the rainbow? What causes the funny color patterns when oil is spilled on top of water? Why does diamond appear so "bright" and exhibit sparkling colors in the sunlight? How does a winemaker know the sugar content of a grape? How does the human eye focus light and how might we correct near- and farsightedness? These questions, and many others, can be answered by a basic understanding of optical phenomena and its uses.

This chapter is your introduction to the field of optical mineralogy and the polarized light microscope (PLM), which is often called a petrographic microscope by those interested in studying the textural relationships of minerals in rocks. Please be patient with your microscope, and realize that learning to use it is like riding a bicycle: it takes a lot of practice. The rewards of understanding a polarizing light microscope are many: the confidence to be able to identify nearly any mineral or solid on the basis of a few grains, and to gain insights into the relationships between coexisting minerals in a thin section (that's the study of petrology!). As an added bonus, you'll always be able to find employment in very diverse fields from forensic science, to art history, to environmental sciences, to pharmaceuticals, to clinical pathology if you only know how to use a polarizing light microscope to identify and characterize particles.

So sit down in front of your microscope, and let's get going.

M.D.D.

Introduction

In this chapter, we tie together the previous two chapters by interweaving them with optics, resulting in two related disciplines: optical crystallography and optical crystal chemistry. The first of these explains how arrangements of atoms in space will interact with light as a function of the symmetry of the material; thus the use of the word "crystallography." So **optical crystallography** basically describes the physical interaction of light with a material with little regard for the actual composition of that material. In optical crystallography, we'll be using "crystallographic" terms we learned in Chapter 4 such as "symmetry" and "crystal systems." **Optical crystal chemistry**, on the other hand, explains how differences in chemistry affect the properties of the material (e.g., what causes color, what affects the refractive index of a mineral, etc.). So here we'll be using the material we learned in Chapter 3, such as "bonding types" and "cations and anions." Optical crystal chemistry addresses the issue of why two identical arrangements of atoms, say halite (NaCl) and sylvite (KCl), which have the same crystal structure, can have different refractive indices: halite = 1.544 and sylvite = 1.490. Also, we can use the theories we'll develop in optical crystal chemistry to explain why muscovite ($KAl_2(AlSi_3O_{10})(OH)_2$) is colorless, and biotite ($K(Mg,Fe)_3(AlSi_3O_{10})(OH)_2$) is a darker color. However, we'll need to call upon optical crystallography to explain why the color of biotite changes depending on the direction in which the light moves through it.

The integration of optical crystallography and optical crystal chemistry is the field of **optical mineralogy**. Often the terms optical crystallography and optical mineralogy are incorrectly used synonymously; as we'll see in our following discussion, optical crystallography is really a subset of optical mineralogy. While a web search on optical mineralogy will bring up thousands of hits and optical crystallography hundreds, no hits are (currently) found for optical crystal chemistry. Thus we are coining this new term, which is effectively a new subdiscipline in geology, to help you better understand how we use crystallography and crystal chemistry to explain the optical properties of minerals.

In this chapter, we'll present a brief overview of what we think are the highlights of the subject at hand: in this case, optical mineralogy. We'll mostly stick with things we can actually see. Later in the book there will be more chapters dedicated to optical mineralogy that will provide the theory behind what we "see" in this chapter. We promise that there will be some overlap between the material presented in this chapter and in the Chapters 16–19, in keeping with our spirit of spiral learning.

Macroscopic View of Polarized Light and Minerals

Before we actually sit down at the microscope, let's begin with a "macroscopic" viewpoint. "Macro" refers to something we can normally see with the naked eye, but it's usually a close-up view. (Some of you might enjoy looking at close-up photos of flowers: these types of "close-ups" are called macrophotography.) So let's look first at the interaction of light with minerals at a larger scale, so as to remove some the "magic" of the microscope. In Figure 5.1, we use a collection of several familiar and unfamiliar items to see how light interacts with minerals. In fact, we can make a simple polarizing light microscope (PLM) out of a hand lens (that you hopefully already acquired and learned to use as a budding geologist!) and the two pieces of gray filter-like material that are sheets of polarizers. These are sheets of Polaroid plastic that absorb light in all but one direction within a plane.

To start, notice how the glass slide is transparent when placed on one polarizer (Figure 5.1a), but when a second polarizer is placed over it, the entire field of view is dark in the area where the two polarizers cross (Figure 5.1a). If we insert a piece of muscovite between the two polarizing sheets, something very different happens (Figure 5.1b).

The trick here is that the polarizers only allow light to travel in one direction, and we have crossed those directions so that no light will come through. In microscopy we call this condition **cross-polarized light** (much more on that later.) Notice in Figure 5.1b that the muscovite between the sheets takes on some unique colors—these are called interference colors, and we'll study them in much, much more detail later. If we rotate the muscovite between the sheets of polarizers, it goes dark at some point; in fact, we find that it goes dark at every 90° of rotation. These interference colors and "light and dark" conditions are really at the heart of how all minerals, except those belonging to the isometric system, behave when they are placed between polarizers.

Next we can move from the macro-world of the piece of muscovite to the micro-world of mineral grains in thin slabs of polished rock called **thin sections**; these are about 1/3 the thickness of a human hair. In Figure 5.2a, we repeat the example

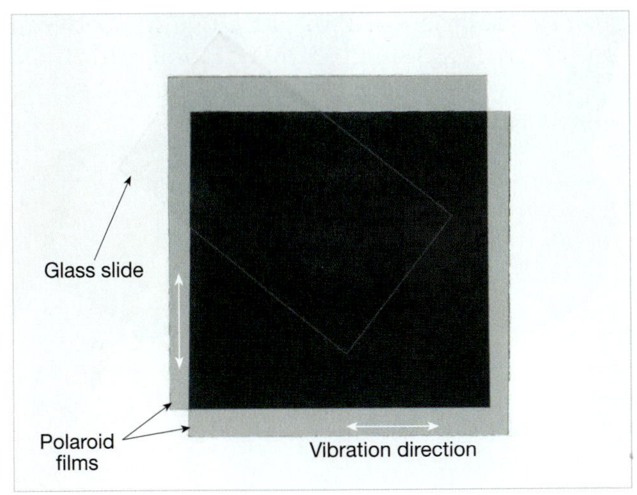

a.

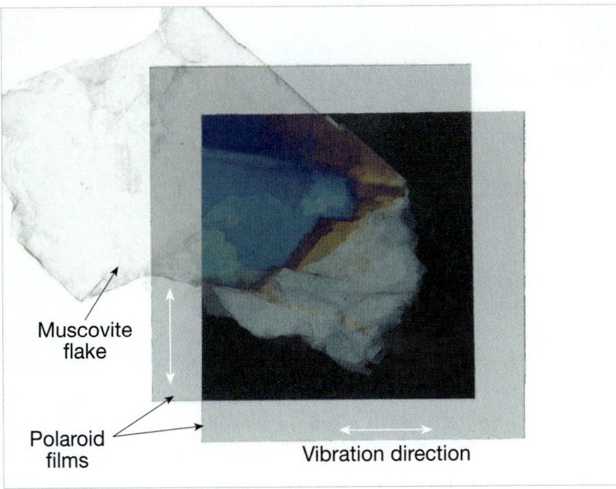

b.

Figure 5.1. a. *Two pieces of 50×50 mm sheets of Polaroid film placed on a light table and arranged where their vibration directions are perpendicular to each other to simulate cross-polarized light in a PLM. Where they do not overlap, light is still transmitted, although of less intensity. Also, note the glass slide sticking out from the polarizers in the upper left corner. No light appears from the glass slide as it enters the polarizer sandwich.* ***b.*** *The same set up as in Figure 5.1a, except a flake of muscovite (of uneven thickness) replaces the glass slide. In this case, not only is light transmitted, but the light takes on interesting colors based on the thickness of the muscovite flake.*

of Figure 5.1, except this time we place a thin section of a granite between the crossed polarizers. We can now see lots of little bright and dark intergrown objects that are individual minerals grains in this igneous rock. Although we can see these grains, it would be nice to "magnify" them a little

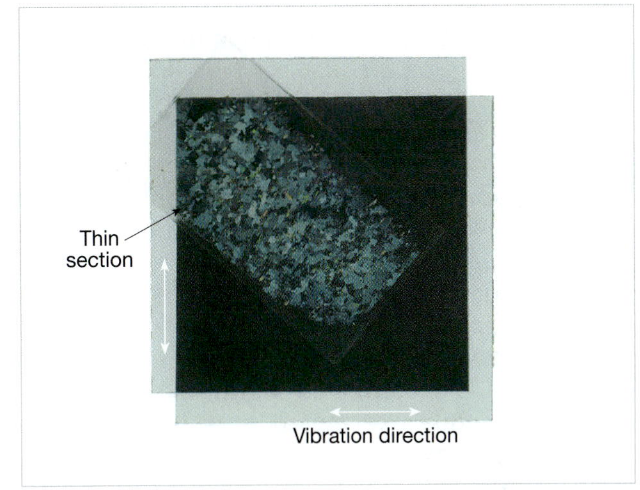

a. Setup with thin section.

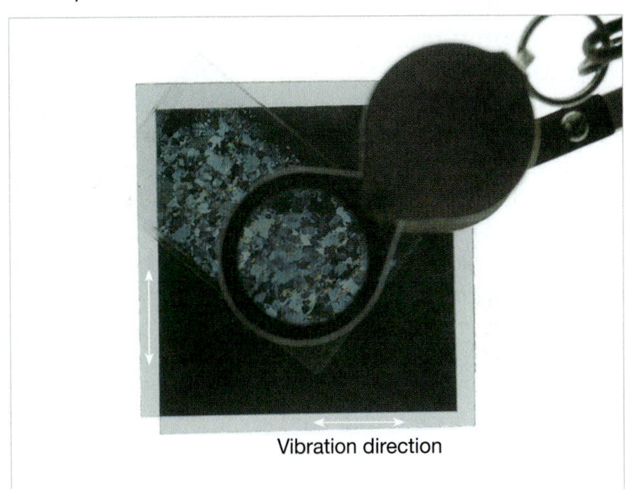

b. Low-power lens.

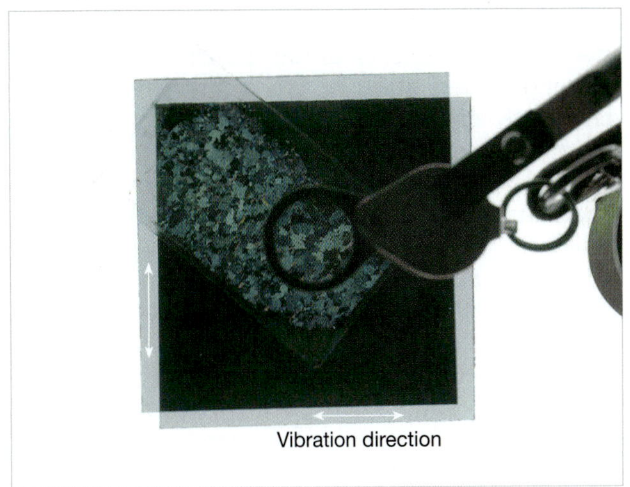

c. High-power lens.

Figure 5.2. a. *The same setup as in 5.1b, except the muscovite flake is replaced by a thin section of an igneous rock. Individual minerals grains in the rock show up as bright and dark specks.* ***b.*** *The same setup as in Figure 5.2a, except a low-power (7×) hand lens is placed over the thin section to simulate the lower power lens on a PLM.* ***c.*** *The same setup as in Figure 5.2b except a higher power hand lens (14×) is placed over the thin section to get a "higher magnification" of the sample. In essence, a PLM is really just two polarizers with some magnification, as simulated in this figure.*

to get a closer look. We could take a 7× hand lens as shown in Figure 5.2b and use it to magnify the image to better see the individual grains. We could also take a 14× hand lens to increase the magnification to better discern the grains. In reality, this is the basis for a PLM: two polarizers (one below and one above the sample) and a means to magnify the image. Of course, there are lots of parts on the microscope, as we'll discuss later, but the basic setup of a PLM is shown in Figure 5.2.

Making a Thin Section

Now that you have some understanding of what we'll be looking at with a microscope, let's see how and why we make thin sections! Geologists usually study rocks and their associated minerals by using **thin sections**, which are polished pieces of rock or mineral that can be placed on a microscope stage. For **reflected light** microscopy of opaque minerals like metals or sulfides, where we are not interested in seeing through the minerals, simple slabs with polished surfaces are used. **Transmitted light** microscopy is used to study doubly-polished pieces of rock thinned to 30 μm.

To make a thin section (Figure 5.3), the sample of interest is first cut into a roughly 1 cm thick slab using a rock saw. The outline of the desired section (usually either rectangular, ~27 × 46 mm, or a 1″ round) is drawn on the slab, and then the desired shape is either drilled out (in the case of a circle) or obtained by a combination of cutting and grinding (for a rectangle); this cut rock is called a **billet**. For a transmitted light slide, an extra step is now added: the resultant rock is polished on one side using various, increasingly finer grits of abrasive paper on a polisher. Then the rock is glued to a glass slide with epoxy or some kind of adhesive before it is trimmed off with a special thin-sectioning saw designed for this purpose. The section is either polished for study under reflected light, or ground down to the 30 μm thickness, again using increasingly finer-grained grit paper and, ultimately, a mixture of SiC or Al_2O_3 grit and water. For a really finely polished surface that is suitable for analytical work, polishing solutions with various sizes of diamond or aluminum oxide particles (commonly 15, 9, 3, 1, 0.3, and 0.05 μm) are sometimes used.

If you have a chance to make at least one thin section on your own, we highly recommend it. This is not a simple procedure; in fact, it's something of an art form. It is also incredibly time-consuming. You may be asking yourself the question: "How do they know when it's 30 μm thick?" You

Cut a piece of rock into a rectangle

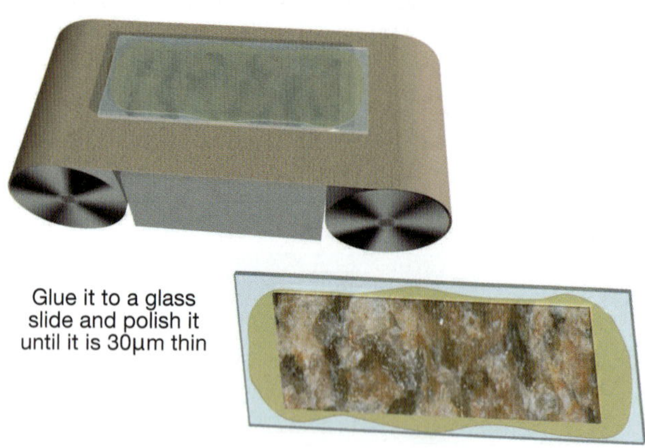
Glue it to a glass
slide and polish it
until it is 30μm thin

Figure 5.3. We've already seen a thin section in Figure 5.2. This illustration (and animation on the MSA website) shows how to make one. Basically we cut a piece of rock into a rectangle (usually with a diamond-coated saw blade), glue it to a glass slide, and then cut, grind, and polish it until it is 30 μm thick, or thin! At this "thinness" most minerals are transparent as shown in Figure 5.2.

could measure it, of course, with a micrometer, though subtracting out the thickness of the epoxy and glass slide can be tricky. More commonly, we use an optical property called retardation to determine the thickness of the slide. Making your own thin section will help you appreciate all the effort that went into creation of the samples you will soon be looking at in your class! To learn more about this process, check out the discussions in Hutchinson (1974) and Humphries (1992).

Meet Your Microscope

In order to see how light interacts with minerals, most of you will be using a **polarizing light microscope** (PLM). Figure 5.2 gave us the basics for a PLM, but as already stated, there's a lot more to it! First off, the PLM may superficially resemble the microscopes used in high school biology class, but a polarizing light microscope is a special design customized for the study of minerals, which is the result of a long development.

Following the discovery of polarized light in 1808 (Malus, 1809; Lima-de-Faria, 1990), many scientists quickly observed the effects of polarized light on minerals such as tourmaline (Biot, 1815). Preparation of thin sections, as invented by William Nicol, was first described in 1831 by Witham, and later by Nicol himself (1834). The first polarizing light microscope was designed between 1838 and 1859 (Needham, 1958), and an article by Sorby (1858) is probably the first application of the microscope to the study of thin sections of rocks and minerals. By the late 1800s, the polarizing light microscope became commercially available, and by the end of that century, it was widely used. Some of these historical papers make very interesting reading, as does the Special Publication by the *Mineralogical Record* on the polarizing light microscope (Kile, 2003). Kile presents an excellent discussion of the development of the PLM, as well as some high-quality photographs of his personal collection of microscopes and accessories, which exceeds that of any museum collection in the world!

You will often see the polarizing light microscope referred to as a **petrographic microscope** in the geological community. While the names may seem synonymous, there is an important difference. The term petrographic light microscope implies that its use is restricted only to the study of petrography (i.e., the texture of rocks). The broader scientific community uses the term polarizing light microscope because this tool is used to study many other things, such as single crystals of minerals, human hair, explosives, drugs, etc. (the list is endless!). Entire fields of study, like forensics science and pathology, use the polarizing light microscope, so get used to thinking of it as such, and expand your future employment options!

To describe the microscope, let's examine the two different scopes shown in Figure 5.4. The one on the left is a newer model and the one on the right a little older. However, they both have the same basic parts, which are located in the same basic areas along the light path. Your microscope will probably be superficially different from these, but realize that they all have the same basic controls (just like a car!).

For each microscope, we have labeled the most important parts, following the light path through the scope and into your eye. At this point we've chosen to list the following eight parts of the scope.

1. **Light source.** A light bulb is "buried" somewhere in the base of scope and light emerges here.
2. **Substage optics.** As the name implies, this is a set of optical elements located below the stage. Usually there are three optical components here: (a) a lens system to focus the light on the sample, (b) an iris diaphragm to control the intensity of the light (just like the iris in your eye), and (c) the lower polarizer, which makes the light all vibrate in the same direction.
3. **Rotating stage.** The stage rotates on PLMs so we can study the effect of light at different angles in minerals.
4. **Focusing knob(s).** For the newer PLMs, the focus is achieved by raising and lowering the stage with this knob. Often there are two knobs, one for coarse and the other for fine focusing.
5. **Objective lenses.** There are usually 3–5 lenses of different magnifications on modern PLM's. The magnification of the scope can be changed

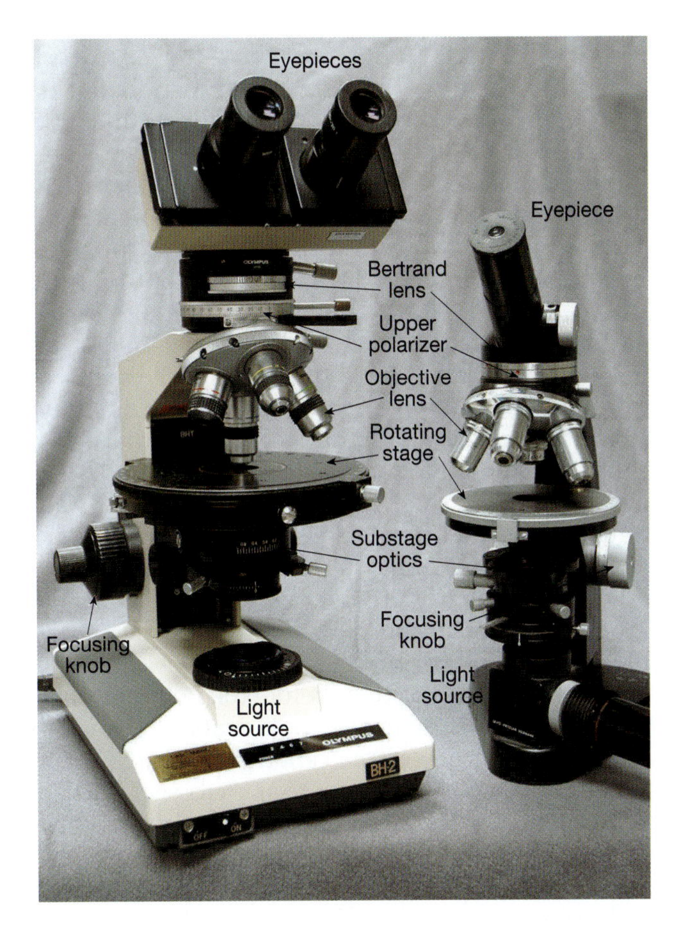

Figure 5.4. Photographs of two PLM's. The one on the left is bit more modern (build in the 1990s) and the one on the right is older (built in the 1970s); however both perform the same functions and have the same basic parts. For each scope we've labeled eight of the most important parts, which are described in detail in the text. Notice how, even though the scopes look different, the parts are in similar locations. Although these scopes may differ from yours, use them to locate the major components.

by rotating the different lenses into the light path.
6. **Upper polarizer.** The upper polarizer is just above the lens, and can be inserted and removed from the light path.
7. **Bertrand lens.** This is a special lens used to view interference figures (more on that later).
8. **Eyepiece.** This forms the image for your eye.

As seen above, a polarizing microscope includes several specializations that make it different from a standard microscope. Of these, probably the most important are rotating stage, the lower polarizer (in the substage optics) and the upper polarizer (in the body tube). The polarizers take light that is vibrating in all possible directions and act to block out all the waves except those that are vibrating in a single plane (Figure 5.5). Thus, when light travels through one polarizer, it is said to be **plane-polarized**.

The lower polarizer is usually fixed at and coincides with what could be called the east-west direction. The upper polarizer is perpendicular to the lower polarizer, oriented in the north-south direction. This east-west and north-south language dates back to the time before electricity. Microscopes used a mirror to gather ambient outdoor light from a north-facing window; thus the north-south direction in the microscope would parallel a north-south direction on land, while the east-west direction in the scope corresponds to the east-west land directions. This early method of lighting the microscope also explains why PLMs often have a blue filter over the light source—to emulate the northern skylight.

The upper polarizer can be inserted and removed, so that images are viewed in either: (1) **plane-polarized light** (PPL) from the lower polarizer, with the upper polarizer removed, or (2) **cross-polarized light** (XPL) with both the upper and lower polarizers inserted at right angles to each other. In the old days before the Polaroid plastic sheets were invented, polarization was performed by calcite prisms called **Nicols** after their inventor (Nicol, 1839). The term "crossed polars" is still sometimes called "**crossed Nicols**" by some of the older professors. If there is nothing on the microscope stage, use of both the lower and upper polarizer will not allow light to pass (Figure 5.6).

The eyepiece contains perpendicular crosshairs aligned in the same directions as the lower and upper polarizers; this allows for angular measurements relative to what is viewed on the rotating stage below. Just below the eyepiece is the **Bertrand lens**, which can also be moved in and out. It magnifies and focuses the focal plane of the image produced by the objective, allowing you to create interference figures. The Bertrand lens is used in conjunction with convergent light, which is usually produced in the substage assembly by insertion of a special, removable lens that converges light waves toward the center of the stage (Figure 5.7).

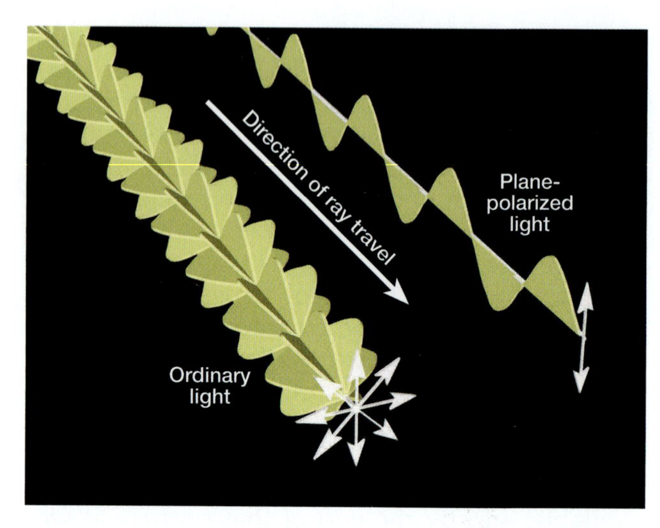

Figure 5.5. Ordinary light ray (left) and a polarized light ray (right). They are both traveling toward us, as indicated by the ray directions; and both are also vibrating perpendicular to the direction of travel.

Your polarizing light microscope can be customized through use of a slot for insertion of various interchangeable, removable **accessory plates**. Each of these accessory plates contains a thin slice of a specially cut mineral (quartz in modern microscopes, and gypsum and mica in older scopes) with a specific thickness and orientation. As we'll

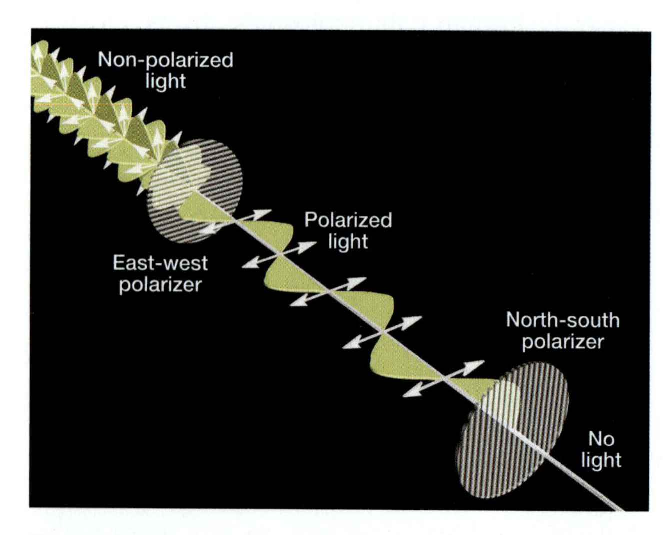

Figure 5.6. An unpolarized light ray is first plane-polarized in an east-west direction (as is the case for the lower polarizer of a PLM), then travels some distance before reaching a north-south polarizer (i.e., one that is perpendicular to the north-south direction, the case of the upper polarizer in a PLM), and then is completely blocked.

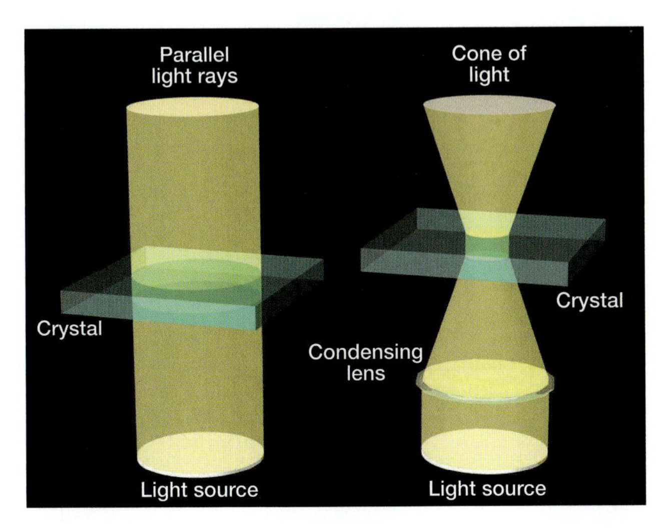

Figure 5.7. *The substage condenser assembly can either focus the light into parallel rays (left) or a cone of light (right). These different sorts of illumination will travel in different ways through a solid placed on the scope stage.*

show at the end of this chapter, these accessory plates are used to slow down, or **retard**, light waves passing through them, and they aid greatly in mineral identification and characterization.

Try to find all these named parts on the microscope you'll be using for this class. Remember that microscopes are like people—they can and will look different from the drawings and photos we have. Like people, all microscopes have the same essential parts, which look different for various individuals. Please treat your microscope as you would treat a friend: with care! The average student polarizing light microscope costs approximately $5,000, and if you drop it, usually some of the parts will break and your instructor will be very unhappy!

Wave Theory of Light

So now you have a microscope, and have glimpsed your first tantalizing view of a psychedelic mineral under crossed polars. To understand this really cool phenomenon, we need to back up a bit to review some theory of light and optics. For this purpose, we will use wave theory, which is based on the electromagnetic spectrum (Figure 5.8). All light that is visible to the human eye is in the energy range where the wavelength of light is about 400–700 nm (where nm refers to nanometer, and there are 1×10^9 (or 1 billion) nanometers in a meter!); the cutoff points of the range are somewhat arbitrary based on an individual's ability to sense light. Different colors correspond to specific wavelength ranges, and the sensitivity of our eyes is highest in the middle of that range (around 550 nm). For this reason, many communities are now re-painting their red fire trucks and warning road signs to yellow-green (~560 nm), in order for the color to be as noticeable as possible!

The electromagnetic spectrum shows the contributions from both electric and magnetic fields as they pass through liquids, solids, or gasses. In minerals, the electric interaction of light is very strong and the magnetic interaction is weak, so we can think of these light waves as electric vectors. A plot of a single wave of plane-polarized light shows the magnitude of these electric vectors (Figure 5.9). The wave is described using the following terms:

The **propagation direction** (or ray path) of the wave is from left to right; this is the direction in which the light wave is traveling.

The **vibration direction** of the wave is up and down (i.e., perpendicular to its ray path).

The **wavelength** (λ) is the distance between two points in the same relative location on adjacent waves.

The **amplitude** (A) of the wave describes the maximum extent of the electric vector in the up and down direction; this relates to the intensity of the light.

The **frequency** of the wave is the number of wavelengths of light that pass a point during a given interval of time.

When a light wave enters a solid (from air), its velocity is reduced as a function of the chemical makeup of the mineral—this is the realm of optical crystal chemistry. Also, the speed of light through a mineral will often vary as a function of orientation; thus most minerals will have more than "one speed." The study of how these speeds vary as a function of direction is relatively well understood, because it falls in the realm of optical crystallography.

Optical Classes and Refractive Index

Let's begin with the six crystal systems we discussed in the last chapter on crystallography, and review the axes in each system:

isometric	$a = b = c$
tetragonal	$a = b \neq c$
hexagonal	$a = b \neq c$
orthorhombic	$a \neq b \neq c$
monoclinic	$a \neq b \neq c$
triclinic	$a \neq b \neq c$.

These are subdivided into three distinct groups, from which you can make three different kinds of shapes, which turn out to be the three **optical classes**:

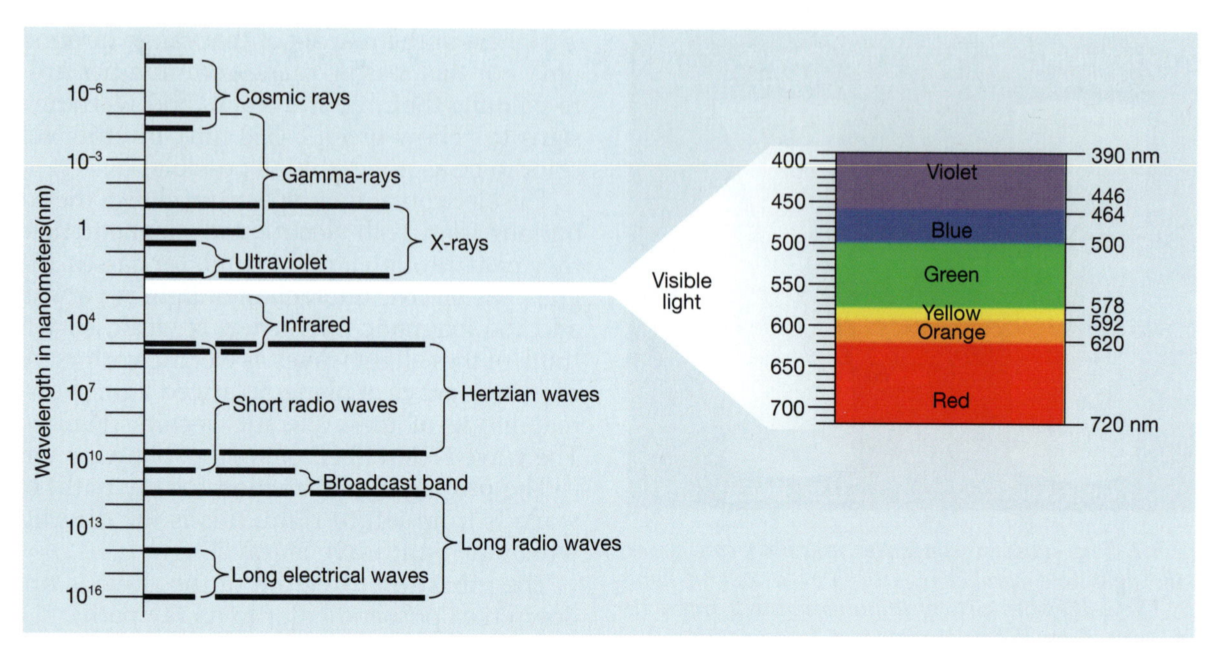

Figure 5.8. The electromagnetic spectrum covers the entire wavelength range of electromagnetic waves from zero to infinity. Visible light is only a small part of this, from about 400–700 nm. For more information, see Chapter 7.

isotropic	$a = b = c$
uniaxial	$a = b \neq c$
biaxial	$a \neq b \neq c.$

Each shape/optical class has a three-dimensional surface that defines the speed of light in a material in any direction. This shape is called an **optical indicatrix**. Imagine an infinite number of light waves radiating out from the center of a crystal. The length of each vector is proportional to the **refractive index**, n, of light that vibrates in that direction (recall the vibration direction is perpendicular to the propagation direction), where

$$n = \frac{\text{velocity of light in a vacuum}}{\text{velocity of light in a material}}.$$

For example, the speed of light in a vacuum is about 300 million m/s. Light slows down to pass through water at a speed of about 225 million m/s. So the refractive index of water would be about 1.33. The majority of silicate minerals have refractive indices between 1.5 and 1.7. Some common materials and their refractive indices are given in Table 5.1.

The velocity of a light wave depends on the *density* (really the density of the electrons) of the medium through which it travels as well as the *distribution* of that density (i.e., how the electrons are distributed, which is related to how the atoms are bonded). The first of these should be intuitive: the denser the material, the "harder" it is for anything to pass through it. The charge distribution

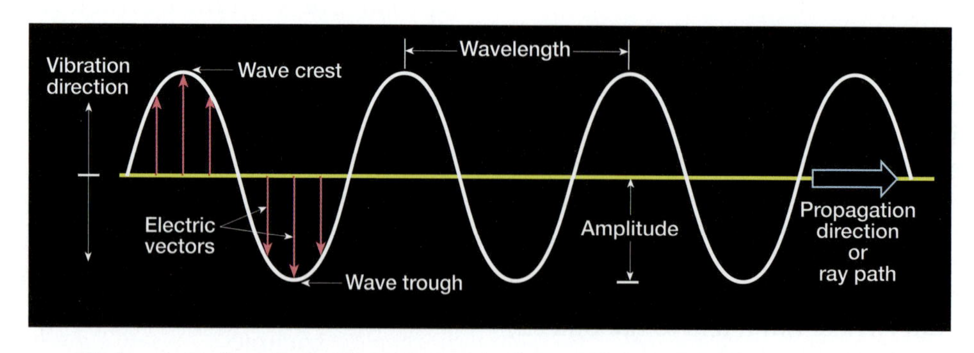

Figure 5.9. A wave of light can be described using this terminology, which is the same vocabulary used to describe waves in the ocean. Note that the vibration direction of the light wave is perpendicular to the propagation direction, or the direction of travel. This is really important to remember, because only the vibration direction "sees" the atoms in the mineral. So when we look down the PLM, we are looking parallel to the propagation direction and the vibration direction is parallel to the scope stage.

Table 5.1. Refractive Indices of Various Gemstones and Other Materials at Room Temperature*

	Gemstones	
Agate 1.544–1.553	Emerald 1.560–1.605	Moonstone 1.518–1.526
Alexandrite 1.746–1.755	Fluorite 1.434	Morganite 1.585–1.594
Almandine 1.75–1.83	Garnet, Grossular 1.72–1.80	Obsidian 1.50
Amber 1.539–1.545	Garnet, Andradite 1.88–1.94	Opal 1.430–1.460
Amethyst 1.532–1.554	Garnet, Demantoid 1.880–1.9	Pearl 1.53–1.69
Andalusite 1.629–1.650	Garnet, Mandarin 1.790–1.8	Peridot 1.635–1.690
Apatite 1.632–1.42	Garnet, Pyrope 1.73–1.76	Quartz 1.544–1.553
Aquamarine 1.567–1.590	Garnet, Rhodolite 1.740–1.770	Ruby 1.757–1.779
Axinite 1.674–1.704	Garnet, Tsavorite 1.739–1.744	Sapphire 1.757–1.779
Beryl 1.57–1.60	Garnet, Uvarovite 1.74–1.87	Spinel 1.708–1.735
Chalcedony 1.544–1.553	Iolite 1.522–1.578	Star Ruby 1.76–1.773
Citrine 1.532–1.554	Jade, Jadeite 1.64–1.667	Tanzanite 1.690–1.700
Clinohumite 1.625–1.675	Jade, Nephrite 1.600–1.641	Topaz 1.607–1.627
Coral 1.486–1.658	Jet 1.660	Topaz, Imperial 1.605–1.640
Chrysoberyl, Catseye 1.746–1.755	Kunzite 1.660–1.676	Tourmaline 1.603–1.655
Danburite 1.627–1.641	Labradorite 1.560–1.572	Zircon 1.777–1.987
Diamond 2.418	Lapis Lazuli 1.50–1.55	Zirconia, Cubic 2.173–2.21
	Metals	
Aluminum 1.39	Mylar 1.65	Silver 1.35
Copper 2.43	Nickel 1.08	Titanium 2.16
Gold 0.47	Platinum 2.33	
	Common Liquids	
Acetone 1.36	Honey, 21% water content 1.484	Rum, White 1.361
Alcohol, Ethyl (grain) 1.36	Ice 1.309	Shampoo 1.362
Alcohol, Methyl (wood) 1.329	Milk 1.35	Sugar Solution 30% 1.38
Beer 1.345	Oil, Clove 1.535	Sugar Solution 80% 1.49
Carbonated Beverages 1.34–1.356	Oil, Lemon 1.481	Turpentine 1.472
Chlorine (liq) 1.385	Oil, Neroli 1.482	Vodka 1.363
Cranberry Juice (25%) 1.351	Oil, Orange 1.473	Water (0° C) 1.33346
Glycerin 1.473	Oil, Safflower 1.466	Water (100° C) 1.31766
Honey, 13% water content 1.504	Oil, vegetable (50° C) 1.47	Water (20° C) 1.33283
Honey, 17% water content 1.494	Oil of Wintergreen 1.536	Whisky 1.356
	Common Transparent Materials	
Eye, Aqueous humor 1.33	Glass, Arsenic Trisulfide 2.04	Glass, Flint, 71% lead 1.805
Eye, Cornea 1.38	Glass, Crown (common) 1.52	Glass, Fused Silica 1.459
Eye, Lens 1.41	Glass, Flint, 29% lead 1.569	Glass, Pyrex 1.474
Eye, Vitreous humor 1.34	Glass, Flint, 55% lead 1.669	Lucite 1.495

*With thanks to Robin Wood, who assembled this compilation; used by permission.

effect is more subtle, but it can be explained with a suitable analogy, as follows:

Imagine that you are with a team of rescuers looking for a lost skier who has ventured off into the woods. Because it is a dark, moonless evening, the search teams march in a straight line with people spaced every few meters apart. The teams split into two groups on either side of the mountain. One team (W) heads for the west side, where stiff prevailing winds prevent trees from growing large. That forest is a mess of tiny, fallen trees that inhibit the progress of the line of rescuers. The other team (S) goes to the south side, where a tall canopy of old growth trees occurs because of increased sunshine and less wind; it is relatively easy for the searchers to walk through the forest on the south side. The mass of the forest (i.e., the "pounds" of trees) is the same on both sides of the mountain, but its distribution is different. So even though team S and team W enter the forest walking at the same rate, team S gets to the bottom of the mountain way ahead of team W. (Let's hope the skier got lost on the south side!)

The moral of the story is that different distributions of charge give rise to varying refractive indices. So, for example, covalently-bonded solids will generally have higher refractive indices because of their overlapping charge distributions (lots of tiny, fallen trees) than ionically-bonded solids, in which the charge is closely associated with the ions (big trees with open spaces between them). For instance, diamond has a much higher refractive index than halite, 2.418 compared to 1.544. So when you think about refractive index, think of those rescuers marching through the woods in the dark: the more tiny, fallen trees there are (high density) and how widely they are distributed (lots of tiny trees vs. a few big trees), the slower they can walk (high refractive index!).

How do we measure refractive index, *n*? There are several ways that we'll consider in detail in Chapter 18, but for now, a simple explanation will suffice. Let's imagine that you have a colorless crystal of the mineral species fluorite, which has a refractive index of 1.434 (there are no units for refractive index because it's a ratio). If you drop your fluorite crystal in a glass of water ($n = 1.33$), you can still see the crystal because the light waves bend when they pass between media with different values of refractive index, and reflection also occurs at interfaces between media of different refractive indices. Now imagine that you want to make the crystal disappear: to do this, its refractive index must match that of the water. By dissolving table sugar in the water, you can increase the water's refractive index until it matches that of the fluorite (about 55% sugar by weight). (There are many kinds of sugars. Herein we are using sugar to be synonymous with the most common type of sugar—common table sugar, or $C_{12}H_{22}O_{11}$, which is known to organic chemists as sucrose.) As you add the sugar, the fluorite crystal will become fainter and fainter, until you can no longer see it. At that point, the refractive index of the fluorite and the sugar water are the same.

The lesson here is that we "see" objects based upon the reflection of light from the interface between the object and the air (or whatever medium) around it. If the object and the air have the same refractive index, then the object will be invisible; neither reflection nor refraction will occur. We'll exploit this method (making a solid of unknown refractive index disappear in a liquid of known refractive index) later when we discuss how we can measure the refractive index of a material.

This dependence of refractive index on sugar content is well-studied, because refractive index is used by vintners (the folks who make wine) to measure the sugar content of a grape's juice while it is still on the vine (Table 5.2); then they pick the grape when its sugar content is at the desired level. The same technique is used to determine harvest times for other fruits such as pineapples and sugar cane. This method is called the **Brix** (Br) scale after its inventor Adolf Brix, and it is used in many different industries where the sugar contents of solutions are measured. Some food processing examples include the manufacture and / or processing of carbonated drinks (5–15% Brix), milk (6–17% Brix), apples (11–18% Brix), honey (58–92% Brix), and even olive oil (70–75% Brix).

As mineralogists, we use refractive index to identify minerals. The database that can be purchased separately on the MSA website lists the refractive index values for each included mineral. Notice that the minerals with low *n* values include many F- and Cl-rich phases, which have ionic bonds! At the other extreme, covalently-bonded diamond has one of the highest refractive indices of all minerals.

Table 5.2. Brix Scale for Sucrose

*% Brix**	*Refractive Index*
0.0	1.3330
5.0	1.3403
10.0	1.3479
15.0	1.3557
20.0	1.3639
25.0	1.3723
30.0	1.3811
35.0	1.3902
40.0	1.3997
45.0	1.4096
50.0	1.4200
55.0	1.4307
60.0	1.4418
65.0	1.4532
70.0	1.4651
75.0	1.4774
80.0	1.4901
85.0	1.5003
90.0	1.4826
95.0	1.5181

* % Brix is the percentage of sugar, by weight, in a sugar solution. This table is based on Table 109 of *NBS Circular 440*.

To identify the refractive index of an unknown mineral in the laboratory, we use a glass slide and a set of liquids with known refractive indices (we now know that we could make some of these liquids using sugar!), making use of the (infamous) concept of the **Becke line** as mentioned in the introduction to this chapter. First you crush up the mineral into tiny grains, which tend to be pieces that are thick in the center and thin on the edges. This makes them act like tiny lenses. When you put mineral grains on the glass slide and surround them with liquid, waves of light passing through the crystal-plus-liquid will bend because they are passing between media with different values of refractive index. If the refractive index of the mineral is *higher* than that of the liquid, light rays will converge toward the center of the grain. If the refractive index of the mineral grain is *lower*, then the light rays will diverge towards the edge of the grain.

To view this phenomenon, put your slide on a polarizing light microscope, making sure that the polars are not crossed. Use a medium-powered objective to focus on the edge of a grain where it is surrounded by liquid. Turn the light down to about half the brilliance you used to find the crystal. Then slowly increase the distance between the sample and the objective by turning the focusing knob to lower the stage. You will see two faint, thin lines appear along the edge of the grain: one light and one dark; these are the Becke lines (Figure 5.10).

In Figure 5.10a, a grain (left image in focus) of $n = 1.50$ is placed in a liquid with $n = 1.48$, thus $n_g > n_l$ and the bright Becke line goes into the grain as the stage is lowered (right out-of-focus image). In Figure 5.10b, grains of $n = 1.50$ are immersed in a liquid with $n = 1.55$ (so $n_l > n_g$); for this case, the bright Becke line goes into the liquid as the stage is lowered. If the Becke lines are colored instead of just light and dark (Figure 5.10c), then it means that your liquid has a refractive index that is very close to that of your crystal. The grain in Figure 5.10c is immersed in a liquid with $n = 1.50$, thus $n_g = n_l$.

The rule is that the lighter Becke line moves toward (and into) the medium with higher refractive index (Figure 5.10a and b) *as long as you are lowering the stage.* This effect is quite subtle, but don't give up until you can reproducibly see the movement of the Becke line. The still images in Figure 5.10 do not do this justice; check the videos of these on the MSA website for better views. Better yet, try to observe these yourself (we don't want any of our students faking it…) Realize that this really is an empowering skill. Once you master the use of the Becke line, you can identify a wide range of minerals with that one tool alone.

Another way we can judge the closeness of a mineral relative to a liquid's refractive indices is by observing **relief**, which is the amount by which a mineral (or other material) stands out from its surroundings, be they a refractive index liquid, mounting epoxy, or adjacent minerals in the thin section of a rock. If you put two materials with different refractive indices next to each other, the relationship of their n values can be described as **high relief** (if they have very different n's), **moderate relief** (if the boundary is visible, but not conspicuous, i.e., the n values are slightly different), or **low relief** (with very similar values of n).

The earlier example of a fluorite crystal disappearing in a glass of sugar water is a demonstration of very low or no relief. However, relief will only tell us *how different* the n values of the solid and the liquid are, while the observation of the Becke line movement will tell us which one is higher. Look at the images in Figure 5.10 and notice how the relief changes for each of the three cases: Figure 5.10a shows low relief (the difference between the grain and liquid is 0.02), Figure 5.10b shows high relief (the difference between the grain and the liquid is 0.05), and for Figure 5.10c, the case where the grain and the liquid match, you basically cannot see the grain, so it has very, very low relief.

Reflection vs. Refraction: Snell's Law

We see relief because light waves bend when they pass through materials of different refractive index and are also reflected at the interface. This phenomenon is described by **Snell's Law**. You may already be familiar with the results of Snell's Law: have you ever dived for a penny on the bottom of the swimming pool? When a light ray moves through materials with different refractive indices, it is bent—so that penny may seem to be in a different location (before you jump in) than it actually is because the light ray first travels in water ($n = 1.33$) and then air ($n = 1.00$). If you've ever tried spearing a fish, the same principle applies: you usually miss. The only way around this problem is to either spear straight down, or, based on experience, compensate for the bending of the light rays and aim high, or is it low? After the next section you should be able to figure this out.

Whenever a light ray passes between materials with different n, it will be reflected off the interface and usually *refracted* (bent) into the second material. When the light wave is *reflected*, it bounces off the interface at exactly the same angle from which it came (Figure 5.11). When the light ray passes through the interface into the second medium (at any angle other than perpendicular), it will be refracted. The amount of bending depends on how extreme the difference is between the two media. We can describe this by using the relationship between two media with refractive indices of n_i and n_t. The incoming (incident) light ray makes an angle of θ_i and the transmitted ray an angle of θ_t with respect to perpendicular (Figure 5.11). Here i represents the incident light ray, and t represents the transmitted ray:

$$n_i \sin \theta_i = n_t \sin \theta_t.$$

If the incoming ray (n_i) is traveling in a higher refractive index medium than the other medium ($n_i > n_t$), the wave will bend up toward the interface. If $n_i < n_t$, then the ray bends away from the interface (Figure 5.11a), toward the perpendicular (Figure 5.11b). If $n_i = n_t$, there is no bending! So the next time you dive for pennies in a swimming pool, try to dive from directly over the penny so you can accurately pinpoint its location. This relationship also explains why the Becke line is so useful: as you lower the microscope stage, the light waves (which form the bright Becke line) bend toward the medium with the higher refractive index.

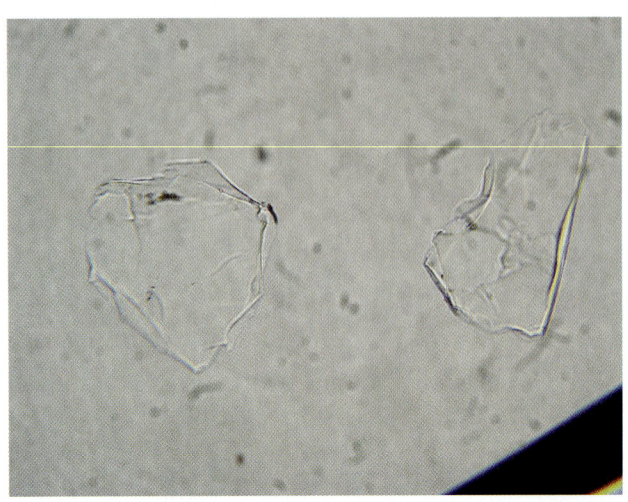

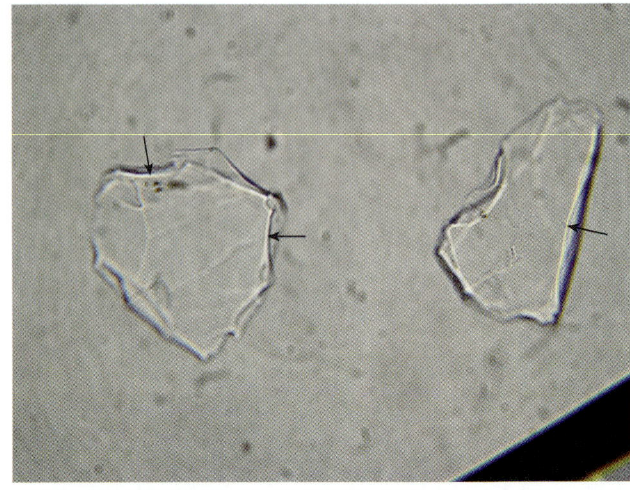

a. The grain is immersed in a liquid with n = 1.48; when the stage is lowered, the light Becke line moves into the grain. Low relief.

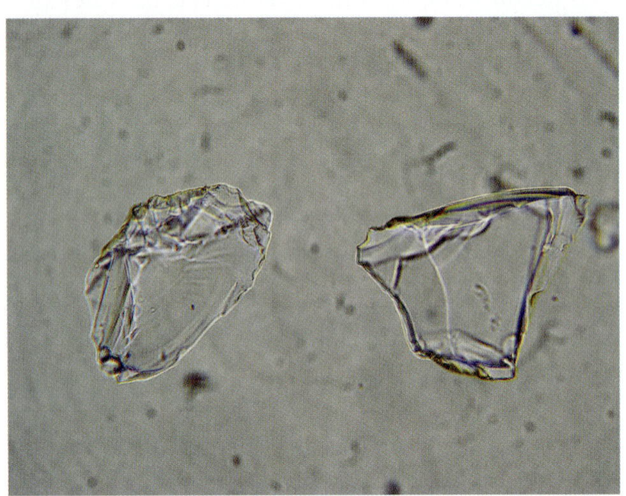

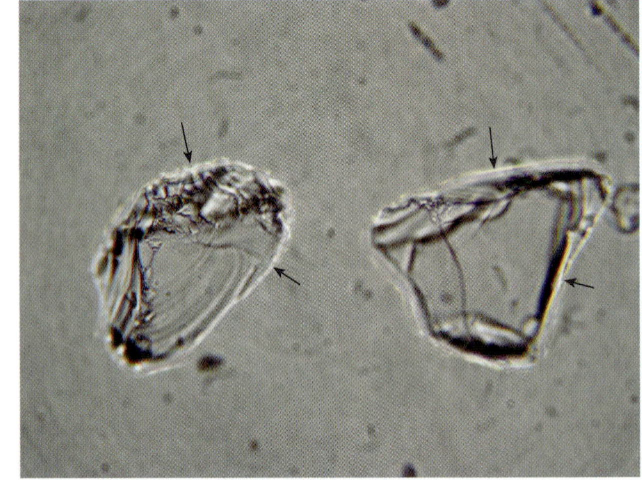

b. The grain is immersed in a liquid with n = 1.55; when the stage is lowered, the Becke line moves into the liquid. High relief.

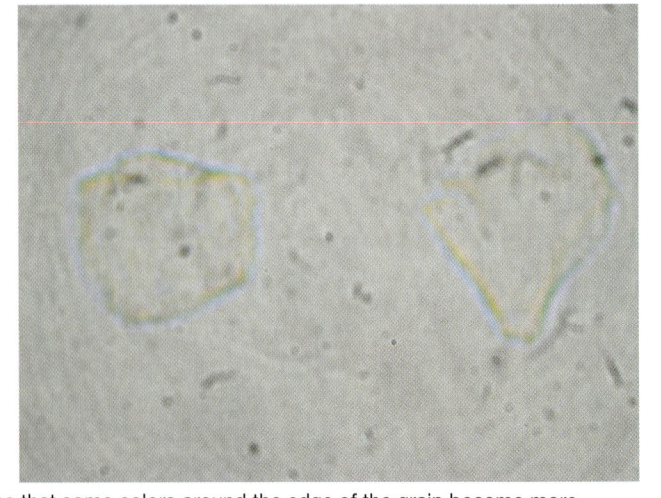

c. Pieces of glass with n = 1.50 immersed in a liquid of n = 1.50. Notice that some colors around the edge of the grain become more intense as the stage is lowered. Also notice that the closer the liquid and grain are to matching, the harder it is to see the grain. Very low relief.

Figure 5.10. Becke lines are optical phenomena that occur when a grain is slightly out of focus. They can be used to help estimate differences in refractive index between a liquid of known refractive index and a grain of unknown. In the three following images, a grain of n = 1.50 *is immersed in different liquids and photographed in focus (left) and with the grain lower and slightly out-of-focus (right; see also the set of movies on the MSA website corresponding to each of the three images). Recall that the light Becke line moves into the material of higher refractive index.* ***a.*** *The grain is immersed in a liquid with* n = 1.48*; when the stage is lowered, the light Becke line moves into the grain.* ***b.*** *The grain is immersed in a liquid with* n = 1.55*; when the stage is lowered, the Becke line moves into the liquid.* ***c.*** *The grain is immersed in a liquid of* n = 1.50*. Notice that some colors around the edge of the grain become more intense as the stage is lowered. Also notice that the closer the liquid and grain are to matching, the harder it is to see the grain.*

Wavelength and Refractive Index

Refractive index is not constant for any material, because it varies as a function of the wavelength of light that passes through it. Figure 5.12 shows a plot of refractive index (on the y axis) vs. wavelength (x axis) of visible light for a typical piece of glass. Notice how n is higher at low wavelengths (the blue-violet end of the spectrum) and decreases systematically as you move toward longer wavelengths at the red end of the spectrum (700 nm). **Dispersion** is the term we use for the change in any optical property as a function of wavelength. So here we talk of dispersion of the refractive index.

Most of the time we look at minerals with natural, white light, which is a mixture of all the visible light wavelengths. You can see the effect of wavelength on refractive index by using a glass prism (Figure 5.13). The width of the rainbow you generate with a prism depends on three variables: the angle of incidence of the white light beam, the refractive index (and dispersion) of the material from which the prism is made, and the angle between the faces of the prism. The result is that the shorter, blue-violet wavelengths of light (with higher n) are refracted (or bent) more than the longer redder wavelengths, producing a rainbow of colors. Think about this the next time you see a rainbow in the sky, and ponder what causes the rainbow to form.

Visual Representations of Refractive Index

So far, we've discovered that interfaces between materials of different refractive indices can cause bending, or refraction, of light waves—a property that is useful when comparing materials with different values of n. We've also established that refractive index, which is a function of the velocity at which light passes through a material, depends on electron density as well as charge distribution. In minerals, or any crystalline material, these properties may vary also as a function of orientation. In other words, you may have a higher electron density in one direction of a crystal than in another. Mineralogists use the differences in refractive index at various orientations as a diagnostic property. We represent these differences using an imaginary three-dimensional surface in which the axial lengths are proportional to refractive index. This 3-D surface is called an **optical indicatrix**, as noted above.

Consider first the **isotropic** case where all the vectors are the same length (Figure 5.14a). The **indicatrix** is defined as the surface connecting the tips of all the vectors of light radiating out from a common point within the crystal. If all the vectors are the same length, then the indicatrix will be a sphere. If the mineral is isometric, then all properties are the same in all directions in the crystal,

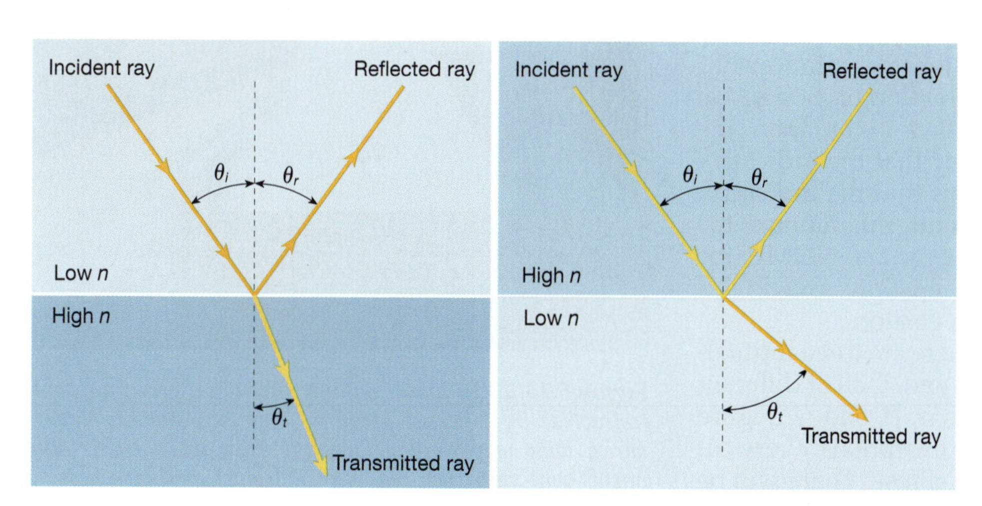

Figure 5.11. When a light ray is incident on the interface between two media of different refractive index two things occur: (1) the ray is reflected at the same angle, and (2) it is refracted. The image on the left shows what happens when a ray moves from low to high refractive index while the one on the right shows a ray moving from high to low refractive index.

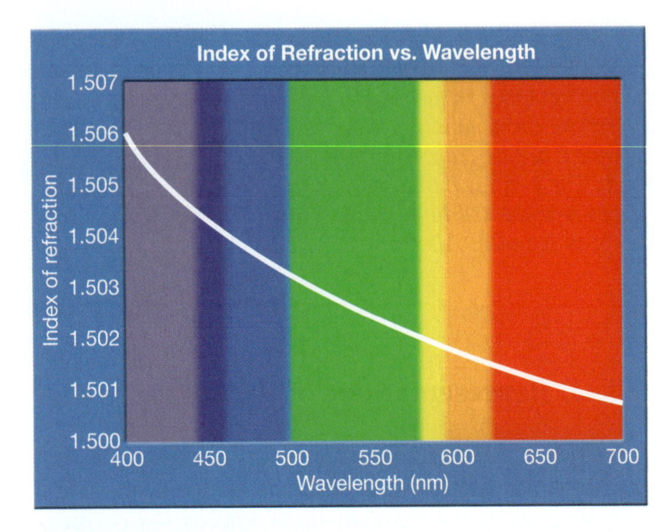

Figure 5.12. Schematic illustration of the change in refractive index with wavelength. This property is called dispersion of the refractive index. In general, the refractive index of a material decreases as the wavelength increases.

which is why the sphere has equal radii in all directions. Any way you cut through a sphere, as long as your cut passes through the center of the sphere, you will get a circle. The radius of that circle is proportional to the refractive index (i.e., the bigger the radius, the bigger the refractive index).

What about the **uniaxial indicatrix**? Now we have two axes that are the same length (*a* and *b*), and one that is different (*c*). There are two possible shapes that satisfy this constraint: one is shaped like a lemon, and one like a flying saucer. Each of these shapes will have a cross section that is circular if you slice it horizontally, but elliptical if you slice it in any other direction (Figure 5.14b and c). The radius of that circle is given the symbol ω (omega) for light vibrating perpendicular to *c*. The refractive index for light vibrating parallel to the *c* axis is called ε (epsilon). If you cut the uniaxial indicatrix through the center but at some random direction, you will always get an ellipse in which **one** direction is ω.

There are two possible shapes (signs) that arise from the two possibilities for the relationship between ε and ω: either $\varepsilon > \omega$ (lemon), or $\varepsilon < \omega$ (flying saucer). Note that if $\varepsilon = \omega$, then it's isotropic! When $\varepsilon > \omega$, we say that the indicatrix is positive, while $\varepsilon < \omega$ means that the indicatrix is negative (Table 5.3). Useful mnemonics for remembering these distinctions are NOME for **n**egative, **o**mega **m**ore than **e**psilon, and POLE for **p**ositive, **o**mega **l**ess than **e**psilon.

For the biaxial indicatrix, there are three mutually perpendicular vectors X, Y, and Z, all of different lengths, which we'll define as α, β, and γ (Figure 5.14d and e). With exactly two exceptions, every cut through the center will be an ellipse. There will be two special angles where the cut will have a circular cross section (Figure 5.14d and e). In this system, the three possible refractive indices have the symbols α (smallest value of *n*), β (intermediate *n*), and γ (largest *n*). The sign of the biaxial figure depends on the relationships between these values. If β is closer to α, then the mineral is positive, and if β is closer to γ, it is negative. Note that when β is equidistant from α and γ, it's not really either.

We define the **optic axis** to be the vector that is perpendicular to the circular section. For the case of uniaxial minerals, the optic axis is, by definition, the same direction as the *c* axis. In the biaxial case, there are two circular sections (Table 5.3), and so there are two optic axes. The acute (less than 90°) angle between them is called **2*V***, and that direction (either X or Z) is the **acute bisectrix**. The direction that bisects the obtuse is called the **obtuse bisectrix** (and is either X or Z depending on the mineral's sign).

A different way of explaining positive (+) and negative (–) for biaxial minerals is to use the location of optic axes relative to the X and Z directions, as just alluded to above. If the acute angle between the two optic axes is bisected by Z (or γ), then the mineral is positive (Figure 5.14d). Alternatively, if the acute bisectrix is X (or α), then the mineral is negative. It is possible that β will be exactly between α and γ—and that situation results in $2V = 90°$.

Next we can relate the refractive indices to the *a*, *b*, and *c* axes and the X, Y, and Z vector set. In the isotropic systems, all directions have a refractive index of *n*, so that *n* can be anywhere with respect to *a*, *b*, and *c*. In a uniaxial system, the *a* axis is the same unit length as the *b* axis, so the plane contain-

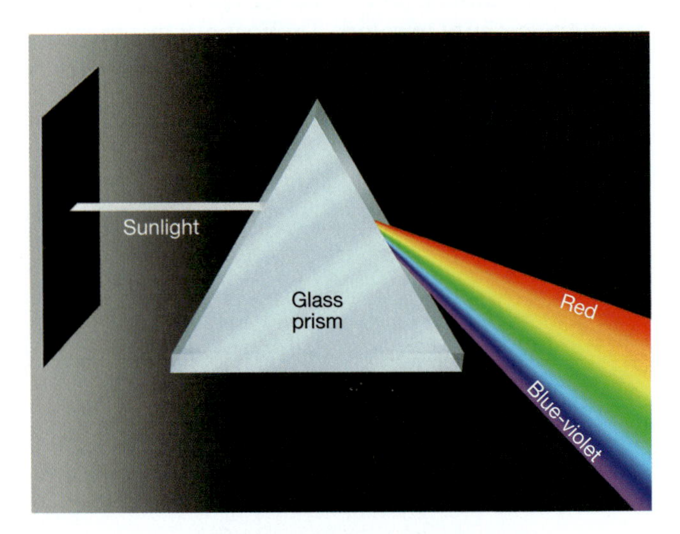

Figure 5.13. Light waves passing through a prism are bent (refracted) by different amounts because refractive index varies with wavelength. Therefore, blue-violet light (400 nm) is bent more than red light (700 nm), as shown here.

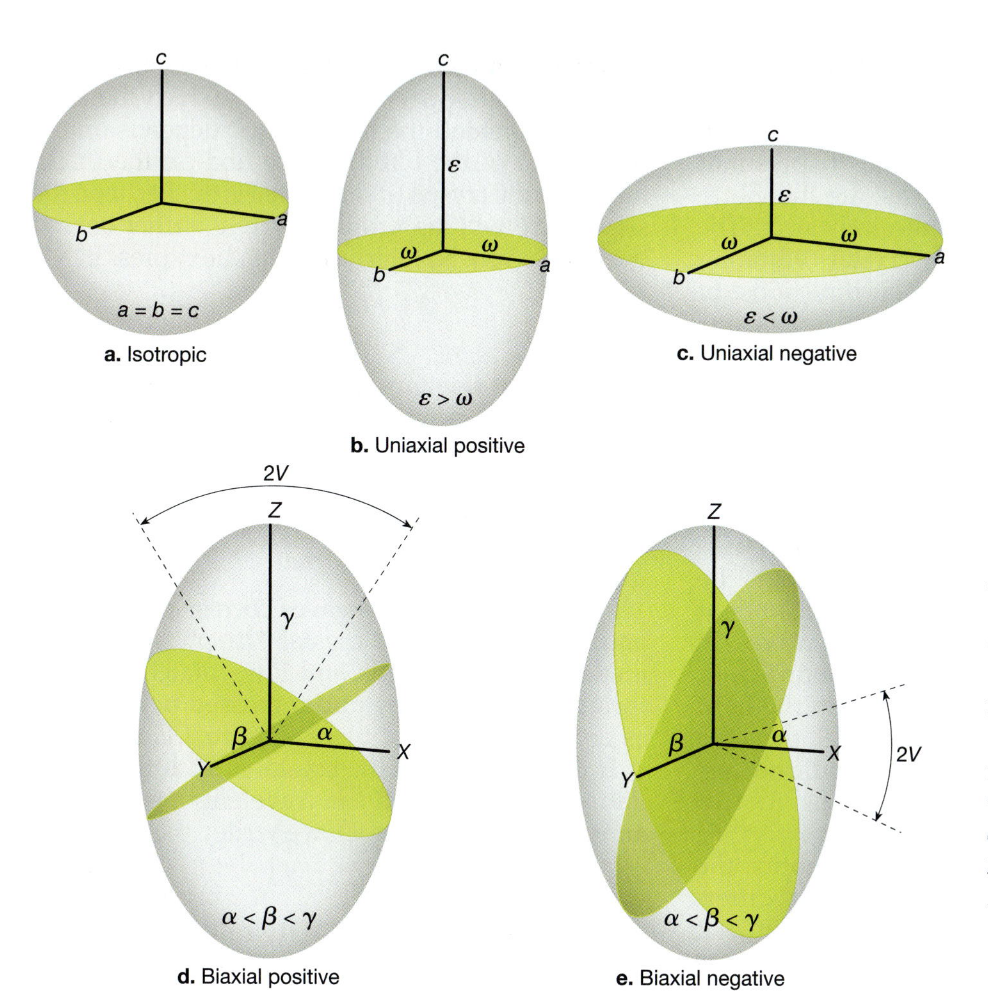

Figure 5.14. The optical indicatrices for crystals with isotropic (Figure 5.14a), uniaxial positive and negative (Figure 5.14b and c), and biaxial positive and negative (Figure 5.14d and e) cases. In Figure 5.14a, the refractive index is the same in all directions. In Figure 5.14b and c, the circular section has a radius of ω (omega), and the refractive index for light vibrating parallel to the c axis is called ε (epsilon). We say that the indicatrix is positive if ε > ω , and negative if ε < ω. In Figure 5.14d and e, there are three possible refractive indices: α (smallest value of n), β (intermediate n), and γ (largest n). Biaxial positive figures have β closer to α, and in biaxial negative figures, β is closer to γ. The acute angle between the optic axes in both cases is called 2V. (Images modified from Bloss 1999.)

ing them is a circle. The radius of that circle is ω (omega). Perpendicular to that circle is the *c* axis, which coincides with ε (epsilon). All the vectors that lie between ε and ω are given the designation ε' (Figures 5.14b and c). Finally, for biaxial systems, the refractive indices α, β, and γ correspond to *X*, *Y*, and *Z*. However, there are six possible optical orientations, or six different ways in which the *X*, *Y*, and *Z* sets can relate to the *a*, *b*, and *c* axes for the orthorhombic case. For the monoclinic, case the *b* axis corresponds to either *X*, *Y*, or *Z*, and for the triclinic case, usually none of the crystallographic axes corresponds to *X*, *Y*, or *Z*.

Birefringence and Retardation

Given what we just learned about refractive index and the optical indicatrix, it should be apparent that different directions in any crystal, other than ones belonging to the isometric system, will have different refractive indices. This occurs because atoms are not arranged uniformly in all directions within a crystal, but are often distributed differently in three-dimensional arrangements (as we discussed in Chapters 3 and 4). What actually happens when a ray of light passes through such an arrangement?

Table 5.3. Isotropic, Uniaxial, and Biaxial Indicatrices, Simplified

Crystal System	*Optical Class*	*Axes*	*# of circular sections*	*# of optic axes*
isometric	isotropic	n	infinite	infinite
tetragonal	anisotropic uniaxial	$\varepsilon > \omega$ (+) $\omega > \varepsilon$ (–)	1	1
hexagonal				
orthorhombic	anisotropic biaxial	β closer α (+) β closer γ (–)	2	2
monoclinic				
triclinic				

When we make a thin section and look at a single crystal, we are essentially looking at a 2-D "slice" that is represented by a 2-D slice of the indicatrix. When light enters that crystal, it splits into two different vectors (n and N) that are the vibration directions in the slice of the indicatrix. The light then travels through the crystal, vibrating parallel to those two perpendicular directions. In an **isotropic** mineral, n and N are the same, because the refractive index is the same in all directions. So the two different vectors emerge from the crystal vibrating in exactly the same way as they were before (Figure 5.15). If you insert the upper polarizer while viewing an isotropic mineral, it will appear black (as we'll shortly see, it is really exhibiting something called zero retardation).

In an **anisotropic** mineral, the n represents the direction with a smaller refractive index, so light travels faster in that direction (the *fast* wave). In the N direction, the wave's motion is retarded, or slowed down (the *slow* wave). So when the light emerges out the other side of the crystal, the slow wave, N, lags (or is retarded) behind the fast wave, n. (Figure 5.16). This brings us to two important terms:

Birefringence, δ, is the difference (notice the Greek symbol is the lower-case "d" for difference) in refractive index of the two directions, or $N - n$. If your cut of the indicatrix happens to be a place where it forms a circle, then there is no birefringence, because $N = n$. Because refractive index is unitless, so too is birefringence. Birefringence can be used as a diagnostic physical property of minerals.

Retardation, Δ, is the distance (notice the Greek symbol is the upper-case "D") by which the slow wave lags behind the fast wave. It depends on the magnitude of the birefringence and on the **thickness**, t, of the crystal (if the crystal is thicker, then the slow wave will lag proportionately farther behind the fast wave!). So retardation can be expressed as:

$$\Delta = t\,(N - n), \text{ or}$$
$$\Delta = t\,\delta.$$

When the thickness of the crystal is such that retardation is equal to the wavelength of light (here assuming monochromatic light—light of one wavelength) passing through it (or an i, integer multiple of that wavelength), we could write:

$$\Delta = i\,\lambda.$$

In this situation, the waves corresponding to N and n will leave the crystal "in phase," which means that their highs and lows match. Viewed through a microscope under crossed polars, such as crystal will appear black (Figure 5.17).

If the thickness of the crystal is such that retardation is equal to half the wavelength of light (or an integer multiple of it), we write:

$$\Delta = (i + {}^{1}/_{2})\,\lambda.$$

The n and N waves emerge from the crystal completely out of phase (Figure 5.18), and when viewed under crossed polars, the mineral will be the same color as the monochromatic light used to illuminate the crystal. Intermediate thickness of crystal will produce intermediate intensities between black and colored.

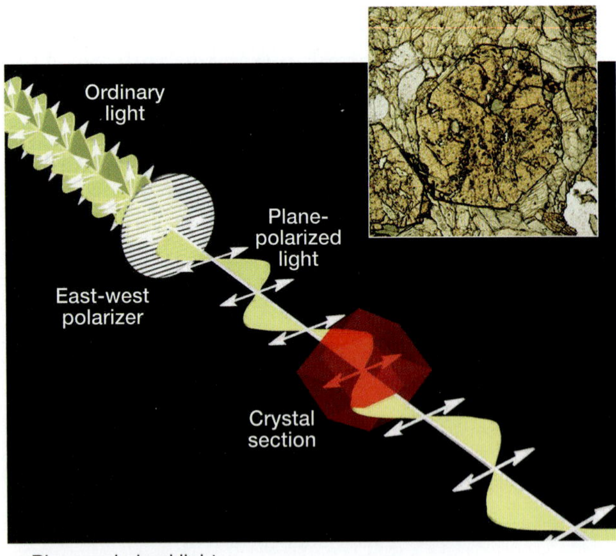

a. Plane-polarized light.

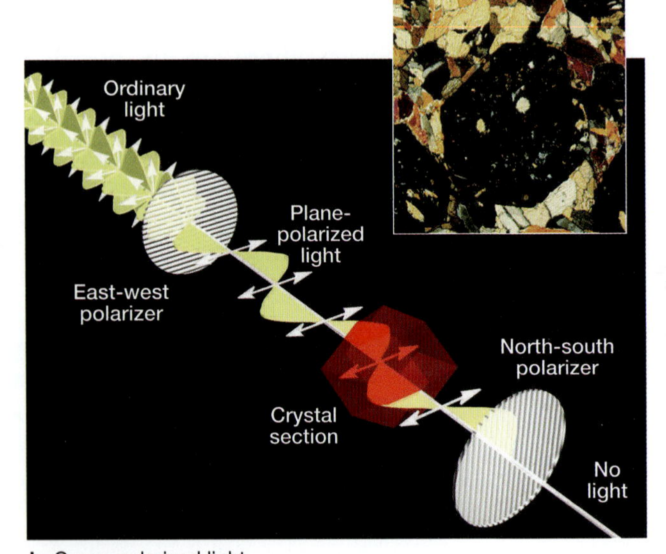

b. Cross-polarized light.

Figure 5.15. Interaction of polarized light with an isometric mineral. Notice how the plane-polarized light passes through the crystal unchanged, but when the polars are crossed, no light is transmitted.

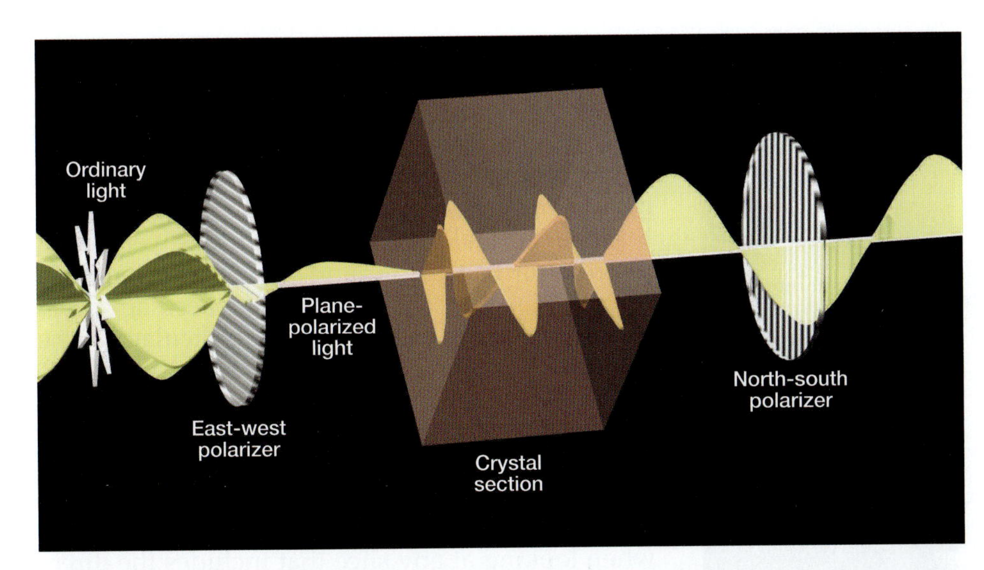

Figure 5.16. *Interaction of polarized light with an anisotropic mineral. Notice how the light wave is split into two different waves when it travels through the crystal, and how a portion of the light is transmitted by the upper polarizer.*

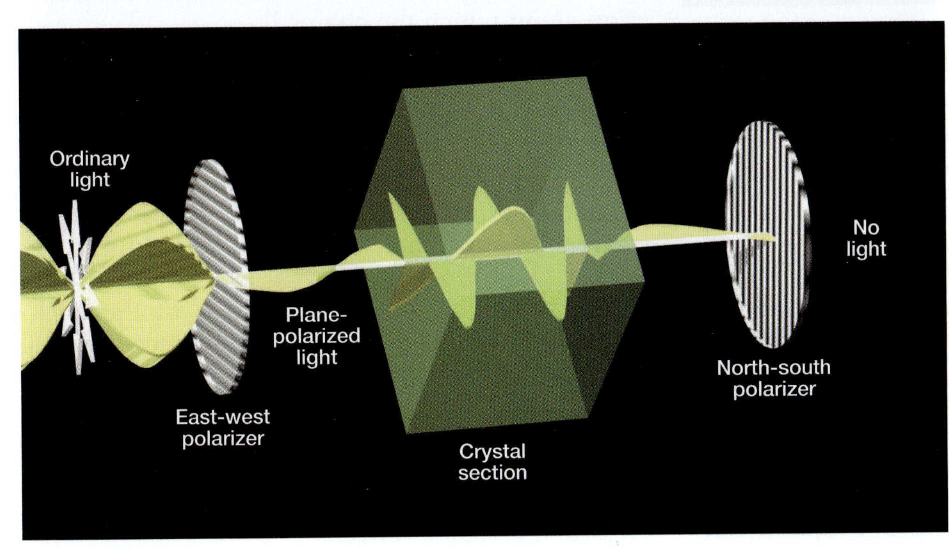

Figure 5.17. *Retardation, Δ, is a measure of how much the slow wave lags behind the fast wave. In some special cases, the thickness of the crystal is such that the retardation is an integer multiple of the wavelength of light (but only when monochromatic radiation is used). So although the two waves change velocity when they are traveling in the crystal, they emerge from the crystal in phase. Then they recombine to be in the same orientation as when they entered the crystal, thus no light makes it through the upper polarizer.*

This variation in color is best understood by first using a single (monochromatic) wavelength of light as used above. Consider the crystal wedge shown in Figure 5.19. The crystal will appear black under crossed polars when its thickness is an integer multiple of the monochromatic illumination. So for red light (λ = 700 nm), it would be black at 700, 1400, 2100, 2800 nm, etc., and bright red at 1050, 1750, 2450, 3150 nm, etc. For violet light (λ = 400 nm), the crystal will be black at 400, 800, 1200, 1600 nm and bright violet at 600, 1000, 1400, 1800 nm.

Your polarizing light microscope probably isn't equipped to pass monochromatic light through your sample. Instead, it uses polychromatic light, which has many different wavelengths (as demonstrated in the prism example earlier in this chapter) covering the whole range of visible light colors. When all these wavelengths are summed up, you get light with a range of colors at different wavelengths (Figure 5.20; we will return to how these colors are formed in more detail in Chapter 16 and their uses).

A **Michel-Lévy chart** (or interference color chart) is a graphical way of showing the relationships between Δ, δ, and t (the back cover of this book). The y axis is the thickness of the crystal, and the bottom x axis is retardation (Δ, sometimes called **path difference**). Diagonal lines from the lower left corner to the upper right and left edges are lines of varying birefringence (δ). Notice how the colors become less saturated ("whiter" or more pale) as you go from left to right on the chart. There are three apparent sets of color series on the chart, which are labeled **first-order**, **second-order**, and **third-order** from left to right.

So far, we've been talking about birefringence as an abstract phenomenon, so let's put it in the context of real mineral structures to make it easier to understand. Consider three mineral group examples from our Big Ten Mineral list: feldspars, micas, and pyroxenes. From Chapter 1, recall that

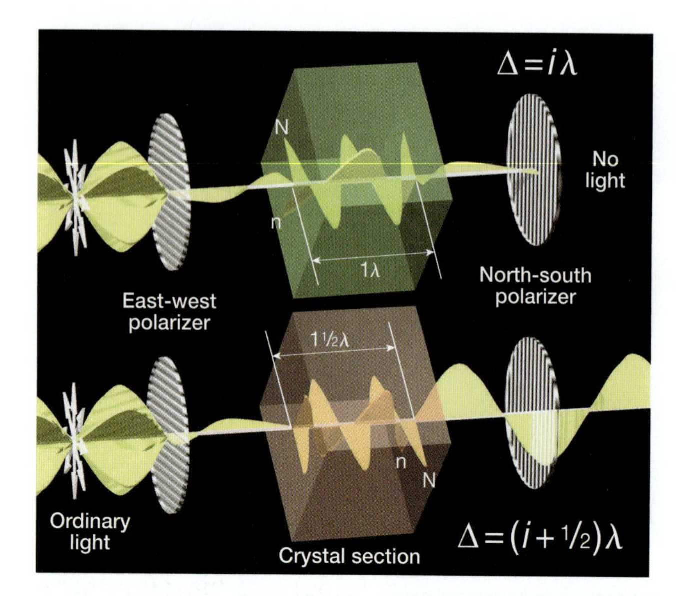

Figure 5.18. *In contrast with Figure 5.17, here* N *lags* ½ *wavelength behind* n. *When the waves recombine at the upper polarizer, they are rotated 90° from the way they were when they entered the crystal; thus a portion of the light is transmitted at the upper polarizer.*

feldspar is a framework silicate, which means that all the corners of every Si and Al tetrahedron are shared. This creates a 3-D tetrahedral network, so the structure has very similar electron densities and charges in all directions. So in feldspars, there are not dramatic differences between orientations, and so the difference between *N* and *n* is small. As a result, feldspars have very low birefringence. What type of birefringence do you think quartz would have?

When you reduce the framework tetrahedra into layer silicates, or micas, you have a different situation because now only three of the four corners of the tetrahedra are shared. For minerals like biotite and muscovite, there are roughly similar atomic and electronic densities within the sheets (i.e., the *a-b* plane), so if you are looking at a mica section where the light vibrates only in the sheet, the retardation will be low. On the other hand, if you happen upon a cut of the indicatrix where one of the refractive indices represents light vibrating *perpendicular* to the sheets in the *a-b* plane (i.e., along *c*), then you see very high retardation because *N* and *n* are so different in magnitude. Can you venture a guess at which direction (i.e., parallel or perpendicular to the sheets) would have the greatest refractive index?

Finally, consider the case of chain silicates like pyroxene, which share only two of four corners of their tetrahedra. Usually in these minerals, the largest refractive index and highest δ occur along the chains. So maximum retardation will occur when looking at any slice that includes the direction along which the chains lie.

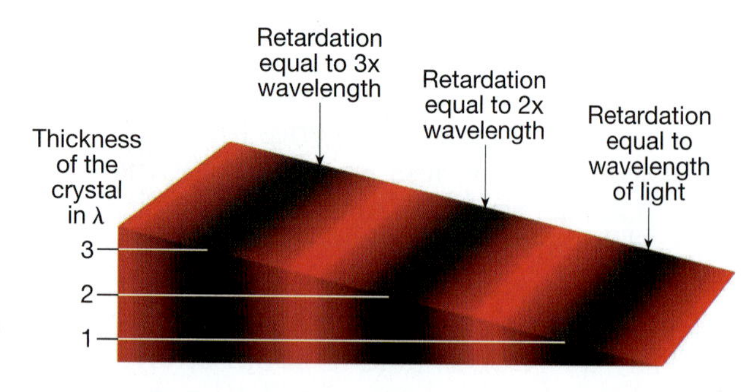

Figure 5.19. *If you have a wedge-shaped crystal illuminated with red light, the changes in thickness will result in color changes in cross-polarized light because of the relationship* $\Delta = t\,(N - n)$ *between retardation,* Δ*, and thickness* (t). *The birefringence,* δ*, of the crystal* (N – n) *remains constant.*

Interference Figures

Although retardation is useful in identifying minerals, it frequently gives non-unique identifications because many minerals have similar values of Δ, and retardation varies as a function of orientation. So it's helpful to have another technique that can help distinguish between uniaxial and biaxial minerals, give information on optic sign and 2*V*, and help determine the orientation of the mineral. For this purpose, we use an **interference figure**, which is produced when we examine light that travels in many directions within a crystal at once.

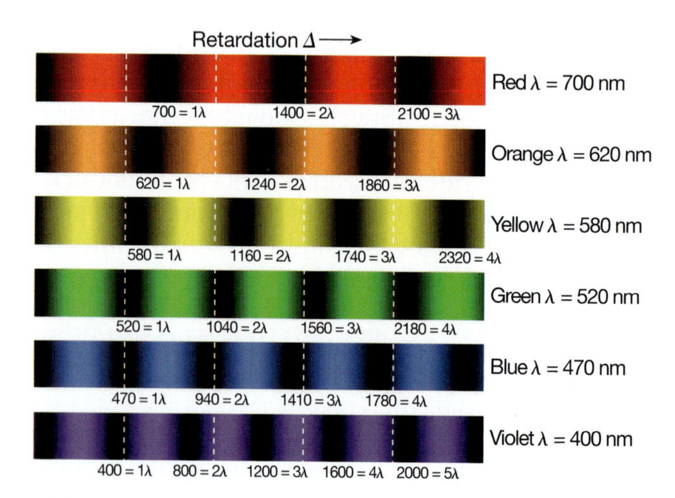

Figure 5.20. *The light source on your microscope contains multiple wavelengths of light, and each of these will have variable retardation (path difference) as thickness changes. When all these wavelengths are added together, the result is a band containing sets of colors: first-order, second-order, third-order, etc. corresponding to increasingly lighter colors as you move from left to right.*

Normally we only observe light that travels straight through a crystal. Historically we've called this type of illumination **orthoscopic**, to imply that the light rays are normal (i.e., orthogonal) to the microscope stage (Figure 5.7). However, we can change the light incident upon the crystal into the shape of a cone, for what we call **conoscopic** illumination, by inserting the substage condensing lens. This allows us to image multiple ray paths through the crystal.

Figure 5.21a shows a piece of fluorescing glass placed on the stage of a PLM. The orthoscopic ray path is shown in Figure 5.21b, and the rays are seen to be parallel; thus they would only illuminate one plane through the crystal. In Figure 5.21c, the substage condensing lens is shown to focus the rays in a cone. Thus these rays would travel along different paths through the crystal and "see" different cuts through the indicatrix.

Next we can replace the glass with an oriented crystal of quartz cut with its *c* axis perpendicular to the microscope stage. To "see" the interference figure, we place over the sample a ping-pong ball cut in half with a piece of polarizer glued to its bottom (Figure 5.22a and b). This method of projected interference figures is called a **Quirke hemisphere**, after the description by Quirke (1938). Medenbach et al. (1977) give a good discussion of the uses of this technique. Now we see the 3-D projection of the interference figure for quartz, a uniaxial mineral. Next we replace the quartz with a piece of the biaxial mineral muscovite to see a biaxial interference figure (Figure 5.22c). On the microscope, we really are looking at a 2-D projection of this 3-D phenomenon. Compare these two interference figures to those for a uniaxial (Figure 5.23a) and biaxial mineral (Figure 5.23b) photographed in the microscope.

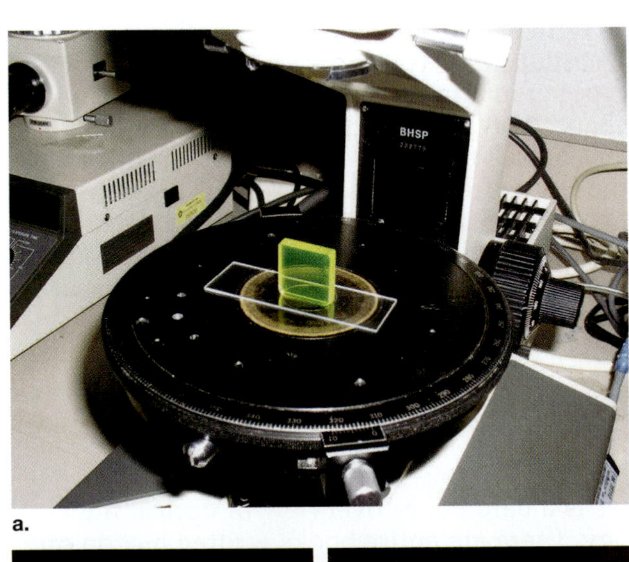

a.

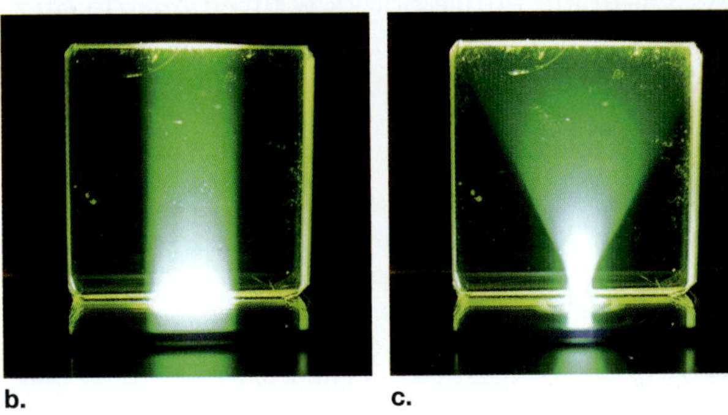

b. c.

Figure 5.21. Orthoscopic and conoscopic illumination shown using a piece of fluorescing glass. ***a.*** *The piece of glass is placed on the stage of a PLM (the lenses have been removed for this demonstration).* ***b.*** *The glass is now illuminated with a parallel bundle of rays from the substage, resulting in orthoscopic illumination.* ***c.*** *Next the glass is illuminated with a cone of rays, for conoscopic illumination.*

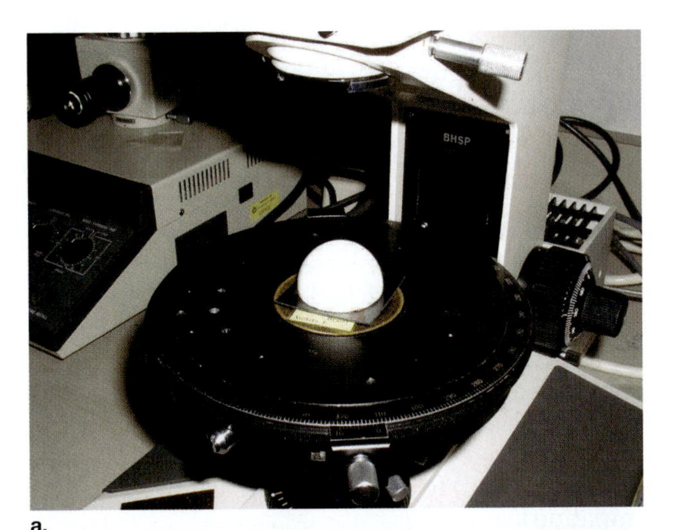

a.

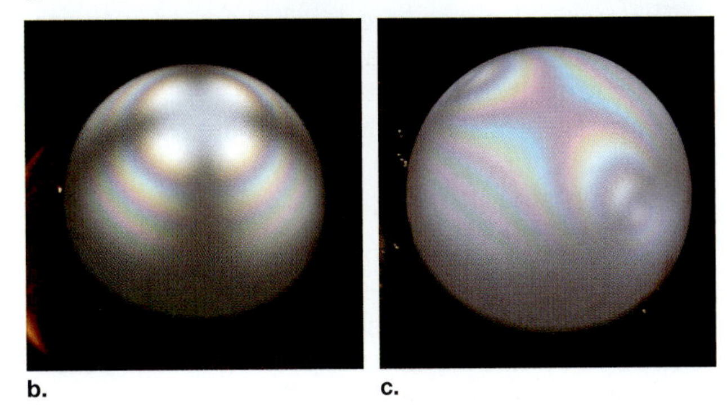

b. c.

Figure 5.22. Illustrations of interference figures projected onto a ping-pong ball. ***a.*** *A photograph of the setup used for the projection. A ping-pong ball is cut in half and glued onto a polarizer (the polarizer replaces the upper polarizer of the PLM).* ***b.*** *Next, a piece of quartz cut with its* c *axis perpendicular to the stage is placed below the ping-pong ball/polarizer combination. Its interference figure is projected onto the spherical surface of the ping-pong ball.* ***c.*** *A muscovite flake replaces the quartz and produces a biaxial interference figure.*

*Figure 5.23. Photomicrographs of: **a.** a centered optical axis interference figure for a uniaxial mineral and **b.** a centered acute bisectrix interference figure for a biaxial mineral. Note how these look just like a 2-D version of the 3-D ping-pong ball projections in Figure 5.22.*

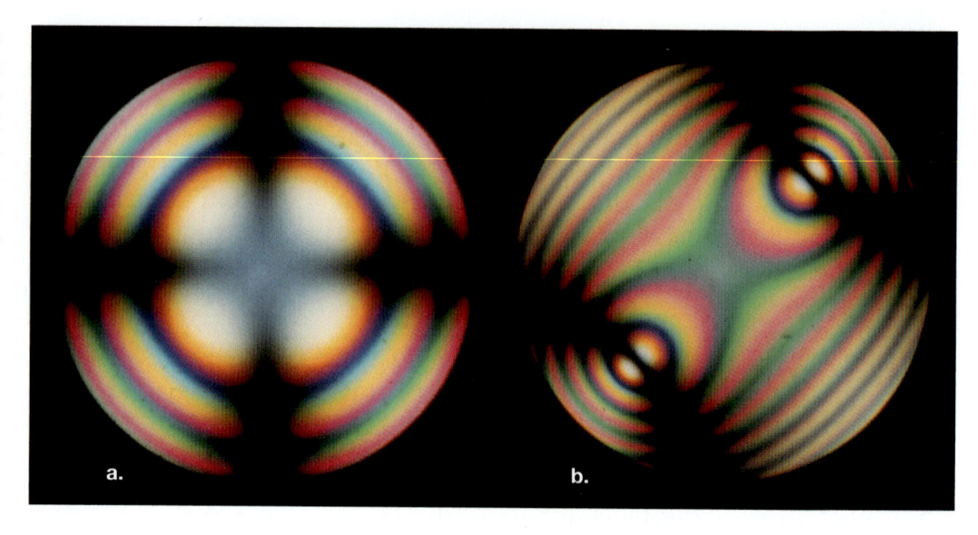

There is a much simpler way to view interference figures without the aid of a microscope or ping-pong balls. This relates back to the image of muscovite placed between two polarizers that we saw back in Figure 5.1b at the start of the chapter. If you place a large single crystal (e.g., muscovite) between crossed polars and bring the three sandwiched pieces as close to your eye as possible, you will simulate illumination in conoscopic light. Figure 5.24a shows the setup for this example. Figure 5.24b is the close view of calcite (a uniaxial mineral) sandwiched between two polarizers showing a uniaxial interference figure. Figure 5.24c shows the up-close view of muscovite showing a biaxial interference figure. Check out the movies on the MSA website showing how the images change as we go from far (i.e., orthoscopic illumination) to near-view (i.e., conoscopic illumination).

The normal procedure to observe an interference figure with your microscope will be to use the Bertrand lens to help image the 3-D rays and the substage condenser to illuminate the mineral with a cone of light. To observe an interference figure, use the following steps:

1. Using a low power objective, choose a large grain with the lowest retardation possible, which increases the chance of looking down the optic axis so a circular section of the indicatrix will be near-parallel to the stage.
2. Switch to the highest power objective you have, and refocus on the grain of interest.
3. Be sure the polars are crossed, and open the substage diaphragm to let in as much light as possible.
4. Switch to conoscopic illumination.

If you are successful, you will see your very own interference figure. Optic axes in uniaxial figures look like black crosses (Figures 5.22b, 5.23a, 24b) that point N-S and E-W and do not change much as the stage is rotated. Optic axes of biaxial figures have curved black lines (Figures 5.22c, 5.23b, 24c) that enter and exit the view as you rotate the stage, changing from nearly straight lines to bending ones (check out the MSA website). Once you learn to distinguish different types of interference figures (more of this in later chapters), you can distinguish between uniaxial and biaxial minerals, estimate $2V$ on the basis of the curvature of the isogyres (which are the dark lines in the interference figure), and even determine the optic sign (+ vs. –) by using the accessory plates. These more advanced topics will be covered in detail in Chapter 17.

Concluding Remarks

Needless to say, what is provided in this chapter is a bare bones overview of optical mineralogy. There are entire books written just on optical crystallography! This chapter is meant to introduce you to some of the wonders of optical

*Figure 5.24. A demonstration to show interference figures by using only two polarizers and a centimeter-sized, oriented crystal. **a.** Materials necessary for the demonstration, which include two polarizers, samples, and a light source (here a light table, although you can hold the samples up next to your eye instead of moving your eye to the light table). **b.** The uniaxial interference figure produced by placing a camera against the polarizers with the calcite (the thin section in Figure 5.24a) between them. **c.** The biaxial interference figure produced by placing a camera against the polarizers with muscovite between them. (Please check out the movies of this on the MSA website. For them, the camera starts about one foot from the samples and records until it makes contact and then moves back out.)*

mineralogy, and get you started with some of the nomenclature and methods. Hopefully this will get you interested in learning more about it. Chapters 16–19 will expand on what's been introduced here, explain the theory of how light interacts with minerals, and discuss more methods to use to identify and characterize minerals.

a.

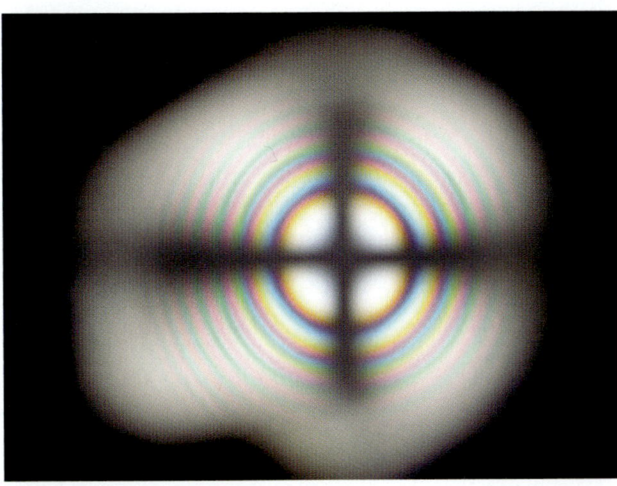

b.

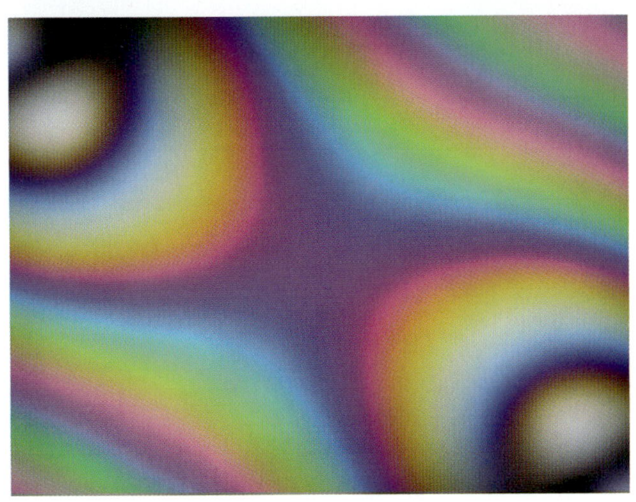

c.

To gain a better understanding of the importance of learning and teaching optical mineralogy in the 21st century, we encourage you to read the article "The polarized light microscope: should we teach the use of a 19th century instrument in the 21st century?" by Gunter (2004).

References

Bloss, F.D. (1999) Optical Crystallography. Mineralogical Society of America, Washington D.C. 239 pp.

Browning, P. (1996) UKESCC: Optical mineralogy module, Terra Nova, 8, 386–389.

Biot, J.-B. (1815) Sur un mode particulier de polarization qui s'observe dans la tourmaline. Annales de Chemie et de Physique, 94, 191–199.

Gunter, M.E. (2004) The polarized light microscope: should we teach the use of a 19th century instrument in the 21st century? Journal of Geoscience Education, 52, 34–44.

Humphries, D.W. (1992) The Preparation of Thin sections of Rocks, Minerals, and Ceramics. Oxford University Press, Oxford, 83 pp.

Hutchinson, C.S. (1974) Laboratory Handbook of Petrographic techniques. John Wiley and Sons, Inc., New York, 527 pp.

Kile, D.E. (2003) The Petrographic Microscope: Evolution of a mineralogical research instrument. Special Publication No. 1, The Mineralogical Record, Tucson, AZ, 96 pp.

Lima-de-Faria, J. (ed.) (1990) Historical Atlas of Crystallography, Kluwer Academic Publishers, Boston, 158 pp.

Malus, E.L.(1809) Sur une porpriété de la lumière réfléchie. Mémoirs de Physique et de Chimie de la Société d'arcueil, 2, 143–158.

Medenbach, O., Weinrich, H., and Medenbach, K. (1977) The demonstration of interference figures of crystals with the projection hemisphere according to Quirke. Scientific and Technical Information, E. Leitz, Inc., 7, 12–14.

Needham, G.H. (1858) The Practical Use of the Microscope. Charles C. Thomas, Springfield, Illinois, 493 p.

Nicol, W. (1834) Observations on the structure of recent and fossil coniferea. Edinburgh New Philosophical Journal, 16, 137–158.

Nicol, W. (1839) Notice concerning an improvement in the construction of the single vision prism of calcareous spar. Edinburgh New Philosophical Journal, 27, 332–333.

Quirke, T.T. (1938) Direct projection of optic figures. The American Mineralogist, 23, 595–605.

Sorby, H.C. (1858) On the microscopical structure of crystals, indicating the origin of minerals and rocks. Quarterly Journal of the Geological Society of London, 14, 453–500.

Chapter 6

Systematic Mineralogy

My love affair with the mineral tourmaline began in college, when my parents bought their retirement home on the coast of Maine. The first time I visited the house, I rushed down to the beach to inspect the rocks, as I was at that point a newly-declared geology major who had just finished taking mineralogy! I was pleased to find growing in the dikes and crevasses of the rocks along the coast there beautiful sprays of a mineral I could easily recognize: tourmaline!

A few years later, I had the good fortune to take a field trip to the localities in southwestern Maine and northern New Hampshire where gem tourmaline was first mined in the U.S. Carl Francis, Bill Metropolis, and Mike Wise introduced me to the many types of occurrences of gem tourmaline, and I got to see the interior of a pegmatite with zones of different (and beautiful!) minerals. The first time you put your head in a newly-blasted gem pocket, you can't help but be hooked! I returned from that field trip with my backpack full of micas as big as dinner plates, and with visions of tourmalines dancing in my head. I soon learned that the chemistry of tourmaline, which is hard to dissolve in acid, was not well-studied by early mineralogists who used wet chemistry (analysis of minerals in solution discussed in Chapter 9) to determine mineral composition. So I set out to study the chemistry of tourmaline using modern analytical techniques.

Since then, I have devoted much of my research time to studying the chemistry and structure of various species of this fascinating mineral group. While doing this, I've learned that tourmaline has a fascinating history. The Saint Wenceslas king's crown, used in the coronations of Czech kings, was made between 1346 and 1387, and its centerpiece is a 39.5 x 36.5 x 15 mm, bright red tourmaline (though it was thought to be a ruby until 1998!) (Hyršl and Neumanova, 1999). An egg-sized, bright pink Burmese rubellite tourmaline pendant in the Treasure Room of the Kremlin dates back to the reign of Emperor Rudolph II (1575–1612) (Glas, 2003). In the 1700s, Dutch East Indian sailors used the pyroelectric property of heated tourmaline crystals to remove ash from their pipes (Glas, 2003). In 1820, tourmaline was first discovered at Mount Mica in Paris, Maine, and this important North American find was mined in earnest beginning in the 1860s, sparking a national interest in gem tourmaline (King, 2000). In World War II, tourmaline was used for pressure gauges in submarines (Heinrich, cited in Simmons, 2003), and today it has the same use in aircraft. The variable chemistry, color, and occurrences of the endlessly fascinating tourmaline group minerals continue to be the subject of extensive research today.

In this chapter, we recap the Big Ten minerals we introduced in Chapter 1, and spiral outward to look at several important silicate mineral groups, including tourmaline. As we embark on this systematic overview of rock-forming minerals, remember that behind every mineral lies a unique history and, in many cases, important industrial applications. Minerals are not just aesthetically pleasing specimens, but they shape our lives in many ways!

M.D.D.

Introduction

There are 78 different mineral classes (yikes!), which is a huge number to learn all at once! Luckily, there are only 27 silicate groups, which constitute 92% of the Earth's crust, and they are all based upon one simple building block: the SiO_4 tetrahedron. Recall that O is the most abundant element in the Earth's crust and Si is the second; thus we'd expect many minerals to contain these two. In this chapter, we will summarize the commonly-occurring members of some of these classes, touching only on what we consider to be the most important and/or interesting representatives. Our goal here is to continue the theme of Chapter 1 by demonstrating how all the silicate structures are related.

You will recall that the $Si^{4+}O^{2-}{}_4$ tetrahedron has a net charge of –4. Because it's not electronically neutral, the SiO_4 tetrahedron has to combine with other elements to make up a neutral solid (a requirement for being a mineral). It can accomplish this either by sharing one or more of the oxygen atoms on its corners, or by linking to other, positively-charged cations in between the tetrahedra. Some of these various possibilities are summarized in Table 6.1, which gives you an idea of the relationships between the mineral groups. We'll begin with the minerals in which the tetrahedra are completely linked (with all four corners shared), and then progressively **depolymerize** (break apart) the tetrahedra until we get to the groups with no shared corners. We begin in this manner because quartz and feldspar belong to this group, and between them they compose almost two-thirds of the minerals in the Earth's crust! At the end of this chapter, we'll also sneak in a look at calcite, which is a non-silicate, and explain why it made our Big Ten list.

Framework Silicates

For framework silicates, the ratio of Si to O is 1:2 because each O^{2-} anion bonds to two Si^{4+} cations (Figure 6.1). Quartz is the first mineral we'll discuss in this group. The formula for quartz, SiO_2, is also one of the easiest to remember! Because quartz makes up about 12% of the Earth's crust, it is the most commonly-occurring mineral species there. It is composed of SiO_4 tetrahedra in which all four corners are shared with other tetrahedra. There are at least five additional polymorphs of SiO_2 found in nature: cristobalite, tridymite, coesite, stishovite, and moganite. These different polymorphs form under different geological conditions (i.e., pressures and temperatures), so finding one of these polymorphs may tell you something about how a rock formed!

In general, quartz is the most stable silica polymorph at the Earth's surface. Cristobalite and tridymite form at high temperatures, so they often occur in association with volcanic rock, while coesite and stishovite occur at high pressures and are often associated with subduction zones and high-pressure metamorphic rocks. Each of these structures (except stishovite) shares the characteristic of having completely-linked SiO_4 tetrahedra, though the arrangements vary; we'll explain this in more detail in Chapter 22. But for now, imagine one of those sponges with the big holes in it that you use to wash your car: the sponge can be stretched, pushed and twisted into a number of different shapes. In that same way, the SiO_2 framework can take on many different shapes, but they all (except for stishovite) share the same basic framework: every Si^{4+} atom is linked to four O^{2-}, and every O^{2-} is linked to two Si^{4+} (Figure 6.1). In stishovite, the Si is surrounded by six oxygens, but it only forms at the very high pressures often associated with meteorite impact craters. Recall that back in Chapter 1, we mentioned that higher pressures often cause increases in coordination number.

The structure of quartz is shown in Figure 6.1a and b, where there are two different projections with the atoms represented in "ball and stick" form. In this type of representation, the atoms are shown as "balls" and the bonds that connect them are the "sticks." Figure 6.1c and d are the same projections as Figure 6.1a and b, except now the structure is shown in a polyhedral representation. In this case, the individual tetrahedra are shown to make it easier to visualize the assemblage of polyhedra in the structure. In Figure 6.1e, the center section of Figure 6.1a is removed and enlarged to show how each O^{2-} anion of the central tetrahedron is shared between two Si^{4+} cations. Figure 6.1f rotates the image by 90° to show the sharing better. Finally, Figure 6.1g and h are polyhedral representations of Figure 6.1e and f. Don't forget to check out the color movies of these static images on the MSA website.

Si^{4+} is not the only ion that can occupy tetrahedra composed of four oxygen atoms. Al^{3+} also commonly occupies tetrahedral coordination polyhedra, usually substituting for Si^{4+}. If you have four formula units of SiO_2 (which would be Si_4O_8) and you substitute in one Al^{3+} atom, you get $(AlSi_3O_8)^{1-}$. The ratio of tetrahedral cations (Si^{4+} and Al^{3+}) to anions (O^{2-}) is still 1:2, so you still have a framework structure. You do need to

Table 6.1. Important Rock-Forming Silicate Minerals

Mineral Class	*Mineral Group*	*Representative Species*	*Chemical Formula*	*Number of tetrahedral corners shared**
framework silicates (tectosilicates)	silica polymorph	quartz	SiO_2	4
	feldspar	albite	$NaAlSi_3O_8$	
		orthoclase	$KAlSi_3O_8$	
		anorthite	$CaAl_2Si_2O_8$	
	zeolite	natrolite	$Na_{16}Al_{16}Si_{24}O_{80}\cdot 16H_2O$	
layer silicates (phyllosilicates)	kaolinite-serpentine	kaolinite	$Al_2Si_2O_5(OH)_4$	3
		antigorite	$Mg_3Si_2O_5(OH)_4$	
	pyrophyllite-talc	pyrophyllite	$Al_2Si_4O_{10}(OH)_2$	
		talc	$Mg_3Si_4O_{10}(OH)_2$	
	mica	muscovite	$KAl_2AlSi_3O_{10}(OH,F)_2$	
		phlogopite	$KMg_3AlSi_3O_{10}(OH,F)_2$	
	smectite	montmorillonite	$(Na,Ca_{0.5})_{0.33}(Al,Mg)_2(Si,Al)_4O_{10}(OH)_2\cdot nH_2O$	
	chlorite	chlorite	$(Mg,Fe,Al)_3(Si,Al)_4O_{10}(OH)_2(Mg,Fe,Al)_3(OH)_6$	
double chains (inosilicates)	amphibole	tremolite	$Ca_2Mg_5Si_8O_{22}(OH)_2$	2 & 3
		grunerite	$Fe^{2+}{}_7Si_8O_{22}(OH)_2$	
single chains (inosilicates)	pyroxene	enstatite	$Mg_2Si_2O_6$	2
		diopside	$CaMgSi_2O_6$	
ring silicates (cyclosilicates)	tourmaline	schorl	$NaFe^{2+}{}_3Al_6(BO_3)_3Si_6O_{18}(OH)_4$	2
disilicates (sorosilicates)	epidote	epidote	$Ca_2(Al,Fe)Al_2O(SiO_4)(Si_2O_7)(OH)$	0 & 1
orthosilicates (nesosilicates)	olivine	fayalite	$Fe^{2+}{}_2SiO_4$	0
		forsterite	Mg_2SiO_4	
	andalusite (aluminosilicates)	andalusite	Al_2SiO_5	0
		sillimanite		
		kyanite		
	garnet	pyrope	$Mg_3Al_2(SiO_4)_3$	0
		almandine	$Fe^{2+}{}_3Al_2(SiO_4)_3$	
		andradite	$Ca_3Fe^{3+}{}_2(SiO_4)_3$	

*includes Al and Si in tetrahedral sites

make room for a monovalent cation to achieve charge balance. The addition of this cation creates the feldspar group minerals:

albite, $Na^{1+}AlSi_3O_8$,
orthoclase, $K^{1+}AlSi_3O_8$, and
anorthite, $Ca^{2+}Al_2Si_2O_8$.

When two Al^{3+} substitute for Si^{4+} (as in the formula of anorthite), a divalent cation is needed to balance the charge.

Figure 6.1i shows the structure of the feldspar group mineral albite in two different projections. Notice that each tetrahedron shares all four corners with another tetrahedron; the darker colored tetrahedra contain Si and the lighter ones, Al. There are more Si tetrahedra than Al tetrahedra, as indicated by the chemical formula. The spheres in the structure represent the Na atoms.

The three feldspar compositions are commonly represented graphically on a nifty plot called a ternary diagram. Although we'll discuss this in more detail later, ternary plots are so useful in explaining relationships between mineral compositions that we'll introduce them here.

First, imagine that either Na^{1+} or K^{1+} can occur in the feldspar, and that they can occur in any range from 0 to 1 total cations. We could represent this by drawing a line with those cations at either end, like this:

Na^{1+}------------K^{1+}

You could then show any composition between 100% Na^{1+} and 100% K^{1+} by putting a dot somewhere along this line. To help do this, you might use a line with tick marks, like this:

0% **K** 20 40 60 80 100% **K**
A **B**
100% **Na** 80 60 40 20 0% **Na**

So a composition with 55% Na and 45% K would lie at point A, and a composition of 15% Na and 85% K would lie at B.

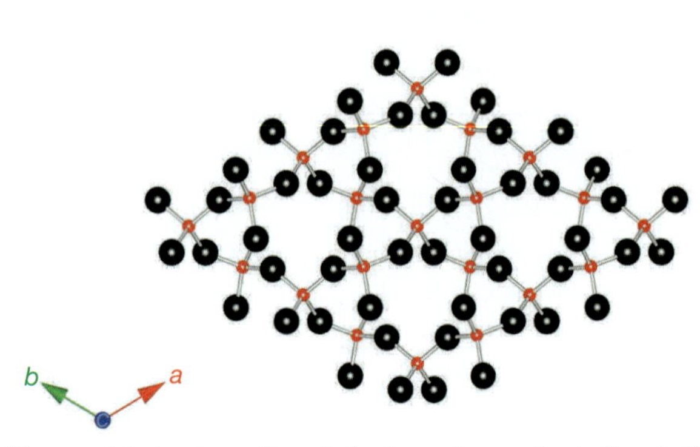

a. The crystal structure of quartz is shown here using ball and stick representation, where the atoms are the balls and the bonds are the sticks. The larger dark spheres represent oxygen and the smaller spheres, silicon. Observation of the center Si atom in the projection shows that the Si tetrahedron shares all of its oxygens with adjacent Si atoms to form this framework silicate. While quartz has the simplest formula of any silicate, its structure is somewhat complex.

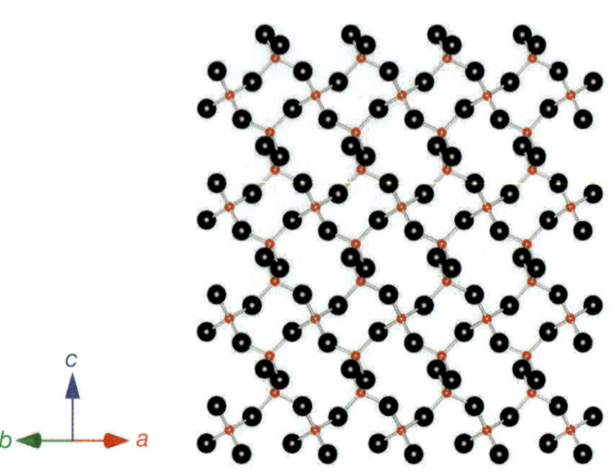

b. The same representation of the structure of quartz as in **a** except rotated by 90°. While this projection at first looks more complex than what is shown in **a**, careful observation again shows that each Si tetrahedron shares all of its corners.

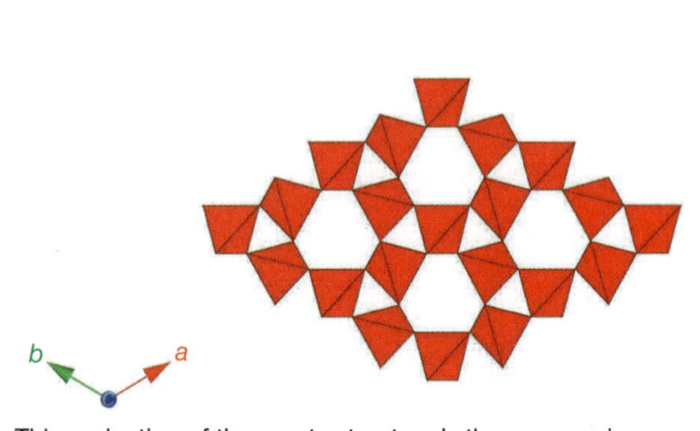

c. This projection of the quartz structure is the same as in **a**, except now the structure is shown in polyhedral form. In this type of representation, the O atoms are connected to form a polyhedron, which in this case is a tetrahedron.

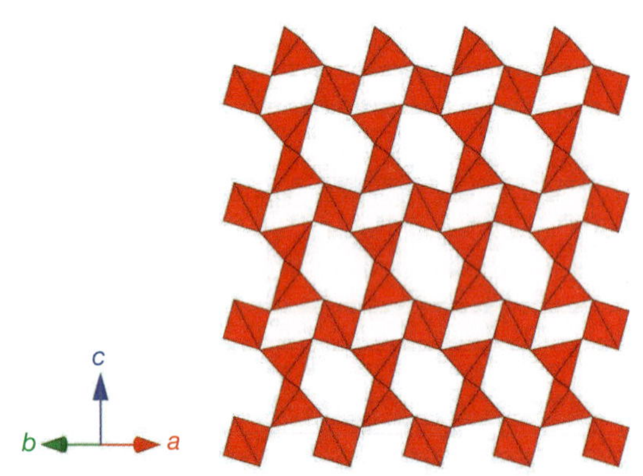

d. This projection is the same polyhedral representation of the quartz structure shown in **c**, but rotated by 90°.

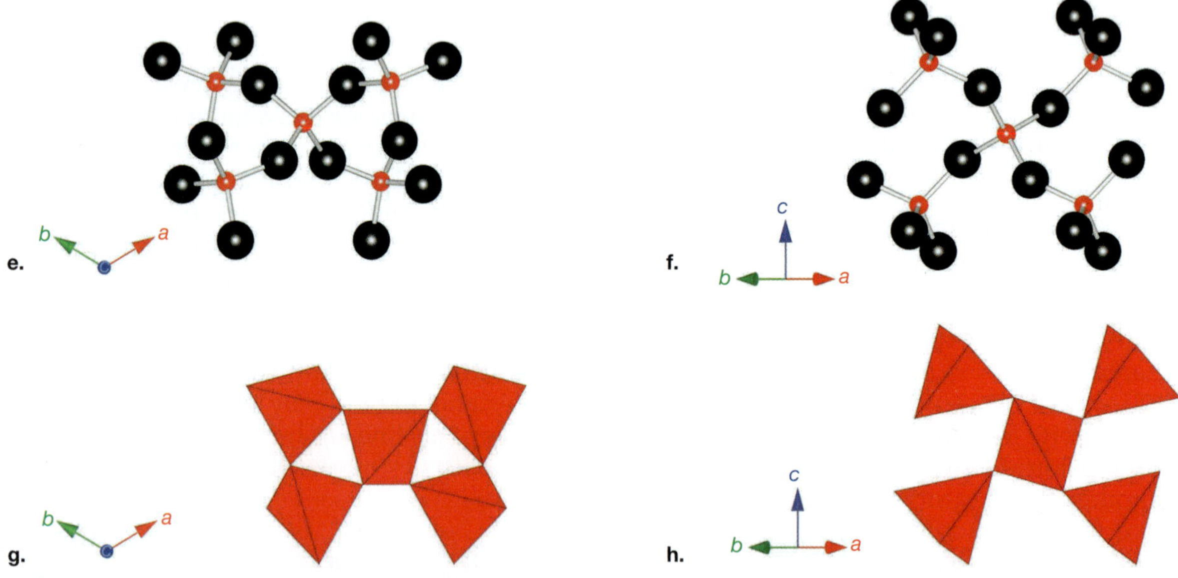

e–h. In these four structure drawings, the central tetrahedron from **a** has been removed and is shown bonded to its four nearest neighbors. **e** and **f** are the ball and stick representations that are in the same orientations as **a** and **b**, and **g** and **h** are the polyhedral representations as in **c** and **d**. These expanded views give a much clearer indication of how the central Si tetrahedron links to its four neighbors.

Figure 6.1. *Framework silicates have structures in which every tetrahedron shares all of its corners. This figure shows examples of three different framework silicates: quartz, the feldspar group mineral albite, and the zeolite group mineral natrolite.*

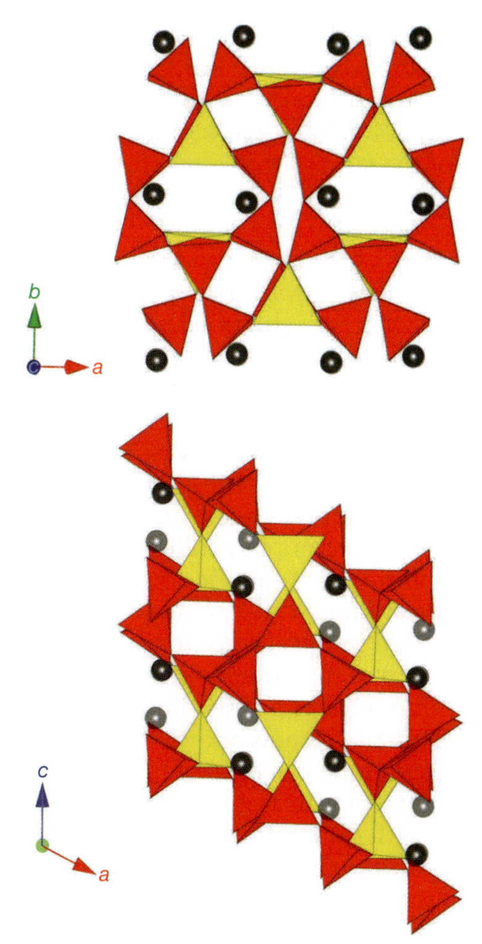

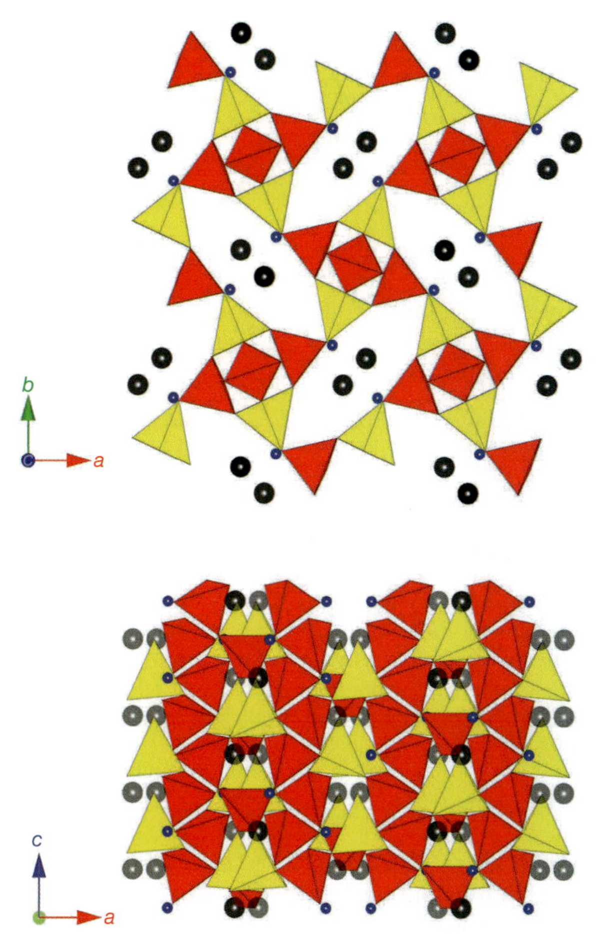

i. The crystal structure of the feldspar group mineral albite, $NaAlSi_3O_8$. The red tetrahedra are filled with Si while the yellow tetrahedra are occupied by Al, and the transparent black sphere is Na, which is bonded to the oxygens in the structure (the bonds are excluded to simplify the drawing). The *a-b* projection on the top shows the Na atoms occupying voids created in the structure. In the *a-c* projection on the bottom, you can also see the Na atoms in the voids. For both projections, note that each tetrahedron shares all four of its corners to form this framework structure, and there are three times more Si than Al tetrahedra.

j. The crystal structure of the zeolite group mineral natrolite, $Na_{16}(Al_{16}Si_{24}O_{80})\cdot16H_2O$. The red tetrahedra are filled with Si, while the yellow tetrahedra are filled with Al. The transparent black sphere is Na, which is bonded to the oxygens in the structure and to the small blue spheres that represent water molecules (those bonds are excluded to simplify the drawing). The top projection is on the *a-b* plane and is looking down the channels that are the hallmark of zeolites. Residing in these channels are the cations (Na here) required for charge balance and H_2O molecules that basically fill the extra space in the channels. In the *a-c* projection on the bottom, you can see that there are more Si than Al tetrahedra. Going back and looking at the structure of quartz (Figure 6.1c) and albite (Figure 6.1i) and comparing them to natrolite, which do you think would have the lowest and highest density?

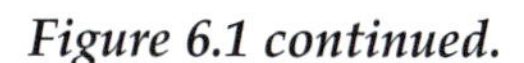
Figure 6.1 continued.

We could also draw similar lines for the other compositional ranges in feldspar.

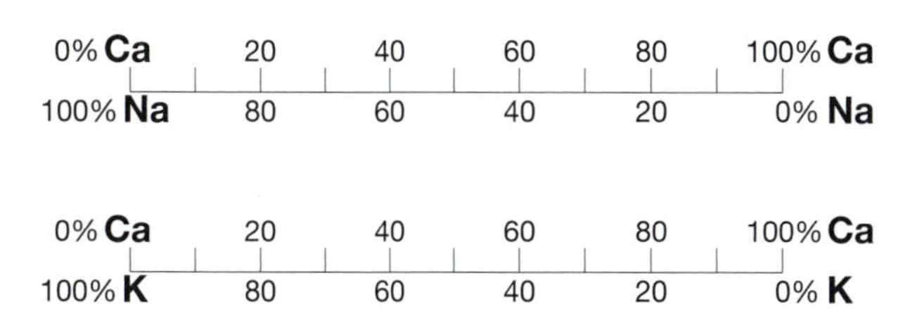

Now imagine that we want to put all three of these on the same plot in order to represent the three variables Ca, Na, and K all at once. So we'll put each of these lines along one edge of a triangle! Figure 6.2 shows a **ternary plot** ("ternary" means three) of the feldspar composition space, with the composition of $Na_{65}K_{10}Ca_{25}$ represented. Each of the three corners of a ternary plot represents 100% of that component (just as the ends of our 2-D line represented pure end-members) and 0% of the other two components.

The lines on the plot represent 10% increments of the components. The line along the bottom side of the triangle, which represents 0% Ca (or anorthite), shows a continuum between 100% Na (albite) and 100% K (potassium) feldspar; feldspars that plot in this region are commonly called the **alkali feldspars** (the first column in the periodic table are the alkali elements). Other horizontal lines, going up toward the pure Ca (anorthite) at the top of the triangle, represent compo-

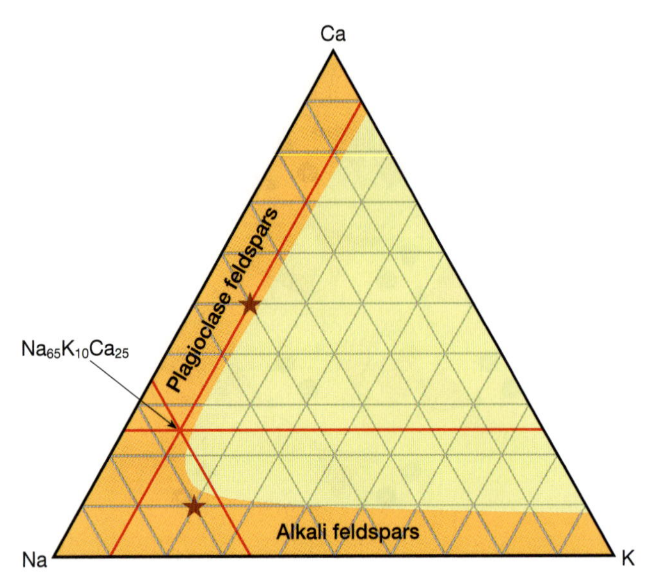

Figure 6.2. A ternary diagram is a way to show the relative contributions of three components in any system where the components can be expressed as adding up to 100%. This plot shows the K, Na, and Ca contents of feldspars. The shaded region represents compositions that actually occur in nature.

sitions that are 10%, 20%, 30%, etc., Ca added in. Similarly, lines parallel to the Na-K side represent varying amounts of the Ca end-member (anorthite), and lines parallel to the K-Ca side represent varying amounts of Na, or albite. The feldspars that plot along the Na-Ca side are referred to as the **plagioclase feldspars**.

See if you can work out the compositions of the locations described by the two stars. Hint: the numbers are expressed in percentages, so all three must add up to 100%! Please take the time to familiarize yourself with this plot; it will continue to reappear, not only in mineralogy, but in many different aspects of geology, from sedimentology to metamorphic petrology, where there is a need to represent compositions with three different components at the same time.

Ternary diagrams are also handy for indicating which compositional ranges are naturally-occurring. The shaded areas of Figure 6.2 represent compositions that are found in nature. Any other compositions are not found in real rocks. This tells us that there are some combinations of cation size and charge that cannot create repeatable, three-dimensional arrays that are stable at room temperature and pressure.

The last framework silicate mineral group that we will discuss here is the **zeolites**, which have great industrial importance and are geologically widespread. Consider the formula for natrolite, which is $Na_{16}(Al_{16}Si_{24}O_{80})\cdot 16H_2O$. First, take a long look at this formula and compare it to albite, which should look similar but without the water. Second, examine Figure 6.1j and convince yourself that this is a framework silicate (hint: check the ratio of the sum of the tetrahedral cations (i.e., $Al^{3+} + Si^{4+}$) to oxygen and see if it is 1:2). In albite, there were one Na^{1+} and one Al^{3+}; here there are 16 Na^{1+} and 16 Al^{3+}, the same ratio.

Zeolites contain molecular H_2O that occurs in channels in their structures (Figure 6.1j). Because the H_2O is neutral in charge, the amount of water present is actually dictated by how much is needed to fill the voids in the channel. In Figure 6.1j, you will see that Na^{1+} atoms also occur in those same channels, and are necessary for charge balance. In future chapters, we'll learn that the channel cations and waters in zeolites can play many different roles, from removing heavy metals like lead from drinking water, to serving as a means of refrigeration in desert environments.

Optically, zeolite minerals have a lot in common with the feldspars and silica polymorphs—each one is just a framework with similar properties in all directions. Because of this characteristic, all the framework silicates have very low birefringence, and will usually be easily recognizable as the "gray guys" when viewed in a thin section in cross-polarized light.

In summary, the formulas of all of the framework silicates share one fundamental constant: the ratio of tetrahedral cations (Si^{4+} or Al^{3+}, or any other cation) to oxygens is always 1:2. This occurs because the basic SiO_4 unit shares all its corners. Each tetrahedron consists of one Si^{4+} cation and half of four oxygen anions ($4 \times 1/2 = 2$), for a 1:2 ratio of tetrahedral cations to oxygens (Figure 6.1). Thus, a formula unit of a framework silicate is always recognizable if you total up the number of tetrahedral Si and Al cations and find that it is half the number of oxygens.

Layer Silicates

Also known as **phyllosilicates** (after the Greek word *phyllon*, for leaf), the layer silicates comprise about 10% of the Earth's crust and always form in the presence of hydrogen. Their basic structure involves sheets of tetrahedra; each one shares three of its four oxygens with an adjacent tetrahedron (Figure 6.3). All the unshared oxygen atoms point away from the plane of tetrahedra, and are termed **apical** oxygens. The basic building block of a layer silicate includes two Si^{4+} cations (or Al^{3+} cations, as in the framework silicates), three oxygens within the outlined basic building block, and four oxygens shared with neighboring build-

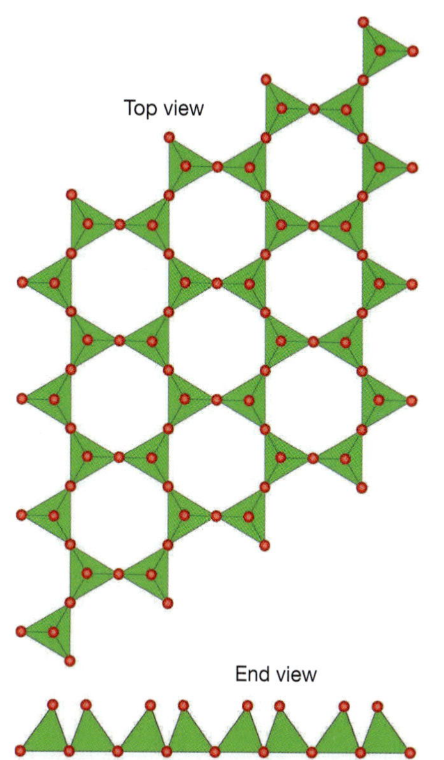

a. The layer silicates are based on sheets of tetrahedra, shown here from the top and the side. The unbonded apical oxygens form bonds to the octahedral sheets as shown in the following figures. The tetrahedral sheets can occur on one or both sides of the octahedral sheets depending on their orientation (i.e., whether they point up or down as shown in this figure in end view).

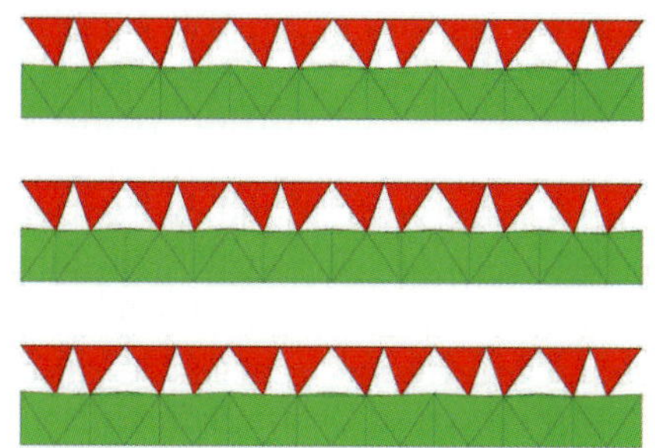

b. Crystal structure of the 1:1 layer silicate of the serpentine group antigorite, $Mg_3Si_2O_5(OH)_4$. This group of layer silicates is composed of one sheet of octahedra (green) linked to one sheet of tetrahedra (red). The layers are held together by H-bonding (not shown) because H^{1+} cations bond to the exposed side of the octahedral sheet (i.e., the side not bonded to the tetrahedral sheets).

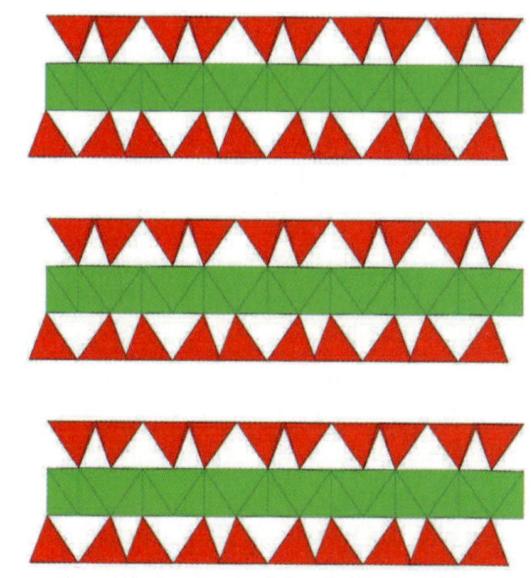

c. Crystal structure of the 2:1 layer silicate talc, $Mg_3Si_4O_{10}(OH)_2$. Talc is composed of two sheets of tetrahedra with a single sheet of octahedra between them; the octahedral sheet is formed by the unbonded apical oxygens on the tetrahedral sheet. The 2:1 layers are held together by weak electrostatic forces.

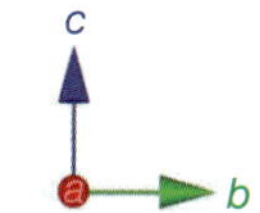

The *b-c* projection is used for Figures **b** through **f**.

d. Crystal structure of the 2:1 layer silicate series biotite, $K(Fe,Mg)_3(AlSi_3O_{10})(OH)_2$. In biotite, one Al^{3+} substitutes for one Si^{4+} in the tetrahedral sheets; this causes a charge imbalance that is satisfied by an interlayer cation, which is K^{1+} in this case. The K atom (black) strongly bonds the 2:1 layers together, so the micas are held together much more tightly than the layers in antigorite or talc.

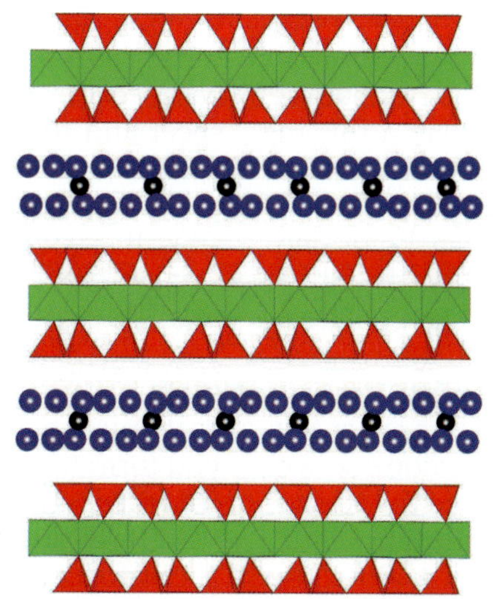

e. Crystal structure of the 2:1 layer smectite clay. This group of clays is similar to the micas but instead there is an inner layer charge of approximately −0.5 instead of −1. This intermediate charge causes a portion of the interlayer sites to be filled with monovalent cations, and H_2O (blue) then fills the voids. Thus these clays can expand or contract depending on the water content. Note that in this structure, the layers are farther apart than in biotite, and the water molecules are on either side of the interlayer cations.

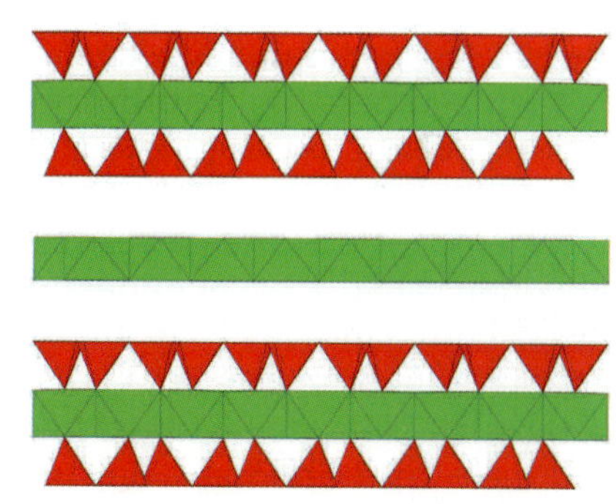

f. Crystal structure of the chlorite group minerals. These minerals are one more twist on the theme of stacking tetrahedral and octahedral layers. As in biotite, chlorite group minerals have a −1 interlayer charge, but here the charge balance is compensated by an octahedral sheet between the 2:1 layers.

Figure 6.3. *Layer silicates display many different structures, but they are all based upon combinations of tetrahedral and octahedral sheets with or without interlayer cations, and (in the case of chlorite only) an octahedral sheet in the interlayer. This series of structure drawings shows examples of five different structures based upon different stacking sequences of tetrahedral and octahedral sheets, all of which are projected perpendicular to the layers to show the differences in stacking sequences.*

ing blocks (Figure 6.4). This gives a ratio of two tetrahedral cations to (3 + (4 × 1/2)) oxygens. So you can always identify a layer silicate by its formula, which will have a 2:5 ratio of tetrahedral cations to oxygens (Table 6.1).

Each of the layer silicates has a sheet of tetrahedra as its fundamental structural unit, and also a sheet of octahedra that bond to it. Interestingly, the geometric arrangement of oxygens in a sheet of octahedra is similar to the geometrical arrangement of the apical oxygens (the unbonded ones, Figure 6.3) in a sheet of tetrahedra (more on this in Chapter 22). So a tetrahedral sheet can bond to an octahedral sheet by sharing these unbonded oxygens. Thus, we can build different layer silicates simply by varying how we stack the tetrahedral (T) and octahedral (O) sheets together. You can think of the T and O sheets as pieces of bread and ham in a sandwich if you like or as different layers in a piece of plywood.

In a formula unit, the Si^{4+} tetrahedral cations have a charge of 2 × 4+, or 8+, while the oxygen atoms have a charge of 5 × 2–, or 10– so the charge on Si_2O_5 is 2–. Thus, as with the feldspars and zeolites, additional cations must be brought in for charge balance. The many different groups of layer silicates result from permutations on how and where the additional cations are found and the fact that Al^{3+} can substitute for Si^{4+} in the T layer. Some of the charge balance is satisfied by the addition of H^{1+} atoms bonded to oxygens in the octahedral layer. Thus, these $(OH)^{1-}$ groups do play a role in charge satisfaction, unlike the H_2Os in zeolites.

Members of the **kaolinite-serpentine** group (Figure 6.3b) are composed of T-O stacks of one tetrahedral and one octahedral sheet, with only a weak electrostatic charge between the layers; these are often called 1:1 clay minerals (kind of an open-faced ham sandwich!). In the serpentine mineral species antigorite ($Mg_3Si_2O_5(OH)_4$), all three of the octahedral sites are occupied by Mg^{2+} atom, so we call them **trioctahedral**. The "tri" refers to the fact that there are three octahedra in the structure. Kaolinite ($Al_2Si_2O_5(OH)_4$) has the same structure but with only two of the three octahedral sites occupied by Al^{3+} (the other one is empty, or "vacant"); this is the **dioctahedral** case (two octahedra occupied). In both cases, the three octahedral sites have a total charge of 6+. For the trioctahedral sites, this is accomplished by three cations of 2+ charge, while for the dioctahedral group, it is accomplished by two cations of 3+ charge.

These minerals are very soft (2–4 on the Mohs scale) because the sheets are neutral in charge, and so the only thing holding one T-O to another T-O is a very weak electrostatic charge of the hydrogen bond.

Pyrophyllite ($Al_2Si_4O_{10}(OH)_2$) and **talc** ($Mg_3Si_4O_{10}(OH)_2$) (Figure 6.3c) are also soft (1–2 on the Mohs scale) because their layers are also electrostatically charge-balanced. These minerals differ from kaolinite-serpentines because they are composed of tetrahedral-octahedral-tetrahedral T-O-T stacks (like ham sandwiches made with *two* pieces of bread). Because there are two tetrahedral sheets for every *one* octahedral sheet, these are called 2:1 clay minerals. These minerals maintain the 2:5 ratio of Si:O, and also come in trioctahedral (talc) and dioctahedral (pyrophyllite) forms.

What happens when you substitute Al^{3+} for Si^{4+} in the tetrahedral site, as we did for the feldspars? Now the T-O-T "sandwiches" are no longer electrostatically neutral, but have a net negative charge that must be compensated. This is accomplished by substitution of interlayer cations, which creates the mica group minerals if only 1+ unit of charge is needed. **Phlogopite** ($KMg_3AlSi_3O_{10}(F,OH)_2$) and **muscovite** ($KAl_2AlSi_3O_{10}(OH,F)_2$) (Figure 6.3d) are the trioctahedral and dioctahedral results of this substitution. The hardness of the micas is higher than that of the Si_4O_{10} silicates, because the charge on the interlayer cation (in this case, K^{1+}) holds the sheets together (Figure 6.3d), just as mayonnaise holds a sandwich together! Notice how there are three Al atoms in the formula for muscovite but we write them separately as Al_2Al. Why? This is done because the first two Al's are in the octahedral sheet and the remaining one is in the tetrahedral sheet. When we write chemical formulas, we tend to list the larger cations (or in this case, the larger sites; those with bigger CN's) first, followed by the smaller cations. In many ways, there is a "grammar" to writing in chemical formulas, and once you understand it, you can start to "read" the formula.

If you replace even less than one Si^{4+} with Al^{3+}, you form the smectite clay minerals like **montmorillonite** ($(Na,Ca_{0.5})_{0.33}(Al,Mg)_2(Si,Al)_4O_{10}(OH)_2 \cdot nH_2O$) (Figure 6.3e); again there are dioctahedral and trioctahedral species of this mineral group and the preceding formula is idealized. But the important part here is that you substitute less than one Al^{3+} for a Si^{4+}. For instance, the tetrahedral sheet can become $Al_{0.5}Si_{3.5}O_{10}$ to make a –0.5 inter layer charge. When this happens, some of the interlayer cations sites are vacant, and water molecules occupy some of this space, in a similar manner to filling the voids in a zeolite.

If all of these possibilities seem confusing, remember this simple rule: if the charge on the T-O or T-O-T layers is 0 or less than 1, it's a clay mineral. If the charge is 1, it's a mica, and monovalent cations like K^{1+} fill the sites between the layers and bond them together tightly. When the charge is less than 1, the layers are less strongly bonded. In order for a clay to contain water molecules, the charge must be greater than 0 (when it's 0, there is no room for water to enter between the layers). All the layer silicates can be distinguished using this rule about layer charge except the members of one group: the chlorites, in which a whole new sheet lies in between the T-O-T layers to accommodate charge balance. In chlorite group minerals like **chamosite** $(Fe^{2+},Mg,Fe^{3+})_5AlSi_3AlO_{10}(OH,O)_8)$, a positively-charged sheet of edge-sharing octahedra, with a formula of $(Mg, Fe^{2+})(OH)_6$, is found between the T-O-T layers. Thus the structure is composed of T-O-T layers alternating with O sheets (Figure 6.3f). Don't worry about how complex these last two formulas are. What's really important is the concept that there are systematics for how the T and O layers fit together and for the impact of charge balance on the contents of the interlayers.

All of the layer silicates share the characteristic of having perfect cleavage in one direction, the direction parallel to the T and O sheets. The optical properties are also similar within the group. The speed of light is pretty much the same in any direction within the *a-b* plane, which is the plane of the layers. Any direction perpendicular to the *a-b* plane will have a very different density of atoms and charge distribution, and so the speed of light (and the refractive index) will be distinct from that within the plane of the layers. So the birefringence of layer silicates will be greatest when the crystals are oriented with the layers lying on their edges.

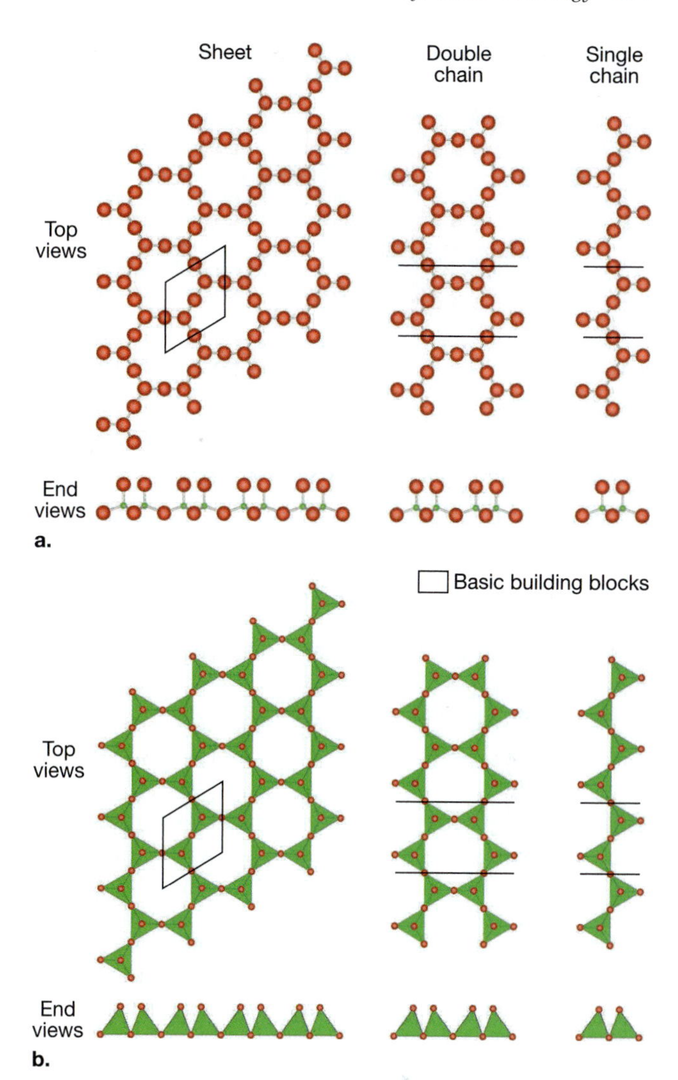

*Figure 6.4. The layer, double-chain, and single-chain silicates are all closely (and systematically) related. **a.** Each of the three is shown in ball and stick representation. Each top view shows the basic building blocks outlined for each: sheet (Si_2O_5), double chain (Si_4O_{11}), and single chain (Si_2O_6). The basic building block is the smallest piece of the polymerized silica groups needed to represent the entire structure. If you can either recall these or derive them from the structures, then you can usually figure out the major silicate group to which a mineral would belong. **b.** Polyhedral representation of the Si^{4+} tetrahedra.*

Chain Silicates: Amphiboles and Pyroxenes

What happens if less than three corners of Si tetrahedra are shared in a repeatable, three-dimensional array? Imagine what would happen if you could take out a tiny pair of scissors and cut the bonds in a sheet of tetrahedra such as those found in the layer silicates. Figure 6.4 shows that you *could* cut them into wide strips, consisting of double chains of tetrahedra, where every four Si tetrahedra would share eleven oxygen atoms, forming the double chains that characterize mineral species in the **amphibole** group (center image Figure 6.4). If you cut very narrow strips, each Si tetrahedron would share two corners, and the result would be the single chains of tetrahedra that characterize the **pyroxene** group minerals (right image Figure 6.4).

The amphibole structure is composed of adjacent chains of corner-sharing tetrahedra that form double chains (Figure 6.5). Because so many tetrahedral corners are no longer shared, the fundamental unit is now Si_4O_{11} (outlined in Figure 6.4), which has a net charge of –6. Thus, the amphibole structure needs several extra cations (usually 7–8 of them depending on their charges) to make the total structure electrostatically neutral. So the chains link up, just as in the micas, with one chain pointing down and the other chain pointing up (Figure 6.3c), with octahedral (C) sites for five

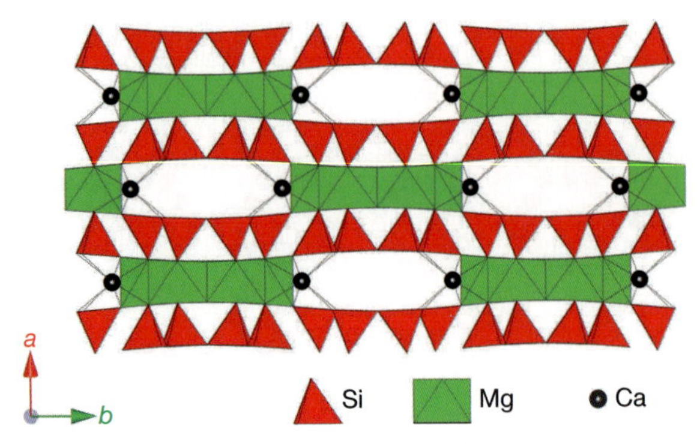

Figure 6.5. Crystal structure of the amphibole mineral tremolite, $Ca_2Mg_5Si_8O_{22}(OH)_2$, projected down its c axis (i.e., down the Si chains), which looks down the double chains seen in Figure 6.4. When the double chains point toward each other (as in the center of the image), their apical oxygens form a strip of octahedra (similar to what is found in the layer silicates). The smaller cations in the amphibole structure can reside in these octahedra. Large sites are created on the sides of the octahedral strips and larger cations can fit here. In this case, Ca^{2+} is shown bonded into the edge of the strip as a black sphere.

cations in between, as shown for the amphibole mineral tremolite in Figure 6.5. Two (B) cations are found at the edges of the strips of octahedra, and have 8-coordination (shown in Figure 6.5 as spheres bonded into the structure). When needed to maintain charge balance, large (A) cations can occupy the big holes between back-to-back tetrahedral rings; this is the same "site" as the cations between the layers in the layer silicates (Figure 6.3d). Just as in the feldspars and layer silicates, the need to maintain charge balance comes from substitution of Al^{3+} for Si^{4+}. As in the micas, amphiboles have hydroxyl $(OH)^{1-}$ ions bound to the octahedral sites. So the amphiboles have a double chain structure and a general formula of $A_{0-1}B_2C_5Si_8O_{22}(OH)_2$.

Because there are so many possible substitutions in this general formula, there are (currently) approximately 80 species in the amphibole group. The most common of these are the variations involving the cations Ca^{2+}, Fe^{2+}, and Mg^{2+}. Because there are three components, we can again use the ternary plot as a way of representing these amphiboles. Figure 6.6a shows this ternary plot, which has anthophyllite ($Mg_7Si_8O_{22}(OH)_2$), ferroanthophyllite ($Fe_7Si_8O_{22}(OH)_2$), and a hypothetical pure-Ca^{2+} amphibole at the top. In practice, the most Ca^{2+} you can put in an amphibole structure is two cations per formula unit, so the ternary plot ends 2/7 up from the bottom. But this plot still can be used to show a lot of different amphiboles.

The pyroxene group minerals (of which there are currently 21 species) also relate directly to the same ternary diagram used for the amphiboles (Figure 6.6b). However, there are fewer sites to occupy because the linked tetrahedra form only a single chain. Thus for pyroxenes, a formula unit is represented by an Si_2O_6 unit, which has a net charge of 4– (outlined in Figure 6.4). So we need only two divalent cations to satisfy the charge imbalance by occupying the octahedral sites between the tetrahedral chains.

Notice that the pyroxene structure (Figure 6.7) appears to be denser overall than the amphibole structure because there are no tetrahedral rings (compare Figure 6.5 to Figure 6.7). This density difference will aid in distinguishing between pyroxenes and amphiboles under an optical microscope (higher density = higher refractive

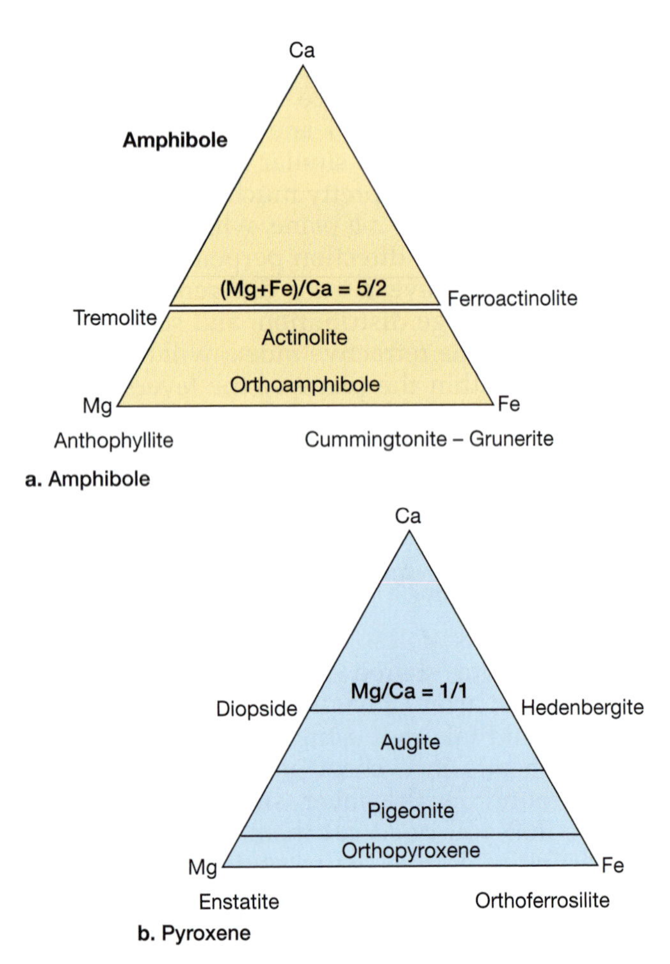

Figure 6.6. Simplified ternary diagrams for the amphibole (Figure 6.6a) and pyroxene (Figure 6.6b) systems. The main difference between the two mineral groups is that amphiboles have six cation sites, while pyroxenes have only two; this results from the different chain widths. Each is based on three atomic end-members with Ca^{2+}, Mg^{2+}, and Fe^{2+}. There are also species names at some of the corners and some intermediate compositions.

index, with some caveats discussed elsewhere), as will the difference in their cleavages (Figure 6.8). Because amphiboles have double chains and pyroxenes have single chains, they cleave differently because they contain different planes of weakness. These chain silicates will break through the cavities (which are places where there are no bonds) and the cavity width is controlled by the chain width (Figure 6.8). Both amphiboles and pyroxenes are elongated parallel to the silicate chains because these chains are the "backbone" of their structures, and their resultant crystal shape will often be elongated. So we once again see that the physical properties (in this case, morphology, cleavage directions, and angles) are directly related to the crystal structures.

Interestingly, there are other minerals with structures based upon chains of tetrahedra that do not belong to the mica, amphibole, or pyroxene groups. These phases are commonly called **biopyriboles**. They are composed of various permutations and combinations of single chains, double chains, and triple chains (effectively pieces of pyroxene, amphibole, and micas structures, respectively). These phases usually require extremely high magnification (such as a transmission electron microscope; see Chapter 9) to distinguish the chains.

Some mineral species consisting of regular patterns of chain widths have been identified. For example, the mineral species jimthompsonite has the formula $(Mg,Fe^{2+})_{10}Si_{12}O_{32}(OH)_4$, and is three chains wide (Figure 6.9a), while chesterite $((Mg,Fe^{2+})_{17}Si_{20}O_{54}(OH)_6)$, is based on double and triple chains (Figure 6.9b). There is some similarity between these formulas and those of the amphiboles. In fact, it might be worth your while to see if you can figure out the basic building block for a triple chain in the same manner we found those for a sheet, double chain, and single chain (Figure 6.4). Mineral structures with chain repetitions from one up to sixty units wide have been reported! The biopyriboles generally form as weathering or alteration products to or from (you guessed it) pyroxenes, amphiboles, and mica.

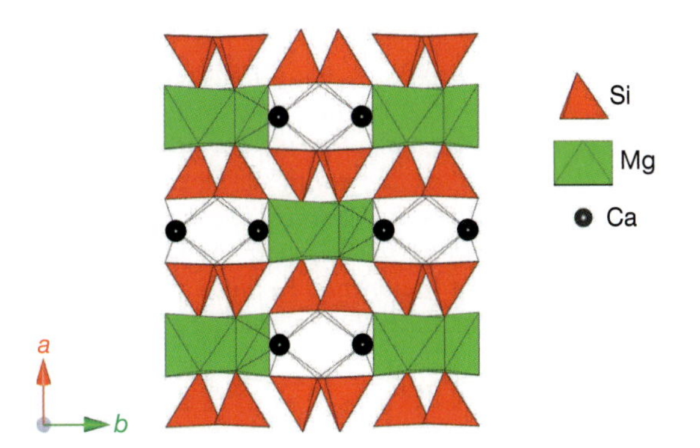

Figure 6.7. Crystal structure of the pyroxene mineral diopside, $CaMgSi_2O_6$, projected down the c axis. The red polyhedra represent the end-on view of the single chains of silicate tetrahedra, while the green polyhedra represent the Mg^{2+} octahedra. The oxygens for these octahedra are the unbonded apical oxygens in the chains, similar to the way that the octahedral sheet is formed in a layer silicate (by the unbonded apical oxygens in the sheet). The black sphere represents the larger Ca^{2+} atom, which is too big to fit in the octahedron formed between the chains. Compare this structure to that of tremolite (Figure 6.5).

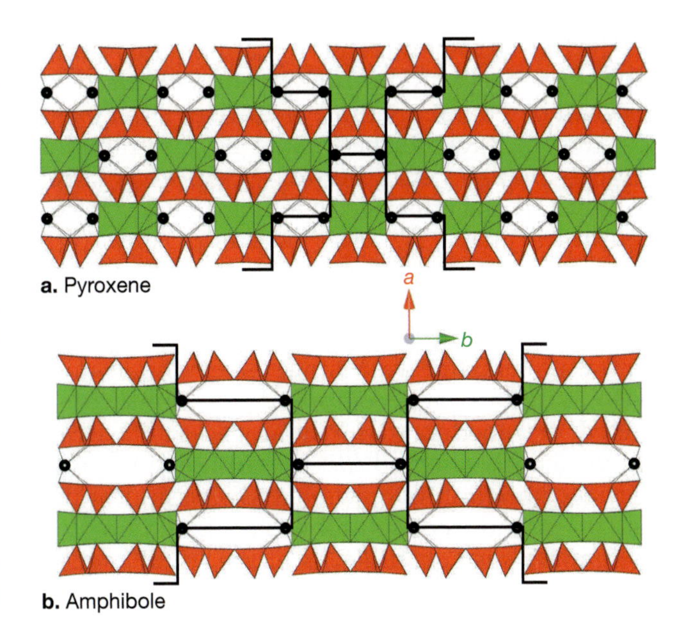

Figure 6.8. Projection of the pyroxene and amphibole structures down the c axis (i.e., down the Si chains). Both minerals are elongated parallel to these chains and have perfect cleavage imparted by the chains. In each case, planes of weakness go through the center of the cavities formed below the base of the chains. In pyroxenes this imparts a near-90° cleavage while in amphiboles it imparts a near-60°/120° cleavage.

Ring Silicates

What happens if we take a chain of tetrahedra and bend it around to make a ring? This gives rise to the group of structures called **ring silicates**, or **cyclosilicates**. In nature, Si^{4+} tetrahedra can form rings composed of 3, 4, 6, 8, 9 and even 12 members, but we will focus here on the most common group of these: the 6-membered, Si_6O_{18} minerals. This formula has a net charge of 12–, so five or six cations need to be added for charge balance. In these minerals, the rings are cross-linked by octahedra and other types of tetrahedra that bind the structure together.

The beryl group minerals (including the mineral species **beryl** ($Al_2Be_3Si_6O_{18}$), are composed of six-membered tetrahedral rings that are linked together by AlO_6 octahedra and BeO_4 tetrahedra (Figure 6.10a). **Cordierite** ($(Mg,Fe^{2+})_2(Al_2Si)[Al_2Si_4O_{18}]$) resembles beryl except that the tetrahedra are linked by Mg^{2+} octahedra and Al_2Si tetrahedral groups (Figure 6.10b). As a result, all of the corners of each Si^{4+} tetrahedron are shared. So why aren't beryl and cordierite considered to be framework silicates? Strictly speaking, framework silicates are composed of completely-linked tetrahedra that contain *only* Si^{4+}. But we've already allowed Al^{3+} to "count" in tetrahedral coordination in our above discussion of the feldspars and zeolites. However, in nature we commonly see substitutions of Be^{3+} in Si^{4+} tetrahedra, so the "Si only" definition of a framework silicate seems pretty arbitrary. In beryl, the three Be^{3+} and six Si^{4+} in tetrahedral coordination result in a 1:2 ratio of tetrahedral cations to O^{2-}. In cordierite, there are four Al^{3+} and five Si^{4+} in tetrahedral coordination giving, again, a ratio of 1:2. In fact, many workers now consider beryl and cordierite to be framework rather than ring silicates. These minerals even have open channels at the centers of their rings where larger cations or molecules can reside—just as with the zeolite group minerals (which are considered framework silicates). Compare the illustration in Figure 6.10 with Figure 6.1, and decide for yourself if beryl and cordierite should be ring silicates or framework silicates!

The highly-linked, dense structures of beryl and cordierite give these minerals great hardness

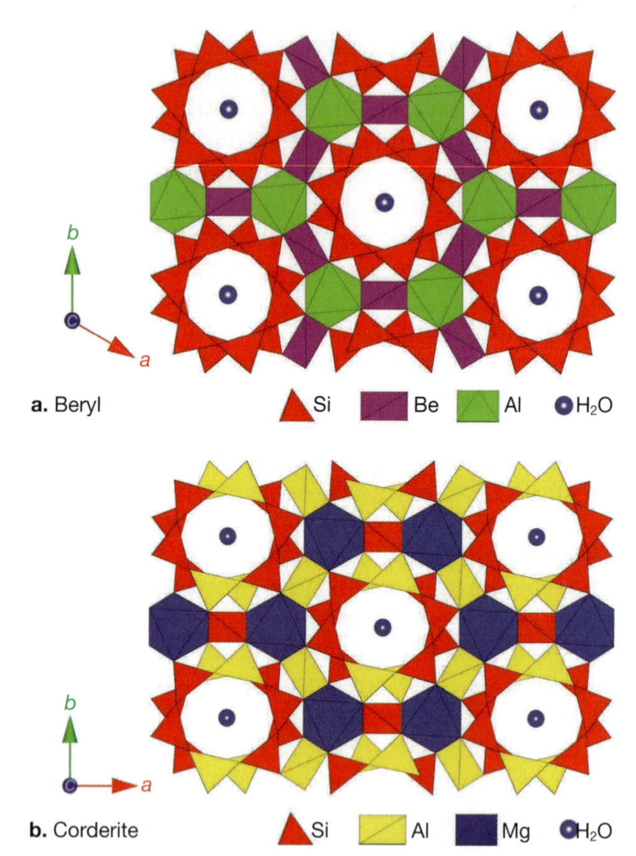

Figure 6.10. *The minerals species beryl, $Al_2Be_3Si_6O_{18}$, and cordierite $(Mg,Fe^{2+})_2(Al_2Si)[Al_2Si_4O_{18}]$, have structures that are based on rings of tetrahedra. Although the two species are generally classified as ring silicates, some recent workers have suggested that they should be framework silicates because all the corners of all tetrahedra in the structure are shared.* ***a.*** *In beryl, the red tetrahedra contain Si^{4+} and compose the 6-membered rings. Purple and elongated, the Be^{3+} tetrahedra occur between the rings and share edges with the green Al^{3+} octahedra.* ***b.*** *In cordierite, four of the six tetrahedra in the rings contain Si^{4+} while the two yellow ones contain Al^{3+}. Al^{3+} also occurs in the yellow colored, elongated tetrahedra (the ones that contain Be^{3+} in beryl) that share edges with blue Mg^{2+} octahedra (the ones that contain Al^{3+} in beryl). Beryl and cordierite are termed "isostructural" because their structures are basically the same; the main difference is in the cations that occupy the polyhedra. Small amounts of cations or water may reside in the centers of the 6-membered rings, as indicated by small blue spheres.*

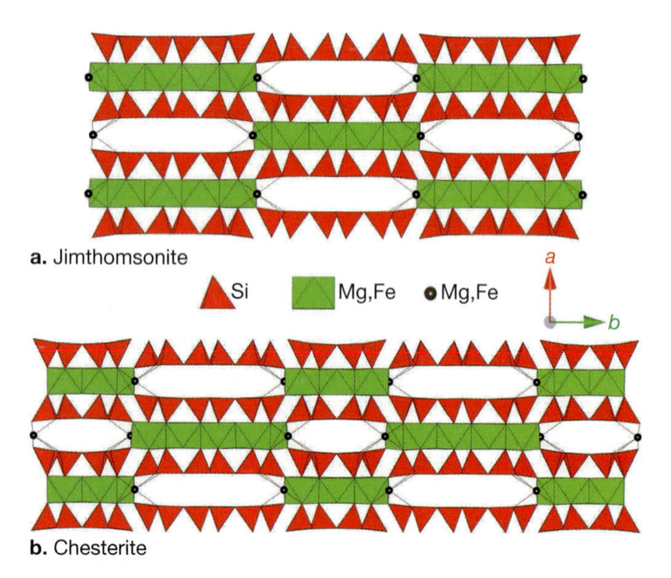

Figure 6.9. *The crystal structures of two minerals with more complicated chain repetitions than just single or double chains.* ***a.*** *Jimthompsonite, $(Mg,Fe^{2+})_{10}Si_{12}O_{32}(OH)_4$, is composed of three laterally-connected Si^{4+} tetrahedra chains, making it a triple chain silicate. (Check out the movie to get a better view of the chains.) When projected down the c axis (i.e., down the chains), the structure is very similar to that of an amphibole (Figure 6.5), but the chains and cavities are wider.* ***b.*** *Chesterite, $(Mg,Fe^{2+})_{17}Si_{20}O_{54}(OH)_6$, is based on double and triple chains. In this projection (again down the chains), notice that the middle layer resembles the triple chain in jimthompsonite, while the upper and lower layers resemble those in an amphibole (Figure 6.5).*

and luster. These characteristics make them very attractive as gemstones. The gem variety of cordierite is called iolite, which has the attractive trichroic color combination of blue to colorless to yellow (Figure 2.11). There are many named varieties of colored beryl, which are created by the substitution of various cations into the structure. The pale blue color of aquamarine arises from Fe^{2+}, while Fe^{3+} produces the golden yellow color of the heliodor variety. Mixtures of Fe^{2+} and Fe^{3+} create darker blue beryls. Green (emerald) beryl can be caused by Cr^{3+} or V^{3+}, while morganite results when Mn^{2+} (pink) or Mn^{3+} (red or green) is present. Blue color can also be caused by radiation damage. Note that beryl and cordierite also tend to have low birefringence, just like the framework silicates, because their tetrahedra are linked at all four corners.

Tourmaline group minerals are true ring silicates with complex structures (Figure 6.11) and chemical formula. Their general formula includes contributions from 9-coordinated W cations (like Na^{1+} or Ca^{2+}), and 6-coordinated X cations (like Mg^{2+}, Fe^{2+}, Al^{3+}) and Y cations (Al^{3+}, Fe^{3+}, Mn^{3+}), along with B^{3+} in 3-coordinated sites and Si in 4-coordinated sites. There are 14 specific end-members in the tourmaline group, and many variations on them. A simplified version of its general formula is $WX_3Y_6(BO_3)_3Si_6O_{18}(O,OH,F)_4$. Are you dizzy yet? Notice that tourmalines have the basic Si_6O_{18} ring structure that is typical of ring silicates.

Tourmaline is highly pleochroic because the plane of the rings has a very different charge density than the direction perpendicular to it—and there are no cross-linking tetrahedra, as we found in beryl, to make the structure framework-like. The color of tourmaline also varies over a wide range as a function of cation substitutions: blue from Fe^{2+} or Cu^{2+}; green from Mn^{2+}, combinations of Fe^{2+} with Ti^{4+}; or Cr^{3+} or V^{3+} alone; amber from Fe^{2+}–Ti^{4+} alone; pink and red from Mn^{3+} or radiation; brown from Mn^{2+} with Ti^{4+}; and black from Fe, Mn, and Ti in multiple valence states (see the Mineral Spectroscopy web site at Caltech for more information). You might begin to notice here that similar cations impart similar colors to minerals—you'll be learning more about this soon in Chapter 7!

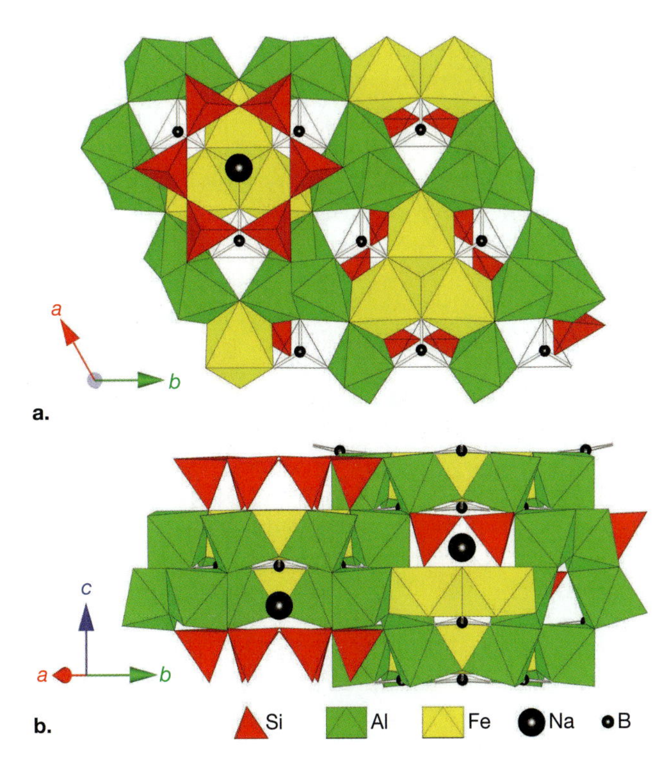

*Figure 6.11. Crystal structure of the tourmaline group mineral schorl, $NaFe_3Al_6(BO_3)_3(Si_6O_{18})(OH)_4$. **a.** Projection of the structure down c onto the a-b plane, showing the 6-membered Si^{4+} tetrahedral ring composed of red tetrahedra. The small black spheres represent B^{3+} 3-coordinated to oxygens. The green octahedra contain Al^{3+} while the yellow ones house Fe^{2+}. Larger black spheres represent Na^{1+} located in the centers of the six-membered rings. **b.** Projection of the structure with c vertical, showing how the rings occur on top of each other and are linked together by the octahedra.*

Disilicates

A large number of diverse minerals form when SiO_4 tetrahedra join up in pairs to form Si_2O_7 units (Figure 6.12); these are also sometimes called **sorosilicates**. By far the most common of the disilicates (di- means two, in reference to the pairs of tetrahedra) are the members of the **epidote** group, particularly the epidote species, $Ca_2(Al,Fe)Al_2O(SiO_4)(Si_2O_7)(OH)$, which forms in low temperature metamorphic rocks. In the epidote group, alternating edge-sharing octahedral chains of $(Fe^{3+},Al)O_6$ and $(Fe^{3+},Al)O_4(OH)_2$ form sheets by linking with Si_2O_7 groups and isolated SiO_4 tetrahedra (Figure 6.12). In fact, the external morphology of the minerals in this group is controlled by these edge-sharing chains of octahedra in the same way that tetrahedral chains controlled the morphology of the chain silicates. Large spaces (9- or 10-coordinated) between the chains are filled by Ca^{2+} or other large cations, such as Ce^{3+}, Pb^{2+}, and Sr^{2+}. Bonding *within* sheets formed by the chains is very strong, but the sheets are held together only by the bonds to Ca^{2+} anions. So epidote group minerals have one direction of perfect cleavage, and are, as stated above, elongated parallel to the octahedral chains.

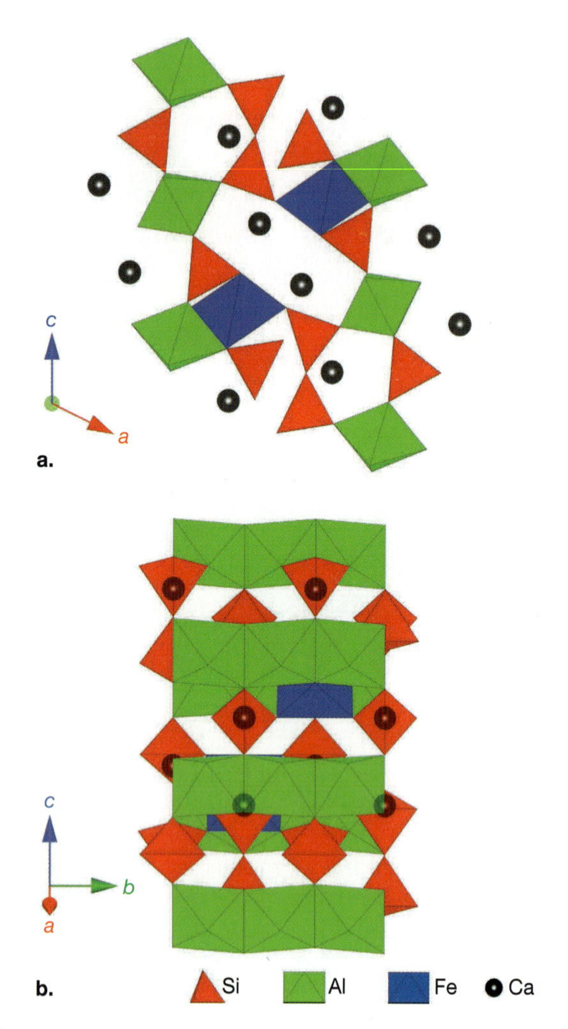

Figure 6.12. Crystal structure of epidote, $Ca_2(Al,Fe)Al_2O(SiO_4)(Si_2O_7)(OH)$, with its isolated SiO_4 tetrahedra alternating with Si_2O_7 units. This is the most common representative of the disilicate, or sorosilicate, class. ***a.*** *A projection down the* b *axis onto the* a-c *plane shows the isolated SiO_4 groups (single red tetrahedra) and the Si_2O_7 groups (double red tetrahedra). The green octahedra contain Al^{3+}, while the blue ones contain Fe^{3+}. The dark spheres represent Ca^{2+} atoms, which bond to 9–10 oxygen atoms in the structure (bonds not shown).* ***b.*** *A projection on the* b-c *plane shows the edge-sharing chains of Al^{3+} octahedra that are the major building units in epidote group minerals.*

Orthosilicates

The orthosilicates (also called **nesosilicates**) represent the opposite end of the polymerization spectrum from framework silicates: orthosilicates have structures composed of isolated SiO_4 tetrahedra, with other cations in between, as required for charge balance. Because the SiO_4 unit has a net charge of 4–, typically two divalent (i.e., 2+ charge) cations are needed for charge balance, though other variations are possible. Here we confine our discussion to three important rock-forming representatives of this group: the olivine and garnet groups, which contain the Si^{4+} tetrahedra in combinations with two cations; and the aluminosilicates, which add additional oxygen atoms (c.f., Al_2SiO_5) and have Al^{3+} in different coordination sites.

In the **garnet** group minerals, the two cations needed to charge balance the SiO_4 tetrahedra occupy two different sites: a large 8-coordinated X site, and a smaller 6-coordinated Y site. The resulting general formula is $X_3Y_2Si_3O_{12}$ (Figure 6.13) in naturally-occurring samples. Other cations such as Al^{3+}, Ga^{3+}, Ge^{4+}, Y^{3+}, and Fe^{3+} are sometimes substituted for the Si^{4+} in synthetic (laboratory-grown) garnets, which have many commercial uses (e.g., YAG, short for yttrium aluminum garnets, lasers are used in surgery to cut tissue, and to erase tattoos!). This mineral group has six ideal end-members:

pyrope, $Mg_3Al_2Si_3O_{12}$,
almandine, $Fe^{2+}{}_3Al_2Si_3O_{12}$,
spessartine, $Mn^{2+}{}_3Al_2Si_3O_{12}$,
uvarovite, $Ca_3Cr^{3+}{}_2Si_3O_{12}$,
grossular, $Ca_3Al_2Si_3O_{12}$, and
andradite, $Ca_3Fe^{3+}{}_2Si_3O_{12}$.

The first three of these are often called the **pyralspite** garnets using the first letters of each species, while the second group is commonly given the name **ugrandite** garnets. The pyralspite group minerals all have Al^{3+} as the Y cation and the ugrandite group minerals have Ca^{2+} as the X cation though solid solutions among all six species do exist.

All the garnets share the same *isometric* structure, which is (therefore) symmetrical in all three crystallographic orientations. The three different types of polyhedra are linked together by sharing edges: the 8-fold site shares ten of its edges, the octahedra share six edges, and the tetrahedra share two opposite edges (Figure 6.13). The sharing makes for a well-bonded, hard structure (H = 6.5–7.5), and garnets are relatively dense with high refractive indices. Garnets are used in many industries as abrasives; they are even used to give denim blue jeans that "well-worn" look!

The **olivine** structure is also composed of isolated SiO_4 tetrahedra bridged together by divalent cations. However, in olivine group minerals, both the divalent cations occupy 6-coordinated sites that form two chains of edge-sharing octahedra (Figure 6.14). There are nine mineral species in this group, but two of them are by far the most important in natural occurrences: **forsterite** (Mg_2SiO_4) and **fayalite** ($Fe^{2+}_2SiO_4$). Their structures are nearly identical, and there is a continuous range of substitution of Mg^{2+} and Fe^{2+} between them. Because they commonly contain Fe^{2+}, olivines are nearly always greenish brown ("olive green").

As we discussed in Chapter 1 (Figure 1.7), there are three **aluminosilicate** minerals with exactly the same composition but different structures. All of them have isolated SiO_4 tetrahedra and one Al^{3+} atom in an octahedral site. They differ because of the coordination environment of the second Al^{3+} atom, which can be 6-, 5-, or 4-coordinated:

kyanite: $^{VI}Al^{VI}Al^{IV}SiO_5$
andalusite: $^{VI}Al^{V}Al^{IV}SiO_5$
sillimanite: $^{VI}Al^{IV}Al^{IV}SiO_5$.

All of these polymorphs of Al_2SiO_5 are based upon edge-sharing chains of octahedra occupied by Al^{3+}, and their crystals are usually elongated along *c* as a result, just as in epidote. The linkages between the octahedral chains and the coordination of the extra Al^{3+} distinguish the different species.

Kyanite has chains of Al^{3+} octahedra linked together by sharing edges with other Al^{3+} octahedra and corners with SiO_4 tetrahedra (Figure 6.15a). In **andalusite** (Figure 6.15b), half of the Al^{3+} cations occupy 5-coordinated sites that alternate with Si^{4+} tetrahedra to form double chains adjacent to the Al^{3+} octahedral chains along the *c* direction. The **sillimanite** (Figure 6.15c) structure has half its Al^{3+} in tetrahedral sites that form 4-membered rings with Si^{4+} tetrahedra. Viewed from this perspective, sillimanite could be considered to be a ring silicate! From a different angle, it can be seen that sillimanite has two adjacent chains: one of Al^{3+} octahedra, and one of alternating Si^{4+} and Al^{3+} tetrahedra. The double chain that is formed is also reminiscent of the double chain structure of amphiboles!

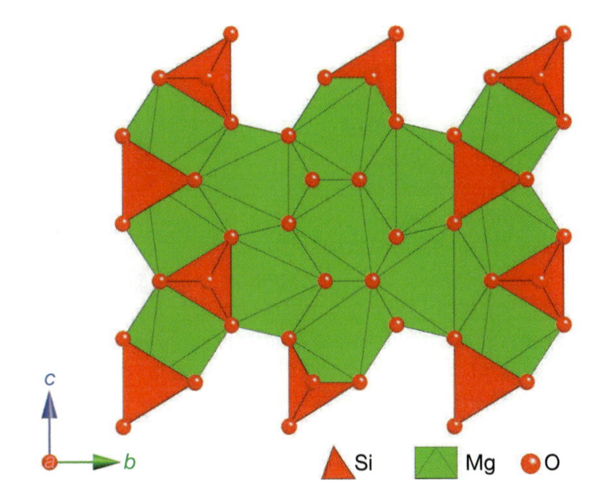

Figure 6.14. Crystal structure of the olivine species forsterite (Mg_2SiO_4). The red polyhedra represent isolated Si tetrahedra in the structure, and the green polyhedra are Mg-containing octahedra that share edges. Polyhedra-defining oxygen atoms at the corners of each polyhedra are also shown.

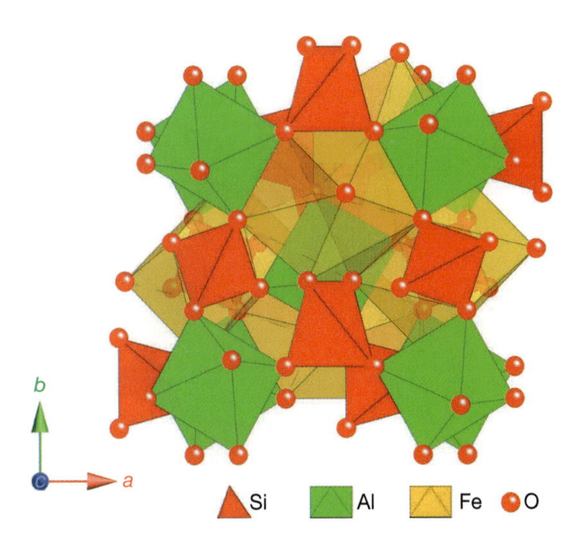

***Figure 6.13.** Crystal structure of the garnet mineral almandine, $Fe^{2+}{}_3Al_2Si_3O_{12}$. There are three different polyhedral sites in almandine: the red, isolated Si-containing tetrahedra (i.e., Si^{4+} has a CN = 4), the green Al-containing octahedra (i.e., Al^{3+} has a CN = 6), and a larger transparent 8-coordinated, irregular-shaped polyhedra that house the Fe^{2+} cation (i.e., Fe^{2+} has a CN = 8). The oxygen atoms are also shown, plotting at the edges of each polyhedron. Observation of the structure (especially the movie on the MSA website) will reveal that the Si^{4+} tetrahedra do not bond to other tetrahedra; thus garnets are orthosilicates. Also notice that many of the polyhedral edges are shared. Edge-sharing as compared to corner-sharing greatly increases the density of minerals.*

The important thing to remember about the aluminosilicates is that these structures actually reflect the minerals' formation conditions, as is the case really for all minerals at some level. These relationships can be viewed on a pressure vs. temperature phase diagram (Figure 6.16). Each Al_2SiO_5 polymorph is found at a different pressure and temperature in metamorphic rocks. The kyanite structure is the densest of the three minerals, and forms at higher pressures. Sillimanite is found at high temperatures, while andalusite is characteristic of low temperature, low pressure metamorphism. In fact, you could predict the density difference of these samples by observing the structures in Figure 6.16. Thus a geologist is lucky to find any of the aluminosilicates in field occurrences, for they can tell much about the metamorphic history of the rocks in which they occur.

Non-Silicates

Because of the large amounts of Si and O in the Earth's crust, silicate minerals are by far the most important, which is why we've chosen them to

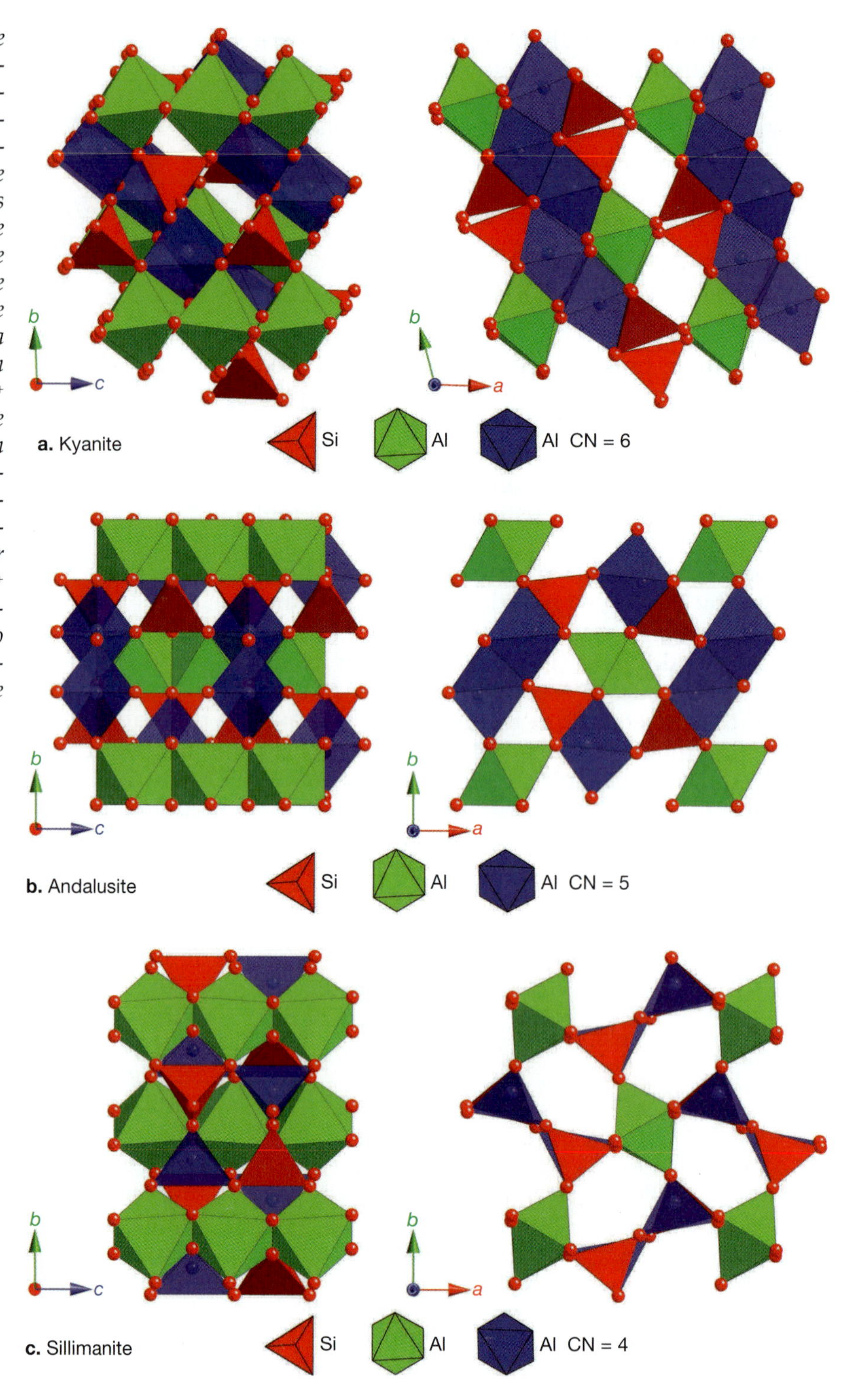

Figure 6.15. *Crystal structures of the three Al_2SiO_5 polymorphs, the aluminosilicates. All have isolated Si^{4+} tetrahedra so they are classified as orthosilicates, and all are based on edge-sharing chains of Al^{3+} octahedra. The major difference between the three is the coordination of the second Al^{3+} site in the structure, which is 6 for kyanite (Figure 6.15a), 5 for andalusite (Figure 6.15b), and 4 for sillimanite (Figure 6.15c). This site is shown as a blue semi-transparent polyhedron in each of the structures while the Si^{4+} tetrahedra and Al^{3+} octahedra are shown in red and green. Also, each structure is shown projected on the* b-c *plane with the chains of edge-sharing octahedra parallel to* c *(left column) and on the* a-b *plane to better show the coordination of the extra Al^{3+} (right column). Based on these drawings, which of the three polymorphs do you think would have the lowest density (hint: which has the most "free space")?*

dominate our Big Ten mineral list (and this chapter). However, you will recall from Chapter 1 that one important non-silicate made our list: **calcite**. Why? It's the most commonly-occurring non-silicate mineral, and it has an important job in stabilizing Earth's atmosphere and keeping it habitable for humans.

When planets in our solar system first formed, gases such as water (H_2O), carbon dioxide (CO_2) and even ammonia (NH_3) were dissolved in their interiors, and later released into the atmosphere by volcanism and augmented by asteroid and cometary impacts. These gases formed the early atmospheres of terrestrial planets, and helped to heat their surfaces via the mechanism called the **greenhouse effect**. Visible wavelengths of sunlight pass through the atmosphere and are absorbed by the surface, then radiated back out

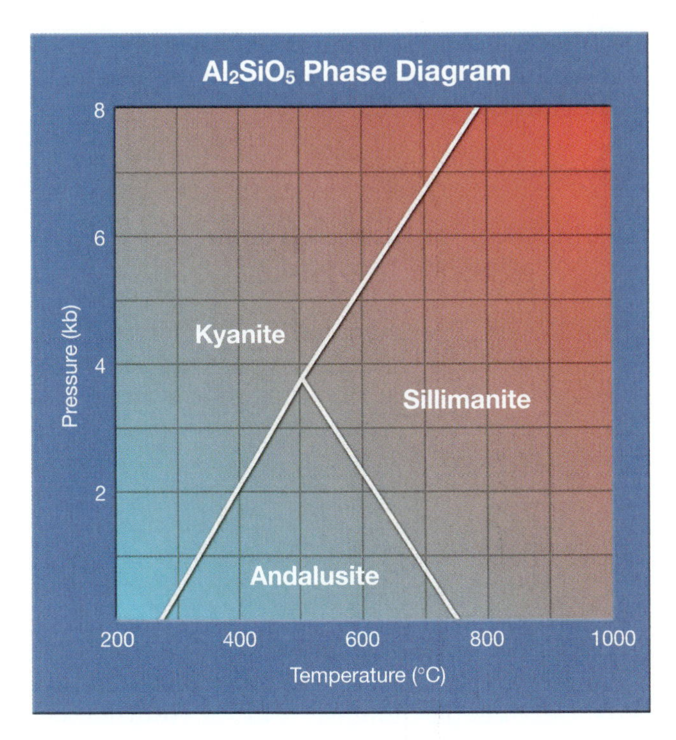

Figure 6.16. This phase diagram for the composition Al_2SiO_5 plots temperature on the x axis against pressure on the y axis. The pressure and temperature ranges for which kyanite, andalusite, and sillimanite are stable can be read from this type of a diagram. For instance, kyanite is the high-pressure polymorph and finding it in a rock would indicate the rock formed at high pressure and low temperatures, possibility in a subduction zone.

to the atmosphere in the form of infrared radiation (heat). NH_3 and CO_2 (along with several others) are called greenhouse gases because they absorb radiation in the infrared part of the spectrum, and don't let heat out. Ammonia doesn't last long when confronted by sunlight (it dissociates into N^{3-} and H^{1+}), but H_2O and CO_2 persist and build up in the atmosphere. Because a planet is heated by radiation from both the Sun and the atmosphere, the net effect of this build-up is to warm the surface.

Venus experienced what is often called a runaway greenhouse effect: it was so close to the Sun that its H_2O effectively boiled off, and the resultant build-up of CO_2 in its atmosphere raised surface conditions to ~740K and 90 bars of pressure. What saved Earth (and perhaps Mars as well) from this fate? CO_2 is extremely soluble in liquid water, and it is gradually dissolved in rivers, lakes, and oceans (CO_2 is especially soluble in saltwater!), eventually forming $CaCO_3$, or calcium carbonate. So on Earth, the carbonate is constantly recycled thanks to plate tectonics, and the oceans store the CO_2 in solution and in the form of the minerals calcite and aragonite. This keeps the CO_2 out of the atmosphere, where it could cause excessive global warming. On Mars, many workers think that there must be CO_2 buried somewhere, but its form and location are currently unknown.

In carbonate minerals, $(CO_3)^{2-}$ groups are the main building blocks, just as SiO_4 tetrahedra form the basis for silicates. Calcite is composed of sheets of flat triangles of O^{2-} with C^{4+} in the middle: the CO_3 groups. These alternate along the *c* axis of the structure with sheets of Ca^{2+} in 6-fold coordination with O^{2-} anions in the CO_3 groups (Figure 6.17), forming a framework similar to that of the feldspars. So this simple mineral structure holds great importance because it locks up CO_2 and keeps it out of Earth's atmosphere!

Concluding Remarks

One of the main goals of this chapter was to reintroduce you to the silicate minerals at a bit high-

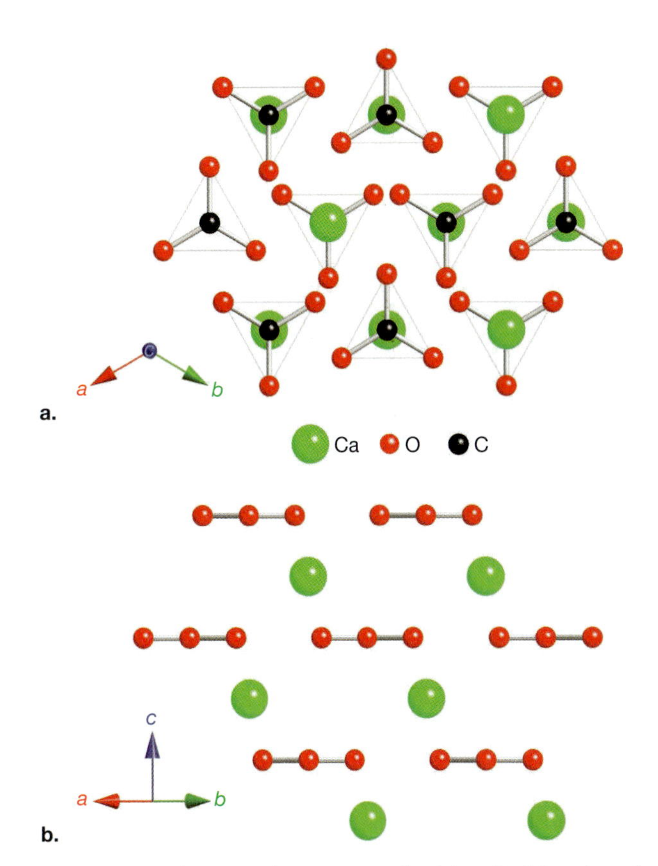

*Figure 6.17. The crystal structure of calcite, $CaCO_3$ viewed in two different orientations: **a.** Looking down the c axis, the carbon atoms (small black spheres) are surrounded by three oxygens (red spheres) at the corners of a triangle. The green spheres represent Ca^{2+} cations and they bond into the O^{2-} anions (bonds not shown). **b.** A view of the structure rotated 90° from a, showing a side projection of the CO_3 groups.*

er level than was done in Chapter 1. We have begun here to stress the crystal structures of the minerals, especially the silicates. Mineral structures are what set mineralogy apart from chemistry. Lack of knowledge about these structures has caused many misunderstandings about mineralogy. It's very useful to think of the silicates minerals as being first built of silicate tetrahedra polymerized in various ways. Then if charge balance is needed, other cations are added. After tetrahedra, octahedra are the most common building blocks of minerals, as you've started to see in this chapter. For example, there are several permutations of how the sheets of tetrahedra and octahedra could link together in the layer silicates. So what's important to know now is that minerals are often assembled from a very *few basic building blocks*, and nature has put these together in many seemingly complex ways. We hope that you are starting to see the systematics in some of these. We'll return in more detail to these and some other silicate mineral groups in Chapter 22. Finally, as the calcite example illustrated, there are also many important minerals in the Earth that are not silicates, and we'll turn our attention to them in Chapter 23.

References

Glas, M. (2003) The Kremlin's Carbuncle. extraLapis English No. 3 Tourmaline, 62–63.

Hyršl, J., and Neumanova, P. (1999) Eine neue gemmologische Untersuchung der Sankt Wenzelskrone in Prag. Zeitschrift der Deutschen Gemmologischen Gesellschaft, 48, 29–36.

King, V. (2000) Mount Mica: The beginnings of Maine mineral production, Mineralogy of Maine, 2, Mining History, Gems and Gemology, 83–127.

Simmons, W.B. (2003) The Tourmaline Group. extraLapis English No. 3 Tourmaline, 10–15.

Chapter 7

Chemistry of the Elements

The majority of people who are attracted to the earth sciences start out with some things in common: a love of the outdoors, appreciation and respect for nature, awe at the time scale of geologic events, and, ultimately, the fact that understanding how the Earth works is really fascinating! If you are one of these people, then you might be thinking that the field of chemistry has very little to do with what you're really interested in. If you've taken a high school or college level chemistry course, you might be wondering how it could possibly be relevant to geology?

The problem is that chemists view things a little differently than geologists. Your chemistry course probably focused on the study of gases and liquids. I remember spending countless (clueless) hours in chemistry labs measuring invisible gases and mixing liquids to make pretty colors for reasons I didn't really understand. Most introductory chemistry courses do not spend much time on crystalline matter because reaction times are slow and fundamental chemical concepts are more easily demonstrated in liquids and gases. Even at upper levels, in courses like thermodynamics, chemists tend to concentrate on non-solids. On the other hand, as mineralogists we really want to understand how crystalline materials behave. Thus, there may seem to be a fundamental disconnect between what chemists teach and what mineralogists want to know.

However, there is a lot of important information in that chemistry class that geologists need to know. There is nothing wrong with sitting in your chemistry class and thinking about how the course material might apply to geological systems and processes. Pay extra-close attention whenever your professor mentions oxygen, the most abundant element in the Earth's crust. Work up the courage to talk with your chemistry professor and let her or him know that you're a geology major, and want to understand how their course applies to the study of geological materials—you might find that the professor will start to use geological examples in lectures!

The truth is that there are very few aspects of geology that do not require some basic knowledge of chemistry, physics, biology, or math. In fact, entire subdisciplines of our field focus on the intersections between geology and other fields: geochemistry, geophysics, biogeochemistry, mathematical geology, etc. Once you have a sense of what kind of geology interests you most, you might find it worthwhile to take some allied science or math courses in support of your geology curriculum.

The sciences of chemistry, physics, biology, astronomy, and geology are becoming more and more interdependent. To succeed in any of these fields, it is now necessary to understand something about related sciences. Sometimes this is not an easy task. Each field has its own culture (by now, you've probably noticed that the geologists at your school are a little different than the other scientists!), and its own conferences, practices, and literature. Old traditions die hard, and it is often still difficult to bridge the vocabulary gap between disciplines—not to mention the psychological barriers between these fields. But breaking down those barriers can lead to profound advances. Geology is already inherently an interdisciplinary field, so accepting early on that you have something to learn from chemists, biologists, physicists, and mathematicians will help speed you on your way to making a contribution to understanding the Earth and how it works!

M.D.D.

Introduction

So, you want to build a mineral structure, or more appropriately, you want to understand mineral structures that occur in nature? In this chapter, our goal is to explore the chemical elements that make up minerals, from start to finish. Here we will develop this topic from first principles, starting with the Big Bang and the creation of the elements in the universe, and then moving to the whys and ways in which elements combine to form minerals with a wide range of compositions. This subject matter spans the range from cosmology to fundamental quantum mechanics and advanced inorganic chemistry—but we will focus our attentions strictly on those aspects that affect the formation of minerals. Our "spin" on this topic may vary somewhat from how your other professors present this information because of our interdisciplinary approach and specific application. But you may be surprised to find that this chapter will help you understand these topics when you meet them in other classes. We hope you will finally understand why chemistry is commonly a required course for a degree in geology, and a prerequisite for a mineralogy class.

The Big Bang

For the purposes of our mineralogy class, we are interested in the Big Bang as the precipitating event for the formation of the elements nature uses to make minerals! The origin of the universe is a fascinating topic in its own right, and is the subject of an entire discipline called cosmology, which is the study of the structure of the universe. Here, we need only summarize the Big Bang succinctly by saying that sometime about 14–15 billion years ago, there was an immense, cataclysmic explosion that produced all of the matter that exists in the universe today. Several different types of matter and energy were initially produced: photons (light), leptons (matter in the form of electrons, positrons, and muons), baryons (which are protons, antiprotons, neutrons, and antineutrons), neutrinos, and even dark matter. Temperatures after the Big Bang were initially too hot for these fundamental particles to stick together, but after a few minutes, the temperature of the universe cooled to a "low" enough temperature (about a billion degrees) for protons and neutrons to stick together and make nuclei. Successive cooling allowed the formation of hydrogen (H) and helium (He) in a 3:1 ratio that can be both predicted and actually observed. Small amounts of lithium (Li) and beryllium (Be) may also have formed then, but there was not enough energy left to make any other elements.

This brings us to a brief discussion of the three fundamental particles (Table 7.1) that are brought together by these reactions. The tiny nucleus of an atom, with a radius of about 10^{-5} Å, is a dense compaction of neutrons and protons. The "cloud" of electrons that surrounds it is significantly larger (roughly 1 Å) and lighter. The properties of the atom are generally described with the following symbols:

Z = atomic number

P = the number of positively-charged protons, ranging up to at least 106

N = the number of neutrally-charged neutrons

$A = P + N = Z + N$ = mass number, ranging from 1–260, or greater

In all atoms except for H, the nucleus is made up of protons and neutrons (H is just a single proton), and the entire atom must have a charge of zero, so the number of protons equals the number of electrons. For $Z<20$ the number of protons and neutrons are similar, but for $Z>20$ there are usually more neutrons than protons.

There is a simple notation for describing all these aspects of an atom, along with its ionic charge ($x\pm$), as follows:

$${}^{A}_{Z}E^{x\pm}$$

So for any element, the number and type of fundamental particles can be conveniently described by using this shorthand. Understanding these symbols will help you puzzle out the scary reactions in the following paragraphs!

After the Big Bang, the processes involved in nuclear fusion in star formation allowed the fundamental particles to begin to come together to build other, heavier elements. For example, the burning of helium into carbon occurs by what is known as the triple alpha process (so-called because a helium nucleus, with two protons and two neutrons, is called an **alpha particle**). "Burning" as used here is really nuclear fusion, the ultimate source of all energy in the universe. The two-stage reaction is as follows:

$${}^{4}_{2}\mathrm{He} + {}^{4}_{2}\mathrm{He} \Longleftrightarrow {}^{8}_{4}\mathrm{Be}$$

$${}^{8}_{4}\mathrm{Be} + {}^{4}_{2}\mathrm{He} \Rightarrow {}^{12}_{6}\mathrm{C} + \gamma$$

where γ is a high-energy photon, or **gamma ray**.

Once you reach this point, you can make lots of heavier elements if you have the thermonuclear energy of a star to assist you in the process. For

Table 7.1. Fundamental Subatomic Particles and their Characteristics (after Gray, 1994)

Characteristic	Electron	Proton	Neutron
symbol	e^-, ${}^{0}_{-1}e$, β^-	p, p^+, ${}^{1}_{1}p$, ${}^{1}_{1}H^+$	n, ${}^{1}_{0}n$, N
absolute mass[a]	0.0005486 amu	1.00757 amu	1.00893 amu
absolute mass	9.109×10^{-31} kg	1.672×10^{-27} kg	1.674×10^{-27} kg
relative mass[b]	1.000	1836.12	1838.65
absolute charge[c]	-1.602×10^{-19} C	1.602×10^{-19} C	0
relative charge	−1	+1	0

[a]1 amu = 1.6606×10^{-24} g
[b]expressed relative to the resting mass of an electron
[c]coulomb (symbolized C) is the dimensionless standard unit of electric charge in the International System of Units

example, carbon atoms can combine with helium to make oxygen by the reaction:

$${}^{4}_{2}He + {}^{12}_{6}C => {}^{16}_{8}O + \gamma$$

The guiding principle in these reactions is that the heavier the element, the higher the temperature that will be required to create it. So only light elements can be generated by the catalytic carbon cycle, which occurs at "low" temperatures near 18 million K; here are some sample reactions (ν_e is the symbol for a neutrino, and β^+ is a positron):

$${}^{12}_{6}C + {}^{1}_{1}H => {}^{13}_{7}N + 1.95 \text{ MeV}$$

$${}^{13}_{7}N => {}^{13}_{6}C + \beta^+ + \nu_e + 2.22 \text{ MeV}$$

$${}^{13}_{6}C + {}^{1}_{1}H => {}^{14}_{7}N + 7.54 \text{ MeV}$$

$${}^{14}_{7}N + {}^{1}_{1}H => {}^{15}_{8}O + 7.35 \text{ MeV}$$

$${}^{15}_{8}O => {}^{15}_{7}N + \beta^+ + \nu_e + 2.75 \text{ MeV}$$

$${}^{15}_{7}N + {}^{1}_{1}H => {}^{12}_{6}C + {}^{4}_{2}He + 4.96 \text{ MeV}$$

$${}^{15}_{7}N + {}^{1}_{1}H => {}^{16}_{8}O + 12.13 \text{ MeV}$$

$${}^{16}_{8}O + {}^{1}_{1}H => {}^{17}_{9}F + 0.60 \text{ MeV}$$

$${}^{17}_{9}F => {}^{17}_{8}O + \beta^+ + \nu_e + 2.76 \text{ MeV}$$

$${}^{17}_{8}O + {}^{1}_{1}H => {}^{14}_{7}N + {}^{4}_{2}He + 1.19 \text{ MeV}.$$

You will note that some (the ones with the "MeV" in the equations) of these reactions produce energy, in the form of heat, which, in turn, causes the star to produce heat and light. As soon as the star attains a temperature of about 100 million K, you have enough energy to create heavier elements by way of the **alpha process**, which just keeps adding ${}^{4}_{2}He$ to the already formed elements as follows:

$${}^{12}_{6}C + {}^{4}_{2}He => {}^{16}_{8}O + \gamma$$

$${}^{16}_{6}C + {}^{4}_{2}He => {}^{20}_{10}Ne + \gamma$$

$${}^{20}_{10}Ne + {}^{4}_{2}He => {}^{24}_{12}M + \gamma$$

$${}^{40}_{20}Ca + {}^{4}_{2}He => {}^{44}_{22}Ti$$

$${}^{44}_{20}Ca + {}^{4}_{2}He => {}^{48}_{22}Ti$$

$${}^{48}_{22}Ti + {}^{4}_{2}He => {}^{52}_{24}Cr$$

$${}^{52}_{24}Cr + {}^{4}_{2}He => {}^{56}_{26}Fe.$$

Elements with values of Z up to 56 can be created by this mechanism. These reactions mostly release energy, keeping the temperature high inside a star. Eventually, though, you run out of the energy to build anything bigger than iron (in fact, the iron-making reaction requires energy rather than creating it), so you need additional processes to make heavier elements.

In order to build heavier elements, we need a process other than simple nuclear fusion in a star. If there are extra neutrons floating around, it is possible for an atomic nucleus to absorb a neutron and create a new element one atomic mass unit heavier than before. Astronomers call this type of neutron capture either ***s*-process** (*s*low neutron addition) or ***r*-process** (*r*apid neutron addition), depending on the time scale at which it happens (which in turn depends on how many free neutrons occur in the star). This occurs when stars have exhausted other fuel and explode (as supernova). This energy allows an element to capture a free neutron, in order to produce heavier nuclei that cannot be made by fusion reactions. The source of the free neutrons is the reaction of a H or He atom with one of the heavier elements in the star as follows:

$${}^{4}_{2}He + {}^{13}_{6}C => {}^{16}_{8}O + n, \text{ or}$$

$${}^{16}_{8}O + {}^{16}_{8}O => {}^{31}_{16}S + n.$$

Once the free neutrons have been produced, they can be captured by other heavy nuclei to produce elements further up the periodic chart (i.e., larger atomic numbers, those higher than Fe). We should note that another mechanism involving the creation of new isotopes by the addition of protons has also been suggested (***p*-process**), but it is as yet poorly understood. For excellent discussions of these mechanisms, we recommend Shu's (1982) astronomy textbook as well as a nice little book by Silk (2001) entitled *The Big Bang*.

The end result of all these reactions is the creation of the elements we see represented by the periodic chart; we'll have more to say about that shortly. For now, you may be wondering how cosmologists have figured all this out!? We have several different types of information on the chemical composition of the universe. These include:

1. **Spectroscopic studies of stars and the interstellar medium.** Light from the surfaces of stars, when analyzed spectroscopically, allows identification of the elements emitting radiation in the form of light. When excited by high temperatures, each element emits (or absorbs) a characteristic energy of light, resulting in bright or dark lines at characteristic wavelengths. In this way, light spectra "fingerprint" the elements in stars. The relative abundances of elements can then be estimated from the intensity of the spectral lines. This field of study greatly overlaps with optical mineralogy.
2. **Radio waves** are used by radio astronomers to study the hydrogen contents of galaxies and interstellar matter at a wavelength of 21 cm.
3. **Meteorites** are pieces of the solar system that fall to Earth, formed through many different processes on planetary bodies. Most of them come from asteroids (small, rocky, minor planets), though a few probably come from comets (small ice and dust bodies). A very few meteorites even come from the Moon (>29 have been found) or Mars (>28 at this writing). Meteorites are the only tangible evidence we have of the substance from which our solar system (and the universe) is made. There are several types: **iron meteorites** are composed of iron and nickel (similar to type M asteroids) and probably represent the cores of terrestrial bodies; stony irons are mixtures of iron and rocky material (type S asteroids); **chondrites** are rocky meteorites that represent the bulk compositions of terrestrial planets; and **achondrites** (like the Mars and Moon meteorites) look like basalts we have here on Earth. From these samples, we can estimate the compositions of rocky bodies in our solar system.
4. **Cosmic rays and the solar wind** have been carefully analyzed by satellites and numerous spacecraft. Although the origin of cosmic rays is a mystery, they do appear to come from the direction of supernova remnants. They consist mostly of protons, alpha particles, and nuclei of carbon, nitrogen, and oxygen atoms, moving at nearly the speed of light. The solar wind is a stream of mostly ionized hydrogen and helium (i.e., the nuclei of these elements with the electrons stripped away) that is emitted from the Sun at velocities greater than 350 km/s. Both cosmic rays and the solar wind tell us much about the fundamental particles involved in the origin of the solar system.
5. **Remote sensing and *in situ* studies of planetary exteriors.** Since the 1960s, both the U.S. and other countries (predominantly the former Soviet Union) have sent orbiters and landers to a number of bodies in our solar system, including more than 40 missions to Venus and Mars. Remote-sensed (from satellites) spectra of the transition elements (i.e., elements with $Z = 21$ to 30 and 39 to 48) are extremely useful in fingerprinting the minerals present on these surfaces, and more advanced analytical techniques are possible on the landers.
6. **The composition of the Earth and the Moon.** Of course, the best source of information about our universe is our own planet, and the samples we collected from the Moon.

Assessment of these elemental abundance data leads us to an understanding of the composition of the universe, the cosmic abundances shown in Table 7.2. As is consistent with the Big Bang theory, hydrogen is by far the most abundant element: 93% of all atoms, or 75% of all matter by weight. Helium is the next most abundant: about 7% of the total atoms, and 24% of the total weight. There is a general decrease in abundance with increasing atomic number, with all the elements beyond helium making up less than 1% of the universe. In fact, these abundances decrease almost exponentially! The abundance of elements with $Z > 50$ are exceedingly low, and above that point their abundances do not vary much with increasing atomic number. There is also an intriguing trend called the **Oddo-Harkins rule**, which says that elements with even atomic numbers are more abundant than those with odd ones. Can you figure out why? Finally, notice that the abundance of iron is noticeably higher than that of other atoms with similar Z's.

These results make a useful contrast to the composition of Earth, which is shown in Table 7.3. For the purposes of our mineralogy class, we are going to be most concerned with elements in the crust (as discussed in Chapters 1 and 3). Notice that the Big Eight elements (O, Si, Al, Fe, Mg, Ca, Na, and K) decrease in abundance in the upper mantle relative to the crust. This occurs because the high pressures and temperatures that occur in the mantle restrict the number of minerals that can be stable there to a very small number. The upper mantle is primarily composed of olivine, pyroxene, spinel, and garnet group min-

Table 7.2. Abundances of Elements in the Solar System expressed in terms of number of atoms relative to 10^6 atoms of Si (after Anders and Grevasse, 1982)

Atomic Number	Element	Symbol	Elemental Abundance*	Atomic Number	Element	Symbol	Elemental Abundance*
1	hydrogen	H	2.72×10^{10}	53	iodine	I	9.0×10^{-1}
2	helium	He	2.18×10^{9}	54	xenon	Xe	4.35×10^{0}
3	lithium	Li	5.97×10^{1}	55	cesium	Cs	3.72×10^{-1}
4	beryllium	Be	7.8×10^{-1}	56	barium	Ba	4.36×10^{0}
5	boron	B	2.4×10^{1}	57	lanthanum	La	4.48×10^{-1}
6	carbon	C	1.21×10^{7}	58	cerium	Ce	1.16×10^{0}
7	nitrogen	N	2.48×10^{6}	59	praseodymium	Pr	1.74×10^{-1}
8	oxygen	O	2.01×10^{7}	60	neodymium	Nd	8.36×10^{-1}
9	fluorine	F	8.43×10^{2}	61	promethium	Pm	0
10	neon	Ne	3.76×10^{6}	62	samarium	Sm	2.61×10^{-1}
11	sodium	Na	5.70×10^{4}	63	europium	Eu	9.72×10^{-2}
12	magnesium	Mg	1.075×10^{6}	64	gadolinium	Gd	3.31×10^{-1}
13	aluminum	Al	8.49×10^{4}	65	terbium	Tb	5.89×10^{-1}
14	silicon	Si	1.00×10^{6}	66	dysprosium	Dy	3.98×10^{-1}
15	phosphorus	P	1.04×10^{4}	67	holmium	Ho	8.75×10^{-2}
16	sulfur	S	5.15×10^{5}	68	erbium	Er	2.53×10^{-1}
17	chlorine	Cl	5.240×10^{3}	69	thullium	Tm	3.86×10^{-2}
18	argon	Ar	1.04×10^{5}	70	ytterbium	Yb	2.43×10^{-1}
19	potassium	K	3.770×10^{3}	71	lutetium	Lu	3.69×10^{-2}
20	calcium	Ca	6.11×10^{4}	72	hafnium	Hf	1.76×10^{-1}
21	scandium	Sc	3.38×10^{1}	73	tantalum	Ta	2.26×10^{-2}
22	titanium	Ti	2.400×10^{3}	74	tungsten	W	1.37×10^{-1}
23	vanadium	V	2.95×10^{2}	75	rhenium	Re	5.07×10^{-2}
24	chromium	Cr	1.34×10^{4}	76	osmium	Os	7.17×10^{-1}
25	manganese	Mn	9.510×10^{3}	77	iridium	Ir	6.60×10^{-1}
26	iron	Fe	9.00×10^{5}	78	platinum	Pt	1.37×10^{0}
27	cobalt	Co	2.250×10^{3}	79	gold	Au	1.86×10^{-1}
28	nickel	Ni	4.93×10^{4}	80	mercury	Hg	5.2×10^{-1}
29	copper	Cu	5.14×10^{2}	81	thallium	Tl	1.84×10^{-1}
30	zinc	Zn	1.260×10^{3}	82	lead	Pb	3.15×10^{0}
31	gallium	Ga	3.78×10^{2}	83	bismuth	Bi	1.44×10^{-1}
32	germanium	Ge	1.18×10^{2}	84	polonium	Po	~0
33	arsenic	As	6.79×10^{0}	85	astatine	At	~0
34	selenium	Se	6.21×10^{1}	86	radon	Rn	~0
35	bromine	Br	1.18×10^{1}	87	francium	Fr	~0
36	krypton	Kr	4.53×10^{1}	88	radium	Ra	~0
37	rubidium	Rb	7.09×10^{0}	89	actinium	Ac	~0
38	strontium	Sr	2.38×10^{1}	90	thorium	Th	3.35×10^{-2}
39	yttrium	Y	4.64×10^{0}	91	protactinium	Pa	~0
40	zirconium	Zr	1.07×10^{1}	92	uranium	U	9.00×10^{-3}
41	niobium	Nb	7.1×10^{-1}	93	neptunium	Np	~0
42	molybdenum	Mo	2.52×10^{0}	94	plutonium	Pu	~0
43	technicium	Tc	0	95	americium	Am	0
44	ruthenium	Ru	1.86×10^{0}	96	curium	Cm	0
45	rhodium	Rh	3.44×10^{-1}	97	berkelium	Bk	0
46	palladium	Pd	1.39×10^{0}	98	californium	Cf	0
47	silver	Ag	5.29×10^{-1}	99	einsteinium	Es	0
48	cadmium	Cd	1.69×10^{0}	100	fermium	Fm	0
49	indium	In	1.84×10^{-1}	101	mendelevium	Md	0
50	tin	Sn	3.82×10^{0}	102	nobelium	No	0
51	antimony	Sb	3.52×10^{-1}	103	lawrencium	Lr	0
52	tellurium	Te	4.91×10^{0}				

erals. Take a look at the compositions of these minerals, and see if you can explain why the abundance of K in the upper mantle is so low.

More interesting comparisons can be made with respect to the lunar data. Most scientists currently believe that the Moon formed when a large object

Table 7.3. Chemical Composition of the Earth's and other Planets (in ppm) (after Burns, 1993)

Element	Solar System (Cosmic)[1]	Earth Average Crust[2]	Earth Oceanic Crust[2]	Earth Upper Mantle[3]	Total Earth[2]	Lunar High Ti Basalt[4]	Lunar Green Glass[4]	Venus Vega 2 Lander[5]	Mars SNC Meteorites[6]	Mars Pathfinder Lander[7]
O	466,000	466,000	409,310	432,900	295,300	405,380	418,830	437,300	422,100	
Si	34,370	277,200	231,000	213,900	15,200	176,400	210,900	212,800	238,000	289,790
Al	2,805	81,300	84,700	17,800	10,900	46,850	39,700	84,700	36,530	56,100
Ti	140	4,400	9,000	900	500	77,940	2,280	1,200	5,400	4,200
Mg	31,945	20,900	46,400	244,600	12,700	50,890	105,520	69,340	56,100	12,060
Fe	61,510	50,000	81,600	63,200	346,300	153,130	155,460	59,810	150,800	92,770
Mn	640	950	1,000	1,000	2,200	2,090	2,010	770	4,030	-
Cr	860	100	270	3,100	2,600	3,030	-	-	1,575	-
Ni	3,540	75	135	700	15,800	2	153	-	36	-
Na	1,615	28,200	20,800	2,500	5,700	2,670	965	14,840	9,870	19,290
Ca	2,995	36,300	80,800	19,200	11,300	76,430	60,710	53,570	70,710	52,170
K	180	25,900	1,250	200	700	415	249	830	1,410	5,800
S	20,205	260	-	-	19,300	-	-	18,800	-	-
Cl	225	130	-	-	-	-	-	$\leq$4,000	-	200

Data from [1]Anders and Grevasse (1989); [2]Mason and Moore (1982) p. 46; [3]Liu and Bassett (1986) p. 234; [4]Taylor (1982): Apollo 17 high-K basalt 74275 and Apollo 15 green glass 15426; [5]Surkov et al. (1984, 1986); [6]Smith et al. (1984) ; [7]McSween et al. (1999), calculated sulfur-free rock. Cosmic abundances are given relative to O = 466,000ppm.

(another, smaller "protoplanet") hit the juvenile Earth about 4.5 billion years ago. The impact blew off the crust and mantle of the newly-evolved Earth, and the debris went into orbit around the earth and eventually aggregated into the Moon. The iron core of the impactor probably melted on impact and merged with the iron core of the Earth (Hartman and Davis, 1975). As a result, the Earth has more than twice as much Fe than the Moon. How do you think this affects the relative density of the Earth vs. the Moon? The Moon is also dry because its H was lost in the giant heat and pressure of the impact, so even though there's lots of O in the rocks on the moon, there's no H to make H_2O.

Finally, you can compare the data from Mars and Venus to the crustal composition of the Earth. Based on these data, would you expect Mars and Venus to have dramatically different mineralogy than the Earth? Do any elements stand out as being different on those two planets relative to the Earth? Do you think this is an artifact of how the data were collected? Such questions are among those currently being faced by planetary mineralogists. Keep your eye on the series of missions to Mars and Venus in the next decades to learn their answers...

Atomic Structure

Now that we see how the elements were created, let's look closely at how they are constructed. As we learned earlier in Chapter 3, atoms are made up of three types of fundamental particles: protons, neutrons, and electrons. The nucleus at the center of an atom is made up of all the protons and neutrons, except in the case of the hydrogen atom, which has just a single proton. Electrons can occur at any distance from the nucleus, and their precise locations at any given time are difficult to describe (and measure) because the electrons are always in motion.

Imagine that you're watching a pig race, where the pigs move around a track. You take a photo of the pig you want to win—and the photograph can tell you exactly where the pig was at the time you snapped your photo. However, you have no way of telling how fast your pig was moving when the picture was taken. To determine speed, you

Figure 7.1. The Heisenberg Uncertainty Principle tells us that you can't determine both the position and momentum of a particle precisely and simultaneously. In this image, we know where the pigs are (Bear Creek, Montana), but we can't tell how fast they are going. Photo by Randall Smith.

would need a speedometer on your pig, and strapping a speedometer on the pig would no doubt make it behave in a different manner (Figure 7.1). This analogy explains why electrons are so tricky to deal with. All of the current methods we have to determine their velocities change their position, and conversely, any method that measures their positions changes their velocities.

So it was a real breakthrough in 1923 when Louis de Broglie suggested that electrons (and all types of matter) could be treated as *waves* rather than particles. He assumed that any particle (be it an electron or an Idaho potato) has a wavelength (λ) that is equal to Planck's constant ($h = 6.6262 \times 10^{-27}$ ergs) divided by its momentums (p):

$$\lambda = \frac{h}{p}$$

where p is also equal to the mass of the particle times its velocity (i.e., $p = mv$). This formulation is useful because it lets us define electrons as waves. So instead of visualizing electrons zipping around a nucleus, we can consider waves, which are mathematically more convenient to describe. Following de Broglie's work, Werner Heisenberg introduced in 1926 his Principle of Indeterminancy (the so-called **Heisenberg Uncertainty Principle**), which states that it is impossible to determine both the position and momentum of a particle precisely and simultaneously. Mathematically it is given as

$$\Delta p \Delta x \geq \hbar / 2$$

where Δp is the uncertainty in momentum, Δx is the uncertainty in position, and $\hbar$ (pronounced "h bar") is the reduced Planck's constant $h/2\pi$.

The point of all this math is that it's hard to describe electrons if you only think of them as particles! The wavelength viewpoint of electronic structure led Erwin Schrödinger in 1926 to develop an equation that described the distribution of electrons in three-dimensional space using de Broglie waves. (Interestingly, he conceived of this idea while sequestered in the Swiss Alps with a Viennese mistress.) The resultant **Schrödinger equation** looks pretty scary:

$$\partial^2\psi/\partial x^2 + \partial^2\psi/\partial y^2 + \partial^2\psi/\partial z^2 +$$

$$[(8\pi^2 m)h^2](E - V)\psi = 0.$$

where ψ is a wave function, m is the mass of the electron, E is the total energy of the atom, and V is the potential energy. Perhaps a simpler way to conceptualize this is to consider it in a form that defines the wave function, ψ, in terms of four variables:

radius $R(r)$, which is the distance of the electron from the nucleus,

$\Theta(\theta)$, which describes the angle of rotation of the electron relative to the z axis,

$\Phi(\phi)$, which is the angle of rotation of the electron around the z axis, and

ψ_s, which is the spin function that describes the direction that the electron is spinning.

Mathematically this can be written as:

$$\psi(r, \theta, \phi) = R(r)\ \Theta(\theta)\ \Phi(\phi)\ \psi_s.$$

This simpler form allows us to visualize how this equation can be used to describe the wave function, so we know where the electrons are located (Figure 7.2). Each of these four functions can be expressed as a quantum number, and together they represent a solution to the Schrödinger equation that describes the size, shape, and orientation in space of the orbitals around a nucleus.

- n is called the **principal quantum number**, and it describes the average distance of the electron from the nucleus, or the size of the electron cloud. It ranges in value from 1 to ∞, and is the most important factor in determining the energy of the electron (Figure 7.3). As the value of n increases, the energies of the electrons also increase. Each value of n is associated with the name of a shell around the nucleus: K, L, M, N, O, etc., as seen in Table 7.4.

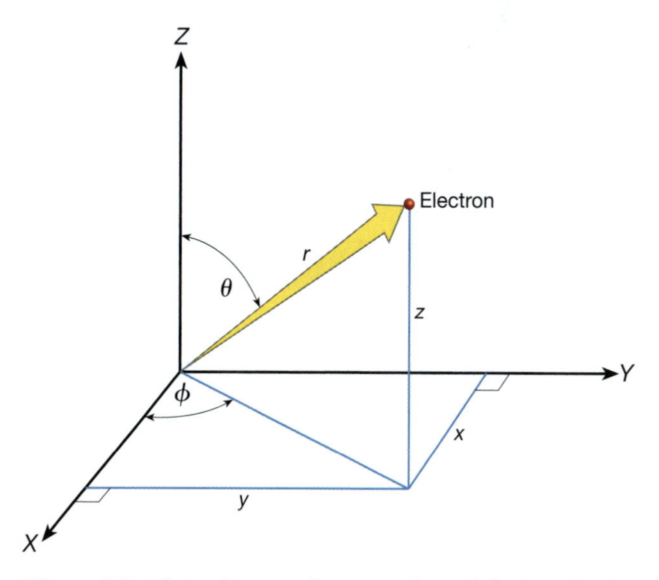

Figure 7.2. The polar coordinates r, θ, *and* ϕ *describe the location of an electron relative to a polar coordinate system. Mathematically, you can derive the relationships between these variables, which include* x = r *sin* θ *cos* ϕ, y = r *sin* θ *sin* ϕ, *and* z = r *cos* θ. *After Gray (1994).*

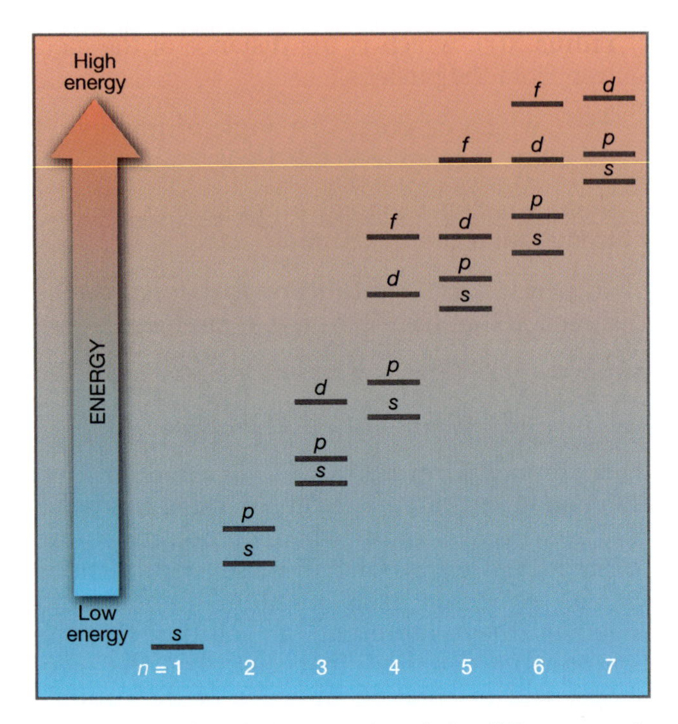

Figure 7.3. *The relative energies of the different atomic orbitals are shown here schematically. This diagram allows prediction of which orbitals will be filled first by electrons, according to the Aufbau Principle.*

• l is the **azimuthal quantum number** (sometimes it is called the secondary quantum number), and it defines the actual shape of the orbital. More precisely, it is a measure of the angular momentum of the electron in the orbital. It has only integer values, but they are limited by the value of n associated with each l, so l ranges from 0 up to $n - 1$. Each integer value is associated with a letter: *s* is used for the $l = 0$ orbital, *p* for the $l = 1$, *d* for the $l = 2$, and so on (Table 7.4). So for example, the K shell ($n = 1$) has only a single value for l: $l = 0$, so there is only one type of l orbital (*s*) associated with $n = 1$. In the L shell, where $n = 2$, l can equal 0 and 1, so there are two types of orbitals, *s* and *p*. In the M shell, $l = 0$, 1, and 2 and there are three orbitals, *s*, *p*, and *d*.

• m_l is the **magnetic quantum number**, which shows how the orbital is oriented relative to the other orbitals. It indicates the direction of maximum extension of the electron cloud around the nucleus, and it occurs in both the θ and ϕ directions shown in Figure 7.2. It ranges from +l to –1, and determines the number of each type of orbital. For example, when $l = 2$, m_l can be –2, –1, 0, 1, or 2, so there are five different types of *d* orbitals.

• m_8 (also called *s*, or m_s) is the spin quantum number. In each orbital, there are two possible opposite directions for an electron to spin: clockwise or counterclockwise. These are assigned values of either +1/2 or –1/2. (This is why the student's beds had to face in opposite directions in the Atomic Apartment of Chapter 3.)

Each combination of *n*, *l*, and m_1 describes a unique probability distribution—the electron clouds that are illustrated in Figure 7.4. Furthermore, no two electrons will ever have the same four quantum numbers (this is known as the **Pauli Exclusion Principle**).

Why is the Schrödinger equation so important? Because it allows us to calculate the probability of finding a particle (in this application, an electron) at a certain position. Schrödinger originally applied this equation to a single particle; for that purpose, this nasty mathematically expression can be solved analytically. However, there are several ways of approximating the solution to this equation for more than one particle (i.e., for

Table 7.4. Atomic Structures as a Function of Quantum Numbers

Shells	*n*	*l = 0 (s orbitals)*	*l = 1 (p orbitals)*	*l = 2 (d orbitals)*	*l = 3 (f orbitals)*	*l = 4 (g orbitals)*
K	1	$1s^2$				
L	2	$2s^2$	$2p^6$			
M	3	$3s^2$	$3p^6$	$3d^{10}$		
N	4	$4s^2$	$4p^6$	$4d^{10}$	$4f^{14}$	
O	5	$5s^2$	$5p^6$	$5d^{10}$	$5f^{14}$	$5g^{18}$
P	6	$6s^2$	$6p^6$	$6d^{10}$	$6f^{14}$	$6g^{18}$
Q	7	$7s^2$	$7p^6$	$7d^{10}$	$7f^{14}$	$7g^{18}$
Number of orbitals		1	3	5	7	9
Number of electrons in orbital		2	6	10	14	18

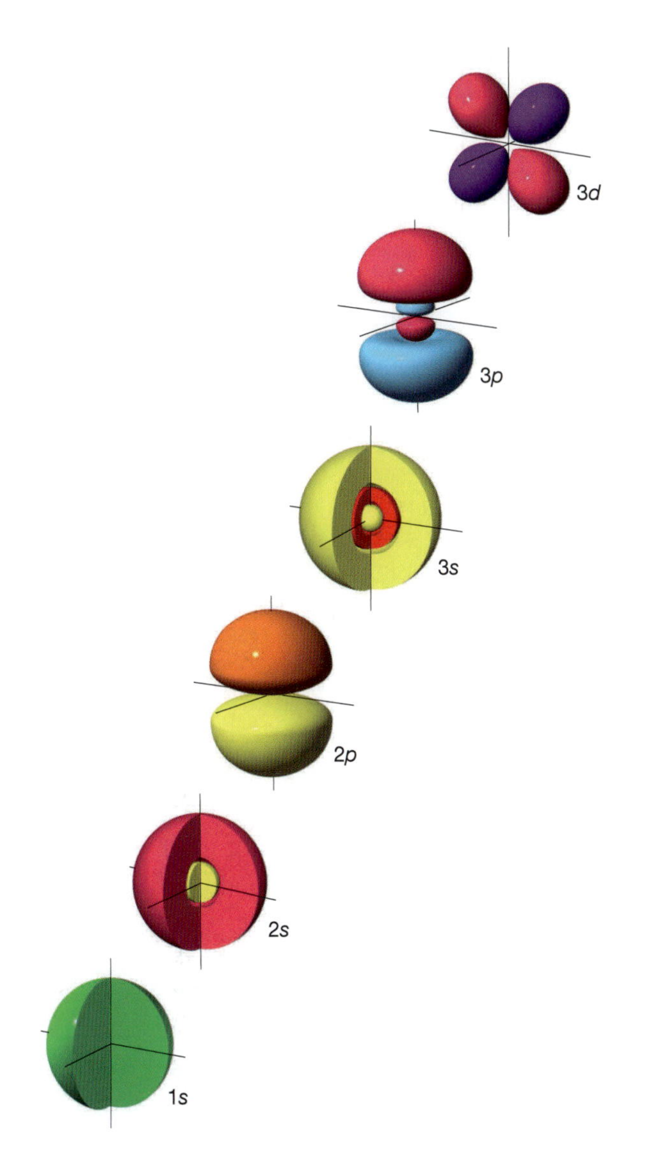

Figure 7.4. Probability distribution functions for s, p, *and* d *orbitals.*

elements other than hydrogen!). Mineralogists use many methods to try to understand the atomic and molecular structures of naturally-occurring crystals, including iterative techniques, something called self-consistent field theory (SCF), and a modification of SCF called the Hartree-Fock approximation. Explanations of all these techniques are beyond the scope of this textbook, but their mineralogical applications are fascinating and worth pursuing in further studies. Their main goal is to predict what minerals would form given a certain set of elements at a given pressure and temperature. This is sort of the Holy Grail of mineralogy!

One more essential characteristic of electrons is needed to allow us to build the periodic chart. The **Aufbau Principle** states that electrons will tend to occupy the lower energy shells and orbitals before the higher energy ones. In Chapter 3, we discussed the fact that students in our Atomic Apartment Complex always occupy rooms on the lowest floors first, because it takes less energy to get to them. Thus we see that electrons, like students, do not wish to exert any more energy than is needed to get them to their orbitals (the beds of the Atomic Apartment). The Aufbau Principle helps us determine the relative order in which the orbitals will be filled, which is $1s<2s<2p<3s<3p<4s<3d<4p<5s<4d<5p<6s$, and so on (Figure 7.3).

So now that we understand the relative energies of orbitals, and the rules for how electrons will fill them, we can construct the periodic chart. As we noted at the beginning of this chapter, each element on the periodic chart has a certain number of protons, neutrons, and electrons. A unique number of protons (its atomic number) defines each element, and thus each element has a characteristic configuration in its atomic state (Table 7.5). The periodic chart (Figure 7.5) is nothing more than a convenient way to arrange the elements to show how they are related by their principal quantum numbers. It was predicted by Dmitri Mendeleev in 1869, even before we understood the structure of an atom. Rows on the periodic chart are called **periods**, and each period begins with the filling of a new outer shell and ends when the shell is full. So the first period, which corresponds to the K shell, has atoms with electrons only in 1*s* orbitals (H and He). The 2nd period, which represents the L shell, has atoms with electrons in the $2s^2$ and $2p^6$ orbitals, so there are eight atoms in that period. The 3rd period corresponds to the M shell, and filling of the $3s^2$ and $3p^6$ orbitals, but the 4th period begins to get complicated by the filling of the $3d^{10}$, $4s^2$, $4p^6$.

Columns on the periodic chart are called **groups**, because they often have similar physical properties. For example, elements numbered 21–30, 39–48 and 72–80 are called transition elements, and they are particularly important to mineralogists (and anyone who likes colored gemstones), because they cause color (see discussion later in this chapter). Other important group names are indicated in Figure 7.5.

Armed with this knowledge of the periodic chart and the structure of elements, there is only one more concept we need before we can start putting atoms together. **Valence electrons** occupy the outermost, highest energy orbitals of an atom. The atom will be most stable if the valence electrons are in one of three configurations: all orbitals are empty, exactly half full, or completely full. Why? Just as our Atomic Apartment Complex was not "balanced" unless all the rooms

Table 7.5. Electronic Configurations and Bonding Information for Chemical Elements

Z	Atom	Orbital electronic configuration	Ionization Energy (eV)	Electron Affinity (eV)	Electronegativity
1	H	$1s^{1}$	13.598	0.754	2.1
2	He	$1s^{2}$	24.587	<0	0
3	Li	$1s^{2}2s^{1}$	5.392	0.61805	0.98
4	Be	$1s^{2}2s^{2}$	9.322	<0	1.57
5	B	$1s^{2}2s^{2}2p^{1}$	8.298	0.2771	2.04
6	C	$1s^{2}2s^{2}2p^{2}$	11.26	1.26293	2.55
7	N	$1s^{2}2s^{2}2p^{3}$	14.534	-.072	3.04
8	O	$1s^{2}2s^{2}2p^{4}$	13.618	1.4611215	3.44
9	F	$1s^{2}2s^{2}2p^{5}$	17.422	3.3993	3.98
10	Ne	$1s^{2}2s^{2}2p^{6}$	21.564	<0	0
11	Na	$1s^{2}2s^{2}2p^{6}3s^{1}$	5.139	0.5479303	0.93
12	Mg	$1s^{2}2s^{2}2p^{6}3s^{2}$	7.646	<0	1.31
13	Al	$1s^{2}2s^{2}2p^{6}3s^{2}3p^{1}$	5.986	0.4411	1.61
14	Si	$1s^{2}2s^{2}2p^{6}3s^{2}3p^{2}$	8.151	1.3855	1.9
15	P	$1s^{2}2s^{2}2p^{6}3s^{2}3p^{3}$	10.486	0.74653	2.19
16	S	$1s^{2}2s^{2}2p^{6}3s^{2}3p^{4}$	10.36	2.07712	2.58
17	Cl	$1s^{2}2s^{2}2p^{6}3s^{2}3p^{5}$	12.967	3.6173	3.16
18	Ar	$1s^{2}2s^{2}2p^{6}3s^{2}3p^{6}$	15.759	<0	0
19	K	$1s^{2}2s^{2}2p^{6}3s^{2}3p^{6}4s^{1}$	4.341	0.501471	0.82
20	Ca	$1s^{2}2s^{2}2p^{6}3s^{2}3p^{6}4s^{2}$	6.113	0.024551	1
21	Sc	$1s^{2}2s^{2}2p^{6}3s^{2}3p^{6}4s^{2}3d^{1}$	6.54	0.1882	1.36
22	Ti	$1s^{2}2s^{2}2p^{6}3s^{2}3p^{6}4s^{2}3d^{2}$	6.82	0.0791	1.54
23	V	$1s^{2}2s^{2}2p^{6}3s^{2}3p^{6}4s^{2}3d^{3}$	6.74	0.5251	1.63
24	Cr	$1s^{2}2s^{2}2p^{6}3s^{2}3p^{6}4s^{2}3d^{4}$	6.766	0.6661	1.66
25	Mn	$1s^{2}2s^{2}2p^{6}3s^{2}3p^{6}4s^{2}3d^{5}$	7.435	<0	1.55
26	Fe	$1s^{2}2s^{2}2p^{6}3s^{2}3p^{6}4s^{2}3d^{6}$	7.87	0.1634	1.83
27	Co	$1s^{2}2s^{2}2p^{6}3s^{2}3p^{6}4s^{2}3d^{7}$	7.86	0.6611	1.88
28	Ni	$1s^{2}2s^{2}2p^{6}3s^{2}3p^{6}4s^{2}3d^{8}$	7.635	1.1561	1.91
29	Cu	$1s^{2}2s^{2}2p^{6}3s^{2}3p^{6}4s^{2}3d^{9}$	7.726	1.2281	1.9
30	Zn	$1s^{2}2s^{2}2p^{6}3s^{2}3p^{6}4s^{2}3d^{10}$	9.394	<0	1.65
31	Ga	$1s^{2}2s^{2}2p^{6}3s^{2}3p^{6}4s^{2}3d^{10}4p^{1}$	5.999	0.302	1.81
32	Ge	$1s^{2}2s^{2}2p^{6}3s^{2}3p^{6}4s^{2}3d^{10}4p^{2}$	7.899	1.22	2.01
33	As	$1s^{2}2s^{2}2p^{6}3s^{2}3p^{6}4s^{2}3d^{10}4p^{3}$	9.81	0.813	2.18
34	Se	$1s^{2}2s^{2}2p^{6}3s^{2}3p^{6}4s^{2}3d^{10}4p^{4}$	9.752	2.020693	2.55
35	Br	$1s^{2}2s^{2}2p^{6}3s^{2}3p^{6}4s^{2}3d^{10}4p^{5}$	11.814	3.3653	2.96
36	Kr	$1s^{2}2s^{2}2p^{6}3s^{2}3p^{6}4s^{2}3d^{10}4p^{6}$	13.999	<0	0
37	Rb	$1s^{2}2s^{2}2p^{6}3s^{2}3p^{6}4s^{2}3d^{10}4p^{6}5s^{1}$	4.177	0.485922	0.82
38	Sr	$1s^{2}2s^{2}2p^{6}3s^{2}3p^{6}4s^{2}3d^{10}4p^{6}5s^{2}$	5.695	0.052066	0.95
39	Y	$1s^{2}2s^{2}2p^{6}3s^{2}3p^{6}4s^{2}3d^{10}4p^{6}5s^{2}4d^{1}$	6.38	0.3071	1.22
40	Zr	$1s^{2}2s^{2}2p^{6}3s^{2}3p^{6}4s^{2}3d^{10}4p^{6}5s^{2}4d^{2}$	6.84	0.4261	1.33
41	Nb	$1s^{2}2s^{2}2p^{6}3s^{2}3p^{6}4s^{2}3d^{10}4p^{6}5s^{2}4d^{3}$	6.88	0.8933	1.6
42	Mo	$1s^{2}2s^{2}2p^{6}3s^{2}3p^{6}4s^{2}3d^{10}4p^{6}5s^{2}4d^{4}$	7.099	0.7461	2.16
43	Tc	$1s^{2}2s^{2}2p^{6}3s^{2}3p^{6}4s^{2}3d^{10}4p^{6}5s^{2}4d^{5}$	7.28	0.552	1.9
44	Ru	$1s^{2}2s^{2}2p^{6}3s^{2}3p^{6}4s^{2}3d^{10}4p^{6}5s^{2}4d^{6}$	7.37	1.052	2.2
45	Rh	$1s^{2}2s^{2}2p^{6}3s^{2}3p^{6}4s^{2}3d^{10}4p^{6}5s^{2}4d^{7}$	7.46	1.1378	2.28
46	Pd	$1s^{2}2s^{2}2p^{6}3s^{2}3p^{6}4s^{2}3d^{10}4p^{6}5s^{2}4d^{8}$	8.34	0.5578	2.2
47	Ag	$1s^{2}2s^{2}2p^{6}3s^{2}3p^{6}4s^{2}3d^{10}4p^{6}5s^{2}4d^{9}$	7.576	1.3027	1.93
48	Cd	$1s^{2}2s^{2}2p^{6}3s^{2}3p^{6}4s^{2}3d^{10}4p^{6}5s^{2}4d^{10}$	8.993	<0	1.69
49	In	$1s^{2}2s^{2}2p^{6}3s^{2}3p^{6}4s^{2}3d^{10}4p^{6}5s^{2}4d^{10}5p^{1}$	5.786	0.32	1.78
50	Sn	$1s^{2}2s^{2}2p^{6}3s^{2}3p^{6}4s^{2}3d^{10}4p^{6}5s^{2}4d^{10}5p^{2}$	7.344	1.22	1.96
51	Sb	$1s^{2}2s^{2}2p^{6}3s^{2}3p^{6}4s^{2}3d^{10}4p^{6}5s^{2}4d^{10}5p^{3}$	8.641	1.075	2.05
52	Te	$1s^{2}2s^{2}2p^{6}3s^{2}3p^{6}4s^{2}3d^{10}4p^{6}5s^{2}4d^{10}5p^{4}$	9.009	1.97083	2.1

Table 7.5. (continued)

Z	*Atom*	*Orbital electronic configuration*	*Ionization Energy (eV)*	*Electron Affinity (eV)*	*Electronegativity*
53	I	$1s^{2}2s^{2}2p^{6}3s^{2}3p^{6}4s^{2}3d^{10}4p^{6}5s^{2}4d^{10}5p^{5}$	10.451	3.05914	2.66
54	Xe	$1s^{2}2s^{2}2p^{6}3s^{2}3p^{6}4s^{2}3d^{10}4p^{6}5s^{2}4d^{10}5p^{6}$	12.13	<0	2.6
55	Cs	$1s^{2}2s^{2}2p^{6}3s^{2}3p^{6}4s^{2}3d^{10}4p^{6}5s^{2}4d^{10}5p^{6}6s^{1}$	3.894	0.471630	0.79
56	Ba	$1s^{2}2s^{2}2p^{6}3s^{2}3p^{6}4s^{2}3d^{10}4p^{6}5s^{2}4d^{10}5p^{6}6s^{2}$	5.212	0.144626	0.89
57	La	$1s^{2}2s^{2}2p^{6}3s^{2}3p^{6}4s^{2}3d^{10}4p^{6}5s^{2}4d^{10}5p^{6}6s^{2}5d^{1}$	5.58	0.53	1.1
58	Ce	$1s^{2}2s^{2}2p^{6}3s^{2}3p^{6}4s^{2}3d^{10}4p^{6}5s^{2}4d^{10}5p^{6}6s^{2}4f^{1}5d^{1}$	5.47		1.12
59	Pr	$1s^{2}2s^{2}2p^{6}3s^{2}3p^{6}4s^{2}3d^{10}4p^{6}5s^{2}4d^{10}5p^{6}6s^{2}4f^{3}$	5.42		1.13
60	Nd	$1s^{2}2s^{2}2p^{6}3s^{2}3p^{6}4s^{2}3d^{10}4p^{6}5s^{2}4d^{10}5p^{6}6s^{2}4f^{4}$	5.49		1.14
61	Pm	$1s^{2}2s^{2}2p^{6}3s^{2}3p^{6}4s^{2}3d^{10}4p^{6}5s^{2}4d^{10}5p^{6}6s^{2}4f^{5}$	5.55		1.13
62	Sm	$1s^{2}2s^{2}2p^{6}3s^{2}3p^{6}4s^{2}3d^{10}4p^{6}5s^{2}4d^{10}5p^{6}6s^{2}4f^{6}$	5.63		1.17
63	Eu	$1s^{2}2s^{2}2p^{6}3s^{2}3p^{6}4s^{2}3d^{10}4p^{6}5s^{2}4d^{10}5p^{6}6s^{2}4f^{7}$	5.67		1.2
64	Gd	$1s^{2}2s^{2}2p^{6}3s^{2}3p^{6}4s^{2}3d^{10}4p^{6}5s^{2}4d^{10}5p^{6}6s^{2}4f^{7}5d^{1}$	6.15		1.2
65	Tb	$1s^{2}2s^{2}2p^{6}3s^{2}3p^{6}4s^{2}3d^{10}4p^{6}5s^{2}4d^{10}5p^{6}6s^{2}4f^{9}$	5.86		1.1
66	Dy	$1s^{2}2s^{2}2p^{6}3s^{2}3p^{6}4s^{2}3d^{10}4p^{6}5s^{2}4d^{10}5p^{6}6s^{2}4f^{10}$	5.93		1.22
67	Ho	$1s^{2}2s^{2}2p^{6}3s^{2}3p^{6}4s^{2}3d^{10}4p^{6}5s^{2}4d^{10}5p^{6}6s^{2}4f^{11}$	6.02		1.23
68	Er	$1s^{2}2s^{2}2p^{6}3s^{2}3p^{6}4s^{2}3d^{10}4p^{6}5s^{2}4d^{10}5p^{6}6s^{2}4f^{12}$	6.101		1.24
69	Tm	$1s^{2}2s^{2}2p^{6}3s^{2}3p^{6}4s^{2}3d^{10}4p^{6}5s^{2}4d^{10}5p^{6}6s^{2}4f^{13}$	6.184		1.25
70	Yb	$1s^{2}2s^{2}2p^{6}3s^{2}3p^{6}4s^{2}3d^{10}4p^{6}5s^{2}4d^{10}5p^{6}6s^{2}4f^{14}$	6.254		1.1
71	Lu	$1s^{2}2s^{2}2p^{6}3s^{2}3p^{6}4s^{2}3d^{10}4p^{6}5s^{2}4d^{10}5p^{6}6s^{2}4f^{14}5d^{1}$	5.43		1.27
72	Hf	$1s^{2}2s^{2}2p^{6}3s^{2}3p^{6}4s^{2}3d^{10}4p^{6}5s^{2}4d^{10}5p^{6}6s^{2}4f^{14}5d^{2}$	6.65	0	1.3
73	Ta	$1s^{2}2s^{2}2p^{6}3s^{2}3p^{6}4s^{2}3d^{10}4p^{6}5s^{2}4d^{10}5p^{6}6s^{2}4f^{14}5d^{3}$	7.89	0.3221	1.5
74	W	$1s^{2}2s^{2}2p^{6}3s^{2}3p^{6}4s^{2}3d^{10}4p^{6}5s^{2}4d^{10}5p^{6}6s^{2}4f^{14}5d^{4}$	7.98	0.8158	2.36
75	Re	$1s^{2}2s^{2}2p^{6}3s^{2}3p^{6}4s^{2}3d^{10}4p^{6}5s^{2}4d^{10}5p^{6}6s^{2}4f^{14}5d^{5}$	7.88	0.152	1.9
76	Os	$1s^{2}2s^{2}2p^{6}3s^{2}3p^{6}4s^{2}3d^{10}4p^{6}5s^{2}4d^{10}5p^{6}6s^{2}4f^{14}5d^{6}$	8.7	1.12	2.2
77	Ir	$1s^{2}2s^{2}2p^{6}3s^{2}3p^{6}4s^{2}3d^{10}4p^{6}5s^{2}4d^{10}5p^{6}6s^{2}4f^{14}5d^{7}$	9.1	1.5658	2.2
78	Pt	$1s^{2}2s^{2}2p^{6}3s^{2}3p^{6}4s^{2}3d^{10}4p^{6}5s^{2}4d^{10}5p^{6}6s^{2}4f^{14}5d^{8}$	9	2.1282	2.28
79	Au	$1s^{2}2s^{2}2p^{6}3s^{2}3p^{6}4s^{2}3d^{10}4p^{6}5s^{2}4d^{10}5p^{6}6s^{2}4f^{14}5d^{9}$	9.225	2.308633	2.54
80	Hg	$1s^{2}2s^{2}2p^{6}3s^{2}3p^{6}4s^{2}3d^{10}4p^{6}5s^{2}4d^{10}5p^{6}6s^{2}4f^{14}5d^{10}$	10.437	<0	2
81	Tl	$1s^{2}2s^{2}2p^{6}3s^{2}3p^{6}4s^{2}3d^{10}4p^{6}5s^{2}4d^{10}5p^{6}6s^{2}4f^{14}5d^{10}6p^{1}$	6.108	0.22	2.04
82	Pb	$1s^{2}2s^{2}2p^{6}3s^{2}3p^{6}4s^{2}3d^{10}4p^{6}5s^{2}4d^{10}5p^{6}6s^{2}4f^{14}5d^{10}6p^{2}$	7.416	0.3648	2.33
83	Bi	$1s^{2}2s^{2}2p^{6}3s^{2}3p^{6}4s^{2}3d^{10}4p^{6}5s^{2}4d^{10}5p^{6}6s^{2}4f^{14}5d^{10}6p^{3}$	7.289	0.9461	2.02
84	Po	$1s^{2}2s^{2}2p^{6}3s^{2}3p^{6}4s^{2}3d^{10}4p^{6}5s^{2}4d^{10}5p^{6}6s^{2}4f^{14}5d^{10}6p^{4}$	8.42	1.93	2
85	At	$1s^{2}2s^{2}2p^{6}3s^{2}3p^{6}4s^{2}3d^{10}4p^{6}5s^{2}4d^{10}5p^{6}6s^{2}4f^{14}5d^{10}6p^{5}$		2.82	2.2
86	Rn	$1s^{2}2s^{2}2p^{6}3s^{2}3p^{6}4s^{2}3d^{10}4p^{6}5s^{2}4d^{10}5p^{6}6s^{2}4f^{14}5d^{10}6p^{6}$	10.748	<0	0
87	Fr	$1s^{2}2s^{2}2p^{6}3s^{2}3p^{6}4s^{2}3d^{10}4p^{6}5s^{2}4d^{10}5p^{6}6s^{2}4f^{14}5d^{10}6p^{6}7s^{1}$	0		0.7
88	Ra	$1s^{2}2s^{2}2p^{6}3s^{2}3p^{6}4s^{2}3d^{10}4p^{6}5s^{2}4d^{10}5p^{6}6s^{2}4f^{14}5d^{10}6p^{6}7s^{2}$	5.279		0.89
89	Ac	$1s^{2}2s^{2}2p^{6}3s^{2}3p^{6}4s^{2}3d^{10}4p^{6}5s^{2}4d^{10}5p^{6}6s^{2}4f^{14}5d^{10}6p^{6}7s^{2}6d^{1}$	5.17		1.1
90	Th	$1s^{2}2s^{2}2p^{6}3s^{2}3p^{6}4s^{2}3d^{10}4p^{6}5s^{2}4d^{10}5p^{6}6s^{2}4f^{14}5d^{10}6p^{6}7s^{2}6d^{2}$	6.08		1.3
91	Pa	$1s^{2}2s^{2}2p^{6}3s^{2}3p^{6}4s^{2}3d^{10}4p^{6}5s^{2}4d^{10}5p^{6}6s^{2}4f^{14}5d^{10}6p^{6}7s^{2}5f^{2}6d^{1}$	5.88		1.5
92	U	$1s^{2}2s^{2}2p^{6}3s^{2}3p^{6}4s^{2}3d^{10}4p^{6}5s^{2}4d^{10}5p^{6}6s^{2}4f^{14}5d^{10}6p^{6}7s^{2}5f^{3}6d^{1}$	6.05		1.38
93	Np	$1s^{2}2s^{2}2p^{6}3s^{2}3p^{6}4s^{2}3d^{10}4p^{6}5s^{2}4d^{10}5p^{6}6s^{2}4f^{14}5d^{10}6p^{6}7s^{2}5f^{4}6d^{1}$	6.19		1.36
94	Pu	$1s^{2}2s^{2}2p^{6}3s^{2}3p^{6}4s^{2}3d^{10}4p^{6}5s^{2}4d^{10}5p^{6}6s^{2}4f^{14}5d^{10}6p^{6}7s^{2}5f^{6}$	6.06		1.28
95	Am	$1s^{2}2s^{2}2p^{6}3s^{2}3p^{6}4s^{2}3d^{10}4p^{6}5s^{2}4d^{10}5p^{6}6s^{2}4f^{14}5d^{10}6p^{6}7s^{2}5f^{7}$	6		1.3
96	Cm	$1s^{2}2s^{2}2p^{6}3s^{2}3p^{6}4s^{2}3d^{10}4p^{6}5s^{2}4d^{10}5p^{6}6s^{2}4f^{14}5d^{10}6p^{6}7s^{2}5f^{7}6d^{1}$	6.02		1.3
97	Bk	$1s^{2}2s^{2}2p^{6}3s^{2}3p^{6}4s^{2}3d^{10}4p^{6}5s^{2}4d^{10}5p^{6}6s^{2}4f^{14}5d^{10}6p^{6}7s^{2}5f^{9}$	6.23		1.3
98	Cf	$1s^{2}2s^{2}2p^{6}3s^{2}3p^{6}4s^{2}3d^{10}4p^{6}5s^{2}4d^{10}5p^{6}6s^{2}4f^{14}5d^{10}6p^{6}7s^{2}5f^{10}$	6.3		1.3
99	Es	$1s^{2}2s^{2}2p^{6}3s^{2}3p^{6}4s^{2}3d^{10}4p^{6}5s^{2}4d^{10}5p^{6}6s^{2}4f^{14}5d^{10}6p^{6}7s^{2}5f^{11}$	6.42		1.3
100	Fm	$1s^{2}2s^{2}2p^{6}3s^{2}3p^{6}4s^{2}3d^{10}4p^{6}5s^{2}4d^{10}5p^{6}6s^{2}4f^{14}5d^{10}6p^{6}7s^{2}5f^{12}$	6.5		1.3
101	Md	$1s^{2}2s^{2}2p^{6}3s^{2}3p^{6}4s^{2}3d^{10}4p^{6}5s^{2}4d^{10}5p^{6}6s^{2}4f^{14}5d^{10}6p^{6}7s^{2}5f^{13}$	6.58		1.3
102	No	$1s^{2}2s^{2}2p^{6}3s^{2}3p^{6}4s^{2}3d^{10}4p^{6}5s^{2}4d^{10}5p^{6}6s^{2}4f^{14}5d^{10}6p^{6}7s^{2}5f^{14}$	6.65		1.3
103	Lr	$1s^{2}2s^{2}2p^{6}3s^{2}3p^{6}4s^{2}3d^{10}4p^{6}5s^{2}4d^{10}5p^{6}6s^{2}4f^{14}5d^{10}6p^{6}7s^{2}5f^{14}6d^{1}$			

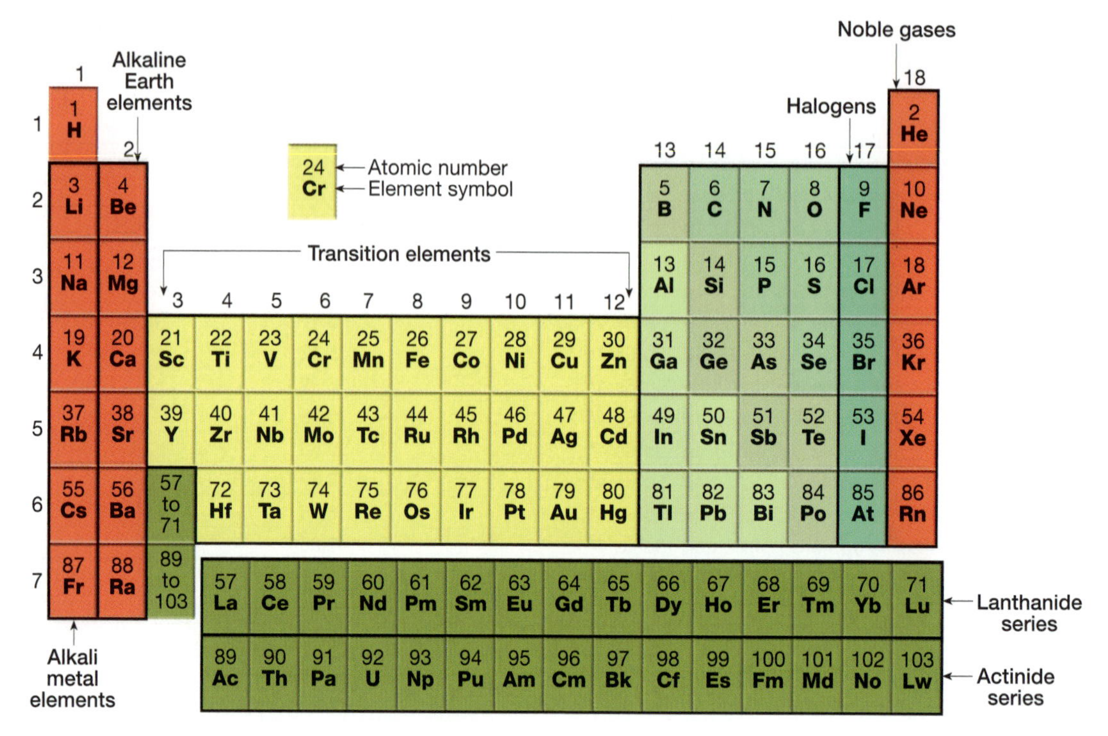

Figure 7.5. *The periodic chart of the elements.*

on the uppermost floor were occupied equally, so too an atom achieves its lowest energy state (i.e., is happiest) when one of these special configurations is achieved. For this reason, many atoms do not like to remain in a neutral state, but will gain or lose electrons in order to become more stable. A good geological example of this phenomenon is the titanium atom (Ti). Its atomic configuration is $1s^2 2s^2 2p^6 3s^2 3p^6 4s^2 3d^2$. On Earth, titanium is most often found in the valence state Ti^{4+}, which indicates that it has lost (given away) four electrons. This empties the 3*d* and 4*s* orbitals, greatly reducing the energy of the structure. The ability of atoms to give away or gain electrons from other atoms not only stabilizes their structures, but gives them an electrostatic charge, making it possible for ions to stick together.

Size

Changes in the number of electrons will also affect the size of the atom, which is also dependent on the number of protons present (i.e., the size of the nucleus). Recall from previous chapters that an atom that gives away electrons is called a **cation**, and has a positive charge. Conversely, an atom that gains electrons is a negatively-charged **anion**. As a general rule, atoms with large numbers of electrons have larger radii that those with smaller numbers of electrons. If an ion gives away electrons, then the difference in charge between the positively-charged nucleus and the electron cloud increases, and the electrons are drawn in close to the nucleus, making the whole ion smaller. This is certainly the case with the Ti^{4+} atom (Table 3.4). To get a better understanding of this relationship, use the data in Table 3.4 to plot the ionic radius of atoms in the first column (group) of the periodic chart when they are in, say, a +1 valence state. Similarly, check out the change in ionic radius as you move across a period (row) from Na^{1+}, to Mg^{2+}, Al^{3+}, Si^{4+}, P^{5+}, and S^{6+}.

In order to build mineral structures, we will need to know the relative sizes of the atoms when they are in their preferred valence states. The size of a cation or anion depends on many factors, including:

- the number of electrons and protons
- the number of surrounding ions (coordination number)
- charge imbalance between protons and electrons
- the degree of covalency in the bond (next chapter)
- the amount of polarization (described below)
- spin state
- pressure and temperature.

The data presented in Table 3.4 give a sense of how ionic radii increase in size as you move to higher atomic numbers (more protons) and change the charge on the ions. A few aspects merit further explanation. **Polarization** refers to how an atom responds to external electrical forces by reshaping or redistributing its charge. It may cause probability distributions of electrons to be skewed toward electrical poles. Polarization tends to shorten bonds on small cations with high charge (like Ti^{4+} and Si^{4+}), large anions with high charge, and non-inert gases (those which are lacking electrons from their normally-full outer orbitals); these are called **Fajan's Rules**. **Spin state** indicates whether the electrons are spinning in the same or opposite directions. Atoms with **high spin** have all the electrons in the outermost shell spinning in the same direction, while atoms with **low spin** have outer shell electrons spinning in opposite directions. Consult Table 3.4 to see which spin state results in a smaller ion.

It is also worth noting that the values given in Table 3.4 were difficult to determine uniquely. Remember from Chapter 3 how we calculated ionic radii? Return again to the halite structure pictured in Figure 1.8. In that illustration, we use balls to show the Na^{1+} and Cl^{1-} ions, but in reality, these are diffuse clouds of electrons that touch each other. Using an X-ray diffractometer, it is a comparatively simple matter to measure the distance between Na^{1+} and Cl^{1-} (see Chapter 15 for this example). However, is it non-trivial to assess the proportions of that distance that are due to Na^{1+} vs. Cl^{1-}. You might then measure the structure of sylvite, which contains K^{1+} and Cl^{1-}, and from that, estimate the difference in ionic radius between Na^{1+} and K^{1+}. You could keep on making structures containing Cl^{1-} bonded to other cations and eventually arrive at a value for all of the cations' radii bonded to Cl^{1-}. Ultimately, in order to get absolute rather than relative numbers for your ionic radii, you would have to make an assumption about the size of Cl^{1-}.

So historically there have been several different versions of ionic radii tables, largely differing in the assumed values for the radii of O^{2-} and F^{-}. The radii presented in Table 3.4 are those generally accepted for use in mineralogy.

Color in Minerals

Given the above description of atomic structure and size, we can now undertake to explore the most conspicuous relationship between chemistry and a physical property of all minerals: color! Why do minerals have color? When is that color diagnostic, and when is it likely to fool you? What does the color of a mineral tell you about its composition? The answers to these questions can be found in the fascinating world of mineral spectroscopy! The following section is derived from a lab exercise in the *Teaching Mineralogy* workbook (Brady et al., 1997).

What is color, exactly? The color perceived by the human eye can be thought of as the sum of the wavelengths that reflect or transmit off an object, with the dominant color being the dominant wavelength of light, or the sum of all the wavelengths reaching the eye. For the most part, color results from the interaction of light waves with electrons (Nassau, 1980) in a range of energies that can be perceived by the human eye. Most humans can only distinguish about 200 gradations of color (hues) in the small region of visible light from about 400–700 nm in wavelength. Your brain creates these colors by using three sets of cones in your eyes, kind of like the red, green and blue electron guns used by your old color television set:

black / white
red / green
yellow / blue.

These wavelengths of visible light make up only a small portion of the electromagnetic spectrum (Figure 7.6). Entire fields of spectroscopy (including subdisciplines of physics, astronomy, biology, and chemistry) are dedicated to the study of various regions of the electromagnetic spectrum. We'll have more to say about how mineralogist and geologists use various wavelengths of energy to analyze minerals in Chapter 9!

Studies of color can often be confusing because spectroscopists use a variety of units to explain their work, due to the fact they often work in differing fields of study. Unfortunately each field tends to use a unique vocabulary to describe what is being measured. The units fall into two categories: those expressed in terms of the energy of light (cm^{-1}, eV, and kJ, to name a few) and those in units of wavelength (μm, nm, or Å). Because the different disciplines tend to use different units, Table 7.6 gives some handy conversion factors for all these units. Most often in the study of color in minerals, the two types of units are wavenumbers (cm^{-1}) and wavelengths (usually expressed as nm). To convert from nm to cm^{-1}, take the inverse of nm and multiply by 10^7. Table 7.7 relates the energies of the visible through infrared regions to their colors.

There are at least fifteen causes of color; extended discussions of each of them are given in an excellent book by Nassau (1983). These are summarized in Table 7.8. Note that all but the first of these are created by interactions of electrons with electromagnetic radiation in the range of visible light. Fortunately, in order to under-

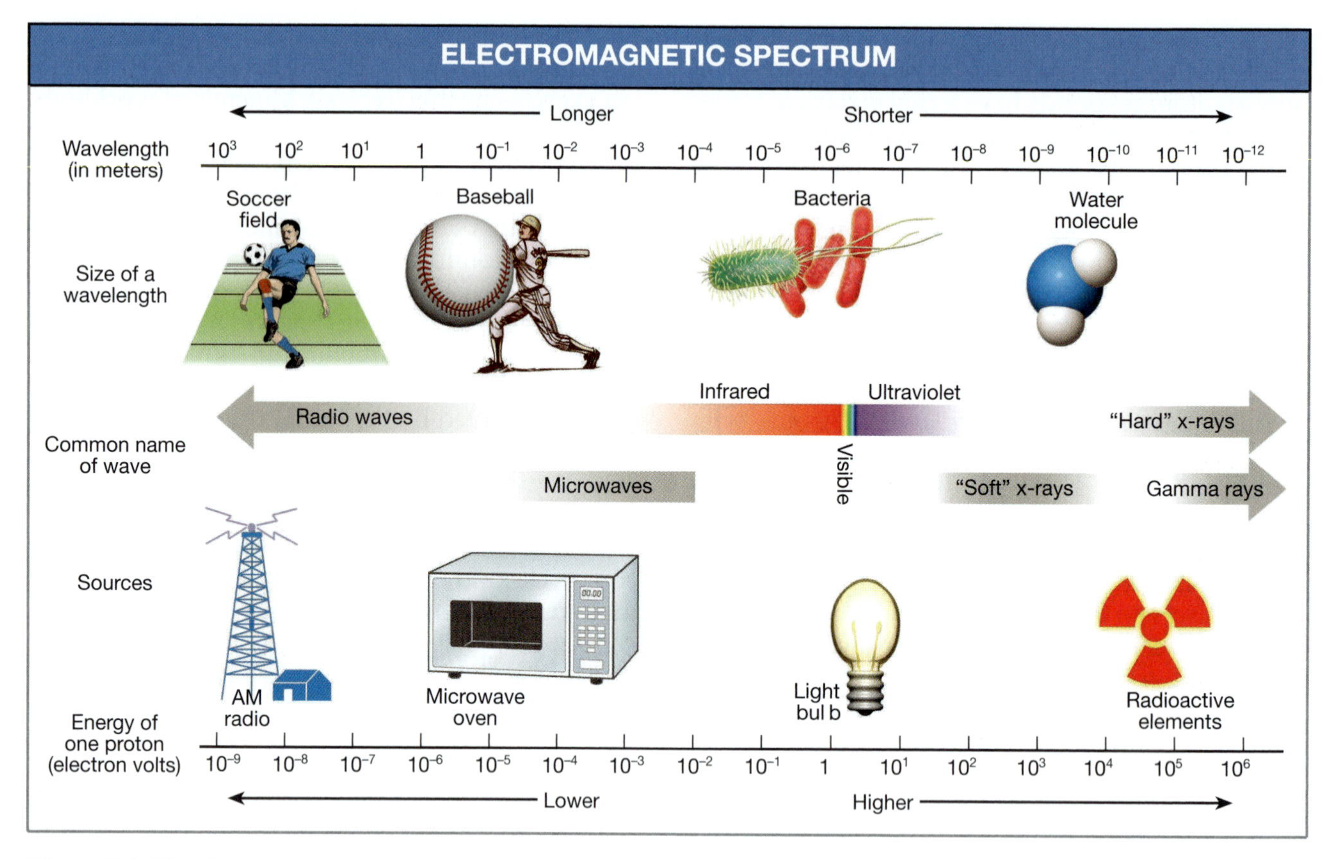

Figure 7.6. *The electromagnetic spectrum.*

stand color in minerals, we only need to examine a few of the items in Nassau's compilation; to learn more, check out his book!

Let's begin our study of colors by concentrating on the transition elements. A transition element is a metal ion with a partially-filled *d* or *f* shell. Generally this definition includes not only neutral ions with those characteristics, but also ions in their commonly-occurring valence states. The so-called first-series transition elements (Z = 21–30) are those with full shells of electrons below 3*p*, and those with incompletely-filled 3*d* orbitals. Another way of saying this is that they have a configuration of:

$1s^2 2s^2 2p^6 3s^2 3p^6 3d^{10-n} 4s^{1 \text{ or } 2}$;

where *n* is an integer from 0 to 10. How many electrons would an atom have in order for one or more to occupy a 3*d* orbital?

Recall from earlier in this chapter that each of the five 3*d* orbitals has a distinctive shape, but they fall into two main groups. Look again at Figure 3.3 and see if you can figure out why we classify them into two distinct groups on the basis of the relationship of their lobes to the *x*, *y*, and *z* axes: the t_2 or t_{2g} orbitals (d_{xy}, d_{yz}, and d_{xz}) and the *e* or e_g orbitals ($d_{x^2-y^2}$ and d_{z^2})?

These weird terms t_2, t_{2g}, *e*, and e_g come from group theory symmetry notation used by chemists to describe the shape of the orbitals. The *e* means that there are two orbitals per principal quantum number (so called *two-fold degeneracy*),

Table 7.6. Conversion table for energy units used in mineral spectroscopy (after Burns, 1993)

	10,000 cm^{-1}	*1 kJ*	*1 kcal*	*1 eV*	*1 mm*
cm^{-1}	10^4	83.59	349.5	8,066	10,000
kJ	119.66	1.0	4.1835	96.49	119.66
kcal	28.59	0.239	1.0	23.06	28.59
eV	1.24	0.01036	0.04336	1.0	1.24
mm	1.0	119.66	28.59	1.24	1.0

Table 7.7. Units Used for the Study of Color

Wavelength Units		*Energy Units*	
Nanometers (nm)	*Ångstroms (Å)*	*Wavenumbers (cm^{-1})*	*Resultant Color*
300	3,000	33,333	ultraviolet
400	4,000	25,000	violet
450	4,500	22,222	blue-violet
500	5,000	20,000	blue-green
550	5,500	18,182	green
600	6,000	16,667	orange
700	7,000	14,286	red
800	8,000	12,500	near-infrared
900	9,000	11,111	near-infrared
1000	10,000	10,000	near-infrared
1500	15,000	6,667	infrared
2000	20,000	5,000	infrared
2500	25,000	4,000	infrared

while the *t* indicates three orbitals (*three-fold degeneracy*). The subscript "2" indicates that the sign of the wave function doesn't change with rotation around the axes diagonal to the Cartesian axes, and "*g*" means that the wave function does not change sign if inverted through the center of the atom. For more information, consult a chemistry text!

If an ion is isolated (i.e., as when floating around unbonded in outer space), all these 3*d* orbitals are perfectly shaped. In this situation the 3*d* orbitals all have the same energy; in this case we say the orbitals are all **degenerate**. Thus, electrons filling the 3*d* orbitals of such an ion would have an equal probability of being located in any of the five orbitals. Graphically, we illustrate this by a drawing like this:

— — — — —

Each individual line symbolizes one of the five 3*d* orbitals, and all the orbital levels lie along the same horizontal line, meaning that they all have the same energy.

In an isolated ion, a 3*d* orbital with ten electrons (all would have the same energy) would be drawn schematically like this:

⇅ ⇅ ⇅ ⇅ ⇅

Each electron is represented by an arrow. The up or down direction of the arrow indicates whether the electron is orbiting the nucleus in a clockwise or counterclockwise direction.

But how would electrons be distributed among the 3*d* orbitals if there are fewer than ten of them? According to **Hund's first rule**, electrons don't pair up until each of the available orbitals contains an electron. So for example, an element with five 3*d* electrons would have one electron in each orbital.

What they don't stress enough in chemistry classes is the fact that these simple, equal energy orbital configurations only work for isolated ions. What happens when a transition metal is incor-

Table 7.8. Examples of the fifteen causes of color (adapted from Nassau, 1987)

Number	*Phenomenon*	*Examples*
1	incandescence (release of thermal energy)	flames, lamps, carbon arc, limelight; and the white color of the Sun, which results from 5700 °C emissions
2	gas excitations (excitations of specific atoms)	vapor lamps, lightning, auroras, gas lasers
3	vibrations and rotations (rock and roll of atoms)	blue ice, iodine, blue gas flames
4	crystal field transitions (where the chromophore* is a major element)	pink rhodochrosite, olivine, almandine, and spessartine
5	crystal field transitions (where the chromophore* is a minor element)	ruby, emerald, chrysoberyl, rubellite, morganite
6	molecular orbitals in organics (electrons belong to several atoms)	dyes, including hair dyes and indigo blue (which is extracted from sea shells), bioluminescence of fireflies
7	intervalence charge transfer (electrons shared by adjacent atoms)	blue sapphire, magnetite, kyanite, vivianite, aquamarine
8	energy bands in metals	copper, silver, gold, brass, iron metal
9	energy bands in semiconductors	silicon, galena, zinc, cadmium, and vermillion paint pigments
10	doped semiconductors	blue (B-doped) and yellow (N-doped) diamond
11	color centers	amethyst, smoky quartz, blue and yellow topaz, zircon, citrine
12	dispersive refraction	rainbows, halos, stars in gemstones
13	scattering	blue sky, red sunsets, blue moon, blue eyes, butterflies
14	interference	oil slicks, soap bubbles, coatings on camera lenses
15	diffraction	opal, moonstone, diffraction gratings, most liquid crystals

*a chromophore is an element that causes color

porated into a mineral structure? You already know that most mineral structures contain cations in 4-, 6-, 8-, or 12-coordinated polyhedra, with oxygen as nearest neighbors. The anions surrounding the transition metals in a mineral structure do not form a perfectly spherical, even distribution of charge around the transition metal. Instead, the charge is unevenly distributed, with the negative charge being concentrated in the vicinity of the anions at the polyhedral corners. The effect of this non-spherical charge distribution is to destroy the degeneracy, so that the energies of all the orbitals are no longer equal. Put simply, the electrons in those orbitals are repelled by the negative charge of the neighboring oxygen ions. Electrons in the orbitals that are close to the oxygen neighbors are repelled more strongly than those that are farther away from the oxygens. Thus, the energies of the formerly equivalent 3*d* orbitals change—they effectively split. Some will have higher energies (those close to the bonding oxygens) and some will have lower energies (those whose lobes point to areas that are oxygen free). The total energy stays the same, however.

Let's examine how this actually happens by considering an octahedral coordination polyhedron. The term *octahedral coordination* is somewhat confusing. It refers to the shape of a polyhedron surrounding the cation; an octahedron has eight sides, thus the name. But the internal cation only has a coordination number of six, because there are six oxygens at the corners of the octahedron (Figure 7.7). The *x*, *y*, and *z* axes in this figure correspond to those in Figure 7.4. The $d_{x^2-y^2}$ and d_{z^2} (e_g) orbitals of the central transition element, which are oriented *along* the *x*, *y*, and *z* axes, point directly toward the neighboring oxygen anions. The repulsion raises their energy compared to the three t_{2g} orbitals, which lie *between* the axes. The sum of the energies of the five orbitals is the same as for the isolated degenerate case, so the energy of the other three orbitals must be lowered to compensate for the increase in the e_g orbitals.

The opposite situation is found in *tetrahedral coordination* (Figure 7.7), because the oxygens on the four corners of the tetrahedral are now closest to the t_2 orbitals. These relationships are shown for five different coordinations in Figure 7.8. The term *tetrahedral coordination* again refers to the shape of the polyhedron surrounding the central cation. In this case, there are both four faces on the polyhedron and four oxygens at its corners.

The difference in energy between the lowest orbitals and the highest orbitals is called **crystal field splitting** and is represented by the symbol Δ (or sometimes, 10 *Dq*). Crystal field splitting energies are dependent on many factors, including :

1. the symmetry and coordination number of the coordination polyhedra,
2. the valence state of the cation,
3. the strength of its bond with the surrounding anions,
4. the distance between the cation and the surrounding anions,
5. pressure, and
6. temperature.

Of these, 1 and 2 are extremely important and very useful (although 3–6 have their utilities for specific problems—see Burns, 1993, for more information). Many techniques for mineral analysis can tell you *what* is in your mineral, but not *where* the cations are located (i.e., which kinds of coordination polyhedra they are in) nor *which* valence states they have (see Chapter 9). For example, an electron microprobe analysis of a mineral might tell you that your sample contains the transition metal iron (Fe), but it cannot tell you which sites the Fe atoms occupy, nor how much of the Fe is Fe^{2+} or Fe^{3+}. Crystal field theory can help us use absorption spectra to provide this much-needed information. It also gives us an alternative to Pauling's rules (see that section later in this chapter) for explaining cation coordination.

The energies of the Δ values vary according to the coordination number, as shown in Figure 7.8. The amount of separation, or Δ, between energy levels can be expressed mathematically as:

$$\Delta_o : \Delta_c : \Delta_d : \Delta_t = 1 : -8/9 : -1/2 : -4/9.$$

These ratios correspond to the magnitude of splitting between t_{2g} or t_g orbitals and e_g or e orbitals. The minus sign implies that the relative stabilities of the two orbitals are reversed for octahedral vs. the other types of coordination polyhedra. In other words, in octahedral coordination, the e_g orbitals are highest in energy, while in tetrahedral, **cubic** (represented above by Δ_c, where the metal atom is at the center of a cube with eight oxygens at its corners), or **dodecahedral** coordination (represented above by Δ_d, where the metal atom is at the center of a dodecahedron with 12 oxygens at its corners), the t_{2g} or t_g orbitals are higher in energy. This relationship tells us that the splitting for a given transition metal will be largest when it is in octahedral coordination, and smallest when it is in tetrahedral coordination. The Δ values also represent the amount of energy that will be needed to move an electron from the low energy orbitals to the high energy ones. Thus, the Δ values obtained from real absorption spectra of minerals can be

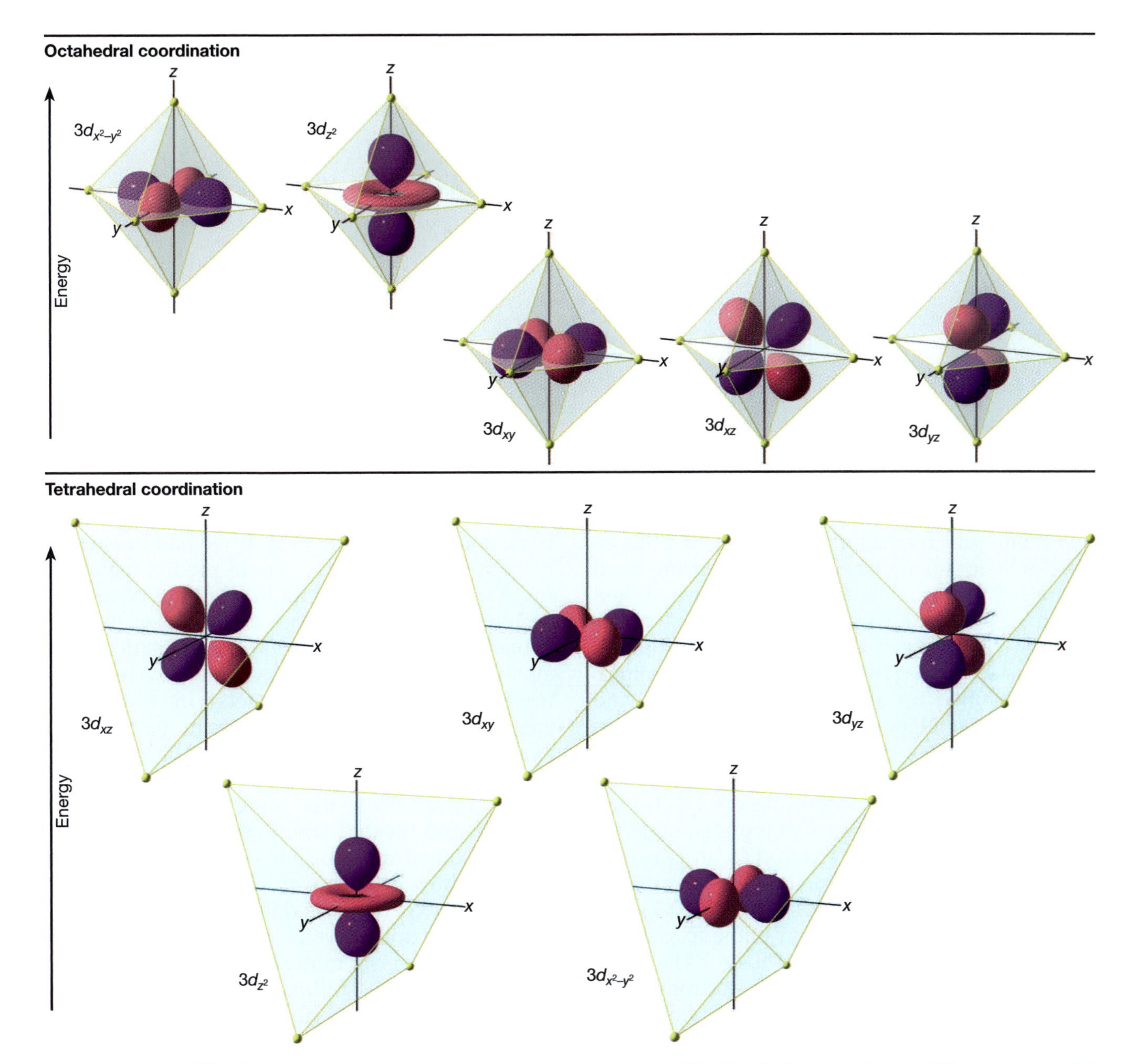

Figure 7.7. Simplified boundary surface representations of electron probability distributions for the five different types of 3d *orbitals in octahedral and tetrahedral coordination. When the atoms are placed in coordination polyhedra with neighboring anions (such as oxygen), these electron distributions become somewhat distorted. Electrons that occupy lobes of the distributions close to neighboring anions are repulsed by them, increasing their energy. The energies of electrons in lobes that are not proximal to anions must decrease to maintain electronic stability.*

used to determine which coordination environment a transition metal is occupying.

How does this concept actually work in practice? The concept of **crystal field stabilization energy** (CFSE) is derived by weighting the contributions of electrons in various orbitals. It describes the total change in energy between the perfectly symmetrical state and the coordinated state. In Figure 7.9, A represents the energies of the transition metal $3d$ orbitals in a free cation (outer space). B shows the energy of the orbitals once the cation is placed in a site surrounded by anions; the orbital energies decrease due to repulsions between anions and electrons *other than* those in the $3d$ orbitals. C shows the change in orbital energy due to the repulsion between anions and the $3d$ electrons in a case where the anions are distributed in a sphere, and D shows the splitting of the $3d$ orbital energy levels in an octahedral crystal field (Burns, 1993).

It's actually quite easy to calculate CFSE, which then allows you to predict which transition metals

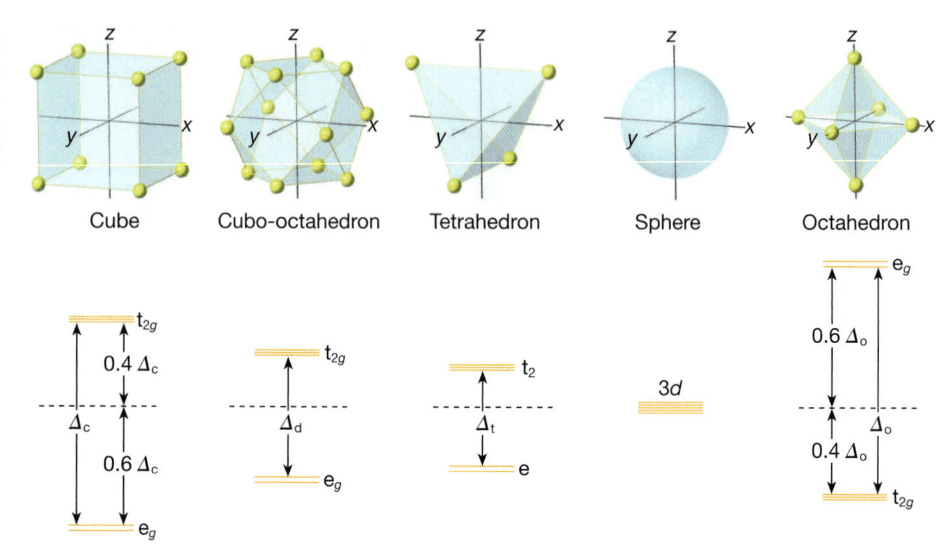

Figure 7.8. Crystal field splitting of 3d orbitals in cubic (Δ_c), dodecahedral (Δ_d), tetrahedral (Δ_t), spherical, and octahedral (Δ_o) coordination polyhedra. Adapted from Burns (1985).

will prefer which kinds of sites. Think of CFSE as the algebraic sum of the energies of electrons in all orbitals. Orbitals with energies that are shifted down (i.e., have lower energy levels—electrons like low energy!) as a result of crystal field splitting are considered negative. Orbitals with energies that are shifted up are considered positive, and likewise have higher energies. For example, in octahedral coordination, each electron in a t_{2g} orbital has a lower energy than it would have in an isolated polyhedron. Therefore, every t_{2g} electron stabilizes the structure by an amount equal to 0.4 Δ_o. It follows that every electron in an e_g orbital would destabilize the structure by an amount equal to 0.6 Δ_o. The multiples of 0.6 and 0.4 come from simple algebra. The energy of the six possible electrons in the t_{2g} orbitals ($6 \times 0.4 = 2.4$) is compensated by the energy of the four electrons in the e_g orbitals ($4 \times 0.6 = 2.4$). The total of stabilizing and destabilizing effects should be zero if all the orbitals are full. High values for CFSE indicate that a cation will be energetically preferred in a coordination polyhedron with that coordination type.

To better understand this calculation, take a close look at Table 7.9, which shows the values of Δ for octahedral coordination. Notice that there are two sets of data: one for high spin (electrons don't pair up—the double rooms of the Atomic Apartment only have one student in them) and one for low spin (electrons pair up before going to higher energy orbitals—two students pair up in a room). As an interesting exercise, try constructing this same table for the case of *tetrahedral* coordination.

How do these theoretical CFSE values compare to actual experimental data? To answer that question, we'd need to know: (1) the CFSE value in a liquid, or some kind of site with no nearest neighbors, and (2) the value of Δ. If we had those two pieces of information, we could predict the value for Δ in a transition metal in any kind of coordination polyhedra to test our theory. Does it work? Table 7.10 shows data for various transition metals in aqueous solution; these values were measured by dissolving each metal in water and acquiring its absorption spectrum to find the peak. The right-hand column in Table 7.10 then shows the experimental results of absorption spectra measured on that same metal, this time in a compound (hexahydrate) where the cation will be in octahedral coordination. You can be the judge of how well this crystal field theory predicts these experimental data!

But how does crystal field theory help us understand color in minerals? When the energies of 3*d* orbitals are split, it becomes possible for electrons to move back and forth between orbitals

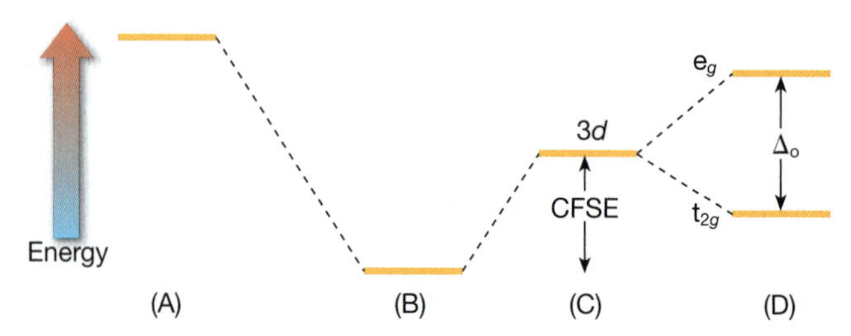

Figure 7.9. (A) represents the energies of the transition metal 3d *orbitals in a free cation (outer space). (B) shows the energy of the orbitals once the cation is placed in a site surrounded by anions; the orbital energies decrease due to repulsions between anions and electrons* other than *those in the* 3d *orbitals. (C) shows the change in orbital energy due to the repulsion between anions and the* 3d *electrons in a case where the anions are distributed in a sphere, and (D) shows the splitting of the* 3d *orbital energy levels in an octahedral crystal field. After Burns (1993).*

Table 7.9. High spin electron configurations for 3d Orbitals in Octahedral Coordination

Cation	Number of electrons in 3d	High Spin Electronic configuration					Unpaired electrons	CFSE
		t2g			eg			
Ca^{2+}, Sc^{3+}, Ti^{4+}	0						0	0
Ti^{3+}	1	↑					1	0.4 Δ_o
Ti^{2+}, V^{3+}	2	↑	↑				2	0.8 Δ_o
V^{2+}, Cr^{3+}, Mn^{4+}	3	↑	↑	↑			3	1.2 Δ_o
Cr^{2+}, Mn^{3+}	4	↑	↑	↑	↑		4	0.6 Δ_o
Mn^{2+}, Fe^{3+}	5	↑	↑	↑	↑	↑	5	0
Fe^{2+}, Co^{3+}, Ni^{4+}	6	↑ ↓	↑	↑	↑	↑	4	0.4 Δ_o
Co^{2+}, Ni^{3+}	7	↑ ↓	↑ ↓	↑	↑	↑	3	0.8 Δ_o
Ni^{2+}	8	↑ ↓	↑ ↓	↑ ↓	↑	↑	2	1.2 Δ_o
Cu^{2+}	9	↑ ↓	↑ ↓	↑ ↓	↑ ↓	↑	1	0.6 Δ_o
Cu^{1+}, Zn^{2+}, Ga^{3+}, Ge^{4+}	10	↑ ↓	↑ ↓	↑ ↓	↑ ↓	↑ ↓	0	0
Cation	**Number of electrons in 3d**	**Low Spin Electronic configuration**					**Unpaired electrons**	**CFSE**
		t2g			**eg**			
Ca^{2+}, Sc^{3+}, Ti^{4+}	0						0	0
Ti^{3+}	1	↑					1	0.4 Δ_o
Ti^{2+}, V^{3+}	2	↑	↑				2	0.8 Δ_o
V^{2+}, Cr^{3+}, Mn^{4+}	3	↑	↑	↑			3	1.2 Δ_o
Cr^{2+}, Mn^{3+}	4	↑ ↓	↑	↑			4	1.6 Δ_o
Mn^{2+}, Fe^{3+}	5	↑ ↓	↑ ↓	↑			5	2.0 Δ_o
Fe^{2+}, Co^{3+}, Ni^{4+}	6	↑ ↓	↑ ↓	↑ ↓			4	2.4 Δ_o
Co^{2+}, Ni^{3+}	7	↑ ↓	↑ ↓	↑ ↓	↑		3	0.8 Δ_o
Ni^{2+}	8	↑ ↓	↑ ↓	↑ ↓	↑ ↓		2	1.2 Δ_o
Cu^{2+}	9	↑ ↓	↑ ↓	↑ ↓	↑ ↓	↑	1	0.6 Δ_o
Cu^{1+}, Zn^{2+}, Ga^{3+}, Ge^{4+}	10	↑ ↓	↑ ↓	↑ ↓	↑ ↓	↑ ↓	0	0

when energy is added. Such energy is usually in the range of the energies associated with visible light. These transitions, of course, must be associated with energy gain or loss to the atom. Thus an electron could jump from a lower energy level to a higher one by absorbing light with an energy equal to Δ. In other words, only a very specific wavelength of light would be absorbed by any particular transition element in a particular site in a mineral. Transition metals are particularly important in this regard, because small energy differences between the split orbitals in certain sites often correspond to those in the visible and near-infrared region of the electromagnetic spectrum. You can see this by comparing the energies of visible light in Table 7.7 to the Δ values (in cm^{-1}) in Table 7.10.

In a laboratory, we can pass light through a mineral one wavelength at a time to look for absorption peaks (more on this topic will follow in Chapter 9). Peaks in the resultant spectrum will represent wavelengths that are absorbed by the sample, and troughs correspond to the transmitted wavelengths.

For example, Figure 7.10 shows the spectrum of an unusual blue elbaite (tourmaline group) from São José da Batalha, Paraíba, Brazil (Rossman et al., 1991). In this plot, wavelength is shown on the x axis, and the y axis shows the amount of absorbance that occurs at each wavelength. The magnitude of the absorbance is a function of the thickness of the crystal, which in this case is 0.20 mm. There are two lines corresponding to spectra taken at two different angles to the crystal: one in which light is vibrating parallel to the c axis, and one with light vibrating perpendicular to the c axis. The two spectra are different because tourmaline is a uniaxial mineral; this means that the densities of the two atoms in the two orientations can be different. The maximum absorption in the visible region occurs at two peaks with energies of 10,870 cm^{-1} and 14,290 cm^{-1} (about 920 and 700 nm). If you compare this spectrum to the colors in Table 7.7, you'll see that absorption occurs in the red to infrared region of the visible light spectrum. Sunlight passing through this elbaite has its red wavelengths absorbed because they are the right energy to cause electrons to jump between split energy levels of the Cu^{2+} atoms. Transmitted light that passes through the elbaite crystal is in the range from 400–600 nm, corresponding to blue light—and

Table 7.10. Δ_o and CFSE for Octahedral Coordination (after Burns, 1993)

Cation	Electronic Structure	Electronic Configuration	Δ aqueous	Δ_o	Calculated CFSE (cm^{-1}) hexahydrate
Ti^{3+}	$3d^1$	$(t_{2g})^3$	18,950	0.4 Δ_o	7,580
V^{3+}	$3d^2$	$(t_{2g})^2$	19,100	0.8 Δ_o	15,280
V^{2+}	$3d^3$	$(t_{2g})^3$	12,600	1.2 Δ_o	15,120
Cr^{3+}	$3d^3$	$(t_{2g})^3$	17,400	1.2 Δ_o	20,880
Cr^{2+}	$3d^4$	hs $(t_{2g})^3(e_g)^1$	13,900	0.6 Δ_o	8,340
Mn^{3+}	$3d^4$	hs $(t_{2g})^3(e_g)^1$	21,000	0.6 Δ_o	12,600
Mn^{2+}	$3d^5$	hs $(t_{2g})^3(e_g)^2$	7,800	0	0
Fe^{3+}	$3d^5$	hs $(t_{2g})^3(e_g)^2$	13,700	0	0
Fe^{2+}	$3d^6$	hs $(t_{2g})^{34}(e_g)^2$	9,400	0.4 Δ_o	3,760
Co^{3+}	$3d^6$	ls $(t_{2g})^6(e_g)^3$	18,600	2.4 Δ_o	44,640
Co^{2+}	$3d^7$	hs $(t_{2g})^5(e_g)^2$	9,300	0.8 Δ_o	7,440
Ni^{2+}	$3d^8$	$(t_{2g})^6(e_g)^2$	8,500	1.2 Δ_o	10,200
Cu^{2+}	$3d^9$	$(t_{2g})^6(e_g)^3$	13,000	0.6 Δ_o	7,800

thus the sample appears blue. Cu is also the coloring agent in the two Cu-carbonate minerals azurite (which is blue) and malachite (which is green). Can you speculate why the same element would cause two different colors in minerals with essentially the same chemistry? (A hint: it has to do with the structure and not the chemistry). You might check out the structures of these two minerals in the mineral database to help you explore the difference as well as Figure 23.23.

For completeness, it is worth mentioning two other causes of color, numbers 7 and 11 in Table 7.8, because they also occur frequently in miner-

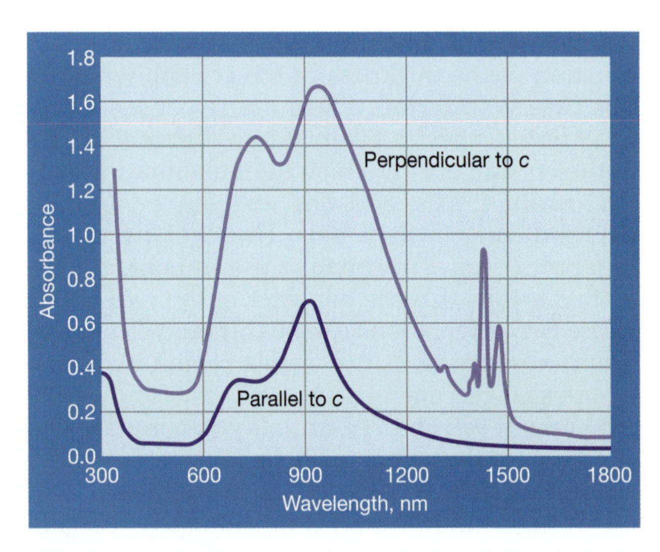

Figure 7.10. *Polarized absorption spectrum of the Brazilian elbaite from São José da Batalha, Paraíba, 0.20 mm thick. Strong absorptions in the red-infrared regions allow light to be transmitted in the blue regions. This tourmaline is blue due to* Cu^{2+}. *Data from the Mineral Spectroscopy web site at Caltech.*

als. Color produced by charge transfer processes is similar to crystal field splitting. Instead of electrons jumping *between orbitals* within the same atom, electrons jump *between atoms*. These intervalence transfers of charge can create very intense colors from very small numbers of shared electrons. In sapphire, for example, Fe^{2+} and Ti^{4+} ions in adjacent sites (ones that would normally be occupied by Al) pass an electron back and forth. At one instant, the charge is distributed as Fe^{2+} and Ti^{4+}, and the next instant it is Fe^{3+} and Ti^{3+}. Very small amounts of Fe^{2+} and Ti^{4+} in an otherwise pure corundum crystal (Al_2O_3), even down to parts per million (ppm) or *parts per billion* (ppb), can still make a sapphire crystal appear blue. In the spectrum shown in Figure 7.11, small peaks at 25,680 cm^{-1} and 22,220 cm^{-1} (389 and 450 nm) represent Fe^{3+}, and the broad bands spanning the range 17,800–14,200 cm^{-1} (562–704 nm) represent intervalence charge transfer peaks.

Color can also be derived from something called a **color center** in a crystal structure. These occur when materials with otherwise perfect structures trap electrons or atoms in metastable sites. The traps can be atomic vacancies (like the electrons filling F vacancies that cause fluorite to be purple), substitutions of trace amounts of color-causing atoms for non-transition metals (as when Fe^{3+} substitutes for Al^{3+} or Si^{4+} in plagioclase feldspar), or just locations in the crystal lattice where a minor charge deficiency provides a place for an electron to rest (as in diamond).

In many cases, heat or another form of energy such as radiation can provide enough energy for an electron to "escape" from its trap, resulting in a color change. Heat and radiation treatments are frequently used in the gemstone industry to

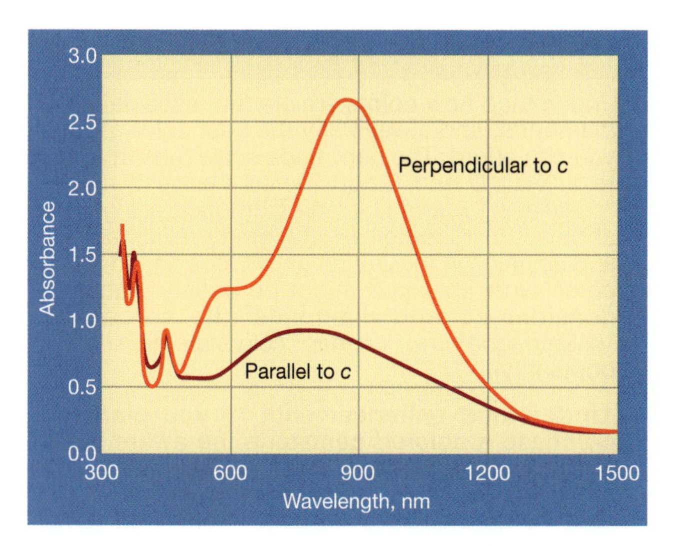

Figure 7.11. Polarized absorption spectra of sapphire from an unknown locality (probably in Australia), normalized to 3.00 mm thick. Sapphire is a variety of the mineral species corundum. As with Figure 7.10, strong absorptions in the red-infrared regions allow light to be transmitted in the blue regions. All the sharp features are from Fe^{3+} and the broad features which dominate the spectrum are from intervalence charge transfer between Fe^{2+} and Ti^{4+} and between Fe^{2+} and Fe^{3+}. Data from the Mineral Spectroscopy web site at Caltech.

change the color of gemstones. For example, heat treatment can turn ugly brown zircons into gemmy blue ones. The apatite shown in Figure 7.12 has probably been heat-treated to change it from green to blue.

To learn more about color in minerals, read the book by Roger Burns (1993), or visit George Rossman's Mineral Spectroscopy web site at the California Institute of Technology. The spectra shown in this chapter come from the latter extensive collection. Most of the spectra printed there can be downloaded and printed out as images or saved as ASCII text files. Check out your favorite mineral!

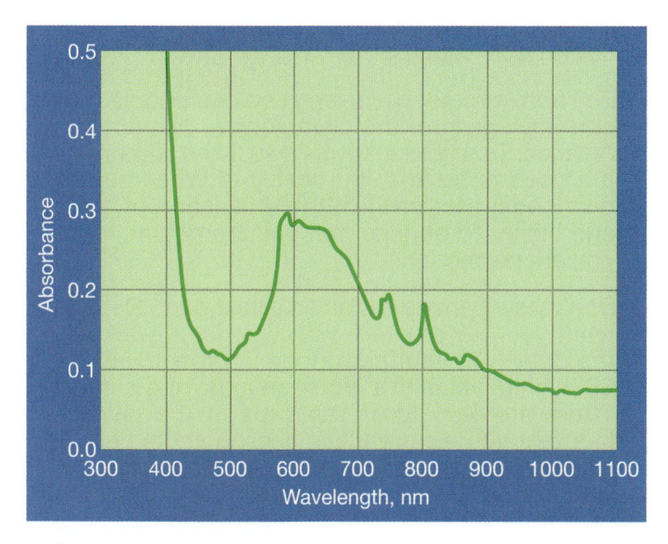

Figure 7.12. Unpolarized absorption spectrum of sky-blue apatite from a neodymium-rich carbonatite rock in Kenya. This sample was most likely heat-treated to turn it from green to blue. 2.53 mm thick. Data from the Mineral Spectroscopy web site at Caltech.

Chemical Substitution

In the sections above, we have considered the constraints imposed by size and coordination on substitution and incorporation of elements into minerals. As seen in the preceding discussion of color in minerals, there are many instances where one element substitutes for another in a mineral structure. These circumstances are nicely explained by **Goldschmidt's Rules of Substitution**, first suggested in 1937.

Rule #1. The ions of one element can extensively replace those of another in ionic crystals if their radii differ by less than about 15% and they have the same charge. Of course, the substitution depends on the amount of ionic bonding, the concentrations of the elements involved, and the temperature (and pressure). Crystals forming at high temperatures are more tolerant of substitutions than those forming at low temperatures, because the structures are somewhat expanded. Elements that occupy the same site in a crystal structure are said to be **diadochic**. If one element is present in very low concentrations (a so-called **trace element**) and the other is a major element, then the trace element is said to be **camouflaged** in the structure. In this context, camouflage means just the same reason you wear camouflage gear when you go hunting: the trace element (you) is hidden in the structure (the woods). A common example of camouflage in minerals is the substitution of Rb^{1+} (1.47 Å) for K^{1+} (1.33 Å) in the feldspar structure. This is useful because some Rb is radioactive and decays to Sr; that reaction can be used to determine the age of the feldspar!

Rule #2. Ions whose charges differ by one unit substitute readily for each other provided electrical neutrality is maintained. Ions with charges that differ by greater than one generally do not substitute. This idea brings us to the notion of **coupled substitutions**, which are pairs of elements substituting simultaneously in order to provide charge balance. For example, consider the pyroxene mineral species aegirine, which has a formula of $Na^{1+}Fe^{3+}Si^{4+}{}_2O_6$. Substitution of Ca^{2+} for the Na^{1+} makes sense on the basis of size, but leaves the structure with excess charge. So the

A FEW WORDS ABOUT BUYING A GEMSTONE

Several years ago, I was lucky enough to attend a lecture by Fred Feldmesser, who is one of the foremost experts of gemstones in the Boston area. He showed me a natural blue diamond that cost more than my entire net worth at the time, but I've never forgotten it. He also passed along some tips about buying gemstones, which I found highly interesting and useful. Since then, I also had the opportunity to work with George Rossman (California Institute of Technology), the man who probably knows more than anyone in the world about color in gemstones (and the science behind how gemstone colors can be artificially "enhanced"). From them, I gleaned the following advice, which may prove useful to you...

1. Match the gemstone with the appropriate use for that stone, e.g. as a ring, earrings, etc. Pay attention to hardness and fracture when choosing gemstones because usage determines whether a "softer" or "harder" stone is required. For example, when I got engaged we wanted my ring to include a garnet (Peter has worked extensively on garnet-bearing rocks), a diamond, and a tourmaline (one of my favorites). But our gemologist refused to let us consider those stones for something that would have to endure the rough treatment of everyday wear, as a ring on my finger would.
2. "Cut" doesn't mean what you think. Most people have the erroneous idea than gemstone facets are cut onto the stones...this is not the case. Facets are polished or ground on to the stones using polishing wheels; the arm that holds the stones can be adjusted to various angles for polishing. Therefore "cut" refers to the shape of the stone. The "cut" shape is often dictated by the natural growth forms. Diamonds are often polished to be round, whereas tourmaline is often an 'emerald' or long cut. This is because diamonds form in cubes and tourmaline grows in long prisms. These cuts allow the maximum amount of the stone to be used.
3. Beware the swindled stone. Gemstones are priced by their weight; for example, <1.0 ct., 1–1.999 ct., 2–2.999 ct., etc., and the price goes up in those steps as the size increases. So a high quality diamond might sell for $1000/ct. if it's less than 1 ct., $3000/ct. if it's in the size range between 1 and 2, and $9000/ct if it's 2–3 ct. The price increase is usually non-linear for diamonds because larger stones are more rare. This means that the person who is faceting the gemstone has a lot of financial incentive to end up with a finished product that's just over 1 or 2 or 3 cts., even if that cut will make an ugly ("swindled") stone! It's estimated that over 80% of commercially-available stones may be swindled (Batcha, 1991). So people "in the know" shop for stones with weights like 0.9 or 1.9 cts.—those are likely to have been cut for appearance, not for weight. Don't let your ego get in the way when you buy a gemstone!
4. Understand how colors are graded, especially in diamonds, and always buy the best color grade you can afford. The gem-trade scale runs from D to Z, with D and E considered the best, most "colorless." F and G grade diamonds are good choices for those of us who aren't millionaires, and anything in the J–Z range will look tinged with color. Note that you might purposely want to consider colored diamonds (the so-called "champagne" and "cognac" colors are gaining in popularity).
5. Understand enhancements. If you plan to purchase a colored gemstone, be aware that probably >95% of the commercially-available stones have been "enhanced" in some way. Treatments include dyes, irradiation, heating in various different kinds of gases, soaking in oil, and even laser surgery! The majority of these effects are long-lasting. The color-enhanced stones are certainly very attractive, and they are far more affordable than naturally-colored stones of the same species. When buying treated stones, ask your jeweler what treatments were used, and how stable they are. However, if you want 'color created by nature', you may want to stick with peridots (olivine), garnets, and tourmalines, or make sure that you buy a naturally colored stone from a reputable dealer.
6. Consider unusual stones. Trends in the jewelry industry dictate prices on gemstones, as with any other commodity (supply and demand). Because you're a geologist, you may want to consider some of the less popular but gorgeous gemstones that aren't as trendy. For example, at this writing it is still possible to buy beautiful iolite (the gem variety of the mineral species cordierite) at low cost. Diopside (a mineral species in the pyroxene group) is also available. Note that these stones are relatively soft and therefore best in necklaces and earrings. Ask your jeweler for ideas–she may have some good suggestions.
7. Buy from a reputable dealer. Look for people with certification from the Gemological Institute of America. Trust your instincts here. If the price looks too good to be true, it probably is! When buying diamonds, ask for the GIA certification of color and clarity to be sure that you are getting what you are paying for.
8. Don't view gemstones as an investment. Despite what your jeweler tells you, when you walk out the door with your new stone, it probably loses 75% of its value (it's worse than buying a car!). There are very large markups in the jewelry industry, so unless you are dealing at the below-wholesale level, you're unlikely to be making any true investment. For this reason, don't go into debt to buy a gem. Buy gemstones because you like them!

substitution of Ca^{2+} for Na^{1+} is usually accompanied by another substitution, such as Fe^{2+} for Fe^{3+}, making the new mineral hedenbergite: $Ca^{2+}Fe^{2+}Si^{4+}{}_2O_6$.

Rule #3. When two different ions can occupy a particular position in a crystal lattice, the ion with the higher ionic potential forms a stronger bond with anions. The charge/radius ratio is called the **ionic potential** of an atom. There are two related terms to this rule. Element **capture** occurs when a minor element preferentially enters a structure due to a *high* ionic potential. Substitution of an element with *low* ionic potential (either low charge or low radius, or both) is called **admission**. So in the biotite structure, which has the formula $K(Mg,Fe^{2+})_3(AlSi_3O_{10})(OH,F)_2$, substitution of Ba^{2+} (1.55 Å) for K^{1+} (1.63 Å) would be considered capture, while substitution of Li^{1+} (0.82 Å) for Mg^{2+} (0.80 Å) would be admission.

Rule #4 (actually written by Ringwood in 1955 to explain problems with rules #1–3): Even when size and charge are similar, the ion with the lower electronegativity will form stronger bonds, and thus be preferentially concentrated when growing crystals from a melt. This idea unites Pauling's Rules of Bonding (Chapter 8) and Goldschmidt's, so we can explain quite effectively how elements behave in crystallizing melts.

We can define a **crystal-liquid distribution coefficient**, *D* as

$$D = C_c / C_1$$

where C_c is the concentration of the element in a crystal, and C_1 is the concentration of the same element in the coexisting liquid. This formulation describes substitutions quantitatively:

- When $D > 1$, you have element capture, so the element prefers to be in the crystal.
- When $D < 1$, you have element admission, so the element prefers to be in the liquid.
- When $D = 1$, you have camouflage, so the element will go either in the liquid or in the crystal.

Distribution coefficients are useful to describe the partitioning of elements between crystals of a mineral forming at equilibrium from a silicate melt. Note that *D*'s are (sometimes highly) temperature dependent, so we can sometimes use them as geothermometers to tell us the temperature of formation or equilibration of minerals.

There are, however, some notable situations where Goldschmidt's and Ringwood's rules do not accurately predict distribution coefficients, especially for the transition metals. For example, species from the mineral group olivine, $(Mg,Fe^{2+})_2SiO_4$, are often the first phases to crystallize from cooling silicate-rich liquid. Goldschmidt's rules would predict that Mg^{2+} would be the preferred cation to occupy the octahedral sites in olivine.

For Ni^{2+} (radius = 0.72Å) 2/0.72 = 2.78

For Mg^{2+} (radius = 0.65Å) 2/0.65 = 3.08

However, it was noted as early as the 1960s that Ni^{2+} was often preferentially incorporated into the first-crystallizing olivines. Based on Goldschmidt's rules, *this should not occur*. So the question of why Ni enrichment was found in magmatic olivines was hotly debated until crystal field theory came along.

If you have experimental data for the crystal field energy of a cation in both octahedral and tetrahedral coordination ($CFSE_o$ and $CFSE_t$), then the difference between those values is a parameter called the **Octahedral Site Preference Energy** (OSPE). OSPE means just what it says: it is a measure of how much that cation prefers being in octahedral coordination relative to tetrahedral coordination. You can do the same comparison for any two types of coordination polyhedra, but the tetrahedral vs. octahedral comparison is most useful because silicates contain mostly octahedral or tetrahedral sites. This is a very powerful concept, and it explains some behaviors that even Goldschmidt did not understand.

Imagine that you have a silicate magma that has arisen from the mantle. Although it is a liquid, the magma is held loosely together by short-range bonds between cations and anions, forming coordination polyhedra of all types, but especially 4-, 5-, and 6-coordinated sites. For most silicate liquids, the first phase to crystallize will incorporate transition metals into an octahedral site, such as those found in olivine or pyroxene. We can describe this scenario by the reaction (Burns, 1993):

$${}^4M^+ + {}^5M^+ + {}^6M \rightarrow {}^6M_{crystal},$$

where M is a transition metal. So the magnitude of the OSPE describes a measure of the relative preferences of cations in a magma to leave the magma behind and occupy octahedral sites in silicate minerals. To see how this works, examine the data in Table 7.11. These results allow us to determine the relative orders of uptake of cations into a crystallizing silicate mineral with an octahedral site (Burns, 1993):

$Ni^{2+} > Cr^{2+} > Cu^{2+} > Co^{2+} > Fe^{2+} > Mn^{2+}$,

Ca^{2+}, and Zn^{2+} and

$Cr^{3+} > Mn^{3+} > Co^{3+} > V^{3+} > Ti^{3+} > Fe^{3+}$, Sc^{3+}, Ga^{3+}.

Now we can understand the Ni^{2+} partitioning behavior: Ni^{2+} has a high preference for occupying octahedral sites. So the minute that octahedral sites in the crystallizing olivines become available, Ni^{2+}

Table 7.11. Δ_o and CFSE for Octahedral Coordination (after Burns, 1993)

Cation	Number of 3d electrons	Octahedral CFSE (kJ/mole)	Tetrahedral CFSE (kJ/mole)	Octahedral Site Preference Energy, OSPE (kJ/mole)
Ca^{2+}, Sc^{3+}, Ti^{4+}	0	0	0	0
Ti^{3+}	1	–87.4	–58.6	–28.8
V^{3+}	2	–160.2	–106.7	–53.5
Cr^{3+}	3	–224.7	–66.9	–157.8
Cr^{2+}	4	–100.4	–29.3	–71.1
Mn^{3+}	4	–135.6	–40.2	–95.4
Mn^{2+}, Fe^{3+}	5	0	0	0
Fe^{2+}	6	–49.8	–33.1	–16.7
Co^{3+}	6	–188.3	–108.8	–79.5
Co^{2+}	7	–92.9	–61.9	–31.0
Ni^{2+}	8	–122.2	–36.0	–86.2
Cu^{2+}	9	–90.4	–26.8	–63.7
Zn^{2+}, Ga^{3+}, Ge^{4+}	10	0	0	0

Data from McClure (1957) and Dunitz and Orgel (1957)

will partition out of the melt and into the olivine. This is just one of countless examples of why we must understand the crystal structures of materials before we can predict how they will behave.

Phase Diagrams

As we have been discussing ionic radii and chemical substitutions, one recurring theme is the idea that pressure and temperature make a difference in how elements are incorporated into mineral structures. It is also desirable to be able to predict which minerals with what compositions will crystallize out of a liquid. These issues are critical to geologists, for whom crystallization (and recrystallization) temperatures and pressures are often of paramount importance in understanding the geological history of an area. Many minerals only form over a limited range of conditions, so when we find them in rocks, we know something about how and where the rocks might have formed.

Phase diagrams give us a graphical method for representing the relationships between temperature, pressure, and composition. They provide the key to interpretation of the genesis of ore deposits, igneous, and metamorphic rocks, and they can also be used to predict which mineral species would be stable in the human lung! Although we will revisit these topics later in Chapter 20, we present here a brief overview of how phase diagrams can be useful in understanding mineral chemistry and structure.

Phase diagrams are used to summarize the conditions at which a material of a given composition is stable as a solid, liquid, or gas. Let's look first at a P-T diagram that describes the relationship between pressure and temperature for a familiar mineral: ice! In Figure 7.13, you will see that the y axis is pressure, the x axis is temperature, and the material is always the same composition: H_2O. The regions of the diagram each represent one of water's phases: ice, liquid, or vapor. The lines between the different regions represent combinations of pressure and temperature where changes occur:

- the line between the solid and liquid phases represents melting vs. freezing,
- the line between solid and gas represents sublimation vs. deposition, and
- the line between liquid and gas represents vaporization vs. condensation.

The graph also has some special points. Notice that there is a point where all three lines come together: this **triple point** is a special pressure and temperature at which water, ice, and vapor can all coexist. The point at which the liquid-gas lines ends is called the **critical point** (374°C and 217 bars); as the temperature increases beyond it, water and vapor become indistinguishable.

This phase diagram may seem irrelevant to mineralogy, but as a planetary mineralogist, I can tell you that I think about this phase diagram a lot! Consider the case of Mars, where the search for evidence of life perseveres with ongoing missions. Our bias is that if we can find evidence for *liquid* water on the martian surface (either now or in the geologic past), then that will be the place where life forms might have evolved. Currently, surface pressures on Mars average about 5.5 millibars—just below the triple point than 6.1 millibars. There

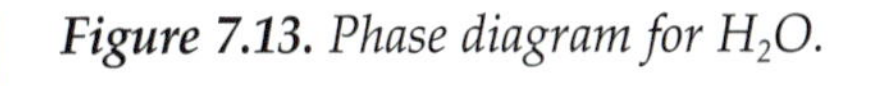

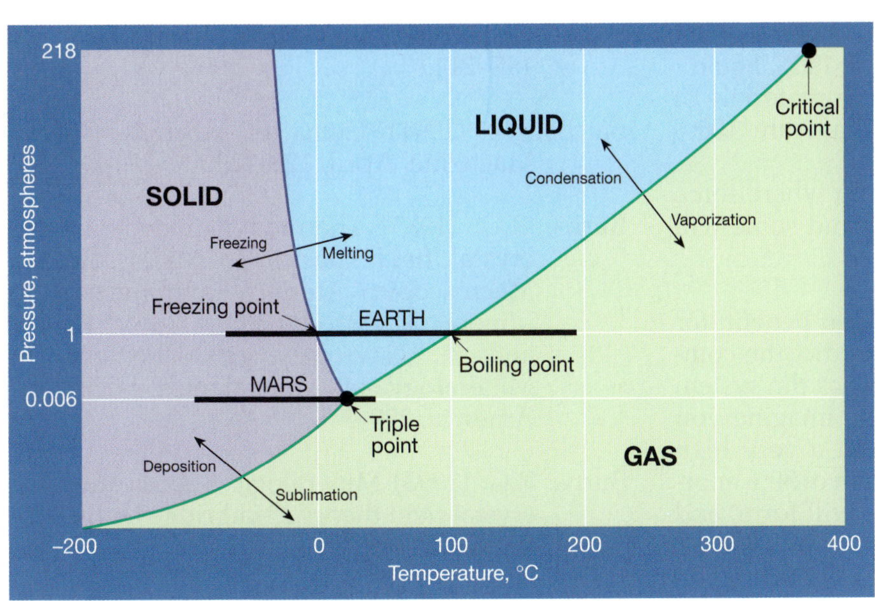

Figure 7.13. *Phase diagram for H_2O.*

may be some low-lying areas where the pressure gets as "high" as 12.4 millibars, like the Hellas basin. The problem is that the boiling temperature there is only about +10°C. If the surface temperature got any higher, the water would boil away. At the landing sites of the Viking missions to Mars, daytime temperatures ranged from –17 to (rarely) 27°C by day, and down to –107°C at night (and that was in the summer!). In most places on Mars, if you were to pour a beaker of water on the surface, it would turn into vapor before it even had a chance to freeze. Take a look at the phase diagram, and ask yourself the question: what would it take to make liquid water stable on Mars today? This is an issue addressed by scientists interested in making conditions on Mars suitable for human habitation…the process of terraforming. Check out the book *The Case for Mars* by Robert Zubrin to learn more about this fascinating topic!

The previous example showed a phase diagram where pressure and temperature varied while composition was held constant. Because the pressure at the surface of the Earth is a fairly constant one bar, geologists often hold the pressure variable constant, and use phase diagrams to express relationships between composition and temperature. A good example is the phase diagram for the plagioclase sub-group of the feldspar group (Figure 7.14). This type of diagram has three regions divided by two lines.

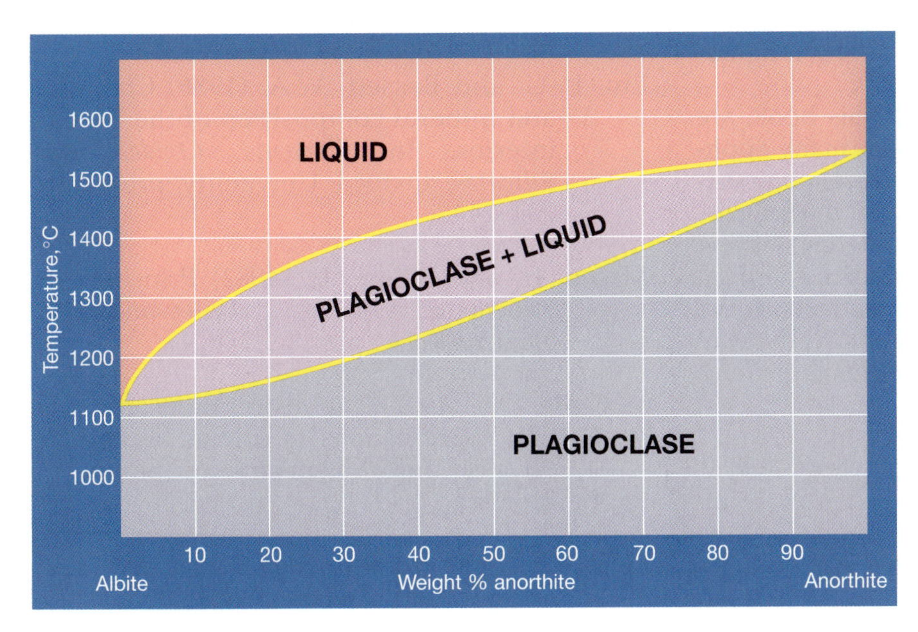

Figure 7.14. *Phase diagram for the plagioclase sub-group of the feldspar group.*

- The liquid region shows where materials of all compositions in the diagram are 100% liquid.
- The solid region shows where materials of all compositions in the diagram are 100% solid.
- The region in between tells you where some of the material will be liquid and some will be solid.

This type of diagram has an added benefit: for any given composition, you can predict the compositions of the liquids and crystals as the system changes temperature. For example, imagine you have a magma that is cooling from a very high temperature, and it has a composition of 50% anorthite. At ~1460°C, the first crystals will form, and their composition can be predicted by drawing a horizontal line through the crystals+liquid region on the plot. At one end of this line is the liquid composition (which is still 50% anorthite because only a volumetrically insignificant number of feldspar crystals has grown so far). At the other end of the line, you can read the composition of the feldspar crystals, which is around 90% anorthite. As you continue to cool the magma, more crystals will grow, and the composition of both crystals and liquid change continuously. Eventually the last little pocket of melt, with a composition of only 10% anorthite, will cool, and those crystals will be 50% anorthite. Below 1220°C the whole system will be a solid that is 50% anorthite.

Another planetary example will explain why this type of phase diagram might be useful. On the moon, most feldspars are about 90% anorthite. The *only* way you can get feldspars with such calcium-rich compositions is to start out depleted in Na and to cool from a *very hot* magma. So if you want to understand the origin of the moon, you have to come up with a mechanism that produces enough heat to form magmas with temperatures well above 1400°C.

As we work our way through this book, we will revisit phase diagrams and their many applications as we look at different mineral compositions. They are one more piece of the puzzle, along with ionic charge, size, and distribution coefficients, that allows us to predict chemical behavior of minerals in geologic contexts. If we are lucky, we can understand and predict the way minerals will form under different geologic (and industrial) conditions.

References

Anders, E. and Grevasse, N. (1989) Abundances of the elements: meteoritic and solar. Geochimica et Cosmochimica Acta, 53, 197–214.

Batcha, B. (1991) Gems of wisdom. Boston Magazine, April, 126–129.

Burns, R.G. (1985) Thermodynamic data from crystal field spectra. In Microscopic to Macroscopic: Atomic Environments to Mineral Thermodynamics. (S.W. Kieffer and A. Navrotsky, eds.) Reviews in Mineralogy, 14, Mineralogical Society of America, 277–316.

Burns, R.G. (1993) Mineralogical applications of crystal field theory. Cambridge University Press, 551 pp.

Dunitz, J.D., and Orgel, L.E. (1957) Electronic properties of transition element oxides. II. Cation distribution amongst octahedral and tetrahedral sites. Journal of Physics and Chemistry of Solids, 3, 318–333.

Goldschmidt, V.M. (1937) The principles of distribution of chemical elements in minerals and rocks. Journal of the Chemical Society of London, 1937, 655–673.

Gray, H.B. (1994) *Chemical bonds; An introduction to atomic and molecular structure.* University Science Books, Mill Valley, California, 232 pp.

Hartmann, W.K. and Davis, D.R. (1975) Satellite-sized planetesimals and lunar origin. Icarus, 24, 504–515.

Liu, L.-G. and Bassett, W.A. (1986) Chemical and mineral composition of the Earth's interior. In *Elements, Oxides, and Silicates*, Oxford University press, pp. 234–244.

Mason, B. and Moore, C. (1982) Principles of Geochemistry, 4th ed., J. Wiley and Sons, New York, 344 pp.

McClure, D.S. (1957) The distribution of transition metal cations in spinels. Journal of Physics and Chemistry of Solids, 3, 31–317.

McSween, H.Y. Jr., Murchie, S.L., Crisp, J.A., Bridges, N.T., Anderson, R.C., Bell, J.F. III,

Britt, D.T., Brueckner, J., Dreibus, G., Economou, T., Ghosh, A., Golombek, M.P., Greenwood, J.P., Johnson, J.R., Moore, H.J., Morris, R.V., Parker, T.J., Rieder, R., Singer, R.B., and Wänke, H. (1999) Chemical, multispectral, and textural constraints on the composition and origin of rocks at the Mars Pathfinder landing site. Journal of Geophysical Research, E, Planets, 104, 8679–8715.

Nassau, K. (1980) The causes of color. Scientific American, October, 124–154.

Nassau, K. (1983) The physics and chemistry of color: The fifteen causes of color. Wiley, New York, 454 pp.

Origin of the Earth and Moon (1998) LPI Contribution No. 957, Lunar and Planetary Institute, Houston, 555 pp.

Ringwood, A.E. (1955) The principles governing trace element distribution during magmatic crystallization. Geochimica et Cosmochimica Acta, 7, 189–202.

Rossman, G.R., Fritsch, E., and Shigley, J.E. (1991) Origin of color in cuprian elbaite from Sao Jos, de Batalha, Paraíba, Brazil. American Mineralogist, 76, 1479–1484.

Shu, F.H. (1982) Physical Universe: An Introduction to Astronomy, University Science Books, 584 pp.

Silk, J. (2001) The Big Bang, 3rd ed.. W.H. Freeman, NY, 512 pp.

Smith, M.R., Laul, J.C., Ma, M.S., Huston, T., Verkouteren, R.M., Lipschutz, M.E., and Schmitt, R.A. (1984) Petrogenesis of the SNC (Shergottites, Nakhlites, Chassignites) meteorites. Implications for their origin from a large dynamic planet, possibly Mars. Journal of Geophysical Research, 89, B612–630.

Surkov, Y.A., Barsukov, V.L., Moskalyeva, L.P., and Kharyukova, V.P. (1984) New data on the composition, structure, and properties of Venus rock obtained by Venera 13 and Venera 14. Journal of Geophysical Research, 89, B393–402.

Surkov, Y.A., Moskalyeva, L.P., Kharyukova, V.P., Dudin, A.D., Smirnov, S.Y., and Zaitseva, S.Y. (1986) Venus rock composition and the Vega 2 landing site. Journal of Geophysical Research, 91, E215–218.

Taylor, S.R. (1982) Planetary Science: A Lunar Perspective. Lunar and Planetary Institute, Houston, 481 pp.

Chapter 8

Bonding and Packing in Minerals

This chapter and the preceding one seem to be focusing on "rules," which in some ways is a bit unfortunate. To me, anything listed as a "rule" might seem like it could be broken; this may come from my high school/motorcycle phase, when I thought it was much cooler to ride around on a motorcycle than go to school! Of course, the price to pay for that was having to attend my senior year in high school twice! I reversed this trend as an undergraduate in college and only missed 3–4 classes in four years. However, I still distrust the use of the word "rule."

In reality the word "rule" as applied to both Goldschmidt and Pauling really refers to some conclusions they drew by observing the natural world. In these cases, they were looking at the structure of minerals (i.e., they looked at the data and they interpreted what they saw to arrive at some conclusions). In science and other fields, data are often constant but our interpretations of the data change with time. For instance, when people first noticed that mountain ranges existed as somewhat linear features on the Earth, they interpreted this as proof that when the Earth cooled, it got smaller (i.e., its radius decreases). Then as it shrank, ridges had to form—the mountain chains! Don't laugh too hard—this was being taught in physical classes as recently as the mid-1960s. At that time, continental drift was mentioned but it had not yet received widespread acceptance as one component of plate tectonics.

As a student, I thought what was presented as interpretation was always the truth. This was especially true in chemistry classes, which are sometimes a prerequisite for a mineralogy course. Then when I hit mineralogy I learned, for example, that we really didn't know the ionic radius of an atom or that there really is no such thing as a pure ionic bond. This, of course, frustrated me because I started questioning much of what I had been "taught."

A few decades later as a professor I realize that, from my view, there are two basic issues that we face when teaching and learning about the natural physical world: (1) the data more or less stay the same with time, but our interpretations change; and (2) we as scientists need to assign words to things that occur in nature. As our interpretations change, they become more refined, and we collect more data. Often times we find yesterday's "rules" get broken and we may be tempted to criticize those who came before and created these "incorrect" rules. But it's only from them that we learn. If those "rules" had never existed in the first place, we would never have been able to refine them—interpretations are also evolving and we must evolve with them, or go extinct. Likewise, we as humans often lack a precise language to describe the natural world with words; however, we can often define it in terms of mathematics. We usually learn the inadequate words first, and become frustrated when we find the true meaning cannot really be described with words. This process describes the way we learn, and is the basis for the spiral learning methods we are using in this book. So in this chapter we'll build upon some of the great observations by those who have gone before us, and refine some of our vocabulary to better understand (we hope) how atoms are held together in minerals.

M.E.G.

Introduction

The goal of the preceding chapter was to give you a sense of the structure, size, and composition of the chemical elements. If we were building a house, we might say that this information constituted a study of the building materials. Now that we know what kind of building materials (e.g., boards and bricks) we are going to use to make minerals, and what sizes they come in (e.g., 2×4's and 1×4's), we can turn to examining the forces (nails and mortar) that hold atoms together in their symmetric configurations. To help understand these, we will rely heavily on a set of "rules" that are really just observations of the ways atoms bond to each to form polyhedra, and, in turn, how these polyhedra join together to form crystal structures. They were developed by Linus Pauling, one of the few people to win two Nobel prizes: one in chemistry and the other in peace. Pauling studied fairly simple, ionically bonded crystalline materials with high symmetry, and noticed some unifying themes in their structures. From these observations, he proposed a set of five "rules" as a general way in which we might better understand the arrangement of atoms in crystalline materials. But before we study Pauling's Rules, we need to revisit, at a higher level, the way atoms are bonded together.

The Forces that Bind

Bonding is the mechanism that is responsible for many of the important physical properties of minerals, such as hardness, electrical conductivity, solubility, crystallization temperature, refractive index, color, etc.

Broadly speaking, there are two types of forces that hold things together: gravitational forces and electrostatic ones. Newton's universal law of gravity explains the attractive force that is experienced by two bodies as a function of the masses of the bodies and their of separation, and is written as:

$$F = G\frac{m_1 m_2}{r^2} \qquad \text{Equation 8.1}$$

where F = attractive force of the between the two bodies, m_1 and m_2 = the masses of the two bodies, r = the distance between the two bodies, and G = the universal gravitation constant, which is 6.67×10^{-11} N·m²/kg².

Although Equation 8.1 applies to objects of all size (our solar system, for example), the mass of a typical atomic particle is small, about 10^{-27} kg² (Table 7.1). You can use any distance you want, but atomic bonds are on the order of 1 to 3Å. Armed with these numbers, you can calculate the force of gravity that holds two particles together. Try this calculation using an Fe atom as an example (Table 7.1) with a separation of distance of 2Å.

On the other hand, electrostatic attractions are governed by Coulomb's law, which is strikingly similar in construction to Equation 8.1:

$$F = \frac{1}{4\pi\varepsilon_0}\frac{q_1 q_2}{r^2} \qquad \text{Equation 8.2}$$

where F = attractive force of the between the two bodies, q_1 and q_2 = the charges on the two bodies, r = the distance between the two bodies, and ε_0 = the vacuum permittivity constant, which is 9.0×10^9 N· m²/C².

A typical unit of charge (q_1 or q_2) for a particle is 1.6×10^{-19} coulombs (Table 7.1). Now you can calculate the value of the electrostatic force, F, for two Fe atoms 2Å apart. How big is the difference in magnitudes of the gravitational and electrostatic forces? Based on this comparison, it should be apparent why we consider electrostatic attractions to be the main attractive force that holds minerals together.

All types of bonds thus depend fundamentally on the strength of electrostatic attractions between particles. However, historically chemists gave names to certain types of bonds, and for better or worse, those names have stuck with us. As we introduced in Chapter 3, bonds in minerals are broadly called ionic, covalent or some mixture of the two. These broad categories also embrace other types of bonding that have been given special names: metallic, van der Waals, and hydrogen bonding, but in reality these are just different expressions used to say "electrostatic bonds." Here we revisit the types of bonding from a more fundamental perspective than in Chapter 3.

The electrostatic attractions between atoms can be understood (and, in fact, *predicted*) on the basis of their **electronegativity**, which is the ability of an atom to attract electrons toward it in a bond. One way of defining electronegativity is in terms of two atomic properties that we can measure: ionization energy and electron affinity; both are usually expressed as electron volts (Table 8.1).

Ionization energy (I) is the energy of the reaction by which one electron is removed to infinity from the outer electron shell of the atom (i.e., the energy to make an neutrally charged atom a cation) and written as:

$$I = E\,(X \rightarrow X^{+} + e^{-}) \qquad \text{Equation 8.3}$$

where I = ionization energy, and $E\,(X \rightarrow X^{+} + e^{-})$ = energy required to remove an electron an, e^{-}, atom, X, to infinity. In this reaction, the atom X is

Table 8.1. Bonding Characteristics for Chemical Elements									
Z	Atom	Ionization Energy (eV)	Electron Affinity (eV)	Electronegativity	Z	Atom	Ionization Energy (eV)	Electron Affinity (eV)	Electronegativity
1	H	13.598	0.754	2.1	53	I	10.451	3.05914	2.66
2	He	24.587	<0	0	54	Xe	12.13	<0	2.6
3	Li	5.392	0.61805	0.98	55	Cs	3.894	0.471630	0.79
4	Be	9.322	<0	1.57	56	Ba	5.212	0.144626	0.89
5	B	8.298	0.2771	2.04	57	La	5.58	0.53	1.1
6	C	11.26	1.26293	2.55	58	Ce	5.47		1.12
7	N	14.534	−.072	3.04	59	Pr	5.42		1.13
8	O	13.618	1.4611215	3.44	60	Nd	5.49		1.14
9	F	17.422	3.3993	3.98	61	Pm	5.55		1.13
10	Ne	21.564	<0	0	62	Sm	5.63		1.17
11	Na	5.139	0.5479303	0.93	63	Eu	5.67		1.2
12	Mg	7.646	<0	1.31	64	Gd	6.15		1.2
13	Al	5.986	0.4411	1.61	65	Tb	5.86		1.1
14	Si	8.151	1.3855	1.9	66	Dy	5.93		1.22
15	P	10.486	0.74653	2.19	67	Ho	6.02		1.23
16	S	10.36	2.07712	2.58	68	Er	6.101		1.24
17	Cl	12.967	3.6173	3.16	69	Tm	6.184		1.25
18	Ar	15.759	<0	0	70	Yb	6.254		1.1
19	K	4.341	0.501471	0.82	71	Lu	5.43		1.27
20	Ca	6.113	0.024551	1	72	Hf	6.65	0	1.3
21	Sc	6.54	0.1882	1.36	73	Ta	7.89	0.3221	1.5
22	Ti	6.82	0.0791	1.54	74	W	7.98	0.8158	2.36
23	V	6.74	0.5251	1.63	75	Re	7.88	0.152	1.9
24	Cr	6.766	0.6661	1.66	76	Os	8.7	1.12	2.2
25	Mn	7.435	<0	1.55	77	Ir	9.1	1.5658	2.2
26	Fe	7.87	0.1634	1.83	78	Pt	9	2.1282	2.28
27	Co	7.86	0.6611	1.88	79	Au	9.225	2.308633	2.54
28	Ni	7.635	1.1561	1.91	80	Hg	10.437	<0	2
29	Cu	7.726	1.2281	1.9	81	Tl	6.108	0.22	2.04
30	Zn	9.394	<0	1.65	82	Pb	7.416	0.3648	2.33
31	Ga	5.999	0.302	1.81	83	Bi	7.289	0.9461	2.02
32	Ge	7.899	1.22	2.01	84	Po	8.42	1.93	2
33	As	9.81	0.813	2.18	85	At		2.82	2.2
34	Se	9.752	2.020693	2.55	86	Rn	10.748	<0	0
35	Br	11.814	3.3653	2.96	87	Fr	0		0.7
36	Kr	13.999	<0	0	88	Ra	5.279		0.89
37	Rb	4.177	0.485922	0.82	89	Ac	5.17		1.1
38	Sr	5.695	0.052066	0.95	90	Th	6.08		1.3
39	Y	6.38	0.3071	1.22	91	Pa	5.88		1.5
40	Zr	6.84	0.4261	1.33	92	U	6.05		1.38
41	Nb	6.88	0.8933	1.6	93	Np	6.19		1.36
42	Mo	7.099	0.7461	2.16	94	Pu	6.06		1.28
43	Tc	7.28	0.552	1.9	95	Am	6		1.3
44	Ru	7.37	1.052	2.2	96	Cm	6.02		1.3
45	Rh	7.46	1.1378	2.28	97	Bk	6.23		1.3
46	Pd	8.34	0.5578	2.2	98	Cf	6.3		1.3
47	Ag	7.576	1.3027	1.93	99	Es	6.42		1.3
48	Cd	8.993	<0	1.69	100	Fm	6.5		1.3
49	In	5.786	.32	1.78	101	Md	6.58		1.3
50	Sn	7.344	1.22	1.96	102	No	6.65		1.3
51	Sb	8.641	1.075	2.05	103	Lr			
52	Te	9.009	1.97083	2.1					

converted into a positively-charged ion, X^+, which is called a cation. The value of I shows a strong tendency to increased from left to right in any horizontal row (period) of the periodic table. Compare, for example the ionization energies of Na, Mg, Al, and Si in Table 8.1.

Electron affinity (E_A) is defined by the energy change involved in the reaction that makes a neutrally charged atom a an anion, and is written as:

$$E_A = E\,(X + e^- \rightarrow X^-) \qquad \text{Equation 8.4}$$

where E_A = electron affinity, and $E\,(X + e^- \rightarrow X^-)$ = energy required to attract an electron to an atom, X, from infinity. Affinity (which means an attraction) is a good term to use for this process, which basically reflects how well an atom can attract other electrons. Values of E_A also tend to increase from left to right on the periodic table; compare O^{2-} and F^-. Small numbers indicate that a less stable negative ion is formed.

We can combine Equations 8.3 and 8.4 to express the balancing act between cations and anions. We define electronegativity as:

$$X = \frac{I + E_A}{a} \qquad \text{Equation 8.5}$$

where X = electronegativity, and I = ionization energy (given in Equation 8.3), E_A = electron affinity (given in Equation 8.4), and a = conversion factor. Electronegativity is sometimes referred to as "electron withdrawing power."

The definition of electronegativity uses the Pauling scale (developed in 1932); two years later Mulliken proposed using a scale that involves the *average* of $E_A + I$, but it is less commonly used. For both scales, the higher the electronegativity (X) of an atom, the more it will pull its electron density in toward the nucleus. So if you have two adjacent atoms with different values of X, the one with the higher X will attract the electron density away from the atom with the smaller X. This redistribution of electron density results in a negative charge on the atom with the higher X (i.e., it becomes an anion), and a positive charge on the atom with the smaller X (i.e., it becomes a cation).

Pauling electronegativity as defined above uses an empirical scale in which a value of 4.0 is assigned to fluorine, which is the most electronegative element. Francium is the least electronegative element (aside from the noble gases), and is assigned a value of $X = 0.7$. As with ionization energy and electron affinity, electronegativity increases from left to right on the periodic table and decreases from top to bottom. Thus the highest electronegativities are toward the upper right of the periodic table. Metals are the least electronegative of the elements.

Electronegativity is a very useful concept because it allows us to predict the type of bonding that will occur between two different atoms. If two elements have very different electronegativities, then electron transfer takes place and ionic bonding occurs by electrostatic attraction. On the other hand, if two elements have very similar electronegativities, then covalent bonding occurs by *sharing* rather than by transfer of electrons. There is really a continuum between ionic and covalent bond types that could be calculated by substituting the values of X for bonded elements in Equation 8.5.

We can apply Pauling's scale of electronegativity to some elements we know will be important in mineral structures. Using the values in Table 8.1, we find that:

The Si^{4+}-O^{2-} bond has about 50% ionic character.
The Mg^{2+}-O^{2-} bond has about 70% ionic character.
The Fe^{2+}-O^{2-} bond has about 60% ionic character.
The Fe^{2+}-S^{2-} bond has about 15% ionic character.

In fact, these estimates are all probably somewhat low, as suggested by certain physical properties, but this method of understanding bond types does at least give a good ballpark estimate of what type of bonding to expect in any given situation. The historical names applied to bonding can now be defined as:

a. Bonding between atoms with extreme differ ences in electronegativity will be **ionic**.
b. Bonding between atoms with high, but equal electronegativities will be **covalent**.
c. Bonding between atoms with intermediate difference between electronegativities will be intermediate between ionic and covalent bonds.
d. Bonding between atoms with low, but equal electronegativities will be **metallic**.

A Dog-Gone Good Analogy for Chemical Bonding

Before we get to a formal definition of the bonding types, let's consider an excellent analogy posted on the Ithaca City School District ChemZone web site (Science Joy Wagon, Lodi, NY). It uses the natural attraction between dogs and bones as an analogy for the attraction between atoms and their electrons.

Metallic bonds. (Figure 8.1) These bonds are best imagined as a room full of puppies who have plenty of bones to go around and are not possessive of any one particular bone. The bones and the puppies circulate freely. Such an arrangement allows electrons to move through a substance with little restriction. The model is often described as kernels of "atoms floating in a sea of electrons."

Covalent bonds. (Figure 8.1) Covalent bonds can be thought of as two or more dogs with equal attraction to the bones. Since the dogs (atoms) are identical, then the dogs share the pairs of available bones evenly. Since one dog does not have more of either bone than the other dog, the charge is evenly distributed among both dogs.

Ionic bonds. (Figure 8.1) Ionic bonding can be best imagined as one big greedy dog stealing another dog's bone. If the bone represents the electron that is up for grabs, then when the big dog gains an electron, he becomes negatively-charged and the little dog who lost the electron becomes positively-charged. The two ions (that's where the name *ionic* comes from) are attracted very strongly to each other as a result of the opposite charges. The two dogs will stay close together because the little dog is so attracted to his lost bone.

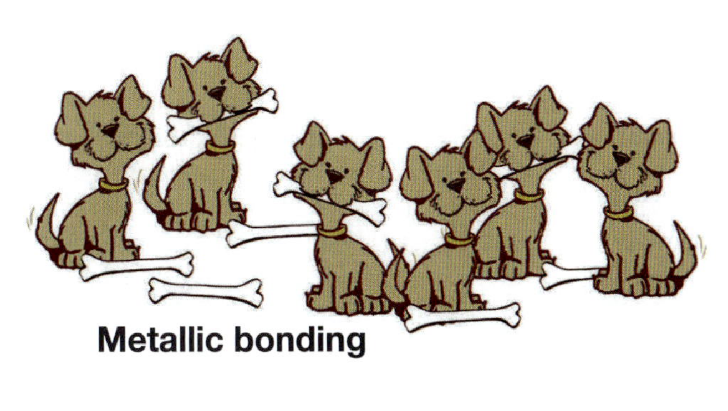

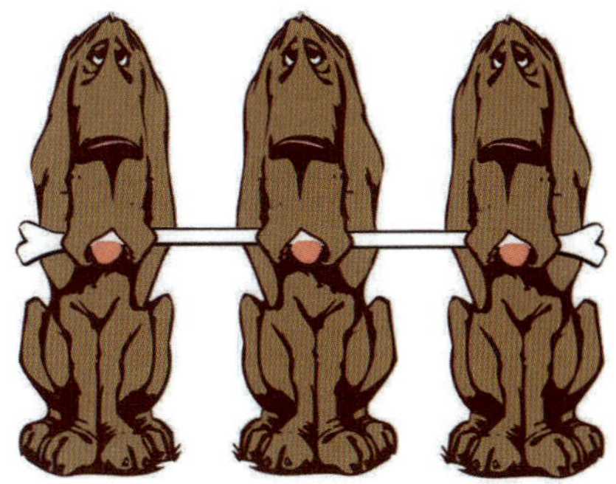

Figure 8.1. A useful analogy for bonding is to consider a roomful of dogs (atoms) with at least one bone (electron) for each dog. In metallic bonding, all the puppies are mellow because there are plenty of bones to go around. At any given time, each dog can be satisfied with a bone. In covalent bonding, dogs (atoms) of equal strength play with equal bones (electrons), which are shared equally. In ionic bonding, a big dog takes a bone away from a little dog, who then follows the big dog around because he's so attracted to his (former) bone (electron).

Metallic Bonds

Now we can get back to a more formal explanation of bonding. **Metallic bonds** are formed between atoms with low and similar (equal or nearly equal) electronegativities. Unlike a covalent bond, the attraction extends outward in all directions from a given central atom. As a result, a metallic bond can form at any point of contact between neighboring metal atoms. This means that pure metals, including the minerals copper (Cu), silver (Ag), gold (Au), platinum (Pt) and taenite (found in meteorites), usually have crystal structures in which their atoms are packed together as closely as possible. From our discussion in Chapter 3, we know that when all the atoms in a structure are the same size, each atom touches twelve neighbors, and such arrangements minimize the pore space between atoms. This close packing arrangement leads to the detachment of electrons from individual atoms so that they can disperse among all the neighboring atoms and are freely mobile.

Of course, we all know that metals are good conductors of electricity. What is it about the sea of electrons surrounding metal atoms that makes this happen? The metal atoms have lost all their valence electrons, which effectively makes them into cations (these are called kernels). The attractions between those positively-charged kernels and the negatively-charged electrons in the surrounding sea make the structure very strong. The delocalization of electrons effectively makes the bonding forces *stronger* over greater distances than would be the case for other types of bonding. Note that the higher the magnitude of positive charge on the metal's nucleus, the greater the strength of the resultant metallic bond.

When you combine this strength with the fact that the bonds are equal in all directions, you get a combination that imparts some key physical characteristics to a metal. Imagine a box full of marbles surrounded by water. No matter how you push on and rearrange the marbles in the box, the water will always follow them. As the marbles move to new places in the box, the bonds (= the water) will not break, because the atoms can slide around while the bonds that hold them extend in all directions. So, minerals with metallic bonds have high tensile strength, ductility, and malleability. Movement of electrons through the structure is unimpeded, and it is this fluid flow of charge that makes metals good conductors of electricity. Note that the process of metal fatigue, which you hear about in the aircraft industry and nuclear power plants, is caused by the introduction of defects that cause a metal to lose its fluid nature and become rigid.

Metals are also excellent thermal conductors, which means they can conduct heat in the form of kinetic energy. How does this work? Application of heat to one side of a metal block will make all the metal atoms (kernels) vibrate. They are free to vibrate because they are so loosely held into the crystal structure. As they heat up more and vibrate more, they begin to bump into adjacent metal kernels, which in turn begin to vibrate. Eventually the entire metal "crystal" is full of vibrating metal atoms, and thus kinetic energy passes easily from one atom to another. In a more typical mineral structure, the atoms are held too rigidly for these vibrations to occur.

Most metals are difficult to work with in their pure form because they are so soft. Nearly pure 99.99% gold, for example, is called 24 carat gold, but you only rarely find jewelry made from this because it is too soft to wear well. For long wear and good looks, 18 carat gold, which is 75% gold and 25% other stronger metals is preferred, though it is very expensive. The type of alloy used to mix with gold depends on the purpose: jewelry manufacturers select alloy materials for both appearance and strength, while dental fillings are alloyed for their ability to cast easily, seal well at the edge of your tooth, and avoid flaking. Wiring manufacturers select their alloys for strength as well, but they have to be careful to avoid changing the metal's properties. Introducing a non-metallic element as an alloy into a metal will reduce its ductility, malleability, electrical, and thermal conductivity.

We still need to address the relationship between the metallic bonds and the structures that their atoms form. Metals represent a somewhat special case because all of the atoms in the structure are the same size, or very close to the same size. This scenario leads to atoms that are packed together in such a way that pore space is minimized; this close packing is particularly common in metals. In fact, we'll see shortly that the term **close packing** is used to describe the different ways spheres of near equal size arrange themselves. Such tight arrangements lead to detachment of electrons from individual atoms so that they are freely dispersed among all neighboring atoms, facilitating the conductive properties discussed above. Atoms in metals tend to have one of three basic arrangements (Table 8.2).

A **body-centered cubic** (bcc) lattice has motifs at the corners and in the center of a cube (Figure 8.2). If each motif is an atom, then each corner atom touches the central atom along the body diagonal of the cube, but the corner atoms do not touch each other. This structure has a packing efficiency that is about 68% higher than if the atoms were only located at the corners of the cube. A good example of a close-packed cubic structure that is based on a bcc lattice is the mineral kamacite (Figure 8.2) (usually $Fe_{0.9}Ni_{0.1}$), which is found in iron meteorites.

The most efficient way (74% efficient!) of packing atoms together and minimizing the amount of pore space between them is to use 12-coordination. There are two ways to do this: **hexagonal close packing** (hcp) and **cubic close packing** (ccp). Each of these types of closest packing consists of single layers of close-packed atoms (one in the center of a hexagon). Only the way in which these layers are stacked distinguishes the two.

Both begin with a layer of close-packed atoms, which we will label A in Figure 8.3. This is the equivalent of placing six oranges around a central orange, and then continuing the pattern in all directions horizontally. Now place another layer that is exactly the same as the first one on top. In Figure 8.3, you will notice that we have labeled the holes between the oranges as "B voids" and "C voids." In a hexagonal closest-packed structure, a second A layer is placed on top of the first one, but the centers of the atoms nestle into and lie above the B voids. The next layer in an hcp structure is another A layer, then a B layer etc. No atoms ever lie over the C voids in an hcp structure. In fact, if you look directly down on the structure, you can see tiny channels running straight through the hcp structure. The mineral species osmium (Os, Ir, Ru) crystallizes in this form.

Table 8.2. Packing Types of Metals

Body-centered cubic lattice type	*Hexagonal close packing*	*Cubic close packing*
Fe	Be	Cu
Cr	Mg	Ag
Mo	Zn	Au
W	Cd	Al
Ta	Ti	Ni
Ba	Zr	Pd
	Ru	Pt
	Os	Pb
	Re	

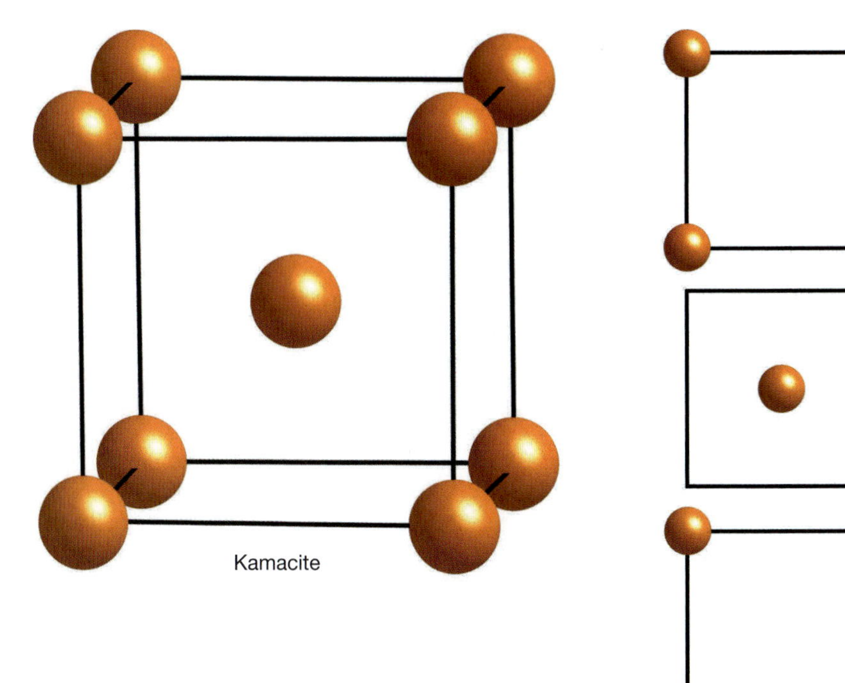

Figure 8.2. The exotic mineral kamacite, which is found in metal meteorites, is a cubic structure based on a body-centered cubic lattice type. Fe or Ni atoms occupy the corners of the cube, with an additional Fe or Ni atom in the center of the cube. Because the Fe and Ni atoms do not favor a particular site, they are represented here as the same color. The figure shows a perspective view of the mineral on the left and three cuts through the mineral on the right.

Another type of cubic close-packed (ccp) structure is based on a face-centered cubic lattice type. *Three* layers of hexagonally-arranged atoms are stacked in the order A (the original starting layer), B (over the B voids) and C (over the C voids). This structure then repeats ABCABC continuously. This type of packing has a high amount of symmetry, containing 2-fold and 4-fold symmetry axes, and it is represented by a number of metallic elements. It also is the basic unit for stacking sulfur, fluorine, and oxygen atoms in many different minerals, especially oxides.

To summarize, hcp is a stacking of A and B layers in the sequence...ABABAB... while ccp is a stacking of layers in the sequence ...ABCABCABC... Given hcp arrangements of spheres, you can see *through* the array, while this is not the case for the ccp array. Interestingly, what would seem to be lower symmetry (hcp) has a simpler stacking sequence than the higher symmetry ccp. We'll

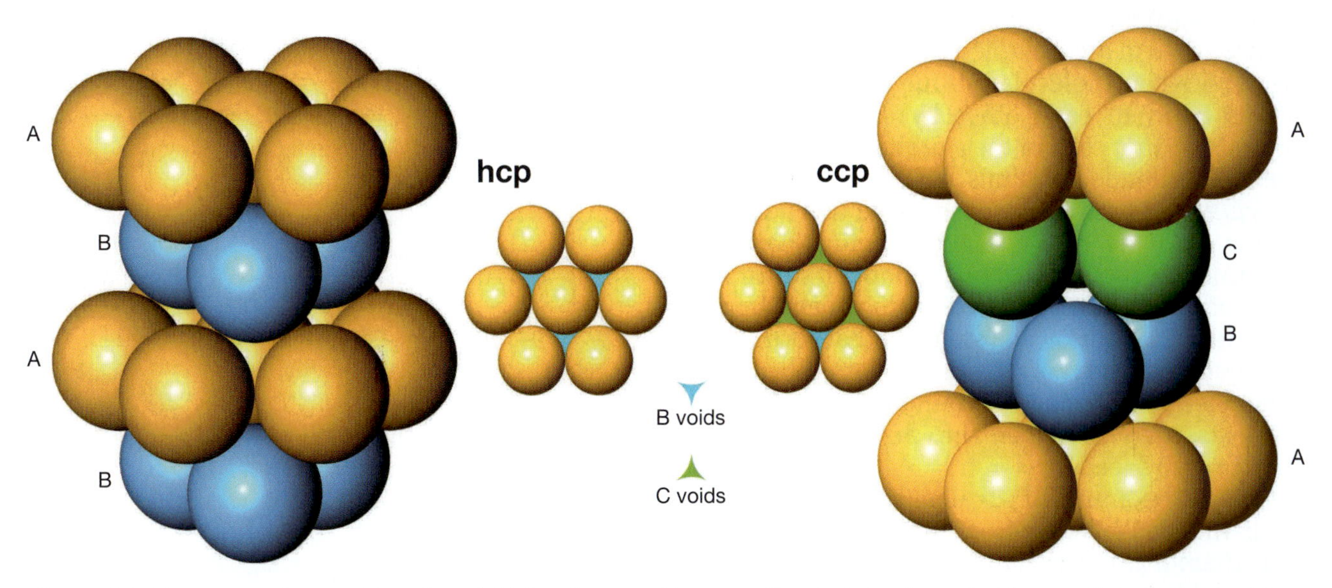

Figure 8.3. Hexagonal close packing (hcp, left) involves stacking two identical closely-packed layers of similarly-sized ions on top of each other; notice that the B layer is slightly offset from the A layer. The B ions lie directly over the B voids in the A layer. Cubic close packing (ccp, right) stacks three layers, each one slightly offset from the other two, in an ABC sequence.

return to layer stacking again when we discuss layer silicates, because they too will have different symmetries based on stacking orders.

Covalent Bonds

When atoms of identical and high electronegativities are brought close together, bonding occurs by sharing of outer electrons. Each atom must have one or more singly occupied orbital or unpaired electrons. The two atoms must approach closely enough to permit considerable overlap of their respective, singly-filled orbitals (i.e., overlap of their electron clouds). When the overlap exceeds some limiting value, two electrons of opposite spins (one from a singly-filled orbital of each atom) become paired in a single orbital. The new orbital has a molecular character, in that it is related to both nuclei. The result is **covalent bonding**, which is inherently directional.

Covalent bonds result from the interactions between spins (not charges) of electrons on adjacent atoms. The energies of covalent bonds depend on the geometrical shapes of the orbitals that describe the distributions of electrons around the atoms. The number of covalent bonds that an atom can form is usually limited by the number of electron with unpaired spins.

Consider the case of molecular sulfur, which has the configuration of $1s^2 2s^2 2p^6 3s^2 3p^4$. If a sulfur atom is going to bond with another sulfur atom, then we would predict a covalent bond (equal and high electronegativities because we are bringing two of the *same* atoms together). Also, recall that there are only three different *p* orbitals, so two of them are occupied by only a single electron. So when molecular sulfur is formed, the S atoms bond to each other by sharing outer *p* orbitals, so that each of the *p* orbitals is fully occupied. The bonded orbitals have a configuration that is equally related to each of the two adjacent atoms, and the result is a connecting covalent bond. The S_8 molecule of orthorhombic sulfur is a ring of eight atoms linked by covalent bonds (Figure 8.4). Most of the minerals that form with covalent bonds involve mixtures of sulfur with other elements commonly found in hydrothermal deposits, such as copper, zinc, arsenic, cadmium, mercury, and lead.

Another example of covalent bonding is found in diamond. We know from Mohs hardness scale that diamond is the hardest known mineral, so we would guess that it must be strongly bonded, and that these strong bonds are propagated in all directions in the structure. The diamond structure (Figure 8.5) resembles that of a cube, with carbon atoms at each of the eight corners and in the middle of each of six faces. Each atom is held to its neighbor by a single covalent bond to form a giant molecule whose size is that of the crystal itself. Each carbon atom is bonded to four others (in tetrahedral coordination), with each C-C-C bond at the tetrahedral angle of 109.28°. This three-dimensional network of covalent bonds explains why diamond is so very hard—you can't scratch the diamond without breaking four bonds for each atom!

Intermediate Covalent-Ionic Bonding

What happens when there are some electronegativity differences between bonded atoms, but they are not extreme? Consider the group of sulfide minerals, each of which is based upon combinations of cations with the sulfur anion. Sulfur has an electronegativity of 2.58, putting it squarely in the middle of the range for common mineral-forming elements. So any other cation in combination with sulfur creates bonds that are neither covalent nor ionic, but somewhere in between.

The crystal structure of sphalerite, ZnS, is shown in Figure 8.6. Each atom is coordinated or bonded to four atoms of the other element. The structure is based on two arrays of atoms, each having a face-centered cubic (fcc) lattice type. Atoms are located at the corners and in the cen-

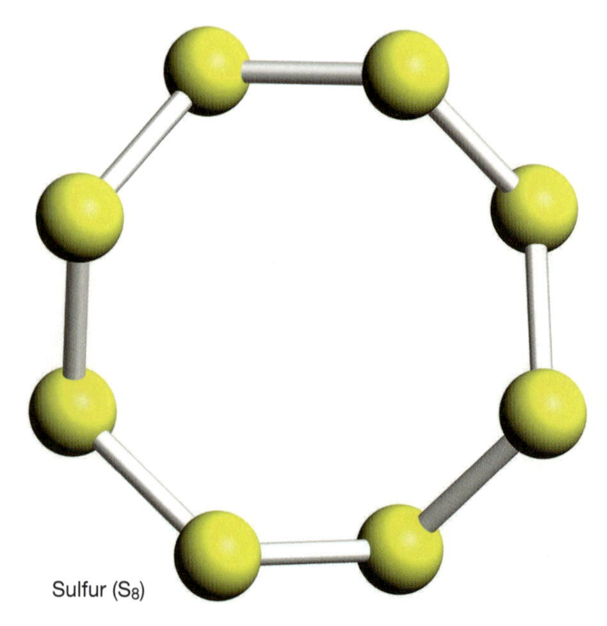

Figure 8.4. *One of the molecular forms of sulfur is S_8, which is composed of 8 sulfur atoms in a covalently-bonded slightly puckered ring. See the movie on the MSA website to aid in visualizing the puckering in the ring.*

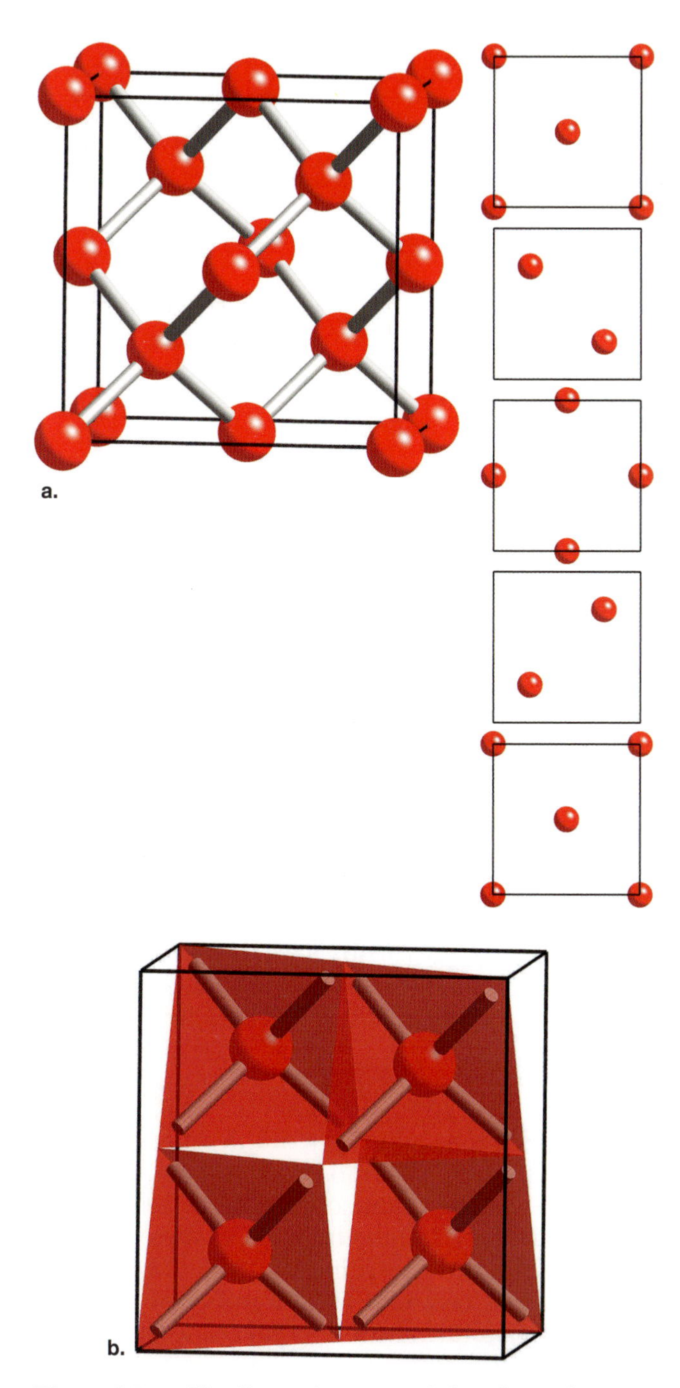

Figure 8.5. ***a.*** *The diamond structure is based on a face-centered cubic lattice type. Each carbon atom tightly (covalently) bonded to four others carbons. A carbon is at the center of a tetrahedron of four other carbon atoms, just as a Si^{4+} cation is typically surrounded by four O^{2-} anions. The figure shows a perspective view of the unit cell (outlined in black) of diamond (above) and five cuts through the mineral (right) starting at the bottom and working to the top.* ***b.*** *A polyhedral representation of diamond showing the central C atom surrounded by four C atoms (not shown) forming the tetrahedron.*

ters of the faces of each cube. The sphalerite structure is simply two face-centered arrangements of atoms that are offset by 1/2 along the body diagonal of the cube. This is a fairly efficient way of packing, and the atoms are close enough together such that most of the bonding is covalent.

Another form of ZnS is the mineral wurtzite, which is hexagonal rather than cubic. Its structure looks just like the hexagonal close packed metals we looked at above, except that now the atoms are alternating Zn^{2+} and S^{2-}. As with sphalerite, the wurtzite structure can be visualized as two interpenetrating arrays of Zn^{2+} and S^{2-}, though in this mineral the arrays are hcp. Each atom is again in tetrahedral coordination with the four surrounding atoms (Figure 8.6).

If you look down on the basal plane of the tetrahedra in the sphalerite and wurtzite structures, their similarities and differences become apparent (Figure 8.6). The only real difference between hexagonal close packing and cubic close packing is the position of the third layer of spheres. A slight shift in their relative positions causes the entire structure to change. If sphalerite is heated to high temperatures (for example, by hydrothermal fluids), the increased energy of the atoms gives them enough energy to make the shift and make wurtzite.

There is one more interesting relative of ZnS with a related structure: $CuFeS_2$, which is chalcopyrite (Figure 8.7). Begin by stacking two unit cells of sphalerite on top of each other. Now substitute Cu^{2+} and Fe^{2+} in alternating sites. This new structure is tetragonal, and very similar to that of sphalerite. Other minerals with structures derived from sphalerite include stannite (Cu_2FeSnS_4) and bornite (Cu_5FeS_4). By now you should be getting the idea that once you understand one mineral structure, you can mix and match elements to make new structures. In minerals, atoms are mixed and matched on the basis of parameters such as temperature, pressure, and availability (bulk composition). So, you begin to see how much fun mineralogists can have, determining and understanding crystal structures. Understanding crystal structures at an atomic level can be very important for the geologist who is interested in interpreting geological conditions of formation!

We can't close our discussion of intermediate covalent-ionic sulfides without mentioning a few more simple sulfide structures, both of which are based on the simple model of the cubic halite structure. Galena, PbS, is our most important source of lead for commercial purposes. Galena has the halite structure, with Pb^{2+} replacing Na^{1+} and S^{2-} replacing Cl^{1-} (Figure 8.8). Its relative, pyrite (FeS_2), has Fe^{2+} atoms located at the Na^{1+} positions in the cubic close-packed cell, based on a face-centered lattice type while pairs of covalently-

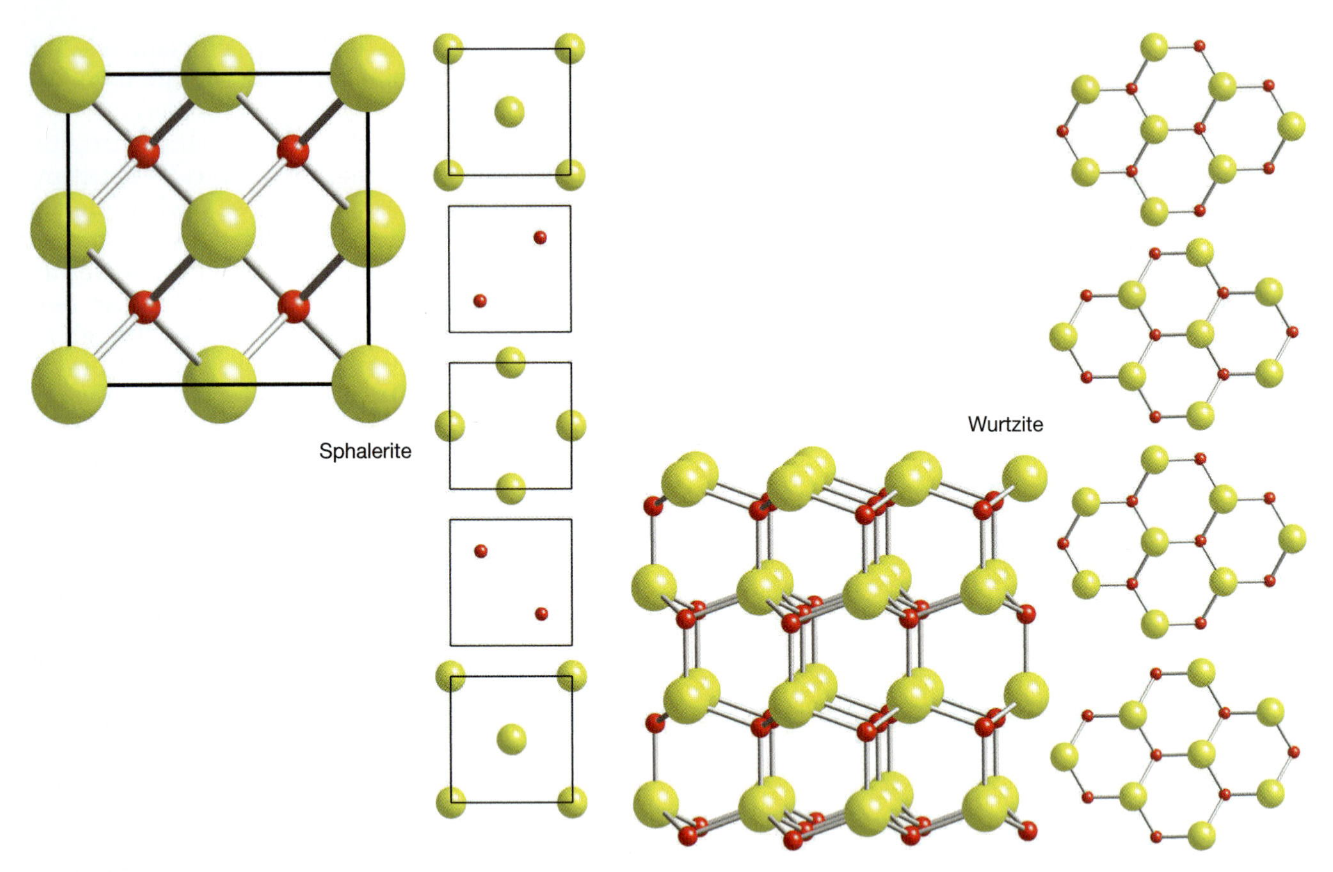

Figure 8.6. *Views of sphalerite (left) and wurtzite (right), two polymorphs of ZnS. The smaller Zn^{2+} atoms are at the center of a tetrahedron formed by four S^{2-} atoms. If we replaced the Zn^{2+} and S^{2-} atoms with C, we would return to the diamond structure of Figure 8.5. Sphalerite has a cubic close-packed, or ccp structure, while the wurtzite structure is based on hexagonal close packing. Thus they have different crystal structures even though they have the same chemical formula. Below each perspective drawing are 2-D layers cut through each mineral to show the atom arrangements on different layers.*

bonded S^{2-} atoms (that look like dumbbells) occupy the Cl^{1-} positions (Figure 8.8) with the center of the dumbbell bar passing through the location of Cl^{1-} atoms of halite. Notice that the dumbbells are not all in the same orientation, but vary in a way that preserves the symmetry elements of this isometric mineral. This structure allows pyrite to assume many interesting crystal forms, including striated cubes, pyritohedra, octahedra, and twinned pyritohedra.

Other sulfide structures are based on hexagonally close-packed (hcp) sulfur atoms with smaller cations in between. They are typical of the many inorganic crystal structures that can be regarded as closest packings of large ions, with smaller ion in the interstices (voids, or crevasses) in between them. Usually the larger ions are anions that are either in hexagonal or cubic closest packings. Earlier, we described these closest packings as consisting of close-packed layers of spheres, with the spheres of the upper layer nestling into the indentations between the lower ions. If you look closely at an hcp or ccp array, you'll notice that two types of interstices are created: ones where six anions come together (octahedral, or O sites) and ones where four anions come together (tetrahedral, or T sites) (Figure 8.9). Mixing and matching of atoms fitting into these "voids" creates a large number of different possibilities, and nature has taken advantage of many of them.

In a cubic closest-packed array of atoms, the unit cell contains eight T sites and four O sites. The four O sites include one at the center of the cell and 1/4 of each O site on each of its twelve edges (since each edge is shared by four unit cells). If you look again at the galena structure, you will see that all the O sites are filled by Pb^{2+}, while the corners anions are all S^{2-}. Similarly, the sphalerite structure is a close-packed array of S^{2-} with only 1/2 the T sites filled.

In hexagonal close packed arrays of ions, there are twelve T sites and six O sites, the locations of which are shown in Figure 8.9. Understanding this type of packing is very useful for describing the structures of other sulfides. In **nickeline**

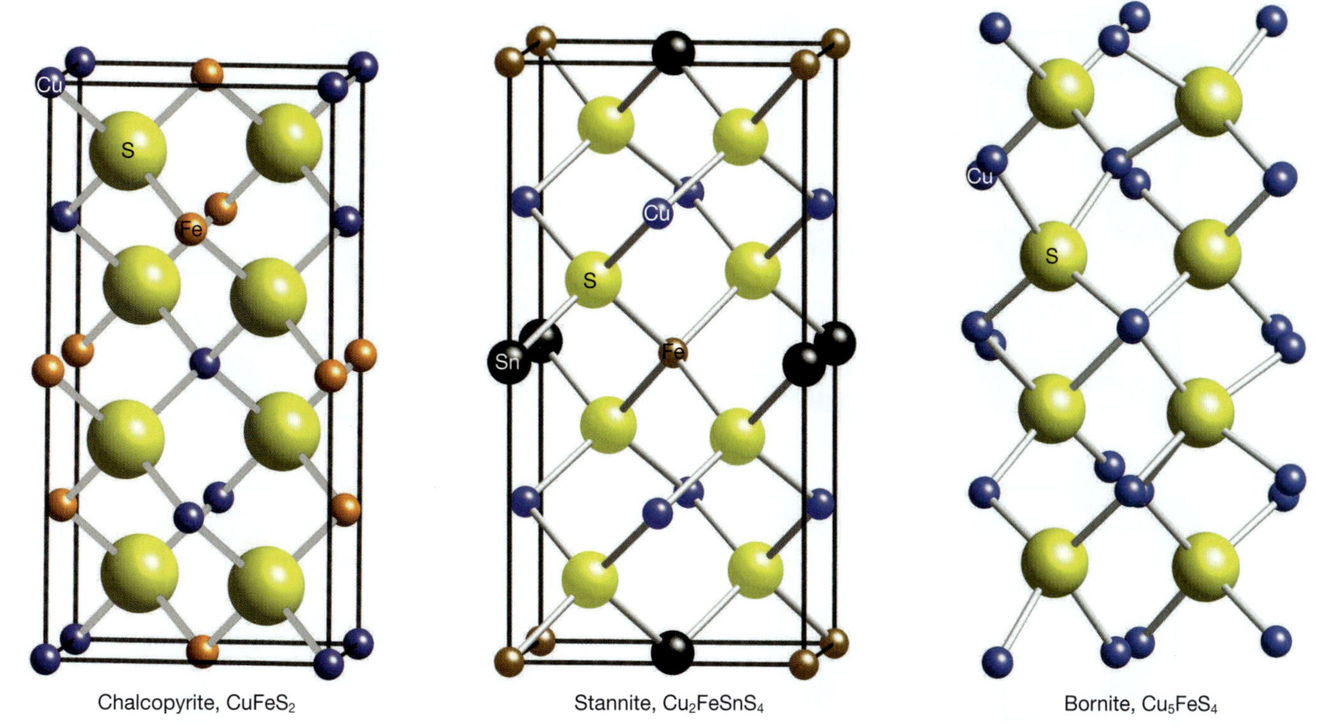

Figure 8.7. *Drawings of the chalcopyrite, $CuFeS_2$ (left), stannite, Cu_2FeSnS_4, (middle), and bornite, Cu_5FeS_4 (right) structures show relationships to the sphalerite structure in Figure 8.6. All have a small cations in tetrahedral coordination with four larger yellow S^{2-} anions. Chalcopyrite is two unit cells of sphalerite stacked on top of each other; the small blue spheres represent Cu^{2+} cations and the orange spheres are Fe^{2+} cations. In stannite, there are three types of atoms in tetrahedral coordination with S^{2-}; the small blue spheres are Cu^{2+}, the small brown spheres are Fe^{2+}, and the slightly larger black spheres represent Sn^{2+}. In bornite, Cu^{2+} and Fe^{2+} cations are disordered (i.e., they are not in a systematic arrangement as in chalcopyrite) and the structure is slightly distorted compared to chalcopyrite).*

(NiAs, formally known as niccolite), As^{3-} anions are hexagonally close-packed, and Ni^{3+} cations fill all the octahedral sites (Figure 8.10). Each As^{3-} atom is also surrounded by six Ni^{3+} atoms and each Ni^{3+} atoms is surrounded by six As^{3-} atoms.

If we replace the Ni^{3+} and As^{3-} in nickeline with Fe^{2+} and S^{2-}, respectively, we get the mineral **troilite**, FeS, which is found in metal meteorites (Figure 8.11). It is closely related to the mineral pyrrhotite, $Fe_{1-x}S$, which is a non-stoichiometric

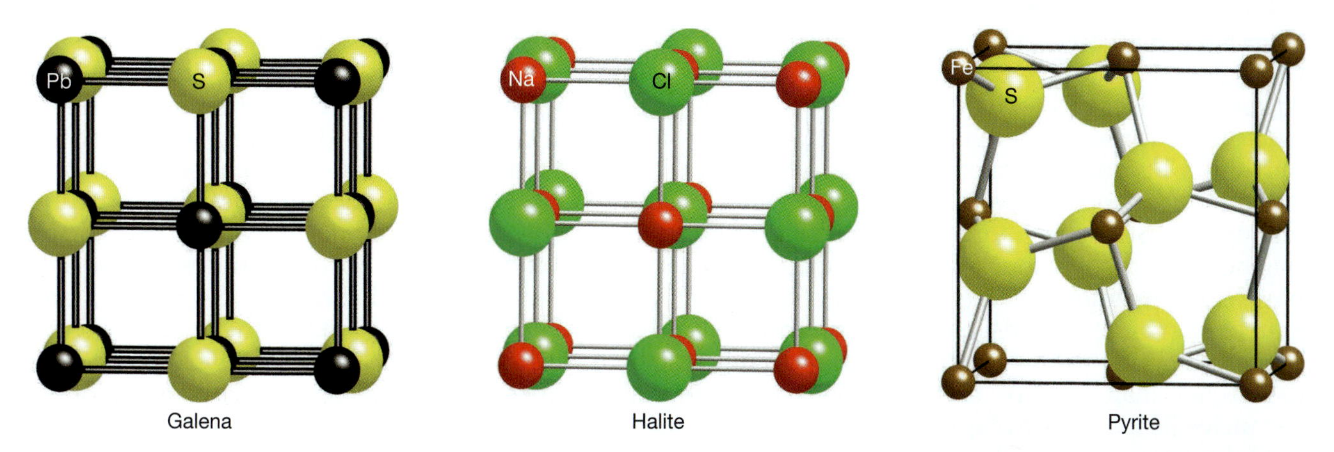

Figure 8.8. *Drawing of galena, PbS, on the left, has the same fcc lattice type as halite (middle), with Pb^{2+} replacing Na^{2+} and S^{2-} replacing Cl^{1-} (the larger spheres are S^{2-} and Cl^{1-}, and the small spheres represent Pb^{2+} and Na^{1+}). Its relative, pyrite, FeS_2 (right) has Fe^{2+} cations located at the Na^{1+} positions in the faces of a cubic cell, while pairs of covalently-bonded S^{2-} anions (that look like dumbbells) occupy the Cl^{1-} positions. The center of the imaginary S^{2-} dumbbells passes through the former Cl^{1-} position. (Look closely at the center of the outlined unit cell for pyrite and compare it to the center Cl^{1-} anion in halite.)*

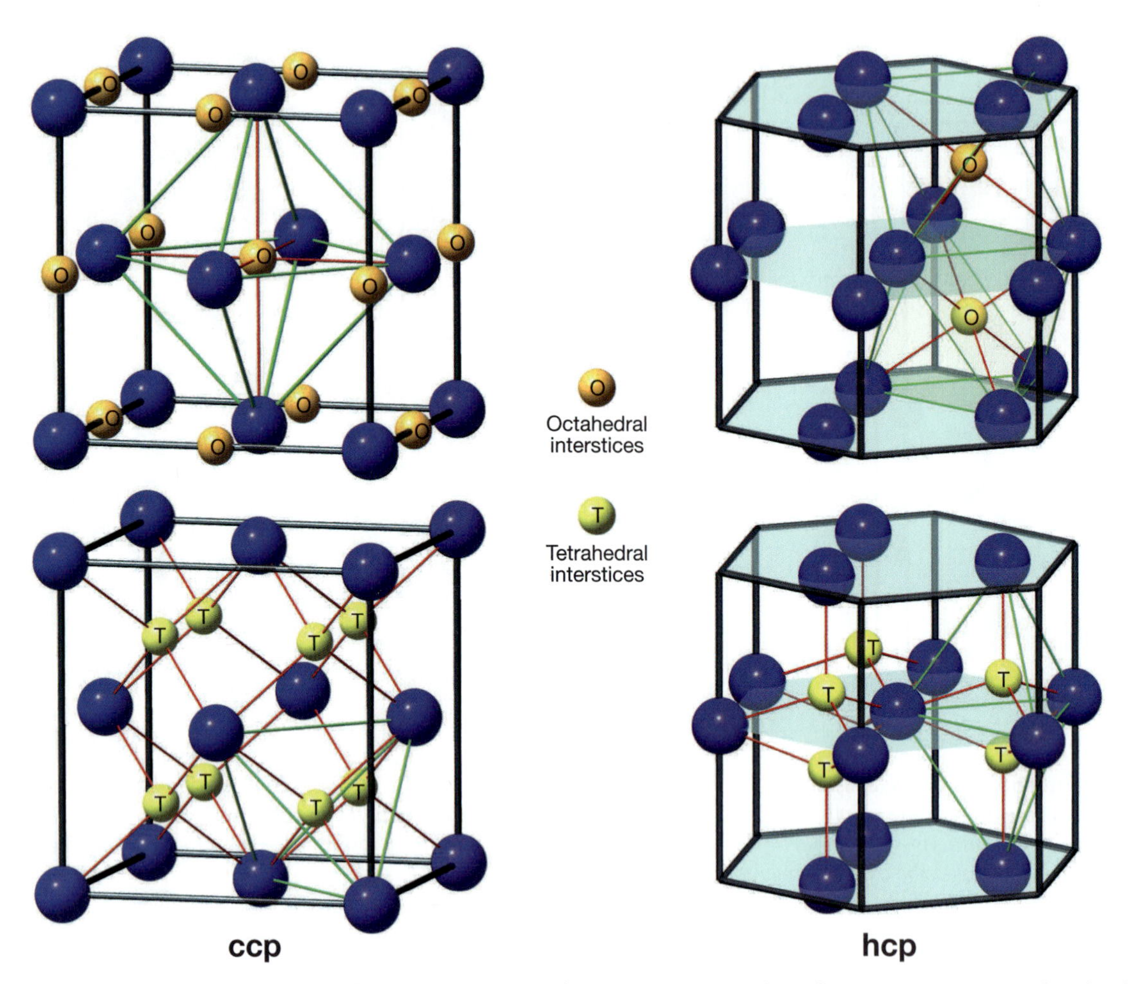

Figure 8.9. In every close-packed array, there are two types of interstices: ones where four atoms come together (T, for tetrahedral) and ones where six atoms come together (O, for octahedral). In a cubic closest-packed array of atoms, the unit cell contains eight T sites and four O sites. In hexagonal close-packed arrays of ions, there are 12 T sites and 6 O sites.

sulfide in which some of the octahedral sites are vacant (Figure 8.11). If all the iron were Fe^{2+}, then the mineral would not charge balance (and that is not allowed!). So some of the Fe^{2+} oxidizes to form Fe^{3+}, and you end up with a neutral formula of $Fe^{2+}_{(1-3x)}Fe^{3+}_{(2x)}\square_{(x)}S$ (the $\square$ represents the vacancy, the empty site). The Fe^{2+} and Fe^{3+} coordinate to behave like ferrimagnets, and the pyrrhotite becomes magnetic. In fact, pyrrhotite is the next most common magnetic mineral after magnetite. So when you see specimens of pyrrhotite in the lab, touch several different samples with a magnet, and see which ones are most and least magnetic. This physical property will tell you whether the pyrrhotite is close to troilite (if it's not magnetic) or has a lot of structural vacancies (if it's highly magnetic)!

Molybdenite (MoS_2) is another sulfide composed of hcp S^{2-} atoms, this time with Mo^{4+} filling half the octahedral sites (Figure 8.12). The S^{2-} layers form covalent bonds to the Mo^{4+}, but are too far away from the adjacent S^{2-} atoms to form strong covalent bonds. Thus, molybdenite forms a sheet structure (with the strong bonds in the plane of the sheet), so it can be used as a lubricant with the high conductivity of a metal. This gray mineral is easily confused with graphite in hand sample. Both are extremely soft because of weak bonding between the sheets of their structures.

Finally, pentlandite, $(Fe,Ni)_9S_8$, often occurs with pyrrhotite (you can tell them apart because the pentlandite is non-magnetic). Its structure would be complicated to explain if you didn't understand packing! But we can describe it simply by saying that it is based on S^{2-} anions in an fcc array (Figure 8.12). Half of the available tetrahedral sites are filled by Fe^{2+} or Ni^{2+}, and half of the octahedral sites are occupied by Fe^{2+} or Ni^{2+}.

Ionic Bonds

When two atoms with extreme differences (>2) in electronegativity come together, the result is

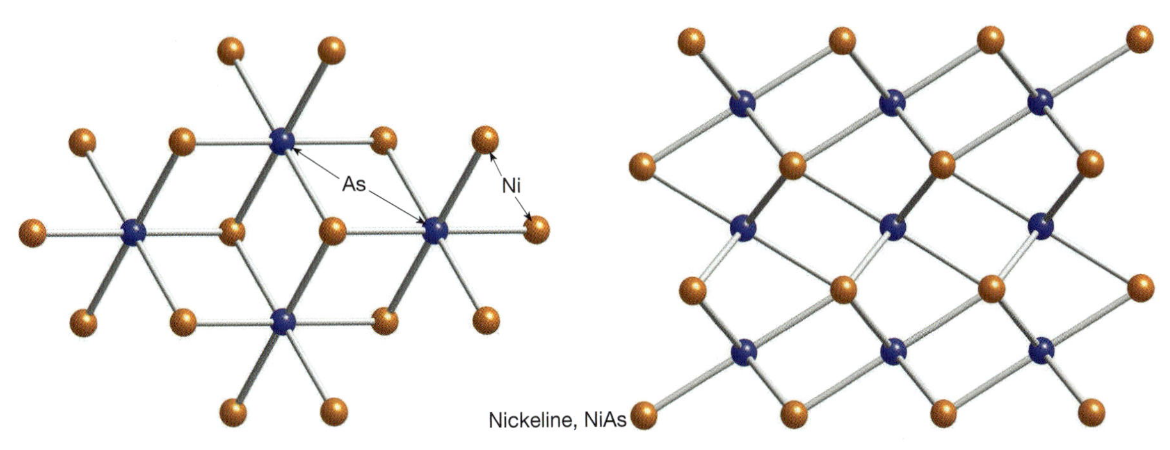

Figure 8.10. The structure of ***nickeline*** *(NiAs),is a hexagonally close-packed array of two ions with similar sizes: As^{3-} anions (small blue spheres) and Ni^{3+} cations (small brown spheres) filling all the octahedral sites. The drawing on the left is projected perpendicular to the close-packed layers and shows the hexagonal nature of the sheets. The drawing on the right is projected parallel to the close-packed layers. It is helpful to look at the movie on the MSA website to see the 3-D version of this structure.*

termed an **ionic bond**, which is most common in non-opaque minerals. Ionic bonds involve a complete transfer of electron(s) from a cation (with a low electronegativity) to an anion (high electronegativity). The bigger the difference in charge between the two atoms, the bigger their attractions—and the stronger the bond! Remember that the charge of the two atoms (q_1 and q_2) in Coulomb's Law is in the *numerator* of the force equation, and so the charges have a first-order effect on the strength of the bond. This also means that if the bonded ions have big radii, the force of attraction between them will be weaker.

Here's an example of how this works. Consider the ever-versatile salt (halite) structure, which we met earlier (Figure 8.8). From Chapter 1, we know that Na has a charge of 1+, so the charge on Cl must be 1–. The electronegativities of Na^{1+} and Cl^{1-} are 0.93 and 3.16, respectively, suggesting that they bond with a strongly ionic character. The Na^{1+}-Cl^{1-} bonds involves the transfer of a single electron.

Now compare NaCl with the mineral periclase, which has the identical structure but a formula of MgO (Figure 8.13). In periclase, two electrons are transferring instead of one. So based on Coulomb's law (here it is again):

$$F = \frac{1}{4\pi\varepsilon_0}\frac{q_1 q_2}{r^2},$$

it follows that there will be *four times* the net attractive force between $Mg^{2+}O^{2-}$ relative to

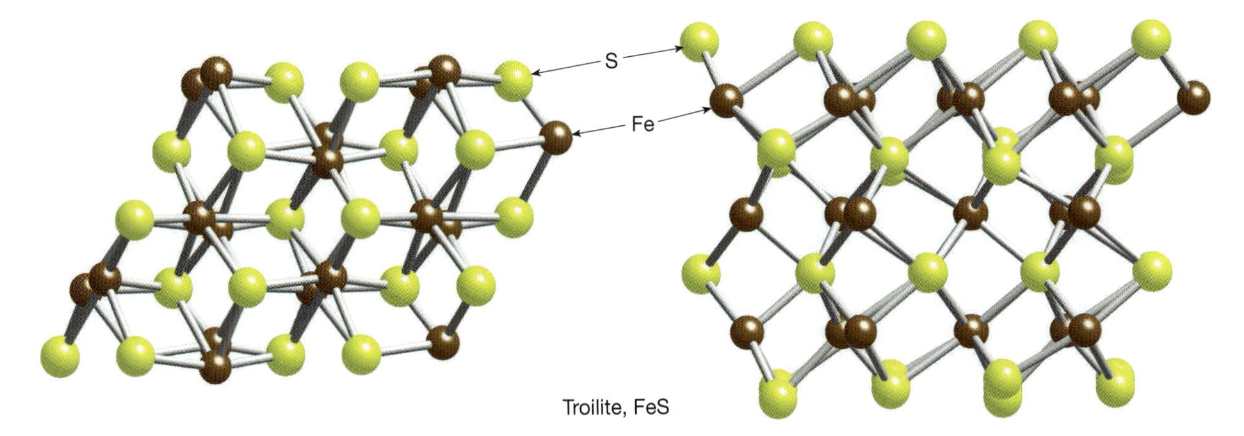

Figure 8.11. The troilite (FeS) structure is almost identical to the nickeline structure, with Fe^{2+} (brown spheres) and S^{2-} (yellow spheres) replacing the Ni^{3+} and As^{3+}, except that the structure is slightly "deformed" (i.e., the atoms have moved from their "ideal" positions in the hcp array). To better compare trolite to nickeline, the left image shows a projection perpendicular to the close-packed layers of S^{2-}. The right image is projected parallel to the layers as in Figure 8.10 for nickeline. Troilite has a similar structure to pyrrhotite, $Fe^{2+}_{1-x}S$, but some of the octahedral sites (i.e., Fe sites, the brown spheres) are vacant.

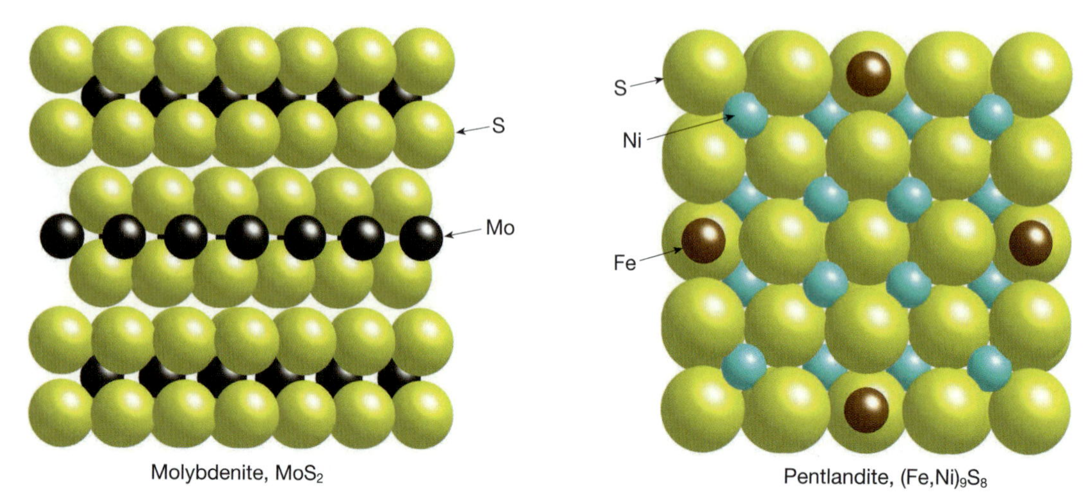

Figure 8.12. Molybdenite, MoS_2, (left) is composed of hcp S^{2-} atoms (larger yellow spheres), this time with Mo^{4+} (smaller black spheres) filling half the octahedral sites. (Here both of these structures are shown with larger atoms than in the other drawings to better show the packing.) Like graphite, molybdenite forms sheets with strong bonds in one plane and weak bonds perpendicular to it. Pentlandite, $(Fe,Ni)_9S_8$, (right) is based on S^{2-} atoms (larger yellow spheres) in a fcc array. Half of the available tetrahedral and octahedral sites are filled by Fe^{2+} (smaller brown spheres) or Ni^{2+} (smaller blue spheres).

$Na^{1+}Cl^{1-}$. Indeed, this is reflected in their respective physical properties (Table 8.3).

Finally, note that in ionic bonding, the respective ions are assumed to be charged spheres (even though we know from an earlier discussion that their shapes are far more complicated than that). This means that ionic bonds have no directional properties—the strength of the bond is assumed to be the same in all directions.

Minerals that are held together by ionic bonds form somewhat predictable crystal structures. This realization became part of a formalism developed by Linus Pauling to explain how atoms arrange themselves in ionic crystals. This set of **Pauling's Rules** works extremely well in explaining most minerals even though they really have a combination of ionic and covalent bonds. So we'll go through each of Pauling's rules to learn some basic principles about crystal structures:

Rule #1. Around every cation, a coordination polyhedron of atoms forms, in which the cation-anion distance is determined by the radius sums, and the coordination number is determined by the radius ratio. This rule assumes that atoms behave like solid balls (recall the example

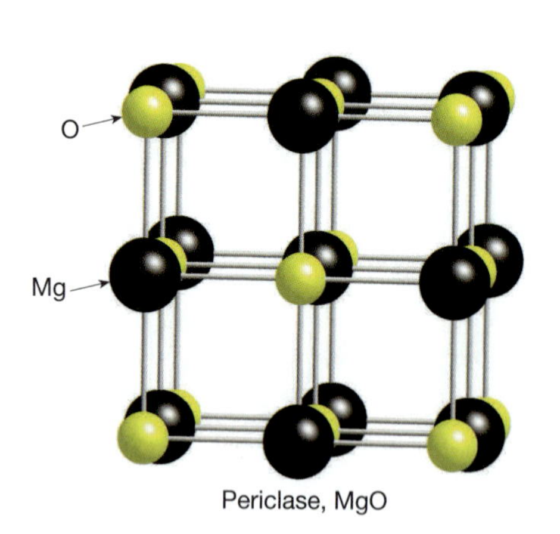

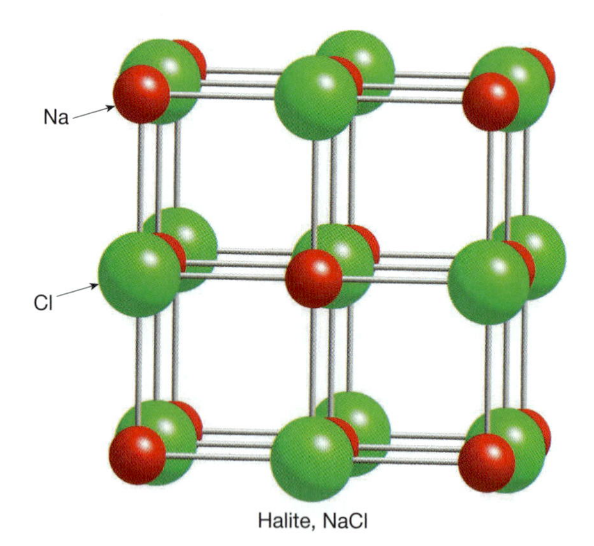

Figure 8.13. View of periclase, MgO, (left) and halite (right) drawn to scale, showing how the Mg^{2+}-O^{2-} bonds are shorter than the Na^{1+}-Cl^{1-} bonds. The net attractive force between Mg^{2+} and O^{2-} in periclase is four times stronger than the attraction between Na^{1+} and Cl^{1-} in the halite structure because the charges are higher.

Table 8.3. Comparison of Properties of NaCl and MgO

Mineral	*NaCl*	*MgO*
Melting point (°C)	801	2800
Hardness	2.5	6.5
Bond Energy (kJ/mole)	640	1000
Density (g/cc)	2.17	3.58
Unit cell size (Å)	5.64	4.21

using oranges and an apple from Chapter 3), and they form bonds when the balls are just touching. The length of the cation-anion bond is then simply the sum of the radii of the cation and anion.

The coordination number of the central cation is determined by the ratio of its size to that of each of the surrounding ions. Five basic types of coordination polyhedra are commonly found in minerals, as shown in Figure 8.14: 3-, 4-, 6-, 8- and 12-coordinated. We learned in Chapter 3 that the size of the hole between these surrounding atoms increases with coordination number. Now we can quantify that relationship, and use it to predict the types of coordination polyhedra that different elements will prefer by calculating radius ratios.

We begin with the simple case of a trigonal (3-coordinated) site shown in Figure 8.15. We will call the radius of the larger anion R_A, and the radius of the cation, R_C. Based on the figure, we see that a right triangle is formed with one long side equal in length to R_A, and a hypotenuse with a length of $R_A + R_C$. So trigonometry allows us to calculate the R_C as follows (remembering that in a right triangle, cos = adjacent/hypotenuse):

$$\cos 30^\circ = R_A \;/\; (R_A + R_C)$$
$$0.866 = R_A \;/\; (R_A + R_C)$$
$$0.866\,(R_A + R_C) = R_A$$
$$0.866\,R_A + 0.866\,R_C = R_A$$
$$0.866\,R_C = R_A - 0.866\,R_A$$
$$0.866\,R_C = 0.134\,R_A$$
$$R_C \;/\; R_A = 0.155$$

This means that any arrangement of anion-cation in which the size of the central cation is exactly 15.5% the size of the surrounding three anions will yield a geometry where all three anions touch each other and the central cation exactly. If the cation is smaller, it will still fit, but if it's bigger than 15.5% of the anion, then the cation won't be "comfortable" in that site.

For six-coordination, the calculation is similar (Figure 8.16). Again, the hypotenuse has a length of $\frac{1}{2} R_A + \frac{1}{2} R_C$, while the shorter side of the right triangle has a length of R_A. So the calculation becomes

$$\cos 45^\circ = R_A \;/\; (R_A + R_C)$$
$$0.707 = R_A \;/\; (R_A + R_C)$$
$$0.707\,(R_A + R_C) = R_A$$
$$0.707\,R_A + 0.707\,R_C = R_A$$
$$0.707\,R_C = R_A - 0.707\,R_A$$
$$0.707\,R_C = 0.293\,R_A$$
$$R_C \;/\; R_A = 0.414.$$

Shape formed by anions	Triangle	Tetrahedron	Octahedron	Cube	Cuboctahedron
Number of corners	3	4	6	8	12
Number of sides	2	4	8	6	14
Shape of polyhedron					

Figure 8.14. Because we know the sizes of most cations and anions fairly well, we can predict the types of structural units they will form using Pauling's Rule. These five types of coordination polyhedra are the most common in minerals.

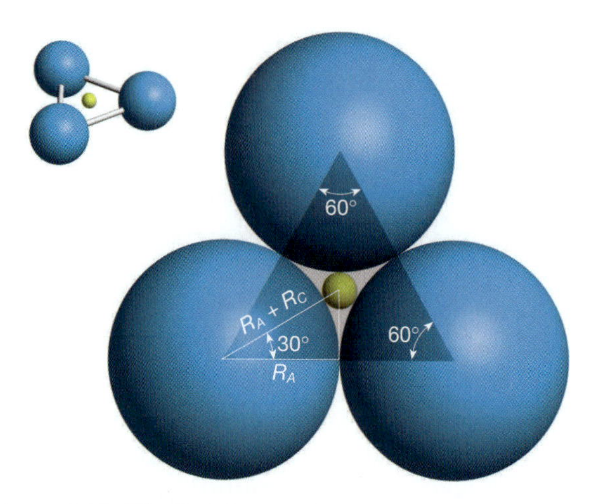

Figure 8.15. The limiting conditions for triangular coordination can be calculated in terms of the radius of the large anion (R_A) vs. the radius of the smaller cation (R_C). Note from the drawing that the three balls define an equilateral triangle. A smaller right triangle, with one angle equal to 30°, allows the relationship between R_A and R_C to be calculated using simple geometry (see text).

It's a good exercise to see if you can derive the other limiting parameters for different types of coordination, as summarized in Table 8.4. Who cares about radius ratios? This is actually a very powerful formulation that allows us to predict what types of coordination polyhedra will be formed in mineral structures.

For example, consider each of the following minerals, and predict the coordination polyhedra on the basis of radius ratios: NaCl (halite), MgO (periclase), PbS (galena), CaF_2 (fluorite), TiO_2(rutile) MnO_2 (pyrolusite), SnO_2 (cassiterite), and Al_2O_3 (corundum). If you apply Pauling's Rule #1 correctly, your prediction will be correct for each of these phases.

Rule #2. An electronic structure will be stable to the extent that the sum of the strengths of the electrostatic bonds that reach an anion equals the charge on that anion. To understand this rule, we define the strength of an atomic bond to be the ratio of charge to coordination number. So for example, the periclase structure is composed of equal numbers of Mg^{2+} and O^{2-} atoms. Every Mg^{2+} cation is bonded to six O^{2-} anions, and every O^{2-} is bonded to six Mg^{2+} atoms (Figure 8.13). So each individual bond has a strength of $-2/6$, or $-1/3$, and the sum of all six charges is 2–, which is the charge on the O^{2-}.

How about the fluorite structure? In CaF_2, each Ca^{2+} cation is surrounded by eight F^{1-} ions in cubic coordination. Each of Ca^{2+}-F^{1-} bonds has a strength of $2/8$, or $1/4$. So the total charge on the Ca ion is two (Figure 8.17).

Rule #3. The sharing of edges, and particularly faces, by two anion polyhedra decreases the stability of a crystal. The idea behind this rule is that cations repel other cations, and so they wish to be as far apart in a crystal structure as possible. This repulsion is particularly high if the cations are highly charged and the coordination number is small. Consider the fundamental building block of silicate minerals, the SiO_4 tetrahedron (Figure 8.18). Corner-sharing results in the maximum possible distance between Si^{4+} cations, with a "screening" O^{2-} between them. If the two tetrahedra share edges, the Si^{4+} atoms are 58% closer than before. If they share faces, they are 38% closer—*too close for the comfort* of these small, highly-charged Si^{4+} cations. For this reason, Si tetrahedra in silicate structures tend to share only corners and rarely edges or faces!

Rule #4. In a crystal that contains different cations, those with high charge and low coordination numbers tend not to share elements of their coordination polyhedra. This one can be

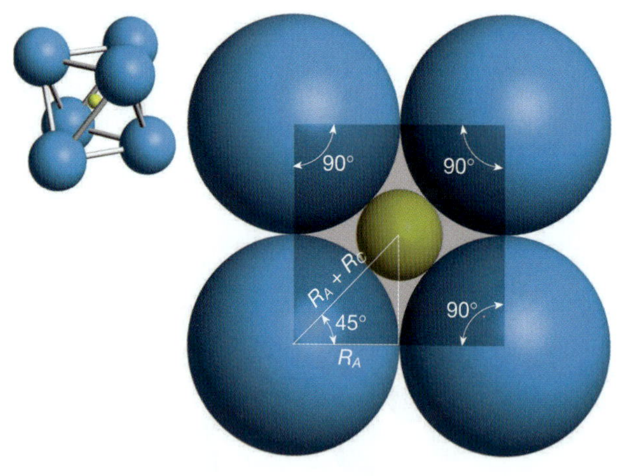

Figure 8.16. The limiting conditions for octahedral coordination can also be calculated in terms of the radius of the large anion (R_A) vs. the radius of the smaller cation (R_C). See text.

Table 8.4. Radius Ratios and Coordination Numbers

Coordination Number	Polyhedron	Radius Ratio Range	Example*
3	triangle	0.155–0.220	C in calcite
4	tetrahedron	0.220–0.414	Si in quartz
5	trigonal bipyramid	-	Al in andalusite
6	octahedron	0.414–0.645	Mg in forsterite
8	square antiprism	0.645–0.730	Ca in garnet
8	cube	0.730–1.00	Ca in fluorite
12	dodecahedron	1.00	native gold

*Thanks to Joe Smyth and his web site: http://ruby.Colorado.edu/~smyth/G30103.html

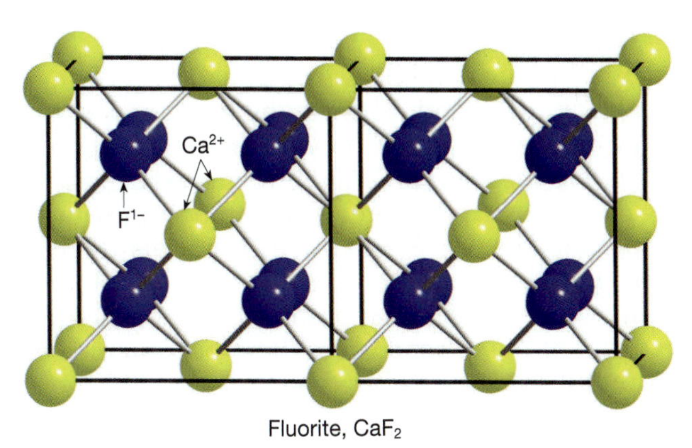

Figure 8.17. In fluorite, CaF_2, each Ca^{2+} cation (the large yellow sphere in the center of the drawing) is surrounded by eight F^{1-} ions (blue spheres) in cubic coordination. (Two unit cells are required to show the coordination of Ca^{2+} with F^{1-}, because Ca^{2+} occurs on the edges of the unit cell.) Each of the eight Ca^{2+}-F^{1-} bonds has a strength of 2/8, or 1/4, for a total charge of eight on the Ca^{2+} cation.

restated as "Rule #3 is especially true if the coordination number is low." So if you have a mineral with multiple cations present, the highly-charged, small cations won't want to share edges or faces. So the other cations with lower charge and higher coordination numbers will be the ones to share faces.

Two good mineralogical examples will serve to illustrate this. In the kyanite structure shown in Figure 1.7, the Si^{4+}-O^{2-} tetrahedra share only corners, while the Al^{3+} octahedra share edges. Figure 8.19 shows the perovskite ($Ca^{2+}Ti^{4+}O_3$) structure. The rule would predict that the Ti^{4+} polyhedra would share only corners, and indeed Ti^{4+} occupies corner-sharing octahedra. On the other hand, the Ca^{2+} cations in this dense structure share faces, and are 12-coordinated.

Rule #5, The Law of Parsimony. The number of essentially different kinds of constituents in a crystal tends to be small. This law says that nature likes to make things simple, so any given mineral will tend to be made up of a very small number of sites, and many different elements can substitute into them. Many geologists take this to imply that while chemistry may vary, the structures are conserved. This is why igneous rock types are all poly-mineralic!

A few final comments are needed on the subject of Pauling's Rules. First, and most obviously, we know that atoms are not hard, round spheres, but rather complicated probability distributions of electrons around a nucleus. So the radii involved in Pauling's Rules are not as rigid as they seem. Second, we know from our previous discussions that minerals bond with a continuum of bond types ranging from purely covalent to purely ionic. The more covalent bonding is involved in a mineral structure, the less Pauling's Rules will apply. Third, most coordination polyhedra are irregularly-shaped—not the perfect geometrical constructions that Pauling envisaged. Finally, there are many cations that occur in two or more different types of coordination polyhedra (Al^{3+}, Fe^{3+}, and Ca^{2+}, to name a few). Although Pauling's Rules do have these significant problems, they do provide an amazingly good way to predict how cations and anions will bond and arrange themselves in minerals, particularly those with predominantly ionic bonds.

Van der Waals Bonds

Most mineral structures are bonded by some combination of ionic and covalent bonding. In some cases, there is charge synchronization or polarization of charge, resulting in a weak distribution of positive or negative charge. These weak bonds are called **van der Waals bonds** (or sometimes, London Dispersive Forces). They common-

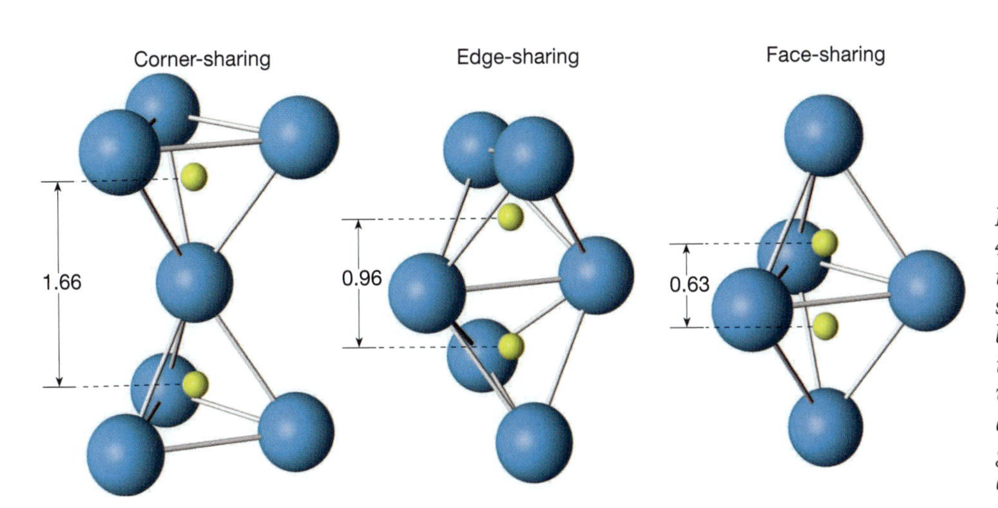

Figure 8.18. Pauling's 3rd and 4th rules explain that coordination polyhedra do not like to share edges or (worse) faces because such sharing increases the closeness of the cations within those polyhedra. These drawings show how the cations get closer together as the amount of sharing increases.

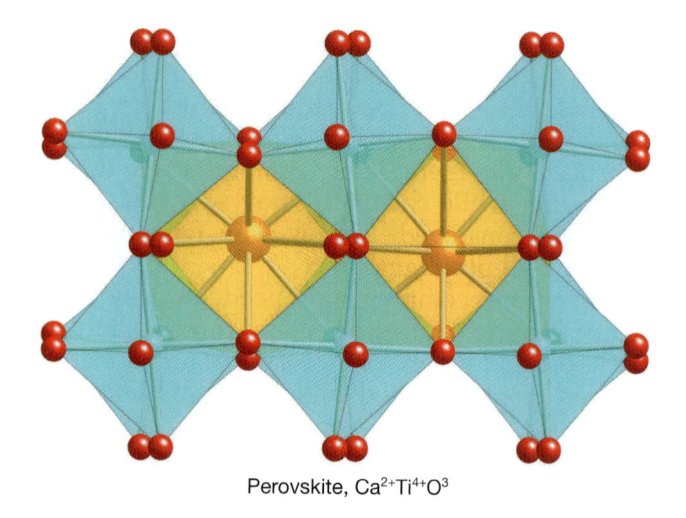

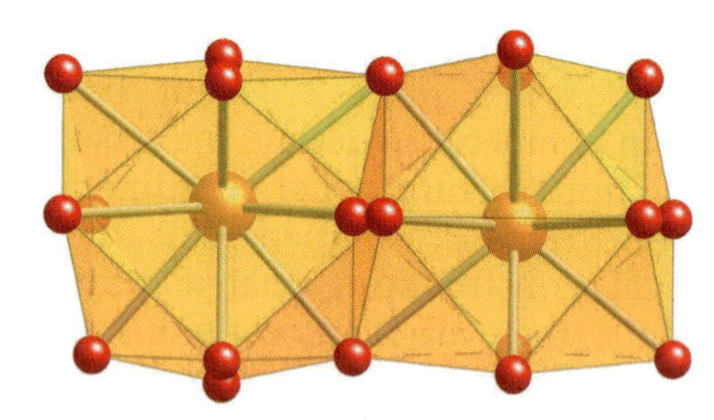

Figure 8.19. The structure of perovskite, $Ca^{2+}Ti^{4+}O_3$ (above) is composed of Ti^{4+} octahedra on the edges and 12-coordinated Ca^{2+} in the center. The O^{2-} atoms (small red spheres) are also shown to define the polyhedra. On the lower image, the octahedra have been removed to better show the Ca^{2+} site. This structure nicely illustrates Pauling's 4th rule. Ti^{4+} cations occupy corner-sharing octahedra, while Ca^{2+} cations share faces, and are 12-coordinated.

ly occur between neutral, nonpolar molecules such as the noble gases (helium, neon, argon, etc.). These weak charges occur as the result of random motions of electrons. For example, at any random moment, the electron density on one side of an atom is anomalously high, and at that same instant, the electron density nearest that point in an adjacent atom is anomalously low. In other words, the electron clouds are synchronized, giving rise to very weak electric multipoles, which can hold adjacent atoms together. These attractions are so weak that materials with van der Waals bonding tend to have extremely low melting and boiling point temperatures, which is why noble gasses are gases at room temperature.

An interesting mineralogical example of this type of bonding is found in the mineral graphite, which was once used to make pencils because it is very soft. Consider again the element carbon, which has the configuration $1s^22s^22p^2$ with two singly-occupied $2p$ orbitals. By analogy with sulfur, carbon might be thought to form divalent bonds. However, some carbon can be quadravalent, so each carbon atom is held by bonds of equal energy to four neighbors in tetrahedral coordination, as in diamond. This can be explained as follows: In the process of bond formation, carbon is excited to the state $1s^22s^12p^3$ by transferring one of its $2s$ electrons to the previously unoccupied third $2p$ orbital. The one $2s$ and three $2p$ orbitals are then mixed (hybridized) to form a new set of four equivalent sp^3 orbitals pointing towards the four vertices of the tetrahedron. These bonds hold the diamond structure tightly together as noted above (see also Figure 3.13).

But carbon can also form covalent bonds involving sp^2 hybrids, as is the case for graphite (Figure 8.20). In its structure, three orbitals form coplanar lobes at 120° apart, leaving the 4th electron in the p orbital lying out of the plane. This p orbital overlaps sideways to form what are called π-bonds, so that the 4th electron becomes delocalized, or detached from its atom. This 4th electron can then help form a van der Waals bond. So in graphite, the carbon atoms have a hexagonal sheet structure, with very strong covalent bonds within the sheets. But the sheets are held to other sheets by only the weak forces of van der Waals bonds. Because the sheets cleave so easily, graphite makes a good lubricant. Graphite also is a good conductor of electricity *along* the sheets because of the delocalized electrons. The observed difference in properties along the sheets (where the covalent bonds are) vs. perpendicular to them (van der Waals bonds) makes graphic dramatically anisotropic, or is **anisodesmic**. As we'll see in a later chapter, some sheet silicates, most notably talc, have layers that are held together with van der Waals bonds.

Hydrogen Bonds

A specialized form of electrostatic attraction happens when a H atom bonds with a highly electronegative anion like F or O. In such a case, the single electron on the hydrogen atom will be so strongly attracted to the anion that it ends up confined to one of its orbitals, leaving the lone proton on the hydrogen atom exposed. The proton can then form bonds with other negative charges, forming a highly directional hydrogen bond. The H_2O^0 molecule is said to be a dipole with a negative side on the oxygen atom and a positive side on the H^{1+} cation (Figure 8.21).

The resultant force is strong because of the large difference in electronegativity between the H atom and the large anion. This type of bond is found in 1:1 layer sheet silicates that contain hydroxyl $(OH)^{1-}$ in their structures (i.e., kaolin-

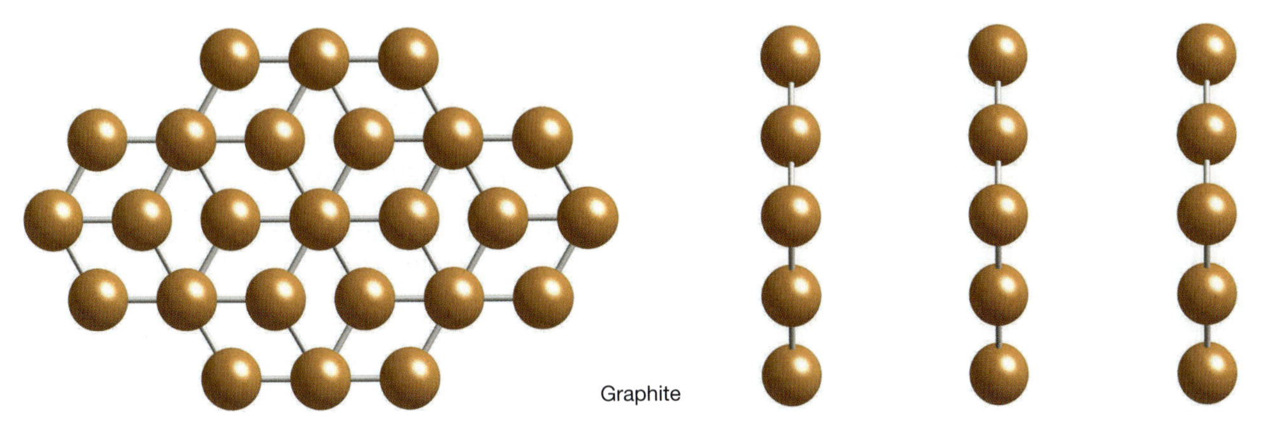

Figure 8.20. *The structure of graphite viewed perpendicular to the sheets composed of 6-membered rings of C (left) and parallel to those sheets (right). As in molydenite (Figure 8.12), the graphite structure is strongly bonded within sheets, but weakly bonded perpendicular to those sheets. This property makes graphite well suited for pencil leads and lubricants.*

ite), as well as in zeolites where the H's in structural H_2O bond to the framework oxygens and the O's to the channel cations.

Hydrogen bonding also helps us understand the interesting properties of water and ice (which is a mineral). As you know from using ice cube

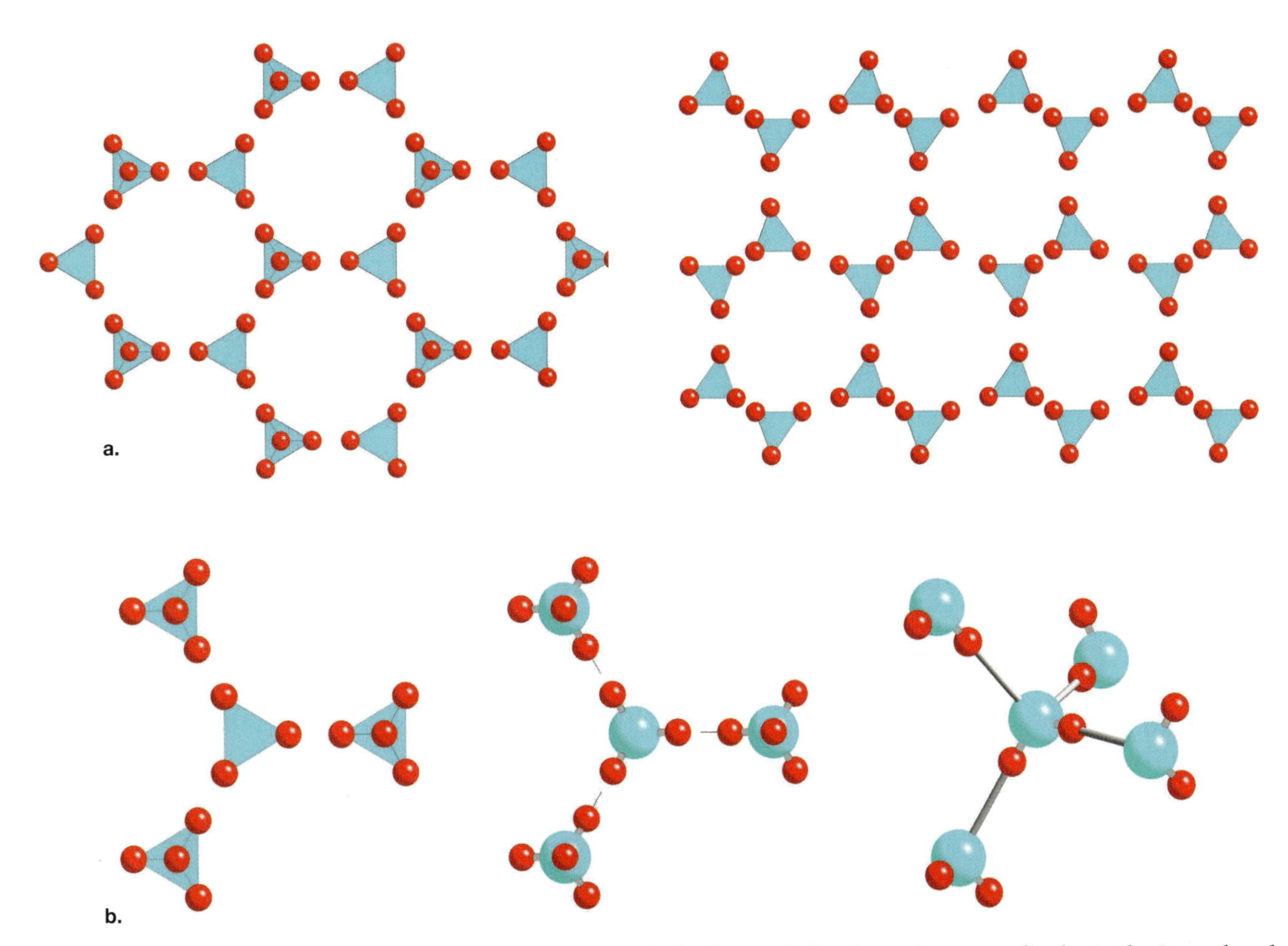

Figure 8.21. a. *The time-averaged structure of ice is composed of HO_4 tetrahedra that point perpendicular to the 6-membered rings.* ***b.*** *A smaller section of the ice structure. The middle drawing shows the ball and stick representation of the drawing on the left. On the right is a snapshot of the actual bonding in ice. The central O^{2-} anion has two short bonds and two long bonds to H^{1+}; these long bonds represent H-bonding. The short and the long bonds are constantly switching, so each O^{2-} has two short bonds and two long bonds in any possible orientations.*

trays in your freezer, ice is unique because it expands upon freezing. The structure of ice is shown in Figure 8.21. The time-averaged structure of ice can be envisioned as sheets of polymerized HO_4 tetrahedra with half of the tetrahedra pointing up and half pointing down. As shown in Figure 8.21b, the H^{1+} cations in H_2O ice form a weak bond between H^{1+} cations and O^{2-} anions of adjacent water molecules. In ice, each O^{2-} anion is surrounded by four H^{1+} cations on the axes of a tetrahedron. Two of the H^{1+} cations are bonded tightly to the O^{2-} with an approximate distance of 0.8 Å with ionic bonds, while the other two H^{1+} cations are H-bonded at a distance of about 1.9 Å. Below 4°C, decreasing kinetic energy causes the H_2O molecules to move closer together. Such close proximity allows hydrogen bonding to take place, forming the hexagonal ice structure. The hydrogen bonding forces the water molecules into an open crystal structure that occupies more space than liquid water. When water is in the liquid phase, the molecules are constantly forming and breaking hydrogen bonds, which have a half-life of only about 10^{-11} seconds! So what really distinguishes water from ice is the lifetime of the hydrogen bonds.

References

Regents of the University of California (1996) Conversation with Nobel laureate Linus Pauling. http://globetrotter.berkeley.edu/conversations/Pauling/pauling7.html

Chapter 9

Chemical Analysis of Minerals

The Merriam-Webster Dictionary I have in my office defines a mineralogist as a scientist who deals with minerals. So, in previous centuries, that moniker described someone who could pick up a rock anywhere in the world and identify the minerals (or at least claim to identify them!) that were present on the basis of their physical properties. Most middle school and high school classes still perpetuate that tradition. As you are reading this, somewhere in the world, hapless students sit patiently at their desks scratching minerals with nails and fingernails and swearing that they will never, ever, be interested in the study of minerals. Come on, admit it: what was your least favorite lab exercise in your Introductory Geology class?

This is not to say that hand sample mineral identification is unimportant. For a geologist in the field, minerals must be identified on the basis of their appearance and association (that's a fancy term describing groups of minerals that usually coexist with one another). The most impressive field mineralogist I ever worked with was Charles Guidotti of the University of Maine. Charlie could pick up a rock and identify all the minerals present with a high degree of accuracy; in five minutes he could tell you things about mineralogy and chemistry that would take me five weeks to confirm back in the lab. He did all this after sprinting to the top of the mountain and balancing precariously next to the most inaccessible outcrop (where the best minerals are, of course). Although Charlie considered himself to be a petrologist, he developed a highly skilled eye for physical characteristics of minerals with years of practice, and he understood the relationships between physical properties and chemistry. I have always aspired to emulate his acuity at field mineralogy.

But for now, I'll excuse myself to run to the lab for a few spectroscopic analyses...many of which are the subject of this chapter. Fifty years ago, the only tools generally available to geologists were the optical microscope, the X-ray diffractometer, and a wet chemical analysis in the laboratory (so-called "wet" because the rocks had to be dissolved in solutions for study). In the 1960s, the introduction of the electron microprobe, which could analyze major elements with remarkable precision and reproducibility, revolutionized geochemistry, mineralogy, and petrology, and led to incredible progress during the last half century. While I was in graduate school, ion microprobe analyses began to make it possible to analyze lighter elements successfully. As I write this, measurement of elemental abundances and ratios at micrometer scales (picogram scale masses) is now routine.

A few elements remain intractable to in situ *microscale analyses in thin sections, however. In general, the lighter the element, the more difficult it is to analyze it at μm scales. Multivalent elements also pose a problem; at present, only the bulk techniques of Mössbauer spectroscopy and wet chemistry can provide accurate and reproducible data on Fe^{2+} and Fe^{3+} (though other techniques are "in the pipeline"). I am hopeful that in the lifetime of this textbook, integration of multiple types of technologies will allow us to characterize every aspect of the crystal chemistry and atomic arrangements of minerals at every scale. It's going to be an exciting time for mineralogists—enjoy the ride!*

M.D.D.

Introduction

The two preceding chapters were focused on how atoms are themselves constructed, and how they arrange themselves and bond together. Now we go on to the interesting issue of mineral analysis: How do we know what elements are present in a mineral? The answer to this question is not always an easy one. Despite the fact that we read in the popular press about how technology makes it possible to analyze anything at almost any scale, routine analyses of some minerals remain problematic.

Consider the species that make up the mineral group tourmaline (Figure 9.1). They share the general formula of $WX_3Y_6(BO_3)_3Si_6O_{18}(O,OH,F)_4$, where:

W = Ca, K, and/or Na;
X = Al, Fe^{2+}, Fe^{3+}, Li, Mg, and/or Mn^{2+};
Y = Al, Cr^{3+}, V^{3+}, and/or Fe^{3+}.

No less than three (and more commonly four or five!) different analytical techniques are needed to fully characterize the chemical composition of a tourmaline sample—a lengthy, expensive process on instrumentation that is difficult to find. So until very recently, the extent of chemical variation in this mineral group was poorly understood.

Historically, the development of new analytical techniques has repeatedly revolutionized different aspects of the field of mineralogy. Before the 1960s, mineralogists had three main tools to use in identifying minerals: X-ray diffraction, the optical microscope, and the technique of wet chemistry, which involves dissolving a mineral in acid and analyzing precipitates from that liquid. This process is tricky, and it requires a lot of sample for a single analysis, so you only get an idea of the averaged chemical composition of large samples. But wet chemistry has one big thing going for it: just about every element can be analyzed in this way, no matter what its position on the periodic chart, and the method can also determine the oxidation states of cations and anions (see details below).

Figure 9.1. The mineral tourmaline is one of the most difficult minerals to analyze completely, because it contains a range of elements from very low on the periodic chart through very high (or heavy). This is a crystal of elbaite, which has a formula of roughly $Na(Li,Al)_3Al_6(BO_3)_3Si_6O_{18}(OH)_4$—though the composition of this particular zoned crystal is far more complicated.

The 1960s saw the introduction of electron beam instruments (and color TV's), which used focused beams of electrons to obtain chemical analyses on small spots on polished minerals and minerals in thin sections. Suddenly it became possible to analyze a single mineral grain, and to see how chemistry changed on very small scales. Overnight, we gained the ability to understand how elements are zoned within individual mineral grains, and how the chemistries of coexisting minerals change as a function of pressure, temperature, and interactions with fluids. The entire field of metamorphic petrology was reborn at this point. The only downside to this revolution was the fact that the electron beam instruments analyze X-rays produced by the samples, so they are inherently limited to the analysis of elements whose electrons can be changed, or excited, through bombardment by high energy electrons or X-rays (to this day, the technique still doesn't work very well on low energy X-rays, that is elements of low atomic number). This limitation forced geologists to focus only on the elements that could be conveniently analyzed by that technique, though it wasn't a very tight constraint: pretty much everything heavier than Na (Z = 11) on the periodic chart could be well-characterized.

Subsequent decades saw the introduction of the ion microprobe, which made it possible to study elements lighter than Na on micro-scales, and the application of spectroscopic techniques to the study of multivalent elements. Improvements in electron microscopes now make it possible to determine light elements with them. Today, it is truly possible to analyze every element in a mineral if you have enough time, money, and patience.

So the goal of this chapter is to provide an overview of the various analytical techniques used by modern mineralogists. Our coverage of techniques is meant to be a leaping-off point for further investigations into mineral analysis. As such, it will be far from comprehensive, yet we aspire to cover (and provide you with some good references for) the methods you are likely to encounter or have access to somewhere nearby. You'll quickly discover that many of the methods are (confusingly) designated with acronyms and abbreviations—so we are introducing you to what is commonly referred to as the alphabet soup of analytical techniques!

Analysis of Minerals

Several different methods are used to obtain the composition of minerals and solids. In general, these can be divided into two broad groups: (1) spectroscopic methods that rely on the fact that each element's configuration will produce a characteristic energy signature, and (2) assorted methods of observation that include empirical techniques developed over the history of mineralogy. In the following sections, we will briefly introduce the most important of these. However, it should be pointed out at the onset that mastering or understanding these methods might require an entire semester-long course, often at the graduate level. But don't fear this. *You don't have to understand all the details of how a car works to drive one.* Many of the computer-aided instruments are even easier to learn to operate than driving a car. We are hopeful that your institution might have some of this instrumentation, so that you might be able to see, or better yet, to use it.

Figure 9.2. *To perform any type of wet chemical analysis on a mineral, the mineral must first be dissolved in a liquid. Then different methods are used to remove the dissolved solids, usually in the form of oxides, from the liquids.*

Wet Chemical Analysis

Historically, chemical analyses of minerals were performed by dissolving them in acids (Figure 9.2). This was not always an easy task—nor was analyzing the resultant solution using various techniques of analytical chemistry. You may well try some of these methods in the laboratory of your introductory chemistry class. Wet chemical analyses were extremely time-consuming and difficult to perform, but the work of the best wet chemists is still admired. Many of the pre-1960 mineral data sets remain the best, most complete sources of chemical analyses on the composition of complex minerals (though this statement is true only for the work of the best analysts!). Very few researchers still use wet chemical methods today because of their difficulty, and the relative ease of more modern techniques.

There are three basic types of wet chemical analyses, all of which involve an initial step of dissolving the samples (usually ground into a fine powder) in acids of various types. In **gravimetric analyses**, the element of interest is precipitated as a solid compound and its weight is compared to the weight of the starting sample to arrive at a weight percent of that element compared to the whole sample. A slight twist on this method was employed for the silicate minerals. For these minerals, the chemist would precipitate oxides, and as we will discuss in the next chapter, this is the reason some researchers still refer to silicate minerals by their weight percent oxides. For instance, if you dissolve the pyroxene group mineral enstatite, $Mg_2Si_2O_6$, you would then precipitate out MgO and SiO_2 by different reactions. Then given the weights of MgO and SiO_2 and the total weight of the sample, the weight percent of each oxide could be calculated. For a more complex mineral like biotite ($K(Mg,Fe)_3(AlSi_3O_{10})(OH)_2$), each of the following oxides would precipitate (and be weighed) individually: K_2O, Fe_2O_3, FeO, MgO, Al_2O_3, SiO_2, leaving behind some H_2O. Other then the fact that this method was tedious, it also required gram-weight, pure mineral samples. If the mineral contained impurities (a very common phenomenon), the results would be in error.

Volumetric analyses use titration, which is the process you may remember from chemistry class. Carefully-measured volumes of a different, reactive liquid (of known concentration) are added to the dissolved sample until a reaction occurs. The concentration of the unknown in solution can then be calculated based upon the volume of the reactive solution that needed to be added. Finally, in **colorimetric analysis** a reactive reagent is added to the dissolved sample solution, causing a color change (Figure 9.3). The intensity of the color can then be related to the concentration of the unknown.

Most wet chemists used multiple techniques and reagents to ensure sensitivity to different elements, so you can see why wet chemical analysis of minerals can be a complicated process for something like tourmaline! To learn more about wet chemical methods, see the good summary in Jeffrey and Hutchison (1981).

"Water Content" of Minerals and Hydrogen Extraction

Water in minerals is currently one of the hottest research areas in mineralogy, so it is important to understand how we represent water in chemical formulas. H_2O is the oxide of H, and H_2O occurs in minerals. When two H's bond to one O, the H^{1+} is occurring in a water molecule. However, H^{1+} more often occurs in minerals bonded to only one O^{2-} and not two. In this case, $(OH)^{1-}$ occurs, which chemists term hydroxide. The confusion results if we represent the composition of an $(OH)^{1-}$-bearing mineral in terms of its weight percent oxide; then it would appear the mineral contains water when it really contains $(OH)^{1-}$.

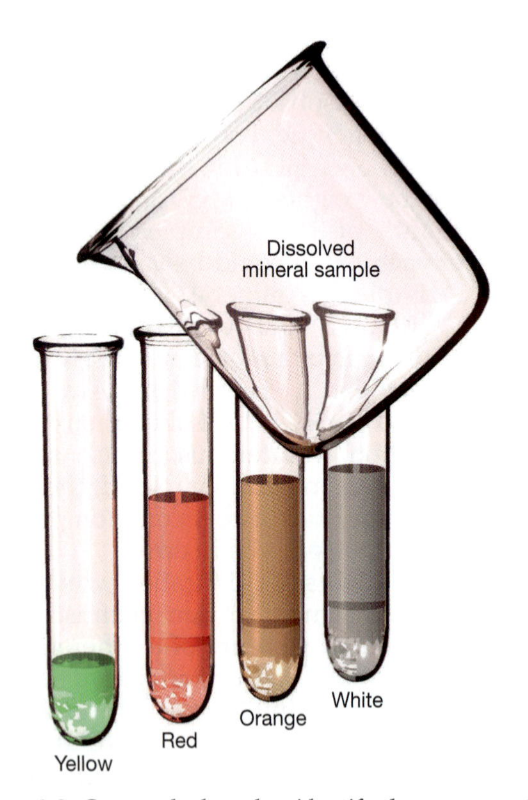

Figure 9.3. *One method used to identify the components in the liquid is based on color changes. Here, each color change would represent a different oxide or element.*

For example, the left column below shows the name, chemical formula, and proportion of oxides in three common feldspars (which hopefully are firmly implanted in your memory by now). Notice that for each of them, SiO_2 appears as one of the oxides, yet we know that feldspar doesn't contain quartz; of course it does contain SiO_2 as one of its oxide components. In the right column are three minerals with slightly more complex formulas than the feldspars. Each one could be chemically derived from the feldspar by the addition of $(OH)^{1-}$ or H_2O groups. Notice that muscovite contains no H_2O in its chemical formula, though it appears to when written in terms of oxides. Chabazite, a zeolite mineral, does contain H_2O as an electrically neutral molecule as shown in the formula, but that H_2O has no charge. When it occurs in a mineral, it appears at the end of the formula separated by a "·." Finally, montmorillonite is a clay mineral that contains both $(OH)^{1-}$ and H_2O, as seen in its chemical formula, but when written as oxides, the $(OH)^{1-}$ is no longer apparent.

orthoclase	muscovite
$KAlSi_3O_8$	$KAl_2AlSi_3O_{10}(OH,F)_2$
$\frac{1}{2}\ K_2O,\ \frac{1}{2}\ Al_2O_3,\ 3\ SiO_2$	$\frac{1}{2}\ K_2O,\ 1\frac{1}{2}\ Al_2O_3,\ 3\ SiO_2\ 1\ H_2O$
anorthite	chabazite
$CaAl_2Si_2O_8$	$Ca(Al_4Si_8O_{24})\cdot 12H_2O$
$1\ CaO,\ 1\ Al_2O_3,\ 2\ SiO_2$	$2\ CaO,\ 2\ Al_2O_3,\ 8\ SiO_2,\ 12\ H_2O$
albite	montmorillonite
$NaAlSi_3O_8$	$NaAl_4\ (AlSi_7O_{20})\ (OH)_4\cdot 3\ H_2O$
$\frac{1}{2}\ NaO\ \frac{1}{2}\ Al_2O_3, 3\ SiO_2$	$\frac{1}{2}Na_2O,\ 2\frac{1}{2}Al_2O_3,\ 7SiO_2,\ 5H_2O$

So, many "hydrous" minerals such as micas and amphiboles contain $(OH)^{1-}$. Others do contain molecular H_2O (gypsum and many of the zeolite group minerals), but you cannot distinguish between $(OH)^{1-}$ and H_2O on the basis of a chemical analysis alone.

Historically, there were several non-spectroscopic methods for determining the water content of minerals. Most of them relied on the assumption that when a mineral is heated, the water will leave the structure. Thus, if a sample is weighed before and after heating, the weight percent water can be determined by the weight loss. While these methods proved useful for some applications, they were often semi-quantitative because the heating was insufficient to liberate all the hydrogen (and sometimes, another easily volatilized element such as carbon was present).

A non-spectroscopic technique that does remain in common usage is **hydrogen extraction** (Figure 9.4), which is used for the analysis of hydrogen content of minerals (and for extracting H for isotropic analysis). The procedure begins by placing a known amount of the material of inter-

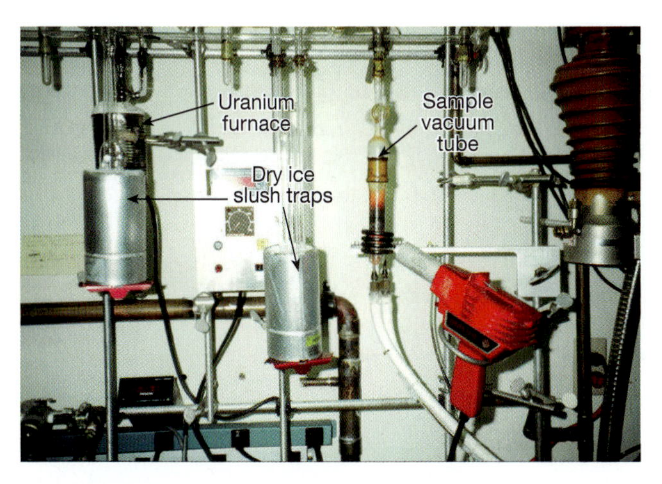

Figure 9.4. Uranium extraction involves a complex series of steps designed to release hydrogen from a mineral structure and isolate it in a volume where it can be measured. Heat from the induction coils wrapped around the glowing sample vacuum tube causes hydrogen gas to be liberated from the mineral inside the tube. The gas is then passed through traps containing liquid nitrogen and a methanol/dry ice mixture to freeze and thaw the hydrogen and condense out all the other gases. The resultant volume of hydrogen gas is measured using a mercury pump and a calibrated volume.

est into a molybdenum crucible, which is heated overnight in a vacuum to drive off any adsorbed atmospheric water from the surfaces of the grains. Then, using an induction furnace, the samples are heated to very high temperatures to release the $(OH)^{1-}$ and H_2O from inside the crystals in the form of gas. The released gases, which at this point include other volatiles such as CO_2, are then passed through a series of liquid nitrogen and methanol-dry ice slush traps to separate water molecules from other condensable and non-condensable gases. Water vapor is then passed over a hot (>750 °C) furnace containing uranium where the water is reduced to H_2 gas ($U + 2H_2O = UO_2 + 2H_2$). A mercury Toepler pump is used to collect the hydrogen vapor in a calibration volume where it can be measured. Calibration of the mercury volume is done using distilled water as a standard.

This method yields very accurate data for the H contents of mineral separates. However, the method is extremely tedious and takes 2–24 hours per sample. More automated approaches include quantitative high temperature carbon reduction, in which the hydroxyl groups in sheet silicates and amphiboles are driven off at 1450 °C in a He gas stream and then reduced to H_2 by carbon reduction. The H_2 is carried into an isotope ratio mass spectrometer for measurement of D/H. Alternatively, lasers can be used as the heat source. While these approaches can analyze very small volumes of minerals and give isotopic information, the uranium reduction technique remains the gold standard for accurate stoichiometric analysis of H.

Introduction to Spectroscopic Methods

Many of the modern analytical techniques utilize spectroscopic methods. They make use of characteristic wavelengths and energies that are produced as electrons or neutrons move from one energy level to another within atoms or are related to the vibrational modes of crystal lattices. X-rays and gamma rays (γ-rays) are often used to analyze minerals, because their wavelengths are about the same size as the individual atoms in mineral structures (Figure 9.5).

Electrons are excited by energy that is just the right wavelength to make them move between the energy levels, or shells, in an atom. As you will recall, electrons can occupy *s, p, d*, and higher energy orbitals. Figure 9.6 shows the relative energies of the different shells around the nucleus (refer back to Table 7.4) in graphical form. The K levels represent the orbitals with quantum number $n = 1$, and are the lowest in energy. As you go further away from the nucleus, the energy of the electrons in the shells increases from K to L to M to N shells, etc. In many spectroscopic methods, energy of various wavelengths is passed through a sample. Electrons lying in the path of that energy flux will absorb some of the energy. The amount of absorption is proportional to the atomic number and density of the element that is absorbing the energy.

Before we explain how we use the energies emitted by electron transition in elements, it is worthwhile to explain some of the terminology used in spectroscopy. We can describe the electromagnetic spectrum in one of three ways: (1) energy, (2) wavelength, and (3) frequency. These three are related by the following equation:

$$\lambda = \frac{K}{E} \text{ or eV} = \frac{K}{\gamma} \qquad \text{Equation 9.1}$$

where λ = wavelength of radiation in Å, E = energy in eV, and K = 12,399.7, which is the conversion derived from Planck's constant, and

$$\lambda = \frac{c}{f} \qquad \text{Equation 9.2}$$

where c = speed of light, 2.9979 x 10^{18} nm/sec, f = frequency in sec^{-1}, λ = wavelength in Å. In Equation 9.1, we see there is an inverse relationship between energy (here given in eV) and wavelength (here given in Å), and we can use this equation to convert between the two. Thus the

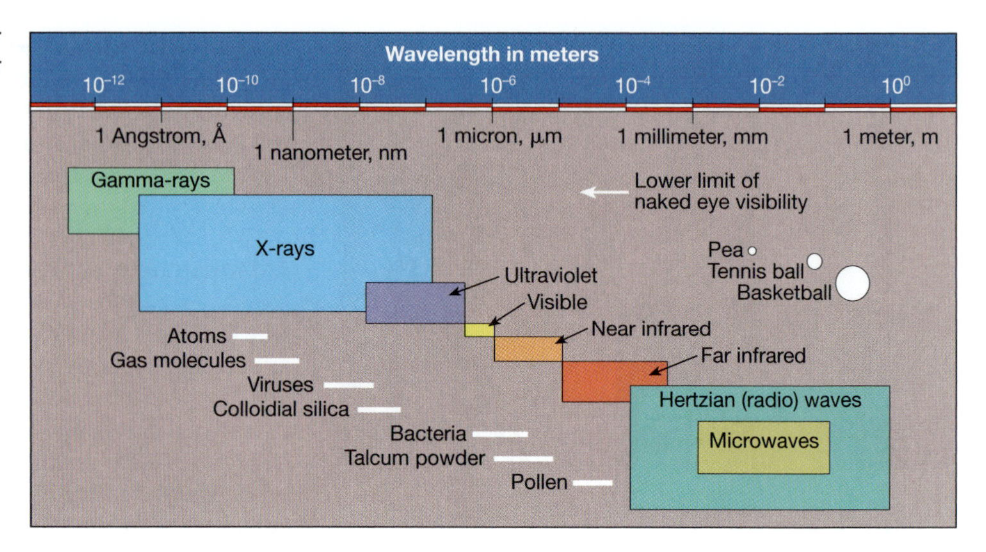

Figure 9.5. Comparison of wavelengths (i.e., sizes!) of various materials.

shorter the wavelength, the higher the energy. For example, X-rays that have a short wavelength (on the order of a few Å's) have high energy. Compare this to visible light, which has a longer wavelength (on the order of 1,000s of Å) and be glad the visible light has a lower power, or the sun would fry us (even more)! Equation 9.2 relates wavelength to frequency, and in turn, Equation 9.1 can be used to relate frequency to energy. Frequency is less often used in mineralogical measurements than energy and wavelength. Table 9.1 gives the relationships (worked out with these two equations) relating wavelength (in various units) to energy and frequency.

As energy increases (and wavelength decreases), one of the electrons in the lowest K shell may gain enough energy to be ejected from the atom (Figure 9.6). The critical energy at which this happens is characteristic of each element. When a low energy electron is emitted, electrons at higher levels will drop down into the vacated position, giving off unneeded energy (or radiation) as they do (Figure 9.7). This release of energy is called **fluorescence**. The concentration of any element is proportional to the number of X-rays given off in this process. Because the energy levels are specific and unique for each element, identification of the characteristic X-ray wave-

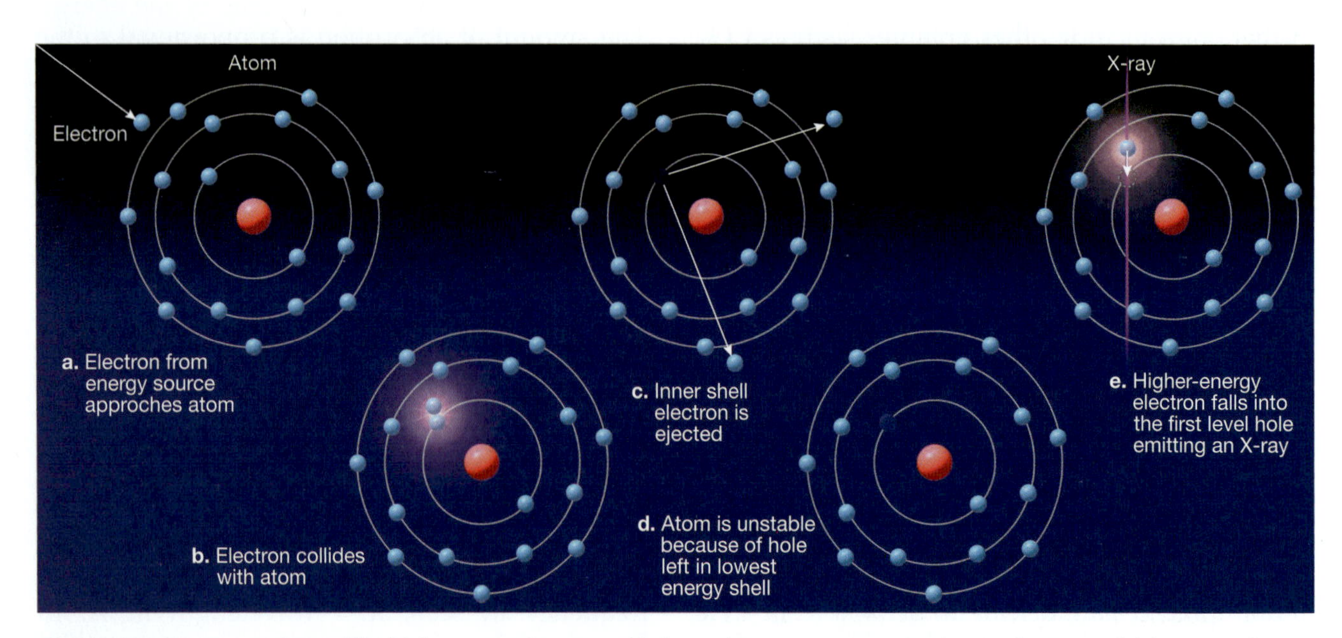

Figure 9.6. X-rays are formed by high-energy electron collisions. An energy source emits an electron, which approaches an atom. It collides with an atom, causing one of the two lowest-energy, inner shell electrons to be ejected. The resultant atom is unstable because it has a hole in its lowermost, lowest-energy shell, so a higher-energy electron falls down into that first level to plug the hole, emitting an X-ray of a characteristic wavelength.

Table 9.1. Relationships between wavelength, energy, and frequency for electromagnetic radiation (values calculated by use of Equations 9.1 and 9.2).

Å	nm	um	mm	cm	cm^{-1}	eV	Frequency
1	0.1	0.0001	0.0000001	0.00000001	100,000,000	12,399.7	3×10^{18}
10	1	0.001	0.000001	0.0000001	10,000,000	1,240	3×10^{17}
100	10	0.01	0.00001	0.000001	1,000,000	124	3×10^{16}
1,000	100	0.1	0.0001	0.00001	100,000	12.4	3×10^{15}
10,000	1,000	1	0.001	0.0001	10,000	1.24	3×10^{14}
100,000	10,000	10	0.01	0.001	1,000	0.124	3×10^{13}
1,000,000	100,000	100	0.1	0.01	100	0.0124	3×10^{12}
10,000,000	1,000,000	1,000	1	0.1	10	0.00124	3×10^{11}
100,000,000	10,000,000	10,000	10	1	1	0.000124	3×10^{10}

lengths can be used to identify which elements are present in the sample being analyzed (Table 9.2). We'll be presenting examples of how these are useful in the following sections as we discuss each technique.

Inductively-Coupled Plasma (ICP) and Atomic Absorption Spectrometry (AAS)

These methods really derive from the virtually forgotten methods of blowpipe analysis, in which a mineral sample was powdered and blown into a flame and the resulting colors observed. Many elements give off distinct colors as they burn which are related to transitions of their electrons. For instance, Na burns yellow (that's why some street lights have a yellow cast). Modern instruments measure the wavelengths of the electron transitions in a quantitative manner, whereas the blowpipe and color changes were more qualitative. But, nevertheless, the elements present could be determined.

Both Inductively-Coupled Plasma (ICP) and Atomic Absorption Spectrometry (AAS) analyses depend on dissolving the sample in a liquid, just as discussed above for wet chemical methods (Figure 9.8). Usually this is done at high temperatures using a flux (something that helps break down) and/or involves some type of acid, so that all the elements break free of the mineral structure to drift around in an aqueous liquid. Next the liquid containing the dissolved sample is sprayed to form an aerosol. The aerosol is sprayed into a flame (AAS) that heats the sample to 2000°C (Figure 9.9), or into a plasma (ICP) that heats the sample to 6000°C.

For **atomic absorption analysis**, a cathode lamp on one side of the flame emits a beam of light only at particular wavelengths that can be absorbed by the element of interest. That element, which is "burning" in the flame, absorbs some of light. A monochrometer and detector on the other side of the flame measure the reduction in light intensity at one of the diagnostic wavelengths, as is caused by absorption by that element in the flame. By comparison with a standard, concentrations of each element can be determined.

An **inductively-coupled plasma** spectrometer (Figure 9.10) works slightly differently. In this method, the aerosol is swept by argon gas into the excitation region of a plasma that is formed by using a radio frequency source to electrically heat the argon gas. The atoms are excited and ionized by the hot plasma, and each emits light of wavelengths that are characteristic of each element. The intensity of light emitted at a given wavelength varies with the concentration of the

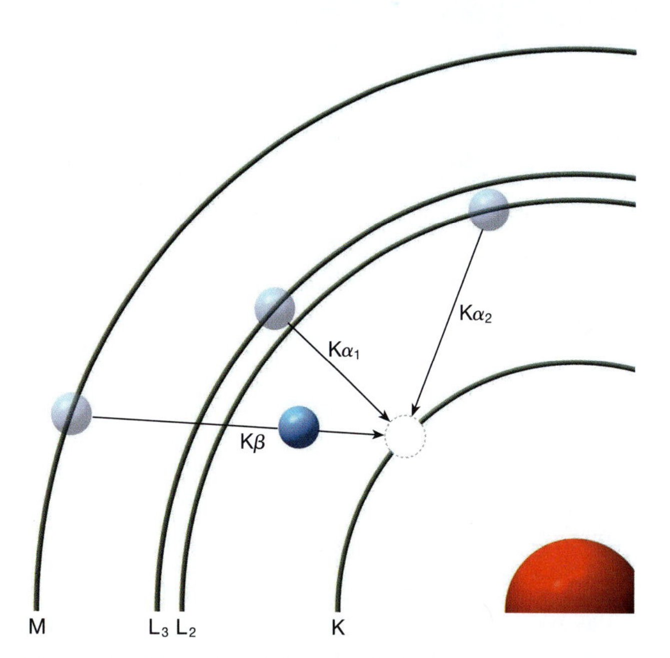

Figure 9.7. *Energies of electron shells with various transitions labeled. The differences in energy between the various energy levels are characteristic of different elements, as listed in Table 9.2.*

Table 9.2. Photon energies of principal K-, L-, and M-shell emission lines, in eV, and $K\alpha_1$ in Å (after Kortright and Thompson, 2000).

Element	$K\alpha_1$	$K\alpha_2$	$K\beta_1$	$K\beta_2$	$L\alpha_1$	$L\alpha_2$	$L\beta_1$	$L\beta_2$	$K\alpha_1$ in λ
3 Li	54.3								228.4
4 Be	108.5								114.3
5 B	183.3								67.6
6 C	277								44.8
7 N	392.4								31.6
8 O	524.9								23.6
9 F	676.8								18.3
10 Ne	848.6	848.6							14.6
11 Na	1040.98	1040.98	1071.1						11.9
12 Mg	1253.60	1253.60	1302.2						9.89
13 Al	1486.70	1486.27	1557.45						8.34
14 Si	1739.98	1739.38	1835.94						7.13
15 P	2013.7	2012.7	2139.1						6.16
16 S	2307.84	2306.64	2464.04						5.37
17 Cl	2622.39	2620.78	2815.6						4.73
18 Ar	2957.70	2955.63	3190.5						4.19
19 K	3313.8	3311.1	3589.6						3.74
20 Ca	3691.68	3688.09	4012.7	341.3	341.3	344.9			3.36
21 Sc	4090.6	4086.1	4460.5	395.4	395.4	399.6			3.03
22 Ti	4510.84	4504.86	4931.81	452.2	452.2	458.4			2.75
23 V	4952.20	4944.64	5427.29	511.3	511.3	519.2			2.50
24 Cr	5414.72	5405.509	5946.71	572.8	572.8	582.8			2.29
25 Mn	5898.75	5887.65	6490.45	637.4	637.4	648.8			2.10
26 Fe	6403.84	6390.84	7057.98	705.0	705.0	718.5			1.94
27 Co	6930.32	6915.30	7649.43	776.2	776.2	791.4			1.79
28 Ni	7478.15	7460.89	8264.66	851.5	851.5	868.8			1.66
29 Cu	8047.78	8027.83	8905.29	929.7	929.7	949.8			1.54
30 Zn	8638.86	8615.78	9572.0	1011.7	1011.7	1034.7			1.44
31 Ga	9251.74	9224.82	10264.2	1097.92	1097.92	1124.8			1.34
32 Ge	9886.42	9855.32	10982.1	1188.00	1188.00	1218.5			1.25
33 As	10543.72	10507.99	11726.2	1282.0	1282.0	1317.0			1.18
34 Se	11222.4	11181.4	12495.9	1379.10	139.10	1419.23			1.10
35 Br	11924.2	11877.6	13291.4	1480.43	1480.43	1525.90			1.04
36 Kr	12649	12598	14112.0	1586.0	1586.0	1636.6			0.98
37 Rb	13395.3	13335.8	14961.3	1694.13	1692.56	1752.17			0.93
38 Sr	14165	14097.9	15835.7	1806.56	1804.74	1871.72			0.88
39 Y	14958.4	14882.9	16737.8	1922.56	1920.47	1995.84			0.83
40 Zr	14775.1	15690	17667.8	2042.36	2039.9	2124.4	2219.4	2302.7	0.84
41 Nb	16615.1	16521.0	18622.5	2165.89	2163.0	2257.4	2367.0	2461.8	0.75
42 Mo	17479.34	17374.3	19608.3	2293.16	2289.85	2694.81	2518.3	2623.5	0.71
43 TC	18367.1	18250.8	20619.0	2424	2420	2538	2674	2792	0.68
44 Ru	19279.2	19150.4	21656.8	2558.55	2554.31	2683.23	2836.0	2964.5	0.64
54 Rh	20216.1	20073.7	22732.6	2696.74	2692.05	2834.41	3001.3	3143.8	0.61
46 Pd	21177.1	21020.1	233818.7	2838.61	2833.29	2990.22	3171.79	3328.7	0.59
47 Ag	22162.92	21.990	24942.4	2984..31	2978.21	3150.94	3347.81	3519.59	0.56
48 Cd	23173.6	22984.1	26095.5	3133.73	3126.91	3316.57	3528.12	3716.86	0.54
49 In	24209.7	24002.0	27275.9	3286.94	3279.29	3487.21	3713.81	3920.81	0.51
50 Sn	25271.3	25044.0	28486.0	3443.98	3435.42	3662.8	3904.86	4131.12	0.49
51 Sb	26359.1	26110.8	29725.6	3604.72	3595.32	3843.57	4100.78	4347.79	0.47
52 Te	27472.3	27201.7	30995.7	3769.33	3758.8	4029.58	4301.7	4570.9	0.45
53 I	28612.0	28317.2	32294.7	3937.65	3926.04	4220.72	4507.5	4800.9	0.43
54 Xe	29779	29458	33624	4109.9					0.42

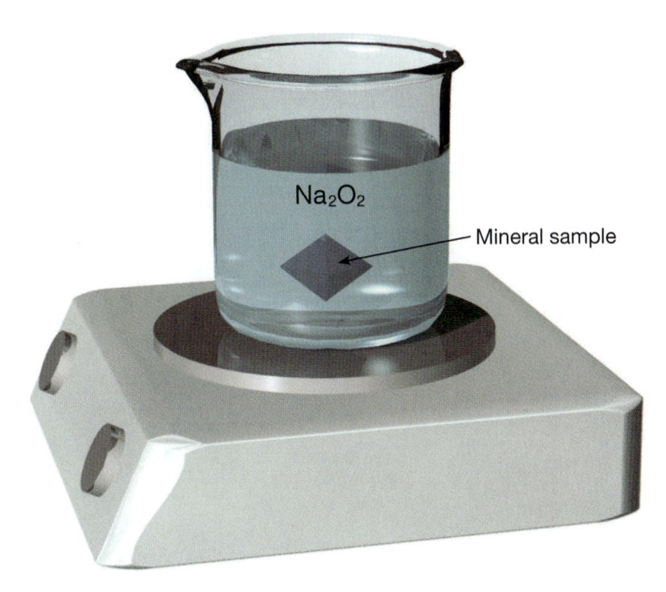

Figure 9.8. For both ICP and AAS, the solid samples must be dissolved into a liquid just as in wet chemical methods. However, for ICP and AAS the liquid is "burned" and the colors that are given off can then be related to the elements in the solution.

element. This technique can be used to detect even light elements (i.e., those with low atomic numbers) at the parts per billion level (ppb). A variation on this instrument, called an ICP-MS (inductively-coupled mass spectrometer), can even measure isotopes of various elements in solution.

Sample preparation issues are the main complications for AAS or ICP analysis. These methods work best for elements with atomic numbers greater than Li, B, and Na—because those elements are commonly used in the dissolution process. For those elements, it is possible (though not always easy) to use different solvents to dissolve the minerals, such as KOH for Li, NaOH for B, KF_2 for Be, etc. Special Zr crucibles and furnaces must be used for such analyses. Commonly, the most difficult and time-consuming part of AAS or ICP is sample dissolution, which can take anywhere from half an hour to several days! Another problem is keeping the elements in solution. Once the sample is in solution, the actual analysis may take only a few minutes, and nearly any element can be analyzed.

To learn more about these methods, there are several good books, including Welz and Sperling (1999) and Lajunen (1997). Your library probably has a book on this topic, which is commonly covered in analytical chemistry courses.

X-ray Fluorescence (XRF)

This method uses an X-ray beam of fixed wavelength (such as Cu Kα) to bombard a sample with X-rays and then determine what elements are present based on emitted wavelengths of secondary X-rays (Figure 9.11). The tube works by using a high voltage power supply to accelerate electrons toward a metal target, which then generates X-rays. The type of metal in the target in turn affects the energies of the released X-rays. When X-rays of just the right energy hit your unknown

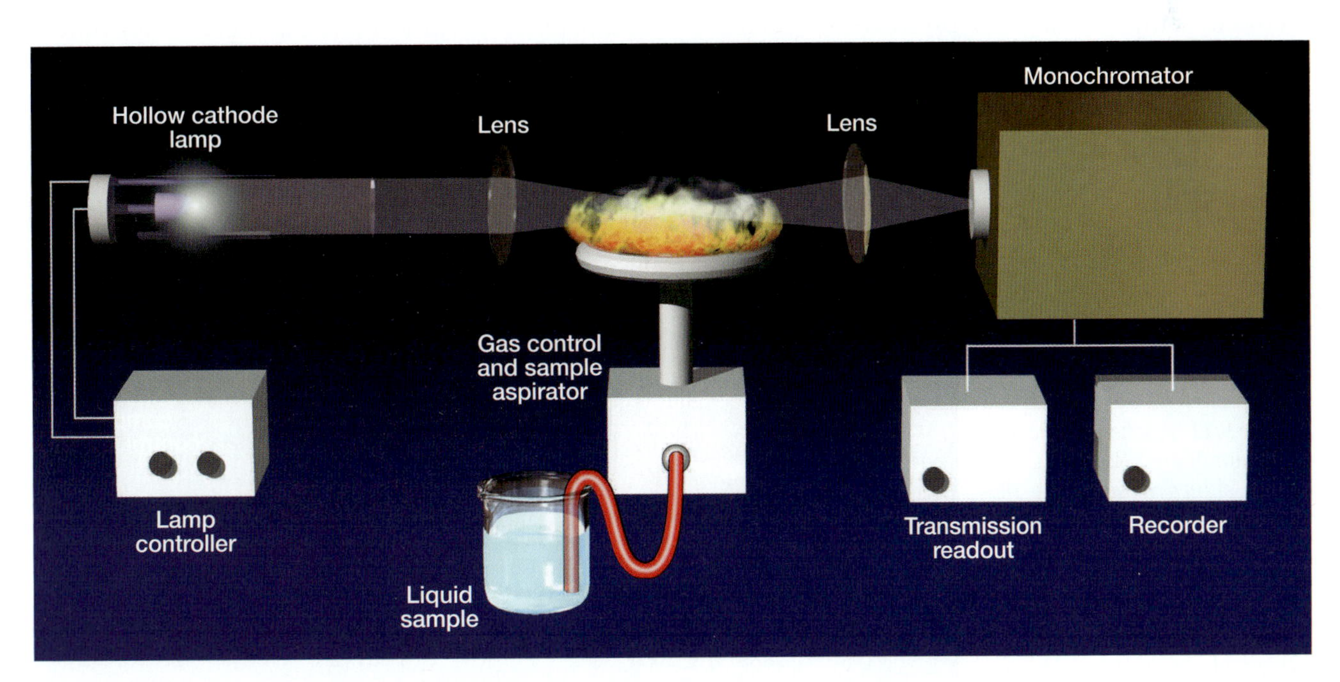

Figure 9.9. This illustration shows a schematic diagram of AAS.

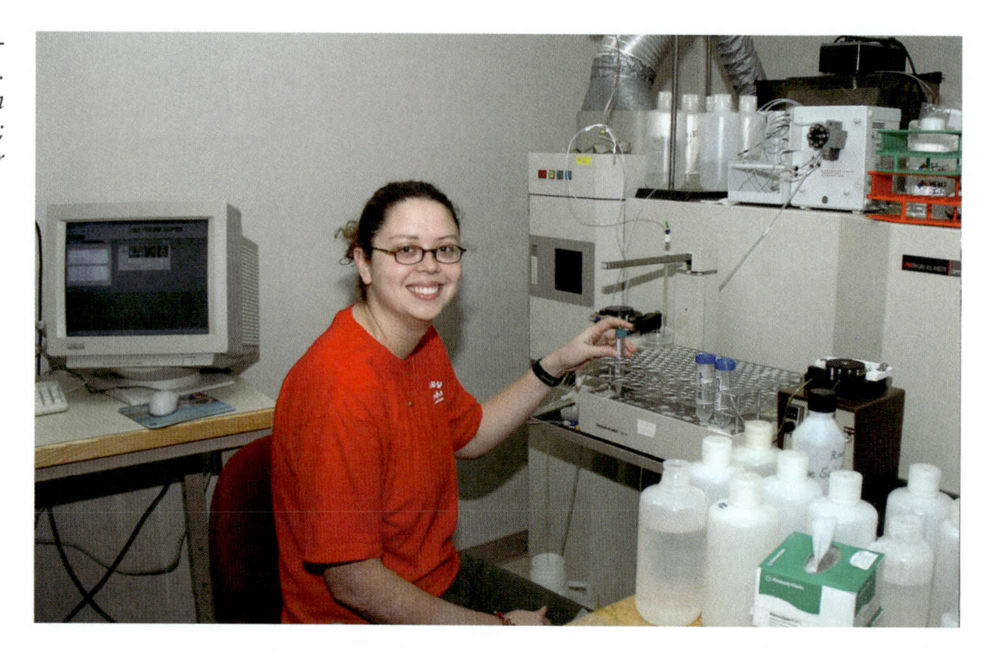

Figure 9.10. *The ICP apparatus at the University of Idaho. The ICP, like most modern equipment, is fully automated; a computer system is used for data collection and analysis.*

sample, they cause the atoms in the sample to eject inner shell electrons, as noted above. Photons are emitted when outer shell electrons fall down to take the place of the ejected one (fluorescence). The energy of the photon is characteristic of the element present, and depends on the energy difference between the outer shell and the inner-shell electrons (Figure 9.7). Diffraction gratings are used to collect the photons generated by the fluorescence of each element in the sample. The gratings or crystals are set up to diffract only the X-rays of interest to the analysis, and the crystal and detector can be rotated to the correct angle for this diffraction to occur. Analysis of a multi-element sample will thus produce a spectrum with distinct peaks for each element and its associated transitions.

X-ray intensities are measured for each element in an unknown (Figure 9.12) sample and compared against intensities of the same elements in a standard. Concentrations are calculated using this relationship, by determining the proportionality constant that relates the concentration of each element to the intensity of its X-rays:

$$C^{\text{element}}_{\text{standard}} = K \text{ Intensity}^{\text{element}}_{\text{standard}},$$

where $C^{\text{element}}_{\text{standard}}$ is the concentration of the element in the standard, $\text{Intensity}^{\text{element}}_{\text{standard}}$ is the intensity of the fluorescence peak at the appropriate energy (see Table 9.2), and K is the proportionality constant. Once a value of K has been determined, then it can be applied to the analysis of unknown minerals where the peak intensity can be meas-

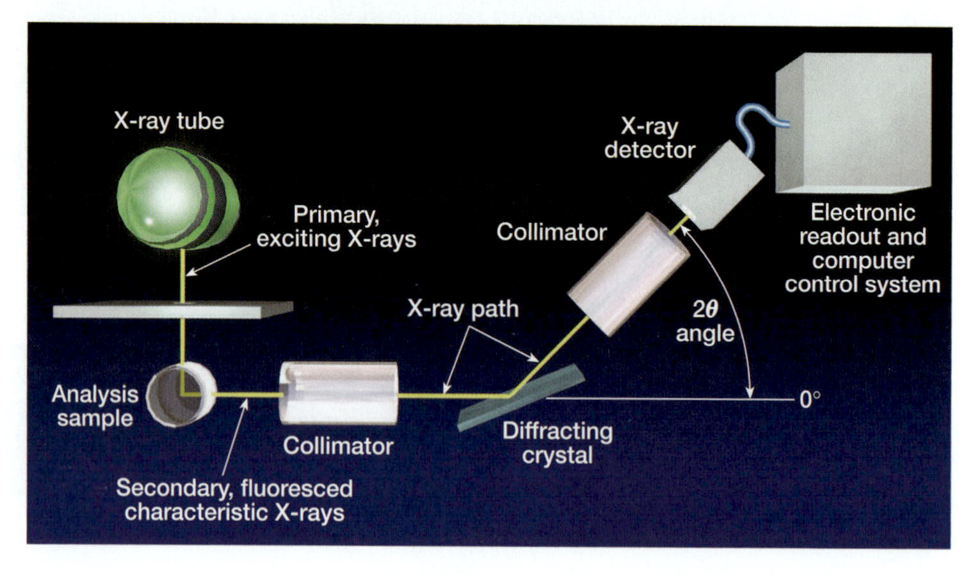

Figure 9.11. *A schematic representation of an XRF. An X-ray source shines X-rays onto the sample. In turn, some of these X-rays are diffracted into a detector. This configuration is similar to that of an X-ray diffractometer, which will be explained in more detail in Chapter 15.*

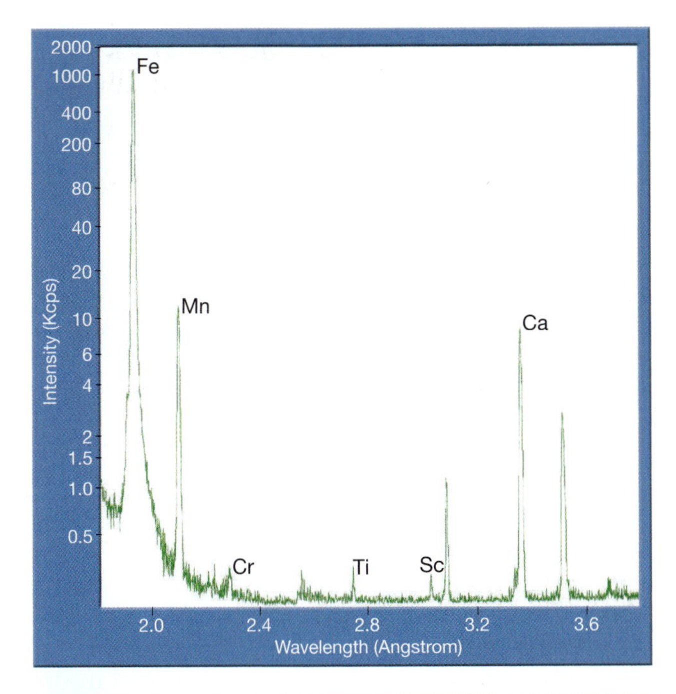

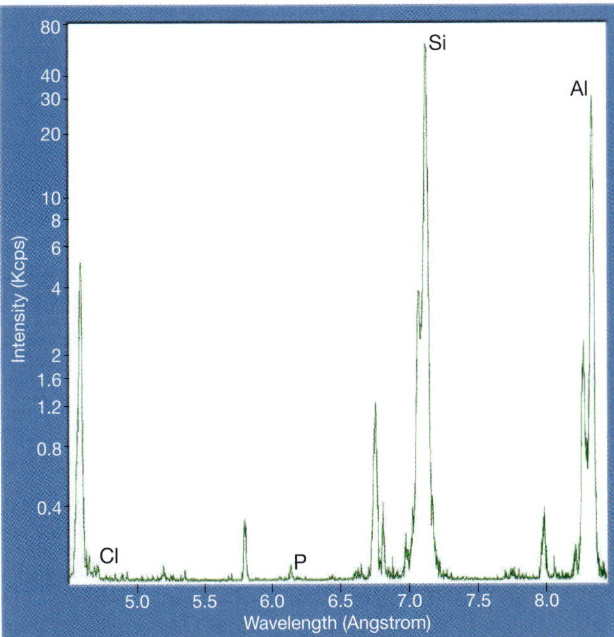

Figure 9.12. XRF spectra of a garnet. Peaks represent electron transitions. Each transition set can be matched to an element. Some of the peaks are identified. Notice that wavelength is plotted along the horizontal axis. Atoms with higher atomic numbers plot at lower wavelengths.

ured, so the concentration can be calculated (c.f. Jenkins, 1999). In practice, this formulation only really works when the concentrations in the standards are roughly the same as those in the unknowns. In other situations, multiple standards and complex calculations to solve for all the interfering effects are required.

Samples are usually prepared for analysis by powdering and then melting them into glass pellets (Figure 9.13), sometimes with the help of a flux to make them melt at a lower temperature (see Buhrke et al., 1997). Once the pellets have been made and a standard has been analyzed, the actual analysis of an unknown takes only a few minutes. A typical XRF unit can measure all the elements in the periodic table that are heavier than oxygen, and detect them at levels as low as parts per million (ppm); light elements are difficult to analyze with this technique. Although the XRF technique was first developed for use on geological samples, advances in technology have also made this a useful tool for many industrial applications, such as quality control! Analyses can also be made on microscopic scales (Janssens et al., 2000).

Electron Microprobe and the Scanning Electron Microscopy

Both electron microprobes (EPMA) and scanning electron microscopes (SEM) combine the capability to image minerals at very small scales with the ability to analyze chemical compositions at those same scales. While the two methods are similar, there are some slight differences. Both instruments can cost $100,000–$700,000 and fill an entire room (Figures 9.14 and 9.15). EPMA requires greater care in sample preparation (the samples must be flat and polished), while angular mineral grains are often analyzed with the SEM. In general, the SEM will yield less accurate and precise chemical data than the EPMA. However, the SEM can be a great aid in identification of minerals, while the EPMA is usually used to characterize minerals. These methods are the most commonly-used tools for determining the chemical composition of minerals. They represent nondestructive techniques capable of rapidly (typically less than 1 minute) analyzing samples on spots as small as a few microns, and sample preparation is very simple.

Electron beam instruments operate somewhat like the XRF with one important difference: they use a beam of electrons, rather than X-rays, to excite the atoms in a specimen. The advantage is that an electron beam can readily be focused to micron spot sizes, whereas this is difficult to do with X-rays. The electrons in the beam are typically accelerated from an electron gun to energies from 15–20 KeV (Figure 9.17). The beam interacts with the mineral sample and excites some of the electrons in the mineral to energy levels above their ground state (which is the lowest energy

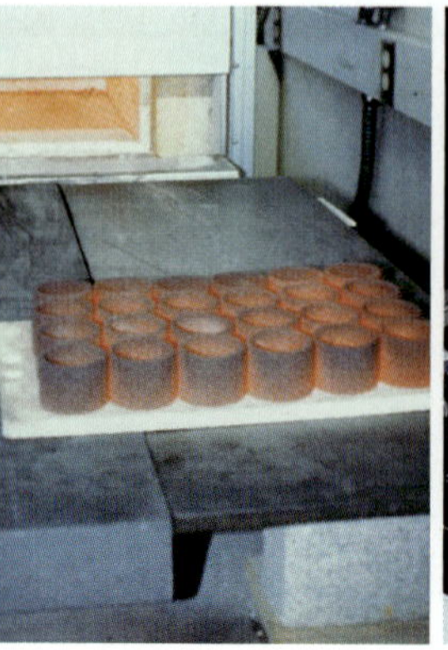

Figure 9.13. *The sample preparation for XRF analysis requires that the mineral be melted into a glass; its surface is then polished. The melting is required to homogenize the sample.*

state). When the excited electrons fall back to their ground states, they fluoresce and emit X-rays (this process should sound familiar from above). Due to the laws of quantum mechanics, the emitted X-rays have very specific wavelengths, or energies, depending upon which elements produced them. The intensity of energy given off at any specific wavelength depends on the concentration of the element in the sample.

Scanning electron microscopes yield high-quality images (Figure 9.16) of minerals as well as semi-quantitative chemistry. The images are created by scanning back and forth over an area of interest and measuring a signal at each pixel. The resultant images look like optical microscopic images, but they're not! The meaning of the image depends on the signal that is recorded by the SEM. **Secondary-electron images** are most sensitive to surface topography and commonly image the morphology of fine-grained minerals. **Backscattered-electron (BSE) images** show contrast in average atomic number and are useful to distinguish elemental distribution of minerals (Figure 9.16). Heavier atoms (e.g., Mn or Fe) tend to backscatter electrons more efficiently and thus, they appear brighter in BSE images than atoms with lower atomic numbers (e.g., Mg or Si), which appear darker. As a result, BSE images convey both morphological and compositional information. SEM's used by mineralogists are normally equipped with **energy-dispersive** (see below) **spectrometers (EDS)** to allow them to analyze X-rays from the sample. An example of an EDS spectrum for a garnet obtained on an SEM is shown in Figure 9.17. Notice how the peaks are plotted as a function of energy. Compare this to the XRF output (Figure 9.11) for a garnet, where the peaks are plotted as a function of wavelength. The combination of morphological information obtained from imaging and compositional data gained from BSE and EDS is extremely powerful in mineral identification.

Electron microprobes (EPMA) for **e**lectron **p**robe **m**icro**a**nalysis are very similar to SEM's and can also create nice images of minerals particularly elemental maps (Figure 9.19). However, they are optimized for analyzing the X-rays that are emitted from minerals (and other materials, of course). The electron gun on an EPMA system can be up to 1000 times brighter than on an SEM, resulting in an electron beam that generates a lot of fluorescence! The wavelength and intensity of the emitted X-rays are analyzed with one of the two types of detectors. **Energy dispersive spectrometers** (EDS) can simultaneously count the number of X-ray photons emitted from a mineral and analyze the energy of each photon, thereby identifying the element that emitted that photon. In contrast, **wavelength-dispersive spectrometers** (WDS) have crystals that are individually tuned to the wavelength of the X-rays emitted by an element of interest (Figure 9.18). Elements from beryllium (Z = 4) to uranium (Z = 92) can be analyzed under optimal conditions. Although this may seem like a more cumbersome method,

PRECISION VS. ACCURACY

In everyday usage, the terms "precision" and "accuracy" are often taken to mean the same thing. However, to an analyst, these terms have very specific, distinct meanings.

Precision effectively means the same thing as reproducibility. It is a measure of how well you can perform an analysis in repeated operations. You can think of it as the degree of perfection in your analysis.

Accuracy describes the difference between the true (or accepted) answer, and the one that you have measured.

To summarize, precision reflects how well you can do the analysis, and accuracy reflects the relationship between your answer and the "truth." A good analyst strives for both high precision and high accuracy.

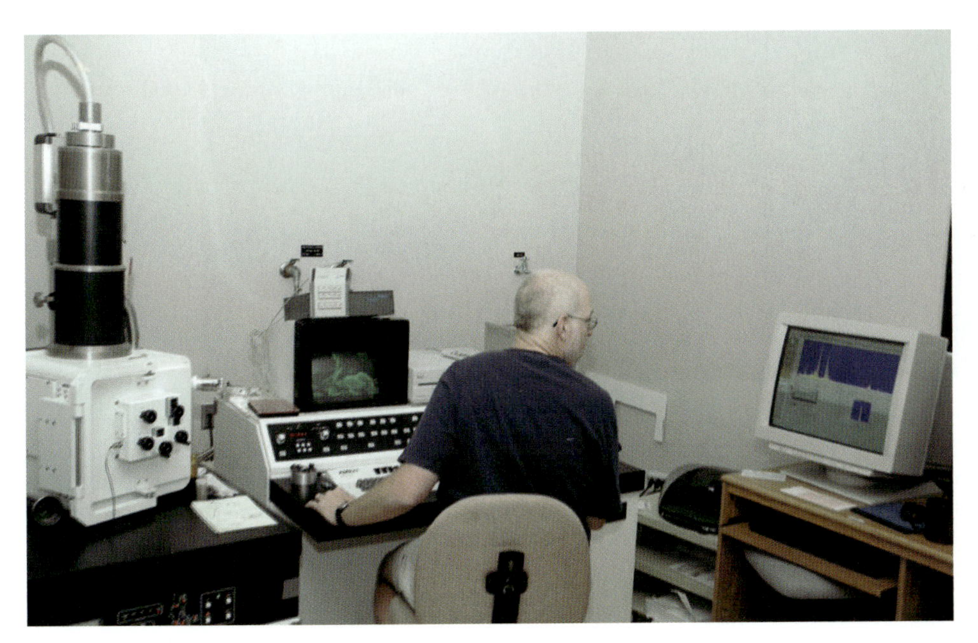

Figure 9.14. A photograph of the SEM at the University of Idaho. The apparatus consists of the electron gun and sample chamber on the left, the viewing screens and required electronics in the center, and a computer to control the systems and collect and process the data. An image of a sample is on the center screen and its EDS spectrum appears on the computer monitor.

the precision and detection limits for WDS are significantly better than for EDS. For more information on SEM and electron microprobe analysis, consult Reed (1993, 1996).

Ion Microprobes

Secondary ion mass spectrometry (SIMS) is used for the study of elements that the electron microprobe can't detect, either because they are too light (low on the period table) or too low in abundance. SIMS is also (and primarily) used for the study of stable and radiogenic isotopes. The SIMS technique uses a beam of ions generated by a duoplasmatron or an alkali metal source. This produces a beam of primary ions such as Cs^+, O_2^+, or O^-, with relatively low energies (0.5–20 KeV). These can be focused to a few μm diameter spot on the surface of a sample. Material is removed or "sputtered" from the surface of the sample by the bombardment of ions (Figure 9.20). A small fraction of the sputtered atoms is ionized by this bombardment (typically 10^{-2} to 10^{-4}).

Sputtered species are emitted in various excited states, as singly and multiply charged positive

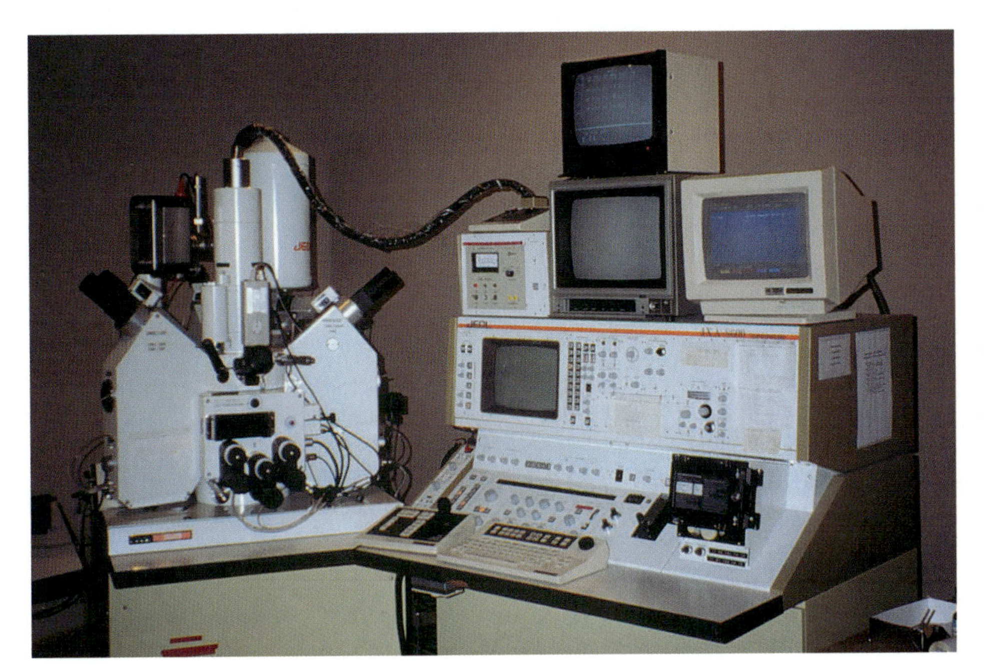

Figure 9.15. A photograph of an electron microprobe, which is similar to the SEM in Figure 9.14. The electron column, optics, and detectors are on the left and the various monitors and electronics are on the right. Notice how much more complicated the area around the electron column appears when compared to the SEM; this is the big difference in the two instruments. The microprobe has a more complex set of detectors that provides more accurate and precise chemical data.

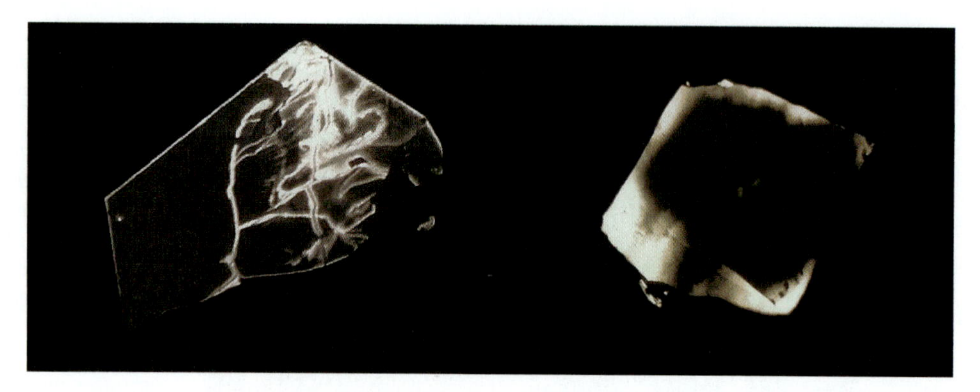

Figure 9.16. Backscattered electron images (BSE) obtained from an SEM. The samples are single crystals of the zeolite group mineral heulandite, $(Na,K)Ca_4[Al_9Si_{27}O_{72}]\cdot 24H_2O$. The lighter areas are portions of crystal in which Pb has diffused around the edges in the top grain and along the cracks in the lower one. Because Pb has a much higher atomic number (and thus more electrons) than the other elements normally occurring in the zeolite, more electrons are backscattered to the detector and image appears brighter were Pb occurs. (Images modified from Gunter et al. (1994) American Mineralogist, 79, 675–682.)

and negative ions, and as ionized clusters of atoms. The ionized species, or secondary ions, are accelerated into the entrance slit of a mass spectrometer, and then counted by a detector after mass selection.

While ion microprobe analysis is a very sensitive technique for light element analysis in geological materials, standardization of the measurements is difficult because of the lack of a satisfactory theoretical model for sputtering from minerals. In the absence of adequate theoretical models, "quantification" is mainly empirical, involving comparison between elemental intensity ratios in the unknown mineral and a standard of similar composition. This reliance on suitable standards for accurate calibration is, however, a fundamental limitation of ion microprobe analysis.

For elemental analysis, the ion microprobe is 1–4 orders of magnitude more sensitive than the electron microprobe, depending on the element and the matrix of other elements surrounding it. In contrast to the electron microprobe, the light elements H, B, Be, and Li are easily detected with this technique (Hervig, 1996 and 2002; Ihinger et al., 1994). Many other trace elements are easily studied by this technique, which can measure elemental abundances at the part-per-million or lower levels for all elements with the exception the noble gases (Reed, 1980; Zinner and Crozaz, 1986; Shimizu and Hart, 1982).

Proton-Induced Emission Spectroscopy

Spatially resolved concentrations of major, minor, and trace elements can be determined using nuclear methods such as proton-induced gamma-ray (PIGE) and X-ray (PIXE) emission spectroscopy. Because an incident proton beam is used, PIXE and PIGE have low background signals relative to the electron microprobe and better overall sensitivities, especially for the lighter (lower atomic number) and trace elements. Both PIGE and PIXE can be performed on either macroscopic powders or on microscope slides with very finely-focused (1 μm diameter) beams. Minimum detection limits can be as good as 5 ppm with relative standard deviations on the

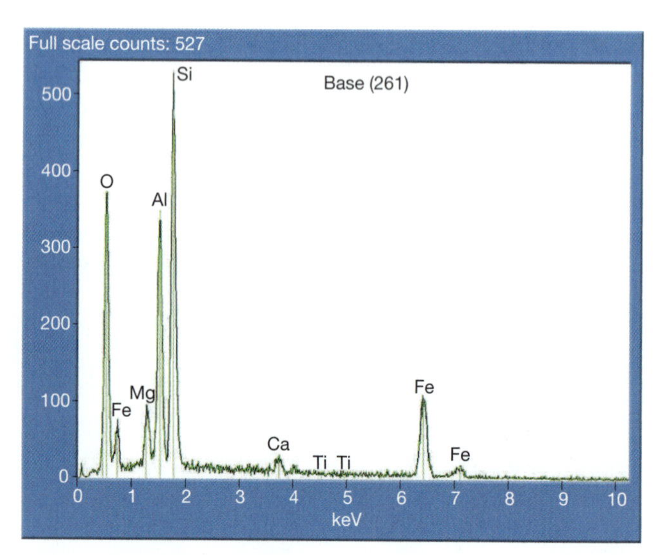

Figure 9.17. EDS output of an Idaho star garnet obtained from an SEM. The horizontal axis is plotted in keV (or 1000s of eV). The most intense peaks in this figure result from Kα transitions. See if you can match up the values in Table 9.2 with the elements in this output. Based on the EDS, the major elements in the garnet are Al, Si, and Fe; thus in this case the garnet species is an almandine, $Fe_3Al_2(SiO_4)_3$. However, other elements occur in smaller amounts, such as Ca, Ti, and Mg. Compare this output with that from the XRF (Figure 9.12).

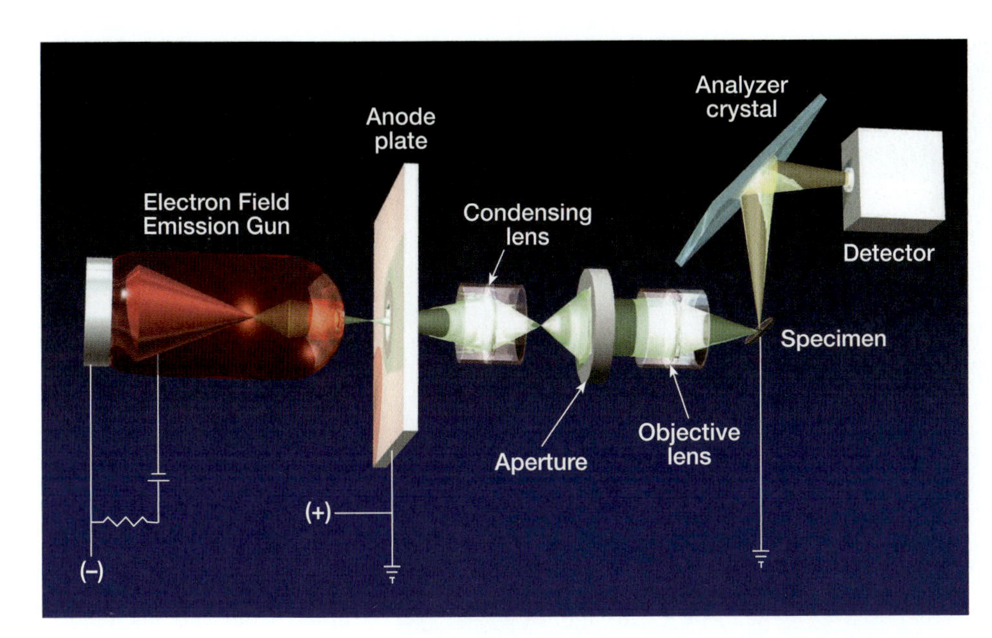

Figure 9.18. An electron microprobe uses an electron field emission gun instead of an X-ray tube to produce X-rays, which are accelerated and focused to a small spot on a thin section. Most electron microprobes have at least three detectors, so that multiple elements can be analyzed simultaneously.

order of a few percent, even on light elements. As with an electron microprobe and SEM, the proton beam can be scanned across a surface to make a map of the concentration of each element of interest.

These techniques can rapidly and simultaneously analyze many elements with very high sensitivities. PIXE is used to analyze major elements from Na to Ca and trace elements with higher atomic numbers than calcium, while PIGE is used for light elements such as H, Li, Be, B, and F. These techniques are based upon the detection of the X-rays or prompt gamma-rays that are emitted following bombardment with a high energy charged particle. Characteristic X-ray emissions by major and trace elements are studied with PIXE using methods similar to those for electron microprobe analysis; in PIGE, γ-ray emissions are used. The energy of the X-ray or γ-ray identifies the element or isotope in the sample, and the intensity of the radiation is used to determine the amount of the element or isotope in the sample.

The beam of protons is supplied by a Van De Graff accelerator, usually operated at an energy of around 2.5-3.5 MeV. The beam is focused onto the sample by a system of electrostatic or magnetic lenses. Detected counts are plotted against energy as with many of the previous techniques we've discussed, and concentrations are determined by comparisons with standards. A classic reference on this technique is the volume edited by Johannson and Campbell (1988); see also Robertson and Dyar (1996). Because these techniques require use of an accelerator, their applications have been somewhat limited.

Neutron Activation Analysis

Neutron activation analysis techniques rely on the fact that most stable atoms can be made radioactive if they are bombarded with enough energy, usually in the form of neutrons, and less commonly, protons, deuterons, or high energy radiation. Once the atoms become radioactive, γ-rays of characteristic energies will be given off by each of the radioactive isotopes. Most commonly, a nuclear reactor is used to create a thermal flux

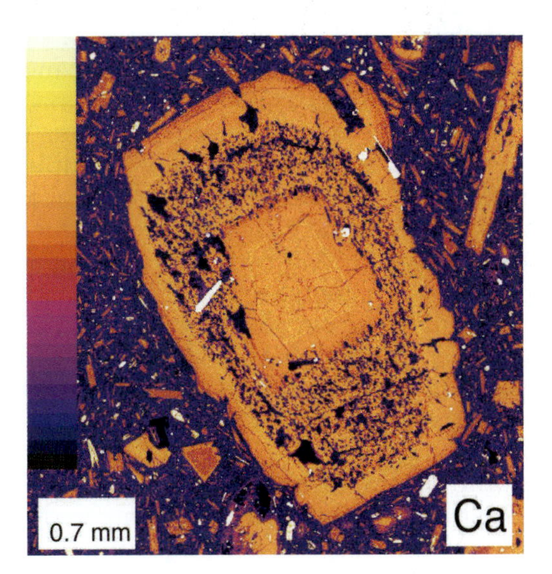

Figure 9.19. X-ray compositional map of relative abundance of Ca in a plagioclase feldspar crystal from the Atascosa Lookout Lava Flow in Arizona. Color scale on left indicates that higher concentrations of Ca correspond to lighter colors. Photo courtesy of Sheila Seaman, University of Massachusetts.

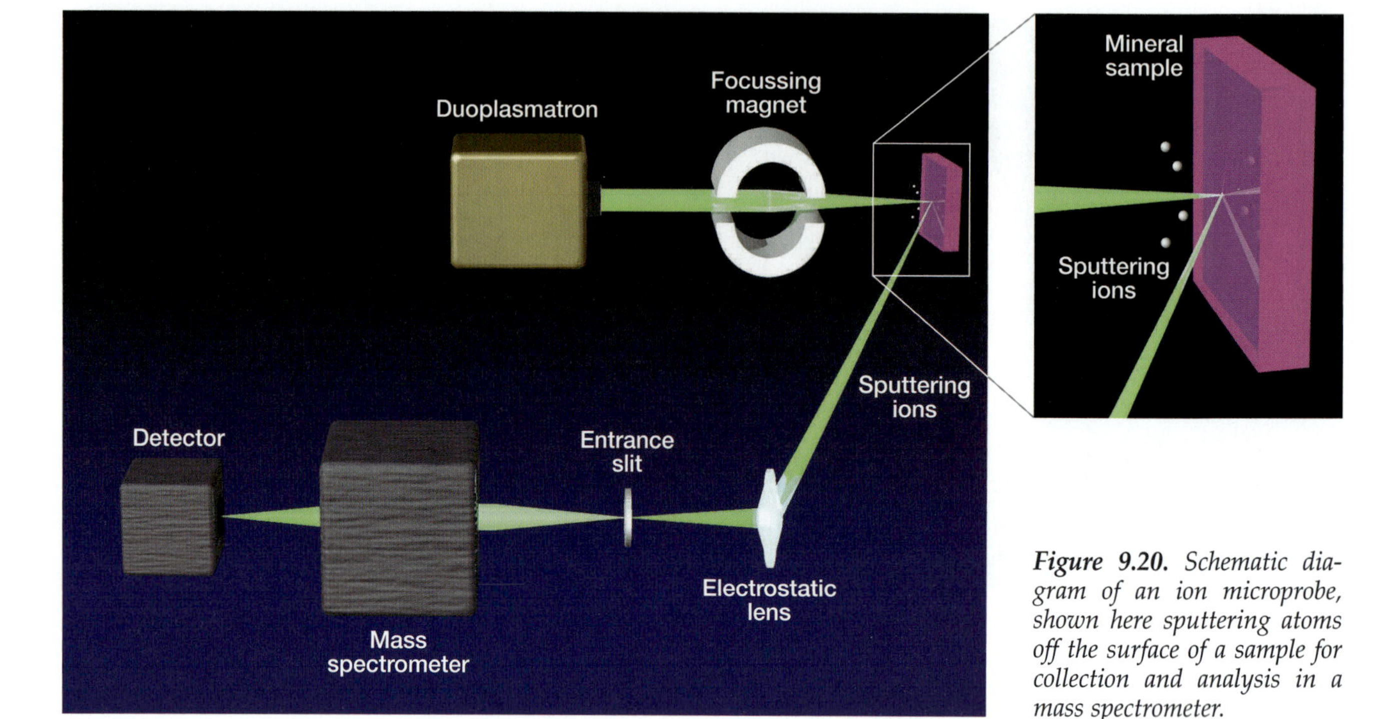

Figure 9.20. Schematic diagram of an ion microprobe, shown here sputtering atoms off the surface of a sample for collection and analysis in a mass spectrometer.

of neutrons. Neutrons work best at interacting with other nuclei because they are neutral in charge and won't be repelled by the charged nucleus. Samples and standards with known concentrations are left in the reactor flux for a period of time ranging from seconds to weeks. When a sample emerges from the reactor, most of its elements (about 70% of all elements can be analyzed via this technique) will have been transformed into another isotope with one unit of a higher mass. If the new isotopes are radioactive, most of them will begin to decay, giving off gamma rays. The number of γ-rays produced at various energies is detected and measured. For example, the element ^{23}Na becomes ^{24}Na when bombarded with neutrons (Figure 9.21). It has a 15 hour half-life and decays to ^{24}Mg by emitting two gamma (γ) photons with energies of 2.76 and 1.38 MeV. The gamma spectrum that results is shown in Figure 9.21, and the areas of the peaks at those energies are proportional to the amount of Na that was present in the original sample before irradiation.

As with several other techniques that we have discussed, the results must be compared with analyses of a standard to derive quantitative concentrations. Typically the standards and the unknowns are irradiated together, allowed to decay for the same length of time, and then counted on the same detector for the same length of time. When these conditions are satisfied, the concentration of elements in the unknown is simply:

$$C_{\text{sample}} = C_{\text{standard}} \frac{W_{\text{standard}} A_{\text{sample}}}{W_{\text{sample}} A_{\text{standard}}}$$

where C_{standard} and C_{sample} are the concentration of the element in the standard and sample, W is the weight of the standard and sample, and A is the activity (or number of counts registered) for the standard and sample.

There are several different "flavors" of neutron activation analysis, depending on the half-lives of the nuclides of interest, which can range from less than a second to several years. In **prompt gamma-ray neutron activation analysis** (PGNAA), the detection and counting of the radioactive nuclides takes place during the irradiation. A beam of neutrons is passed through a special window in the reactor, and the detector is positioned very close to the sample (Figure 9.22). This method is useful for the study of elements with high **neutron capture cross-sections** (i.e., those in which the probability of a reaction occurring between a neutron and a nucleus is large, such as B, Cd, Sm, and Gd), those with very fast decay times, and elements that have very weak gamma rays that are hard to detect.

If counting takes place after irradiation, then several variations on analysis are possible. These are all grouped under the general term **delayed gamma-ray neutron activation analysis** (DGNAA). The most common of these is called **instrumental neutron activation analysis** (INAA), because the analysis is accomplished purely with instrumental methods and does not require human intervention. Samples are irradiated, then removed from the reactor so the radiation can be counted over a period of days. This method uses from 1–500 mg of sample, and can determine up to 30 or 40 elements in a single run (Figure 9.23). INAA is both accurate and precise, yielding results in the ppm and ppb range for most elements (Table 9.3). For these reasons, INAA is particularly useful for analysis of rare earth elements and many trace elements (e.g., Cr,

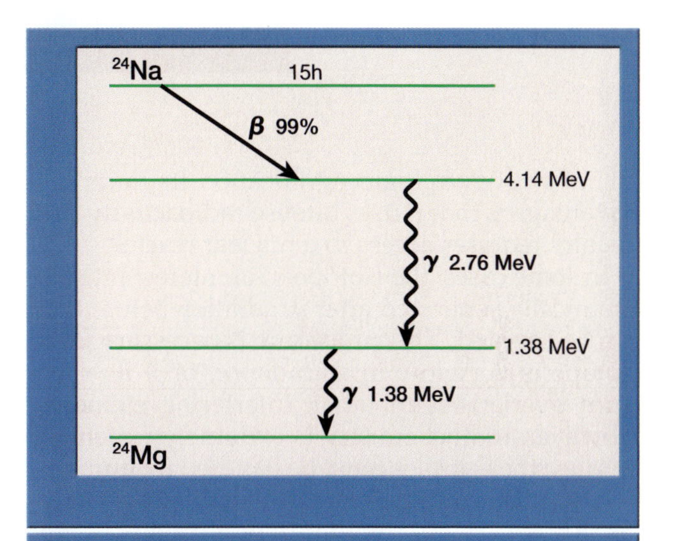

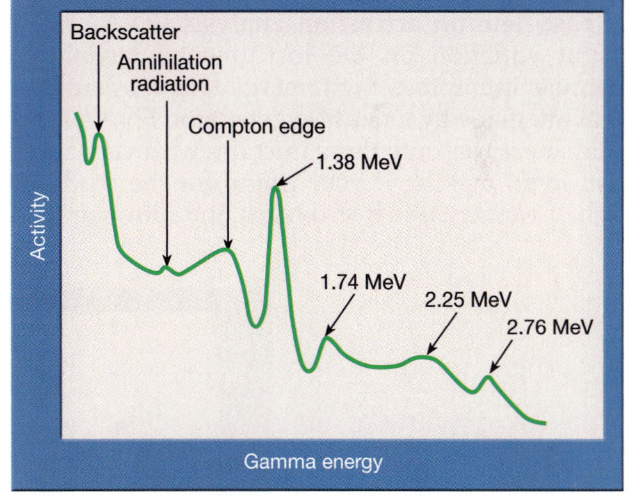

Figure 9.21. *When ^{25}Na, which is stable in natural samples, is bombarded with neutrons in a nuclear reactor, it turns into ^{24}Na, which is* unstable *and has a half-life of 15 hours. ^{24}Na then decays to ^{24}Mg by emitting two gamma (γ) photons with energies of 2.76 and 1.38 MeV (top), which are then seen in the gamma spectrum (bottom). Other peaks are also present, and are due to known phenomena such as pair production, which gives rise to the peaks at 2.25 and 1.74 MeV. (Pair production occurs when a high energy photon comes close to a nucleus, and the incoming photon is suddenly replaced by a pair of particles: an electron and a positron.)*

RADIOACTIVE DECAY: A QUICK SUMMARY OF USEFUL TERMINOLOGY

Geochemists often take advantage of the process of radioactive decay to analyze minerals. So it's helpful to acquaint yourself with a few of the terms used to describe radioactive decay.

Recall from Chapter 7 that although the volume of the atom is mostly in the electron cloud, most of the mass is in nucleus. Because of this, the nucleus mostly determines the basic properties of the atom, which are described with the following terminology:

Z is the number of protons, or atomic number of the element

N is the number of neutrons in the nucleus

A is the mass number, which is the sum of *Z* + *N*.

In the nomenclature for describing atoms, the "*A*" is usually given as a superscript before the symbol for the element.

So an element bases its "identity" on the number of protons it has, but electrons and neutrons can vary. A ***nuclide*** is an element that occurs naturally with a variable number of neutrons. For example, ^{12}C has 6 protons and 6 neutrons, while ^{13}C has 6 protons and 7 neutrons. There are several terms associated with the possible permutations of *Z*, *N*, and *A*:

An ***isotope*** has the same *Z*, but different values of *N* and *A* (e.g., ^{54}Fe, ^{56}Fe).

An ***isobar*** has the same *A*, but different *Z* and *N* (e.g., ^{56}Fe, ^{56}Co).

An ***isotone*** has the same *N*, but different *Z* and *A* (e.g., ^{56}Fe, ^{57}Co, ^{58}Ni).

Although most atomic nuclei are stable (>99.99%!), there are more than 1700 nuclides that are known to be unstable, which means they are in the process of changing into something else by radioactive decay. There are three main ways this happens:

In ***alpha decay*** (α-decay), two protons and two neutrons are given off as an α particle (that's a ^{4}He nucleus, by the way). Examples of this are ^{235}U => ^{231}Th and 1^{147}Sm => ^{143}Nd.

In ***beta-minus decay*** (β^- decay), a neutron turns into a proton, an electron, and something called an antineutrino, and the latter two leave the nucleus. The remaining atom has one more proton than it started out with, so it moves up on the periodic chart, though the value of *A* stays the same. Examples of this are ^{234}Th => ^{234}Pa and ^{87}Rb => ^{87}Sr.

Electron capture occurs when the nucleus "captures" an electron, dragging it into the nucleus where it combines with a proton to form a neutron (and a neutrino, which is ejected from the nucleus). In the resultant element, *Z* decreases by one, but *A* stays the same because a new neutron has been created. Geologists use the decay of ^{40}K to ^{40}Ar to determine the ages of rocks. Electron capture is also exploited by studies of Fe in mineral using the Mössbauer effect, and the reaction ^{57}Co => ^{57}Fe.

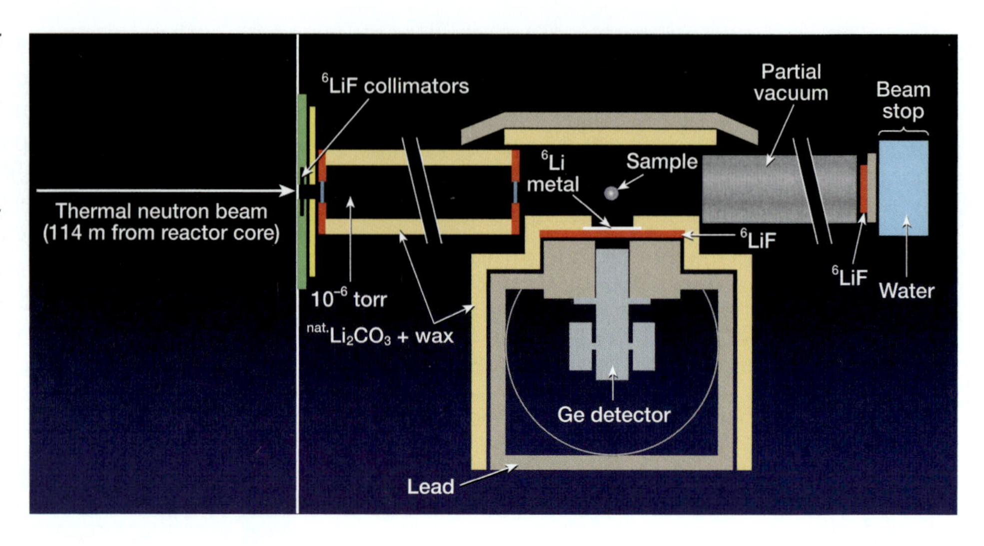

Figure 9.22. The technique of prompt gamma neutron activation analysis uses a detector that can count gamma rays while the sample is being irradiated. The sample sits in a teflon bag directly in the path of the neutron source, and emitted gamma-rays are counted at a detector that sits as close as 75 cm away (at a 90° angle).

Ni, V) with petrologic significance. Its main disadvantage is that it uses intense radioactivity, and usually requires access to a nuclear reactor.

In some cases, the isotope(s) of interest must be chemically separated after irradiation before they can be counted. This procedure is necessary if the isotope is very low in abundance, or if its spectrum overlaps with other, interfering elements. However, radiochemistry is extremely expensive because the samples must be handled by humans, and so **radiochemical neutron activation analysis** (RNAA) is less commonly performed.

Fast neutron activation analysis (FNAA) is a slight variation on the instrumental technique, because it employs high energy ("fast") neutrons that are made by a neutron generator. FNAA typically measures only those nuclides with very short half-lives, making it very useful for the study of lighter elements such as oxygen and nitrogen. The samples must be "counted" immediately (within seconds) after irradiation. To accomplish this, samples are placed in an apparatus that passes them through the neutron flux source (Figure 9.24) and then immediately back to the counting apparatus, usually through a tube (Figure 9.25).

FNAA sensitivities for most elements are about 500 ppm, depending on the element to be analyzed. It is the preferred method for precise oxygen (^{16}O has a half-life of 7.1 s) determinations, but it is useful for other light elements such as ^{19}F ($t_{1/2}$ = 26.9 s), ^{31}P ($t_{1/2}$ = 2.24–2.50 m), and ^{14}N ($t_{1/2}$ = 9.97 m) as well. FNAA was also used to study the first lunar samples.

Neutron activation analyses of all types are often considered the most sensitive techniques for analysis of various elements, and results of these measurements are often used as standards for other techniques. The web site for the

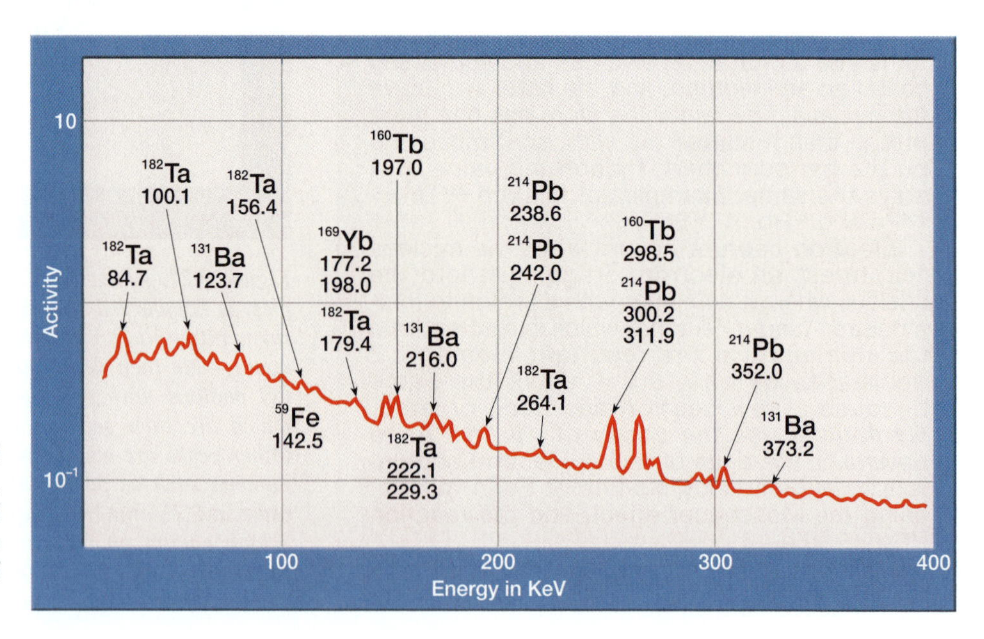

Figure 9.23. In instrumental neutron activation analysis, many elements can be analyzed from the same spectrum, as shown in this spectrum plotted as a function of energy.

Table 9.3. Estimated detection limits for INAA using decay gamma rays. Assuming irradiation in a reactor neutron flux of $1x10^{13}$ n cm^{-2} s^{-1}.	
Sensitivity (picograms)	*Elements*
1	Dy, Eu
1–10	In, Lu, Mn
10–100	Au, Ho, Ir, Re, Sm, W
100–1000	Ag, Ar, As, Br, Cl, Co, Cs, Cu, Er, Ga,Hf, I, La, Sb, Sc, Se, Ta, Tb, Th, Tm, U, V, Yb
1000–10,000	Al, Ba, Cd, Ce, Cr, Hg, Kr, Gd, Ge, Mo, Na, Nd, Ni, Os, Pd, Rb, Rh, Ru, Sr, Te, Zn, Zr
10,000–100,000	Bi, Ca, K, Mg, P, Pt, Si, Sn, Ti, Tl, Xe, Y
100,000–1,000,000	F, Fe, Nb, Ne
10,000,000	Pb, S

University of Missouri Research reactor gives an excellent summary of the details of these techniques, and is highly recommended. Other good source of information include these books: Alfassi (1990, 1994, 1998), Alfassi and Chung (1995), Ehmann and Vance (1991), and Nargolwalla and Przybylowicz (1973).

Mössbauer Spectroscopy

Highly specialized methods are needed to determine the oxidation states of elements in minerals (and in other materials). One such method is the Mössbauer effect, which is widely used in mineralogy to examine the valence state of iron, which

Figure 9.24. Dr. William Ehmann of the University of Kentucky was one of the pioneers of the technique of fast neutron activation analysis. He's showing me the neutron flux source (don't worry—it's turned off!) in the sub-basement of the Chemistry Department at the University of Kentucky. Notice the concrete blocks and other heavy shielding that are necessary to contain neutrons.

Figure 9.25. Schematic of the fast neutron activation analysis set-up at the University of Kentucky. Samples are enclosed in polyethylene vials and then "shot" into a flexible pipe using compressed air. They travel several floors down in the building to reach the neutron flux apparatus, in the basement (Figure 9.24), where they stay for 30 seconds to be spun (to avoid preferred orientation effects) and irradiated. The "hot" samples then return back upstairs in about 6 seconds to the counting apparatus, where they are counted for another 20 or more seconds.

is found in nature as Fe^0 (metal), Fe^{2+}, and Fe^{3+}. Because the valence state of iron reflects the amount of oxygen present when rocks crystallize, a Mössbauer instrument was included on each of the Mars Exploration Rovers.

The Mössbauer effect, as generally applied to the study of minerals, relies on the fact that ^{57}Fe, which is a decay product of ^{57}Co, is unstable. ^{57}Fe decays by giving off a gamma ray (γ-ray). If a nucleus gives off radiation or any other form of energy (in this case, in the form of a γ-ray), the nucleus must recoil (or move) with an equal and opposite momentum to preserve its energy (E), in the same way that a gun (by analogy, the nucleus) recoils when a bullet (the γ-ray) is fired out of it. We describe this general case in terms of energy by saying that:

$$E_{\gamma\text{-ray emission}} = E_{\text{transition}} - E_{\text{recoil}},$$

where $E_{\gamma\text{-ray emission}}$ = the energy of the emitted γ-ray, $E_{\text{transition}}$ = the energy of the nuclear transition, E_{recoil} = the energy of the recoil. Sometimes the nucleus absorbs the energy of the γ-ray and it doesn't recoil (instead, the entire structure, rather than just the nucleus, absorbs the energy). The variable f, called the **recoil-free fraction**, indicates the probability of this happening. The Mössbauer effect occurs because in solids, the value of f is high enough that recoil-free γ-ray absorption is possible. Thus an atom of ^{57}Co can decay to ^{57}Fe, which gives off a γ-ray, and may be absorbed without recoil by a nearby ^{57}Fe, which happens to have just the right splitting between the energy levels in its nucleus to absorb it. This scenario will only happen if the decaying Co atom is surrounded by the same atoms as the absorbing Fe. If the receiving Fe atoms are in a different matrix (say, in a mineral) than in the emitter, then no absorption can occur.

A Mössbauer spectrometer overcomes this restriction by moving the emitting source back and forth so that the γ-rays being given off will be Doppler-shifted and span a range of energies, in order to maximize the chance that they can be absorbed by the receiving Fe atoms in a sample (Figure 9.26). The x axis in a Mössbauer spectrum is usually plotted in units of mm/s, which is the velocity of the source as it moves toward and away from the sample; 1 mm/s equals 4.8×10^{-8} eV. Spectra are usually calibrated against the spectrum of metallic iron, which provides a constant value for zero velocity and a reference gradient for all other velocities.

The resultant spectrum is defined by two parameters: (1) **isomer shift** (δ), which arises from the difference in s electron density between the source and the absorber, and (2) **quadrupole splitting** (Δ), which is a shift in nuclear energy levels that is induced by an electric field gradient caused by nearby electrons. Graphically, quadrupole splitting is the separation between the two component peaks of a doublet, and isomer shift is the difference between the midpoint of the doublet and zero on the velocity scale (Figure 9.27). Mössbauer parameters are temperature-sensitive, and this characteristic is sometimes exploited by using lower temperatures to improve peak resolution and induce interesting magnetic phenomena.

If the electrons around the Fe atom create a magnetic field, as in the case of magnetite, then the energy levels in the Fe nucleus will split to allow six possible nuclear transitions, and a sextet (six-peak) spectrum results. The positions of the peaks in the sextet define what is called the **hyperfine splitting** (H_{int} or B_{Hf}, depending on the units used) of the nuclear energy levels.

The combination of isomer shift and quadrupole splitting (along with the hyperfine field, in the case of magnetic phases) is used to identify the valence state and site occupancy of Fe in a given site and individual mineral (Figure 9.28). In some cases, Mössbauer spectrometers are also used to identify minerals. This application is limited, however, by the fact that many different minerals can have site geometries that are the same, such that their Mössbauer spectra and the resultant peak parameters will also be the same. For example, the spectra of amphibole and pyroxene group minerals are all very similar, so you could not tell these minerals apart by their Mössbauer spectra alone!

Mössbauer spectroscopy is a topic that is frequently covered in quantum mechanics courses, so it is likely that your school may even have a Mössbauer apparatus in the Physics department. The Mössbauer effect may also be used to study other isotopes with long-lived, low-lying excited nuclear energy states such as ^{99}Ru, ^{151}Eu, ^{155}Gd, ^{193}Ir, ^{195}Pt and ^{197}Au. However, among all the elements, the isotope with the strongest recoil-free resonant absorption is ^{57}Fe, and for this reason the vast majority of Mössbauer studies are done using ^{57}Fe. For more information on application of this technique to the study of minerals, consult the classic reference by Bancroft (1973) or the more recent updates by Hawthorne (1988) and Dyar et al. (2006).

Visible and Infrared Spectroscopy

In Chapter 7 we discussed the differences in the energies of orbitals that allow electrons to jump between orbitals, absorbing energy and causing

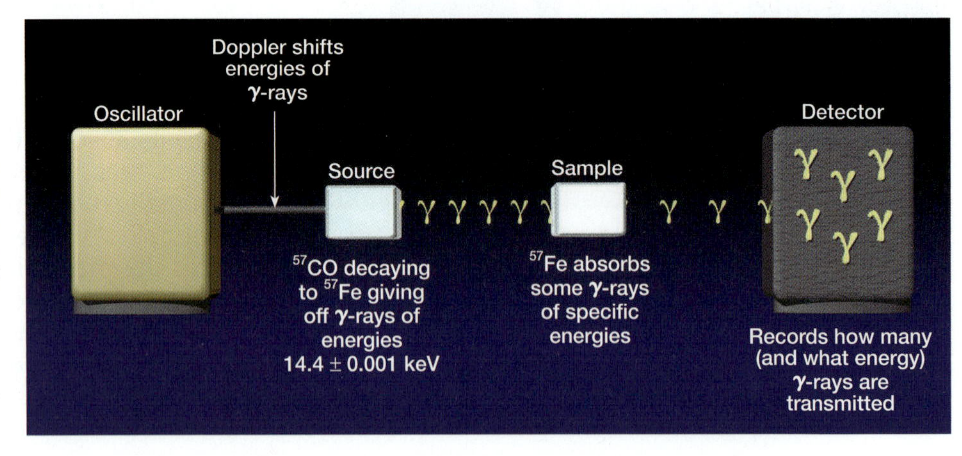

Figure 9.26. A Mössbauer spectrometer works by oscillating a radioactive source in order to change the energies of the γ-rays that are being given off (ever so slightly!). The resultant range of γ-ray energies spans a large enough range to maximize the chance that the γ-rays will be absorbed by the receiving Fe atoms in the sample.

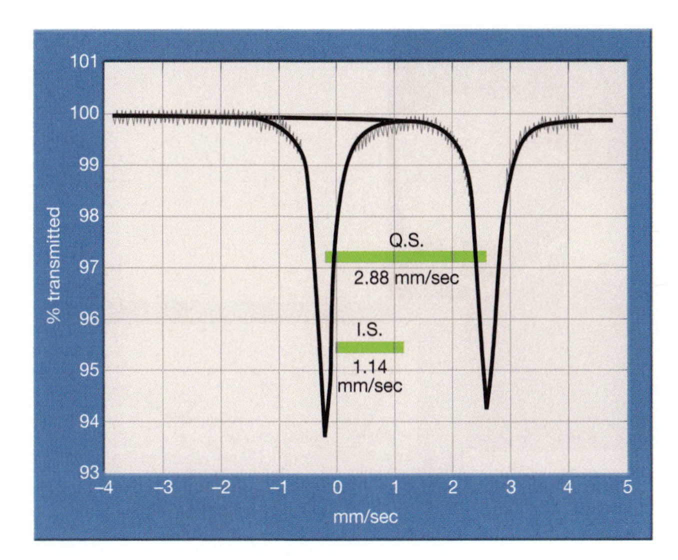

Figure 9.27. *The Mössbauer spectrum of an iron atom in a single site in a mineral structure will give rise to a doublet if nearby electrons do not induce a magnetic field. The doublet can be described using two chemical parameters that relate to the structure of the nucleus of the iron atom. Quadrupole splitting is defined as the separation between the two component peaks of a doublet, and isomer shift is the difference between the midpoint of the doublet and zero on the velocity scale.*

color. The spectra in Figures 7.10–7.12 were acquired with an optical spectrometer, which measures the wavelength range from about 250 nm (ultraviolet) up to about 3000 nm (mid-infrared). Infrared spectrometers cover the range from 1000 nm (10,000 cm^{-1}) up to 100,000 nm (100 cm^{-1}) in the far-infrared. The unit, cm^{-1} commonly used in infrared studies, is a reciprocal of wavelength and is proportional to the energy of the light.

The spectrometers that obtain these spectra use a beam of light generated by either a tungsten-halogen (visible, near-IR) or a deuterium (UV) light source. A series of mirrors, gratings, and slits focuses the beam on two holders: one occupied by the mineral sample, and the other remaining empty as a baseline or reference beam (Figure 9.29). The light beam passes through both the sample and the reference, while the wavelength changes. Some of the wavelengths are absorbed by the sample, and will show up as absorption bands in the resulting spectrum. For most applications, the light is polarized, or constrained to vibrate in only one direction, as it passes through the sample (Figure 9.30). This allows us to study absorptions in different directions within the crystal that result from different electron densities and structural arrangements along different orientations. Among other uses, visible region spectra can determine the valence state and coordination environments of transition metals; Fe, Ti, V, Cr, Mn, Ni, and Cu are commonly studied using this technique. Depending on the mineral and the geometry of the site occupied by these cations, absorption will occur at characteristic wavelengths (Table 9.4).

Infrared spectroscopy works in basically the same way, except that in mineralogy, spectra are used to identify phases, determine the amount and type of H-bearing components (H_2O, $(OH)^{1-}$) and carbon species (CO_2, CO_3^{2-}, C-H) that may be present, and to obtain fundamental thermodynamic parameters. Analysis of H-bearing components is possible because the bonds between H and O atoms are strongly polar, and will move when infrared photons are absorbed. IR spectra are acquired on a **Fourier Transform Infrared** (FTIR) spectrometer. These instruments measure two interferograms, one of light passing through the sample and the other without the sample. A computer analyzes the interferograms and determines what wavelengths were removed by the sample.

Different populations of ions within a single sample can be quantified if the molar absorption coefficient, ε, for the ion in the mineral being studied is known. When ε is known, the **Beer-Lambert** law can be used to calculate the concentration, C, of the element of interest in a sample:

$$\log\frac{I_0}{I} = \varepsilon C d,$$

where d = the thickness of the sample, ε = the molar absorption coefficient, C is the concentration of the species of interest, I_0 = the intensity of

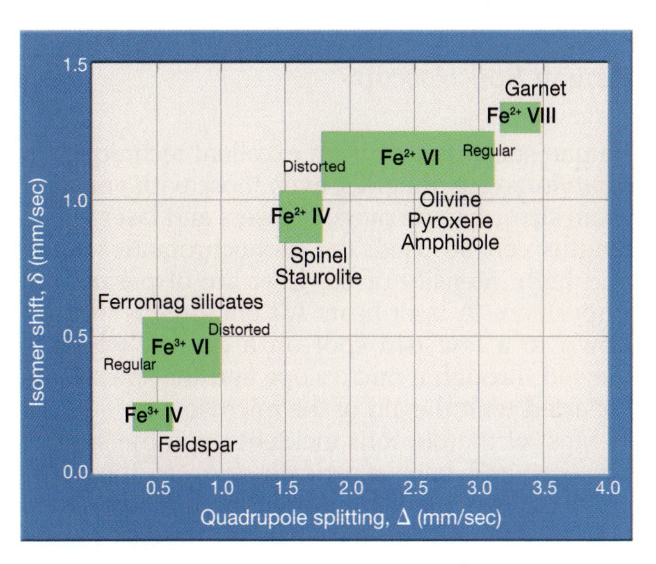

Figure 9.28. *Once the isomer shift and quadrupole splitting have been determined for a Mössbauer doublet, they can be used to understand the valance state and coordination type of the Fe atom by using this plot, which is adapted from Burns and Solberg (1990).*

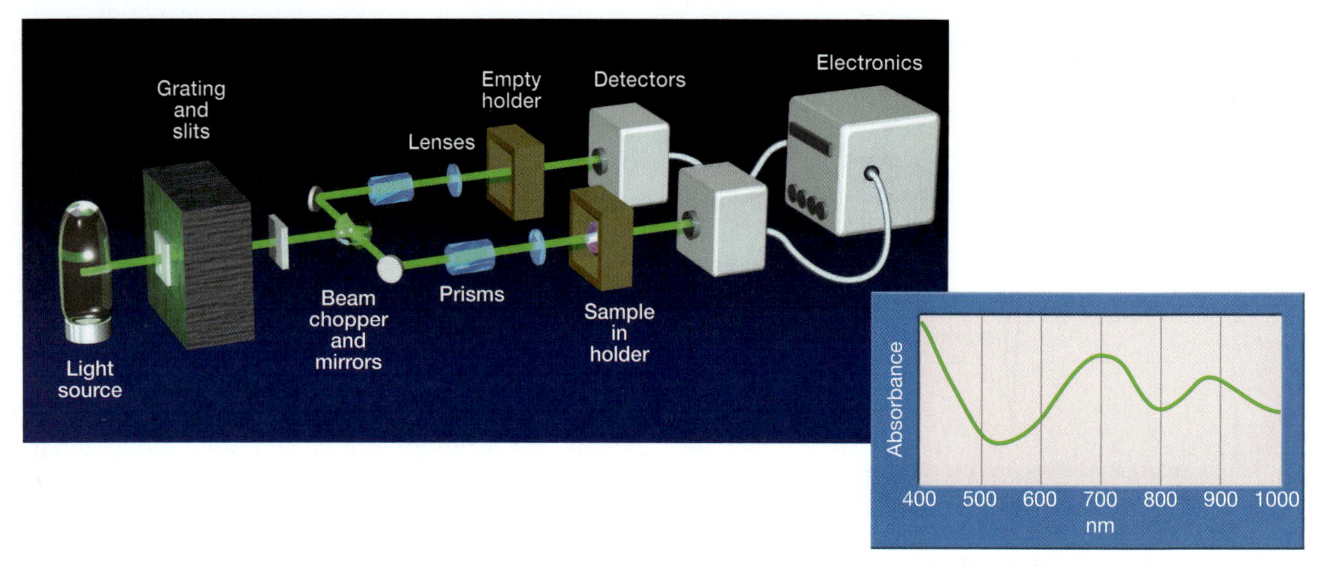

Figure 9.29. *A visible region spectrometer uses a series of mirrors, gratings, and slits to focus its beam on two holders: one occupied by the mineral sample, and the other remaining empty as a baseline or reference beam. The light beam passes through both the sample and the reference, while the wavelength changes. Some of the wavelengths are absorbed by the sample, and will show up as absorption bands in the resulting spectrum.*

light entering the sample, and I = the intensity of light as it emerges from the sample.

FTIR is also very useful for studying Si^{4+}-O^{2-} bonds, and SO_4 and PO_4 species (among others). Unlike visible region spectroscopy, it is very useful for identification of non-Fe-bearing minerals (especially silicates) and is nicely complementary to Raman spectroscopy. More extensive explanations of these techniques can be found in Farmer (1973), Rossman (1988a, b), Burns (1993), and McMillan and Hoffmeister (1988).

Raman Spectroscopy

Raman spectroscopy is an excellent technique for *identification* of minerals, even those with very fine grain sizes. A wide range of lasers and laser wavelengths can be used; the monochromatic nature and high intensity of the laser are of paramount importance. A laser beam (of photons) is focused down to a few μm spot on a crystal, which is viewed through a microscope and the spectra are collected with the aid of the microscope.

Most of the photons incident upon the sample are scattered back elastically (i.e., at the same wavelength (and frequency) they initially had—the wavelength of the incident laser) (Figure 9.31). This is called **Rayleigh scattering**. About one in ten million photons, however, scatters back at a different (typically longer) wavelength, typically a longer wavelength than that of the incident photons. This inelastic scattering is called Raman scattering. The difference in energy between the incident light (E_i) and the Raman scattered light (E_s) is the amount of energy needed to make the molecule vibrate (E_v), and is called the **Raman shift** and defined as:

$$E_v = E_i - E_s.$$

Table 9.4. Energies of Common Absorptions in Minerals

Wavelength (nm)	*Species*
500	Fe^{2+}
550–750	Fe^{3+}
900–1000	Fe^{2+}
1400	OH stretch
1700	CO_3 overtones
1800–2000	Fe^{2+}
1900	H_2O stretch and bend
2200	CO_3 overtones
2300–2500	cation-OH motions
2350	CO_2 stretch
2700–2900	OH stretch
2800	H_2O stretch
6100	H_2O bend
7000	CO_3 stretch
8500	SO_4 stretch
9400	PO_4 stretch
29,000	OH
30,000	H_2O
34,000	CH
59,000	H_2O
62,000	CO_3 stretch

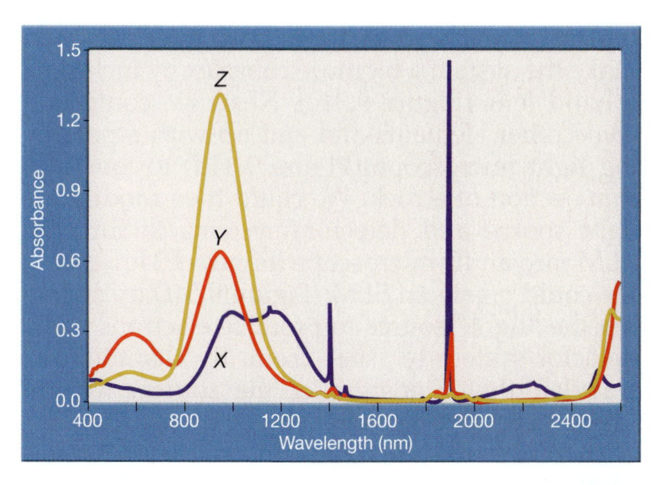

Figure 9.30. *Polarized absorption spectrum of cordierite from Manatowage, Ontario, Canada. Spectra were acquired with the beam polarized along the X, Y, and Z optical directions (i.e., the α, β, and γ refractive index directions). The color of cordierite comes from its Fe^{2+} and Fe^{3+} content. In this sample, Fe^{2+} in the octahedral site causes the pair of absorption bands near 1000 and 1300 nm that have almost no effect on the color because they are in the near-infrared. The strong absorption band in the Z direction is also from Fe^{2+}, but in a different site. The color is predominantly due to the Fe^{2+}-Fe^{3+} intervalence charge transfer band near 600 nm, which results when two adjacent Fe cations share an electron. Sharp bands near 1400 and 1900 nm are from water molecules in channels that run parallel to the c axis of the crystal. (Spectra and description courtesy of George Rossman.)*

The most important point is that the energy/frequency difference between the incoming and the Raman-scattered photons is the exact same energy/frequency as a vibration in the sample's molecules. Raman spectra therefore are plotted in terms of intensity of scattered photons vs. amount of shift in frequency that occurred between the incoming and outgoing photons (Figures 9.32 and 9.33).

The same sample may have many different vibrational motions within its molecules, which may give rise to multiple Raman shifts and multiple spectral peaks. In this way, a Raman spectrum may be used as a "fingerprint" to identify a mineral or other compound, somewhat analogous to the identification of minerals by X-ray diffraction, except the Raman spectra relate to smaller scale molecular clusters (a few Å) while X-ray diffraction relates to slightly larger scale phenomena (a few 10's of Å).

Among the petrologically and geochemically useful aspects of this analytical technique are its ability to identify and quantify chemical species (e.g., distinguishing carbonate from elemental carbon from carbon monoxide), to distinguish molecular and crystalline structure in polymorphs (e.g., 8-fold vs. 6-fold rings of elemental sulfur, graphite vs. diamond, calcite vs. aragonite), and to determine degree of crystallinity of materials (e.g., silica glass vs. chert vs. quartz). The microprobe configuration of a Raman spectrometer allows for excellent spatial imaging and analysis of sample areas as small as 1 μm in diameter. For instance, within a three-phase fluid inclusion 15 μm in diameter, it is possible to identify anhydrite (as distinguished from the compositionally identical, except hydrated form, gypsum), determine the concentration of sulfate dissolved in the coexisting liquid brine, and determine the molar proportions of methane and carbon dioxide in the coexisting gas phase. Raman microprobe analysis has been applied both to natural samples (to better characterize them) and to experimental analogs (to better understand and track the controlling variables in geologic systems). For more information, consult McMillan and Hofmeister (1988) and Pasteris (1998).

Common Themes

There are many similarities between the different types of analytical methods we described. Here we would like to point those out explicitly. After the chapter on diffraction (Chapter 15), we'll also highlight the similarities between some of the diffraction methods used to determine the crystal

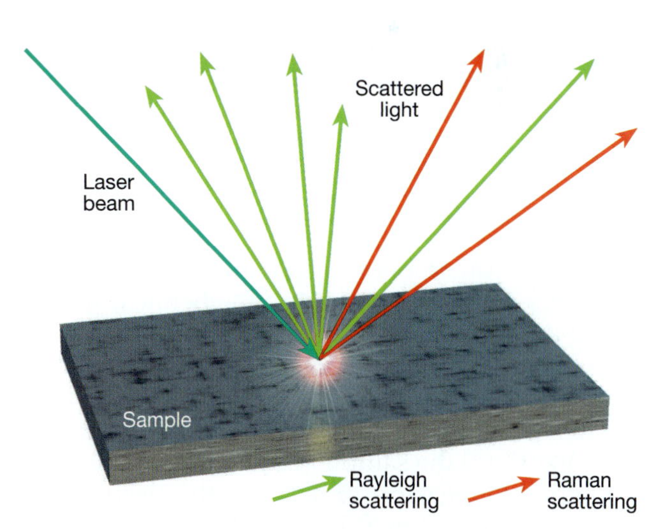

Figure 9.31. *In a Raman experiment, photons (from a laser) are focused onto a surface at a non-vertical angle. The vast majority of the photons is scattered back elastically at the initial wavelength and frequency; this process is known as Rayleigh scattering (green arrows in figure above). About one in ten million photons scatters back inelastically at a different wavelength, typically a longer wavelength than that of the incident photons, in what is called Raman scattering (red arrows).*

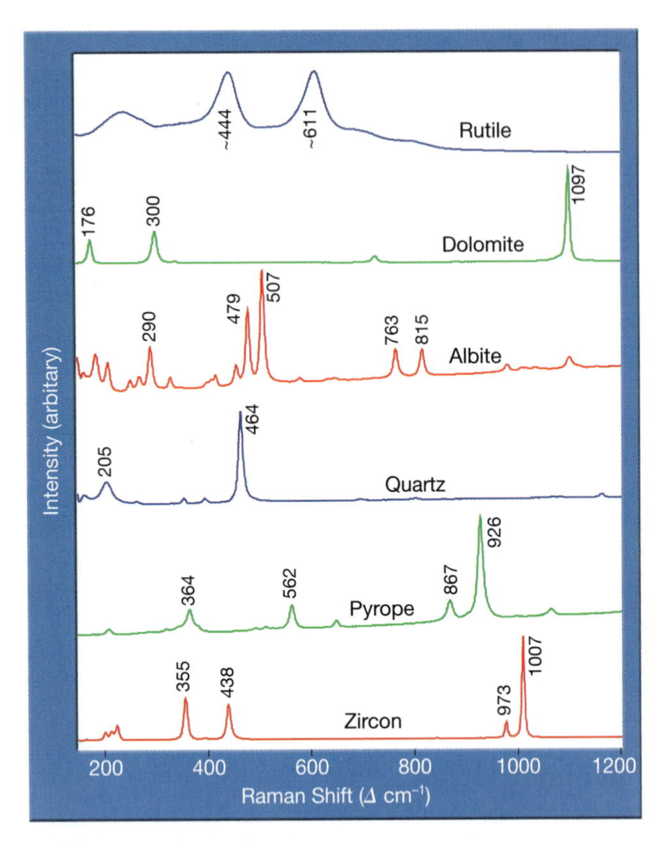

Figure 9.32. *Raman spectra of individual mineral grains. The number of peaks relates to the symmetry of the mineral structure, and the positions of the peaks reflect the bond energies within the chemical functional groups (e.g., silicate, carbonate, oxide). All the major peaks in the four bottom spectra arise from vibrational motions within the SiO_4 tetrahedra, whereas the major peak in the dolomite spectrum arises from vibrations with the CO_3 trigonal planar groups. The Raman peak positions therefore can be used to identify minerals, as well as other crystalline compounds, amorphous materials, dissolved (aqueous) species, and gas species. Each spectrum required 1–2 minutes. Figure courtesy of Brigitte Wopenka, Washington University of St. Louis.*

structures of minerals and the spectroscopic methods used to determine the chemical makeup of minerals as discussed in Chapter 19.

In general, all types of spectroscopic instruments can be thought of in terms of three to four systems: (1) a source of electromagnetic radiation (i.e., light), (2) the sample (i.e., a mineral), and (3) a method to measure, or observe, the electromagnetic radiation after it interacts with the sample (i.e., a detector). A fourth system may also be used between the source and the sample and/or after the sample and the detector. Of course, sometimes the sample might need special preparation. Figure 9.34a is a schematic of the basic three-component system. The simplest example of a real system is shown in Figure 9.34b, where the light source is the sun, the sample is just a mineral, and the detector is your eye. We could make the system a bit more complex by including a hand lens (Figure 9.34c). Next, we could add some other elements and end up with a polarizing light microscope (Figure 9.34d) to look at a thin section of a rock. We could then modify the light source and detectors systems to turn the PLM into an IR-microscope (Figure 9.34e). Lastly, we could create an SEM (Figure 9.34f) by changing the "light" source to produce electrons and a detector system to "see" them. By changing the wavelength (or energy) of the source, we can probe several different aspects of the mineral. With X-rays and electrons, we can see at much smaller resolutions than with visible light. In summary, analytical techniques share many similarities, and a good basic understanding of any technique will serve you well when you begin to explore others. Don't be intimidated by the complex technologies presented here. Instead, try to focus on the interesting information that can be learned from them about minerals!

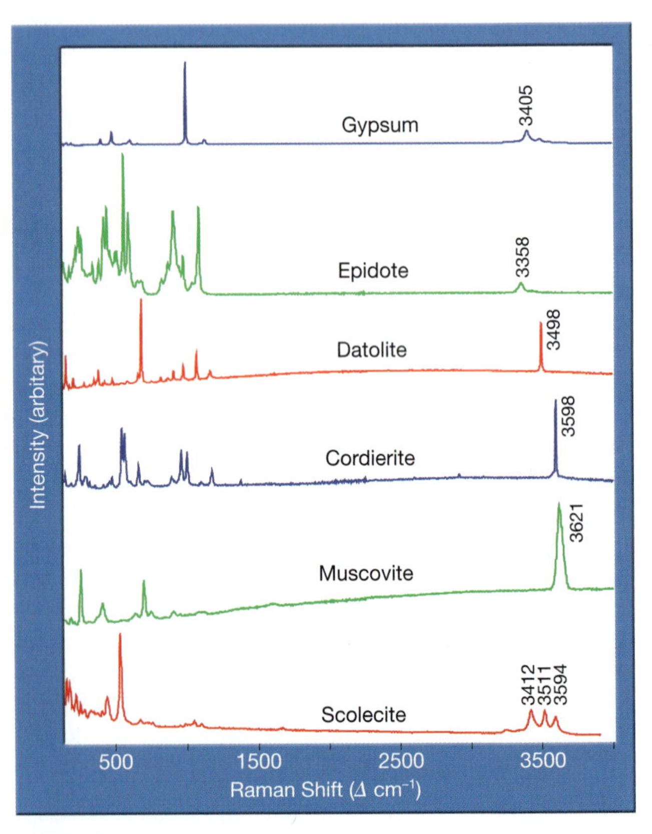

Figure 9.33. *Raman spectra can also be used to characterize the hydroxyl $(OH)^{1-}$ in a mineral and the nature of its hydrogen-bonding. All of the bands shown in the region from about 3400 to 3600 cm^{-1} indicate the stretching vibration between O^{2-} and H^{1+} in $(OH)^{1-}$ or H_2O. If the $(OH)^{1-}$ or H_2O occupies more than one type of (symmetrically distinguishable) site, then the spectrum will have multiple peaks. Figure courtesy of Brigitte Wopenka, Washington University, St. Louis.*

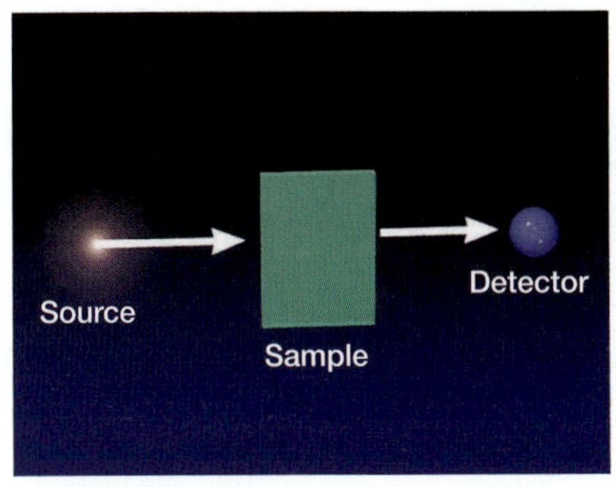

a. A general schematic showing a source, sample, and detector.

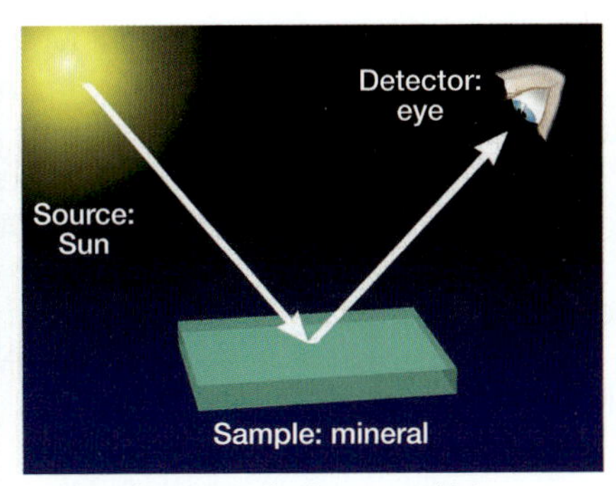

b. Here the source is the sun, which produces sunlight that reflects off a mineral sample and enters the human eye.

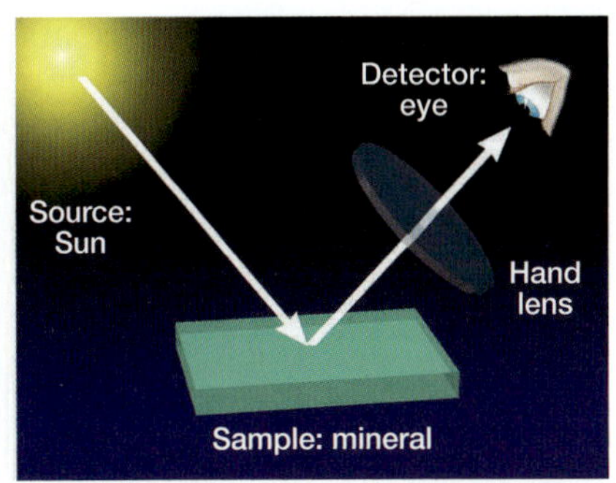

c. This system is a slight modification of the previous one where a hand lens is added between the sample and the eye to magnify the image.

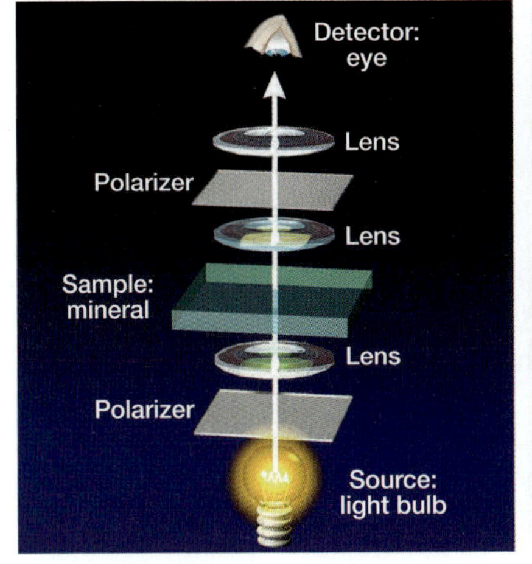

d. We can modify the above system by replacing the sun with an electric light source, adding some optics above and below the sample, and still using the eye as a detector. Once this is done, we have basically made the polarizing light microscope (PLM) that was introduced in Chapter 5.

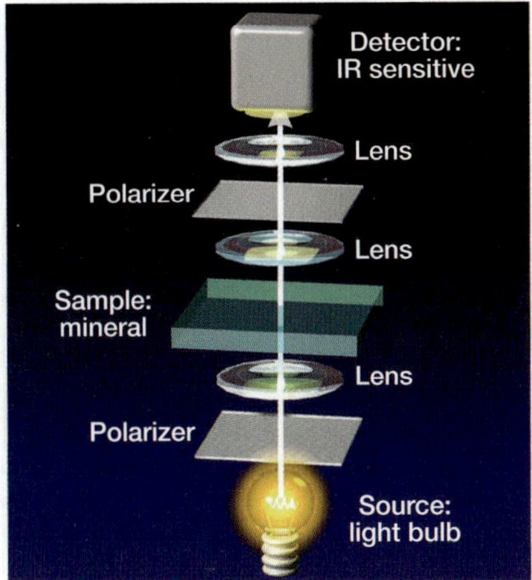

e. Next we can turn the PLM system into an IR-spectrometer by replacing the human eye with a detector system that can be scanned over the IR wavelength region.

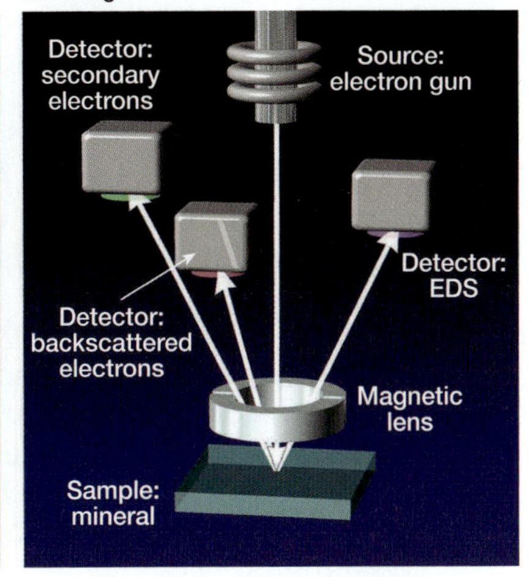

f. As the last example, we can use an illumination system that uses electrons instead of visible light. The electrons are focused onto the sample. As they interact with the sample, they produce both electrons that can be used to image the sample and electrons of different energies that are "seen" by another detector that will give us the chemistry of the sample.

Figure 9.34. This series of schematic illustrations is meant to show the basic components of a spectroscopic system and how that system can evolve from simple to complex. All of these systems are composed of at least three elements: (1) a source of radiation, (2) the sample, and (3) a detector. Some of them also have components added between the source and the samples and the sample and the detectors.

Acknowledgments. This chapter has greatly benefited from useful reviews of various sections by Peter Crowley (Amherst College), Charles Douthitt (ThermoElectron Corporation), Michael Glascock (University of Missouri), Richard Hervig (Arizona State University), Kurt Hollocher (Union College), Jill Pasteris (Washington University at St. Louis), J. Micheal Rhodes (University of Massachusetts), J. David Robertson (University of Missouri), George Rossman (California Institute of Technology), and Martha Schaefer (Louisiana State University). We appreciate the use of figures contributed by Sheila Seaman, University of Massachusetts and Brigitte Wopenka. Useful information on Raman spectroscopy was obtained from the web site of Kaiser Optical Systems, Inc.

References

Alfassi, Z.B. (1990) Activation Analysis, vol. 1 and 2. CRC press, Boca Raton, FL, 192 and 632 pp.

Alfassi, Z.B. (1994). Chemical Analysis by Nuclear Methods. John Wiley and Sons: New York, NY, 576 pp.

Alfassi, Z.B. (1998). Instrumental Multi-Element Chemical Analysis. Kluwer Academic Publishers: Dordrecht, the Netherlands, 520 pp.

Alfassi, Z.B., and Chung, C. (1995) Prompt Gamma Neutron Activation Analysis, CRC Press, Boca Raton, FL, 244 pp.

Bancroft, G.M. (1973) Mössbauer spectroscopy: an introduction for inorganic chemists and geochemists. Wiley and Sons, New York, 251 pp.

Bigeleisen, J., Perlman, M.L., and Prosser, H.C. (1952) Conversion of hydrogenic materials to hydrogen for isotopic analysis. Analytical Chemistry, 24, 1356–1357.

Buhrke, V.E., Jenkins, R., and Smith, D.K. (1997) A practical guide for the preparation of specimens for X-ray fluorescence and X-ray diffraction analysis. John Wiley and Sons, New York, 360 pp.

Burns, R.G. (1993) Mineralogical Applications of Crystal Field Theory, 2nd ed. Cambridge University Press, Cambridge, 551p.

Burns, R.G. and Solberg, T.C. (1990) ^{57}Fe-bearing oxide, silicate, and aluminosilicate minerals, in: Spectroscopic Characterization of Minerals and their Surfaces, L.M. Coyne, S.W.S. McKeever and D.F. Blake, eds. 415, pp. 262–283, American Chemical Society, Washington D.C.

Dyar, M.D., Agresti, D.G., Schaefer, M., Grant, C.A., and Skulte, E.C. (2006) Mössbauer spectroscopy of earth and planetary materials. Annual Reviews of Earth and Planetary Science, 34, 83–125.

Ehmann, W.D. and Vance, D.E. (1991). Radiochemistry and Nuclear Methods of Analysis. John Wiley and Sons: New York, NY, 531 pp.

Farmer, V.C. (1974) The Infra-Red Spectra of Minerals. Mineralogical Society, London, 539 pp.

Haswell, S.J. (1991) Atomic Absorption Spectrometry; Theory, Design and Applications. Elsevier, Amsterdam.

Hawthorne, F.C. (1988) Mössbauer spectroscopy. Reviews in Mineralogy, 18, 255–340.

Hervig, R.L. (1996) Analyses of geological materials for boron by secondary ion mass spectrometry. Reviews in Mineralogy, 33, 789–802.

Hervig, R.L. (2002) Beryllium analyses by secondary ion mass spectrometry. Reviews in Mineralogy, 50, 319–332.

Holdaway, M.J., Dutrow, B.L., Borthwick, J., Shore, P., Harmon, R.S., and hinton, R.W. (1986) H content of staurolite as determined by H extraction line and ion microprobe. American Mineralogist, 71, 1135–1141.

Ihinger, P.D., Hervig, R.L., and McMillan, P.F. (1994) Analytical methods for volatiles in glasses. Reviews in Mineralogy, 30, 67–121.

Janssens, K.H.A., Adams, F.C.V., and Rindby, A. (2000) Microscopic X-ray fluorescent analysis. John Wiley and Sons, New York, 420 pp.

Jeffrey, P.G., and Hitchison, D. (1981) Chemical methods of rock analysis, 3rd ed. Oxford, New York, 379 pp.

Jenkins, R. (1999) X-ray fluorescence spectrometry, 2nd ed. Wiley-Interscience, 232 pp. Kortright, J.B., and Thompson, A.C. (2000) X-ray data Booklet, Section 1.2. X-ray Emission Energies. Lawrence Berkeley National Laboratory, http://xdb.lbl.gov/Section1/Table_1-2.pdf.

Lajunen, L.H.J. (1997) Spectrochemical analysis by atomic absorption and emission. Royal Society of Chemistry, 254 pp.

McMillan, P.F., and Hofmeister, A.M. (1988) Infrared and Raman spectroscopy. Reviews in Mineralogy, 18, 99–160.

Nargolwalla, S.S. and Przybylowicz, E.P. (1973) Activation Analysis with Neutron Generators, John Wiley, New York, 662 pp.

Pasteris, J.D. (1998) The laser Raman microprobe as a tool for the economic geologist. Reviews in Economic Geology, 7, 233–250.

Petersen, O.V., Francis, C.A., Dyar, M.D., and Rosing, M.T. (2002) Dravite from Qârusulik, Ameralik Fjord, southern West Greenland: a forgotten classic tourmaline occurrence. extraLapis English, 3, 42–46.

Reed, S.J.B. (1980) Ion microprobe analysis—a review of geological applications. Mineralogical Magazine, 53, 3–24.

Reed, S.J.B. (1993) Electron microprobe analysis. 2nd ed. Cambridge University Press, New York, 326 pp.

Reed, S.J.B. (1996) Electron microprobe analysis and scanning electron microscopy in geology. Cambridge University Press, New York, 201 pp.

Robertson, J.D., and Dyar, M.D. (1996) Nuclear methods for analysis of boron in minerals. Reviews in Mineralogy, 32, 805–820.

Rossman, G.R. (1988a) Optical spectroscopy. Reviews in Mineralogy, 18, 207–254.

Rossman, G.R. (1988b) Vibrational spectroscopy of hydrous components. Reviews in Mineralogy, 18, 193–206.

Shimizu N. and Hart S.R. (1982) Applications of the ion probe to geochemistry and cosmochemistry. Annual Reviews of Earth and Planetary Science, 10, 483–526.

Vocke, R.D. (1997) Atomic weights of the elements 1997. Pure And Applied Chemistry 71, 1593–1607.

Welz, B., and Sperling, M. (1999) Atomic Absorption Spectrometry, 3rd ed. Wiley and Sons, 965 pp.

Zinner E. and Crozaz, G. (1986) A method for the quantitative measurement of rare-earth elements in the ion microprobe. International Journal of Mass Spectrometry and Ion Processes, 69, 17–38.

Chapter 10

Mineral Formulas

If you look in old cookbooks, you'll sometimes see that all the recipes are given in terms of weight instead of volume. This always drives me crazy. I even found the following recipe for s'mores in a recent magazine advertisement:

- *1 15 oz box of graham crackers*
- *15 chocolate bars, 1.55 oz each = 23.25 oz*
- *1 10 oz pkg. of large marshmallows*

If I'm a scout leader and I'm trying to figure out if this recipe will feed my hungry troop of 30 kids, I don't want to know the ingredients by weight. I want to know how many s'mores it will make! So I have to open the packages and count up what's inside them in order to convert from weight to "objects."

- *A 15 oz box of graham crackers contains 30 rectangles, and each rectangle makes one s'more.*
- *Each 1.55 oz chocolate bar makes 2 squares, so 15 bars, or 23.25 oz. of chocolate makes 30 squares.*
- *A 10 oz package of marshmallows contains 30 marshmallows, and you put 3 marshmallows on each s'more.*

So now I know I can make one s'more for each child (hoping that none of the graham crackers will be crushed; otherwise, I'm in trouble!). If I knew the weight of each ingredient in terms of weight per serving, I could actually calculate it, like this:

Ingredient	Package weight	Weight percentage	Weight per serving	Proportion of ingredients
Graham crackers	15 oz	100× 15/48.25= 31.1%	0.5 oz	31.1/0.5= 62.20
Chocolate bars	23.25 oz	100× 23.25/48.25= 48.2%	0.775 oz	48.2/0.775= 62.19
Marshmallows	10 oz	100× 10/48.25 = 20.7%	0.333 oz	20.7/0.333= 62.16
Total	48.25 oz			

This would tell me that given the package sizes I have on hand, I should come out with a 62:62:62 ratio, which is a basically 1:1:1 ratio, of graham crackers:chocolate bars:marshmallows. In order to tell how many servings I have, I'd need to divide the weight of a package by the weight per serving. All of these things I can do, even when I've been up all night with a hyperactive puppy, but it all seems a bit complicated just to feed some girl scouts!

So what does this have to do with mineralogy? In the previous chapter, we saw that many analytical techniques measure mineral compositions in the form of weight percent. But what we really want to know is how many of each cation will make up a mineral formula. So we have to go through calculations like the one above. If you don't already know how to use a spreadsheet, this chapter should motivate you to learn—it will make your life so much easier (and not just for this class!).

M.D.D.

Introduction

In the preceding chapter, we surveyed many of the analytical techniques used to determine the chemical composition of minerals. Different types of analyses will report data in several different formats: as weight percent oxides, as weight percent of elements, as parts per million, etc. The key is to transform these data into the chemical formulae of minerals, using crystal chemical insights to decide where each element resides in the structure. So there are two parts to the process: (1) convert from weights to atoms, in the process known as **mineral formula recalculation**, and (2) decide where each atom will go in the structure. The first of these tasks is a straightforward calculation once mastered, and will comprise the first half of this chapter. The second task can be a very difficult undertaking. To determine the cation **site occupancies** (i.e., the location of the atoms at particular places in the structure, sort of like their street addresses), we'll need to combine our chemical data with other types of mineralogical information, such as X-ray diffraction data that tell us the sizes of the coordination polyhedra, and spectroscopic data that may indicate which atoms are next to each other (sort of like finding out who your neighbors are).

Mineral Formula Calculation

To calculate chemical formulas, we'll need to reintroduce ourselves to some old terms and learn some new ones. Let's begin with **atomic weight**, which is the average mass of the atom relative to $1/12$ the mass of ^{12}C in its neutral and electronic ground state (Table 10.1). ^{12}C is used as a reference, and is defined as weighing exactly 12 **atomic mass units**, or amu's. The **molecular weight** is the weight of all the atoms in a molecule (such as H_2O). In this text, we will use the word "weight" as the commonly-used term for what might more properly be termed "mass," in order to maintain consistency with most scientific publications. It is calculated as the sum of the atomic weights of all its constituents. For example, the molecular weight of H_2O is twice the weight of H plus the weight of one O or, (2×1.00794) + 15.9994, which equals 18.01528 amu.

The term **mole** is strictly defined by the International Union of Pure and Applied Chemistry (IUPAC) as:

> "The mole is the amount of substance of a system which contains as many elementary entities as there are atoms in 0.012 kilogram of ^{12}C. When the mole is used, the elementary entities must be specified and may be atoms, molecules, ions, electrons, other particles, or specified groups of such particles."

So one mole of anything contains as many "entities" as there are in 12 grams of ^{12}C (or 0.012 kilogram), which happens to be 6.022×10^{23} atoms. The number 6.022×10^{23} is called **Avogadro's Number** and has the symbol *N*. A different way of saying this is that one mole of anything will have 6.022×10^{23} entities. This means that a mole can also be thought of as the amount of an element that will have a weight equal to its atomic weight. So, a mole of H would weigh 1.00794 grams and a mole of O would weigh 15.9994 grams. Think of it as a "chemist's dozen:" a dozen donut holes would weigh less than a dozen donuts and a dozen birthday cakes would weigh even more! But in all three cases there are a dozen, just as for the chemist a mole will be of 6.022×10^{23} of whatever the particle of interest might be.

Let's begin formula recalculation by assuming that you have an analysis of a mineral. Depending on what type of mineral it is, and what instrumentation you used to analyze it, one of two scenarios is possible: (1) your analysis has a value for every element in your mineral, or (2) your analysis is cast in terms of the weight percent of oxygen, because oxygen can't easily be analyzed, and it is given to you in the form of weight percent oxides. We'll consider these cases separately because they turn out to be rather different.

Complete Chemical Analysis (Minerals without Oxygen)

If you are working on a mineral that doesn't contain oxygen (admittedly, a rare thing!), then you'll probably have an analysis that gives the weight of each element in terms of weight percent of the total mineral. This is typically done for sulfide minerals and other ores of economic importance. Some examples of how to calculate a chemical formula given the weight percent of the elements present are given in Table 10.2A.

Let's imagine you are an exploration geologist for a mining company. You go into the field and bring back an assortment of interesting ore minerals, and you send them all out for analysis. The results come back to you in the form shown in columns 1 and 2 of Table 10.2B: the name of each element analyzed, and the amount present by weight. Weights, which are units of mass, are not very convenient because they don't translate easily to mineral proportions. So you want to convert from mass units (i.e., weights) to atomic units (i.e., numbers). For instance, we are used to writing the formula for water as H_2O (the numbers of each elements) rather than as $H_{2.00588}O_{15.9994}$ (the weights of each element present).

A good analogy for this problem is to consider a packed school bus about to depart for an away soccer game. On the bus are 10 cheerleaders and 20 soccer players. The coach climbs on the bus and asks "How many cheerleaders are here?" and some smart alecky student replies: "20 weight percent, coach!" Well, that's not what the coach wants to know—he needs to know if there are 10 cheerleaders on the bus! So he does a quick calculation in his head based on the fact that the bus holds 30 students, and then tells the bus driver to start driving. How does he convert from weight percent cheerleaders to the number of cheerleaders?

First, he makes some assumptions: the average weight of each cheerleader is about 45 kg and the average weight of a soccer player is about 90 kg, or twice that of a cheerleader. So 10 cheerleaders would weigh a total of $10 \times 45 = 450$ kg, and 20 soccer players weigh a total of $20 \times 90 = 1800$ kg. The total weight of students on the bus is 2250 kg.

Table 10.1. An alphabetical list of important elements in minerals, giving their atomic and molecular weights (i.e., oxide weight) in amu.

For each element, a conversion factor is provided to convert oxide weight percent to atomic weight percent (by multiplying the weight percent oxide by the appropriate conversion factor), the most common requirement in mineralogy. You can also convert in the other direction, from atomic weight percent to oxide weight, by multiplying atomic weight percent by 1/conversion factor. (See text for a full discussion.)

Element	Atomic weight	Oxide	Molecular weight	Conversion factor	Element	Atomic weight	Oxide	Molecular weight	Conversion factor
Al	26.981538(2)	Al_2O_3	101.9613	0.5293	Mn	54.938049(9)	MnO_2	86.9368	0.6319
As	74.9216(2)	As_2O_3	197.8414	0.7574	Mn	54.938049(9)	Mn_3O_4	228.8117	0.7203
Au	196.96655(2)	Au_2O	409.9325	0.9610	Mo	95.94(1)	MoO_3	143.9382	0.6665
B	10.811(7)	B_2O_3	69.6202	0.3106	Na	22.9898(2)	Na_2O	61.9790	0.7419
Ba	137.327(7)	BaO	153.3264	0.8957	Nb	92.906(2)	Nb_2O_5	265.8090	0.6990
Be	9.012182(3)	BeO	25.0116	0.3603	Nd	144.24(3)	Nd_2O_3	336.4782	0.8574
C	12.0107(8)	CO_2	44.0095	0.2729	Ni	58.6934(2)	NiO	74.6928	0.7858
Ca	40.078(4)	CaO	56.0774	0.7147	O	15.9994(3)			
Ce	140.116(1)	CeO_2	172.1148	0.8141	P	30.973762(4)	P_2O_5	141.9445	0.4364
Ce	140.116(1)	Ce_2O_3	328.2302	0.8538	Pb	207.2(1)	PbO	223.1994	0.9283
Cl	35.4527(9)				Pr	140.90765(2)	Pr_2O_3	329.8135	0.8545
Co	58.9332(9)	CoO	74.9326	0.7865	Rb	85.4678(3)	Rb_2O	186.9350	0.9144
Cr	51.9961(6)	Cr_2O_3	151.9904	0.6842	S	32.066(6)	SO_3	80.0642	0.4005
Cs	132.90545(2)	Cs_2O	281.8103	0.9432	Sb	121.76(7)	Sb_2O_3	291.5182	0.8354
Cu	63.546(3)	CuO	79.5454	0.7989	Sc	44.95591(8)	Sc_2O_3	137.9100	0.6520
Dy	162.5(3)	Dy_2O_3	372.9982	0.8713	Si	28.0855(3)	SiO_2	60.0843	0.4674
Er	167.26(3)	Er_2O_3	382.5182	0.8745	Sm	150.36(3)	Sm_2O_3	348.7182	0.8624
Eu	151.964(1)	EuO	167.9634	0.9047	Sn	118.71(7)	SnO_2	150.7088	0.7877
Eu	151.964(1)	Eu_2O_3	351.9182	0.8636	Sr	87.62(1)	SrO	103.6194	0.8456
Fe	55.845(2)	FeO	71.8444	0.7773	Ta	180.9479(1)	Ta_2O_5	441.8928	0.8190
Fe	55.845(2)	Fe_2O_3	159.6882	0.6994	Tb	158.92534(2)	Tb_2O_3	365.8489	0.8688
F	18.9984023(5)				Th	232.0381(1)	ThO_2	264.0369	0.8788
Ga	69.723(1)	Ga_2O_3	187.4442	0.7439	Ti	47.867(1)	TiO_2	79.8658	0.5993
Gd	157.25(3)	Gd_2O_3	362.4982	0.8676	Ti	47.867(1)	Ti_2O_3	143.7322	0.6661
Ge	72.61(2)	GeO_2	104.6088	0.6941	Tm	168.93421(2)	Tm_2O_3	385.8666	0.8756
H	1.00794(7)	H_2O	18.0153	0.1119	U	238.0289(1)	UO_2	270.0277	0.8815
Hf	178.49(2)	HfO_2	210.4888	0.8480	U	238.0289(1)	U_3O_8	842.0819	0.8480
Ho	164.93032(2)	Ho_2O_3	377.8588	0.8730	V	50.9415(1)	V_2O_5	181.8800	0.5602
K	39.0983(1)	K2O	94.1960	0.8301	W	183.84(1)	WO_3	231.8382	0.7930
La	138.9055(2)	La_2O_3	325.8092	0.8527	Y	88.90585(2)	Y_2O_3	225.8099	0.7874
Li	6.941(2)	Li_2O	29.8814	0.4646	Yb	173.04(3)	Yb_2O_3	394.0782	0.8782
Lu	174.967(1)	Lu_2O_3	397.9322	0.8794	Zn	65.39(2)	ZnO	81.3894	0.8034
Mg	24.305(6)	MgO	40.3044	0.6030	Zr	91.224(2)	ZrO_2	123.2228	0.7403
Mn	54.938049(9)	MnO	70.937449	0.7745					

Values from Vocke (1999) or calculated from them. The number in parentheses after the atomic weight gives the uncertainty in the last digit. Conversion factors are calculated as follows: Multiply the element's weight by the number of times it appears in the oxide formula, then divide that number by the molecular weight of the oxide. See examples in text.

So weight of the cheerleaders divided by the total weight of students on the bus would be: 450/2250 which is really 20 weight percent cheerleaders. No doubt you can come up with many other examples!

Now let's use a mineralogical example. Let's say you think you have a sample of bornite, which has the formula Cu_5FeS_4. You want to compare a bornite analysis with one of your unknown minerals, so you'd like to know the weight percent of each element present. To convert from the number of atoms in a molecule to weight percent, you multiply the number of each atom present times its atomic weight, which tells you the total weight in a mole (the weight of the bus when it is full, i.e., the weight of the cheerleaders and soccer players) of bornite. Then the weight percent of each element is the weight of that element divided by the total weight of the molecule times 100. This calculation is shown in detail in Table 10.2A.

Table 10.2. Sample calculations showing how to convert between atomic weight percent and chemical formulas for non-oxygen bearing minerals.*

A) This example shows how to find the weight percent of an element in a mineral, given the chemical formula for the mineral. The atomic weight (column B) of each element is multiplied by the number of atoms in the chemical formula (column C); this gives the weight of each element (column D), and these summed give the total weight of the mineral's formula (bottom of column D). Lastly, to find the weight percent of the element (column E), divide its weight by the total weight of the elements in the mineral formula (bottom of column D).

B) This example shows how to convert weight percent elements to a chemical formula; the reverse procedure of Table 10.2A. The weight percent of each element (column B), is divided by its atomic weight (column C), to yield its atomic proportion, or its "subscript" in the chemical formula (column D). The next step is to write the formula in some chosen, standard manner by selecting a "base" number of atoms in the formula. In column E, each atom number (from column D) is divided by the sum of the atoms (bottom of column D) and the ratio of the atoms found. Lastly (column F), to simplify the formula and write it with the standard number of atoms, the values from column E are multiplied by a standard value (selected for each mineral), and given at the bottom of column E, to arrive at the atoms in the formula in column F.

10.2A: To convert from a mineral formula to the weight % of its individual elements:
Example: bornite, Cu_5FeS_4

A: Element	*B: Atomic Weight*	*C: Atoms in formula*	*D: Weight of that element in this formula*	*E: Weight % of that element*
Cu	63.546	5	5 × 63.546 = 317.730	100 × 317.73/501.839 = 63.313
Fe	55.845	1	1 × 55.845 = 55.845	100 × 55.845/501.839 = 11.128
S	32.066	4	4 × 32.066 = 128.264	100 × 128.264/501.839 = 25.559
			Σ = 501.839	Σ = 100.000

10.2B: To determine a mineral formula from the weight % of the elements:
Example: pyrite, ideally FeS_2

A: Element	*B: Weight % of that element*	*C: Atomic Weight*	*D: Atomic proportion*	*E: Number of those elements present*	*F: Clear fractions*
Fe	46.55	55.845	46.55/55.845 = 0.834	0.834/2.501 = 0.333	0.333 x 3= 1
S	53.45	32.066	53.45/32.066 = 1.667	1.667/2.501 = 0.667	0.667 x 3 = 2
	Σ = 100.00		Σ = 2.501	3 atoms	FeS_2

Example: skutterudite, ideally $CoAs_{2.5}$

A: Element	*B: Weight % of that element*	*C: Atomic Weight*	*D: Atomic proportion*	*E: Number of those elements present*	*F: Clear fractions*
Co	17.95	58.933	17.95/58.933 = 0.305	0.305/1.425 = 0.214	0.214 x 3.5 = 0.75
Ni	5.96	56.693	5.96/56.693 = 0.105	0.105/1.425 = 0.074	0.074 x 3.5 = 0.26
As	76.09	74.922	76.09/74.922 = 1.016	1.016/1.425 = 0.713	0.074 x 0.713 = 2.49
	Σ = 100.00		Σ = 1.425	3.5 atoms	$Co_{0.75}Ni_{0.26}As_{2.49}$

Example: pentlandite, ideally $(Fe,Ni)_9S_8$

A: Element	*B: Weight % of that element*	*C: Atomic Weight*	*D: Atomic proportion*	*E: Number of those elements present*	*F: Clear fractions*
Fe	32.56	55.845	32.56/55.845 = 0.583	0.583/2.223 = 0.262	0.262 x 17.16 = 4.50
Ni	34.21	56.693	34.21/56.693 =0.603	0.603/2.223 = 0.271	0.271 x 17.16 = 4.66
S	33.23	32.066	33.23/32.066 = 1.036	1.036/2.223 = 0.466	0.466 x 17.16 = 8
	Σ = 100.00		Σ = 2.223	17 atoms	$Fe_{4.50}Ni_{4.66}S_8$

Example: kesterite, ideally $Cu_2(Zn,Fe)SnS_4$

A: Element	*B: Weight % of that element*	*C: Atomic Weight*	*D: Atomic proportion*	*E: Number of those elements present*	*F: Clear fractions*
Zn	10.38	65.390	10.38/65.390 = 0.159	0.159/1.756 = 0.090	0.090 x 8.3 = 0.750
Fe	2.95	55.845	2.95/55.845 = 0.053	0.053/1.756 = 0.030	0.030 x 8.3 = 0.250
Cu	26.89	63.546	26.89/63.546 = 0.423	0.423/1.756 = 0.241	0.241 x 8.3 = 2.000
Sn	32.65	118.71	32.65/118.71 = 0.275	0.275/1.756 = 0.157	0.157 x 8.3 = 1.300
S	27.14	32.066	27.14/32.066 = 0.846	1.846/1.756 = 0.482	0.482 x 8.3 = 4.000
	Σ = 100.00		Σ = 1.756	8.3 atoms	$Cu_2Zn_{0.75}Fe^{2+}_{0.25}Sn_{1.3}S_4$

*analyses from webmineral.com

More often, as a mineralogist you will be doing the calculation the other way: you will need to convert from weight percent (relative weights of cheerleaders and soccer players) to formula units (students on the bus). Luckily, you know the weights of each element (student) very precisely (Table 10.1), though you have to decide how many atoms (students) are in a formula unit (the bus).

To do this calculation, you first divide the weight % of each element by its atomic weight (see Table 10.2B). This gets you away from weights, and into atomic proportions (i.e., numbers of elements). Then total up the number of atom proportions of all the elements present. The percentage of any element present is its atomic proportion divided by the total number of atoms. Once you have the atoms in proportions that add up to 100, then you can clear the fractions to end up with the right total number of atoms (students on the bus), and give your formula using the lowest possible multiples, hopefully as integers.

In Table 10.2B, we show several different examples of this calculation. For each mineral, we start out with an idea of how many atoms are going to be present in a formula unit: 3 for pyrite, 3.5 for skutterudite, 17 for pentlandite, and 8 for kesterite. When we get to the point of clearing the fractions, we find that there are some minor problems with our analyses, and our formulas don't turn out exactly like the ideal compositions of the mineral species. That's OK, because we know that nature rarely makes perfectly stoichiometric minerals! Also, because of the covalent bonding in sulfide minerals, the charges on the ions in these sulfide minerals are somewhat variable, so we can't use charge balance to tell us if we've got the right proportions. The final calculated formulas are given in the lower right corner for each analysis. The best way to make these calculations is by using a spreadsheet, which we highly encourage you to do! Finally, don't be concerned if the numbers in your analysis don't add exactly to 100 wt% because, as we discussed in the last chapter, we are using (imperfect) experimental methods to determine the elemental content of the minerals.

WHY OXIDES?

You've probably asked yourself why we present chemical compositions of minerals in weight percent oxides and then convert them to mole fractions, or atoms per formula unit (apfu) for chemical formulas? That is a very good question, and it has an historical explanation.

In the old days (pre-1960s), chemical analysis of minerals (and solids in general), was done by a method we call wet chemistry. Possibly you've had the misfortune to do some of this sort of work in a chemistry lab. The process goes something like this (Chapter 9). You obtain a solid, carefully weight it, and dissolve it in a liquid. Then you precipitate solids from this liquid, and carefully weigh each of these solids. Well, these solids are individual oxides from the original multi-oxide solid you dissolved.

So you continue this process until you've precipitated all of the oxides from the liquid. Then you dry and weigh each of the oxides. Because you know the total weight of the original solid, you divide the weights of the individual oxides by the total weight of the starting material and arrive at the weight percent oxides for each of the individual oxides.

Our current electron beam instruments, on the other hand, directly determine the concentrations of the individual atoms based on the wavelength, or energy, of their electron transitions. However, even the state of the art instruments still can't analyze oxygen to high precision. So we usually just analyze the cations (e.g., Si or Mg) and add to each one enough oxygen to make the oxide by charge balance (e.g., SiO_2 or MgO), and compare the result against a calibration standard. We run our instruments to give the results of our analyses in terms of weight percent oxides. This makes it easy to check the quality of an analysis, which ideally should total to 100%, but it makes it hard on the students, and researchers, because we have to convert the oxide weight percents into apfus. In the old days chemical formulas for minerals were given in oxide units (i.e., the number of oxides in that mineral) as shown in Figure 10.1, but with modern instrumentation, this practice is no longer necessary!

Formula Recalculations Based on Oxides

Because oxygen is by far the most common element in the Earth's crust, it is most likely that you will be studying minerals that contain oxygen. Ironically, as you have noticed from the analytical descriptions in the last chapter, oxygen is one of the most difficult elements to analyze accurately, particularly at microscales. *So it is very rare that you will have an analysis for the most abundant element in your sample.*

In reality, there are several ways in which we could list the chemical composition of a mineral, especially a silicate or any oxygen-bearing mineral. For example, take a look at the labels on the mineral photographs in Figure 10.1. These are old samples! In the old days (early 1900s), mineral formulas were often given in terms of the number of oxide units they contained. Thus the enstatite in the photo contains one unit of MgO and one of

SiO_2. These "oxide units" were used because wet chemical methods measured the elements as oxides, and when the structures of minerals were unknown, it was logical to assume that they may have just been made of oxides. Nowadays, instead of writing $MgO{\cdot}SiO_2$ for the formula of enstatite, we would write $MgSiO_3$, which is just the sum of those two oxides, or even $Mg_2Si_2O_6$.

How about the other mineral in the photo, nephelite (which we now refer to as nepheline)? Its oxide unit formula is written as: $3Na_2O{\cdot}K_2O{\cdot}4Al_2O_3{\cdot}9SiO_2$. Multiplying out these terms and expressing them as a "modern" chemical formula would give $K_2Na_6Al_8Si_9O_{34}$. For an exercise, check to be sure that this formula is charge balanced, and, assuming the Al^{3+} is in tetrahedral coordination, decide the silicate class to which this mineral belongs. What would you guess would be its retardation in thin section? If you ponder this formula a bit longer, you might come to the conclusion that it is similar to that of the feldspars (i.e., a framework silicates with K^{1+} and Na^{1+}). You're not alone in this observation! In fact, this mineral belongs to a group of minerals called the feldspathoids, and the suffix "oid" means similar!

Table 10.3 shows several different ways that the composition of a mineral might be reported. There are five minerals in this table representing five of the major silicate groups; thus they all contain Si and O. The extra cation needed for charge balance is Mg^{2+}, and the amphibole mineral anthophyllite and the sheet silicate talc also contain H^{1+}. In Table 10.3A, the compositions of the five minerals are listed with their weight percent (commonly abbreviated as wt%) oxides and oxide units, similar to the way the chemical formulas are expressed in Figure 10.1. In Table 10.3B, each mineral's chemistry is represented by the wt% of each element and the numbers of atoms in the chemical formula. This latter term is often called **atoms per formula unit** and abbreviated apfu, which we will use herein to designate the subscripts in a chemical formula. The data in Table 10.3 can be used to make several points: (1) O, the element we can't measure, is the most abundant one in these silicate minerals, (2) you can check your ability to convert between oxide wt%, oxide units, element wt%, and, most importantly, apfu, and (3) you may notice a couple of (interesting?) trends in the table, including how the percentages of Mg decrease and those of Si and O increase as the silicate tetrahedra become more polymerized.

Given the conversion factors presented in Table 10.1, it is possible to go back and forth from wt% oxide to wt% element. These factors were calculated by dividing the atomic weight of the cation in the oxide by the atomic weight of the oxide. To convert from wt% oxide to weight percent element, multiply the number in the right-hand column of Table 10.1 (listed under the column tilted "Conversion Factor") times the wt% oxide, to obtain the wt% of that element. For example:

55.49 wt% SiO_2 x 0.4674 = 25.94 wt% Si,
18.61 wt% MgO x 0.6994 = 11.22 wt% Mg,
and
25.90 wt% CaO x 0.7147 = 18.51 wt% Ca.

Once you have the wt% of each element in your mineral, you can go back and forth between the elemental weights and the mineral formula numbers. Also, once we have the elemental weights, we *could* use the procedure explained in Table 10.2 to calculate a formula, except for one small problem: while our wt% oxide values sum to 100, our elemental weights do not total to 100 wt%, because we didn't analyze for oxygen. In fact, we would find that there's 44.33 wt% O in the above sample, by summing the wt% of Si + Mg + Ca = 55.67%, and subtracting that from 100% (i.e., 100-55.67 = 44.33). So we need a slight variation on this method to get around that problem.

In Table 10.4, we show the same idealized pyroxene analysis as above in terms of wt% oxides. Moving from left to right in the table from columns A to G, you can work through a series of calculations to convert wt% oxides to apfu, and then base the chemical formula on a set number of O. The number of O atoms for each mineral group is chosen to relate to the polymerization ratios of the silicates as explained in Chapters 6 and 22.

Next, let's move on to a more complex example that involves O^{2-} and other anions. In the pyroxene example, we could calculate the number of O^{2-} anions based on charge balance somewhat easily, but when other anions (e.g., F^- or Cl^-) occur, the calculation becomes a bit more complex. The method we are presenting here is somewhat different from what has been traditionally taught (as you'll see later in the chapter), but it has two huge benefits: (1) the calculations make more mineralogical sense, and (2) our method handles these other anions more easily.

Table 10.5 gives the raw data for a real chemical analysis of the amphibole species kaersutite. The compositions of Cl and F and all the cations, except H, were obtained by electron probe microanalysis (EPMA). The relative amounts of Fe^{2+} and Fe^{3+} were measured by Mössbauer analysis and H was analyzed by hydrogen extraction. All of these techniques were explained in the previous chapter. To perform the calculation, we use the method summarized in Table 10.4 and listed below with some minor variations:

1. Convert from wt% oxide to wt% element by use of the conversion factor.
2. Find the apfu values.
3. Determine the total positive change.
4. Calculate how much oxygen is needed to balance the charge on everything else. To do this, take the sum of the charges and divide it by –2 (because oxygen has a charge of –2).
5. Now you have the proportions of all the atoms in your mineral. The numbers are all odd fractions, so you'd like to convert them to something with integers in it. So, you check the ideal formula for amphibole, and learn that F + Cl + O is supposed to equal 24. Now you can calculate an "O-factor" by which to multiply all your apfu values, so as to make the sum of Cl + F + O to come out to 24. You do this by this formula:

$$\text{O-factor} = \frac{\text{number of anions}}{\text{charge of O} + \text{charge on F} + \text{charge on Cl}}$$

$$\text{O-factor} = \frac{24}{2.6544 + 0.00 + 0.1166} = 9.0015$$

6. Once you have the O-factor, multiply it times each of your apfu values and you will have your mineral formula expressed in apfu based on a chosen, standard number of oxygens.

To do this in a spreadsheet, consult Table 10.6, which has all the formulas needed to do the calculations. You can adapt this basic spreadsheet to use on any mineral simply by adding in rows for additional elements, and changing the "24" in cell F18 to whatever oxygen basis is desired. You could also write your own spreadsheet, which will help you understand the calculation. To do this, use the data in Table 10.3 and 10.4 to check your answers for less-complicated systems, then advance to the complex analysis in Table 10.5 that we just performed.

Figure 10.1. *Close-up photographs of two old hand samples of minerals. The chemical formulas on their labels are given in terms of oxide units (i.e., the number of oxides in the formula) rather than in terms of apfu, as is normally done now. For instance, we would write the formula for enstatite as $Mg_2Si_2O_6$ (or sometimes, $MgSiO_3$), while in this photo, it is written as $MgO{\cdot}SiO_2$. If we added these two oxides together, we'd get $MgSiO_3$, the current apfu representation. See if you can figure out what the formula would be for nephelite (which we now call nepheline). Once you've determined the formula, see if you can guess its silicate class (assuming Al^{3+} is in tetrahedral coordination).*

Table 10.3.

In the following two tables, five minerals are listed along with their chemical formulas. In A, the oxide percent and oxide units are given for each. In B, the atomic wt% and atoms in the formula unit (apfu) are given for each. You will notice that these minerals represent five silicate classes, and the "charge-balancing" cation is always Mg.

A. Minerals expressed in terms of wt% oxide and numbers of oxide units

Name	Chemical formula	MgO	SiO_2	H_2O	MgO	SiO_2	H_2O
forsterite	Mg_2SiO_4	57.29	42.71	-	2	1	-
enstatite	$Mg_2Si_2O_6$	40.15	59.85	-	2	2	-
anthophyllite	$Mg_7Si_8O_{22}(OH)_2$	36.13	61.56	2.31	7	8	1
talc	$Mg_3Si_4O_{10}(OH)_2$	31.88	63.37	4.75	3	4	1
quartz	SiO_2	-	100	-	-	1	-

B. Minerals expressed in terms of wt% elements and atoms per unit formula (apfu)

Name	Chemical formula	Mg	Si	O	H	Mg	Si	O	H
forsterite	Mg_2SiO_4	34.55	19.96	45.49	-	2	1	4	-
enstatite	$Mg_2Si_2O_6$	24.21	27.98	47.81	-	2	2	6	-
anthophyllite	$Mg_7Si_8O_{22}(OH)_2$	21.79	28.78	49.18	0.26	7	8	24	2
talc	$Mg_3Si_4O_{10}(OH)_2$	19.23	29.62	50.62	0.53	3	4	12	2
quartz	SiO_2	-	46.74	53.26	-	-	1	2	-

Table 10.4. A sample calculation showing the step-by-step procedure (actually column-by-column!) to convert from weight percent in oxides to atoms in the chemical formula (apfu) for the pyroxene mineral diopside.

A. This is the cation and its charge when it is associated with an oxide.
B. This is the wt% of the oxide (often provided from a chemical analysis of the mineral, but here calculated from the chemical formula).
C. In this column we convert from wt% oxide to wt% of the cation by multiplying the wt% of the oxide by its conversion factor (Table 10.1).
D. The atomic weight of each of the three cations is given here. These values come from Table 10.1.
E. Here we find the atoms per formula unit (apfu′) by dividing the weight percent of the element (column C) by its atomic weight (column D). This apfu′ value is almost our answer but not quite. We want to express the chemical formula in terms of a whole, set number of O, so there is one more step!
F. The goal of this column is to determine the amount of positive charge associated with the cations in the formula; once this is done, we can charge balance the formula by adding the correct number of O's. To do this, we multiply the charge on the cation (column A) by the number of atoms present (i.e., the apfu′ values in E). Then we sum those charges (bottom of F). Next, we divide that number by 2 (the charge on O) which will give us the number of O's to associate with the cation apfu′ values. This value of 2.7707 is also the number of O's in the chemical formula, or O apfu′. What remains is to set the O value to some chosen, standard number and scale the apfu′ values accordingly. In this example, we have 2.7707 O apfu′, and we want it to be 6. So we need to find a value of x to satisfy the equation $2.7707x = 6$, so $x = 6/2.7707 = 2.1655$, which we term the O-factor.
G. Finally, we convert the apfu′ values in column E to apfu values for the set number of O atoms by multiplying them all by 2.1655.

A: Cation	*B: Wt% of oxide*	*C: Wt% of element*	*D: Atomic Weight*	*E: apfu′*	*F: Charge on apfu′*	*G: apfu (based on 6 O)*
Si^{4+}	55.49	55.49 × 0.4674 = 25.9360	28.0855	25.9360/28.0855 = 0.9235	4 × 0.9235 = 3.6940	0.9235 × 2.1655 = 1.9998
Mg^{2+}	18.61	18.61 × 0.6030 = 11.2218	24.305	11.2218/24.305 = 0.4617	2 × 0.4617 = 0.9234	0.4617 × 2.1655 = 0.9998
Ca^{2+}	25.90	25.90 × 0.7147 = 18.5107	40.078	18.5107/40.078 = 0.4620	2 × 0.4620 = 0.9240	0.4620 × 2.1655 = 1.0005
					Sum of this column = 5.5414	
O^{2-} calculated, not measured!					Charge on oxygen = ½ x 5.5414 = 2.7707	2.7707 × 2.1655 = 6.0000
					O-factor: 6/(2.7707) = 2.1655	

Calculated formula: $CaMgSi_2O_6$

The Trouble with Iron

You may have noticed in the preceding examples that iron is always given as two numbers: one for Fe^{2+}, in terms of its oxide $Fe^{2+}O$, and one for Fe^{3+}, in terms of its oxide $Fe_2^{3+}O_3$. That's because, as we noted previously in this book, Fe is an element that commonly occurs in two different valence states. There are many other multivalent elements, but fortunately, they are usually minor in abundance (so assuming the wrong valence state doesn't matter that much). The most commonly-occurring of these are Cr^{2+}/Cr^{3+}, Ti^{3+}/Ti^{4+}, Mn^{2+}/Mn^{3+}, and $V^{2+}/V^{3+}/V^{4+}/V^{5+}$. If you happen to be studying a mineral with significant amounts of any of these elements, it's worthwhile to invest some time into researching which of these valence states is the proper one to use (see below).

So, your analysis probably only has FeO. Why? Chemical analyses obtained with an electron microprobe WDS (wavelength-dispersive X-ray spectrometer) or SEM-EDX (energy dispersive X-ray spectrometer) do not have the resolution to distinguish between the oxidation states of elements. If you only have an electron probe or SEM analysis, your analysis will usually be given as total FeO because Fe^{2+} is generally more abundant in terrestrial rocks than Fe^{3+}. The other common multivalent elements are generally given as Cr^{2+}, Ti^{4+}, Mn^{2+}, and V^{5+}.

However, if you really want to get a good formula for your mineral, you might try to approximate how much Fe^{3+} is present. If you look in the mineralogical literature, you'll find that several people have proposed methods for estimating how much Fe^{3+} is present based on a complicated set of assumptions relating to charge balance (e.g., Droop, 1987). *These don't work very well, and should be avoided, particularly in minerals containing* Si^{4+}. A better idea is to consult the literature on your mineral of interest, and see if someone else has measured Fe^{3+} and Fe^{2+} in samples from similar types of rocks. You can use their results to "guesstimate" the amount of Fe^{3+} and Fe^{2+} in your sample. A good source of such analyses is the series by Deer, Howie, and Zussman on rock-forming minerals (your library probably has them). If you really want to know the Fe^{3+} and Fe^{2+} in your sample, you need to find someone who does either wet chemical analysis or Mössbauer spectroscopy, and arrange for them to analyze your sample.

Mössbauer data are usually reported as the percentage of total Fe atoms that are Fe^{3+} (%Fe^{3+}). To use these data, you need the %Fe^{3+} to convert

Table 10.5.

A more complex example of determining the chemical formula of a mineral; the (real) data for this example came from three different types of analysis: oxide data are from an electron microprobe (all the elements except H), H was determined with H extraction (explained in Chapter 9), and the Fe^{3+}, Fe^{2+} values were obtained with Mössbauer spectroscopy (also explained in Chapter 9). The data are for the amphibole group mineral kaersutite, which has an ideal formula of $NaCa_2(Mg, Fe^{2+})_4Ti(Si_6Al_2)O_22(OH)_2$. See if you can work through this example; it might help to refer back to Table 10.4. See the explanation in the text on how to deal with the additional anions Cl and F in this data set.

Cation	*Wt% of oxide*	*Wt% of element*	*Atomic Weight*	*apfu'*	*Charge on apfu'*	*apfu (based on 24 O)*
Si^{4+}	38.91	38.91 × 0.467 = 18.188	28.086	18.188/28.085 = 0.647	4 × 0.647 = 2.590	0.647 × 9.001 = 5.829
Al^{3+}	14.64	14.64 × 0.529 = 7.748	26.982	7.748/26.981 = 0.287	3 × 0.287 = 0.862	0.287 × 9.001 = 2.585
Ti^{4+}	4.84	4.84 × 0.599 = 2.902	47.867	2.902/47.867 = 0.060	4 × 0.060 = 0.243	0.060 × 9.001 = 0.546
Fe^{2+}	7.88	7.88 × 0.773 = 6.125	55.845	6.125/55.845 = 0.109	2 × 0.109 = 0.219	0.109 × 9.001 = 0.987
Fe^{3+}	4.48	4.48 × 0.699 = 3.133	55.845	3.133/55.845 = 0.056	3 × 0.056 = 0.168	0.056 × 9.001 = 0.505
Cr^{3+}	0.03	0.03 × 0.684 = 0.021	51.996	0.021/51.996 = 0.000	3 × 0.000 = 0.001	0.000 × 9.001 = 0.004
Mg^{2+}	12.16	12.16 × 0.603 = 7.334	24.305	7.334/24.305 = 0.301	2 × 0.301 = 0.604	0.301 × 9.001 = 2.716
Mn^{2+}	0.17	0.17 × 0.720 = 0.132	54.938	0.132/54.938 = 0.002	2 × 0.002 = 0.005	0.002 × 9.001 = 0.022
Ca^{2+}	11.02	11.02 × 0.714 = 7.876	40.078	7.876/40.078 = 0.196	2 × 0.196 = 0.393	0.196 × 9.001 = 1.769
Na^{1+}	2.61	2.61 × 0.741 = 1.936	23.000	1.936/22.000 = 0.084	1 × 0.084 = 0.084	0.084 × 9.001 = 0.758
K^{1+}	1.66	1.66 × 0.830 = 1.378	39.100	1.378/39.100 = 0.035	1 × 0.035 = 0.035	0.035 × 9.001 = 0.317
F^{1-}	0.22	0.220	18.998	0.22/18.998 = 0.012	–1 × 0.011 = –0.012	0.011 × 9.001 = 0.104
Cl^{1-}	0.01	0.010	35.452	0.001/35.452 = 0.003	–1 × 0.003 = –0.000	0.000 × 9.001 = 0.003
H^{1+}	1.05	1.05 × 0.111 = 0.118	1.008	0.118 / 1.008 = 0.117	1 × 0.116 = 0.117	0.116 × 9.001 = 1.049
					Sum of this column = 5.309	
O^{2-} calculated, not measured!					Charge on oxygen = ½ × 5.038 = 2.654	2.654 × 9.001 = 23.894
					O-factor: 24/(2.654-0.000–0.116) = 9.002	

Calculated formula: $Na_{0.758}Ca_{1.768}K_{0.317}Mg_{2.716}Fe^{2+}_{0.987}Fe^{3+}_{0.505}Ti_{0.545}Cr_{0.003}Mn_{0.021}Si_{5.829}Al_{2.584}O_{22.844}(F_{0.104}Cl_{0.002}(OH)_{1.049})$!

the wt% of total FeO from your microprobe analysis, as supplied by your microprobe analysis, to reflect the fact that some of the iron is FeO and some is Fe_2O_3, as follows:

$$\text{wt\%FeO} = \frac{\text{wt\%FeO}_{\text{microprobe}}}{\left(1 + \frac{\%\text{Fe}^{3+}}{100 - \%\text{Fe}^{3+}}\right)}$$

and

$$\text{wt\%Fe}_2\text{O}_3 = \frac{\text{wt\%FeO}_{\text{microprobe}} \times \%\text{Fe}^{3+} \times 79.86}{71.844 \times 100}$$

These factors are derived from the ratios of the atomic weights of Fe and O as FeO and Fe_2O_3. Now you can substitute your corrected wt% FeO and wt% Fe_2O_3 into your analyses, and do the recalculation the same way as above.

There's a related conversion you might want to make when dealing with Fe. On occasion your sample might be all Fe^{2+} or Fe^{3+}, but the analysis might be expressed in the wrong form. So you may want to convert between FeO and Fe_2O_3. For example, let's say you have your elemental data in wt% FeO and want to convert it into wt% Fe_2O_3. These conversions utilize the molecular weights of FeO and Fe_2O_3 and a conversion factor:

$$\text{wt\%Fe}_2\text{O}_3 = \text{wt\%FeO}\left(\frac{159.6888}{2 \times 71.8444}\right) = \text{wt\%FeO} \times 1.111.$$

As the above equation indicates, to convert FeO wt% to Fe_2O_3 wt%, you would just multiply the FeO values times 1.111. To convert Fe_2O_3 to FeO, you would multiply the Fe_2O_3 values time 0.9001 (the inverse of 1.111).

What about Hydrogen?

It's also quite unlikely that you will have access to a good analysis for the hydrogen content of your mineral. Even though H is very important in helping us understand mineral formation conditions, H is very hard to analyze. But if you are studying a mineral like a mica or an amphibole that contains H, what do you do? In this situation, you can make the assumption that the amount of OH + F + Cl for that mineral is the ideal value. So you would calculate your "OH, F, Cl factor" based on the ideal number of oxygens, as follows (see Table 10.5).

$$\text{O, F, Cl factor} = \frac{\text{O factor} - 0.5 \times \text{ideal H}}{\text{Charge on oxygen proportion} - 0.5 \times (\text{F} + \text{Cl})}$$

Please realize that this assumption, too, is flawed—nature often varies the H content of a mineral in order to charge balance other substitutions in the structure, resulting in vacancies in the OH site. At least, this method gives you a good approximation to work with.

Formula Recalculations Based on Oxides

Many geologists calculate mineral formulas directly from the oxide wt%, using the methods displayed in Table 10.7 and 10.8. We think the method we described earlier (Table 10.4 and 10.5) is a lot easier to understand, especially when dealing with minerals that contain other anions than O^{2-} (i.e., F^{1-} and Cl^{1-}). However, because the oxide recalculation method has been used by so many people, and will be the one your professor learned, we'll include an explanation here.

First, consider that what we want to know is:

$$\frac{\text{atoms of element}}{\text{formula of mineral}} = \frac{\text{moles of element}}{\text{moles of mineral formula.}}$$

To get there, we follow four steps, similar to the ones we've already done, as given in Tables 10.7–10.9. The biggest difference is that in the previously-described method, we determine the O

Table 10.6.

An example of how to program the calculations to determine the chemical formula for a mineral given the raw data from chemical analysis into a spreadsheet. A) Is modeled after the data in Table 10.5, and B) is similar except it shows the example when no H data are available.

A. Same example as in Table 10.5, assuming you have a wt% H_2O analysis							
1	*A: Cation*	*B: Wt% of oxide*	*C: Wt% of element*	*D: Atomic Weight*	*E: apfu'*	*F: Charge on apfu'*	*G: apfu*
2	Si^{4+}	38.91	=B2*0.4674	28.0855	=C2/D2	=4*E2	=E2*F18
3	Al^{3+}	14.64	=B3*0.5293	26.9815	=C3/D3	=3*E3	=E3*F18
4	Ti^{4+}	4.84	=B4*0.5993	47.867	=C4/D4	=4*E4	=E4*F18
5	Fe^{2+}	7.88	=B5*0.773 0	55.845	=C5/D5	=2*E5	=E5*F18
6	Fe^{3+}	4.48	=B6*0.6993	55.845	=C6/D6	=3*E6	=E6*F18
7	Cr^{3+}	0.03	=B7*0.6842	51.9961	=C7/D7	=3*E7	=E7*F18
8	Mg^{2+}	12.16	=B8*0.6030	24.305	=C8/D8	=2*E8	=E8*F18
9	Mn^{2+}	0.17	=B9*0.7203	54.938	=C9/D9	=2*E9	=E9*F18
10	Ca^{2+}	11.02	=B10*0.7147	40.078	=C10/D10	=2*E10	=E10*F18
11	Na^{1+}	2.61	=B11*0.7419	22.9898	=C11/D11	=1*E11	=E11*F18
12	K^{1+}	1.66	=B12* 0.8301	39.0983	=C12/D12	=1*E12	=E12*F18
13	F^{1-}	0.22	=B13	18.998	=C13/D13	=-1*E13	=E13*F18
14	Cl^{1-}	0.01	=B14	35.4527	=C14/D14	=-1*E14	=E14*F18
15	H^{1+}	1.05	=B15*0.1119	1.008	=C15/D15	=1*E15	=E15*F18
16						=SUM(F2:F15)	
17	O^{2-} calculated, not measured!					=0.5*F16	=F17*F18
18						=24/(F17+F14+F13)	
B. Same example as in Table 10.6, but you have NO wt% H_2O analysis							
15	H^{1+}						=4-G13-G14
16						=SUM(F2:F15)	
17	O^{2-} calculated, not measured!					=0.5*F16	
18						=22/-(F17- 0.5*(F14+F13))	

apfu values by charge balance; the older method does this but not as directly or obviously.

To illustrate the older method, let's consider an analysis of the pyroxene mineral diopside, which has 55.49 wt% SiO_2, 18.61 wt% MgO, and 25.90 wt% CaO. (Recall we've already seen this mineral in Table 10.4.)

Step 1. Divide the weight percentage of each oxide by the molecular weight of the oxide, to obtain its molecular proportion (Table 10.7). The goal of this step is to convert our analysis from a wt% oxide basis to a molar basis. (In Table 10.3, we used the term oxide unit to be synonymous with molar basis.) Once we perform this first step, we have effectively recast our formula in terms of moles (numbers of oxide units) per unit weight, so we could, right now, write our mineral formula as:

$$(CaO)_{0.46185}(MgO)_{0.46166}(SiO_2)_{0.92353}.$$

In spite of the many digits, you might see that the number of moles of CaO and the number of moles of MgO are nearly identical, and both are

Table 10.7.

Sample calculation of a different approach to convert weight percent oxides into a chemical formula. This example is for the pyroxene group mineral diopside, which was also used as an example in Table 10.4. Compare the two approaches and see which you think makes more sense and is easier to perform.

Step 1. Divide the weight percentage of each oxide by the molecular weight of the oxide, to obtain its molecular proportion (Table 10.1).

$$\frac{\text{wt\% SiO}_2}{\text{molecular weight of SiO}_2} = \frac{55.49}{60.085} = 0.92353\ \frac{\text{moles of SiO}_2}{\text{100g of mineral}},$$

$$\frac{\text{wt\% MgO}}{\text{molecular weight of Mg}} = \frac{18.61}{40.311} = 0.46166\ \frac{\text{moles of MgO}}{\text{100g of mineral}}, \text{ and}$$

$$\frac{\text{wt\% CaO}}{\text{molecular weight of CaO}} = \frac{25.90}{56.079} = 0.46185\ \frac{\text{moles of CaO}}{\text{100g of mineral}}.$$

Step 2. Multiply the molecular proportion of each oxide by the number of oxygens in the oxide formula. For example, MgO has 1 oxygen, SiO_2 has 2 oxygens, and Al_2O_3 would have 3.

$$\frac{\text{moles of SiO}_2}{\text{100g of mineral}} = \frac{\text{oxygen units of SiO}_2}{\text{molecular proportion of SiO}_2} = 0.92353 \times 2 = 1.84705\ \frac{\text{oxygen units of SiO}_2}{\text{100g of mineral}},$$

$$\frac{\text{moles of MgO}}{\text{100g of mineral}} = \frac{\text{oxygen units of MgO}}{\text{molecular proportion of MgO}} = 0.46166 \times 1 = 0.46166\ \frac{\text{oxygen units of MgO}}{\text{100g of mineral}}, \text{ and}$$

$$\frac{\text{moles of CaO}}{\text{100g of mineral}} = \frac{\text{oxygen units of CaO}}{\text{molecular proportion of CaO}} = 0.46185 \times 1 = 0.46185\ \frac{\text{oxygen units of CaO}}{\text{100g of mineral}}.$$

Step 3. Multiply the "oxygen number" of each oxide by a normalization constant that is equal to the number of oxygens in the desired formula divided by the sum of the "oxygen numbers." Therefore, each number is multiplied by 6/2.7705 = 2.16567:

$$\frac{\text{oxygen units of SiO}_2}{\text{100g of mineral}} = \frac{\text{100g of mineral}}{\text{6 oxygen units}} = 1.84705 \times 2.16567 = 4.00\ \frac{\text{oxygen units of SiO}_2}{\text{6 oxygen units of mineral}},$$

$$\frac{\text{oxygen units of MgO}}{\text{100g of mineral}} = \frac{\text{100g of mineral}}{\text{6 oxygen units}} = 0.46166 \times 2.16567 = 1.00\ \frac{\text{oxygen units of MgO}}{\text{6 oxygen units of mineral}}, \text{ and}$$

$$\frac{\text{oxygen units of CaO}}{\text{100g of mineral}} = \frac{\text{100g of mineral}}{\text{6 oxygen units}} = 0.46185 \times 2.16567 = 1.00\ \frac{\text{oxygen units of CaO}}{\text{6 oxygen units of mineral}}.$$

Step 4. Multiply the "normalized oxygen numbers" of each oxide by the number of cations per oxygen in the oxide formula. So MgO would be multiplied by 1/1, SiO_2 is multiplied by 1/2, Al_2O_3 would be multiplied by 2/3, K_2O by 2/1, etc.

$$\frac{\text{oxygen units of SiO}_2}{\text{6 oxygen units of mineral}} = \frac{\text{moles of Si}}{\text{oxygen units of SiO}_2} = 4 \times 0.5 = 2.00\ \frac{\text{moles of Si}}{\text{6 oxygen units of mineral}},$$

$$\frac{\text{oxygen units of MgO}}{\text{6 oxygen units of mineral}} = \frac{\text{moles of Mg}}{\text{oxygen units of MgO}} = 1 \times 1 = 1.00\ \frac{\text{moles of Mg}}{\text{6 oxygen units of mineral}}, \text{ and}$$

$$\frac{\text{oxygen units of CaO}}{\text{6 oxygen units of mineral}} = \frac{\text{moles of Ca}}{\text{oxygen units of CaO}} = 1 \times 1 = 1.00\ \frac{\text{moles of Ca}}{\text{6 oxygen units of mineral}}.$$

approximately half the number of moles of SiO_2. In this case, we can see "by inspection" that multiplying this formula times some constant, in this case the inverse of 0.46166, would give us a nice tidy formula:

$$(CaO)_{1.0}(MgO)_{1.0}(SiO_2)_{2.0} \text{ or } CaMgSi_2O_6.$$

Unfortunately, it is not always possible to "clear the fractions" by observation. So further processing is usually required to obtain a formula.

Step 2. Multiply the molecular proportion of each oxide by the number of oxygens in the oxide formula. For example, MgO has one oxygen, SiO_2 has two oxygens, and Al_2O_3 would have three

Table 10.8.

This table shows a different way to write out the procedure in Table 10.7 and gives yet one more example of calculating the chemical formula of a mineral based on actual microprobe data. In this case, we are using the mica group mineral muscovite as an example. In the upper section of the table, H_2O data are not available (the usual case) and in the lower section of the table, H_2O data are known. Notice that the formulas really don't change much between the two case (except that H is given for one), and the number of O's in the formulas also differ: 22 for the non-H_2O case and 24 for the H_2O case.

To determine a mineral formula from the wt% of the elements, expressed as oxides, in a hydrous mineral where you don't know the H_2O content. Example: Muscovite from western Maine, Ra-d37-66

Oxide	*Wt% of oxide*	*Molecular Weight*	*Molecular proportion*	*Atomic proportion*	*Anions based on 22 oxygens*	*Cations based on 22 oxygens*
SiO_2	45.48	60.0843	45.48/60.08 = 0.7570	2 × 0.7570 = 1.5140	1.5140 × 7.9387 = 12.019	12.019/2 = 6.009
Al_2O_3	37.74	101.9613	37.74/101.9613 = 0.3701	3 × 0.3701 = 1.1104	1.1104 × 7.9387 = 8.815	8.815/1.5 = 5.877
TiO_2	0.37	79.8658	0.37/79.87 = 0.0046	2 × 0.0046 = 0.0093	0.093 × 7.9387 = 0.074	0.074/2 = 0.037
FeO	0.74	71.8444	0.74/71.84 = 0.0103	1 × 0.0103	0.0103 × 7.9387 = 0.082	0.082/1 = 0.082
Fe_2O_3	0.39	159.6882	0.39/159.69 = 0.0024	3 × 0.0024 = 0.0073	0.0073 × 7.9387 = 0.058	0.058/1.5 = 0.039
MgO	0.50	40.3044	0.50/40.30 = 0.0124	1 × 0.0124 = 0.0124	0.0124 × 7.9387 = 0.098	0.098/1 = 0.098
Na_2O	0.75	61.9790	0.75/61.9790 = 0.0121	1 × 0.0121 = 0.0121	0.0121 × 7.9387 = 0.096	0.096/0.5 = 0.192
K_2O	8.99	94.1960	8.99/94.1960 = 0.0954	1 × 0.0954 = 0.0954	0.0954 × 7.9387 = 0.758	0.758/0.5 = 1.515
Sums	94.96			Σ = 2.7712		
Factor				22/2.7712 = 7.9387		

Calculated formula: $K_{1.515}Na_{0.192}Mg_{0.098}Fe^{2+}_{0.082}Fe^{3+}_{0.039}Ti_{0.037}Al_{5.877}Si_{6.009}O_{20}(OH)_4$

To determine a mineral formula from the wt% of the elements, expressed as oxides, in a hydrous mineral where you DO know the H_2O content (same mineral as above, only you now have an H_2O analysis!)

Oxide	*Wt% of oxide*	*Molecular Weight*	*Molecular proportion*	*Atomic proportion*	*Anions based on 24 oxygens*	*Cations based on 24 oxygens*
SiO_2	45.48	60.0843	45.48/60.08 = 0.7570	2 × 0.7570 = 1.5140	1.5140 × 7.9706 = 12.067	12.067/2 = 6.034
Al_2O_3	37.74	101.9613	37.74/101.9613 = 0.3701	3 × 0.3701 = 1.1104	1.1104 × 7.9706 = 8.851	8.851/1.5 = 5.901
TiO_2	0.37	79.8658	0.37/79.87 = 0.0046	2 × 0.0046 = 0.0093	0.093 × 7.9706 = 0.074	0.074/2 = 0.037
FeO	0.74	71.8444	0.74/71.84 = 0.0103	1 x 0.0103	0.0103 × 7.9706 = 0.082	0.082/1 = 0.082
Fe_2O_3	0.39	159.6882	0.39/159.69 = 0.0024	3 × 0.0024 = 0.0073	0.0073 × 7.9706 = 0.058	0.058/1.5 = 0.039
MgO	0.50	40.3044	0.50/40.30 = 0.0124	1 × 0.0124 = 0.0124	0.0124 × 7.9706 = 0.099	0.099/1 = 0.099
Na_2O	0.75	61.9790	0.75/61.9790 = 0.0121	1 × 0.0121 = 0.0121	0.0121 × 7.9706 = 0.096	0.096/0.5 = 0.193
K_2O	8.99	94.1960	8.99/94.1960 = 0.0954	1 × 0.0954 = 0.0954	0.0954 × 7.9706 = 0.761	0.761/0.5 = 1.521
H_2O	4.32	18.0153	4.32/18.0153 = 0.2397	1 × 0.2397 = 0.2397	0.2397 × 7.9706 = 1.911	1.911/0.5 = 3.823
Sums	99.28			Σ = 3.0110		
Factor				24/3.0110 = 7.9706		

Calculated formula: $K_{1.521}Na_{0.193}Mg_{0.099}Fe^{2+}_{0.082}Fe^{3+}_{0.039}Ti_{0.037}Al_{5.901}Si_{6.034}O_{20}(OH)_{3.823}$

(Table 10.7). The sum of all the resulting "oxygen numbers", or atomic proportions (which we called apfu′ above), is 1.84705 + 0.46166 + 0.46185 = 2.7705. We want this sum to be 6, not 2.7705, because the formula of diopside has 6 oxygens.

Step 3. Multiply the "oxygen number" of each oxide by a normalization constant that is equal to the number of oxygens in the desired formula divided by the sum of the "oxygen numbers." Therefore, each number is multiplied by 6/2.7705 = 2.16567 (Table 10.7). This isn't quite finished, because the numbers are expressed as oxygen units rather than as moles, so we need one more step. (This step is similar to finding the total charge needed on the number of oxygens as explained earlier.)

Step 4. Multiply the "normalized oxygen numbers" of each oxide by the number of cations per oxygen in the oxide formula. So MgO would be multiplied by $^1/_1$, SiO_2 is multiplied by $^1/_2$, Al_2O_3 would be multiplied by $^2/_3$, K_2O by $^2/_1$, etc. (Table 10.7). This gives you what we've called the apfu. This calculation is much easier to *do* than to explain, so if all the words just provided don't make sense to you, ignore them and look at the Tables (10.8 and 10.9) instead. Some people learn better with words, and other with numbers—pick your learning style!

Graphical Depictions of Mineral Chemistry

While we are on the subject of mineral chemistry, one final concept is worth mentioning again: the idea that mineral chemistries can be represented *graphically*, by illustrations. We discussed this back in Chapter 6, but it is such an important concept that we'll go over it again here. Let's begin with a simple example, where we have two elements that can freely substitute for each other: Mg^{2+} and Fe^{2+}. Imagine a situation where these two cations completely fill a site, so that the number of formula units of $Mg^{2+} + Fe^{2+} = 1$. No other components are present. This could be, for example, the carbonate mineral group, with the two relevant species magnesite and siderite ($Mg^{2+}CO_3$ and $Fe^{2+}CO_3$, respectively). We would call these pure idealized starting components **end-members**. Mixtures of them would represent intermediate compositions between them.

You have a sample with the following formula: $Mg^{2+}_{0.5}Fe^{2+}_{0.5}CO_3$. Imagine that you want to draw a diagram illustrating the 1:1 relationship between Mg^{2+} and Fe^{2+}. So you draw a line with Mg^{2+} at one end, and Fe^{2+} at the other, and put an "X" exactly in the middle, to represent a composition that is halfway between the ends, like this:

Mg^{2+} ----------X-------------Fe^{2+}

A composition that is all Mg^{2+} would look like this:

Mg^{2+} X ------------------------Fe^{2+}

and a composition that is all Fe^{2+} would look like this:

Mg^{2+} ------------------------X Fe^{2+}.

These are the simple cases. Now what would happen if your sample is $Mg^{2+}_{0.9}$ and $Fe^{2+}_{0.1}$? It should plot nearly at the Mg^{2+} end, because it is so close to being all Mg^{2+}. So its plot would look like this:

Mg^{2+} ---X -------------------- Fe^{2+}

A plot of $Mg_{0.1}Fe^{2+}_{0.9}CO_3$ would look like this:

Mg^{2+} ---------------------X---Fe^{2+}

These plots are easy to visualize because in our example, the two components total to exactly one, which we can think of as 100%. So we can draw our line in a more quantitative way, like this:

0% Fe^{2+}	20	40	60	80	100% Fe^{2+}
100% Mg^{2+}	80	60	40	20	0% Mg^{2+}

and we can plot anything we want by thinking of its composition in terms of the percentages of the two cations that are at the ends of the solid solution. Another possibility is to think of the composition in terms of its molecular percentages, on a line like this:

0% $FeCO_3$	20	40	60	80	100% $FeCO_3$
100% $MgCO_3$	80	60	40	20	0% $MgCO_3$

In other words, we are now plotting the composition in terms of its molecular end members, siderite and magnesite, instead of just using Fe^{2+} and Mg^{2+}. There may be situations where one or the other approach is favored, but both work to uniquely illustrate any composition along this two-component solid substitution.

What if you wanted to plot these in terms of *wt% oxides*? Now this analysis gets trickier. First, you need to know that one formula unit of magnesite weighs 84.31 g, while a formula unit of siderite weighs 115.86 g. To plot these on a line, we need to make them total to 100%, as follows:

$$84.31\ g + 115.86\ g = 200.17\ g$$

$$\frac{100 \times 84.31}{200.17} = 42.12\%\ MgCO_3 \text{ by weight, and}$$

$$\frac{100 \times 115.86}{200.17} = 57.88\% Fe^{2+}CO_3 \text{ by weight.}$$

So if we plotted our composition by molecular wt%, it would look like this:

Table 10.9.

One last example showing how to handle the halogens (F and Cl), in this older method of calculating chemical formulas. In this case, the raw data are from Petersen et al. (2002), and the mineral example is a tourmaline.

Oxide	Wt% of oxide	Molecular Weight	Molecular proportion	Atomic proportion	Anions based on 31 oxygens	Cations based on 31 oxygens
SiO_2	36.42	60.0843	36.42/60.08 = 0.6062	2 × 0.6062 = 1.2124	1.2124 × 9.8097 = 11.893	11.893/2 = 5.947
Al_2O_3	33.14	101.9613	33.14/101.9613 = 0.3250	3 × 0.3250 = 0.9751	0.9751 × 9.8097 = 9.565	9.565/1.5 = 6.377
TiO2	0.33	79.8658	0.33/79.87 = 0.0041	2 × 0.0041 = 0.0083	0.083 × 9.8097 = 0.081	0.082/2 = 0.041
FeO	3.12	71.8444	3.12/71.84 = 0.0434	1 × 0.0434 = 0.0434	0.0434 × 9.8097 = 0.426	0.426/1 = 0.426
Fe_2O_3	0.58	159.6882	0.58/159.69 = 0.0036	3 × 0.0036 = 0.0109	0.0109 × 9.8097 = 0.107	0.107/1.5 = 0.071
MgO	8.77	40.3044	8.77/40.30 = 0.2176	1 × 0.2176 = 0.2176	0.2176 × 9.8097 = 2.135	2.135/1 = 2.135
MnO	0.03	228.817	0.03/228.817 = 0.0004	1 × 0.0004 = 0.0004	0.0004 × 9.8097 = 0.004	0.004/1 = 0.004
ZnO	0.01	81.3894	0.01/81.3894 = 0.0001	1 × 0.0001 = 0.0001	0.0001 × 9.8097 = 0.001	0.001/1 = 0.001
Cr_2O_3	0.02	151.9904	0.02/151.9904 = 0.0001	3 × 0.0001 = 0.0004	0.0004 × 9.8097 = 0.004	0.004/1.5 = 0.003
Na_2O	2.21	61.9790	2.21/61.9790 = 0.0357	1 × 0.0357 = 0.0357	0.0357 × 9.8097 = 0.350	0.350/0.5 = 0.700
CaO	0.66	56.0774	0.66/56.0774 = 0.0118	1 × 0.0118 = 0.0118	0.0118 × 9.8097 = 0.115	0.115/1 = 0.115
K_2O	0.01	94.1960	0.01/94.1960 = 0.0001	1 × 0.0001 = 0.0001	0.0001 × 9.8097 = 0.001	0.001/0.5 = 0.002
B_2O_3	10.80	69.6202	10.80/69.6202 = 0.1551	3 × 0.1551 = 0.4654	0.4654 × 9.8097 = 4.565	4.565/1.5 = 3.044
H_2O	3.18	18.0153	3.18/18.0153 = 0.1765	1 × 0.1765 = 0.1765	0.1765 × 9.8097 = 1.732	1.732/0.5 = 3.463
F	0.08	18.9984	0.08/18.9984 = 0.0042	1 × 0.0042 = 0.0042	0.0042 × 9.8097 = 0.041	0.041/1 = 0.041
Sums	99.36			Σ = 3.1623		
F, Cl correction				3.1623 – (½ 0.0042) = 3.1602		
Factor				31/3.1602 = 9.8097		

Calculated formula: $K_{0.002}Ca_{0.115}Na_{0.700}Zn_{0.001}Mn_{0.004}Mg_{2.135}Fe^{3+}{}_{0.071}Fe^{2+}{}_{0.426}Ti_{0.041}Al_{6.377}(B_{3.004}O_9)Si_{5.947}O_{18}(OH_{3.463}F_{0.041})$

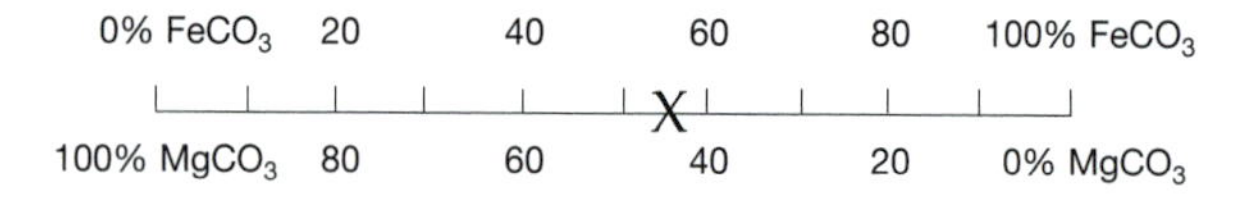

Notice that you only need to specify the percentage of one of the end members to completely describe the composition. This happens because there are only two components in the system, and they are required to sum to 100. So for the olivine group minerals that contain both Mg^{2+} and Fe^{2+}, if you have a sample that is 25% forsterite (the Mg^{2+} end-member), it must be 100 – 25 = 75% fayalite (the Fe^{2+} end-member). Sometimes we describe this scenario by just using the abbreviation for the mineral species name, followed by its percentage: that composition would be either Fo_{25} or Fa_{75}. You will often see this terminology used when the mineral of interest only has two end-members.

To practice your understanding of these drawings, try plotting the following compositions on this line: Fo_{10}, Fa_{45}, Fo_{70}, Fo_{55}, and Fa_{25}. Which two of these expressions are identical?

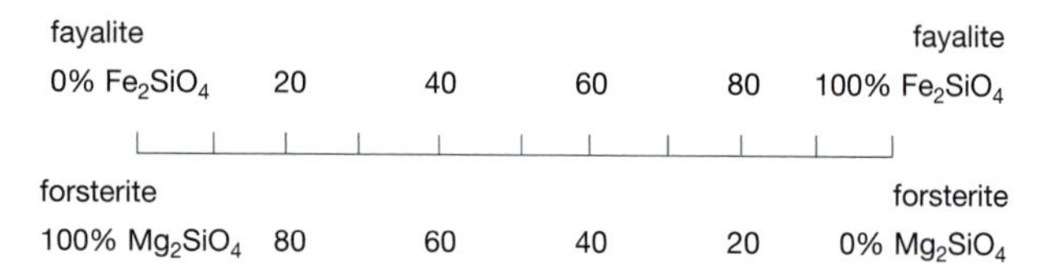

Most minerals have more than two end-members, so they need more than two-dimensional space to represent them. While it is impractical to use graphical representations for systems with four or more components, three-dimensional systems are common. For visualization of a three-component system, we use a special kind of plot called a **ternary diagram**. Again we have to make all our components add up to 100%, only this time, there are three of them. To uniquely specify a location on this plot, you need to specify two of the three components, and they all have to add up to 100%.

A ternary diagram for the olivine group system fayalite-forsterite-tephroite (the latter is Mn_2SiO_4) is shown in Figure 10.2. Each of the three corners of a ternary plot represents 100% of that component and 0% of the other two components. The lines on the plot represent 10% increments of each component (just as the ends of our line represented pure end-members). The bottom line, which represents 0% tephroite, is the exact same Mg^{2+}-Fe^{2+} line that we presented above. Other horizontal lines, moving up toward the pure Mn^{2+} tephroite at the top of the triangle, represent compositions that are 10%, 20%, 30%, etc. tephroite with variable Mg^{2+} and Fe^{2+}. Similarly, lines parallel to the Mg^{2+}-Mn^{2+} side represent varying amounts of the Fe^{2+} end-member, and lines parallel to the Fe^{2+}-Mn^{2+} side represent varying amounts of Mg^{2+}, or Mg_2SiO_4. Note that we are using molecular proportions here, not molecular weights! You can practice plotting various per-mutations of $(Mg,Fe,Mn)_2SiO_4$ on this diagram. Notice that the composition with equal amounts of Mg^{2+}, Fe^{2+}, and Mn^{2+} plots at the very center and would be $Fo_{33}Fa_{33}Tep_{33}$ (excusing the round-off error).

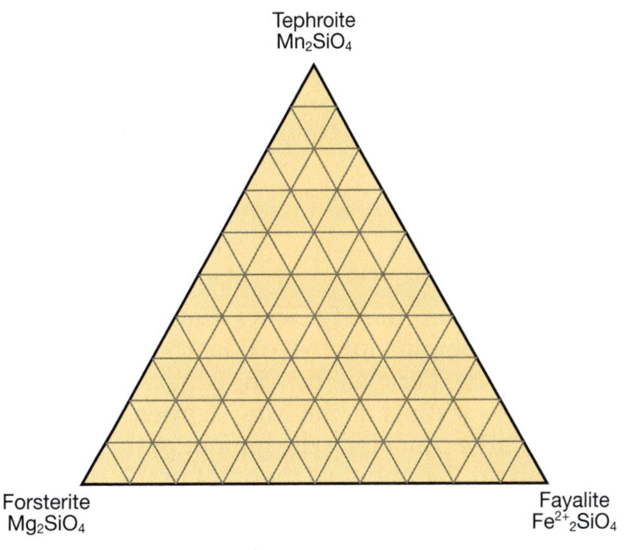

Figure 10.2. Ternary plot representing the olivine group end-members forsterite (Mg_2SiO_4), fayalite ($Fe_2{}^{2+}SiO_4$) and tephroite (Mn_2SiO_4). These sorts of diagrams are used to represent any combination of the three end-members (i.e., we could plot $(Mg^{2+},Fe^{2+},Mn^{2+})_2SiO_4$ values). See the text for a more thorough discussion.

As described in Chapter 6, ternary diagrams are commonly used to describe compositions in the pyroxene, amphibole, and feldspar groups. First, let's revisit the feldspar diagram (Figure 10.3). Most of the interior of the feldspar ternary plot is labeled **miscibility gap**, and it represents a range where naturally-occurring minerals do not exist. You can think of a miscibility gap as the opposite of a solid solution! Only on the left side and along the bottom of the feldspar ternary do natural compositions actually occur. Feldspars along the left side, in the region between $NaAlSi_3O_8$ and $KAlSi_3O_8$, are sometimes called the **alkali feldspars**, while those along the bottom (between $NaAlSi_3O_8$ and $CaAl_2Si_2O_8$) are the **plagioclase feldspars**. (Ask your professor if $NaAlSi_3O_8$ is a plagioclase or alkali feldspar—even we would like to know the answer to that one!)

The pyroxene ternary diagram uses two pyroxene species along the bottom ($Mg_2Si_2O_6$ and $Fe_2Si_2O_6$) and the pyroxenoid species wollastonite ($CaSiO_3$) on the top of the triangle (Figure 10.4). There are no naturally-occurring pyroxene species in the upper half of the triangle (i.e., above the line connecting diopside and hedenbergite), but this format makes it easy to specify common pyroxene compositions in the bottom half. Wollastonite is not a pyroxene but, as the name implies, is similar to a pyroxene (recall that "oid" means similar).

Finally, it is sometimes useful to describe a mineral by specifying the oxides that compose it, using a range of end-members that will define a number of related mineral species. One example is the MgO-SiO_2-CaO ternary plot shown in Figure 10.5, which is used to explain ultramafic metamorphic rocks that are commonly so low in Fe that their bulk composition can be expressed with three components. Let's make a list of some minerals that contain combinations of MgO, SiO_2, and CaO:

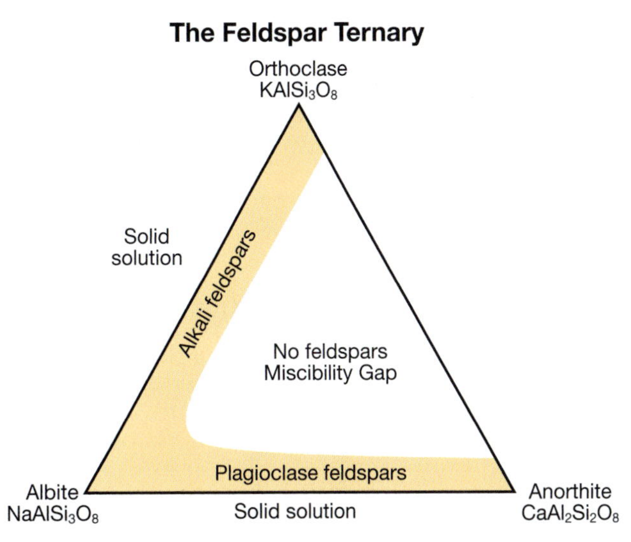

Figure 10.3. Feldspar ternary diagram showing the extent of solid solution between the feldspar minerals albite, anorthite, and orthoclase. Feldspar is not *stable for a large range of compositions that lie in the region labeled "miscibility gap."*

MgO	periclase
CaO	lime
SiO_2	quartz
Mg_2SiO_4	forsterite
$Mg_2Si_2O_6$	enstatite
$CaMgSi_2O_6$	diopside
$CaSiO_3$	wollastonite
$Ca_7Mg(SiO_4)_4$	bredigite
$Ca_2MgSi_2O_7$	åkermanite

To plot them on the MgO-SiO_2-CaO ternary, you need to separate out the oxide components in each formula. Periclase has one MgO and no CaO or SiO_2, so it plots at the MgO corner. Forsterite and enstatite plot along the bottom of the ternary. What about the minerals that plot in the middle?

Let's work through the example of åkermanite, $Ca_2MgSi_2O_7$. Its formula can be broken down in terms of oxide units:

2 CaO
1 MgO
2 SiO_2.

To plot this composition, we have to make the components add up to 100%, as we did for the two-component example above:

$$2\ CaO + 1\ MgO + 2\ SiO_2 = 5$$

$$\frac{100 \times 2}{5} = 40\%\ CaO,$$

$$\frac{100 \times 1}{5} = 20\%\ MgO,\ and$$

$$\frac{100 \times 2}{5} = 40\%\ SiO_2.$$

To plot the line of compositions that are 40% CaO, count in four lines, or 40% worth, from the MgO-SiO_2 edge of the plot, or you can count in six lines, or 60% worth, from the CaO corner. Now plot the line of compositions representing 20% MgO, and the intersection of those two lines will uniquely define our åkermanite. This example is plotted for you on Figure 10.5. You can even check to see if you got this correct by checking that the third oxide, SiO_2, plots where it should. For fun, try plotting the remaining Ca-Mg-Si minerals on the diagram yourself! The real power of these sorts of diagrams will be evident when you take igneous petrology. Petrologists will use the oxide weight of the rock to make such plots. In fact, they refer to the chemical analysis of a rock (mixture of minerals), often done by XRF or ICP, as a **whole-rock analysis**. From those chemical data, they'll do everything from deriving naming systems for the rocks based on their chemistry to helping explain how volcanic rock compositions change over the age of a magma system.

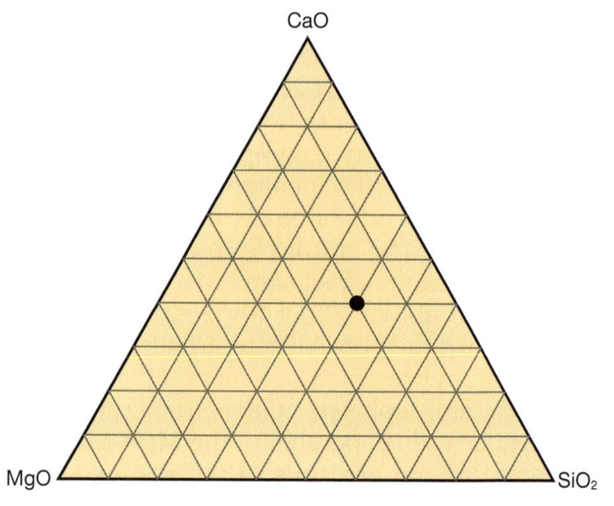

Figure 10.5. Ternary plot of the system MgO-SiO_2-CaO. The composition of åkermanite ($Ca_2MgSi_2O_7$) is calculated in the text and is CaO = 40%, MgO = 20%, and SiO_2 = 40%. These data are then plotted on the ternary diagram. Their point of intersection locates the composition of åkermanite in MgO-SiO_2-CaO space. (See the text for details on how to plot these data.)

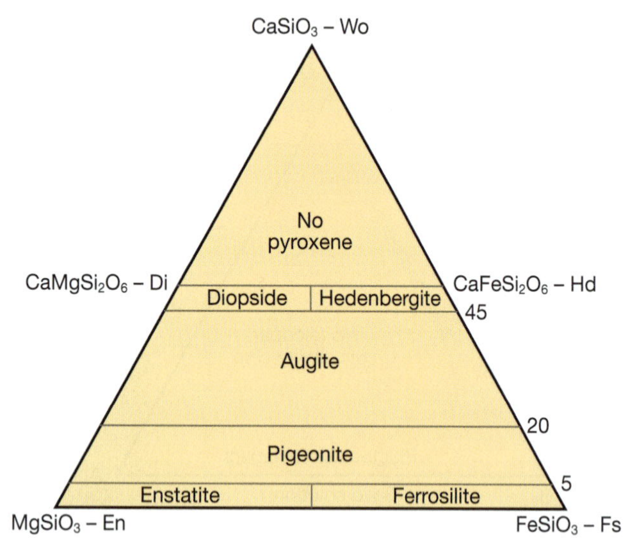

Figure 10.4. Ternary plot of the compositional space where pyroxenes (and pyroxenoids, which lie above the diopside-hedenbergite line) are represented. The different composition fields have mineral names associated with them as shown.

Compositional Variation in Minerals

Another concept follows from the above discussion, and is useful in describing mineral compositions. This is the realization that minerals have variable compositions, and usually deviate (sometimes, significantly) from their ideal, published formulas. While this often tends to frustrate students, as well as professional scientists outside the fields of mineralogy and geology, we must realize we are dealing with natural systems, and these systems often contain small amounts of many other elements. As we discussed in Chapters 1, 3 and 6, cations will occupy sites in minerals if their sizes and charges are appropriate. So if multiple elements with similar sizes and charges are "available" for incorporation into the structure when a mineral is crystallizing, they are likely to substitute for each other, and create compositions that are intermediate between ideal formulas. Some common examples of these substitutions are given in Table 10.10. Simple substitutions involve one-for-one replacement of one element by another element. These include Fe^{2+} for Mg^{2+}, F^{1-} for Cl^{1-}, Al^{3+} for Fe^{3+}, Na^{1+} for Ca^{2+}, etc. as shown in Table 10.10.

One of the most important and common substitutions in silicate minerals is Al^{3+} for Si^{4+}. However, charge balance is not maintained in this substitution, so it can only occur when the Al^{3+}-for-Si^{4+} substitution is charge-compensated by another substitution elsewhere in the mineral. For example, in the plagioclase feldspars, the loss of charge from the Al^{3+}-for-Si^{4+} substitution is charge balanced by substitution of Ca^{2+} for Na^{1+}. Another method of charge balancing is for a cation to be added somewhere in the structure. For instance, K^{1+} can be incorporated into the (previously vacant) interlayer site of a sheet silicate when one Al^{3+} substitutes for one Si^{4+} in the tetrahedral layer. Understanding these substitutions will help you not only understand why mineral compositions vary, but it will also help you to understand the systematics of mineral chemistries.

Table 10.10.

Examples of common substitutions that occur naturally in many minerals. (Many more substitutions can occur, but we are limiting them here to the major elements in the Earth's crust plus Cl, F, and H. Also the symbol "□" is used to represent a vacancy in the structure (i.e., a crystallography site that is empty).)

$Si^{4+} \leftrightarrow Al^{3+}$
$Al^{3+} \leftrightarrow Mg^{2+}$
$Mg^{2+} \leftrightarrow Fe^{2+}$
$Na^{1+} \leftrightarrow K^{1+}$
$Na^{1+} \leftrightarrow Ca^{2+}$
$K^{1+} \leftrightarrow Ca^{2+}$
$Na^{1+}Si^{4+} \leftrightarrow Ca^{2+}Al^{3+}$
$Na^{1+}Al^{3+} \leftrightarrow 2Mg^{2+}$
$3Mg^{2+} \leftrightarrow \square + 2Al^{3+}$
$OH^{1-} \leftrightarrow Cl^{1-} + \square$
$OH^{1-} \leftrightarrow F^{1-} + \square$
$OH^{1-} \leftrightarrow O^{2-} + \square$
$F^{1-} \leftrightarrow Cl^{1-} \leftrightarrow \square$

Assigning Cations to Structural Sites

We began this chapter by pointing out that there are two parts to writing the chemical formula for a mineral: (1) determining the chemical composition of the mineral in terms of the atoms in its formula, which we termed apfu, and (2) being able to place these atoms in the correct sites (e.g., bonded in an appropriately sized coordination polyhedral). We'll now take a more thorough look at the second of these issues.

Site occupancy questions set the study of minerals and crystalline materials apart from the study of chemistry of liquids, gases, or non-crystalline materials (i.e., this is really the guts of mineralogy)! So far, you should have learned that cations with different sizes fit in varying types of coordination polyhedra (you will recall that as size increases, so does coordination number). If a mineral has cations with vastly different sizes (i.e., big Ca^{2+} vs. tiny Si^{4+} in a garnet), it should be clear which sites they would occupy. If we think about similarly-sized cations (i.e., Mg^{2+} and Fe^{2+} in sheet or chain silicates or Si^{4+} and Al^{3+} in a framework silicate), it might be more difficult to determine site occupancy. As we are about to learn, site occupancy can be difficult to determine, and may require integration of *chemical* (e.g., compositional data obtained from EPMA), *structural* (e.g., size of coordination polyhedral obtained by X-ray diffraction), and *spectroscopic* (e.g., oxidation state of state of Fe from Mössbauer spectroscopy) information.

Fortunately, this problem can be greatly simplified through an understanding of mineral structures. A general overview of site occupancy will show the similarities of sites in many minerals and demonstrate that many site assignments can be made based on structural knowledge and good old-fashioned mineralogical common sense. The following discussion presents a simplified version of where to put cations in common rock-forming minerals. We encourage you, however, to consider each mineral group more carefully by consulting the discussion of nomenclature for each one, as presented in Chapters 21–23.

Garnets and olivines. As a first example, consider the site assignments for the garnet mineral andradite, $Ca_3Fe_2(SiO_4)_3$. There are three cations with sizes of $Ca^{2+} > Fe^{3+} > Si^{4+}$. Now take a look at the structure drawing of a garnet (Figure 10.6). Notice there are three polyhedra with different coordination numbers (CN): a large distorted polyhedron with CN = 8 (translucent yellow in the figure), an octahedron with CN = 6 (green) and a tetrahedron, CN = 4 (red). Given the three cations Ca^{2+}, Fe^{3+}, and Si^{4+}, where would you predict they would fit in the garnet? You'd put the biggest cation, Ca^{2+}, in the biggest site and the smallest cation, Si^{4+}, in the smallest site. This leaves the intermediate site (CN = 6) for the intermediate-sized cations (Fe^{3+}); thus the site assignments would be Ca^{2+} in the CN = 8 site, Fe^{3+} in the CN = 6 site, and Si^{4+} in the CN = 4 site.

Let's try another orthosilicate group, the olivines, and use the same three elements. Olivine group minerals have the general formula $(Ca^{2+},Fe^{2+},Mn^{2+},Mg^{2+})_2SiO_4$, so we'll consider two different olivines: fayalite, Fe_2SiO_4 (Figure 10.7a), and monticellite, $CaFeSiO_4$ (Figure 10.7b). The olivine structure has three different polyhedra: two octahedral sites, which are called M1 and M2 (this "M" nomenclature is used to denote a "metal" or cation site in a mineral without regard to what cation actually occupies the site) and the isolated tetrahedral sites. So it's a safe bet to assign Si^{4+} to the tetrahedral site, but what about the M sites? Recall that the olivine structure is based on two edge-sharing octahedral chains that run parallel to the *c* axis (Figure 10.7); the M1 octahedron is the green one, while the M2 site is translucent yellow. In fayalite (Figure 10.7a), both of these octahedra appear to be about the same size; in monticellite (Figure 10.7b), one of them is larger, and somewhat distorted from a perfect octahedron. What's your guess on where Ca would go in monticellite? Hopefully you'd put Ca in the larger, distorted site. Thus in fayalite, M1 and M2 are about the same size (Figure 22.41), but in monticellite, M2 > M1 because M2 stretched to house the larger cation. (Recall the house analogy from the start of the chapter, as a family would grow in size, they'd need to enlarge the house!)

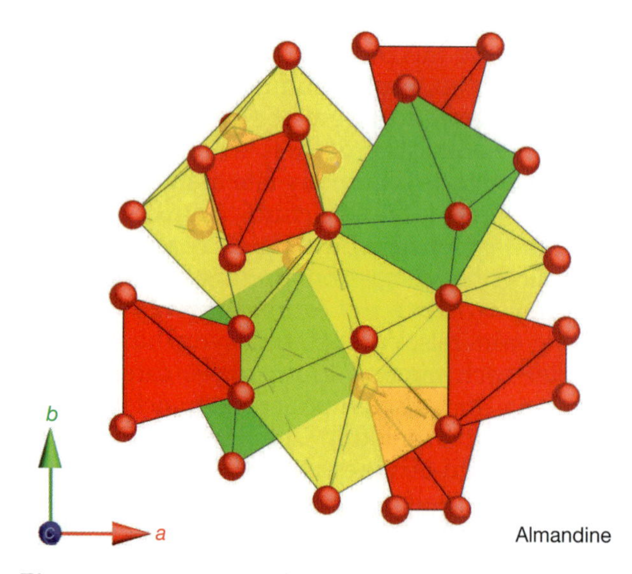

Figure 10.6. A portion of the crystal structure of a garnet, showing its three coordination polyhedra. From largest to smallest these are: a distorted CN = 8 site (translucent yellow), an octahedron CN = 6 (green), and a tetrahedron CN = 4 (red). Large cations like Fe^{2+} would enter the CN = 8 site, intermediate cations like Al^{3+} would enter the CN = 6 site, while small cations like Si^{4+} would reside in the CN = 4 site. The oxygen anions defining the corners of each polyhedron are shown as small spheres.

The situation for site assignments gets more complicated when you think about including Mg^{2+} and Mn^{2+} in the M1 and M2 sites. Fe^{2+}, Mg^{2+}, and Mn^{2+} are all about the same size, so their respective site occupancies (i.e., M1 vs. M2) cannot be predicted simply on the basis of charge or ionic radius. To solve the puzzle of where these cations lie in any olivine, it is necessary to get additional information, usually from X-ray diffraction or spectroscopic studies.

Pyroxenes. Next, let's take a look at the issue of site occupancy in the pyroxenes and finally learn why we write $MgSiO_3$ as $Mg_2Si_2O_6$. The pyroxenes are based on single chains of Si tetrahedra, and the unbonded apical oxygens in these chains form octahedra between them, while another octahedral site lies beneath the bases of the chains. Portions of the structures of enstatite, $Mg_2Si_2O_6$ (Figure 10.8a), and diopside, $CaMgSi_2O_6$ (Figure 10.8b), show these octahedral sites and the tetrahedral chains. There are again two octahedral sites: the M1 site formed by the apical oxygens between the chains (green), and the M2 sites below the bases of the chains and next to M1. M2 is larger than M1 (Figure 22.28). (In fact, you may be seeing a trend here in that, in general, as the number of the M-site increases, so does its size). In enstatite, both the M1 and M2 sites contain Mg^{2+}, while in diopside, M2 contains the larger Ca^{2+} cation and Mg^{2+} is in M1. Thus, when site assignments are made for the pyroxenes, large cations like Ca^{2+} and Na^{1+} can only enter M2 because M1 is too small. Notice that M2 has some room to expand laterally into the cavity created below the base of the tetrahedral chains. Now we can see why the pyroxene formula is written as $Mg_2Si_2O_6$: because there are two octahedral sites, and the Ca^{2+} content (as shown in Figure 10.4) cannot exceed one apfu because Ca^{2+} can only go in M2.

Amphiboles. Now we are ready to undertake the site assignments in amphiboles, which some folks call "garbage can" minerals. They have earned this lovely name because they can contain lots of different elements; the many substitutions are possible because amphiboles have several sites of different sizes! So if your professor insists on using this name, you need to think of garbage cans ranging in size from the small wastebasket in your dorm room (for the small cations like Si^{4+} and Al^{3+}) to the large dumpster outside (for the large cations like K^{1+} and Na^{1+}) as well as inter-

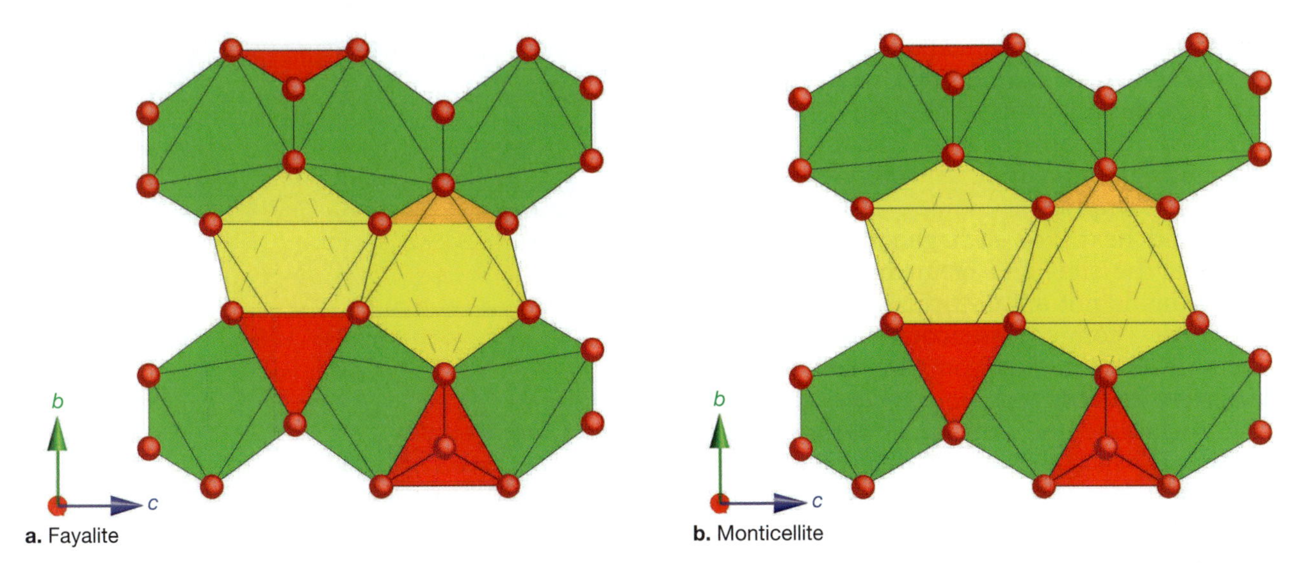

Figure 10.7. Portions of the crystal structure of two olivine minerals. The larger M2 octahedra are translucent yellow and sandwiched between the smaller M1 octahedra which are green, and the Si^{4+} tetrahedra are red; small spheres represent O and define the three coordination polyhedra. ***a.*** The structure of the olivine group mineral fayalite, Fe_2SiO_4 ; Fe^{2+} occupies both of the octahedral coordination polyhedra. ***b.*** The structure of the olivine group mineral monticellite, $CaFeSiO_4$; Ca^{2+} has enlarged the M2 site while the Fe^{2+}-containing M1 site remains nearly the same size as in fayalite.

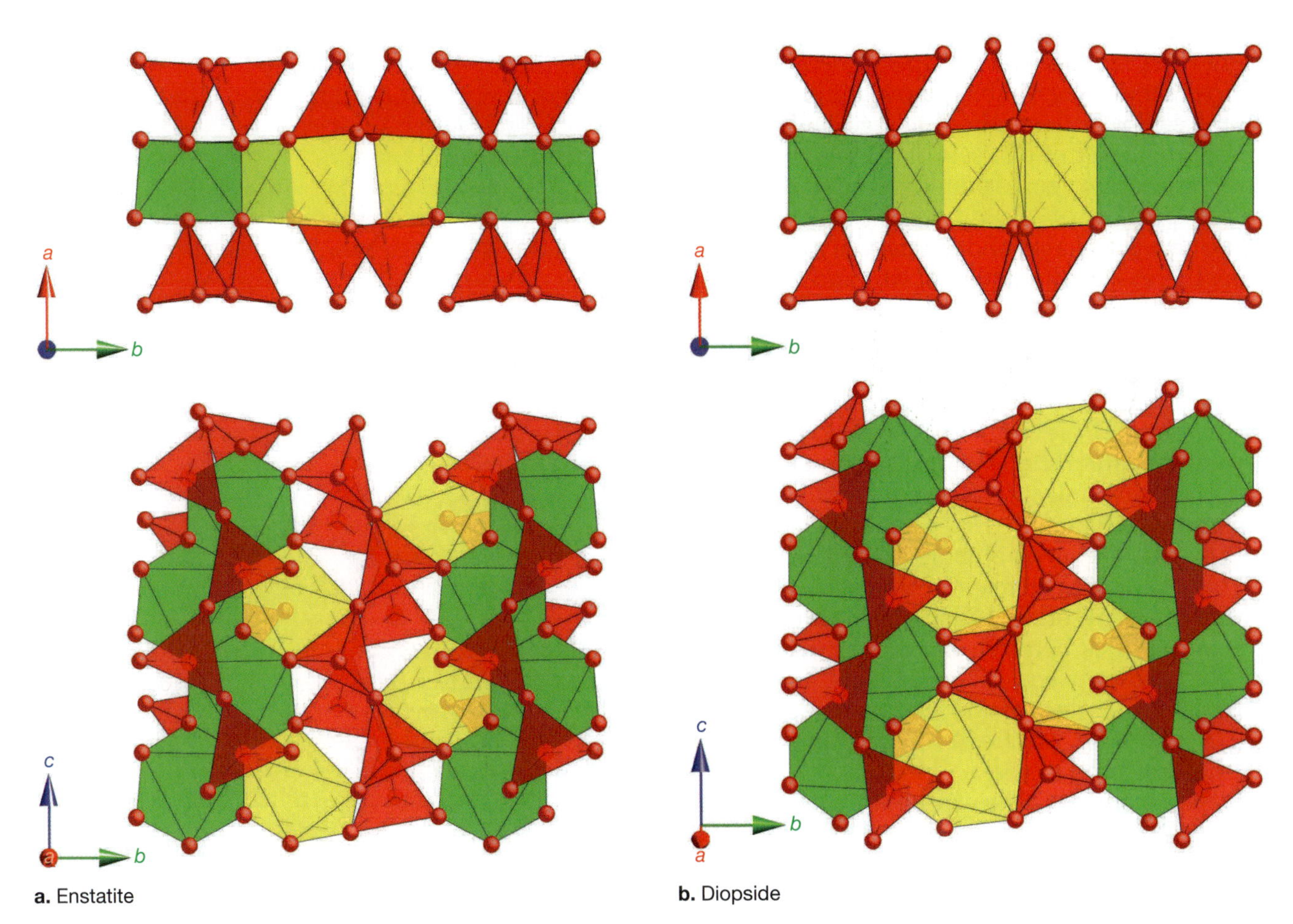

Figure 10.8. Portions of the crystal structure of two pyroxene mineral species. The larger M2 polyhedron is translucent yellow, the smaller M1 octahedron is green, the Si^{4+} tetrahedra are red, while small red spheres are O^{2-} anions and define the corners of the polyhedra. The upper projection is looking down the chains, while the lower projection is a top view perpendicular to the chains. ***a.*** The structure of enstatite, $Mg_2Si_2O_6$; both M2 and M1 contain Mg^{2+}. ***b.*** The structure of diopside, $CaMgSi_2O_6$; in this case, the larger cation Ca^{2+} occupies the larger M2 polyhedron while Mg^{2+} is in M1. Notice how the M2 sites have expanded into the cavity in the center of the structure.

mediate-sized cans (for cations like Mg^{2+} and Fe^{2+}) like the ones your parents have at home.

To put this in context, examine Table 10.11, which lists the general formulas for the above examples (garnets, olivines, and pyroxenes) as well as for amphiboles, micas, and feldspars, to be covered next. For each general formula, the table lists the sites and elements that can fit in them. (This is not an all-inclusive list, but gives the most commonly-encountered elements, which are also the most common elements in the Earth's crust!) For amphiboles, there are two separate formulas: the first, A_{0-1} $M4_2M3M2_2M1_2Si_8O_{22}(OH)_2$, lists the four M sites individually, while the second, A_{0-1} $B_2C_5T_8O_{22}(W)_2$, combines M1-3 (because these sites are of similar size) into the C site and M4 into the B site.

Figure 10.9 shows portions of three different amphibole group minerals in two different orientations. Figure 10.9a shows a portion of the structure of the amphibole group mineral anthophyllite, $Mg_7Si_8O_{22}(OH)_2$. The upper drawing shows a projection down the double chains. On its left and right sides are octahedral strips formed by the apical oxygens of the double chains pointing inward. In the center of the figure is a large cavity formed under the bases of two sets of double chains with their apical oxygens pointing outward. In the lower portion of the drawing, we are looking down on the chains and octahedral strip. Notice how there is an alternation of two and three green octahedra. The two-wide strips are the two M1 sites. The three-wide strip has two M2's on the outside with an M3 sandwiched in the middle. (Notice how these numbers fit the ratio in the general formula.) Next, there is a polyhedron in a light transparent shade that occurs on the edges of the M1 octahedral strip protruding into the cavity; this is the M4 site.

The M4 site in amphibole should remind you of the M2 site in pyroxene, while the M1-3 sites are reminiscent of the M1 site in pyroxene. Given the similarities between the two structures, you might predict where the cations are going to fit, based on the observation that the size of M4 > M1-3; again, M1-M3 are lumped together because they are of similar size.

Next, consider the structure of tremolite, $Ca_2Mg_5Si_8O_{22}(OH)_2$, as shown in Figure 10.9b. Note how similar it is to anthophyllite when projected down the *c* axis. When it's rotated to look down on the octahedral layer, notice how the occupancy by Ca^{2+} has expanded the M4 site. In Figure 10.9b for tremolite, we've also shown the H atom, which bonds off an O^{2-} in an octahedron located below the center of the ring formed by the above tetrahedral chains. Finally, a less common amphibole called richterite, $Na(CaNa)Mg_5Si_8O_{22}(OH)_2$, is shown in Figure 10.9c. The main thing to notice here is the addition of Na^{1+} in the A site, which is located in center of the cavity formed by basal oxygens in the double chains. Here the M4 site contains one Ca^{2+} and one Na^{1+}, so Na^{1+} occupies two sites in this amphibole.

Table 10.11.
General formulas and site assignments for garnets, olivines, pyroxenes, amphiboles, micas, and feldspars. (Many more cations can enter these minerals, but we are listing the major elements in the Earth's crust plus Cr, Mn, Ti and the anions Cl and F. More details for each mineral group will be given in Chapter 21.)

Garnets: $A_3B_2(SiO_4)_3$ where, A > B, and A = Ca^{2+}, Mg^{2+}, Fe^{2+}, Mn^{2+} B = Al^{3+}, Fe^{3+}, Cr^{3+}
Olivines: $M2M1SiO_4$ where, M2 > M1, and M2 = Ca^{2+}, Mg^{2+}, $Fe^{2+/3+}$, Mn^{2+} M1 = Mg^{2+}, $Fe^{2+/3+}$, Mn^{2+}
Pyroxenes: $M2M1Si_2O_6$ where, M2 > M1, and M2 = Ca^{2+}, Na^{1+}, Mg^{2+}, Fe^{2+} M1 = Mg^{2+}, $Fe^{2+/3+}$, Mn^{2+}, Al^{3+}, Cr^{3+}, Ti^{4+}
Amphiboles: $A_{0-1}M4_2M3M2_2M1_2Si_8O_{22}(OH)_2$ or $A_{0-1}B_2C_5T_8O_{22}(OH)_2$ where, A > M4 > M3 ~> M2 = M1, or A > B > C > T, and A = K^{1+}, Na^{1+} B = Na^{1+}, Ca^{2+}, Mg^{2+}, Fe^{2+}, Mn^{2+} C = Mg^{2+}, $Fe^{2+/3+}$, Mn^{2+}, Al^{3+}, Cr^{3+} T = Si^{4+}, Al^{3+}, Ti^{4+} OH = $(OH)^{1-}$, F^{1-}, Cl^{1-}, O^{2-}
Micas: $IM_{2-3}vac_{0-1}T_4O_{10}(OH)_2$ where, I > M > T, and I = K^{1+}, Na^{1+}, Ca^{2+} M = Al^{3+}, Mg^{2+}, $Fe^{2+/3+}$, Ti^{4+} T = Si^{4+}, Al^{3+}, Fe^{3+} OH = $(OH)^{1-}$, F^{1-}, Cl^{1-}, O^{2-}
Feldspars: $M_{1-2}T_4O_8$ where, M > T, and M = K^{1+}, Na^{1+}, Ca^{2+} T = Si^{4+}, Al^{3+}, Fe^{3+}

These three examples should lend insights into site assignments and the general formulas for amphibole group minerals. Basically there are five types of sites in this mineral: the A site (CN = 12), the B site or M4 (CN = 6–8), the C site or M1-3 (CN = 6), the T site (CN = 4), and the OH site, which can either be a H^+ bonded to an O^{2-}, or Cl^{1-} or F^{1-} replacing the O^{2-} in the octahedral sheet. Once you have calculated the number of cations for each formula unit in an amphibole, you need to assign them to these five sites. The process is based on what we've learned in the pyroxenes and some good old mineralogical common sense. First, the A site is the only place in the structure that could house atoms as large as K^{1+}. Na^{1+} could fit in either A or M4. M4 would be the only one of the M sites big enough to house Ca^{2+} (the A site is too big, and M1-3 are too small); this is why no more than two Ca^{2+} atoms can enter the structure. Smaller cations like Mg^{2+}, Fe^{2+}, and Al^{3+} would fit just fine in M1-4, while the smallest cations, like Si^{4+} and Al^{3+} would fit in the T sites. This would leave $(OH)^{1-}$, Cl^{1-}, and F^{1-} in the last site.

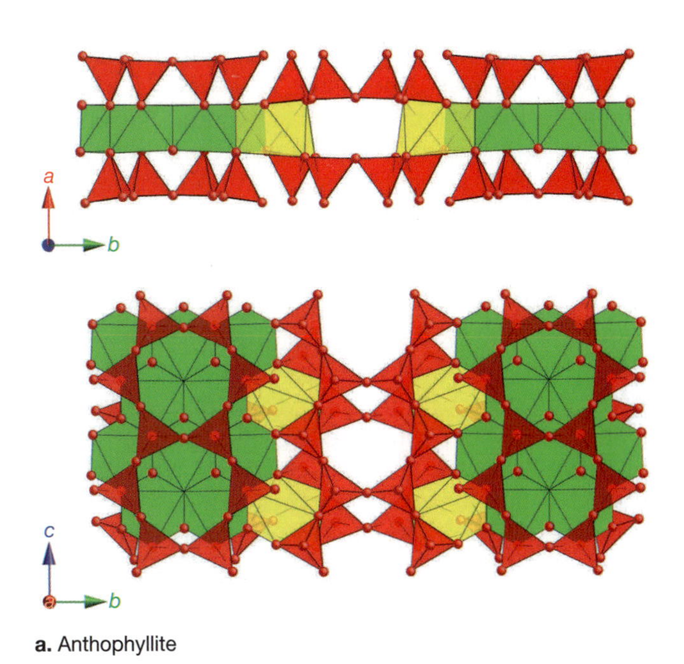

a. Anthophyllite

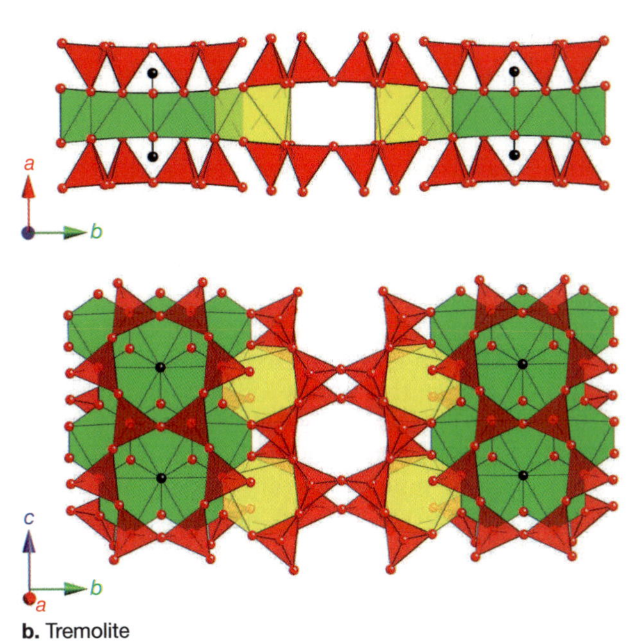

b. Tremolite

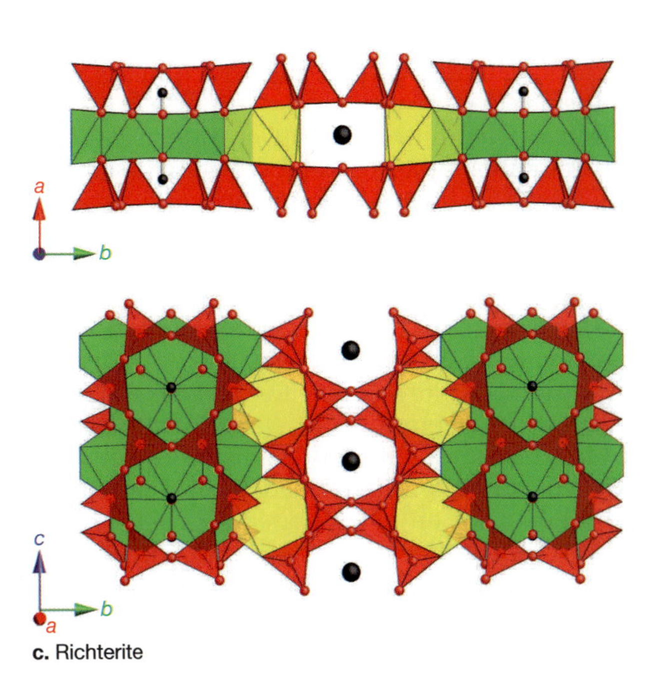

c. Richterite

Figure. 10.9. *Portions of the crystal structures of three amphibole group minerals. The double chains of silicate tetrahedra (translucent red) point inward to form the octahedral strips that contain the M1-3 sites (green). The larger M4 site (translucent yellow) is at the edge of these strips, and intrudes into the cavity created between the bases of another set of chains whose apical oxygens point upward, while small red spheres are O atoms and define the corners of the polyhedra. Two projections are shown, the top one looking down the chains and the lower one looking perpendicular to the chains.* ***a.*** *The structure of anthophyllite, $Mg_7Si_8O_{22}(OH)_2$; Mg^{2+} occupies all four of the M sites.* ***b.*** *The structure of tremolite, $Ca_2Mg_5Si_8O_{22}(OH)_2$. In this structure, Ca^{2+} replaces the Mg^{2+} from A in the M4 sites. Notice how M4 is bigger (seen best in the bottom projection perpendicular to the chains) now that Ca^{2+} occupies it. H^{1+} atom positions (small black sphere) have been added to show how H^{1+} bonds to an oxygen in the octahedral layer that is below the six-membered rings in the tetrahedral layer. That O^{2-} also is the only O^{2-} in the octahedral layer that is not bonded to the tetrahedral layer.* ***c.*** *The structure of richterite, $Na(CaNa)Mg_5Si_8O_{22}(OH)_2$. In this structure, Na^{1+} occupies the A site in the cavity created by the basal oxygens of the silicate chains. This site is 12-coordinated to six oxygens above and six below, both sets being in the six-membered rings formed by the double chains (best seen in the bottom projection).*

Thus, there is an established method to assign elements into the sites in an amphibole (and most other minerals, we'll discuss this more when we turn our attention to the individual mineral groups in Chapter 22 and 23). For the amphiboles, it proceeds as follows:

1. Fill the T site until it totals 8, by first using Si^{4+}, then Al^{3+}, then Ti^{4+}.
2. Fill the C site with five cations by first using excess Al^{3+} and Ti^{4+} from step 1, and then (in this order) Cr^{3+}, Fe^{3+}, Mn^{3+}, Mg^{2+}, Fe^{2+}, Mn^{2+}.
3. Sum the B site to 2 by first using excess Mg^{2+}, Fe^{2+}, Mn^{2+} from step 2, and then Ca^{2+}, then Na^{1+}.
4. Finally, assign excess Na^{1+} from step 3 and any K^{1+} to the A site.

Notice that this is an idealized method that assumes there are no vacancies in the T or C sites; this may not be the case for naturally-occurring amphiboles. But it gives a good first approximation for cation locations.

Table 10.12 is an example of the results obtained for this method for richterite. It should be stressed that this is the only correct way to assign the appropriate species name to an amphibole: first to perform a detailed chemical analysis and then to make the site assignments as described above. This kind of systematic approach also extends to most silicate minerals. While this entire exercise may seem to be of interest only to a mineralogist, we'll return to its importance later in Chapter 21, because of the health effects of amphibole-asbestos and the need to clearly understand the mineral species involved in different settings. Also, many minerals have been, and continue to be, misidentified because researchers have not taken the time to perform careful chemical and structural analysis of them.

Micas. The micas have fewer cation sites than the amphiboles (i.e., less of a variation in garbage can sizes!), so their site occupancies are also simplified. A general formula for the micas can be written as $IM_{2-3}\square_{0-1}T_4O_{10}(OH)_2$. ($\square$ refers to a **vacancy** in the structure, which means that one of the sites, in this case an octahedron, does not contain a cation. This symbol was picked years ago and is intended to indicate an empty site because it looks like an empty box). Recall that micas are 2:1 layer silicates with one octahedral sheet sandwiched between and formed by the unbonded apical oxygens on two tetrahedral sheets (Figure 10.10). Substitution of one Al^{3+} into one of every four Si^{4+} sites (i.e., Si_4O_{10} becomes $AlSi_3O_{10}$) causes a charge imbalance on these 2:1 layers, which is offset by the charge on the interlayer (I) cation.

Figure 10.10a shows three views of the structure of the mica group mineral phlogopite, $KMg_3(AlSi_3O_{10})(OH)_2$. The top projection is parallel to the tetrahedral and octahedral sheets. The middle view is a projection from the top showing the H^{1+} ion located in the center of the six-membered rings bonded into the octahedral sheet, just as in the amphiboles. The bottom projection shows the K^{1+} ion centered in the six-membered rings. So there are four types of sites in this mineral: the interlayer sites with CN = 12 (six O's in the tetrahedral layer above and six below), M sites with CN = 6 (there are two of them, M2 and M1, whose size is similar but which differ in their association with the $(OH)^{1-}$ groups), the T sites with CN = 4, and the OH sites. Large cations such as K^{1+} and Na^{1+} occupy the I site, intermediate size cations like Mg^{2+}, Fe^{2+}, and Al^{3+} are the most common occupants of the M sites, the smaller cations like Si^{4+} and Al^{3+} most commonly occupy the T sites and $(OH)^{1-}$, Cl^{1-} and/or F^{1-} reside in the OH site. Notice that Al^{3+} occurs only in the tetrahedral layer.

Figure 10.10b shows different projections of the trioctahedral mica series biotite, $KMg_2Fe(AlSi_3O_{10})(OH)_2$. In this idealized illustration, Mg^{2+} occupies the M2 sites and Fe^{2+} occupies the M1 sites, but these cations are basically interchangeable for each other (see also phlogopite in Figure 22.17). The dioctahedral mica species, muscovite, $KAl_2(AlSi_3O_{10})(OH)_2$ (Figure 10.10c) has vacant M1 sites, because the greater charge on Al^{3+} relative to Mg^{2+} and Fe^{2+} requires that only two of the three octahedra be filled. Notice that Al^{3+} occurs in both the tetrahedral and octahedral sheets. The O-H bond is no longer perpendicular to the octahedral sheet, but tilts slightly into the vacant octahedral site.

The preceding examples demonstrate that there are many similarities among the cation sites in the pyroxene, amphibole, and mica groups; the nomenclature is summarized in Table 10.13. These systematics, which change as we go from one chain (pyroxenes) to two chains (amphiboles) to many chains (i.e., layer silicates), should help you to understand how the three groups are related. Explicitly, note how the interlayer cation sites in the layer silicates are basically the same as the A cation sites in the amphiboles: both occur between the bases of 6-membered rings of tetrahedra, and both have CN = 12. This site is missing from the pyroxene group because the single chains don't form 6-membered rings. Along the same lines of reasoning, the rings provide space in micas and amphiboles for an H^{1+} ion to bond onto the octahedral sheet and penetrate into the rings, but this "ringed" site, and thus H^{1+}, is lacking in the pyroxenes.

Table 10.12.
Example of site assignment for the amphibole group mineral richterite based on first obtaining the apfu's from a detailed chemical analysis.

apfu			
Si	7.97	T-Si	7.97
Al	0.05	T-Al	0.03
Ti	0.01	sum T	8.00
Fe^{2+}	0.19		
Fe^{3+}	0.34	C-Al	0.02
Mg	4.43	C-Ti	0.01
Mn	0.04	C-Fe^{2+}	0.19
Ca	1.12	C-Fe^{3+}	0.34
Na	1.17	C-Mg	4.43
K	0.19	C-Mn	0.01
H	1.63	sum C	5.00
F	0.37		
		B-Mn	0.03
		B-Ca	1.12
		B-Na	0.85
		sum B	2.00
		A-Na	0.32
		A-K	0.19
		sum A	0.51

After site assignments, the formula is (where the superscripted A, B, C, and T represent the sites):

$^{A}(K_{0.19}Na_{0.32})\ ^{B}(Na_{0.85}Ca_{1.12}Mn_{0.03})\ ^{C}(Mn_{0.01}Mg_{4.43}Fe^{3+}_{0.34}Fe^{2+}_{0.19}Ti_{0.01}Al_{0.02})\ ^{T}(Al_{0.03}Si_{7.97}O_{22})(OH)_{1.63}F_{0.37}$

Because the micas are made of an infinite number of side-linked chains, they don't really have an equivalent to the M4 site in amphiboles and M2 site in pyroxenes; both of these are quite similar, and occur on the edge of the octahedral sheet. The M sites in mica are similar to M1-3 in the amphiboles and M1 in pyroxenes, forming from the unbonded apical oxygens of the chains. Finally, all of the T sites are basically the same in these mineral groups; they mainly house Si^{4+} and Al^{3+}. From all these comparisons, it is evident that amphiboles share some characteristics with the micas and some with the pyroxenes; thus amphiboles have more sites with different sizes (Table 10.13) than either micas or pyroxenes. The pyroxenes, amphiboles, and micas make up about 20% of the Earth's crust. Given the similarities of these minerals at the atomic scale, it might come as no surprise that they often form from each other (i.e., amphiboles can form from the pyroxenes by weathering). If you understand these minerals at the atomic level, it will really help you understand how they behave in nature as they form and alter in rocks.

Feldspars. The last site assignment example we will show involves Al^{3+}/Si^{4+} distribution in the tetrahedral sites of the feldspars. This is one of the most important substitutions in all silicate minerals, because it can be related to the geological processes by which the feldspars form. The reason for its geological importance results from the phenomenon of order/disorder in minerals. When an element is **ordered**, it consistently occupies the same site throughout the mineral (usually because the pressure and temperature conditions conspire to create a thermodynamically stable location). A **disordered** element (or elements) would be randomly distributed among different sites in the mineral. In other words, all the sites are similarly low in energy, and thus no cation has a preference for any single site. Feldspars that form in deep-seated plutons (i.e., have cooled for long periods of time) have an ordered array of Si^{4+} and Al^{3+} in their tetrahedral sites, while feldspars associated with volcanic rocks (i.e., have cooled rapidly) have a disordered array of Si^{4+} and Al^{3+} over the tetrahedral sites. There is also a gradation of ordering between these two extremes which, if determined, can provide useful information on the formation of a igneous rock body.

As an example, let's take a look at a portion of the structure of the feldspar mineral albite, $NaAlSi_3O_8$ (Figure 10.11). This figure shows the

Table 10.13.
CN, site equivalences, site names, and site occupancies for the micas, amphiboles and pyroxenes.

CN	*Micas*	*Amphiboles*	*Pyroxenes*	*Cations*
12	I	A	–	K^{1+}, Na^{1+}
6–8	–	M4	M2	Na^{1+}, Ca^{2+}, Mg^{2+}, Fe^{2+}
6	M	M1–3	M1	Mg^{2+}, $Fe^{2+/3+}$, Al^{3+}
4	T	T	T	Al^{3+}, Si^{4+}
–	OH	OH	–	$(OH)^{1-}$, Cl^{1-}, F^{1-}

"crankshaft chain" formed by rings of four tetrahedra that run parallel to the *a* axis in the mineral. Figure 10.11a shows an ordered albite; in this case, one tetrahedron out of every four is fully occupied by Al^{3+} (yellow translucent shape) and all the others by Si^{4+} (red shape). In the disordered albite (Figure 10.11b), the Si^{4+} and Al^{3+} are equally distributed among all the sites (so they all have the same shading), thus there is a 25% chance of finding an Al^{3+} in any one tetrahedron. Notice in these two figures that the Al^{3+} tetrahedra are bigger than the Si^{4+} ones because Al^{3+} is slightly larger than Si^{4+}. The locations of the Al^{3+} atoms can thus be detected by X-ray diffraction experiments that can be used to precisely determine the bond distances.

Finally, compare the four-membered rings for the disordered vs. ordered cases. In the ordered sample, the ring is a little distorted because the lone Al^{3+} tetrahedron is slightly larger. This lowers the symmetry of the mineral, and that loss of symmetry can be measured by X-ray diffraction or observed by using a polarizing light microscope. Thus, it is possible (even with a simple thin section) to determine the geological conditions under which a feldspar formed.

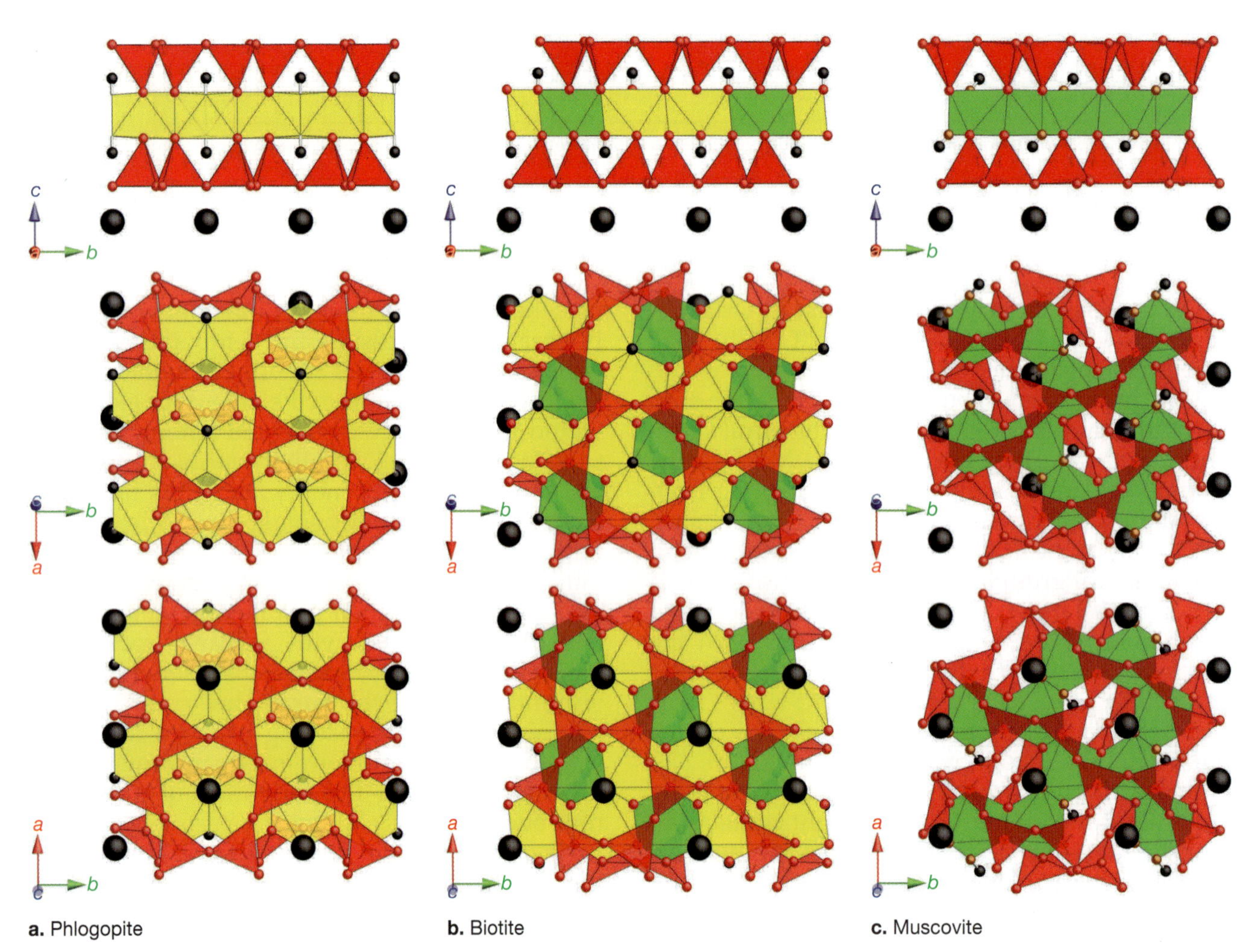

Figure 10.10. *Portions of the structure of three micas. The solid yellow octahedra represent the M2 site and the translucent yellow octahedra represent the M1 site. The large black sphere is the interlayer cation K^{1+}, the small black sphere represents H^{1+}, and small red spheres are O^{2-} ions and define the corners of the polyhedra. The upper projection is parallel to the layers while the middle and lower projections are perpendicular to the layers. The middle projection is looking down from the top to see the location of the H^{1+} ions (small black spheres) and the lower one is looking up from the bottom to see the location of the K^{1+} interlayer cation (large black spheres).* ***a.*** *In phlogopite, $KMg_3(AlSi_3O_{10})(OH)_2$, all the octahedral sites are filled with Mg^{2+}. Note in the middle projection that H^{1+} is bonded to the octahedral layer and points upward into the six-membered tetrahedral ring. In the lower projection, K^{1+} resides in the interlayer beneath the six-membered rings, and bonds to six oxygens in the sheet shown and six in a sheet below it (not shown).* ***b.*** *In members of the biotite series, $KMg_2Fe(AlSi_3O_{10})(OH)_2$, Fe^{2+} has replaced Mg^{2+} in the M1 site.* ***c.*** *In muscovite, $KAl_2(AlSi_3O_{10})(OH)_2$, only two Al^{3+}'s are needed for charge balance of the $3Mg^{2+}$'s in phlogopite. This, in turn, causes every third octahedral site to be empty, causing the O-H bond to deviate from being perpendicular to the octahedral sheet and deflect toward the octahedral vacancy. See also Figure 22.17.*

Other mineral groups. The preceding examples have shown the logic behind cation site assignments in some of the major rock-forming mineral groups. You can use the same type of reasoning to decipher the site assignments of cations in any other mineral. Several good mineralogical references will help you understand these relationships. We recommend the many volumes on *Rock-Forming Minerals* by Deer, Howie, and Zussman, as well as the *Reviews in Mineralogy and Geochemistry* series published by the Mineralogical Society of America and the Geochemical Society. Two other excellent references contain succinct summaries of site assignments in all mineral species: *Dana's New Mineralogy* (currently in its 8th edition, by Gaines et al., 1997) and *Strunz Mineralogical Tables* (currently in the 9th edition, by Strunz and Nickel, 2001). The principles that govern site assignments of cations will empower you to understand and predict many geologically important phenomena.

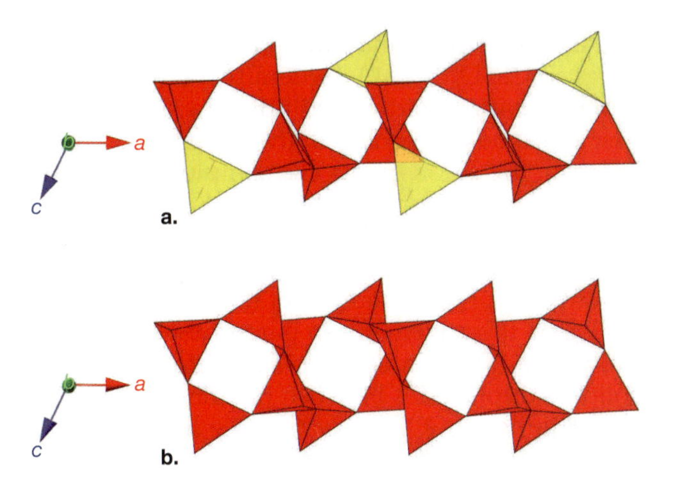

Figure 10.11. *Portions of the albite structure show the crankshaft chains composed of four-membered rings of tetrahedra that are near parallel to the a crystallographic axis. Recall that albite has the formula* $NaAlSi_3O_8$*, so a four-membered ring could have one* Al^{3+} *and three* Si^{4+} *tetrahedra.* ***a.*** *In this illustration, the* Al^{3+} *(translucent yellow) and* Si^{4+} *(red) tetrahedra are said to be* ordered. *This means that the* Al^{3+} *occurs in the same tetrahedron in all the rings. The* Al^{3+} *tetrahedra are slightly bigger than the* Si^{4+} *tetrahedra, and they distort the ring slightly by enlarging it.* ***b.*** *In this chain, the* Al^{3+} *(and* Si^{4+}*) are randomly distributed among the four sites in the ring (i.e., there is a 25% chance of finding* Al^{3+} *and a 75% chance of finding* Si^{4+} *in any tetrahedral site). The rings are no longer distorted. Thus the sample in Figure 10.11a would have a lower symmetry than the one in Figure 10.11b.*

The "Grammar" Rules for Mineral Formula

Now that we know how to calculate a mineral formula, and where to locate the cations in the various sites of each important mineral group, we must add a final note about use of notation. As mentioned in several places so far in this book, there is an unwritten set of rules for writing chemical formulas for minerals. Like any set of rules, they are not always followed! However, when they are used properly, it makes it much easier for students to "read" mineral formulas correctly and gain an understanding of the underlying crystal structures of the minerals of interest. If the formula is not written in these styles, then the systematics become much harder to see and frustration can easily take over!

As an example, consider the formula for muscovite, which is sometimes written as $KAl_3Si_3O_{10}(OH)_2$. We prefer to write it as $KAl_2(AlSi_3O_{10})(OH)_2$. To a chemist, this would seem like wasted typing to write the Al_3 from the first formula as Al_2 and Al in the second. But this notation shows the there are two different structural sites for the Al: the Al_2's are in octahedral coordination while the single Al is in a tetrahedral site. By writing the formula in this manner, it is evident that the ratio of tetrahedral cations to oxygen is 4:10, cluing you into the fact that this is a layer silicate. We could then write the formula for the biotite series $K(Fe,Mg)_3(AlSi_3O_{10})(OH)_2$ on the board right below the formula for muscovite to show that the only difference in these two formulas is the replacement of Al_2 by $(Fe,Mg)_3$.

Another example is the formula for magnetite, which is sometimes written as Fe_3O_4. If we know that the charge on oxygen is –2, what's the charge on the iron? Based on the way the formula is written, it works out to be –2.667, which is rather strange! To clarify this, the formula for magnetite is often written as $Fe^{2+}Fe^{3+}_2O_4$ to show that iron in magnetite, like Al^{3+} in muscovite, is in two structural sites and in this case, even has two separate charges. Hopefully these examples will motivate you to observe and understand the following loose set of rules.

For simple two-element compounds like Na^+Cl^-, $Ca^{2+}F^{1-}_2$, $Pb^{2+}S^{2-}$, we list the cation first, followed by the anion; this is the convention for all minerals in groups that are based on anions. The practice also continues when we move to groups of minerals that are based on anionic complexes, for example, $Pb(SO_4)$, $CaCO_3$, $Mg(OH)_2$, and $CaSO_4$. When minerals contain molecular water (i.e., H_2O groups), and not just OH groups (as in the hydroxides), the water is added to the end of the formula and separated from it by a dot "·" as in $CaSO_4 \cdot 2H_2O$. Generally a chemical formula is written starting with the largest elements first and then proceeding to the smaller ones, ending with the anion or anionic complex. For example, the formula for cryolite is Na_3AlF_6. Finally, as we've shown for the case of the silicate minerals, we represent solid solutions (or elements that can substitute for each other in the same site) by enclosing those elements in parentheses and separating them with a comma. For instance, for the solid-solution series between magnesite ($MgCO_3$) and siderite ($FeCO_3$), we write $(Mg,Fe)CO_3$.

So, let's build on these rules and add a couple more for the silicate minerals. You may be getting tired of seeing so much emphasis placed on the silicates, but recall that they compose 93% of the Earth's crust, with quartz and the feldspars composing almost two-thirds of it! An added benefit of following a set of rules for the silicates is that you can figure out which silicate class a mineral belongs to by observing its Si to O ratio, as has been stated many times. For example, look at the formula for the amphibole group mineral actinolite: $Ca_2(Mg^{2+},Fe^{2+})_5Si_8O_{22}(OH)_2$. It follows all the rules stated above (i.e., largest cation on the left moving to smaller ones to the right, the substitutions for (Mg^{2+},Fe^{2+}), and anionic complex written next), but there are two new things to observe in this formula: (1) we can tell which silicate class this mineral belongs to based on the Si:O ratio of 8:22, and (2) if $(OH)^{1-}$ occurs in a silicate mineral, it comes after the silicate anionic complex; this is also the spot for other anions or anionic complexes to occur in the structure. Here's another formula for the silicate mineral group scapolite, which you've not seen before: $(Na,Ca)_8(AlSiO_4)_6(SO_4,S,Cl)_2$. Notice how it follows our rules. To which silicate class does this belong?

Next let's show an example of a silicate that doesn't (usually) follow the rules: andalusite, Al_2SiO_5. You may recall that andalusite is an orthosilicate, that would mean it should contain SiO_4, but it has SiO_5. So what gives? Another way to write this formula is Al_2OSiO_4, but why did we add that O^{2-} in before the SiO_4 group? This seems to be breaking our rule of putting the anions, or anionic complexes, at the end of the formula. While we ponder that formula, let's look at this one: $Ca_2(Al,Fe)Al_2O(SiO_4)(Si_2O_7)(OH)$. You've seen it before; it's epidote. It has isolated SiO_4 groups and Si_2O_7 groups, which are two tetrahedra sharing a single oxygen, but again it has an extra O^{2-} in the middle. For one last example with a minor modification, let's take the formula for the 2:1 layer layer silicate talc, $Mg_3Si_4O_{10}(OH)_2$, and remove its H^{1+} atoms to yield $Mg_3Si_4O_{10}(O)_2$. Taking a look at the two different "O"s in this formula will solve our problem of why "O's" end up in the middle of the formula. The "O" in Si_4O_{10} is part of the tetrahedral sheet and the basic building block for layer silicates. The other O^{2-} is in the octahedral sheet, and does not bond to the tetrahedral sheet. Yes, the O's in the Si_4O_{10} layer do bond to the octahedral sheet (in fact they make it) but they also bond to the Si in the tetrahedral sheet. So the extra O^{2-} is part of the octahedral sheet exclusively and we could write this formula as $Mg_3O_2Si_4O_{10}$.

This analogy could be extended to the amphibole group minerals. For instance, tremolite is normally written as $Ca_2Mg_5Si_8O_{22}(OH)_2$, but if we remove all the H^{1+}, we would have $Ca_2Mg_5Si_8O_{22}(O)_2$ and then rewriting would yield $Ca_2Mg_5O_2Si_8O_{22}$. Take a look at the amphiboles in Figure 10.9, and you can see that the O^{2-} associated with H^{1+} in the octahedral layer does not bond to any O^{2-} in the tetrahedral layer. Thus it is not included in the Si_8O_{22} portion of the formula, but is normally added at the end with the H^{1+}. Now going back to Al_2OSiO_4 for andalusite, that extra O^{2-} belongs to the octahedral chain in andalusite and does not share a bond to Si^{4+}. In a similar manner, the extra O^{2-} in epidote does not bond to any Si^{4+} atoms, but bonds into octahedra. In fact, notice that epidote really has one anion (the O^{2-}) and two anionic complexes in its formula and structure.

General formula. Several terms are used as modifiers for mineral formulas, and we want to end this section by describing them and giving some examples. The terms **general**, or generalized, and **ideal** or idealized all mean the same thing: they indicate a simplified formula for a mineral. No silicate mineral, except maybe the silica polymorphs, ever really has an ideal formula, because minerals form in nature and nearly always incorporate some other elements, at least in small amounts. In a general formula, we idealize the elements present by using integers for the apfu values instead of using the more precise decimal values and including the minor elements that they would contain in nature. The general formulas, especially for the silicates, allow us to identify the silicate group of each mineral and, in turn, to possibly predict some of its physical properties based on its crystal structure.

Here are several examples of general formulas for minerals followed by actual chemical analyses:

andalusite: Al_2SiO_5
$(Al_{0.95}Fe_{0.03}Mn_{0.02})Al_{1.01}Si_{0.99}O_5$

almandine: $Fe_3Al_2(SiO_4)_3$
$(Fe_{2.12}Mg_{0.61}Ca_{0.10}Mn_{0.04})Al_{2.04}Si_{3.04}O_{12}$

tremolite: $Ca_2Mg_5Si_8O_{22}(OH)_2$
$K_{0.01}(Na_{0.03}Ca_{1.97})(Mg_{4.52}Fe_{0.47}Mn_{0.01})$
$(Al_{0.03}Si_{7.94}O_{22})(OH)_2$

vermiculite: $(Ca,Mg)_{0.35}(Mg,Fe,Al)_3[(Al,Si)_4O_{10}]$
$(OH)_2 \cdot 8H_2O$
$(Ca_{0.05}Mg_{0.38})(Mg_{1.92}Fe^{2+}_{0.08}Fe^{3+}_{0.46}Ti_{0.11}Al_{0.22})$
$(Al_{1.28}Si_{2.72}O_{10})(OH)_2 \cdot 4.43H_2O$

plagioclase : $(Ca,Na)(Al,Si)_3O_8$ or
$Ca_xNa_{1-x}Al_{1+x}Si_{3-x}O_8$
$Ca_{0.69}Na_{0.30}Al_{1.66}Si_{7.94}O_8$

Notice how we've enclosed many of the cation and anion groups in parentheses. These elements all occupy the same site in the mineral, so it's easy to see what substitutes for what in the formulas. We've included two general formulas for plagioclase: the second one, $Ca_xNa_{1-x}Al_{1+x}Si_{3-x}O_8$, is written in such a way as to show the coupled substitutions. For instance, when $x = 0$, the formula would be that of albite, $NaAlSi_3O_8$. When $x = 1$ we would have the end-member anorthite, $CaAl_2Si_2O_8$. We could represent all the intermediate compositions with the ranges of x from 0 to 1.

End-member and intermediate compositions. Earlier in this chapter, we used the term **end-member** to describe idealized compositions at either end of a solid-solution series. For instance, in the pyroxenes, we could write the solid-solution series $Ca(Mg,Fe)_2Si_2O_6$, which is represented in the middle of Figure 10.4. One end-member would be $CaMgSi_2O_6$ and the other $CaFeSi_2O_6$. The former is diopside and the latter is hedenbergite, but what about the intermediate compositions, say $CaMg_{1.5}Fe_{0.5}Si_2O_6$? Usually we use something called the "50% rule" to name minerals that occur with complete solid solutions. So a $Ca(Mg,Fe)_2Si_2O_6$ pyroxene would have to be called diopside if its composition were in the range of $CaMgSi_2O_6$ to $CaMg_{1.0}Fe_{1.0}Si_2O_6$ and hedenbergite if it was in the range of $CaFeSi_2O_6$ to $CaMg_{1.0}Fe_{1.0}Si_2O_6$.

Now consider the amphibole analogs of the pyroxenes just mentioned: tremolite, $Ca_2Mg_5Si_8O_{22}(OH)_2$, and ferroactinolite, $Ca_2Fe_5Si_8O_{22}(OH)_2$. A third amphibole of intermediate composition between these two is called actinolite. Tremolite occupies the composition range from $Ca_2Mg_5Si_8O_{22}(OH)_2$ to $Ca_2Mg_{4.5}Fe_{0.5}Si_8O_{22}(OH)_2$ (only 10% of the Mg, Fe range), actinolite from $Ca_2Mg_{4.5}Fe_{0.5}Si_8O_{22}(OH)_2$ to $Ca_2Mg_{2.5}Fe_{2.5}Si_8O_{22}(OH)_2$ (the 10 to 50% Fe range) and ferroactinolite from $Ca_2Mg_{2.5}Fe_{2.5}Si_8O_{22}(OH)_2$ to $Ca_2Fe_5Si_8O_{22}(OH)_2$ (the 50 to 100% Fe range). We'll deal more with mineral nomenclature when we treat each group in detail in Chapters 21 to 23.

Concluding Remarks

There is no doubt that representing minerals with chemical formulas is one of the central themes of mineralogy. Mineral formulas can be your friends if you take the time to understand what they are telling you! Both the chemical relationships, such as charge balance, and the structural relationships, where cations of similar size substitute for one another, come into play when understanding formulas. A basic understanding of the "grammar" rules for formulas will help you gain insights into the crystal structure of a mineral by correctly "reading" them.

However, the mess created by having to convert wt% oxides to chemical formulas adds unnecessary confusion to an already complex subject. Although we'll never get away from having to learn chemical formulas, we could most certainly stop reporting mineral chemistry in terms of wt% oxides, and in turn alleviate the need for students to convert oxides to formula units. This would make the sections of this chapter somewhat less necessary and the authors and future mineralogy students much less confused!

Acknowledgments. We appreciate the inspiration provided by the table for elemental site assignments from James Wittke, Northern Arizona University. Parts of this chapter were based upon excellent handouts on mineral formula recalculations (John Brady, Smith College) and chemical substitutions (Stephen A. Nelson, Tulane University), with their permission. The reader is referred to their excellent web sites at their respective institutions for more complete information.

References

Droop, G.T.R. (1987) A general equation for estimating Fe^{3+} concentrations in ferromagnesian silicates and oxides from microprobe analyses using stoichiometric criteria. Mineralogical Magazine, 51, 431–5.

Peterson, O.V., Francis, C.A., Dyar, M.D., and Rosing, M.T. (2002) Dravite from Qârusulik, Ameralik Fjord, southern West Greenland: a forgotten classic tourmaline occurrence. extraLapis English, 3, 42–46.

Vocke, R.D. Jr. (1999) Atomic weights of the elements 1997 (technical report). Pure and Applied Chemistry, 71(8), 1593–1607.

Chapter 11

Introduction to Symmetry

The mineral collections of the National Museum of Natural History in the Smithsonian Institution contain some of the most spectacular examples of huge, gem-quality crystals in the world. If you bypass the crowds around the Hope Diamond and the Logan Sapphire, you may see the mammoths of the collection, which are among my favorites: two amazing topaz crystals weighing 31.8 kg (more than my Siberian Husky) and 50.4 kg (what I weighed in high school)! Each of these is a homogeneous, naturally-occurring crystalline solid with a constant chemical composition. How did these minerals grow to be so big with their atoms arranged in such an amazing, nearly-infinite series of perfectly repeated positions?

To understand the answer to this big question, we have to think of minerals as built from the ground up. First, we need to understand which repeatable arrangements of atoms are possible in three-dimensional space. This is akin to figuring out how many different shapes of building blocks you can construct out of marshmallows and toothpicks. If you vary the lengths of the toothpicks and the angles between them, you can reproduce the 230 space groups, which represent all the specific ways in which 3-D periodic structures can have symmetry. Once you understand the possible arrangements of "toothpicks and marshmallows," then you can make the structures different by changing the size and type of atom, or element, at each corner. Every known mineral can be described by knowing the geometric arrangement of its atoms, and the "identity" of each atom.

In this chapter, we begin with the most abstract concept in this book, which is the repetition of space. At this point we are not concerned with the types of "marshmallows" at the corners of the blocks, but only the possible arrangements of "toothpicks" of different lengths and at different angles to each other. These arrangements make patterns that are familiar to all of us. As you read, I hope you will begin to become conscious of the world of patterns around you. By the end of this chapter, you will never look at floor tiles the same way again!

M.D.D.

Introduction

How many repeatable patterns can you make in two dimensions? We all know that if you draw four equally-spaced dots at the corners of a square on a piece of paper, you can draw two more spots next to it to make another square, and then two more. Eventually you will cover a paper with a pattern in which, if it is properly drawn, every spot is at the corner of four identical squares (Figure 11.1). In this pattern, every dot is an equal distance from its closest four neighbors, and if you draw lines between the dots, they will be at 90° to each other. This is just *one* of the 17 possible patterns you can draw in two-dimensional space. If you are clever and have a lot of time to waste, you might be able to arrive at all 17 possible ways to arrange your dots by trial and error. It would take a lot of straws and marshmallows, however, to figure out all the ways in which 3-D structures can have periodic symmetry. An easier way is to discover them through the tools of the field known as **crystallography**.

Geometrical crystallography is generally regarded as one of the few "closed" sciences. It was brought to formal completion about 1908, which is to say that by then, all 230 3-D arrangements could be mathematically derived from elementary principles. However, as noted by M.J. Buerger in his penultimate text on this subject,

> "the fundamental ideas concerning crystal symmetry are inherently simple. They require no complicated mathematics, and can be understood by anyone whose background includes high school mathematics. In fact, they fall into the category of an interesting kind of geometry"
>
> (*Elementary Crystallography*, MIT Press, 1978).

One of the aspects of crystallography that makes it particularly appealing is its vocabulary, which while slightly antiquated to modern ears, gives highly specific meanings to its terminology. These classical terms are used here because they are historically correct and rather endearing.

Operations in Two Dimensions

An ideal crystal is constructed by the infinite repetition in space of identical **structural units**. These "units" may be repetitions of the same atom (e.g., gold, silver, or carbon) or a group of atoms corresponding to the chemical composition or formula of a mineral. Each repeated unit is

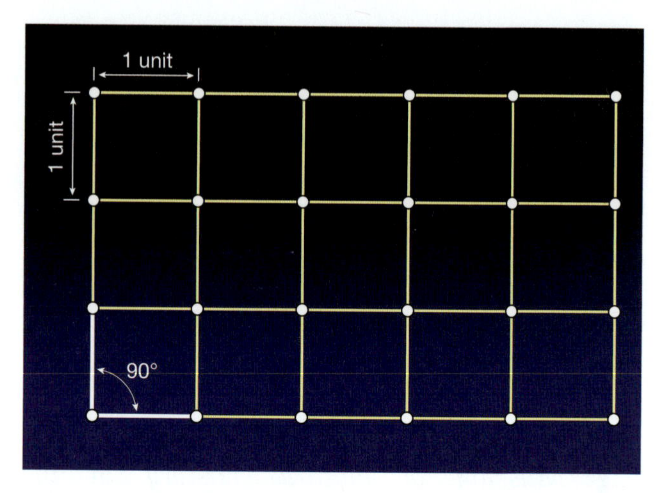

Figure 11.1. If you begin with one point in space, and repeat it at regular distances, you make 2-D pattern. In this drawing, the spaces among all the points are identical, and lines drawn between the points would be at right angles to each other.

called a **motif** or **basis**. Most crystals are made up of groups of atoms; water is composed of H_2O molecules, and the mineral **periclase** is made from MgO molecules. In extreme examples, such as proteins (which are not strictly crystalline), you can have up to 10^4 atoms in one motif. Each molecule is a copy or repetition of some arbitrarily-chosen motif molecule. The words "structural unit," "motif" and "basis" are, for our purposes, used interchangeably as well as broadly to refer to anything than can be indefinitely repeated in a pattern.

The repetition of motifs in a pattern is governed mathematically and is constrained by the number of dimensions you wish to operate in. As we already know, in 2-D there are only 17 available ways of repeating a general motif; in 3-D, there are 230. To begin building patterns, we need to explore the different types of operations, or ways of repeating space. The term **operation** was originally defined as any motion that brings the original motif into the same motif elsewhere in the pattern (Figure 11.2). Think of a pattern that is drawn with a stencil, in which every motif in the pattern is repeated at regular intervals in two dimensions. An operation can be thought of as any movement of the stenciled motif from one position to another in the pattern.

A related concept to operations is the property of **symmetry**. A pattern is said to have symmetry if an operation can be performed causing points in the motif to be moved, but with the result that the appearance of the pattern remains the same.

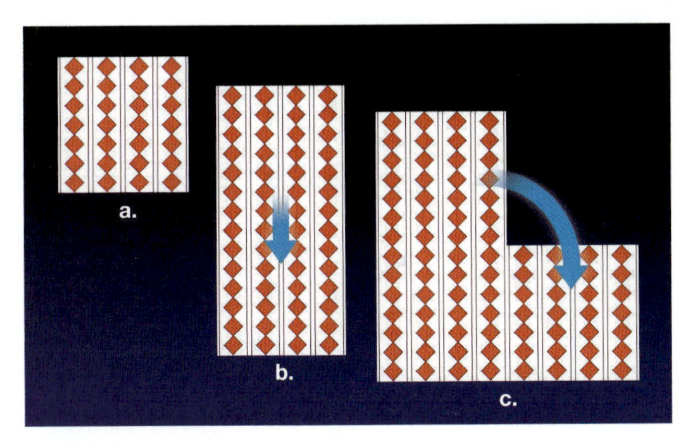

Figure 11.2. If you copy this pattern and pick it up, you can move it over so that the motifs again line up. The action of moving the pattern is called an operation, and there are five operations in 2-D space.

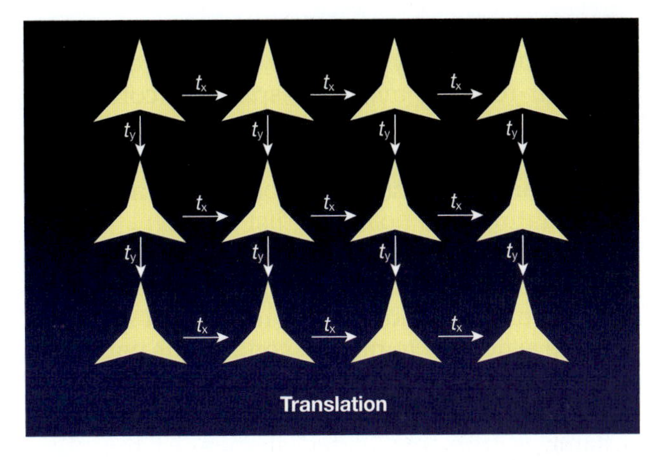

Figure 11.3. Translation is an operation that simply generates a pattern at intervals equal to some constant distance, or translation direction, t.

A better way to understand operations and symmetry is to look at the five operations that are possible in two dimensions.

1. **Translation.** Translation is an operation that generates a pattern at regular (identical) intervals (Figure 11.3). The translation direction is usually given the designation t. In multidimensional space, the translations have names that indicate which direction they move in: t_x, t_y, and t_z. If the translation directions don't line up with the Cartesian axes, then they're just designated t_1, t_2, t_3, etc.
2. **Rotation.** Rotation is an operation that generates a pattern by rotating the motif through an angle around an imaginary axis (Figure 11.4). The rotation angle is usually given the designation α or sometimes g, and the general symbol for any rotation is called n. All objects in two dimensions have what is called the **identity operation**, which means that they can be rotated 360° and still look the same. A different term for this is to say that these objects possess **1-fold rotational** symmetry, or the ability to be rotated $2\pi/1$ or 360° and still look the same. Other basic types of rotations include:

2-fold	$2\pi/2 = 180°$ rotation
3-fold	$2\pi/3 = 120°$ rotation
4-fold	$2\pi/4 = 90°$ rotation
6-fold	$2\pi/6 = 60°$ rotation

As you increase the value of n, you progress to "higher" symmetry, which means that the pattern possesses more than one type of rotational symmetry. For example, any object with 2-fold symmetry also has 1-fold symmetry. This means that if you rotate the object 180°, it will be exactly the same as it was before you rotated it, but if you rotate another 180°, (a total of 360°) it will still look exactly the same as before. Similarly, any object with 6-fold symmetry also has 1-fold, 2-fold, and 3-fold symmetry inherent in it. One of the elegant things about the science of crystallography and symmetry is its simplicity: if a pattern has 6-fold symmetry, then the other rotational symmetries can just be assumed without specifying them.

Without translation, rotations are kind of like revolving doors: they rotate in place, and

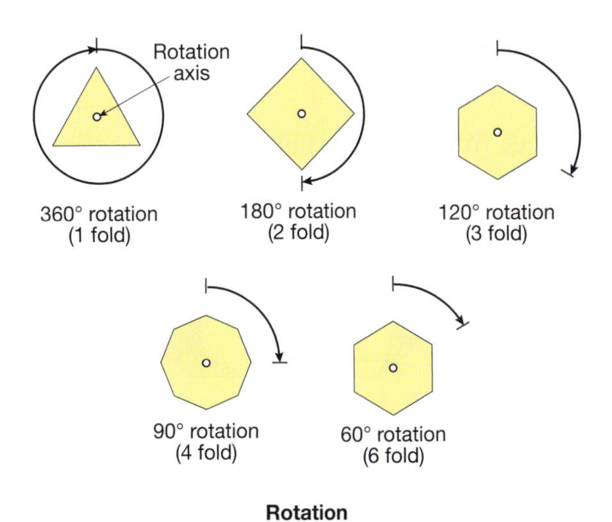

Figure 11.4. Each of these objects possesses a different type of rotational symmetry. This means that each object can be rotated around an imaginary axis (sticking up out of the page), and after each rotation, the object will look exactly the way it did before. Patterns combining rotation and translation can only be made using 1-fold, 2-fold, 3-fold, 4-fold, and 6-fold rotation axes.

don't go anywhere. In order to create crystals with rotational symmetry, we need to be able to repeat rotating motifs in space by combining rotation with translation. This combination imposes restrictions on the values that are possible, and explains why 5-fold, 7-fold and higher rotations are not in the list above.

One way of understanding this problem is to visualize a pattern made up of 5-sided shapes (Figure 11.5). No matter how hard you try, there is no way to make a continuous pattern of 5-sided shapes (where all sides are the same length) without leaving gaps in between. The same is true for 7-sided shapes. At the other extreme, try to make a pattern with eight sides of equal length. Notice that this pattern has all the attributes of 4-fold symmetry combined with translation. Thus, there is no need to define a rotational symmetry with, say, 8-fold or 9-fold rotation, because all of the characteristics of these patterns can in fact be more simply explained using "lower" symmetry such as 4-fold or 3-fold.

The limitations imposed by the need to combine the operations of rotation and translation are more simply explained by a mathematical derivation. Start with a simple translation between two points, D and D′. Add a rotation (α), and rotate your original translations around point D through that angle. Let the end of that translation be called E. Now rotate through a around D′, and let the end of that translation be called E′ (Figure 11.6). You want distance E – E′ to be an integer multiple (where m is the integer) of distance D-D′, such that the length of

$$E - E' = mt.$$

Notice that this equation can't be satisfied for 5-fold or 7-fold rotation (Figure 11.7).

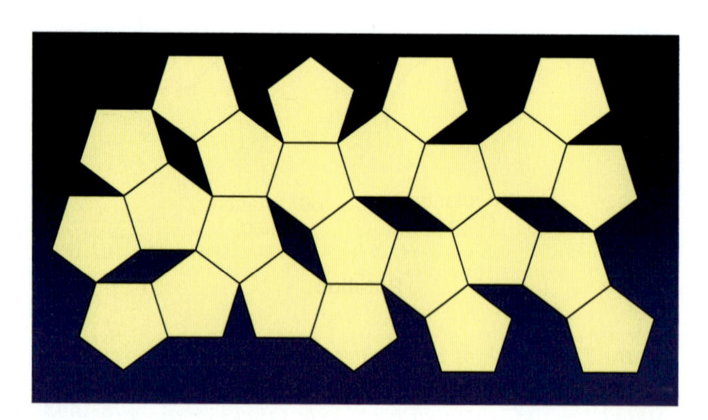

Figure 11.5. *It is impossible to draw a pattern in which the motifs are all regularly-shaped five-sided objects. Gaps between the shapes inevitably result as the pattern is repeated in two dimensions.*

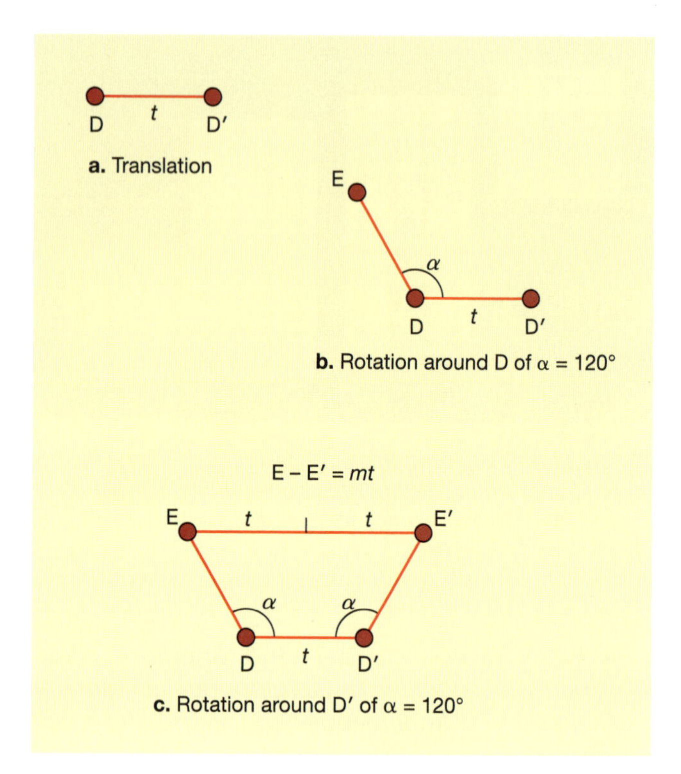

Figure 11.6. *When the combination of rotation with translation results in the creation of new points with spacings that are even integer multiples of the original translation distance, then that rotation can be used to build patterns in two- (and three-) dimensional space. In this case, a value of 120° is used to repeat at translation and form a lattice based on 3-fold rotation.*

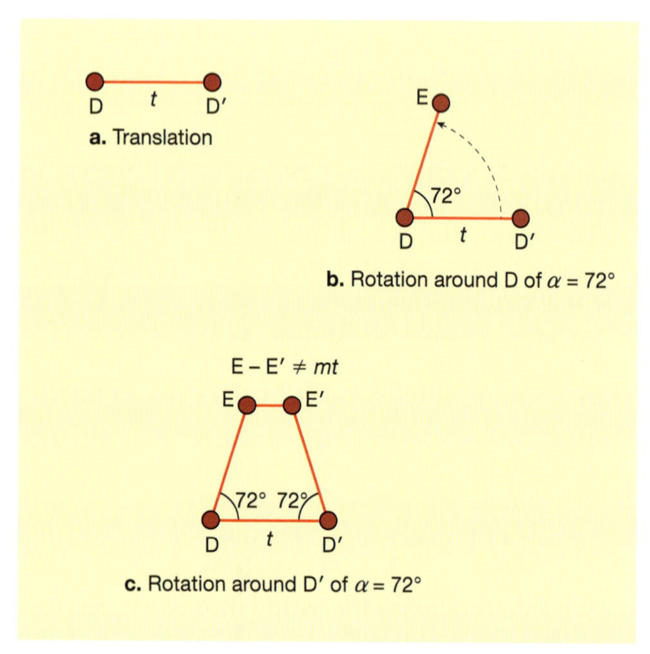

Figure 11.7. *When a rotation of 1/5 of 360° is combined with translation, it is impossible to generate a new set of points with a spacing that is an integer multiple of the original translation distance. This is why 5-fold symmetry is not commonly found in nature.*

From Figure 11.6, it can also be seen that distance E – E′ can be described as a function of α and t, because

$$E - E' = t - 2t \cos \alpha \,.$$

By combining those two equations, you get

$$mt = t - 2t \cos \alpha$$

$$m = 1 - 2 \cos \alpha$$

$$\cos \alpha = \frac{1 - m}{2}$$

Only a few of the solutions to this equation yield values of α that are permitted by trigonometry, as seen in Table 11.1. Thus, due to the restriction that translation and repetition must be combined to build arrays or patterns, only five different rotation axes are compatible with translational symmetry.

3. **Reflection**. The third operation that can be used to repeat a motif in space is called a reflection, or mirror plane. It is usually given the symbol m. This concept is intuitively obvious because the common usage of the term "reflection" means the same thing as the formal symmetry term. When something is reflected across a plane (such as your bathroom mirror), the resultant image is identical to the original one, except that every part of the original image has been translated

   an equal and opposite direction across and perpendicular to the plane of the mirror

   (Figure 11.8). If you hold your hands out in front of you and imagine a vertical mirror plane perpendicular to your face, then your hands can be thought of as mirror images of each other.
4. **Inversion**. The fourth operation in two-dimensional space is called **inversion**, and the symbol for it is i. It is similar to reflection except that inversion translates everything an equal and opposite distance through a point rather than through a plane (Figure 11.9).

Table 11.1. Possible Combinations of Rotation and Translation in 2D				
m	$\cos \alpha$	α	Angle (°)	n-fold axis
–2	3/2	no solution	—	—
–1	1	0	0° or 360°	∞ or 1-fold
0	1/2	π/3	60°	6-fold
1	0	π/2	90°	4-fold
2	–1/2	2 π/3	120°	3-fold
3	–1	π	180°	2-fold
4	–3/2	no solution	—	—

5. **Glide**. The last operation in two dimensions has two steps: reflection followed by translation. It is called a **glide** plane (Figure 11.10), and the symbol for it is the letter g. It combines a reflection across a mirror plane (every part of the motif is reflected an equal and opposite distance perpendicular to and across a plane) with a translation along a line parallel to the mirror plane. This is the pattern you make with your bare feet in the sand when you walk along the beach!

Operations 1, 2, and 4 are fundamentally different from operations 3 and 5 in two-dimensional space, a subtle point that is best shown through

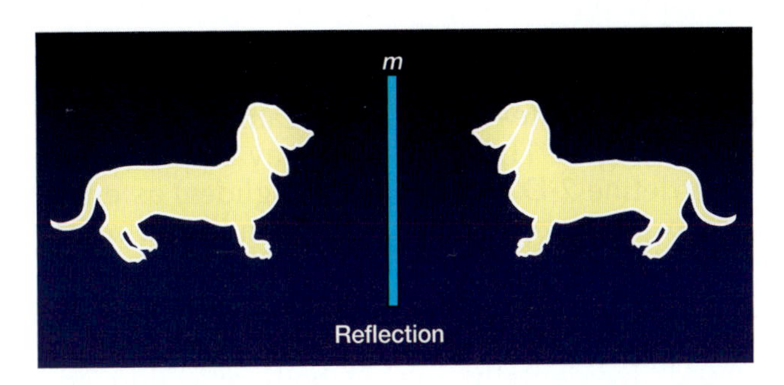

Figure 11.8. When a motif is reflected, every part of the motif moves an equal and opposite distance across and perpendicular to a "mirror" plane.

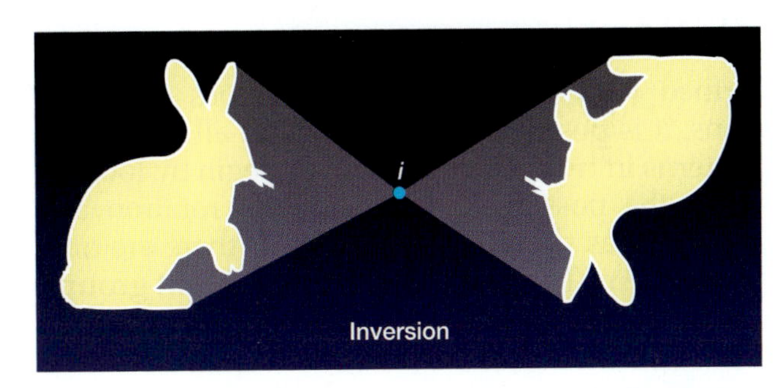

Figure 11.9. The operation of inversion involves projecting every part of a motif an equal and opposite distance through a point in space.

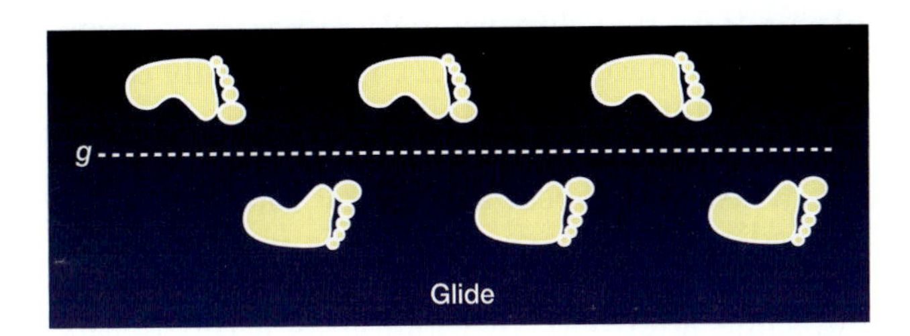

Figure 11.10. A glide plane is another type of operation in two dimensions. It is a combination of reflection across a mirror plane, followed by translation parallel to the mirror.

illustrations (Figure 11.11). In the cases of translation, rotation, and inversion, the images that are created by the operations are identical in appearance to the original motifs for 2-D. If you make a pattern using a stencil, each image is identical to every other image, and if the images are directly superposed, they are indistinguishable. These are called **congruent** operations. In other words, if you place two figures over each other and they correspond exactly, then that repetition results from a congruent operation.

If you make a pattern with your hand prints using alternately your right hand and your left hand, you now have a pattern that cannot be so simply reproduced. There is no way (short of turning the paper over) to make the print of your left hand match the print of your right hand. The right-hand print is a reflection of the left-hand print. Thus, in two dimensions the operation of reflection creates motifs that cannot be superposed (Figure 11.11). Another way of saying this is to define 2-D reflections (and glides) as **enantiomorphous** operations. This term in fact comes from the concept of "handedness"; the shape of your left hand is identical to that of your right hand except that they cannot be directly superposed. Incidentally, in 3-D both reflection and inversion produce enantiomorphous patterns.

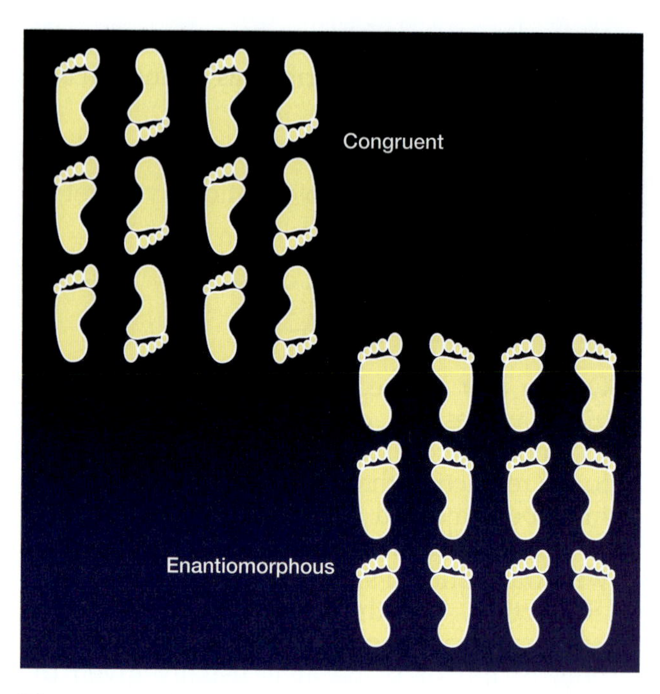

Figure 11.11. *Congruent operations in 2-D, such as the one on the left, repeat a motif via translation, rotation, translation plus rotation, or inversion. In all cases, any motif can be moved "on top of" any other motif, and they will exactly correspond. This is not true for reflection in 2-D. The pattern shown on the right combines reflection and translation. No movement of the motif in this picture can make it correspond to every other motif.*

Combinations of Operations: Planar Point Groups

Armed with the preceding five types of operations, it is possible to build any and all repeatable patterns in two dimensions. Let's begin by looking at all the possible combinations of rotation and reflection as shown in Figure 11.12; there are only ten of them! These are called the ten **point groups**, and they are defined as the group of operations that repeat an object around a fixed point. In other words, by using rotation and reflection, you can make ten different patterns around a point that repeat a motif but leave the center point unmoved. In some books you will see them called **planar point groups** to make it clear that they apply to planar (i.e., two-dimensional) rather than three-dimension space (Table 11.2). We can easily derive the ten point groups, as follows.

Begin with the question of what space looks like when you repeat a motif by rotating it 360° around an imaginary point. This is the identity operation again, because the motif looks exactly the same after 360° of rotation as it did before. This point group has the symbol 1.

Point groups with the symbols 2, 3, 4, and 6 do the same thing: they take a motif, rotate it 180°, 120°, 90°, or 60° around an imaginary central point, and repeat it each time. Point group *m* is a little different. It does not rotate the motif around an imaginary central point, but it reflects it an equal and opposite distance across a plane that contains that point. Thus, the first six point groups are either *only rotation*, or *only reflection*.

The last four point groups represent combinations of rotation and reflection. If you add a mirror plane (another way of saying that you add reflection) to a 2-fold rotation, you get the group

Table 11.2. Planar Point Groups: Ten Possible Combinations of Rotation and Reflection

Symbol	*Type of rotation or reflection*
1	360° rotation only
2	180° rotation only
3	120° rotation only
4	90° rotation only
6	60° rotation only
m	reflection only
2*mm*	180° rotation plus reflection
3*m*	120° rotation plus reflection
4*mm*	90° rotation plus reflection
6*mm*	60° rotation plus reflection

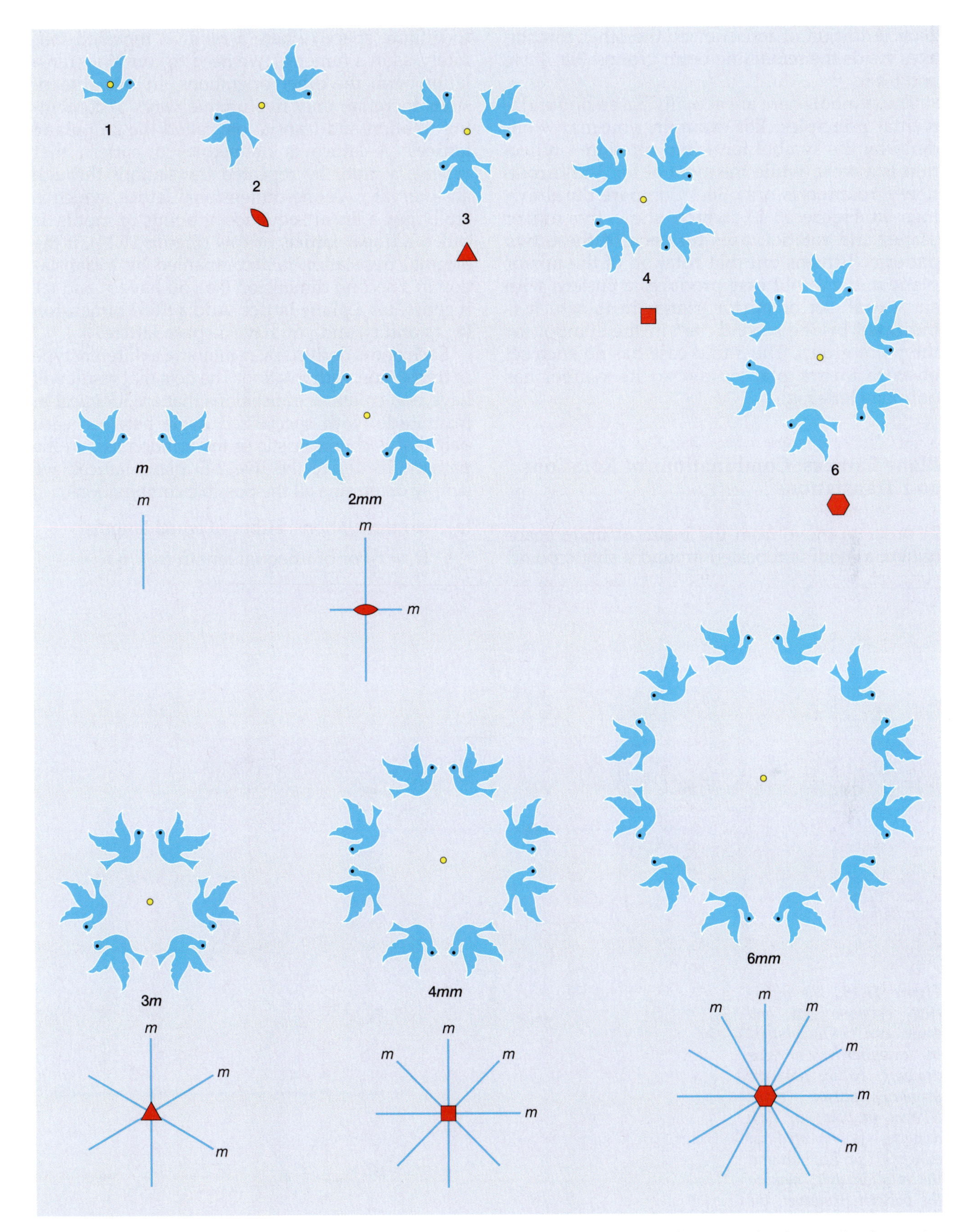

Figure 11.12. *All ten possible combinations of rotation and reflection around a central (imaginary) point, or the plane point groups, are shown here. Above, the ten groups are shown along with the locations of the imaginary points around which the motifs are rotated or reflected. Below, the symmetry notations, or symbols, for each of the two groups are given. Note that the symbol for a 2-fold rotation axis is a lozenge, for 3-fold a triangle, for 4-fold a box, and for 6-fold a hexagonal shape.*

2*mm*. Addition of reflection to the other rotation axes yields the remaining point groups: 3*m*, 4*mm*, and 6*mm*.

The symbols here are actually based on fundamental principles. For example, you may wonder why the symbol for 4-fold rotation + reflection is a 4*mm*, while the symbol for 3-fold rotation + rotation is only 3*m*? Compare the drawings in Figure 11.13, which show the mirror planes and rotation axes for each of these two patterns. It turns out that rotation of the mirror plane in the 4-fold case produces a pattern with an "extra" set of mirror planes in it, which is indicated by the second "*m*" in the symbol for the point group. The 3-fold case has no such set of extra mirror planes, and so its symbol has only one letter *m*.

Plane Lattices: Combinations of Rotations and Translations

In order to move from the realm of finite space (where a motif is repeated around a single point) to infinite space (where a motif is repeated infinitely, as in a mineral), we need to combine translation with the other operations. In two dimensions there are only five unique ways of combining rotation and translation, called the 2-D **plane lattices**. A lattice is defined as a pattern that repeats a motif by repeated translations through an interval, t. A one-dimensional lattice, which is really just a linear sequence of points or motifs, is called a **linear lattice**, or row (Figure 11.14). If the original translation is accompanied by a translation in a second dimension (so you have t_1 and t_2), it generates a **plane lattice**. Add a third dimension (t_1, t_2, and t_3) and you have a **space lattice.**

Each one is derived by combining a different type of translation with rotation. The nets that result will have non-collinear translations that are identical in magnitude, with specialized angle between each pair that is characteristic of the rotation axis in the pattern. To derive the five 2-D plane lattices, we simply determine all the possible combinations:

(a) Two translations, either of equal lengths ($t_1 = t_2$) or of unequal length ($t_1 = t_2$)

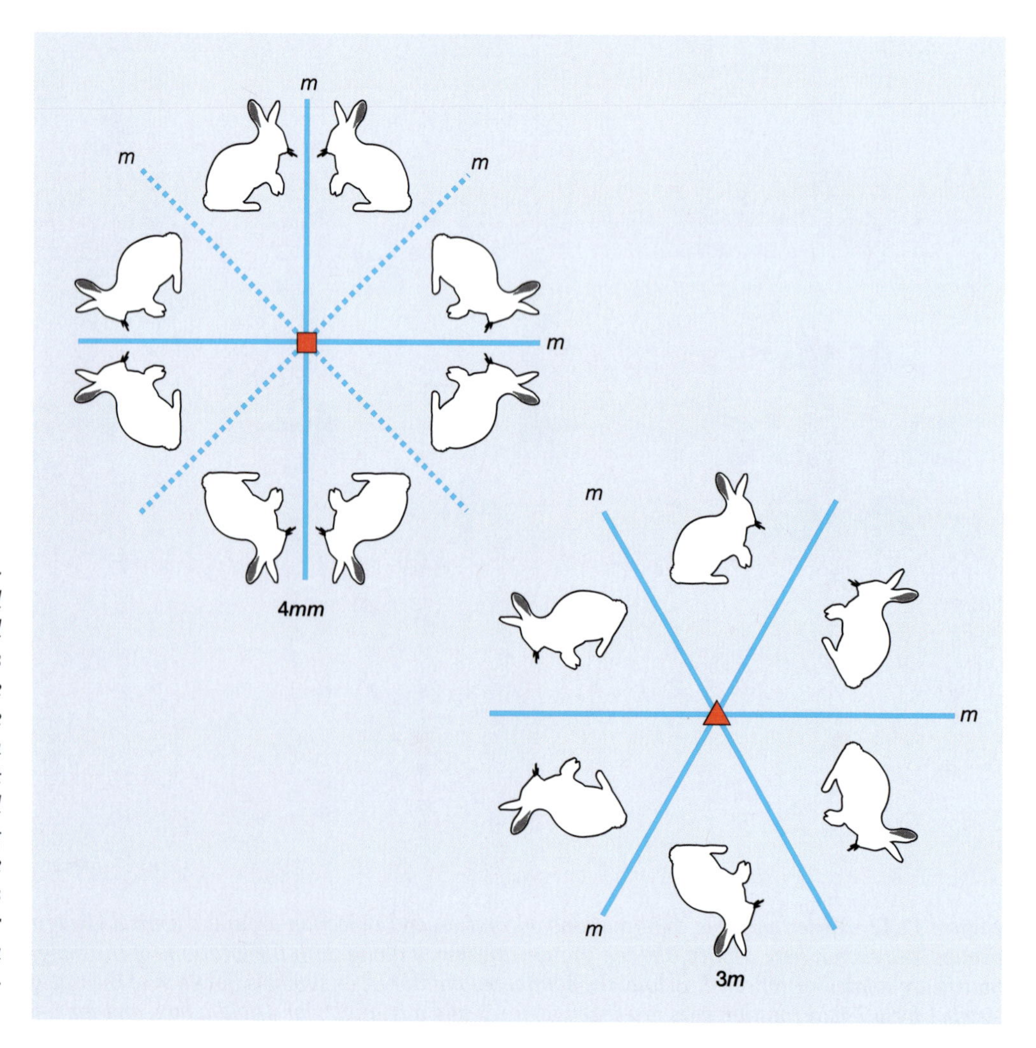

Figure 11.13. *The difference between 3m and 4mm can be understood by examination of this drawing. Notice that the primary mirror plane (shown in bold) in each drawing is repeated at either 90° or 120° around the rotation axis, making the pattern. However, the pattern for 4mm also has an inherent second mirror plane in it (shown as a dashed line), while the pattern for 3m does not.*

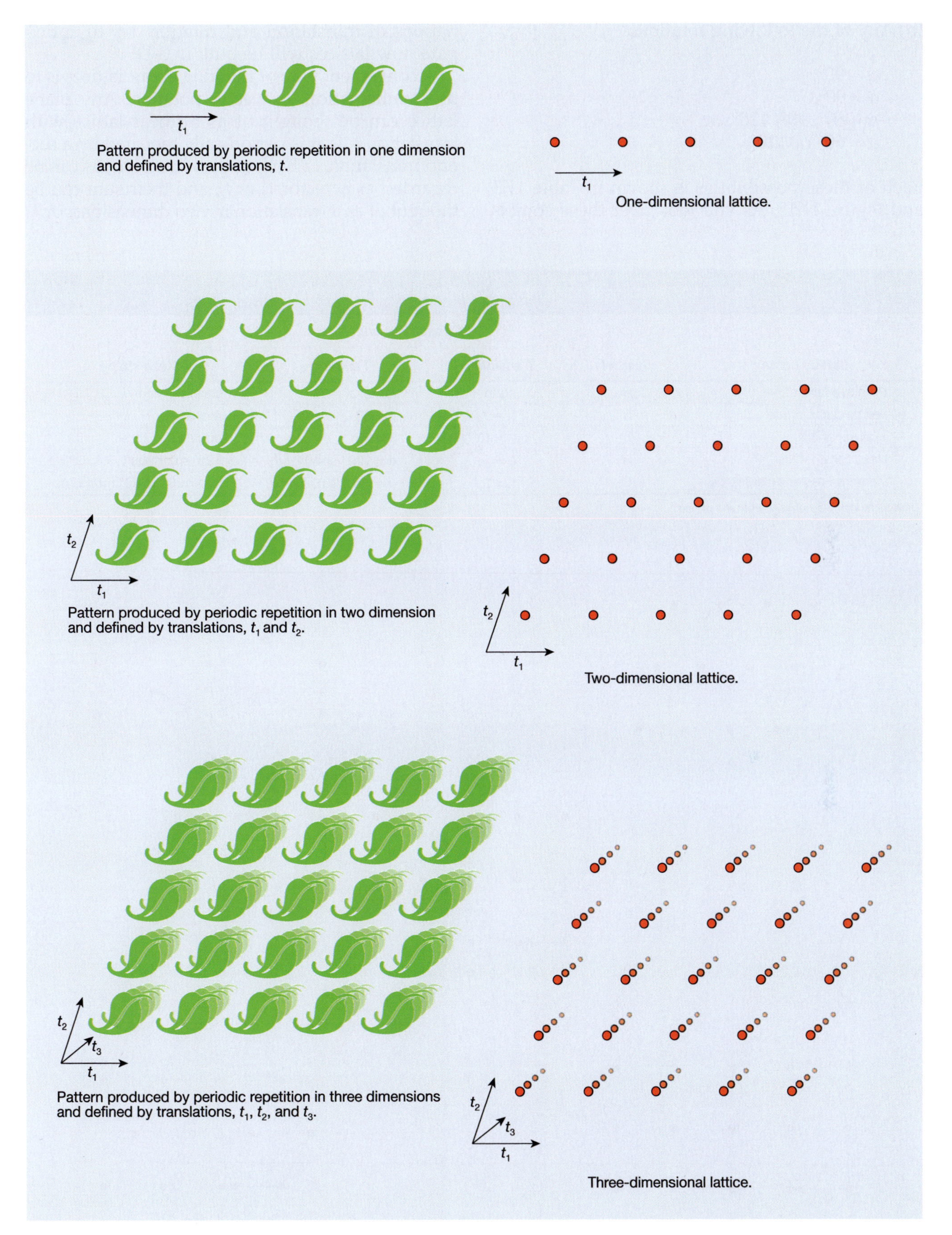

Figure 11.14. *Lattices are patterns created by repetition of a motif or point in one (linear lattice), two (plane lattice), or three (space lattice) dimensions. Each is produced by a periodic repetition of the motif at intervals designated* t_1 *(for 1-D),* t_1 *and* t_2 *(for 2-D), and* t_1 *,* t_2 *, and* t_3 *for 3-D.*

(b) Any of the following rotations:

$\alpha = 90°$,
$\alpha \neq 90°$,
$\alpha \neq 60°, 90°, 120°$, or
$\alpha = 60°$ or $120°$.

Each of these possibilities is shown in Table 11.3 and Figure 11.15. As you look over these combinations of translation and rotation, try to anticipate how lattices will be built in 3-D!

A convenient way of thinking about lattices is to use mathematical, or vector notation. Any plane lattice can be thought of as a linear lattice with translation t_1 that is periodically repeated by a second translation, t_2. Each of these translations can be regarded as a vector, $\vec{t_1}$ or $\vec{t_2}$, and their sum can be thought of as a translation in two dimensions, or $\vec{T}$:

Table 11.3. Two-Dimensional Lattices: Combinations of Translation and Rotation

		Constraints on:		
Lattice name	*Symbol*	*Translation*	*Rotation*	*Mesh shape*
Primitive lattice*	*P*	none	none; a ≠ 90°	parallelogram
Rectangular lattice	*R*	$t_1 \neq t_2$	$\alpha = 90°$	rectangle
Square lattice	*S*	$t_1 = t_2$	$\alpha = 90°$	square
Hexagonal lattice	*H*	$t_1 = t_2$	$\alpha = 120°$ or $60°$	120° rhombus
Centered rectangular lattice	*C*	$t_1 = t_2$	$\alpha \neq 60°, 90°, 120°$	non-120° or 60° rhombus

*Other terms for this are oblique lattice and clinonet.

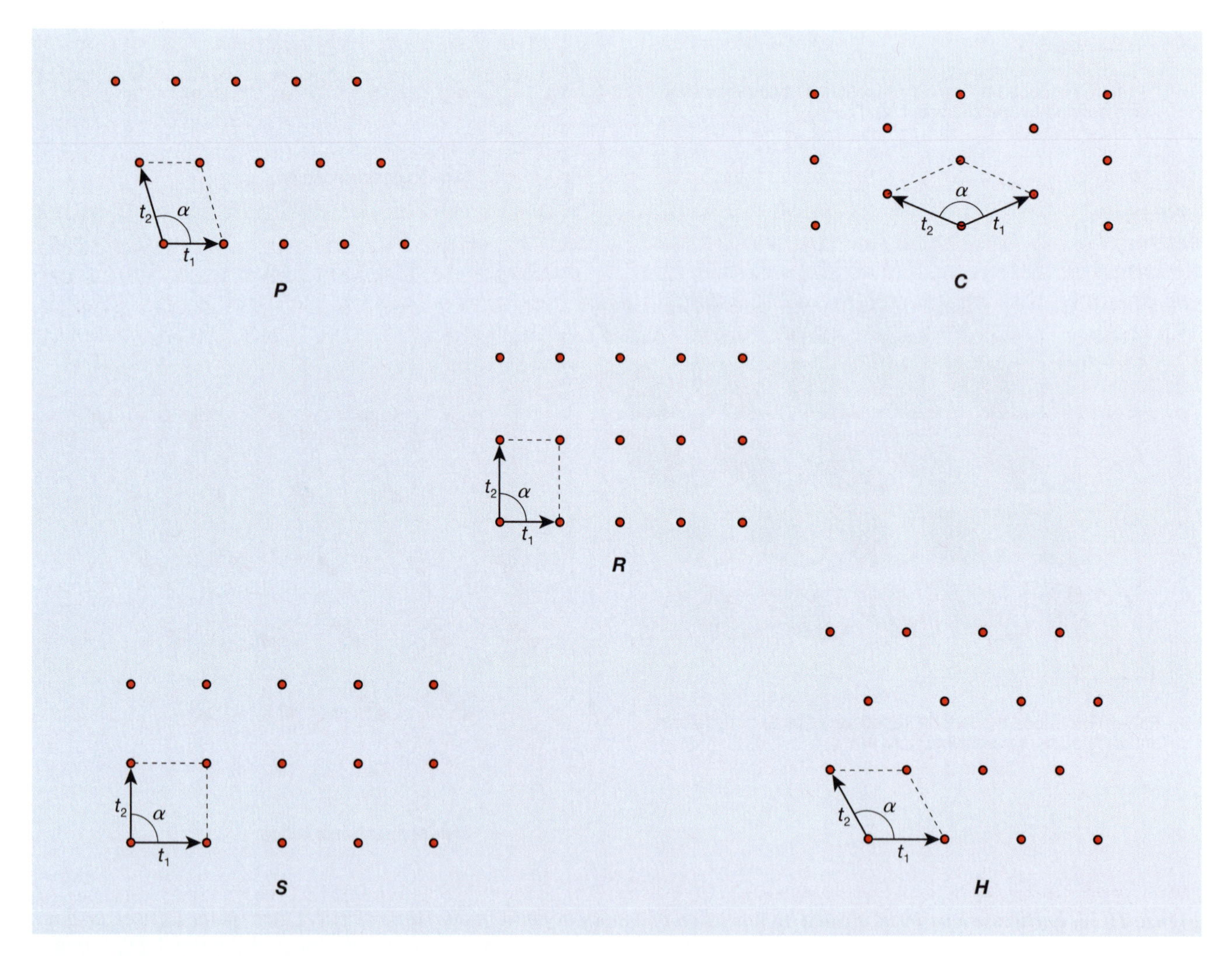

Figure 11.15. *The five 2-D plane lattices represent combinations of rotation and translation.*

$$\vec{T} = \vec{t}_1 + \vec{t}_2.$$

If you choose any point in the lattice as your origin, you can describe the location of any other lattice point by using a vector such as $\vec{T}$. For example, in Figure 11.16 a new lattice point can be described as the linear sum of vectors $\vec{t}_1$ and $\vec{t}_2$ by the equation

$$\vec{T} = 3\vec{t}_1 + 2\vec{t}_2.$$

This can be generalized to describe any point in a 2-D lattice by the translation

$$\vec{T} = u\vec{t}_1 + v\vec{t}_2,$$

where u and v are integer multiples of your original unit vectors in either direction. In 3-D, this is

$$\vec{T} = u\vec{t}_1 + v\vec{t}_2 + w\vec{t}_3.$$

If we let the symbol K stand for a collection of vectors, then the complete set of vectors that make up the entire lattice can be described in 3-D as

$$K\{u\vec{t}_1 + v\vec{t}_2 + w\vec{t}_3\}.$$

The set of points that makes the lattice is sometimes given the designation

$$T = K\{\cdot u\vec{t}_1 + v\vec{t}_2 + w\vec{t}_3 \cdot\}.$$

The dots in the equation represent the points located at the end of each vector.

Any pair of translations (also known as **conjugate translations**) t_1 and t_2 will define a parallelogram called the **primitive cell**. Every primitive cell of the lattice has the same area, and each primitive cell is associated with exactly one lattice point (Figure 11.17). This leads us to the definition of a **unit cell**, which is the smallest portion of the lattice that contains all of the symmetry of the pattern as a whole, and is therefore entirely representative of the structure as a whole.

You can also have **non-primitive cells**, which are also called **multiple cells** (Figure 11.18). These are cells that contain one or more lattice points per cell, usually one point total from the corners of the cell, and additional lattice points in the interior. When we describe minerals, it will sometimes be easier to use multiple cells to gain a full understanding of a particular structure, so it is a useful concept to remember.

So far, we have covered two permutations of combinations of operations: rotation + reflection (the ten plane point groups) and rotation + translation (the five plane lattices). In order to derive the complete set of possible patterns in two-dimensional space, we can now move on to **plane groups**, which are combinations of translation, rotation, reflection, and glide.

Plane Groups = Translation + Rotation + Reflection + Glide

As explained at the beginning of this chapter, there are only 17 possible patterns you can draw in two dimensions. These are called **plane groups**, although the term **two-dimensional space groups** is also sometimes used. These patterns can be generated by two types of combinations of operations: **symmorphic** plane groups, which are combinations of translation + rotation + reflection, and **non-symmorphic** groups, which combine translation + rotation + glides (Table 11.4). A few examples should illustrate how these groups are derived; for a complete explanation, see the text by O'Keefe and Hyde (1996) or Buerger (1978).

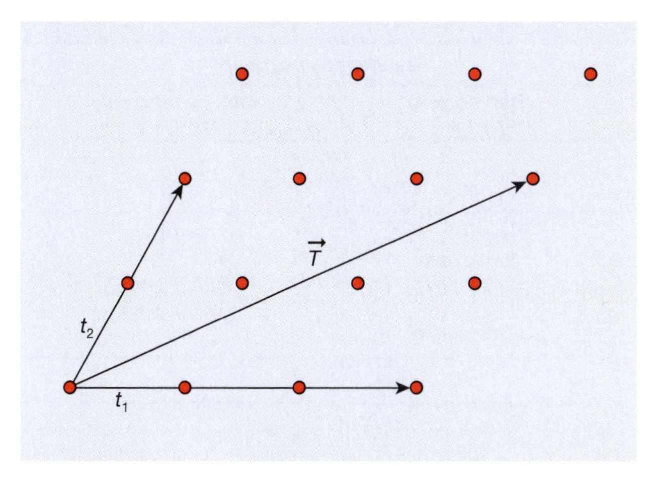

Figure 11.16. *Any point in a two-dimensional lattice can be described as a linear sum of the translations in the two directions* t_1 *and* t_2.

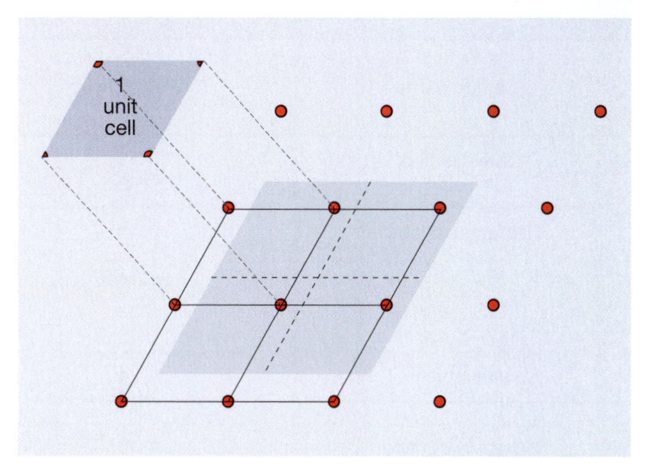

Figure 11.17. *For any given lattice, a unit cell can be defined as the smallest portion of the lattice that can be repeated indefinitely to form it. The unit cell by definition contains only one lattice point per cell.*

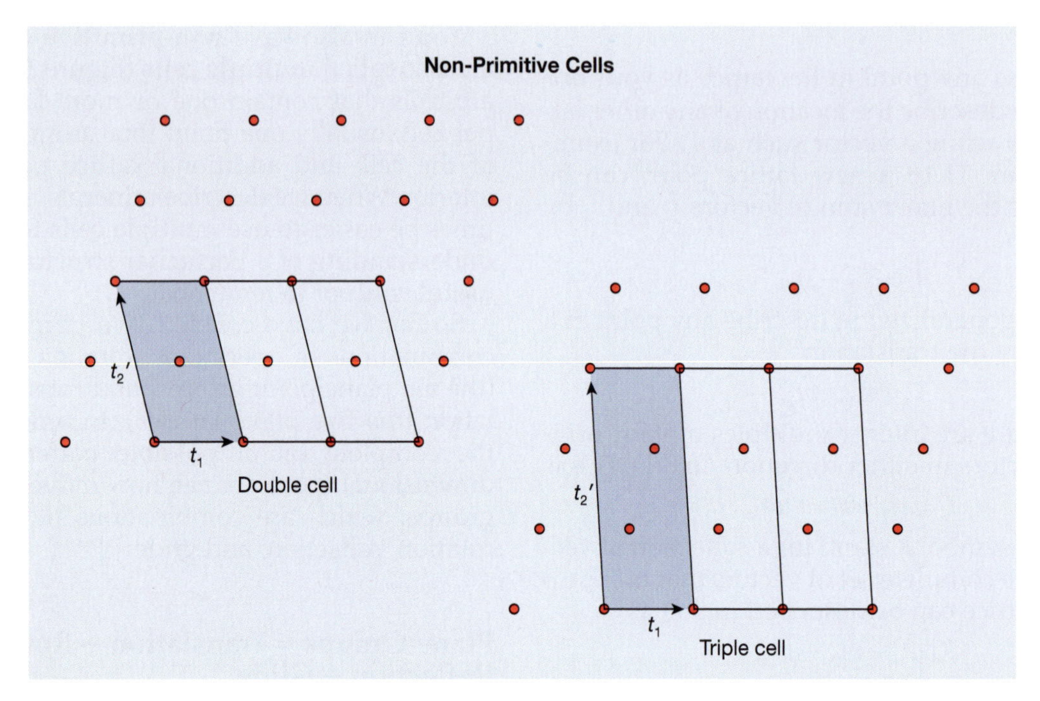

Figure 11.18. *Non-primitive, or multiple cells, contain more than one lattice point per cell.*

What do you get if you combine the oblique lattice type *p* with a 6-fold rotation axis and a mirror plane? This is like saying that you will take an oblique lattice (*p*) and add at each corner a *6mm* plane point group; the new symbol is *p6mm*. Remember that the plane point groups can only repeat space around a *point*; you need the addition of *translation* to make the pattern infinitely repeatable! The results are shown in Figures 11.19, 11.20 and 11.21; this pattern has all the symmetry of *6mm*, but it can be repeated indefinitely in two dimensions.

Similarly, the combination of an oblique lattice (*p*) with a mirror plane (*m*) gives you plane group *pm*, while the combination of a centered rectangular lattice (*c*) with a mirror gives *cm*.

The terminology gets a little more complicated for the case of *p* with *2mm* when you attempt to replace the mirrors with glide planes. Note that this combination yields two subtly different patterns

Table 11.4. Plane Groups Created by Combinations of Two-Dimensional Operations

Lattice name	*Point group*	*Resultant plane group**	
		Symmorphic $t+r+n$	*Non-symmorphic* $t+r+g$
Primitive lattice	1 2	*p*1 *p*2	
Rectangular lattice	*m* 2*mm*	*pm* (*p*1*m*1) *cm* (*c*1*m*1) *p*2*mm* (*pmm*) *c*2*mm* (*cmm*)	*pg* (*p*1*g*1) *p*2*mg* (*pmg*) *p*2*gg* (*pgg*)
Square lattice	4 4*mm*	*p*4 *p*4*mm* (*p*4*m*)	 *p*4*gm* (*p*4*g*)
Hexagonal lattice	3 3*m* 6 6*mm*	*p*3 *p*3*m*1 *p*31*m* *p*6 *p*6*mm* (*p*6*m*)	

*Different applications use different symbols for the plane groups; alternate names are given in parentheses.

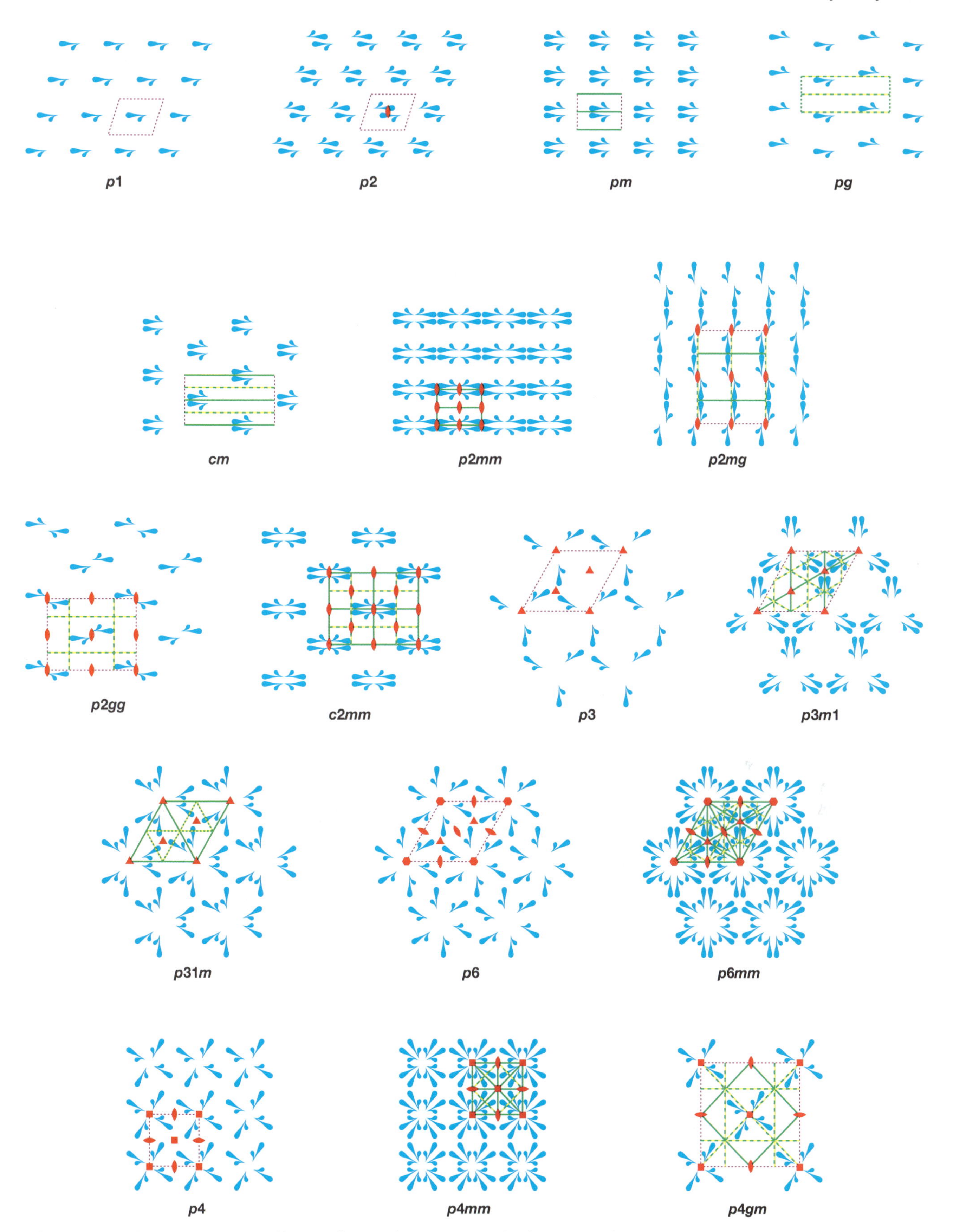

Figure 11.19. *Patterns corresponding to the 17 plane groups, also known as the 2-D space groups, are shown here. The arrangement of motifs in these patterns is a "traditional" one that dates to some of the original crystallographers; most recently, simpler versions of these figures can be found in Buerger (1978).*

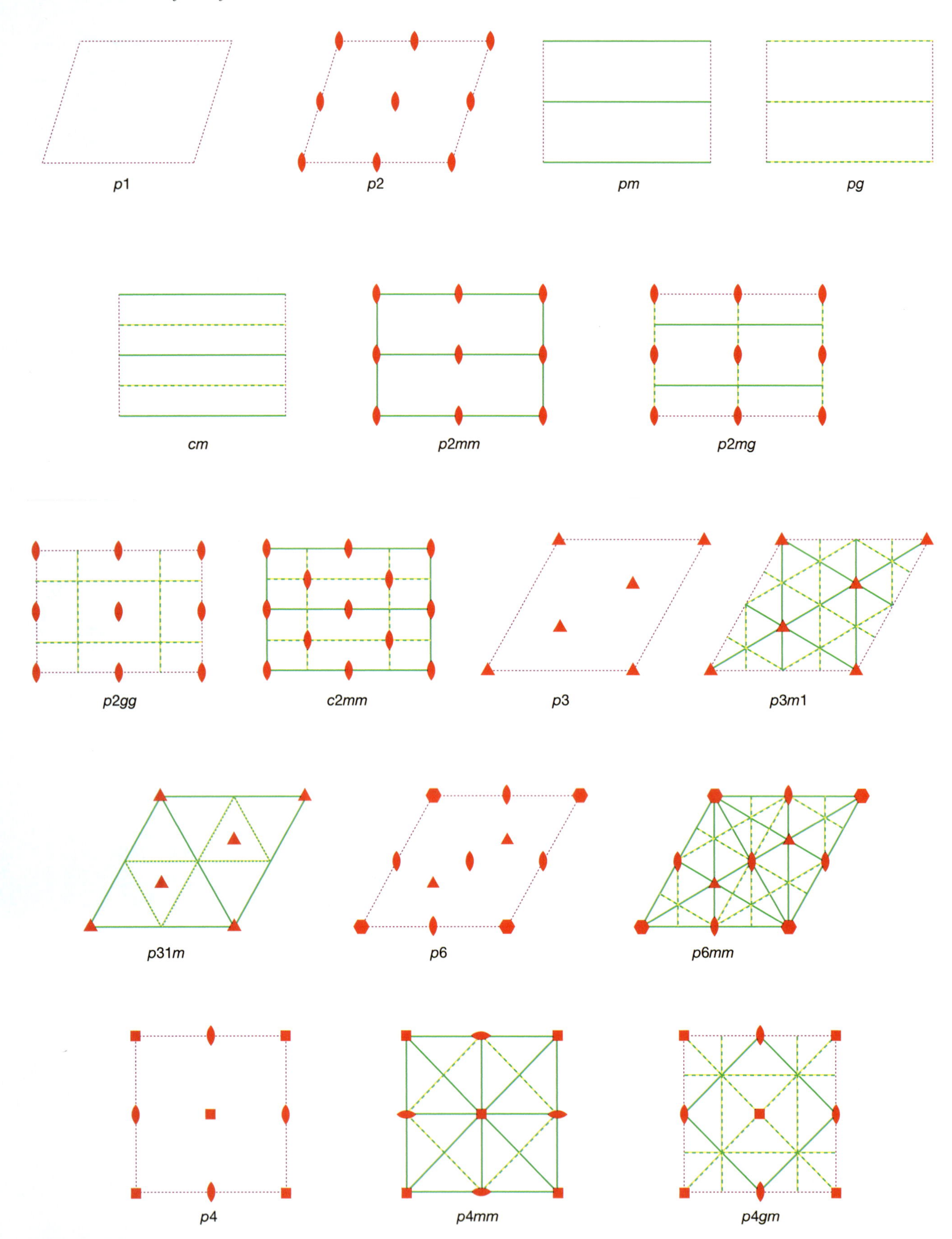

Figure 11.20. *The symmetry elements that correspond to each of the 17 plane groups. Dotted lines are used to outline the cells, solid lines represent mirror planes perpendicular to the page, and dashed lines represent glide planes, also perpendicular to the page.*

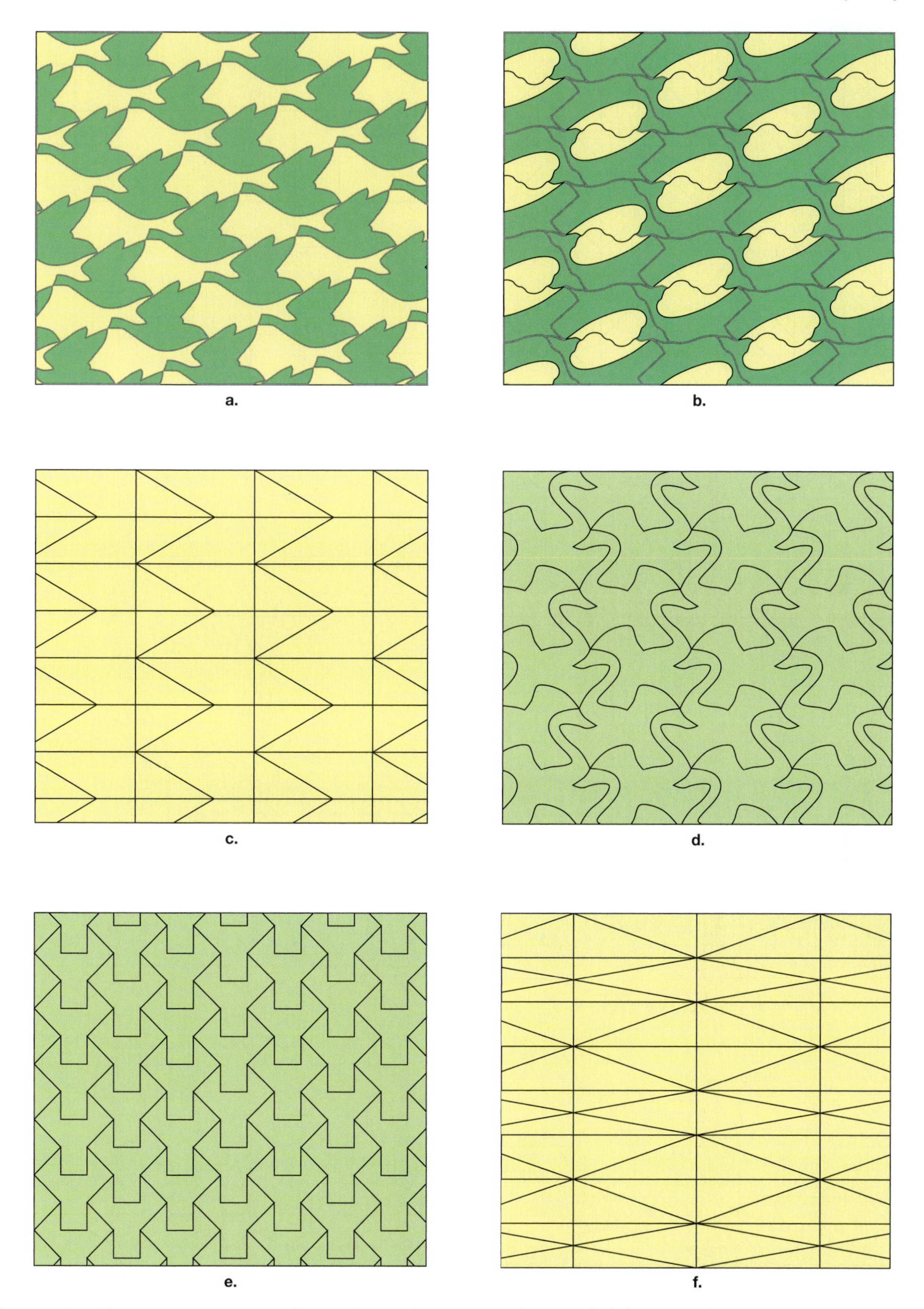

Figure 11.21. *More patterns corresponding to the 17 plane groups, deliberately left unlabeled to provide a convenient exercise in pattern recognition (hints are on the MSA website, however). For ease of use, these patterns can be enlarged on a copier or scanner. Note that many of these are traditional quilt patterns (patterns include pineapple,* ***n.****, spools,* ***o.****, and tumbling blocks,* ***q.****).*

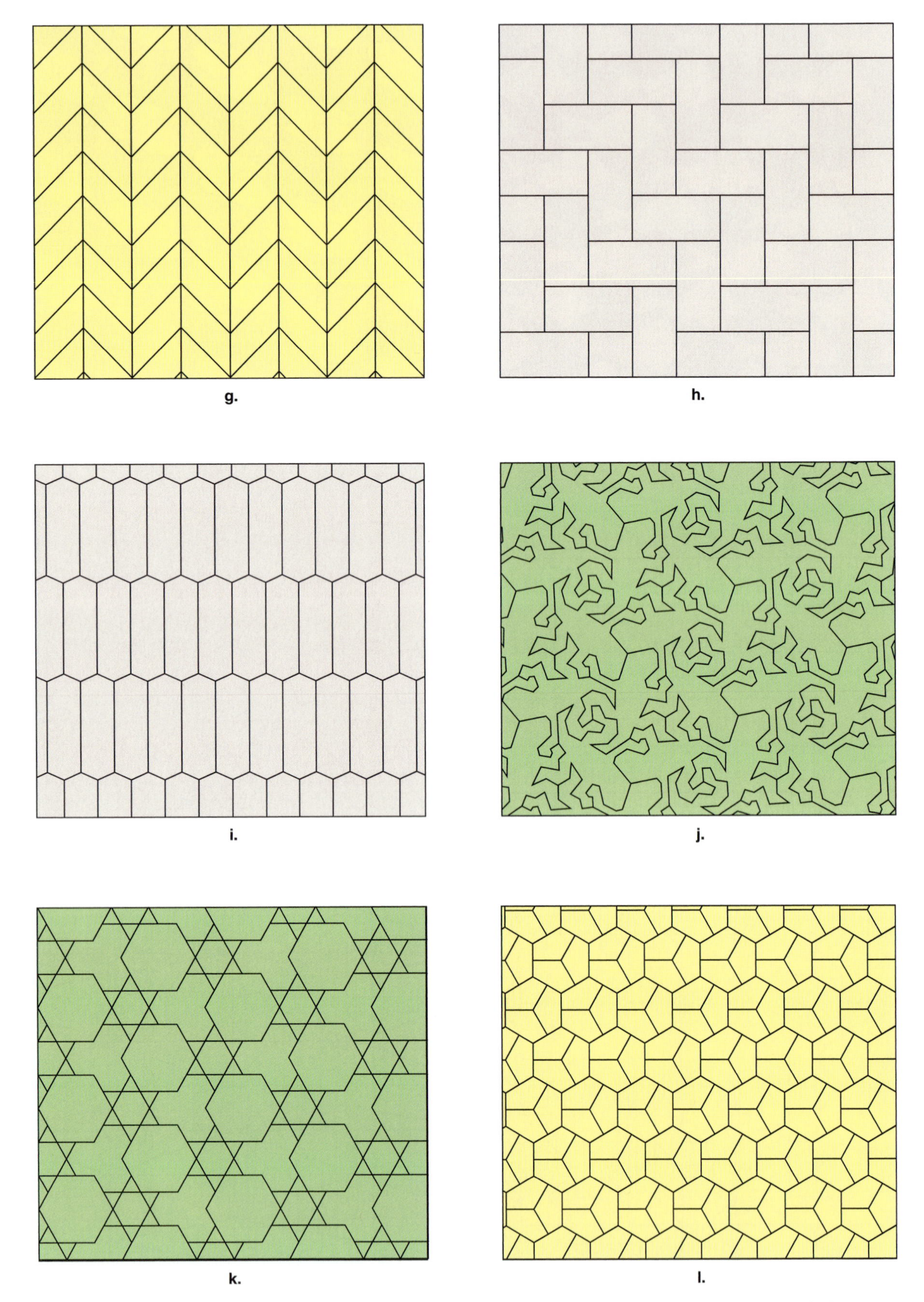

Figure 11.21. (continued) *The Escher patterns and the crescent wrench design are adapted from the book* Contemporary Quilts, *by Kay Parker (1981, The Crossing Press, Trumansburg, New York). The elongated hexagon pattern is based on a quilt design by Kathryn Kuhn and the divided hexagon pattern was designed by Edith Zimmer; both are shown in…*

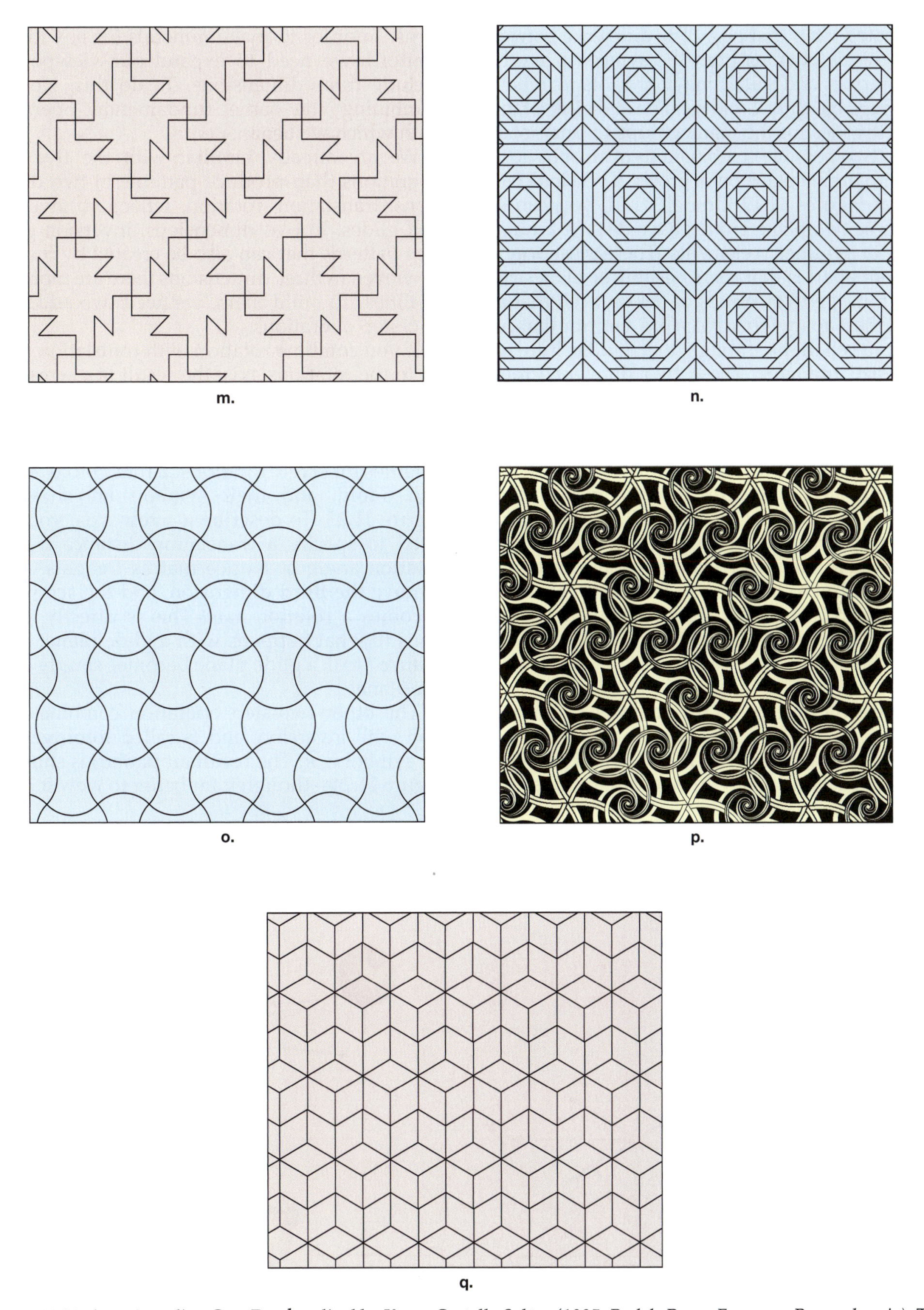

Figure 11.21. (continued)...One Patch, *edited by Karen Costello Soltys (1995, Rodale Press, Emmaus, Pennsylvania). The pencil point,* ***c.****, and hang glider,* ***f.****, patterns are from* Rectangular Quilt Blocks, *by Jean Roesler (1982, Wallace-Homestead Book Co., Des Moines, Iowa). The* p6 *pattern is from Edwards (1967).*

designated *p2mg* and *p2gg*. The difference between these two is obvious from the symbol: *p2mg* has both mirrors and glides inherent in its symmetry, while *p2gg* has only 2-fold rotation and glide planes.

The trickiest of the plane groups to understand are the two possible combinations of an oblique lattice with the 3*m* plane point group. If you look carefully at Figures 11.19 through 11.21, you can see that these two plane groups have distinct arrangements of symmetry elements. The easiest way to tell them apart is that *p*3*m*1 has mirror planes running from the two corners of the unit cell that are the furthest apart, while *p*31*m* has a mirror that cuts the unit cell right through the middle into two equal-size triangles. These two different groups arise from the fact that coordinate systems (i.e., adding *x*, *y*, and sometimes *z* axes to our symmetry drawings) are sometimes used to describe the locations of atoms within a unit cell. In systems where the angles are 90°, the axes are easily picked to line up with the edges of the unit cell. However, two different coordinate systems are possible for symmetries associated with 3-fold rotation: *x* and *y* axes can be picked to lie between the mirror planes, or along them. This gives rise to two different types of *p* + 3*m* symmetries (Figure 11.22).

Operations in Three Dimensions

So far, we have limited our discussion to operations that result largely in two-dimensional patterns. In order to make minerals (or any kind of matter!), we need to expand our viewpoint to include three dimensions. To do this, we keep combining the same fundamental operations with which we began.

We are already familiar with the five basic operations than produce patterns in two dimensions: translation, rotation, reflection, inversion, and glides. In two dimensions, inversion generates patterns that can also be created by rotation. However, in three dimensions there are more possibilities. To build them, we need two additional two-step operations.

If you combine rotation with translation parallel to the rotations axis, the result is a **screw axis** (Figure 11.23), so called because it creates a pattern that repeats like the threads on a screw. Its symbol is usually given as n_m. There are four possibilities for screw rotations: 2-fold, 3-fold, 4-fold, and 6-fold, and their symbols are shown in Figure 11.23. To describe a screw axis, you only need to specify a translation distance *t* and a rotation angle α. Notice that as *t* equals 0, you remove the third dimension, and the screw axis becomes a rotation axis. This is directly analogous to what happens with a glide plane: if you reduce *t* to 0, a glide plane becomes simply a mirror plane!

The other two-step operation combines rotation with inversion, and is called **rotoinversion**. Its symbol is $\overline{n}$. The resultant pattern is shown in Figure 11.24—though it far better to view it on the

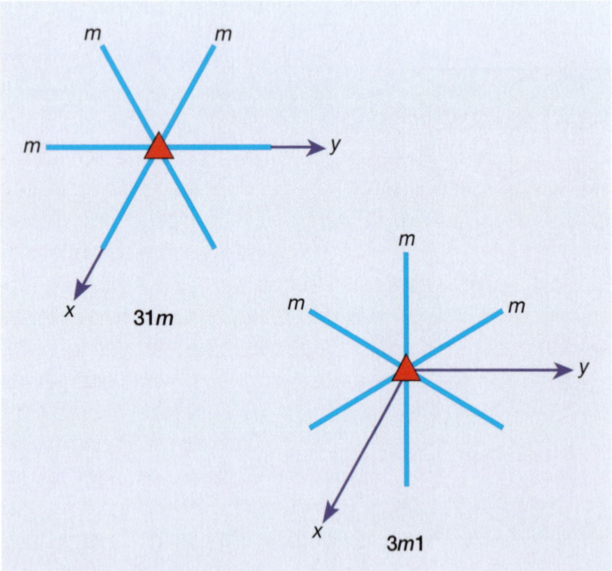

Figure 11.22. The difference between the p31m *and* p3m1 *plane groups results from the fact that* x *and* y *coordinate axes can be assigned to these patterns in two different orientations as shown above (after O'Keefe and Hyde, p. 8).*

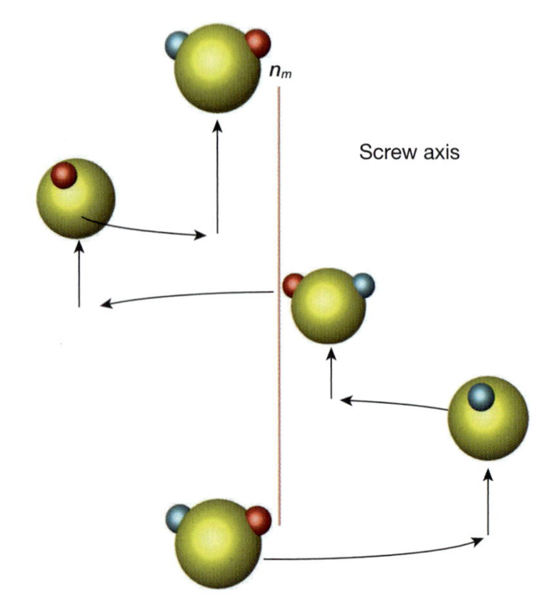

Figure 11.23. A combination of rotation and translation parallel to the rotation axis is called a screw axis, and given the symbol n_m. *The rotation can be either in a clockwise or counterclockwise direction, giving rise to left-handed or right-handed screw axes.*

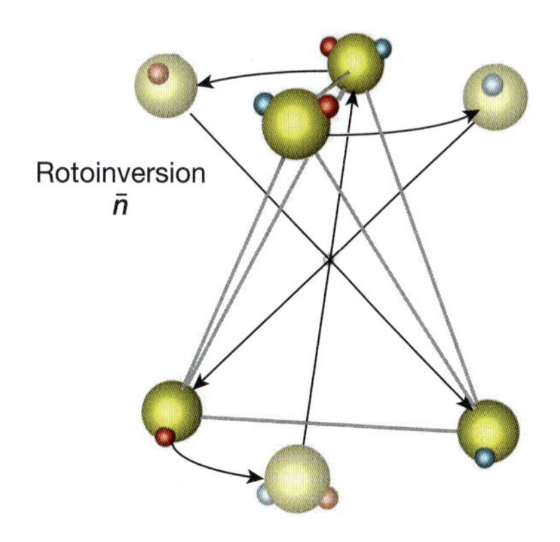

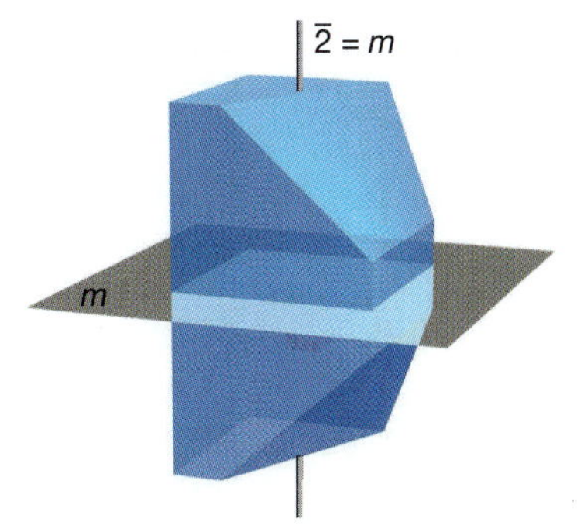

2-fold Rotoinversion. The operation of 2-fold rotoinversion involves first rotating the object by 180° then inverting it through an inversion center. This operation is equivalent to having a mirror plane perpendicular to the 2-fold rotoinversion axis. A 2-fold rotoinversion axis is symbolized as a 2 with a bar over the top, and would be pronounced as "bar 2." Because this is the equivalent of a mirror plane, *m*, the bar 2 is rarely used.

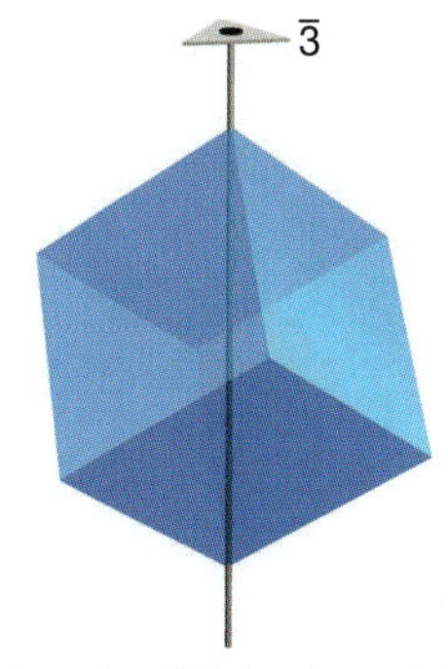

3-fold Rotoinversion. This involves rotating the object by 120° ($^{360}/_{3}$ = 120), and inverting through a center. A cube is good example of an object that possesses 3-fold rotoinversion axes. A 3-fold rotoinversion axis is denoted as $\bar{3}$ (pronounced "bar 3"). Note that there are actually four axes in a cube, one running through each of the corners of the cube. If you hold one of the $\bar{3}$ axes vertical, then there are three faces on top, and $\bar{3}$ identical faces upside down on the bottom that are offset from the top faces by 120°.

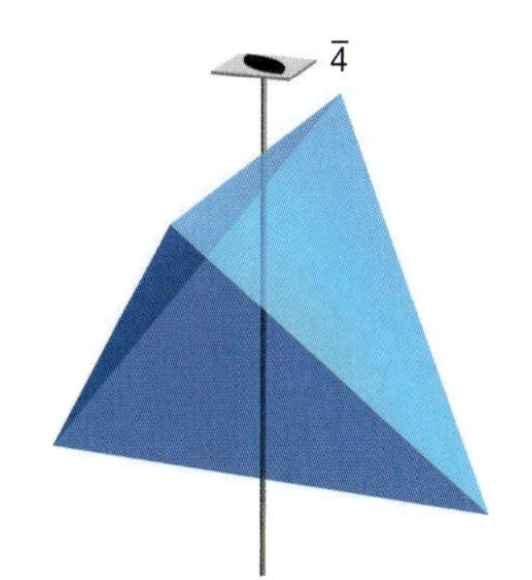

4-fold Rotoinversion. This involves rotation of the object by 90° then inverting through a center. A 4-fold rotoinversion axis is symbolized as $\bar{4}$. An object possessing a 4-fold rotoinversion axis will have two faces on top and two identical faces upside down on the bottom, if the $\bar{4}$ axis is held in the vertical position.

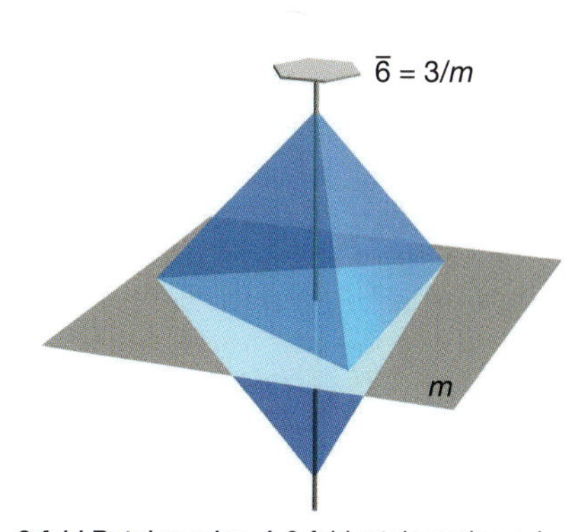

6-fold Rotoinversion. A 6-fold rotoinversion axis ($\bar{6}$) involves rotating the object by 60° and inverting through a center. Note that this operation is identical to the combination of a 3-fold rotation axis perpendicular to a mirror plane.

Figure 11.24. *Rotoinversion, which is given the symbol $\bar{n}$, represents a combination of rotation and inversion, producing an enantiomorphous motif. The resultant combinations often have other inherent symmetry. For example, a 1-fold rotoinversion axis is the same as a center of symmetry. Objects with $\bar{2}$ symmetry (rotate 180°, then invert through inversion center) have symmetry that is the same as a mirror plane, so the $\bar{2}$ symbol is not usually used. A common example of $\bar{3}$ is the familiar shape of a cube, in which each of the four body diagonals is a 3-fold rotoinversion axis. When you hold a cube on opposite corners, you can see that the top three faces are repeated on the bottom half, only offset by 120°. Shapes with $\bar{4}$ symmetry (rotate 90°, then invert through inversion center) similarly have two faces on top, with the same two faces repeated upside down on the bottom when the 4-fold rotoinversion axis is vertical. Finally, $\bar{6}$ symmetry (rotate 60°, then invert through inversion center) turns out to have the same result as the combination of a 3-fold axis that is perpendicular to a mirror plane. For this reason, $\bar{6}$ symmetry is the same as 3/m symmetry, with the latter term generally preferred.*

MSA website. This gives us seven total operations, which is all we need to build three-dimensional space.

The seven operations also make it possible to revisit the notion of congruent vs. enantiomorphous operations (handedness) in a more quantitative context, broadened to 3-D. Let's begin with the simple case of two dimensions, *x* and *y*, and use Figure 11.25 to illustrate the difference between congruent and enantiomorphous operations. The rabbit shown in A has his head up in the +*y* direction, and his tail toward the +*x* direction. If the direction of the *x* axis is reversed, as seen in image B, the head still points up in the +*y* direction (that remains unchanged), but the tail now points to the left, in the –*x* direction. Rabbits A and B are enantiomorphs: neither translation nor rotation can bring rabbit A to lie over rabbit B and overlap him perfectly. We see the same relationship when we reverse only the direction of the *y* axis: rabbit A and rabbit D are also enantiomorphs. Thus, in two dimensions, as we learned earlier in the chapter, reversing one direction results in the enantiomorphous operation of reflection. In other words, reversal of +*x* to –*x* is the same thing as reflection across *y*, and reversal of +*y* to –*y* is the same as reflection across *x*.

But what is the relationship between our starting rabbit A and rabbit C in Figure 11.25? Rabbit C represents a reversal in the direction of both the *x* and the *y* axes. You could think of this as an inversion, because certainly, every point on Rabbit A has been projected an equal and opposite direction across the origin into rabbit C. However, this demonstrates an important point about 2-D symmetry: reversal of two axes generates a congruent image that can be overlain on the original image with only a 180° rotation.

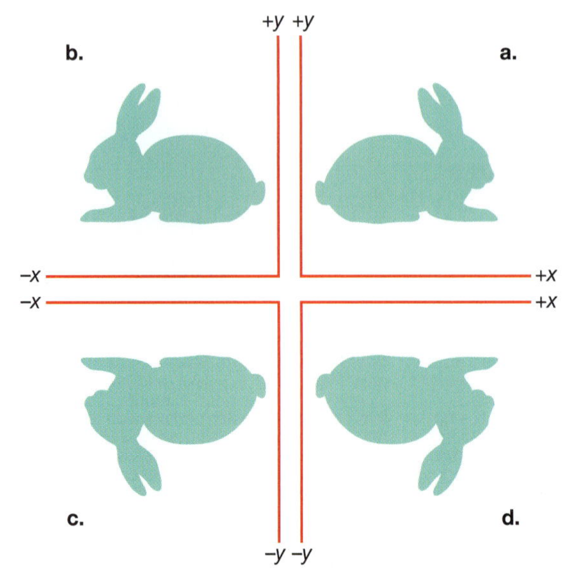

Figure 11.25. In two dimensions, there are two axes, and thus there are four possible combinations of orientations for the x *and* y *axes: +*x *and +*y *(**a**), –*x *and +*y *(**b**), –*x *and –*y *(**c**), +*x *and –*y *(**d**). This figures shows what happens to a 2-D image of a rabbit when the axes are reversed. Reversal of the direction of one axis (as in Figure 11.25 b and d) leads to an enantiomorphous motif—one that cannot be perfectly overlapped with the original rabbit in Figure 11.25a. Reversal of both axes results in a congruent rabbit (**c**) that is identical to the original image, though rotated by 180°.*

Now on to three dimensions! We can reverse one, two, or three directions, resulting in the eight possible combinations shown in Figure 11.26. The results are simplified with the example in Figure 11.27, where we show the four possible permutations: our original coordinate set, with one, two, and three axis directions reversed. The original figure is an inflated balloon with some text written on it (I use this as a demonstration in my class). If you reverse only one direction (in this case, the *x* direction), you are reflecting through the plane formed by the other two axes. The resultant figure is an enantiomorph relative to the original. If you reverse two directions (here, *x* and *y*), it is the same as rotating by 180° around the un-reversed axis, and the result is a congruent figure to our original balloon. Finally, if you reverse all three axes, you are performing the operation of inversion, and you have an enantiomorph relative to the original. We can conclude that in 3-D, either reflection or inversion creates an enantiomorph, while reflection creates a congruent figure. These basic relationships will help us later on, when we try to understand the types of symmetry that result from performing various operations on a crystal.

Combinations of Operations in 3-D: Crystal Classes

In the 2-D case, we described ten plane point groups that repeat a motif around a fixed point using only rotation and reflection, without the operation of translation. There are 32 analogous **point groups** (sometimes called **crystallographic point groups**) that do exactly the same thing: repeat a motif around a single point, leaving that point unmoved. To create the 32 point groups (also commonly called **crystal classes**), we combine rotation and reflection, but add inversion and rotoinversion to the mix. The resultant possible combinations are given in Table 11.5. The first column names the classes base on rotational (*n*) symmetry alone: 1, 2, 3, 4, and 6. Each of those classes has an equivalent in rotoinversion space, labeled $\bar{1}$, $\bar{2}$, $\bar{3}$, $\bar{4}$, or $\bar{6}$. When you combine the characteristics of rotation and rotoinversion, you end up with the classes in the fourth column, which can be more simply described as combinations of *n*-fold rotation exactly perpendicular to a mirror plane: 2/*m*, 3/*m*, 4/*m*, and 6/*m* (though note that 3/*m* is the same as $\bar{6}$, and thus doesn't count as a distinct class either!). If you combine rotation with reflection across a plane that contains the rotation axis (i.e., *m* and the *n*-fold axis are coplanar), you create the classes 2*mm*, 3*m*, 4*mm*, and 6*mm*. You can also add the mirror plane to combinations that lie between two other axes. So possible combinations of *n*, *m*, *i*, and $\bar{n}$ create the remaining classes: 2/*m*2/*m*2/*m*, 4/*m*2/*m*2/*m*, 6/*m*2/*m*2/*m*, 23, 2/*m*$\bar{3}$, 432, $\bar{4}$ 3*m*, and 4/*m*$\bar{3}$2/*m*. We'll have a lot more to say about these in the following chapter. For now, the important conclusion is that in 3-D, those four operations create 32 possible ways to repeat space around a point

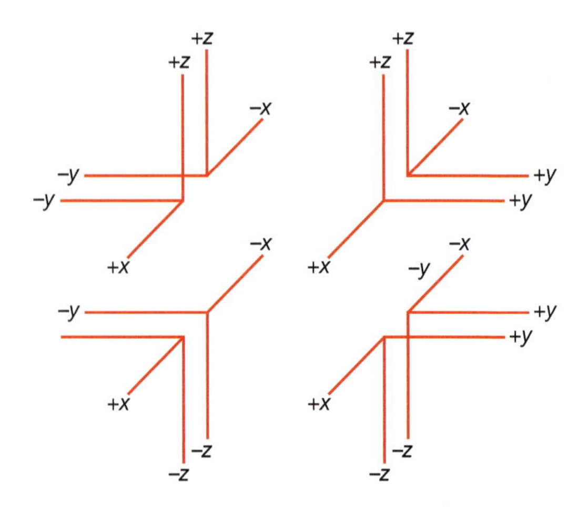

Figure 11.26. In three dimensions, there are eight possible arrangements of the x, y, *and* z *axes in the + and – directions, which are shown here. Adapted from Buerger (1978).*

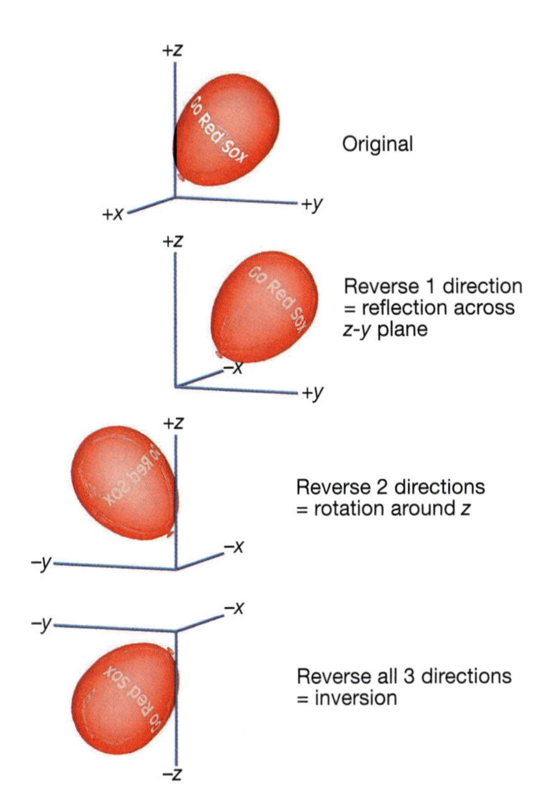

Figure 11.27. *Fortunately, the eight possibilities show in Figure 11.26 can be simplified because there are only four permutations: the starting motif (top), and reversals of one, two, or three axis directions. Reversal of one direction is the same as reflection across the plane of the other two, un-reversed axes, and creates an enantiomorphous figure. Reversal of two axes corresponds to rotation around the un-reversed axis, and results in a congruent motif. Reversal of three axes is the same as inversion, and makes an enantiomorphous figure.*

(Table 11.6). Note that the table also includes the name of the shape associated with each crystal class. These may seem like somewhat antiquated terms ("tetartoid"), but some students will find these names, often learned in geometry classes, to be useful in visualizing the classes, which is why they are included.

Space Lattices

Earlier in this chapter (Table 11.3), we developed the vector notation for describing lattices as combinations of vector translation in one, two, and three dimensions, and discussed the primitive and non-primitive lattices that could result. There were five possible combinations of translation and rotation that gave rise to the five 2-D lattices. In 3-D, the addition of the third translation direction creates additional possibilities, but there are still only five different space lattices (seven if you count the face-centered lattices, *A*-, *B*-, and *C*-, separately—see Table 11.7). We can derive them in the same way as for the plane lattices, by combining:

(a) three translations, with lengths that are equal ($t_1 = t_2 = t_3$), unequal ($t_1 \neq t_2 \neq t_3$), or something in between (e.g., $t_1 \neq t_2 = t_3$), and

(b) Any of the following rotations:

$$\alpha = \beta = \gamma = 90°$$
$$\alpha = \gamma = 90° \neq \beta$$
$$\alpha = \beta = 90°,\ \gamma = 120°$$
$$\alpha \neq \beta \neq \gamma$$

It is convenient to describe the resultant 3-D lattice types as the "edges" of shapes associated with each of the crystal systems (remember them from Chapters 1 and 4?), starting with the least symmetry and working our way up. So we begin with the triclinic system, which corresponds to the 1 and $\bar{1}$ point groups. There are three translations of unequal lengths, and three axes inclined to one another. This arrangement produces a perfectly general oblique net, or parallelopiped, and is only occurs with primitive lattice types.

The monoclinic system also has three translations of unequal lengths, though now only one axis is inclined relative to the other two (that are at right angles to each other). This system is

Table 11.5. Operations that create the 32 Space Point Groups, or Crystal Classes

Operations	*Classes Associated with Different Rotational Symmetries*				
Rotation (*n*) only	1	2	3	4	6
Rotoinversion ($\bar{n}$) only	$\bar{1}$ (=1)	$\bar{2}$	$\bar{3}$	$\bar{4}$	$\bar{6}$ (=3/*m*)
Multiple *n* with no mirror planes (*m*)		222	32	422	622
One mirror plane (*m*) coplanar with *n*-fold axis		2*mm*	3*m*	4*mm*	6*mm*
One mirror plane (*m*) perpendicular to *n*-fold axis		2/*m*	3/*m* (=$\bar{6}$)	4/*m*	6/*m*
$\bar{n} + n + m$			$\bar{3}$ 2/*m*	$\bar{4}$ 2*m*	$\bar{6}$ 2*m*
Three *n*-fold axes perpendicular to *m*		2/*m*2/*m*2/*m*		4/*m*2/*m*2/*m*	6/*m*2/*m*2/*m*
Others (isometric class only)		23 2/*m*$\bar{3}$		432, $\bar{4}$3*m*, 4/*m*$\bar{3}$ 2/*m*	

Table 11.6. Summary of Symmetry Inherent in the 32 Space Crystal Classes

Mirror Planes	*Rotation Axes*				*Center of Symmetry*	*Symbol*	*Name of Shape*	*System*
	2-fold	*3-fold*	*4-fold*	*6-fold*				
						1	pedion	triclinic: 1-fold rotation ± inversion
					1	$\bar{1}$ (=1)	pinacoid	
1						*m*	dome	monoclinic: 2-fold rotation and/or *m*
	1					2	sphenoid	
1	1				1	2/*m*	prism	
2	1					2*mm*	orthorhombic pyramid	orthorhombic: 3 2-fold rotations and/or 3*m*
	3					222	orthorhombic disphenoid	
3	3				1	2/*m*2/*m*2/*m*	orthorhombic dipyramid	
		1				3	trigonal pyramid	hexagonal: 1 3-fold or 1 6-fold rotation axis
		1			1	$\bar{3}$	rhombohedron	
3		1				3*m*	ditrigonal pyramid	
	3	1				32	trigonal trapezohedron	
3	3	1			1	$\bar{3}$ 2/*m*	trigonal scalenohedron	
1		1				$\bar{6}$ (=3/*m*)	trigonal dipyramid	
				1		6	hexagonal pyramid	
1				1	1	6/*m*	hexagonal dipyramid	
4	3	4				$\bar{6}$ 2*m*	ditrigonal dipyramid	
6				1		6*mm*	dihexagonal pyramid	
	6			1		622	hexagonal trapezohedron	
7	6			1	1	6/*m*2/*m*2/*m*	dihexagonal dipyramid	
	1					4	tetragonal disphenoid	tetragonal: 1 4-fold or $\bar{4}$ axis
			1			$\bar{4}$	tetragonal pyramidal	
1			1		1	4/*m*	tetragonal dipyramid	
2	3					$\bar{4}$ 2*m*	tetragonal scalenohedron	
4			1			4*mm*	ditetragonal pyramid	
	4		1			422	tetragonal trapezohedron	
5	4		1		1	4/*m*2/*m*2/*m*	ditetragonal dipyramid	
	3	4				23	tetartoid	isometric: 4 3-fold rotation axes
3	3	4			1	2/*m* $\bar{3}$	diploid	
6	3	4				$\bar{4}$ 3*m*	hextetrahedron	
	6	4	3			432	gyroid	
9	6	4	3		1	4/*m* $\bar{3}$ 2/*m*	hexoctahedron	

derived from the classes with a single 2-fold axis (2, *m*, and 2/*m*). This shape is effectively a right prism with a parallelogram base; it comes in both primitive and *C*-centered types.

The orthorhombic system contains classes with three orthogonal 2-fold axes, and is thus compatible with the 222, 2*mm*, and 2/*m*2/*m*2/*m* point groups. This comes in six of the seven possible lattice types: primitive, body-centered, side-centered (*A*-, *B*-, and *C*-), and face-centered. The resultant shape is a rectangular parallelepiped.

The tetragonal system contains classes with 4 or $\bar{4}$ axes. Thus, it has two axes of equal length and one of unequal length, along with all angles of 90°. These have the shape of a square prism, which can be either primitive or body-centered in shape. Tetragonal symmetry is compatible with primitive and body-centered lattice types.

Hexagonal crystals belong to classes with a single 3, $\bar{3}$, 6, or $\bar{6}$ axis. They can be thought of as cells that make up 1/3 of a hexagonal prism, where the *c* axis is the 3, $\bar{3}$, 6, or $\bar{6}$ axis. Now the translation lengths are such that $a = b \neq c$, while $\alpha = \beta = 90°$ and $\gamma = 120°$. This system is found in primitive or rhombohedral lattice types.

Finally, the isometric crystal system is composed of four 3-fold axes; the angles between *a*, *b*, and *c* are always 90°. Cubic systems are compatible with primitive, body-centered, and face-centered lattice types.

The resultant 14 Bravais Lattice types are given in Table 11.8 and will be explained (and illustrated) in much greater detail in Chapter 12 (if you can't wait, peek ahead to Figures 12.48, 12.49, 12.50, 12.51, and 12.52). For now, it is only necessary to explain that as before, some lattice

Table 11.7. 3-D Space Lattices

Lattice Type	Symbol	Lattice Points	Location of lattice points >1	Crystal systems
Primitive lattice	*P*	1	none	triclinic, monoclinic, orthorhombic, tetragonal, hexagonal, isometric
Body-centered lattice	*I*	2	One at center of cell at intersection of body diagonals	orthorhombic, tetragonal, isometric
A-centered	*A*	2	One at center of *A* (100) face	
B-centered	*B*	2	One at center of *B* (010) face	orthorhombic
C-centered	*C*	2	One at center of *C* (001) face	
Face-centered	*F*	4	Three, one each at centers of *A*, *B*, and *C* faces	orthorhombic, isometric
Rhombohedral	*R*	3	Two, along long body diagonal of cell	hexagonal

types are only compatible with certain types of rotations. For example, rhombohedral lattices are not compatible with 4-fold rotation, and body-centered lattices do not occur with solely 1-fold or $\bar{1}$ rotations. This brings us full circle back to the six crystal systems we explained in Chapter 1.

Space Groups

If you take the 3-D lattice types we have just described and combine them with the 32 point groups, there are actually only 65 possible combinations. But you'll notice that the point groups in 3-D are only combinations of rotation, reflection, inversion, and rotoinversion. What about screw axes and glides? It is possible to replace the rotation axes in each point group with screw axes, and to substitute glide planes for all the mirror planes. The net result is 230 unique space groups, which represent all possible combinations of the seven symmetry operations (t, n, $\bar{n}$, m, i, g, and n_m). These are listed, along with their associated point groups, in Table 11.9.

Summary

The goal of this chapter was to present the development of two- and three-dimensional symmetry operations in a conceptual framework. We hope that this background will prepare you for the more mathematical approach, grounded in stereographic representations, that follows in the next two chapters. With this pro-

Table 11.8. Bravais Lattices (14) related to Crystal Systems (adapted from Buerger, 1978)

Bravais Lattices (14)	Lattice Symbol	Crystal Systems (6)	Translations and rotations	Point Groups (32)
primitive	*P*	triclinic	$a \neq b \neq c$ $\alpha \neq \beta \neq$	1, $\bar{1}$
primitive side-centered	*P* C	monoclinic	$a \neq b \neq c$ $\alpha = \beta = 90°$ or $\alpha = \beta = 90°$	2, *m*, 2/*m*
primitive side-centered body-centered face-centered	*P* *C* *I* *F*	orthorhombic	$a \neq b \neq c$ $\alpha = \beta = \gamma = 90°$	222, 2*mm*, 2/*m*2/*m*2/*m*
primitive body-centered	*P* *I*	tetragonal	$a = b \neq c$ $\alpha = \beta = \gamma = 90°$	4, $\bar{4}$, 4/*m*, 422, 4*mm*, $\bar{4}$2*m*, 4/*m*2/*m*2/*m*
primitive	*P*	hexagonal	$a = b \neq c$ $\alpha = \beta = 90°$, $\gamma = 120°$	6, 3/*m*, 6/*m*, 622, 6*mm*, $\bar{6}$2*m*, 6/*m*2/*m*2/*m*
rhombohedral	R			3, $\bar{3}$, 32, 3*m*, $\bar{3}$ 2/*m*
primitive body-centered face-centered	*P* *I* *F*	isometric	$a = b = c$ $\alpha = \beta = \gamma = 90°$	23, 2/*m*$\bar{3}$, 432, $\bar{4}$3*m*, 4/*m*$\bar{3}$ 2/*m*

Table 11.9. The 230 space groups, their numbers, and associated point groups:
***Triclinic*:**
1: *P*1 (1)
$\bar{1}$: $P\bar{1}$ (2)
***Monoclinic*:**
2: *P*2 (3), $P2_1$ (4), *C*2 (5)
m: *Pm* (6), *Pc* (7), *Cm* (8), *Cc* (9)
2/*m*: *P*2/*m* (10), $P2_1/m$ (11), *C*2/*m* (12), *P*2/*c* (13), $P2_1/c$ (14), *C*2/*c* (15)
***Orthorhombic*:**
222: *P*222 (16), $P222_1$ (17), $P2_12_12$ (18), $P2_12_12_1$ (19), $C222_1$ (20), *F*222 (21), *I*222 (22), $I2_12_12_1$ (23)
***mm*2**: *Pmm*2 (25), $Pmc2_1$ (26), *Pcc*2 (27), *Pma*2 (28), $Pca2_1$ (29), *Pnc*2 (30), $Pmn2_1$(31), *Pba*2 (32), $Pna2_1$ (33), *Pnn*2 (34), *Cmm*2 (34), $Cmc2_1$ (35), *Ccc*2 (37), *Amm*2 (38), *Abm*2 (39), *Ama*2 (40), *Aba*2 (41), *Fmm*2 (42), *Fdd*2 (43), *Imm*2 (44), *Iba*2 (45), *Ima*2 (46)
2/*m* 2/*m* 2/*m*: *P*2/*m* 2/*m* 2/*m* (47), *P*2/*n* 2/*n* 2/*n* (48), *P*2/*c* 2/*c* 2/*m* (49), *P*2/*b* 2/*a* 2/*n* (50), $P2_1/m\ 2/m\ 2/a$ (51), $P2/c\ 2_1/n\ 2/a$ (52), $P2/m\ 2/n\ 2_1/a$ (53), $P2_1/c\ 2/c\ 2/a$ (54), $P2_1/b\ 2_1/a\ 2/m$ (55), $P2_1/c\ 2_1/c\ 2/n$ (56), $P2/b\ 2_1/c\ 2_1/m$ (57), $P2_1/n\ 2_1/n\ 2/m$ (58), $P2_1/m\ 2_1/m\ 2/n$ (59), $P2_1/b\ 2/c\ 2_1/n$ (60), $P2_1/b\ 2_1/c\ 2_1/a$ (61), $P2_1/n\ 2_1/m\ 2_1/a$ (62), $C2/m\ 2/c\ 2_1/m$ (63), $C2/m\ 2/c\ 2_1/a$ (64), *C*2/*m* 2/*m* 2/*m* (65), *C*2/*c* 2/*c* 2/*m* (66), *C*2/*m* 2/*m* 2/*a* (67), *C*2/*c* 2/*c* 2/*a* (68), *F*2/*m* 2/*m* 2/*m* (69), *F*2/*d* 2/*d* 2/*d* (70), *I*2/*m* 2/*m* 2/*m* (71), *I*2/*b* 2/*a* 2/*m* (72), $I2_1/b\ 2_1/c\ 2_1/a$ (73), $I2_1/m\ 2_1/m\ 2_1/a$ (74)
***Tetragonal*:**
4: *P*4 (75), $P4_1$ (76), $P4_2$ (77), $P4_3$ (78), *I*4 (79), $I4_1$ (80)
$\bar{4}$: $P\bar{4}$ (81), $I\bar{4}$ (82)
4/*m*: *P*4/*m* (83), $P4_2/m$ (84), *P*4/*n* (85), $P4_2/n$ (86), *I*4/*m* (87), $I4_1/a$ (88)
422: *P*422 (89), $P42_12$ (90), $P4_122$ (91), $P4_12_12$ (92), $P4_222$ (93), $P4_22_12$ (94), $P4_322$ (95), $P4_32_12$ (96), *I*422 (97), $I4_122$ (98)
4*mm*: *P*4*mm* (99), *P*4*bm* (100), $P4_2cm$ (101), $P4_2nm$ (102), *P*4*cc* (103), *P*4*nc* (104), $P4_2mc$ (105), $P4_2bc$ (106), *I*4*mm* (107), *I*4*cm* (108), $I4_1md$ (109), $I4_1cd$ (110)
$\bar{4}2m$: $P\bar{4}2m$ (111), $P\bar{4}2c$ (112), $P\bar{4}2_1m$ (113), $P\bar{4}2_1c$ (114), $P\bar{4}m2$ (115), $P\bar{4}c2$ (116), $P\bar{4}b2$ (117), $P\bar{4}n2$ (118), $I\bar{4}m2$ (119), $I\bar{4}c2$ (120), $I\bar{4}2m$ (121), $I\bar{4}2d$ (122)
4/*m* 2/*m* 2/*m*: *P*4/*m* 2/*m* 2/*m* (123), *P*4/*m* 2/*c* 2/*c* (124), *P*4/*n* 2/*b* 2/*m* (125), *P*4/*n* 2/*n* 2/*c* (126), $P4/m\ 2_1/b\ 2/m$ (127), $P4/m\ 2_1/n\ 2/c$ (128), $P4/n\ 2_1/m\ 2/m$ (129), $P4/n\ 2_1/c\ 2/c$ (130), $P4_2/m\ 2/m\ 2/c$ (131), $P4_2/m\ 2/c\ 2/m$ (132), $P4_2/n\ 2/b\ 2/c$ (133), $P4_2/n\ 2/n\ 2/m$ (134), $P4_2/m\ 2_1/b\ 2/c$ (135), $P42/m\ 2_1/n\ 2/m$ (136), $P4_2/n\ 21/m\ 2/c$ (137), $P4_2/n\ 21/c\ 2/m$ (138), *I*4/*m* 2/*m* 2/*m* (139), *I*4/*m* 2/*c* 2/*m* (140), $I4_1/c\ 2/m\ 2/d$ (141), $I4_1/a\ 2/c\ 2/d$ (142)
***Hexagonal*:**
3: *P*3 (143), *P*31 (144), *P*32 (145), *R*3 (146)
$\bar{3}$: $P\bar{3}$ (147), $R\bar{3}$ (148)
32: *P*312 (149), *P*321 (150), $P3_112$ (151), $P3_121$ (152), $P3_212$ (153), $P3_221$ (154), *R*32 (155)
3*m*: *P*3*m*1 (156), *P*31*m* (157), *P*3*c*1 (158), *P*31*c* (159) *R*3*m* (160) *R*3*c* (161)
$\bar{3}2/m$: $P\bar{3}\ 1\ 2/m$ (162), $P\bar{3}\ 1\ 2/c$ (163), $P\bar{3}\ 2/m\ 1$ (164), $P\bar{3}\ 2/c\ 1$ (165), $R\bar{3}\ 2/m$ (166), $R\bar{3}\ 2/c$ (167)
6: *P*6 (168), $P6_1$ (169), $P6_5$ (170), $P6_2$ (171), $P6_4$ (172), $P6_3$ (173)
$\bar{6}$: $P\bar{6}$ (174)
6/*m*: *P*6/*m* (175), *P*63/*m* (176)
622: *P*622 (177), $P6_122$ (178), $P6_522$ (179), $P6_222$ (180), $P6_422$ (181), $P6_322$ (182)
6*mm*: *P*6*mm* (183), *P*6*cc* (184), *P*63*cm* (185), *P*63*mc* (186)
$\bar{6}m2$: $P\bar{6}m2$ (187), $P\bar{6}c2$ (188), $P\bar{6}2m$ (189), $P\bar{6}2c$ (190)
6/*m* 2/*m* 2/*m*: *P*6/*m* 2/*m* 2/*m* (191), *P*6/*m* 2/*c* 2/*c* (192), $P6_3/m\ 2/c\ 2/m$ (193), $P6_3/m\ 2/m\ 2/c$ (194)
***Isometric*:**
23: *P*23 (195), *F*23 (196), *I*23 (197), $P2_13$ (198), $I2_13$ (199)
2/*m* $\bar{3}$: $P2/m\ \bar{3}$ (200), $P2/n\ \bar{3}$ (201), $F2/m\ \bar{3}$ (202), $F2/d\ \bar{3}$ (203), $I2/m\ \bar{3}$ (204), $P2_1/a\ \bar{3}$ (205), $I2_1/a\ \bar{3}$ (206)
432: *P*432 (207), $P4_232$ (208), *F*432 (209), $F4_132$ (210), *I*432 (211), $P4_332$ (212), $P4_132$ (213), $I4_132$ (214)
$\bar{4}3m$: $P\bar{4}m$ (215), $F\bar{4}3m$ (216), $I\bar{4}3m$ (217), $P\bar{4}3n$ (218), $F\bar{4}3c$ (219), $I\bar{4}3d$ (220)
4/*m* $\bar{3}$ 2/*m*: $P4/m\ \bar{3}\ 2/m$ (221), $P4/n\ \bar{3}\ 2/n$ (222), $P4_2/m\ \bar{3}\ 2/n$ (223), $P4_2/n\ \bar{3}\ 2/m$ (224), $F4/m\ \bar{3}\ 2/m$ (225), $F4/m\ \bar{3}\ 2/c$ (226), $F4_1/d\ \bar{3}\ 2/m$ (227), $F4_1\ \bar{3}\ 2/c$ (228), $I4_1/m\ \bar{3}\ 2/m$ (229), $I4_1/a\ \bar{3}\ 2/d$ (230)

gression, we aspire to move your understanding of symmetry and crystallography up several "levels" along the spiral learning curve as we discussed in the Preface.

The important message for most geologists to get from this chapter is that a few operations can be used to derive all the possible combinations of space in three dimensions. This groundwork will allow us to go on to understand three-dimensional arrangements of atoms, which will then lead to mineral structures! So symmetry represents perhaps the most fundamental underpinning of the study of minerals.

References

Bernal, I., Hamilton, W.C., and Ricci, J.S. (1972) Symmetry. A Stereoscopic Guide for Chemists. W.H. Freeman and Co., San Francisco, 182 pp.

Boisen, M.B., and Gibbs, G.V. (1990) Mathematical Crystallography. Reviews in Mineralogy, vol. 15, Mineralogical Society of America, Washington, D.C., 406 pp.

Buerger, M. J. (1978) Elementary Crystallography. MIT Press, Cambridge, Massachusetts, 528 pp.

Cotton, F.A. (1990) Chemical Applications of Group Theory. 3rd Ed., John Wiley and Sons, New York, 295 pp.

Edwards, E.B. (1967) Pattern and Design with Dynamic Symmetry. Dover Publications, Inc., New York, 122 pp.

Gruenbaum, B., and Shepard, G.C. (1986) Tilings and Patterns. W.H. Freeman, New York, 700 pp.

Kennon, N.F. (1978) Patterns in crystals. John Wiley and Sons, New York, 197 pp.

MacGillavry, C.H. (1965) Symmetry Aspects of M. C. Escher's Periodic Drawings, Oosthoek, Utrecht, 84 pp.

O'Keefe, M., and Hyde, B.G. (1996) Crystal Structures I. Patterns and Symmetry. Mineralogical Society of America, Washington, D.C., 453 pp.

Smith, J.V. (1982) Geometrical and Structural Crystallography. John Wiley and Sons, New York, 450 pp.

Terpstra, P. (1955) Introduction to the Space Groups. J.B. Wolters, Groningen, Djakarta, 160 pp.

Chapter 12

Symmetry

If you go to the library and check out any of the older mineralogy textbooks, you'll quickly see that the material covered in this chapter was usually the first chapter in the book. This is how I was first introduced to mineralogy. Although I liked it, I had to read the first chapter in the book (the current edition is Bloss, 1994) 6–7 times before I got it—but I did get it. Now it's a bit humbling to try and cover this material in my own words, and to bring his amazing illustrations to 3-D animations. Of course, the fact that I spent seven years working with and learning from Bloss while getting my Ph.D. has been a big help in furthering my understanding of crystallography and his methods of explaining and illustrating it. Many of the illustrations in this chapter rely heavily on those in his book—because it is the way I learned that material!

There is no doubt that 3-D symmetry is one of the most important aspects of science, because it does nothing less than explain the atomic structure of materials! However, it is perhaps one of the hardest subjects to master (and to teach!). Symmetry lectures are often presented out of context (i.e., the student sees no real application for what they are supposed to be learning). Students are asked, often for the first time in their academic careers, to think in three dimensions. This necessary skill is essential to understanding not only the crystal structure of a mineral, but almost every aspect of geology (i.e., folded rocks, stratifications, the shape of a fossil—the list is endless)!

There are many things professors can do to help the students visualize 3-D objects and in turn manipulate them in their minds. These include (1) building and using 3-D models in class, (2) making 3-D projections on a 2-D chalkboard or overhead, and (3) using computer animations. Our illustrations should help accomplish number 2, while the MSA website serves number 3. For other good ideas, we recommend the website from the 2004 NAGT workshop "Teaching Geoscience with Visualizations: Using Images, Animations, and Models Effectively" for examples and ongoing discussion of the use of visualizations and animations in teaching earth sciences.

To me, crystallography is what sets mineralogists apart from many other of the physical scientists of the world. While chemistry, physics, and material science may use crystallography, the other disciplines rarely require the in-depth understanding that is needed by a well-trained mineralogist. Chemists tend to deal with much simpler systems than mineralogists—they often go into the lab and grow their compounds with known amounts of only a few elements—a condition that rarely occurs in nature. Both physicists and material scientists tend to deal with materials of fairly high symmetry—isometric and hexagonal, while mineralogists must deal with the common silicates, which are mainly of low symmetry—monoclinic and triclinic.

While crystallography may be (and really is!) quite foreign to the student who entered geology to look at rocks and hang out outdoors, crystallography is really the "guts" of mineralogy!

M.E.G.

Introduction

There are four chapters in this book dedicated to symmetry, after the very brief introduction in Chapter 1. Chapter 4 was a brief overview of the more important aspects of crystallography, especially the crystal systems. In Chapter 11, we described the seven symmetry operations and described how mixing and matching them can result in the 230 space groups. In this chapter, we move up the spiral learning curve to understand symmetry using stereographic projections. Finally, Chapter 13 will repeat much of the material in this chapter using a mathematical approach. No other subject in this book is given such thorough and repetitive coverage, but the reason for this is simple—this stuff is challenging to understand, but is fundamental to mineralogy. Mastery of 3-D symmetry is also needed to go on to the crystal structures of minerals, which are presented in Chapter 14.

There are two types of symmetry that we will discuss in this chapter: point symmetry and space symmetry. **Point symmetry** was originally used to describe the external morphology of well-formed crystals. From them, it was discovered that external symmetry could be observed and internal symmetry inferred. For instance, Figure 12.1 shows a pair of well-formed halite and quartz crystals with markedly different external symmetry, from which it was postulated that their internal structures were also different. However, it was not until the advent of X-ray diffraction that we could directly determine the internal arrangement of atoms and determine their **space symmetry**. Figure 12.1 also shows the main structural units that, in turn, control the external morphology of the crystals. Thus the internal and external symmetries of materials are related, with the internal symmetry controlling the external symmetry. So as mineralogy has evolved over the past century, we now spend more time studying the internal atomic arrangements of minerals, and less time studying their external morphology, because by knowing the former, we can predict the latter.

Both point and space symmetry are very common in everyday life. The mirror you look into every morning uses a type of point symmetry (Figure 11.8), because it keeps an entire plane of points fixed (i.e., the mirror's surface). Your footprints in the sand or snow (Figure 11.10), however, represent a type of symmetry that does not keep any points fixed—you are walking. To relate your tracks, you need two symmetry operations: a mirror plane (you have right and left feet) and translation (your stride). The mirror plane is then transformed by the translation and the resulting space symmetry operation, as we learned in Chapter 11, is a glide plane.

Another type of point symmetry we see daily is based on rotation. For instance, we can use point symmetry to generate the leaves of a 4-leaf clover (Figure 12.2) based on the point symmetry operation of rotation; the center point of the clover stays fixed. Translations can also be added to rotations to allow them to repeat space in 3-D. Consider the jar of olives in Figure 12.3a. From the top view, it appears as if the olives have 4-fold symmetry, though they look a little different when viewed from the side. What happens in the jar is that two opposite olives are rotated 180° and translated upward; this type of symmetry operation is a screw axis, which we first saw in Chapter 11, and is very common in minerals. Now look at the other olive jar (Figure 12.3b) and think about what kind of symmetry it shows.

Stereographic Projections

One of the most powerful techniques for gaining insight into the angular relationships of 3-D geometric features (i.e., lines and planes) is the technique of **stereographic projections**, which we introduced in Chapter 4. In keeping with the concept of spiral learning, we will again discuss the method here and build on it from the earlier discussion.

Stereographic projections provide a way to envision 3-D space in 2-D, and show the relationships between angular data (see Figure 4.6). As an example, we used the latitude and longitude of some major (and not so major) cities and plotted them on a globe. We then projected those locations onto the equatorial plane of the Earth, the plane that passed through the equator and was perpendicular to the N and S poles. Figure 12.4a shows a projection similar to the one we used in Chapter 4, including a line and a plane. Position "P" on the upper hemisphere can be projected down to the equatorial plane by connecting its upper hemisphere position to the "South" pole in the diagram. Then the point at which the line crosses the equatorial plane would be its projection point. To represent the plane also shown in Figure 12.4a, we would have to trace an infinite number of lines (only a couple of which are shown) to trace out the plane's projection onto the equatorial plane. The resultant 2-D stereographic projections of the point and the plane are shown in Figure 12.4b. This method effectively reduces 3-D space to 2-D space. The result is that things that are 2-D in 3-D space (e.g., a line) would plot as 1-D points on the stereographic projection. Similarly, 3-D objects such as planes would be represented as 2-D lines on the stereographic projection.

Next, we need to be able to specify locations on the stereographic projections in a quantitative way. Recall that in Chapter 4 we used latitude and longitude as our polar coordinate set. The longitude values increased clockwise around the circumference of the projection (i.e., along the equator) with 0° starting in England. The latitude values were defined to be 0° at the equator and increase in both a northerly and southerly direction; thus they would be 90° at both the North and South poles.

In mineralogical applications, we use a similar coordinate system, but with different names and a slight modification. We replace the latitude term with the Greek letter ϕ (phi) and define ϕ to be 0° on the right hand side of the projection (Figure 12.5) and to increase in a clockwise fashion just as longitude did. The difference between the two systems is how we measure the other angle. For latitude, we started at the equator and increased

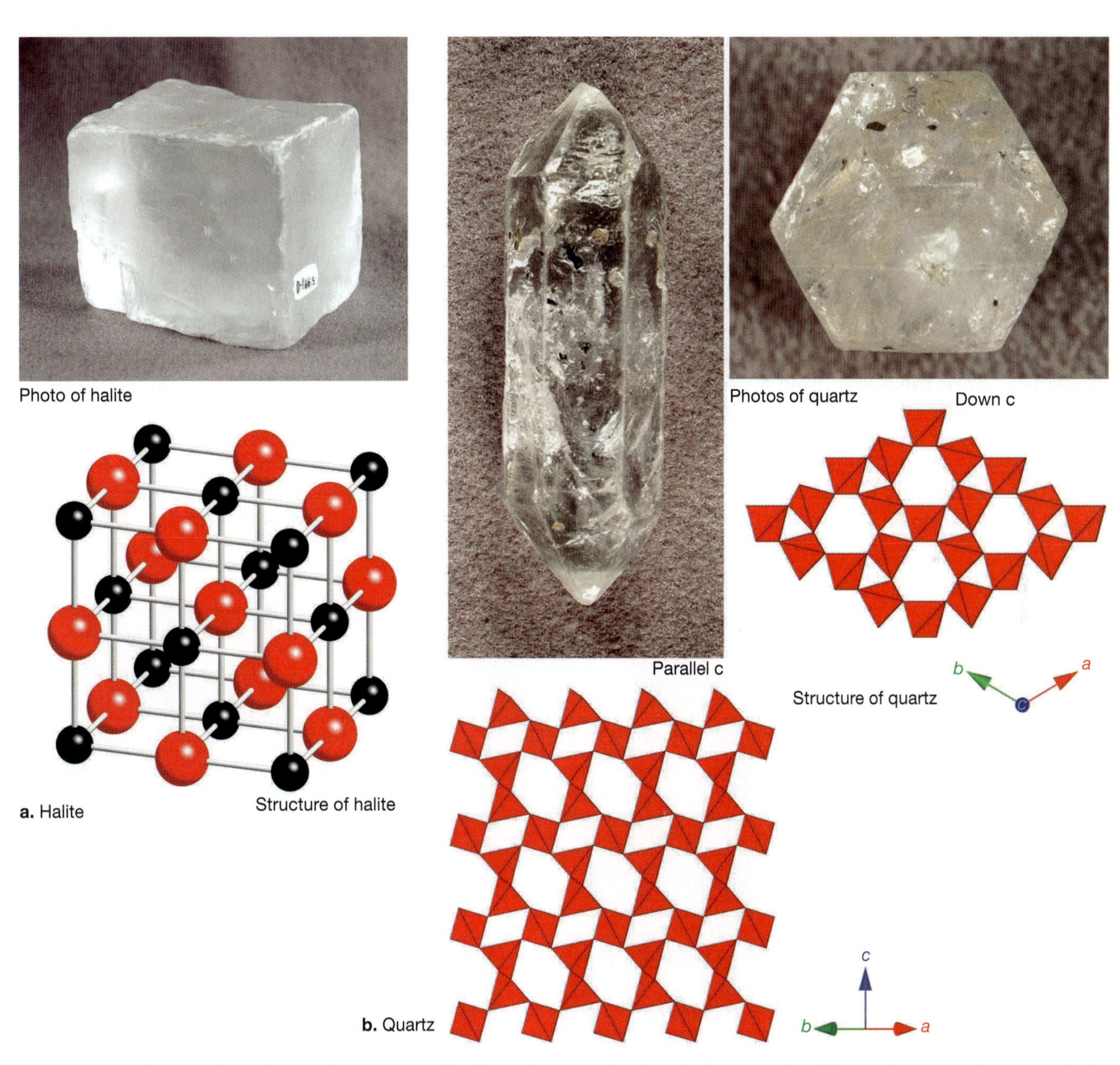

Figure 12.1. *Photographs and crystal structures of halite and quartz showing their external and internal symmetries, respectively.* ***a.*** *Halite occurs in the form of a cube and belongs to the isometric crystal system. The internal arrangement the atoms in halite clearly controls its external symmetry.* ***b.*** *Quartz, which belongs to the hexagonal crystal system, is shown in two orientations with the* c *axis parallel (center) and vertical (right); the symmetry becomes apparent projected down* c*. This structure helps explain why quartz grows elongated parallel to* c*, because it is based on the spiraling chains of linked tetrahedra parallel to* c*, and has hexagonal symmetry projected down* c*.*

*Figure 12.2. Photographs of 3- and 4-leaf clovers exhibiting the symmetry operations of rotation and reflection. **a.** The 3-leaf clover has a 3-fold axis, so the leaves repeat every 120°. There are also three diagonal mirror planes at 120° that relate to the 3-fold rotation axes. **b.** The 4-leaf clover has a 4-fold axis (i.e., the leaves repeat every 90°) and it also has four diagonal mirror planes that intersect at 90°, conforming to the symmetry of the rotation axis.*

the values inward, while for the stereographic projections the opposite is used. Here we replace the latitude term with the Greek letter ρ (rho) and measure it from the North pole outward. In other words, a latitude of 20° would equal a ρ value of 70°. Figure 12.5 is a 3-D projection defining the ϕ and ρ polar coordinate system. It shows that the ϕ and ρ coordinates for the point "P" in Figure 12.4 would be $\phi = 45°$ and $\rho = 60°$.

The next task is to define a 2-D grid system that we can use to plot the ϕ and ρ coordinates. This is done by "contouring" the equatorial plane with these two coordinates. The ϕ values are easy; they are just the angular readings around the circle starting with 0° on the east (Figure 12.4). The ρ values are measured from the North pole. We can envision planes passing through the sphere (as in Figure 12.4) and then being projected to the equatorial plane (Figure 12.6). The angle the plane makes with the North pole ($\phi = 0°$ direction) increases from 0° to 90°. The planes in Figure 12.7 are drawn every 10° in bold and every 2° in lighter shading. These planes are called **great circles** to distinguish them from the smaller arcs used to denote the ϕ angles.

To actually plot some points, let's take some of the latitude and longitude data from a few of the cities we used back in Chapter 4; they're given in Table 12.1. The ϕ values are synonymous with longitude, but we need to convert the latitude data to ρ by taking the complement of it (i.e., ρ = 90 – latitude). Thus two new columns appear in Table 12.1: one for ϕ and the other ρ for each of the cities. It is best to begin by estimating where these points will plot. Start by drawing a circle with a radius of 5 inches and label the ϕ values as in Figure 12.8. Next, estimate a ϕ value of 73° around the circumference of the circle for South Hadley, Massachusetts. Now move out from the center about a little over halfway, which would be the ρ angle of 48°. This is the approximate loca-

*Figure 12.3. Examples of screw axis symmetry in jars of olives in two different projections. **a.** Looking down into the jar, it appears as if the olives are in 4-fold symmetry. However, the side view shows that the olives are stacked in a way that really is 2-fold symmetry. Each pair is related to the pair above or below by a 90° rotation. So this symmetry is a combination of a rotation and a translation, which is termed a screw axis. **b.** This jar of olives appears to have 6-fold symmetry when viewed from the top. The side view shows a similar type of symmetry as in Figure 12.3a; here each layer of three olives is rotated 60° and then translated up a later. Thus each layer of olives is related to the one above and below by rotation and translation.*

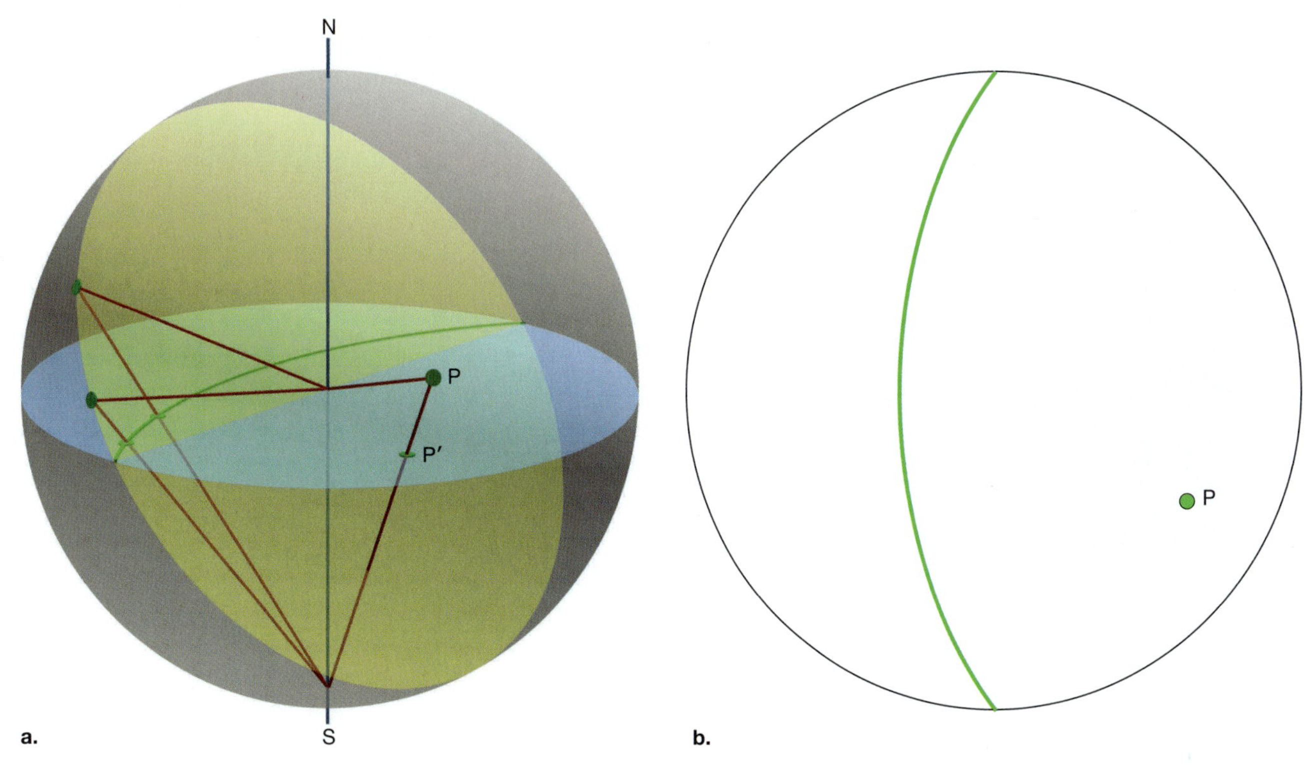

Figure 12.4. *The concept behind stereographic projection.* ***a.*** *A sphere (e.g., the Earth) with its equatorial plane added (i.e., the imaginary plane that would pass through the center of the Earth and intersect it along its equator). A point "P" is plotted on the outer surface of the sphere and projected onto the equatorial plane as a point. An inscribed plane would project onto the equatorial plane as an arc. (Modified from Figure 4.2, Bloss, 1994.)* ***b.*** *The equatorial plane "removed" from Figure 12.4a and projected down the North pole direction, showing the point "P" and the arc in Figure 12.4a. This is the projection surface that we'll use for stereographic projections.*

tion for South Hadley, so label this point SH. Next plot Moscow, Idaho (use MO), Bern (use BE), and Kyoto (use KY). Your plot should look something like the one shown in Figure 12.8.

Now that we've approximated the location of the four cities, let's plot them precisely. Make a copy of the Wulff net given in Figure 12.7 and find yourself some tracing paper (the kind you can see through) and a thumbtack. (If you want to get fancy, you can mount the stereonet onto a thin piece of cardboard, as shown in Figure 12.9.) Now poke the thumbtack up through the back of the stereonet at its center; you might want to put a piece of tape over the back of it to hold it in place (and keep a pencil eraser stuck onto the thumbtack while you're carrying this in your backpack). Next, trace out the circumference of the underly ing stereonet, and mark the $\phi = 0°$ with a small line as in Figure 12.9 (this whole process is animated on the MSA website). Now we're ready to plot the cities!

Using the thumbtack as a rotation point, we will rotate the paper on top of the stereonet in order to plot points and measure the angles between them. Let's plot South Hadley first. Put a mark at $\phi = 73°$, then rotate the stereonet counter-clockwise as shown in Figure 12.10 until the mark at 73° aligns with the $\phi = 0°$ (i.e., the E point on the net). Now count out from the center 48°; this is the ρ value for South Hadley. This rotation step is necessary so we could count out the ρ angle. The only place we can correctly measure ρ is along one of the great circles that project as a straight line (i.e., the E-W or N-S lines on the stereonet). Now plot the remaining three cities; they should look like those shown in Figure 12.10. For kicks, overlay the plot you made from Figure 12.8 over the one in Figure 12.10 and see how close your estimates come to the precise plots.

Table 12.1. Values of latitude, longitude, and ϕ, ρ for selected cites in the northern hemisphere. They are plotted by latitude and longitude in Figure 4.6, and by ϕ and ρ in Figures 12.8 and Figure 12.10.

	Latitude	*Longitude*	ϕ	ρ
South Hadley	42°	73°	73°	48°
Moscow	47°	117°	117°	43°
Bern	47°	353°	353°	43°
Kyoto	33°	224°	224°	57°

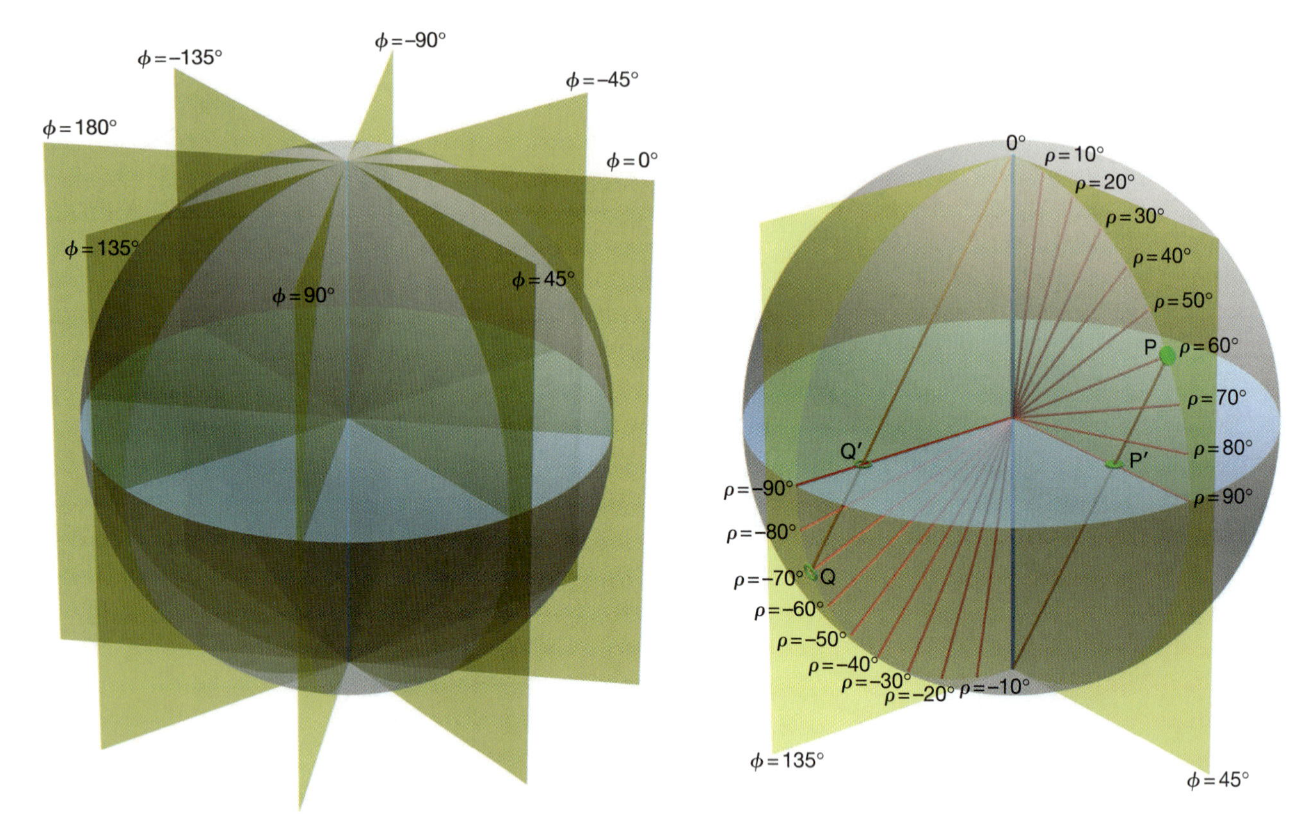

Figure 12.5. *The polar coordinate system used in stereographic projections. The ϕ angles range from 0° to 360°, or 0° to 180° and 0° to –180°, and are plotted around the circumference of the projection. The ϕ values are similar to longitude. The ρ values are measured out from the North pole, so they are the compliments of longitude. Point "P" is shown plotted as a solid circle, while point "Q" is shown plotted as a hollow circle because it plots on the lower hemisphere. (It might help to think that the hollow circle would represent a open man-hole cover and if you walked over it, you'd fall in (i.e., to the lower hemisphere).) (Modified from Figure 4.3, Bloss 1994.)*

One of the major uses of stereographic plots is to graphically determine the angular relationships between lines and planes. (We can also calculate these values, and a method to do that is given in the next chapter.) To determine the angles between two points (say, Moscow and South Hadley), you'll need to do some more paper rotation. Rotate the paper until the two towns rest on the same great circle (as in Figure 12.10). Now count off the number of degrees. Next, measure the angles between Moscow and Bern, and Moscow and Kyoto. You should get the answers listed on the projection in Figure 12.11.

Finally, let's plot some crystallographic data and measure the angles between some points. Table 12.2 gives the ϕ and ρ coordinates for a series of points. Plot these ten points and make sure your results look like those in Figure 12.11. Point number 5 might present somewhat of a challenge because ρ = 105° (i.e., it's greater than 90°). Think about what this means. This would be a vector that makes a 105° angle with the vertical, thus it would project on the lower hemisphere (when that happens, we use a hollow circle). So we'd count in 15° (= 105° – 90°) to plot this point. Now measure angles between points 1 and 4, 1 and 7, etc., as listed in the Table 12.2 and make sure you get the same results. You'll notice that the angles between 1 and 4 equal the angles

Table 12.2. Stereographic coordinates (i.e., ϕ and ρ) for a series of example points plotted in Figure 12.11, as well has the angles between them.

Point	ϕ	ρ
1	0°	90°
2	44°	90°
3	63°	90°
4	90°	90°
5	90°	105°
6	90°	15°
7	–	0°
8	–44°	90°
9	–63°	90°
10	30°	60°

angles between:
1:4 = 90°, 1:7 = 90°, 1:6 = 90°, 1:5 = 90°, 5:7 = 105°, 2:8 = 88°, 3:9 = 126°, 6:10 = 53°

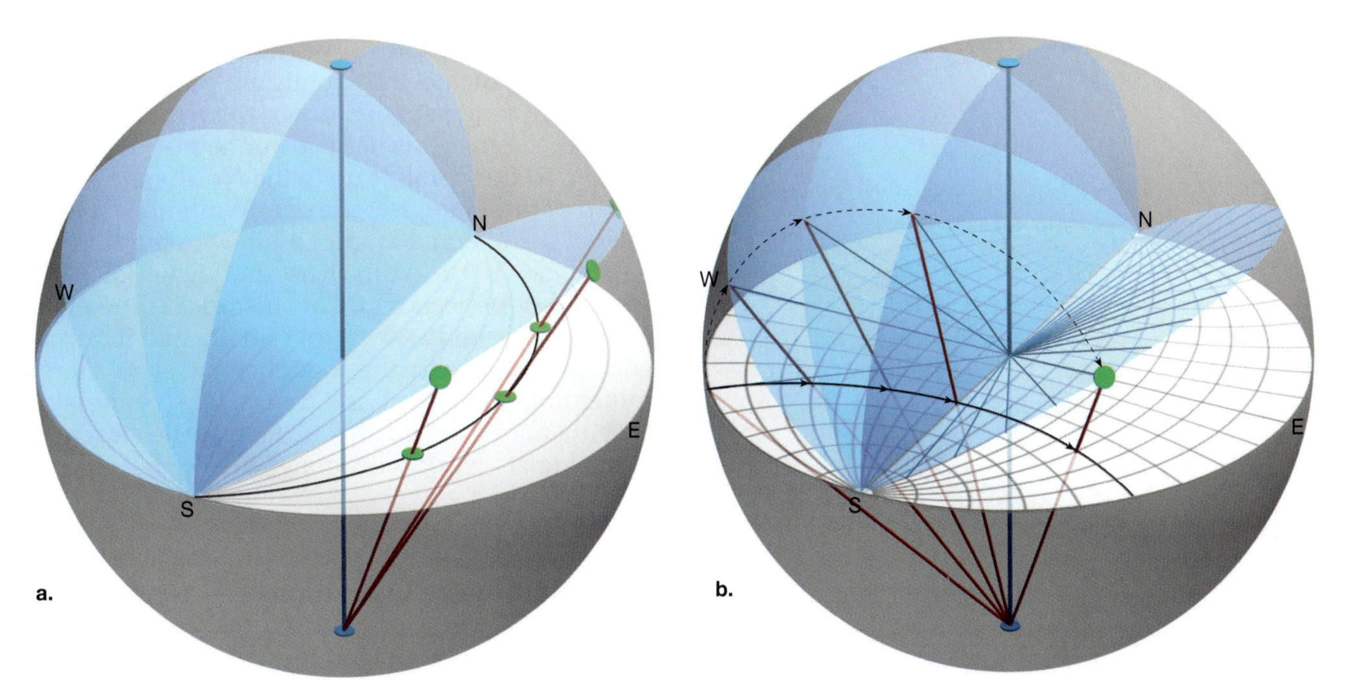

Figure 12.6. Description of how the plotting grids are made for stereographic projections. **a.** The great circles are produced by projecting a family of planes that are "hinged" on the N-S line as shown above. As the ϕ angle is changed from 0° to 90°, the planes move from vertical to horizontal. A plane with a ϕ angle of 60° is given as an example. (Modified from Figure 4.6, Bloss, 1994.) **b.** The small circles are produced by setting a fixed angle of ρ within one plane and then projecting it as that plane is rotated about the N-S direction. (Modified from Figure 4.7, Bloss, 1994.)

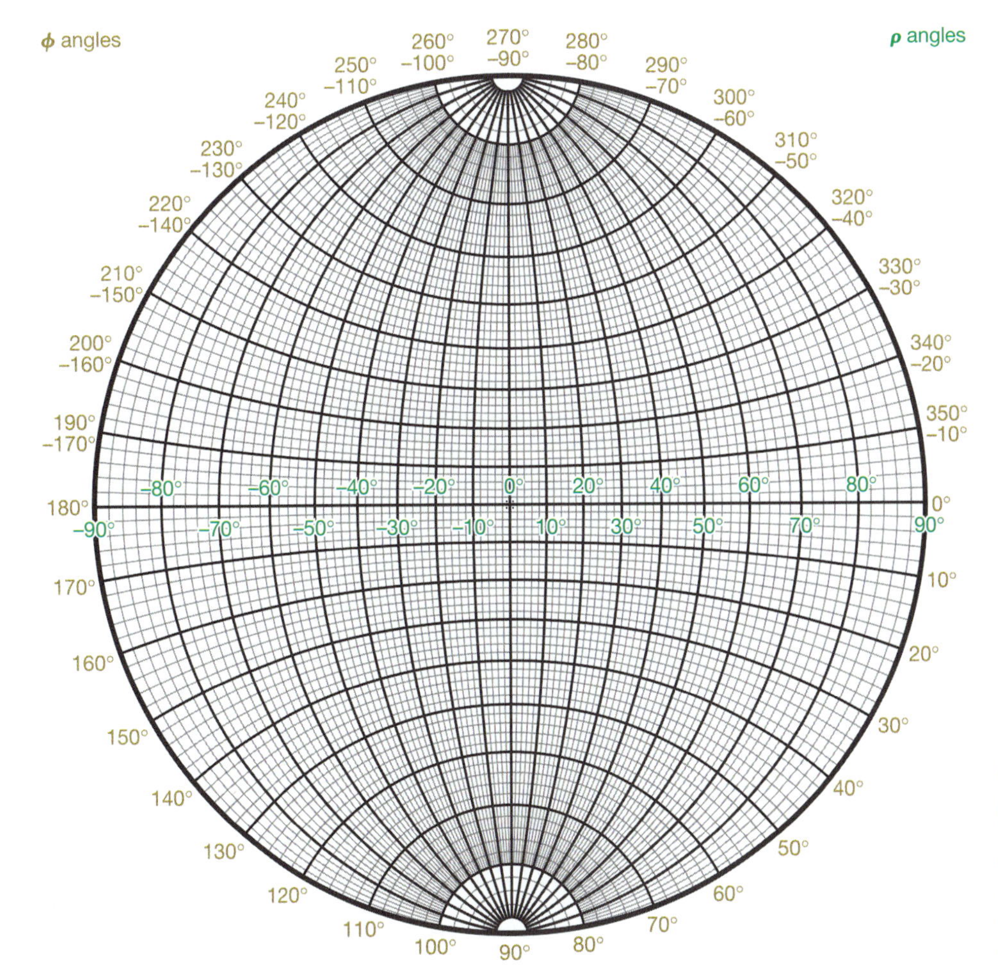

Figure 12.7. The two sets of projections shown in Figure 12.6 result in a stereonet with the ϕ and ρ coordinate system, with ϕ measured around the circumference and ρ measured outward from the center.

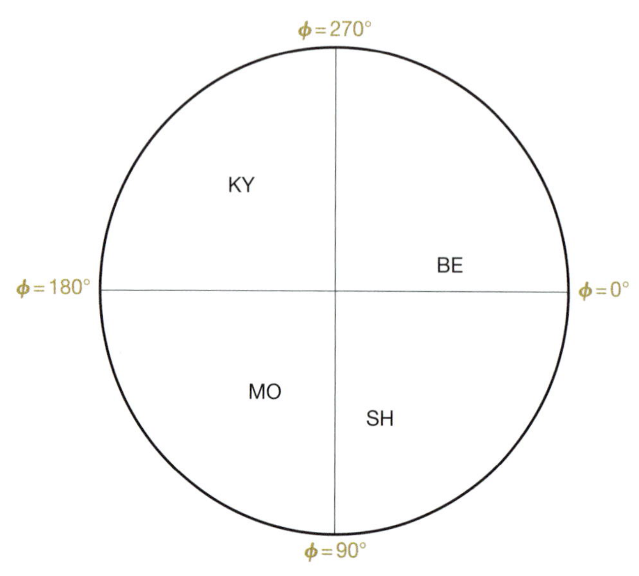

Figure 12.8. The approximate locations of four cities, South Hadley Massachusetts (SH), Moscow Idaho (MO), Bern Switzerland (BE), and Kyoto Japan (KY) as estimated from the ϕ *and* ρ *coordinates in Table 12.1.*

between 1 and 7, 1 and 6, and 1 and 5, and all of them are 90°. Thus points 4, 5, 6, and 7 all occur in a plane that is 90° away. You'll be seeing lots of stereoplots in this chapter, so make sure you understand this section before moving on. Also, you can check out the animations on the MSA website, which should help you visualize the leap from 3-D to 2-D space.

One final comment seems necessary to add, because many of you who are taking this class are also studying structural geology. For historical reasons, mineralogists and structural geologists use stereo projections in different ways. Mineralogists use an "upper hemisphere" projection, which means that points are projected from anywhere in the sphere, down to the South pole. Structural geologists use a lower hemisphere projection, so they project up to the North pole. Note also that mineralogists use only Wulff nets, which project using equal angles, while structural geologists use both Wulff nets and Schmidt nets (equal *area* projections). This historical difference confused me endlessly until I finally had it pointed out to me, and the logic helped even more: mineralogists work with things (e.g., minerals) that are above the Earth's surface, while structural geologists work with things below the surface (e.g., bedrock).

Point Symmetry Operations

Now that you understand how to represent these relationships graphically, we can return to the subject at hand: representing points in space in 3-D. As you recall from the previous chapter, point symmetry operations are those that leave at least one point fixed in space. These are the 3-D expansion of the ten planar point groups (rotation plus reflection) that we discussed in the previous chapter. The point symmetry operations are those that keep a point, a line, or a plane fixed:

1. a center of symmetry , or **inversion** (labeled as "*i*") keeps only a point in space fixed (i.e., all other parts of space are reproduced by it);
2. **rotation** axes (labeled as *n*-fold, where *n* is some integer), keep a line in space fixed;
3. a **mirror plane** (labeled as "*m*") keeps a plane in space fixed; and

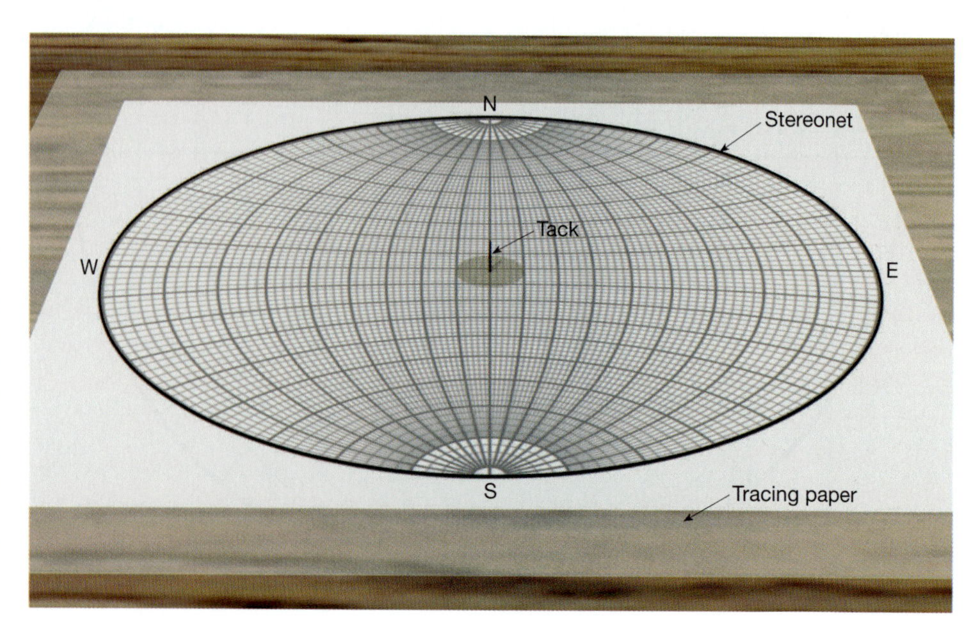

Figure 12.9. The standard plotting setup used in stereographic projections. First, make a copy of the stereonet in Figure 12.7. Then stick a thumb tack through its center point from the rear. (You might want to glue the stereonet on a thin piece of stiffer paper first to make it more durable.) Next a piece of tracing paper is placed over the stereonet, its center impaled by the thumb tack, and trace the stereonet's circumference onto the tracing paper.

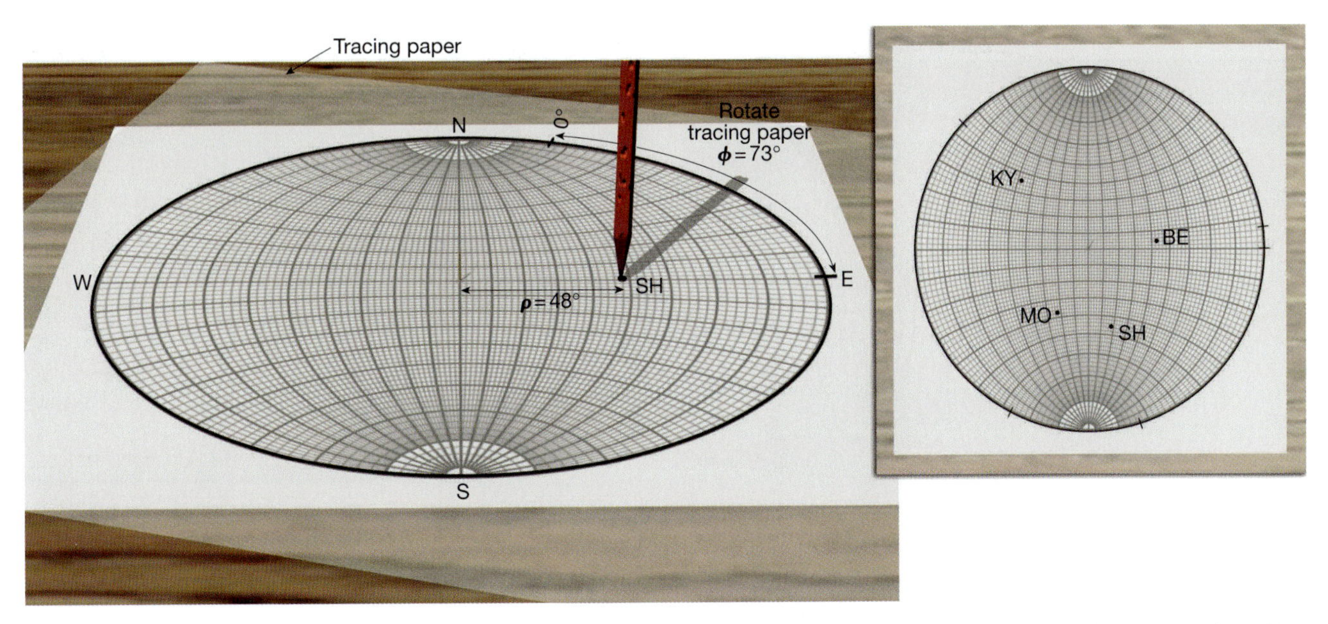

Figure 12.10. *Once you've got the setup in Figure 12.9 ready, you can precisely plot the cities from Table 12.1 and see how close we came to our estimates in Figure 12.8. First, place a mark on the tracing paper at $\phi = 0°$; this is needed because we are going to be rotating the paper to plot the cities and we'll need to return to 0° when we are done. Next, plot South Hadley with $\phi = 73°$ and $\rho = 48°$. The ϕ value is going to cause South Hadley to plot in the lower right quadrant of the stereonet. Now find $\phi = 73°$ on the stereonet's circumference and place a small mark over it on the tracing paper. Rotate the $\phi = 73°$ mark counterclockwise to the E-W (i.e., $\phi = 0°$) direction. (It is necessary to rotate the paper to this direction so we can measure ρ; ρ values can only be measured along the E-W straight great circle or the N-S straight great circle.) Next, count out 48° from the center of the stereonet to for the ρ value; at that point place a solid circle and label it "SH." Now repeat the process for Moscow and label it MO, Bern (BE), and Kyoto (KY). Lastly, compare this to Figure 12.8; the results should be similar.*

4. **rotoinversion**, a uniquely 3-D operation, combines inversion with rotation to keep a point in space fixed.

We'll review these only briefly, because we just discussed them in the previous chapter.

Mirror planes. A mirror plane might be the most commonly encountered "symmetry generator." We usually see one every morning that produces a mirror image of us by reflection. In this type of symmetry operation only two images exist—the original and the reflected one (Figures 11.8 and 12.12). Also note that these two images have a different handedness (i.e., a right paw appears as a left paw in the mirror, because the mirror flips the points). Reflection projects every point an equal and opposite direction across a plane.

Rotation axes. The example of a 3- or 4-leaf clover was already given in the introduction; all that is required is one leaf and then symmetry produces the rest (Figure 12.2). In fact, as we discussed in the previous chapter, there are exactly five types of rotation axes that are of interest in mineralogy: 1-, 2-, 3-, 4-, and 6-fold (Figure 12.13). Some mineralogists also call these a full-turn, 1/2 turn, 1/3 turn, 1/4 turn, and 1/6 turn to indicate their rotation angle (i.e., 360°, 180°, 120°, 90°, and 60° respectively).

Center of symmetry. To determine if an object has a center of symmetry, or **inversion center**, we must first locate its geometric center. Next draw a line from the center outward to some point on the object. Then extend this line in the opposite direction (from the center) and look at the tip of each line (Figure 12.14). If they are identical, the object has a center of symmetry (as in Figure 12.15a, c, e). What is really going on here is that a point in the object is projected though the object center's to an equidistance point. This is the equivalent of reflection (where we projected an equal and opposite distance across a plane), except that inversion is a projection of an equal and opposite direction *through a point* instead of *across a plane*. If this type of symmetry is performed on a collection of points, for instance the right hands in 3-D (Figure 12.15), the left hands are the symmetrically-related objects and the material has a center of symmetry.

Rotoinversions. Rotoinversion is a combination of a rotation and an inversion. (A center of symmetry really is an inversion of points through the center of an object; and it's the inversion that changes the handedness.) So we will need to examine what new symmetry operations occur when we add a center of symmetry to a 1-fold, 2-fold, 3-fold, 4-fold, and 6-fold axis.

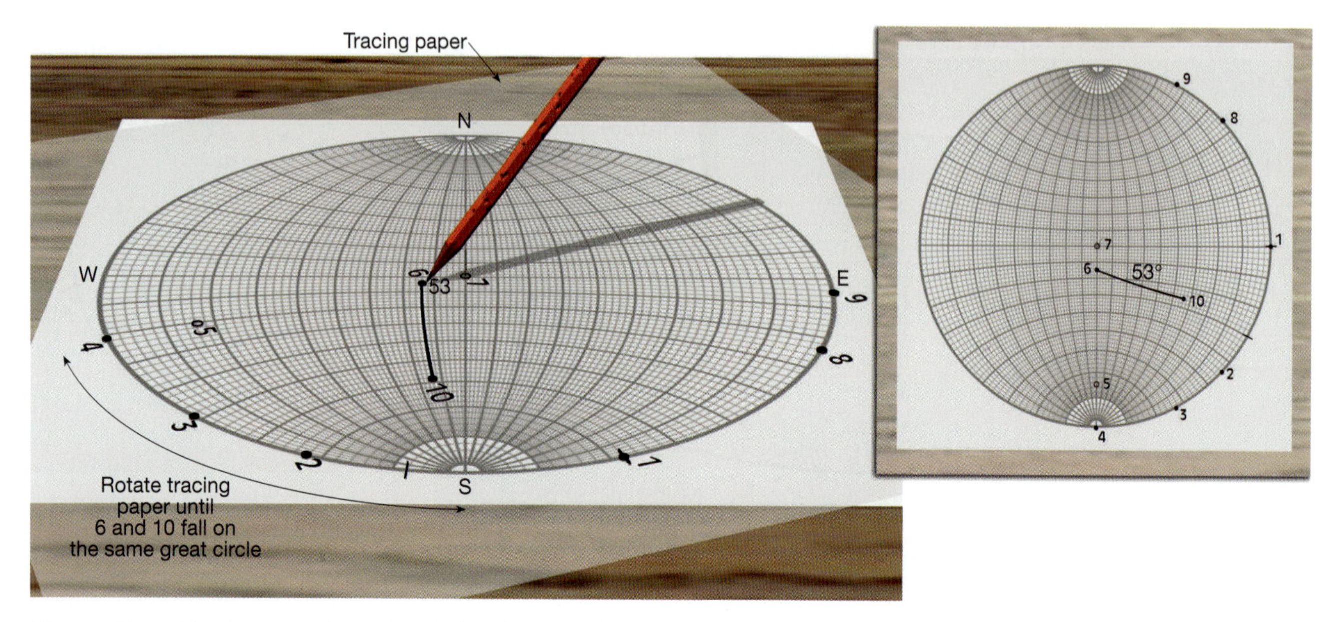

Figure 12.11. Plot the ten points given in Table 12.2, labeling them 1 to 10. Check to see if they look like the points in this figure. What is the angle between some of these points? Just by observation, what is the angle between 1 and 4? 1 and 7? 1 and 6? 1 and 5? Those angles are all 90°. Next, let's measure the angle between 6 and 10. To do this, you must place them on the same great circle (i.e., you must rotate the tracing paper until 6 and 10 fall on the same great circle). Once this is done, count the degrees (along the great circle) to see the angular relationship between the two points. Now check the answer at the bottom of Table 12.2. Finally, find the angles between 2 and 8 and 3 and 9 and make sure they agree with those in Table 12.2.

Figure 12.15 shows graphically what occurs when we add a center of symmetry to our five types of rotation axes. In Figure 12.15a, a center of symmetry is added to a 1-fold axis, which is a rotation of 360°. First the right hand (R1) is rotated 360° and then inverted (each point is projected an equal and opposite distance) through the center of the sphere to produce a second left hand (L1). The "new" symmetry element is the center of symmetry, $\bar{1}$, which is read "bar 1."

Figure 12.12. "Mirror" image of Karen and Dennis Tasa's dog. Take a good look and make sure to see how well this image reproduced the real dog on the left into the mirror on the right.

What happens when we add a center of symmetry to a 2-fold axis (Figure 12.15b)? We first rotate the right hand (R1) 180°, then invert it through the center to make the second left hand (L1). The "new" symmetry element could be labeled a $\bar{2}$, but the hand in Figure 12.15b is really just reflected across a plane that would be normal to the rotation axis, and the right hand (R1) is a mirror image of the left hand (L1). As we noted in Chapter 11, a 2-fold axis plus a center of symmetry is the same thing as a mirror plane.

Let's add a center of symmetry to a 3-fold axis and see what happens. Start with a right hand (R1) (Figure 12.15c), rotate it 120° and then invert it through the center of the sphere to create a left hand (L1). We are going to keep performing symmetry operations until we end up with the hand we started with (i.e., R1). Next we take L1 rotate in 120° and invert it through the center to create a right hand (R2). The process is continued until six hands are created (Figure 12.15c). The resultant rotoinversion axis is termed a $\bar{3}$, and is really the first "new" symmetry type created by adding a centers of symmetry to rotation axes. In Figure 12.15c, there are three right hands on the upper hemisphere and three left hands on the lower hemisphere. The upper and lower sets of hands are offset by 60°.

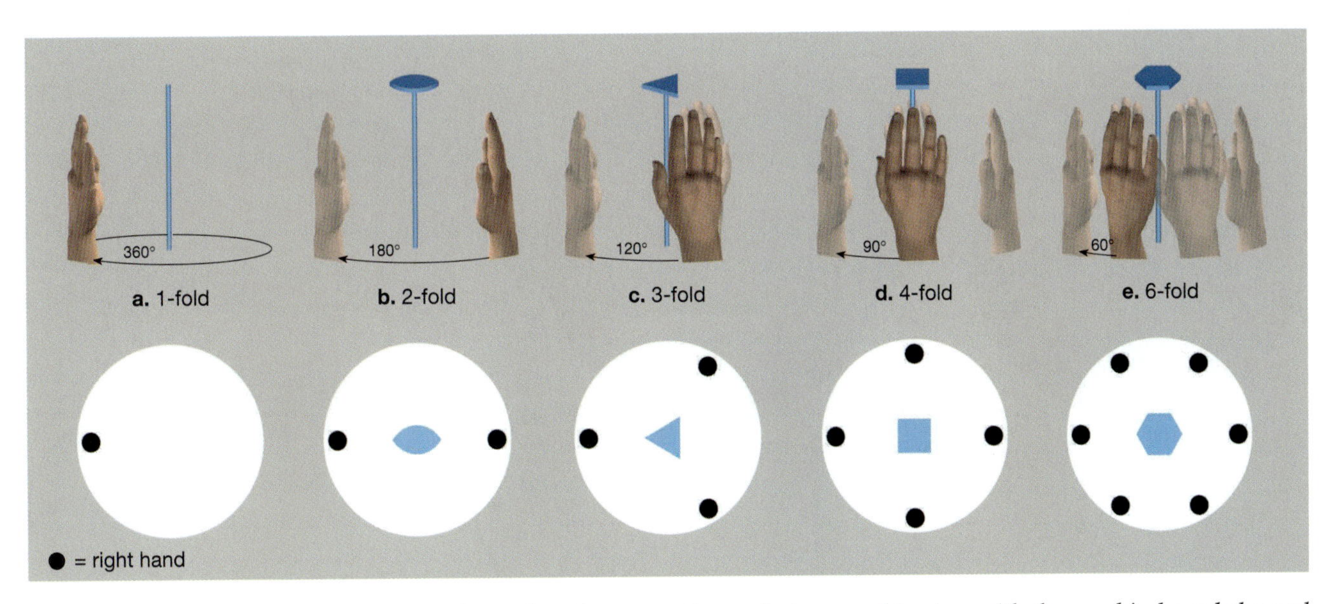

*Figure 12.13. The five rotation axes are shown here in perspective and stereographic view with the graphical symbols used to denote them and a series of symmetry-generated right hands. The number of hands corresponds to the "n" in n-fold, and also to the number of corners on the graphical symbols. Lastly, all of the angular relationships for them are shown on the stereographic projections. **a.** 1-fold: Creates only one symmetry point, which is equivalent to a 360° rotation. **b.** 2-fold: Two points are created at 180° increments with a 2-pointed graphical symbol. **c.** 3-fold: Three points are created at 120° increments and the graphical symbol is a triangle. **d.** 4-fold: Four points are created at 90° increments and the graphical symbol is a square. **e.** 6-fold: Six points are created at 60° increments and the graphical symbol is a hexagon.*

For a 4-fold axis, we start with right hand (R1) (Figure 12.15d), rotate 90°, and invert it through the center of the sphere forming a left hand as (L1) on the lower hemisphere. Next, we repeat the 90° rotation with L1, and invert in through the center to form a right hand as R2. Lastly, R2 is rotated 90° and inverted to form L2. If L2 is again rotated 90° and inverted through the center, we'd end up with R1 again. This type of axis is termed a $\bar{4}$.

The details of the $\bar{6}$ axis (Figure 12.15e) will be left to you to figure out, with a few hints. How do you interpret what the meaning of the $\bar{6}$ symbol is? Think about the difference between the $\bar{3}$ and $\bar{6}$ as we discussed in Chapter 11. Can you think of a combination of two of the above symmetry elements that might do the same thing as a $\bar{6}$? "Reflect" on this for a moment and think about 3-fold axes, and notice that the upper hemisphere is full of right hands, while the lower has left hands.

To summarize this section, there are four different types of point symmetry: inversion, reflection, rotation, and rotoinversion. Table 12.3 summarizes the 12 possible types of point symmetry. If we mix and match the operations of reflection, rotation, inversion, and rotoinversion, there are 32 different possibilities in 3-D space. These are the **point groups**, sometimes called **crystal classes**, which are collections of point symmetry operations and listed by crystal system (Table 12.4).

Groups of Point Symmetry Operations

Most minerals, or 3-D objects, contain more than one type of point symmetry. To appreciate this, consider again the 2-D representation of our 3- and 4-leaf clovers (Figure 12.16a and b). The 4-leaf clover had a 4-fold rotation axis. Do you see any other types of symmetry in Figure 12.16a?

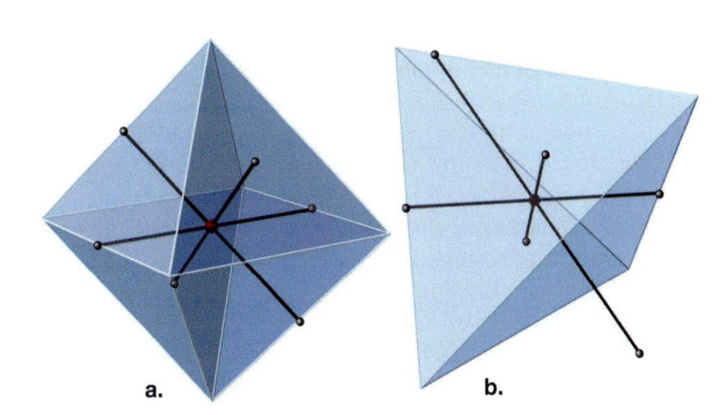

*Figure 12.14. Geometric shapes with and without a center of symmetry. **a.** An octahedron has a center of symmetry. To test this, note any point on it and then locate a point that lies in an equal but opposite direction for a line passing through the center (as shown for a couple of points in the figure). **b.** A tetrahedron lacks a center of symmetry. To prove this to yourself, select a point and then project it an equal distance back through the center and see if a similar point occurs on the opposite side (as shown for three points in the figure).*

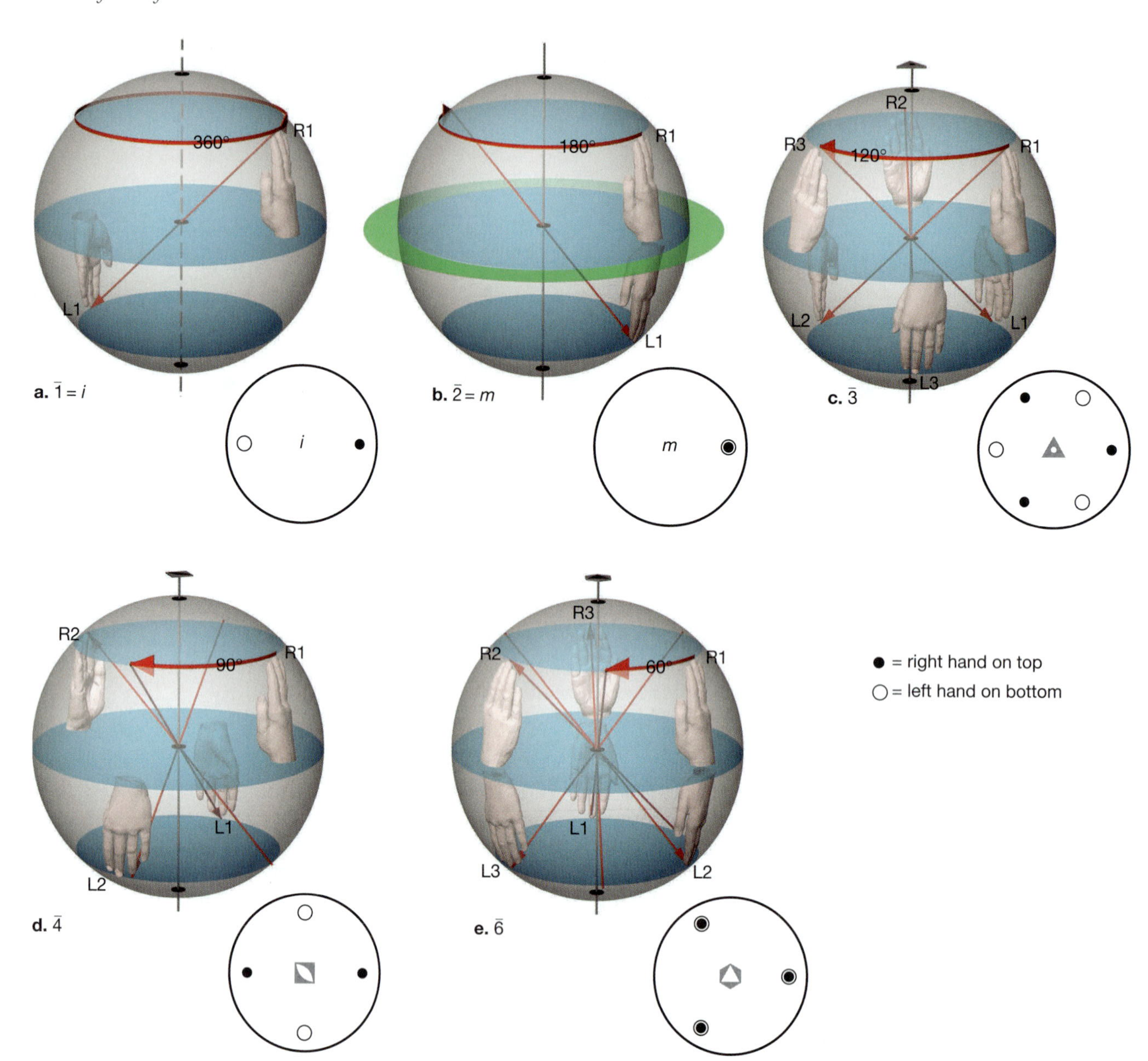

Figure 12.15. *Combinations of a center of symmetry with the five different types of rotation axes to create rotoinversion axes (modified from Figure 1.7, Bloss, 1994). All the rotations are in the clockwise direction.* ***a.*** *1-fold axis plus center of symmetry: The right hand (R1) in the upper hemisphere is rotated 360° and then inverted through the center of the sphere to from the left hand (L1) on the lower hemisphere. This is the same thing as a center of symmetry.* ***b.*** *2-fold axis plus center of symmetry: The right hand (R1) in the upper hemisphere is now rotated 180° and inverted through the center to form the left hand (L1) in the lower hemisphere just below R1. This is the same as a mirror plane perpendicular to the rotation direction, so we term this type of symmetry as a mirror plane.* ***c.*** *3-fold axis plus center of symmetry: For this example, the rotations are in 120° increments. A right hand (R1) on the upper hemisphere is rotated 120° and inverted through the center to form a left hand (L1) on the lower hemisphere. Next L1 is rotated 120° and inverted through the center to form R2 on the upper hemisphere. Then R2 rotates 120° and is inverted through the center to form L2 on the lower hemisphere. The process is repeated until all six hands—three rights on the upper hemisphere and three lefts on the lower hemisphere—are formed. This procedure creates a new symmetry element that is called* $\bar{3}$*, and pronounced "bar 3." Its graphical representation is the same as for a 3-fold axis except a dot is placed in its center to indicate the inversion and, in this case, a center of symmetry.* ***d.*** *4-fold axis plus center of symmetry: Starting again with R1 in the upper hemisphere, but this time rotating 90° and inverting arrives at L1 on the lower hemisphere. L1 then rotates 90°, is inverted, and forms R2 in the upper hemisphere. Lastly, R2 rotates 90°, is inverted, forming L2 in the lower hemisphere. This operation creates a* $\bar{4}$ *axis. Its symbol is that of a 4-fold axis with an inscribed 2-fold axis. The symbol makes sense because there is a 2-fold imbedded in the* $\bar{4}$*.* ***e.*** *6-fold axis plus center of symmetry: The above process is repeated, except this time the rotations are only 60°. See if you can follow the six hands from R1 to L1 to R2 to L2 to R3 to L3. This operation forms a* $\bar{6}$ *axis with the same symbol as a 6-fold except with a 3-fold symbol embedded. Reflect on any other types of symmetry in this image.*

Table 12.3. Point symmetry operations, names, and symbols.
With respect to a plane
m (*m* = mirror plane)
With respect to a line
1-fold
2-fold ⬬
3-fold ▲
4-fold ■
6-fold ⬢
With respect to a point
i (*i* = center of symmetry)
n-folds plus a center of symmetry
i
1-fold + *i* = *i*
2-fold + *i* = *m*
3-fold + *i* = $\bar{3}$ fold
4-fold + *i* = $\bar{4}$ fold
6-fold + *i* = $\bar{6}$ fold

Table 12.4. The 32 point groups listed by crystal system, from low to high symmetry, with the number of point groups in each crystal system given in parenthesis. At the bottom of the table are some of their commonly used abbreviations.
Triclinic (2): 1, $\bar{1}$
Monoclinic (3): 2, *m*, 2/*m*
Orthorhombic (3): 222, *mm*2, 2/*m* 2/*m* 2/*m*
Tetragonal (7): 411, $\bar{4}$11, 4/*m*11, 422, 4*mm*, $\bar{4}$2*m*, 4/*m* 2/*m* 2/*m*
Hexagonal (12): 311, $\bar{3}$11, 321, 3*m*1, $\bar{3}$2/*m*1, 611, $\bar{6}$11, 6/*m*11, 622, 6*mm*, $\bar{6}$*m*2, 6/*m* 2/*m* 2/*m*
Isometric (5): 231, 2/*m* 31, 432, $\bar{4}$3*m*, 4/*m* $\bar{3}$ 2/*m*
Commonly used abbreviations: * 1-fold axes left out as follows: 411 = 4, $\bar{4}$11 = $\bar{4}$, 4/*m* 1 1 = 4/*m* 311 = 3, $\bar{3}$11 = $\bar{3}$, 3*m*1 = 3*m*, $\bar{3}$ 2/*m* 1 = $\bar{3}$ 2/*m* 611 = 6, $\bar{6}$11 = $\bar{6}$, 6/*m* 1 1 = 6/*m* 231 = 23, 2/*m* $\bar{3}$1 = 2/*m* $\bar{3}$
Highest symmetry point groups have left out elements: 2/*m* 2/*m* 2/*m* = *mmm* 4/*m* 2/*m* 2/*m* = 4/*mmm* 6/*m* 2/*m* 2/*m* = 6/*mmm* 2/*m* $\bar{3}$ 1 = *m* $\bar{3}$ 4/*m* $\bar{3}$ 2/*m* = *m* $\bar{3}$*m*

How about a horizontal mirror plane? How about a vertical mirror plane? Or two diagonal mirror planes? The 4-leaf clover has (at least) four mirror planes along with the 4-fold axis (Figure 12.16a). The clover also has a 2-fold axis, which is implied by the presence of the 4-fold axis.

What if we wanted to assign an axis set to the 4-leaf clover and then use this axis set to describe its geometry? There are many axis sets we could pick. First, we need to pick an origin, which we will select to make our lives as easy as possible. So we make the origin at the center of the clover and the axes perpendicular (Figure 12.16a). Notice how this axis set would "fit" the 4-fold symmetry, and all those mirror planes. The 4-fold axis would take the *a* axis and move it to the *b* axis. Because these axes are equivalent, the symmetry "fits."

Now, examine the 3-leaf clover (Figure 12.16b). What type of symmetry operation(s) does it have and what type of axis set could we create to fit? The 3-leaf clover has a 3-fold axis. It also has three mirror planes that bisect each clover (Figure 12.16b). Notice that the mirror planes "fit" the symmetry of the 3-fold axis (i.e., they are located at intervals of 60° to each other, just as the mirror planes for the 4-fold were at increments of 45°). As for the axes choice, let's use the same origin at the center. The angles between the axes must be some increment of 60°. In fact, 60° would work well, but an angle of 120° (Figure 12.16b) was chosen in the days of old, so that's what we will use.

The clover is a 2-D example. When you add the third dimension, the situation becomes more complicated. A mineralogical example of 4- and 3-fold symmetry is shown in Figure 12.17, with four Cl^{1-} atoms surrounding a one Na^{1+} from halite and three O^{2-} anions surrounding one C^{4+} from calcite. Notice how you could also add the mirror planes and axis sets to each of these crystal structures.

So each of the 32 point groups can contain many different types of symmetry. In the following section, we will describe the characteristics of each point group in terms of the combinations of symmetry elements it possesses. While we are doing this, keep in mind that each of these 32 point

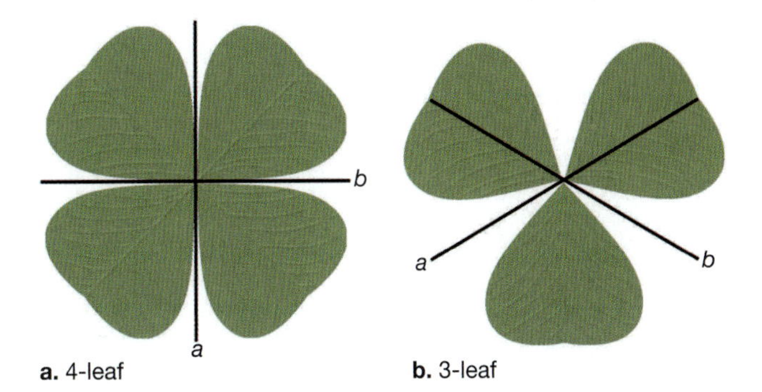

Figure 12.16. *Choice of basis vector sets (i.e., crystal systems) to match symmetry.* ***a.*** *As discussed earlier, a 4-leaf clover has a 4-fold axis that generates a leaf every 90°. We could "define" the system geometrically with a vector set whose two axes,* a *and* b*, are 90° apart, thus "fitting" 4-fold symmetry. The origin is placed at the center of the leafs.* ***b.*** *For the case of a 3-leaf clover, two axes are chosen with an origin at the center. In order to "fit" the symmetry of a 3-fold axes, the axes are at 120° to each other.*

groups is consistent with one of the six **crystal systems** we've been talking about since Chapter 1. So let's do a systematic "fitting" of each axis set to 3-D objects of the correct geometry (Figure 12.18).

Relating Point Symmetry Operations to Crystal Systems

Isometric system. We'll start with the isometric system because it is more familiar to us, even though this system has the most combinations of symmetry operations. Our standard basis vector set for an isometric crystal system is shown in Figure 12.19a, where $a = b = c$, and all the interaxial angles are 90° (i.e., $\alpha = \beta = \gamma = 90°$). First, let's place a 4-fold axis parallel to *a* (Figure 12.20a). Because all of these axes are equivalent, that's the same thing as saying "what goes on along *a* must go on along both *b* and *c*." So we must have 4-fold axes along all three directions. Then, 3-fold axes running from opposite corners of the cube bisect the *a*, *b*, and *c* axes (i.e., makes the same angle with all three of them). The 3-fold axes would map these onto one another. For instance, a 120° ccw rotation would take +*a* to +*b*, +*b* to +*c* and +*c* to +*a* (Figure 12.20b). The 4-fold axis that is parallel to *c* would create three other 3-fold axes, for a total of four 3-folds (Figure 12.20b), given that the first 3-fold axis bisects the positive ends of each of the three crystallographic axes. This is the defining characteristic of the isometric system: the presence of four 3-fold axes. There are also several 2-fold axes (Figure 12.19b) and mirror planes that occur in this system. Figure 12.19c is a stereographic representation of these symmetry elements, showing the angular locations of all the symmetry. Collectively, they represent the $4/m\ \bar{3}\ 2/m$ point group, which has the highest symmetry of the five point groups in the isometric system. We'll get to the others shortly; for now, you can guess at what the symbols in the point group mean!

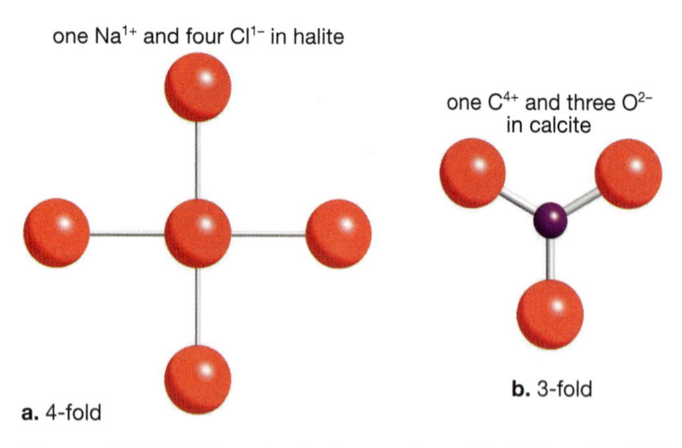

*Figure 12.17. Mineralogical examples of 4-fold and 3-fold symmetry. **a.** In halite, a small Na^{1+} atom is surrounded by six larger Cl^{1-} atoms and a 4-fold axis occurs, looking down the O^{2-} anions as shown. **b.** In calcite, a smaller C^{4+} atom is surrounded by three larger O^{2-} anions.*

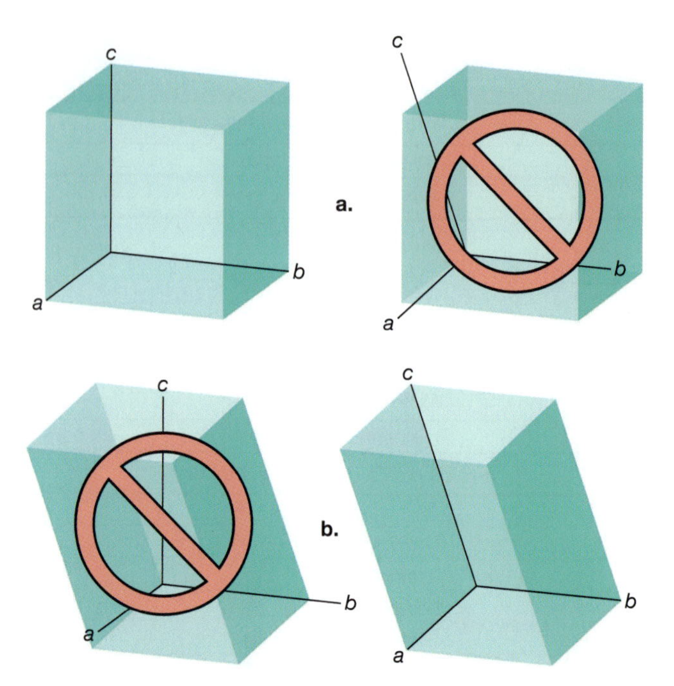

***Figure 12.18.** Examples of fitting axes sets to wooden blocks of different shapes. **a.** For the case of a wooden cube we can fit an isometric axes set to it, but not a triclinic set. **b.** For the case of a block of wood cut to fit the triclinic axes set, an isometric axes set would no longer "fit," but a triclinic axes set would.*

Figure 12.19c also shows several solid and hollow representations. The 48 points shown would result from plotting one point on the stereographic projection and allowing symmetry operations to produce the others. For instance, there are six points around each one of the 3-fold axes. Compare these to the $\bar{3}$ axes shown in Figures 12.15c, 12.19b, and 12.20b and see if you can rationalize how those points got there. (Hint: think about what the $\bar{3}$ axis would do, while noting that there is a mirror plane perpendicular to it). Figure 12.20 shows how the 4-, 3-, and 2-fold axes for this point group fit to the symmetry of the point group by relating the symmetrically-equivalent crystallographic axes to each other.

By this point you'll notice there are 4-, 3-, and 2-fold axes and mirror planes going all sorts of directions. We need a better way to keep track of all of their orientations than saying, for instance, "the 4-fold axis that is parallel to the *c* axis," or "the 3-fold axis that makes an equal angle to each of the 3 perpendicular crystallographic axes." So before we continue with the remaining crystal systems, let's devise a better method to discuss

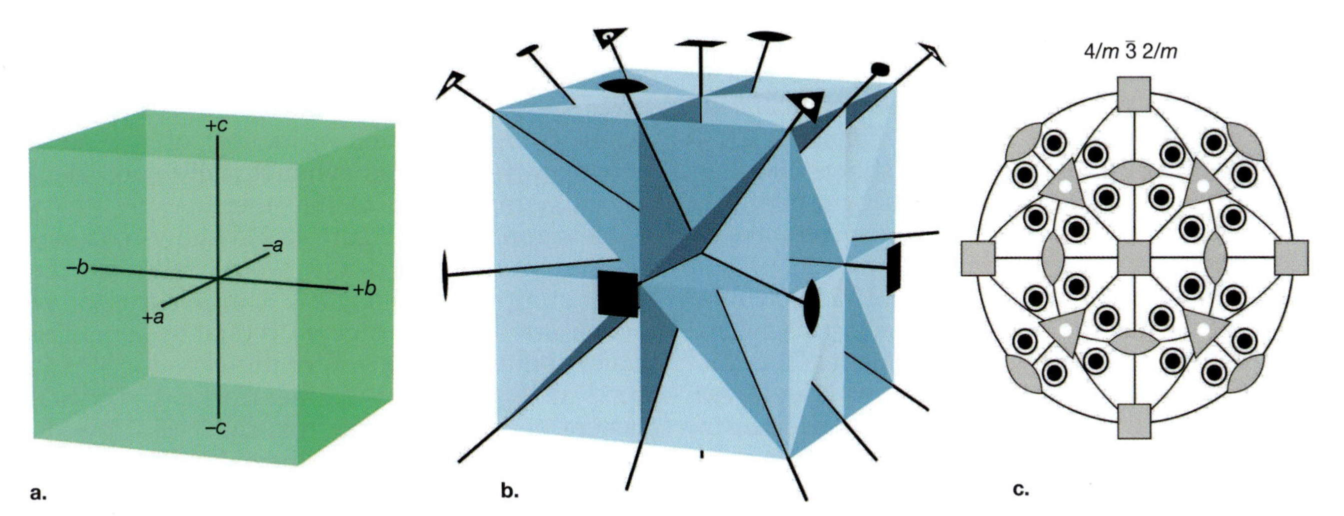

Figure 12.19. Geometric representation of the isometric crystal system and its symmetry operations in perspective and stereographic views. ***a.*** *The axis set with a = b = c and α = β = γ = 90° is first inscribed inside a cube-shaped block of wood. The axis system fits the shape of the cube.* ***b.*** *The cube from Figure 12.19a, with its symmetry elements added. There are three 4-fold axes, all perpendicular to the six faces, four 3-fold axes that are along the body diagonals, six 2-fold axes along each edge, and nine mirror planes.* ***c.*** *A stereographic representation of the symmetry elements in Figure 12.19b. Also included are the 48 points that would be created by the symmetry operations for this point group.*

the orientation of the many point group operations that collect themselves together to form a crystal system.

Figure 12.21a shows the familiar Cartesian coordinate system (*x*, *y*, *z*) with crystallographic nomenclature (*a*, *b*, *c*). We can represent any direction in this coordinate system as a vector, and then mathematically define the vector's direction by use of the coordinate system. First we write each crystallographic axis in terms of the other three. For instance, *a* could be written 1*a*, 0*b*, 0*c* or in vector notation [100]. Likewise, *b* would be [010], and *c* would be [001]. Then we could superscript the symmetry so as to show its orientation. So to represent "a 4-fold axis that is parallel to the *c* axis," we could write $^{[001]}4$ (Figure 12.21b). Likewise, [111] would denote a direction that bisects the three crystallographic axes. To represent "the 3-fold axis that makes an equal angle to each of the three perpendicular crystallographic axes," we would write $^{[111]}3$ (Figure 12.21b). Next, the 2-fold axes that bisect *a* and *b* would be written $^{[110]}2$

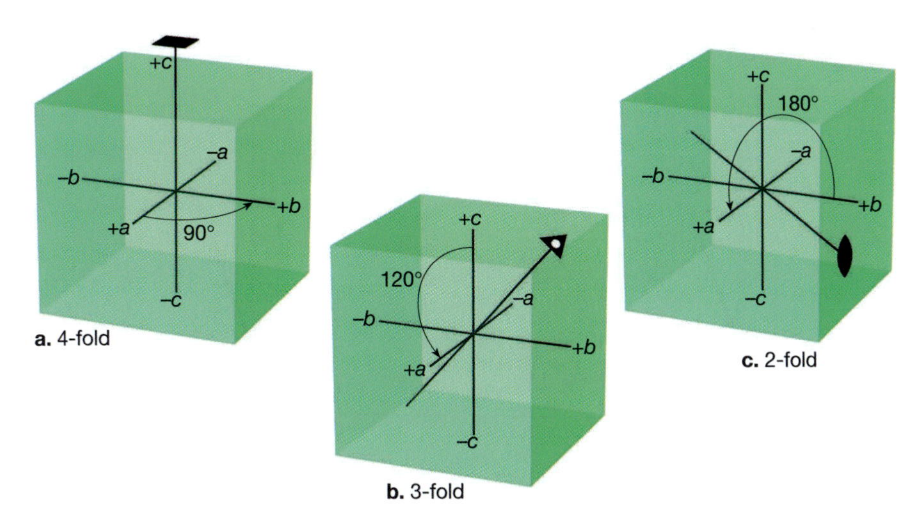

Figure 12.20. Examples of how the 4-fold, 3-fold, and 2-fold axes "fit" to the isometric crystal system axis set. ***a.*** *The 4-fold axes parallel to one crystallographic axes (e.g., c) would rotate +a into +b. Because a is equivalent to b, the symmetry operation "fits."* ***b.*** *The 3-fold axis bisecting a, b, and c would rotate +a to +b, +b to +c, and +c to +a. Because a, b, and c are equivalent, this symmetry operation "fits" the isometric axis set.* ***c.*** *Any of the 2-fold axes would rotate the positive end of a crystallographic axis to its negative end.*

(Figure 12.21b). Finally, recall that we define a plane by its normal (i.e., a line that is perpendicular to it). Thus "a mirror plane perpendicular to the b axis" would be written $^{[010]}m$ (Figure 12.21b).

Tetragonal case. Figure 12.22a shows the basis vector set for the tetragonal crystal system with $a = b \neq c$ and $\alpha = \beta = \gamma = 90°$. In this system, the 4-fold axis is parallel to c, which we write as $^{[001]}4$. If we operate on the a axis (Figure 12.22a) with a $^{[001]}4$, we see that the $+a$ axis goes to $+b$; this means that the a and b axes are equivalent. Thus the symmetry fixes the equivalences of a and b as well as the 90° for α, β, and γ. We also see in Figure 12.22a that the $^{[111]}3$ of the isometric system is lost in the tetragonal system because c is no longer equivalent to a and b. Objects with these point symmetries will belong to the tetragonal crystal systems. In Figure 12.22b, we see that there must be a $^{[100]}2$ that would relate $+c$ to $-c$ by a 180° rotation. If a is a 2-fold axis, then so is b (because of the equivalence of a and b).

So far, the block in Figure 12.22a has a $^{[001]}4$, $^{[100]}2$, and $^{[010]}2$. Do you see any other possible 2-fold axes? If you had a block of wood in the shape shown in Figure 12.22a, you can twirl it around [110] and indeed see that it is a 2-fold axis. If there is a $^{[110]}2$ axis, then the $^{[001]}4$ would create a $^{[\bar{1}10]}2$ (here the $\bar{1}$ means "along the $-a$ direction). At this point we now have one 4-fold and four 2-fold axes. Are we done? There is a mirror plane perpendicular to each of the five rotation axes; thus we have a $^{[001]}m$, $^{[100]}m$, $^{[010]}m$, $^{[110]}m$, and $^{[\bar{1}10]}m$. These elements combine to make the $4/m\ 2/m\ 2/m$ crystal class, which has the highest symmetry of the seven crystal classes in the tetragonal system.

To summarize, Figure 12.22c shows these point group symmetries on a stereographic plot. Figure 12.23 shows how the 4- and 2-fold axes for this point group fit to the symmetry of the point group by relating the symmetrically equivalent crystallographic axes to one other.

Hexagonal case. Figure 12.24a shows our axis set and wooden block that would represent the point groups with a single 3, $\bar{3}$, 6, or $\bar{6}$ axis, all of which are in the hexagonal crystal system. The major symmetry element for this class is a $^{[001]}6$; a and b are equivalent and differ from c. The crystal system derives its name from the $^{[001]}6$. How do these two conditions "fit" together? Because the symmetry in the $^{[001]}6$ relates a to b, then γ must be some multiple of 60°. Generally, γ is chosen to be 120° (Figure 12.24a), although it could have just as easily been chosen to be 60°. So the γ angle is the major difference between the tetragonal and hexagonal systems.

What other types of symmetry elements do you see in the block (Figure 12.24a)? Because there is a $^{[100]}2$ axis, there must also be a $^{[010]}2$ at 120° away, which is the γ angle. The 2-fold axes relate the $+c$ and $-c$ ends of the $^{[001]}6$ (Figure 12.25b). There are also 2-fold axes every 60° perpendicular to the $^{[001]}6$ that are produced by the inherit symmetry of the $^{[001]}6$. Thus there would be 360/60 2-fold axes, or 6 of them, and their orientations would be (going counterclockwise from a): $^{[100]}2$, $^{[210]}2$, $^{[110]}2$, $^{[120]}2$, $^{[\bar{1}10]}2$, and $^{[\bar{1}20]}2$. Now for the first time we see "2's" in our vector notation. They just mean that the vector has twice the component in the "2" direction; thus, we could write [210] as $2a$, $1b$, $0c$. Finally, there is a mirror plane

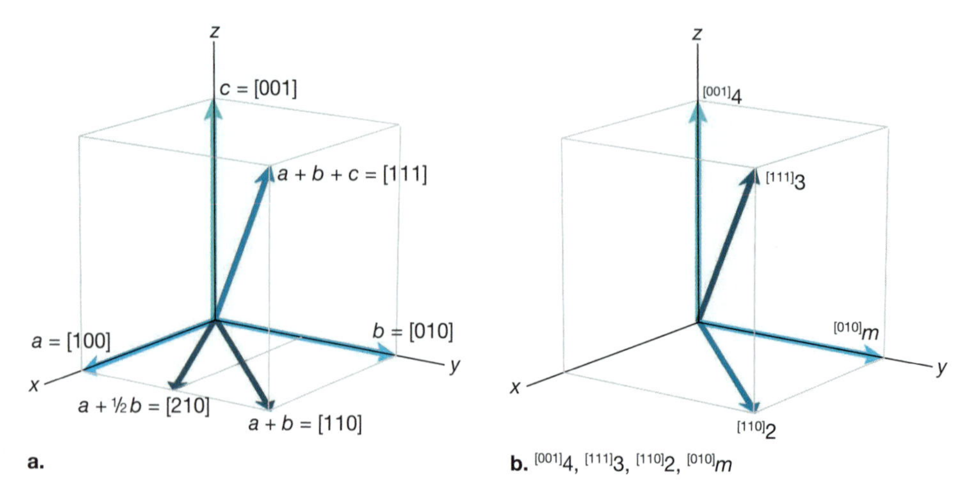

***Figure 12.21.** Naming directions (e.g., crystallographic axes) in minerals. **a.** We can assign numbers to directions by using vector notation. For instance, we use the notation [uvw] where* u *is the projection along* a, v *is the projection along* b, *and* w *is the projection along* c. *Thus* a = *[100],* b = *[010], and* c = *[001] is shown above, along with some other directions.* ***b.** This example shows the notation used to denote the direction of one of the 4-fold axes, one 3-fold axis, one 2-fold axis, and one mirror plane for the example shown in Figure 12.20.*

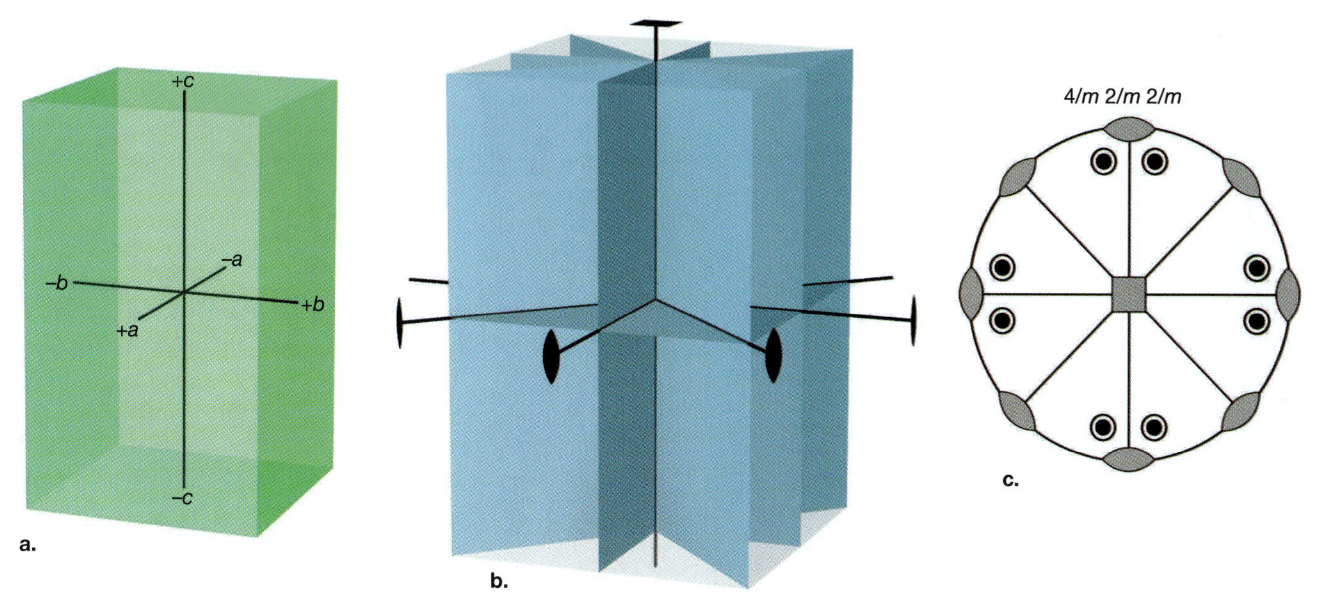

Figure 12.22. *Geometric representation of the tetragonal crystal system and its symmetry operations in perspective and stereographic views.* ***a.*** *The axis set for the tetragonal crystal system,* a = b ≠ c *and* $\alpha = \beta = \gamma = 90°$*, inscribed inside of a block of wood.* ***b.*** *The block/axis set combination from Figure 12.22a, with symmetry added. There are five rotation axes: a,* $^{[001]}4$*, and four 2-fold axes,* $^{[100]}2$*,* $^{[010]}2$*,* $^{[110]}2$*, and* $^{[\bar{1}10]}2$*. There are also five mirror planes,* $^{[001]}$m*,* $^{[100]}$m*,* $^{[010]}$m*,* $^{[110]}$m*, and* $^{[\bar{1}10]}$m*, all of which are perpendicular to the five rotation axes.* ***c.*** *The stereographic projection shows the orientation of the point symmetry operations as well as a set of points that their symmetry would create.*

perpendicular to each of the seven rotation axes, so we have: $^{[001]}m$, $^{[100]}m$, $^{[210]}m$, $^{[110]}m$, $^{[120]}m$, $^{[110]}m$, and $^{[\bar{1}20]}m$. Collectively, these symmetry elements make up the 6/*m* 2/*m* 2/*m* crystal class, which has the highest symmetry of the twelve classes in the hexagonal crystal system. Figure 12.25 shows how the 6- and 2-fold axes for this point group fit to the symmetry of the point group by relating the symmetrically equivalent crystallographic axes to each other.

Before we move on to the next crystal system, compare the hexagonal system in Figure 12.24 with the tetragonal system in Figure 12.22. They really are very much alike. The difference in the γ angle is the difference between the 6-fold rotation angle of 60° and the 4-fold rotation angle of 90°. What makes them similar is the equivalence of *a* and *b*, which will cause the physical properties for both systems to be the same in the *a-b* plane, and allow them to differ in directions not in that plane.

Orthorhombic case. What happens when you keep the 90° angles, but do not allow any 4-fold axes? Effectively, we are removing the constraints placed on the cell parameters, so that we arrive at the case where $a \neq b \neq c$, with no required symmetry equivalence for these three directions. The crystal derives its name from the 90° angles between axes, which is called orthogonal. So where does this relaxation in symmetry lead?

Take a look at the wooden block in Figure 12.26a, and the inscribed orthorhombic axis set. Because the 3-D axis set is orthogonal, and the positive and negative ends of each axis are related to each other by a 180° rotation, there are many 2-fold axes. This can be seen in Figure 12.27. Thus +*a* is symmetrically related to –*a* by a

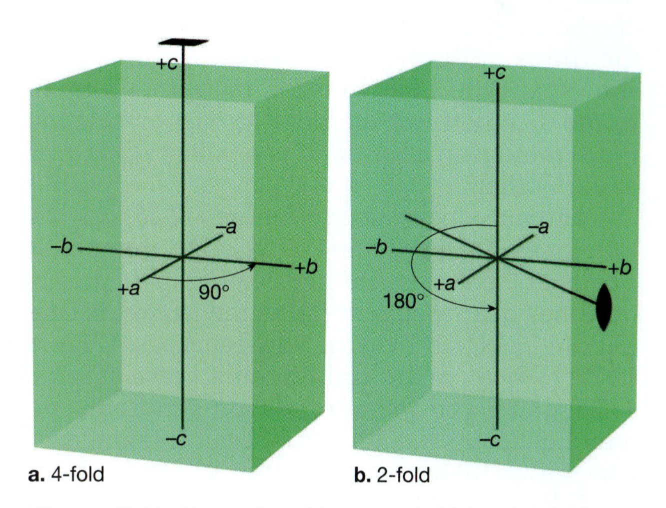

Figure 12.23. *Examples of how the 4-fold and 2-fold axes "fit" to the tetragonal crystal system axis set.* ***a.*** *The* $^{[001]}4$ *relates* +a *to* +b *with a 90° rotation; thus the axis set fits the crystal system in that* a *and* b *are symmetrically equivalent.* ***b.*** *The three 2-fold axes that are perpendicular to* c *would relate* +c *to* –c *with a 180° rotation as well as relating the positive and negative ends of* a *and* b.

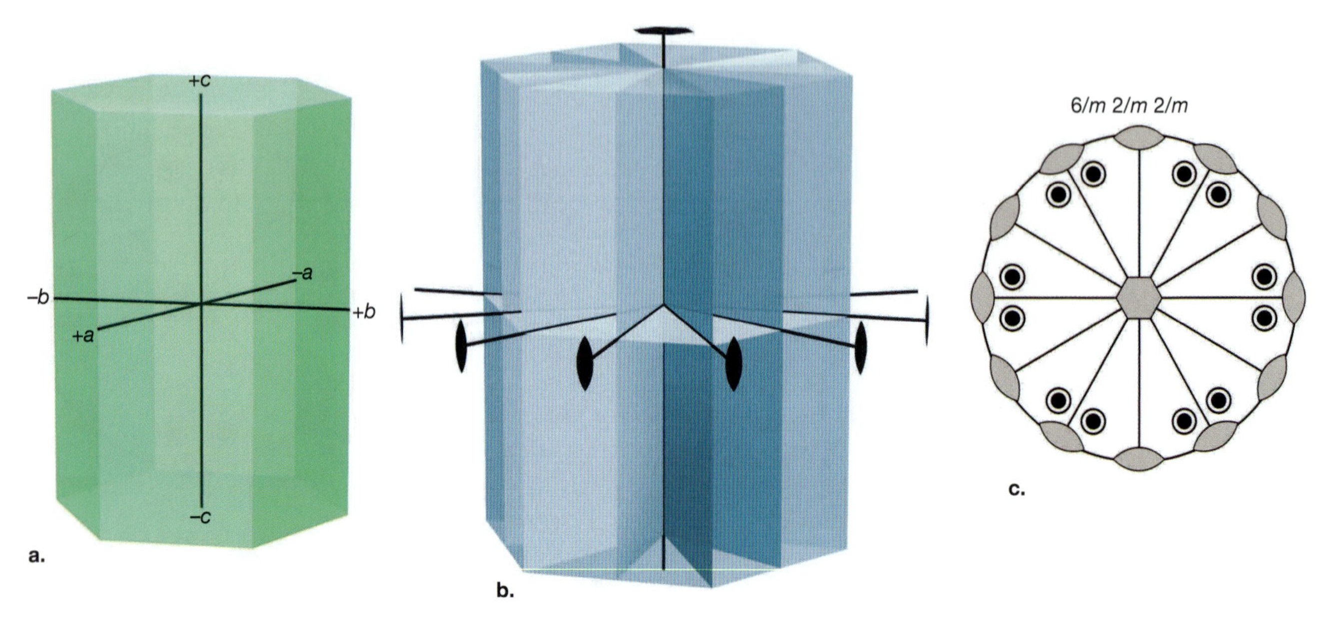

Figure 12.24. *Geometric representation of the hexagonal crystal system and its symmetry operations in perspective and stereographic view.* ***a.*** *A hexagonal block is shown with the hexagonal axis set (i.e.,* a = b ≠ c *and* α = β = 90° *and* γ = 120°) *inscribed. The difference between this axis set and that for the tetragonal case is the* γ *angle between* a *and* b. ***b.*** *The point symmetry elements have been added to the block / axis set from Figure 12.24a. There are one 6-fold axis, six 2-fold axes, and seven mirror planes normal to each of the rotation axes.* ***c.*** *A stereographic projection of the seven rotation axes and seven mirror planes in this point group, along with the 24 symmetrically generated points.*

180° rotation about *c*, so we have a [001]2. The same logic would produce the other two 2-fold axes: a [100]2 and a [010]2. These three mutually-perpendicular 2-fold axes impose the symmetry requirements of $\alpha = \beta = \gamma = 90°$, while not requiring equivalence of *a*, *b*, and *c*. There are also mirror planes perpendicular to the three 2-fold axes, giving us [100]*m*, [010]*m*, and [001]*m*. This 2/*m* 2/*m* 2/*m* crystal class has the highest symmetry of any of the three orthorhombic classes. Figure 12.27 shows how the 2-fold axes for this point group fit to the symmetry of the point group by relating the symmetrical equivalent crystallographic axes to each other.

Compare this system (Figure 12.26) to the tetragonal system (Figure 12.22). The number of points in the corresponding stereographic projections has decreased from 48 in the isometric case (Figure 12.19c) to 16 in the tetragonal case (Figure 12.22c), to 8 in the orthorhombic case (Figure 12.26c). Can you see a correlation between symmetry elements (i.e., the number of rotation axes and mirror planes) and those little dots and circles? While you're looking backward, take a quick peek forward to the monoclinic case (Figure 12.28c) and the triclinic case (Figure 12.30c) to see if your prediction holds there.

Monoclinic case. Now we will relax the constraint on interaxial angles in addition to those on axis lengths. "Mono" means "one" and "clinic" will refer to "inclined," or a non-90° angle. So the monoclinic crystal system has one non-90° angle, while the other two are still required to be 90°. We

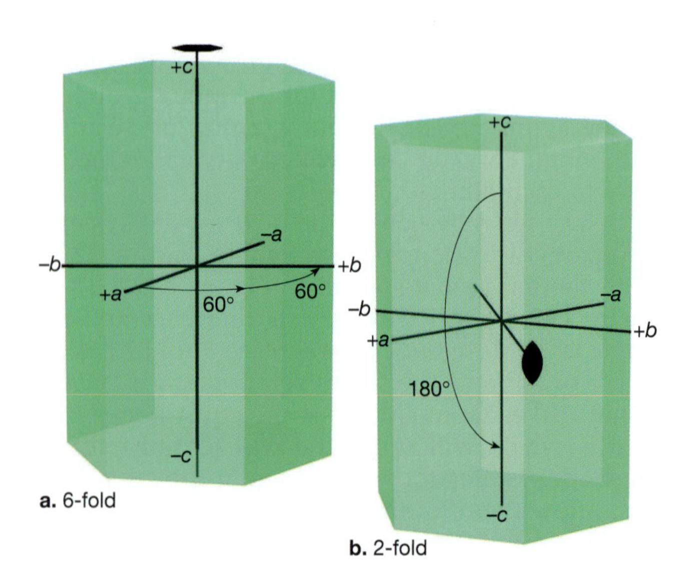

Figure 12.25. *Examples of how the 6-fold and 2-fold axes "fit" to the hexagonal crystal system axis set.* ***a.*** *The 6-fold axis relate* +a, *or [100], to [110] with a 60° rotation. The [110] would also be related to* +b, *or [010], by another 60° rotation. Thus* a, b, *and [110] are symmetrically equivalent by the operation of the 6-fold parallel to* c. ***b.*** *All of the 2-fold axes are perpendicular to the* c *axis so a 180° rotation on any of them would relate* +c *to* –c. *Any of these rotations would bring symmetrically equivalent directions in the* a-b *plane into coincidence.*

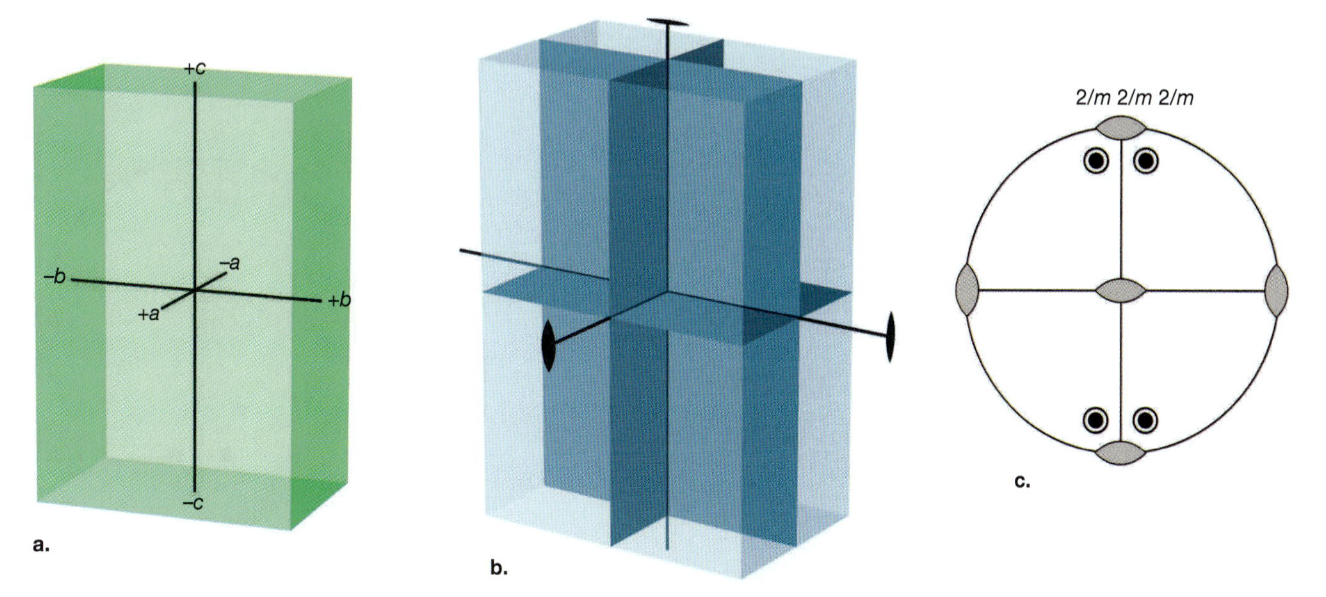

*Figure 12.26. Geometric representation of the orthorhombic crystal system and its symmetry operations in perspective and stereographic views. **a.** The crystallographic axis set for orthorhombic minerals (a ≠ b ≠ c and α = β = γ = 90°) inscribed inside a wooden block. **b.** The block from Figure 12.26a with three 2-fold axes coinciding with the three mutually-perpendicular crystallographic axes. As usual, a mirror plane is perpendicular to each of three 2-fold axes **c.** Stereographic projection of the six symmetry elements for this point group, as well as the eight points they would generate.*

generally assign the non-90° to β. So we use $\alpha = \gamma = 90°$ and $\beta \neq 90°$ to describe the interaxial angles for the monoclinic system.

Given these conditions, what type of symmetry operations could we "fit" into such a basic vector set? Look at the *a*, *b*, and *c* axes in Figure 12.28a, and see if any (or all) of them could be 2-fold axes. You'll find that in the monoclinic system, only *b* can be a 2-fold axis; a 180° rotation around *b* takes +*a* to −*a* and +*c* to −*c*. If you try a 180° rotation about *c*, you'll find that +*b* goes to −*b* but +*a* does not go to −*a*. This happens because *a* does not make a 90° angle with *c*. So [010]2 is our only rotation axis, and this is the defining characteristic shared by all three monoclinic crystal classes.

How about the mirror planes? As usual, there is a mirror plane perpendicular to the 2-fold, and we would denote it [010]*m*. There is also a center of symmetry (Figure 12.28c); this combination of elements gives rise to the 2/*m* crystal class (one of the three monoclinic crystal classes). In fact, all of the above examples contain a center of symmetry, which always arises when you have a mirror plane perpendicular to a rotation axis. Figure 12.29 shows how the 2-fold axis and mirror plane for this point group fit to the symmetry of the point group by relating the symmetrical equivalent crystallographic axes to each other.

Triclinic case. The triclinic case has the lowest symmetry, and thus the fewest constraints of all the crystal systems; it is often referred to as a per-

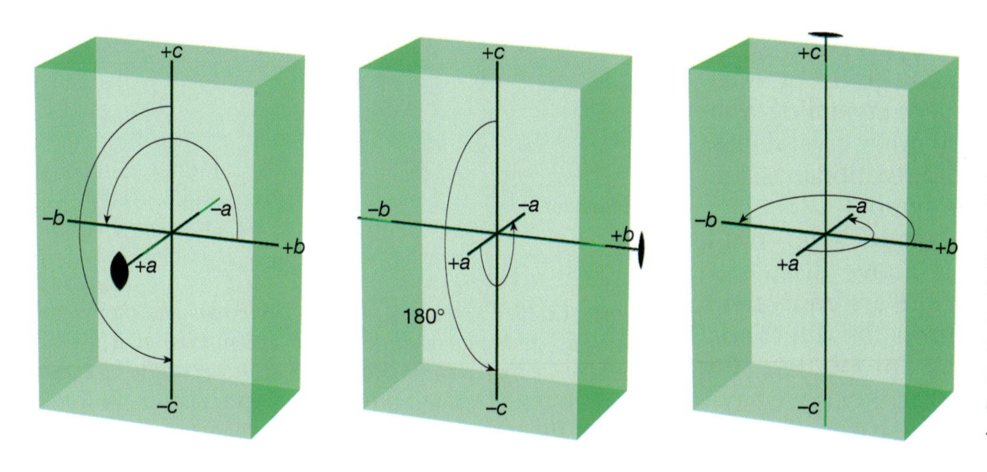

Figure 12.27. Examples of how the 2-fold axes "fit" to the orthorhombic crystal system axis set. Each of the three 2-fold axes would relate the + and − ends of the three crystallographic axes. Unlike the above systems, no two axes are symmetrically equivalent, but their + and − ends still are.

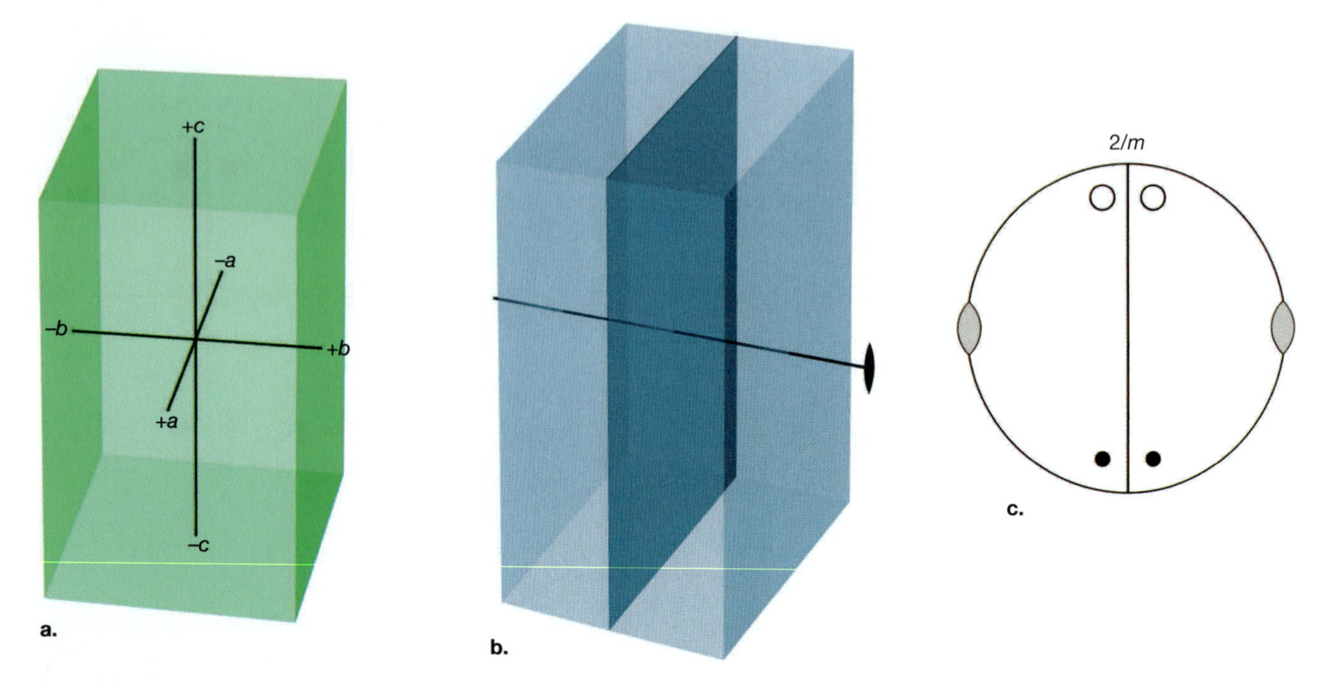

*Figure 12.28. Geometric representation of the monoclinic crystal system and its symmetry operations in perspective and stereographic views. **a.** The axis set for the monoclinic system ($\alpha \neq$ b $\neq$ c and $\alpha = \gamma = 90°$, $\beta \neq 90°$) inscribed into a block with fitting symmetry. **b.** The two point symmetry operations added to the block from Figure 12.28a. For the monoclinic system, there is one 2-fold axis parallel to* b *with a mirror plane normal (perpendicular) to* b. ***c.** A stereographic projection showing the two symmetry elements and the four symmetry-generated points.*

fectly general parallelopiped. As in the previous two systems, there is no equivalence of the crystallographic axes ($a \neq b \neq c$). As you may have guessed, "tri" means three, so for this system we have three non-90° angles. Following from the logic of the monoclinic case, no 2-fold axes can occur unless we have an axis set that contains some 90° relationships; and this system does not (Figures 12.30 and 12.31). So is there any type of symmetry that can "fit" this type of axis set? The stereographic plot in Figure 12.30c shows two points that are related by a center of symmetry. They relate the positive end of each axis to its negative end. That is the sole symmetry element for this crystal class, for which the symbol is simply $\bar{1}$: a center of symmetry. Figure 12.31 shows how the center of symmetry for this point group fits to the symmetry of the point group by relating the symmetrically equivalent crystallographic axes to each other. The only other crystal class in the triclinic system has 1-fold rotation axis, and is designated simply with the symbol 1.

One final point must be made here: You'll notice that somewhere in the above discussion we've started saying that a and b are equivalent, writing $a = b$. Symmetry controls the directional dependence of the physical properties. For instance, in an isometric mineral, when we say $a = b = c$, we really mean that all these directions are symmetry equivalent (e.g., their physical properties must be the same in all of these directions). When we say $a \neq b \neq c$ in an orthorhombic mineral, we are *not* saying that symmetry forces these directions to be different. Symmetry forces only

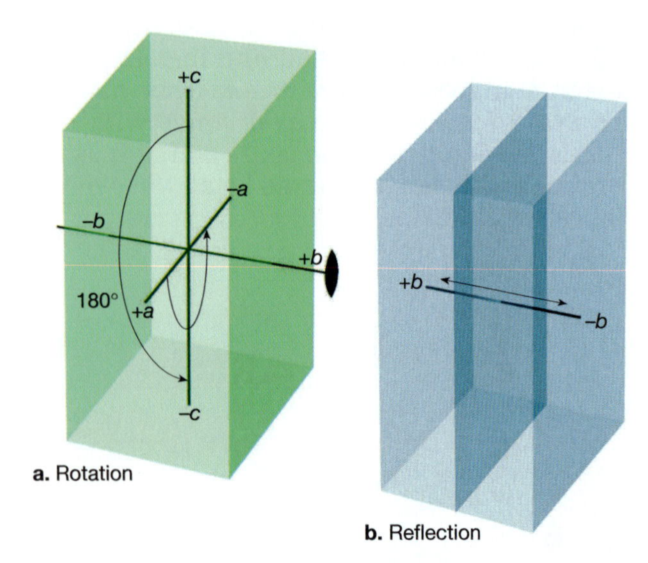

Figure 12.29. Example of how the 2-fold axis and the mirror plane "fit" to the monoclinic crystal system axis set. The $_{[010]}2$ *would relate the* +a *and* +c *to* –a *and* –c*, respectively. Likewise, the* $_{[010]}$m *would relate* +b *to* –b*. However, neither the 2-fold axis nor the mirror planes can occur in any other directions.*

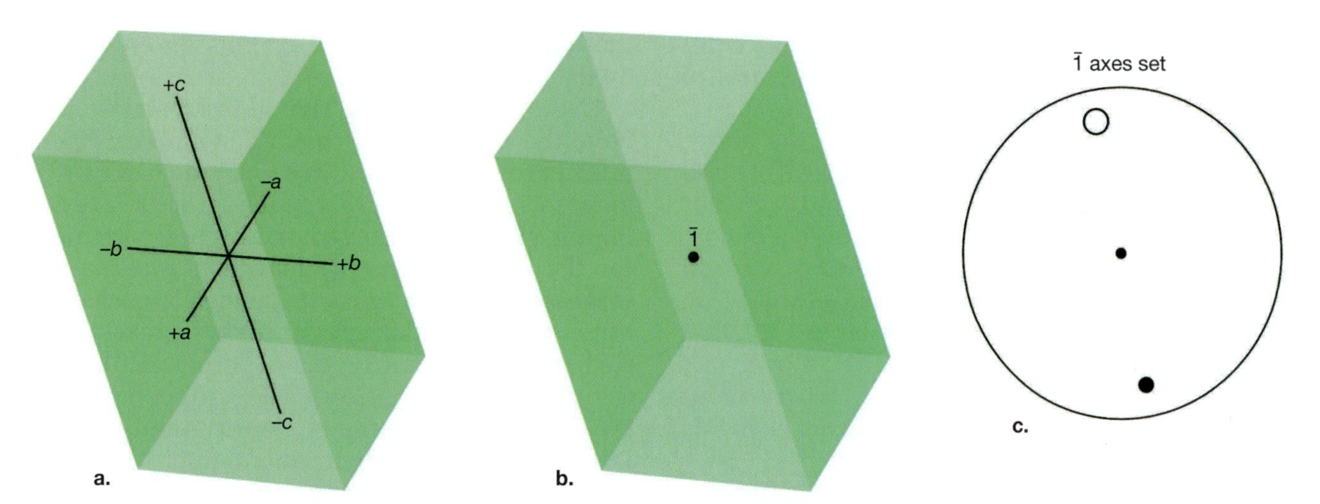

Figure 12.30. Geometric representation of the triclinic crystal system, and its symmetry operations in perspective and stereographic views. ***a.*** *The triclinic basis vector set* (a ≠ b ≠ c *and* α ≠ β ≠ γ ≠ 90°) *inscribed inside a block fitting its symmetry.* ***b.*** *The block from Figure 12.30a with a center of symmetry added; this is the only symmetry operation for this basis vector set.* ***c.*** *A stereographic projection showing the center of symmetry and the two points it would generate.*

equivalence, but not non-equivalence, although non-equivalence is usually the case. Symmetry will dictate, for instance, that the speed of light *must* be the same in all directions for a mineral belonging to the isometric systems, but does *not* dictate that the speed of light differs for a mineral in the orthorhombic system; however, it usually does. Also, there is no reason why the lengths of *a*, *b*, and *c* cannot be the same in an orthorhombic minerals, however they *must* be the same in an isometric one. Typically we write the "=" or "≠" sign to relate *a*, *b*, and *c* and let it go with that. But we're really talking about symmetry equivalence or non-equivalence. In the case of equivalence, all the directional physical properties (e.g., speed of light, hardness, color, etc.) *must* be the same, while in the case of non-equivalence they can (and usually do) vary, although they are not required to on the basis of symmetry.

Naming Point Groups

In the previous sections, we used the proper terms to indicate the point groups we were discussing, but left the explanation of these terms until now. In each case, we highlighted the point group with the highest symmetry (i.e., the most symmetry operations). Point group names are based on the types and orientations of the different point group operations, and each crystal system has a different naming scheme, which may at first seem like a lousy way to create a nomenclature. However, it will become apparent during our discussion that the terminology can be useful. It allows you to look at a point group symbol and tell three things: (1) the crystal system, (2) the types of symmetry operations, and (3) their orientations.

In this section, we will deliberately reverse the order of discussion, in order to start with low symmetry and build upon it. Thus we begin with the single symmetry element for the triclinic case and build up to the 48 elements of the isometric crystal system. Table 12.4 gives a preview of the crystal systems and associated point groups.

Triclinic case. Naming the directions for the triclinic case is trivial because there are none—the sole symmetry element for our triclinic example (Figure 12.31) was a center of symmetry, or a point in space. Because a point does not have a direction, we don't need to worry about naming its direction. So if we collect the symmetry ele-

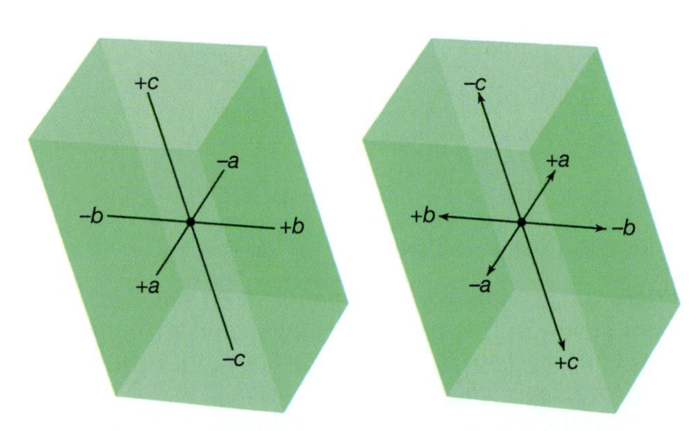

Figure 12.31. Example of how the center of symmetry in the triclinic crystal system axis set fits. The center of symmetry would relate the + and – end of each axis; and fit this lowest-symmetry basis vector set.

ments for the triclinic case we have only a center of symmetry, which is indicated by a $\bar{1}$ axis of rotation for the $\bar{1}$ point group.

Monoclinic case. Referring back to the above monoclinic case (Figure 12.28), we see we have two symmetry operations: a 2-fold axis and a mirror plane. The 2-fold axis is parallel to *b*, so we denote it as a $^{[010]}2$. The mirror plane is in the *a-c* plane with its normal coinciding with *b*, so we call it a $^{[010]}m$. When a mirror plane is perpendicular to a rotation axis, we write it as the $2/m$ (read it "2 over *m*") point group. We don't really need to worry about stating the direction for these symmetry elements because they are both in the *b* direction.

However, there is a slight historical complication. In the past, some researchers assigned these symmetry elements to the *c* direction, which in turn caused the γ angle be non-90° and the β angle to be 90° (Figure 12.32). When the *c* axis is assigned the symmetry direction, it is termed the **first setting** and when *b* is it is termed the **second setting**. This is mentioned here for two reasons: (1) at some point you might run into a monoclinic mineral in the first setting, which is unlikely unless you are really using old literature, and (2) more importantly, to show you that decisions about how to assign things change with time. As we'll see when we discuss the orthorhombic, tetragonal, and hexagonal systems, it would have been nice if we still used the first setting for the monoclinic system, because the "*c*" axis will be

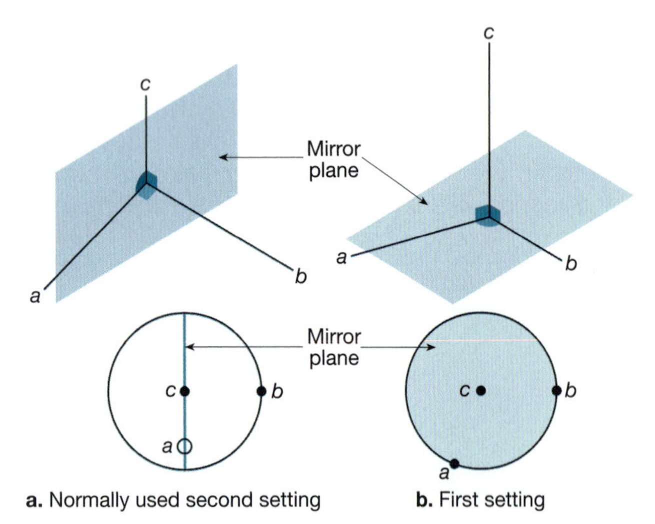

Figure 12.32. Naming convention used for the monoclinic point groups. ***a.*** *The commonly used second-setting for the monoclinic system. In this case the symmetry elements are oriented along* b, *thus* $\alpha = \gamma = 90°$, *while* $\beta \neq 90°$. ***b.*** *In the first setting, the symmetry elements are oriented along* c, *thus* $\alpha = \beta = 90°$, *while* $\gamma \neq 90°$. *The first setting, while making more sense (because in the remaining point groups the* c *axis has the highest symmetry elements assigned to it) is rarely used.*

the one that consistently exhibits different symmetry than the others.

Orthorhombic case. The other crystal systems have the possibility of having symmetry operations in more than one direction, which may often be associated with the crystallographic axes. In Figure 12.26, we showed the highest symmetry for the orthorhombic system, $2/m\ 2/m\ 2/m$. For this case, a 2-fold axis corresponded to each of the three perpendicular crystallographic axes. Also, each axis had a mirror plane perpendicular to it (Figure 12.26b and 12.26c). The symbol represents the collection of symmetry elements and provides information about their orientation. The directional naming scheme employed for the orthorhombic system assigns the *a*, *b*, and *c* directions respectively to the three distinct entries in the point group. So we could write $^{[100]}2/m\ ^{[010]}2/m\ ^{[001]}2/m$ to explicitly define the directions. This is never done; it is always assumed that the reader understands there are orientation directions assigned to each of three entries.

This relationship between the symbols and the orientation directions is critical to understanding point groups of lower symmetry, which can and do occur within each of the six crystal systems because of differences in internal composition. For example, there are two other point groups in the orthorhombic system: 222 and $mm2$. The 222 symbol means there are three 2-fold axes that coincide with the three perpendicular crystallographic axes and $mm2$ means that there are mirror planes perpendicular to two of the axes and a 2-fold axis parallel to the third axis. We could expand $mm2$ to be $^{[100]}m\ ^{[010]}m\ ^{[001]}2$. Table 12.5 gives the relationships between the symmetry directions and crystallographic axes and Figure 12.33 shows a graphical representation of them. Now you should see why we need to designate the directions for these three entries. If there is a difference in the entries for the symbol, the *c* axis will be the one to differ, and it will be the axis of highest symmetry for the remainder of the crystal systems.

Tetragonal case. For the tetragonal crystal system, the action revolves around the sole 4-fold axis. The 4-fold axis coincides with the *c* crystallographic axis, so we could write it as $^{[001]}4$. Also, for the example given above there was a $^{[001]}m$ (i.e., a mirror plane perpendicular to *c* as in Figure 12.22b). The other symmetry directions are all perpendicular to the [001] direction. Thus the symmetry elements in Figure 12.22b resulted in the point group $4/m\ 2/m\ 2/m$. In this case, the first entry coincides with the *c* axis (for the orthorhombic case it was the last entry that coincided with *c*; see Figure 12.33 and Table 12.5). The change in the order was done (I think) to emphasize the

Table 12.5. Method and logic of assigning symmetry operations in the point group symbols listed by crystal system.
triclinic: only one possible symmetry element (1 or $\bar{1}$), thus only one symmetry element in name
monoclinic: the single symmetry element coincides with the *b* crystallographic direction, or [010]
orthorhombic: named for symmetry operations along the *a*, *b*, *c* directions, or [100], [010], [001], respectively
tetragonal: named for the symmetry operation along the *c*, *a* (= *b*), *a* + *b*, directions or [001], [100], [110], respectively
hexagonal: named for symmetry operation along the *c*, *a* (= *b*), 2*a* + *b* directions or [001], [100], [210], respectively
isometric: named for symmetry operation along the *c* (= *a* and *b*), *a* + *b* + *c*, *a* + *b* directions or [001], [111], [110], respectively

importance of the *c* axis, because it is the one with the highest symmetry. But what about the other two entries? Would it make sense for them to be, say the *a* and *b* directions? Probably not, since *a* and *b* are equivalent in this example.

So let's again pick a lower symmetry point group; in this case $\bar{4}2m$. As previously stated, the first entry is along [001] and the next entry is along [100], indicating that it is also along [010] because *a* and *b* are equivalent. Thus, in Figure 12.34 the 2-fold axis coincides with both *a* and *b*. Finally, to define the orientation of the mirror plane, we use its normal to the mirror plane (the one marked [110]), which projects midway between the *a* and *b* axis in Figure 12.34. This direction is also symmetrically-equivalent to the [$\bar{1}$10] direction that contains the other mirror planes. So we could explicitly write the point group $\bar{4}2m$ as $^{[001]}\bar{4}\ ^{[100]}2\ ^{[110]}m$. Table 12.5 gives the naming sequence and Figure 12.34 is a graphical representation for $\bar{4}2m$ point group.

Hexagonal case. The naming system for the hexagonal system is like that for the tetragonal case, with the exception that not only are the [100] and [010] directions symmetrically-equivalent, but so too is the [110] (i.e., saying [100], [010], and [110] are symmetrically equivalent would be like writing $a = b = a + b$). So returning to our above hexagonal example (Figure 12.24), there were one 6-fold axis parallel to *c*, six 2-fold axes normal to *c*, and a mirror plane normal to each of the rotation axes, which gave us the point group $6/m\ 2/m\ 2/m$. The sequence of naming is that [001] corresponds to the first entry, [100] (and [010] and [110]) all correspond to the second entry, and the last entry is the [210] and symmetrically-equivalent [120] direction, so we could write $^{[001]}6/m\ ^{[100]}2/m\ ^{[210]}2/m$. Table 12.5 gives the naming sequence and the orientations and Figure 12.35 shows them graphically for point group $\bar{6}m2$.

Isometric case. If you examine the cube in Figure 12.19 to see which crystallographic axes correspond to 4-fold axes; you'll notice that all of them do! So it would make little sense to assign the symmetry operations in our point group symbol the same way here as we did for the orthorhombic, tetragonal, or hexagonal systems above. For the isometric case, the first symbol in our point group will be for the symmetry operation that coincides with all of the crystallographic axes (i.e., [100], [010], [001]) because they are all symmetrically equivalent.

What about the second symbol? If you look back on the discussion of relating the crystallographic axes to the point symmetry operations you'll see the second symmetry operation we talked about was the 3-fold axes that equally bisect the *a*, *b*, and *c* directions. There are four of these symmetrically equivalent directions: [111], [$\bar{1}$11], [$\bar{1}\bar{1}$1], and [1$\bar{1}$1]. So the second term in this point group is along [111]. Finally, there were six 2-fold axes located midway between two of the crystallographic axes. For instance, the [110] direction would plot halfway between *a*, or [100], and *b*, [010]. So the last symbol for the isometric system is for the six symmetrically equivalent 2-fold directions: [110], [$\bar{1}$10], [101], [011], [$\bar{1}$01], [0$\bar{1}$1]. So we could collect all of the symmetry elements together form Figure 12.25b and write $4/m\ \bar{3}\ 2/m$ or explicitly write $^{[001]}4/m^{[111]}\bar{3}^{[110]}2/m$. Table 12.5 lists the naming sequence for this crystal system and a lower symmetry example, $\bar{4}3m$, as shown in Figure 12.36.

The 32 Point Groups: Organizing Collections of Point Groups into Crystal Systems

In the above section, we correlated each crystal system to a collection of point group operations,

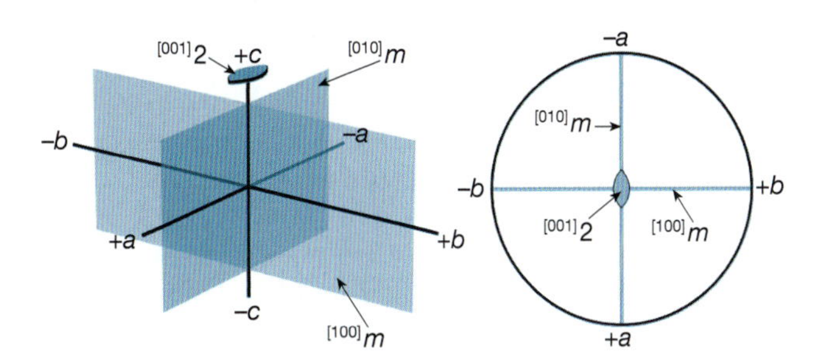

Figure 12.33. Naming convention used for the orthorhombic point groups. Orthorhombic point groups have three terms in their symbols. The first term corresponds to the a *axis, the second to* b, *and the last to* c. *All three axis directions are used because they are not symmetrical equivalent. Thus point group* mm2, *shown above, could be fully written as* $^{[100]}$m $^{[010]}$m $^{[001]}$2. *The axis set and symmetry elements are shown in perspective view and in stereographic projection.*

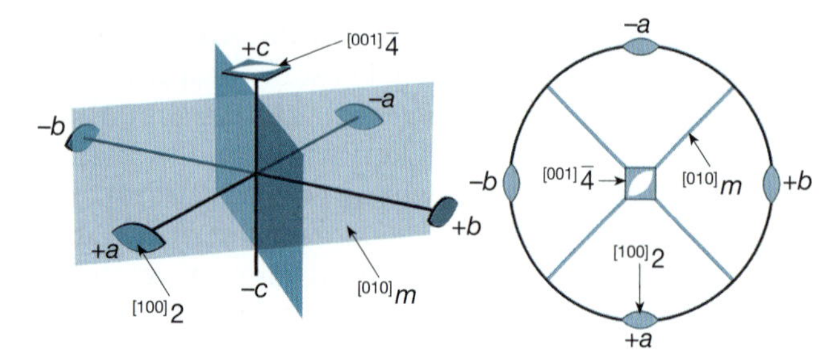

Figure 12.34. Naming convention used for the tetragonal point groups. For the tetragonal system, the first symbol represents the symmetry coinciding with c*, the second with* a *(which is equivalent to* b*), and the third along* a + b*. Thus the above example,* $\bar{4}$2m*, could be written as* [001]$\bar{4}$ [100]2 [110]m*. The axis set and symmetry elements are shown in perspective view and stereographic projection above.*

and talked about how they are named. So far we've discussed only a few of the 32 point groups. Now all that remains is to show how symmetry is lowered from the "high symmetry" point groups discussed above to form the remaining point groups.

But first, how can symmetry be lowered? Figure 12.37 is a non-mineralogical example of this. In this figure, it appears as if we are looking down a 2-fold axis with a vertical mirror plane. However, a 180° rotation of the photograph would show that we really are *not* looking at a 2-fold axis. In fact, what we thought was 2/*m* symmetry is actually just *m* symmetry. We could happily return 2/*m* symmetry to this image by replacing the top frown with a smile. Hopefully, after reading this your face follows the 2/*m* pattern and not the *m* pattern.

For the following 32 point groups we are going to use still images in the book and animate them on the MSA website. The animations are required to show how the symmetry operations change in 3-D. Students (or their professors) are encouraged to build 3-D models of each of the axes sets shown in these 3-D animations; holding an object in the hand and comparing it to the animation will help to "see" these symmetry.

Triclinic point groups. Now on to some more mineralogical examples of how symmetry can be lowered. Figure 12.38a redraws the triclinic axis set from Figure 12.30a except with spheres on the end of each axis. The spheres are the same size for each of the three axes. This setup would represent the triclinic point group $\bar{1}$. Now what would happen if the size of one of the balls changes, so that the one on the +*c* end is a different size than on the –*c* end (Figure 12.38b)? This would destroy the center of symmetry, leaving no symmetry at all. In this situation, crystallographers use a 1, which really means a 1-fold axis (i.e., a 360° rotation). So there are only two point groups in the triclinic system: $\bar{1}$ and 1. A stereographic representation of them is shown below each axis set in Figure 12.38, illustrating both their symmetry elements and the points that would be produced.

Monoclinic point groups. Let's again begin by taking the monoclinic axis set shown in Figure 12.28 and placing atoms of identical size at the end of each axis (Figure 12.39a); this would represent the 2/*m* case. Now if we change the size of the atom at +*b*, the mirror plane is destroyed, and we would create the new, lower symmetry point group 2 (Figure 12.39b). If we return to the 2/*m* case again, but this time change the size of the atom at +*a*, this would destroy the 2-fold axes, but preserve the mirror plane and create point group *m* (Figure 12.39c). Notice (in the stereographic projections in Figure 12.39) that 2/*m* had a center of symmetry that is lost in both 2 and *m*.

Orthorhombic point groups. Figure 12.40a shows the orthorhombic axes from Figure 12.26 with identical atoms added to the ends of the axes; this symmetry represents the point group 2/*m* 2/*m* 2/*m*. We can lower the symmetry of 2/*m* 2/*m* 2/*m* by changing the size of the atom at +*c*, causing the [100]2 and [010]2 to be destroyed, as well as [001]*m*. Thus, 2/*m* 2/*m* 2/*m* is converted to the lower symmetry point group *mm*2, and loses its center of symmetry (Figure 12.40b).

We'll have to use a bit more creativity to form the next point group, and many of the others that follow by adding other spheres to the axes set. Also, to see these you'll really need to refer to the MSA website for the animations. Let's bond two other atoms to the atoms at +*a* and –*a*. In a mineral, these might represent H atoms bonding to an oxygen atom. These added H's destroy the three mirror planes but maintain the three 2-fold axes,

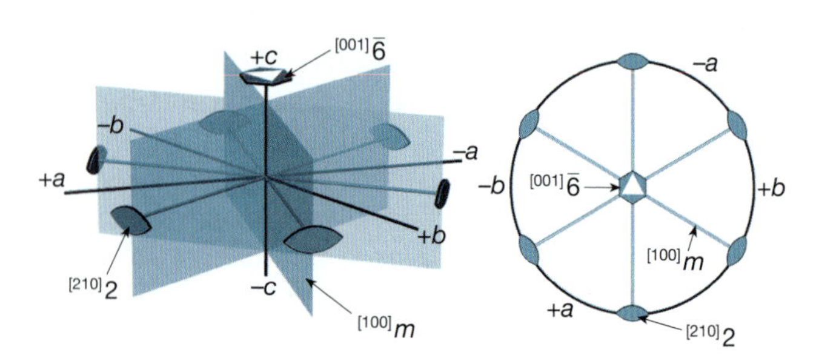

Figure 12.35. Naming convention used for the hexagonal point groups. For the hexagonal system, the first symbol represents the symmetry coinciding with c*, the second with* a *(which is equivalent to* b *and* a + b*), and the third along [210]. Thus, in the above example* $\bar{6}$m2 *could be written as* [001]$\bar{6}$ [100]m [210]2*. The axis set and symmetry elements are shown in perspective view and in stereographic projection.*

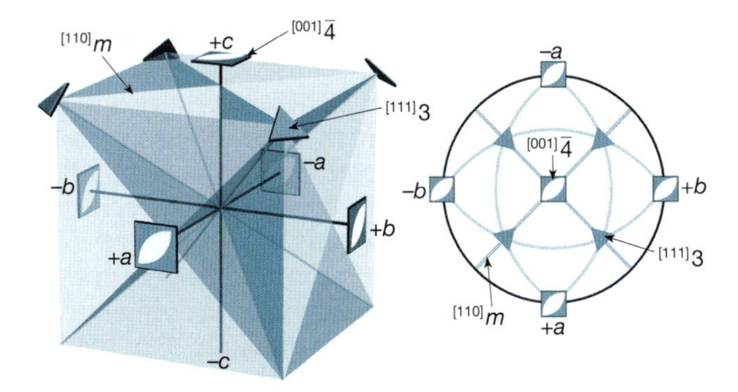

Figure 12.36. Naming convention used for the isometric point groups. For the isometric system, the first symbol represents the symmetry coinciding with c *(that is equivalent to* a *and* b*), the second to [111], and the third with [110]. Thus, in the above example, could be written as* $^{[001]}\bar{4}$ $^{[111]}3$ $^{[110]}$m*, which is normally shortened to* $\bar{4}3$m*. The axis set and symmetry elements are shown in perspective view and in stereographic projection.*

so the new point group is 222 (Figure 12.40c). The center of symmetry is also maintained for 222.

Tetragonal point groups. There are going to be a total of seven point groups in this system, and they can all be formed by starting with the highest symmetry tetragonal point group, $4/m$ $2/m$ $2/m$, and changing the atom clusters as done above. So we begin with the axis set in Figure 12.22 and again add atoms to the end of the axes (Figure 12.41a). If we change the size of the atom at $+c$, this destroys $^{[001]}m$, the 2-fold axes (i.e., $^{[100]}2$, $^{[110]}2$, $^{[010]}2$, and $^{[\bar{1}10]}2$), and the center of symmetry, while preserving the other mirror planes that pass through c (i.e., $^{[100]}m$, $^{[110]}m$, $^{[010]}m$, and $^{[\bar{1}10]}m$. This creates the new point group $4mm$ (Figure 12.41b).

To create the next point group, 422, we'll start with $4/m$ $2/m$ $2/m$ and remove all the mirror planes, which will remove the center of symmetry as well. Figure 12.41c shows how this was done. We added a cluster of two new atoms at $+a$. The symmetry will reproduce this atom pair at $+a$ to $+b$, $-a$, and $-b$, but in doing so, all of the mirror planes are destroyed while the rotation axes remain. This results in the 422 point group, in which no mirror planes exist; its stereographic representation is shown in Figure 12.41c.

The next point group, $4/m$, can be created from $4/m$ $2/m$ $2/m$ by adding a single small atom (Figure 12.41d) to each of the larger atoms in the a-b plane. In doing so, the $^{[001]}4$ and $^{[001]}m$ symmetry, as well as the center of symmetry, are conserved but the other mirror planes and the 2-fold axes are destroyed. In reality, we should call this point group $4/m$ 1 1, because we know that there are 1-fold axes in the [100] and [110] directions (the 2nd and 3rd entries for the tetragonal crystal systems). However the 1's are always left out of this (and all future) point group symbols, so we simply use $4/m$. These adopted "short cuts" in writing most certainly add to the confusion, so feel free to start a new trend and fully write these out!

Next, if we change the size of the atom at $+c$ in $4/m$ (Figure 12.41d), we create the new point group 411, which is usually written as the number 4 (Figure 12.41e). The remaining two point groups in this system involve $\bar{4}$ axes along c instead of along a 4-fold axis. By adding two atom clusters that differ in orientation (Figure 12.41f), we create the point group $\bar{4}2m$.

Finally, we can destroy the $^{[100]}2$ and the $^{[110]}m$ of $\bar{4}2m$ by removing one of the two black spheres at the end of a and b axes, which creates $\bar{4}11$ or just $\bar{4}$ (Figure 12.41g). It's worth a final look at Figure 12.41 to note how the number of points in the stereographic drawings has decreased as the symmetry has decreased from $4/m$ $2/m$ $2/m$ to $\bar{4}$.

Hexagonal point groups. This group is really broken into two separate groups: there will be seven point groups that have a 6 or $\bar{6}$ along c as their major symmetry element and five point groups that will have 3 or $\bar{3}$ along c as their major symmetry element. Sometimes (especially in Europe), these subgroups are split into the hexagonal and trigonal systems, respectively. More often, they are grouped together (as herein) because they have a common basic vector set (i.e., $a = b \neq c$ and $\alpha = \beta = 90°$ and $\gamma = 120°$).

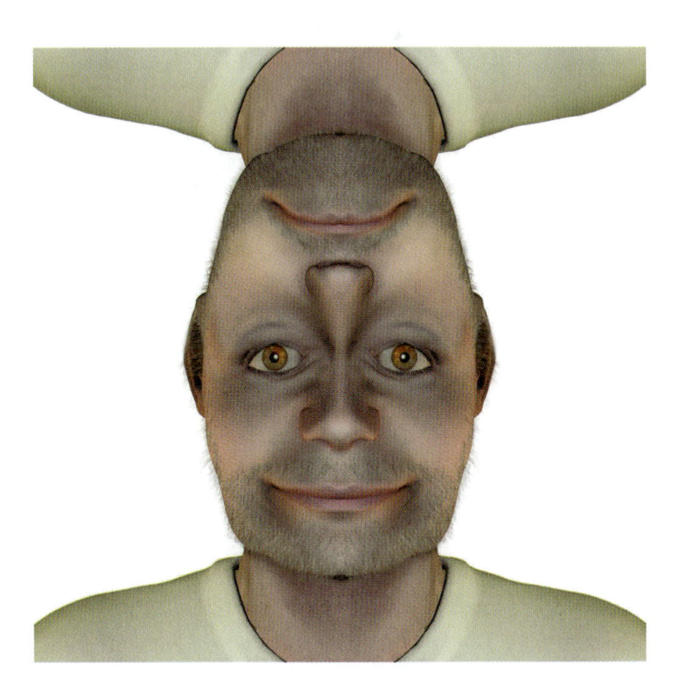

Figure 12.37. An example of a "unique" type of symmetry. At first, the image appears to have both a mirror plane and a 2-fold axis. In fact it only has a vertical mirror plane. Rotate the image 180° to see why.

The axis set for the point group 6/*m* 2/*m* 2/*m* is shown again in Figure 12.42a with the addition of single, same-sized "atoms" located at the end of the axes. Notice that we've now added the [110] direction to the sketch. We can form the first lower symmetry point group by changing the size of the atom at +*c*; this destroys all the 2-fold axes, the $^{[001]}m$, and the center of symmetry, but it preserves the other mirror planes and the $^{[001]}6$ to form the new point group 6*mm* (Figure 12.42b).

If we remove all the mirror planes from 6/*m* 2/*m* 2/*m*, the result is the point group 622 (Figure 12.42c), this is accomplished by placing atom clusters on the ends of the *a* and *b* axes. Continuing in the same way as for the tetragonal case, we create two point groups 6/*m* (Figure 12.42d) and 6 (Figure 12.42e) by adding a single atom on the end of *a* and *b* axes and in the *a-b* plane; thus, lowering the symmetry in the same way we did to form 4/*m* and then making the atom at +*c* smaller to form 4, except that we are using the hexagonal basis set and not the tetragonal basis set. The last two point groups here are based on $\bar{6}$ instead of 6-fold axes: $\bar{6}m2$ (Figure 12.42f) and $\bar{6}$ (Figure 12.42g) were created by changing the orientation of the atoms clusters at the end of axes in the *a-b* plane.

Next we will tackle the subset of the hexagonal crystal systems that are based on 3 or $\bar{3}$ axes parallel to [001]. There are three point groups based on 3-fold axes: 32, 3*m*, and 3, which could be written in the non-abbreviated versions as 321, 3*m*1, and 311. There are two point groups based on a $\bar{3}$ axis $\bar{3}2/m$ and $\bar{3}$, which could be written as $\bar{3}2/m$ 1 and $\bar{3}11$. To describe each of these in detail, we'll use the same basis vector set as was used for the hexagonal system (Figure 12.42), but there will be a set of three $\bar{3}$'s replacing the 6 or $\bar{6}$. Figure 12.43 shows these five point groups represented by the "stick axes" and their stereographic projections. See if you can figure out how each of these axis sets conforms to the point symmetry operations.

Isometric point groups. If you're keeping track, you know that we have gone through 27 of the 32 point groups so far; thus there must be five remaining point groups that belong to the isometric system. We have already discussed the 4/*m* $\bar{3}2/m$, the highest symmetry point in this or any system (Figure 12.19). What remains is to see how the symmetry of this point group can be lowered to create the remaining four point groups.

For this group, the four 3- or $\bar{3}$-fold axes really define the group; they are located along [111], [$\bar{1}$11], [$\bar{1}\bar{1}\bar{1}$], and [1$\bar{1}\bar{1}$]. The first term in the point group (i.e., the symmetry element oriented along *a*, *b*, and *c*) can be either a 4-fold or a 2-fold axis. There are three point groups: 4/*m* $\bar{3}2/m$, 43*m*, and 432, that have a 4-fold axis *along* the crystallographic axes and there are two point groups that have a 2-fold axis *parallel* to the crystallographic axes: 2/*m* and 23, which could also be written as 2/*m* $\bar{3}$1 and 231. However, note that all of these have 3-fold or $\bar{3}$ axes on their second position. Starting with 4/*m* $\bar{3}2/m$ (Figure 12.44a), we can reduce the symmetry to create $\bar{4}3/m$ (Figure 12.44b) and 432 (Figure 12.44c). Next we can further reduce the symmetry to create 2/*m* $\bar{3}$ (Figure 12.44d) and 23 (Figure 12.44e).

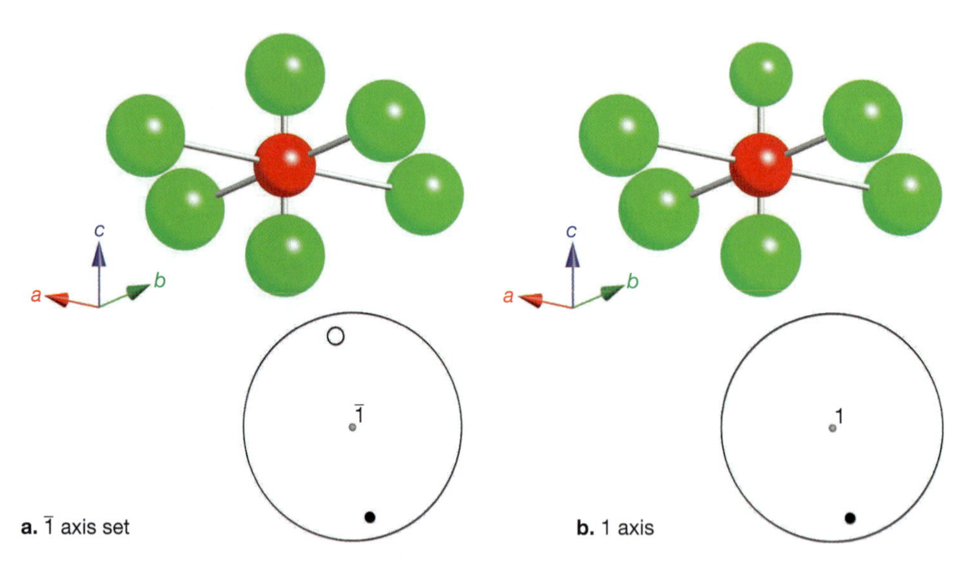

Figure 12.38. *The two triclinic point groups. For each point group, the upper image represents an axis set that conforms to the point group symmetry. Below is a stereographic projection of the symmetry elements and a set of points that would be created by them.* ***a.*** *$\bar{1}$: Shows a center of symmetry because the "+" end of each crystallographic axes can be related to its "–" by inversion through the center. The stereographic projection shows the two points created by a center of symmetry.* ***b.*** *1: The center of symmetry is lost in this point group by replacing the "atom" at +c with a smaller "atom." In the stereographic projection, only one point appears. It should be clear in the stereographic projection that the center of symmetry is destroyed in 1 as compared with $\bar{1}$.*

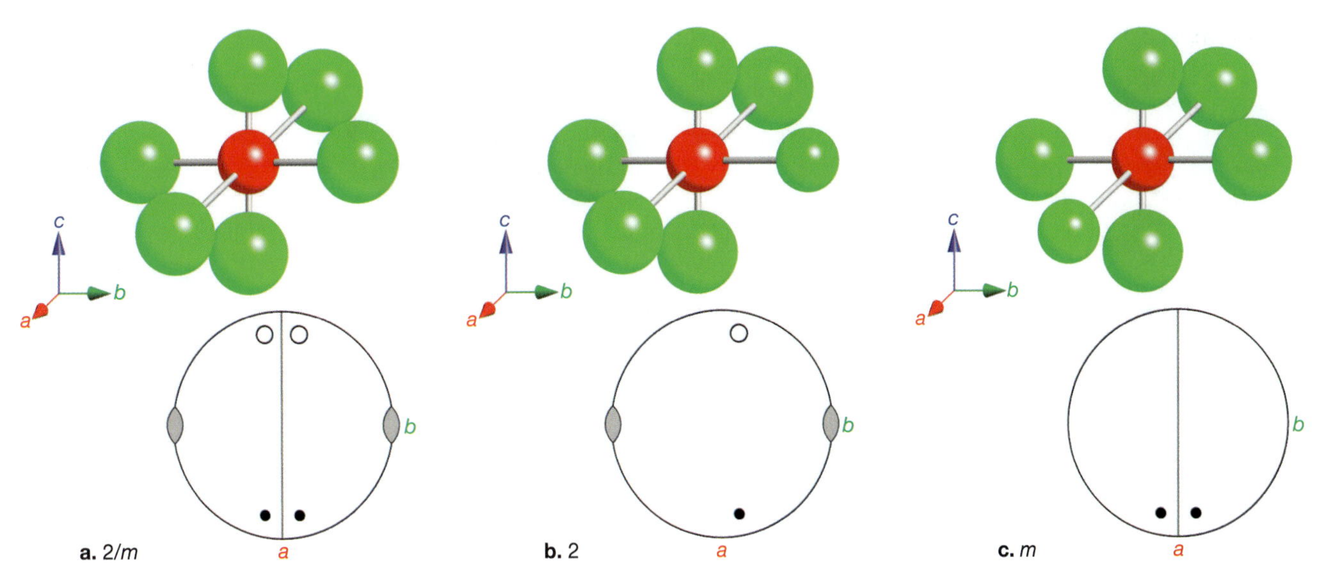

Figure 12.39. *The three monoclinic point groups. For each point group, the upper image represents an axis set that conforms to the point group symmetry. Below is a stereographic projection of the symmetry elements and a set of points that would be created by them.* ***a.*** 2/m*: This highest symmetry point group for the monoclinic system has a* [010]2 *and a* [010]m*. The stereographic plot shows the position of the two symmetry elements and four points conforming to* 2/m *symmetry, which also has a center of symmetry.* ***b.*** 2*: The mirror plane in* 2/m *is destroyed by changing the size of atom at* +b*. On the stereographic plot, this also destroys the center of symmetry.* ***c.*** m*: The 2-fold axis from* 2/m *is destroyed by changing the atom type at* +a*, which also destroys the center of symmetry.*

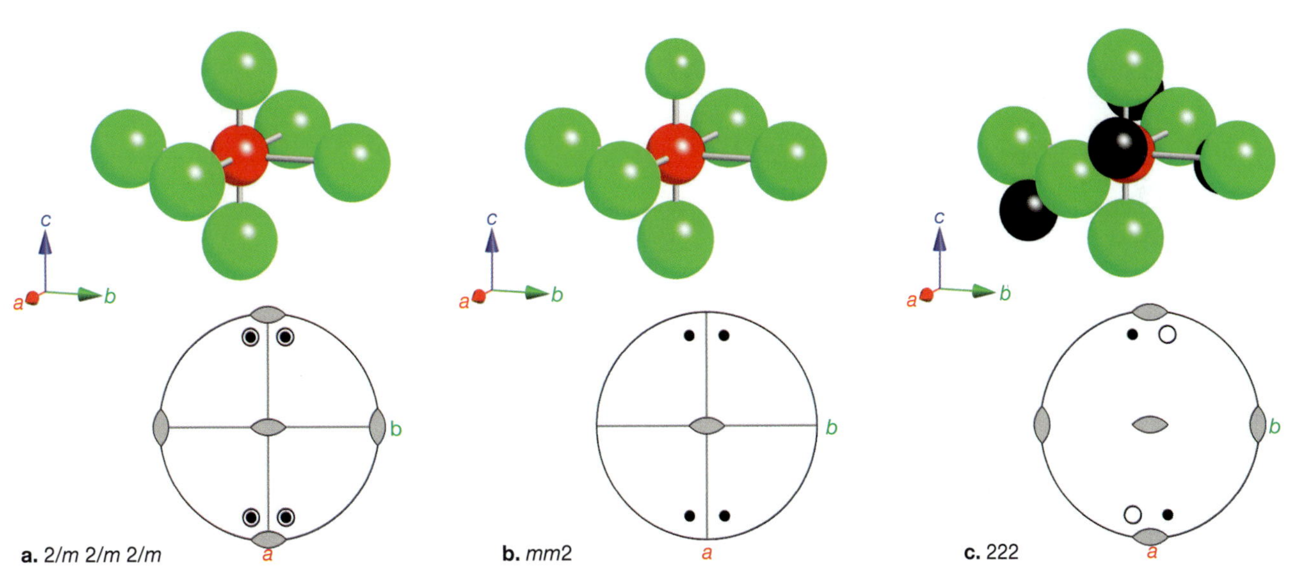

Figure 12.40. *The three orthorhombic point groups. For each point group, the upper image represents an axis set that conforms to the point group symmetry. Below is a stereographic projection of the symmetry elements and a set of points that would be created by them.* ***a.*** 2/m 2/m 2/m*: This is the orthorhombic point group with the highest symmetry. Three mutually-perpendicular 2-fold axes and mirror planes coincide with the crystallographic axes in the order* a, b, c*. The point group has a center of symmetry, and there are eight symmetry related points on the stereographic projection.* ***b.*** mm2*: By changing the size of the atom at* +c*, the 2-fold axes perpendicular to* c *(i.e.,* [100]2 *and* [010]2*) are lost, as well as the* [001]m*. However, the* [100]m*,* [010]m*, and* [001]2 *are preserved. A center of symmetry is absent, and there are only four symmetry generated points.* ***c.*** 222*: To create this point group, additional atoms have been added to the atoms at the ends of* +a *and* –a *(the smaller, darker ones). These atoms destroy all the mirror planes in* 2/m 2/m 2/m*, as well as the center of symmetry, but the three mutually-perpendicular 2-fold axes are preserved. This point group has the same number of generated points as* mm2*, but their placement should differ on the stereographic projection.*

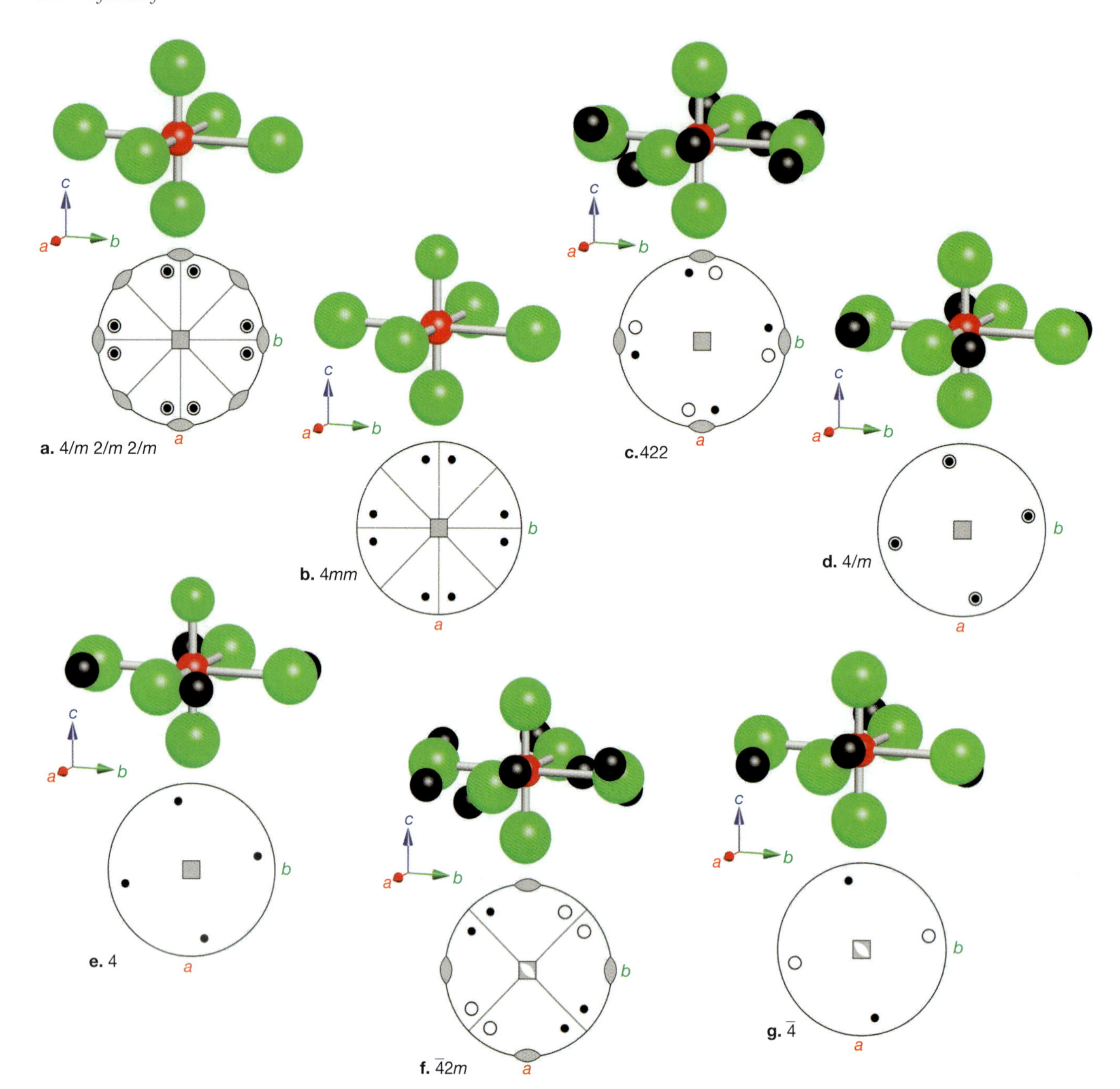

Figure 12.41. *The seven tetragonal point groups. For each point group, the upper image represents an axis set that conforms to the point group symmetry. Below is a stereographic projection of the symmetry elements and a set of points that would be created by them.* ***a.*** 4/m 2/m 2/m*: This is the point group in the tetragonal system with the highest symmetry. The symmetry elements in the point group correspond to the [001], [100], and [110] directions. This point groups also has a center of symmetry, and generates sixteen symmetrically related positions.* ***b.*** 4mm*: By reducing the size of the atom that plots at +c, the 2-fold axes perpendicular to* c *(i.e.,* $^{[100]}2$*,* $^{[110]}2$*,* $^{[010]}2$*, and* $^{[\bar{1}10]}2$*) are lost, as well as the* $^{[001]}$m *and the center of symmetry. This also reduces the number of symmetry points to eight all plotting on the upper hemisphere.* ***c.*** 422*: The symmetry of* 4/m 2/m 2/m *can be reduced to* 422 *by adding a couple of smaller atoms to the existing atoms at the ends of both* a *and* b*. This destroys all the mirror planes and the center of symmetry. Again creating eight points as in Figure 12.41b, except the points alternate between the upper and lower hemispheres.* ***d.*** 4/m*: To create this point group, only one atom is added at the ends of both* a *and* b*, and it is located in the* a-b *plane to preserve the* $^{[001]}$m*. The addition of these atoms destroys all the other symmetry elements in* 4/m 2/m 2/m *except for the* $^{[001]}4$*,* $^{[001]}$m*, and the center of symmetry. Again eight points are created, but this time four on the upper and four on the lower hemisphere.* ***e.*** 4*: Point group* 4 *can be created from* 4/m *by changing the size of the atom plotting at +c; this also destroys the center of symmetry. Four symmetry points at 90° all occurring on the upper hemisphere.* ***f.*** $\bar{4}2$m*: Point group* $\bar{4}2$m *is very similar to* 422 *above; two atoms are again added to the atoms that plot at the ends of* a *and* b*. However, the symmetry of the* $^{[001]}\bar{4}$ *differs from the* $^{[001]}4$ *in* 422*. Take a moment and compare these two symmetry types by looking at the stereographic plots.* ***g.*** $\bar{4}$*: By loss of the* $^{[100]}2$ *and* $^{[110]}$m *in* $\bar{4}2$m*, the two extra smaller dark atoms reduce to one to satisfy the sole symmetry element in this point group.*

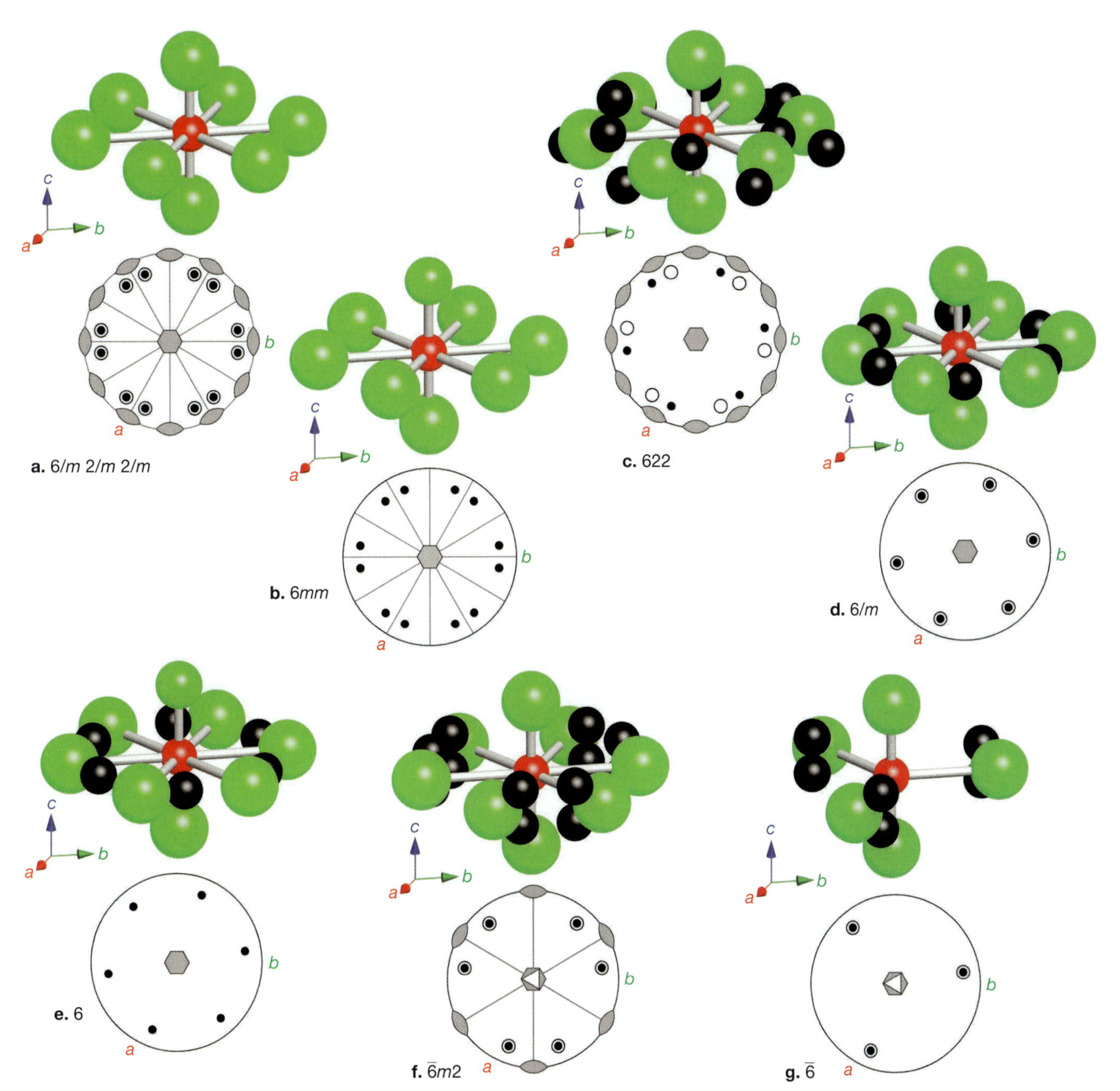

Figure 12.42. *The seven hexagonal point groups based on a 6- or* $\overline{6}$*-fold axis. For each point group, the upper image represents an axis set that conforms to the point group symmetry. Below is a stereographic projection of the symmetry elements and a set of points that would be created by them.* ***a.*** 6/m 2/m 2/m: *The highest symmetry point group for this system. The symmetry elements in the point group symbol coincide with the [001], [100], and [210] directions. This point group also has a center of symmetry, and creates 24 symmetry related points has shown in the stereographic projection.* ***b.*** 6mm: *This point group is formed from* 6/m 2/m 2/m *by simply changing the size of the atom that plots at* +c. *This destroys all the 2-fold axes perpendicular to* c *(because* +c *and* –c *are not longer symmetry related) as well as the* [001]m *and the center of symmetry; now only 12 points are created all plotting on the upper hemisphere.* ***c.*** 622: *To create this point group from* 6/m 2/m 2/m, *two atoms plotting above and below the* a-b *plane are added to the ends of the* a, b, *and [110] directions. This also destroys the center of symmetry. Note the similarity between this point group and 422.* ***d.*** 6/m: *This point group is created by just adding one extra atom in the* a-b *plane (note the similarity of this to 4/m above). The* 6/m *does contain a center of symmetry.* ***e.*** 6: *Point group 6 can be created from* 6/m *by changing the size of the atom at* +c, *thus destroying the* [001]m *because the atom sizes differ at* +c *and* –c. ***f.*** $\overline{6}$m2: *This point group is very similar to 622 above, except the operation of the* $\overline{6}$ *creates a different arrangement of the extra atoms than the 6-fold axis.* ***g.*** $\overline{6}$: *To create* $\overline{6}$ *from* $\overline{6}$m2, *the* [100]m *and* [210]2 *are removed. Once this has been done, there is a reduction of the extra atoms to arrive at* $\overline{6}$, *which has a center of symmetry.*

In summary, Table 12.4 lists the 32 point groups and their associated crystal systems. Table 12.6 shows the number of minerals occurring in each point group. The highest symmetry point group for each crystal system has the most minerals. Over one-third of the known minerals are monoclinic, so you should expect lots of monoclinic examples in this book, with orthorhombic and hexagonal coming in second and third. As a generalization, more complicated structures tend to have lower symmetry, and most minerals are somewhat complicated because they often contain several elements of vastly different sizes and charges (i.e., O^{2-}, Si^{4+}, Al^{3+}). These materials tend to have lower coordination numbers and form more complex structures. Simple compounds with few elements of like size and charge (i.e., Na^{1+}, K^{1+}, Cl^{1-}, F^{1-}) have higher symmetry because they generally have higher coordination numbers and form less complicated structures.

Determining crystal systems from point groups. One of the goals of this section is to be able to look at a point group and determine the crystal system. This statement could have gone on to say "without memorizing all 32 point groups and the six crystal systems to which they belong." The trick to determining the crystal system from the point group is based on the knowledge that each crystal system uses a different sequence to compose the point group, so what at first seemed like a lousy system might actually pay off now. Let's go back and review the differ-

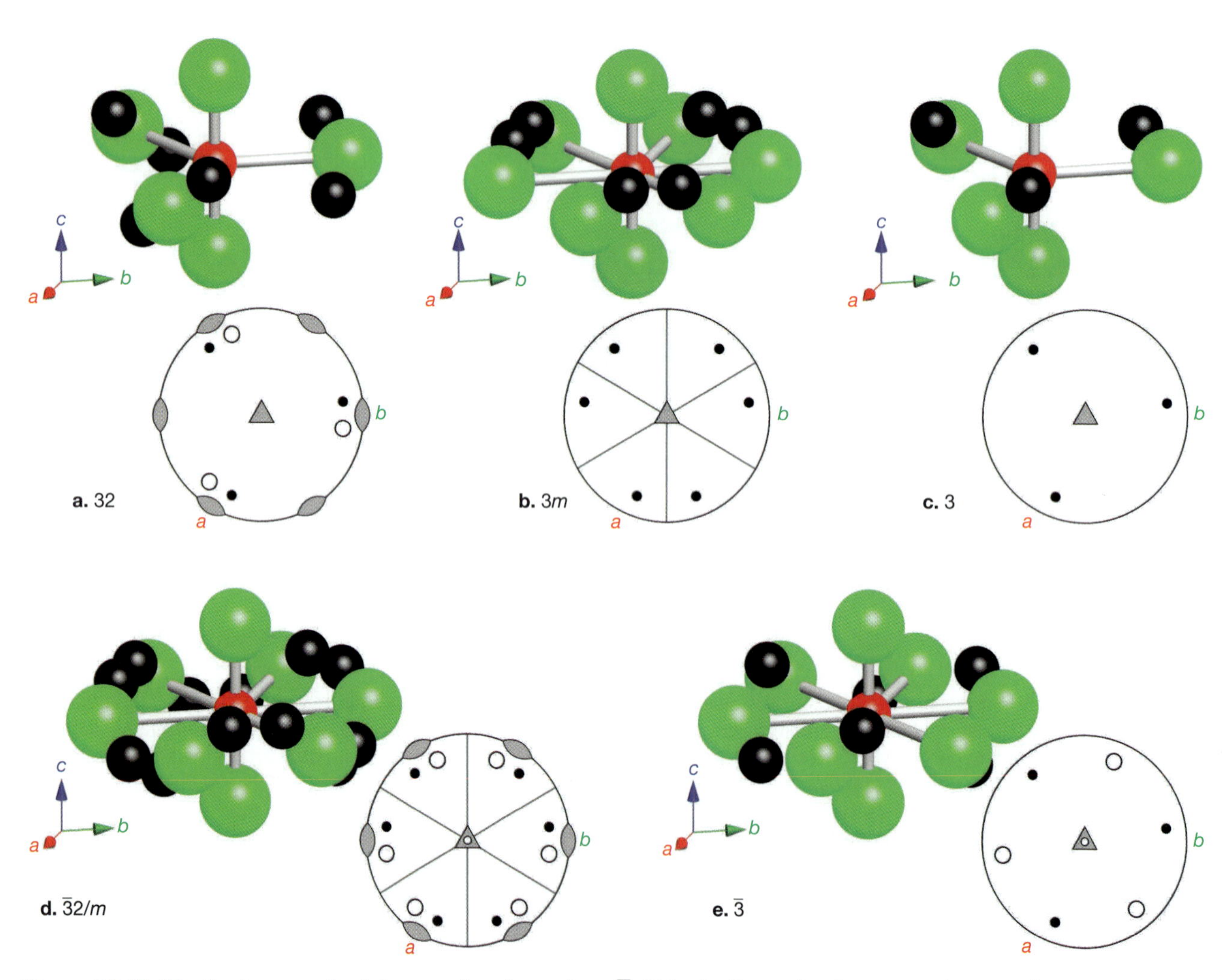

Figure 12.43. The five hexagonal point groups based on a 3- or $\bar{3}$-fold axis. For each point group, the upper image represents an axis set that conforms to the point group symmetry. Below is a stereographic projection of the symmetry elements and a set of points that would be created by them. ***a.*** *32: This point group could really be called 321, to represent the symmetry directions of a* $^{[001]}3$, $^{[100]}$m, *and* $^{[210]}1$. *The axis set has extra atoms added on the ends of* a *and* b *to represent this point group.* ***b.*** 3m: *The difference between* 3m *and 32 above is the replacement of the* $^{[100]}2$ *with a* $^{[100]}$m, *as shown in the perspective and stereographic plots.* ***c.*** *3: To form 3, only one extra atom is added at the ends of the* a *and* b *axes.* ***d.*** $\bar{3}2$/m: *In* $\bar{3}2$/m, *atoms are plotted every 60°, which is different from the above three point groups for which they plot at 120°. The replacement of the 3 with a* $\bar{3}$ *adds these extra atoms.* ***e.*** $\bar{3}$: *The* 2/m *symmetry of* $\bar{3}2$/m *is removed in this point group by a reduction in the size of the atoms added to the axis ends.*

ent naming methods in Table 12.5 and the different point groups given in Table 12.4. Notice that the triclinic and monoclinic systems only have one entry for symmetry; for triclinic systems it is either 1 or $\bar{1}$ and for monoclinic, either *m*, 2, or 2/*m* as the single entry. Next, the orthorhombic system has three entries of which only 2's and *m*'s occur, and only three point groups: 222, *mm*2, and 2/*m* 2/*m* 2/*m*. For the tetragonal system, the first entry will be either a 4 or $\bar{4}$; for the hexagonal system, either a 3, $\bar{3}$, 6, or $\bar{6}$.

This method could cause someone to conclude incorrectly that an isometric point group like 432 was actually tetragonal. The solution to this is to realize that the second entry is also important when the first character is either 4 or $\bar{4}$. The identifying characteristic of the isometric system is its four 3 or $\bar{3}$'s, and these are the second entry in the

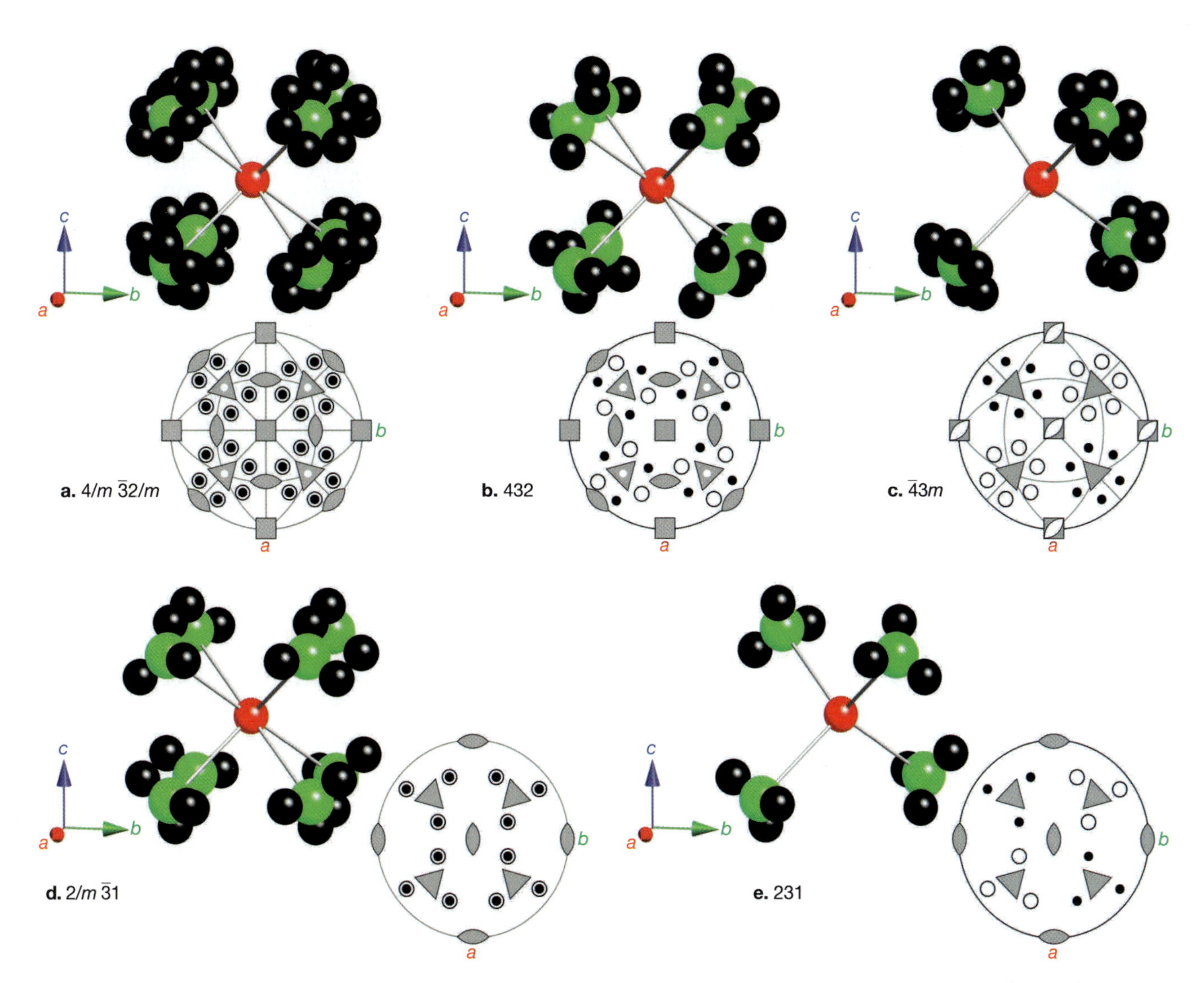

Figure 12.44. *The five isometric point groups. For each point group, the upper image represents an axis set that conforms to the point group symmetry. The axes sets are changed for this group as compared to the other five crystal systems; for this group the [111] set of directions (i.e., the four 3 and $\bar{3}$ directions) are used to show how the symmetry changes between the point groups. Below is a stereographic projection of the symmetry elements and a set of points that would be created by them.* ***a.*** 4/m $\bar{3}$2/m*: This is the highest symmetry isometric point group (and highest symmetry of all 32 point groups). For the isometric system, the first symmetry element in the point group coincides to the three symmetrical equivalent crystallographic axes (i.e.,* a, b, *and* c). *The second entry corresponds to the body diagonal (i.e., [111]) and the last to [110]. Note that 48 points symmetry related points are generated in the stereographic projection.* ***b.*** *432: Look carefully at the atom cluster and the stereographic projections and convince yourself there are no mirror planes for this point group as compared to* 4/m $\bar{3}$2/m *above. Note also that the 48 points from* 4/m $\bar{3}$ 2/m *are reduced to 24 by loss of the mirror planes.* ***c.*** $\bar{4}$3m*: The same atom set as used in Figure 12.44b is used to form* $\bar{4}$3m. *The only difference is that the set of point symmetry operations produces a very different atom cluster. Notice how this cluster is similar to a tetrahedron.* ***d.*** 2/m$\bar{3}$1*: This point group is somewhat similar to 432, but the* ${}_{[001]}4$ *is replaced with a* ${}_{[001]}2$/m, *the* ${}_{[111]}3$ *with a* ${}_{[111]}\bar{3}$, *and the* ${}_{[110]}2$ *with a 1-fold axis. Look carefully to see how these two differ.* ***e.*** *231: The last point group with the lowest symmetry appears more similar to* $\bar{4}$3m *but with fewer atoms, to reduce the symmetry.*

Table 12.6.
The number of minerals for each crystal system and point group as of 2001.
(Data from Dana's New Mineralogy (Gaines et al., 1997).

Crystal system	*Number*	%
triclinic	284	9.0
1	47	1.5
$\bar{1}$	238	7.5
monoclinic	1098	34.8
2	63	2.0
m	104	3.3
2/*m*	931	29.5
orthorhombic	592	18.8
222	105	3.3
*mm*2	41	1.3
2/*m*2/*m*2/*m*	446	14.1
tetragonal	272	8.6
4	4	0.1
$\bar{4}$	8	0.3
4/*m*	58	1.8
$\bar{4}2m$	37	1.2
4*mm*	2	0.1
422	19	0.6
4/*m*2/*m*2/*m*	144	4.6
hexagonal	580	18.4
3	27	0.9
$\bar{3}$	60	1.9
3*m*	78	2.5
32	31	1.0
$\bar{3}2/m$	151	4.8
6	2	0.1
$\bar{6}$	22	0.7
6/*m*	42	1.3
$\bar{6}m2$	19	0.6
6*mm*	21	0.7
622	24	0.8
6/*m*2/*m*2/*m*	103	3.3
isometric	328	10.4
23	20	0.6
$2/m\bar{3}$	48	1.5
43*m*	54	1.7
432	5	0.2
$4/m\bar{3}2/m$	201	6.4

point group symbol. So a 23 would be isometric and a 422 would be tetragonal.

One last bit of nomenclature is important before we depart from the story of crystal systems. Some books and papers will use a nomenclature other than the *a*, *b*, and *c* used herein to denote the crystallographic axes. They adopt the method that when two axes are equal (i.e., equivalent by symmetry), they rename them as a_1 and a_2 instead of *a* and *b*. In such as system, *a*, *b*, and *c* are used for triclinic, monoclinic, and orthorhombic, a_1, a_2, and *c* are used for tetragonal and hexagonal, and a_1, a_2, and a_3 are used for isometric. We have avoided this usage for two important reasons: (1) it seems to confuse students, and (2) it is not commonly used in the mineralogical research literature, so when students see *a*, *b*, and *c* later, they get confused. Often people who use this type of nomenclature also use a_1, a_2, a_3, and *c* for the hexagonal system (Figure 12.45). Not only is this confusing, it is wrong. The crystallographic axes really form a basis vector set that allows us to define the geometry of the mineral—either the external geometry of the faces and axes or the internal geometry of the atoms. A basis vector set needs only as many vectors as the dimension of space we are interested in. Extra vectors, like a_3 for the hexagonal case, could be written as a linear combination of the others (in this case $a_3 = -a_1 + a_2$), and are useless to use; they only cause (more) confusion for the students.

Unit Cells

The **unit cell** is defined as the literal, as well as figurative, basic building block for a mineral, or any crystalline material. Given the unit cell of a crystalline material, we can produce any size mineral by simply replicating the unit cell. For example, the unit cell of halite is shown in Figure 12.46a. The 5.63 Å size cube, and its contents, are all we need to know to replicate any size piece of halite we choose; Figure 12.46b "grows" the structure to eight unit cells. A piece of table salt that is approximately 0.5 mm on a side would contain 10,000,000 unit cells along one of its edges (Figure 12.46), so it would contain 10,000,000 x 10,000,000 x 10,000,000 (or 1 x 10^{21}) unit cells!

In biological terms, the unit cell can be thought of as a single cell. In chemistry, a unit cell is like a molecule, which is the smallest part of a compound that is replicated to produce larger portions. The unit cell is really just a 3-D object and is described by the lengths of three non-coplanar axes (i.e., the by-now familiar crystallographic axes: *a*, *b*, and *c*) and the angles between those axes (i.e., the likewise familiar interaxial angles, α, β, and γ). The axis lengths and interaxial angles are called the **cell parameters**, because they define the unit cell. At this point we are only concerned with the geometry of the unit cell (i.e., its 3-D shape). The symmetry of the unit cell (i.e., its point group, discussed above and its space group, discussed below) and the content of the unit cell (i.e., the atoms present and their positions) are also important and considered later in this chapter and in Chapters 13 and 14.

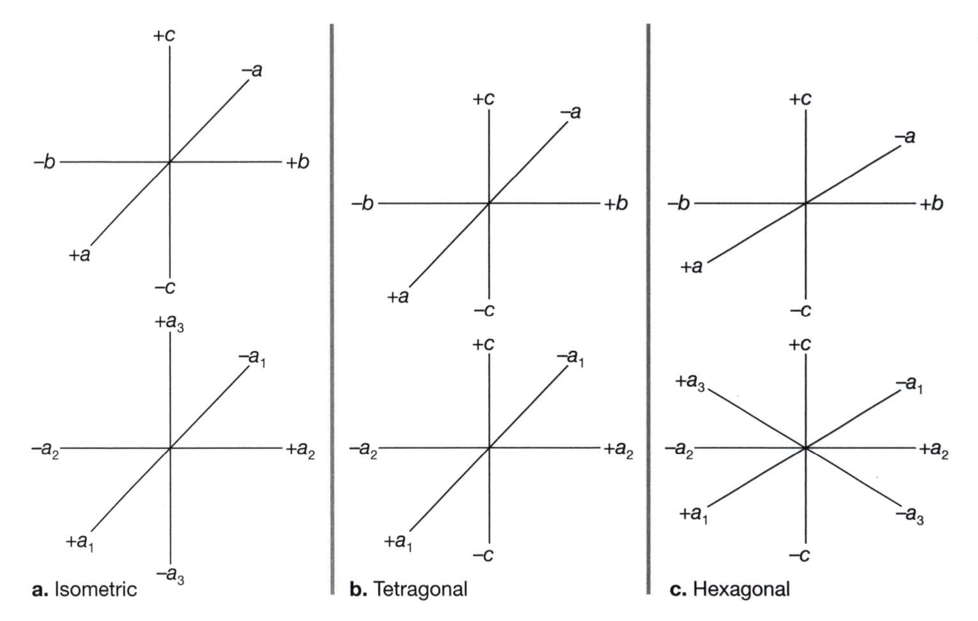

Figure 12.45. *Different naming systems are sometimes used for the axes sets of the isometric, tetragonal, and hexagonal crystal systems, whereby symmetrically equivalent axes are noted with the same letter but with the addition of subscripts.* ***a.*** *Isometric: Here* a, b, *and* c *are used to designate the three symmetry equivalent axes for the isometric system. Some authors use* a_1, a_2, *and* a_3 *in the hope that students will realize the* a*'s are all the same.* ***b.*** *Tetragonal: For the tetragonal system, instead of using* a = b, *some authors use* a_1 *and* a_2 *(i.e., they replace our* a *with* a_1 *and* b *with* a_2*).* ***c.*** *Hexagonal: Even more confusion for the hexagonal system is sometimes caused by adding a third (unnecessary)* a *axis. In this case,* $a_3 = a_1 + (-a_2)$*. Whenever one axis can be written as a linear combination of the other two, it serves no (useful) purpose in defining the basis vector set of a mineral.*

Bravais Lattices

Strictly speaking, the point groups do nothing but repeat space around a *single object*—a point, line, or plane, as noted above. That's because they only combine four symmetry operations: rotation, reflection, inversion, and rotoinversion. (Really only three because rotoinversion is a combination of a rotation and an inversion.) In order to generalize the point groups and create a full, repeatable, 3-D space, we need to add in the operation of translation (and the resultant two-step operations involving translation) to the mix. As with the 2-D case, we do this by first considering the possibilities for lattices, which are arrangements of translations and point symmetry operations.

In general, we can think of the unit cell as a 3-D object, often called a **parallelopiped**, or an object with a closed form like a cube, rectangle, etc. Based upon our discussion so far of crystal systems, you might conclude that there would be at least six different 3-D shapes—one for each of

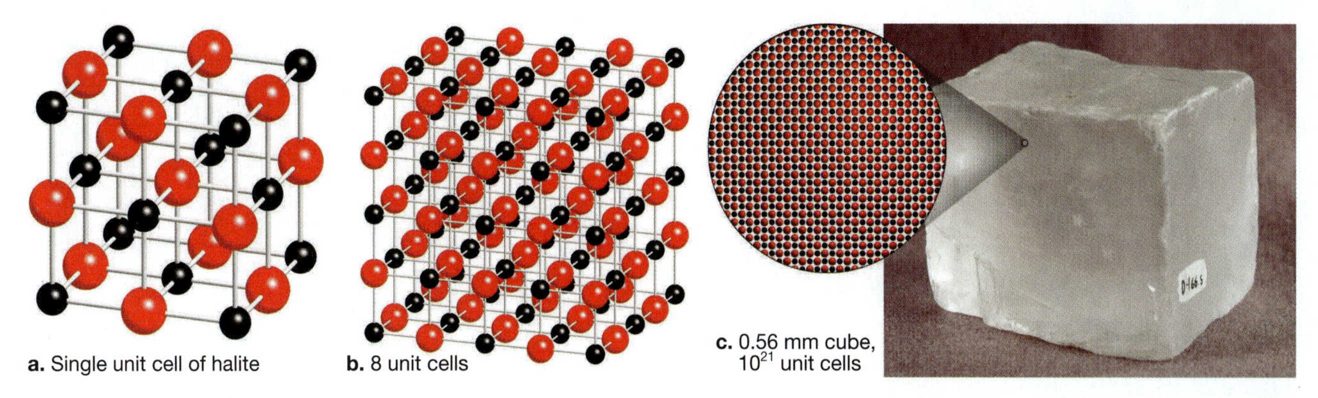

Figure 12.46. *The concept of the unit cell of a mineral. The unit cell is the smallest 3-D portion of a mineral that is required to replicate the mineral. Its size and shape are defined by the mineral's cell parameters (i.e.,* a, b, c *and* α, β, γ*).* ***a.*** *A single unit cell of the isometric mineral halite with* a = b = c = *5.6Å and* $\alpha = \beta = \gamma = 90°$*.* ***b.*** *Eight unit cells of halite creating a piece of halite that measures 11.2Å on each side.* ***c.*** *A 0.56* mm *piece of halite would contain 10,000,000 unit cells of halite on a side.*

the six crystal systems. (You can refer back to Figure 1.12 to see these six basic shapes.) In 2-D, a lattice represented a shape in a plane. In 3-D, a lattice represents the generalized shapes (i.e., scale independent) of the 3-D objects that would be created by the six crystal systems (see also the six blocks of wood in Figure 1.12, and the six perspective views shown in Figures 12.19, 12.22, 12.24, 12.26, 12.28, and 12.30).

For instance, the unit cell *shape* for the isometric minerals halite and sylvite would be the same. What differs is their "scale": in halite a = 5.63 Å (Figure 12.47a) and in sylvite, a = 6.28 Å (Figure 12.47b). Thus both of them have an isometric lattice, but unit cells of different sizes. The scalings of the a, b, and c axes would also extend into the other crystal systems, and they change in systems where there are no constraints on the interaxial angles. The unit cells for halite and sylvite are fairly simple; they are highly symmetrical and contain only a few atoms. Common silicate minerals (i.e., the ones that make up the rocks in the Earth) are much more complex. Figure 12.47c shows the crystal structure of the monoclinic mineral muscovite with its unit cell outlined in black. The cell is much larger than in halite, with $a = 5.19$, $b = 9.00$, and $c = 20.10$ Å. As you can see, it contains many more atoms. And it is also a different shape with $\alpha = \gamma = 90°$ and $\beta = 95°$.

To introduce the 3-D lattices, we'll make lattices for each crystal system, but then add another variable to the story of lattices (Table 12.7). Lattices are our templates for how atoms, or groups of atoms, repeat in space. In defining the lattice of a material, crystallographers seek to define the highest symmetry that describes a material, not the lower symmetry cases.

Table 12.7. Possible Bravais lattices and associated crystal systems for the "standard settings"

Crystal system: lattices
Triclinic: *P*
Monoclinic: *P*, *C*
Orthorhombic: *P*, *C*, *I*, *F*
Tetragonal: *P*, *I*
Hexagonal: *P*, *R*
Isometric: *P*, *I*, *F*

Going back to the six shapes corresponding to the six crystal systems in Figures 12.19, 12.22, 12.24, 12.26, 12.28, and 12.30, we are now going to redraw them all a little differently. This time we will put a lattice point on the corner of each of the blocks (Figure 12.48). In the most general case, we can always represent the lattice of a material with atoms, or groups of atoms, at these lattice points. These lattices are termed **primitive**, because they are the simplest. We use the symbol "*P*" as their abbreviation.

Other lattice types occur when crystallographers select a higher symmetry for the material, just like the centered 2-D lattice in Figure 11.15c. In 3-D, there are four other types of lattices and they are denoted by names that describe the geometry of the extra points (i.e., those not located at the lattice corners as in the primitive case). For the face-centered case, these are logically denoted by the letter "*F*." There are six extra lattice points: one on each of the six faces (Figure 12.49).

Side-centered lattices are a variant of the *F* type. They contain two extra lattice points locat-

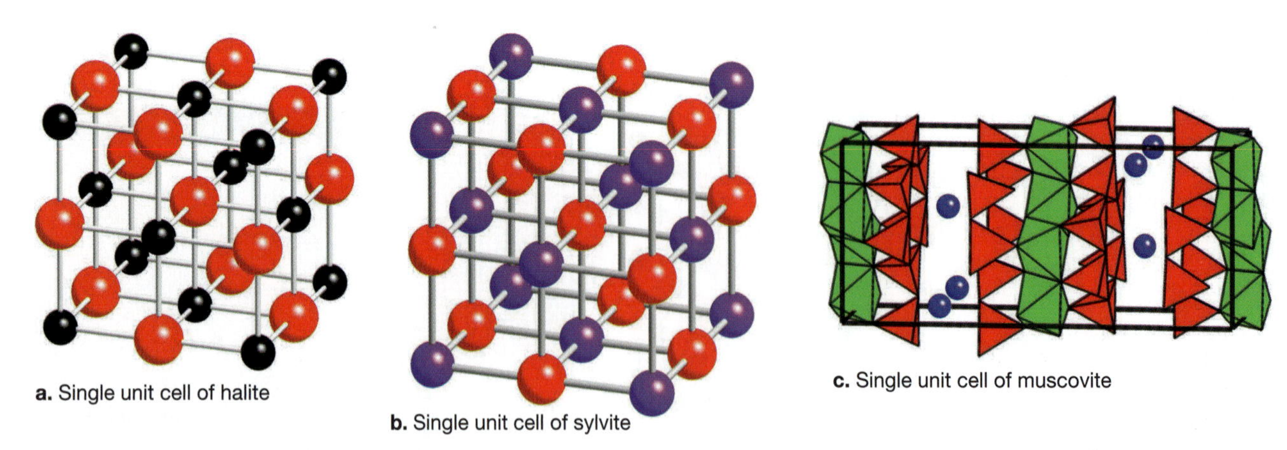

Figure 12.47. *Three different sizes and two shapes of unit cells for three different minerals.* ***a.*** *Halite,* a = *5.6Å with* Na^{1+} *and* Cl^{1-} *ions.* ***b.*** *Sylvite,* a = *6.3Å with* K^{1+} *and* Cl^{1-} *ions. Even though sylvite has the same shape as halite, it is bigger. The size increase is because the larger* K^{1+} *substitutes for the smaller* Na^{1+}. ***c.*** *Muscovite, a monoclinic mineral with* a = *5.19,* b = *9.00,* c = *20.10Å and* $\alpha = \gamma = 90°$ *with* $\beta = 95°$. *Notice how the unit cell is much bigger for muscovite than for halite or sylvite, and thus it contains many more ions. This structure is fairly typical of the silicate minerals that compose the majority of the Earth's crust.*

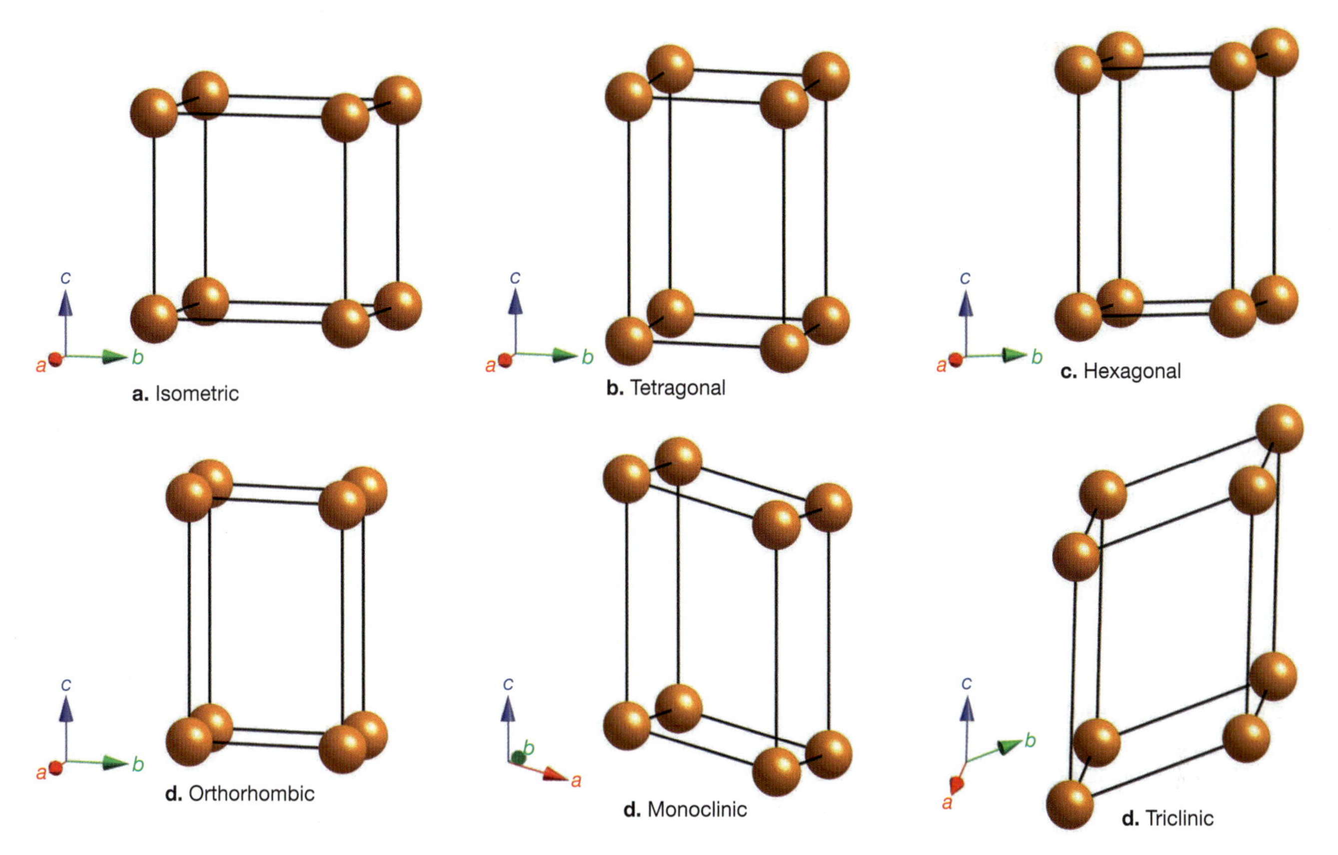

Figure 12.48. *The six primitive (P) unit cells. Only one lattice point occurs at each corner and the shape is determined by the crystal system.*

ed on two symmetrically opposite faces (Figure 12.50). The names we apply to these lattice are ***A*-centered**, ***B*-centered**, and ***C*-centered**, based upon the locations of the extra points. For example, the letter "*C*" refers to the situation where the extra points are located on the *a-b* face (Figure 12.50). This is the same sort of the odd-man-out style used to designate the interaxial angles in regards to the crystallographic axes. It would have made much more sense to call a "*C*" lattice an "*a-b*" lattice; however this mistake was made 100s of years ago. Part of learning crystallography, or other any mature (i.e., well-developed) science, is the need to appreciate the historical development of the nomenclature. Try not to get too upset with your professor—he/she had nothing to do with developing the nomenclature!

The next type of lattice has a fairly simple concept, but its abbreviation is bit foreign (pun intended!). The isometric, tetragonal, and orthorhombic crystal systems (Figure 12.51) all have a lattice type named "*I*." The extra lattice point is in the center of the lattice. We call these lattices "*I*" or **body-centered**. But where did the "*I*" come from? It's German for "innenzentrierte" which means "body-centered."

The last lattice type is infrequently used for the hexagonal crystal system. Figure 12.48c shows a *P* **hexagonal** lattice, whose cell parameters are those of the hexagonal system (i.e., $a = b \neq c$, and $\alpha = \beta =$ 90° and $\gamma = 120°$). There is also a lattice type termed **rhombohedral**, or *R* for short, that has a different set of cell parameters with $a = b = c$ and $\alpha = \beta = \gamma$ (Figure 12.52). The constraints here are that all of

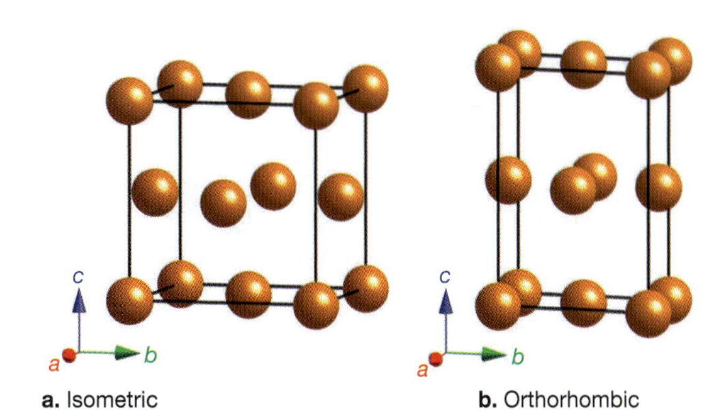

Figure 12.49. *The two face-centered (F) lattices. One lattice point occurs on each corner and another lattice point is centered on each of the six faces. Because of symmetry constraints, only the isometric and tetragonal cells have face-centered lattices.*

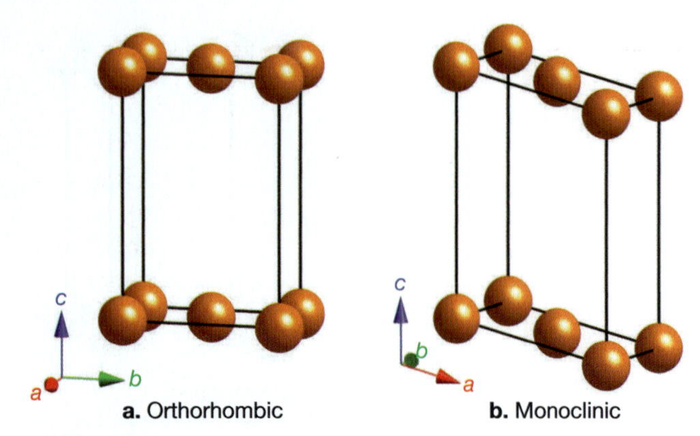

Figure 12.50. The side-centered lattices. A lattice point occurs on each corner and on a single pair of opposite faces. The name given to the lattice depends on the face where the extra lattice is located. These would either be "A," "B," or "C," if the face was the b-c *plane, the* a-c *plane, or the* a-b *plane, respectively. Due to symmetry constraints, only the orthorhombic and monoclinic systems have side-centered lattices.*

the crystallographic axes and interaxial angles are equal, but they are not constrained to be any certain angle. This is something that we've not seen before! The good news is that the *R* lattice is seldom used, and it can be can be converted to a *P* hexagonal lattice. Figure 12.53a shows both an *R* lattice and an *R* lattice converted to a *P* hexagonal lattice. A better view can be seen in the crystal structure of corundum in Figure 12.53b,c. The two structures are the same, but all that differs is the outline of the unit cell. So it is our coordinate choice that changes, not the structure of the material.

You might notice that certain crystal systems do not have all the lattice types. For instance, there are no *I* or *F* monoclinic lattices. Why? To answer this question, ponder the symmetry elements associated with each crystal system, and decide if those elements would allow, for instance, a face-centered orthorhombic lattice.

Before we leave this section we want to try and dispel what is a common misconception about lattice types. Often students and professors think that lattice points represent only atoms. That's a reasonable assumption based on the way we tend to show them (as in Figures 12.48 to 12.52). However, lattice points rarely are indicated by individual atoms, and are far more commonly found to be located relative to collections of atoms. The lattice points are simply the "points" around which these groups repeat space, so they are really symmetrically equivalent points. In some cases, there may not even be an atom present at a lattice point; the atoms are instead arrayed around the lattice point. Figure 12.54a shows the crystal structure of diamond. What lattice type do you think it has (hint: look at the positions of the same atoms)? Diamond has an *I*-centered lattice.

Now look at the isometric garnet mineral pyrope in Figure 12.54b; what lattice type do you think it has? By looking at the series of images that strip away some of the "extra" atoms, you can see that garnet has similar structural units (in this case octahedra) located at the corners and center of its unit cell.

Finally, study the structure of two different pyroxenes, diopside (Figure 12.55a) and enstatite (Figure 12.55b). Again removing the "extra" atoms and paying attention to the orientation of their crystallographic axes, the structural units can be seen to be centered on the *a-b* plane for diopside at the edge of the unit cell; thus, making it a monoclinic *C* lattice. Stripping the extra atoms from enstatite reveals a *P* lattice (Figure 12.55b). You might notice that we've left out the face-centered lattice here. But you've seen it several times with the "much used" structural drawing of halite (refer back to Figure 12.46).

Naming Planes and Lines in Crystals

It is useful (perhaps critical) to assign names to directions in minerals. There are two types of

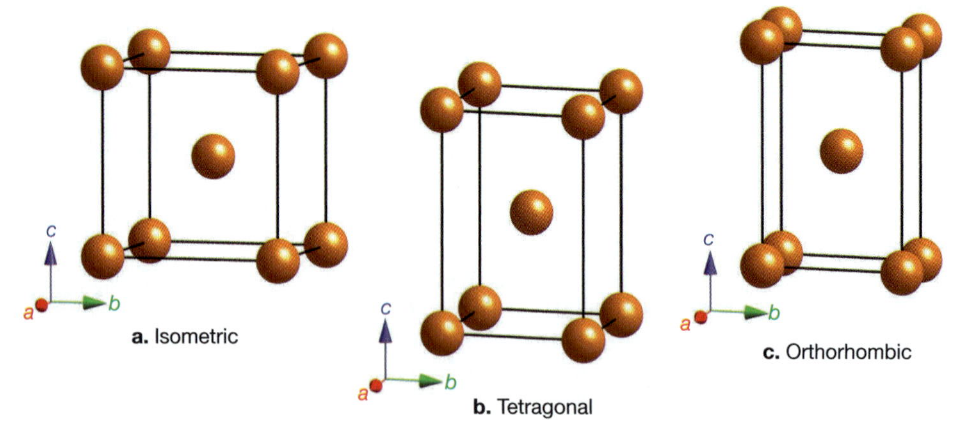

Figure 12.51. The three body-centered (I) lattices. A lattice point occurs on each of the six corners plus one extra lattice point is located in the center of the unit cell. Based on symmetry constraints, only the isometric, tetragonal, and orthorhombic systems can have this lattice type.

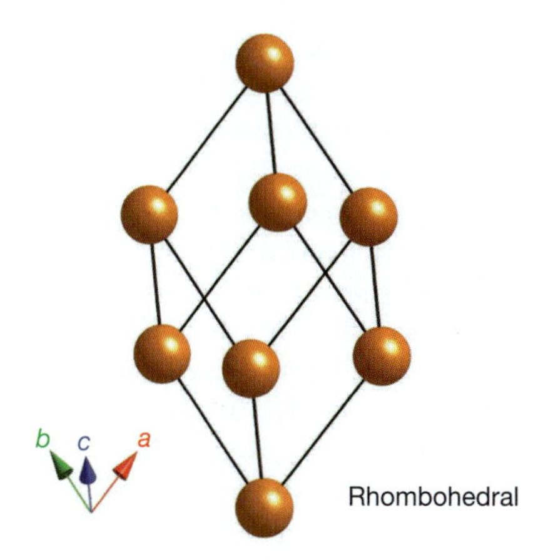

Figure 12.52. *The rhombohedral* (R) *lattice has cell parameters* a = b = c *and* α = β = γ, *where the single length or angle can vary. For this type of lattice, there are eight lattice points that occur at the eight corners.*

directional features to name: *planes*, which internally represent layers of atoms in the mineral and externally faces of the mineral, and *lines*, which represent such things as the crystallographic axes of a mineral's unit cell or the direction of elongation of a mineral.

We have already devised a method to name lines, and we have used it extensively in our discussion of point symmetry. Figure 12.21 shows the axis set for the isometric crystal system. The three crystal axes are labeled *a*, *b*, and *c* as usual, and can be numerically expressed as $+a$ = [100], $+b$ = [010], and $+c$ = [001]. We could also write $-a$ = [$\bar{1}$00], $-b$ = [0$\bar{1}$0], and $-c$ = [00$\bar{1}$]. We could then write directions not coinciding with the crystallographic axes as linear combinations of *a*, *b*, and *c*. For instance, the body diagonal [111] could be written as $a + b + c$ and the vector that bisects *a* and *b* would be $a + b$ or [110]. A vector that had twice the component along *a* as *b* and a zero component along *c* would be $2a + 1b + 0c$, or [210], etc. (Figure 12.21). Historically, this method of naming directions in crystals was derived from defining axes based upon the intersection of two prominent faces on a mineral (Figure 12.56). The term **zone axis** was applied to this direction, but we can generalize these vectors as [*u v w*], where *u*, *v*, and *w* are integers that give the component of the axis in the *a*, *b*, and *c* coordination system. Although it is not of apparent significance at this point, note that we use bracket (i.e., "[]") to denote *vectors* in minerals.

We have also discussed the orientations of mirror planes in the point groups. We used a perpendicular line to designate the orientation of these planes, and we'll use a similar method to denote *planes* in minerals. Planes in minerals are represented using parentheses, "()", not brackets. So when you see a number like (100) enclosed in "()," realize that it's a plane.

We also use three numbers to designate the orientation of a plane. We can generalize a plane's orientation to be (*hkl*), where *h* will give us information about the plane relative to the *a* axis, *k* is relative to the *b* axis, and *l* to the *c* axis. The number tells us if the plane intersects that axis; if a "0" appears, that means the plane doesn't interest that axis. So if the symbol is (200), we immediately know that the plane doesn't intersect the *b* or *c* axis. When a number appears (and it will usually be a small positive or negative number like 1–4), it describes where the plane cuts the crystallographic axis. We need to refer back to the unit cell concept here. Recall that the unit cell is a geometric 3-D shape defined by the lengths of the crystallographic axes and the interaxial angles. When we see a 2, for instance, as in (002), we know that this plane cuts the *c* axis in half (Figure 12.57a). If we see a 1, as in (010), then the plane cuts the end of the axis (Figure 12.57a). If we see two 1's, as in (110), we know that the face cuts *a* and *b* and is parallel to *c* (Figure

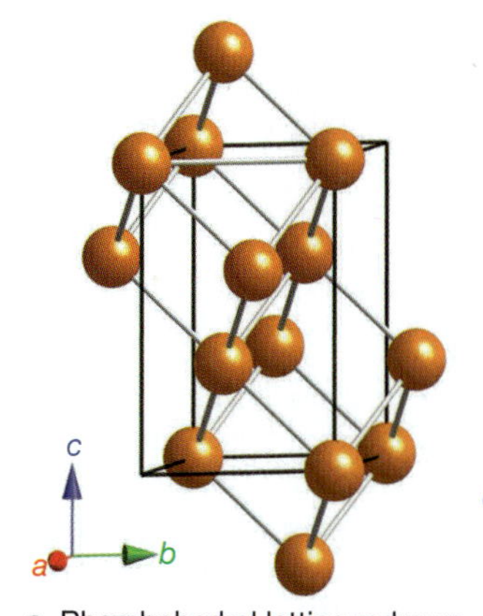

a. Rhombohedral lattice redrawn in a hexagonal *P* lattice

b. Corundum based on *R* lattice

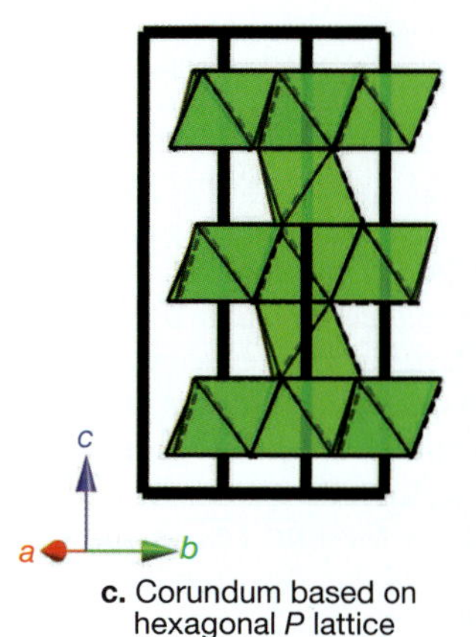

c. Corundum based on hexagonal *P* lattice

Figure 12.53. *A mineral that can be assigned an* R *lattice can also be assigned a* P *lattice.* ***a.*** *The rhombohedral* (R) *lattice from Figure 12.52 recast into a hexagonal* P *lattice.* ***b.*** *For example, corundum (see center image) is indexed on an* R *lattice (left) and a* P *hexagonal lattice (right). The atomic arrangement stays the same, but the system we use to describe it changes.*

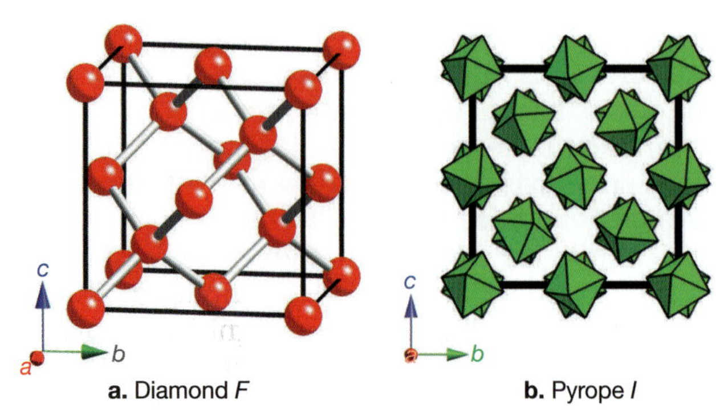

*Figure 12.54. Mineralogical examples of lattice types. **a.** Diamond is one of the simplest examples of a face-centered lattice. Carbon atoms are located at the corners of the unit cell, and on the center of each face. **b.** Garnet group minerals, like the pyrope ($Mg_3Al_2Si_3O_{12}$) shown here, present a slightly more complex example of a body-centered lattice than that shown in Figure 12.48. For this case, it helps to remove some the "extra" cations (the 4-coordinated Si^{4+} and the 8-coordinated Mg^{2+}). When just the Al^{3+} octahedra are left in the structure, it is apparent that they occur on the corners of the lattice. However it also looks like they occur on the faces, but if you carefully look you'll notice they are in a different orientation. And there is one in the center that is in the same orientation as the ones on the corner; it's hard to see it without removing the ones on the face.*

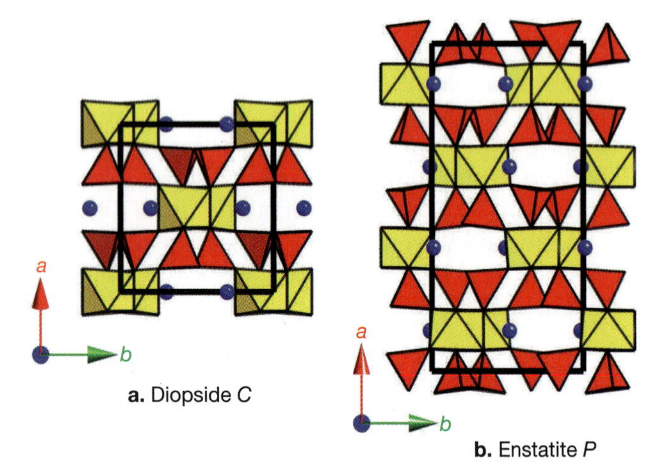

Figure 12.55. Examples of C *side-centered and* P *lattices in pyroxenes. **a.** Diopside ($CaMgSi_2O_6$) is a monoclinic pyroxene. Similar atomic arrangements occur around each corner and on the* a-b *plane, resulting in a* C *lattice. No actual atoms lie on the corners, or are centered on the* a-b *plane, but groups of atoms are centered around those lattice points. Thus actual atoms do not need to occur at the lattice points to define a lattice type. **b.** Enstatite ($Mg_2Si_2O_6$) is an orthorhombic pyroxene. Comparison of its structure with that of diopside shows a doubling of the unit cell in the* c *direction and a* P*, instead of* C*, lattice type. Again, notice no actual atoms lie on the corners of the unit cell.*

12.57a). Figure 12.57b shows three more examples: (100) cuts *a* but is parallel to *b* and *c*, (010) cuts *b* but is parallel to *a* and *c*, and (001) cuts *c* but is parallel to *a* and *b*. These three planes would define three of the faces of the unit cell and the distance from the origin of the unit cell to each face would represent the lengths of the *a*, *b*, and *c* axes accordingly. Finally, Figure 12.58 shows a series of planes that cut all three axes, but at different lengths with their associated names. So to generalize, we can write (*hkl*) for a plane that cuts the *a* axis *h* times, the *b* axis *k* times, and the *c* axis *l* times.

Before we knew the internal structure of minerals, mineralogists spent considerable time studying and measuring the angular relationships of growth and cleavage faces. They devised the now-archaic method of Weiss parameters for naming faces. Gibbs (1997) shares this view and states "one avoids using Weiss parameters which have not only outlived their usefulness but should be purged from the mineralogical literature." In the approach presented above, we are only concerning ourselves with how planes intersect the unit cell. This in turn will be the basis for mineral identification and characterization by diffraction methods (discusses in Chapter 15). To name an external face on a mineral, we simply use the (*hkl*) value that parallels that face as obtained from the unit cell. Because the physical properties of minerals are directly related to their crystal structures, we can often predict by looking at the planes in a crystal structure draw-

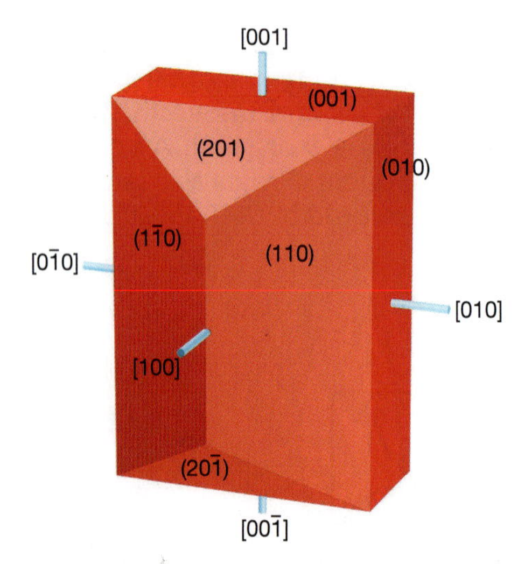

Figure 12.56. When two cleavage or growth faces of a mineral intersect, their intersection forms a line historically termed a zone axis. This zone axis can be labeled [uvw], where u *indicates its component on the* a *axis,* v *its component on the* b *axis, and* w *its component on the* c *axis. This [uvw] nomenclature has evolved into the method described in Figure 12.21 to name directions in minerals.*

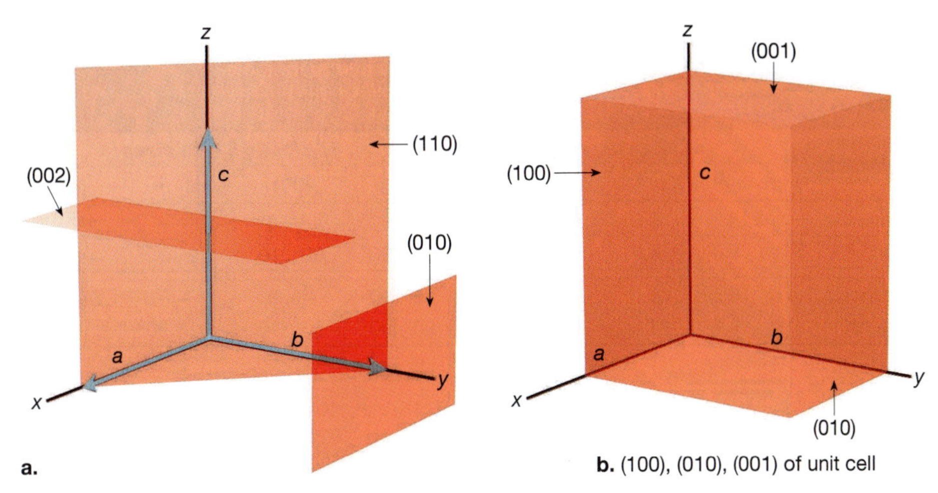

Figure 12.57. Faces and planes (either growth, cleavage, or defined by layers of atoms) in minerals are named by the method of Miller indices. These are similar to the method of naming directions in that we use three integers to refer to the three crystallographic axes. The general term (hkl) *is used to define a plane's orientation, where* h *tells us how the plane intersects* a, k *how it intersects* b, *and* l *how it intersects* c. *When an* hkl *value equals zero, the plane does not intersect the respective crystallographic axis. When* hkl *values are non-zero, the plane cuts the respective axis or axes.* **a.** *In the above figure, (010) would cut* b *at its end while lying parallel to* a *and* c*; (110) would intersect* a *and* b *at the ends while being parallel to* c, *and (002) would be parallel to* a *and* b *but cut* c *in two pieces (thus the 2).* **b.** *We only deal with the unit cell when we use the* (hkl) *nomenclature. When a plane cuts at 1, that is the end of the axis, and larger numbers represent cutting the axes into smaller parts (as shown in Figure 12.57a above and in Figure 12.58). Thus, the (100), (010), and (001) would outline the unit cell. This "inverse" relationship will make more sense when we study how X-rays are diffracted by minerals.*

ing where such external properties as cleavage or hardness will be affected by the internal arrangement of the atoms.

Figure 12.59a shows our familiar cube ($4/m\ \bar{3}2/m$) with the six faces and crystallographic axes labeled. In Figure 12.59b, the cube has been placed inside a sphere (the one used for stereographic projections) and the poles (i.e., normals) to each plane are extended to the edge of the sphere. All the poles to the planes (i.e., the (*hkl*) values) and the crystallographic axes coincide. Figure 12.59c shows the stereographic projection of the face poles and axes. Figure 12.60a is the same setup but this time for an octahedron, which has the same point group symmetry as a cube, though all of its faces intersect the crystallographic axes (Figure 12.60b); a stereographic plot (Figure 12.60c) is also shown. The last isometric example shows a tetrahedron (Figure 12.61), which has $\bar{4}3m$ symmetry and lacks a center of symmetry (Figure 12.61c).

Next let's look at some lower symmetry examples of common orthorhombic and monoclinic minerals using the pyroxene and amphibole groups as examples. Table 12.8 gives the cell parameters and calculated ϕ and ρ coordinates (based on the cell parameters) for select faces for enstatite, diopside, anthophyllite, and tremolite. A set of perspective drawings for each crystal is shown in Figures 12.62 and 12.63, along with a 2-D projection on the *a-b* plane showing the lengths of *a* and *b*, the (110) or (210) plane, and associated stereographic projections of their crystallographic axes and selected (*hkl*)'s.

In general, crystals of the pyroxene and amphibole groups are elongated parallel to their *c* crystallographic axes (the *c* dimension for both is approximately 5.3 Å). Also, both groups tend to form prism-shaped crystals with prominent cleavage faces that cut both the *a* and *b* axes and are parallel to *c* (Figure 12.62 and 12.63). We can

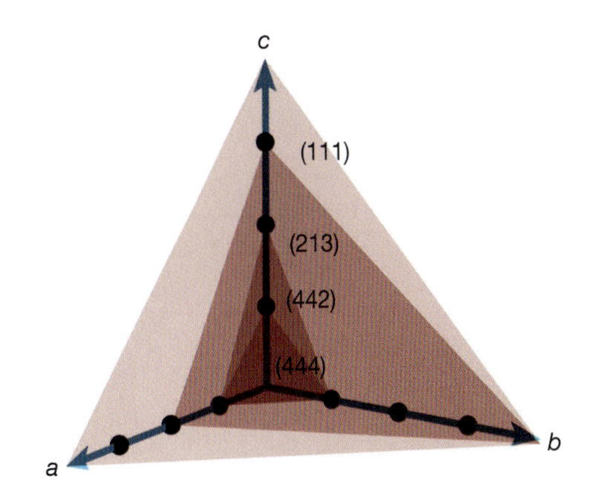

Figure 12.58. A series of faces that intersect (i.e., cut) the a, b, *and* c *axes at different positions. The black dots on the axes represent 1/4 distances along each axis. Notice how each of these planes intersect all three axes, thus the* (hkl) *values are all non-zero. Also notice how (111) and (444) are parallel to each other and the larger the* (hkl) *planes are closer to the origin.*

Table 12.8.
Crystallographic data for pyroxenes and amphiboles.
Examples of both monoclinic and orthorhombic species are given. These data are used for the plots in Figures 12.62 (for pyroxenes) and 12.63 (for amphiboles).

Mineral	Cell parameters				ϕ, ρ for (hkl) face poles			
	a(Å)	b(Å)	c(Å)	β°	(100)	(001)	(110)	(210)
Enstatite	18.2	8.8	5.2	90°	90°, 90°	---, 0°	----	44°,90°
Diopside	9.8	8.9	5.3	106°	90°, 90°	90°, 16°	42°, 90°	-----
Anthophyllite	18.5	18.0	5.3	90°	90°, 90°	---, 0°	----	63°, 90°
Tremolite	9.9	18.1	5.3	105°	90°, 90°	90°, 15°	61°, 90°	-----

generalize this type of cleavage as (*hk*0), meaning that the faces cut *a* and *b* but are parallel to *c*. There are both orthorhombic and monoclinic varieties of pyroxenes and amphiboles. The difference is the length of their *a* axes. (Compare the length of *a* for the monoclinic and orthorhombic versions of each mineral in Table 12.8. What's the relationship?). Finally, notice how the *b* length is the same for both pyroxenes and doubled for the amphiboles. These relationships show how monoclinic and orthorhombic minerals will differ in the angular relationships between their planes and crystallographic axes.

Next we are going to plot the face poles for the (100) and (001) faces, which sometime occur for these minerals. Let's start with the mineral enstatite, an orthorhombic pyroxene. The (*hk*0) faces are labeled as (210) in Figure 12.62a because in the projection on the *a*-*b* plane, the plane cuts *a* in half and intercepts *b* at its end. Next, some basic trigonometry can be used to find the angles (210) makes with the *a* and *b* axes (because we know the lengths of *a* and *b*). We can then find the cleavage angles and determine the ϕ and ρ values so as to plot the (210) face on stereonet. The cleavage angles are 88° and 92°. Relating these to the different faces, (210):(2$\bar{1}$0) = 88° and (210):($\bar{2}$10) = 92°, and the (210) face would plot at $\phi = 44°$ and $\rho = 90°$. It would have plotted exactly at 45° if the *a* axis had been exactly twice as long as the *b* axis. So Figure 12.62a shows the stereographic plot of the *a*, *b*, and *c* crystallographic axes and the (100), (001), (210), ($\bar{2}$10), ($\bar{2}\bar{1}$0), and (2$\bar{1}$0) faces. You might notice that the face pole for (100) and (001) coincides with the *a* and *c* axes.

In the above discussions, we've been listing all four faces (i.e., (110), ($\bar{1}$10), ($\bar{1}\bar{1}$0), (1$\bar{1}$0) and (210), ($\bar{2}$10), ($\bar{2}\bar{1}$0), (2$\bar{1}$0)) for the two types of cleavage in the pyroxenes and amphiboles. Listing all of these faces each time becomes somewhat tiresome to type and read. Fortunately, mineralogists have developed a shorthand method to describe a serious of symmetrically related faces. To do this, we enclose an *hkl* value in curly brackets like this "{}," instead of parentheses, so {110} would represent the set of four symmetrically faces (i.e., (110),

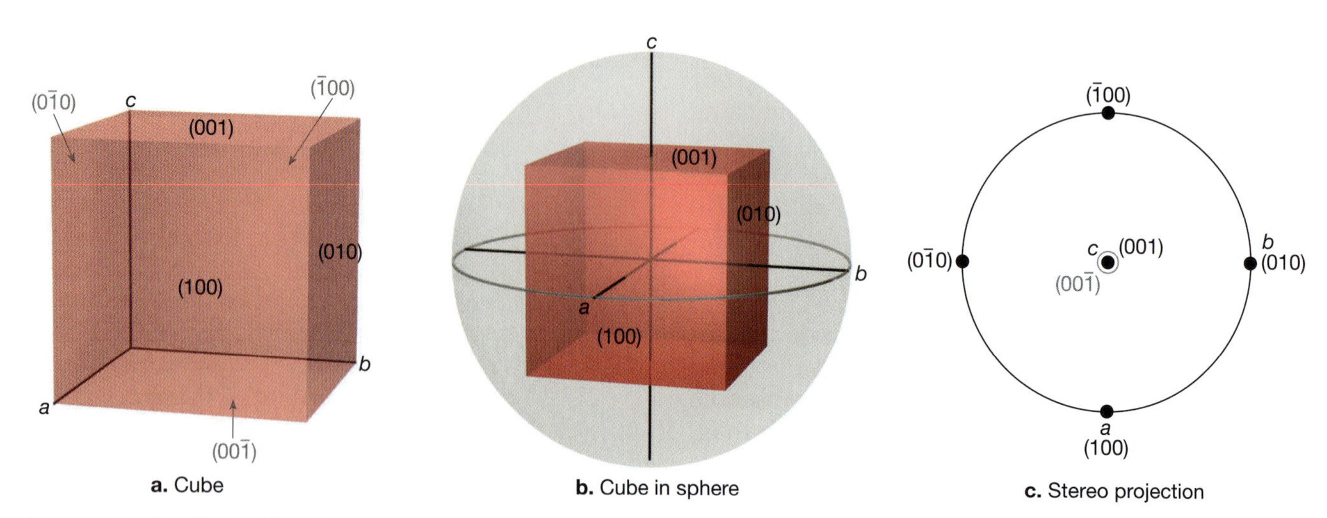

Figure 12.59. The (hkl) values for the six faces of a cube and their stereographic projections. ***a.*** *A cube with its six faces labeled and its associated isometric axis set.* ***b.*** *The cube in Figure 12.59a, placed inside a sphere with the normals to each face (and the crystallographic axes) extended to intersect with the sphere's surface.* ***c.*** *Stereographic projection of the six faces and crystallographic axes. Notice how (100) coincides with* a, *(010) coincides with* b *and (001) coincides with* c.

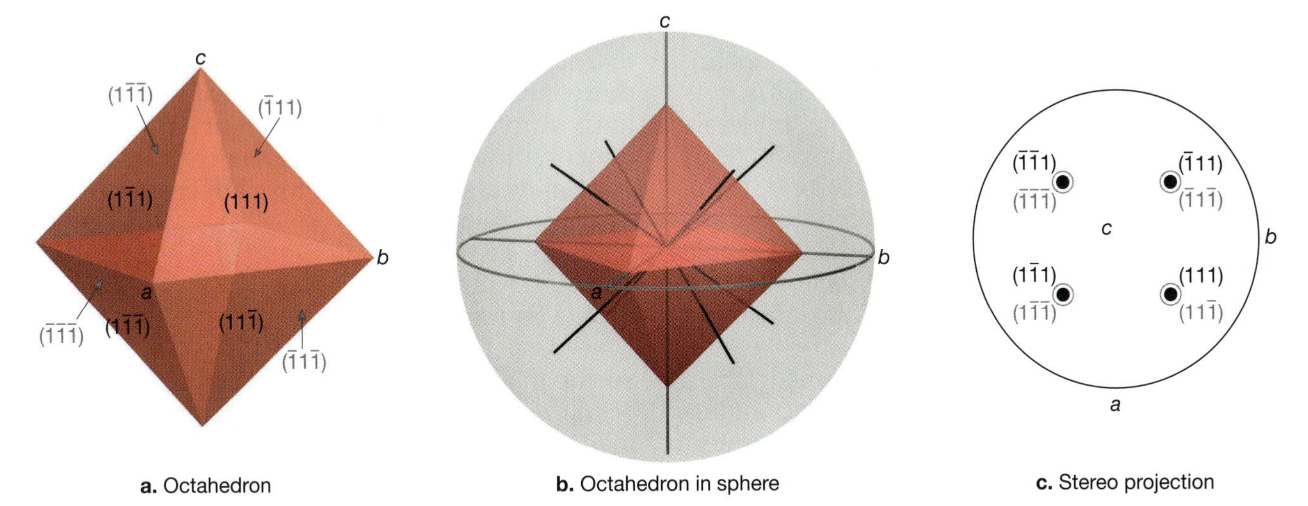

Figure 12.60. *The (hkl) values for the eight faces of an octahedron and their stereographic projections.* ***a.*** *An octahedron with its associated isometric axis set and the (hkl) values labeled for its eight faces.* ***b.*** *The octahedron placed inside a sphere with the normals to its faces and crystallographic axes extended to intersect with the sphere's surface.* ***c.*** *A stereographic projection of the eight faces and crystallographic axes. The {111} faces plot midway between the three crystallographic axes (i.e., they make the same angle with each of them).*

($\bar{1}$10), ($\bar{1}\bar{1}$0), 1$\bar{1}$0), and {210} would represent (210), ($\bar{2}$10), ($\bar{2}\bar{1}$0), (2$\bar{1}$0) set of symmetrically related faces.

Next, we repeat the above calculations and plots for diopside, a monoclinic pyroxene. The (*hk*0) face is (110), thus it cuts both *a* and *b* at their endpoints (Figure 12.62b). In reality, the (210) cleavage in the orthorhombic pyroxene is the same plane through the structure as the (110) plane in the monoclinic pyroxene. Why? (Hint: take a look at the *a* cell lengths in Table 12.8. Next notice that the (110):(1$\bar{1}$0) = 84° and (110):($\bar{1}$10) = 96°, similar, but slightly different from enstatite, and (110) plots at $\phi = 42°$ and $\rho = 90°$.) Compare these to the stereographic projections of *a*, *b*, and *c*, and the (100), (001), and {110} in Figure 12.62.

What is the relationship between *a* and *c* and (100) and (001)? They no longer coincide as they did for the orthorhombic case. Recall that the "0's" in the (*hkl*) nomenclature mean that a face is parallel to an axis and a non-zero integer means that it cuts the indicated axis. Just because a plane is parallel to one axis (e.g., (100) is parallel to *b* and *c* for diopside) does not mean that it is perpendicular to another axis (e.g., (100) is not perpendicular to *a*, it just passes through *a*). For this

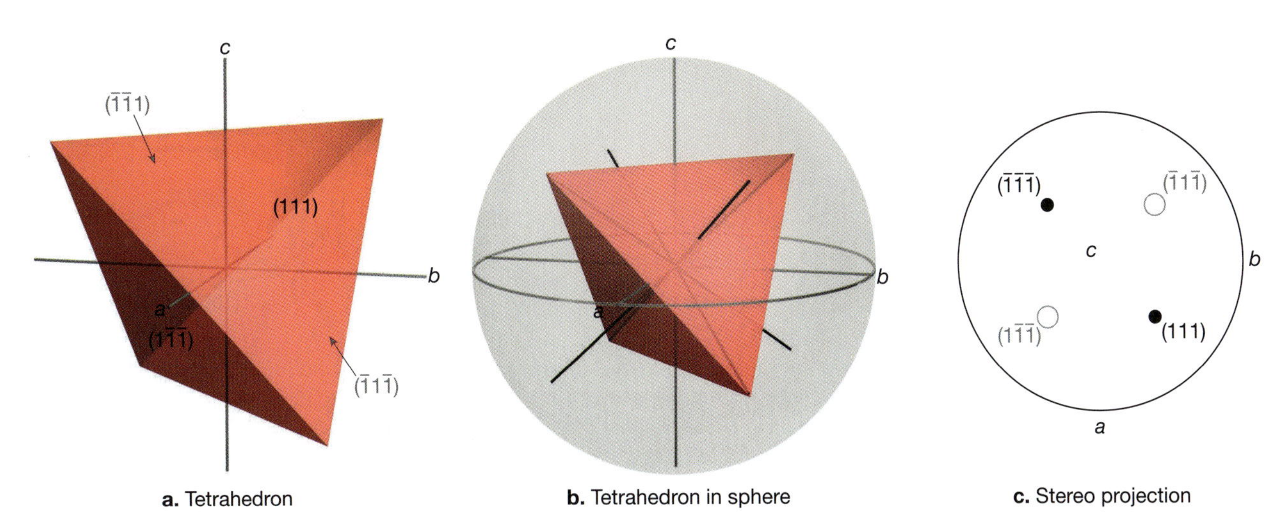

Figure 12.61. *The (hkl) values for the four faces of a tetrahedron and their stereographic projection.* ***a.*** *A tetrahedron with its four faces labeled and its associated crystallographic axes.* ***b.*** *The tetrahedron from Figure 12.61a, placed inside a sphere with the normals to each face and the crystallographic axes extended to the surface of the sphere.* ***c.*** *A stereographic projection of the faces and crystallographic axes for the tetrahedron.*

reason, the face pole to (100) is not symmetry-required to coincide with the *a* axis for monoclinic minerals like diopside (Figure 12.62b). It is symmetry-required for orthorhombic minerals like enstatite (Figure 12.62a). For diopside, (001) and *c* do not coincide, while (010) and *b* do. Figure 12.63 shows the crystal sketches and plots for an orthorhombic and monoclinic amphibole as was just done for the pyroxenes. Why are the cleavage angles different for the pyroxenes and amphiboles?

We will be using (*hkl*) and [*uvw*] nomenclature throughout our discussions of minerals to understand the angular relationships between a mineral's axes and faces, to help identify minerals, and to understand how mineral surfaces may react in different environments. For example, in a human lung, amphiboles expose different surfaces when they occur as a cleavage fragment vs. an asbestos fiber, and this affects their toxicity. The list of such examples is endless!

Space Symmetry

So far in this chapter, we have been examining combinations of four operations **point groups** are made up of rotation, reflection, inversion, and

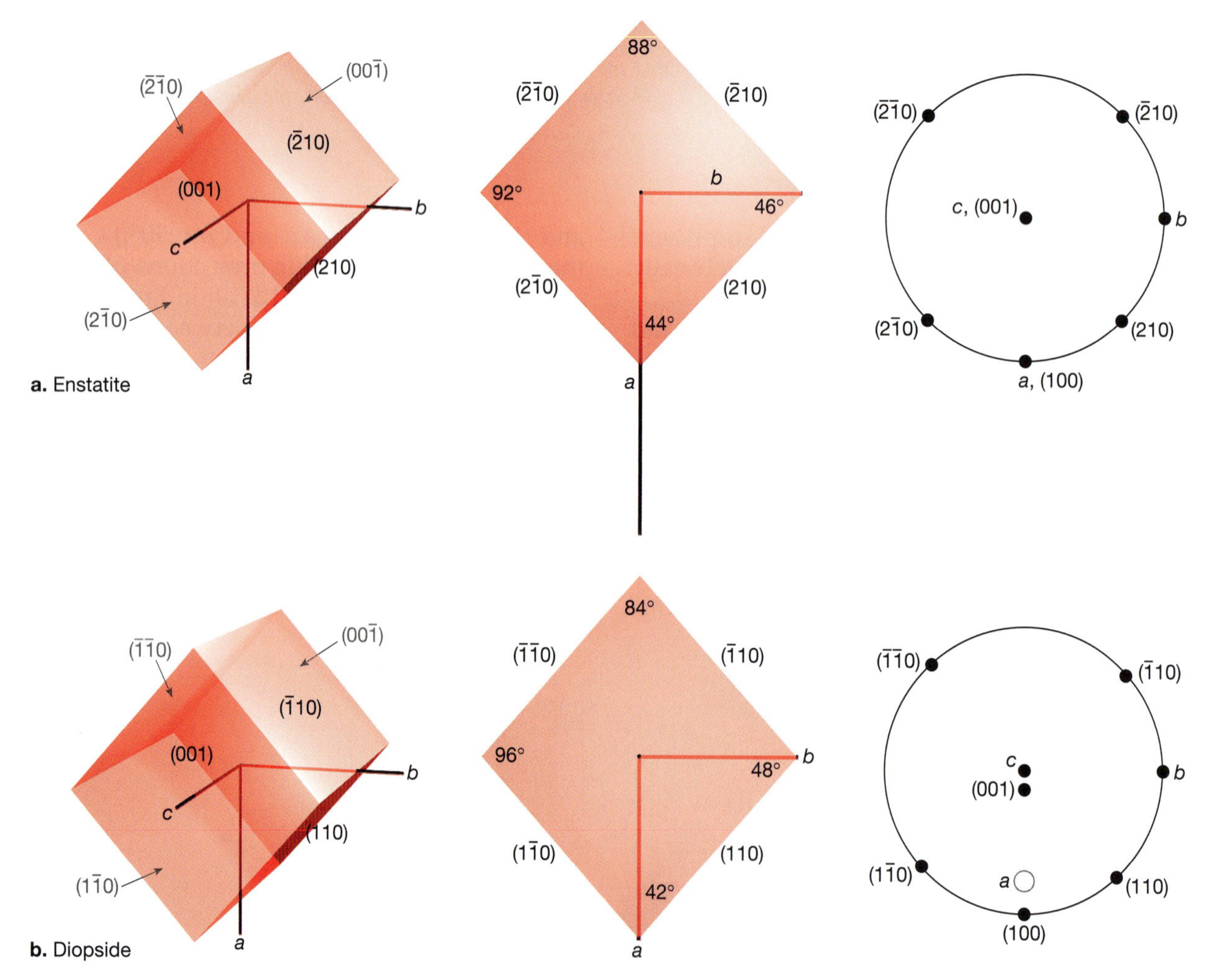

Figure 12.62. *Morphological sketches and stereographic projections of orthorhombic and monoclinic pyroxenes.* ***a.*** *The morphological sketch (left) of enstatite, an orthorhombic pyroxene, shows that the mineral is elongated parallel to* c *and has four cleavage faces parallel to* c. *A projection down* c *onto the* a-b *plane (center) shows the (210) cleavage plane cutting* a *in half and* b *at its end. Based on the lengths of* a *and* b *in Table 12.8, the cleavage angles as shown can be calculated. At right is a stereographic projection of the crystallographic axes and the four {210} face poles along with the (100) and (001) planes. The* a *and* c *axes correspond to (100) and (001) face poles.* ***b.*** *A similar set of three sketches as in Figure 12.62a, but now for the monoclinic pyroxene diopside. The length of* a *is half that of enstatite. Thus the {210} faces in enstatite are now {110}. The cleavage angles of the two pyroxenes are similar, and vary only by a few degrees. Another major difference is that the* a *and* c *directions no longer coincide with the face poles for (100) and (001) respectively. This occurs because of the non-90° angle between* a *and* c. *Thus, (100) is parallel to* c *but its face pole no longer coincides with* a.

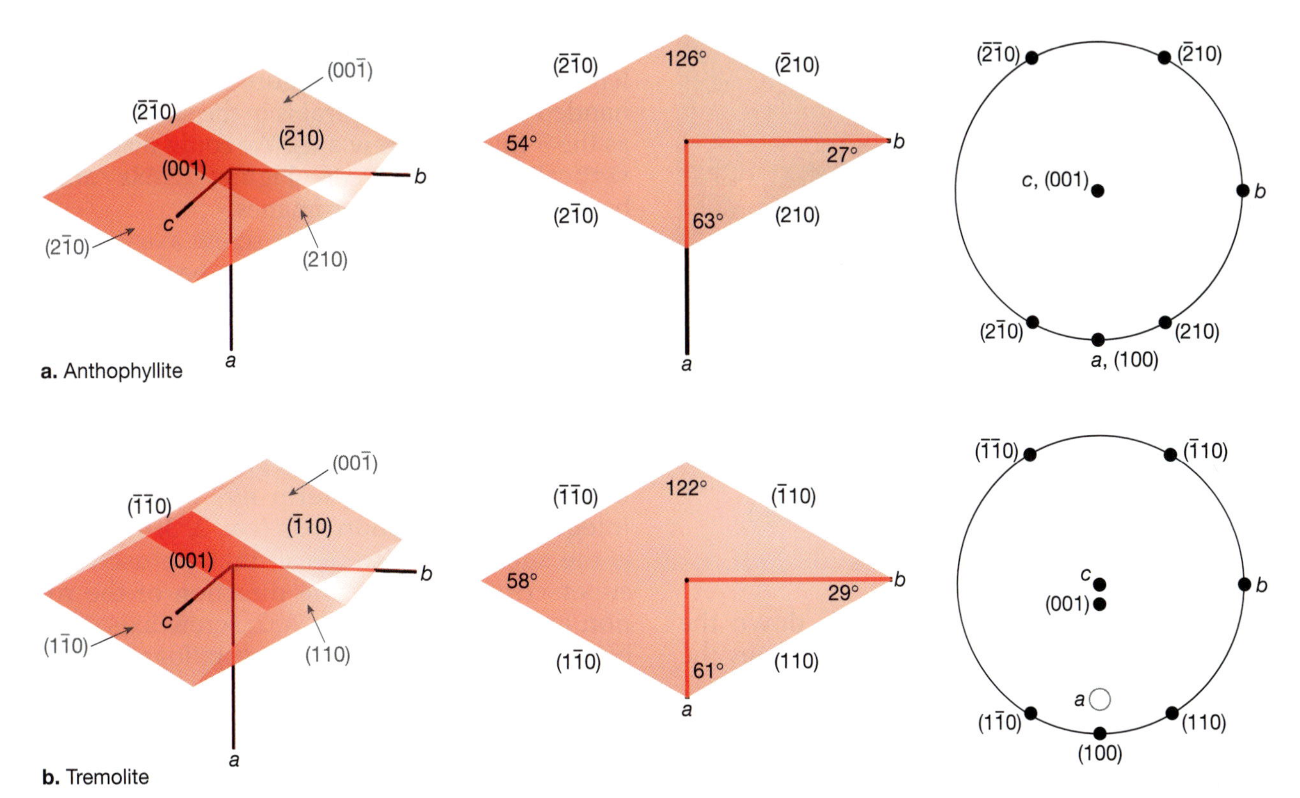

Figure 12.63. *Morphological sketches and stereographic projections of orthorhombic and monoclinic amphiboles.* ***a.*** *A series of sketches for anthophyllite (an orthorhombic amphibole), similar to those for the pyroxenes shown in Figure 12.62. The external morphology (left) of amphiboles is similar to that of pyroxenes (i.e., they are elongated parallel to* c *and have well-formed cleavage parallel to* c*). However, the* b *axis is doubled in amphibole (a consequence of going from a single chain to a double chain) while* a *and* c *remain basically the same. This doubling of* b *is seen in the* a-b *projection (center), with the resulting change in cleavage angles for the {210} family of planes. Note (right) the changes in the stereographic plots for the cleavage faces and crystallographic axes.* ***b.*** *Tremolite, a monoclinic amphibole, follows the same trends as the orthorhombic / monoclinic pyroxenes—the* a *axis is half the length in the monoclinic variety when compared to the orthorhombic one. Thus, the {210} planes in anthophyllite become {110} planes for tremolite (center), as seen in the* a-b *projection. In the stereographic projection (right), neither (100) and* a *nor (001) and* c *coincide. We'll see later how we can use these angular relationships to distinguish orthorhombic and monoclinic amphiboles.*

rotoinversion, and **lattices** which are combinations of these plus translations. Ultimately we want to arrive at all 230 possible combinations of all the symmetry operations, which will represent 230 so-called space groups. To accomplish this, we need the final two symmetry operations that we discussed in Chapter 11: **screw axes** and **glide planes**.

Screw axes. A screw axis is nothing more than a rotation axis (i.e., a 2-, 3-, 4-, or 6-fold axis) with a translation parallel to the rotation axis. A point is translated some distance parallel to the rotation axis, then rotated to create a new point. Screw axes are less common in nature than rotation axes. Figure 12.3 showed a jar of olives that exhibits a screw axis; the olives are moved/translated up to toward the top of the jar before being rotated around. Figure 12.64 shows a strange type of screw axis whereby the cat's right front paw is related to his right rear paw by a 2-fold rotation accompanying by a translation of his body length; you can conclude from this example that screw axes preserve handedness (or in this case, pawedness!).

Graphically, we can relate rotation axes to their accompanying screw axes. Figure 12.65a shows a 2-fold axis in perspective, with two points shown resulting from the 180° rotation. If we translate the two points from the 2-fold axis by moving them "one repeat" parallel to the 2-fold, this will simulate translating points from one unit cell into the next. In Figure 12.65b, the 2-fold axis is a 2-fold screw axis. In this case, the first point is rotated 180° as usual, but then it is translated parallel to the rotation axis "half a repeat" to create the new point. If this operation is continued for several translations and the points are traced out in 3-D space, a "screw" effect can be seen. Mineralogical examples of both a 2-fold axis and 2-fold screw axis are shown in the structure of natrolite (Figure 12.66a); both are parallel to *c*.

Figure 12.64. *A photo of one our cats in a strange pose. He has twisted his back in such a way that his body exhibits "screw axis" symmetry. His front right paw can be related to his right rear paw by a rotation of 180° and a translation of his body length. (Note: a screw axis doesn't change pawedness.)*

Figure 12.66b isolates the 2-fold axis down the tetrahedral chains and Figure 12.66c isolates the 2-fold screw axis down the center of the channel, where the translation is 1/2 along *c* (i.e., 1/2 the length of the *c* crystallographic axis).

A new twist is added when we move to the case of 3-fold screw axes because the rotations can be either clockwise (cw) or counterclockwise (ccw). Figure 12.67 shows perspective drawings for a 3-fold axis with translations added to rotations in both directions. It begins in the same way as the 2-fold screw axis in Figure 12.65, except the rotation is 120° instead of 180°. So we rotate 120° *counterclockwise*, then translate, to create a new point. Next, we rotate again by 120°, do another 1/3 translation, and create the third point. We could continue doing this to make spirals as is done in Figure 12.68. However, unlike the case for the 2-fold screw axis, we could rotate 120° *clockwise* and then do the 1/3 translations—this variation is shown in Figure 12.67c. Note that these two variations produce two different types of screw axes, depending on the direction of the spiral. These two axes are mirror images of each other and as such have a different handedness.

A mineralogical example of these two axes is shown in Figure 12.68 for quartz. On the left side of the drawing (Figure 12.68a) is right-handed quartz and on the right is left-handed quartz (Figure 12.68b). Notice how these two spiral in different directions. To aid in seeing these spirals we can use the cork screw from a Swiss Army knife (Figure 12.68c). That corkscrew (and for that matter, most thread on screws and bolts) is an example of a right-handed screw axis (Figure 12.68c), and its photo shows the same direction of spiraling. Playing some tricks with digital photography, we can create a mirror image of the knife and corkscrew in Figure 12.68c to produce a left-handed corkscrew (Figure 12.68d). The left-handed corkscrew's spiral is in the same direction as the 3-fold left screw axis. We could also draw a vertical plane between Figure 12.68a and b because they are mirror images.

A system is used to name screw axes and represent them with symbols. We begin with the rotation axis symbol, and add a subscript to its right to describe the type of screw. The little subscript holds two pieces of information: the amount of translation and its rotation direction (i.e., right- vs. left-handed or, synonymously, a ccw or cw rotation). We can generalize the translation amount by the term *n*/*r*, where *r* is the integer for the rotation axis (i.e., 2, 3, 4, 6) and *n* is the subscript. Also, recall that we used a left superscript to denote the direction of the rotation axis. For instance a $^{[010]}2_1$ screw axis would be oriented parallel to the *b* crystallographic axis and would have a translation of 1/2 , or half of the parallel to *b*. So we can rename our 3-fold right screw axis (Figures 12.67b and 12.68a) as a 3_1 and our 3-fold left screw axis (Figures 12.67c and 12.68b) as a 3_2.

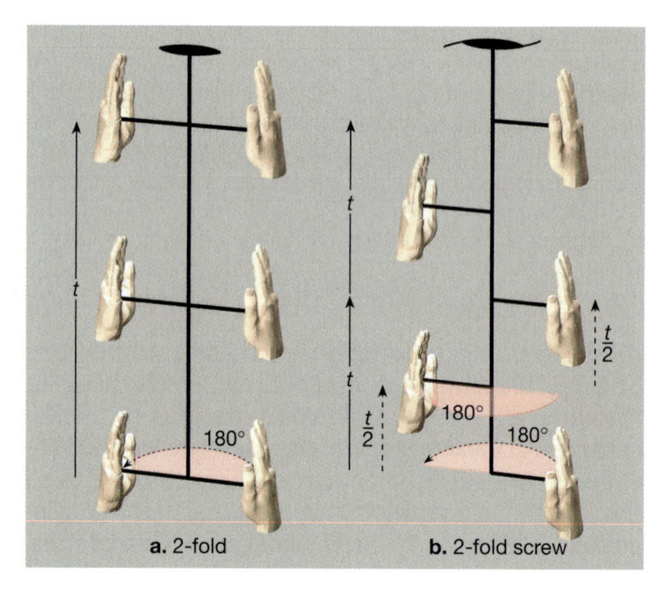

Figure 12.65. *Examples of 2-fold and 2-fold screw axes shown in perspective and their effect on hands (modified from Figure 7.4, Bloss, 1994).* ***a.*** *A 2-fold axis would relate the right hand to another right hand in the same plane after a 180° rotation. We could translate this pair of hands the length of the crystallographic axes, parallel to the 2-fold axis, to produce another set of hands in the adjacent unit cell.* ***b.*** *For a 2-fold screw axis, the 180° rotation is performed first, followed by a translation of half the cell length parallel to the 2-fold screw axis. Then a new symmetry point is created within the same unit cell. The process could be continued to create more hands by first a 180° rotation and then a 1/2 translation.*

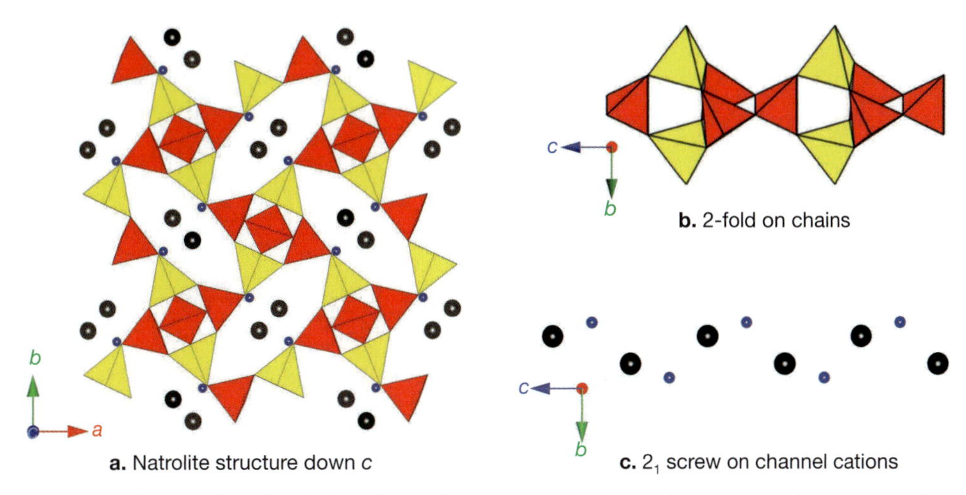

Figure 12.66. Mineralogical examples of 2-fold and 2-fold screw axes in the zeolite group mineral natrolite. ***a.*** *Natrolite projection down its* c *axis onto its* a-b *plane.* ***b.*** *A 2-fold axes parallel to* c, *and centered down the tetrahedral chains, is isolated from the structure shown in Figure 12.66a. Compare the tetrahedrons parallel to the [001]2 with the hands shown in Figure 12.65a.* ***c.*** *A 2-fold screw axis parallel to* c, *and centered down the channels, is isolated from the structure shown in Figure 12.66a. Again, compare the atoms parallel to this [001]2 screw to those in Figure 12.65b.*

Next we move on to the 4-fold axis. A $^{[001]}4_1$ (Figure 12.69b) would be a 4-fold screw axis parallel to *c* with a 1/4 translation parallel to *c*, while a $^{[001]}4_2$ (Figure 12.69c) would also be parallel to *c*, but have a half translation parallel to *c*. Last is a $^{[001]}4_3$ (Figure 12.69d). Based on the above description of translations, we would write 3/4 as the translation component for this screw axis. In

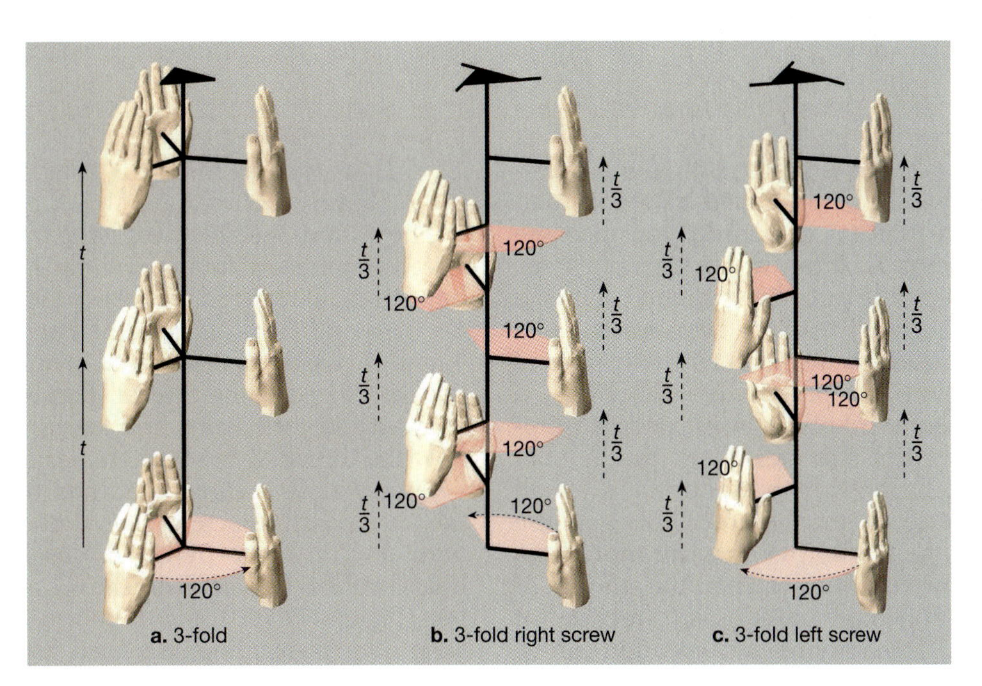

Figure 12.67. Examples of 3-fold and 3-fold screw axes shown in perspective, along with their effect on hands (modified from Figure 7.5, Bloss, 1994). ***a.*** *A 3-fold axis creates three right hands in the same plane. This set of hands can translate parallel to the 3-fold axis into the next unit cell.* ***b.*** *A 3-fold screw axis first performs a 120° ccw rotation and then translates the hand 1/3 of the distance along the crystallographic axis parallel to it before creating the second hand. Next, another 120° ccw occurs, accompanied by a 1/3 translation to create the third hand. All three of these would be in the same unit cell. The process could be continued to create atoms in the next unit cell.* ***c.*** *A slightly different* twist *on the 3-fold screw axis shown in Figure 12.67b. This time, the 120° rotation is clockwise instead of counterclockwise. While this still creates a 3-fold screw, the rotation is in a different direction. These two different rotation directions are mirror images of each other. To help show that, we've used a left hand for the cw rotation direction for this and future screw axes.*

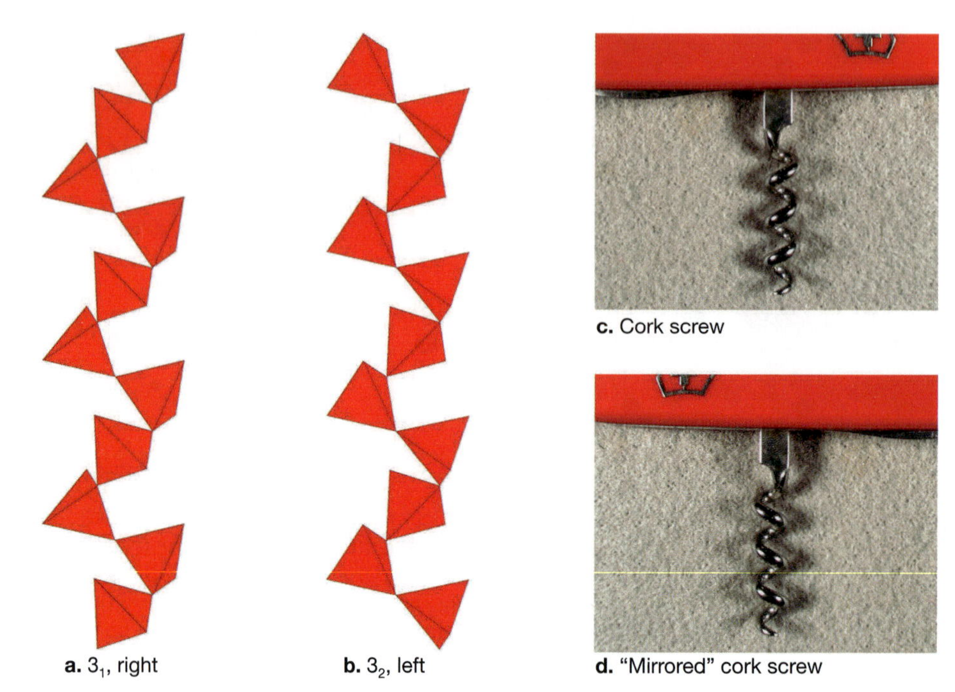

Figure 12.68. *A mineralogical and a physical example of a 3_1 (i.e., right-handed) screw axis that results from ccw rotations combined with a 3_2 (i.e., left-handed) screw axis, which results from cw rotations.* ***a.*** *The [001]3_1 screw axis of quartz shows how the silicate tetrahedra are related by a 120° ccw rotation and 1/3 translations along* c. ***b.*** *The [001]3_2 screw axis of quartz shows how the silicate tetrahedra are related by a 120° cw rotation and a 1/3 translation along* c. *Quartz can form with either right-handed or left-handed screw axes; both occur in nature with no apparent geological reason for their different handednesses. The 3_1 and 3_2 are mirror images of each other.* ***c.*** *A corkscrew is an example of a right-handed screw type of thread. Notice how it spirals in the same direction as the 3_1 screw in quartz.* ***d.*** *This image is the mirror reflection of the knife and corkscrew shown in Figure 12.68c. The spiral of the corkscrew in this image is the same as in a 3_2 screw. Think about how you would have to twist the cork screw in this image to get it to go into a cork as compared to the "normal" cork screw in Figure 12.68c.*

Figure 12.69d you see a point at 3/4 along the 4-fold axis that would be produced by a ccw rotations (as is the normal case), but you also see points at 1/4 and 1/2. A better way to look at a 4_3 (as well as the 3_2 and the upcoming 6_4 and 6_5) is as a cw rotation associated with the translation. Then we could treat the 3/4 translation as $1 - 3/4 = 1/4$.

Notice that in the perspective drawings, a slightly modified symbol has been used to represent the screw axis. We start with the symbol used for the rotation axis, but add barbs (small straight lines) to them (Figure 12.66c). These barbs are like the subscripts; they do more than identify a screw axis. They tell you the amount of translation in and the handedness (direction) of the screw axis. The amount of translation can be found by observing the number of barbs on the symbol. When a barb occurs on each corner of the symbol, then a translation occurs at every rotation increment. For instance, for the 4_1 symbol (Figure 12.69b), a 1/4 translation occurs every 90°, while for the 4_2, which only has two barbs, a translation only occurs every 180° (Figure 12.69c). The orientation of the barbs differs between the 4_1 (Figure 12.69b) and the 4_3 (Figure 12.69d), as well as for the 3_1 (Figure 12.67b) and the 3_2 (Figure 12.67c).

The 4_1 and 3_1 screw axes have translation associated with a ccw rotation, while 4_3 and 3_2 would have them associated with cw rotation directions. Thus, 4_1 and 3_1 have the same handedness, as do 4_3 and 3_2. Lastly, 4_1 and 4_3 and 3_1 and 3_2 are mirror images of each other. Recall Figure 12.68, which showed the cork screw and its mirror image and how the threads run in different directions.

Figure 12.70a shows a normal 6-fold axis and the five possible screw axes that are associated with it. The same trends discussed above also hold for the 6-fold screws. The 6_1 is the simplest to see (Figure 12.70b) and involves a 60° rotation with a 1/6 translation, repeated by 60° rotations and 1/6 translations. The 6_5 screw (Figure 12.70c) is its mirror image, with a different handedness than the 6_1 screw. For the 6_2 screw (Figure 12.70d) and its mirror image the 6_4 (Figure 12.70e), a 120° rotation is following by a 1/3 translation. The 6_3 screw (Figure 12.70f), which is somewhat similar to a 4_2 or 2_1 screw, shows the translation component is 1/2.

In summary, we've taken the four rotation axes (2-, 3-, 4-, and 6-fold) and added a translation component to them to make the eleven screw axes. Table 12.9 gives a list of symbols for all of the screws and their translational components. The screw axes are defined with a subscript to their right side that indicates the amount of translation and the handedness of the rotation. The values for translation refer to a distance in the unit cell of a mineral. For instance, Figure 12.71 shows a unit cell for a tetragonal mineral with $a = b = 5$ Å and $c = 10$ Å projected onto the *a*-*b* plane. Let's assume we have a $^{[001]}4_1$ axis and a $^{[010]}2_1$ axis (note these axes can occur at different locations in the unit cell). The $^{[001]}4_1$ axis would translate a point 2.5Å along the *c* axis (1/4 of its length), while the $^{[010]}2_1$ axis would translate a point 2.5Å along (half its length). If the $^{[001]}4_1$ axis was a $^{[001]}4_2$ axis instead, then the translation along [001] would be 5Å or half the length of *c*. Just in case you're wondering what "use" all of this might be, it turns out that handedness of molecules and handedness of mineral surfaces are helping explain the origin of life. Also drug companies spend billions of dollars annually to produce drugs with the correct handedness.

Glide planes. Glide planes are mirror planes plus a translation; instead of reflecting an object across a mirror plane, the object is first translated before the reflection occurs. Footsteps in the snow (Figure 12.72a) are the classic example of a glide plane. The right foot is "reflected" into the left foot by an imaginary plane running down the middle of the "steps" only after a translation (i.e., step) occurs. Compare this glide plane to a mirror plane (Figure 12.72b) where the feet are reflected back and forth (this took some fancy stepping to make!). Unlike screw axes, which can have several different amounts of translations, a glide plane always has a translation of either 1/2 (i.e., half the repeat distance of the unit cell) or 1/4 (i.e., 1/4 the repeat distance of the unit cell). The translation can be in just one direction (say, along the *a* axis), or along two directions (e.g., there could be a 1/2 translation along *a* followed a 1/2 translation along *b*). So we'll need to find a way to keep track of the translation direction(s), as well as the plane in which the translation occurs.

Consider first the symbols used for the different kinds of glide planes. Glide planes are represented by single letters (just as a mirror plane is just represented by the single letter "*m*"). There are really three kinds of glide planes: (1) simple

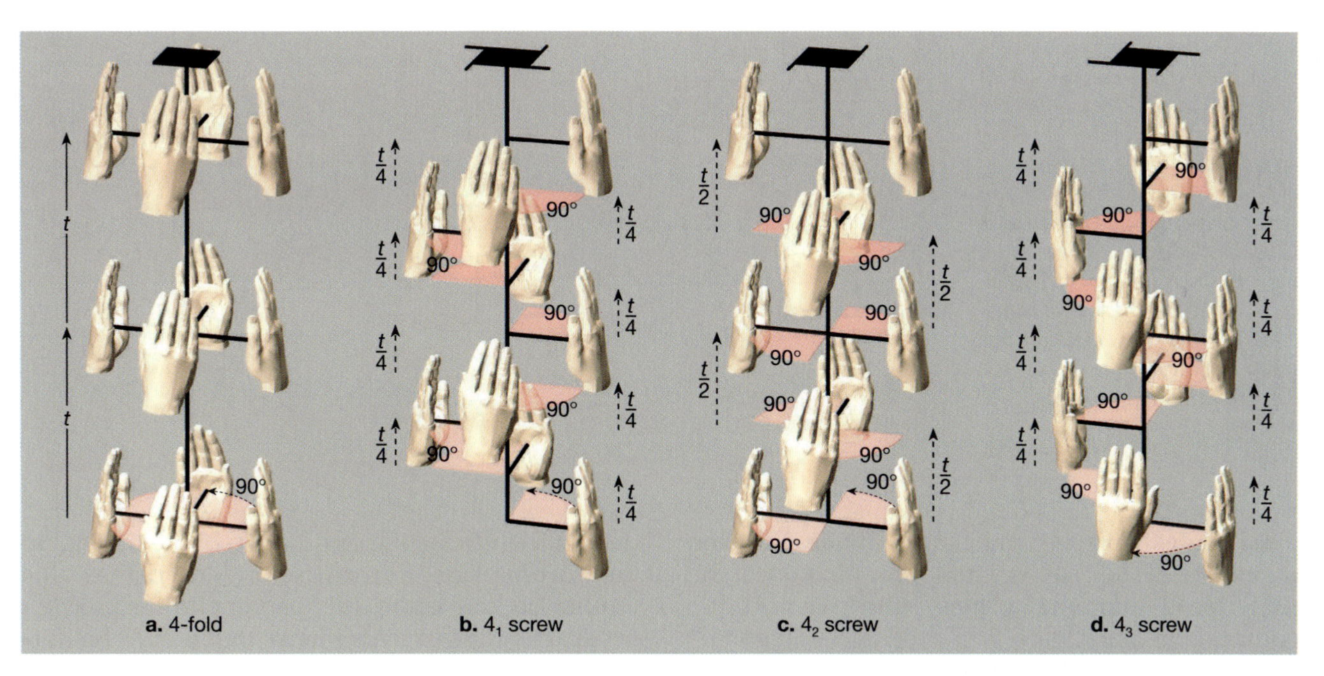

Figure 12.69. Examples of 4-fold and 4-fold screw axes shown in perspective and their effect on hands (modified from Figure 7.6, Bloss, 1994). ***a.*** *A normal 4-fold axis creates four hands in the same plane. This group of hands can be translated parallel to the rotation axis to create another set of hands in the next unit cell.* ***b.*** *A 4_1 screw would first rotate the right hand by 90° ccw and then translate it 1/4 parallel to the rotation axis before creating the next hand. This is followed another 90° ccw rotation and 1/4 translation to create the next hand, and finally one more 90° rotation and 1/4 translation to create the fourth hand. All of these hands would be in the same unit cell and would form a right-handed spiral of right hands similar to the 3_1 screw shown in Figure 12.67b.* ***c.*** *A 4_2 screw differs slightly from the screw axes we've seen so far. It would create two hands in the same plane at 180° from one another. Next, this pair would be translated half way along the screw axis with a rotation of 90°. The olive jar in Figure 12.3a is an example of a 4_2 screw.* ***d.*** *The 4_3 screw is similar to the 4_1 screw except the rotations are cw, thus forming a left-handed spiral similar to the 3_2 screw shown in Figure 12.67c. The 4_1 and 4_3 screws are mirror images of each other.*

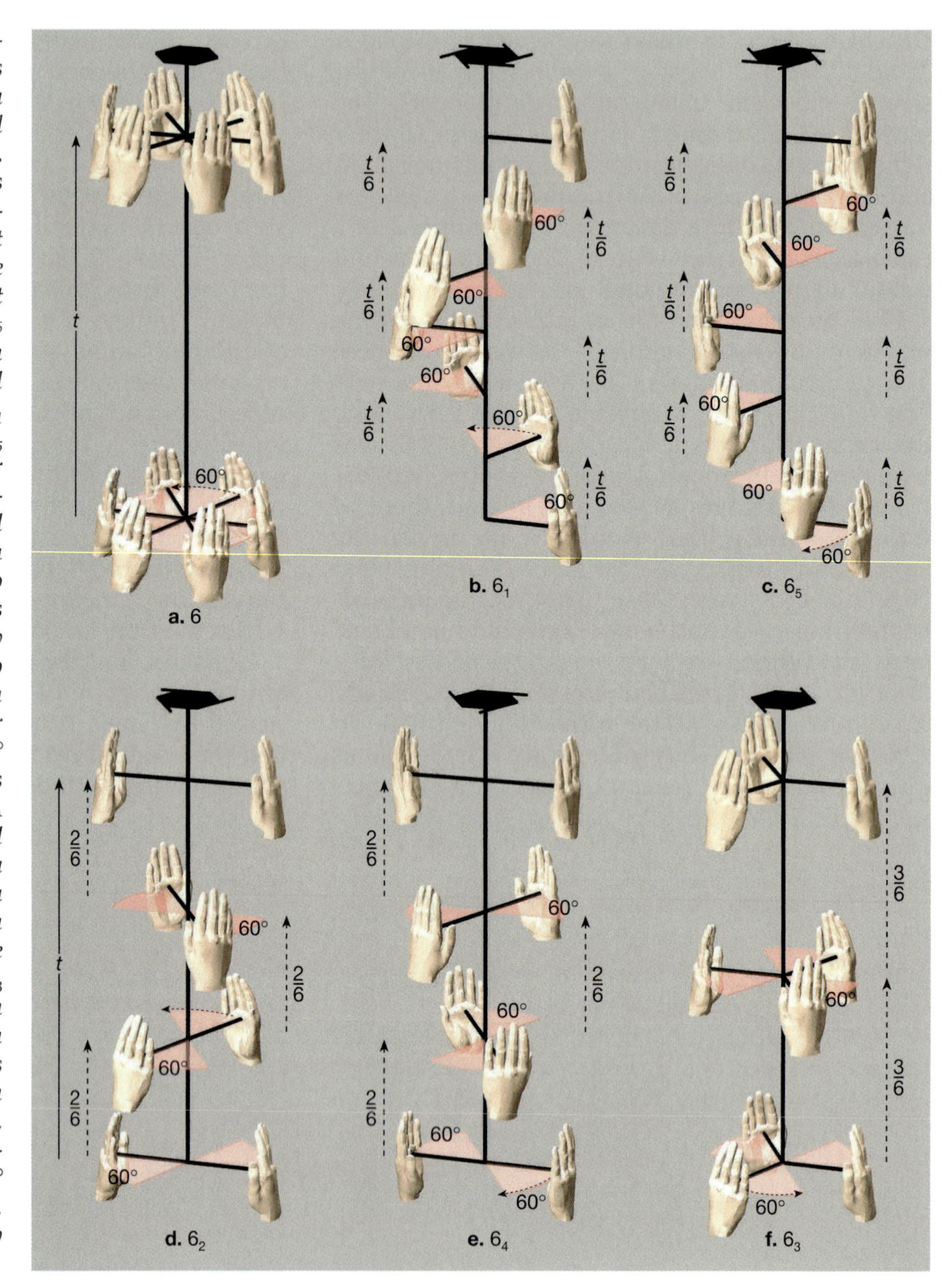

Figure 12.70. *Examples of 6-fold and 6-fold screw axes shown in perspective along with their effect on hands (modified from Figure 7.7, Bloss, 1994).* ***a.*** *A 6-fold axis creates six hands in the same plane at 60° increment. We can translate the set parallel to the axis to create more atoms in adjacent unit cells.* ***b.*** *A 6_1 screw has one 1/6 translation per 60° ccw rotation to create the right-handed spiral above. Compare this to the 4_1 screw and the 3_1 screw.* ***c.*** *The 6_5 screw has one 1/6 translation per 60° cw rotation to create the left-handed spiral above. The 6_1 and 6_5 are mirror images of each other. Compare the 6_5 screw to the other left-handed screws (i.e., 3_2 and 4_3).* ***d.*** *A 6_2 screw has a translation of 2/6 or 1/3. Two hands are created by the rotation in each plane. In turn, this pair is translated 1/3 and rotated 60° before repeating. The process is continued as shown above.* ***e.*** *A 6_4 screw is the left-handed equivalent of the 6_2. Again, a pair of hands is created in each plane with a 1/3 translation between them, the difference being a cw rotation.* ***f.*** *A 6_3 screw would have a translation of 1/2 and a rotation of 120°. In any one plane, this screw axis looks like a 3-fold axis, based on the 120° rotation. However, each triple set of hands is translated 1/2 and then rotated 60° with respect to the lower set. The jar of olives in Figure 12.3b contains a 6_3 screw.*

axial glides with a 1/2 translation in one direction along one of the three crystallographic axes of the mineral, (2) **diagonal glides**, with two 1/2 glide components parallel to two axes in a "diagonal" direction, and (3) **diamond glides**, which are just like diagonal glides except the translation is 1/4 in both directions. In a glide plane, the translation direction is restricted to a plane that would contain two of the crystallographic axes and be perpendicular to the third. Thus the translation can take place along either of the directions *in* the plane (Figure 12.73), but *not normal* to it. Recall that when we wanted to distinguish the orientation of mirror planes, we used the superscript to the left to indicate the orientation of the normal to the mirror plane. In the case of glide planes, this would be the direction where gliding *couldn't* occur. The three mirror planes that would be normal to the *a*, *b*, and *c* axes would be $^{[100]}m$, $^{[010]}m$, and $^{[001]}m$, respectively. Thus gliding could not take place in the *a* direction for $^{[100]}m$, the *b* direction for $^{[010]}m$, or the *c* direction $^{[001]}m$.

Consider first the three types of **axial glides**, which are represented by *a*, *b*, or *c* to indicate the direction of translation. Thus an *a* glide has translation in the *a* direction, a *b* glide in the *b* direction, and a *c* glide in the *c* direction. The *a* glide needs to have translation in the *a* direction, so we could

Table 12.9. Space group operations, names, and symbols

With respect to a line: screw axes	*Symbol*
2_1 = 2 fold + 1/2 translation parallel to a crystallographic axis	
3_1 = 3 fold + 1/3 ccw translation parallel to a crystallographic axis	
3_2 = 3 fold + 1/3 cw translation parallel to a crystallographic axis	
4_1 = 4 fold + 1/4 ccw translation parallel to a crystallographic axis	
4_2 = 4 fold + 1/2 translation parallel to a crystallographic axis	
4_3 = 4 fold + 1/2 cw translation parallel to a crystallographic axis	
6_1 = 6 fold + 1/6 ccw translation parallel to a crystallographic axis	
6_2 = 6 fold + 1/3 ccw translation parallel to a crystallographic axis	
6_3 = 6 fold + 1/2 translation parallel to a crystallographic axis	
6_4 = 6 fold + 1/3 cw translation parallel to a crystallographic axis	
6_5 = 6 fold + 1/6 cw translation parallel to a crystallographic axis	
With respect to a plane: glide planes	
a = mirror + 1/2 translation parallel to the *a* crystallographic axis	
b = mirror + 1/2 translation parallel to the *b* crystallographic axis	
c = mirror + 1/2 translation parallel to the *c* crystallographic axis	
d = mirror + 2 1/2 translation parallel two separate crystallographic axes	
n = mirror + 2 1/4 translation parallel two separate crystallographic axes	

have a $^{[010]}a$ or $^{[001]}a$ (Figure 12.74a). A *b* glide can have translation in the *a* or the *c* direction, $^{[100]}b$ or $^{[001]}b$ (Figure 12.74b). A *c* glide needs translation in the *c* direction, $^{[100]}c$ or $^{[010]}c$ (Figure 12.74c).

Note that there is again a sort of odd-man-out relationship here. If we think of the crystallographic directions (*a*, *b*, *c*) in terms of their vector notations ([100], [010], [001]) then the glide planes $^{[100]}[100]$, $^{[010]}[010]$, and $^{[001]}[001]$ cannot exist mathematically. But $^{[010]}[100]$ and $^{[001]}[100]$ can occur and would be $^{[010]}a$ and $^{[001]}a$; the same is true for $^{[100]}b$, $^{[001]}b$, $^{[100]}c$, and $^{[010]}c$. The confusing part (at least one of them!) is that the letters we use to represent axial glides really tell us the *direction* of translation. They do not directly describe the plane where they lie in the mineral, but they do tell us *where they are not*. For instance, an *a* glide cannot be perpendicular to the *a* axis, a *b* glide cannot be perpendicular to the *b* axis, and a *c* glide cannot be perpendicular to the *c* axis. Figure 12.75 shows an example of a $^{[010]}c$-glide in diopside.

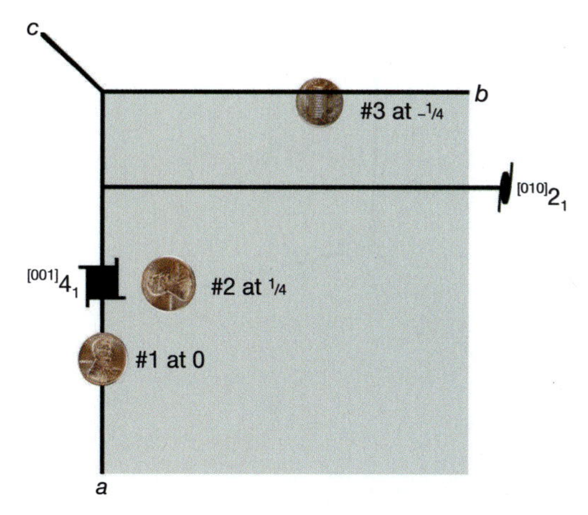

Figure 12.71. *A projection of a tetragonal unit cell* (a = b = 5, c = 10Å) *onto the* a-b *plane. Included in the projection is a* $^{[001]}4_1$ *(normal to the projection) and a* $^{[010]}2_1$ *(in the page and parallel to* b*). The* $^{[001]}4_1$ *would rotate an atom (penny number 1) by 90° ccw and translate it to* 1/4 c *(i.e., 2.5Å) along* c *resulting in a new atom (penny number 2). The* $^{[010]}2_1$ *would take the atom (penny number 2) rotate it 180° and translate it* 1/2 b *(i.e., 2.5Å) to produce a third atom (penny number 3).*

Figure 12.72. *Physical examples of a glide plane and a mirror plane.* ***a.*** *This photograph shows my footprints in the snow. As I walk, my right foot and left foot are translated by my stride. If you placed a line between my right and left feet, you would see something similar to a mirror plane, except that a translation occurs between the reflections. So my right foot translates parallel to the mirror plane before it is reflected. In crystallography, this type of symmetry element is called a glide plane.* ***b.*** *Here I've just gone back and added some extra steps that would convert the glide plane in Figure 12.72a to a mirror plane. The reflection changes the handedness of an object, or in this case the "footedness!"*

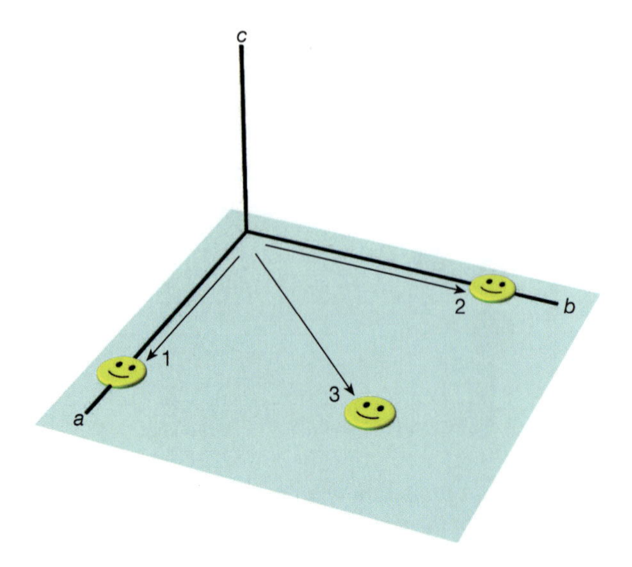

Figure 12.73. *Glide planes are based on mirror planes with a translation parallel to the mirror plane. For example, a* [001]m *would have* a *and* b *in the plane of the mirror. Thus, translation could occur parallel to* a *(as for vector 1), parallel to* b *(as for vector 2), or parallel to a combination of* a *and* b *(as for vector 3). Translations can only occur in the plane parallel to the mirror plane. This is critical to understanding the nomenclature associated with mirror planes.*

The next types of glides are those that have translations in two directions; these are the **diagonal** (translations of $1/2$) and the **diamond** (translations of $1/4$) glides. Unfortunately, both of these glide types start with the same letter. The customary method is to use "*n*" for the diagonal glide and "*d*" for the diamond glide. The same rules about translation directions being confined to planes and naming the orientation apply here as did for the axial planes. Thus, a [100]*n* glide would have translations confined to the (100) plane, and it would glide $1/2$ in both the *b* and *c* directions (Figure 12.76a). A [001]*d* glide (Figure 12.76b) would have $1/4$ translations confined to the (001) plane in both the *a* and *b* directions. Figure 12.77 shows an example of a diamond glide in diamond.

To summarize (Figure 12.78), glide planes are mirror planes with a translational component *in* the plane of reflection. Translations of $1/2$ and $1/4$ are possible. Axial glides have a $1/2$ translation parallel in one direction, diagonal glides have two perpendicular $1/2$ translations, and diamond glides have two perpendicular $1/4$ translations.

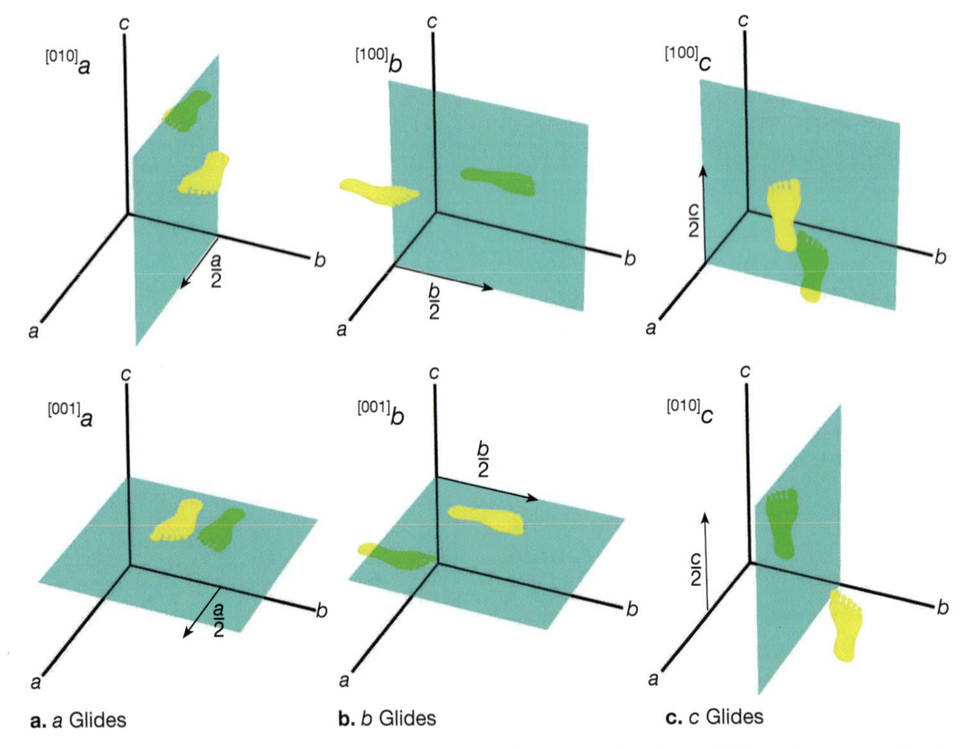

Figure 12.74. *Examples of the six possible types of axial glides that occur for three differently-oriented mirror planes:* [100]m, [010]m, *and* [001]m. *These are all based on $1/2$ translations within the plane parallel to two possible crystallographic directions. For each case, the glide direction replaces the "*m*" in the symbol. The arrows show the direction of translation. The right foot is the original point and is translated and reflected into a left foot.* ***a.*** *By definition,* a *glides have $1/2$ translation in the* a *direction. Thus* [010]a *and* [001]a *can exist, but* [100]a *cannot, because the translation would have to be perpendicular to the mirror plane, which is not allowed (refer back to Figure 12.73).* ***b.*** *For* b *glides, the translation is $1/2$ parallel to* b. *So* [100]m *with a $1/2$ translation parallel to* b *would become* [100]b *and* [001]m *with a $1/2$ translation parallel to* c *would become a* [001]b. ***c.*** *Finally, $1/2$ translations could occur parallel to* c *for the case of* [001]m. *For this case, we would arrive at* [100]c *and* [010]c.

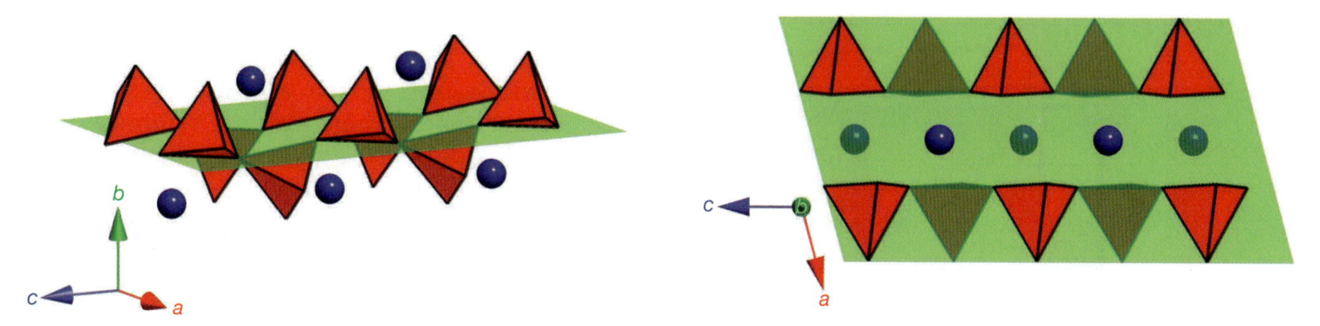

Figure 12.75. A mineralogical example of [010]c *in diopside with the* Mg^{2+} *(i.e., M1 sites) removed, the* Ca^{2+} *cations (i.e., M2 sites) shown as spheres, and portions of two tetrahedral chains. Recall* [010]c *is composed of a mirror plane perpendicular to* b *with a* ½ *translation in the* c *direction. The image on the left is a perspective view and shows how the* Ca^{2+} *cations are related to one another with a translation parallel to* c *and then a reflection across the green plane (which represents* [010]c*). The view on the right is projected perpendicular to the glide plane and shows the* Ca^{2+} *cations alternately above and below the glide plane and translated parallel to* c.

The glides are named for the direction and amount of translation. Finally, in Table 12.9 is a list of the glide planes.

Space Groups

With the addition of screw axes and glide planes, we now have the full complement of symmetry operations that exist in 3-D, so we can mix and match them to create all 230 of their possible combinations, which are called **space groups**. Each of these has its own symbol and number, from 1 to 230. The 230 space groups are listed by crystal system and point group in Table 11.9.

To start, let's look at the anatomy of a space group symbol. The space group symbols are a combination of the lattice type and one of the 32 point groups. The possible number of space groups grows as we allow for the translational symmetry that creates screw axes and glide planes. Consider the space groups $P2/m\ 2/m\ 2/m$ (Figure 12.79a) and $P2_1/b\ 2_1/a\ 2/m$ (Figure 12.79b). Both of these are orthorhombic and have a P lattice (the letter representing the lattice type always proceeds the symmetry operators). $P2/m\ 2/m\ 2/m$ contains only point group operations while $P2_1/b\ 2_1/a\ 2/m$ contains both point and space group operations.

A bias in the nomenclature has evolved for use of space group symbols. Crystallographers often abbreviate space group symbols, as well as point group symbols discussed above. For instance, $P2/m\ 2/m\ 2/m$ is abbreviated as $Pmmm$ and $P2_1/b\ 2_1/a\ 2/m$ as $Pbam$. Although a crystallographer would understand these abbreviations, we will avoid abbreviations of space groups in this text. The amount of time saved in making such abbreviations is an order of magnitude less than the confusion they have caused! Table 11.9 lists the 230 space groups associated with the 32 point groups and six crystal systems. It's our challenge now to briefly discuss these space groups to see how they are created, so we can use them in Chapter 14 to represent crystal structures. (By the way, one of us (M.E.G.) had an entire semester course on space groups as a graduate student, so this is really a *brief* overview!)

Triclinic system. We'll develop space groups from lowest to highest symmetry, and begin with the triclinic system. Recall that there are only two

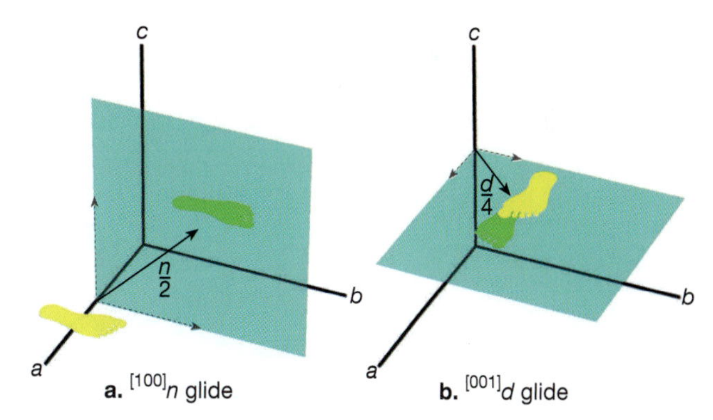

*Figure 12.76. Examples of a diagonal (*n*) glide and diamond (*d*) glide. Both of these glides have translations in two, rather than one direction, as for the axial glides.* ***a.*** *Diagonal (*n*) glides have two* ½ *glide translations parallel to the two crystallographic axes in the plane; the end result of this is a "diagonal" translation (with respect to the two axes) across the plane. The vector 3 in Figure 12.73 would be the result of a* ½ *translation parallel to* a *and a* ½ *translation parallel to* b. *Thus, a* [100]n *would have* ½ *translation in* b *and* ½ *in* c *and transform the right foot into a left foot as above.* ***b.*** *A diamond (*d*) glide is similar to an* n *glide, except that the translations are only* ¼ *in both directions instead of* ½. *So a* [001]d *would be similar to a* [001]n, *though the length of the diagonal translation is less. Thus the* [001]d *would translate* ¼ *in* a *and* ¼ *in* b *and transform the right foot into the left foot as above.*

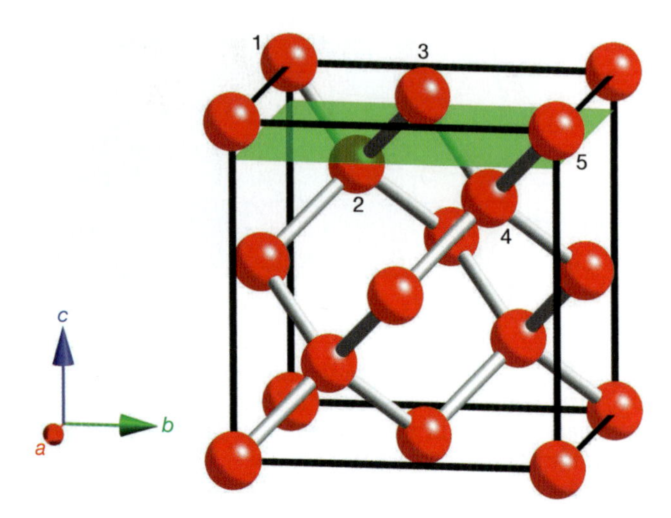

***Figure 12.77.** An example of a diamond glide in diamond. The [001]d causes the C atoms to be translated by 1/4a and 1/4b before being reflected across a plane parallel to (001). The [001]d is shown on the upper portion of the unit cell of diamond. Atom 1 is related to 2 by the operation of the [001]d, as is atom 2 to 3, 3 to 4, and 4 to 5.*

triclinic point groups: 1 and $\bar{1}$. To make space groups from these point groups, we need to associate each of them with a Bravis lattice (Figures 12.48 to 12.51). For the triclinic system, only *P* lattice can occur. Thus, we add the "*P*" to each of our point groups and arrive at *P*1 and $P\bar{1}$ space groups. Are there other types of symmetry elements that could exist within the triclinic system if we allowed for the addition of translational components (i.e., any screw axes or glide planes)? The answer is no, because there are no mirror planes in this systems and there is no such thing as a 1-fold screw axis.

Monoclinic system. This system contains 13 space groups, and there are more minerals from them than from any of the others (Table 11.9). There are two lattice types (Figures 12.48 and 12.50) that can occur in this system, *P* and *C*, where *C* is a side-centered lattice with an extra lattice point on the (001) face. Recall there are three point groups in this system: 2, *m*, and 2/*m*. So we add a *P* and *C* lattice type to each of these three space groups to arrive at *P*2, *C*2, *Pm*, *Cm*, *P*2/*m*, and *C*2/*m*.

Next we need to consider what types of screw axes and glide planes might occur in this system. A 2-fold axis with a 1/2 translation would be a 2_1 screw (more precisely here a $^{[010]}2_1$), so we could form $P2_1$ and $P2_1/m$. There is no $C2_1$ because a $^{[010]}2_1$ and *C*-centered lattice are not compatible

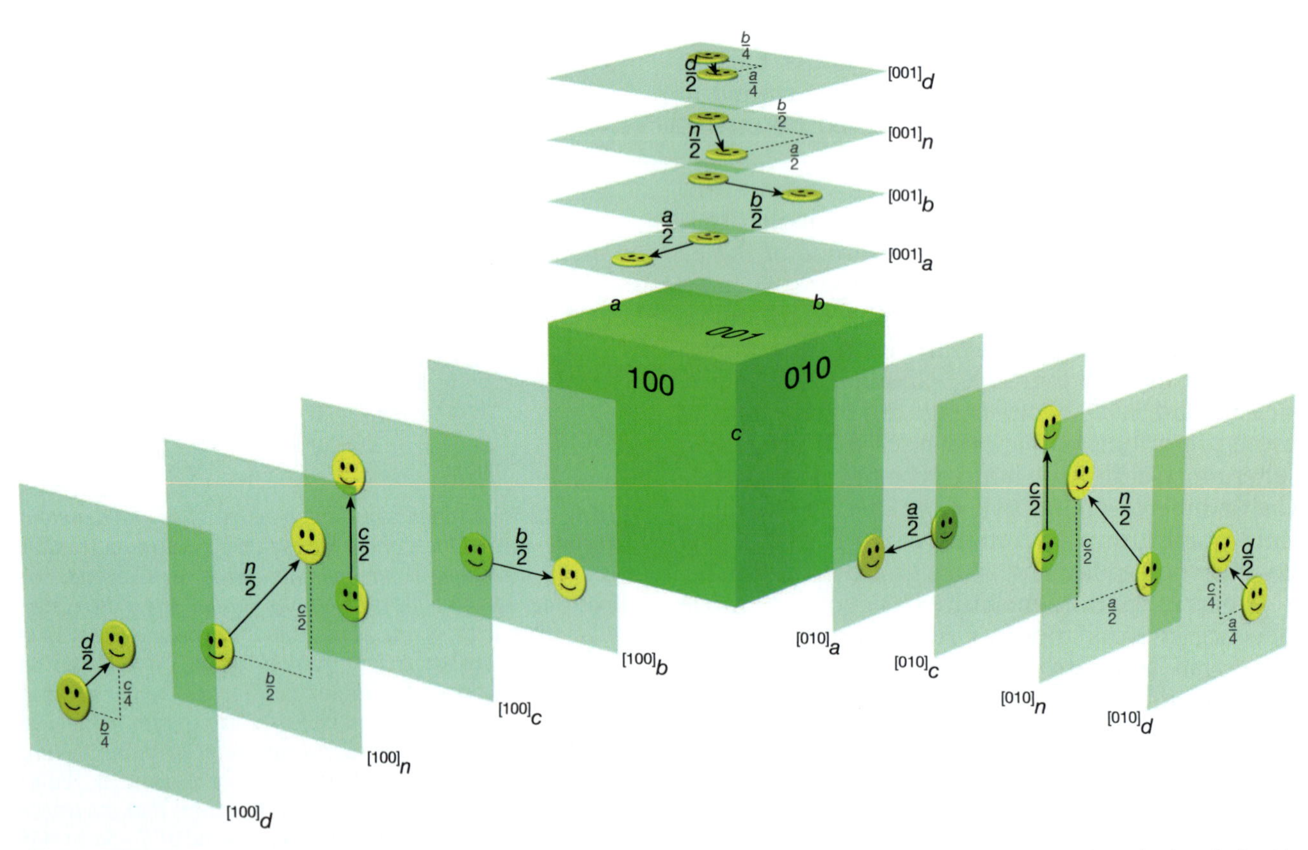

***Figure 12.78.** A summary of the different types of glide planes and how they are symmetry-constrained to the (100), (010), and (001) faces. As shown, there are four possible glide planes for each of the three faces: two axials, a diagonal, and a diamond glide. Adapted from Bloss (Figure 7.9, page 170, 1994).*

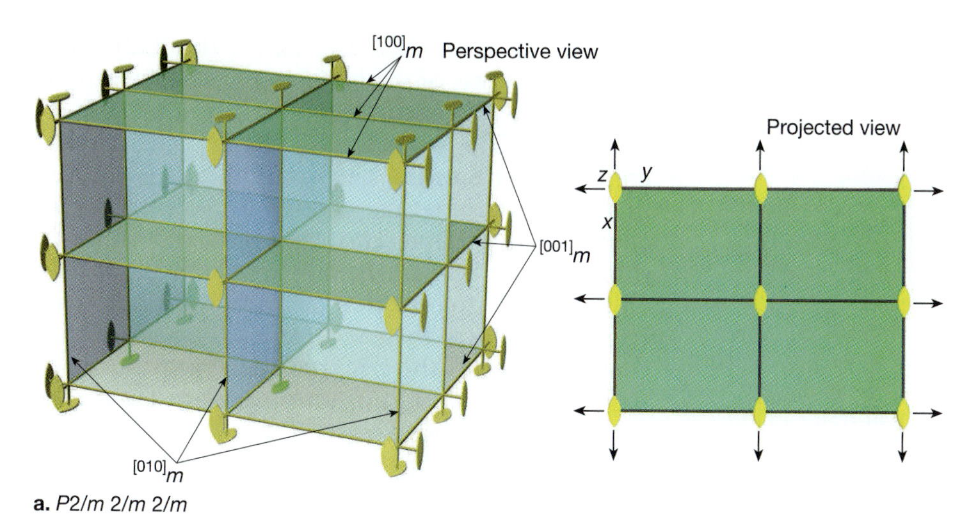

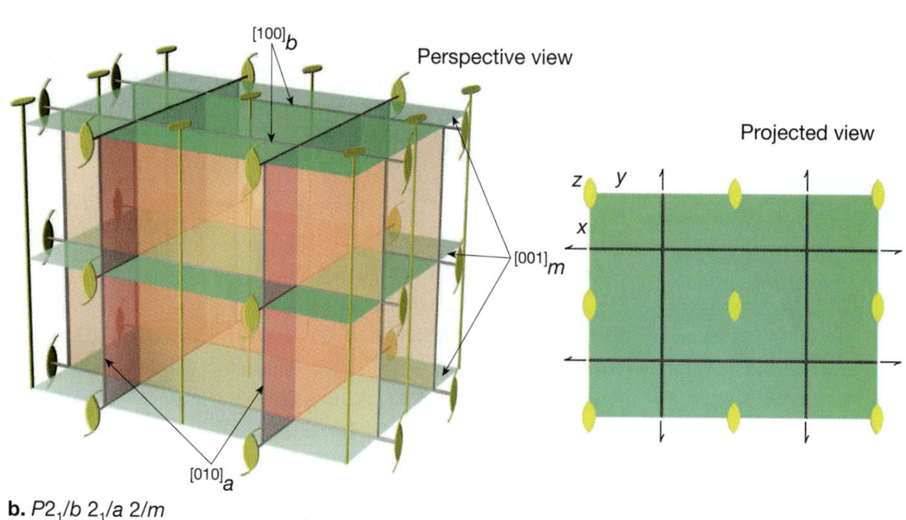

*Figure 12.79. Examples of orthorhombic space groups both in a perspective and projected views. See Tables 12.3 and 12.9 for the meaning of the symbols **a.** The space group P 2/m 2/m 2/m would have 2-fold axes and mirrors coinciding with, and perpendicular to, all three crystallographic axes. **b.** The space group P2_1/b 2_1/a 2/m is similar to P 2/m 2/m 2/m, except two of the 2-fold axes are now 2_1 screws and two of the mirror planes are now axial glide planes.*

(i.e., the two symmetry types cannot exist at the same time.) What kind of glide planes can occur in the monoclinic system? Recall that we have a $^{[010]}m$, which can have glide components in either *a* or *c* (Figure 12.80). Because of the monoclinic symmetry, the glide direction can only be in *c* (Figure 12.80a). By adding the *c* glide, we create *Cc*, *P2/c*, *P*2_1*/c*, and *C2/c*. So we have created the thirteen monoclinic space groups from the symmetry-allowed permutations of the three point groups, two lattice types, and screw axes and glide planes (i.e., the $^{[010]}2_1$ and $^{[010]}c$).

Some of the space groups in Table 12.10 have a "*" next to them. These are the so-called "non-standard" space groups and they occur because some researchers have chosen different orientations for the symmetry elements and crystallographic axes. There are at least ten non-standard space groups for the triclinic case, 30 for monoclinic, and a whopping 165 for the orthorhombic systems (fortunately, none for the other systems because none of the other crystal systems has these none-standard space groups; they are more symmetry-constrained and the choice of symmetry directions cannot be varied). Several of these non-standard space groups are used to describe the feldspars and are extremely common in minerals (Table 12.10). We'll turn our attention to the non-standard space groups after the discussion of space groups is completed.

Orthorhombic system. Because we are using permutations of symmetry elements to compose the space groups, the numbers of possibilities grows with the number of symmetry elements and lattice types. The orthorhombic system only has three point groups (222, *mm*2, and 2/*m* 2/*m* 2/*m*), there are four possible lattice types (Figures 12.48 to 12.51). So we start with twelve space groups before considering screw axes and glide planes. When we add them in, we gain an addition 38 new space groups (Table 11.9). Unlike the monoclinic case, we can have glide planes along (100), (010), and (001) because the crystallographic axes in the orthorhombic system are mutually perpendicular. Also, there are *a*, *b*, *c*, and even *n* glides. What's more, past researchers have selected different ori-

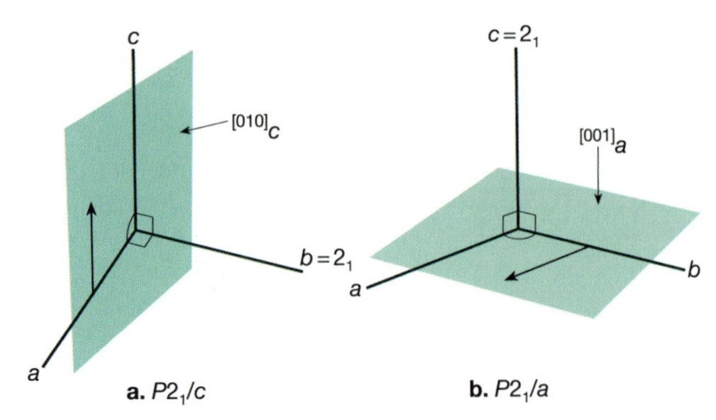

Figure 12.80. *Different choices of axes will result in different space groups. For instance, in the monoclinic minerals either the* b *or* c *axis can be chosen to correspond to the sole symmetry direction.* ***a.*** *The commonly-chosen second setting for monoclinic minerals, where* b *is selected as the symmetry direction, results in the standard space group* $P2_1/c$*, where the* $^{[010]}m$ *is converted into a* $^{[010]}c$ *glide by translation parallel to* c. ***b.*** *In the so-called first setting, where* c *is selected to coincide with the sole symmetry axis, the* $P2_1/c$ *shown in Figure 12.80a would become* $P2_1/a$*, because the* $^{[001]}m$ *would change to* $^{[001]}a$ *with the glide parallel to* a *(as indicated by the arrow). In Figure 12.80a, (the second setting)* $\alpha = \gamma = 90°$ *and* $\beta \neq 90°$*, while for the first setting,* $\alpha = \beta = 90°$ *and* $\gamma \neq 90°$*. The crystal structures of the minerals are not changing, but we are just choosing different basic vector sets to describe them.*

entations for some of the symmetry operations, creating 165 non-standard space groups (Table 12.10, which will be discussed later.)

Tetragonal system. There are seven point groups in this system but only two lattice types, *P* and *I*, which gives us 14 space groups to start. Another 52 space groups are created by the addition of screw axes and glide planes (Table 11.9).

Hexagonal system. Again there are only two lattice types in this system, *P* and *R*. The *R* lattice type can only occur in minerals with a 3- or $\bar{3}$-fold axis parallel to *c*, of which there are five point groups. Incorporation of the *P* and *R* lattice types with the space symmetry operations yields 25 point groups. There is some overlap between these space groups, as pointed out in Figure 12.53 where corundum was generated based on either an *R* or *P* lattice. The seven point groups that start with a 6 or $\bar{6}$ can only have *P* lattices. The addition of translations then yields the 27 space groups shown in Table 11.9.

Isometric system. This system has 3-fold axes that coincide with the [111] direction, making five point groups. Three possible lattice types can occur for this system, *P*, *I*, and *F*. Combinations of the five point groups and three lattice types with translational symmetries result in the 36 remaining space groups (Table 11.9).

Table 12.10 lists the top 20 (of the 230) space groups in terms of the numbers of known minerals to form in each. Interestingly, the sum of these 20 is 53%, so only 9% of the space groups contain over half the known minerals! $P\bar{1}$ occurs in the most minerals, followed by four monoclinic space groups. The space group systems in this table follow fairly closely those shown in Table 12.8, except that triclinic system now comes in second instead of third. But also notice that all (two) of the triclinic space groups are in the top 20 and not far from where the orthorhombic ones are.

Non-standard space groups. As we mentioned above, there are non-standard space groups in the triclinic, monoclinic, and orthorhombic systems. Fortunately, there are no non-standard space groups in the systems with higher symmetry because the symmetry axes are constrained to conform to the crystallographic axes in a fixed manner (i.e., the 4-fold and 6-fold axes of the tetragonal and hexagonal systems always coincide with *c*, while the 3-fold axes of the isometric systems always coincide with [111]).

The non-standard space groups in the triclinic system result from the choice of a different lattice type. Table 12.11 lists the ten non-standard space groups for the triclinic system, which show all the different lattice types except *R*. These differ-

Table 12.10.
The "top 20" space groups and the number and percent of the minerals they contain.
(Data from Dana's New Mineralogy (Gaines et al., 1997)).

Space group	*# minerals*	%
$P\bar{1}$ (2)	221	6.9
$C2/m$ (12)	217	6.8
$C2/c$ (15)	138	4.3
$P2_1/c$ (14)	132	4.1
$P2_1/a$ (14)*	124	3.9
$R\text{-}\bar{3}\ 2/m$ (166)	101	3.2
$P1$ (1)	90	2.8
$F4_1/d\ \bar{3}\ 2/m$ (227)	81	2.5
$P2_1/n$ (14)*	80	2.5
$P2_1/m$ (11)	75	2.3
$P6_3/m\ 2/m\ 2/c$ (194)	66	2.1
$F4/m\ \bar{3}\ 2/m$ (225)	58	1.8
$P2_1$ (4)	44	1.4
$P2_12_12_1$ (19)	42	1.3
$P6_3/m$ (176)	42	1.3
$P2_1/b\ 2_1/n\ 2_1/m$ (62)*	41	1.3
$R\bar{3}$ (148)	38	1.2
$R3m$ (160)	37	1.2
$P2_1/b\ 2_1/c\ 2_1/a$ (61)	32	1.0
$P2_1/a\ 3$ (205)	30	0.9

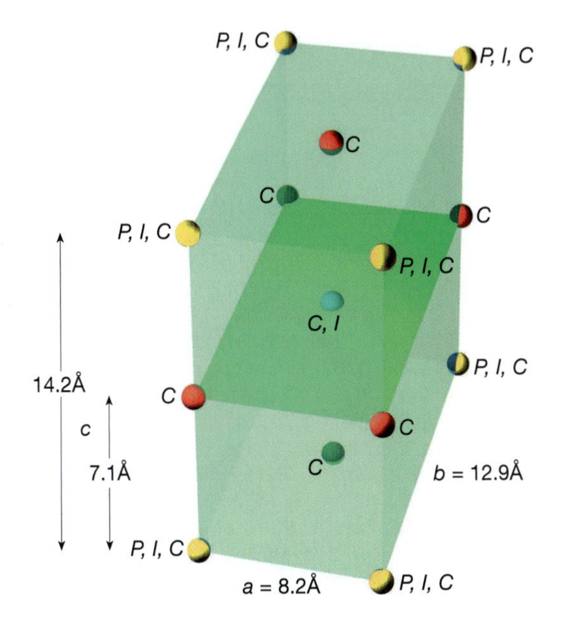

Figure 12.81. *Above three different lattice types are drawn on a unit cell of a feldspar. Three letters are used* (P, I, C) *to mark the different cells, with the letters representing the lattice corners, face, and/or center. Again notice how the* P *lattice repeat in c is twice that of the* C *lattice. (Modified from Figure R.12, Ribbe, 1975.) For the triclinic system, there are non-standard space groups that result from choosing lattices other than* P. *For instance, the standard space group* $P\bar{1}$ *is often used to describe feldspars, though past researchers have (unfortunately) chosen other related lattice types, such as* $I\bar{1}$ *or* $C\bar{1}$. *Note that as in the case of a* P *lattice for enstatite (Figure 12.55b) and a* C *lattice for diopside (Figure 12.55a), the* a *repeat is twice as long for* P *as for* C. *The same condition holds here for the* C *triclinic lattice as compared to the* P *lattice.*

ent lattice types occur by changing the size of the unit cell. For instance, $P\bar{1}$, $I\bar{1}$, and $C\bar{1}$ lattice types have all been used to describe feldspars (Figure 12.81). While $P\bar{1}$ would work to describe all of the structures for the feldspars, earlier researchers unfortunately picked these other types of space groups. Because we often want to compare current structures to those done earlier, we have adopted some of these non-standard space groups. The remaining eight non-standard space groups could be described similarly based on changing the size of the unit cell to incorporate different lattice points.

In the monoclinic system, non-standard space groups result from selection of different symmetry axes and non-standard lattice types. For instance, $P2_1/c$ could be represented by the non-standard space group $P2_1/a$ (Figure 12.80) by choosing a different axis set (i.e., the former has *b* as the symmetry axis while the latter has *c*). Different non-standard lattice types can also be chosen. Table 12.11 shows the 27 non-standard monoclinic space groups that result from selecting *c* instead of *b* as the sole symmetry axis, and allowing for other lattice types.

Because there are six permutations of how to fit the three mutually perpendicular symmetry operation directions with the three mutually perpendicular crystallographic axes, there are 156 possible non-standard space groups in the orthorhombic system! Table 12.11 lists six entries for each of the 57 orthorhombic space groups that result from permutations of the *a*, *b*, and *c* axes. Consider space group number 41. The point group for this space group is *mm*2. Thus we can expect to have mirror planes or glide planes in the first two entries that correspond to the *a* and *b* directions and a 2-fold or 2_1 screw in the last (or *c*) direction. The standard setting for this space group creates *Aba*2, which we could write as $A^{[100]}b\ ^{[010]}a\ ^{[001]}2$. Notice that there are five other possible ways that *a*, *b*, and *c* could fit to this symmetry if we choose the axes differently. For example, if we write $C^{[100]}c\ ^{[010]}2\ ^{[001]}a$, it would imply that *c* and *b* have changed places; this is (hopefully) apparent because the position of the "2" was changed.

When the axis directions are changed relative to the symmetry directions, the other symmetry types change (Figure 12.82). Here, we are changing only the coordinate system we use to describe the minerals. We start with *a*, *b*, and *c* as the three symmetry directions to arrive at *Aba*2. Next, we change the orientation of the symmetry elements to the order *c*, *a*, *b* to arrive at *Cc*2*a*. This says that the symmetry along *a* for *Aba*2 is along *c* for *Cc*2*a*, the symmetry that was along *b* for *Aba*2 is along *a* for *Cc*2*a*, and the symmetry along *c* for *Aba*2 is along *b* for *Cc*2*a*. Yes, this is very confusing! This is why crystallographers have worked to standardize all these orientations. However, we have included these older, non-standard space groups here just in case you ever need to refer to the earlier crystallographic work.

Summary

This chapter had the task of exploring the subject of symmetry at a higher level than Chapter 11. As stated at the start, we really deal with two kinds of symmetry in mineralogy: external, which relates to mineral shapes and internal, which relates to atomic arrangement. External morphology was studied first, before analytical methods were developed that allowed us to explore internal atomic arrangements. Because of this history, some of the nomenclature developed to understand external symmetry has been carried over to internal symmetry. However we now emphasize

Table 12.11.
Non-standard space groups based on different axes choices and orientations.

Triclinic:	*A*1 (1), $A\bar{1}$(2), *B*1 (1), $B\bar{1}$ (2), *C*1 (1), $C\bar{1}$ (2), *I*1 (1), $I\bar{1}$ (2), *F*1 (1), $F\bar{1}$ (2)
Monoclinic:	*Pn* (7), *Pa* (7), *P*2/*n* (13), *P*2/*a* (13), *P*2/*b* (13), $P2_1/n$ (14), $P2_1/a$ (14), $P2_1/b$ (14), *A*2(5), *Aa* (7), *An* (9), *A*2/*m* (12), *A*2/*a* (13) $A2_1/a$ (14), *A*2/*n* (15), *B*2 (5), *Bm* (8), *Bn* (9), *B*2/*m* (12), *B*2/*n* (15), *I*2 (5), *Im* (8), *Ia* (9), *Ib* (9), *I*2/*m* (12), *I*2/*a* (15), *I*2/*b* (15)
Orthorhombic:	$P22_12$ (17), $P2_122$ (17), $P22_121$ (18), $P2_122_1$ (18), *Pm*2*m* (25), *P*2*mm* (25), $Pcm2_1$ (26), $Pb2_1m$ (26), $Pm2_1b$ (26), $P2_1am$ (26), $P2_1ma$ (26), *Pb*2*b* (27), *P*2*aa* (27), *Pbm*2 (28), *Pc*2*m* (28), *Pm*2*a* (28), *P*2*cm* (28), *P*2*mb* (28), $Pbc2_1$ (29), $Pb2_1a$ (29), $Pc2_1b$ (29), $P2_1ab$ (29), $P2_1ca$ (29), *Pcn*2(30), *Pn*2*b* (30), *Pb*2*n* (30), *P*2*na* (30), *P*2*an* (30), $Pnm2_1$ (31), $Pm2_1n$ (31), $Pn2_1m$ (31), $P2_1mn$ (31), $P2_1nm$ (31), *Pc*2*a* (32), *P*2*cb* (32), $Pbn2_1$ (33), $Pn2_1a$ (33), $Pc2_1n$ (33), $P2_1nb$ (33), $P2_1cn$ (33), *Pn*2*n* (34), *P*2*nn* (34), *P*2/*b* 2/*m* 2/*b* (49), *P*2/*m* 2/*a* 2/*a* (49), *P*2/*c* 2/*n* 2/*a* (50), *P*2/*n* 2/*c* 2/*b* (50), $P2/m\ 2_1/m\ 2/b$ (51), $P2_1/m\ 2/a\ 2/m$ (51), $P2/m\ 2/c\ 2_1/m$ (51), $P2/b\ 2_1/m\ 2/m$ (51), $P2/c\ 2/m\ 2_1/m$ (51), $P2_1/n\ 2/n\ 2/b$ (52), $P2/n\ 2/a\ 2_1/n$ (52), $P2_1/n\ 2/c\ 2/n$ (52), $P2/b\ 2/n\ 2_1/n$ (52), $P2/c\ 2_1/n\ 2/n$ (52), $P2/n\ 2/m\ 2_1/b$ (53), $P2/m\ 2_1/a\ 2/n$ (53), $P2/n\ 2_1/c\ 2/m$ (53), $P2_1/b\ 2/m\ 2/n$ (53), $P2_1/c\ 2/n\ 2/m$ (53), $P2/c\ 2_1/c\ 2/b$ (54), $P2_1/b\ 2/a\ 2/b$ (54), $P2/b\ 2/c\ 2_1/b$ (54), $P2/b\ 2_1/a\ 2/a$ (54), $P2/c\ 2/a\ 2_1/a$ (54), $P2_1/c\ 2/m\ 2_1/a$ (55), $P2/m\ 2_1/c\ 2_1/b$ (55), $P2_1/b\ 2/n\ 2_1/b$ (56), $P2/n\ 2_1/a\ 2_1/a$ (56), $P2_1/c\ 2/a\ 2_1/m$ (57), $P2/c\ 2_1/m\ 2_1/b$ (57), $P2_1/b\ 2_1/m\ 2/a$ (57), $P2_1/m\ 2/c\ 2_1/a$ (57), $P2_1/m\ 2_1/a\ 2/b$ (57), $P2_1/n\ 2/m\ 2_1/n$ (58), $P2/m\ 2_1/n\ 2_1/n$ (58), $P2_1/m\ 2/n\ 2_1/m$ (59), $P2/n\ 2_1/m\ 2_1/m$ (59), $P2/c\ 2_1/a\ 2_1/n$ (60), $P2_1/c\ 2_1/n\ 2/b$ (60), $P2/b\ 2_1/n\ 2_1/a$ (60), $P2_1/n\ 2_1/c\ 2/a$ (60), $P2_1/n\ 2/a\ 2_1/b$ (60), $P2_1/c\ 2_1/a\ 2_1/b$ (61), $P2_1/m\ 2_1/n\ 2_1/b$ (62), $P2_1/n\ 2_1/a\ 2_1/m$ (62), $P2_1/m\ 2_1/c\ 2_1/n$ (62), $P2_1/b\ 2_1/n\ 2_1/m$ (62), $P2_1/c\ 2_1/m\ 2_1/n$ (62), $A2_122$ (20), *A*2*mm* (35), $A2_1ma$ (36), $A2_1am$ (36), *A*2*aa* (37), *Am*2*m* (38), *Ac*2*m* (39), *Am*2*a* (40), *Ac*2*a* (41), $A2_1/m\ 2/m\ 2/a$ (63), $A2_1/m\ 2/a\ 2/m$ (63), $A2_1/b\ 2/m\ 2/a$ (64), $A2_1/c\ 2/a\ 2/m$ (64), *A*2/*m* 2/*m* 2/*m* (65), *A*2/*m* 2/*a* 2/*a* (66), *A*2/*b* 2/*m* 2/*m* (67), *A*2/*c* 2/*m* 2/*m* (67), *A*2/*b* 2/*a* 2/*a* (68), *A*2/*c* 2/*a* 2/*a* (68), *B*2212 (20), *Bm*2*m* (35), *Bm*21*b* (36), *Bb*21*m* (36), *Bb*2*b* (37), *Bmm*2 (38), *B*2*mm* (38), *Bma*2 (39), *B*2*cm* (39), *Bbm*2 (40), *B*2*mb* (40), *Bba*2 (41), *B*2*cb* (41), $B2/m\ 2_1/m\ 2/b$ (63), $B2/b\ 2_1/m\ 2/m$ (63), $B2/m\ 2_1/a\ 2/b$ (64), $B2/b\ 2_1/c\ 2/m$ (64), *B*2/*m* 2/*m* 2/*m* (65), *B*2/*b* 2/*m* 2/*b* (66), *B*2/*m* 2/*a* 2/*m* (67), *B*2/*m* 2/*c* 2/*m* (67), *B*2/*b* 2/*a* 2/*b* (68), *B*2/*b* 2/*c* 2/*b* (68), $Ccm2_1$ (36), *Cm*2*m* (38), *C*2*mm* (38), *Cm*2*a* (39), *C*2*mb* (39), *Cc*2*m* (40), *C*2*cm* (40), *Cc*2*a* (41), *C*2*cb* (41), $C2/c\ 2/m\ 2_1/m$ (63), $C2/c\ 2/m\ 2_1/b$ (64), *C*2/*m* 2/*m* 2/*b* (67), *C*2/*c* 2/*c* 2/*b* (68), *Im*2*m* (44), *I*2*mm* (44), *Ic*2*a* (45), *I*2*cb* (45), *Ibm*2 (46), *Im*2*a* (46), *Ic*2*m* (46), *I*2*mb* (46), *I*2*cm* (46), *I*2/*c* 2/*m* 2/*a* (72), *I*2/*m* 2/*c* 2/*b* (72), $I2_1/m\ 2_1/m\ 2_1/b$ (74), $I2_1/m\ 2_1/a\ 2_1/m$ (74), $I2_1/m\ 2_1/c\ 2_1/m$ (74), $I2_1/b\ 2_1/m\ 2_1/m$ (74), $I2_1/c\ 2_1/m\ 2_1/m$ (74), *Fm*2*m* (42), *F*2*mm* (42), *Fd*2*d* (43), *F*2*dd* (43)

internal symmetry because of the plethora of analytical methods we have to study it. Point group symmetry leads to an understating of external morphology, followed by the development of space group symmetry which has its basis in point groups, but allows translations which are critical to form the internal arrangement of atoms.

To link (literally and figuratively) the material presented in the chapter, Figure 12.83 shows a concept map of the relationships between point sym-

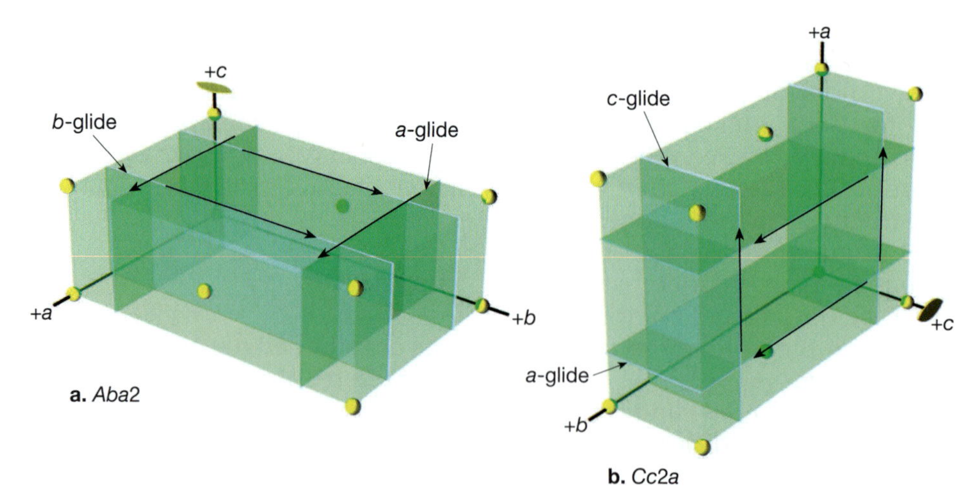

Figure 12.82. *An example of a non-standard space group for the system. (Modified from Figure 13.34, Bloss, 1994.)* ***a.*** *A sketch of the unit cell and major symmetry elements for the standard space group* Aba2 *(#41). This space group is derived from the point group* mm2 *and has an* A *side-centered lattice. The orientation of the symmetry elements for* Aba2 *could be written explicitly as* $A^{[100]}b^{[010]}a^{[001]}2$*, which is the accepted convention for orthorhombic minerals (Table 12.7 and Figure 12.33).* ***b.*** *The non-standard space group* Cc2a *results from a different axis choice. In fact, there are six permutations (= 3!) for assigning the three mutually-perpendicular symmetry operations to the three mutually-perpendicular crystallographic axes. In Figure 12.82a, the symmetry directions were assigned in the order* a, b, c. *For* Cc2a, *they are assigned in the order* b, c, a.

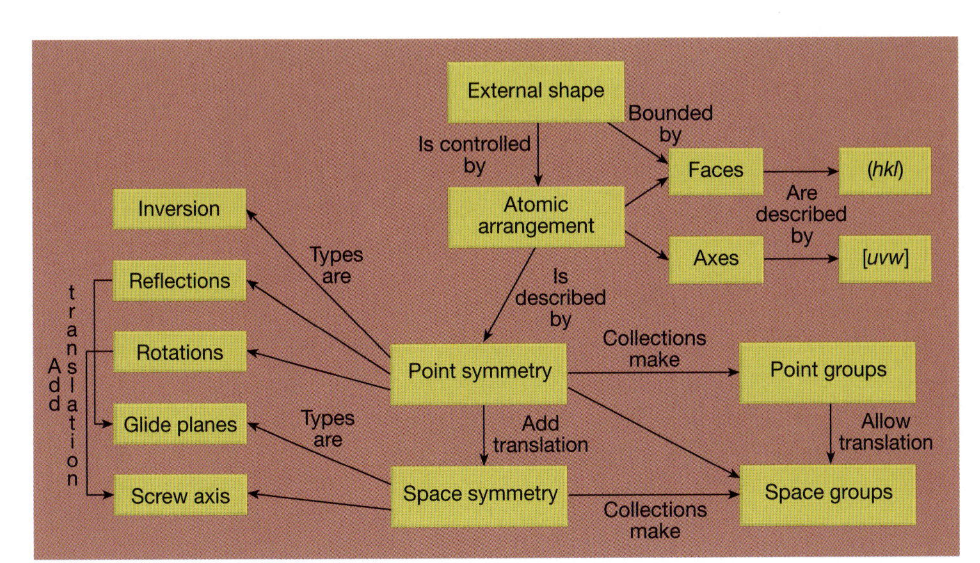

Figure 12.83. A concept map that relates the crystallographic nomenclature that we have developed in this chapter. From an historical perspective it is important to remember that all of this nomenclature has roots in the external shape of a mineral, because this was the first things observed by humans.

metry, crystal systems, point groups, Bravais lattices, (*hkl*)'s, [*uvw*]'s, space group symmetry, and space groups. It has taken centuries for crystallographers to arrive at the crystal systems, point and space groups and all the nomenclature necessary to describe the crystalline state. The effort you spend trying to comprehend this material now will reward you many fold (pun intended) as you pursue your career in geology. While this material may make your head hurt, it will teach you to envision and manipulate 3-D objects in your mind! And this will make you a better geologist!

References

Buerger, M. (1978) Elementary crystallography. MIT Press, Cambridge, Mass., 528 pp.

Bloss, F.D. (1994) Crystallography and crystal chemistry. Mineralogical Society of America, Washington D.C., 545 pp.

Gaines, R.V., Skinner, H.C.W., Foord, E.E., Manson, B., and Rosenweig, A. (1997) Dana's New Mineralogy, 8th Edition. John Wiley and Sons, Inc., 1819 pp.

Gibbs, G.V. (1997) The metrical matrix in teaching mineralogy. In Teaching Mineralogy, editors J.B. Brady, D.W. Mogk, and D. Perkins, III, 201–212.

Hahn, T. (1995) International table for crystallography, volume A, space-group symmetry. Kluwer Academic Publishers, Dordrecht, Holland, 878 pp.

Chapter 13

Mathematical Crystallography

While many might view this as the least important chapter in this book, I view it as the most important for several reasons. Probably the most fulfilling and enjoyable lectures I ever attended were those of Jerry Gibbs' mineralogy class at Virginia Tech when I was a first-year graduate student. I had been assigned as one of the graduate teaching assistants for the course. It began with crystallography and symmetry, so I thought it would be a review from my undergraduate course, but boy, was I ever wrong! Jerry immediately starting discussing symmetry from an approach that centered on representing the symmetry elements (i.e., the rotations and reflections that were introduced in the last chapter) by the use of matrices. Who would have thought that my linear algebra course was more useful in learning mineralogy than my mineralogy course would be! As the grad student TA, I stayed about one lab ahead of what I was teaching the undergrads. I probably learned more than anyone in the course. I'm fairly sure I enjoyed it more than anyone.

Why was the course so meaningful to me? For two main reasons: (1) it was the second "spiral" for me in the spiral learning process—the method we are using in this book (i.e., I had already seen this material before, and on the second spiral its coverage was much more thorough), and (2) it was quantitative—unlike the phenomenological process we went through in the last chapter of showing symmetry in objects and then arm-waving to arrive at the symmetry operations, crystal systems, point groups, and space groups. Here were mathematically-sound methods presented to us as proofs that clearly showed how all of the foundations of crystallography were firmly rooted in provable theories. I loved it! I'm also sure many of the undergraduates, who were seeing the material for the first time, hated it. I think they lacked the mineralogical maturity to see the usefulness of the methods being presented to them, especially without the advantage of the "first spiral" that I'd already been through.

As you'll see if you work through this material, the math behind the crystallography is really very simple. Use of matrix algebra greatly simplifies many complicated calculations. With the ease of matrix manipulation in current spreadsheets, it is very simple to perform these calculations yourself. The material in this chapter really is the underpinning for how we actually represent (i.e., draw projections) of crystal structures—the subject of the next chapter.

I would like to end my comments to the introduction to this chapter with a quote from Jerry Gibbs (1997) in reference to what happens when you fail to use a mathematical approach to crystallography: "the study of the geometry and symmetry of minerals becomes a qualitative chore, providing little meaningful insight into the nature of their properties and their possible uses." *One of the goals of this book is to understand the properties and uses of minerals, which can only be understood by having a thorough understanding of the crystal structure of a material, which in turn can only be accomplished by use of the mathematical methods presented in this chapter. Jerry did not develop these methods in a vacuum. He worked closely with Monte Boisen from the mathematics department. I have tried to distill their book* Mathematical Crystallography *(Boisen and Gibbs, 1990) into this single chapter.*

M.E.G.

Introduction

In the last chapter, we introduced the point group symmetry operations. This was done using the rather classical, traditional approach of studying the external symmetry of 3-D objects. Of course, historically we were concerned with the symmetry of macroscopic 3-D objects long before we thought about the symmetry of the microscopic world. Over the past 100 years, our concerns have shifted to studying the atomic arrangement of crystalline materials. In fact, we've found that external symmetry is quite often controlled by the internal symmetry of the atomic arrangement. Thus, it seems appropriate to refine our methods of understanding symmetry. To that end, the unifying method of mathematical crystallography was developed and recently refined by the use of matrix algebra methods. In this chapter, we will present many of these recent advances. These are based upon work by Boisen and Gibbs (1976) on point group derivation, Boisen and Gibbs (1978) on matrix representation of space group operations, and Gibbs (1997) on using the metrical matrix, here called the metric tensor, in teaching mineralogy. All of this material is given in their book (Boisen and Gibbs, 1990), which should be consulted for a more in-depth treatment of the material presented herein. We also highly recommend the classic text by Nye (1985), *Physical Properties of Crystals*. These methods greatly simplify complex calculations and provide the basis for representing the crystal structures of materials. Also, the availability of modern spreadsheets makes it very simple for students to perform these calculations.

Matrix Representation of Symmetry Operations

Matrices provide a simple method for representing the point group symmetry operations: rotation axes, reflection planes, and the center of symmetry. However, a minimal understanding of linear algebra nomenclature and simple matrix manipulations is required to make use of these methods. For those who feel rusty in this area, please refer to Appendix A at the end of this chapter for a brief introduction to and/or review of the important concepts used in this chapter.

There are two basic approaches to arriving at the matrix representation of a symmetry element: (1) by observation of how the basis vector set changes for the point group operation, and (2) by a more mathematical approach and the use of the general Cartesian rotation matrix. Recall that the **basis vector set** is defined as an axis set with an equal number of axes to the dimension of space of concern (three in our case), where none of those three axes can be written as a linear combination of the other two. Thus in crystallography we have six basis vector sets—the axis sets that define the six crystal systems. We will use the first method, that of observation, in this chapter. However, Appendix B (located at the very end of this chapter) provides a full description of the general Cartesian rotation matrix. We encourage interested students to program this matrix into a spreadsheet and experiment with possible point group operations.

To begin, we'll use an example of $^{[001]}2$ to show how simple it is to find its matrix representation. Figure 13.1 shows a representation of a 3-D axis set (x, y, z) for Cartesian space. We ask the question: what happens to this axis set if we apply a $^{[001]}2$ rotation to it? Stated in words, the x axis would go to negative x, the y axis would go to negative y, and the z axis would stay the same, as seen in Figure 13.1. But we seek a mathematical expression to describe this, so we'll write the original axis set as (x, y, z) and the new axis set as (x', y', z') to represent the effect of the two fold. Then we could say $-x > x'$, $-y > y'$, and $-z > z'$. Next, we could put this into matrix form where each column of the 3×3 matrix would be a mathematical representation of the transformed axis set (x', y', z') written in terms of (x, y, z). Thus, the first column would represent the x' direction, the second column the y' direction, and the third column the z' direction written in terms of the original axes set. In reality, saying all this seems much more confusing then simplifying looking at the resultant matrix! The 3×3 matrix would be:

$$^{[001]}2 = \begin{bmatrix} -1 & 0 & 0 \\ 0 & -1 & 0 \\ 0 & 0 & 1 \end{bmatrix}.$$

By observation of Figure 13.2, make sure that you understand how we arrive at the next two matrices, which represent 2-folds parallel to the x and y axes, respectively:

$$^{[100]}2 = \begin{bmatrix} 1 & 0 & 0 \\ 0 & -1 & 0 \\ 0 & 0 & -1 \end{bmatrix} \quad ^{[010]}2 = \begin{bmatrix} -1 & 0 & 0 \\ 0 & 1 & 0 \\ 0 & 0 & -1 \end{bmatrix}.$$

For each of the three 2-fold axes, the axes coinciding with the rotation direction do not move, and the other two axes have a change in sign. This just means that their positive and negative ends are switched by the 180° rotation.

Think about what happens with a mirror plane. It will keep two axes fixed, and flip the third—which is the axis perpendicular to the mirror plane (i.e., exchange its positive and negative ends). Recall we named the mirror plane based on its perpendicular direction. Now by observation of Figure 13.3, the following three matrix representations, for three different orientations of a mirror plane, should make sense:

$$^{[100]}m = \begin{bmatrix} -1 & 0 & 0 \\ 0 & 1 & 0 \\ 0 & 0 & 1 \end{bmatrix} \; ^{[010]}m = \begin{bmatrix} 1 & 0 & 0 \\ 0 & -1 & 0 \\ 0 & 0 & 1 \end{bmatrix} \; ^{[001]}m = \begin{bmatrix} 1 & 0 & 0 \\ 0 & 1 & 0 \\ 0 & 0 & -1 \end{bmatrix}.$$

Before we move on to more examples, we'd like to offer a caveat or two at this point. As stated earlier in this chapter, the material herein is based on the 460-page book of Boisen and Gibbs (1990), and as such is much condensed. Also, some of the mathematical "purity," especially in notation conventions, is excluded. For instance, we would write $^{[001]}m$ (as above) to represent the matrix representation for the symmetry. More precisely the $^{[001]}m$ is the symmetry operation, while the notation for its matrix representation should be written as: $M_D(^{[001]}\underline{m})$, where M_D represents the basis vector set D. Next by underlining the m we refer to its matrix operation. Although we respect such detailed notation, we've chosen to simplify the expressions in the hope of making this material accessible to math-challenged students. So, the "=" signs are not strictly true, in the abbreviated notations used herein. Those who excel in math should see Boisen and Gibbs (1990) for the more pure presentation of this material.

Finally, we must point out that the choices of (1) writing position vectors as either columns or rows, and (2) whether to pre- or post-multiply by the transformation matrix, are both completely arbitrary. Once these choices are made, however, a single convention must be consistently used.

Let's resume with two more examples that introduce additional terminology: (1) a +90° rotation and (2) a –90° rotation, both parallel to z. In the 180° rotation, we didn't need to concern ourselves with the sense of rotation (i.e., whether it was clockwise (cw) or counterclockwise (ccw)). Crystallographers use a right-handed coordinate system. As such, ccw rotations are defined as "+" and cw rotations as "–." Figure 13.4 shows a $^{[001]}4$ rotation and how the axes move after the 90° ccw rotation. By observation, we could write the resultant 3×3 matrix as:

$$^{[001]}4 = \begin{bmatrix} 0 & -1 & 0 \\ 1 & 0 & 0 \\ 0 & 0 & 1 \end{bmatrix}.$$

To distinguish positive rotations (i.e., ccw) from negative rotations (i.e., cw), we place a –1 superscript to the right of the rotation axis; thus $^{[001]}4^{-1}$ would represent a –90° rotation parallel to z. Figure 13.4 shows the results of a –90° rotation on the axis set. By observation, we could write the following matrix:

$$^{[001]}4^{-1} = \begin{bmatrix} 0 & 1 & 0 \\ -1 & 0 & 0 \\ 0 & 0 & 1 \end{bmatrix}.$$

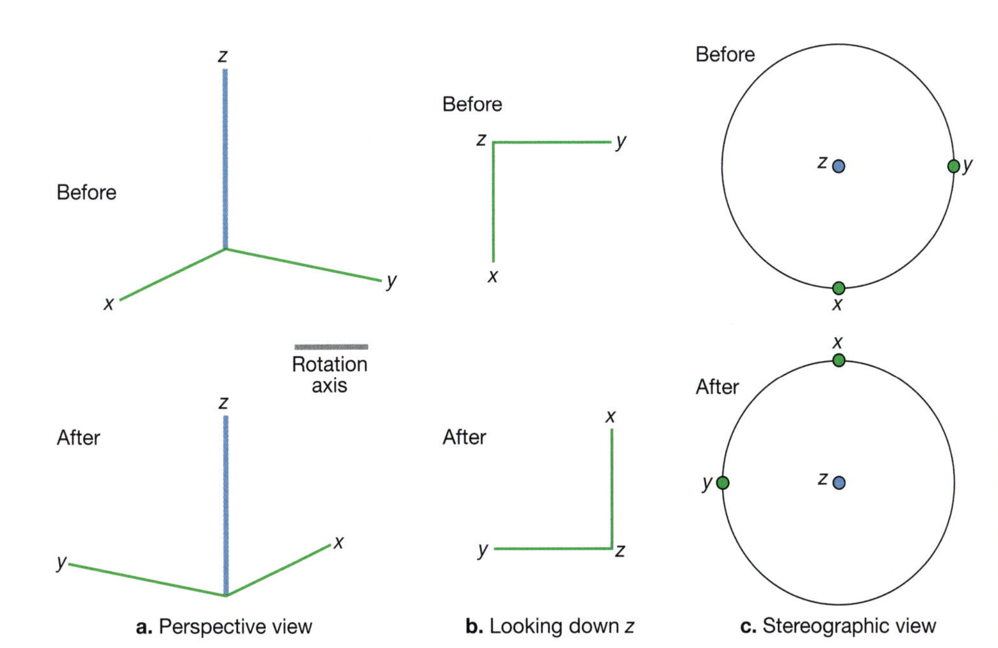

Figure 13.1. *An example of a $^{[001]}2$ rotation (i.e., a 2-fold 180° rotation parallel to z) showing how the axes set is transformed in* ***a.*** *a perspective view,* ***b.*** *the view looking down z, and* ***c.*** *a stereographic projection. Notice how the z axis remains unchanged, the x axis goes to –x, and the y axes go to –y after the 180° rotation.*

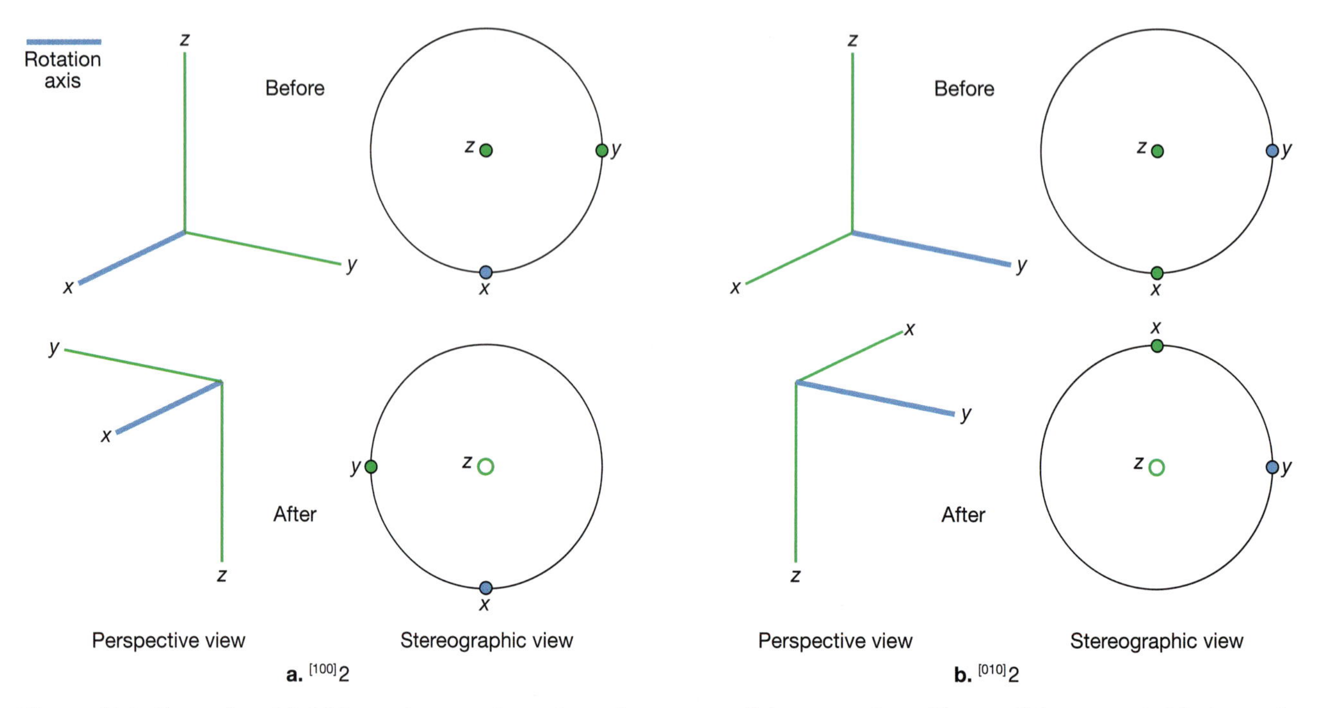

Figure 13.2. *Examples of 2-fold axes in two other orientations:* ***a.*** *parallel to* x, *[100]2, and* ***b.*** *parallel to* y, *[010]2. Notice again how one axis stays the same and the other two rotate about it.*

At this point, we should pause for a few comments. Some good news is that the entries in these matrices representing point group operations will usually be either 0, 1, or –1. They will never be non-integers. In fact, if you experiment with the general rotation Cartesian matrix in Appendix B, you'll find that all of the permissible point group operations (i.e., rotation angles and rotation directions) will result in 0, 1, and –1 values in the matrix. Forbidden operations will result in decimal values. The second point is, as we'll see later in this chapter, that while this matrix method might be a good end in itself (i.e., to better understand how point group operations work), another reason we use it is that we will be able to mathematically and graphically represent crystal structures. This is accomplished by multiplying the atom locations (which are 3×1 matrices) by the 3×3 matrix of the symmetry operations. Finally, there is another property of matrices that represent symmetry operations that needs to be mentioned. For any basis set, a sym-

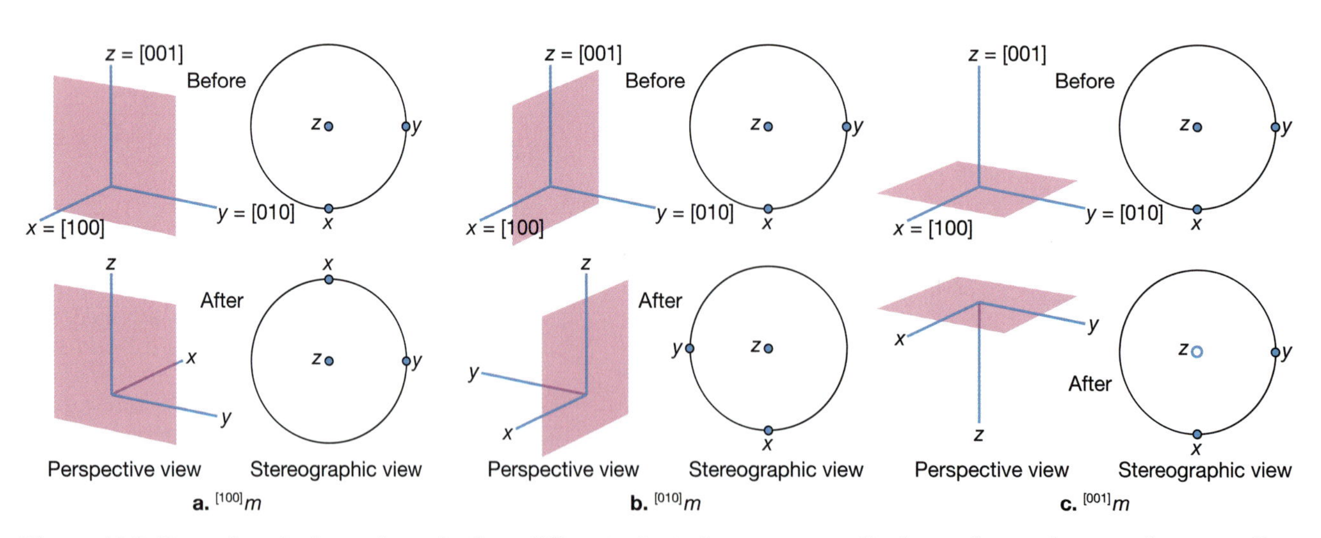

Figure 13.3. *Examples of mirror planes in three different orientations:* ***a.*** *perpendicular to the* x *axis, [100]m,* ***b.*** *perpendicular to the* y *axis, [010]m,* ***c.*** *perpendicular to the* z *axis, [001]m. Recall we define a mirror plane by vector that is perpendicular to it. Notice that the mirror planes keep two axis the same, and flip the third—the one that is normal to it.*

metry operation does not change the size of the transformed object. Therefore, the corresponding matrix has a determinant of +1 or –1.

For example, given an atom that would plot at [1 1 0] on the stereographic projection in Figure 13.5, we could ask the (rhetorical) question "where would this atom go with a $^{[001]}2$ rotation?" Even though you could answer this question by observation, we will need to develop mathematical methods to represent complex crystal structures that do not have such obvious solutions. All that is required is to write the 3×3 matrix for the symmetry operation (here a $^{[001]}2$) and then multiply that matrix by the 3×1 matrix (i.e., a vector) that represents the position of the atom. Doing so yields:

$$\begin{bmatrix} -1 & 0 & 0 \\ 0 & -1 & 0 \\ 0 & 0 & 1 \end{bmatrix} \begin{bmatrix} 1 \\ 1 \\ 0 \end{bmatrix} = \begin{bmatrix} -1 \\ -1 \\ 0 \end{bmatrix}.$$

So the atom plotting at [1 1 0] would be mapped to [–1 –1 0] by the $^{[001]}2$. Even though, as stated, we could have determined this by observation, we now have a mathematical method that is easily programmed into a computer (using a spreadsheet) to make these calculations for us.

Although it might not seem to be required, we also will find a use for the matrix representation of a 1-fold axis (i.e., the symmetry operation that would repeat itself every 360°). This may seem like a complete lack of symmetry, but imagine how hard it would be to do algebra with the number "1"! The 1-fold axis matrix can be written as:

$$1 = \begin{bmatrix} 1 & 0 & 0 \\ 0 & 1 & 0 \\ 0 & 0 & 1 \end{bmatrix}.$$

Hopefully it is apparent that if we multiply an atom position (x, y, z) times this matrix we would simply get (x, y, z) (i.e., the same value or that which would be produced by a 360° rotation). As described in Appendix A, this matrix has a special name in linear algebra. It is the identity matrix, [I].

Another important matrix is the "–1" of linear algebra, which turns out to be the matrix representation of an inversion, and is written as:

$$i = \begin{bmatrix} -1 & 0 & 0 \\ 0 & -1 & 0 \\ 0 & 0 & -1 \end{bmatrix}.$$

Given an element located at the general position (x, y, z), a center of symmetry (i.e., the inversion operations) would map this atom to $(-x, -y, -z)$.

Let's take a closer look at the symmetry contained in $^{[001]}2$. It involves a 2-fold rotation, but this symmetry operation also contains a 1-fold axis. So it contains two symmetry elements and

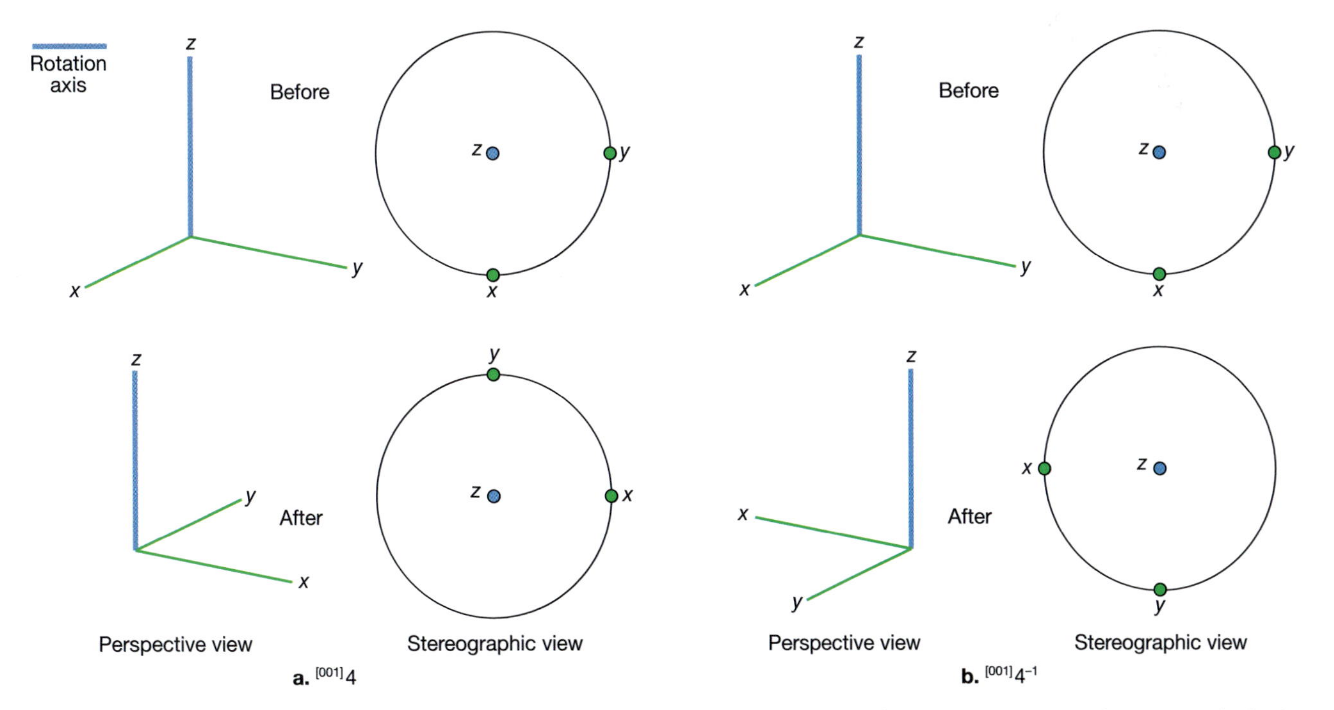

Figure 13.4. Two examples of a 4-fold rotation axis parallel to z. ***a.*** *For this case the rotation is 90° in the counterclockwise (ccw) direction, written as* $^{[001]}4$*; the ccw sense of rotation is defined as positive in crystallography.* ***b.*** *For this case, the rotation is 90° in the clockwise (cw) direction, written as* $^{[001]}4^{-1}$*; the cw sense of rotation is defined as negative in crystallography.*

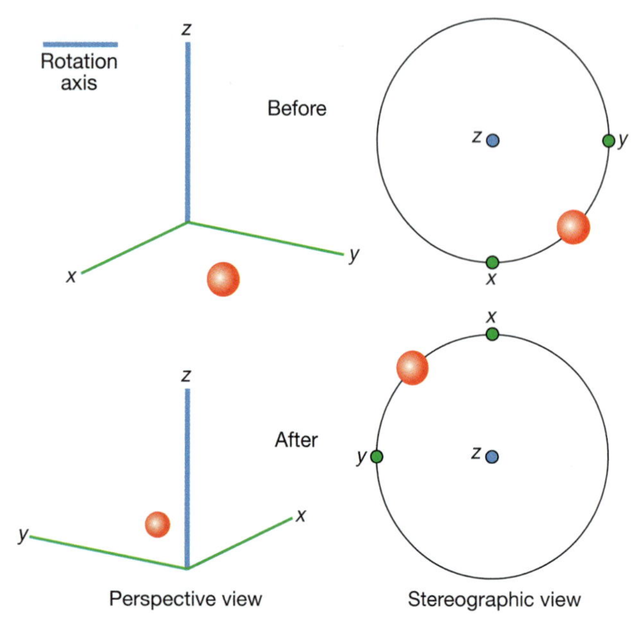

Figure 13.5. *A perspective and stereographic projection of what would happen to an atom plotting at (1, 1, 0) when it is operated upon by a* ${}^{[001]}2$.

would produce two atoms; so we say it has an order of 2. We would write the symmetry elements ${}^{[001]}2 = \{1, 2\}$ to indicate this. Likewise, a ${}^{[010]}m = \{1, m\}$. How about a ${}^{[001]}4$? By observation of Figure 13.6, we see that there are four atoms reproduced by ${}^{[001]}4$, so ${}^{[001]}4$ would have an order of 4. We would expect it to be composed of four symmetry elements that we would write as ${}^{[001]}4 = \{1, 2, 4, 4^{-1}\}$. Stated in words, a ${}^{[001]}4$ contains 1, 4, 2, and 4^{-1} rotations. So if we wanted to know the symmetry position generated for an atom plotting at (1 1 0), as shown in Figure 13.6, we would need to write the four matrices and perform the required matrix multiplications as follows:

$$\overset{1}{\begin{bmatrix}1&0&0\\0&1&0\\0&0&1\end{bmatrix}\begin{bmatrix}1\\1\\0\end{bmatrix}=\begin{bmatrix}1\\1\\0\end{bmatrix}},\quad \overset{4}{\begin{bmatrix}0&-1&0\\1&0&0\\0&0&1\end{bmatrix}\begin{bmatrix}1\\1\\0\end{bmatrix}=\begin{bmatrix}-1\\1\\0\end{bmatrix}},\quad \overset{2}{\begin{bmatrix}-1&0&0\\0&-1&0\\0&0&1\end{bmatrix}\begin{bmatrix}1\\1\\0\end{bmatrix}=\begin{bmatrix}-1\\-1\\0\end{bmatrix}},\quad \overset{4^{-1}}{\begin{bmatrix}0&1&0\\-1&0&0\\0&0&1\end{bmatrix}\begin{bmatrix}1\\1\\0\end{bmatrix}=\begin{bmatrix}1\\-1\\0\end{bmatrix}}.$$

A final example of the different symmetry operations contained within a 6-fold rotation axis parallel to z is: ${}^{[001]}6 = \{1, 6, 3, 2, 3^{-1}, 6^{-1}\}$. Thus the order of ${}^{[001]}6$ is 6, and we would need to write six matrices to determine the effect each rotation axis would have on individual elements. Throughout the above discussion we have used the Cartesian system and labeled the axes as x, y, and z. However, we can also use the a, b, and c axes that define the basis vector sets for the different crystal systems. The resultant matrices will, in turn, relate to these different basis vector sets.

Before we leave this section, it's time for another caveat on mathematical purity. The astute reader might notice that we can mean several things (three to be exact) when we write ${}^{[001]}6$. We could mean: (1) a symmetry operation—a 6-fold axis parallel to c, (2) the matrix representation for a 6-fold axis parallel to c, and (3) a set of six symmetry operations. At the introductory level, as presented in this chapter, we think it's clear which of these we mean based on the context of their use. In Boisen and Gibbs (1990), a notation set is developed to clearly deal with these three meanings. First, they use bold lettering to refer to a symmetry operation; thus, a 6-fold axis parallel to c is written there as: ${}^{[001]}\mathbf{6}$. To denote the matrix representation, they use the symbol $M_D(\alpha)$ to refer to the 3×3 matrix operation of the point isometry α given in the basis vector set D. Thus, the notation to denote the matrix representation for a 6-fold parallel to c in the hexagonal system would be: $M_{hex}({}^{[001]}\underline{6})$. Finally, to denote the set of six point symmetries in the 6-fold parallel to c, they use $M_D(G)$, where G would represent the group of symmetry operations included in the 6-fold or $M_{hex}(6)$. Again we

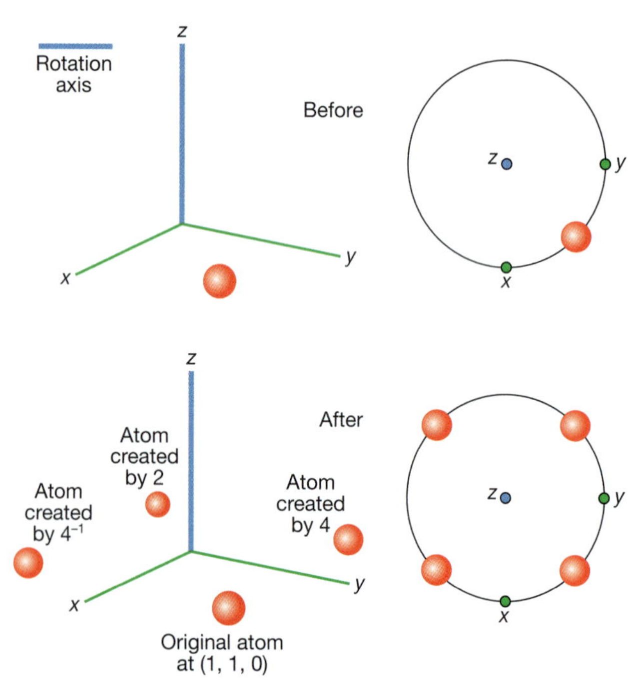

Figure 13.6. *A perspective and stereographic projection of what would happen to an atom plotting at 1, 1, 0 when it is operated upon by a* ${}^{[001]}4$*. In this case we will leave the original atom at 1, 1, 0 and produce three new atoms. Recall that a* ${}^{[001]}4$ *has the following set of symmetry operations {1, 4, 2,* 4^{-1}*}; from this set the "1" would create the atom plotting at 1, 1, 0; the "4" would create the atom 90° ccw from the original atom, the "2" would create the atom at 180° from the original atom, and the "*4^{-1}*" would create the atom at –90° (i.e., cw) from the original atom.*

have avoided using this notation style, and instead believe you can see the meaning of our simplified notation by its context.

Rotoinversion Axes

Mathematically, it is a simple matter to create a rotoinversion axis, given the matrix representation for a rotation axis. Recall that in Chapter 11 we defined a rotoinversion axis as a rotation plus an inversion. These are denoted with a bar written above the rotation symbol. In other words, to find the new point created by say, $\bar{4}$, first invert it through the center and then rotate the point 90° ccw. It is much harder to find these matrices by observation. Creation of these rotoinversions is greatly simplified by matrix manipulations. All that is required is to multiply the point group operation, [PG], times the center of symmetry $[I]^{-1}$, and the resultant matrix will be the rotoinversion. As an example, let's find the ${}^{[001]}\bar{4}$. To do this, we write:

$$^{[001]}\bar{4} = [^{[001]}4]\,[I]^{-1}$$

or,

$$^{[001]}\bar{4} = \begin{bmatrix} 0 & -1 & 0 \\ 1 & 0 & 0 \\ 0 & 0 & 1 \end{bmatrix} \begin{bmatrix} -1 & 0 & 0 \\ 0 & -1 & 0 \\ 0 & 0 & -1 \end{bmatrix} = \begin{bmatrix} 0 & 1 & 0 \\ -1 & 0 & 0 \\ 0 & 0 & -1 \end{bmatrix}.$$

Next we could find the matrix representation for ${}^{[001]}\bar{4}^{-1}$ by multiplying the matrix representation for a ${}^{[001]}4^{-1}$ by $[I]^{-1}$ as follows:

$$^{[001]}\bar{4}^{-1} = \begin{bmatrix} 0 & 1 & 0 \\ -1 & 0 & 0 \\ 0 & 0 & 1 \end{bmatrix} \begin{bmatrix} -1 & 0 & 0 \\ 0 & -1 & 0 \\ 0 & 0 & -1 \end{bmatrix} = \begin{bmatrix} 0 & -1 & 0 \\ 1 & 0 & 0 \\ 0 & 0 & -1 \end{bmatrix}.$$

As another example, we could find what happens if we combine a 2-fold axis and a center of symmetry. We find the product of the ${}^{[001]}2$ with $[I]^{-1}$ to be:

$$^{[001]}\bar{2} = \begin{bmatrix} -1 & 0 & 0 \\ 0 & -1 & 0 \\ 0 & 0 & 1 \end{bmatrix} \begin{bmatrix} -1 & 0 & 0 \\ 0 & -1 & 0 \\ 0 & 0 & -1 \end{bmatrix} = \begin{bmatrix} 1 & 0 & 0 \\ 0 & 1 & 0 \\ 0 & 0 & -1 \end{bmatrix}.$$

You might notice that the combination of a center of symmetry with a 2-fold axis produces a mirror plane, so we would write the set of symmetry operations for ${}^{[001]}\bar{4} = \{1, \bar{4}, m, \bar{4}^{-1}\}$. In fact, there is no such thing as a $\bar{2}$ axis. It is instead called a mirror plane, and denoted by the letter *m*.

In summary, Table 13.1 presents several matrix representation for various symmetry operations in different orientations. Some of these have been explained in the text. The reader is encouraged to derive all of these examples. You could also use the general Cartesian matrix given in Appendix B to determine the matrix representation of any permissible point group operation.

This section presents two major reasons why we would want to express symmetry operations as matrices: (1) to better understand them, and (2) to aid in the representation of crystal structures. The following section will extend reason #2 by adding translation components of symmetry and the resultant space group operations, while the last two sections in this chapter will delve more deeply into reason #1 and point group derivations.

Derivation of Space Group Symmetry Group Operations: Symmetry Propagation of Atoms

One of the ultimate goals of crystallography is to be able to represent the structures of crystalline materials. This is accomplished by determining three basic things: (1) the geometry of the unit cell (i.e., the cell parameters and the interaxial angles), (2) the positional parameters for each atom, and (3) the symmetry of the atoms composing the unit cell (i.e., the lattice type and space group). We'll talk more about #2 in Chapter 14 on representing crystal structures and in Chapter 15 on how we use diffraction to locate the atoms. What we would like to address here is how we can use matrix methods, once given the location of an atom in space, to propagate atoms based on the symmetry of the crystalline material.

In general, we will begin with the positional parameter of an atom, which in general we can represent as a 3×1, or 1×3, vector of the form [*x y z*], where *x* would represent the proportion of the distance the atom plotted along the *a* axis, *y* along the *b* axis, and *z* along the *c* axis. These values are between 0 and 1 for atoms plotting within the unit cell. Given the atom's position, we simply multiply it by the 3×3 matrices that represents the collection of symmetry operations for the space group. As an example (see Figure 13.7), let's say we have a ${}^{[001]}2$ and an atom that plots at [0.1 0.5 0.2]. To determine the atom's new position under the rotation we could write:

$$^{[001]}2 = \begin{bmatrix} -1 & 0 & 0 \\ 0 & -1 & 0 \\ 0 & 0 & 1 \end{bmatrix} \begin{bmatrix} 0.1 \\ 0.5 \\ 0.2 \end{bmatrix} = \begin{bmatrix} -0.1 \\ -0.5 \\ 0.2 \end{bmatrix}.$$

In the general case, we could say that the ${}^{[001]}2$ would create a new atom plotting at (–*x*, –*y*, *z*). Given the (*x*, *y*, *z*) positional parameters, we could determine its new coordinates based on

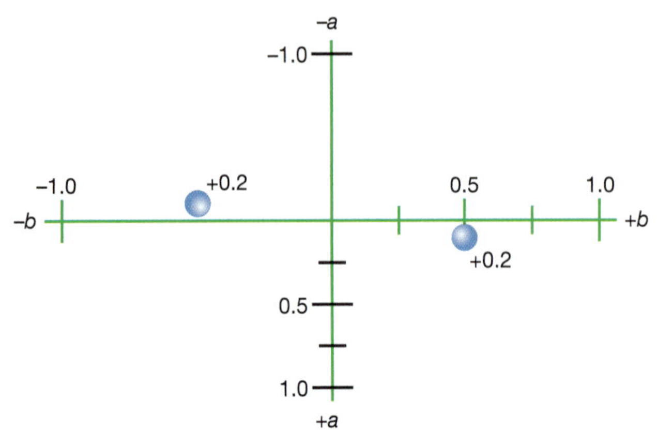

Figure 13.7. *An example of a coordinate system used to plot atoms, and how an atom plotting at 0.1, 0.5, 0.2 would move when operated on by a* $^{[001]}2$. *Notice the axis lengths go from 0 to 1. These lengths are proportioned in this manner for reasons fully explained in the next chapter on representing crystal structures. Because this is a 2-D projection down z, we cannot show the height of the atom in z, so to represent this we write "+0.2" next to the original atom. The* $^{[001]}2$ *creates a new atom, as described by the matrix multiplication in the text, at –0.1, –0.5, 0.2.*

the above matrix, and matrix multiplication. Another example for the more general case, is given below for a $^{[010]}m$:

$$^{[010]}m = \begin{bmatrix} 1 & 0 & 0 \\ 0 & -1 & 0 \\ 0 & 0 & 1 \end{bmatrix} \begin{bmatrix} x \\ y \\ z \end{bmatrix} = \begin{bmatrix} x \\ -y \\ z \end{bmatrix}.$$

Thus $^{[010]}m$ would take all the atoms in the structure and create their symmetry-equivalent atoms by multiplying all the atoms by the above matrix (i.e., in this simple case, really just making the y positional parameter negative.)

The above two examples are point group operations, which by definition leave at least one point in space fixed. However, crystal structures are described by space groups, which include translational symmetry. So we need to add the translational portion to calculate the new positional parameters. Also, in the space lattice the symmetry operation's location must be taken into consideration. The above example for a $^{[001]}2$ and $^{[010]}m$ would only work to locate new atoms in a structure if the symmetry operations pass through the origin of the unit cell. The origin is defined as the 0 points for the three crystallographic axes (Figure 13.8). Thus, we will need to devise a method to consider the location of the point group operation and its position in the lattice.

Equation 13.1 is the general matrix representation for any space group operation as derived by Boisen and Gibbs (1978). The main difference from the above point group operations is the additional column at the end of the matrix that will define the location and translational component of the symmetry operation.

$$\mathrm{SG} = \begin{bmatrix} l11 & l12 & l13 & t1 \\ l21 & l22 & l23 & t2 \\ l31 & l32 & l33 & t3 \\ 0 & 0 & 0 & 1 \end{bmatrix} \qquad \text{Equation 13.1}$$

where l_{nm} = the values for the point group operation, t_n = the values for the location and translational components, and

$$[t] = ([\mathrm{I}] - [\mathrm{PG}])[l] + [d]. \qquad \text{Equation 13.2}$$

In this equation, [I] = 3×3 identity matrix, [PG] = matrix representation of point group operation, $[l]$ = a vector representing the location of the symmetry operation, and $[d]$ = a vector representing the displacement of the symmetry operation.

As an example, we will find the [SG] for a $^{[001]}2_1$ screw axis located at 1/4 a and 1/4 b (Figure 13.9). Our goal will be to determine the twelve elements of [SG] given in Equation 13.1. The point group component of $^{[001]}2_1$ is $^{[001]}2$, and already determined above, thus:

$$\mathrm{SG} = \begin{bmatrix} -1 & 0 & 0 & t1 \\ 0 & -1 & 0 & t2 \\ 0 & 0 & 1 & t3 \\ 0 & 0 & 0 & 1 \end{bmatrix}.$$

Now we need to find the values for $[t]$. To accomplish this, we solve Equation 13.2, by letting $[l]$ = [1/4

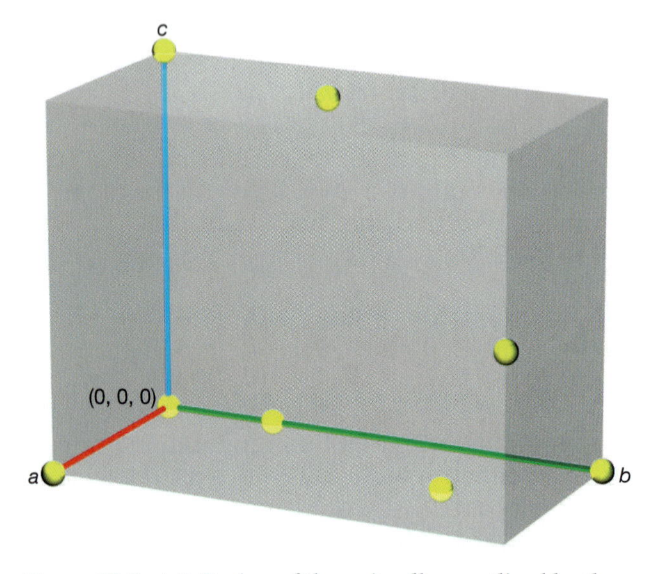

Figure 13.8. *A 3-D view of the unit cell, as outlined by the* a, b, *and* c *axes of an orthorhombic mineral* a, b, *and* c *are different lengths and the interaxial angles are all 90°. This drawing is an extension of Figure 13.7 to 3-D. Also shown are several atoms plotting at different positions. Notice the origin of the unit cell is 0, 0, 0.*

¼ 0], the location of ${}^{[001]}2_1$, and $d = [0\ 0\ ½]$, the direction and amount of the displacement of a ${}^{[001]}2_1$ (i.e., the displacement direction is parallel to the c axis, and the translation for a 2_1 screw is ½, see Chapter 12 for details). Thus we use Equation 13.2 becomes:

$$\begin{bmatrix} t1 \\ t2 \\ t3 \end{bmatrix} = \left(\begin{bmatrix} 1 & 0 & 0 \\ 0 & 1 & 0 \\ 0 & 0 & 1 \end{bmatrix} - \begin{bmatrix} -1 & 0 & 0 \\ 0 & -1 & 0 \\ 0 & 0 & 1 \end{bmatrix} \right) \begin{bmatrix} 1/4 \\ 1/4 \\ 0 \end{bmatrix} + \begin{bmatrix} 0 \\ 0 \\ 1/2 \end{bmatrix} = \begin{bmatrix} 1/2 \\ 1/2 \\ 1/2 \end{bmatrix}.$$

Now [SG] can be written as, for the general case of an atom located at x, y, z:

$$[SG] = \begin{bmatrix} -1 & 0 & 0 \\ 0 & -1 & 0 \\ 0 & 0 & 1 \\ 0 & 0 & 0 \end{bmatrix} \begin{bmatrix} 1/2 \\ 1/2 \\ 1/2 \\ 1 \end{bmatrix} \begin{bmatrix} x \\ y \\ z \\ 1 \end{bmatrix} = \begin{bmatrix} -x + 1/2 \\ -y + 1/2 \\ z + 1/2 \\ 1 \end{bmatrix}$$

Next we could calculate the new location of an atom plotting at (0.3, 0.5, 0.1) as follows:

$$\begin{bmatrix} -1 & 0 & 0 \\ 0 & -1 & 0 \\ 0 & 0 & 1 \\ 0 & 0 & 0 \end{bmatrix} \begin{bmatrix} 1/2 \\ 1/2 \\ 1/2 \\ 1 \end{bmatrix} \begin{bmatrix} 0.3 \\ 0.5 \\ 0.1 \\ 1 \end{bmatrix} = \begin{bmatrix} 0.2 \\ 0.0 \\ 0.6 \\ 1 \end{bmatrix}.$$

Table 13.1. Examples of symmetry operations and their matrix representations broken down by rotation axes and orientations. (Use of the equal signs below refer to the assumed basis vector sets associated with each crystal system and their symmetry operations as explained in the text.)

	1-fold $= \begin{bmatrix} 1 & 0 & 0 \\ 0 & 1 & 0 \\ 0 & 0 & 1 \end{bmatrix}$ *center of symmetry (i)* $= \begin{bmatrix} -1 & 0 & 0 \\ 0 & -1 & 0 \\ 0 & 0 & -1 \end{bmatrix}$
mirror planes	${}^{[100]}m = \begin{bmatrix} -1 & 0 & 0 \\ 0 & 1 & 0 \\ 0 & 0 & 1 \end{bmatrix}$, ${}^{[010]}m = \begin{bmatrix} 1 & 0 & 0 \\ 0 & -1 & 0 \\ 0 & 0 & 1 \end{bmatrix}$, ${}^{[001]}m = \begin{bmatrix} 1 & 0 & 0 \\ 0 & 1 & 0 \\ 0 & 0 & -1 \end{bmatrix}$, ${}^{[110]}m = \begin{bmatrix} 0 & -1 & 0 \\ -1 & 0 & 0 \\ 0 & 0 & 1 \end{bmatrix}$
2-folds	${}^{[100]}2 = \begin{bmatrix} 1 & 0 & 0 \\ 0 & -1 & 0 \\ 0 & 0 & -1 \end{bmatrix}$, ${}^{[010]}2 = \begin{bmatrix} -1 & 0 & 0 \\ 0 & 1 & 0 \\ 0 & 0 & -1 \end{bmatrix}$, ${}^{[001]}2 = \begin{bmatrix} -1 & 0 & 0 \\ 0 & -1 & 0 \\ 0 & 0 & 1 \end{bmatrix}$, ${}^{[110]}2 = \begin{bmatrix} 0 & 1 & 0 \\ 1 & 0 & 0 \\ 0 & 0 & -1 \end{bmatrix}$
3-folds	${}^{[001]}3 = \begin{bmatrix} 0 & -1 & 0 \\ 1 & -1 & 0 \\ 0 & 0 & 1 \end{bmatrix}$, ${}^{[001]}3^{-1} = \begin{bmatrix} -1 & 1 & 0 \\ -1 & 0 & 0 \\ 0 & 0 & 1 \end{bmatrix}$, ${}^{[111]}3 = \begin{bmatrix} 0 & 0 & 1 \\ 1 & 0 & 0 \\ 0 & 1 & 0 \end{bmatrix}$
$\bar{3}$ folds	${}^{[001]}\bar{3} = \begin{bmatrix} 0 & 1 & 0 \\ -1 & 1 & 0 \\ 0 & 0 & -1 \end{bmatrix}$, ${}^{[001]}\bar{3}^{-1} = \begin{bmatrix} 1 & -1 & 0 \\ 1 & 0 & 0 \\ 0 & 0 & -1 \end{bmatrix}$, ${}^{[111]}\bar{3} = \begin{bmatrix} 0 & 0 & -1 \\ -1 & 0 & 0 \\ 0 & -1 & 0 \end{bmatrix}$
4-folds	${}^{[100]}4 = \begin{bmatrix} 1 & 0 & 0 \\ 0 & 0 & -1 \\ 0 & 1 & 0 \end{bmatrix}$, ${}^{[010]}4 = \begin{bmatrix} 0 & 0 & 1 \\ 0 & 1 & 0 \\ -1 & 0 & 0 \end{bmatrix}$, ${}^{[001]}4 = \begin{bmatrix} 0 & -1 & 0 \\ 1 & 0 & 0 \\ 0 & 0 & 1 \end{bmatrix}$
$\bar{4}$ folds	${}^{[100]}\bar{4} = \begin{bmatrix} -1 & 0 & 0 \\ 0 & 0 & 1 \\ 0 & -1 & 0 \end{bmatrix}$, ${}^{[010]}\bar{4} = \begin{bmatrix} 0 & 0 & -1 \\ 0 & -1 & 0 \\ 1 & 0 & 0 \end{bmatrix}$, ${}^{[001]}\bar{4} = \begin{bmatrix} 0 & 1 & 0 \\ -1 & 0 & 0 \\ 0 & 0 & -1 \end{bmatrix}$
6-folds	${}^{[001]}6 = \begin{bmatrix} 1 & -1 & 0 \\ 1 & 0 & 0 \\ 0 & 0 & 1 \end{bmatrix}$, ${}^{[001]}6^{-1} = \begin{bmatrix} 0 & 1 & 0 \\ -1 & 1 & 0 \\ 0 & 0 & 1 \end{bmatrix}$
$\bar{6}$ folds	${}^{[001]}\bar{6} = \begin{bmatrix} -1 & 1 & 0 \\ -1 & 0 & 0 \\ 0 & 0 & -1 \end{bmatrix}$, ${}^{[001]}\bar{6}^{-1} = \begin{bmatrix} 0 & -1 & 0 \\ 1 & -1 & 0 \\ 0 & 0 & -1 \end{bmatrix}$

Figure 13.9. *The same unit cell as in Figure 13.8 (shown both in perspective view and plotting looking down* c *but with the addition of a* $^{[001]}2_1$ *located at* 1/4a *and* 1/4b*). The matrix representation in the text shows what would happen to a point plotted at (0.3, 0.5, 0.1) by the screw axis.*

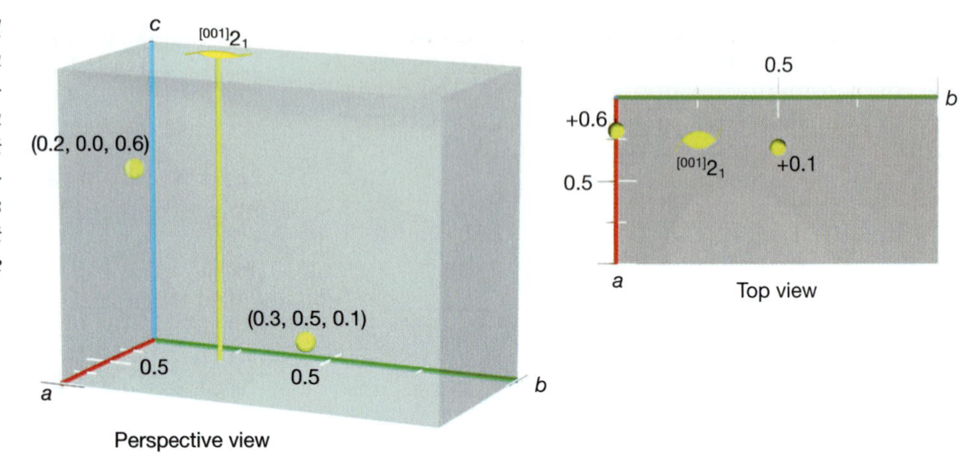

Thus the symmetry operation represented by a $^{[001]}2_1$ located at 1/4*a* and 1/4*b* would take an atom located at 0.3, 0.5, 0.1 and create at new atom at 0.2, 0.0, 0.6 as shown Figure 13.9.

The next example is for a glide plane, in this case $^{[010]}a$. First, take a mirror plane perpendicular to *b*, which we represent as $^{[010]}m$. In this example, the mirror plane will be plotted at 1/4 the distance along *b* (Figure 13.10), so the vector [l] = [0 1/4 0]. Next we need to define the displacement of the $^{[010]}a$ glide, which would be 1/2 along *a*, so *d* = [1/2 0 0]. We could have also made the mirror plane into a *c*-glide by defining the displacement as 1/2 along *c*, or as an *n*-glide, which would make [*l*] = [1/2 0 1/2] (i.e., displacement along both *a* and *c*). All that remains is to construct [SG] based on Equations 13.1 and 13.2 as follows: First we add the point group operation of a $^{[010]}m$:

$$[\mathrm{SG}] = \begin{bmatrix} 1 & 0 & 0 & t1 \\ 0 & -1 & 0 & t2 \\ 0 & 0 & 1 & t3 \\ 0 & 0 & 0 & 1 \end{bmatrix}.$$

Next we solve Equation 13.2 given [PG], [*l*], and [*d*] for the $^{[010]}a$ glide plane:

$$\begin{bmatrix} t1 \\ t2 \\ t3 \end{bmatrix} = \left(\begin{bmatrix} 1 & 0 & 0 \\ 0 & 1 & 0 \\ 0 & 0 & 1 \end{bmatrix} - \begin{bmatrix} 1 & 0 & 0 \\ 0 & -1 & 0 \\ 0 & 0 & 1 \end{bmatrix} \right) \begin{bmatrix} 0 \\ 1/4 \\ 0 \end{bmatrix} + \begin{bmatrix} 1/2 \\ 0 \\ 0 \end{bmatrix} = \begin{bmatrix} 1/2 \\ 1/2 \\ 0 \end{bmatrix}$$

Substituting the vector [*t*] into [SG] we arrive at:

$$[\mathrm{SG}] = \begin{bmatrix} 1 & 0 & 0 & 1/2 \\ 0 & -1 & 0 & 1/2 \\ 0 & 0 & 1 & 0 \\ 0 & 0 & 0 & 1 \end{bmatrix} \begin{bmatrix} x \\ y \\ z \\ 1 \end{bmatrix} = \begin{bmatrix} -x + 1/2 \\ -y + 1/2 \\ z \\ 1 \end{bmatrix}$$

Next we could calculate the new location of an atom plotting at (0.2, 0.3, 0.4) as follows:

$$\begin{bmatrix} 1 & 0 & 0 & 1/2 \\ 0 & -1 & 0 & 1/2 \\ 0 & 0 & 1 & 0 \\ 0 & 0 & 0 & 1 \end{bmatrix} \begin{bmatrix} 0.2 \\ 0.3 \\ 0.4 \\ 1 \end{bmatrix} = \begin{bmatrix} 0.7 \\ 0.2 \\ 0.4 \\ 1 \end{bmatrix}.$$

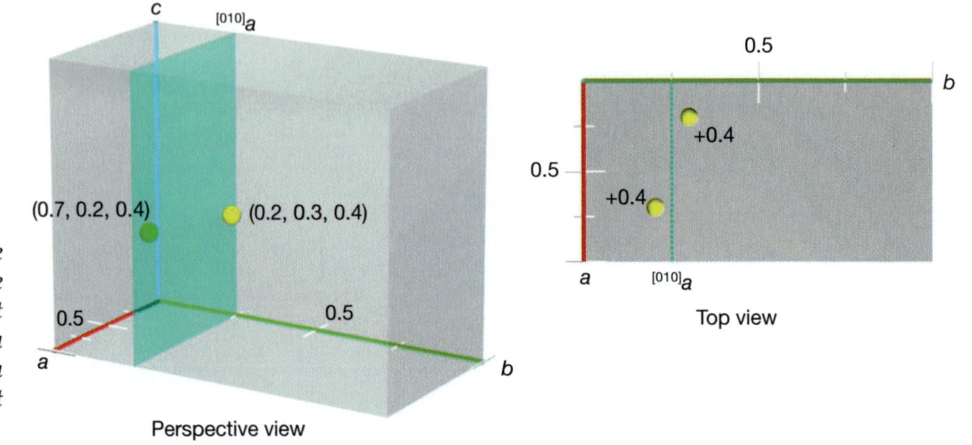

Figure 13.10. *This example uses the same unit cell as in the preceding two figures but shows what happens to an atom plotting at (0.2, 0.3, 0.4) when it is operated on by a* $^{[010]}$a *that is located at* 1/4b.

Thus the symmetry operation represented by an $^{[010]}a$ located at $1/4b$ would take an atom located at (0.2, 0.3, 0.4) and create a new atom at (0.7, 0.2, 0.4) as shown Figure 13.10. It is left to the reader to show that the simpler cases of a $^{[001]}2$ located at the origin and a $^{[010]}m$ passing through the origin would create general points at $(-x, -y, z)$ and $(x, -y, z)$, respectively.

The above two examples show that given the position of an atom in a unit cell, we can find the other atoms that would be generated based on the space group of that mineral. While the above examples might seem trivial, recall that space groups can have as many as 192 symmetry operations. That means we would need to write 192 matrices to determine all the symmetry generated positions! Also, crystal structures of minerals commonly have 5–10 unique atom locations. So to represent a crystal structure, we must make use of computers. The computer programs are written based on the matrix manipulations presented in this chapter. In the next chapter we will put these calculations to work representing crystal structures of minerals. We will use the methods developed here to represent the crystal structures of seven different minerals from the six different crystal systems. We'll also make reference to the *International Tables for Crystallography* (Hahn, 1995) which contains detailed descriptions of all 230 space groups. The methods presented above can be used to determine the symmetry equivalent positions listed for each of the space groups.

The Metric Tensor

Use of the metric tensor as described in Boisen and Gibbs (1990) and Gibbs (1997) provides a simple, elegant method to perform complex and tedious crystallographic calculations. There are six different crystal systems and therefore, six different basis vector sets. Typically in the more traditional approach an equation is given for each of the different crystal systems to perform some desired calculation (Bloss 1994, p. 347). For example, Equation 13.3a below would be used to calculate the volume of the unit cell for materials belonging to the isometric system, 13.3b the tetragonal system, 13.3c the orthorhombic system, 13.3d the hexagonal system, 13.3e the monoclinic system, and 13.3f the triclinic system:

$$\text{volume} = a^3 \qquad \text{Equation 13.3a}$$

$$\text{volume} = (a^2)\,(c) \qquad \text{Equation 13.3b}$$

$$\text{volume} = (a)\,(b)\,(c) \qquad \text{Equation 13.3c}$$

$$\text{volume} = (a^2)\,(c)\sin 60^\circ \qquad \text{Equation 13.3d}$$

$$\text{volume} = (a)\,(b)\,(c)\sin\beta \qquad \text{Equation 13.3e}$$

$$\text{volume} = (a)\,(b)\,(c)\,(1 - \cos^2\alpha - \cos^2\beta - \cos^2\gamma + 2(\cos\alpha\cos\beta\cos\gamma))^{1/2}, \qquad \text{Equation 13.3f}$$

where a, b, c, are the lengths of the crystallographic axes defining the unit cell and α, β, γ are the interaxial angles.

Equation 13.3f is the most general form of the equation used to calculate volume for a triclinic material and, in turn, would work for all six crystal systems. Observation of the above equations shows that as the interaxial angles become equal to 90°, the sin 90° term simplifies to 1's and the cos term to 0's. Historically, researchers (and mineralogy texts) have selected the "appropriate" equation for the desired crystal system. In general this leads to considerable confusion for a newcomer to the field of crystallography (i.e., being given the choice of six different equations to calculate the volume of a mineral). Another example of the use of six equations for the calculation of *d*-spacings is given in Chapter 15. In all fairness to past authors, in the pre-computer and calculator days it was very difficult to make even simple calculations. However there is *no* reason today not to use the most general forms of these equations and avoid the confusion factor. It is quite simple to use a calculator or program the equations into a spreadsheet.

The metric tensor greatly simplifies routine crystallographic calculations by serving as a unifying "tool" for many types of calculations. The metric tensor, [MT], is mathematically defined as follows:

$$[\text{MT}] = \begin{bmatrix} a \cdot a & a \cdot b & a \cdot c \\ b \cdot a & b \cdot b & b \cdot c \\ c \cdot a & c \cdot b & c \cdot c \end{bmatrix} \begin{bmatrix} a^2 & ab\cos\gamma & ac\cos\beta \\ ab\cos\gamma & b^2 & bc\cos\alpha \\ ac\cos\beta & bc\cos\alpha & c^2 \end{bmatrix}. \qquad \text{Equation 13.4}$$

Mathematically, the metrical tensor is a collection of dot products of a basic vector set as shown in Equation 13.4, which can also be written in terms of the cell parameters. In the later form it is a simple matter to compose this matrix, given the cell parameters and interaxial angles of a mineral. A spreadsheet can be written to determine the metric tensor based on input of the cell parameters and interaxial angles, and in turn expand that spreadsheet to perform the calculations described below. As in the volume equations above, the off-diagonal terms of [MT] would take on zero values as the interaxial values became equal to 90°, as is the case for minerals of higher symmetry. However, we will employ the metric tensor written in its most general form. This section draws heavily on similar calculations modified slightly from Gibbs (1997).

A numerical example of a metric tensor will be helpful to illustrate its construction and is used throughout the following examples. Students and instructors are strongly encouraged to duplicate the following set of calculations by construction of a spreadsheet to perform each one. Also, accuracy of the student's work can be checked with the following numerical examples. We will select the most general case, a triclinic mineral, anorthite. Anorthite will also be used in the next chapter dealing with representations of crystal structures.

Based on the structure refinement by Angel et al. (1990), the cell parameters and interaxial angles for anorthite are: $a = 8.175$, $b = 12.873$, $c = 93.11$Å, $\alpha = 93.11$, $\beta = 115.89$, and $\gamma = 91.28°$. Thus [MT] can be written based on Equation 13.4:

$$[MT] = \begin{bmatrix} 66.806 & -2.3508 & -50.5808 \\ -2.3508 & 165.7141 & -9.8963 \\ -50.5808 & -9.8963 & 200.7889 \end{bmatrix}.$$

Volume Calculations

To determine the volume of a unit cell of any crystal system, you simply calculate the determinant of the metric tensor. (If you can't remember what the determinant is, or how to calculate it, see the Appendix A of this chapter.) The determinant is a programmed function in most modern spreadsheets. The volume for anorthite is simply the square root of the determinant of the above [MT], which is 1,337.90 Å^3.

Bond Distances

Whether determining the crystal structure of a material or empirically measuring ionic radii, it is important to be able to calculate the bond distances between two atoms. While this is a fairly straightforward calculation in Cartesian space, it is very difficult in the general triclinic class. Use of the metric tensor, again, greatly simplifies this calculation. To make the calculation, you need to know the cell parameters and interaxial angles to compose the metric tensor and the location (i.e., positional parameters, discussed in more detail in Chapter 14) of the two atoms of interest. Given that information, the following matrix is used:

$$d(1-2)^2 = [PP]\,[MT]\,[PP]^t \qquad \text{Equation 13.5}$$

where $d(1 - 2)^2$ = the square of the distance between atom 1 and atom 2, PP = $[x2 - x1 \;\; y2 - y1 \;\; z2 - z1]$ = a 1×3 matrix with each element equal to the difference between the x, y, and z positional parameters of atom 1 and atom 2.

There are three matrices in Equation 13.5. [PP] is a 1×3 matrix, [MT] is a 3×3 matrix, and $[PP]^t$ is a 3×1 matrix. The multiplication proceeds from left to right, with the result being a 1×1 matrix (i.e., a scalar). In Cartesian space, [MT] would simply be the identity matrix. So to determine the distance between two points in Cartesian space, you would write $d(1-2)^2 = [PP][PP]^t$.

The following example will be used to calculate the bond distance of Si-O bonds and Al-O bonds in the anorthite structure. We'll use Si1 and O1 to demonstrate the calculation; their positions are:

Si1 = (0.009, 0.1589, 0.1039), O1 = (0.0236, 0.1246, -0.0043).

Thus,

[PP] = [0.0236 – 0.009 0.1246 – 0.1589 –0.0043 – 0.1039] = [0.0146 –0.0343 –0.0182].

Substituting [PP] and $[PP]^t$ into Equation 13.5 yields,

$$[0.0146\; -0.0343\; -0.1082]\begin{bmatrix} 66.8306 & -2.3580 & -50.5808 \\ -2.3508 & 165.7141 & -9.8963 \\ -50.5808 & -9.8963 & 200.7889 \end{bmatrix}\begin{bmatrix} 0.0146 \\ -0.0343 \\ -0.1082 \end{bmatrix} = 2.6488$$

The product of multiplying these three matrices is 2.6488, and its square root is 1.628. We would then write that d(Si1 – O1) = 1.628Å. Equation 13.5 can be programmed into a spreadsheet. Then it becomes a simple task to determine other bond distances for a structure by inputting the new positional parameters. For example, the position of O25 is (0.1833, 0.1053, 0.1910), which also bonds to Si1. We could determine its bond distance to be 1.622, or d(Si1 – O25) = 1.622 Å. Given these bond distances, if we assume a radius of O^{2-}, say 1.32 Å we can determine that the radius of Si^{4+} would be 0.31 Å (based on the O1 bond) and 0.30 Å (based on the O25) bond. In fact, this is the manner in which ionic radii have been determined. Researchers have taken many of the crystal structures that have been determined over the years and calculated bond distances for cations, and then averaged them to arrive at the ionic radii presented in Table 3.4 in Chapter 3.

Another example will help demonstrate the point. Recall that anorthite has both Si^{4+} and Al^{3+} tetrahedra. A quick glance at Table 3.4 in Chapter 3 shows that Al^{3+} has a greater ionic radius than Si^{4+}. We could check this by calculating the bond distance for Al-O in the anorthite structure. Given Al1 = (0.0069, 0.1609, 0.6111), O19 = (0.0215, 0.2910, 0.6485), O27 = (0.2136, 0.1025, 0.6840) from the structure data in Table 13.7, we would use

Equation 13.5. We would find that d(Al7 – O19) = 1.715 Å, and d(Al1 – O27) = 1.762 Å. Thus Al^{3+} would have a radius of 0.40 Å (based on the O19 bond) and 0.44 Å (based on the O27 bond). We see that we could arrive at many different radii of Al depending on which bond we used. This is why many bonds are averaged to arrive at the ionic radii. Regardless, you can see that Al^{3+} is about 0.1 Å larger than Si^{4+}, which is in good agreement with the data given in Table 3.4.

Bond Angles

Another common crystallographic calculation is to determine the bond angle between three bonded atoms. For example, the angle between O-Si-O in an ideal tetrahedron would be 109.5°. The calculation of bond angles, much like bond distances, is very tedious in non-Cartesian systems, but use of the metric tensor again simplifies the calculation. To determine the bond angle, we again need the cell parameters and interaxial angles of the mineral of interest so that its metric tensor can be constructed. Then we will require the positional parameters of three atoms. In the general case, if we number the atoms 1, 2, 3 (with the central atom as atom 1) we would have the following equation to determine the bond angle:

bond angle = $\cos^{-1}$ ([PP31] [MT] $[PP21]^t$ / (d(1 – 2) d(1 – 3)) Equation 13.6

where [PP31] = [$x3 - x1$ $y3 - y1$ $z3 - z1$] = a 1×3 matrix with the difference of the positional parameters for atom 3 and atom 1, and $[PP21]^t$ = [$x2 - x1$ $y2 - y1$ $z2 - z1$] = a 3×1 matrix with the difference of the positional parameters for atom 2 and atom 1.

The following example will be used to calculate an O-Si-O bond angle in anorthite. We'll use again Si1 = (0.009, 0.1589, 0.1039), O1 = (0.0236, 0.1246, –0.0043), and O25 = (0.1833, 0.1053, 0.1910). Thus,

[PP31] = [0.1833 – 0.0090 0.1053 – 0.1589 0.1910 – 0.1039] = [0.1743 –0.0536 0.0871]

[PP21] = [0.0236 – 0.009 0.01246 – 0.1589 –0.0043 – 0.1039] = [0.0146 –0.0343 –0.1082].

From the above example, substituting [PP31], $[PP21]^t$, d(Si1 – O1) = 1.628 A°, and d(Si1 – O25) = 1.622 A° into Equation 13.6 yields:

O25-Si1-O1 = $\cos^{-1}$([0.1743 –0.0536 0.0871] [MT] [0.0146 –0.0343 –0.1082]^t / (1.628)(1.622).

Solving the above using the [MT] for anorthite yields O1-Si1-O25 = 101.8°. As an exercise, you could also solve the O19-Al1-O27 angle given the positions and the bond distances used in the bond distance calculation and find that O19-Al1-O27 = 111.1°. Thus, both of these angles show that there is a slight deviation from the ideal tetrahedral bond angle of 109.5°.

d-spacings and Reciprocal Space

Table 15.1 lists the six different equations that are commonly used to calculate the d-spacings of minerals. Much like the above example for the volume calculations, d-spacing calculations are greatly simplified by use of the metric tensor and matrix multiplication. D-spacings refer to the distance, usually given in Å, between lattice planes within a crystalline material. One of their main uses is to aid in the identification of unknowns by use of powder X-ray diffraction; every crystalline material has a unique set of d-spacings. An in-depth discussion of their use will be given in Chapter 15. However, one important point to make here is that when X-rays are diffracted by a crystalline material, they produce a pattern that is related to the arrangement of atoms in the material. The X-ray pattern is "inversely" related to the structure. In Chapter 15 we'll talk of direct space (this is the space we live in) and reciprocal space (this will be the pattern produced by X-ray diffraction). To determine the structure of a material in direct space, we must interpret the diffraction pattern in reciprocal space and mathematically relate the two.

In matrix form, the d-spacing for a particular (h k l) plane can be found with the following equation:

$1 / d^2 = [h\ k\ l]\ [MT]^*\ [h\ k\ l]^t$ Equation 13.7

where [h k l] = the Miller indices for the plane of interest [MT]* = the inverse of [MT].

So far, the numerical examples we have given are in direct space. Calculations involving d-spacings, however, are in reciprocal space because they correspond to X-ray diffraction. To move between these two types of space (really just two separate vector basis systems) we use the metric tensor. The metric tensor as it was defined in Equation 13.4 represents direct space. However, by taking the inverse of the metric tensor, we arrive at the following:

$$[MT]^{-1} = [MT]^* = \begin{bmatrix} a^{*2} & a^*b^*\cos\gamma^* & a^*c^*\cos\beta^* \\ a^*b^*\cos\gamma^* & b^{*2} & b^*c^*\cos\alpha^* \\ a^*c^*\cos\beta^* & b^*c^*\cos\alpha^* & c^{*2} \end{bmatrix}.$$

Equation 13.8

The "*" used in Equation 13.8 represents the reciprocal cell parameters and interaxial angles (discussed in Chapter 15) and is read as "star." So to arrive at [MT]* we simply take the inverse of [MT]. For the anorthite example, this would yield:

$$[MT]^* = \begin{bmatrix} 0.0185 & 0.0005 & 0.0047 \\ 0.0005 & 0.0061 & 0.0004 \\ 0.0047 & 0.0004 & 0.0062 \end{bmatrix}.$$

For any mineral we could form the metric tensor, then find its inverse, [MT]*, to determine the reciprocal axes and angles. Given the entries in [MT]* for anorthite above, we could calculate, based on the relationships given in Equation 13.8, the values for anorthite as: $a^* = 0.1361$, $b^* = 0.0779$, $c^* = 0.0786$ 1/Å, $\alpha^* = 87.6$, $\beta^* = 64.0$, $\gamma^* = 84.9°$.

Given [MT]*, we can then select different values of (h k l) and determine their d-spacings by use of Equation 13.7 as follows. For instance, say we want to find d(1 0 0) for anorthite. We would solve:

$$1/d^2 = [1\ 0\ 0] \begin{bmatrix} 0.0185 & 0.0005 & 0.0047 \\ 0.0005 & 0.0061 & 0.0004 \\ 0.0047 & 0.0004 & 0.0062 \end{bmatrix} \begin{bmatrix} 1 \\ 0 \\ 0 \end{bmatrix} = 0.0185.$$

Then, solving for d would yield a d-spacing for (1 0 0) = 7.3448 Å. If we performed this calculation with a mineral that had interaxial angles of $\alpha = \beta = \gamma = 90°$, d(1 0 0) would have equaled a. For a mineral of lower symmetry, this is not the case; a = 8.175 Å for anorthite. As in the previous examples, this calculation can be programmed into a spreadsheet. As another example, we could find that d(3 7 2) = 1.3127 Å.

Angles Between Two Axes

To determine the angle between two directions, the inner or dot product is used (see Appendix A for a description). To determine the angle between two axes for any crystal system, we would write the matrix formula:

$$\text{angle} = \cos^{-1}([u1\ v1\ w1]\ [MT]\ [u2\ v2\ w2]^t\ /\ |u1\ v1\ w1|\ |u2\ v2\ w2|) \qquad \text{Equation 13.10}$$

where [$u1$ $v1$ $w1$], [$u2$ $v2$ $w2$] = the numerical values for the two axes, and $|u1\ v1\ w1|$, $|u2\ v2\ w2|$ = the magnitude of each axes. The magnitude of the axes can be written in matrix form as:

$$|u\ v\ w|^2 = [u\ v\ w]\ [MT]\ [u\ v\ w]^t \qquad \text{Equation 13.11}$$

For a Cartesian system, [MT] would equal the identity matrix, and thus is normally excluded from the calculation, because most workers, except for crystallographers, only work in the Cartesian system.

As an example let's determine the angle between [1 0 0] and [0 1 0] for anorthite. First, calculate the magnitude for each as follows:

$$|1\ 0\ 0|^2 = [1\ 0\ 0] \begin{bmatrix} 66.8306 & -2.3508 & -50.5808 \\ -2.3508 & 165.7141 & -9.8963 \\ -50.5808 & -9.8963 & 200.7889 \end{bmatrix} \begin{bmatrix} 1 \\ 0 \\ 0 \end{bmatrix} = 66.83$$

so, $|1\ 0\ 0| = 8.175$. Likewise, we could calculate $|0\ 1\ 0| = 12.873$. Next, we substitute this magnitude in along with the values for u, v, and w, and [MT] into Equation 13.10.

$$\text{angle} = \cos^{-1} \left([1\ 0\ 0] \begin{bmatrix} 66.8306 & -2.3508 & -50.5808 \\ -2.3508 & 165.7141 & -9.8963 \\ -50.5808 & -9.8963 & 200.7889 \end{bmatrix} \begin{bmatrix} 0 \\ 1 \\ 0 \end{bmatrix} / (8.175)(12.873)\right) = 91.28°$$

The length of [1 0 0] is equal to the a axis, the length of [0 1 0] is equal to the b axis, and the angle between them, γ, was correctly calculated to be 91.28°. For a more general case, the angle between [4 1 8] and [1 9 3] to be 67.55° would be found.

Angles Between Two Planes

The dot product is used in a similar manner to Equation 13.9 to calculate the angle between two faces, given their (h k l) values and the associated metric tensor. We define a plane by a direction that is perpendicular to it, so in reality we are finding the angle between the face poles. However, unlike the calculation to find the angle between two axes, the inverse metric tensor is used (because we are dealing with the perpendiculars to planes and not axes). The resultant equation is:

$$\text{angle} = \cos^{-1}([h1\ kl\ 11]\ [MT^*][h2\ k2\ l2]^t / |h1\ kl\ 11|\ |h2\ k2\ l2|) \qquad \text{Equation 13.11}$$

where [$h1$ $k1$ $l1$], [$h2$ $k2$ $l2$] = the numerical values for the two planes, and $|h1\ k1\ l1|$, $|h2\ k2\ l2|$ = the magnitude of each plane perpendicular. The magnitude of the plane's perpendicular can be written in matrix form as:

$$|h\ k\ l|^2 = [h\ k\ l]\ [MT]^*\ [h\ k\ l]^t. \qquad \text{Equation 13.12}$$

As an example, calculate the angle between the face poles for (1 0 0) and (0 1 0) for anorthite. First, determine the magnitude for each as follows:

$$|1\ 0\ 0|^2 = [1\ 0\ 0] \begin{bmatrix} 0.0185 & 0.0005 & 0.0047 \\ 0.0005 & 0.0061 & 0.0004 \\ 0.0047 & 0.0004 & 0.0062 \end{bmatrix} \begin{bmatrix} 1 \\ 0 \\ 0 \end{bmatrix} = 0.0185.$$

So, $|1\ 0\ 0| = 0.1361$. Likewise, we could calculate $|0\ 1\ 0| = 0.0779$. Next, we substitute these magnitudes, alone with the values for h, k, and l, and [MT]*, into Equation 13.11:.

$$\text{angle} = \cos^{-1}[1\,0\,0]\begin{bmatrix}0.0185 & 0.0005 & 0.0047\\0.0005 & 0.0061 & 0.0004\\0.0047 & 0.0004 & 0.0062\end{bmatrix}\begin{bmatrix}1\\0\\0\end{bmatrix}/(0.1361)(0.0779)=87.06^\circ$$

The angle between (100) and (010) is γ^*. The magnitudes of (1 0 0) and (0 1 0) are a^* and b^*. Recall these "*" values will be used to describe the geometry of the unit cell as observed in a diffraction pattern (see Chapter 15 for thorough discussion). For a more general case, we would find that the angle between (4 1 8) and (1 9 3) is 57.82°.

Angles Between an Axis and a Plane Normal

The last permutation in this series of angle calculations is to determine the angle between a plane normal and an axis. Again the dot product is used as follows:

$$\text{angle} = \cos^{-1}([hu + kv + lw] \;/\; |h\,k\,l|\;|u\,v\,w|)$$
Equation 13.13

where $(h\,k\,l)$ = the plane of interest, and $[u\,v\,w]$ = the axis of interest.

We have already seen how to calculate the magnitude for $(h\,k\,l)$ and $[u\,v\,w]$. As an example of this calculation, find the angle between (0 0 1) and [0 0 1]. First we find the magnitude of the face pole $|0\,0\,1| = 0.0786$ and then the magnitude of the axis $|0\,0\,1| = 14.17$. Then we substitute the values of h, k, l, and u, v, w as follows:

$$\text{angle} = \cos^{-1}([(0)(0) + (0)(0) + (1)(1)] \;/\; (0.0786)(14.17)) = 26.19^\circ.$$

Note that the magnitude of [0 0 1] is equal to c and the magnitude of the (0 0 1) face pole is c^*. In a crystal system with the interaxial angles equal to 90°, the angle between the face pole to (0 0 1) and the direction [0 0 1] would equal zero.

Derivation of the 32 Crystallographic 3-D Point Groups and Collection into Six Crystal Systems

To me, it was always a mystery that we arrived at exactly 32 3-D point groups. This section is included in this chapter because it would have answered that for me. Much of this material is based on a paper by Boisen and Gibbs (1976), though the same discussion is included in their more recent book (Boisen and Gibbs, 1990). For those interested in a more thorough discussion on this subject, please refer to these two references.

The symmetry operations must map the crystallographic axes onto themselves. An example of this can be seen in Figure 13.11, where the 4-fold rotation axis of a tetragonal crystal would map the symmetry equivalent a axis to the b axis. Also shown in Figure 13.11 is the case where this condition does not hold, when a 4-fold rotation axis would map the non-equivalent a axis of an orthorhombic material onto its b axis.

Proper Point Groups

The **proper point groups** are defined as those that do not contain an inversion center. The five monaxial point groups are represented by the point group symbols: 1, 2, 3, 4, 6, referring to the five possible rotation axes that would map 3-D lattice onto itself. Next the task is to determine the possible combinations of these rotation axes that will make the polyaxial point groups. Fortunately, a theorem exists to aid in the seemingly-endless combinations of these rotation axes in 3-D space. Klein's theorem states that a polyaxial point group can only be made of a combination of non-identity (i.e., the 1-folds are excluded) monaxial point groups such that $1/n1 + 1/n2 + 1/n3 > 1$, where $n1$, $n2$, and $n3$ are the orders of the monaxial point groups. So we are left with solving this inequality for all combinations of 2, 3, 4, and 6.

There are twenty possible combinations given in Table 13.2. Solutions of these possible combinations show only six possible combinations that satisfy Klein's theorem: 222, 322, 422, 622, 332, 432. (The point group 332 is better represented as 231, for reasons explained in Boisen and Gibbs, 1990.) Klein's theorem also states that the order of a polyaxial point group is $(2n1n2n3)/(n1n2 + n1n3 + n2n3 - n1n2n3)$. Thus the order (i.e., number of unique symmetry operations contained in the point group) of 222 = 4, 322 = 6, 422 = 8, 622 = 12, 332 = 12, and 432 = 24. At this point we have eleven of the 32 point groups.

Improper Point Groups

The improper point groups are generated by starting with the proper point groups and performing two separate operations. This will be done using set theory and the union of the proper point group, G, with the set composed of the same proper point, G but with i (the inversion center) added. In set theory, this is written as G U Gi, (pronounced "gugi"). As an example of G U

Figure 13.11. Perspective views of a. a tetragonal unit cell (a and b are equal and c may differ) and b. an orthorhombic unit cell (all axes lengths differ). For the tetragonal unit cell, a 90° rotation maps a to b because they are of equal lengths, while for the orthorhombic unit cell the same 90° rotation still maps a to b, but because a and b are of different lengths as shown in the figure, the 90° rotation would not be symmetry allowable for the orthorhombic cell.

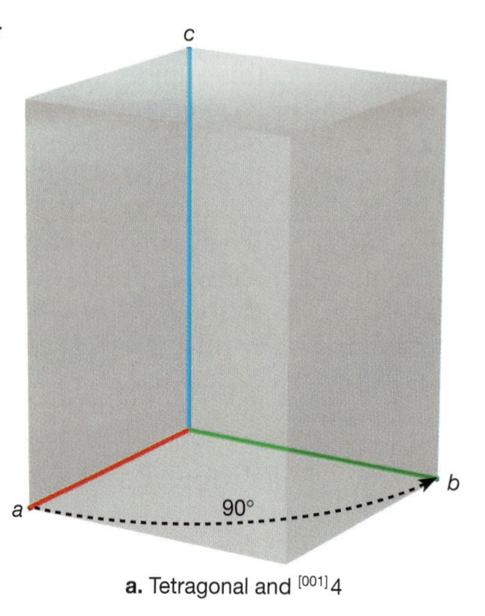

a. Tetragonal and $^{[001]}4$

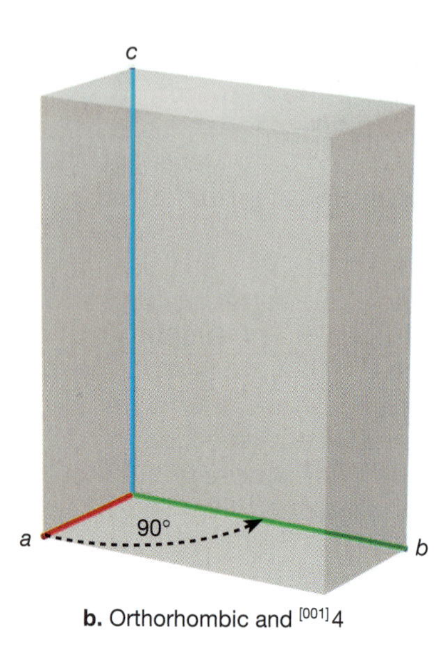

b. Orthorhombic and $^{[001]}4$

G*i* we can take the proper monaxial point 4 and write 4 U 4*i*. The elements of 4 = {1, 4, 2, 4^{-1}} and the elements of 4*i* = {1, 4, 2, 4^{-1}}*i* = {1*i*, 4*i*, 2*i*, $4^{-1}i$} = {*i*, $\overline{4}$, *m*, $\overline{4}^{-1}$}. If we collect the elements, we arrive at 4 U 4*i* = {1, 4, 2, 4^{-1}, *i*, $\overline{4}$, *m*, $\overline{4}^{-1}$}. The order of this new improper point group would be eight and it would be named 4/*m*. We showed in the first section of this chapter the resultant symmetry elements when *i* was combined with either a 2-fold and 4-fold rotation axes. Table 13.3 lists the eleven possible improper point groups that would result from the G U G*i* construction. Note that these are the only point groups that contain a center of symmetry.

To find the remaining point groups, we use a slightly more complicated method to add the inversion center to the eleven proper crystallographic point groups. To form this union, we first need to find a **halving group** for each of the proper crystallographic point groups (see Table 13.3). The halving group in this case refers to the order of the point groups. So before a point group can be a halving group of another, its order must be half. For instance, 2 would be a halving group of 4 and 4 would be a halving group of 422. Table 13.3 contains the halving groups for the 11 proper crystallographic point groups.

The construction type is termed HU(G\H)*i* and read as "hugi."The G\H means to exclude the elements in G that are also in H. As an example, let's use 4 = {1, 4, 2, 4^{-1}} and its halving group 2 = {1, 2}. Thus, we would write 2 U (4\2)*i*, and expanding would yield {1, 2} U ({1, 4, 2, 4^{-1}} \ {1, 2})*i* = {1, 2} U {4, 4^{-1}}*i* = {1, 2} U {4*i*, $4^{-1}i$} = {1, 2} U {$\overline{4}$, $\overline{4}^{-1}$} = {1, 2, $\overline{4}$, $\overline{4}^{-1}$}, so the order of this new group is 4, its called point group $\overline{4}$, and it lacks a center of symmetry. Table 13.3 shows the 10 resultant point groups based on the HU(G\H)*i*, note they all lack a center of symmetry.

In summary, Table 13.4 lists all the 32 point groups with their individual symmetry operations arranged in the order in which they were derived in this section. There are also several more example of G U G*i* and H U (G\H)*i* construction given at the bottom. Students are encouraged to complete the remainder of the set operations.

Table 13.2.
Klein's theorem for 1/*n*1 + 1/*n*2 + 1/*n*3 calculations for each of the 20 possible permutations of monaxial proper rotation axes. Based on this theorem, only those combinations 1/*n*1 + 1/*n*2 + 1/*n*3 > 1 can be a polyaxial point group, thus only 222, 322, 422, 622, 332, and 432 meet this condition.

222 = 1.50	322 = 1.33	422 = 1.25	622 = 1.17
332 = 1.17	432 = 1.08	632 = 1.00	442 = 1.00
642 = 0.92	662 = 0.83	333 = 1.00	433 = 0.92
633 = 0.83	443 = 0.83	643 = 0.75	663 = 0.67
444 = 0.75	644 = 0.67	664 = 0.58	666 = 0.50

Determining the Angles Between the Rotation and Rotoinversion Axes and Grouping the 32 Point Groups into the Six Crystal Systems

Now that we have derived the 32 point groups, we would like to group them into the six crystal systems. This was done based on symmetry arguments in the previous chapter. However, we would like to find a more quantitative grouping method. Recall in the last chapter, Table 12.5 showed the naming system for symmetry operations as a function of rotations axes. Based on equivalence of symmetry directions we found that different conventions were used for each of the six crystal systems. For instance, for 432, the 4 was oriented along [001], the 3 along [111], and the 2 along [110], while for 422, the 4 was oriented along [001], the first 2 along [100] and the last 2 along [110]. But we did not provide a method to determine the angles between these different directions for the polyaxial point groups.

Euler's theorem can be used to calculate the interaxial angles for the polyaxial point groups. Euler's theorem states that given r_1, r_2, and r_3 are three nonidentity rotations with rotations angles of $\rho 1$, $\rho 2$, and $\rho 3$ about rotations axes l_1, l_2, and l_3 and $r_1 \times r_2 = r_3$, then their intersection angles can be found by:

$$\cos(l_1{:}l_2) = (\cos(\rho_1/2)\cos(\rho_2/2) + \cos(\rho_3/2)) \;/\; \sin(\rho_1/2)\sin(\rho_2/2$$

$$\mathrm{cosl}(l_1{:}l_3) = (\cos(\rho_1/2)\cos(\rho_3/2) + \cos(\rho_2/2)) \;/\; \sin(\rho_1/2)\sin(\rho_3/2)$$

Table 13.3.
The 32 point groups and their constructions (modified from Boisen and Gibbs 1977). The number in ( ) represents the order of the point group (i.e., its number of unique symmetry operations). Also, at the bottom of the table are some worked examples to complement those presented in the text.

Proper point groups	Halving group	Improper point groups	
		G U Gi	*H U (G\H)i*
1 (1)	none	$\bar{1}$ (2)	none
2 (2)	1	2/*m* (4)	*m* (2)
3 (3)	none	$\bar{3}$ (6)	none
4 (4)	2	4/*m* (8)	$\bar{4}$ (4)
6 (6)	3	6/*m* (12)	$\bar{6}$ (6)
222 (4)	2	2/*m* 2/*m* 2/*m* (8)	*mm*2 (4)
322 (6)	3	$\bar{3}$ 2/*m* 1 (12)	3*mm* (6)
422 (8)	4	4/*m* 2/*m* 2/*m* (16)	4*mm* (8)
	222		$\bar{4}$2*m* (8)
622 (12)	6	6/*m* 2/*m* 2/*m* (24)	6*mm* (12)
	322		$\bar{6}$2*m* (12)
332 = 231 (12)	none	2/*m* $\bar{3}$ 1 (24)	none
432 (24)	231	4/*m* $\bar{3}$ 2/*m* (48)	$\bar{4}$3*m* (12)

Examples of G U Gi and H U (G\H)i (see text for details)

G U Gi constructions

2	$\{1, 2\} \cup \{1, 2\}i$	$\{1, 2\} \cup \{i, m\}$	$\{1, 2, i, m\}$
4	$\{1, 4, 2, 4^{-1}\} \cup \{1, 4, 2, 4^{-1}\}i$	$\{1, 4, 2, 4^{-1}\} \cup \{i, 4, m, 4^{-1}\}$	$\{1, 4, 2, 4^{-1}, i, \bar{4}, m, \bar{4}^{-1}\}$
222	$\{1, {}^{[100]}2, {}^{[010]}2, {}^{[001]}2\} \cup \{1, {}^{[100]}2, {}^{[010]}2, {}^{[001]}2\}i$		$\{1, {}^{[100]}2, {}^{[010]}2, {}^{[001]}2, i, {}^{[100]}m, {}^{[010]}m, {}^{[001]}m\}$
322	$\{1, {}^{[001]}3, {}^{[001]}3^{-1}, {}^{[100]}2, {}^{[010]}2, {}^{[110]}2\} \cup \{1, {}^{[001]}3, {}^{[001]}3^{-1}, {}^{[100]}2, {}^{[010]}2, {}^{[110]}2\}i$		$\{1, {}^{[001]}3, {}^{[001]}3^{-1}, {}^{[100]}2, {}^{[010]}2, {}^{[110]}2, i, {}^{[001]}\bar{3}, {}^{[001]}\bar{3}^{-1}, {}^{[100]}m, {}^{[010]}m, {}^{[110]}m\}$

H U (G\H)i constructions

G	H			
2	1	$\{1\} \cup (\{1, 2\} \setminus \{1\})i$	$\{1\} \cup \{2\}i$	$\{1, m\}$
4	2	$\{1, 2\} \cup (\{1, 4, 2, 4^{-1}\} \setminus \{1, 2\})i$	$\{1, 2\} \cup \{4, 4^{-1}\}i$	$\{1, 2, \bar{4}, \bar{4}^{-1}\}$
222	2	$\{1, 2\} \cup (\{1, {}^{[100]}2, {}^{[010]}2, {}^{[001]}2\} \setminus \{1, 2\})i$		$\{1, 2, {}^{[100]}m, {}^{[010]}m, {}^{[001]}2\}$
322	3	$\{1, {}^{[001]}3, {}^{[001]}3^{-1}\} \cup (\{1, {}^{[001]}3, {}^{[001]}3^{-1}, {}^{[100]}2, {}^{[010]}2, {}^{[110]}2\} \setminus \{1, {}^{[001]}3, {}^{[001]}3^{-1}\})i$	$\{1, {}^{[001]}3, {}^{[001]}3^{-1}\} \cup \{{}^{[100]}2, {}^{[010]}2, {}^{[110]}2\}i$	$\{1, {}^{[001]}3, {}^{[001]}3^{-1}, {}^{[100]}m, {}^{[010]}m, {}^{[110]}m\}$

Table 13.4. The 32 point groups and their individual elements organized by type.
Proper monaxial point groups
$1 = \{1\}$
$2 = \{1, {}^{[010]}2\}$
$3 = \{1, {}^{[001]}3, {}^{[001]}3^{-1}\}$
$4 = \{1, {}^{[001]}4, {}^{[001]}2, {}^{[001]}4^{-1}\}$
$6 = \{1, {}^{[001]}6, {}^{[001]}3, {}^{[001]}2, {}^{[001]}3^{-1}, {}^{[001]}6^{-1}\}$
Proper cyclic point groups
$222 = \{1, {}^{[100]}2, {}^{[010]}2, {}^{[001]}2\}$
$322 = \{1, {}^{[001]}3, {}^{[001]}3^{-1}, {}^{[100]}2, {}^{[010]}2, {}^{[110]}2\}$
$422 = \{1, {}^{[001]}4, {}^{[001]}2, {}^{[001]}4^{-1}, {}^{[100]}2, {}^{[010]}2, {}^{[110]}2, {}^{[-110]}2\}$
$622 = \{1, {}^{[001]}6, {}^{[001]}3, {}^{[001]}2, {}^{[001]}\text{-}3, {}^{[001]}\text{-}6, {}^{[100]}2, {}^{[010]}2, {}^{[110]}2, {}^{[-110]}2, {}^{[120]}2, {}^{[210]}2\}$
$231 = \{1, {}^{[100]}2, {}^{[010]}2, {}^{[001]}2, {}^{[111]}3, {}^{[111]}3^{-1}, {}^{[-1-11]}3, {}^{[-1-11]}3^{-1}, {}^{[-11-1]}3, {}^{[-11-1]}3^{-1}, {}^{[1-1-1]}3, {}^{[1-1-1]}3^{-1}\}$
$432 = \{1, {}^{[001]}4, {}^{[001]}2, {}^{[001]}4^{-1}, {}^{[100]}4, {}^{[100]}2, {}^{[100]}4^{-1}, {}^{[010]}4, {}^{[010]}2, {}^{[010]}4^{-1}, {}^{[110]}2, {}^{[-110]}2, {}^{[101]}2, {}^{[011]}2, {}^{[0-11]}2, {}^{[-101]}2, {}^{[111]}3, {}^{[111]}3^{-1}, {}^{[-1-11]}3, {}^{[-1-11]}3^{-1},$ $ {}^{[-11-1]}3, {}^{[-11-1]}3^{-1}, {}^{[1-1-1]}3, {}^{[1-1-1]}3^{-1}\}$
Improper point groups with a center of symmetry (G U Gi constructions)
$\bar{1} = \{1, i\}$
$2/m = \{1, {}^{[010]}2, i, {}^{[010]}m\}$
$\bar{3} = \{1, {}^{[001]}3, {}^{[001]}3^{-1}, i, {}^{[001]}\bar{3}, {}^{[001]}\bar{3}^{-1}\}$
$4/m = \{1, {}^{[001]}4, {}^{[001]}2, {}^{[001]}4^{-1}, i, {}^{[001]}\bar{4}, {}^{[001]}m, {}^{[001]}\bar{4}^{-1}\}$
$6/m = \{1, {}^{[001]}6, {}^{[001]}3, {}^{[001]}2, {}^{[001]}3^{-1}, {}^{[001]}6^{-1}, i, {}^{[001]}\bar{6}, {}^{[001]}\bar{3}, m, {}^{[001]}\bar{3}^{-1}, {}^{[001]}\bar{6}^{-1}\}$
$2/m\ 2/m\ 2/m = \{1, {}^{[100]}2, {}^{[010]}2, {}^{[001]}2, i, {}^{[100]}m, {}^{[001]}m, {}^{[001]}m\}$
$\bar{3}\ 2/m\ 1 = \{1, {}^{[001]}3, {}^{[001]}3^{-1}, {}^{[100]}2, {}^{[110]}2, {}^{[010]}2, i, {}^{[001]}\bar{3}, {}^{[001]}\bar{3}^{-1}, {}^{[100]}m, {}^{[110]}m, {}^{[010]}m\}$
$4/m\ 2/m\ 2/m = \{1, {}^{[001]}4, {}^{[001]}2, {}^{[001]}4^{-1}, {}^{[100]}2, {}^{[010]}2, {}^{[110]}2, {}^{[-110]}2, i, {}^{[001]}\bar{4}, {}^{[001]}m, {}^{[001]}\bar{4}^{-1}, {}^{[100]}m, {}^{[010]}m, {}^{[110]}m, {}^{[-110]}m\}$
$6/m\ 2/m\ 2/m = \{1, {}^{[001]}6, {}^{[001]}3, {}^{[001]}2, {}^{[001]}3^{-1}, {}^{[001]}6^{-1}, {}^{[100]}2, {}^{[210]}2, {}^{[110]}2, {}^{[120]}2, {}^{[010]}2, {}^{[-110]}2, i, {}^{[001]}\bar{6}, {}^{[001]}\bar{3}, m, {}^{[001]}\bar{3}^{-1}, {}^{[001]}6^{-1}, {}^{[100]}, {}^{[210]}m, {}^{[110]}m,$ $ {}^{[120]}m, {}^{[010]}m, {}^{[-110]}m\}$
$2/m\ \bar{3}\ 1 = \{1, {}^{[001]}2, {}^{[010]}2, {}^{[010]}2, {}^{[111]}3, {}^{[111]}3^{-1}, {}^{[-1-11]}3, {}^{[-1-11]}3^{-1}, {}^{[-11-1]}3, {}^{[-11-1]}3^{-1}, {}^{[1-1-1]}3, {}^{[1-1-1]}3^{-1}, i, {}^{[001]}m, {}^{[010]}m, {}^{[010]}m, {}^{[111]}\bar{3}, {}^{[111]}\bar{3}^{-1}, {}^{[-1-11]}\bar{3},$ $ {}^{[-1-11]}\bar{3}^{-1}, {}^{[-11-1]}\bar{3}, {}^{[-11-1]}\bar{3}^{-1}, {}^{[1-1-1]}\bar{3}, {}^{[1-1-1]}\bar{3}^{-1}\}$
$4/m\ \bar{3}\ 2/m = \{1, {}^{[001]}4, {}^{[001]}2, {}^{[001]}4^{-1}, {}^{[100]}4, {}^{[100]}, {}^{[100]}4^{-1}, {}^{[010]}4, {}^{[010]}2, {}^{[010]}4^{-1}, {}^{[110]}2, {}^{[-110]}2, {}^{[001]}2, {}^{[011]}2, {}^{[0-11]}2, {}^{[-101]}2, {}^{[111]}3, {}^{[111]}3^{-1}, {}^{[-1-11]}3, {}^{[-1-11]}3^{-1},$ $ {}^{[-11-1]}3, {}^{[-11-1]}3^{-1}, {}^{[1-1-1]}3, {}^{[1-1-1]}3^{-1}, i, {}^{[001]}\bar{4}, {}^{[001]}m, {}^{[001]}\bar{4}^{-1}, {}^{[100]}\bar{4}, {}^{[100]}m, {}^{[100]}\bar{4}^{-1}, {}^{[010]}\bar{4}, {}^{[010]}m, {}^{[010]}\bar{4}^{-1}, {}^{[110]}m, {}^{[-110]}m, {}^{[010]}m, {}^{[011]}m, {}^{[0-11]}m, {}^{[-101]}m,$ $ {}^{[111]}\bar{3}, {}^{[111]}\bar{3}^{-1}, {}^{[-1-11]}\bar{3}, {}^{[-1-11]}\bar{3}^{-1}, {}^{[-11-1]}\bar{3}, {}^{[-11-1]}\bar{3}^{-1}, {}^{[1-1-1]}\bar{3}, {}^{[1-1-1]}\bar{3}^{-1}\}$
Improper point groups which lack a center of symmetry (H U (G\H)i constructions)
$m = \{1, m\}$
$\bar{4} = \{1, {}^{[001]}2, {}^{[001]}\bar{4}, {}^{[001]}\bar{4}^{-1}\}$
$\bar{6} = \{1, {}^{[001]}3, {}^{[001]}3^{-1}, {}^{[001]}\bar{6}, {}^{[001]}m, {}^{[001]}\bar{6}^{-1}\}$
$mm2 = \{1, {}^{[100]}m, {}^{[010]}m, {}^{[001]}2\}$
$3mm = \{1, {}^{[001]}3, {}^{[001]}3^{-1}, {}^{[100]}m, {}^{[110]}m, {}^{[010]}m\}$
$4mm = \{1, {}^{[001]}4, {}^{[001]}2, {}^{[001]}4^{-1}, {}^{[100]}m, {}^{[110]}m, {}^{[010]}m, {}^{[-110]}m\}$
$\bar{4}2m = \{1, {}^{[100]}2, {}^{[010]}2, {}^{[001]}2, {}^{[001]}\bar{4}, {}^{[001]}\bar{4}^{-1}, {}^{[110]}m, {}^{[-110]}m\}$
$6mm = \{1, {}^{[001]}6, {}^{[001]}3, {}^{[001]}2, {}^{[001]}3^{-1}, {}^{[001]}6^{-1}, {}^{[100]}m, {}^{[210]}m, {}^{[110]}m, {}^{[120]}m, {}^{[010]}m, {}^{[-110]}m\}$
$\bar{6}2m = \{1, {}^{[001]}3, {}^{[001]}3^{-1}, {}^{[100]}2, {}^{[010]}2, {}^{[110]}2, {}^{[001]}\bar{6}, m, {}^{[001]}\bar{6}^{-1}, {}^{[210]}m, {}^{[120]}m, {}^{[-110]}m\}$
$\bar{4}3m = \{1, {}^{[100]}2, {}^{[010]}2, {}^{[001]}2, {}^{[111]}3, {}^{[111]}3^{-1}, {}^{[-1-11]}3, {}^{[-1-11]}3^{-1}, {}^{[-11-1]}3, {}^{[-11-1]}3^{-1}, {}^{[1-1-1]}3, {}^{[1-1-1]}3^{-1}, {}^{[001]}\bar{4}, {}^{[001]}\bar{4}^{-1}, {}^{[100]}\bar{4}, {}^{[100]}\bar{4}^{-1}, {}^{[010]}\bar{4}, {}^{[010]}\bar{4}^{-1}, {}^{[110]}m,$ $ {}^{[-110]}m, {}^{[101]}m, {}^{[111]}m, {}^{[0-11]}m, {}^{[-101]}m\}$

$\cos(l_2{:}l_3) = (\cos(\rho_2/2)\cos(\rho_3/2) + \cos(\rho_1/2)) / \sin(\rho_2/2)\sin(\rho_3/2)$

Table 13.5 lists the angular relationships for the proper polyaxial point groups. Notice how the angles change for each thus requiring different crystal system assignments, as also listed in the table.

The directions listed in the Table 13.5 are not those used to describe the basis vector sets for each of the six crystal systems. They do, however, define the non-symmetric (i.e., non-equivalent directions) which are used in Table 12.5 as the naming conventions for the six crystal systems.

Appendix A: Brief Introduction to Linear Algebra and Matrix Manipulation as Applied to Mineralogy

Linear algebra is an extension of algebra to higher dimensions. In algebra, we use single numbers or variables and have a set of rules for their manipulation (i.e., addition, multiplication, division, etc.) In linear algebra, we replace the single variable with a matrix, where the matrix is nothing more than an array of numbers, and develop a set of rules for matrix manipulation (i.e., addition, multiplication, division, etc.). Because we are concerned about describing 3-D space in mineralogy, matrices provide powerful tools by which to accomplish that goal.

A matrix is first defined based on its dimension, where the dimension is the size of the array. The first number is the number of rows in the matrix and the second is the number of columns.

Table 13.5. The angles (written as $l_1{:}l_2$ and read as the angle between l_1 and l_2, for example) between the three rotation axes (l_1, l_2, and l_3) as determined by the Euler equations as presented in the text. (Note 332 could be rewritten as 231 then its angles would be the same as 432.) Based on these angles we see the need for the six different crystal systems.

Axes set			
$l_1\ l_2\ l_3$	$l_1{:}l_2$	$l_1{:}l_3$	$l_2{:}l_3$
2 2 2	90°	90°	90°
3 2 2	90°	90°	120°
4 2 2	90°	90°	45°
6 2 2	90°	90°	30°
3 3 2	70.53°	54.74°	54.74°
4 3 2	54.74°	45°	35.26°

For instance, a 3×3 (read "3 by 3") matrix would have three rows and three columns, while a 1×3 matrix would have one row and three columns. The 3×3 matrix would have nine elements in the array while a 1×3 matrix would have three. In the two matrices below, each element is subscripted with its row and column number. When we denote a matrix by an algebraic symbol, we enclose that symbol in brackets to denote that the symbol represents a matrix.

$$[X]\text{, an example } 3\times3 \begin{bmatrix} x_{11} & x_{12} & x_{13} \\ x_{21} & x_{22} & x_{23} \\ x_{31} & x_{32} & x_{33} \end{bmatrix}$$

$$[Y]\text{, an example } 3\times1 \begin{bmatrix} y_{11} \\ y_{21} \\ y_{31} \end{bmatrix} \text{ or } \begin{bmatrix} y_1 \\ y_2 \\ y_3 \end{bmatrix}$$

The subscripts in the above matrices have a special meaning. In general, we would write an element in a matrix as x_{rc}, where r is the row number and c is the column number. Thus, x_{23} above would represent the element occurring in the 2nd row and the 3rd column.

A square matrix is one in which the number of rows equals the number of columns. There are some special terms we use in dealing with square matrices. The diagonal elements are those for which $r = c$. For instance, the diagonal elements of [X] above would be x_{11}, x_{22}, and x_{33}. The other elements, those where $r \neq c$, are termed the off-diagonal elements. A square matrix, such as a 3×3, is said to be symmetric if $x_{12} = x_{21}$, $x_{13} = x_{31}$, and $x_{23} = x_{32}$; this is basically a mirror plane running parallel to the diagonal.

In mineralogy we will mainly use two sizes of matrices: a 3×3 matrix that will commonly represent a symmetry operation, and a 3×1 matrix that will, usually, represent the position of an atom in space. Multiplication of a symmetry operation matrix and an atom's position will then determine the location of a new atom that was produced by the symmetry operation. But before we can use these methods, we must know some very basic rules of linear algebra.

Multiplication of Matrices

Matrices are multiplied on a term-by-term basis. Thus, matrices must be conformable for multiplication. By **conformable**, we mean that the number of columns of the first matrix must be equal to the number of rows of the second matrix. The product will have the row size of the first matrix and the column size of the second. Thus the order of multiplication matters in linear algebra, while it does not in algebra. For example, a 3×3 matrix [A] (see below) could be post-multiplied (this means [B] comes after [A], thus the term "post") by a 3×1 matrix [B] with the product a 3×1 matrix. The order of multiplication (i.e., [B] [A]) could not be reversed. If two matrices are conformable for multiplication, they are multiplied and summed as shown in [A] [B] below:

$$[A]=\begin{bmatrix} a11 & a12 & a13 \\ a21 & a22 & a23 \\ a31 & a32 & a33 \end{bmatrix} \quad [B]=\begin{bmatrix} b_{11} \\ b_{21} \\ b_{31} \end{bmatrix} \quad [A][B]=\begin{bmatrix} a_{11}b_{11}+a_{12}b_{21}+a_{13}b_{31} \\ a_{21}b_{11}+a_{22}b_{21}+a_{23}b_{31} \\ a_{31}b_{11}+a_{32}b_{21}+a_{33}b_{31} \end{bmatrix}$$

The 3×3 matrix often represents a symmetry operation and the 3×1 matrix the position of an atom in a crystal. A numerical example of the multiplication of a 3×3 matrix times a 3×1 matrix would be:

$$[C]=\begin{bmatrix} -1 & 0 & 0 \\ 0 & -1 & 0 \\ 0 & 0 & 1 \end{bmatrix} \quad [D]=\begin{bmatrix} 0.25 \\ 0.50 \\ 0 \end{bmatrix} \quad [C][D]=\begin{bmatrix} -0.25 \\ -0.50 \\ 0 \end{bmatrix}$$

In the above example, [C] might represent a 2-fold axis parallel to *c*, which can be written as $^{[001]}2$, and [D] would represent an atom that occurs at ¼ along *a*, ½ along *b*, and 0 along *c*. The product of [C][D] would represent the atom's location after it was operated on by the 2-fold axis.

We will also find it necessary to multiply a 3×3 matrix, [E], times a 3×3 matrix, [F]. In this case the resultant will also be a 3×3 matrix. Again, the multiplication is done on a term-by-term basis summing each product to find the result as below:

$$[E]=\begin{bmatrix} e11 & e12 & e13 \\ e21 & e22 & e23 \\ e31 & e32 & e33 \end{bmatrix} \qquad [F]=\begin{bmatrix} f11 & f12 & f13 \\ f21 & f22 & f23 \\ f31 & f32 & f33 \end{bmatrix}$$

$$[E][F]=\begin{bmatrix} e_{11}f_{11}+e_{12}f_{21}+e_{13}f_{31} & e_{11}f_{12}+e_{12}f_{22}+e_{13}f_{32} & e_{11}f_{13}+e_{12}f_{23}+e_{13}f_{33} \\ e_{21}f_{11}+e_{22}f_{21}+e_{23}f_{31} & e_{21}f_{12}+e_{22}f_{22}+e_{23}f_{32} & e_{21}f_{11}+e_{22}f_{23}+e_{23}f_{33} \\ e_{31}f_{11}+e_{32}f_{21}+e_{33}f_{31} & e_{31}f_{12}+e_{32}f_{22}+e_{33}f_{32} & e_{31}f_{13}+e_{32}f_{23}+e_{33}f_{33} \end{bmatrix}.$$

A numerical example of the multiplication of a 3×3 matrix, [G], times a 3×3 matrix, [H], would be:

$$[G]=\begin{bmatrix} 0 & -1 & 0 \\ 1 & 0 & 0 \\ 0 & 0 & 1 \end{bmatrix} \quad [H]=\begin{bmatrix} -1 & 0 & 0 \\ 0 & -1 & 0 \\ 0 & 0 & -1 \end{bmatrix} \quad [G][H]=\begin{bmatrix} 0 & 1 & 0 \\ -1 & 0 & 0 \\ 0 & 0 & -1 \end{bmatrix}.$$

In the above example, [G] might represent a 4-fold axis parallel to *c*, and [H] represents a center of symmetry. The product of these two matrices is a 4-fold rotoinversion.

Transpose of a Matrix

The transpose of a matrix is nothing more than switching the rows and columns of a matrix. The superscript "*t*" used to the right of a matrix to denote its transpose. For instance, the transpose of a 3×2 matrix would be a 2×3 matrix and the transpose of a 3×1 matrix results in 1×3 matrix, which could be written as:

$$\begin{bmatrix} x1 \\ x2 \\ x3 \end{bmatrix}^t = [x1 \ \ x2 \ \ x3].$$

The transpose of a 3×3 matrix would be written as:

$$\begin{bmatrix} x11 & x12 & x13 \\ x21 & x22 & x23 \\ x31 & x32 & x33 \end{bmatrix}^t = \begin{bmatrix} x11 & x21 & x31 \\ x12 & x22 & x32 \\ x13 & x23 & x33 \end{bmatrix}.$$

Inversion ("Division") of Matrices

Just as there is an inverse of a scalar number in algebra, there is an inverse of a matrix in linear algebra. To determine the inverse of a 3×3 matrix (only square matrices have an inverse) by hand, might require several hours of laborious calculations. However, by use of a hand-held calculator or computer, the process is vastly simplified. In fact, most spreadsheets have the ability to determine the inverse of a matrix, and it will take longer to input the numbers than to perform the calculation. There are two main reasons we will need to be able to determine the inverse of matrix in crystallography: (1) we need to solve algebraic equations and (2) we use a basis vector set called **direct space** (this is the world around us) to describe the crystal structure of a material. However, we determine the structure of a material with diffraction experiments and their results are in something we call **reciprocal space**. Luckily, direct and reciprocal space are the inverse of each other. Thus, the diffraction experiment results can be converted into a crystal structure by using the inversion property of a matrix.

We'll make extensive use of matrix inversion, but we'll let the computer perform the calculation for us. (For those interested in seeing how one approaches this sort of calculation manually, please refer any linear algebra book.) Briefly, the process of finding the inverse of a square matrix

[A] is to satisfy the mathematically expression $[A][A]^{-1} = [I]$. However the process to satisfy this equation, as already stated, is mathematically tedious and best left to a calculator or computer program.

We'll give four examples of matrix inversion here, with several more given in this chapter. The first will use the identity matrix, [I], and its inverse, $[I]^{-1}$, the "–1" superscript denotes inverse, then give two other numerical examples. We'll also use the example of something called the metric tensor, [MT], and its inverse, $[MT]^{-1}$.

The identity matrix is basically the "1" of linear algebra. It is a square matrix. A 3×3 matrix would be written as:

$$[I] = \begin{bmatrix} 1 & 0 & 0 \\ 0 & 1 & 0 \\ 0 & 0 & 1 \end{bmatrix}.$$

Its inverse, just like the inverse of 1 (i.e., 1/1) in algebra, is the same, and can be written as:

$$[I]^{-1} = \begin{bmatrix} 1 & 0 & 0 \\ 0 & 1 & 0 \\ 0 & 0 & 1 \end{bmatrix}.$$

Again thinking about algebra, the inverse of 2 is 1/2 or 0.5, so it somewhat follows that in linear algebra we would have:

$$\begin{bmatrix} 2 & 0 & 0 \\ 0 & 2 & 0 \\ 0 & 0 & 2 \end{bmatrix}^{-1} = \begin{bmatrix} 0.5 & 0 & 0 \\ 0 & 0.5 & 0 \\ 0 & 0 & 0.5 \end{bmatrix}.$$

However, when the off-diagonal terms are nonzero it becomes more complicated to calculate the inverse. Here's one last example:

$$\begin{bmatrix} 100 & 50 & 9 \\ 5 & 150 & 10 \\ 9 & 10 & 200 \end{bmatrix}^{-1} = \begin{bmatrix} 0.010055 & -0.000306 & -0.000437 \\ -0.000306 & 0.006698 & -0.000321 \\ -0.000437 & -0.000321 & 0.005036 \end{bmatrix}.$$

The metric tensor [MT] is a 3×3 matrix composed of the cell parameters and interaxial angles that define the unit cell of a mineral. It has many uses in standard crystallographic calculations (see examples in this chapter). The metric tensor is written as:

$$[MT] = \begin{bmatrix} a^2 & ab\cos\gamma & ac\cos\beta \\ ab\cos\gamma & b^2 & bc\cos\alpha \\ ac\cos\beta & bc\cos\alpha & c^2 \end{bmatrix},$$

and its inverse can be written as

$$[MT]^{-1} = \begin{bmatrix} a^{*2} & a^*b^*\cos\gamma^* & a^*c^*\cos\beta^* \\ a^*b^*\cos\gamma^* & b^{*2} & b^*c^*\cos\alpha^* \\ a^*c^*\cos\beta^* & b^*c^*\cos\alpha^* & c^{*2} \end{bmatrix}.$$

The cell parameters and interaxial angles have a "*" superscript; see the chapter text for an explanation. Also, both [MT] and $[MT]^{-1}$ are symmetrical matrices in which the diagonal elements are a function of the cell parameters only. The off-diagonal elements are a function of both the cell parameters and interaxial angles.

The Determinant of a Matrix

The determinant of a matrix is a scalar value. Only square matrices have a determinant. The calculation of a determinant, like an inverse, is somewhat tedious. Fortunately a spreadsheet or calculator can be used to find the determinant. The determinant of a 3×3 matrix is calculated as follows:

$$\det \begin{bmatrix} x11 & x12 & x13 \\ x21 & x22 & x23 \\ x31 & x32 & x33 \end{bmatrix} = x_{11}x_{22}x_{33} + x_{12}x_{23}x_{31} + x_{13}x_{21}x_{32} - x_{13}x_{22}x_{31} - x_{12}x_{21}x_{33} - x_{11}x_{23}x_{32}$$

Our main use of the determinant will be to determine the volume of the unit cell. The square root of the determinant of the metric tensor, [MT], is the volume of the unit cell.

Addition and Subtraction of Matrices

Matrices are added or subtracted on a term-by-term basis. Before two matrices are conformable for addition, or subtraction, they must be of the same size. The following would represent an example of the addition of two 3×3 matrices:

$$\begin{bmatrix} a11 & a12 & a13 \\ a21 & a22 & a23 \\ a31 & a32 & a33 \end{bmatrix} + \begin{bmatrix} b11 & b12 & b13 \\ b21 & b22 & b23 \\ b31 & b32 & b33 \end{bmatrix} = \begin{bmatrix} a11+b11 & a12+b12 & a13+b13 \\ a21+b21 & a22+b22 & a23+b23 \\ a31+b31 & a32+b32 & a33+b33 \end{bmatrix},$$

while the subtraction of two 3×1 matrices would be:

$$\begin{bmatrix} x1 \\ x2 \\ x3 \end{bmatrix} - \begin{bmatrix} y1 \\ y2 \\ y3 \end{bmatrix} = \begin{bmatrix} x1-y1 \\ x2-y2 \\ x3-y3 \end{bmatrix}.$$

Dot Products

The dot product is used to calculate the angle between two vectors. The dot product for two vectors, u and v, in a Cartesian system (i.e., 3-D system, with orthogonal basis vectors) is written as:

$$\cos\theta = \frac{u \cdot v}{|u|\,|v|}$$

where $u \cdot v$ (read "u dot v") $= u_1v_1 + u_2v_2 + u_3v_3$ or, in matrix form,

$$\begin{bmatrix} u1 \\ u2 \\ u3 \end{bmatrix} [v1\ v2\ v3].$$

Here $|u|$ and $|v|$ are the magnitudes of u and v (i.e., $\sqrt{u_1^2+u_2^2+u_3^2}$), and θ is the angle between u and v. In 3-D space, the product of the multiplication of a 3×1 matrix times a 1×3 will result in a 1×1 matrix (i.e., a scalar). In reality, a 3×3 matrix is placed in between the u and v vectors. For Cartesian space, we would have:

$$u \cdot v = [u1 \quad u2 \quad u3] \begin{bmatrix} 1 & 0 & 0 \\ 0 & 1 & 0 \\ 0 & 0 & 1 \end{bmatrix} \begin{bmatrix} v1 \\ v2 \\ v3 \end{bmatrix}$$

For the general case for any crystal system, we have:

$$u \cdot v = [u1 \;\; u2 \;\; u3] \begin{bmatrix} a^2 & ab\cos\gamma & ac\cos\beta \\ ab\cos\gamma & b^2 & bc\cos\alpha \\ ac\cos\beta & bc\cos\alpha & c^2 \end{bmatrix} \begin{bmatrix} v1 \\ v2 \\ v3 \end{bmatrix}$$

where a, b, c, α, β, γ are the cell parameters and interaxial angles of a mineral. See the chapter text for examples of this type of calculations.

Cross Products

The cross product determines a vector (w) that is perpendicular to two vectors (u and v). In Cartesian space we can write:

$$w = u \times v = \begin{vmatrix} u2 & u3 \\ v2 & v3 \end{vmatrix}, - \begin{vmatrix} u1 & u3 \\ v1 & v3 \end{vmatrix}, \begin{vmatrix} u1 & u2 \\ v1 & v2 \end{vmatrix} = u2v3-u3v2, u3v1-u1v3, u1v2u2v1$$

The vertical lines in the above equation are the determinants of the 2×2 matrices. We'll find uses for cross products when we deal with direct and reciprocal lattices, as discussed at length in Chapter 14.

Appendix B: The General Cartesian Rotation Matrix and Its Use to Arrive at Matrix Representations for Rotations and Rotoinversions

Earlier in this chapter, we showed how to arrive at the matrix representation of rotations by observing how the basis vector sets behaved under that rotation. Next, we showed how to find the associated rotoinversions by multiplying the rotation operation matrix by the inversion matrix, which represented the inversion center. Basically each of the three columns in a 3×3 matrix represents the three axes, and, in turn, each of the three elements in that column represents the projections of that axis along the others. For example, if we have after a rotation:

"x"	"y"	"z"
0	–1	0
1	0	0
0	0	1

We could see that x went to y, y went to $-x$, and z stayed the same. This process that we have so far treated by observation has its roots in something called the general Cartesian rotation matrix, [CR], which is written as:

$$[\mathrm{CR}] = \begin{bmatrix} \cos(x_r)^2(1-\cos\rho)+\cos\rho & \cos(x_r)\cos(y_r)(1-\cos\rho) - \cos z_r\sin\rho & \cos(x_r)\cos(z_r)(1-\cos\rho) + l_2\sin\rho \\ \cos(x_r)\cos(y_r)(1-\cos\rho) - \cos z_r\sin\rho & \cos(y_r)^2(1-\cos\rho)+\cos\rho & \cos(y_r)\cos(z_r))(1-\cos\rho)-\cos(x_r)\sin\rho \\ \cos(x_r)\cos(z_r)(1-\cos\rho)-\cos(y_r)\sin\rho & \cos(y_r)\cos(z_r)(1-\cos\rho) + \cos(x_r)\sin\rho & \cos(z_r)^2(1-\cos\rho)+\cos\rho \end{bmatrix}$$

where ρ = rotation angle (positive if counterclockwise and negative if clockwise), x_r = angle between the rotation axis and x Cartesian axis, y_r = angle between the rotation axis and y Cartesian axis, z_r = angle between the rotation axis and z Cartesian axis and, based on the law of directional cosines:

$$(\cos x_r^2 + \cos y_r^2 + \cos z_r^2)^{1/2} = 1.$$

No doubt the above matrix looks somewhat foreboding. However, notice that terms in the matrix contain the sin and cos functions. If the angle of rotation, ρ, equal 90°, then the $\sin\rho$ terms simplify to 1 and the $\cos\rho$ terms to 0. Also, if the orientation of the rotation axis is parallel to one of the Cartesian axes, then the cos terms in the equation become either 0 (if the rotation axis is 90° to a Cartesian axis) or 1 (if the rotation axis coincides with a Cartesian axis). Thus the above matrix will simplify considerably for the typical rotations we will encounter in crystallography. Figure 13B.1 shows the angular relationships between the rotation axes and the three mutually perpendicular axes of a Cartesian basic vector set, x, y, and z.

Let's take a rather complex numerical example. Assume we want to make a 45° rotation (i.e., ρ = 45°) around a line that has x_r = 85.4, y_r = 66.6, and z_r = 23.9°. Figure 13B.2 shows the graphical representation of this scenario before and after the rotation. The resultant rotation matrix would be:

$$\begin{bmatrix} 0.7090 & -0.6371 & 0.3023 \\ 0.6558 & 0.7533 & 0.0496 \\ -0.2594 & 0.1631 & 0.9519 \end{bmatrix}.$$

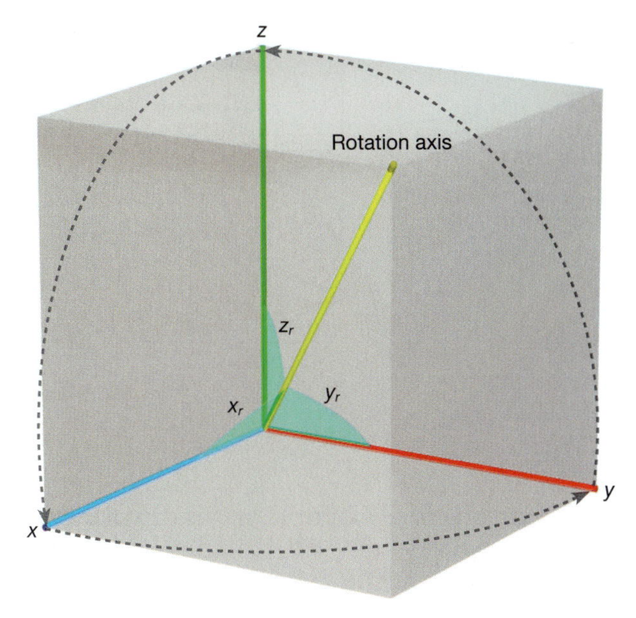

Figure 13.B1. Graphical relationships for a rotation axis in a general orientation in a Cartesian basis vector set. The angles x_r, y_r, and z_r represent the angle the rotation axis makes with each of the three Cartesian axes, x, y, and z, respectively.

Under this rotation, the x axis would have moved from [1 0 0] to [0.7090 0.6558 –0.2594]. What does this tell us about this rotation? First it describes the new locations of the axes. Notice how none of the axes is mapped to one of the others. For example, [1 0 0] did not go to [0 1 0]) as in the first example in this section. This would tell us that it is not an allowable rotation operation.

The next example will show an allowable rotation based on symmetry. Figure 13B.3 shows a rotation axis coinciding with the +z direction (i.e., $x_r = 90$, $y_r = 90$, $z_r = 90°$). Given a 90° positive rotation (i.e., counterclockwise) the rotation matrix becomes:

$$\begin{bmatrix} 0 & -1 & 0 \\ 1 & 0 & 0 \\ 0 & 0 & 1 \end{bmatrix},$$

which was the rather simple example at the beginning of this section. Notice that in the above matrix, only the values of 0, 1, and –1 occur. If other numbers occur, as in the example shown in Figure 13B.2, then this symmetry operation is not allowable (i.e., it would not preserve the symmetry of the material). It is very instructive for a student (or his/her instructor) to write a simple spreadsheet with the general Cartesian rotation matrix and then input different combinations of rotation axis orientations and rotation angles to see which are allowable. Once this is done, you'll get the set of all possible point group operations that we have discussed to date. A slightly more mathematical way to say the above is that the determinant of [CR] must be equal to 1 for the rotation operation to be allowable.

At the same time, you can create the rotoinversions associated with the rotations. All that is required is to post-multiply the resultant [CR] by the 3×3 matrix that represents an inversion (i.e., the inverse of the identity matrix). The determinant of the allowable rotoinversions will be –1.

Non-Cartesian Case

The general Cartesian rotation matrix works for any crystal system that has an orthogonal basis vector set, regardless of the lengths of the vectors. Thus it can be used for the isometric, tetragonal, and orthorhombic crystal systems. It cannot be directly used to determine the rotation matrices for the general case of the hexagonal and monoclinic system, because the matrix is written for Cartesian systems. However, coordinate system

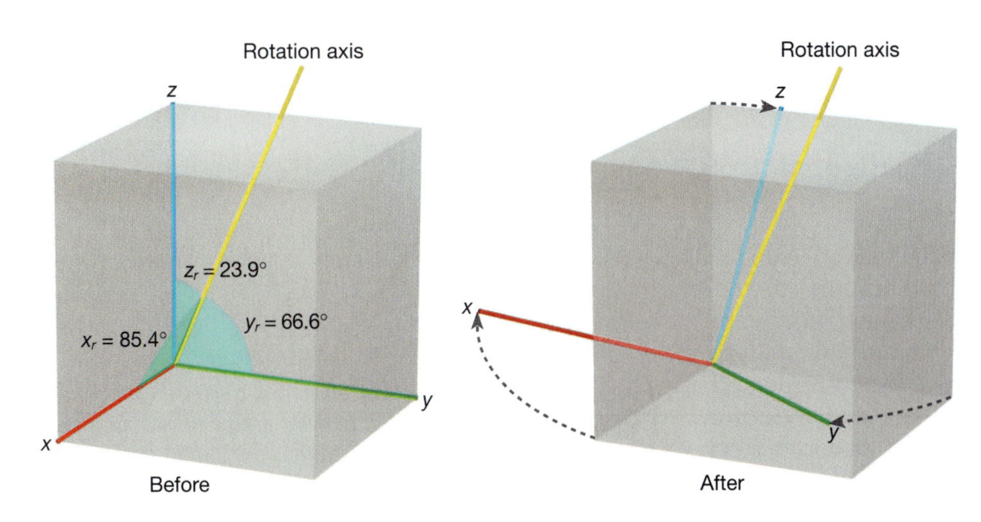

Figure 13B.2. Graphical relationship for a rotation axis directed along a line that plots at $x_r = 85.4°$, $y_r = 66.6°$, and $z_r = 23.9°$, before and after a 45° counterclockwise rotation.

Figure 13B.3. Graphical relationships for a rotation axis directed along the +z Cartesian axis (i.e., $x_r = 90°$, $y_r = 90°$, $z_r = 0°$), before and after a 90° counterclockwise rotation.

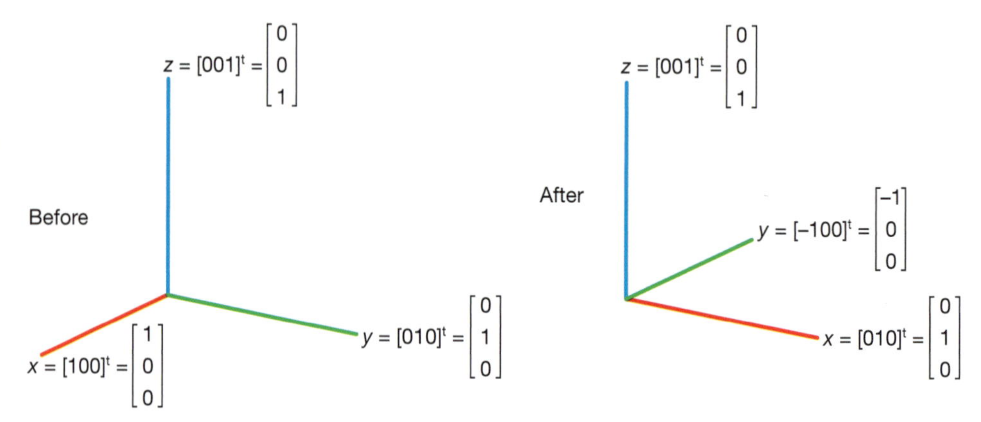

transformation matrices can be written to transform these non-Cartesian systems into a Cartesian system. Then the general Cartesian matrix can be used, and its results transformed back to the non-Cartesian system. While this sounds tedious, the process becomes very straightforward by the use of matrix multiplication. The most difficult aspect is to determine the transformation matrix. This can be done graphically, and is explained below for the hexagonal system (the monoclinic system will be left as an student exercise).

To use the [CR] with the hexagonal system, first a transformation matrix must be found that will convert the hexagonal system to the Cartesian system. (For this conversion we are only concerned with angles.) Figure 13B.4 shows the projection of the axes on the *a-b* plane. To find [TF] we need to express a_c in terms of a_h, b_c in terms of b_h, and c_c in terms of c_h, where the subscripts represent the basis vector set. By observation of Figure 13B.4 we see that *b* and *c* coincide for both systems (we are looking down *c*); however *a* does not. Some basic trigonometry yields that a_c can be written in the hexagonal basis as the vector [cos(120°–90°) sin(90°–120°) 0] or [$\sqrt{3/2}$ –0.5 0] or numerically [0.866 –0.5 0]. So in matrix form, [TF] becomes:

$$[TF] = \begin{bmatrix} 0.8660 & 0 & 0 \\ -0.5 & 1 & 0 \\ 0 & 0 & 1 \end{bmatrix}.$$

What remains is to perform the matrix product of:

$$[TF]^{-1}\,[CR]\,[TF].$$

This converts the hexagonal system into a Cartesian system, performs the rotation calculations in the Cartesian basis, and then reconverts the calculation back to the hexagonal system. Again, no doubt this all seems a bit complicated, but when written out as a series of matrices on a spreadsheet, the calculations can become very straightforward and meaningful.

We'll do one final example of this calculation in the hexagonal system: a 120° rotation parallel to *z* (i.e., a 3rd turn). The following represents the required matrix multiplication:

$$[TF]^{-1}[CR][TF] =$$

$$\begin{bmatrix} 1.1547 & 0 & 0 \\ 0.5774 & 1 & 0 \\ 0 & 0 & 1 \end{bmatrix}\begin{bmatrix} -0.5 & 0.8660 & 0 \\ -0.8660 & -0.5 & 0 \\ 0 & 0 & 1 \end{bmatrix}\begin{bmatrix} 0.8660 & 0 & 0 \\ -0.5 & 1 & 0 \\ 0 & 0 & 1 \end{bmatrix} = \begin{bmatrix} -1 & 1 & 0 \\ -1 & 0 & 0 \\ 0 & 0 & 1 \end{bmatrix}.$$

Thus the matrix representation for a 120° rotation about the +*z* axis would be

$$\begin{bmatrix} -1 & 1 & 0 \\ -1 & 0 & 0 \\ 0 & 0 & 1 \end{bmatrix}$$ as shown in the above calculation.

We could also write the following expression to find the matrix representation for a –3 rotoinversion axis parallel to z (the only difference between this and the above equation is post multiplying the above result by the matrix representation for a center of symmetry):

$$[TF]^{-1}[CR][TF][I]^{-1} = \begin{bmatrix} 1 & -1 & 0 \\ 1 & 0 & 0 \\ 0 & 0 & 1 \end{bmatrix}.$$

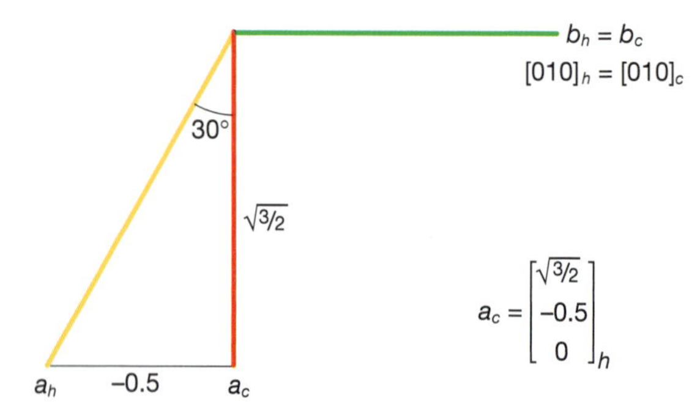

Figure 13B.4. A projection on the a-b *plane (for both a Cartesian,* c, *and hexagonal,* h, *basis vector set) showing the mathematical relation of* a *in the Cartesian system,* a_c, *expressed in terms of the hexagonal system,* a_h.

References

Angel, R.J., Carpenter, M.A., and Finger, L.W. (1990) Structural variation associated with compositional variation and order-disorder behavior in anorthite-rich feldspars. *American Mineralogist*, 75(1–2), 150–162.

Bloss, F.D. (1994) Crystallography and crystal chemistry. Mineralogical Society of America, Washington D.C. 545 pp.

Boisen, M.B., Jr. and Gibbs, G.V. (1990) Mathematical crystallography. Second Edition. Reviews in Mineralogy, 15, Mineralogical Society of America, Washington D.C. 460 pp.

Boisen, M.B., Jr. and Gibbs, G.V. (1978) A method for constructing and interpreting matrix representations of space group operations. Canadian Mineralogist, 16, 293–300.

Boisen, M.B., Jr. and Gibbs, G.V. (1976) A derivation of the 32 crystallographic point groups using elementary theory. American Mineralogist, 61, 145–165.

Gibbs, G.V. (1997) The meterical matrix in teaching mineralogy. In Teaching Mineralogy, editors J.B. Brady, D.W. Mogk, and D. Perkins III, 201–212.

Hahn, T. (1995) International tables for crystallography, volume A, space-group symmetry. Kluwer Academic Pub., Dordrecht, Holland, 878 pp.

Chapter 14

Representation of Crystal Structures

As we've stressed throughout this book, "the physical properties of minerals are directly related to their crystal structures." Thus, we need a method to "show" the crystal structure of a mineral. I think that the main reason we bother to teach crystallography (i.e., symmetry operations, point groups, lattice types, and space groups) is as a means to understand crystal structures. So, this chapter is the culmination of the past three chapters.

The atomic arrangement of atoms is really the guts of mineralogy and what, I think, differentiates a mineralogist from a chemist. For example, we know that there is a vastly different health effect in the lung between crystalline forms of silica and amorphous forms. We even know there are different effects that result from differences within the crystalline species. Thus for many reasons, as discussed throughout this text, crystal structures are important.

But how do we represent a crystal structure? As we'll see, there are two ways to do this. The first deals with the positions of atoms in 3-D space and the symmetry they conform to; we can put this information in a table that's only a few lines in length. Second, we can, by one of many methods we'll explore, attempt to make a 3-D visualization of the structure, either by physically building a model or some type of computer visualization. I have been doing both of these since 1979. All of the graduate students at Virginia Tech who studied crystal chemistry with Jerry Gibbs built physical models (termed ball and stick models) of minerals; I built scapolite. As I recall, it took me about 40 hours to build that model—but what a learning experience! As you'll see in this chapter, at the University of Idaho we still build and use these ball and stick models in teaching and research. If you don't want to spend the time to build these models, you can purchase them, though they are not cheap. The other option is to use computer software to produce graphical images of the minerals; so far you've seen countless ones in this book, with more to come. These drawings are not that difficult to produce with modern computers and menu-assisted software. But the first time I made a computer drawing of a mineral back in 1979, it required: (1) the use of a mainframe computer, (2) a special graphics terminal (back then computer graphics were a lot more complex and required special terminals, and you didn't worry about color schemes—they came out black and white, not even grayscale!), (3) assistance from another student on how to use the program, and (4) 3–4 hours in a darkened lab on a beautiful spring Sunday afternoon in the Blue Ridge Mountains. However, the result was very useful for my graduate research. To this day, I'm proud of that drawing (it's included in this chapter as Figure 14.7d). By looking at it and the other images in that figure, you'll see how far we've come in visualizations of crystal structures since I was a student.

M.E.G.

Introduction

If you've worked your way through the chapters in this book and the figures on the MSA website, you've seen 100's of crystal structure drawings that illustrate many of the important aspects of mineralogy. Just in case you've for some reason jumped to this chapter to start reading, take a look at Figure 14.1, which shows two separate crystal structures, each represented in two ways. Figure 14.1a is about as simple as a structure gets—this is the all-to-familiar halite. Figure 14.1b is the much-more-complex amphibole mineral tremolite. If you look at this figure closely, you can see that the upper images of halite and amphibole are photographs of physical models of the mineral, while the lower images are computer-generated graphical images. (For an idea of scale, the balls in the physical models are 1/2" in diameter.)

Most mineralogists call the physical models **ball and stick models** because they are made of balls (that are used to represent the atoms) and sticks (used to represent the bonds between the atoms). In the education world, physical models are often termed **manipulatives**. Regardless of what you call them, they are very useful in teaching and research dealing with minerals. However, as stated above, you either have to build these models (it would take 2–3 hours to build the halite model and 30–40 to build the tremolite) or buy them (halite would cost about $200 and tremolite at least $2,000). So it would seem to make more sense if we could use computer-generated graphics to represent crystal structures. The lower two images in Figure 14.1 are just that. The newer software packages are much easier to use than a decade ago and are just packed with features that we'll touch in this chapter. For example, notice the lighting on the balls in the ball and stick model in Figure 14.1, then notice how that lighting is mimicked in the lower computer-generated images. However, current education research shows that most of us need the manipulatives (i.e., physical models that we can touch and turn) first to learn to visualize objects in 3-D space. As soon as we are given some of these skills, we can then "see" the 3-D in the computer-generated images. So regardless of how far computer graphics have come, most of us might still resort (at least initially) to physical models.

What information is required to build these models or generate the computer images? Most of you could probably figure out how to build a model of halite out of styrofoam balls and toothpicks, or make a drawing of it on the computer—after all, it's just a square array in 2-D that would turn into a cube in 3-D. But how about tremolite, or the countless other structures you've seen so far in the book? In this book we've used *CrystalMaker* software (CrystalMaker Software Ltd., 1994–2007) for our illustrations. We've used the structure data from files supplied in *CrystalMaker* as well as those in the *American Mineralogist* crystal structure database (Downs and Hall-Wallace, 2003, Clarke and Downs, 2004), and the inorganic crystal structure database (ICSD 2004). *CrystalMaker* is a commercially-available software package; a free demo can be downloaded from its site. Although not used herein, *XtalDraw*, discussed by Bartelmehs et al. (1993), Downs and Hall-Wallace (2003), and Clarke and Downs (2004), is an excellent shareware software package available for Windows-based machines. Thus the software and databases are readily available for students and professors to explore the beauties of crystal structures!

Once you've reached the point of understanding a structure by looking at it, you'll see how the methods presented in this chapter can be used to help us conduct research into the beneficial uses of minerals. Thus, the goals of this chapter are: (1) to show the methods mineralogists use to visualize crystal structures, (2) to work through a detailed example of how to manually produce, draw, and build a ball and stick model for a mineral, and (3) to work through the details of ten crystal structures, giving examples of how to create the structures and to use them to show different crystallographic features of each. We'll return to this same set of ten minerals in the next chapter and use them for examples of how they diffract X-rays.

Visualizations of Crystal Structures

There are several different ways to visualize the crystal structure of a mineral. Broadly speaking, we can view all of these as models of a mineral's atomic arrangement. In the earliest days, these were hand-drawn on paper or hand-made models. With the advent of computer graphics, computer software has replaced the hand-drawn versions. However, it will be worth our while to work through one hand-drawn structure so you'll gain an understanding of what the computer software is actually doing.

Regardless of how we view a structure, we must know the atomic arrangement of the atoms before we proceed: the shape and size of the unit cell and the locations of the atoms within that cell. So the atomic arrangement will be the required starting point for whatever method we chose to use to visualize a structure. Figure 14.2 summarizes several different methods for viewing mineral structures, and the atomic arrangement is central to all of these.

To proceed, we will use two very simple structures, diamond and fluorite—both of which are isometric with face-centered lattices—and examine the different methods of viewing each as summarized in Figure 14.2. Before we start, it should be noted that the goal in this section is to look at these two structures and gain an understanding of how we define the atomic arrangement. Figure 14.3 presents a series of seven different views of the unit cells (outlined in black) of each of these

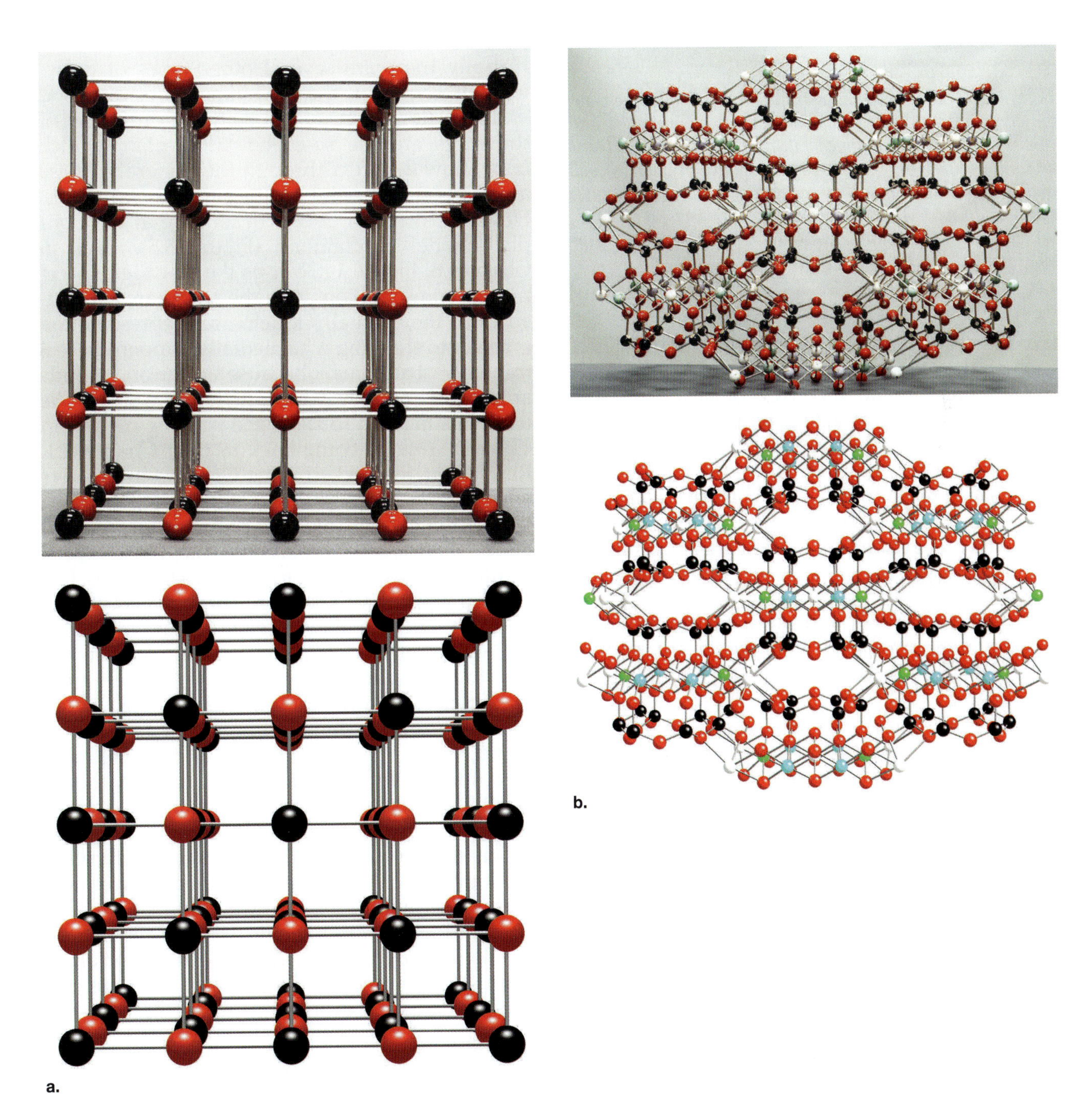

Figure 14.1. *Photographs (top) and computer-generated representations (bottom) of ball and stick models of halite (Figure 14.1a) and tremolite (Figure 14.1b). Halite is one of the simplest mineral structures, while tremolite is a bit more complex. Halite only has two ions in its structure:* Na^{1+} *(black spheres) and* Cl^{1-} *(red spheres), each of which has its own specific sites. Tremolite (ideally* $Ca_2Mg_5Si_8O_{22}(OH)_2$*) has four sites for the large cations (white, green, and turquoise spheres in the very center of the structure), two tetrahedral sites (black spheres), seven oxygen sites (red spheres), and one site for* H^{1+} *(not shown). One of the major goals of this chapter is to explain how we generate and visualize complex structures like tremolite.*

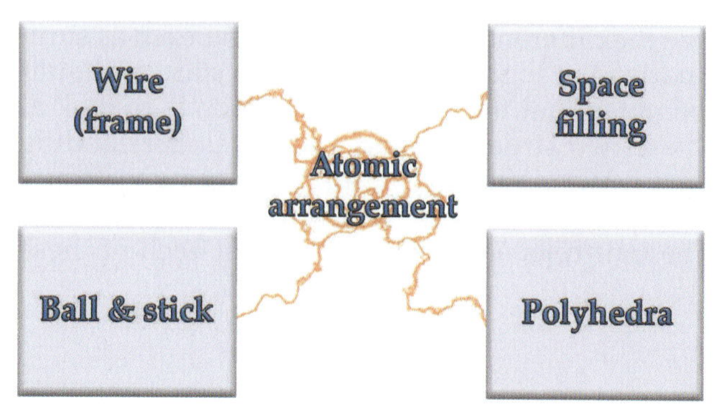

Figure 14.2. Graphical representation of the different ways we can represent the crystal structures of minerals, and some of the associated nomenclature. These can be contrasted with Figure 14.1, which gave examples of physical ball and stick models and their computer generated counter parts, and Figure 14.3, which shows other ways to visualize minerals using two simple structures, diamond and fluorite, as examples.

two minerals (there are associated color movies of each on the MSA website). Figure 14.3 first shows the familiar ball and stick view of each. Note that diamond has one atom, C, while fluorite, CaF_2, has two (with Ca^{2+} shown as the yellow sphere and F^{1-} the blue one). The atomic arrangement is based on three things you've already learned in Chapters 11–13. (1) the size and shape of the unit cell, which is defined by the cell parameters (a, b, c and α, β, γ), (2) the symmetry of the atoms in the unit cell, which is given by the space group, and (3) a list of the unique atoms within the cell; these are the atoms formed by the 1-fold axis symmetry and are termed the atoms of the asymmetric unit.

Both of these minerals are isometric (recall that means $a = b = c$ and $\alpha = \beta = \gamma = 90°$), so the unit cells are cubes, with slightly different sizes: for diamond, $a = 3.6$Å while for fluorite, $a = 5.0$ Å. The space group for diamond is $F4_1/d\,\bar{3}\,2/m$ and that of fluorite is $F4/m\,\bar{3}\,2/m$. Diamond has one unique C atom that plots at 0, 0, 0 in the unit cell; this means it would plot at the origin of the unit cell where all values of a, b, and c are 0. Fluorite has two unique ions: Ca^{2+}, which plots at 0, 0, 0 and F^{1-}, which plots 1.25 Å along the a, b, c direction.

Look at the diamond structure in Figure 14.3a and figure out where these atoms would plot; then ask yourself the question, where did all the other atoms come from in each ball and stick model? The answer is that the other atoms were all generated by the symmetry operations contained in the space groups. In diamond, the atoms produced by the F-centered lattice are C atoms at the corners of the unit cell and centered on each face. Likewise, for fluorite there are Ca^{2+} ions on the corners and centered on each face. But how about the other four C atoms in diamond not on the faces and the four F^{1-} anions not on the faces? They are produced with different symmetry operations. We'll explore the details of how that works in the next section, but first let's look at six other structure representations.

The ball and stick representations are the ones we've used the most throughout the book, but we recognize that they don't work for everyone. Happily there are several other types of drawings, as shown in the other images of diamond and fluorite in Figure 14.3. Left out of this set of drawings are the so-called "floor-plan" which only show 2-D cuts through the minerals. These are rarely used except for the simplest structures. However it might be worth looking at the floor plan model of diamond (Figure 8.5a) back in Chapter 8. Figure 14.3b uses the so-called space-filling representation. In it, the atoms are expanded until they actually touch. This representation is closer to showing what actually happens in real mineral structures, because electron clouds around atoms may well touch and overlap. We used this method to show the packing models in Chapter 3, we'll come back to it in Chapter 21. Recall that we packed large O^{2-} ions into layers and then inserted smaller atoms in the voids. Unfortunately, this method only shows the front layer of atoms and obscures the back layers. While it may work well for simple structures, it is not too helpful for the more complicated atomic arrangements we tend to see in silicate minerals.

Polyhedral models are especially useful when viewing complicated minerals. (We'll see more examples of their use later in this chapter.) Figure 14.3c shows the polyhedral representation of diamond, with C atoms at the center of the tetrahedron formed by C atoms. Recall that polyhedra are 3-D solids with apices (i.e., corners) defined by a chosen set of atoms. For diamond, the polyhedra slightly obscure the structure. In fluorite, the four Ca^{2+} cations form a tetrahedron around a central F^{1-} anion. For the fluorite structure, the tetrahedra have been made translucent so you can see through the structure to the central F^{1-} atom. Note how the Ca^{2+} cations define the corners of the tetrahedrons.

Figure 14.3d is the opposite of the space-filling representations. In this case, the atoms have been excluded altogether and just the bonds are shown. This sort of representation is either termed a **stick figure** (when the lines connecting the central points of the non-shown atoms are big, as is the case here) or a **wire figure** (when the connecting lines are very thin). These representations sometimes find use in complicated zeolite

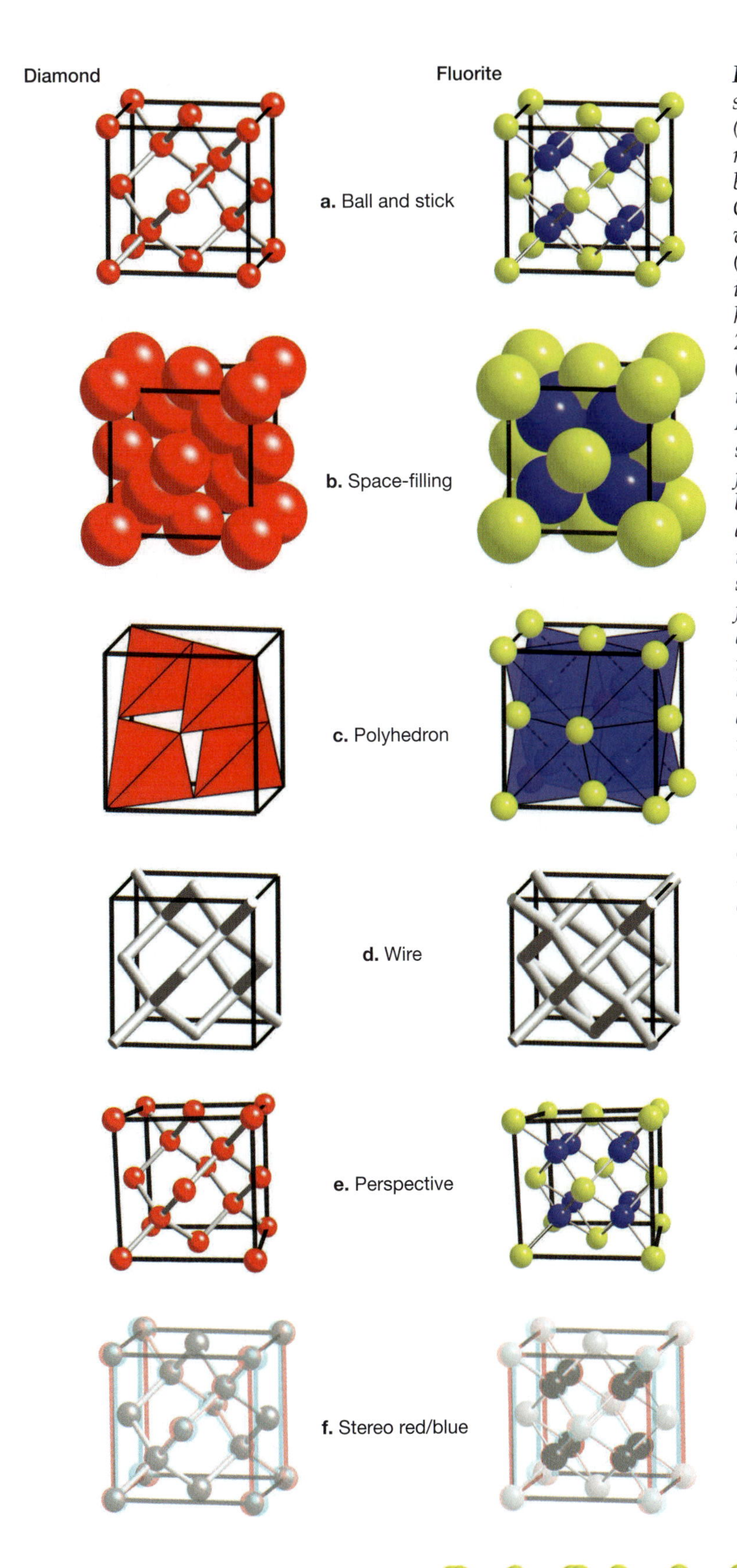

Figure 14.3. *This series of illustrations shows seven different representations of the unit cells (outlined in black) of two simple structures: diamond (left column) composed of tetrahedrally-bonded C atoms and fluorite (CaF_2) composed of Ca^{2+} cations (yellow spheres) forming tetrahedra with F^{1-} anions (blue spheres) at their center (right column). The main goal of this illustration is to show the different methods that researchers have developed to represent crystal structures in 2-D and attempts to show 3-D on 2-D surfaces (i.e., the pages of a book). Movies, and color renditions, of each of these structures are also on the MSA website.* ***a. Ball and stick****: This method of structure visualization derives its name from the fact that atoms are represented as balls and the bonds between them as sticks. Sometimes these are also called "ball and stick."* ***b. Space-filling****: As the name implies, in this method the atoms are scaled to fill space, but because all the space is filled by the atoms, it is difficult to visualize the arrangement of the atoms.* ***c. Polyhedron****: In this method, a 3-D solid (i.e., a polyhedron) is defined based on a central atom bonded to its nearest atoms at the apices of a polyhedron. In the diamond structure, this results in tetrahedra defined by four C atoms (at the apices of a tetrahedron) with a C atom in the center. For fluorite, four Ca^{2+} cations (shown as yellow spheres) form a tetrahedron with F^{1-} in the center. The tetrahedra have been made translucent so that the central F^{1-} anion can be seen.* ***d. Wire*** *(or stick) drawings: In these drawings, the atoms are excluded and a network of bonds (termed wires or sticks) is used to represent the structure.* ***e. Perspective views****: In this type of projection, as well as the next two, an attempt is made to show the 3-D nature of the mineral. Compare these structures to the ball and stick ones above. The distances in front are longer than in the rear of the image, so the size of the unit cell is shorter on the rear of the image.* ***f. Red-blue stereoscopic projections****: To obtain a 3-D image of a mineral (or anything, for that matter), two images are produced in different colors and then they are slightly offset. When viewed with red-blue glasses (red over one eye and blue over the other), the images appear to be 3-D.* ***g. Stereopairs****: Instead of two images printed nearly on top of each other as for the red-blue projections, for this method the two images are printed separated and rotated slightly with respect to each other. When you view the right image with the your right eye and the left image with your left eye, your brain will "see" in 3-D. To see this effect, you must relax your eyes; this is sometimes accomplished with a set of viewing glasses called stereopairs. (Yes, this is exactly how air photos work to see 3-D!)*

structures, but tend to be used more by chemists than mineralogists.

The final three representations all attempt to show the 3-D structure using 2-D projection. Figure 14.3e is a perspective view. This type of view was the sort of thing the art teacher tried to teach you when drawing landscapes with railroad tracks or roads. The road gets smaller in the distance. Compare Figure 14.3e with Figure 14.3a and you can see that the unit cell gets smaller in the rear of the projection. The more atoms in the structure, the better the sense of perspective. For example, the structures of halite and tremolite in Figure 14.1 are less useful that the movie of halite on the MSA website. There is no perspective view used in the movie, so you only see the 2-D image of the front atoms. The atoms behind are covered up. Figure 14.3f looks like it is out of focus. In reality, this is a 3-D red/blue projection like you may have seen at the movies. The structure is drawn slightly offset and with different colors for each of the images. When you view this with a red filter over one eye and a blue one over the other, your mind creates a 3-D image because the eye with the red filter sees the red image and the eye with the blue filter sees the blue image; this "tricks" the brain into seeing 3-D—so get some red/blue glasses and give it a try.

The last set of images (Figure 14.3g) is called a **stereopair** because it uses a pair of images. If you look at these images you'll notice the right and the left ones are slightly offset (i.e., they are rotated with respect to each other). If you can trick your eyes to relax, and without allowing them to rotate in, change your focus to bring the left image into focus in your left eye and the right one into focus in your right eye, your brain will produce a 3-D image. Yes, for those who've seen 3-D images from air photographs, this is the same effect. At the time of this writing, a new type of 3-D imaging that's making the scene in geology departments nation-wide is the so-called "geowall." The geowall produces 3-D images in a 2-D projection in a similar manner as the red/blue idea. Two images are produced and overlaid, but instead of red/blue for the images, they are polarized in different directions. So when you put on a pair of polarized glasses, your brain makes a 3-D image. So we are always seeking newer ways to envision 3-D objects in the geosciences.

One last example is needed before we move on to generating atoms based on symmetry operations. Now we are going to look at three physical models (i.e., ones you could touch) of a 2:1 layer mica. Figure 14.4a is a ball and stick representation of a mica with the interlayer cations in the center of the image. In turn, there is a tetrahedral-octahedral-tetrahedral combination above and below the interlayer cations. Figure 14.4b show the tetrahedral-octahedral-tetrahedral layers as polyhedra. This gives a much clearer representation of them, but still shows the interlayer cations as ball and sticks. Overall, Figure 14.4b does a much better job of representing the structure of the mica. Figure 14.4c is a space-filling model attempting to show the tetrahedral-octahedral-tetrahedral layers. The large spheres represent close-packed sheets of O^{2-} with BB's (i.e., 4 mm steel spheres) in the tetrahedral voids simulating the Si^{4+} cations in tetrahedral coordination. Smaller darker spheres are filling the octahedral voids. These space-filling models don't do a very good job of representing the structure. A better portrayal of this structure was created for the cover of the *Reviews in Mineralogy and Geochemistry* dealing with the micas (volume 46). The image is a perspective view mixing polyhedra (to show the tetrahedral-octahedral-tetrahedral sheets) and light-colored spheres to show the interlayer cations.

Andalusite Story

After the somewhat phenomenological explanation of the diamond and fluorite structures, we'd now like to carefully work through the development of the crystal structure of a mineral. Now we will pay more attention to the symmetry operations that generate the atoms in the unit cell of the minerals. As an example, we'll use andalusite because its structure is simple enough that we can begin to draw it by hand, yet complicated enough that you should see the usefulness of letting a computer do the calculations.

The basic pieces of information needed to generate the andalusite crystal structure are given in Table 14.1. Andalusite is orthorhombic with a = 7.7980, b = 7.9031, c = 5.5566Å and $\alpha = \beta = \gamma = 90°$; this tells us the size and shape of the unit cell. The space group is $P\ 2_1/n\ 2_1/n\ 2/m$ (which is based on the point group $2/m\ 2/m\ 2/m$). It contains eight symmetry operations. The locations (or so-called positional parameters or fractional coordinates) for the seven unique atoms comprising the structure of andalusite (i.e., the ions in the asymmetric unit) are also given.

Before we proceed, you may have noticed that the atomic positions are given in terms of numbers that are all less than one. They are termed **fractional coordinates**, and they are all between 0 and 1. Remember that in fluorite, Ca^{2+} plotted at 0, 0, 0 (the origin of the unit cell) or 0 on the a, 0 on b, and 0 on c and F^{1-} plotted at 1.25Å, 1.25Å, 1.25Å, or 1.25Å along a, 1.25Å along b, and 1.25Å along c. Another way to denote the position of F^{1-} would be to give its fractional coordinate (x, y, z), where: $x = d_a / a$, $y = d_b / b$, $z = d_c / c$ and d_a, d_b, d_c are the distances the plots along a, b, c. Recall that for fluorite, $a = 5.0$Å, so the fractional coordinates for F^{1-} would be 0.25, 0.25, 0.25. In other words, F^{1-} would plot 1/4 of the way along a, b, and c. So instead of giving the distance in Å along each cell direction to plot atoms, the standard is to give the distance in terms of the fraction of each of the axes. Thus, for andalusite Al1 would plot at 0, 0, 0.2419, or, converting to distances, 0Å along a, 0Å along b, and 0.2419×5.5566 Å $= 1.3441$Å along c. While it may seem strange to use fractional coordinates, it simplifies the entire process.

The first step in creating a structural drawing of andalusite is to draw the lattice. Before we can do this, we'll need to pick a direction in which to view the structure, because we are going to be making a 2-D projection of a 3-D image. (This drawing is going to be what is sometimes referred to as a floor-plan as discussed above.)

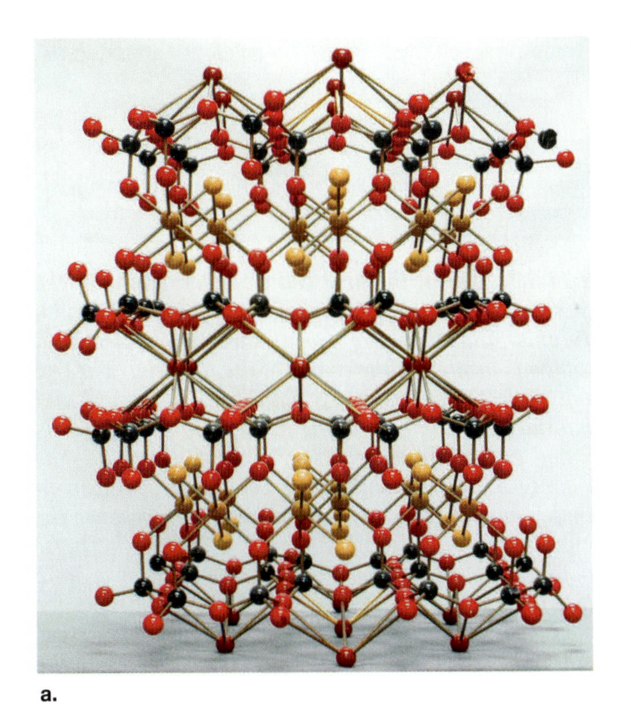

a.

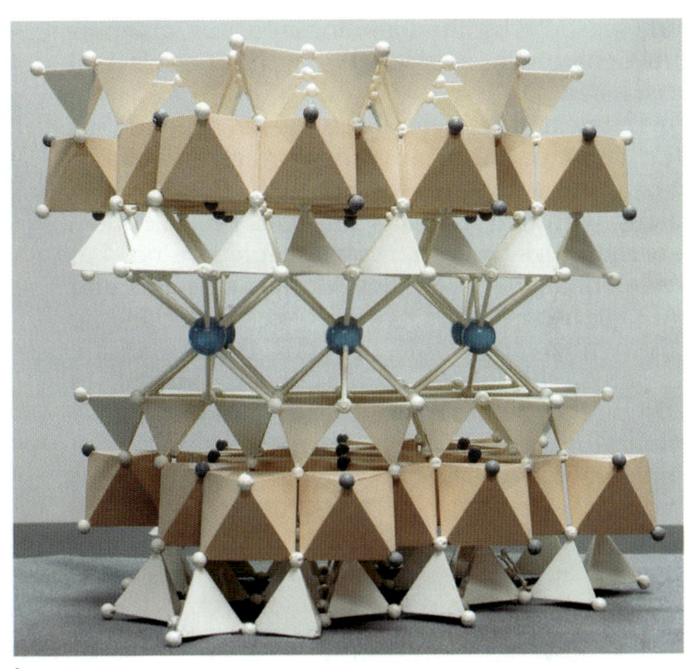

b.

c.

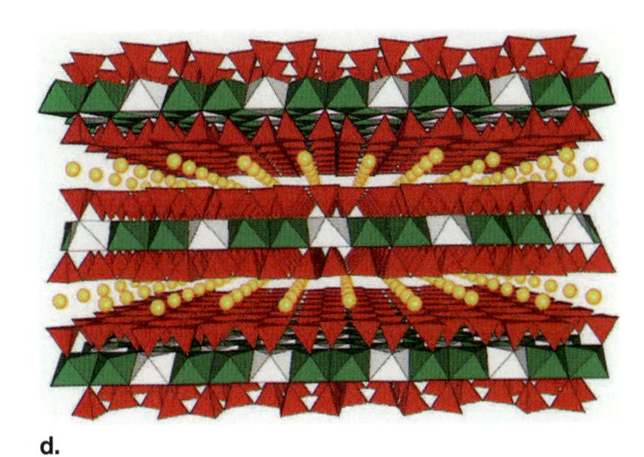

d.

Figure 14.4. *Four separate views of the mica structure projected along the layers (i.e., the* c *axis is vertical).* ***a.*** *A ball and stick model with the interlayer cation layer in the center of model.* ***b.*** *A physical model similar to the ball and stick model, except that tetrahedral and octahedral sheets are shown as polyhedra.* ***c.*** *A space-filling model showing two tetrahedral sheets pointing inward toward the octahedral sheet. The model is composed of clear and orange spheres of the same size, which represent O^{2-} and $(OH)^{1-}$, respectively. The smaller darker spheres fit in the octahedral interstices between the close-packed sheets. Although it is hard to see, BB's (i.e., 4mm diameter metal spheres) are in the tetrahedral interstices formed between the outermost and next layer of clear balls.* ***d.*** *A perspective drawing showing three sandwiches of tetrahedral and octahedral sheets. Try to figure out the relationships between the color scheme of the tetrahedral and octahedral sheets to the book (hint: think about flags).*

Often, structures are viewed with two of the axes in the plane of the page and the third perpendicular to the page. (This would not work for, say the monoclinic and triclinic systems, but works well for the others.) So for reasons that will be apparent later, we are going to chose an *a*-*b* projection (i.e., *a* and *b* are in the plane of the page) with *c* perpendicular to the page. Figure 14.5 shows the *a*-*b* projection, with *a* vertical (pointed down) and *b* horizontal. Also, there are fractional coordinates plotted along *a* and *b*. Although it is somewhat difficult to tell from this projection, the length of the *a* axis is shorter than the length of *b*, as given in Table 14.1. The difference in these lengths is one reason why the ions' locations are given as fractions of the axis lengths and not the actual distances.

The next step is to plot the seven unique ions (i.e., those in the asymmetric unit) given in Table 14.1. First, note that there is only one Si^{4+} ion, along with two Al^{3+}, and four O^{2-}. Because there are multiples of several of these ions, we need to differentiate between them. In this case, Al1 and Al2 are used for the two Al^{3+} sites (i.e., they are given numbers) while the four O^{2-} anions are denoted with letters as OA, OB, OC, and OD to distinguish them. We've already laid out the *a*-*b* plane for andalusite in Figure 14.5, so next the ions are added based on their fractional coordinates (*x*, *y*, *z*). We cannot visually denote *z*, the value in the *c* direction, because we are looking down that direction. Instead, we place the value of *z* in parentheses next to the ion. You can now see how the seven ions are plotted in the upper left portion of the unit cell shown in Figure 14.5. We're going to discuss how those *other* seven ions, those outlined by dots, in the lower right portion of the unit cell got there next.

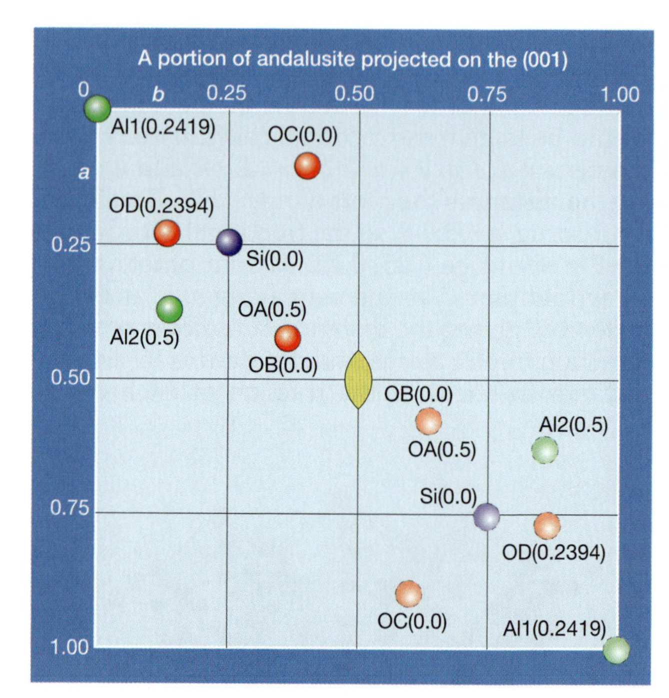

Figure 14.5. *A plot of the unit cell of andalusite with the seven unique ions in the asymmetric unit (those created by a 1-fold axis) and the set of seven new ions created by a* $^{[001]}2$*, which is located at the center of the unit cell. The cell is drawn projected onto the* a-b *plane with* a *vertical and* b *horizontal. Each of the seven ions from Table 14.1 is labeled and plotted based on its* x, y *coordinates (i.e., the values along* a *and* b*), while the z coordinate is given in parentheses next to each one (this is necessary because the projection is down* c*). Next, that set of seven ions is transformed to a new set of symmetry-related ions by a* $^{[001]}2$ *located at 0.5*x *and 0.5*y*. Table 14.2 list the coordinates for the seven new ions. What would happen to an ion that plotted at 0.5*x*, 0.5*y*, 0.5*z*? (Nothing, because it lies on the* $^{[001]}2$ *and therefore, no new ions would be created by the symmetry operation.)*

Table 14.1.
Structural information for andalusite: chemical formula, space group, cell parameters, and fractional coordinates of the seven unique ions in the asymmetric unit (Winter and Ghose 1979).

Al_2SiO_5, $P2_1/n\ 2_1/n\ 2/m$, a = 7.7980, b = 7.9031, c = 5.5566 Å			
Atom (Ion)	*x*	*y*	*z*
Si	0.2460	0.2520	0.0
Al1	0.0	0.0	0.2419
Al2	0.3705	0.1391	0.5
OA	0.4233	0.3629	0.5
OB	0.4246	0.3629	0.0
OC	0.1030	0.4003	0.0
OD	0.2305	0.1339	0.2394

Now it's time to explore the symmetry operations that will take the seven unique ions in the asymmetric unit and fill it. The symmetry operations are included in the space group, which is $P\ 2_1/n\ 2_1/n\ 2/m$. The space group is derived from the orthorhombic point group $2/m\ 2/m\ 2/m$. For the orthorhombic system, the three entries in the space (or point) group are mutually perpendicular and coincide with the minerals *a*, *b*, and *c* axes, respectively.

As a simple first step in generating the remaining ions, let's take the $^{[001]}2$ as an example; we'll perform its symmetry operation graphically and then move on using the mathematical crystallography methods from Chapter 13. We also need to locate the symmetry operation within the unit cell. Some of the operations are located at the origin of the unit cell, while others are located throughout the cell. There will be multiples of $^{[001]}2$ located in different portions of the unit cell. This was discussed some in Chapter 13, and some examples are shown in Figures 13.8 to 13.10.

Look again at Figure 14.5 and notice that the symbol for a 2-fold axis is plotted at $x = 0.5$ and $y = 0.5$; this is the location of our $^{[001]}2$. Next let's explore graphically what the symmetry at this location does to the seven unique ions plotted in the upper left corner of the unit cell using Si and OD as examples. They would be rotated by the 2-fold axis to create a new Si* and OD* atom by a 180° rotation into the lower right portion of the cell. *Note that for the rest of this chapter, we'll refer to atoms rather than ions for simplicity*. Of course, the atoms in most of the minerals discussed here are really cations and anions. In the context used here, we are referring to sites that happen to be named after the atoms/ions that occupy them.

Table 14.2.
The fractional coordinates of the seven symmetry-related sites in the asymmetric unit of andalusite and the set of seven (shown with a "*") created by the operation of a $^{[001]}2$ axis located at the middle of the unit cell. Both sets of ions are plotted in Figure 14.5, projected on the *a-b* plane.

Atom (Ion)	*x*	*y*	*z*
Si	0.2460	0.2520	0.0
Si*	0.7540	0.7480	0.0
Al1	0.0	0.0	0.2419
Al1*	1.0	1.0	0.2419
Al2	0.3705	0.1391	0.5
Al2*	0.6295	0.8609	0.5
OA	0.4233	0.3629	0.5
OA*	0.5767	0.6371	0.5
OB	0.4246	0.3629	0.0
OB*	0.5754	0.6371	0.0
OC	0.1030	0.4003	0.0
OC*	0.8970	0.5997	0.0
OD	0.2305	0.1339	0.2394
OD*	0.7695	0.8661	0.2394

We can take this a step further by looking at the coordinates of the two ions. The original Si cations plot at 0.2460, 0.2520, 0.0 and OD at 0.2305, 0.1339, 0.2394. The Si* created by the 2-fold axis plots at 0.7540, 0.7480, 0.0 and the new OD* plots at 0.7695, 0.8661, 0.2394. The $^{[001]}2$ has no effect on the *z* component for either of the cations. (If you like, you could go back to Chapter 13 and use Equations 13.1 and 13.2 to derive the matrix representation of the $^{[001]}2$ plotting at $x = 1/2$ and $y = 1/2$.) Table 14.2 lists the original coordinates for each of the seven unique sites as well as the coordinates for the seven new sites created by the $^{[001]}2$. So where are we in terms of plotting all the ions in the unit cell? We need to find the *x*, *y*, and *z* values for all seven ions for the collection of symmetry operations in the space. So far we've located the seven ions with one of the symmetry operations, and you might be thinking this is a tedious process. You'd be correct! Next we want to call on what we learned in the last chapter to remove the tedium, but before we can do that we'll need to address all the symmetry operations, and their locations in andalusite lattice.

The space group $P\ 2_1/n\ 2_1/n\ 2/m$ contains eight symmetry operations which are {1, $\bar{1}$, $^{[100]}2_1$, $^{[010]}2_1$, $^{[001]}2$, $^{[100]}n$, $^{[010]}n$, $^{[001]}m$}, so we will need to consider what each of these eight operations would do to the seven unique ions in the structure. Also, as stated, we have to consider the location of each of these in the unit cell by referring to the *International Tables for Crystallography* (Hahn, 1995). An *a-b* projection of the space group $P\ 2_1/n$ $2_1/n\ 2/m$ lattice is shown in Figure 14.6a, with *a* vertical and *b* horizontal, matching the orientation of the *a* and *b* axes in Figure 14.5. This is the standard projection of the lattice and the reason we chose the same projection of our andalusite example. Notice that there are lots of symbols on the lattice. These are named in our figure, but are not in the reference book. The $^{[001]}2$ located at $x = 0.5$ and $y = 0.5$ that we used in the above example is present, but there are also $^{[001]}2$'s at the corners and midpoints of the unit cell. The dot inside each of them refers to the location of centers of symmetry (denoted as $\bar{1}$ in the above symmetry list). There are two $^{[100]}2_1$'s and two $^{[010]}2_1$'s that are located at 1/4 and 3/4 along the unit cell directions and plot parallel to the projection (their symbol is a 1/2 barbed arrow). Likewise, there are two $^{[100]}n$'s and two $^{[010]}n$'s plotting at 1/4 and 3/4 in the unit cell; these glide planes plot as lines because we are looking down on them. The dots and dashes tell us these are *n*-glides (these have two 1/2 translations). If these lines were solid, they would represent mirror planes. If they were just dashed, they would be *b*- or *c*-glides (with only one 1/2 translation). Finally, the $^{[001]}m$ is indicated in the upper left portion of the diagram by two lines at 90° to each other. A key to all these symbols was given in Chapter 12; don't get hung up on the symbols at this point, it's more important to try and under-

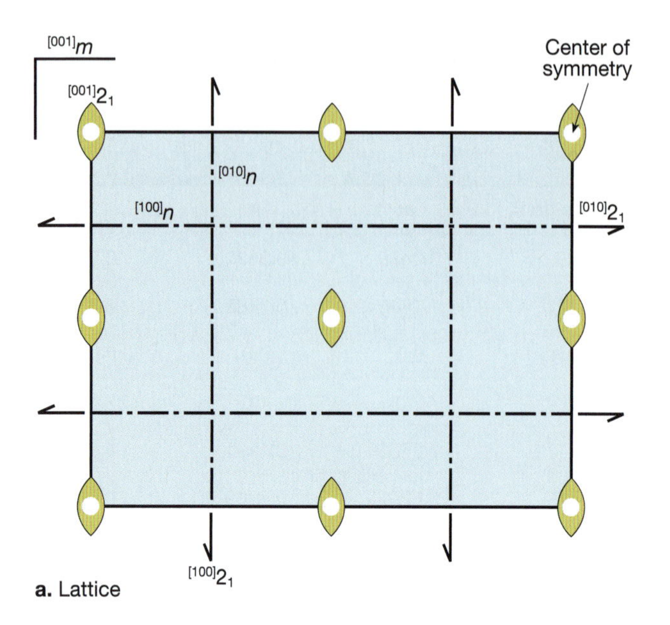

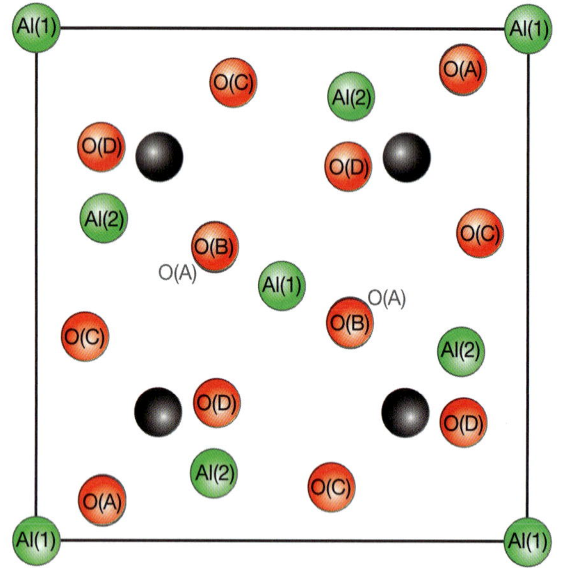

b. *a-b* projection of the 32 ions in the unit cell

Figure 14.6. *An a-b projection of the lattice and unit cell contents of andalusite, with* a *vertical and* b *horizontal.* ***a.*** *A schematic view of the symmetry elements of the space group* P2_1/n 2_1/n 2/m. *The orientations of each symmetry element correspond to entries in the space group symbol. There are more than one of each operation (e.g., there are nine* [001]*2's).* ***b.*** *An* a-b *projection of the contents of the unit cell of andalusite, with all the ions labeled, except for Si (which is the black sphere). In Figure 14.4 the seven ions in the asymmetric unit were plotted along with the seven ions generated by a* [001]*2.However, andalusite has the space group* P2_1/n 2_1/n 2/m *and contains eight symmetry operations: 1,* $\bar{1}$, [100]2_1, [100]n, [010]2_1, [010]n, [001]2, *and* [001]m, *thus each symmetry element will generate new ions.*

stand how the symmetry operations affect the ions—which is explained next.

Now we are ready to populate the entire unit cell. We can do this in one of two ways. We could return to the discussion in Chapter 13 of how to create the symmetry operations (i.e., write a matrix) for each of the eight symmetry operations in $P\ 2_1/n\ 2_1/n\ 2/m$; those matrices are given in Table 14.3. The second method is to again refer to the *International Tables of Crystallography* and look up the so-called coordinates created by each symmetry operation in terms of x, y, and z; these too are given in Table 14.3. Then we either multiply the x, y, and z positions for each of the seven ions in the asymmetric unit by the matrices for each symmetry operations, or we substitute the values of x, y, and z for each ion into the coordinates.

For example, if we take the fractional coordinates for Si (i.e., 0.2460, 0.2520, 0) and operate on them with the first symmetry element in Table

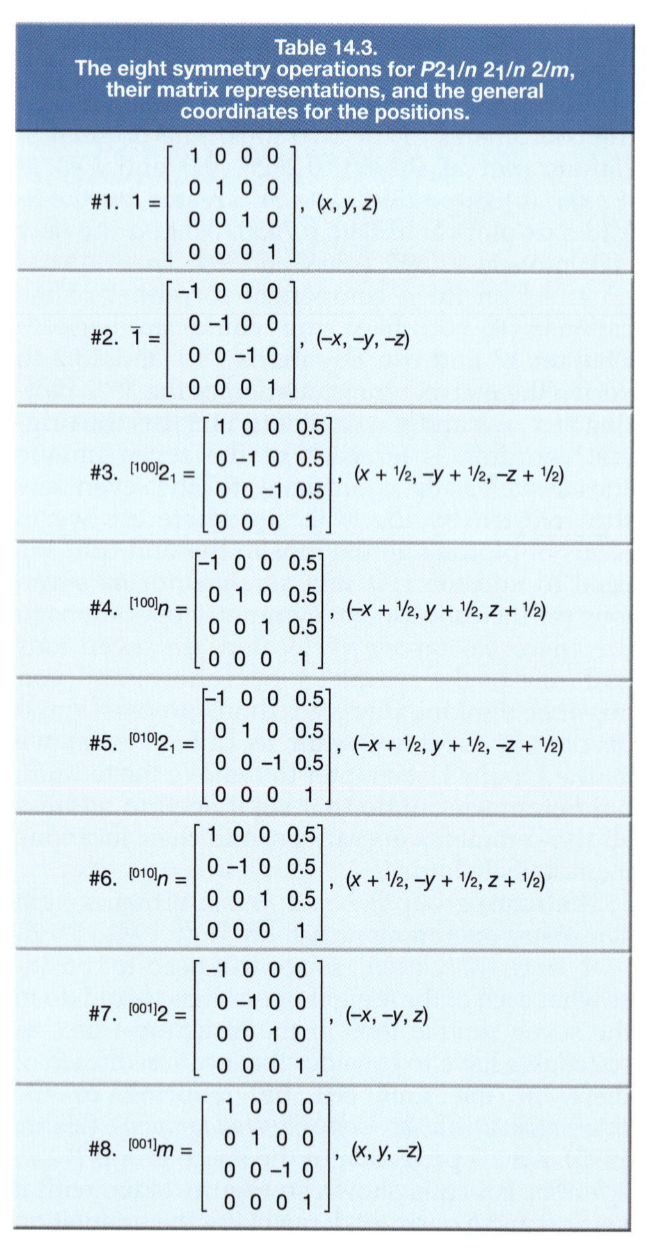

Table 14.3.
The eight symmetry operations for $P2_1/n\ 2_1/n\ 2/m$, their matrix representations, and the general coordinates for the positions.

#1. $1 =$	$\begin{bmatrix} 1 & 0 & 0 & 0 \\ 0 & 1 & 0 & 0 \\ 0 & 0 & 1 & 0 \\ 0 & 0 & 0 & 1 \end{bmatrix}$	(x, y, z)
#2. $\bar{1} =$	$\begin{bmatrix} -1 & 0 & 0 & 0 \\ 0 & -1 & 0 & 0 \\ 0 & 0 & -1 & 0 \\ 0 & 0 & 0 & 1 \end{bmatrix}$	$(-x, -y, -z)$
#3. $^{[100]}2_1 =$	$\begin{bmatrix} 1 & 0 & 0 & 0.5 \\ 0 & -1 & 0 & 0.5 \\ 0 & 0 & -1 & 0.5 \\ 0 & 0 & 0 & 1 \end{bmatrix}$	$(x + 1/2, -y + 1/2, -z + 1/2)$
#4. $^{[100]}n =$	$\begin{bmatrix} -1 & 0 & 0 & 0.5 \\ 0 & 1 & 0 & 0.5 \\ 0 & 0 & 1 & 0.5 \\ 0 & 0 & 0 & 1 \end{bmatrix}$	$(-x + 1/2, y + 1/2, z + 1/2)$
#5. $^{[010]}2_1 =$	$\begin{bmatrix} -1 & 0 & 0 & 0.5 \\ 0 & 1 & 0 & 0.5 \\ 0 & 0 & -1 & 0.5 \\ 0 & 0 & 0 & 1 \end{bmatrix}$	$(-x + 1/2, y + 1/2, -z + 1/2)$
#6. $^{[010]}n =$	$\begin{bmatrix} 1 & 0 & 0 & 0.5 \\ 0 & -1 & 0 & 0.5 \\ 0 & 0 & 1 & 0.5 \\ 0 & 0 & 0 & 1 \end{bmatrix}$	$(x + 1/2, -y + 1/2, z + 1/2)$
#7. $^{[001]}2 =$	$\begin{bmatrix} -1 & 0 & 0 & 0 \\ 0 & -1 & 0 & 0 \\ 0 & 0 & 1 & 0 \\ 0 & 0 & 0 & 1 \end{bmatrix}$	$(-x, -y, z)$
#8. $^{[001]}m =$	$\begin{bmatrix} 1 & 0 & 0 & 0 \\ 0 & 1 & 0 & 0 \\ 0 & 0 & -1 & 0 \\ 0 & 0 & 0 & 1 \end{bmatrix}$	$(x, y, -z)$

14.3 (the 1-fold axis), the cation plots at 0.2460, 0.2520, 0. If we move to the second symmetry element, the center of symmetry or $\bar{1}$, Si^{4+} cations will plot at –0.2460, –0.2520, 0. The third symmetry element, the $^{[100]}2_1$ would plot the Si^{4+} cation at 0.7460, 0.2480. 0.5. We would continue with this method for the remaining five symmetry operations for Si^{4+}, then move on to Al1, Al2, and the 4 O^{2-}. Table 14.4 lists the results. So we have now seen the effect of the collection of symmetry operations in $P\ 2_1/n\ 2_1/n\ 2/m$ on the seven unique elements in the andalusite structure. All that remains is to plot the ions given in Table 14.4 to show the unit cell contents of andalusite, as in Figure 14.6b.

Before we leave the andalusite example, we need to take a closer look at Table 14.4. You'll notice that even though we have eight symmetry operations (listed in Table 14.3), there are only 32 unique atoms generated for the unit cell, and not, as might be expected, 56. Let's explore why this is. In Table 14.3 each of the symmetry operations is listed and numbered. In Table 14.4 each of the eight symmetry operations is applied to each atom, and, for example, eight Si's are written out. Notice how Si #5 (the one generated by the $^{[010]}2_1$) plots at the same location as Si#4 (the one generated by $^{[100]}n$); a similar situation is true for Si#6-#8. So, in reality, we only have four new symmetry-generated Si cations. Likewise, as you look down the table, you see that there are only four ions in the unit cell for each of the other ions, except OD, which has the as-expected eight. What's going on? In the figure caption for Figure 14.5 the question was asked, "what would be the effect of the $^{[001]}2$ located at $x = 0.5$ and $y = 0.5$ on an ion that plotted at 0.5, 0.5, 0.5?" Well the answer is nothing (i.e., no new ion would be generated), because the ion at 0.5, 0.5, 0.5 would lie on the $^{[001]}2$. Thus in the case of andalusite, all the ions except OD must lie on some symmetry operation.

Looking more closely at Table 14.4, we see that Si, Al2, OA, OB, and OC have a duplication of symmetry generators of their eight ions, creating only four unique ones. For each of these five ions, the last four in the list are duplicates and follow the order of symmetry duplications with #5 = #4, #6 = #3, #7 = #2, and #8 = #1. However, this trend doesn't hold for Al1. We can interpret this to mean that Si, Al2, OA, OB, and OC must rest on similar symmetry elements and Al1 a different one, and OD doesn't plot on any symmetry element. We can conclude by saying that OD plots on a *general* position while the other atoms plot at *special* positions.

Table 14.5 is taken from the *International Tables for Crystallography*. There are three entries in the table. The first is the so-called multiplicity of the site (i.e., how many ions would be generated, which equals the number of symmetry operations for this space group), the second is the site symmetry, and the last gives the fractional coordinates for that site. For example, Al1's fractional coordinate is 0, 0, 0.2419. If we look in Table 14.5, we see that the fourth line down has a fractional coordinate of 0, 0, z (meaning x and y are constrained to 0, while z can take on any values). Thus for ions plotted at 0, 0, z (like Al1), the multiplicity is reduced from eight to four. Also, notice that this location is on the $^{[001]}2$. Si, Al2, OA, OB, and OC all have fractional coordinates of the form x, y, 0, corresponding to the second line in Table 14.5. Thus, they plot on a mirror plane (the $^{[001]}m$ one) and their site multiplicity is reduced from eight to four. Finally, the fractional coordinates of OD are 0.2305, 0.1339, 0.2394, so it plots at a general position in the unit cell at a site with 1-fold symmetry, and has a multiplicity of eight.

You might be wondering why the formula for andalusite is Al_2SiO_5 when there are one Si^{4+}, two Al^{3+}, and only four O^{2-} listed in the structure (Table 14.1). From Table 14.4, we see that the original seven ions in the structure generated 32 total ions in the unit cell of andalusite; they are four Si^{4+}, eight Al^{3+}, and twenty O^{2-} (i.e., four each of OA, OB, and OC and eight of OD). Thus we could write a unit cell formula, or the number of ions in the unit cell, as: $Al_8Si_4O_{20}$, which looks a lot like Al_2SiO_5. In Chapter 10, we used the term atoms per formula unit and abbreviated it as apfu. For andalusite, there would be two Al's, one Si, and five O's apfu.

Now it's time for a new term (which I've never seen used before): **atoms per unit cell** or apuc. For andalusite this would be: eight Al^{3+}, four Si^{4+}, and twenty O^{2-}. There is a relationship between apfu and apuc. In the case of andalusite, there are four apfu's in one apuc. The symbol that expresses this relationship is the letter Z, which is the number of apfu in the unit cell. For the case of andalusite, Z = 4.

Building Computer-Based and Physical Model of Minerals

It would be very useful to use a computer for all the calculations required to represent the structure of a mineral. As input, we would use the structure data, like those given in Table 14.1 for andalusite. We might have to enter in the symmetry operations (i.e., data like those given in Table 14.3) in some form, but most of the computer programs either contain or can calculate the space groups, from which the symmetry operations can be easily obtained.

As was noted in the introduction, there are several software packages currently available for generating crystal structure drawings on the computer. Throughout this book we have used *CrystalMaker*; it's an excellent program. Table 14.6a is a partial output from the *CrystalMaker* program for the andalusite data in Table 14.3. The results in this table look very similar to the results we obtained in Table 14.4. The major difference is that there are three more entries given for each

Table 14.4.
This table lists all symmetry-generated sites in the unit cell of andalusite, based on the seven ions in the asymmetric unit of andalusite and the symmetry operations contained in the space group $P2_1/n\ 2_1/n\ 2/m$. Each ion is numbered to correspond to 1 of the 8 operations given in Table 14.2, and each operation is also listed in the second column; the numbers serve to show equivalence of symmetry operations as listed in parentheses in column one.

Atom (Ion)	Symmetry generator	x	y	z	Atom (Ion)	Symmetry generator	x	y	z
Si (#1)	1	0.2460	0.2520	0.0	OA (#5 = #4)	$^{[010]}2_1$	0.0767	0.8629	0.0
Si (#2)	$\bar{1}$	0.7540	0.7480	0.0	OA (#6 = #3)	$^{[010]}n$	0.9233	0.1371	0.0
Si (#3)	$^{[100]}2_1$	0.7460	0.2480	0.5	OA (#7 = #2)	$^{[001]}2$	0.5767	0.6371	0.0
Si (#4)	$^{[100]}n$	0.2540	0.7520	0.5	OA (#8 = #1)	$^{[001]}m$	0.4233	0.3629	0.5
Si (#5 = #4)	$^{[010]}2_1$	0.2540	0.7520	0.5	OB (#1)	1	0.4246	0.3629	0.0
Si (#6 = #3)	$^{[010]}n$	0.7460	0.2480	0.5	OB (#2)	$\bar{1}$	0.5754	0.6371	0.0
Si (#7 = #2)	$^{[001]}2$	0.7540	0.7480	0.0	OB (#3)	$^{[100]}2_1$	0.9246	0.1371	0.5
Si (#8 = #1)	$^{[001]}m$	0.2460	0.2520	0.0	OB (#4)	$^{[100]}n$	0.0754	0.8629	0.5
Al1 (#1)	1	0.0	0.0	0.2419	OB (#5 = #4)	$^{[010]}2_1$	0.0754	0.8629	0.5
Al1 (#2)	$\bar{1}$	0.0	0.0	0.7581	OB (#6 = #3)	$^{[010]}n$	0.9246	0.1371	0.5
Al1 (#3)	$^{[100]}2_1$	0.5	0.5	0.2581	OB (#7 = #2)	$^{[001]}2$	0.5754	0.6371	0.0
Al1 (#4)	$^{[100]}n$	0.5	0.5	0.7419	OB (#8 = #1)	$^{[001]}m$	0.4246	0.3629	0.0
Al1 (#5 = #3)	$^{[010]}2_1$	0.5	0.5	0.2581	OC (#1)	1	0.1030	0.4003	0.0
Al1 (#6 = #4)	$^{[010]}n$	0.5	0.5	0.7419	OC (#2)	$\bar{1}$	0.8970	0.5997	0.0
Al1 (#7 = #1)	$^{[001]}2$	0.0	0.0	0.2419	OC (#3)	$^{[100]}2_1$	0.6030	0.0997	0.5
Al1 (#8 = #2)	$^{[001]}m$	0.0	0.0	0.7581	OC (#4)	$^{[100]}n$	0.3970	0.9003	0.5
Al2 (#1)	1	0.3705	0.1391	0.5	OC (#5 = #4)	$^{[010]}2_1$	0.3970	0.9003	0.5
Al2 (#2)	$\bar{1}$	0.6295	0.8609	0.5	OC (#6 = #3)	$^{[010]}n$	0.6030	0.0997	0.5
Al2 (#3)	$^{[100]}2_1$	0.8705	0.3609	0.0	OC (#7 = #2)	$^{[001]}2$	0.8970	0.5997	0.0
Al2 (#4)	$^{[100]}n$	0.1295	0.6391	0.0	OC (#8 = #1)	$^{[001]}m$	0.1030	0.4003	0.0
Al2 (#5 = #4)	$^{[010]}2_1$	0.1295	0.6391	0.0	OD (#1)	1	0.2305	0.1339	0.2394
Al2 (#6 = #3)	$^{[010]}n$	0.8705	0.3609	0.0	OD (#2)	$\bar{1}$	0.7695	0.8661	0.7606
Al2 (#7 = #2)	$^{[001]}2$	0.6295	0.8609	0.5	OD (#3)	$^{[100]}2_1$	0.7305	0.3661	0.2606
Al2 (#8 = #1)	$^{[001]}m$	0.3705	0.1391	0.5	OD (#4)	$^{[100]}n$	0.2695	0.6339	0.7394
OA (#1)	1	0.4233	0.3629	0.5	OD (#5)	$^{[010]}2_1$	0.2695	0.6339	0.2606
OA (#2)	$\bar{1}$	0.5767	0.6371	0.5	OD (#6)	$^{[010]}n$	0.7305	0.3661	0.7394
OA (#3)	$^{[100]}2_1$	0.9233	0.1371	0.0	OD (#7)	$^{[001]}2$	0.7695	0.8661	0.2394
OA (#4)	$^{[100]}n$	0.0767	0.8629	0.0	OD (#8)	$^{[001]}m$	0.2305	0.1339	0.7606

Table 14.5.
Site multiplicity information for andalusite (Hahn 1995) multiplicity, site symmetry, coordinates

8, 1: x, y, z; $-x, -y, -z$; $x + 1/2, -y + 1/2, -z + 1/2$; $-x + 1/2, y + 1/2, z + 1/2$; $-x + 1/2, y + 1/2, -z + 1/2$; $x + 1/2, -y + 1/2, z + 1/2$; $-x, -y, z$; $x, y, -z$
4, m: $x, y, 0$; $-x, -y, 0$; $-x + 1/2, y + 1/2, 1/2$; $x + 1/2, -y + 1/2, 1/2$
4, 2: $0, 1/2, z$; $1/2, 0, -z + 1/2$; $0, 1/2, -z$; $1/2, 0, z + 1/2$
4, 2: $0, 0, z$; $1/2, 1/2, -z + 1/2$; $0, 0, -z$; $1/2, 1/2, z + 1/2$
2, $2/m$: $0, 1/2, 1/2$; $1/2, 0, 0$
2, $2/m$: $0, 1/2, 0$; $1/2, 0, 1/2$
2, $2/m$: $0, 0, 1/2$; $1/2, 1/2, 0$
2, $2/m$: $0, 0, 0$; $1/2, 1/2, 1/2$

atom that tell the computer where to plot the atoms. The program basically converts the orthorhombic basis vector set into a Cartesian basis vector set to plot the atoms. This is not that hard for the orthorhombic case, but becomes a lot trickier for monoclinic and triclinic. This program, and the others that perform these calculations, have at their hearts the matrices we discussed back in Chapter 13.

Figure 14.6b was a *CrystalMaker* plot of the ions in Table 14.6a. Going back to look at Figure 14.6b, you see it's only a plot of the ions—not that informative. What is next needed is to add things like bonds between the ions, different views, the unit cell, a key for crystallographic axes, etc. to make the diagrams look more realistic and be more informative. This is all done in a new drawing of andalusite shown in Figure 14.7a. For comparison, Figure 14.7b is a photograph of a ball and stick model of andalusite that we built. Also shown in Figure 14.7c is a polyhedral view of the structure in the same orientation. It's really easy with modern computer software to switch between different methods of viewing the structures to optimize whatever you're trying to show. Finally, in Figure 14.7d is a bit of history of how we made structure drawings in the "old days." This is the familiar view of andalusite projected slightly tilted out of the *a-b* plane. (This is also the structure that I talked about taking all afternoon to make back in 1979 in the beginning of this chapter.) The labels denoted the unit cell and ions were added by hand by a draftsperson. All in all, it took over eight hours to produce this one projection. Today it would take more like eight minutes, and it could be in color, and projected in any direction, have movies made from it, etc.

The structure shown in Figure 14.7d was produced with a program called *ORTEP* (an acronym for Oak Ridge thermal ellipsoid plot) and discussed in Johnson (1965). If you look closely, you'll see that the atoms really aren't spheres, but ellipses. The shape of the ellipse relates to the thermal motion of the atom. This program was so widely used in the 1970s and 1980s that some of the older folks call all crystal structure drawings, ORTEP drawings.

While the computer-generated models are now so easy to make, and very useful both in mineralogical teaching and research, in many ways we still prefer the manipulative ball and stick models. You could never build as many as you would need for teaching and research, but you might be lucky to build a few! We think they help in the transition to visualizing the computer drawings. As stated earlier, to build a physical ball and stick model of a mineral like halite (Figure 14.1a), you don't really require anything complicated: some balls, something to connect them, and a way to drill holes at 90° angles (because all the bonds are at right angles). However, when you decide to build a model of the structure of a common silicate mineral like tremolite (Figure 14.1b), mica (Figure 14.3a), or andalusite (Figure 14.7b), things get a bit more complicated. First off, the angles are no longer 90°, so you need some kind of device to hold a ball and allow you to rotate it to any possible angle, and you need a way to determine what those angles are, and how long to cut the "bonds."

Years ago we (Gunter and Downs, 1991) modified a program that ran on a mainframe called *Drill* to aid in the building of ball and stick models. The program requires as input the cell parameters of the mineral, its space group, and the fractional coordinates of the atoms in the asymmetric unit. From that information, the program generates, in 3-D space, the set of atoms in the unit cell. We then tell the program which atom was an anion (i.e., O^{2-}) or a cation (i.e., Al^{3+}, Si^{4+}, Mg^{2+}, etc.) and the maximum bond distance for each atom. Then the program looks in a sphere equal to the maximum bond distance, and if it was a cation, bonds it to any anion, and if it was an anion it bonds to any cation within that sphere. The program outputs the spherical coordinates (recall stereonets in Chapter 12?) to drill the bonds for that ion. Next, you cut rods to the correct length (also given in the output of the program) to join the ions. Table 14.6b is a section of sample output for andalusite. It starts with Si(1), one of four Si's in the unit cell, and then gives the angular coordinates, in terms of ϕ and ρ, to drill four holes into the Si atom. Next you look up in the output the 4 O anions bonded to Si(1). For example, OB(1) is listed further down in the output. You drill it, and join it to Si(1). In the full output, there are 32 ions that need to be drilled (with 3–6 holes each). They are all joined together to make the andalusite shown in Figure 14.7b. To build such a model requires a special drill shown in Figure 14.8. With it, you first drill a 1/8″ hole in the ball at any angle (that's what all the 999's mean for the first entry in Table 14.6b). Then the ball is placed on the rod running to the ϕ angular adjustment, and the remaining holes are drilled. After each hole is drilled, a small sticker is placed next to it indicating what ion the hole bonds to. Yes, it's tedious to build these, and yes it takes a long time (about 20–30 hours for such a model), but once it's done it's a great aid in teaching and learning about mineral structures.

Examples

The last section of this chapter is going to use structural data for ten minerals, given in Table 14.7, to show several important crystal structural concepts. A paragraph or two will be dedicated to each structure and the concepts they have been carefully chosen to show. We'll return to the same set of ten minerals in the next chapter, which deals primarily with X-ray diffraction.

Table 14.6.
Partial sample output from CrystalMaker (a.) and Drill (b.) needed to either draw a structure or build one.

a. CrystalMaker output

Listing of atomic coordinates for first unit cell

Label	*Elmt*	*Fractional Coordinates*			*Orthogonal Coordinates*		
		x	*y*	*z*	*xor*	*yor*	*zor*
Si	Si	0.2460	0.2520	0.0000	1.992	−1.918	−0.000
Si	Si	0.2540	0.7520	0.5000	5.943	−1.981	2.778
Si	Si	0.7460	0.2480	0.5000	1.960	−5.817	2.778
Si	Si	0.7540	0.7480	0.0000	5.912	−5.880	−0.000
Al(1)	Al	0.0000	0.0000	0.2419	−0.000	0.000	1.344
Al(1)	Al	0.5000	0.5000	0.7419	3.952	−3.899	4.122
Al(1)	Al	0.0000	0.0000	0.7581	−0.000	0.000	4.212
Al(1)	Al	0.5000	0.5000	0.2581	3.952	−3.899	1.434
Al(2)	Al	0.3705	0.1391	0.5000	1.099	−2.889	2.778
Al(2)	Al	0.1295	0.6391	0.0000	5.051	−1.010	−0.000
Al(2)	Al	0.8705	0.3609	0.0000	2.852	6.788	−0.000
Al(2)	Al	0.6295	0.8609	0.5000	6.804	−4.909	2.778
O(A)	O	0.4233	0.3629	0.5000	2.868	−3.301	2.778
O(A)	O	0.0767	0.8629	0.0000	6.820	−0.598	−0.000
O(A)	O	0.9233	0.1371	0.0000	1.084	−7.200	−0.000
O(A)	O	0.5767	0.6371	0.5000	5.035	−4.497	2.778
O(B)	O	0.4246	0.3629	0.0000	2.868	−3.311	−0.000
O(B)	O	0.0754	0.8629	0.5000	6.820	−0.588	2.778
O(B)	O	0.9246	0.1371	0.5000	1.084	−7.210	2.778
O(B)	O	0.5754	0.6371	0.0000	5.035	−4.487	−0.000
O(C)	O	0.1030	0.4003	0.0000	3.164	−0.803	−0.000
O(C)	O	0.3970	0.9003	0.5000	7.115	−3.096	2.778
O(C)	O	0.6030	0.0997	0.5000	0.788	−4.702	2.778
O(C)	O	0.8970	0.5997	0.0000	4.739	−6.995	−0.000
O(D)	O	0.2305	0.1339	0.2394	1.058	−1.797	1.330
O(D)	O	0.2695	0.6339	0.7394	5.010	−2.102	4.109
O(D)	O	0.7305	0.3661	0.7394	2.893	−5.696	4.109
O(D)	O	0.7695	0.8661	0.2394	6.845	−6.001	1.330
O(D)	O	0.7695	0.8661	0.7606	6.845	−6.001	4.226
O(D)	O	0.7305	0.3661	0.2606	2.893	−5.696	1.448
O(D)	O	0.2695	0.6339	0.2606	5.010	−2.102	1.448
O(D)	O	0.2305	0.1339	0.7606	1.058	−1.797	4.226

b. Drill output

Central Atom	*Coordinating Atoms*	*rho*	*phi*	*dist*	*Peg length*
Si (1) 1	17 OB (1)	999.99	999.99	1.646	1.1
	21 OC (1)	101.39	.00	1.618	1.0
	25 OD (1)	111.58	241.39	1.630	1.0
	30 OD (6)	111.58	118.61	1.630	1.0
Si (2) 2	18 OB (2)	999.99	999.99	1.646	1.1
	22 OC (2)	101.39	.00	1.618	1.0
	26 OD (2)	111.58	118.61	1.630	1.0
	29 OD (5)	111.58	241.39	1.630	1.0
Al1 (1) 9	14 OA (2)	999.99	999.99	1.827	1.3
	15 OA (3)	85.28	.00	1.827	1.3
	18 OB (2)	96.68	179.73	1.891	1.4
	19 OB (3)	178.02	187.83	1.891	1.4
	25 OD (1)	88.64	90.91	2.086	1.6
	31 OD (7)	90.79	271.43	2.086	1.6
Al2 (1) 5	13 OA (1)	999.99	999.99	1.817	1.3
	22 OC (2)	86.62	.00	1.839	1.3
	23 OC (3)	160.65	.00	1.898	1.4
	25 OD (1)	99.09	233.95	1.814	1.3
	30 OD (6)	99.09	126.05	1.814	1.3
OA (1) 13	5 Al2 (1)	999.99	999.99	1.817	1.3
	10 Al1 (2)	130.67	.00	1.827	1.3
	11 Al1 (3)	130.67	151.85	1.827	1.3
OB (1) 17	1 Si (1)	999.99	999.99	1.646	1.1
	10 Al1 (2)	124.63	.00	1.891	1.4
	11 Al1 (3)	124.63	134.33	1.891	1.4
OC (1) 21	1 Si (1)	999.99	999.99	1.618	1.0
	6 Al2 (2)	123.85	.00	1.839	1.3
	7 Al2 (3)	130.17	180.00	1.898	1.4
OD (1) 25	1 Si (1)	999.99	999.99	1.630	1.0
	5 Al2 (1)	126.46	.00	1.814	1.3
	9 Al1 (1)	111.10	169.82	2.086	1.6

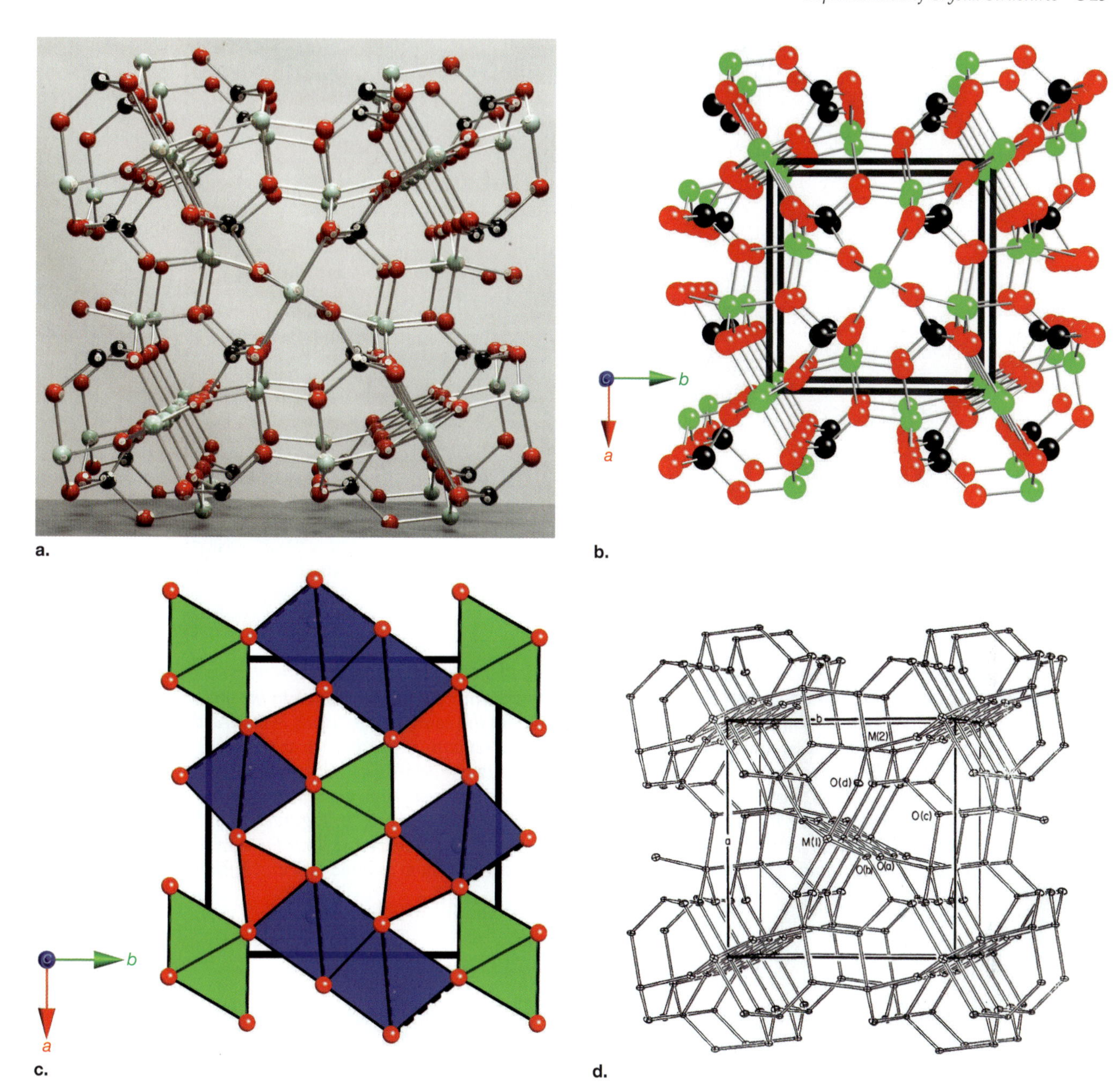

Figure 14.7. Four different representations of the structure of andalusite, all projected down c *with* a *vertical and* b *horizontal. (Green spheres represent Al^{3+}, red spheres O^{2-}, and black spheres Si^{4+}.) The goal of this figure is to show again different ways to represent the same crystal structure as was done in Figure 14.3 for simpler structures and Figure 14.4 for mica. Compare these images to one other as well as those in Figures 14.5 and 14.6 to see how we obtained these models from the seven original atoms in Table 14.1.* ***a.*** *A photograph of a ball and stick model built using the computer program Drill, described later.* ***b.*** *A computer-generated ball and stick drawing with the unit cell outlined, shown in perspective view in an attempt to match the ball and stick model.* ***c.*** *A computer-generated polyhedral drawing with the unit cell outlined (green polyhedra represent 6-coordinated Al^{3+}, blue translucent polyhedral represent 5-coordinated Al^{3+}, and red polyhedral represent 4-coordinated Si^{4+}).* ***d.*** *An ORTEP drawing, tilted slightly out of the* a-b *plane. This was the state-of-the art computer program used for generated images in the early 1980s. ORTEP drawings could represent an atom's position with an ellipsoid that was related to the motion of the atom in the structure (image from Gunter and Bloss, 1982).*

Table 14.7 is meant to show a few trends. First off, there are seven entries for each mineral: mineral species name, crystal system, space group symbol, space group number (these go from 1 to 230, and are often used instead of the symbols), multiplicity (the total number of symmetry operations the space group contains), number of ion sites (the number of unique ions in the asym-

Figure 14.8. *A photograph of the drill and bond cutter used to build ball and stick models. A ball is placed as shown in the drill. The first hole is drilled into the ball and then the ball is mounted on the end of a rod connected to the dial marked "ϕ angle." Now the ball can be rotated about ϕ and, in turn, into a plane parallel to the counter top, which is labeled the "ρ angle." Thus the ϕ and ρ coordinates represent a spherical coordinate system. All that remains is to adjust ϕ and ρ to the desired angle. Also shown near the middle of the photograph is the "bond cutter." It is used to cut desired lengths of aluminum welding rods to fit into the holes drilled into the balls. Finally, the model is assembled. We have used this apparatus to build simple models like halite (Figure 14.1a) to more complex models like tremolite (Figure 14.1b) and andalusite (Figure 14.6a). Mickey even built a tourmaline structure for Darby as a present! (A special thanks to Jerry Gibbs of Virginia Tech for introducing MEG to this method. Refer to Gunter and Downs, 1991, for a more thorough discussion on how to build ball and stick models.)*

Table 14.7.
Crystal structure information for the 10 mineral samples used as examples to explain various aspects of crystal structure visualization and representation.

Mineral	*Crystal system*	*Space group symbol*	*Space group #*	*General site multiplicity*	*# atom sites*	*Z*
halite	isometric	$F\,4/m\,\bar{3}\,2/m$	225	192	2	4
pyrope	isometric	$I\,4_1/a\,\bar{3}\,2d$	230	96	4	8
quartz	hexagonal	$P\,3_2\,2\,1$ or $P\,3_1\,2\,1$	154	6	2	3
cristobalite	tetragonal	$P\,4_1\,2_1\,2$	92	8	2	4
enstatite	orthorhombic	$P\,2_1/b\,2_1/c\,2_1/a$	61	8	10	8
tremolite	monoclinic	$C\,2/m$	12	8	14	2
orthoclase	monoclinic	$C\,2/m$	12	8	8	4
anorthite	triclinic	$P\bar{1}$	2	2	52	4
heulandite	monoclinic	$C\,2/m$	12	8	23	1
Pb-exchanged heulandite	monoclinic	$C\,m$	8	4	43	1

metric unit), and Z (the number of formula units in the unit cell). The first trend is that the symmetry deceases for the first eight minerals in the table. Meanwhile, the multiplicity decreases, the number of atoms increases, and Z decreases. If the multiplicity is high, then you need fewer atoms. For instance, halite (isometric) only has two atoms while anorthite (triclinic) has 52! This is because halite, being isometric, has many, many more symmetry operations than a triclinic mineral like anorthite, (which being $P\overline{1}$ only has two symmetry operations). The last two examples in the table show the same trend for $C2/m$, which contains more symmetry operations that Cm, so we must have a greater number of unique atoms in Cm than $C2/m$ based on a lowering of the symmetry.

Halite (isometric). We have no doubt seen the structure of halite more than any other mineral in this book, so it seems reasonable to examine it here as well. Table 14.8 lists the structural parameters for halite with space group $F\ 4/m\ \overline{3}\ 2/m$. There are only two unique atoms in the asymmetric unit. Figure 14.9 shows the step-by-step construction of this structure from these data. First the unit cell is shown (Figure 14.9a). Next the Na^{1+} cation is added at 0, 0, 0 and then the symmetry generates the Na's at the eight corners of the unit cell and in the center of the six faces (Figure 14.9b). This arrangement of atoms would satisfy this F-centered lattice. Then the Cl^{1-} is plotted at 0.5, 0, 0 and the remaining 12 symmetry-generated Cl^{1-} anions (Figure 14.9c). Finally (Figure 14.9d) the bonds are added between the Na^{1+} and Cl^{1-} ions.

Halite's space group contains 192 symmetry operations (i.e., if an atom occurred in a general position, there would be 192 of them). However, the multiplicity of both Na^{1+} and Cl^{1-} is just four because they plot at special positions. So the unit cell of halite contains four Na^{1+} cations and four Cl^{1-} anions. However, there are 14 Na^{1+} cations and 13 Cl^{1-} anions in Figure 14.9d. What's the difference in these two counts? In Figure 14.9d, eight of the Na^{1+} cations are on the corners of the unit cell and each of these would be shared with eight unit cells, while the other six Na^{1+} cations are on the faces, so they would be shared between two unit cells. If you do the math, then you'll find that are four Na^{1+} cations. Likewise, the number of Cl^{1-} anions can also be determined. There is a 1:1 ratio of Na^{1+} to Cl^{1-}.

Pyrope (isometric). A ball and stick representation of the unit cell contents of the garnet mineral pyrope ($Mg_3Al_2Si_3O_{12}$) is shown in Figure 14.10a. You might think that because pyrope is isometric like halite, its crystal structure would be just as simple, but clearly, it's not. To start with, notice that a = 11.548 Å for pyrope and a = 5.63 for halite, so the volume of the unit cell (i.e., a^3) would be 1,540.0 Å^3 for pyrope and 178.5 Å^3 for halite. Next, look at the fractional coordinates for the four atoms listed in Table 14.9, and notice the multiplicity as compared to the site multiplicity for halite. We see that the unit cell of pyrope contains 24 Mg^{2+}, 16 Al^{3+}, 24 Si^{4+}, and 96 O^{2-}. We could write a formula for the unit cell as $Mg_{24}Al_{16}Si_{24}O_{96}$. Based on this formula, you should notice that this mineral has a Si:O ratio of 1:4 and is an orthosilicate. Also, based on the discussion back in andalusite, we see that O^{2-} plots on a general x, y, z position and thus has a greater multiplicity than the other three atoms that plot on symmetry positions. Note in Table 14.7 that pyrope contains eight formula units in its unit cell (i.e., Z = 8), so we write the formula for pyrope as $Mg_3Al_2Si_3O_{12}$.

Table 14.8.
Structural information for halite: chemical formula, space group, cell parameters, and fractional coordinates of the two unique atoms in the asymmetric unit along with their site multiplicity.

NaCl, $F\ 4/m\ \overline{3}\ 2/m$, a = 5.63 Å				
Atom (Ion)	*x*	*y*	*z*	*Multiplicity*
Na^{1+}	0	0	0	4
Cl^{1-}	0.5	0	0	4

To gain an understanding of the structure of complex minerals like pyrope, it is useful to view the structure as a series of polyhedra, and to start off looking at smaller portions of it. In Figure 14.10b–f, we'll do just that. Figure 14.10b shows only the Al^{3+} octahedra in a portion of the unit cell. In Figure 14.10c, the Mg^{2+} dodecahedra are added, and then in Figure 14.10d, the Si^{4+} tetrahedra are added. Building the structure in this manner out of individual polyhedra gives you a much better understanding of the structure than the ball and stick drawing in Figure 14.10a. In Figure 14.10e and 14.10f, the plotting range for the drawing has been expanded; a larger portion of the unit cell is shown, so more ions must be added. In Figure 14.10f, the Al^{3+} octahedra occur on the corners of the unit cell. If you go back to Figure 14.10b, you'll note that an Al^{3+} octahedron also occurs in the center of the unit cell; thus, pyrope is body-centered.

Quartz (hexagonal) and cristobalite (tetragonal). Quartz and cristobalite are silica polymorphs that each contain one unique Si atom and one unique O^{2-} ion in an asymmetric unit (Tables 14.10 and 14.11). Quartz crystallizes in space group $P3_2\ 2\ 1$ (or $P3_1\ 2\ 1$, because quartz can be either right- or left-handed) and cristobalite in $P4_1\ 2_1\ 2$; thus they both contain screw axes and no

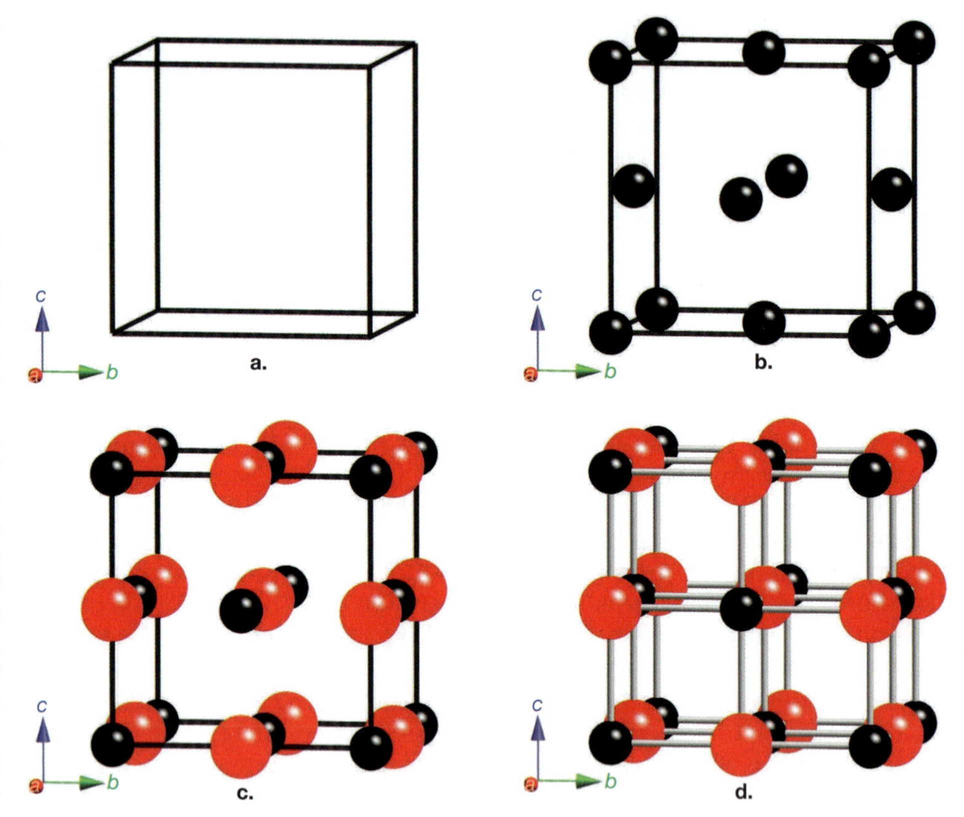

Figure 14.9. The following four images, along with the data in Table 14.8, will work us through the structure of halite, which is cubic with a = *5.63Å, space group* F4/m $\bar{3}$ 2/m. *It contains two different atoms in its asymmetric unit:* Na^{1+} *(small black sphere) and* Cl^{1-} *(larger red-sphere), which plot at 0, 0, 0 and 0.5, 0, 0 within the unit cellrespectively.* ***a.*** *The empty cube-shaped unit cell of halite showing the crystallographic axes.* ***b.*** Na^{1+} *plots at 0, 0, 0 (i.e., at the origin of the unit cell). Halite has an* F *lattice, in which there are identical points at the corners and the faces of the unit cell. Thus the one* Na^{1+} *cation at 0, 0, 0 must plot on the corners of the unit cell as well as at the center of each face.* **c.** *The* Cl^{1-} *anion plots at 0.5, 0, 0 (i.e., 1/2 way along* a *and with zero components along* b *and* c*). The one* Cl^{1-} *anion is in turn operated on by symmetry elements to produce the remaining* Cl^{1-} *anions (plotting at 1, 0, 1/2 and 1, 1/2, 0) on the front lower corner of the unit cell, and it would also generate the three* Cl^{1-} *anions on the upper rear corner of the unit cell plotting around the* Na^{1+} *cation at 0, 1, 1. It was easier to use the matrices, as we did for andalusite, than to describe how the symmetry generates each atom.* ***d.*** *Finally, the bonds between* Na^{1+} *and* Cl^{1-} *are added in this last image to complete the structure. Before we leave this rather simple structure, there is one more concept that should be explained. The formula for halite is NaCl and Z = 4 (i.e., there are 4 formula units per unit cell), so the content of the unit cell could be written as* Na_4Cl_4*. From Table 14.8, it can be seen that the multiplicity of both* Na^{1+} *and* Cl^{1-} *is 4. That means that symmetry will generate four ions of* Na^{1+} *and four of* Cl^{1-} *within the unit cell, but there appear to be lots more ions than that in Figure 14.9d. This quandary can be resolved by noting that each of the* Na^{1+} *cations that plots on the corners of the cell would occur in eight unit cells. Thus each corner* Na^{1+} *would count only 1/8 and there are eight of them for a total of one* Na^{1+}*. The six* Na^{1+} *cations on the faces are shared between two unit cells, so each cell has three* Na^{1+} *cations. Thus, the final count for a single unit cell is four* Na^{1+} *cations. See if you can figure out how we end up with four* Cl^{1-} *anions!*

mirror or glide planes. Figures 14.11a–c and 14.12a–c show ball and stick representations of the unit cell contents of quartz and cristobalite, respectively. Both of these structures contain few atoms in the unit cell, thus the ball and stick representation works well for them.

Figures 14.11d and 14.12d are *a-b* projections of the lattice for quartz and cristobalite. Compare the $^{[001]}3_2$ screw axis in quartz that plots at 1/3 *a* and 2/3 *b* and inspect how it relates to the three Si and three O atoms. Examine the cristobalite structure shown in Figure 14.12a and ask yourself what type of a symmetry operator resides in the cell's center (i.e., *a* = 1/2 and *b* = 1/2)? It may at first appear to be a 2-fold axis, but close examination of Figure 14.12b and 14.12c should convince you that it's a 2_1. The lattice in Figure 14.12d confirms this. Before we leave this example, please spend a few minutes correlating the symmetry operations in these space groups with the symmetry operations shown on the lattices so you can see how this all fits into the structures.

Table 14.9.
Structural information for pyrope: chemical formula, space group, cell parameters, and fractional coordinates of the four unique atoms in the asymmetric unit along with their site multiplicity (Hazen and Finger 1989).

$Mg_3Al_2(SiO_4)_3$, $I\,4_1/a\,\bar{3}\,2d$, a = 11.548 Å

Atom (Ion)	*x*	*y*	*z*	*Multiplicity*
Mg	0.125	0	0.25	24
Al	0	0	0	16
Si	0.375	0	0.25	24
O	0.0341	0.049	0.653	96

Enstatite (orthorhombic) and tremolite (monoclinic). Ball and stick representations of the unit cell contents of the single chain silicate enstatite ($Mg_2Si_2O_6$) and the double chain silicate tremolite ($Ca_2Mg_5Si_8O_{22}(OH)_2$) are shown in Figures 14.13 and 14.14; their structural data are given in Tables 14.12 and 14.13, respectively. The series of three drawings shows first the O^{2-} anions, then the Si^{4+}, and finally the remaining cations (Mg^{2+} in the case of enstatite and Ca^{2+} and Mg^{2+} for tremolite). Enstatite has ten unique ions in its asymmetric unit, each with a multiplicity of eight (i.e., they all occur on general positions) so there would be a total of 80 ions in the unit cell with Z = 8. Tremolite contains 14 unique ions in its asymmetric unit, with most in general positions, and the remainder at special positions, so there are 82 ions in its unit cell and Z = 2.

Table 14.10. Structural information for quartz: chemical formula, space group, cell parameters, and fractional coordinates of the two unique atoms in the asymmetric unit along with their site multiplicity (Kihara 1990).

SiO_2, $P\,3_2\,2\,1$ or $P\,3_1\,2\,1$, a = 4.9137, c = 5.4047 Å				
Atom (Ion)	*x*	*y*	*z*	*Multiplicity*
Si	0.4697	0	0	6
O	0.4133	0.2672	0.1188	6

As discussed in Chapter 12 and as is shown in Figures 12.62 and 12.63, there is a systematic relationship in the unit cell parameters for the chain silicates. In Tables 14.12 and 14.13, note that the *c* cell dimension is similar for both enstatite and tremolite (this is the repeat along the silicate

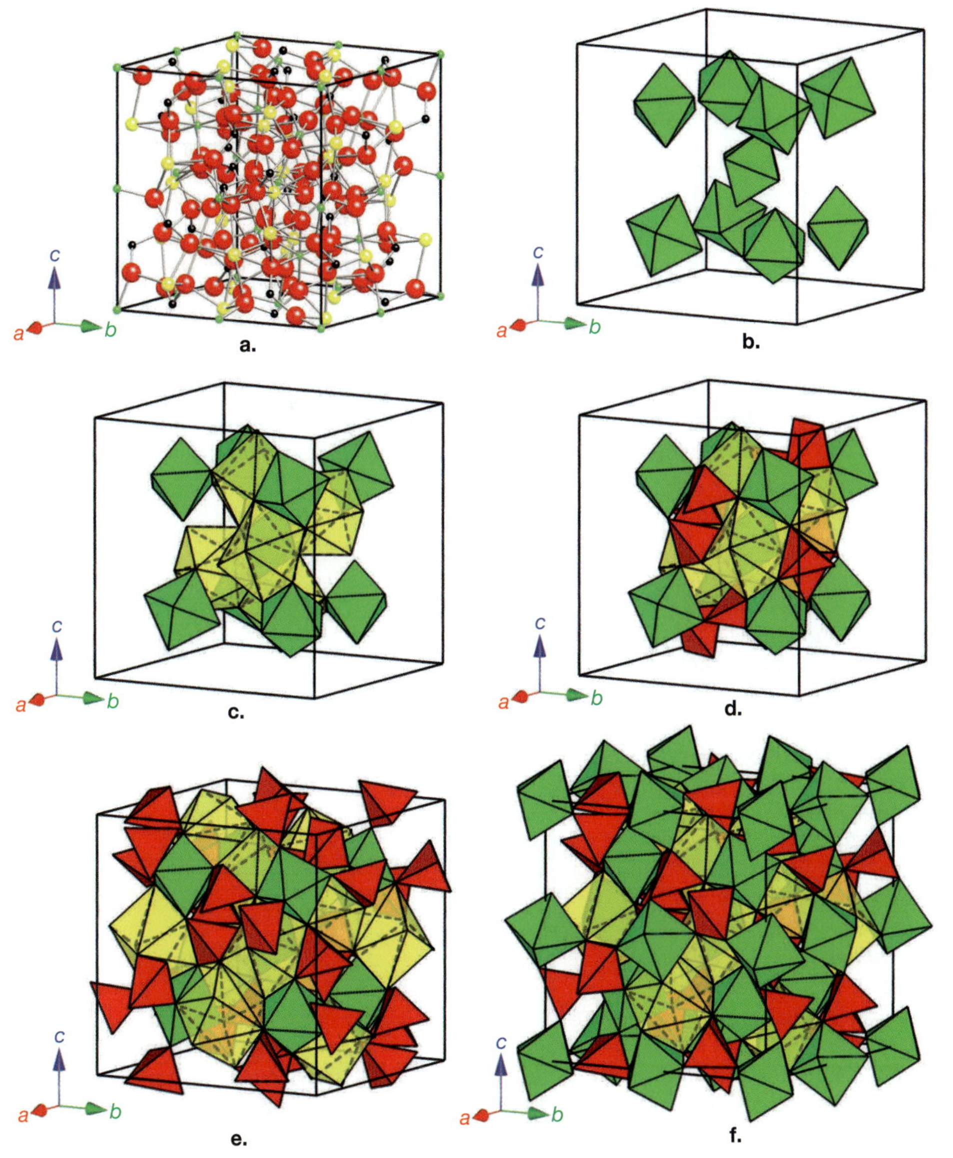

Figure 14.10. *Representation of the crystal structure of the garnet group mineral species pyrope, $Mg_3Al_2(SiO_4)_3$. The structural parameters are given in Table 14.9.* ***a.*** *A ball and stick representation of the unit cell contents. There are 160 ions in the unit cell: 96 O^{2-} (large red spheres), 24 Si^{4+} (small black spheres), 16 Al^{3+} (small green-spheres), and 24 Mg^{2+} ions (intermediate size yellow-spheres). It is difficult (or impossible) to understand this structure based on a ball and stick model of the entire unit cell. About all you can tell from this drawing is that the Al^{3+} cations plot on the corners and midpoints of the unit cell. (As shown in Table 14.9, Al plots at 0, 0, 0.)* ***b.*** *To simplify the structure, the central portion of the unit cell is shown with just the Al octahedra plotted.* ***c.*** *Next the polyhedra formed by the 8-coordinated Mg^{2+} cations are added.* ***d.*** *The polyhedra formed by the 4-coordinated Si^{4+} cations are added.* ***e.*** *Now the cell is expanded (i.e., the range of ions plotted is increased).* ***f.*** *Finally, the cell is expanded even more. Note the similarity in this final plot to Figure 14.10a (i.e., the Al^{3+} cations, here centered in octahedra, occupy the corners and midpoints of the unit cell).*

chains). The *a* axis is also similar for both: ~9 Å for the monoclinic ones and 18 Å for the orthorhombic one. The *b* axis relates to the chain width, doubling from enstatite to tremolite.

Orthoclase (monoclinic) and anorthite (triclinic). Ball and stick representations for the eight unique ions in the asymmetric unit of orthoclase ($KAlSi_3O_8$) and the 52 (yes 52!) ions for anorthite ($CaAl_2Si_2O_8$) are shown in Figures 14.15 and 14.16. Tables 14.14 and 14.15 give the structural parameters for these two feldspar group minerals. The multiplicity for an ion plotting in a general position for *C*2/*m* (the space group for orthoclase) is eight, while that for the space group of anorthite, $P\overline{1}$, is only two; this partially explains why there are so many more ions needed to describe the structures of the lower symmetry anorthite as compared to orthoclase. You'll also notice that *c* is doubled for anorthite. The doubling of *c* was discussed in Chapter 12 and was shown in Figure 12.81 based on the type of lattice chosen.

In this example we've plotted all the unique ions in the asymmetric unit of orthoclase (Figure 14.15a) and anorthite (Figure 14.16a). For orthoclase, it appears that the unit cell is about 1/8 filled and for anorthite it appears half-full. The unfilled 7/8 for orthoclase will be filled by the seven other symmetry-generated ions, while for anorthite, the other half is filled by the center of symmetry (i.e., the $\overline{1}$) axis. A few of the ions have been labeled in the drawing so you can distinguish the ions originating from the symmetry-generated ones.

Table 14.11. Structural information for cristobalite: chemical formula, space group, cell parameters, and fractional coordinates of the two unique atoms in the asymmetric unit along with their site multiplicity (Downs and Palmer 1996).

SiO_2, $P\,4_1\,2_1\,2$, a = 4.9717, c = 6.9222 Å				
Atom (Ion)	*x*	*y*	*z*	*Multiplicity*
Si	0.3004	0.3004	0	4
O	0.2398	0.1044	0.1789	8

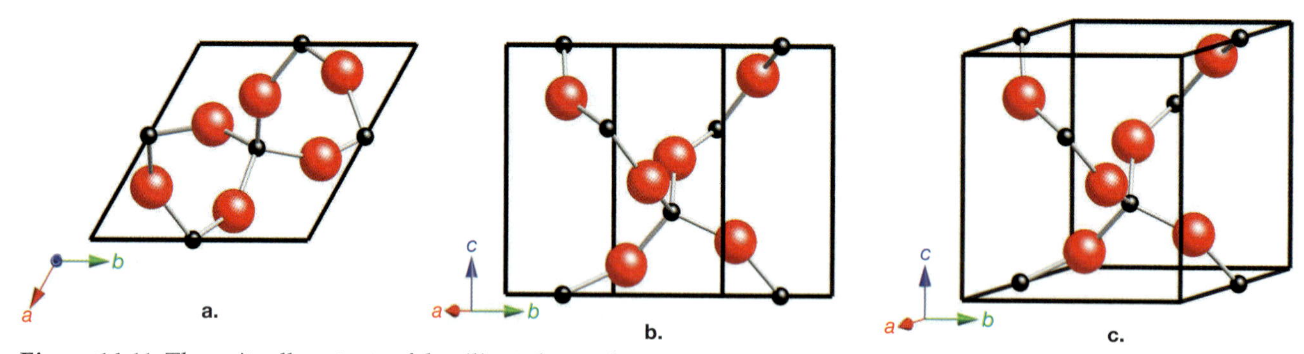

Figure 14.11. *The unit cell contents of the silica polymorph quartz, based on the data in Table 14.10, using an* a-b *projection (large red spheres represent* O^{2-} *and small black ones* Si^{4+}*) of its lattice to show the symmetry elements.* ***a.*** *A projection onto the* a-b *plane showing the hexagonal lattice. All the* Si^{4+} *cations occur on the cells edges and all the* O^{2-} *anions lie within the unit cell.* ***b.*** *A projection down the [210] direction.* ***c.*** *A perspective projection. Based on these three projections, try to determine which of the* Si^{4+} *and* O^{2-} *ions are plotted from the Table 14.10; the others are generated by symmetry (the movie on the MSA website might be helpful for this). Note the relationships among the atom multiplicity, atoms in the unit cell, and the 1:2 ratio of* Si^{4+} *to* O^{2-}*. (Hint: count the number of* Si^{4+} *and* O^{2-} *ions that belong to the unit cell.)* ***d.*** *A projection of the space group* P 3_2 2 1 *onto the* a-b *plane, with its labeled symmetry elements. We could write out the space group as* P $^{[001]}3_2$ $^{[100]}2$ *and* $^{[210]}1$ *to clearly denote the orientation of the rotation axes. Try to locate these in these projections (several of them are labeled, so that should help). For this crystal system, the symmetry must be equivalent along the [100] = a, [010] = b, and [110] = a + b directions. Notice that the* 3_2 *screws occur at the corners of the cell (which fits the* P *lattice) and internally, but not at the lattice's center. There are several* 2_1 *screw axes that lie in the plane of the projection. The numbers next to them and some of the 2-folds indicate the height (in terms of z) where these occur; unlabeled operations occur at z = 0.*

1/6
1/6
1/6
$^{[001]}3_2$
1/3
1/3
$^{[110]}2$
1/3
1/3
2_1
1/3
1/3
$^{[100]}2$
$^{[010]}2$
1/6
1/6
1/6
d.

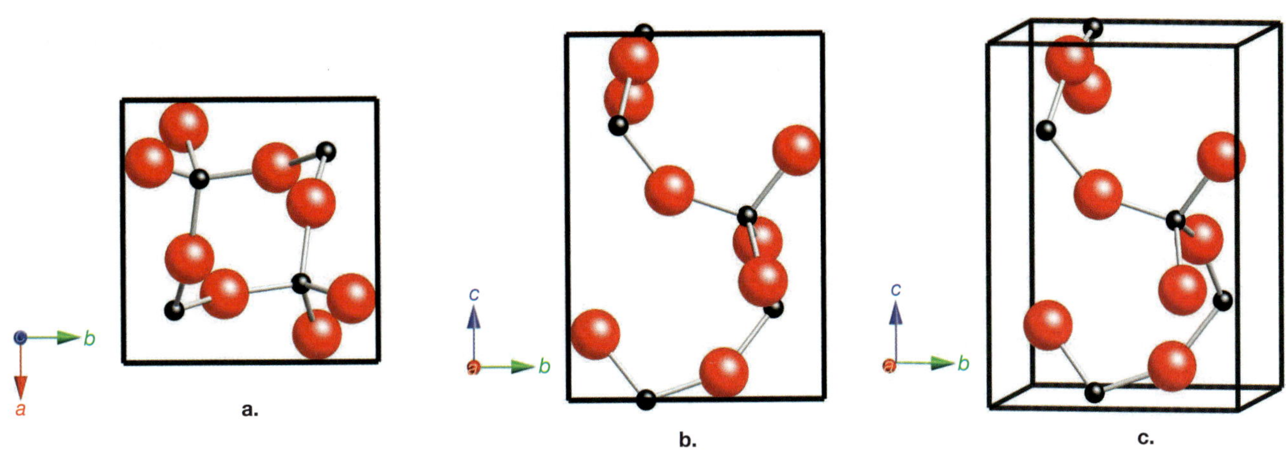

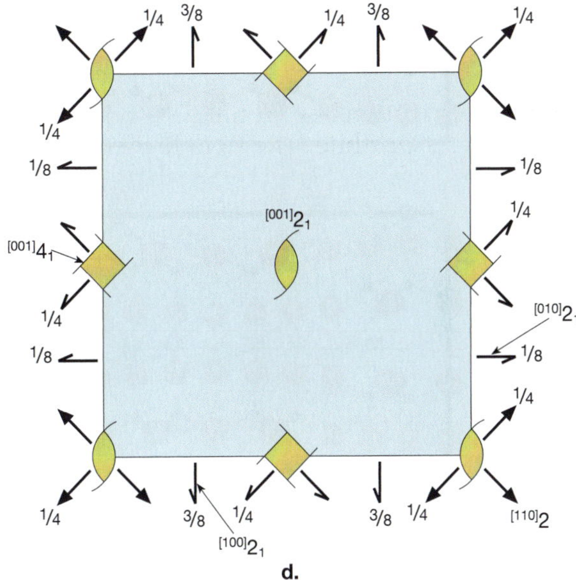

Figure 14.12. *The unit cell contents of the silica polymorph cristobalite, based on the data in Table 14.11 (large red spheres represent oxygen and small black ones silicon), and the* a-b *projection of its lattice showing the symmetry elements.* ***a.*** *A projection onto the* a-b *plane showing the tetragonal lattice.* ***b.*** *A projection down the [100] direction.* ***c.*** *A perspective projection that shows the 4_1 axis and how it operates on the Si^{4+} cations. You can see the five Si^{4+} cations spiraling upward in a counterclockwise direction. Based on these three projections, see if you can determine which of the Si^{4+} and O^{2-} ions are plotted from the Table 14.11; the others are generated by symmetry. The relationships among the atom multiplicity, atoms in the unit cell, and the 1:2 ratio of Si^{4+} to O^{2-} should be noted. (Hint: count the number of Si^{4+} and O^{2-} ions that belong to the unit cell.) Finally, compare this structure to that of quartz.* ***d.*** *A projection of the space group* P $4_1\ 2_1\ 2$ *onto the* a-b *plane, with its symmetry elements labeled. We could write out the space group as* P $^{[001]}4_1\ ^{[100]}2_1\ ^{[110]}2$ *to clearly denote the orientation of the rotation axes. Locate these symmetry operations in these projections (several of them are labeled, so that should help). For this crystal system, the symmetry must be equivalent along the [100] =* a, *[010] =* b. *For this space group, the $^{[001]}4_1$ does not occur at the origin, and the $^{[100]}2_1$ and $^{[010]}2_1$ do not coincide with* a *and* b, *but are parallel to them such that a $^{[001]}2_1$ occurs in the cell's center. As in Figure 14.11d, the numbers next to the symmetry elements refer to the height of the atoms above z; unlabeled operations occur at z = 0. See Table 14.11 for structure data.*

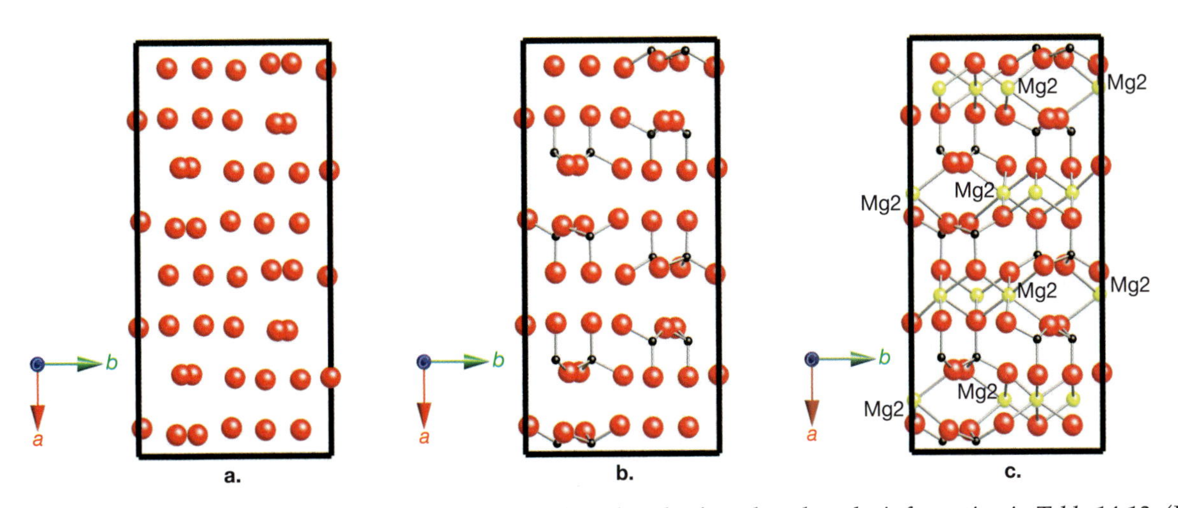

Figure 14.13. *The unit cell of enstatite, $Mg_2Si_2O_6$, projected on the* a-b *plane, based on the information in Table 14.12. (Large red anions represent O^{2-}, intermediate yellow cations are Mg^{2+}, and small black cations are Si^{4+}.) The accompanying movie on the MSA website shows each set of atoms, and associated names, adding to build the structure.* ***a.*** *O^{2-} anions.* ***b.*** *O^{2-} anions with Si^{4+} cations, forming Si^{4+} tetrahedra. In this view, we are looking down the axis of the single chains of Si^{4+} tetrahedra.* ***c.*** *Finally, the 6-coordinated Mg^{2+} cations are added, with Mg2 (the larger site) labeled.*

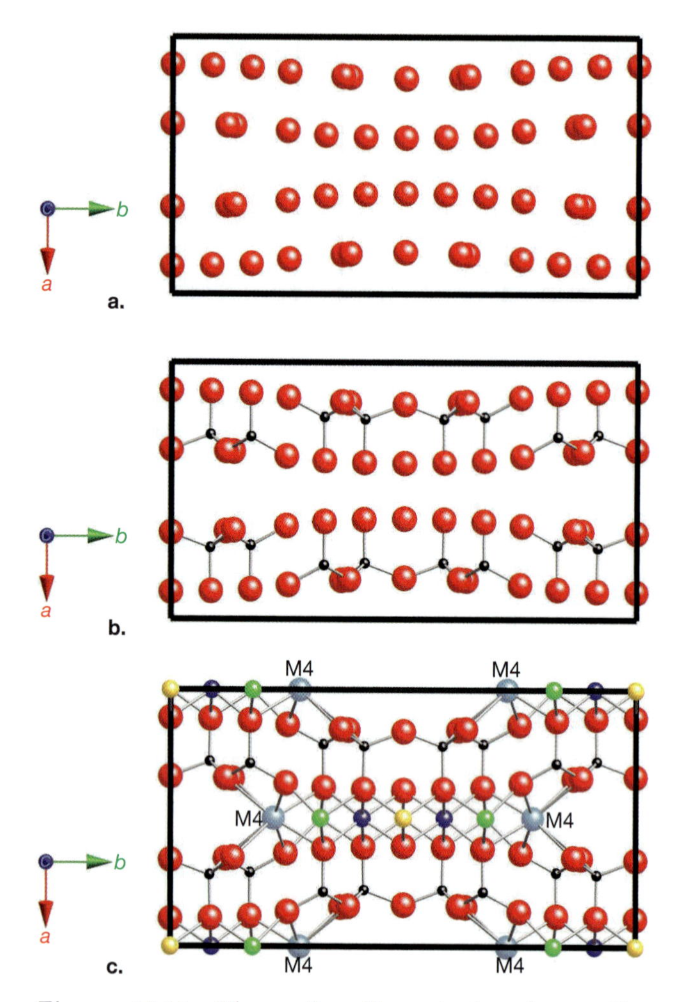

Figure 14.14. The unit cell contents of tremolite, $Ca_2Mg_5Si_2O_8(OH)_2$, projected on the a-b plane, based on the information in Table 14.13. Large red ions represent O^{2-}, small black atoms are Si^{4+}, and M-sites are shown in the center of the last projection. The movie on the MSA website shows each set of ions, and associated names, adding to build the structure. ***a.*** *O^{2-} anions.* ***b.*** *O^{2-} anions with Si^{4+} cations, forming Si tetrahedra. In this view, we are looking down the axis of the double chains of Si tetrahedra.* ***c.*** *Finally, the four M-sites are added to completely the structure, and M4, the largest, is labeled.*

Natural and Pb-exchanged heulandite. The last example moves from the realm of teaching you about structure representations to its use in research projects. In fact, the details of this example are discussed in a full-length research article (Gunter et al., 1994). Herein our goal is to point out some more subtle features about crystal structures. Figure 14.17 shows the crystal structure of a natural (a) and Pb-exchanged (b) zeolite group mineral called heulandite. Zeolites are framework silicates, with both Al^{3+} and Si^{4+} in tetrahedral coordination, represented by the polyhedra in the figure. Tables 14.16a and b give the structural information. The heulandite structure also contains two

Table 14.12.
Structural information for enstatite: chemical formula, space group, cell parameters, and fractional coordinates of the ten unique atoms in the asymmetric unit along with their site multiplicity (Hugh-Jones and Angel 1994).

$Mg_2Si_2O_6$, $P\,2_1/b\,2_1/c\,2_1/a$, a = 18.233, b = 8.8191, c = 5.1802 Å				
Atom (Ion)	*x*	*y*	*z*	*Multiplicity*
Mg1	0.3763	0.6541	0.8663	8
Mg2	0.3769	0.4872	0.3589	8
SiA	0.2715	0.3418	0.0506	8
SiB	0.4736	0.3374	0.7981	8
O1a	0.1835	0.3407	0.034	8
O1b	0.5622	0.3414	0.798	8
O2a	0.3114	0.5029	0.043	8
O2b	0.4324	0.4836	0.69	8
O3a	0.3025	0.2224	0.832	8
O3b	0.4483	0.1951	0.603	8

Table 14.13.
Structural information for tremolite: chemical formula, space group, cell parameters, and fractional coordinates of the 14 unique atoms in the asymmetric unit along with their site multiplicity (simplified from Hawthorne and Grundy 1976).

$Ca_2Mg_5Si_8O_{22}(OH)_2$, $C\,2/m$, a = 9.863, b =18.048, c = 5.285 Å β = 104.79°				
Atom (Ion)	*x*	*y*	*z*	*Multiplicity*
M1	0	0.0883	0.5	4
M2	0	0.177	0	4
M3	0	0	0	2
M4	0	0.2779	0.5	4
T1	0.2799	0.0842	0.2974	8
T2	0.2882	0.1713	0.8056	8
O1	0.1114	0.0857	0.2182	8
O2	0.1187	0.1707	0.7251	8
O3	0.1082	0	0.7154	4
O4	0.3642	0.2482	0.7928	8
O5	0.3467	0.1339	0.0998	8
O6	0.3437	0.1181	0.591	8
O7	0.338	0	0.2921	4
H	0.2088	0	0.7628	4

channels (as is typical of zeolites): an upper, smaller one and a lower, larger one. Within the channels are the exchanged cations and water molecules.

Natural heulandite crystallizes in *C*2/*m* while the Pb-exchanged has space group *Cm*. A general site in *C*2/*m* has a multiplicity of eight, while in *Cm* the multiplicity is reduced to four because *Cm* lacks the [010]2 and center of symmetry of *C*2/*m*. Table 14.16b thus shows more unique atoms than are needed to describe the *Cm* sample than the *C*2/*m* sample in Table 14.16a. In the tetrahedra, there are five unique Si^{4+} cations and ten O^{2-} anions for the natural sample and nine Si^{4+} and 19 O^{2-} anions for the *Cm* sample—basically double the amount in *C*2/*m*. In fact, it would be exactly doubled if it were not for T5, O1, and O5, all of which plot on special positions. Within the unit cell, the natural sample would contain 36 tetrahedral sites and 72 O^{2-} anions for the both of the samples. This defines the framework portion of the structures as $T_{36}O_{72}$. In the channels of the natural samples, there are three distinct sites listed as O's in Table 14.16 after the cations. Based on the multiplicity of all of these atoms we arrive at the formula given in the table.

Something different occurs in these structures than we've seen before. Notice in Figure 14.17a that some numbers have been added. These numbers represent bond distances between atoms. For example, the bond distance between Na^{1+} and 017 is 2.36Å, a reasonable bond distance for Na^{1+} to an anion. In the smaller channel, three bond distances are given for Ca^{2+} and two for water molecules. This would indicate that this mineral has Ca^{2+}-Ca^{2+} bonds; these sorts of bonds (metal to metal) do not normally occur in silicates. So what's up? These Ca^{2+} sites are never occupied at the same time; in other words, the site population for Ca^{2+} is 0.5. A new column for site population has been added in Table 14.16a. Its value is 1 for

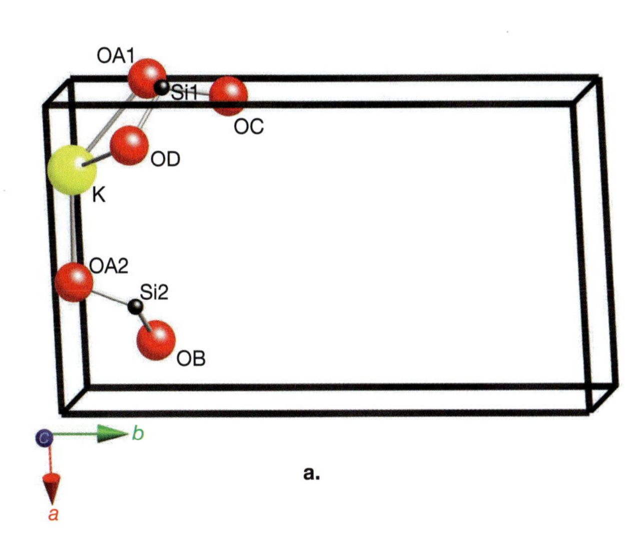

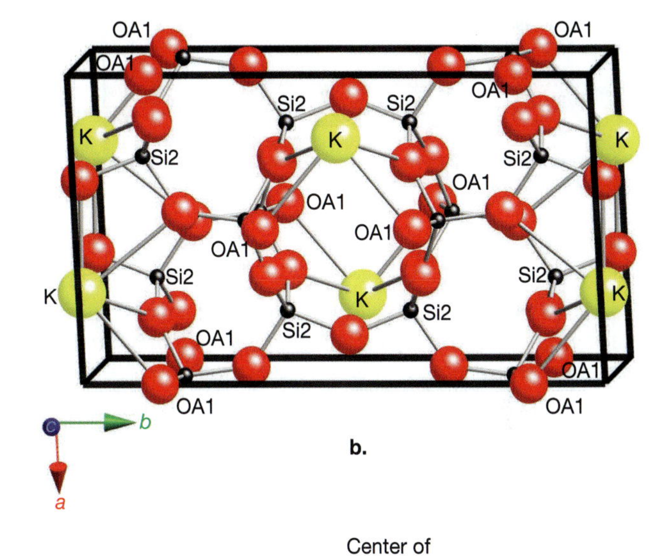

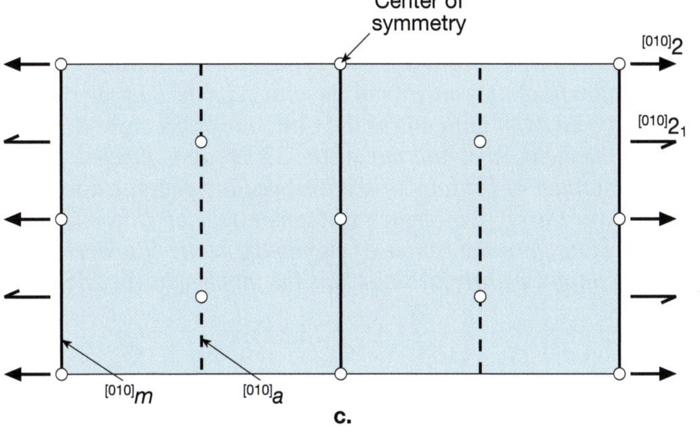

Figure 14.15. *The unit cell of orthoclase,* $KAlSi_3O_8$, *with space group* C2/m. *The projections are offset slightly from the* a-b *plane. Data from the structure are in Table 14.14. (Large yellow spheres represent* K^{1+}, *intermediate red spheres,* O^{2-}, *and small black spheres,* Si^{4+}.) *The associated movies on the MSA website allow you to rotate these structures to explore the symmetry relationships among the eight unique atoms in the asymmetric unit and the symmetry-generated ones.* ***a.*** *The eight unique atoms of the asymmetric unit given in Table 14.14 are plotted in the cell and labeled. Because* C2/m *has 8 symmetry operations, this is basically ⅛ of the contents of the unit cell.* ***b.*** *The remainder of the atoms in the unit cell generated by the* C2/m *space group symmetry are added to the plot. The one* K^{1+}, *one of the two* Si^{4+} *cations (Si2), and one of the five unique* O^{2-} *anions (OA1) are labeled to aid in understanding the symmetry.* ***c.*** *The space lattice representation of* C2/m *showing each of the symmetry elements. How do the atoms in the structure conform to the symmetry operations shown? Note that no atoms plot at the corners of the cell or are centered on the (001) face. Thus, as stated earlier in Chapter 12, atoms are not required to plot at these locations in order for the symmetry around their locations them to conform to C-centering.*

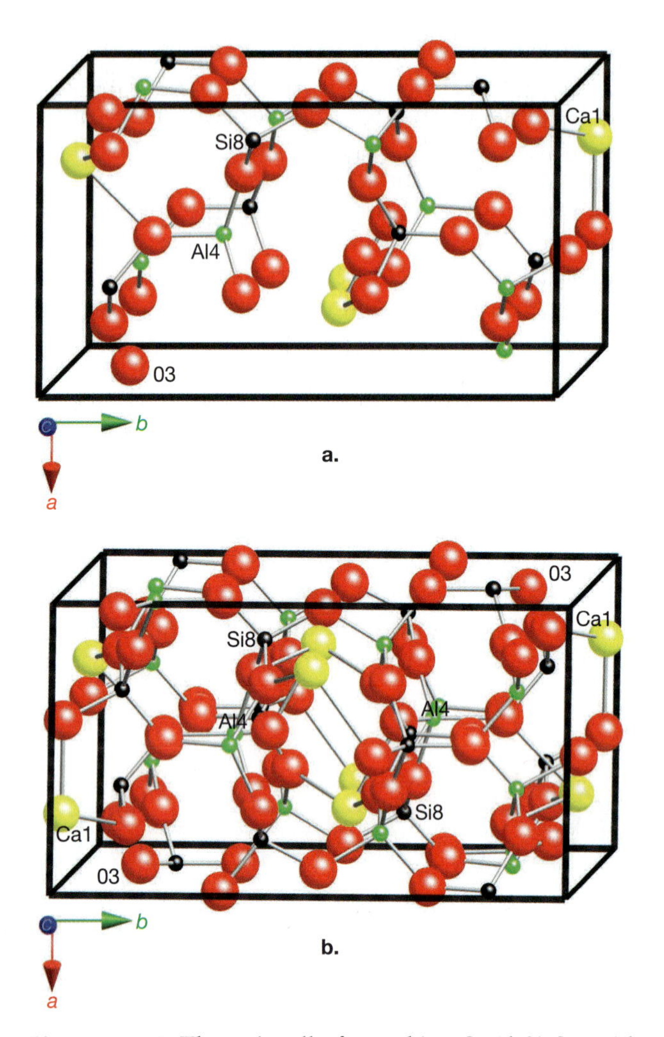

Figure 14.16. The unit cell of anorthite, $CaAl_2Si_2O_8$, with space group $P\bar{1}$. The projections are offset slightly from the a-b plane. Data from the structure are in Table 14.15. (Large yellow spheres represent Ca^{2+}, intermediate red spheres, O^{2-}, small black spheres, Si^{4+}, and small green spheres, Al^{3+}.) The associated movies on the MSA website allow you to rotate these structures to explore the symmetry relationship between the eight unique atoms and the symmetry generated ones. ***a.*** *The 52 unique atoms in the asymmetric unit, given in Table 14.15, are plotted within the unit cell and labeled. Because $P\bar{1}$ contains two symmetry operations, this is basically 1/2 of the contents of the unit cell (the half formed by the 1-fold axis) with one of the Ca^{2+}, one of the eight Si^{4+}, one of the eight Al^{3+}, and one of the 32 O^{2-} ions labeled.* ***b.*** *The remainder of the ions in the unit cell are generated and plotted by the $\bar{1}$ (i.e., center of symmetry) for this space group. How does the center of symmetry relate the labeled ions? It might be helpful to look at the movies on the MSA website.*

most of the atoms except the channel cations. In fact, for Ca^{2+} it's 0.484(4); this value is determined in the crystal structure analysis (more on how that works in Chapter 15), and the "(4)" at the end is the error associated with this value. There are also three channel cations sites that have the labels Na1, Ca2, and K3. Just because the site has an element name associated with it doesn't prevent it from containing other elements! In fact, at the bottom of Table 14.16a you'll see the contents of each of those sites. For example, Na1 contains both Ca^{2+} and Na^{1+}, and K3 contains both K^{1+} and Na^{1+}, while Ca2 contains only Ca^{2+}. Of the five water sites, all are fully occupied except for O17, which is only 43% filled. Why are the five water sites (O13, O14, O16, O17 and O19) are denoted as O's and not H_2O? That's because the exact position of the H^{1+} atoms is not known in the structures (more on this in the next chapter).

In the Pb-exchanged sample (Figure 14.17b), you see a similar distribution of the atoms in the channels to the natural sample. All the bond distances make sense except for the (Pb3′-Pb1′) distance of 1.31Å. Again this might seem to indicate a metal bonding, which doesn't occur in these types of structures. So, as with the Ca^{2+} cations in the natural samples, you'd assume that both of these sites are not fully occupied. From checking Table 14.16b, you'd see that Pb1′ is 40.4% occupied while Pb3′ is 12%, and both cannot be occupied at the same time. Also notice that several of the water sites are not fully occupied. We'll return to these samples in the next chapter in the discussion of crystal structure analysis, so we can explain how we figured out the site occupancies.

Conclusions and Final Thoughts

After looking at hundreds of crystal structure drawings in the previous thirteen chapters, we

Table 14.14.
Structural information for orthoclase: chemical formula, space group, cell parameters, and fractional coordinates of the eight unique atoms in the asymmetric unit along with their site multiplicity (Colville and Ribbe 1968).

$KAlSi_3O_8$, *C* 2/m, *a* = 8.561, *b* = 12.996, *c* = 7.192 Å, *β* = 104.79°

Atom (Ion)	*x*	*y*	*z*	*Multiplicity*
K	0.2838	0	0.1373	4
Si1	0.0095	0.1844	0.2239	8
Si2	0.7089	0.1178	0.3443	8
OA1	0	0.1459	0	4
OA2	0.6346	0	0.2851	4
OB	0.828	0.147	0.2282	8
OC	0.0349	0.3106	0.2607	8
OD	0.1815	0.1258	0.4065	8

Table 14.15.
Structural information for anorthite: chemical formula, space group, cell parameters, and fractional coordinates of the 52 unique atoms in the asymmetric unit along with their site multiplicity (Angel et al. 1990).

$CaAl_2Si_2O_8$, $P\bar{1}$, $a = 8.175$, $b = 12.873$, $c = 14.17$ Å, $\alpha = 93.11$, $\beta = 115.89$, $\gamma = 91.28°$

Atom (Ion)	*x*	*y*	*z*	*Multiplicity*	*Atom (Ion)*	*x*	*y*	*z*	*Multiplicity*
Ca1	0.2652	0.9861	0.0869	2	O7	0.5723	0.9892	0.6370	2
Ca2	0.2683	0.2683	0.5430	2	O8	0.0716	0.4924	0.1383	2
Ca3	0.7735	0.5359	0.5412	2	O9	0.8133	0.1016	0.0798	2
Ca4	0.7637	0.5051	0.0751	2	O10	0.3331	0.5952	0.6053	2
Si1	0.0090	0.1589	0.1039	2	O11	0.8117	0.0967	0.6057	2
Si2	0.5066	0.6560	0.6045	2	O12	0.2857	0.6034	0.0792	2
Si3	0.0058	0.8153	0.6132	2	O13	0.8177	0.8549	0.1444	2
Si4	0.5050	0.3203	0.1103	2	O14	0.2984	0.3554	0.6111	2
Si5	0.6817	0.1034	0.6647	2	O15	0.8105	0.8521	0.6026	2
Si6	0.1707	0.6067	0.1491	2	O16	0.3421	0.3577	0.1342	2
Si7	0.6739	0.8829	0.1877	2	O17	0.0151	0.2792	0.1354	2
Si8	0.1761	0.3793	0.6733	2	O18	0.5095	0.7772	0.6350	2
Al1	0.0069	0.1609	0.6111	2	O19	0.0215	0.2910	0.6485	2
Al2	0.4988	0.6660	0.1129	2	O20	0.5078	0.7962	0.1495	2
Al3	0.9909	0.8153	0.1176	2	O21	0.0	0.6801	0.1034	2
Al4	0.5083	0.3186	0.6214	2	O22	0.5148	0.1799	0.6107	2
Al5	0.6843	0.1132	0.1512	2	O23	0.0093	0.6885	0.6006	2
Al6	0.1910	0.6111	0.6677	2	O24	0.5075	0.1955	0.0993	2
Al7	0.6814	0.8716	0.6724	2	O25	0.1833	0.1053	0.1910	2
Al8	0.1895	0.3776	0.1816	2	O26	0.7012	0.6079	0.6800	2
O1	0.0263	0.1246	−0.0043	2	O27	0.2136	0.1025	0.6840	2
O2	0.4860	0.6236	0.4859	2	O28	0.6930	0.6043	0.2013	2
O3	0.9811	0.1256	0.4836	2	O29	0.2038	0.8732	0.2108	2
O4	0.5179	0.6238	0.9965	2	O30	0.6900	0.3636	0.7317	2
O5	0.5757	0.9906	0.1434	2	O31	0.1742	0.8569	0.7183	2
O6	0.0721	0.4881	0.6346	2	O32	0.6992	0.6992	0.1983	2

have now explained the basis for how they are represented. This chapter built on the concepts of symmetry operations and lattice types that we learned in Chapter 12. Chapter 13 then laid the mathematical groundwork to produce structure representations by turning the symmetry operations into mathematical operations. Thus now we know that only knowledge of a mineral's symmetry, the size and shapes of its unit cell, and the fractional coordinates of the unique atoms in the asymmetric unit are needed to represent its structure. This chapter also showed that there are many options in currently-available computer software for representing structures in such a way as to illustrate important mineralogical concepts.

As stated at the start of this chapter, we believe the main reason to understand symmetry in min-

Table 14.16.
Structural information for a natural (A) and Pb-exchanged (B) heulandite: chemical formula, space group, cell parameters, fractional coordinates unique atoms in the asymmetric unit, site population, site multiplicity, and atoms in the unit cell (apuc) (Gunter et al. 1994).

a. natural heulandite: $Ca_{3.7}K_{0.13}Na_{1.3}(Al_{8.9}Si_{27.1}O_{72})•24.8H_2O$, *C 2/m*, *a* = 17.671(1), *b* = 17.875(7), *c* = 7.412(3) Å, *β* = 116.93(3)°

Atom (Ion)	*x*	*y*	*z*	Site pop.	Site mult.	Apuc	Atom (Ion)	*x*	*y*	*z*	Site pop.	Site mult.	Apuc
T1	0.17940(8)	0.16975(8)	0.0967(2)	1	8	8	O9	0.2103(2)	0.2542(2)	0.1783(6)	1	8	8
T2	0.28859(8)	0.08989(8)	0.5008(2)	1	8	8	O10	0.1155(2)	0.3723(2)	0.3988(6)	1	8	8
T3	0.29172(8)	0.30959(8)	0.2831(2)	1	8	8	Na1	0.1549(3)	0	0.6666(7)	1.108(8)	4	4.43*
T4	0.06460(8)	0.29843(8)	0.4110(2)	1	8	8	Ca2	0.5402(2)	0	0.2031(5)	0.484(4)	4	1.94**
T5	0	0.2131(1)	0	1	4	4	K3	0.2763(5)	0	0.010(1)	0.494(5)	4	1.98***
O1	0.3048(4)	0	0.5466(9)	1	4	4	O13	0.4228(3)	0.0817(3)	0.0307(9)	1	8	8
O2	0.2313(3)	0.1194(2)	0.6132(6)	1	8	8	O14	0.5	0	0.5	1	2	2
O3	0.1833(3)	0.1543(3)	−0.1158(6)	1	8	8	O16	0.0949(1)	0	0.285(3)	1	4	4
O4	0.2385(2)	0.1065(2)	0.2555(6)	1	8	8	O17	0.082(1)	0.005(7)	0.867(3)	0.43(1)	8	3.44
O5	0	0.3259(3)	0.5	1	4	4	O19	0	0.0873	.5	1	4	4
O6	0.0820(2)	0.1587(2)	0.0626(6)	1	8	8							
O7	0.3729(3)	0.2660(3)	0.4508(6)	1	8	8							
O8	0.0089(3)	0.2662(3)	0.1842(6)	1	8	8							

*Na1 = 1.76 Ca + 0.91 Na
** Ca2 = 1.94 Ca
*** K3 = 0.39 Na + 0.13K

b. Pb-exchanged heulandite: $Pb_{4.4}(Al_{8.9}Si_{27.1}O_{72})•16.4H_2O$, *Cm*, *a* = 17.767(3), *b* = 17.917(2), *c* = 7.432(2) Å, *β* = 116.33(2)°

Atom (Ion)	*x*	*y*	*z*	Site pop.	Site mult.	Apuc	Atom (Ion)	*x*	*y*	*z*	Site pop.	Site mult.	Apuc
T1	0.1796(4)	0.1695(3)	0.091(1)	1	4	4	O8	0.0099(9)	0.2643(8)	0.194(2)	1	4	4
T1'	−0.1778(4)	−0.1685(3)	−0.094(1)	1	4	4	O8'	−0.0123(9)	−0.2743(8)	−0.183(2)	1	4	4
T2	0.2874(4)	0.0894(3)	0.493(1)	1	4	4	O9	0.2118(9)	0.2526(8)	0.178(2)	1	4	4
T2'	−0.2839(4)	−0.0903(3)	−0.497(1)	1	4	4	O9'	−0.2109(9)	−0.2522(8)	−0.171(2)	1	4	4
T3	0.2961(4)	0.3081(3)	0.288(1)	1	4	4	O10	0.1187(8)	0.3726(7)	0.401(2)	1	4	4
T3'	−0.2895(4)	−0.3081(3)	−0.282(1)	1	4	4	O10'	−0.118(1)	−0.3750(9)	−0.420(2)	1	4	4
T4	0.0670(4)	0.2991(3)	0.417(1)	1	4	4	Pb1	0.1505(5)	0	0.679(1)	0.318(4)	2	0.64
T4'	−0.6034(4)	−0.3020(3)	−0.411(1)	1	4	4	Pb1'	−0.1480(4)	0	−0.673(1)	0.404(4)	2	0.81
T5	0	0.2160(2)	0	1	4	4	Pb2	0.0381(3)	0.5	0.1979(9)	0.608(3)	2	1.22
O1	0.299(1)	0	0.555(3)	1	2	2	Pb3	0.217(1)	0	0.877(3)	0.144(4)	2	0.29
O1'	0.199(1)	0.5	0.458(3)	1	2	2	Pb3'	−0.2060(8)	0	−0.860(3)	0.12(3)	2	0.24
O2	0.2351(9)	0.1248(8)	0.613(2)	1	4	4	Pb4	−0.011(1)	0.07689(9)	0.667(3)	0.144(6)	4	0.58
O2'	−0.2238(9)	−0.1230(8)	−0.597(2)	1	4	4	Pb5	0.009(3)	−0.100(2)	−0.516(6)	0.069(4)	4	0.28
O3	0.188(1)	0.1508(9)	−0.115(2)	1	4	4	O13	0.4218(8)	0.0873(8)	1.034(2)	1	4	4
O3'	−0.1807(9)	−0.1518(7)	0.177(2)	1	4	4	O14	0.499(3)	0	0.502(7)	1	2	2
O4	0.231(1)	0.1017(9)	0.245(3)	1	4	4	O16	0.077(2)	0	0.268(5)	1.14(5)	2	2.28
O4'	−0.2389(8)	−0.1052(6)	−0.248(2)	1	4	4	O16'	−0.076(2)	0	−0.879(6)	0.88(4)	2	1.76
O5	0	0.3280(6)	0.5	1	4	4	O22	0.017(1)	0.074(1)	0.322(3)	0.76(5)	4	3.04
O6	0.082(1)	0.1623(9)	0.047(2)	1	4	4	O27	0.274(2)	0	−0.004(5)	0.6(4)	2	1.2
O6'	−0.0824(7)	−0.1586(6)	−0.067(2)	1	4	4	O27'	−0.276(2)	0	−0.018(4)	0.86(6)	2	1.72
O7	0.3807(()	0.2654(8)	0.453(2)	1	4	4	O28	0.29(1)	0	0.25(2)	0.22(3)	2	0.44
O7'	−0.369(1)	−0.2632(8)	−0.442(2)	1	4	4							

erals is to lay the foundation for crystal structure representation. One major reason we care about structures is that a basic understanding of them allows us to predict the physical properties of the minerals, which in turn will aid us in their identification. Finally, we think everyone will benefit from a 3-D understanding of minerals. Most of us will see structure drawings again in our professional lives. In the next chapter and the final one in this section, we'll explain how to do something that few people will ever have a chance to do—determine the crystal structure of a mineral. In the process, we will gain knowledge that will aid us in the identification of a mineral!

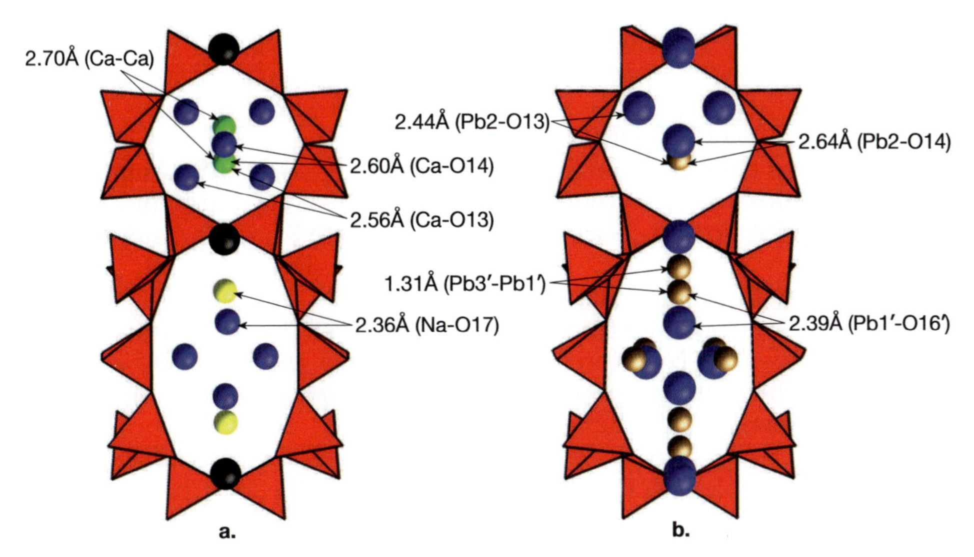

Figure 14.17. *Crystal structure drawings of a portion of the structure of a natural heulandite (Figure 14.17a) and a Pb-exchanged heulandite projected onto the* a-b *plane looking down the major channels; the Al and Si tetrahedra define these channels.* ***a.*** *There are three cations in the natural sample. In the large channel, K^{1+} is represented by the large black sphere and Na^{1+} by the smaller yellow sphere; the remainder of the intermediate-sized spheres represent H_2O. In the smaller channel there are five waters and two Ca^{2+} cations near the center. The bond distances for several cation–waters are also given. A Ca^{2+}-Ca^{2+} distance of 2.70 Å is also given. Simultaneous occupancy of both those sites by Ca^{2+} is too close for these cations; typically only one of these sites is occupied.* ***b.*** *In the Pb-exchanged sample, there are only Pb^{2+} cations (represented by the smaller spheres) and H_2O molecules in the channels. Some bond distances between Pb^{2+} and H_2O atoms have been added to the figure, as well as a distance for Pb3′ – Pb1′. Again these are too close together for comfort of the cations, so these sites are never both occupied at the same time. See Table 14.16b for structure data.*

References

Angel, R.J., Carpenter, M.A., and Finger, L.W. (1990) Structural variation associated with compositional variation and order-disorder behavior in anorthite-rich feldspars. American Mineralogist, 75, 150–162.

Bartelmehs, K.L., Gibbs, G.V., Boisen, M.B, Jr., and Downs, R.T. (1993) Interactive computer software used in teaching and research at Virginia Tech. Geological Society of America Fall Meeting, Boston, A-347.

Bloss, F.D. (1994) Crystallography and crystal chemistry. Mineralogical Society of American, Washington D.C. 545 pp.

Boisen, M.B., Jr. and Gibbs, G.V. (1990) Mathematical crystallography. Second Edition. Reviews in Mineralogy, 15, Mineralogical Society of America, Washington D.C. 460 pp.

Clark, C.M. and Downs, R.T. (2004) Using the American Mineralogist crystal structure database in the classroom. Journal of Geoscience Education, 51, 76–80.

Coville, A.A. and Ribbe, P.H. (1968) The crystal structure of an adularia and a refinement of the structure of orthoclase. American Mineralogist, 53, 25–37.

CrystalMaker Software Ltd. (1994–2007) CrystalMaker: an interactive program for visualization of crystal and molecular structures. CrystalMaker Software Ltd., Oxfordshire, U.K. (http://www.crystalmaker.com).

Downs, R.T. and Hall-Wallace, M. (2003) The American Mineralogist crystal structure database. American Mineralogist, 88, 247–250.

Downs, R.T. and Palmer, D.C. (1996) The pressure behavior of alpha-cristobalite. American Mineralogist, 79, 9–14.

Gibbs, G.V. (1997) The meterical matrix in teaching mineralogy. In Teaching Mineralogy, editors J.B. Brady, D.W. Mogk, and D. Perkins III, 201–212.

Gunter, M.E. and Bloss, F.D. (1982) Andalusite-kanonaite series: Lattice and optical parameters. American Mineralogist, 67, 1218–1228.

Gunter, M.E. and Downs, R.T. (1991) DRILL: A computer program to aid in the construction of ball and spoke crystal models. American Mineralogist, 76, 293–294.

Gunter, M.E., Armbruster, T., Kohler, T., and Knowles, C.R. (1994) Crystal structure and optical properties of Na- and Pb-exchanged heulandite group zeolites. American Mineralogist, 79, 675–682.

Hahn, T. (1995) International table for crystallography, volume A, space-group symmetry. Kluwer Academic Publishers, Dordrecht, Holland, 878 pp.

Hawthorne, F.C. and Grundy, H.D. (1976) The crystal chemistry of the amphiboles: IV. X-ray and neutron refinements of the crystal structure of tremolite. Canadian Mineralogist, 14, 334–345.

Hazen R.M. and Finger L.W. (1989) High-pressure crystal chemistry of andradite and pyrope: Revised procedures for high-pressure diffraction experiments. American Mineralogist, 74, 352–359.

Hollocher, K. (1997) Building crystal structure ball models using pre-drilled templates: Sheet silicates, tridymite, and cristobalite. In Teaching Mineralogy, editors J.B. Brady, D.W. Mogk, and D. Perkins III, 255–282.

Hugh-Jones, D.A. and Angel, R.J. (1994) A compressional study of $MgSiO_3$ orthoenstatite up to 8.5 GPa. American Mineralogist, 79, 405–410.

ICSD (2004) The inorganic crystal structure data base, published by The National Institute of Standards and Technology.

Johnson, C.K. (1965) ORTEP: A FORTRAN thermal-ellipsoid plot program for crystal structure illustrations. ONRL Report 3794, Oak Ridge, Tennessee, Oak Ridge National Laboratory.

Kihara, K. (1990) An X-ray study of the temperature dependence of the quartz structure. European Journal of Mineralogy, 2, 63–77.

Mogk, D. W. (1997) Directed-discovery of crystal structures using ball and stick models. In Teaching Mineralogy, editors J.B. Brady, D.W. Mogk, and D. Perkins III, 283–290.

Winter, J.K. and Ghose, S. (1979) Thermal expansion and high-temperature crystal chemistry of the Al_2SiO_5 polymorphs. American Mineralogist, 64, 573–586.

Chapter 15

Diffraction

Like many students, I have found it somewhat easy to misidentify minerals by just looking at them. Shortly (unfortunately) after completing my undergraduate mineralogy course, I discovered the X-ray lab in our geology department. There were several machines that supposedly had the ability to correctly identify minerals. This naturally interested me, because it offered the hope that for once and for all, I could now identify minerals without guessing! However, back in those days it was rare for undergraduates to use these machines; they were not like the computer-driven X-ray machines of today. Not only did they look complicated with many knobs, but if the knobs were set incorrectly, bad things would happen (for instance over the years I've seen smoke, fires, leaky radiation etc. from these 1970s vintage machines). The instruments of today, as discussed in this chapter, can easily be used by students with only a few minutes of training. Not only can you use them to identify minerals, but X-rays can also determine mineral structures.

X-ray diffraction has played a huge part in the advancement of science in the 20th century. In my mineralogy and introductory geology classes and in casual conversations, I often argue that the most significant scientific discovery of the 1900s was the structure of DNA, which was described using X-ray diffraction. The fact that the discovery took place in 1953, the year I was born, is especially interesting to me. The structure of DNA won a Nobel prize for the discoverers. X-ray diffraction continues to be an integral part of many science disciplines, where it remains the "gold standard" for describing crystalline structures.

Technology has advanced rapidly. From 1950–1980, crystal structure refinements were extremely difficult. Many a Ph.D. student earned a degree on the basis of the XRD structure determination of a single mineral. Advances in computing speed and automated X-ray instrumentation now mean that crystal structure determinations are relatively straightforward. Now even an undergraduate thesis may need to determine several structures!

So unlike when I was an undergraduate, my students don't have to wait until after their mineralogy class is over to see our X-ray lab. I take the students there during the first lab section and show them how to use the instruments to identify minerals. Later in the semester, many of the students will use the instrument for projects that vary from figuring out what's in clumping cat litter (and how it changes when it gets "wet"), to determining the difference between elk antlers and cow horns—after all, I do live in Idaho! So the information in this chapter will give you the keys to making positive mineral identifications—even if you can't identify them in hand samples!

M.E.G.

Introduction

It seems appropriate to begin by defining **diffraction**. The simple definition of diffraction is bending. This might be a bit confusing because refraction also means bending. However, the two terms refer to different phenomena. In *refraction*, the bending occurs when a ray travels between two materials at different speeds and the ray changes direction at the interface (remember Snell's Law from Chapter 5); the change in velocity causes the bending. For *diffraction*, the bending occurs because the wavelength of the electromagnetic wave, be it light or X-rays, is similar to that of the spacing of a periodic array of the material, causing the wave to change direction even though the speed does not change. That periodic array can be anything as simple as scratches on a window of a car, or a screen wire, or as complex as the arrangement of atoms in a mineral.

Unlike some more abstract mineralogical concepts, diffraction is often observed in nature. You have probably observed diffraction on a daily basis, and not even realized it. I hope that after reading this chapter, you'll see it everywhere! In this chapter, I use the approach of first explaining diffraction in the phenomenological sense and then showing you the many places to see diffraction in everyday life—especially after dark. After this explanation, we will then turn to diffraction in crystals, and then to an explanation of diffraction from a mathematical standpoint. Of course, entire books and graduate classes (I have had two of them!) are dedicated to this subject, so what is presented here is meant to be only an introduction.

There are two main uses for diffraction in the study of minerals. One will be for mineral identification, because every mineral species, by definition, has its own distinct crystal structure (i.e., periodic arrangement of atoms). The other use of diffraction is to determine the crystal structure of a material. To achieve either of these, we will need to interpret a diffraction pattern and relate it to an array of atoms that will produce such a pattern.

In this chapter, we will briefly touch on three different types of diffraction: (1) **light diffraction**, (2) **X-ray diffraction**, and (3) **electron diffraction**. Light diffraction is by far the simplest to observe because it can be seen with the eye. X-ray diffraction will require generation of X-rays that will interact with a sample to produce a diffraction pattern. These must be "seen" either on X-ray film (like your dentist uses) or with some electronic device similar to the CCD array in your digital camera. These allow you to make a "picture" using X-rays instead of visible light. To observe electron diffraction, we typically use a transmission electron microscope (TEM). In this instrument, a high-energy beam of electrons shines through a thin sample and is diffracted as it passes. Electron and X-ray diffraction studies may be done either on single crystals or powders. **Single-crystal diffraction** occurs, as the name implies, when the energy is incident on a single crystal in a specific orientation. In **powder diffraction**, the incident energy interacts with many, perhaps hundreds of thousands, of small crystals that theoretically are in all possible orientations. In all cases, diffraction occurs when the wavelength of the incident energy is similar to some physical spacing of the material. As we will shortly see, the wavelength must be smaller than the spacing. Observable diffraction will occur even when the wavelength is 20 to 30 times smaller than the physical spacing.

Figures 15.1 and 15.2 show examples of both single-crystal and powder diffraction for halite (Figure 15.1) and tremolite (Figure 15.2). Figure 15.1a–c shows insets of the crystal structure of halite projected on the (001), (111), and (110) planes. Associated with each of these projections are the diffraction patterns that would result from an X-ray or electron diffraction study. Notice how the diffraction patterns appear very similar to the crystal structures. These are rudimentary examples of how the crystal structures can be determined from the diffraction patterns.

The symmetry of each of the diffraction patterns relates directly to the crystal structure. In Figure 15.1a, you are looking down a 4-fold axis, so the diffraction pattern exhibits a four-fold symmetry. In Figure 15.1b, you are looking down a 3-fold axis, and the diffraction pattern shows 3-fold symmetry. In Figure 15.1c, you are looking down a 2-fold axis, and thus the diffraction pattern shows 2-fold symmetry. There are many mirror planes in both the structures and the associated diffraction patterns. Each simple mineral produces several different diffraction patterns as a function of its orientation. Figure 15.1d is a calculated powder diffraction pattern for halite. Each peak represents diffraction that occurs in a specific plane in the mineral. The peak positions are plotted as $1/d$ (where d is the spacing between planes of atoms) and the units are $Å^{-1}$. Plotting in these strange units gives us some indication of what is to come, in that there is an inverse relationship between distance in the crystal structure and the diffraction pattern. We will see that powder diffraction is the typical method that we use to identify minerals, while single-crystal diffraction is commonly used to determine crystal structures of minerals.

Figure 15.2 is a similar set of calculated diffraction patterns for the more complicated mineral tremolite, and is meant to be compared with the

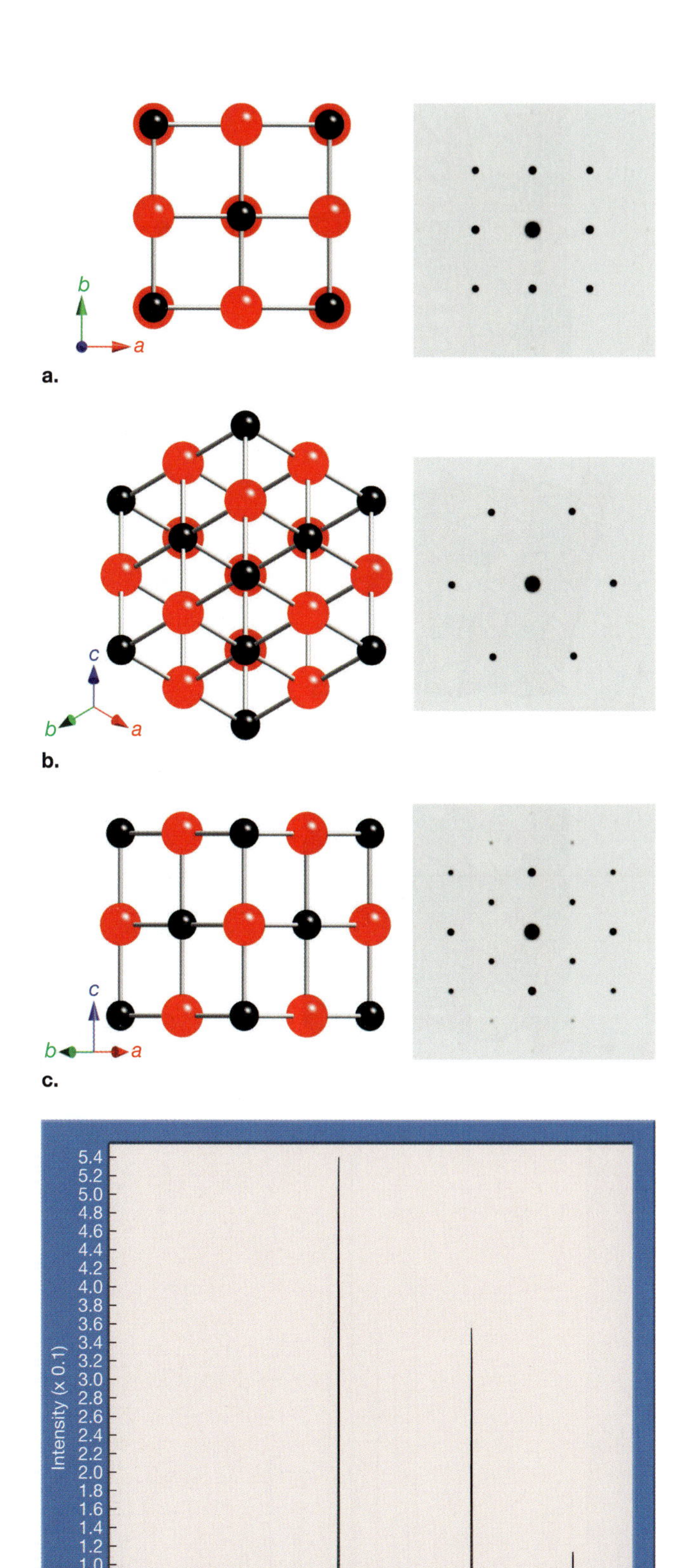

Figure 15.1. *Single-crystal (a–c) diffraction patterns (right) and corresponding sample orientations (left) and powder diffraction pattern (d) for halite. Notice how the diffraction patterns (a–c) change as a function of sample orientation. These diffraction patterns and those shown throughout this chapter (unless noted) are calculated using SingleCrystal (for single-crystal diffraction patterns) and CrystalDiffract (for powder diffraction patterns.)* ***a.*** *Halite projected down [001]4. Note the relationship between the symmetry (4-fold and mirror planes) in the structure and its diffraction pattern.* ***b.*** *Halite projected down [111]3. Again there is a relationship between the symmetry (3-fold and mirror planes) in the structure and the diffraction pattern. Although the image appears to be rotated by 30°, this is actually just the result of diffraction, as explained later in this chapter.* ***c.*** *Halite projected down [110]2. The symmetry (2-fold and mirror planes) relates to the structure and its diffraction pattern. Some diffraction spots (small black dots) are absent on the horizontal and vertical directions.* ***d.*** *A calculated powder diffraction pattern for halite. This pattern represents the summation of all possible single-crystal orientations. Notice the horizontal axis is in 1/*d *(where* d *represents the* d*-spacing in the mineral), giving us a hint that diffraction is somehow relating* d*-spacing in the mineral to 1/*d *spacing in the diffraction pattern (i.e., there is an inverse relationship).*

less complicated structure of halite (Figure 15.1). For tremolite (as for halite) there are three different sets of diffraction patterns associated with three sets of projections of the structure. However, comparing tremolite with halite you quickly see that tremolite is much more complicated (i.e., there are many more spots, and the spots have different intensities at different orientations). In Figure 15.2b, it appears as if the structure and the diffraction pattern have slightly different orientations. This is not the case, and the explanation of this apparent anomaly will be discussed later in the chapter. Figure 15.2d is a calculated (theoretical) powder diffraction pattern for tremolite. It is significantly more complicated (i.e., it has a lot more peaks) than halite. Each of these peaks represents the spacing of different planes in the minerals. Tremolite has many more planes than halite.

Finally (by way of introduction), let's compare halite (NaCl) with its close relative, sylvite (KCl). Both of these minerals have the halite structure, but they contain different cations. K^{1+} is larger than Na^{1+}. As we saw in Figures 15.1d, and 15.2d, these patterns are plotted as a function of $1/d$. Thus, these two diffraction patterns should look similar except for the spacing of the peaks. Figure 15.3 shows single-crystal and powder diffraction patterns of sylvite (a) compared with halite (b). As predicted, the diffraction pattern of sylvite is shifted to lower $1/d$ values (i.e., greater values of d—the planes of atoms are farther apart!). Notice also that the heights of the peaks change between sylvite and halite. *The intensities of the peaks are telling us something about which atoms are present in the structures, while the spacings of the peaks relate to the sizes*

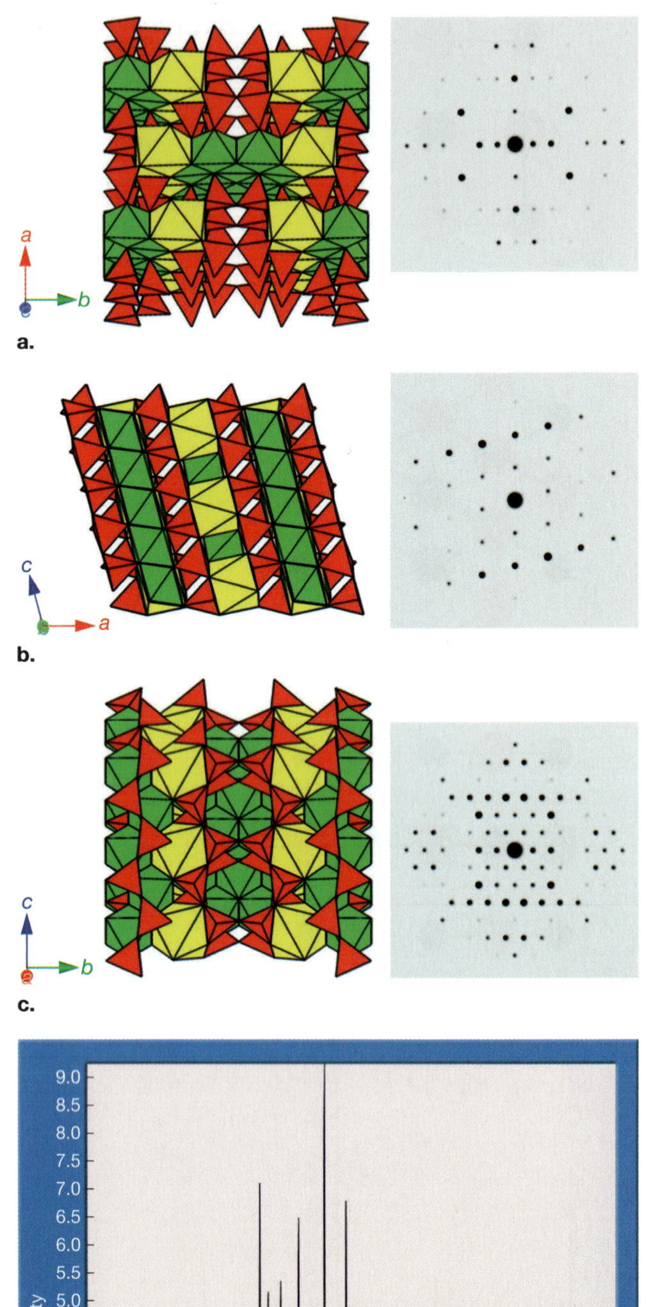

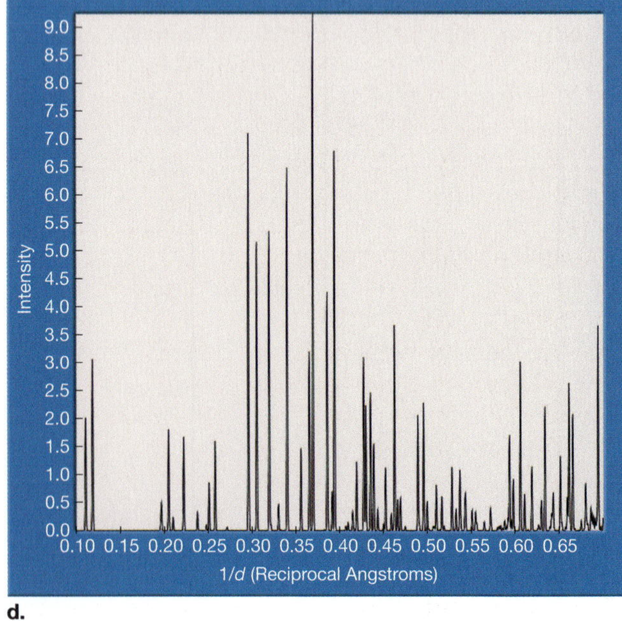

Figure 15.2. *Single-crystal (a–c) diffraction patterns (right) and corresponding sample orientations (left) and powder diffraction pattern (d) for tremolite. Compare this figure to that of halite (Figure 15.1). The more complicated tremolite structure produces more complex diffraction patterns. Note that tremolite is monoclinic (2/*m*) while halite is isometric.* ***a.*** *Tremolite projected onto its (100) plane. Because tremolite is monoclinic,* a *is not perpendicular to the page. The symmetry (i.e.,* $^{[010]}2$ *and* $^{[010]}$m*) is revealed in the diffraction pattern.* ***b.*** *Tremolite projected onto its (010) plane. In this case we are looking down* $^{[010]}2$*, revealing the 2-fold symmetry in the plane. However, notice that the structure and the diffraction pattern appear to be rotated with respect to each other—this is an effect of diffraction and will be discussed later in this chapter (Figures 15.7–15.9).* ***c.*** *Tremolite projected onto its (001) plane. This orientation is similar to that in Figure 15.2a above and shows the* $^{[010]}2$ *and* $^{[010]}$m *symmetry, as well as the* c *axis being tilted slightly out of the plane of the page.* ***d.*** *The powder diffraction pattern for tremolite. Relative to the pattern of halite (Figure 15.1d), this one has many, many more peaks because tremolite's structure is much more complex, and has lower symmetry, than halite. Many more peaks appear at lower 1/*d *values, which correspond to the longer repeats (e.g., cell edges) in tremolite.*

of the structures. You could use these patterns to calculate how much bigger the ionic radius of K^{1+} is than Na^{1+}. In fact, Brady and Boardman (1995) do just that, along with giving other nice teaching examples of the use of powder X-ray diffraction.

These ideas are summarized in Table 15.1, which gives equations relating the *d*-spacings in minerals to their *hkl* values and cell parameters. Notice how the isometric minerals (e.g., halite and sylvite) have simple formulae. But as symmetry decreases, the formulae get much more complicated. This "complication" with decreasing symmetry will be a theme throughout this chapter.

Light Diffraction

As previously stated, light diffraction is all around us and can be observed in many places! Diffraction occurs when light is bent by a periodic pattern. In mineralogy, we are concerned with the periodic patterns made by atoms in minerals. But first, let's look at some other periodic patterns that occur in nature so we can later relate them to the patterns we cannot see with our own eyes.

Figure 15.4a shows rows of small Christmas trees photographed from different angles. In the top image, we are looking parallel to the rows. The next two images show other "rows" that occur from looking at different directions through the trees. These still images do not do this analogy justice, so take a look on the MSA website of the movie of this. As you drive along and look at a row of trees, or corn, or anything in rows, they align themselves in different directions. If you drew lines to connect the trees in different orientations, you would arrive at different parallel planes passing through the trees (Figure 15.4b). In the figure, each series of planes is labeled with three numbers. These should remind you of the Miller indices we learned about in Chapters 12 and 13. Noticed how the spacing between the planes decreases as you go from (110) to (210) to (310) to (410) to (510). If this was a mineral structure (instead of a field of trees!), X-rays would be diffracted off these planes, resulting in the hypothetical unit cell shown in Figure 15.4c.

We can relate these spacings between planes and their hypothetical X-ray diffraction patterns to periodic arrays and light diffraction. It is easy to observe diffraction with point sources of lights (i.e., car headlights or streetlights) shining through a screen window after dark. In Figure 15.5a, the car on the right is viewed through a window screen, while there is no screen on the left side. Figure 15.5b is a photograph taken after dark showing how the car lights on the right are diffracted through the screen compared to the car on the left. This diffraction effect is somewhat subtle and really appears as a smearing of the light away from the headlights. In Figure 15.5c, the screen is moved so it now covers the headlights of both cars, which now show the square diffraction pattern. In Figure 15.5d the screen is removed and neither car's headlights exhibits diffraction. Finally, Figure 15.5e shows the view from the house with the screen completely removed. Although you can't see the wires in the screen in these images, the spreading of light in the vertical direction relates to the screen wire in the horizontal direction, because the diffraction is perpendicular to the periodic array that caused it.

Another very common diffraction pattern occurs in the rain when light shines through a car windshield while the wipers are in action. Unlike the two-dimensional diffraction pattern shown in Figure 15.5, this type of diffraction pattern is only one-dimensional. The next time you drive after dark in the rain with your windshield wipers on, you will notice this diffraction pattern, so drive carefully! Figure 15.6 and the associated movie on the MSA website show this phenomenon. In Figure 15.6a, I have taken a photograph of the light on the outside of my garage while sitting *inside* my car in daylight. In Figure 15.6b, it is now dark and the garage light is turned on, but it is not raining. In Figure 15.6c and d, the light looks streaky, and the effect is different in the two separate images. Why? The streaking occurs perpendicular to the directions of the lines of water made by the windshield wiper. In Figure 15.6c the left hand windshield wiper has just passed between the camera and light, while in

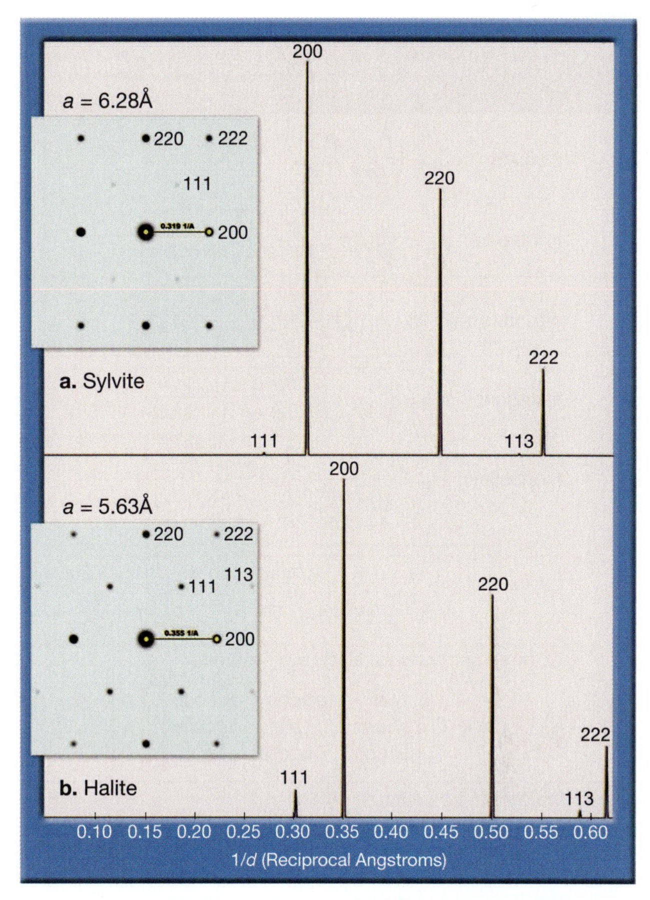

Figure 15.3. *Diffraction patterns for (a) sylvite (KCl) and (b) halite (NaCl). The single-crystal patterns result from diffraction looking down [110]2. The diffraction patterns are "indexed" with their associated* hkl *values (i.e., the lattice planes causing direction are added to the patterns). These two minerals share the same structure, and differ only in the length of the* a *unit cell repeat, which is larger for sylvite because the ionic radius for K^{1+} is greater than that for Na^{1+}. The larger spacing for sylvite shows up in the powder diffraction patterns by shifting the peaks to the left (i.e., smaller 1/*d *values, which correspond to larger* d *values). A similar phenomenon occurs for the single-crystal patterns in that the diffraction spots are closer together for sylvite, indicating that the atoms are farther apart. Notice the lack of a (100) diffraction spot or peak. The (100) peak "should" occur halfway between the center of the single-crystal diffraction pattern and the (200) spot, but it is "absent;" these "absent" spots will help further refine the symmetry elements of a mineral. Also, notice that the intensities of the peaks differ between the two patterns; thus the intensities must be related to the atoms. As is seen in these patterns of two simple compounds, interpretations of diffraction patterns tell us: (1) symmetry (based on arrangements of diffraction spots), (2) atomic spacings (based on distances of diffraction in 1/*d *space, and (3) atomic differences (based on the intensity of the diffraction peaks and/or spots).*

Figure 15.6d, the effect of the right hand windshield wiper is shown. These effects show up best if you look though the area of windshield where

Table 15.1.
Formula, by crystal system, to calculate d_{hkl} values given unit cell parameters and $(h\ k\ l)$ followed by a matrix method (discussed in Chapter 13) that is generalized to all crystal systems and easily programmed.

isometric: $$d_{hkl} = \sqrt{\frac{a^2}{h^2+k^2+l^2}}$$

tetragonal: $$d_{hkl} = \sqrt{\frac{a^2b^2c^2}{h^2b^2c^2+k^2a^2c^2+l^2a^2b^2}}$$

orthorhombic: $$d_{hkl} = \sqrt{\frac{a^2b^2c^2}{h^2b^2c^2+k^2a^2c^2+l^2a^2b^2}}$$

hexagonal: $$d_{hkl} = \sqrt{\frac{3a^2b^2c^2}{4c^2(h^2+hk+k^2)+3l^2ab}}$$

monoclinic: $$d_{hkl} = \sqrt{\frac{b^2\sin^2\beta}{b^2\left(\frac{h^2}{a^2}+\frac{l^2}{c^2}-\frac{2hl\cos\beta}{ac}\right)+k^2\sin^2\beta}}$$

triclinic: $$d_{hkl} = \sqrt{\frac{1-\cos^2\alpha-\cos^2\beta-\cos^2\gamma+2\cos\alpha\cos\beta\cos\gamma}{\frac{h^2}{a^2}\sin^2\alpha+\frac{k^2}{b^2}\sin^2\beta+\frac{l^2}{c^2}\sin^2\gamma+\frac{2hk}{ab}(\cos\alpha\cos\beta-\cos\gamma)+\frac{2hl}{ac}(\cos\alpha\cos\gamma-\cos\beta)+\frac{2kl}{bc}(\cos\beta\cos\gamma-\cos\alpha)}}$$

or, more generally for all crystal systems:

$$\frac{1}{d^2_{hkl}} = [h\ k\ l]\begin{bmatrix} a^2 & ab\cos\gamma & ac\cos\beta \\ ab\cos\gamma & b^2 & bc\cos\alpha \\ ac\cos\beta & bc\cos\alpha & c^2 \end{bmatrix}^{-1}\begin{bmatrix} h \\ k \\ l \end{bmatrix}$$

which can be rewritten as:

$$d_{hkl} = \sqrt{\frac{1}{[h\ k\ l]\begin{bmatrix} a^2 & ab\cos\gamma & ac\cos\beta \\ ab\cos\gamma & b^2 & bc\cos\alpha \\ ac\cos\beta & bc\cos\alpha & c^2 \end{bmatrix}^{-1}\begin{bmatrix} h \\ k \\ l \end{bmatrix}}}$$

the two windshield wipers overlap. You can even see them in daylight if you have an older car and live in an area high in dust. Over time, the quartz particles in the dust scratch the windshield as the wipers move forward and backwards. So even in the daylight you may be able to see the streaking effect as sunlight is being diffracted through the scratches in the windshield.

If you don't own a car, you can simulate this effect by dipping your finger in water and smearing it on glass. Next look at a point light source through the glass. Notice how the diffracted light pattern is perpendicular to the direction you wiped the glass. Look around you and you'll see this effect in many, many places: polished metal, scratched glass, etc.

Now we move on to diffraction in different sizes of screen wire. To do this, we are using something you may have seen in your sedimentary class—sieves. Sieves come in various sizes (the spacings between the wires are different) so that sediments can be separated into different size fractions. A diffraction pattern can be created by shining a laser through theses sieves and projecting these patterns on a distance wall. Figure 15.7a shows such an experimental setup where a green laser shines through a sieve located a few inches away from it and projects a diffraction pattern on a screen ten feet away. Figure 15.7b–d show several diffraction patterns and the sieves that produced them. Notice how the square pattern of the sieve produces a square diffraction pattern, which is similar to the diffraction pattern of halite in Figure 15.1a. In Figure 15.7c, a smaller sieve size is used. We might predict that the diffraction pattern from a smaller sieve would have more closely spaced diffraction spots than a lager sieve. However, observation of the resulting diffraction pattern in Figure 15.7c shows the opposite to be true (i.e., the closer the spacing of the wires, the farther apart the dots are in the diffraction pattern). Also, diffraction occurs perpendicular to the repeating array that forms it (recall the windshield wipers). In Figure 15.7d, the screen in Figure 15.7b has been sheared so the wires are no longer perpendicular. Now the horizontal wires of the sieve produce the vertical lines of diffraction patterns, and inclined diffraction patterns are produced perpendicular to the inclined wires in the sieve. Finally, Figure 15.7e shows two superimposed diffraction patterns resulting from the use of two separate lasers: a red laser with wavelength of 632 nm, as used in the previous figures, and a green laser with wavelength of 532 nm (stop reading now and see if you can figure out why they differ). The red laser, with longer wavelength, is diffracted at greater angles than the green laser. Thus, diffraction is not only related to the spacing of the wires, but is also related to the wavelength of light that is used.

There is one last light diffraction experiment you can do yourself. In Figure 15.8, a series of diffraction gratings has been produced by making closely-spaced lines with a graphical computer program and then printing them on transparencies using a high-resolution laser printer. Johnson (2001) discusses a similar demonstration, but he uses dots instead of lines. In Figure 15.8a, a series of closely-spaced horizontal lines results in a vertical diffraction pattern. In Figure 15.8b, the spacing of the horizontal lines has been increased, thus causing the diffraction spots' spacing to decrease. In Figure 15.8c, the grating used in Figure 15.8b is duplicated and the second grating (with the same spacing) is placed perpendicular to the first. This results in a square diffraction pattern similar to that produced by halite in Figure 15.1a. Figure

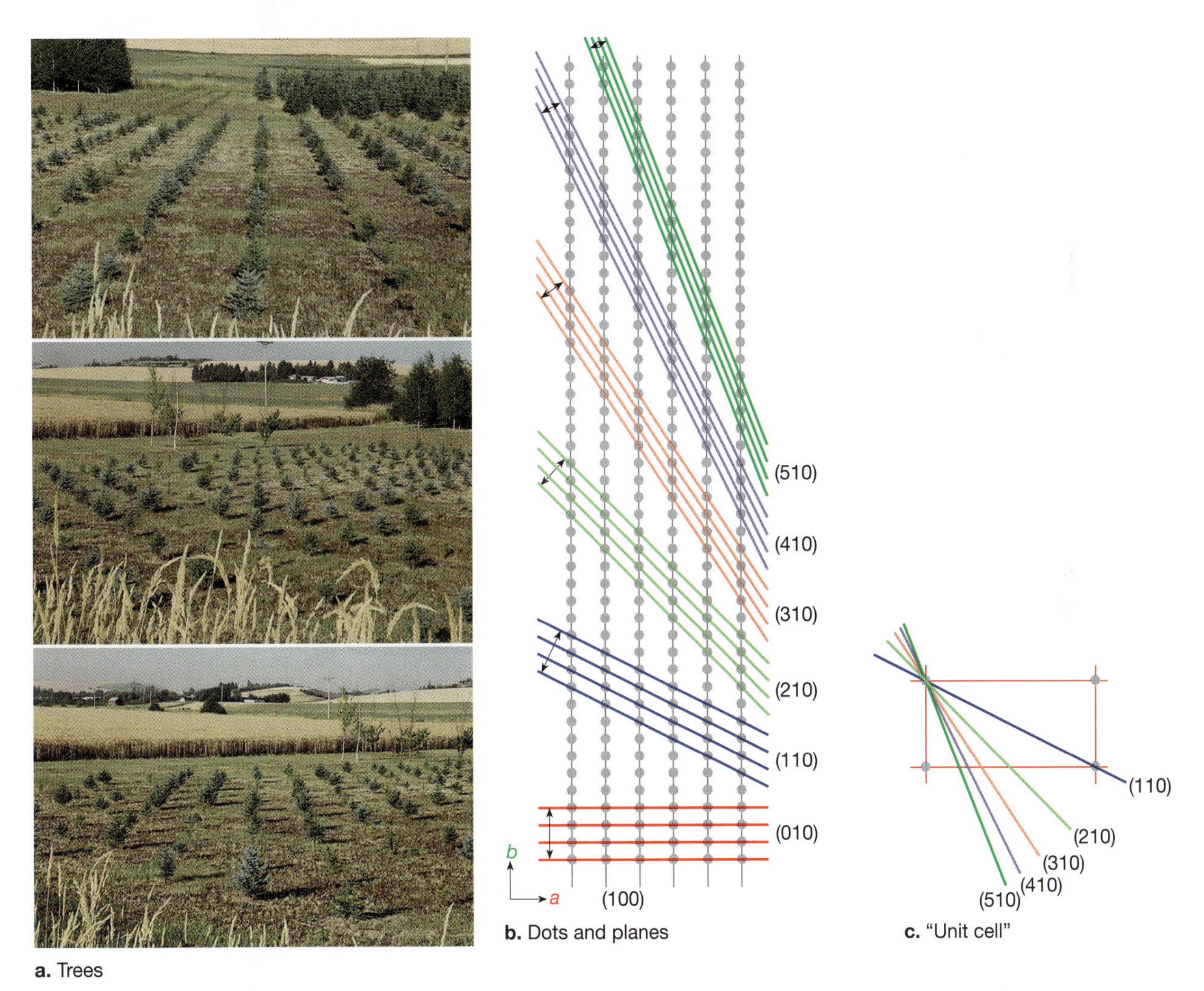

Figure 15.4. Different spacings in rows of planted trees (a) and their correspondence to different spacings in rows (b) of atoms in minerals with an enlarged view (c) of the "unit cell." ***a.*** *The upper image is a view down rows of planted trees and the next two photos show other "rows" that occur as we view the trees from different directions (it might help to notice the relationship between the shadows the trees cast and the trees). You will see this as you drive down the road and look at rows of planted trees, corn, or anything lined up in equally spaced rows. (Check out the movie showing this on the MSA website.)* ***b.*** *An overhead view of a series of rows (vertical) of plants that would correspond to atoms in 2-D. The repeat along* a *(horizontal), would represent the spacing of the "rows" and* b *(vertical) would represent the spacing on the plants within the rows. Next, a series of planes ((110), (210), etc.) has been added. Notice how the spacing changes between these planes (i.e., it decreases as the numbers increase).* ***c.*** *An enlarged view of the single unit cell with the (110), (210), etc. planes placed inside the cell. The distance from the origin of the cell to the various planes decreases as the numbers increase.*

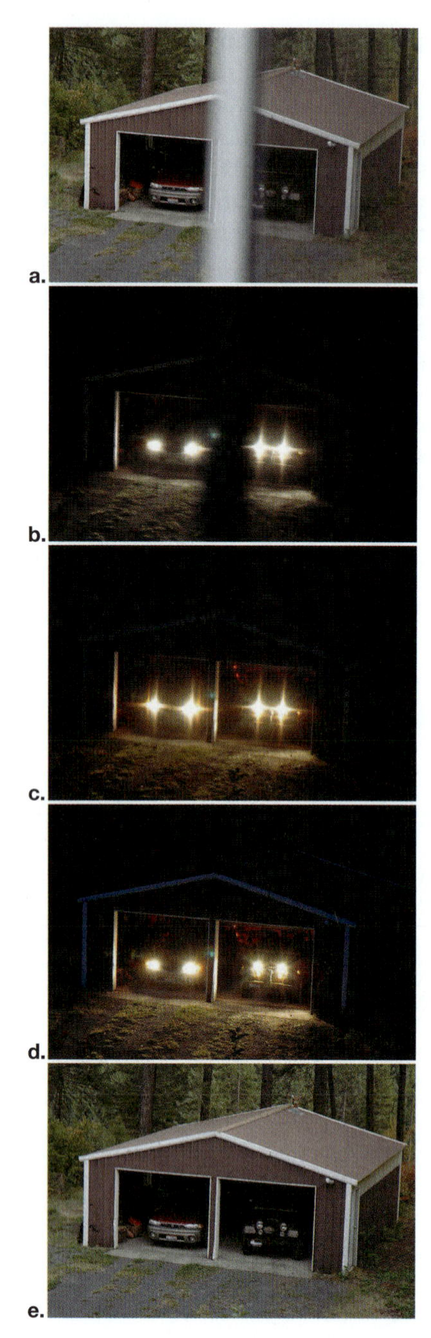

*Figure 15.5. A series of photos showing diffraction of car headlights by a screen window taken from the front porch of my house looking at cars in my garage. **a.** A screen window is placed on the right side in front of the camera lens. The camera is positioned so that the screen is in front of the car on the right, but not the one on the left. Now we wait for dark! **b.** After dark, the headlights of both cars are turned on and diffraction is seen for the headlights on the right, through the screen. **c.** In this photo, the screen is moved to cover both sets of headlights. **d.** Now the screen is removed and no diffraction occurs. **e.** Daylight with no screen.*

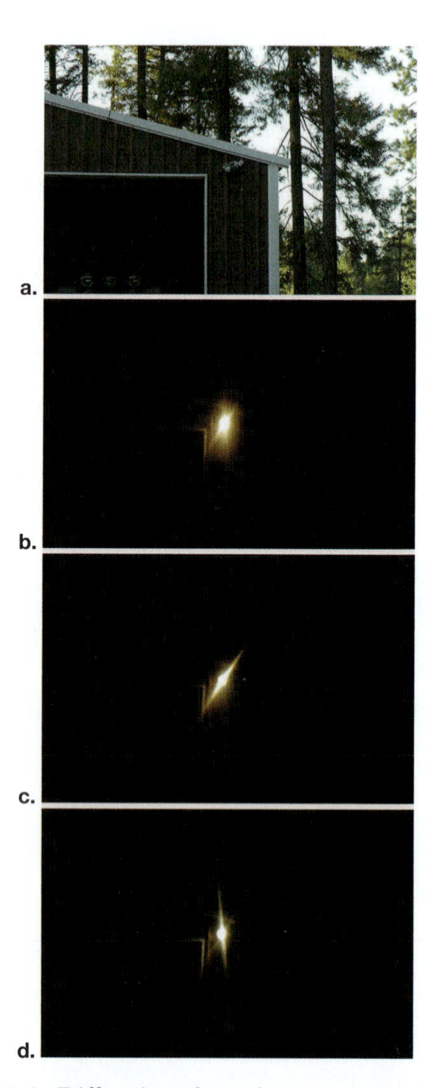

*Figure 15.6. Diffraction through a car windshield after dark, in the rain. **a.** This photograph shows the outdoor light on my garage (same one as in Figure 15.5) at home viewed through my car windshield during daylight. The garage light is centered on the windshield in the area where the cars windshield wipers overlap. **b.** Same image as in Figure 15.6a, except now it is dark and the light is turned on. **c.** Now the windshield wipers are turned on and rain is simulated. Notice the diffraction in one direction—the diffraction streak is perpendicular to the water streaks created by the wiper. **d.** This is the same perspective as in Figure 15.6c, except the other windshield wiper (the one on the passenger's side) has just moved over the light, causing the orientation of the diffraction pattern to change. Now it is perpendicular to the second wiper. (Check out the movies of this on the MSA website—they add the motion.) The next time you are riding (not driving, this is too cool of a distraction!) in a car after dark in the rain, you can clearly see this effect.*

15.8d simulates the orthorhombic diffraction pattern that results when the gratings in Figure 15.8a and b are placed perpendicular to each other. Figure 15.8e uses those same two gratings, but the one that was vertical in Figure 15.8c is now rotated counterclockwise. This counterclockwise rotation causes the horizontal diffraction in Figure 15.8c to be rotated, in a similar manner as the sheared screen in Figure 15.7d.

Reciprocal Lattices and *d*-spacings

All the preceding demonstrations have shown that there is a relationship between the crystal structures of the materials (i.e., the periodic arrangement of the atoms in minerals) and their diffraction patterns. It is very important to understand that shorter distances within a periodic array appear as longer distances in their diffraction patterns. Obtuse angles in the periodic arrays appear as acute angles in the diffraction patterns, while 90° angles in the periodic array appear as 90° angles in the diffraction patterns. These conclusions demonstrate the relationship between the direct lattice and the reciprocal lattice. The **direct lattice** is the crystal structure of the material, while the **reciprocal lattice** is the name given to what is produced in the diffraction pattern. The term reciprocal derives from the fact that there is an inverse relationship between the distances in the crystal structure and the diffraction pattern. We have now seen the phenomenon demonstrated using light waves, and this concept was also explained mathematically in Chapter 13.

Figure 15.9a–d shows a series of lattices for isometric, orthorhombic, hexagonal, and monoclinic crystal systems. To represent the direct lattice, we used the familiar cell parameters (a, b, c and α, β, γ) and for the reciprocal lattices, we use the same letters but with a "star" on each one. The result is the reciprocal lattices parameters (a^*, b^*, c^* and α^*, β^*, γ^*). You might be able to see that there is a geometric relationship between a, b, c and a^*, b^*, c^*. Notice how a^* is perpendicular to the plane defined by b and c; b^* is perpendicular to the plane of a-c; and the c^* is perpendicular to the plane defined by a-b. Likewise a is perpendicular to the plane defined by b^*-c^*; b is perpendicular to the plane defined by a^*-c^* and c is perpendicular to the plane defined by a^*-b^*.

So, for the case of an isometric lattice a, b, and c coincide with a^*, b^*, c^* (Figure 15.9a), but the distances would still be inverted between direct and reciprocal space. This relationship was shown for the diffraction patterns of sylvite and halite in Figure 15.3. A similar situation of coincidence of the *directions* of the direct and reciprocal axes occurs for the orthorhombic system (Figure 15.9b), though now there is an inverse relationship

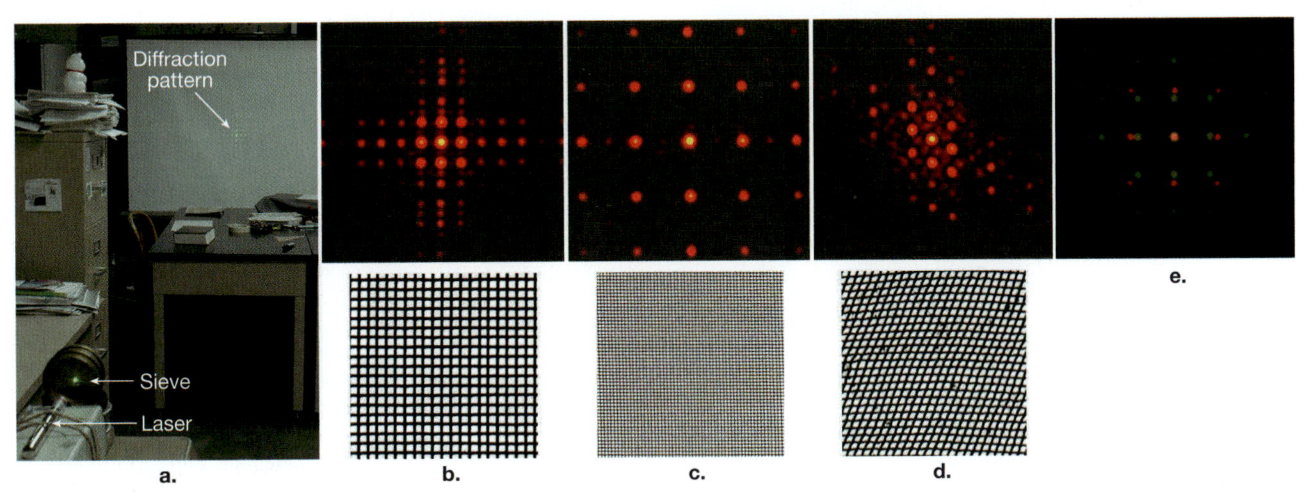

Figure 15.7. A series of photos showing diffraction patterns created by shining a laser through sieves. (Two sieves are used: a 100 mesh (the one with the greater spacing) and a 300 mesh. (the mesh size refers to the number of holes in the sieve per inch.) ***a.*** *The experimental setup shows the laser shining through a sieve with the resultant image projected from about ten feet away on a white screen.* ***b.*** *The upper image shows the diffraction pattern created from the sieve pictured above.* ***c.*** *Now the diffraction spots are farther apart than in Figure 15.7b and surprisingly, the sieves holes are closer together! Recall the inverse relationship between atom spacing and diffractions back in Figures 15.1–15.3.* ***d.*** *In this pattern, the 100 mesh sieve has been distorted to represent a monoclinic lattice. At first glance it would appear that the sieve and pattern have been rotated (just like in Figures 15.1c and 15.2b), but it is the horizontal wires that produce the vertical set of diffraction spots (just like the windshield wipers in Figure 15.6). Notice how the wires that are inclined from vertical produce diffraction spots that are perpendicular to them. The perpendicular relationships of planes to diffraction effects are seen in materials that do not have perpendicular axis sets.* ***e.*** *This is a repeat of the 300 mesh pattern, except now two separate lasers are used to form the pattern—a red laser (λ = 632 nm) and a green laser (λ = 532 nm). The red laser with a longer wavelength produces more widely-spaced diffraction spots.*

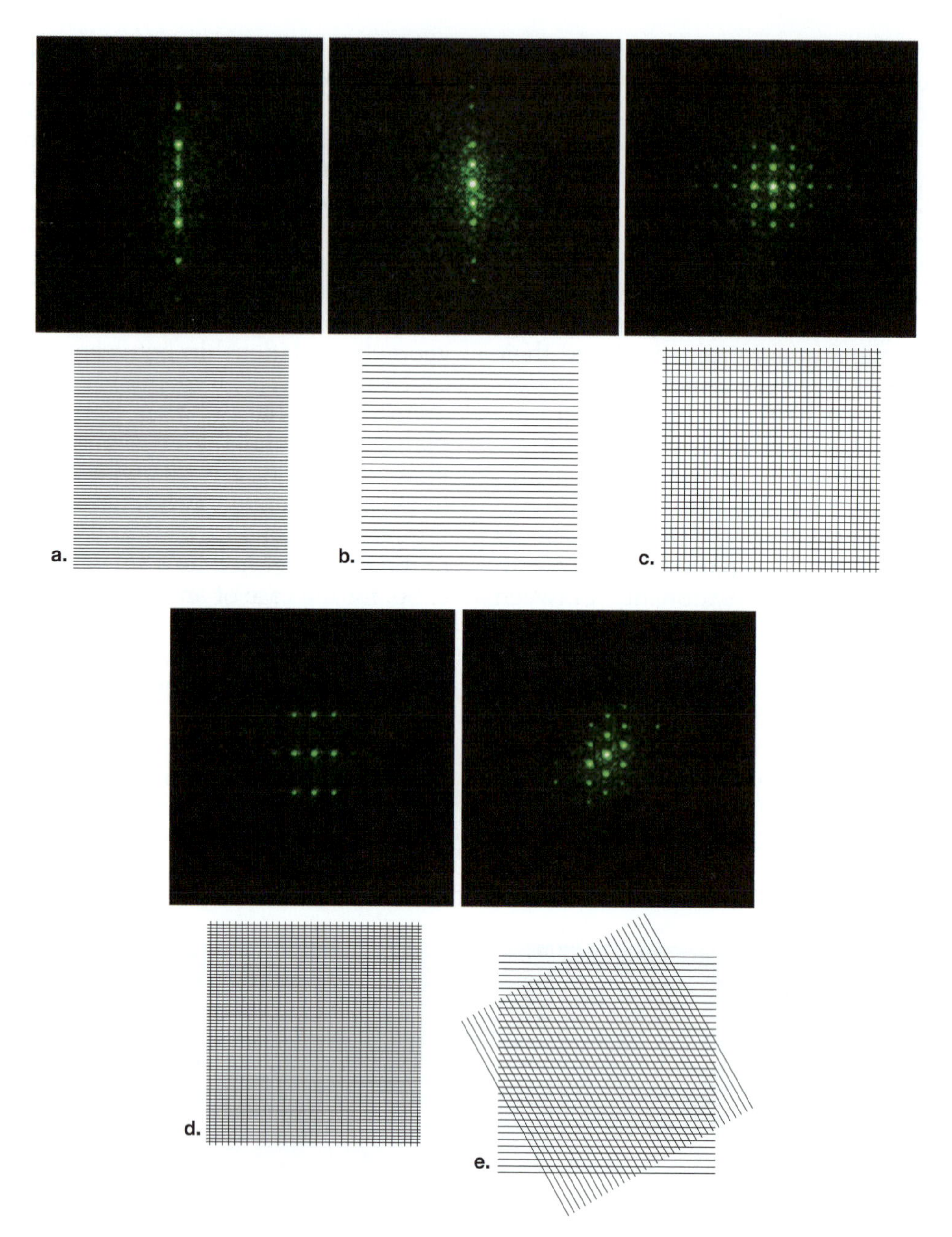

Figure 15.8. A series of diffraction patterns (upper image of each layer), and associated computer-generated gratings (lower image) printed on clear transparencies with a high-resolution laser printer. ***a.*** *A one-dimensional diffraction pattern is formed perpendicular to a closely-spaced set of lines. This setup simulates the windshield wiper effect in Figure 15.6.* ***b.*** *Another one-dimensional diffraction pattern produced at right angles to a closely-spaced set of parallel lines. In this case the lines are farther apart than in Figure 15.8a and thus produce diffraction spots that are closer together.* ***c.*** *A two-dimensional diffraction pattern. To make this grating, the pattern from Figure 15.8b was used with a duplicated copy rotated 90° and placed over it. This simulates the car headlights in Figure 15.5 and the sieves in Figure 15.7a and b.* ***d.*** *A two-dimensional diffraction pattern is produced by taking the grating from Figure 15.8a and placing the one from Figure 15.8b, after a 90° rotation, on top of it. This represents the diffraction pattern from an orthorhombic lattice. The longer repeat in the grating results in closer spacing of points in the diffraction pattern. Again, an inverse relationship exists between the distance between the lines in the grating and the spacings between points in the diffraction pattern.* ***e.*** *The two-dimensional diffraction pattern produced by placing two gratings from Figure 15.8b on top of each other at a non-90° angle. This pattern simulates the distorted sieve shown in Figure 15.7d, except the inclined lines are tilted to the left of vertical, thus producing a different orientation in the rows of diffracted spots.*

between *lengths* of a and b vs. a^* and b^*. Figure 15.9c shows a hexagonal lattice and the associated γ angle of a 120° between a and b. The associated reciprocal lattice again shows the aforementioned relationship of the directions between the direct and reciprocal lattice. Notice how the row of atoms that defines the *a* direction is perpendicular to the b^* direction, while the row of atoms defining the *b* direction is perpendicular to the a^* direction. This is the same phenomenon that we saw in Figure 15.2b in tremolite, which made the structure of tremolite appear rotated from the diffraction pattern. Also notice that in the hexagonal lattice, γ^* becomes the supplement (i.e., they total to 180°) of γ. Finally, for the monoclinic lattices (Figure 15.9d), the lengths of *a* and *b* are shown inverted as a^* and b^*, while the obtuse γ angle becomes an acute angle denoted as γ^* that is the supplement of γ.

"Reflection" of X-rays

We can now begin to explain what causes some of the diffraction effects that we have been observing. To do this, we will relate light reflection from a material's surface to X-ray diffraction by the material. However, while light reflects off of a surface at all angles, X-rays will only "reflect" from the structure of the mineral at certain angles, where these angles relate to the wavelength of the X-ray and the spacings of the planes in the mineral. As an example, refer to Figure 15.10 and the associated movies on the MSA website.

The setup of the experiment shown in Figure 15.10 places a sample of halite in the center of a powder X-ray diffractometer (Figure 15.10a). The X-ray source is on the left, the sample is in the center, and the detector is on the right. Figure 15.10b shows the same setup as in Figure 15.10a, except that the X-ray tube and detector have been rotated to an angle that is labeled θ. The X-ray diffractometer has a mechanical system that simultaneously rotates the X-ray tube and the detector. Figure 15.10c shows the crystal structure of halite that was placed in the diffractometer. The horizontal, parallel lines drawn through

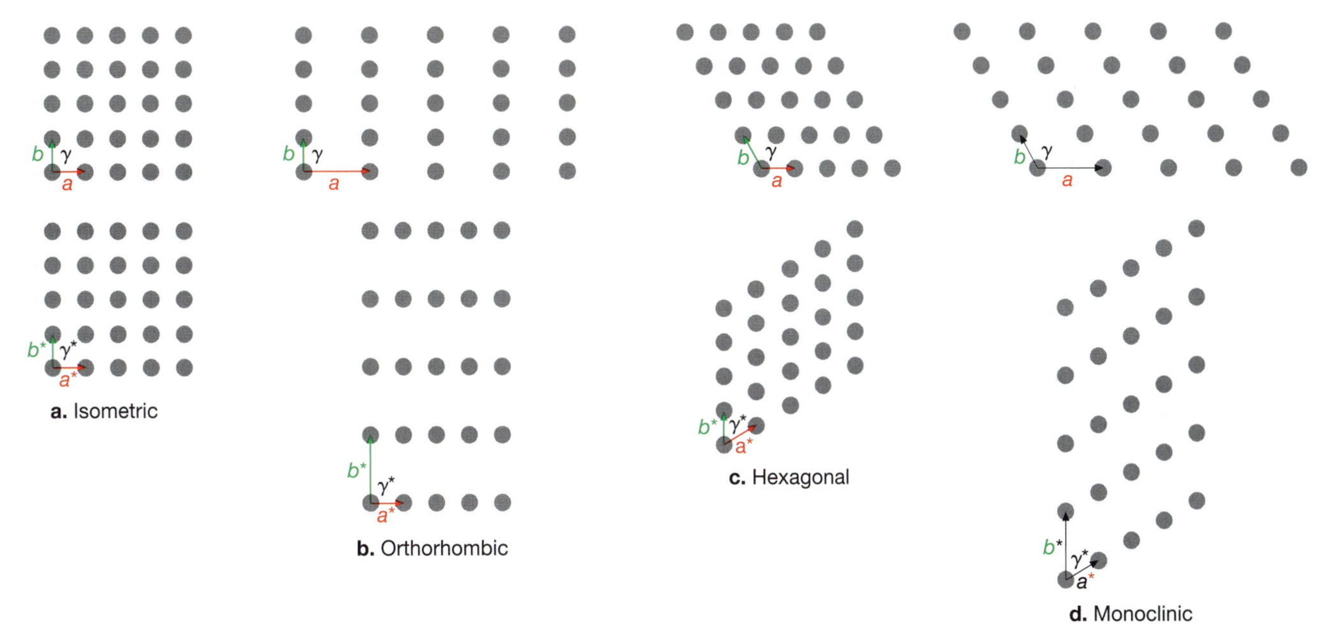

Figure 15.9. *Two-dimensional projections of direct (upper) and reciprocal (lower) lattices for the isometric (a), orthorhombic (b), hexagonal (c), and monoclinic (d) crystal systems. The direct lattices shown here relate to the gratings shown in Figure 15.8, with atoms (small circles) replacing the intersections of the grating lines. The reciprocal lattices are related to the diffraction patterns shown in Figure 15.8. The direct lattices are represented by the all-familiar cell parameters (i.e.,* a, b, c *and* α, β, *and* γ*) and projected on the* a-b *plane (i.e., down* c*). The reciprocal lattices are represented by the "star" cell parameters and listed as* a*, b*, c* *and* α*, β*, *and* γ**. Thus, they are projected on the* a*-b* *plane looking down* c**. While the direct lattice dimensions are fixed, those for the reciprocal lattice vary as a function of the wavelength that is used and the distance between the sample (i.e., the sieve location in Figure 15.7a) and the plane in which the diffraction image is formed (i.e., the screen in Figure 15.7a). Finally, the general observation that should be gained from these projections is that a reciprocal axis (e.g.,* a**) is perpendicular to two other direct axes (e.g.,* b *and* c*).* ***a.*** *For the isometric systems, the geometries of the direct and reciprocal lattices are identical.* ***b.*** *For the orthorhombic systems, the shorter of the two axes in direct space becomes the longer axis in reciprocal space.* ***c.*** *For the hexagonal systems, the lengths of* a *and* b *are similar to those of* a* *and* b**. Here* γ *is the obtuse angle 120°, while* γ* *is the acute angle 60°; thus these two angles are the supplement of each other (i.e., they total to 180°). Thus non-90° angles also change between direct and reciprocal space, as shown in the case of the gratings in Figure 15.7. Diffraction occurs perpendicular to the lattices rows. So* a *is perpendicular to* b* *and* b *is perpendicular to* a**.* ***d.*** *For the monoclinic systems, the lengths of the repeats along* a *and* b *are, as expected, inverted between the direct and reciprocal lattices. Also,* γ *and* γ* *are supplements,* a *is perpendicular to* b**, and* b *is perpendicular to* a**.*

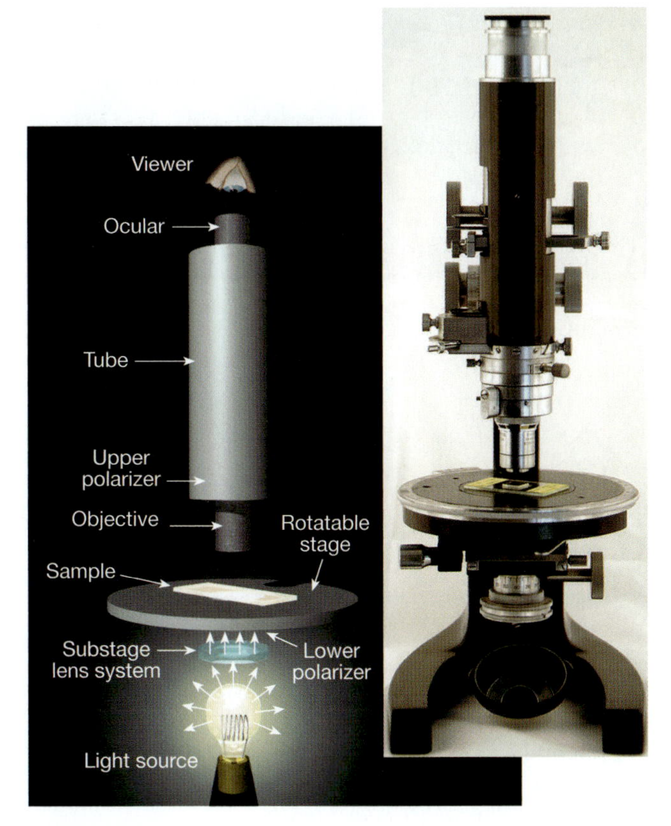

Figure 17.2. A photograph and sketch of a simple microscope. In many ways, a microscope is nothing more than a fancy hand lens (composed of the ocular, tube, objective combination) with a light source (the light bulb, substage lens system combination) and a place to precisely position and focus on a sample (the microscope stage). In a polarizing light microscope (PLM), two polarizers are added: one below the sample and one above it.

diaphragm located in the substage lens system. Like the iris of an eye, its purpose is to control the amount of light reaching the sample.

Next, the light rays pass through a circular hole in the microscope's stage. The sample is placed on the stage over this hole and is illuminated by light rays that have an incident angle of 0° (i.e., they are perpendicular to the sample). The stage is circular (unlike light microscopes you may have used in biology lab), which enables it to rotate, a feature that will be necessary for the study of anisotropic minerals. The stage can also move up and down to focus on the samples and, as we'll see in the next chapter, allow diagnosis of Becke lines that will aid in determining the refractive index of materials.

The next set of optical components includes an objective lens, a tube, and the ocular. These really constitute nothing more than a fancy hand lens: they serve to magnify the sample placed on the stage. Both the objective and ocular magnify the image, and the product of their magnifications equals the total magnification for the microscope. The objectives and the oculars can be changed to increase magnification, but usually only the objectives are swapped in modern microscopes.

Figure 17.3 shows a more modern PLM with its various parts labeled. It will be worthwhile now to go over these parts and their uses in a bit more detail. This PLM has two oculars instead of one (as was seen in Figure 17.2), along with several objective lenses. Other than that, these microscopes are very similar.

This newer microscope has four separate objective lenses (Figure 17.4). Each objective lens is in focus on the thin section below. From left to right, the space between the lens and the sample increases from the 63× power lens in Figure 17.4a to the 4× lens in Figure 17.4d. The overall magnification of the microscope is the power of the objective lens times that of the ocular, and in general, the oculars are 10×. So this particular microscope would have a range of magnification from 630× to 40×.

Along with the magnification, each lens has other characteristics that you will need to under-

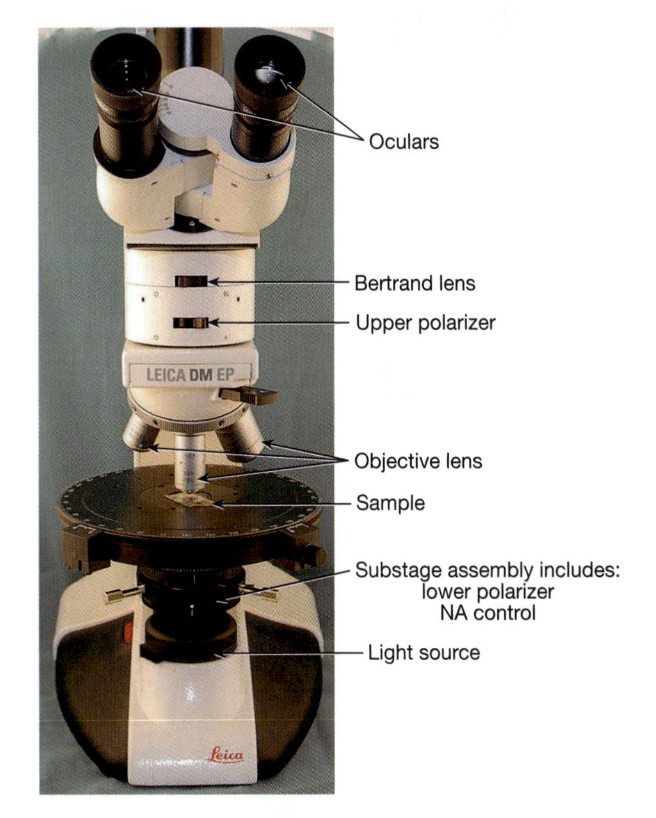

Figure 17.3. A photograph of a modern polarizing light microscope showing a few basic upgrades from the older PLM shown in Figure 17.2. These include (from bottom to top) an electronic light source, a more sophisticated substage system that allows for better control of the incident light, multiple objectives mounted on a rotating turret, the accessory plate used to help determine orientations of mineral grains, a Bertrand lens for obtaining interference figures (e.g., Figure 17.1b), and a set of binocular oculars for ease of viewing.

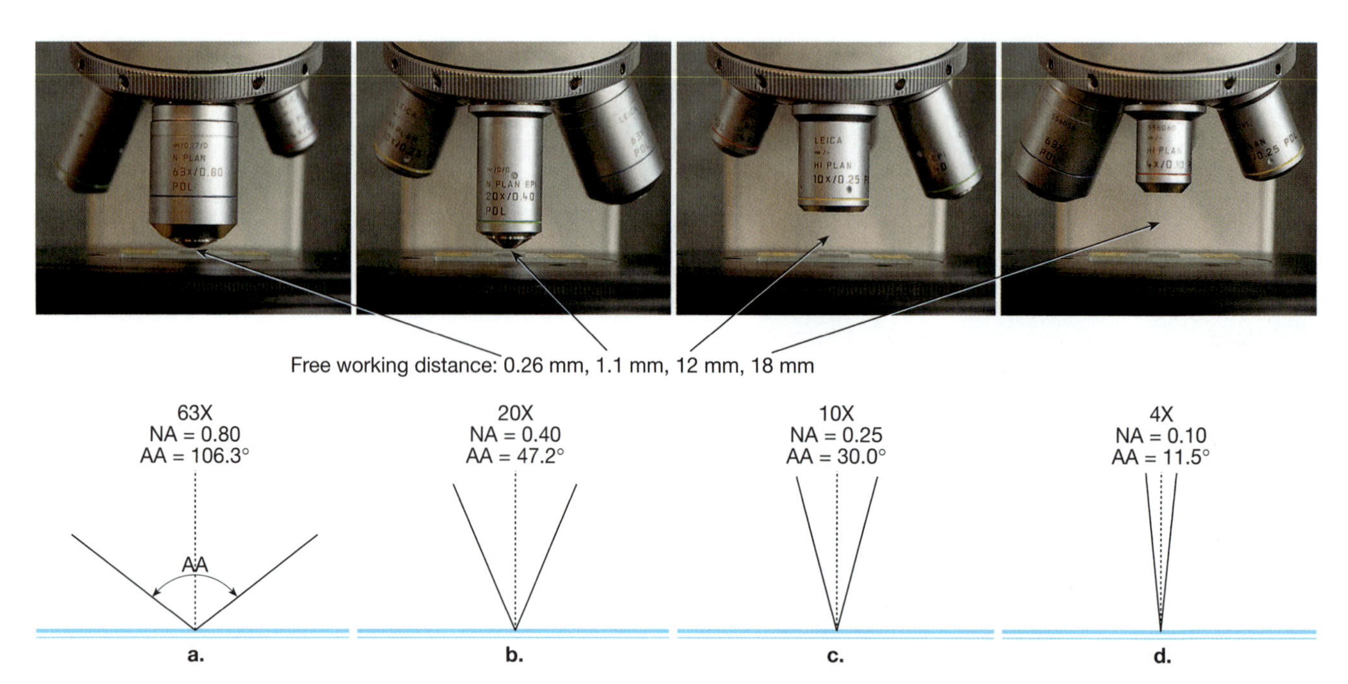

Figure 17.4. *Views of the four separate objective lenses from the PLM in Figure 17.3. For each lens, the top row shows the free working distance between the bottom of the lens and the top of the thin section when the sample is in focus. The bottom row shows the corresponding angular apertures (the maximum angle between the divergent rays that can be admitted to the lens) for each one. The lenses show a progression from high power in Figure 17.4a to low power in Figure 17.4d, where Figure 17.4a is a 63× lens, Figure 17.4b is 20×, Figure 17.4c is 10×, and Figure 17.4d is 4×. The free working distance decreases as the power of the lens increases, and the angular aperture increases with the higher power lenses.*

stand: the **free working distance**, which is the distance from the bottom of the lens to the top of the sample (or cover slide), and the **angular aperture** (*AA*), which is the angle between the two most divergent rays that the lens can accept. These are schematically illustrated below the photograph of each lens. Another useful term is the **numerical aperture** (*NA*) of each objective, which is related to the angular aperture by the following equation:

$$AA = 2\sin^{-1}NA. \qquad \text{Equation 17.1}$$

Both the numerical aperture and the angular aperture of the lens decrease as the magnification of the lens decreases. Figure 17.4 summarizes all of these optical terms for each of the four lenses. Each lens would also produce a different field of view that would decrease as the magnification increases. To determine the precise field of view for any lens, it is necessary to use a stage micrometer (i.e., a tiny ruler on a piece of glass, or even a plastic ruler for low magnifications!). For the microscope illustrated here, the approximate fields of view are 5 mm for the 4× lens, 2 mm for the 10× lens, 1 mm for the 20× lens, and 0.3 mm for the 63× lens. Notice the relationship between the field of view and the magnification. The free working distance decreases as the magnification and *NA* of the objective lens increase.

Each objective lens is brought into the light path of the microscope by a rotating turret (Figure 17.5). On the turret, two small tools are shown inserted into screw holes above the 63× lens. The majority of PLM's manufactured today have fixed rotating stages (i.e., they cannot be moved laterally). Thus, each individual lens must be centered over the fixed center of rotation of the microscope stage. This is accomplished by rotating the stage and observing what happens to a grain. In Figure 17.6, the center of rotation is noted by a small grain at the center of a circle. As the stage is rotated counterclockwise, the large grain that begins at the center of the crosshairs initially appears to orbit around the smaller grain. To center this stage, translate that small grain to the crosshairs (Figure 17.6b) using the centering tools shown at the top of the photograph in Figure 17.5. This makes the center of the microscope stage coincide with the center of the lens, as represented by the intersection of the crosshairs. Notice that the microscope stage shown in Figure 17.5 is calibrated in angles, which can be read off the markings on the rim surrounding the stage. Angular measurements are needed to determine orientations of mineral grains.

Below the stage (the substage), is a dial used to match the angular aperture of light to each of the objective lenses (4×, 10×, 20×, 63×). Figure 17.7 shows what happens when this dial is adjusted

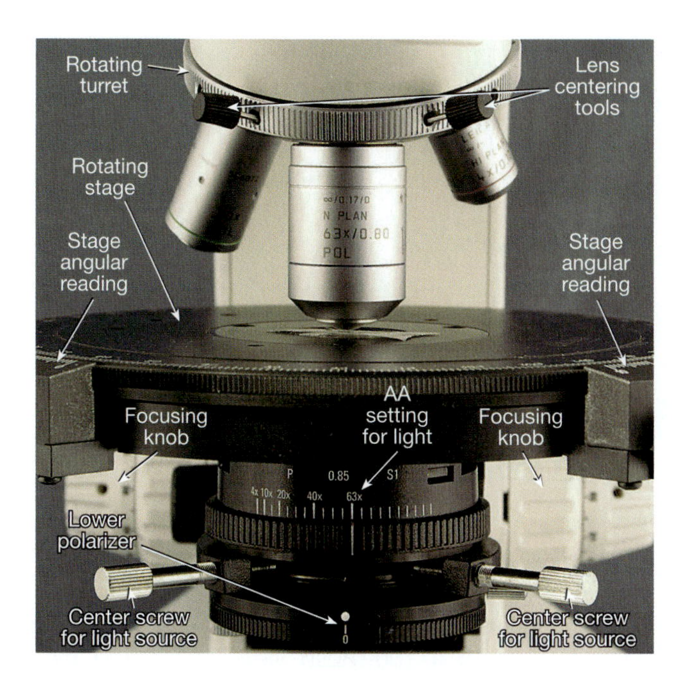

Figure 17.5. Close-up photograph showing several of the most important mechanical aspects of a PLM. Below the microscope stage is the substage assembly, which contains the lower polarizer, the two centering screws for the incident light source, and a control to adjust the angular aperture of the light. Above that is the rotating stage with scales inscribed to allow precise angles to be determined. At the top, a close-up of the rotating turret shows several of the lenses with the centering screws inserted. In the background (slightly out of focus) are the focusing knobs used to raise and lower the stage.

(see also Figure 5.21). In Figure 17.7a, the light rays can be seen to be parallel. Historically, this has been called **orthoscopic illumination.** Figure 17.7b shows the rays emerging from the sample at various angles, as seen in the fluorescing glass placed on top of the sample. Classically, this type of light configuration is called **conoscopic illumination**, because the sample is illuminated by a cone of light. It is used to produce interference figures (introduced in Chapter 5), which aid in determination of optical class, orientation of the mineral, and other optical properties. As noted in Figure 17.4, different objective lenses have different numerical apertures, and thus they will image cones of light with different shapes. This fact becomes important when observing interference figures, which must always be thought of as spherical projections of the three-dimensional space that is imaged on a two-dimensional plane in the microscope. When interpreting interference figures, the larger the area to be imaged, the easier it is to make observations.

Figure 17.8 shows the effect of changing the numerical aperture of the lens, with the resultant different images of interference figures for biaxial (Figure 17.8a, upper row) and uniaxial (Figure 17.8b, lower row) minerals. From left to right, the images show the result of lowering the numerical aperture (from 0.8 to 0.4 to 0.15), successively imaging less and less of the interference figure. In each row, the most useful interference figure is the one on the left, because it shows the most information. For this reason, the highest numerical aperture available on each microscope is generally used when examining interference figures.

At the very bottom of Figure 17.5, a dial marked with a zero indicates that orientation of the lower polarizer. And in it current setting it is in the so-called E-W direction.

Sample Types and Preparations

Several different methods are used to prepare samples for study with PLM. In general, the

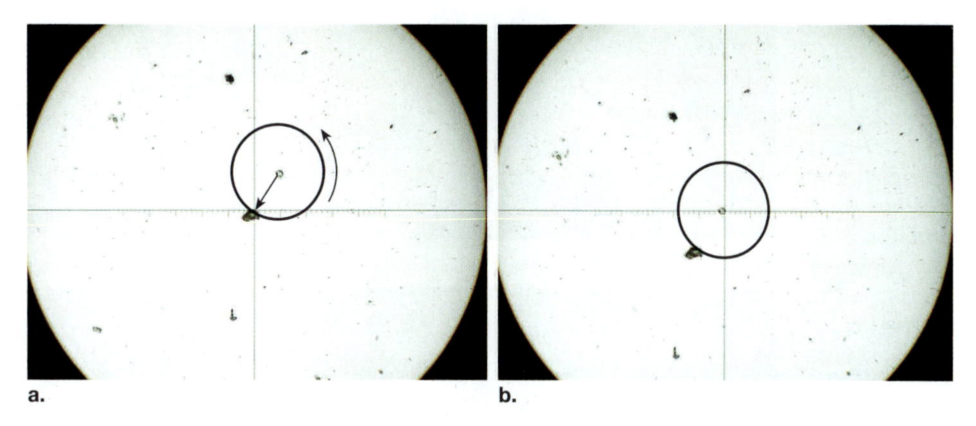

*Figure 17.6. Two photographs of particles in an uncentered (left) and centered (right) polarizing light microscope. **a.** When the stage is rotated counterclockwise, the particle that starts out at the center follows a path shown by the circle. To center the stage, the center of the dark circle needs to be moved (translated) so that it corresponds to the center of the crosshairs. This is accomplished by carefully moving the centering screws (Figure 17.5) to translate the lens in the proper direction. **b.** After centering, the circle is positioned so that its center lies directly on the crosshairs.*

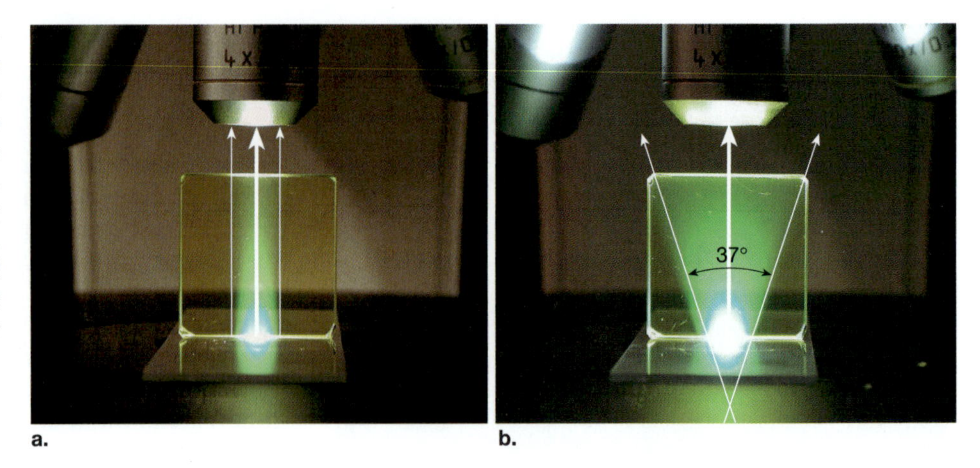

*Figure 17.7. A plate of fluorescing glass is placed on the microscope stage and illuminated in two different ways by adjustment of the substage angular aperture control. **a.** The rays are parallel, creating orthoscopic illumination. **b.** The rays form a cone and diverge through the sample at an angle of ~37°, creating conoscopic illumination.*

method is chosen on the basis of sample type, availability, and what information is being sought. We'll briefly discuss each of the sample preparation methods here and then go on to show examples in this and the next chapter.

Grain mounts (or the immersion method). Refractive index is one of the optical properties often used for identification of mineral grains. A few crystals are placed into a refractive index liquid and inspected to determine the refractive index of the grain relative to the liquid. This is called the immersion method, and it has been used for over a hundred years. Sets of liquids with different refractive indices are commercially available. The sets usually contain liquids ranging from 1.4 to 1.7 in increments of 0.002 or 0.004. Liquids are also available above 1.7, but they are used less frequently. With this method, the optical class of the mineral can also be determined.

A grain mount is made by first crushing and sieving the material to a desired size, usually about 100 μm grains. The crushed material is typically passed through a 100-mesh sieve (with 150 μm holes) but will not pass through a 200-mesh sieve (with 75 μm holes). A few of these grains are then sprinkled onto a glass slide with a drop or two of refractive index liquid already on the slide (Figure 17.9a). If the material's refractive index is truly unknown, then usually a liquid in the middle of the refractive index range is selected (e.g., 1.55). If you already have an idea what the mineral might be and are trying to confirm this, you would choose a liquid that matches the material's refractive index. A glass cover slip is then placed over the grain-liquid combination, and the slide is placed on the microscope for observation. We'll describe this method in greater detail in the next chapter.

Spindle stage. A spindle stage is a slight modification of the immersion method in which the

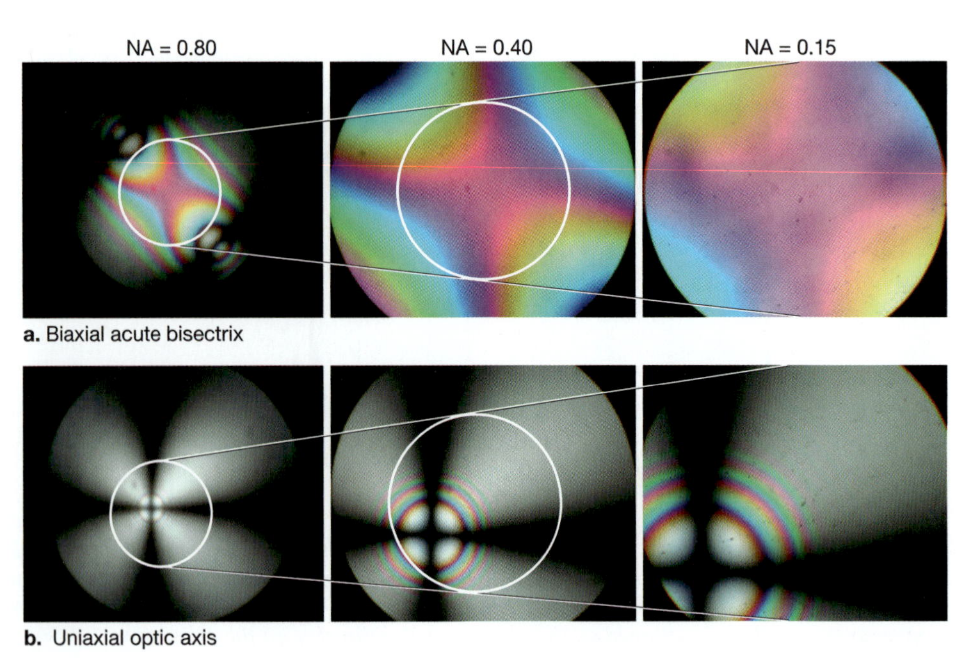

Figure 17.8. Images photographed using conoscopic illumination with three of the objectives shown in Figure 17.4, and different numerical apertures. Figure 17.8a shows a biaxial interference figure, while Figure 17.8b shows a uniaxial figure. From left to right, the numerical aperture decreases from 0.8 to 0.15, so that a smaller portion of the interference figure is seen. The white circles show the part of the figure that is visible in the view through the lower numerical aperture lens to its right. As the numerical aperture decreases (from left to right), smaller portions of the mineral are viewed because the rays are passing through a smaller range of angles.

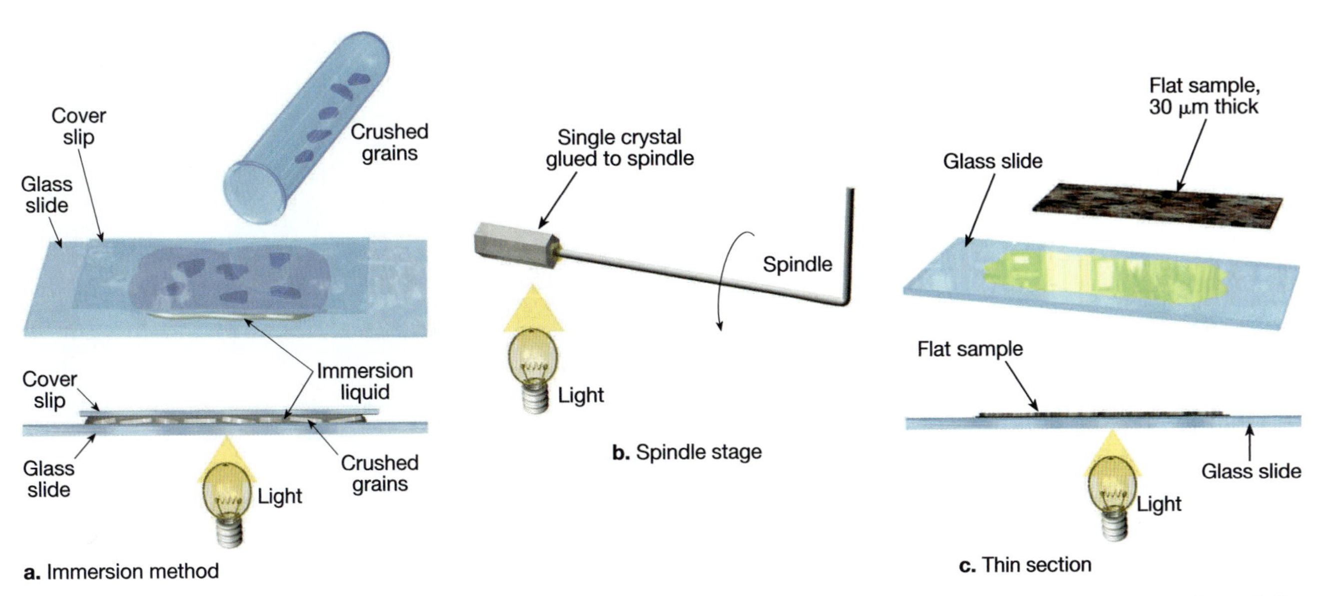

Figure 17.9. Example of three sample preparations used to observe mineral grains and rocks. ***a.*** *A grain mount is made by placing a drop or two of immersion liquid on a glass slide, and then sprinkling a few mineral fragments into the liquid. Next, a glass cover slip is placed over the liquid. This is the setup for the immersion method.* ***b.*** *Individual grains can be mounted on the end of a bent wire, and then the wire can be placed on the microscope stage and rotated so as to see the grain in different orientations. The grain is typically placed into a liquid to aid in optical observation. This setup is the basis for the spindle stage.* ***c.*** *A thin section of a mineral or rock is made by cutting and polishing a piece of minerals or rock to 30 μm thickness.*

same grain can be studied in different refractive index liquids. A single grain is glued onto a bent wire (Figure 17.9b), which can then be placed on a microscope stage and rotated to bring different orientations of the grain into the plane of the microscope stage. This is useful for studying morphologies of grains, and also for orienting grains to measure refractive index. In this chapter, the spindle stage will be used to show how certain optical properties such as retardation and the orientation of the optical indicatrix can be observed in the same grain at different orientations, producing different colors (like those seen in the quartz grain in Figure 17.1d).

Thin sections. Grain mounts and the spindle stage are typically used for identification of individual mineral grains, while thin sections are usually made to study the minerals present in a rock and the relationship of those minerals to one another (i.e., the texture). Thus thin sections are used more by petrologists than mineralogists. A thin section is a rectangular piece of a rock glued onto a glass slide. The rock slab is approximately 20 × 30 mm and with a final thickness of 30 μm and is made by cutting, trimming, and polishing a piece of rock (Figure 17.9c). During this process, the rock slab is glued to a 25 × 45 mm glass slide. Although the process can be automated somewhat, the process of making thin sections is kind of an art because it is quite difficult to make them flat and exactly 30 μm thick. We don't usually measure the 30 μm, but use the retardation color, to arrive at the correct thickness. Most geologists send their rock samples to commercial labs to produce thin sections for them. The cost per thin section is usually around $10–15. Sometimes a glass cover slip is glued on top of the thin section. This protects the sample while viewing it on a microscope, but renders it useless for other types of studies, such as those using an electron microscope for chemical analysis (discussed in Chapter 9).

Generally speaking, refractive indices of minerals cannot be measured in thin section. However, relief can be observed between adjacent materials. A thin piece of rock is mounted onto a glass slide with an epoxy that usually has a refractive index of about 1.55. The epoxy takes the place of the liquid in the immersion method. Thus, a mineral with a refractive index close to 1.55 will exhibit low relief. The Becke line method can also be used if the grain's edge can be seen next to the epoxy or another grain of known refractive index.

Geologists tend to study many more thin sections than grain mounts, because they are usually more interested in the texture of the rock than in mineral identification. Grain mounts are used to help identify mineral fragments. Other, more expensive analytical methods exist for identification. Electron microprobes are used to measure the chemical compositions of minerals, while X-ray diffraction is used to disclose the crystal structure of a mineral. However, both of these methods require the use of complex instruments costing as much as a million dollars. A new PLM and a set of refractive index liquids cost less than $6,000 (used microscopes are even less).

Birefringence, Retardation, and Orientation of *N* and *n*

In Chapter 16, the concepts of birefringence and retardation were discussed at length, with an emphasis on the colors that occurred on a wedge of quartz placed between crossed polarizers (Figure 16.23). When the wedge was illuminated in polychromatic light, a series of different colors was produced by interference (constructive and destructive) of the different wavelengths of polychromatic light. A similar representation is now shown in Figure 17.10a for a single quartz grain viewed with a PLM. The interference colors increase from the thin edge of the grain toward the thicker center. In monochromatic light such as the blue shown in Figure 17.10b, only one color (i.e., blue) is seen, with black isochromes. If the grain is the right thickness for a wave of light to be retarded by one full wavelength behind the other when it emerges from the grain, none of the resultant light will be oriented so it can pass through the upper polarizer, and the black isochrome will be seen. As the thickness increases, the slow wave lags two or three wavelengths behind the fast wave, and second- and third-order black isochromes are seen. If the wavelength of the illuminating light is increased from blue to yellow (Figure 17.10c), a similar phenomenon occurs. Because the yellow light has a longer wavelength than the blue light, the isochromes are now spaced slightly farther apart. The increased spacing is most clearly seen in Figure 17.10d, where monochromatic red light is used; now only two black isochromes appear. By analogy with the contour lines on a topographic map, here the spaces between black lines (isochromes) have a larger "contour interval" as the wavelength of illuminating light is increased. For example, in Figure 17.10b, the spacing is 450 nm of retardation between black lines, while the spacing is 650 nm between the black lines in Figure 17.10d.

When the grains are illuminated in monochromatic light, the black lines represent integer wavelength differences between the slow and the fast ray. In polychromatic light, the integer wavelengths differences are noted by different orders of red (as explained in Chapter 16 and Figure 16.27), as shown in Figure 17.10a starting with first-order red (retardation of 550 nm), second-order red (Δ = 1100 nm), and third-order red (Δ = 1650 nm). The isochromes map out areas of equal retardation. As the thickness of the grain increases, so too does the retardation (Figure 17.11). The birefringence of quartz ($\varepsilon - \omega$) is 0.009. The retardation observed in the microscope would be the product of the birefringence and the grain's thickness. Thus, first-order red (550 nm retardation) would occur at a grain thickness of 61 μm, while

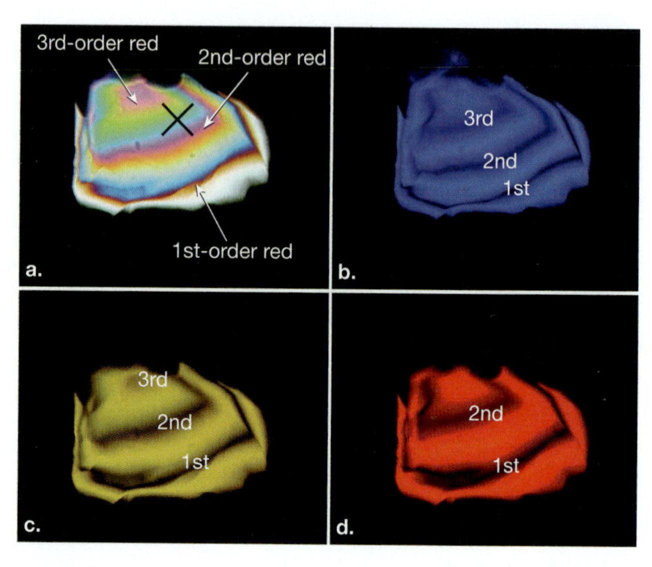

Figure 17.10. The same mountain-shaped crystal shown in Figure 17.1d, illuminated in polychromatic light (Figure 17.10a), and in three different wavelengths of monochromatic light from short (Figure 17.10b, blue ~450 nm) through medium (Figure 17.10c, yellow ~590 nm) to the longest wavelength (Figure 17.10d, red~ 650). ***a.*** *The interference colors should be reminiscent of those seen in the quartz wedge in Chapter 16. The two mutually-perpendicular black lines represent the vibration directions of the crystal. The first-, second-, and third-order red isochromes (lines of equal color) are also labeled.* ***b.*** *The black line nearest the (thin) edge of the crystal corresponds to the thickness where the slow wave lags behind the fast wave by one wavelength (in this case, ~450 nm), so no light is transmitted at the upper polarizer. As the thickness of the crystal increases, second- and third-order black isochromes are seen, representing lags of the slow wave behind the fast wave at thicknesses of two (~900 nm) and three (1350 nm) wavelengths. In the blue areas of the crystal, light is transmitted because the waves are either partially or totally in phase when they exit the crystal. In Figure 17.10c and d, the analogous situation is shown for yellow (~590 nm) and red (~650 nm) light.* ***c.*** *In yellow light, the third-order isochrome is barely seen because the isochromes are farther apart than for blue light.* ***d.*** *The red light has the longest wavelength seen here, so the isochrome spacing is farthest apart, and the crystal isn't thick enough for a third-order isochrome to be visible. The isochromes here are analogous to lines on a contour map of elevation, in that they map out lines of equal retardation (and thickness).*

second-order red (1100 μm) would occur at a thickness of 122 μm, and third-order red would occur at a thickness of 183 μm (Figure 17.11).

Accessory Plates

Mineral identification and characterization sometimes require the ability to distinguish between N (the larger refractive index) and n (the smaller refractive index) for an individual grain in a specific orientation. Accessory plates are used to add (or subtract) either a fixed amount of retardation (flat plate) or a variable amount of retardation (a wedge-shaped plate). The plate of choice is inserted into the light train of the PLM, above the objectives and below the upper polarizer, with its vibration directions at a 45° angle to those of the polarizers (Figure 17.3). The most commonly-used plate produces first-order red retardation, and is made from a 61 μm slab of quartz cut parallel to its optic axis. The N (in this case, the optic axis) is perpendicular to the long direction of the accessory plate, as shown in Figure 17.12. Typically, accessory plates will have the N direction etched on them. The n direction, which is mutually-perpendicular to N, is seldom etched on the plate.

The accessory plates allow use of the N_{plate} to distinguish between N_{grain} and n_{grain} (Figure 17.12). If N_{plate} coincides with N_{grain}, then N_{grain} is further retarded when it passes through the plate, causing the retardation colors to increase (addition). If N_{plate} coincides with n_{grain}, then n_{grain} is less retarded when it passes through the plate, and the retardation colors decrease (subtraction). In practice, the grain's vibration directions are placed at 45° off extinction, causing the vibration direction that is oriented NE-SW to coincide with N_{plate}. As an example, consider the case where insertion of the plate causes the first-order red isochrome to become black (Figure 17.12a,b). This is the case for subtraction, because 550 nm of retardation is subtracted by insertion of the plate, and N_{plate} coincides with n_{grain}. The black background in Figure 17.12a appears as a first-order red background in

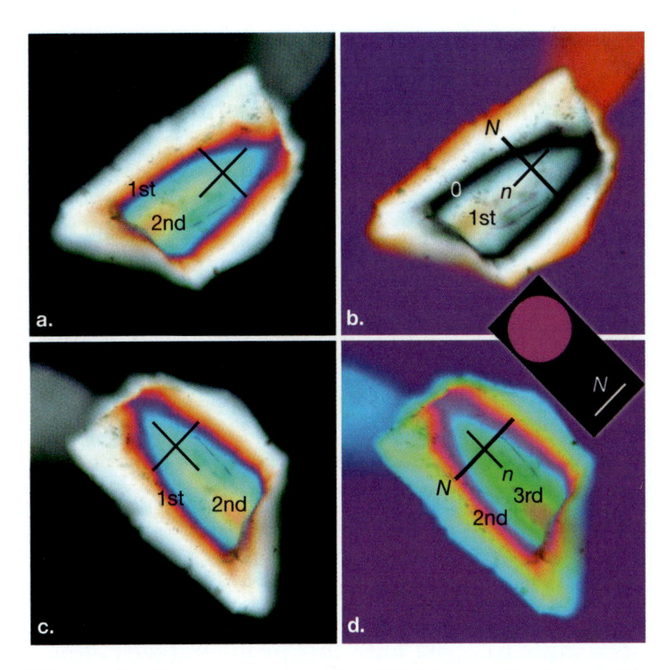

*Figure 17.12. **a.** A quartz grain at 45° from extinction, with the vibration directions and isochromes labeled. The left images have a black background, while the right images have a reddish background. These show how the orientation of* N *(the large refractive index) can be distinguished from the* n *(the small refractive index) direction. The grain shows both first- and second-order red interference colors. **b.** The quartz plate, which produces 550 nm of retardation, has been inserted between the two polarizers, with its* N *refractive index at a 45° angle to the vibration direction of the upper and lower polarizers. Insertion of the quartz plate causes the background to turn from black to first-order red (which looks somewhat like a purplish-red) and also affects the retardation of the isochromes on the grain. In this case, the first-order red isochrome is reduced to zero retardation. The second-order red color is reduced to first-order red. This is the case of subtraction, because the interference colors decrease as a result of the fact that* N_{plate} *is parallel to* N_{grain}*. **c.** The grain has been rotated 90° from its orientation in Figure 17.12a. **d.** Insertion of the quartz plate now causes the interference colors of the grain to increase (for example, first-order red becomes second-order red); this is called addition. Here* N_{plate} *is parallel to* N_{grain}*, so the wave that was retarded in the crystal is further retarded in the plate.*

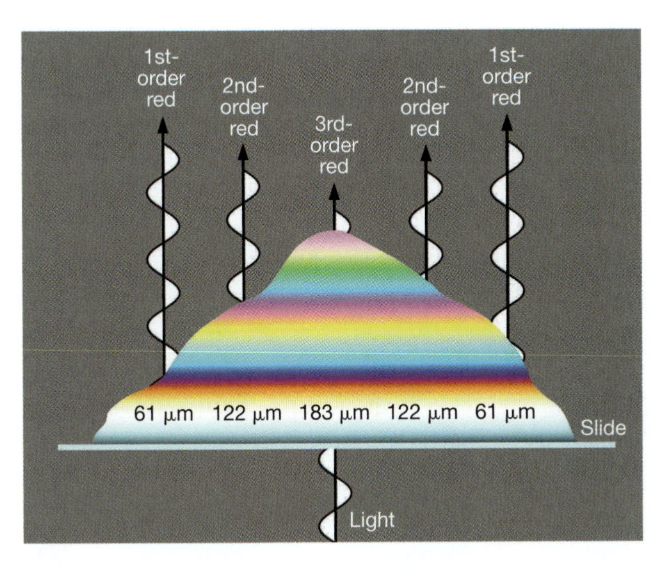

Figure 17.11. A schematic cross section through the quartz grain shown in Figures 17.1d and 17.10, showing the thickness in various locations. Because quartz has a retardation of 0.009, the different thicknesses produce first-, second-, and third-order red when the grain is illuminated with polychromatic light.

Figure 17.12b because the plate was inserted. If the grain is rotated 90°, insertion of the plate causes addition, and all the interference colors increase by 550 nm. Thus first-order red becomes second-order red, and second-order red goes to third-order red. Here, N_{plate} corresponds to N_{grain}.

To aid in visualizing the difference between addition and subtraction, the interference color chart is used (Figure 16.27). Figure 17.13 shows a small portion of that chart with examples of addition and subtraction as would be accomplished by use of a quartz plate (sometimes called a first-order red plate). It is by far the easiest to observe subtraction of first-order red to black than to see its addition from first-order red to second-order red. Likewise, it is difficult to distinguish between subtraction and addition for second-order red changing to either first- or third-order red. Fortunately, there is a wedge-shaped accessory plate to assist with this problem (see below).

A more common problem is distinguishing between subtraction and addition when the retardation of the grain is less than 550 nm. A different style of interference color chart may be used to help visualize this problem (Figure 17.14). This chart is centered around zero thickness, and its retardation colors increase to the right and left of the center line. At the bottom of the chart, an example is shown for the addition and subtraction of 550 nm from first-order red. Moving up the chart, a starting retardation of ~400 nm (first-order yellow) would add to become second-order yellow and subtract to become first-order gray. The next line shows the case for a starting retardation of first-order gray (275 nm), which adds to become second-order green and subtracts to first-order white. The top line is one of the most commonly-observed interference colors: it begins with low first-order gray ~100 nm, adds to become second-order blue, and subtracts to become first-order yellow. It is often worthwhile when making these observations to check for both addition and subtraction in two separate 90° orientations of the grain, as was done in Figure 17.12.

Another commonly-used accessory plate is the quartz wedge, which has variable thickness (Figure 16.18) and thus produces a range of interference colors (Figure 16.23). This was used to explain the formation of interference colors in cross-polarized light. Quartz is cut in the same orientation in the wedge as it is in the quartz plate, with its optic axis parallel to the plate and perpendicular to its long direction. The orientation of N_{grain} is determined again by orienting the grain at 45° off extinction and assessing whether addition or subtraction of interference colors occurs. The quartz wedge becomes more useful when the retardation of the grain is higher, because as discussed above, it can be difficult to distinguish between addition or subtraction for second-order red. When the wedge is inserted, movement of the isochromes can be used to ascertain whether addition or subtraction occurs.

As an example (Figure 17.15), consider a quartz grain with fourth-order red retardation, placed at 45° off extinction. When the quartz wedge is inserted, the isochromes would appear to move out as indicated by the arrows on the photograph. Higher order isochromes would replace lower order isochromes, indicating addition because N_{grain} is parallel to N_{grain}. If the grain is rotated 90° counterclockwise, insertion of the quartz wedge would cause the isochromes to appear to move inward, with low order isochromes replacing higher order ones. This would be the case for subtraction because N_{plate} is parallel to n_{grain}. Regardless of which accessory plate is used, observation of a black isochrome is usually the most definitive indicator of addition vs. subtraction because it only occurs through subtraction. The wedge is also use-

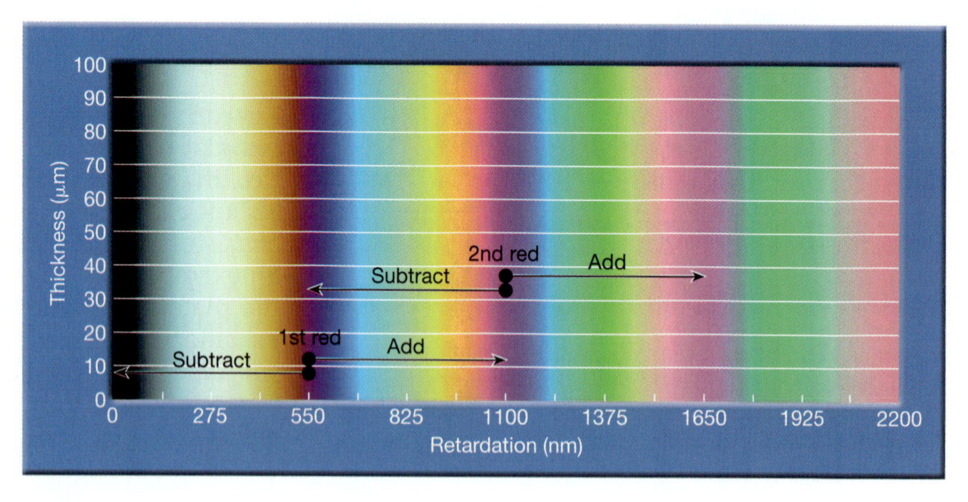

Figure 17.13. The color chart from Figure 16.27 can be used to explain the conditions for addition and subtraction with insertion of the quartz plate. The quartz plate produces 550 nm of retardation, so it effectively changes the retardation by moving horizontally on this chart: adding to increase retardation (N_{plate} parallel to N_{grain}), or subtracting to decrease it (N_{plate} parallel to n_{grain}). Two of the many possibilities for addition and subtraction, as shown in Figure 17.12, are shown.

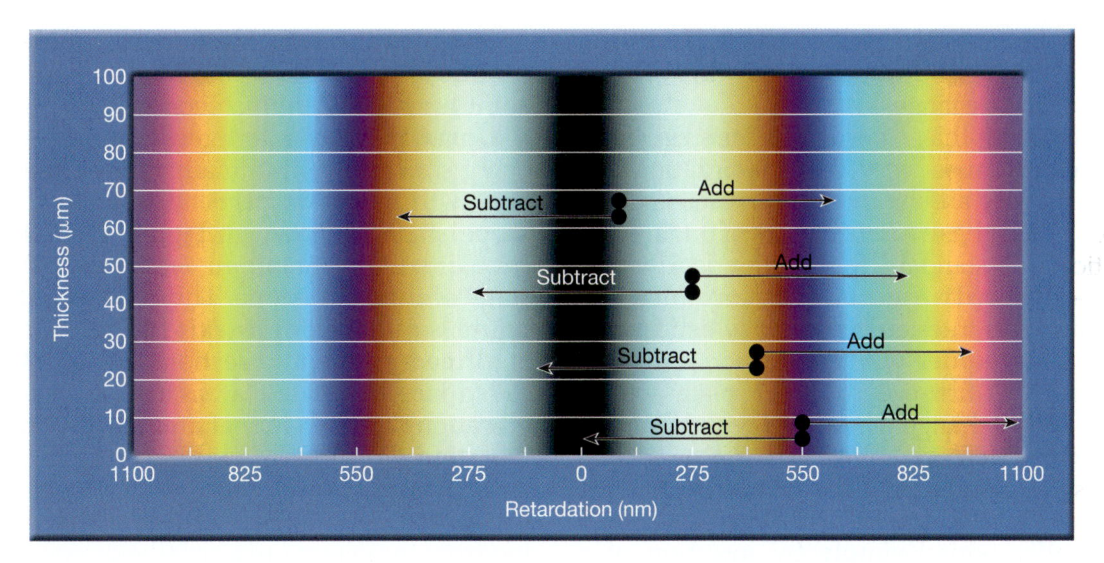

Figure 17.14. A smaller portion of the interference chart from 0 nm to second-order red (1100 nm) is shown on the right, with its mirror image on the left. This depiction is useful for understanding retardation of grains in cases where the retardation is < 550 nm, which is typically the case for minerals in thin section. Four separate examples are shown: low first-order gray, first-order white, first-order yellow, and first-order red. For first-order gray (~100 nm retardation), the addition and subtraction colors are very distinctive: addition results in second-order blue and subtracts to first-order yellow. For first-order white (~275 nm), addition yields second-order green but subtraction results in first-order white. It is often important to check for both addition and subtraction by rotating the grain 90°. (Thanks to Andrew Knudsen for suggesting the idea for this color chart.)

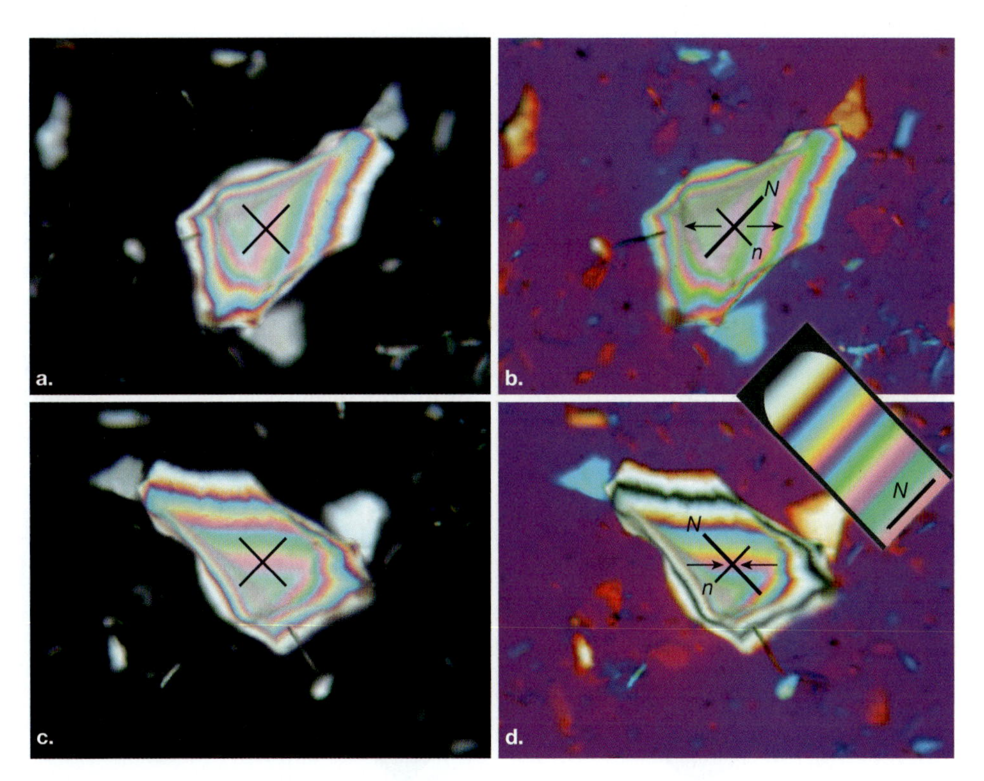

Figure 17.15. A quartz grain with higher retardation than in Figure 17.14. In this case, a quartz wedge is used to determine the orientations of N *and* n. ***a.*** *Quartz crystal rotated to 45° off extinction.* ***b.*** *The quartz wedge has been partially inserted, and the isochromes move outward, as indicated by the arrows. This is the situation where* N_{wedge} *is parallel to* $N_{quartz\ grain}$*, and the higher-order isochromes replace the lower-order isochromes as the wedge is inserted.* ***c.*** *The grain has been rotated 90° from Figure 17.15a.* ***d.*** *The quartz wedge has again been partially inserted. In this case, the isochromes move inward, as indicated by the arrows; low-order isochromes replace high-order isochromes. This represents the case of subtraction, where* N_{wedge} *is parallel to* $n_{quartz\ grain}$.

ful for color blind students because it shows the movements of the isochromes so they do not need to interpret color changes.

Some minerals have very high birefringence, and therefore cause very high retardation. Of those, carbonates are the most common, especially calcite. Calcite has a birefringence of 0.172, so its retardation in a standard thin section is 5160 nm (~eighth-order red). Figure 16.26 explained the formation of this interference color, which is usually called high-order white. Addition and subtraction with a first-order quartz plate would show little color change when used with a grain of such high interference color. This makes it relatively easy to distinguish between calcite and the framework silicates quartz and feldspar (which are first-order white) simply by insertion of a quartz plate. Addition and subtraction for quartz and feldspar would result in changes to second-order blue or first-order yellow, respectively. Addition and subtraction for calcite with this plate would look basically the same (Figure 17.16).

Interference Figures

Interference figures were first seen formed on ping-pong balls in Chapter 5 of this text, where we explained the difference in figures between uniaxial and biaxial minerals. In thin sections of rocks, quartz (uniaxial) is typically distinguished from the feldspars (biaxial) by obtaining and interpreting interference figures. Everything we have learned so far (retardation, addition and subtraction, orientation of N vs. n, the microscope parts, etc.) will all be required to make optimal use of these interference figures, which are produced (in anisotropic minerals) by a PLM by using conoscopic illumination. These are valuable for determining the optical class and orienting minerals. Interference figures can be produced on minerals in grain mounts, on a spindle stage, or in thin section.

Uniaxial. The interference figure produced using conoscopic illumination for a uniaxial mineral with its c axis perpendicular to the microscope stage is shown in Figure 17.17. Here, the circular section containing ω is in the plane of the stage of the microscope. This type of interference figure is called an **optic axis figure**. A black cross is formed by the intersections of two black lines called **isogyres**. These areas are dark because the vibration direction in that area corresponds to the vibration directions of the upper and lower polarizers. Familiar interference colors changing from first-order red to higher orders are seen, increasing from the center outward (the opposite of what is seen with orthoscopic illumination). Insertion of the quartz plate results in addition in the upper right and lower left quadrants, and in subtraction in the upper left and lower right. Insertion of a quartz wedge would show the same trends but with apparent movement. The analogous figures obtained on a grain of tourmaline are shown in Figure 17.17c,d. Insertion of the plates now causes subtraction in the upper right and lower left quadrants, and addition in the upper left and lower right—the opposite of what was observed in quartz. Why?

Recall that there are two separate classes of uniaxial minerals: uniaxial positive ($\varepsilon > \omega$) and uniaxial negative ($\varepsilon < \omega$). The refractive index value ε' is always between ε and ω in a uniaxial mineral. In a uniaxial positive mineral, $\varepsilon' > \omega$, and for a uniaxial negative mineral, $\varepsilon' < \omega$. In Figure 17.17b, N_{grain} must correspond to N_{plate} in the upper right quadrant, while n_{grain} must relate to N_{plate} in the upper right quadrant of Figure 17.17d. The orientations of ω and ε' in these interference figures will determine whether you see addition or subtraction.

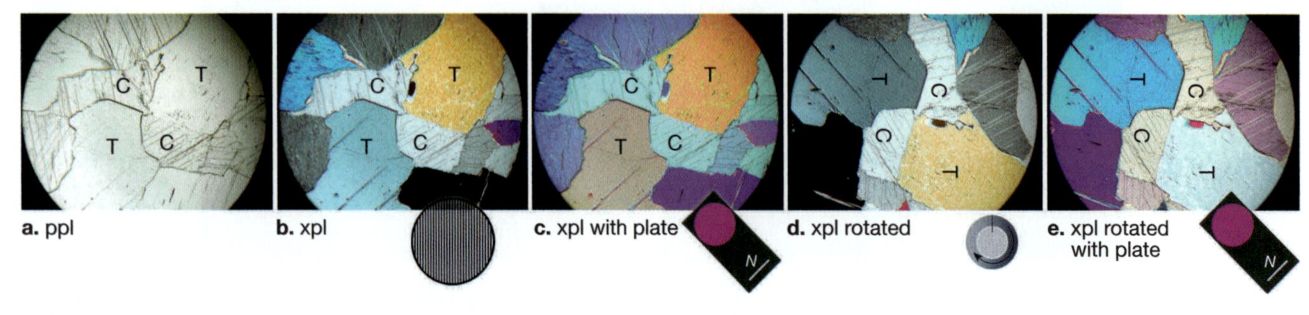

Figure 17.16. *Thin section of rock containing calcite (C) and tremolite (T).* ***a.*** *Plane-polarized light.* ***b.*** *The same view in cross-polarized light. Calcite now appears with grayish (high-order white) interference colors.* ***c.*** *The quartz plate has been inserted, with very little change (from gray to purplish blue) in the retardation color of calcite.* ***d.*** *The sample has been rotated by 90°.* ***e.*** *The quartz plate is inserted. Again, notice the minor change in the color of calcite. Here calcite is exhibiting ~5000 nm of retardation, so that adding or subtracting 550 nm has very little effect on its color. Compare this to the graph shown in Figure 16.26, which illustrates how high-order white colors are created.*

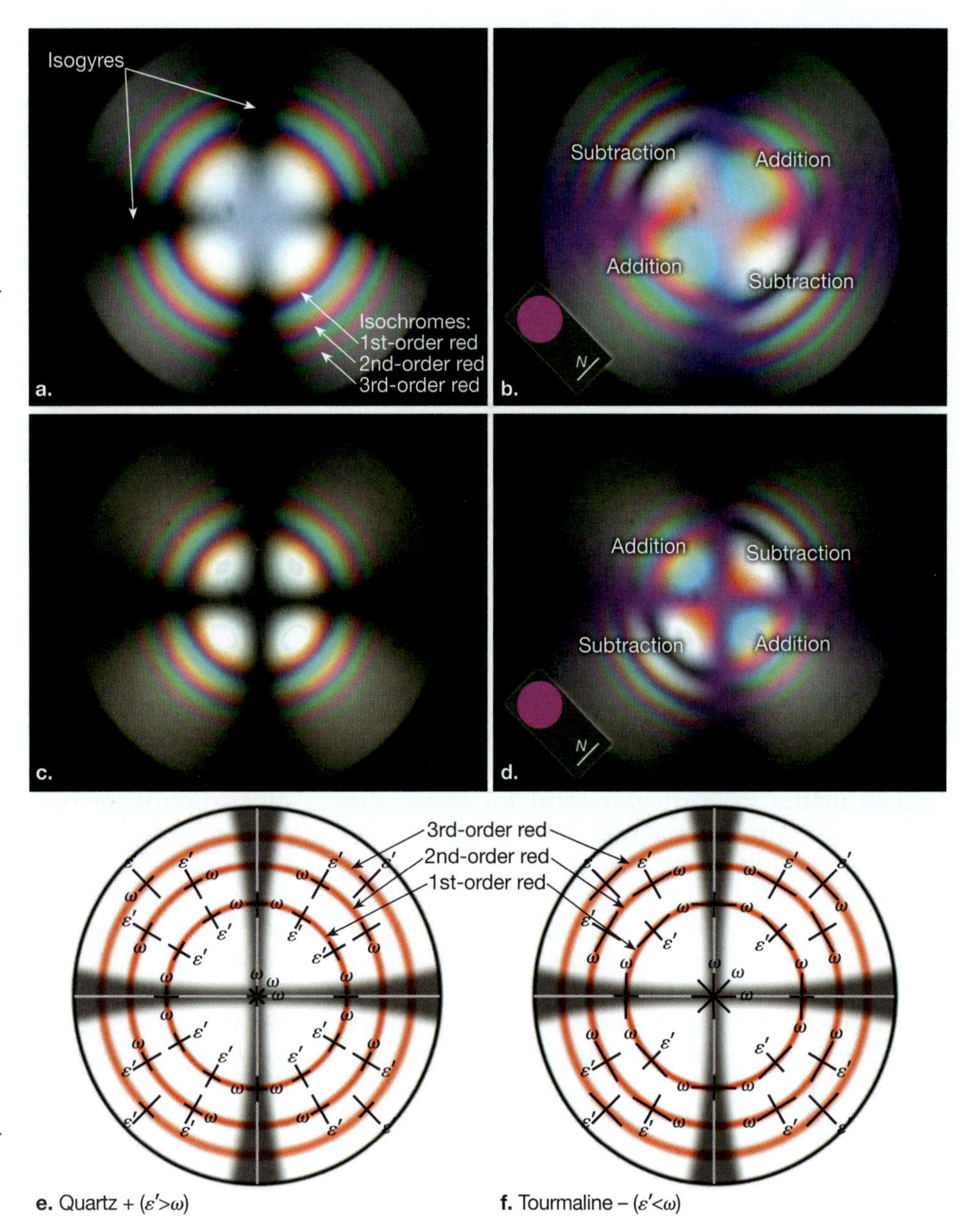

***Figure 17.17.** Views down the optic axis of a uniaxial mineral with conoscopic illumination, along with associated sketches of their vibration directions. **a.** Isochromes form in the interference figure of a quartz grain, increasing in retardation from the center of the grain outwards. This is the opposite of what was seen in the quartz grains in orthoscopic illumination in the previous figures. The dark wedges in the interference figure are called isogyres, representing areas where the vibration directions in the crystal correspond to those in the polarizers. **b.** The same configuration as in Figure 17.17a, but a quartz plate has now been inserted to show areas of addition and subtraction. If a quartz wedge is inserted instead, areas of addition and subtraction can be determined by observing the movement of the isochromes. For example, addition occurs when the colors change from low- to high-order, so the isochromes will appear to be moving inward; again the opposite is seen in orthoscopic illumination. **c.** View down the optic axis of a tourmaline grain. **d.** The same grain as in Figure 17.17c, with a quartz plate inserted. Addition and subtraction occur in separate regions of the interference figure. **e.** Schematic representations of the isochromes, isogyres, and vibration directions of quartz. **f.** Schematic representations of the isochromes, isogyres, and vibration directions in tourmaline. In both cases, the ω vibration direction is seen at the very center. Moving out from the center, the ω vibration direction is tangent to the isochromes. By definition, other vibration directions (in this case, ε′) are always perpendicular to ω, and therefore perpendicular to the isochromes. The vibration directions in the isogyres correspond to the vibration directions of the polarizers, causing the isogyres to be dark. In the upper right quadrant of Figure 17.17e and f, the ε′ direction coincides with the N direction of the accessory plate. In Figure 17.17e, ε′ > ω (so ε′ = N) and insertion of a plate would cause addition in that region. In Figure 17.17f, ε′ > ω so ε′ = n. Overall, notice that the same orientation of the vibration directions ε′ occurs in both the lower left and the upper right. The ε′ vibration direction is rotated by 90° in the upper left and lower right quadrants, resulting in subtraction in those regions in Figure 17.17e and addition in those regions in Figure 17.17f. In might help to use the mnemonic "WITTI" to recall the "ω is tangent to isochrome." "BURP," meaning "blue upper right positive" can be helpful to tell positive from negative uniaxial minerals as long as you remember that the "blue" occurs in the area that was gray.*

Figure 17.17e,f shows schematic illustrations of the vibration directions, isochromes, and isogyres for the cases of a uniaxial positive mineral ($\varepsilon' > \omega$) and a uniaxial negative one ($\varepsilon' < \omega$), respectively. The value of ε' increases from the center outward, causing an increase in retardation. In the areas where the isogyres occur, the ω and ε' vibration directions coincide with those of the polarizers' vibration directions (i.e., they are N-S and E-W). At the very center of the interference figure, the ω vibration direction occurs in all directions.

Returning to determination of the optic sign, notice in Figure 17.17e that ω is tangent to the isochromes, and thus, ω changes orientation in different quadrants of the interference figure. The orientation of ω relative to ε' is the same in the upper right and lower left quadrants, and rotated 90° from that in the upper left and lower right quadrants. Insertion of the quartz plate causes addition in the upper right and lower left, because N_{plate} is parallel to N_{grain} (i.e., you are looking at ε'). In the upper left and lower right quadrants, ω becomes parallel to N_{plate} because ω and ε' are rotated 90°, and subtraction occurs. In Figure 17.17f, $\omega > \varepsilon'$ because the mineral is optically negative. Here, subtraction occurs in the upper right and lower left quadrants, while addition occurs in the upper left and lower right quadrants (the opposite of what was seen in Figure 17.17e).

The mnemonic "WITTI" may be helpful in summarizing these figures: "ω is tangent to the isochrome." If ω corresponds to n, the optic sign of the mineral is positive. If ω corresponds to N, then the mineral is negative.

To understand the formation of interference figures in the PLM, it is helpful to realize that only one ray in the microscope is perpendicular to the mineral in conoscopic illumination. All the others are incident at ever-increasing angles from the center out to the edge of the field of view. Snell's Law (Figure 16.5b) shows that a light ray moving from a low to a high refractive index material is refracted toward the normal to the interface. A cross-sectional view of the analogous phenomena for a series of rays of light incident upon the grain at different angles is shown in Figure 17.18. When a ray of light is incident on an anisotropic material, it is forced to vibrate in two mutually-perpendicular directions (see calcite in Figure 16.13). The two mutually-perpendicular vibration directions correspond to rays with two different refractive indices. Thus, each one incident ray is refracted into two separate rays vibrating in orthogonal directions. Rays incident upon a grain that are near-parallel and emerge from the grain as O-rays can interfere with the E-rays from another ray. In other words, rays that travel in slightly different directions in the crystal can interfere with each other to produce retardation colors. In turn, the amount of retardation will increase from the center of the field of view toward the edge because of two separate phenomena: (1) the difference in ω and ε' (birefringence) increases from the center of the field of view outward, and (2) the rays will travel a greater distance through the crystal as the angle increases, thus increasing the distance traveled by the waves and in turn, retardation.

Figure 17.18b is a perspective view of Figure 17.18a showing cones of equal retardation for the two mutually-perpendicular vibration directions after they pass out of the crystal. The retardation increases outward from the center, and the ω vibration direction is in all cases tangent to each cone while the ε' vibration direction is perpendicular. Each of these labeled cones would in turn produce first-, second-, third-, fourth-, and fifth-orders of red retardation color, each representing the circular isochromes in Figure 17.17. The vibration directions would form dark areas defining the isogyres in the form of a cross because their vibration directions coincide with those of the polarizers.

Uniaxial interference figures do have applications outside of geology! The Inset shows a device called an optical ring sight made from a uniaxial axis figure. This was first used during World War II as a sight on anti-aircraft guns and is used in photography during parachuting. Details of how these images are produced are described in the Inset.

Biaxial interference figures. In the previous section, the optic axis figures for uniaxial minerals had only one optic axis. For the case of biaxial minerals, there are two separate optic axes—thus the prefix "bi." The most useful type of interference figure for this class of minerals will be the one looking down the bisector between the two optic axes. This was seen earlier for the case of muscovite in Chapter 5 (Figure 5.23). Muscovite's cleavage is perpendicular to the bisector between its two optic axes. Thus, it is very easy to obtain this type of interference figure from muscovite and other micas.

As we progress through discussion of interference figures, the interference figure that is observed in the microscope will always be a function of the orientation of a mineral's indicatrix. Often it will be difficult to interpret the interference figures of biaxial minerals. The interference figure for muscovite looking down the bisecting angle between the two optic axes, along with its schematic, is shown in Figure 17.19a and c. This view is termed the **acute bisectrix** (*AB*) because it bisects the acute angle between the two optic axes. The mineral has been rotated 45° from extinction in orthoscopic illumination to obtain this figure. The two optic axes outcrop in the upper left and lower right quadrants in the center of a concentric circle that exhibits first-order red retardation (Figure 17.19a). A series of isochromes also appears, as was seen in a uniaxial optic axis figure. The main difference is that in the biaxial case, the isochromes are no longer circular. With the insertion of an accessory plate, subtraction occurs in the center of the field of view and addi-

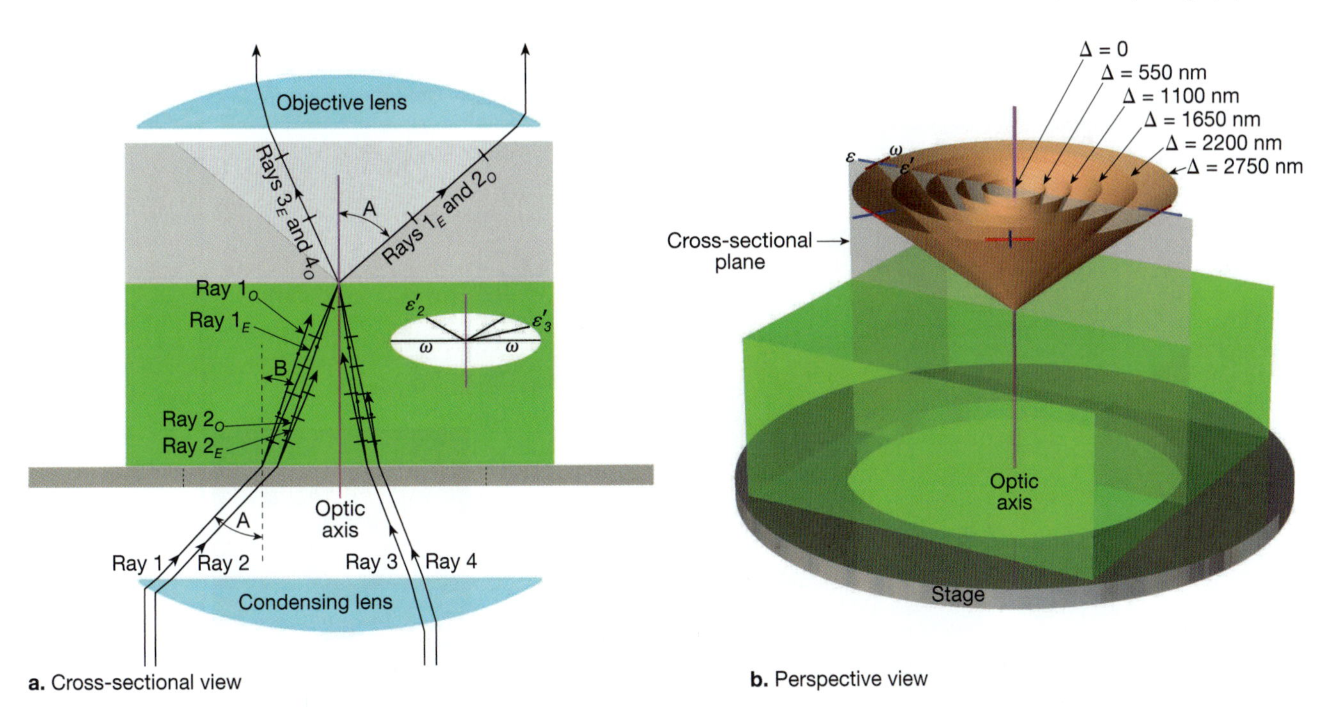

***Figure 17.18.** Formation of interference figures shown in Figure 17.17 for uniaxial negative minerals. An indicatrix is inscribed into the crystal to aid in understanding E- and O-rays. The formation of interference colors, retardation, and E-ray and O-rays is shown as a series of rays interacting at differing angles with the anisotropic material (modified from Bloss 1999, Figure 7.13) **a.** Cross-section view showing a series of rays impinging on the crystal at different angles, as a result of conoscopic illumination. Each ray is broken into an E-ray and an O-ray with mutually-perpendicular vibration directions. The O-ray's vibration direction is perpendicular to this view, and the E-ray's vibration direction is in the plane of the page. The E-ray and the O-ray are associated with different directions in the crystal with different refractive indices, so they are refracted at different angles and retarded differently. The E-ray for Ray 3 arrives at the surface of the crystal at the same time as the O component of Ray 4, so one wave would be lagging behind the other and interference would occur. Because there are a series of rays at different angles impinging upon the crystal, those rays travel along different directions and different thicknesses of the crystal, producing increasing levels of retardation as the angles become more divergent. **b.** Perspective view showing cones of equal retardation (Δ) increasing outward from the center. Here the ω and ε′ vibration directions are shown, where ω is again tangent to the cone. The interference figures shown in Figure 17.17 are an image of the top of these cones.*

tion on the concave side of the two isogyres (Figure 17.19b). At the center of the field of view, $n_{crystal}$ coincides with N_{plate}. On the concave side of the isogyres, $N_{crystal}$ coincides N_{plate}.

In biaxial minerals, there are three principal refractive indices, and the relationship of their magnitudes is $\alpha < \beta < \gamma$. The intermediate values between α and β are α' and the intermediate values between β and γ are γ'. In general, we can also write that $\alpha < \alpha' < \beta < \gamma' < \gamma$. We will see that α, α', β, γ', and γ all appear in the interference figures (Figure 17.19c). Both β and γ may outcrop at the center of the interference figure. In Figure 17.19b, β corresponds to N_{plate}, so subtraction occurs. In all cases, β is normal to an imaginary line joining the two optic axes. On the concave side of the isogyres, β is still parallel to N_{plate}, but γ has now been replaced by α'. Thus, although β was n in the center portion of the interference figure (resulting in subtraction), on the concave side of the isogyre, β becomes N (addition occurs).

In Figure 17.19d, the crystal has been rotated 90° counterclockwise from Figure 17.19a, as seen schematically in Figure 17.19f. Insertion of the accessory plate shows addition over the central portion of the interference figure and subtraction on the concave side of the isogyres (Figure 17.19e). This is just the opposite of what was seen in Figure 17.19b. Observation of the interference figure (Figure 17.19f) now shows γ ($N_{crystal}$) is parallel to N_{plate} in the center of the field of view, while α' ($n_{crystal}$) coincides with N_{plate} on the concave side of the isogyre.

Topaz is also a biaxial mineral, but its sign is opposite that for muscovite. An interference figure for topaz is shown in Figure 17.20a looking down the acute bisectrix as was done for muscovite; the accompanying schematic is shown in Figure 17.20c. The optic axes are now at the edges of the field of view, indicating that $2V$ (the angle between the two optic axes), is greater in topaz than in muscovite. The interference figures for

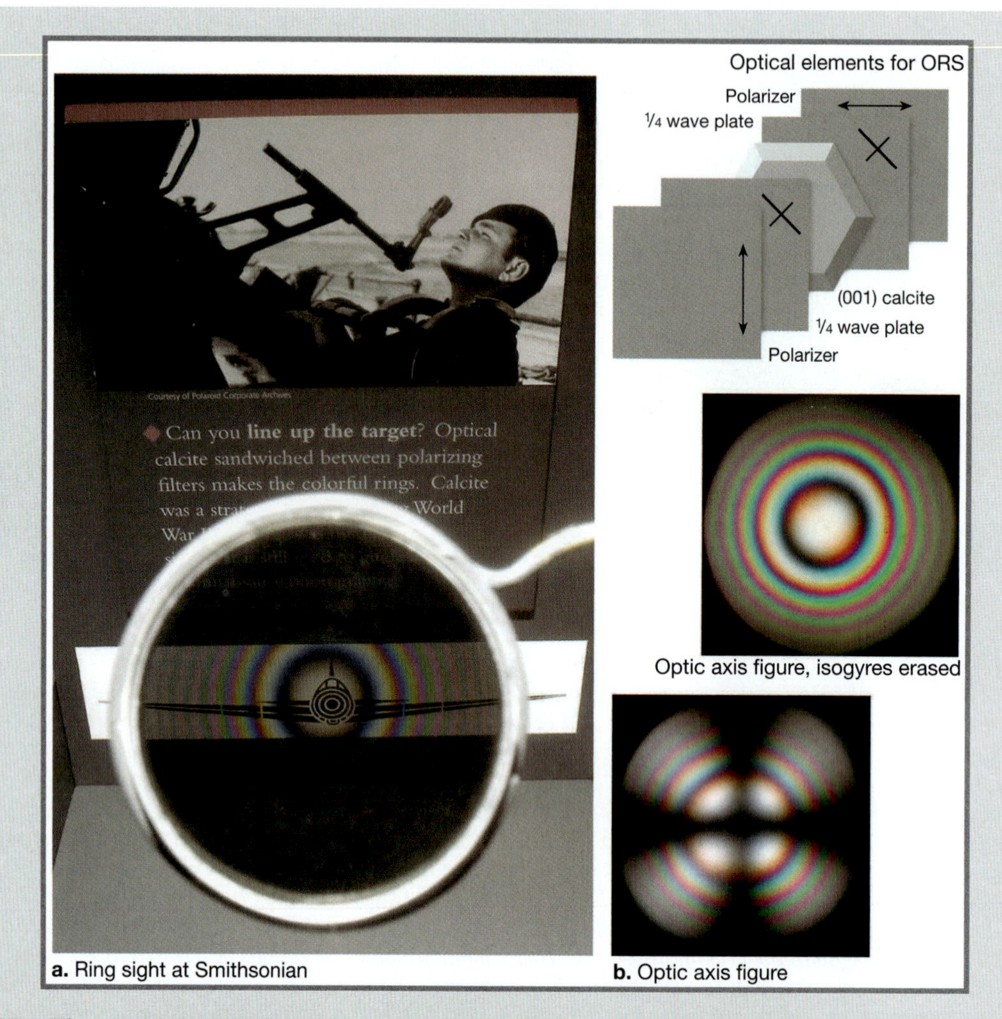

Figure INSET. The optical ring sight (ORS) was developed during World War II to aid anti-aircraft gunners in targeting other planes. The sight lacks the problem of parallax associated with a front and back sight as commonly used for aiming. For a more detained discussion see Gunter (2003). ***a.*** *Photograph of a display discussing the sight's past and present use (set up in 1997) at the Smithsonian Institution, Washington, D.C., showing a modern, commercially-available optical ring sight. (Photo courtesy of Jeff Post and Dane Penland of the Smithsonian Institution, Washington, D.C.).* ***b.*** *A series of figures used to explain development of the ORS. An optic axis figure of calcite is formed when a cone of converging light rays strikes a (001) plate of calcite sandwiched between two cross-polarizers. The isogyres (black cross) of the interference figures can be removed, yielding the ORS, by placing two 1/4-wave retarders (i.e., 125 nm retardation) in front of and behind the calcite, oriented at 45° to the polarizer's vibration directions with their slow directions parallel.*

topaz and muscovite look similar, although fewer isochromes are visible in the topaz figure because it has lower retardation than muscovite. Insertion of the accessory plates shows addition in the center of the interference figure and subtraction on the concave side of the isogyre (Figure 17.20b). Here the orientation of β is the same as it was in muscovite, but α has now replaced γ at the center of the field of view. This causes addition to occur in the center. We can also use this image to define the difference between biaxial positive and negative figures. Topaz is a biaxial positive mineral, and we know this because the acute bisector is equal to the γ vibration direction. Muscovite represents the opposite case where α bisects the acute angle between the optic axes, and this is defined as biaxial negative.

Further 90° counterclockwise rotation of the topaz crystal (Figure 17.20d,f) and insertion of the accessory plate (Figure 17.20e) shows subtraction over the central portion of the interference figure and addition on the concave side of the isogyres. This is just the opposite of what was seen in Figure 17.20b. Observation of the interference figure now shows that in the central portion of the crystal, α ($n_{crystal}$) coincides with N_{plate}, resulting in subtraction. On the concave side of the isogyres, γ' ($N_{crystal}$) coincides with N_{plate}.

If $2V_{mineral} > 60°$, the optic axes fall outside of the field of view, as shown for the centered acute bisectrix interference figure of staurolite in Figure 17.21. The large $2V$ of this mineral can be noted from the symmetry of the isochromes, as compared with the two previously-shown acute bisectrix figures of muscovite and topaz. If an accessory plate is inserted, addition occurs in the center of the field of view (Figure 17.21b). In this orientation, β_{grain} coincides with N_{plate}. Thus, if addition is occurring, α must be the other vibration direction occurring at the center of the field of view. We are looking down γ and so, by definition, the mineral is optically positive. When the optic axes of a centered acute bisectrix figure are positioned in the upper left and lower right quadrants, β will be outcropping in the center of the field of view and coinciding with N_{plate}. Thus, when $\beta = N$, addition occurs, and the mineral is optically positive.

If the staurolite grain is rotated 90° counterclockwise (Figure 17.21c) and the accessory plate is inserted (Figure 17.21d), subtraction occurs over the central portion of the interference figure because now α (i.e., n_{grain}) is parallel to N_{plate}. In the previous three figures, all of the samples were oriented at 45° off extinction. When an acute bisectrix interference figure is rotated to extinction, it may bear a slight resemblance to a centered, uniaxial optic axis figure, forming a cross (Figure 17.22). However, closer observation of these crosses shows that the isochromes are no longer circles around the center of the figure, but are centered around the two separate optic axes, which lie along the horizontal isogyre. Comparing the four different minerals in Figures 17.22, the $2V$ increases from 18° in aragonite, 45° in muscovite, 60° in topaz, to > 60° in staurolite (where the optic axes occur outside of the field of view). The widths of the horizontal and vertical isogyres are different for each of the minerals, but the isogyres containing the optic axes are always thinnest (see also Figure 17.23).

An acute bisectrix figure also changes as the microscope stage is rotated. All the images shown in Figure 17.22 were taken with the crystals at extinction, whereas in Figures 17.19–17.21, all the samples were rotated 45° off extinction. Figure

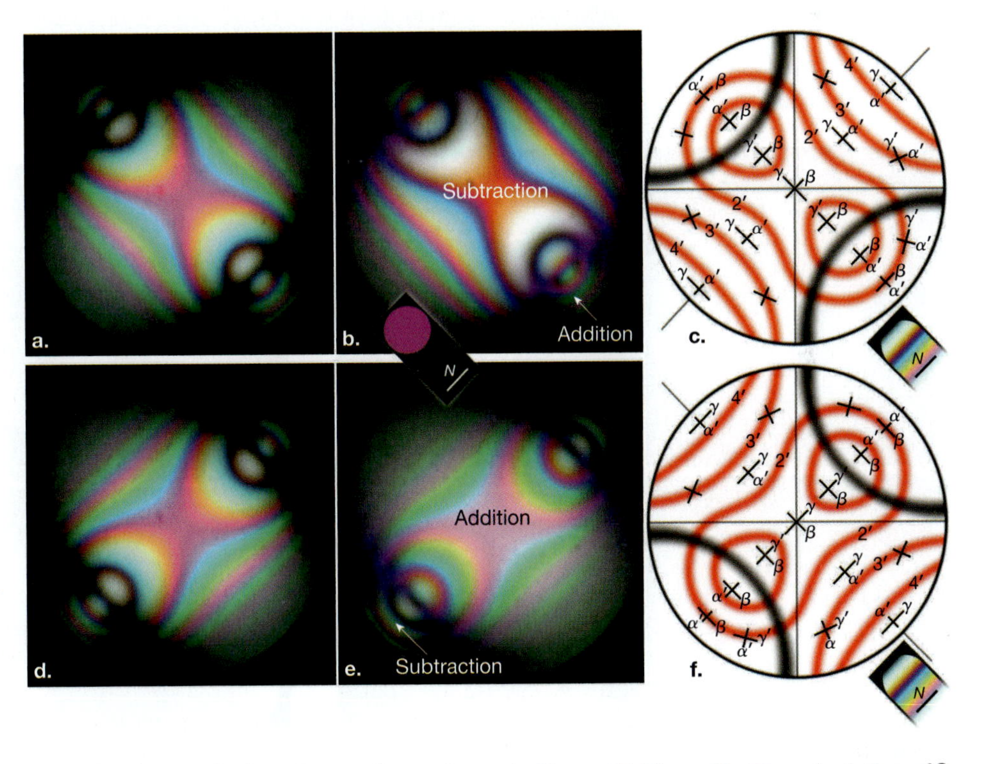

Figure 17.19. *Interference figure for the biaxial mineral muscovite, shown in the same fashion as the uniaxial interference figures (Figure 17.17).* ***a.*** *Photograph of a centered acute bisectrix figure at 45° off extinction. The two optic axes emerge in the upper left and lower right portions of the field of view, and are in the center. The isochromes generally increase in retardation as they move out from the center of the field of view toward the upper right and lower left. Isogyres intersect the optic axes in the upper left and lower right.* ***b.*** *A quartz plate is inserted. Subtraction occurs over the majority of the field of view, with addition occurring in the very lower right-hand corner and the very upper left corner.* ***c.*** *A schematic illustration of the isogryes, isochromes, and associated vibration directions corresponding to the interference figure shown in Figure 17.19a and b. The orientation of β is perpendicular to the line that would join the two optic axes. In the center of the field of view, γ is seen to be perpendicular to β. On the concave side of the isogyres, α′ is perpendicular to β. Thus, when an accessory plate is inserted, addition would occur on the concave side of the isogyres because N_{plate} corresponds to the larger refractive index, N, in the interference figure (β). In the center of the field of view, subtraction would occur because β is now equal to n.* ***d.*** *The muscovite interference figure has been rotated 90° from that in Figure 17.19c.* ***e.*** *Insertion of the quartz plate now causes addition in the center portion of the figure, and subtraction on the concave side of the isogyres.* ***f.*** *A schematic illustration of the vibration directions, isogryes, and isochromes for the crystal shown in Figure 17.19d and e. In the center of the field of view, both β and γ occur. The γ (N) is now parallel to N_{plate}, so addition occurs in the center region. On the concave side of the isogyres, α′ occurs perpendicular to β, so subtraction is seen in these regions when an accessory plate is inserted.*

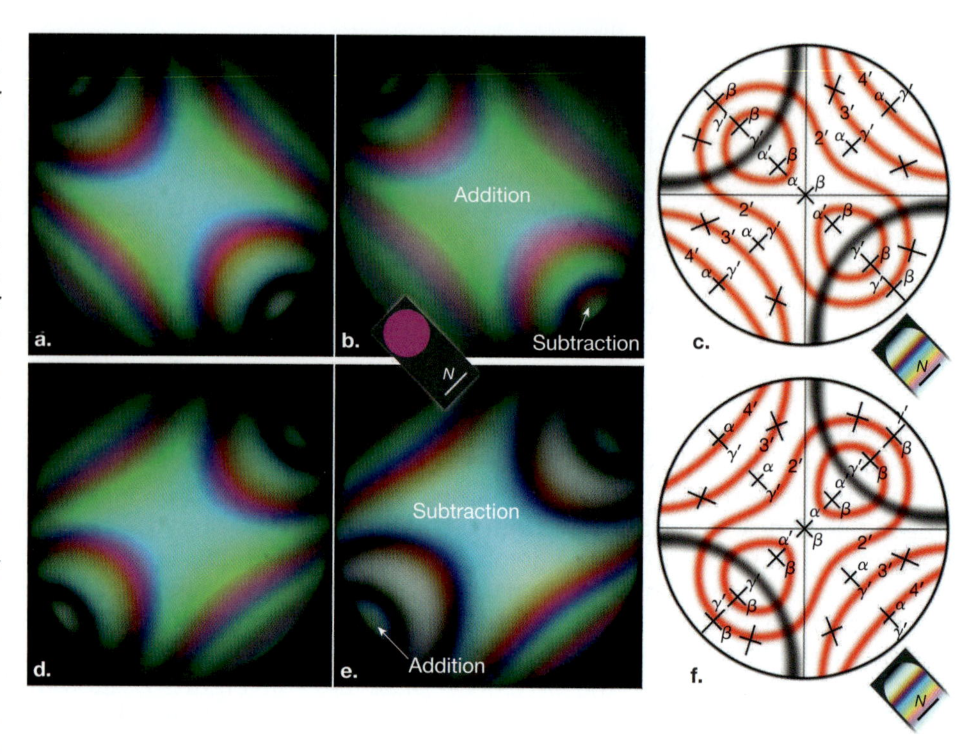

Figure 17.20. *Set of figures analogous to those in Figure 17.19, using topaz instead of muscovite. The optic axes are now near the edges of the fields of view.* ***a.*** *Isochromes and optic axes in the center of each of the isogyres.* ***b.*** *When the accessory plate is inserted, addition is now seen in the center of the field of view, with subtraction on the concave side of the isogyre.* ***c.*** *Schematic sketch of the interference figure like the one shown in Figure 17.19c for muscovite. The difference is that α has replaced γ in the center portion of the field of view, and γ has replaced α on the concave side of the isogyre. This results in addition at the center because N_{plate} is parallel to N_{grain}, which is β.* ***d.*** *Now the stage has been rotated 90° from what was shown in Figure 17.20a.* ***e.*** *Subtraction occurs in the center of the field of view after insertion of an accessory plate.* ***f.*** *Orientations of the vibration directions that correspond to Figure 17.20d and e. Subtraction would occur at the center of the field of view because n_{grain} (here α) is parallel to N_{plate}.*

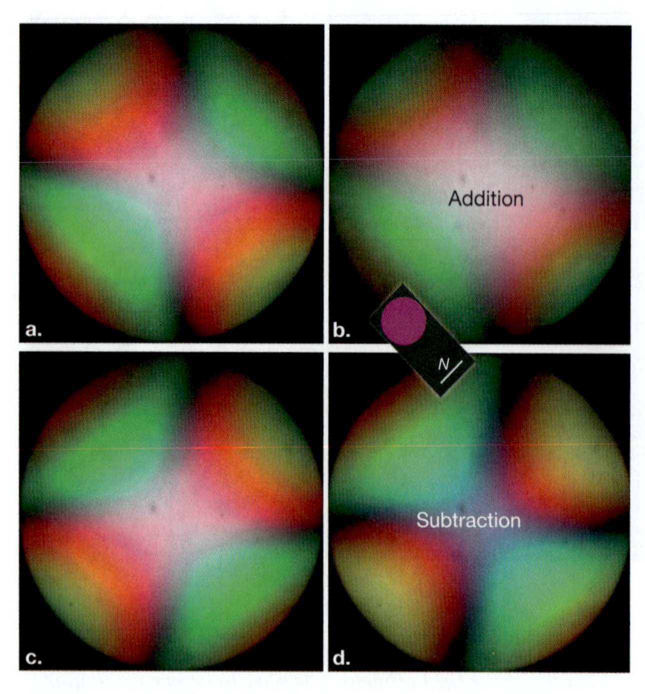

Figure 17.21. *The optic axes are outside of the field of view in this set of images of a staurolite grain.* ***a.*** *The optic axes lie outside the upper left and lower right quadrants.* ***b.*** *Insertion of the quartz plate causes addition over the entire field of view, as was seen in topaz.* ***c.*** *The grain has been rotated 90° from Figure 17.21a.* ***d.*** *Insertion of an accessory plate results in subtraction.*

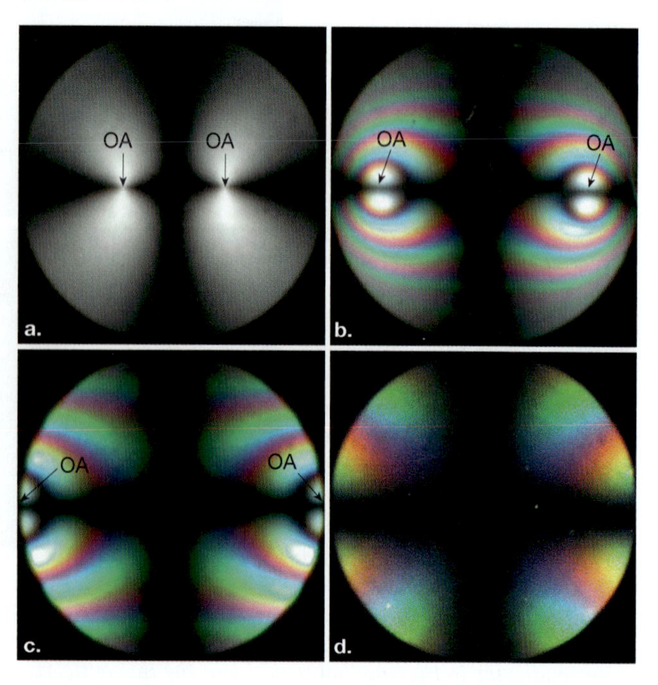

Figure 17.22. *Minerals arranged so the angle between the optic axes increases from 18° to 45°, 60°, and >60°. The optic axes occur at the most narrow part of the isogyres. Note the horizontal isogyre is consistently thinner than the vertical isogyre.* ***a.*** *Aragonite.* ***b.*** *Muscovite.* ***c.*** *Topaz.* ***d.*** *Staurolite. The latter three grains have all been rotated 45° counterclockwise from their orientations in Figures 17.19a, 17.20a, and 17.21a.*

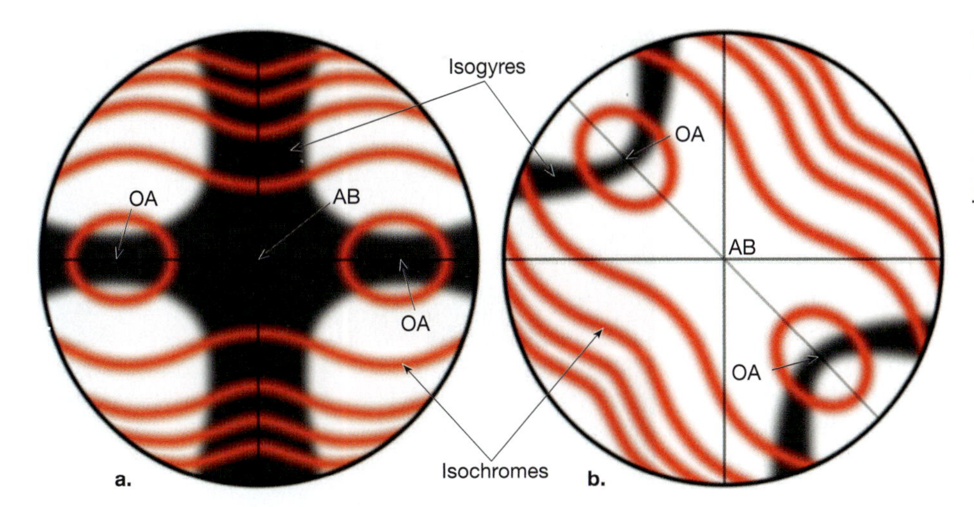

Figure 17.23. *Simplified schematic representation of the isochromes, isogyres, and the when the grain is at extinction (Figure 17.23a) or rotated 45° from extinction (Figure 17.23b).*

17.23 reviews this basic concept, showing a schematic illustration of the crystal placed at extinction and then rotated 45° clockwise. In both cases, we are looking down the centered acute bisectrix figure, with the associated optic axes and isogyres also shown.

A graphical method can be used to determine the orientations of the vibration directions within an acute bisectrix interference figure. Examples of this construction are shown in Figure 17.24, where the figures correspond to the orientations shown in Figure 17.23a and b. The grain is at extinction, with the two optic axes resting on the E-W cross-hair. A point labeled 1 in the field of view is chosen, and then a line is drawn connecting that point to each of the two optic axes. The vibration directions are then found by bisecting the acute angle formed by those two lines, thus showing one of the vibration directions. The second vibration direction is perpendicular to that; this portion of the interference figure is not at extinction. For point 2, a similar construction is made, but now the vibration directions correspond to those of the polarizers, so this area would be at extinction. Further construction of points in Figure 17.24a would then define the isogyres. The same construction is shown in Figure 17.24b, though the crystal has been rotated 45° to match the interference figure in Figure 17.23b.

Another type of interference figure for biaxial minerals occurs when the optic axis is perpendicular to the stage of the microscope. In many

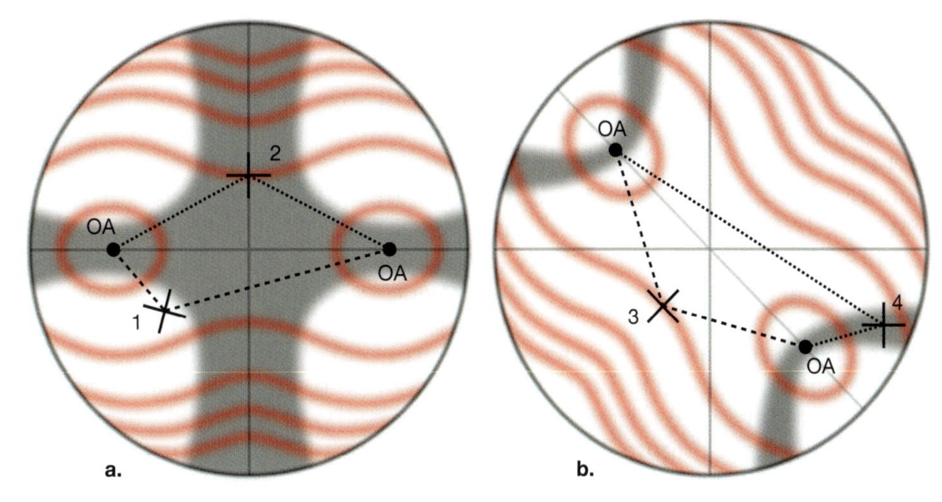

Figure 17.24. *Determining the orientation of the vibration directions within a biaxial interference figure.* ***a.*** *The two optic axes are shown plotting on a horizontal line, as seen in Figure 17.23a. To determine the orientation of the mutually-perpendicular vibration directions at any position on the interference figure, a graphical projection is used. Lines connecting the two optic axes to point 1 are first drawn. The vibration direction is then drawn so it equally bisects the two lines drawn from the optic axes. The orientation of the second vibration direction is then perpendicular to those lines. The directions of these vibrations do not correspond to those of the polarizers, so light would pass through the upper polarizer. If point 2 is selected and a similar construction performed, the vibration direction will correspond to that of the polarizers, so this region of the interference figure would appear dark (so an isogyre would occur in that area).* ***b.*** *The stage, and in turn the optic axes, rotated 45° clockwise from Figure 17.24a so they correspond to the interference figure show in Figure 17.23b. Point 3 is at the same location as point 1, and the vibration directions have now changed so they are at a greater angle to the polarizers, and thus greater light intensity would reach the upper polarizer. The vibration direction of point 4 corresponds to those of the polarizers, so no light is transmitted. This point would be occurring on an isogyre.*

ways, this is the easiest interference figure to find. As in the case for uniaxial minerals, whenever an optic axis is perpendicular to the stage, a circular section is parallel to the stage, and the crystal will exhibit very low retardation. This is demonstrated in the centered optic axis figures in Figure 17.25. In barite ($2V = 37°$), a single curved isogyre moves through the center of the field of view and a series of near-concentric isochromes circles the outcrop of the optic axis. The acute bisectrix is about two-thirds of the way to the edge of the field of view in the upper left quadrant.

The optic sign can also be determined from a centered optic axis figure. In barite (Figure 17.25b), addition occurs on the convex side of the isogyre and subtraction on the concave side, as was previously seen in the acute bisectrix figure of muscovite (Figure 17.19). An olivine sample with a $2V \approx 80°$ is shown in Figure 17.25c; its isogyre appears to be less curved than in barite—an indication of its higher $2V$. This figure can also be used to determine the optic sign because the isochromes are not circular, but are slightly elongated toward the direction of the acute bisectrix (which would outcrop outside of the field of view in the upper left quadrant). Insertion of the accessory plate shows addition on the concave side of the isogyres and subtraction on the convex side, indicating that this olivine is optically positive.

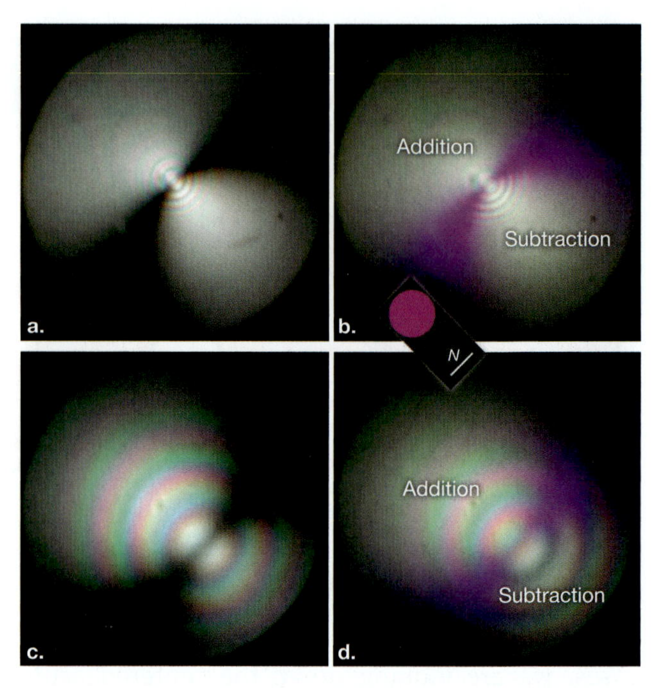

Figure 17.25. *Images of barite (left) and olivine (right) show centered interference figures obtained by looking straight down one of the optic axes of each of these biaxial minerals. The isochromes form almost-concentric circles around the optic axes.* ***a.*** *A portion of a circular isochrome in the upper left has its center outside the field of view; that center is the other optic axis.* ***b.*** *The accessory plate is inserted, showing addition on the convex side of the single isogyre, and subtraction on its concave side (analogous to what was seen in topaz and staurolite).* ***c.*** *In olivine, the other optic axis outcrops well outside the field of view, again in the upper left. The circle formed by the isochromes elongates in the direction of the other optic axis.* ***d.*** *Insertion of an accessory plate shows addition and subtraction in the same areas as in Figure 17.25b.*

The Indicatrix Revisited

When the optical indicatrix was first mentioned in Chapter 1, we explained that the indicatrix was a three-dimensional surface that could be used to model the speed of light in a material. For the simplest case of isotropic minerals, the indicatrix was a sphere with a radius representing the single refractive index, n. For the uniaxial case, the sphere became either an prolate ellipsoid (stretched vertically, with $\varepsilon > \omega$) or a oblate ellipsoid (squashed vertically, with $\omega > \varepsilon$), representing uniaxial positive or negative minerals, respectively (Figure 5.14).

The positive biaxial indicatrix results from taking a positive uniaxial indicatrix and reducing the length along one side (Figure 5.14). To describe the biaxial indicatrix (Figure 17.26a), we use a mutually-perpendicular vector set, with X, Y, and Z representing the three different refractive indices of the mineral: α, β, and γ. To further construct the three-dimensional surface of the biaxial indicatrix, three mutually-perpendicular ellipses are drawn, with their major and minor axes corresponding to the lengths α, β, and γ (Figure 17.26b). Because the length of β is intermediate between those of α and γ, it can be inscribed on the two-dimensional ellipse that has major and minor axes equal to γ and α, respectively. Values within the ellipse between α and β are labeled α', and those between β and γ are labeled γ'.

The biaxial indicatrix can be represented by dissecting the three-dimensional indicatrix with a series of two-dimensional planes. When a mineral rests on the stage of a microscope, its intersection with the biaxial indicatrix will be a two-dimensional plane, with major and minor axes corresponding to the mutually-perpendicular vibration directions. Each of the planes can be described using the lengths of the major and minor axes, which will correspond to α, β, and/or γ.

The first three cuts through the biaxial indicatrix are all ellipses (Figure 17.27). The X-Z plane has lengths α and γ, respectively. This plane is the most important in the biaxial optical indicatrix, and it is called the **optic plane** because it contains the two optic axes. The Y-Z plane has lengths of β

and γ, while the X-Y plane has lengths of α and β. The fourth cut is one of the two circular sections that can be inscribed inside this triaxial ellipsoid, and its radius is β.

Returning to the optical plane, Figure 17.27e shows the lengths inscribed on a slice through the indicatrix. This two-dimensional view is looking down on the edge of the circular sections. This contrasts with the uniaxial case in which there was one circular section and one optic axis perpendicular to it. In the biaxial case, each of the two circular sections has an optic axis perpendicular to it.

The $2V$ angle is also indicated on both slices shown in Figure 17.27e,f. The X direction bisects the acute angle between the two optic axes in Figure 17.27e. By definition, when the acute bisectrix (AB) equals X, the mineral is defied to be biaxial negative. When $AB = Z$, the mineral is optically positive (Figure 17.27f). Biaxial negative and positive minerals can also be explained in terms of the values of α, β, and γ. Optically positive minerals have β closer to α than γ. In fact, if $\beta = \alpha$, then there would be only one circular section, and the mineral would be uniaxial positive. Optically negative minerals have β closer to γ, and likewise if β were to become equal to γ, the mineral would only possess one circular section in which $\omega > \varepsilon$, so the mineral would be uniaxial negative.

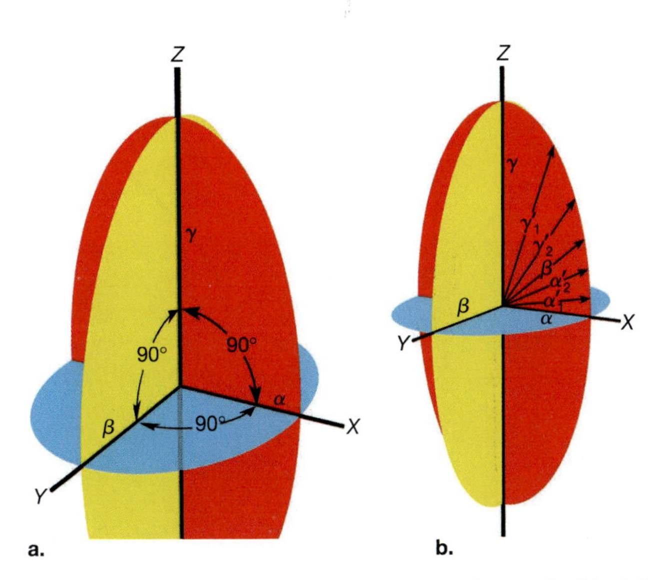

*Figure 17.26. Detail of nomenclature relating to the biaxial indicatrix (modified from Bloss, 1999, Figure 10.1). **a.** The biaxial indicatrix is defined by three mutually-perpendicular directions X, Y, and Z, which have lengths equal to α, β, and γ, where α < β < γ **b.** The surface defined by three mutually-perpendicular ellipses describes the biaxial indicatrix. In the X-Z plane, α corresponds to X (in the horizontal plane) and γ corresponds to Z. Thus, β must lie somewhere on the plane between α and γ. Values of refractive index between α and β are termed α′, and those between β and γ are γ′, so α < α′ < β < γ′ < γ.*

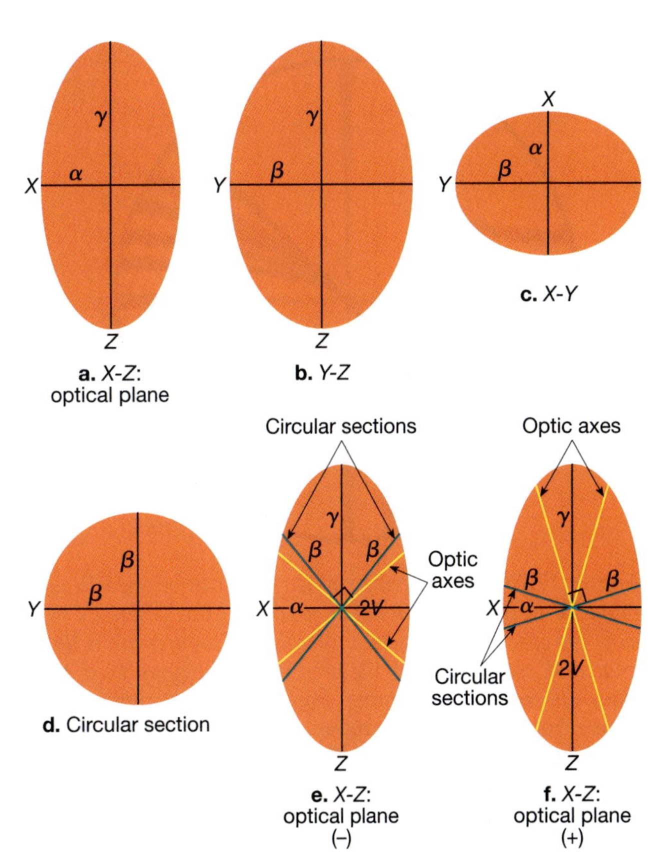

*Figure 17.27. A series of two-dimension planes that cut the three-dimensional biaxial indicatrix. Table 18.1 lists equations relating refractive index values, orientation, and 2V for biaxial minerals. **a.** The X-Z plane is an ellipse with axes equal in length to α and γ. This plane is termed the optic plane because it contains the two optic axes. **b.** The Y-Z plane is also an ellipse, with minor and major axes equal in length to β and γ. **c.** The X-Y plane is an ellipse with minor axis length of α and major axis length of β. **d.** The circular section is a circle that occurs at two special positions within the biaxial indicatrix with radius β. **e.** The optic plane for a negative biaxial mineral showing traces of the circular sections, the two optic axes, and the angle 2V, which is defined as the acute angle between the two optic axes. In this graphical construction, the length β occurs between α and γ. The two optic axes (shown here as dashed lines) are perpendicular to each of the two circular sections. Because the 2V angle is bisected by X or α, by definition this mineral is optically negative. The length of β is closer to γ than to α. **f.** When β is closer to α than to γ, the 2V angle is bisected by Z (γ), defining the mineral to be optically positive.*

Variations in 2*V*. To aid in the identification of biaxial minerals, it is useful to be able to estimate $2V$ on the basis of an interference figure, which is done most easily by inspection of a centered optic axis figure. The curvature of the isogyres increases as a function of $2V$. This was already apparent from the interference figures of barite, which had a $2V$ of 37°, and olivine ($2V = 80°$). Figure 17.28 shows that for a $2V$ of 15°, the second optic axis

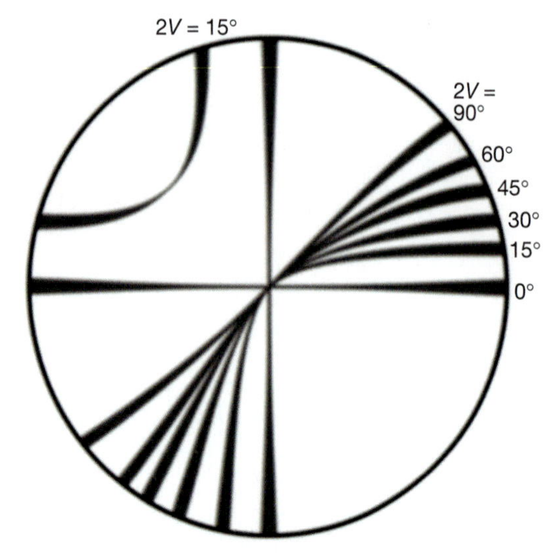

Figure 17.28. *The curvature of isogyres can be used to estimate 2V in figures centered on an optic axis. When 2V = 0, as is the case for a uniaxial mineral, the isogyres are orthogonal and correspond to the crosshairs. Isogyres that arise from values of 2V varying from 0–90° are shown here. Lower values of 2V are more curved than higher values (modified from Bloss, 1999, Figure 11.15).*

outcrops in the upper left quadrant, about halfway to the edge of the field of view. For the case of $2V = 30°$, the second optic axis occurs at the edge of the field of view. When $2V > 30°$, the second optic axis is not seen. These statements assume that the highest numerical aperture lens (objective) on the microscope is used. Typically, this will be an image of ~60° of the interference figure.

When imaging interference figures, conoscopic illumination must be used so the rays are increasingly divergent, radiating out from the center of the crosshairs. To calculate the maximum divergence angles at the edge of the field of view, Snell's law is used to trace the rays through the crystal. Figure 17.29 shows the most divergent rays in a grain, which are labeled AA_g (angular aperture within the grain). When a ray leaves the crystal, it is refracted away from the normal, as shown by the angle θ_t. AA_a, the angular aperture in air, is inscribed on top of the grain; it relates to the numerical aperture of the lens being used.

From Chapter 5, recall that Snell's law is $n_i \sin \theta_i = n_t \sin \theta_t$. Making the necessary substitutions from Figure 17.29 yields:

$$n_g \sin\left(\frac{AA_g}{2}\right) = n_a \sin\left(\frac{AA_a}{2}\right),$$

and because $(AA_a/2) = NA$, we arrive at

$$AA_g = 2\sin^{-1}\left(\frac{NA}{n_g}\right). \qquad \text{Equation 17.2}$$

AA_g is the angle of the most divergent rays in the grain, AA_a is the angle of the most divergent rays in air, n_g is the refractive index of the grain, n_a is the refractive index of air (i.e., 1.0), and NA is the numerical aperture of the objective.

If the refractive index of a grain and the numerical aperture of the lens are known, then AA_g can be calculated. As NA decreases, a smaller area of the indicatrix is seen. Likewise, as n_g increases, a smaller region would be seen. For example, given $NA = 0.8$ and $n_g = 1.6$, the maximum AA_g would be 60°. In other words, if the optic axes outcrop at the edge of the field of view looking down the AB direction (in this microscope configuration), $2V$ would be 60°.

Likewise, the optic axes of minerals with $2V > 60°$ will always occur outside of the field of view in a centered acute bisectrix figure. As was shown in Figure 17.28 for the case of a centered optic axis figure, if $2V = 30°$, then the second optic axis will plot at the edge of the field of view based on Equation 17.2.

To determine the $2V$ of a mineral with a centered acute bisectrix figure, measure the distance from the center of the field of view to the edge, and the distance from the center of the field of view to the one optic axis. Then use:

$$A_o = \sin^{-1}\left(\left[\frac{D_o}{D_e}\right]\left[\frac{NA}{n_g}\right]\right), \qquad \text{Equation 17.3}$$

where A_o is the general angle an object makes from vertical, D_o is the distance to the object, and

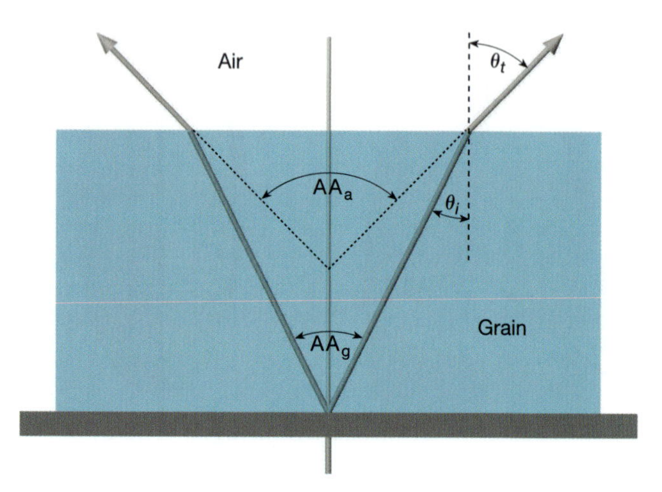

Figure 17.29. *A graphical depiction of the rays traveling through a grain in conoscopic illumination, showing that they will diverge as they leave the grain. The associated angles can be precisely calculated by the application of Snell's Law. More specifically, this illustration shows how the unknown angular aperture within the grain (AA_g) can be determined based on the angular aperture in air (AA_a), which in turn is related to the numerical aperture of the lens and the refractive index of the grain; this relationship is given in Equation 17.1.*

D_e is the distance to the edge of the field. This equation can be used to determine $2V$, which is twice the value of A_o.

To further explain the derivation of Equation 17.3, the image in Figure 17.30b shows a side view of the rays as they move through a mineral grain, illuminated with conoscopic light. The edge of the field of view is D_e, and the distance to the object is labeled D_o. A_o is the actual angle that is calculated from Equation 17.3.

With respect to optical properties such as the indicatrix, there is a continuum from uniaxial through biaxial minerals (Figure 17.31). For example, we could say that a biaxial mineral with $2V = 0°$ would be uniaxial. As $2V$ increases to 30°, the isogyres in the interference figure break apart. As $2V$ increases to 60°, the isogyres are seen at the edge of the field of view.

As a final reminder that the interference figures observed in the microscope are really two-dimension projections of three-dimensional optical phenomena, look again at the ping-pong ball images in Figure 17.32. A centered uniaxial optic axis figure is shown, with a single optic axis emanating from the center of the cross (Figure 17.32a). This image would stay the same even when the stage is rotated. For a centered acute bisectrix figure, we see first the interference figure projected at 45° off extinction on the ping-pong ball (Figure 17.32b) and then rotated to extinction (Figure 17.32c). Notice how the thin isogyre contains the optics axes.

Dispersion

Most optical properties of minerals change as a function of wavelength; this property is called **dispersion**. In general, the refractive indices decrease as the wavelength increases. The amount of dispersion and the rate at which the refractive indices change with wavelength are not necessarily the same for individual refractive index directions in a mineral. Sometimes these changes can cause interesting optical phenomena, as in the case of the mineral brookite when looking down the γ vibration direction (Figure 17.33). At extinction, the grain producing the acute bisectrix figure sort of resembles a centered optic axis figure for a uniaxial mineral. When it is rotated 45° clockwise, it produces a strange figure resembling a centered acute bisectrix figure for a biaxial mineral. Figure 17.33a,b is illuminated in white light, while Figure 17.33c–e is illuminated in monochromatic light with wavelengths of 500, 555, and 650 nm. In light with $\lambda = 500$ nm, a typical acute bisectrix figure is seen, but when the wavelength changes to $\lambda = 555$ nm, the resultant figure resembles a uniaxial optic axis figure because α and β become equal at that wavelength. At longer wavelengths (650 nm), the mineral again exhibits an acute bisectrix figure but with the optic axes in the opposite quadrants from those at 500 nm. This example is a very rare occurrence for minerals, but it shows that optical classes of minerals are indeed a function of refrac-

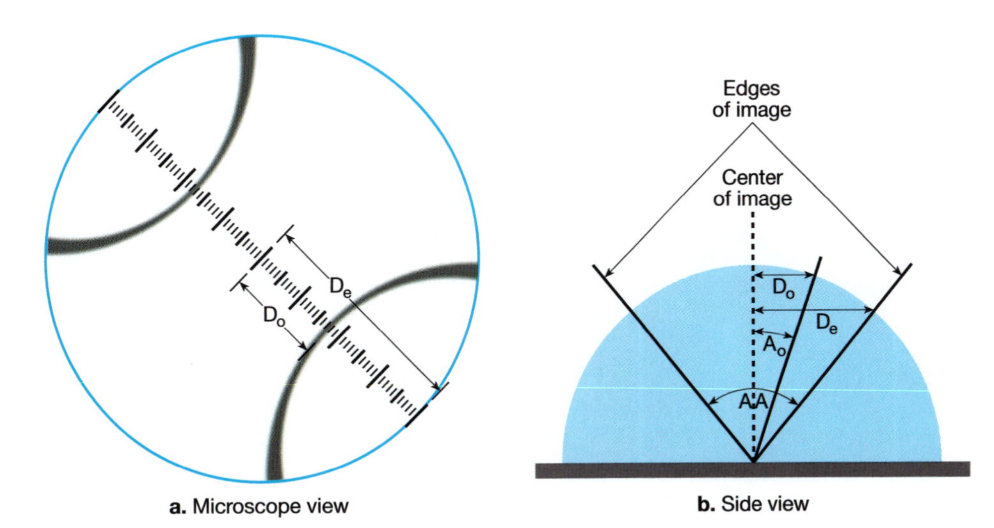

*Figure 17.30. A graphical explanation of how angles in conoscopic illumination can be calculated. **a.** A centered acute bisectrix interference figure. Measurements can be made by using a graduated ocular to determine 2V for mineral. The angle between the ray and the normal ray path on the PLM (i.e., how much it deviates from being normal to the stage) is termed A_o. The distance to the edge of the field of view (D_e) and the distance to the object (D_o) can be measured and applied to Equation 17.2 to determine the angle. In this case, twice that angle ($2A_o$) would be 2V. **b.** A side view of the image in Figure 17.30a, showing that an interference figure is a 2-D image of the 3-D spherical surface whose edge is defined by the angular aperture (AA) of the lens.*

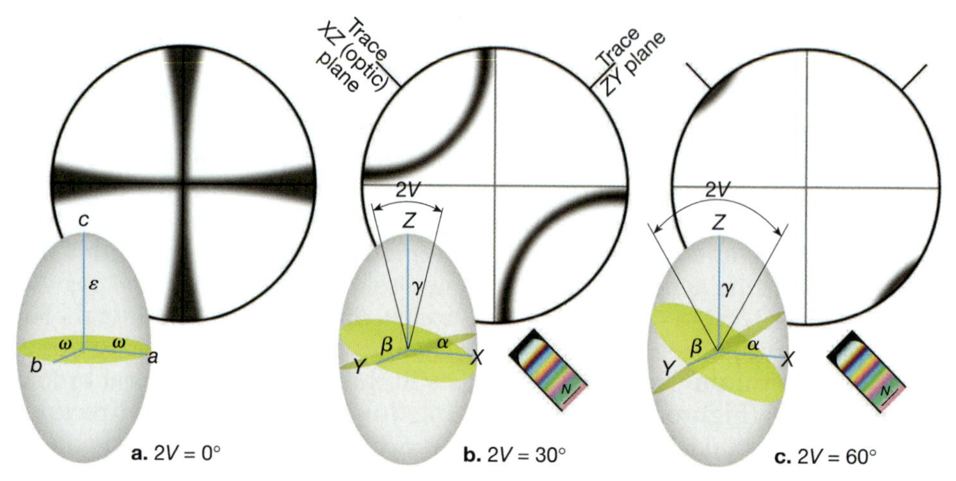

Figure 17.31. Minerals with 2V varying from 0–60°. The upper illustrations show the interference figures, while the lower ones show the corresponding indicatrices that produced the figures. ***a.*** *The uniaxial case where 2V = 0°.* ***b.*** *For 2V ≈ 30°, the figure has broken apart as the isogyres move to the upper left and lower right, forming two separate isogyres.* ***c.*** *2V is now ~60°, and the optic axes are outcropping near the edge of the field of view.*

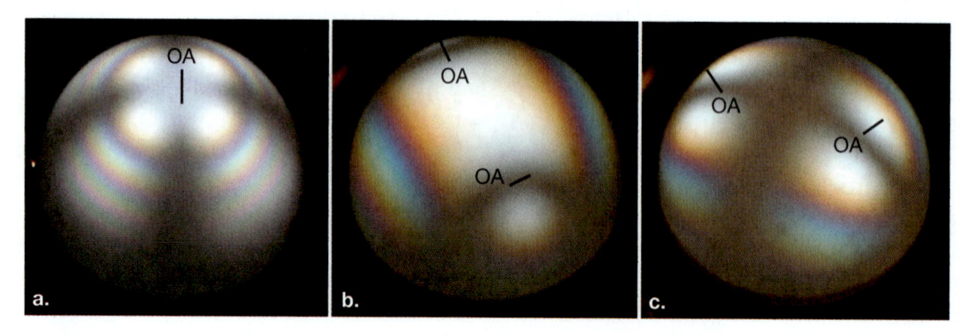

Figure 17.32. The interference figure seen in a microscope is simply a two-dimensional depiction of a three-dimensional surface. These images show interference figures for uniaxial (Figure 17.32a) and biaxial (Figure 17.32b and c) minerals projected onto a ping-pong ball cut in half to rest on top of a microscope stage, with conoscopic illumination. ***a.*** *The optic axis for quartz emanates from where the isogyres cross. As the angular aperture of the microscope lens increases, more of the isochromes can be seen (as seen in Figure 17.8).* ***b.*** *For a biaxial mineral that is rotated off extinction, the optic axis comes from the center of the isogyres.* ***c.*** *The same biaxial grain has been rotated to extinction, and the trace of the optic plane is again shown by the center of the two isogyres.*

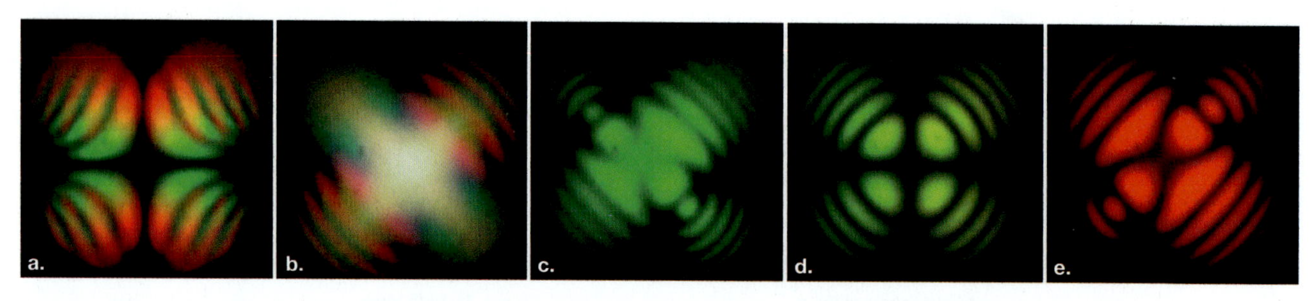

Figure 17.33. The centered acute bisectrix interference figure for brookite is illuminated here at two different orientations relative to polychromatic light (Figure 17.33a and b) and at the same orientation (Figure 17.33c, d, and e) illuminated by different wavelengths of monochromatic light. ***a.*** *The grain positioned at extinction (similar to Figure 17.23a).* ***b.*** *Grain rotated 45° clockwise to match Figure 17.23b.* ***c.*** *Grain in the same orientation as in Figure 17.33b, illuminated by light with a wavelength of 500 nm, and showing the typical acute bisectrix interference figure.* ***d.*** *Grain illuminated by light with 555 nm wavelength, and the resultant figure that resembles a uniaxial optic axis figure because α and β become equal at that wavelength.* ***e.*** *At longer wavelengths (650 nm), a biaxial interference figure is again formed, but the optic axes appear to be in different quadrants than in Figure 17.33c.*

tive indices, and in turn refractive indices are a function of the wavelength of illumination.

The three refractive indices for brookite can be graphed relative to wavelength (Figure 17.34). All three refractive indices decrease as wavelength increases. At low wavelengths, α and β values are very close together, and the value of β changes more rapidly with wavelength than does α. The values of α and β become equal at ~550 nm, causing this orthorhombic mineral to exhibit a uniaxial interference figure. As the wavelength increases past 555 nm, the mineral again becomes biaxial. For this example, it is really more meaningful to discuss the refractive indices in relationship to crystallographic directions than to attempt to assign α and β directions.

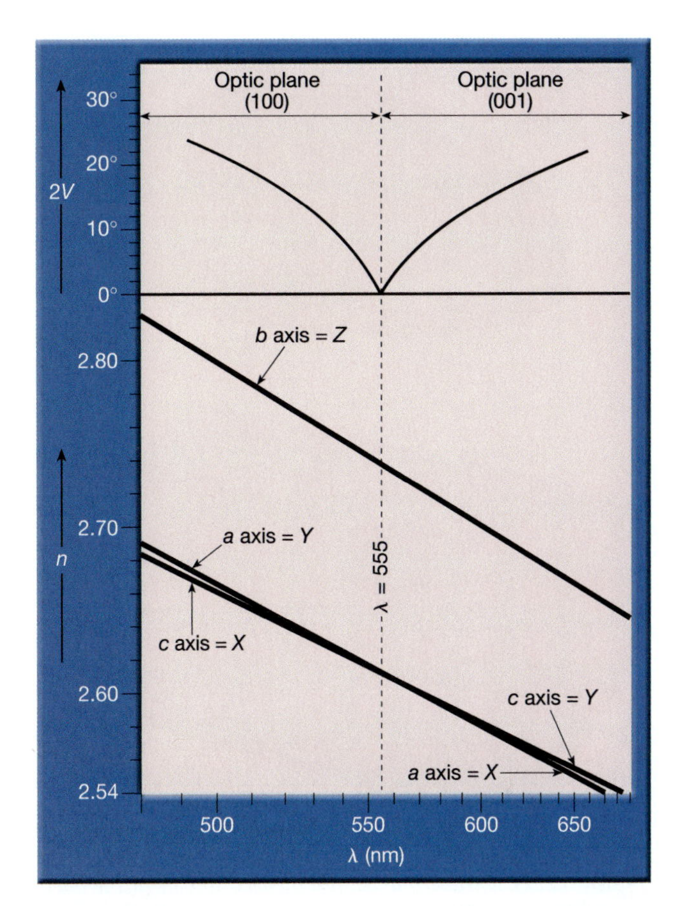

Figure 17.34. A plot of the change in refractive indices as a function of wavelength for brookite. As is typical, refractive index decreases as wavelength increases. The refractive index parallel to the a *axis decreases at a greater rate than that parallel to the* c *axis, causing these two values to become equal at ~555 nm (as seen in the previous figure). Thus, the orthorhombic brookite shows a uniaxial interference figure when illuminated by light of 555 nm wavelength.*

Indicatrix on "Stage"

The appearance of minerals resting on the microscope stage, either as single crystals or thin sections, is dependent on the orientation of the optical indicatrix, so a thorough understanding of the optical indicatrix is needed to help identify minerals with a PLM. While the light travels perpendicular to the stage of the microscope, it vibrates in the plane of the stage (i.e., within the two-dimensional cut of the indicatrix). If a biaxial indicatrix rests on its *Y-Z* plane, the refractive indices β and γ would be measured in orthoscopic illumination, and retardation associated with $\gamma - \beta$ would be observed (Figure 17.35). In conoscopic illumination, a centered acute bisectrix figure for a negative mineral would be seen, arising from a series of rays in mutually-perpendicular vibration directions emerging from the crystal at different orientations. This is analogous to the formation of the isochromes and isogyres for centered uniaxial optic axis figures (Figure 17.18b). Biaxial interference figures are directly analogous to those for uniaxial minerals. When unpolarized light is forced to vibrate in two mutually-perpendicular directions along *N* and *n* (typically α' and γ'), interference occurs when the rays emerge from the crystal, producing retardation that is manifested as isochromes.

In Chapter 16, a series of illustrations (Figure 16.15) shows how light behaves when traveling through a uniaxial material. Figure 17.36 shows an analogous series of constructions for a biaxial mineral. The grain is initially at a random orientation, with none of the principal vibration directions (*X*, *Y*, or *Z*) in the plane of the stage of the microscope. Both rays behave as E-rays traveling through the mineral. When only one principal vibration direction is in the plane of the microscope stage, that ray behaves as an O-Ray; the other randomly-oriented vibration direction is an E-ray. When two principal vibration directions are located in the plane of the stage of the microscope, the third direction is perpendicular to the stage, and both rays traveling through the crystal behave as O-rays. In all cases, the values of α, β, and γ would be associated with O-rays, and α' and γ' would be associated with E-rays. In order to measure a principal refractive index, a crystal must be oriented in this way, with the direction of interest in the plane of the microscope stage; this can be accomplished through use of interference figures.

If a biaxial indicatrix is placed in the center of a mineral and each crystal face is placed on a microscope stage, a projection of the indicatrix can be seen (Figure 17.37). The resultant ellipse represents the vibration direction in the plane of the microscope stage. The top face represents the γ and β vibration directions, the face at the lower right is α and β, and the front face is α and γ. Two

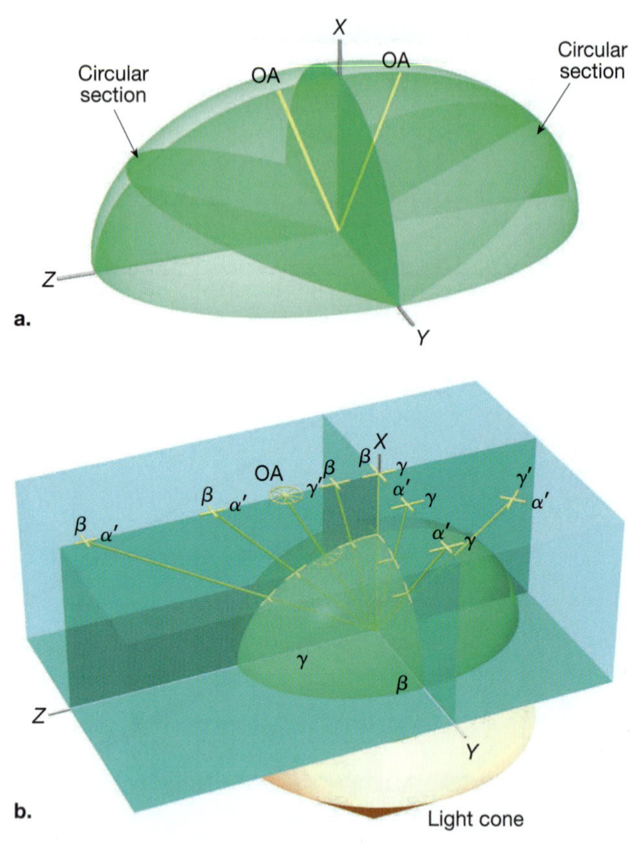

Figure 17.35. *Three-dimensional views of the negative biaxial indicatrix with its Z-Y plane coinciding with the microscope stage.* ***a.*** *The relationships between X, Y, and Z, the two optic axes, and their mutually-perpendicular circular sections.* ***b.*** *Series of light rays emanating out of the biaxial indicatrix when the grain is viewed with conoscopic illumination. This three-dimensional perspective view relates directly to Figures 17.32b and 17.19c (modified from Bloss, 1999, Figure 10.14).*

other faces occur at random; one is cut perpendicular to the optic axis to show a circular section. The retardation for this mineral will be zero (black) when it rests on its circular section, and will be the greatest when resting on the face showing the γ and α values, with the other faces having intermediate retardation values. Changes in orientation that result in changes in retardation are evidenced by different colors.

Interference Figures in Practice

The appearance of a mineral relates directly to its optical indicatrix, so that precise mineral identification is only possible when orientation is constrained. This means that grains in an immersion mount or thin section must be selected so they can provide useful interference figures to discriminate between uniaxial and biaxial minerals, determine optic sign and, for the case of biaxial minerals, estimate the $2V$. A series of photomicrographs of grains in immersion mounts and oriented thin sections is provided here to lend insights into how orthoscopic and conoscopic views are related, and

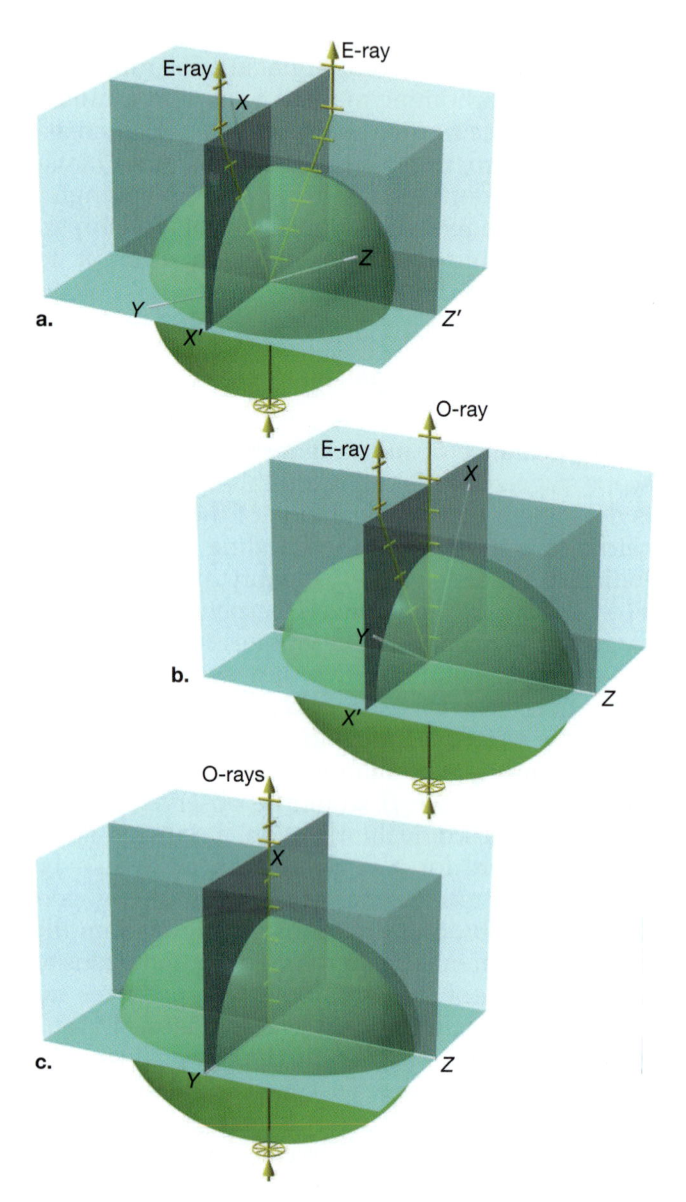

Figure 17.36. *The biaxial indicatrix is shown here resting on different surfaces of a microscope stage with the resultant ray paths for the two mutually-perpendicular vibration directions (modified from Bloss, 1999, Figure 10.11).* ***a.*** *The indicatrix here is in a random orientation, so that none of the principal vibration directions lies in the plane of the microscope stage. The incident unpolarized light waves are forced to vibrate along two mutually-perpendicular directions with neither wave obeying Snell's Law. Thus, both light rays behave as E-rays.* ***b.*** *The Z vibration direction now lies in the plane of the microscope stage. Light coinciding with the Z direction behaves as an O-ray, whereas the other vibration direction behaves as an E-ray.* ***c.*** *Both Y and Z are in the plane of the stage of the microscope, so both waves behave as O-rays.*

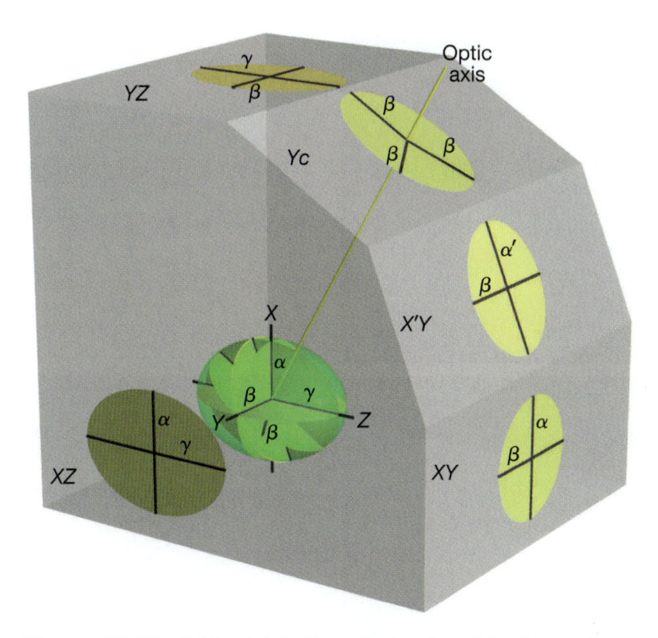

Figure 17.37. A biaxial indicatrix is placed inside a three-dimensional block bounded by planes in several different orientations. Each plane represents a cut through the block. A two-dimensional projection of the ellipsoid is shown on each plane. If the mineral rested on any of those planes, light would be forced to vibrate along those refractive index directions (modified from Bloss, 1999, Figure 10.10).

to demonstrate how the appearance of crystal can change as a function of orientation.

The most useful interference figure for grains in an immersion mount will often come from the largest grain with the smallest retardation (the grayest colors), because these grains will probably have a circular section near-parallel to the stage, and in turn, will produce an optic axis figure. Likewise, the smallest grain in an immersion mount with the highest retardation will produce a flash figure for uniaxial minerals (Figure 17.38) or an optic normal interference figure for a biaxial mineral (where γ and α are in the plane of the microscope stage).

The same idea is illustrated in Figure 17.39 for a different mineral, but in this case, the interference figure of the smallest retardation grain would identify this to be a biaxial mineral. The highest retardation grain produces an optic normal interference figure. Grains in immersion mounts vary in size and thickness, and will tend to produce higher retardation colors, so they will tend to have higher retardation colors than are seen in standard thin sections. Thus, it is typically easier to find and interpret interference figures in grain mounts than in thin sections. Isochromes in grain mounts also tend to be concentric because the grains are often wedge-shaped or pyramidal.

In a thin section, the grains are all the same thickness by definition (30 μm). At this thickness, framework silicates appear with first-order colors, and the other common silicates with higher birefringence (olivine, pyroxene, etc.) can be easily distinguished. An unfortunate consequence of the 30 μm standard thickness is that the grains aren't thick enough to have very many isochromes, so interpretation of interference figures becomes much more difficult. Figures 17.40 and 17.41 illustrate this problem by showing thin sections of oriented quartz grains and the resultant interference figures. When the quartz plate is inserted with the centered optic axis figure, addition occurs in the upper right quadrant, where first-order gray changes to second-order blue. The mnemonic

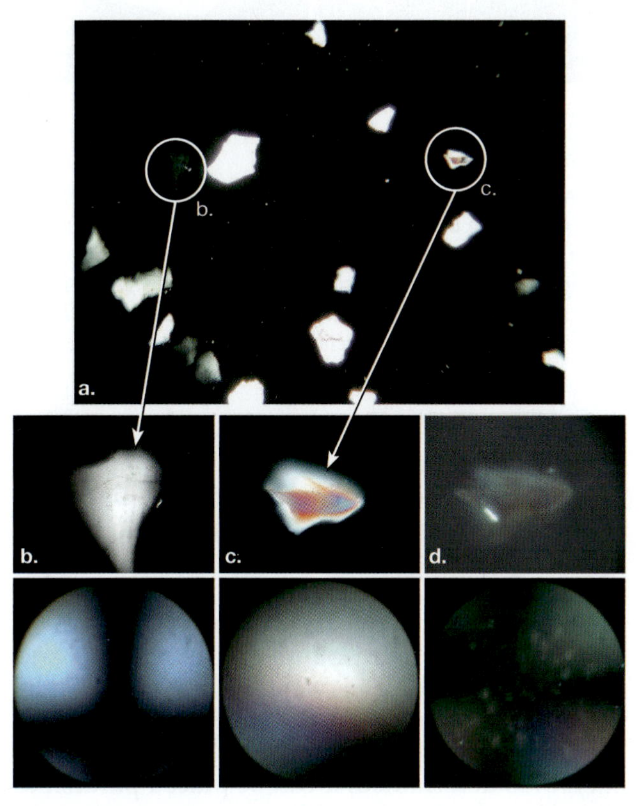

Figure 17.38. *Grain size and retardation can be used to select grains that might produce certain types of interference figures for a uniaxial mineral.* ***a.*** *To obtain optic axis figures, look for the largest grain with the lowest retardation. In this case, the grain circled and labeled "b" is chosen. To obtain a flash figure, look for the smallest grain with the largest retardation: here, the circled grain labeled "c."* ***b.*** *In the upper image, the grain labeled "b" is shown at higher magnification and in orthoscopic illumination. Below it, the corresponding interference figure, obtained using conoscopic illumination, shows that the grain gives a slightly off-centered optic axis figure.* ***c.*** *The "c" grain is shown at higher magnification, slightly off extinction, and in orthoscopic illumination. Below it is the corresponding interference figure.* ***d.*** *Grain "c" is now rotated to extinction in orthoscopic illumination in the upper image, with a conoscopic view of a centered uniaxial flash figure in the lower image.*

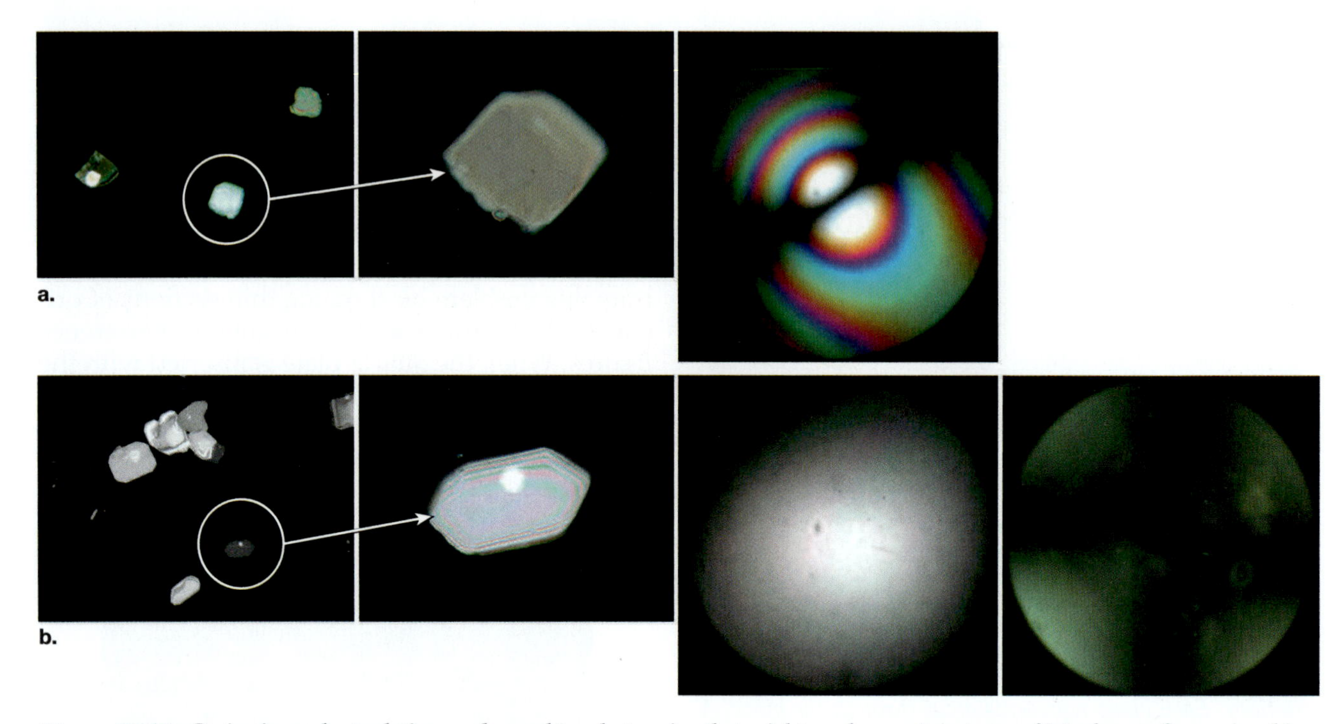

Figure 17.39. Grain size and retardation can be used to select grains that might produce certain types of interference figures in a biaxial mineral. ***a.*** *In the circle is a large grain exhibiting low retardation that is suggestive of a grain oriented so as to produce an optic axis figure. The grain is shown in a higher magnification orthoscopic view in the center. At the right is a conoscopic view of a centered optic axis figure for a biaxial mineral.* ***b.*** *The circled grain has the highest retardation visible in this collection of grains. To the right is the same grain seen at higher magnification, slightly off extinction. The next image to the right shows the interference figure obtained from this grain. At far right is an interference figure of the grain when rotated to extinction, showing a centered, optic-normal figure.*

"BURP" is sometimes used to remember that "**b**lue in the **u**pper **r**ight quadrant is **p**ositive."

Optic sign can also be measured from a flash figure by determining which of the principal vibration directions corresponds to ε (the single optic axis). When the flash figure is rotated in either direction, the isogyres will break up and move toward the edge of the field of view in the quadrants where the optic axis is being rotated. When the vibration direction of the optic axis has been thus determined, it can be placed parallel to N_{plate}, and checked for addition vs. subtraction. In Figure 17.41c, the mineral happens to be optically positive (because this is quartz!).

The optical orientations of minerals relate their crystallographic directions to their optical indicatrices. In isotropic minerals, the optical orientation is meaningless because the *a*, *b*, and *c* crystallographic axes occur within a spherical indicatrix. In uniaxial minerals (tetragonal and hexagonal),

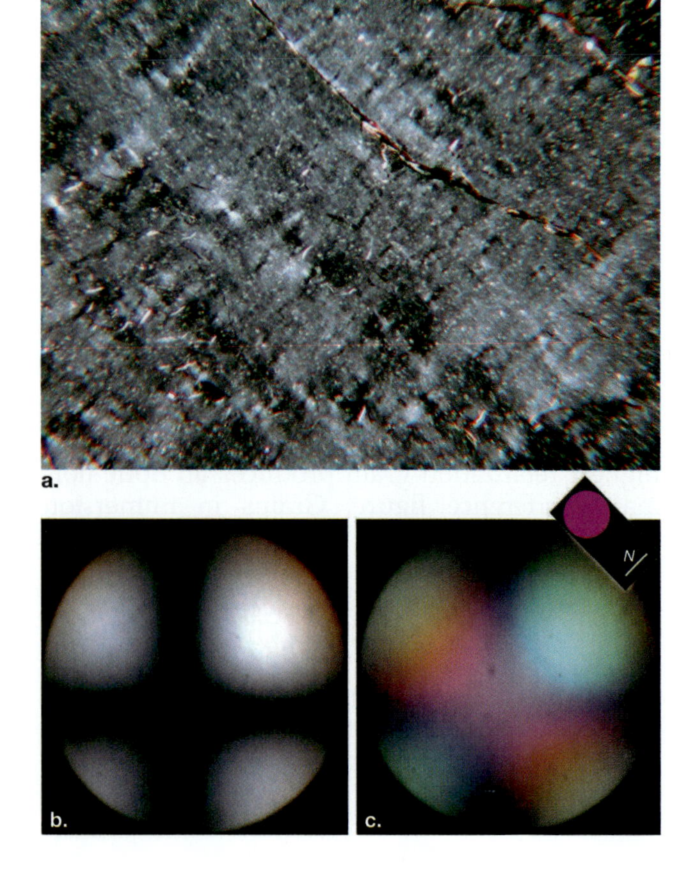

Figure 17.40. An oriented thin section of a quartz grain, specially prepared so that its optic axis is perpendicular to the slide. ***a.*** *The sample in cross-polarized light.* ***b.*** *Interference figure of the sample.* ***c.*** *Insertion of the quartz plate shows addition in the upper right and lower left, leading to the conclusion that this mineral is optically positive.*

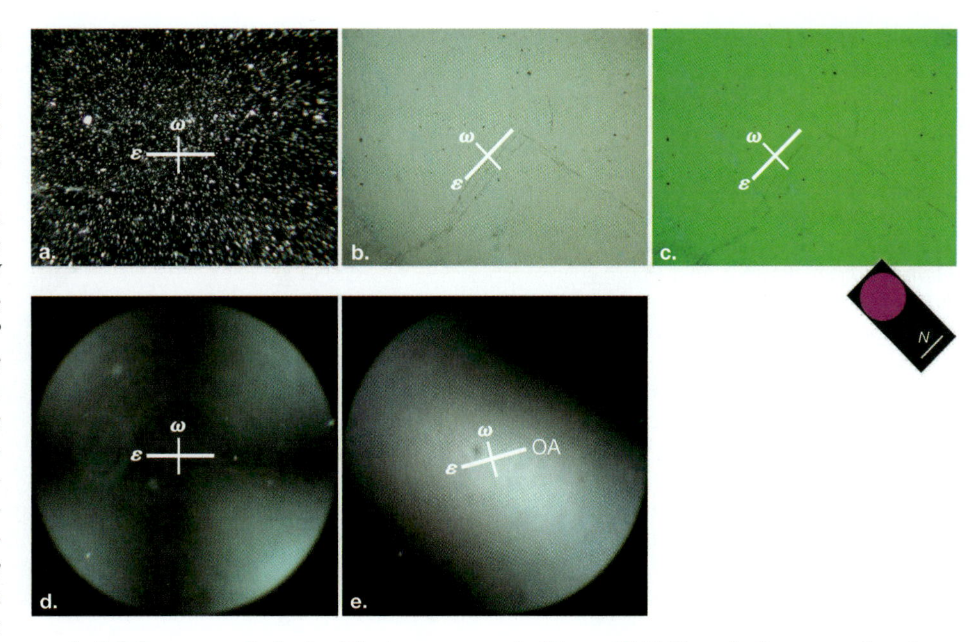

Figure 17.41. An oriented thin section of a quartz grain mounted with its optic axis lying in a plane parallel to the surface of a glass slide. The ε *and* ω *vibration directions are indicated.* ***a.*** *Orthoscopic view of the sample in cross-polarized light, with the optic axis oriented in the E-W direction.* ***b.*** *View of the sample when it has been rotated 45° counterclockwise to produce the 1st-order white retardation color.* ***c.*** *Insertion of the quartz plate shows addition because the interference color increases to 2nd-order green, leading to the conclusion that this mineral is optically positive.* ***d.*** *The interference figure obtained from the grain in the same orientation as in Figure 17.41a.* ***e.*** *The stage has now been rotated slightly counterclockwise. The isogyres seen in Figure 17.41d have broken up and are leaving the field of view toward the upper right and lower left, which is the quadrant in which the optic axis now lies (after the rotation).*

the *c* crystallographic axis coincides with a single optic axis; the *a* and *b* axes occur perpendicular to *c* in the single circular section. So, a PLM can determine the orientation of the *c* crystallographic axis (because it coincides with the optic axis), but it cannot distinguish between or identify the *a* and *b* axes (Figure 17.42a).

In biaxial minerals, the situation is even more complicated. In the simplest case of orthorhombic minerals, the mutually-perpendicular crystallographic axis set (*a*, *b*, *c*) corresponds in one of six possible arrangements to the three mutually-perpendicular directions of the optical indicatrix (*X*, *Y*, and *Z*). In monoclinic minerals, only one of the principal vibration directions corresponds to the *b* crystallographic axis, which is either the 2-fold axis or is perpendicular to a mirror plane. For the triclinic case, none of the crystallographic directions is forced to correspond to any of the optical directions by symmetry.

An orthorhombic example is shown for topaz (Figure 17.42b). In Figures 17.43–17.47, a crystal was cut perpendicular to the *Z*, *X*, and *Y* directions in the optical indicatrix to show a centered

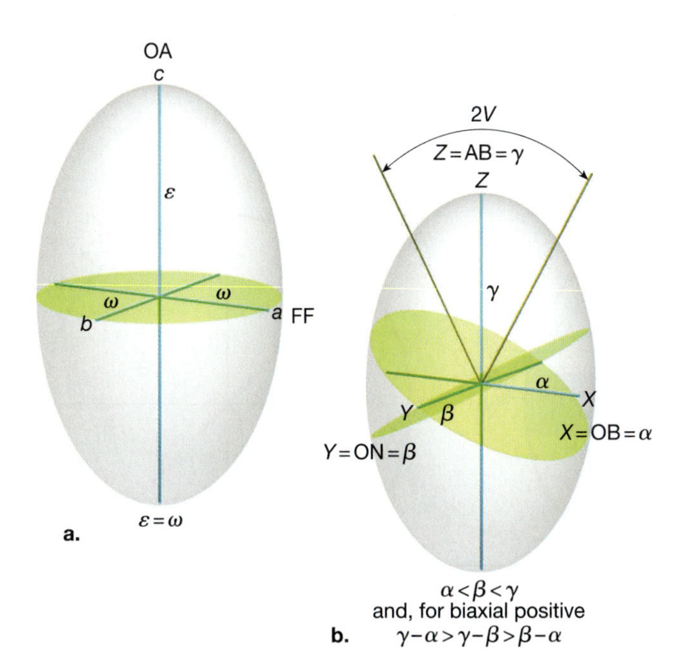

Figure 17.42. The optical orientation of a mineral relates its indicatrix directions to its crystallographic axes. ***a.*** *In quartz (and all uniaxial minerals), the* c *crystallographic direction corresponds to the* ε *vibration direction and optic axis. Looking down the* c *crystallographic axes, an optic axis figure is obtained. The* a *and* b *crystallographic axes are both perpendicular to* c*, and thus are located in the single circular section. When the mineral is viewed perpendicular to the optic axis, a uniaxial flash figure is observed.* ***b.*** *In the orthorhombic mineral topaz, the crystallographic axes form a mutually-perpendicular vector set in which the crystallographic axes directly correspond to the* X, Y, *and* Z *directions in the biaxial indicatrix. This drawing shows that the acute bisectrix for topaz corresponds to the* Z *direction, so that topaz is optically positive. The* X *direction could be defined here as the obtuse bisectrix because it bisects the obtuse angle made by the two optic axes. The* Y *direction is the optic normal because it is perpendicular to the optic plane (defined as the plane which contains the two optic axes). Thus, there are three separate directions that can be viewed in this orthorhombic mineral: the acute bisectrix, the obtuse bisectrix, and the optic normal.*

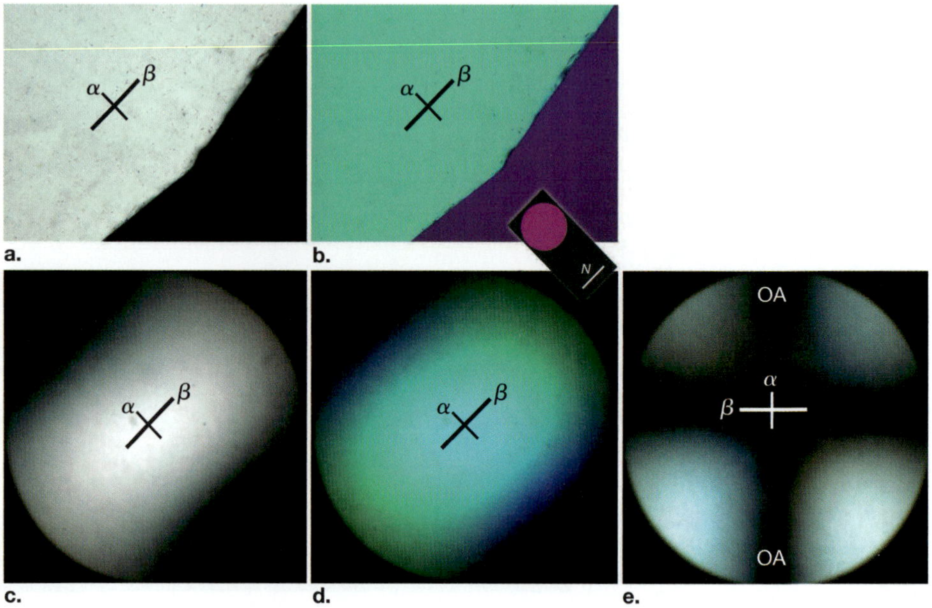

Figure 17.43. *Topaz grain viewed down its acute bisectrix (AB), with α and β in the plane of the microscope stage.* ***a.*** *The grain is rotated 45° off extinction, showing a retardation that corresponds to β – α, which is 1st-order white.* ***b.*** *The quartz plate has been inserted, and the interference colors increase.* ***c.*** *Conoscopic image of the same grain, in the same orientation as in Figure 17.43a. The two optic axes outcrop at the upper left and lower right corners of the field of view.* ***d.*** *When the quartz plate is inserted, the retardation color is the same as in Figure 17.43b.* ***e.*** *Conoscopic view of the grain, rotated to extinction.*

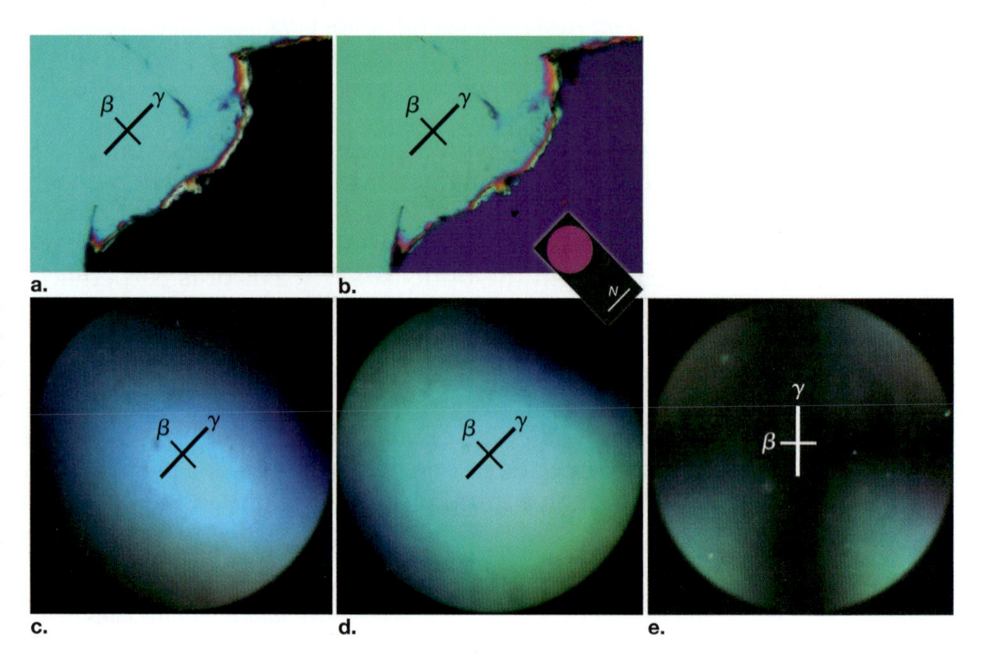

Figure 17.44. *A second oriented topaz grain viewed down its obtuse bisectrix (OB), with γ and β in the plane of the microscope stage.* ***a.*** *The grain is rotated 45° off extinction, showing a 2nd-order blue retardation color that corresponds to γ – β.* ***b.*** *The quartz plate has been inserted, and the interference colors increase to 3rd-order blue-green.* ***c.*** *Conoscopic image of the same grain in the same orientation as in Figure 17.44a.* ***d.*** *When the quartz plate is inserted into Figure 17.44d, the retardation color is the same as in Figure 17.44b.* ***e.*** *Conoscopic view of the grain rotated to extinction.*

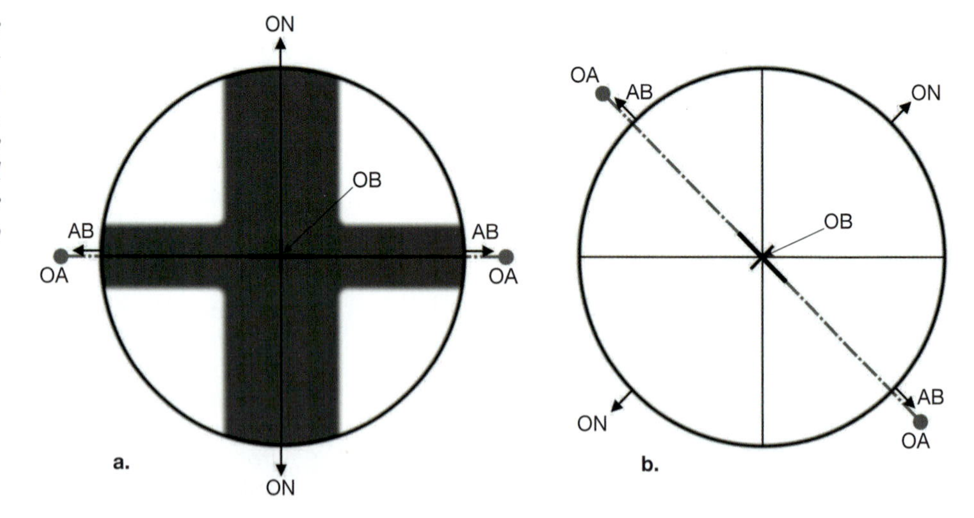

Figure 17.45. *Sketch of the interference figure for a centered obtuse bisectrix figure.* ***a.*** *The isogyres are at extinction, and the locations of the optic axes (OA) are outside the field of view.* ***b.*** *A 45° clockwise rotation causes the isogyres to leave the field of view.*

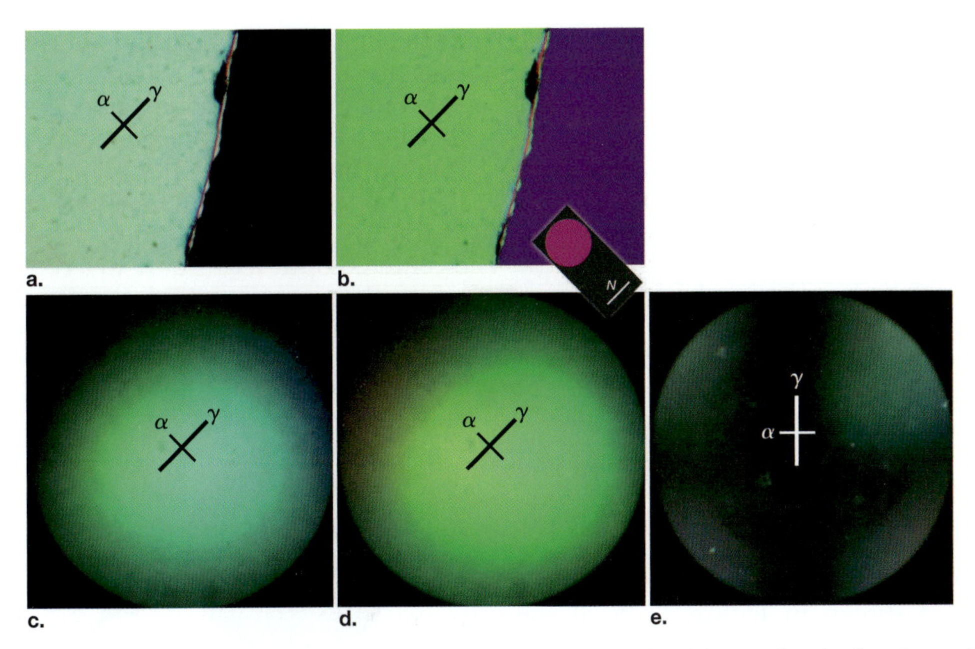

Figure 17.46. *A third topaz thin section viewed down its optic normal (ON), with γ and α in the plane of the microscope stage. The retardation color is as high as it can be for this mineral, because γ and α are the most different in magnitude.* ***a.*** *The grain is rotated 45° off extinction, exhibiting a 2nd-order green retardation color.* ***b.*** *The quartz plate has been inserted, and the interference colors increase to 3rd-order green.* ***c.*** *Conoscopic image of the same grain at the same orientation as in Figure 17.46a.* ***d.*** *When the quartz plate is inserted, the retardation color is the same as in Figure 17.46b.* ***e.*** *Conoscopic view of the grain rotated to extinction. Compare this conoscopic view to those in Figures 17.43e and 17.44e, and notice a progression in the 'fuzziness' of the isogyres from the* AB *to* OB *to* ON *centered interference figures.*

acute bisectrix (*AB*) figure, a centered obtuse bisectrix (*OB*) figure, and an optic normal (*ON*) figure, respectively. Topaz is optically positive, with a $2V \approx 60°$, so β should be closer to α than it is to γ. Therefore, the acute bisectrix will exhibit retardation related to $\beta - \alpha$, the obtuse bisectrix $\gamma - \beta$, and the optic normal $\gamma - \alpha$. The retardation thus increases in this series of images. In the progression from *AB* to *OB* to *ON*, the isogyres become more diffuse, and cover more of the field of view. The isogyres break up and leave the field of view with successively smaller stage rotations.

As a means to quantify this difference, stage rotation can be measured for each of the three types of interference figures. For an acute bisectrix figure with a $2V < 60°$, the isogyres do not

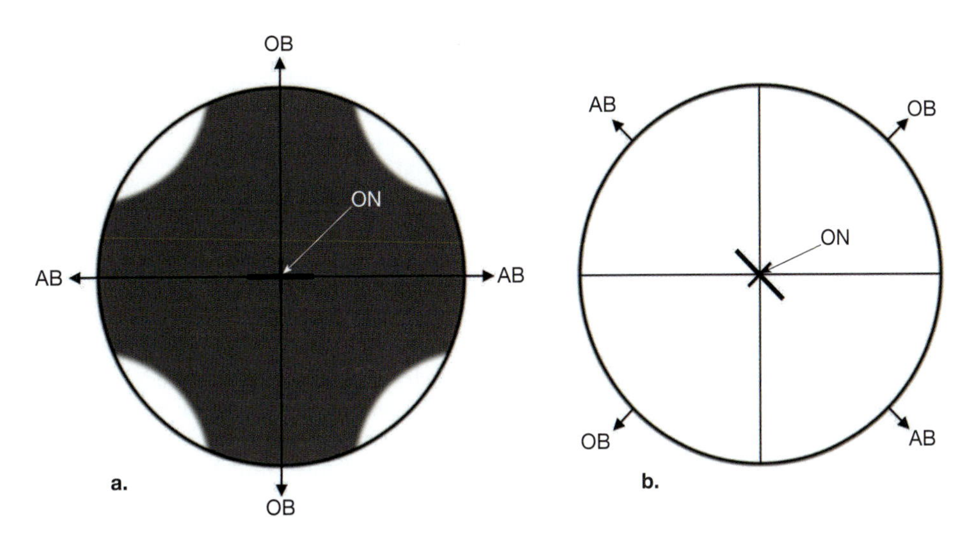

Figure 17.47. ***a.*** *Interference figures for an optic normal figure showing the isogyres at extinction.* ***b.*** *A clockwise rotation causes the isogyres to leave the field of view.*

leave the field of view (assuming use of a 0.8 numerical aperture lens). For $2V > 60°$, the isogyres do leave the field of view. When $2V \approx 80°$, the difference between an *AB* and an *OB* figure can be determined by measuring the amount of stage rotation (*SR*) that is needed to make the isogyres leave the field of view. The following equation (modified from Kamb, 1958) allows $2V$ of the interference figure to be calculated:

$$2V = 2\sin^{-1}\sqrt{\frac{1}{\left[\sin(2SR)\left[\frac{1}{(NA/n_g)^2} - 0.5\right]\right] + 0.5}}\,.$$

Equation 17.4

If $2V < 90°$, then the figure was an acute bisectrix, while if $2V > 90°$, the figure was an obtuse bisectrix. A graphical method for solving this equation is also shown in Figure 17.48. If the equation does not yield an answer, or if the stage rotation does not intersect one of the numerical aperture curves on the graph in Figure 17.48, then the interference figure was either an optic normal for a biaxial mineral or a flash figure for a uniaxial mineral. In either case, the figure could still be used to determine optic sign or to measure refractive indices.

Throughout our discussion of interference figures, we have focused on those that show two principal vibration directions in the plane of the microscope stage or an optic axis perpendicular to the plane to it. Such grains would always be in a specific orientation. However, most grains are in a random orientation and produce interference figures that can be difficult to interpret. When an interference figure bisects the crosshairs and does not coincide with them, the mineral is unequivocally biaxial (Figure 17.49a). If the isogyre both bisects the crosshairs and is parallel to one of them, then the mineral can be either biaxial or uniaxial (Figure 17.49b). If it is uniaxial, its ω vibration direction would be E-W. If it is biaxial, one of its three principal refractive indices (α, β, or γ) would be E-W and a non-principal vibration direction would be N-S.

Orientational Dependence of Images

A spindle stage is useful to further illustrate the orientational dependence of the optical properties because it allows individual crystals to be rotated into different orientations on the PLM. This device was briefly mentioned in Figure 17.9b as a sample preparation method, and will be discussed in more detail in Chapter 18 as a useful way to identify minerals using a PLM.

To show how retardation and interference figures are a function of orientation, consider three separate grains: scapolite, a sphere of corundum (both uniaxial) and sanidine (biaxial). The former two are mounted on the spindle's needle with their *c* crystallographic axes perpendicular to the "spindle." In this orientation, we can move from a centered optic axis figure to a centered flash figure to observe how retardation increases in orthoscopic illumination. Sanidine is mounted with its

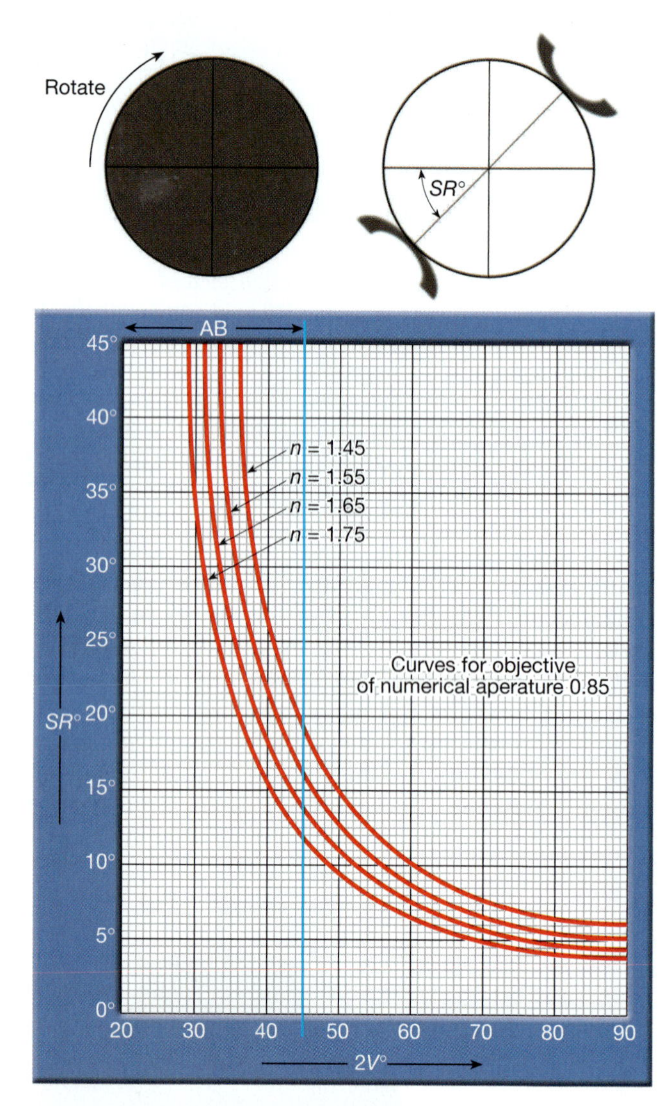

Figure 17.48. Different types of biaxial interference figures can be distinguished by determining the point at which the isogyres leave the field of view with rotation of the stage. In routine conoscopic viewing, the isogyres leave the field of view for acute bisectrix figures when 2V exceeds ~60°. The 2V of an interference figure can be approximated by noting the stage rotation angle where the isogyres just leave the field of view, as indicated here. A graphical solution for Equation 17.4 is shown at right. If 2V < 90°, then the view is of an acute bisectrix figure. If 90° < 2V < 180°, then the view is of an obtuse bisectrix figure. If rotating the stage does not cause the isogyres to intersect any of the curves on the drawing, it is an optic normal figure.

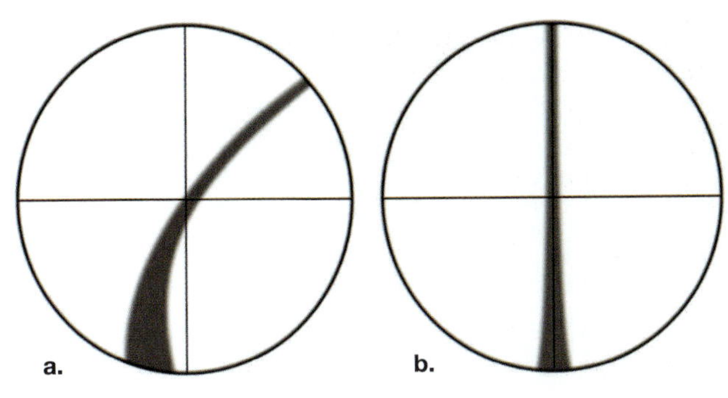

Figure 17.49. *When an interference figure is off-center, it can be difficult to distinguish between uniaxial and biaxial minerals.* ***a.*** *If an isogyre bisects the crosshairs without being parallel to them, then the mineral is biaxial.* ***b.*** *If any isogyre lines up perfectly with a crosshair, then the mineral can be either uniaxial or biaxial.*

optic normal coinciding with the axis of rotation of the spindle, to allow rotation from a centered acute bisectrix to a centered optic axis to a centered obtuse bisectrix figure in conoscopic illumination. In orthoscopic illumination, we will be able to see how the retardation changes.

The series of images of scapolite shown in Figure 17.50 illustrates the changes in retardation and associated interference figures with orientation, while the analogous images for the corundum sphere appear in Figures 17.51 and 17.52. In the latter images, the crystal is spherical so its thickness is the same in all orientations. Thus, any change in retardation is directly related to a change in the orientation of the sample. Corundum is shown at extinction ($S = 0°$), at 45° off extinction ($S = 45°$), and at 90° from extinction ($S = 90°$) with both conoscopic and orthoscopic illumination. S is the symbol that is generally used to represent the angular rotation of the sample around the spindle axis. Thus, S is the angle made by the circular section relative to the plane of the microscope stage. These images show that the retardation is indeed a function of orientation. Figure 17.53 shows a simulation of the corundum sphere with its c axis vertical and its associated retardation colors overlain on the surface of the sphere. These would result if light could emanate from the center of the sphere.

Figures 17.54 and 17.55 show the analogous images for a sanidine crystal at extinction (Figure 17.54), and at 45° off extinction (Figure 17.55). Four separate S settings are shown, where S represents the departure of the acute bisectrix from being perpendicular to the stage. At $S = 0°$, the acute bisectrix is perpendicular to the stage. At $S = 11°$, the AB has been rotated to

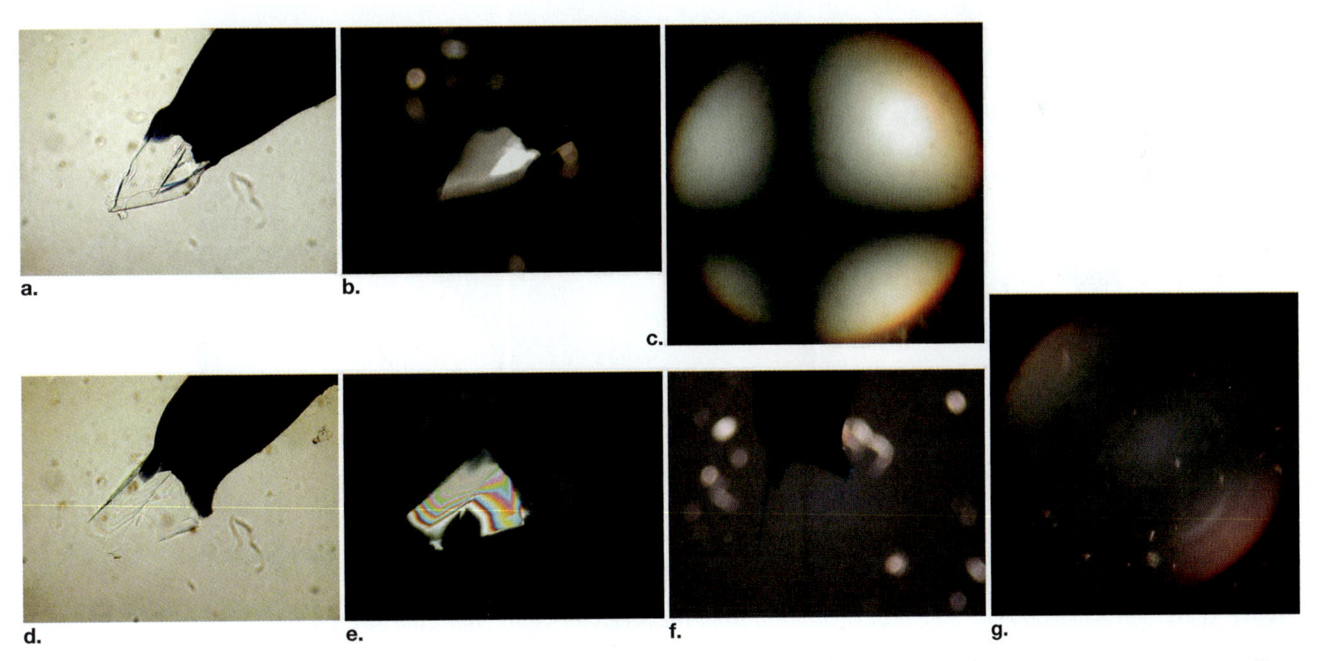

Figure 17.50. *Single crystal of the uniaxial mineral scapolite, mounted on the end of a needle with its optic axis perpendicular to the needle, immersed in a fluid with a near-matching refractive index.* ***a.*** *Plane-polarized light view.* ***b.*** *Cross-polarized light view, showing low retardation color.* ***c.*** *Conoscopic view, showing a near-centered optic axis figure.* ***d.*** *The grain has now been rotated 90° about the needle so the optic axis is in the plane of the microscope stage.* ***e.*** *The crystal in Figure 17.50d, viewed in cross-polarized light, showing very high retardation. This is the same crystal as in Figure 17.50b, just viewed at a different orientation!* ***f.*** *The crystal in Figure 17.50e has now been rotated ~45° counterclockwise to extinction.* ***g.*** *The interference figure produced by the view of the grain in Figure 17.50f.*

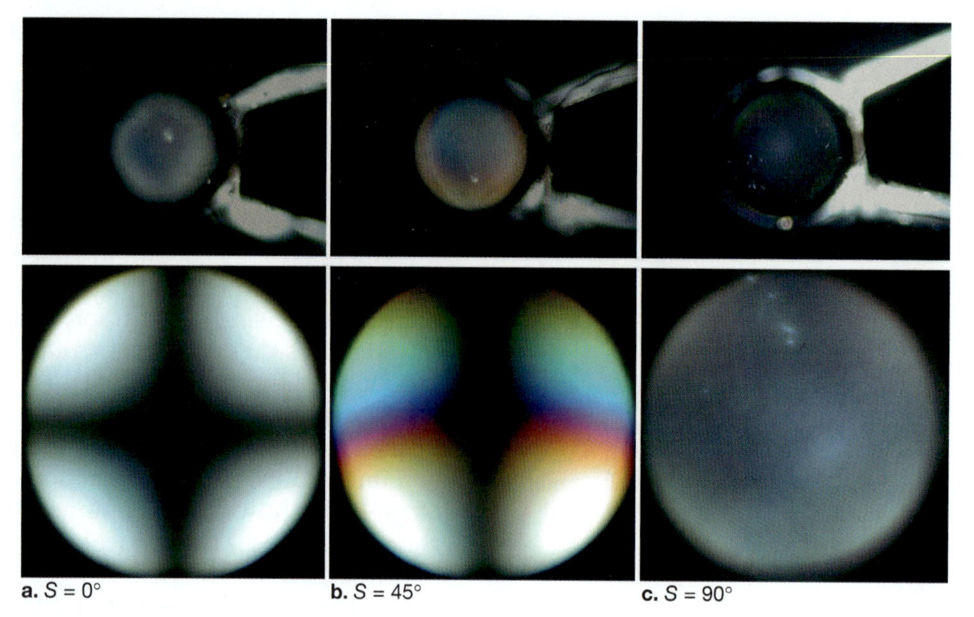

Figure 17.51. *A 300 μm sphere of the uniaxial mineral corundum, mounted with its* c *axis perpendicular to the needle. The upper row of photographs was taken in orthoscopic illumination, while the bottom row was taken in conoscopic illumination. The letter 'S' is used to denote the amount of rotation around the needle.* ***a.*** *View down the optic axis (S = 0°), so a centered optic axis figure is observed.* ***b.*** *The crystal has been rotated by 45° (S = 45°), so the view is half way between a centered optic axis figure and a centered flash figure.* ***c.*** *The crystal has been rotated another 45° (S = 90°) into the plane of the stage of the microscope, resulting in a centered flash figure. For all cases of orthoscopic illumination, the sphere appears dark; in Figure 17.51a this is because it exhibits zero retardation and in Figure 17.51c because the grain is at extinction.*

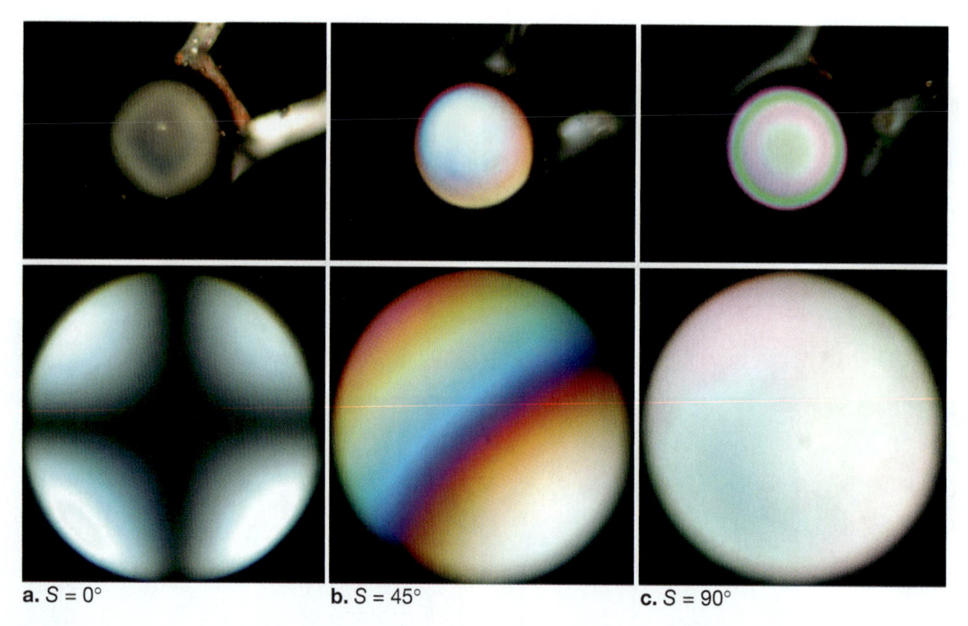

Figure 17.52. *The same 300 μm sphere of corundum from the previous figure, with the microscope stage rotated 45° counterclockwise. Crystals at Figure 17.52b and c are no longer at extinction. The upper row of photographs was taken in orthoscopic illumination, while the bottom row was taken in conoscopic illumination.* ***a.*** *View down the optic axis (S = 0°), so a centered optic axis figure is observed.* ***b.*** *The crystal has been rotated by 45° (S = 45°), so now the optic axis is emerging in the lower right hand corner of the image.* ***c.*** *The crystal has been rotated another 45° (S = 90°), and the optic axis is now in the plane of the microscope stage, oriented NW-SE. These images illustrate the adage that what goes on in the middle of a conoscopic figure is the same as what is seen over the entire field in orthoscopic illumination. The light ray emerging at the center of the field of view in conoscopic illumination is in the same orientation as all the light waves in orthoscopic illumination.*

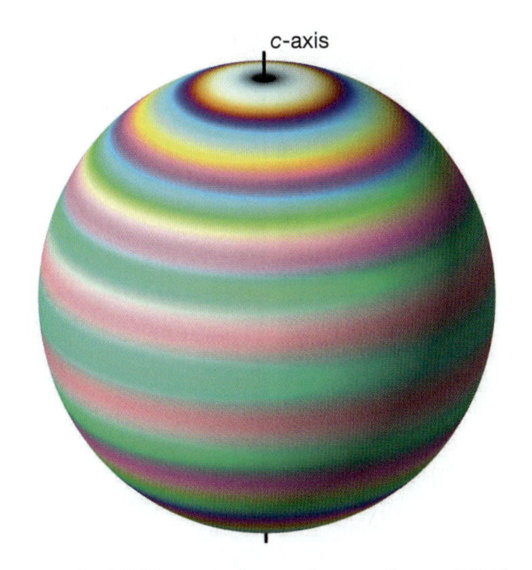

Figure 17.53. A 300 μm sphere of corundum with the c axis vertical, showing the different retardation colors that result from changes in orientation of an illuminated light source. In such a case, light would originate from the center of the sphere and move outward in all directions.

the edge of the field of view, such that the optic axis is at the center of the field of view. $S = 70°$ shows a random orientation, and $S = 90°$ is the obtuse bisectrix. The upper row of photos is in plane-polarized light, to show the change in shape of the crystal as it is rotated. The middle row shows the crystals in cross-polarized light, while the bottom row of photos is in conoscopic illumination.

Other Optical Phenomena

Moving beyond the uses of interference figures and accessory plates, there are several other phenomena that can be viewed in cross-polarized light that are useful in the identification and characterization of minerals.

Sign of elongation. When a crystal is elongated, it is sometimes useful to ascertain the orientation of N or n relative to that elongation.

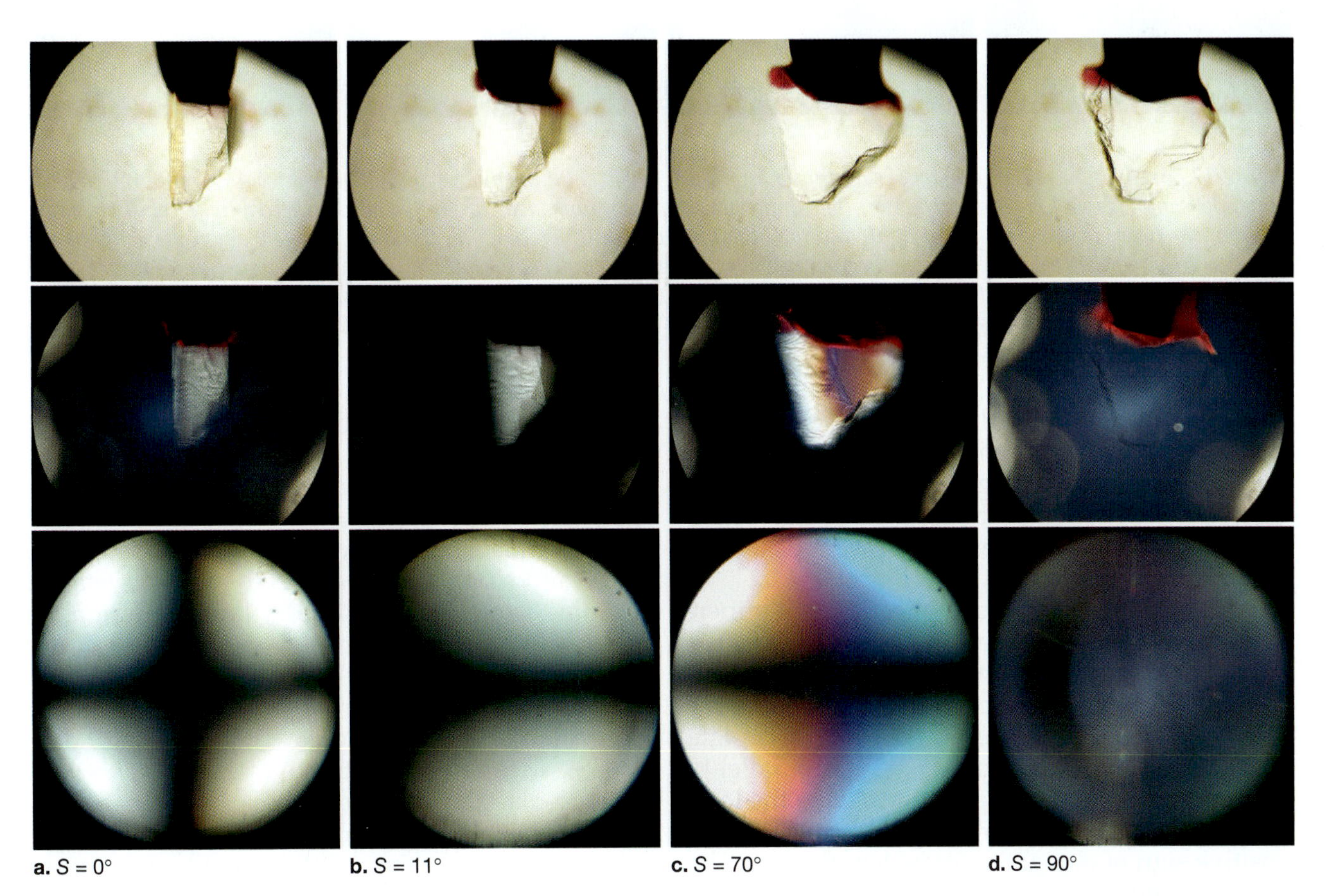

Figure 17.54. A single crystal of the biaxial mineral sanidine (2V = ~22°), mounted with its optic normal parallel to the needle, which places the optic plane perpendicular to the needle. The upper row of photographs was taken in plane-polarized light, the middle row in cross-polarized light, and the bottom row in conoscopic illumination. This set of photographs was taken using a lower numerical aperture lens than in the previous conoscopic photographs. ***a.*** *View down the acute bisectrix direction (S = 0°).* ***b.*** *The crystal has been rotated 11° (S = 11°), placing an optic axis at the center of the field of view.* ***c.*** *The crystal has been rotated to S = 70° resulting in an off-centered figure.* ***d.*** *The crystal has been rotated 90° from Figure 17.54a until the acute bisectrix is in the plane of the microscope stage, resulting in a centered obtuse bisectrix interference figure.*

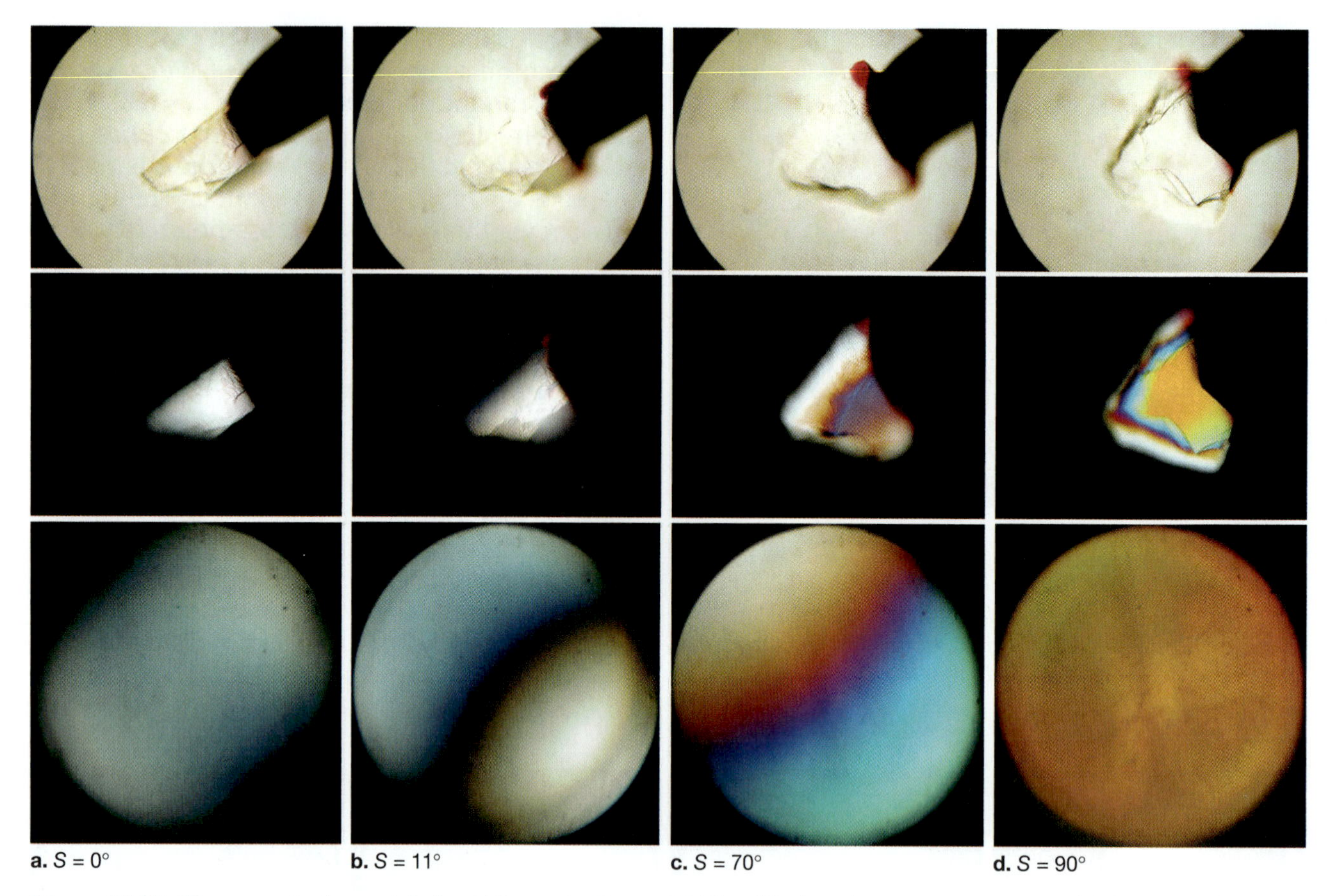

Figure 17.55. The same single crystal of sanidine as in Figure 17.54, with the microscope stage rotated 45° counterclockwise. Thus, the sample is no longer at extinction. The upper row of photographs was taken in plane-polarized light, while the middle row was taken in cross-polarized light, and the bottom row is in conoscopic illumination. ***a.*** *View down the acute bisectrix (S = 0°), and the two optic axes can be seen in the upper left and lower right of the interference figure.* ***b.*** *Centered optic axis figure.* ***c.*** *Off-center interference figure.* ***d.*** *Off-extinction, centered obtuse bisectrix figure. In both the orthoscopic and conoscopic views, retardation greatly increases as the grain is rotated from a centered* AB *to a centered* OB.

Historically, this is called the **sign of elongation** of a mineral, which is defined to be positive when N corresponds to the long direction, and negative when n corresponds to the long direction. These are sometimes called **length slow** and **length fast**, respectively.

To determine the sign of elongation of a mineral, a crystal must be viewed in cross-polarized light, and its long axis must be positioned to be parallel to N_{plate} (see the crystal to the right of the "+" in Figure 17.56a). Insertion of the quartz plate in this case causes addition, so N_{grain} is parallel to N_{plate}, implying that N_{grain} corresponds to the long axis of the grain. The grain would thus have a positive sign of elongation or be length slow. The crystals in Figure 17.56 are all tremolite, and most amphiboles are length slow. The two major asbestiform amphiboles, amosite and crocidolite, are distinguished on the basis of their sign of elongation: crocidolite is length fast, while amosite, like most other amphiboles, is length slow. Notice how the other tremolite grains change in color with insertion of the plate and stage rotation.

The natrolite group minerals natrolite, mesolite, and scolecite (Figures 17.57–17.59) all have similar morphologies with well-developed (110) faces and elongations parallel to their c crystallographic axes (Figure 17.60). Natrolite is length slow, and scolecite is length fast; these are shown as single crystals on a spindle stage viewed in plane-polarized and cross-polarized light with insertion of a quartz plate (Figures 17.57 and 17.58). Mesolite is distinctive (Figure 17.59) in plane-polarized and cross-polarized light with insertion of a quartz plate. In Figure 17.59a–c, the crystal appears length slow, while in Figure 17.59d–f, it appears to be length fast. The two sets of images are different because in the latter set, the crystal has been rotated 90° around the spindle. In mesolite, β is parallel to the long direction. So in the first row of photographs, α is the direction perpendicular to β, so the crystal is length slow. In the second row of photos, γ has

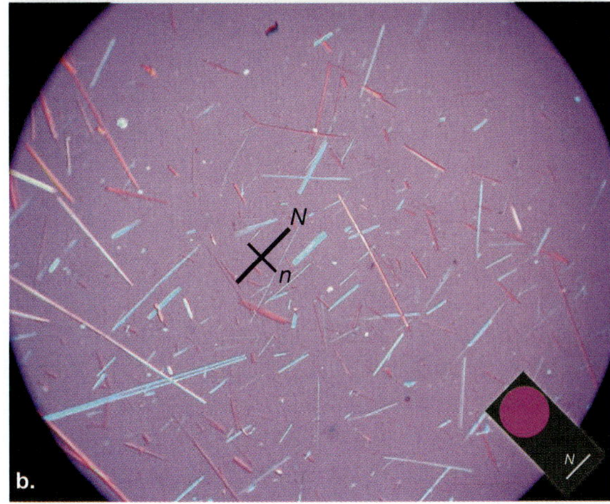

Figure 17.56. *Either the large or the small refractive index can be oriented parallel to the direction of elongation in a mineral grain, and this can be a useful diagnostic property.* ***a.*** *To determine which refractive index is parallel to the long axis of the mineral, rotate the long axis parallel to the* N *direction of the inserted accessory plate, and then either addition or subtraction is observed.* ***b.*** *If addition occurs, then* N *corresponds to its long direction, as is the case for the tremolite grains in the immersion mount shown here.*

been rotated into the plane of the stage of the microscope, and β would become n, showing subtraction. Figure 17.60 summarizes these conclusions by showing the optical orientations for the three minerals. Both natrolite and mesolite are orthorhombic, and scolecite is monoclinic. The change in the sign of elongation within this group is apparent from the fact that Z is parallel to c for natrolite, while Y is parallel to c for mesolite, and X is near-parallel to c for scolecite. Use of the sign of elongation is the simplest way to distinguish among the mineral species within this group of minerals.

Extinction types and angles. Extinction angles relate the orientation of the optical indicatrix of a mineral to its crystallographic directions as well as morphology. Extinction angles are very useful in mineral identification, especially in thin sections. They are most commonly used to distinguish between monoclinic and orthorhombic amphiboles and pyroxenes. Orthorhombic minerals exhibit parallel extinction, while monoclinic minerals can exhibit inclined extinction. True determination of extinction type requires knowledge of the optical orientation of the mineral as well as its orientation. Extinction angles are only meaningful when they are obtained from crystals that exhibit centered interference figures, such that two principal vibration directions are in the plane of the microscope stage. Also, the minerals must show some morphologic linear feature. The extinction angle is then defined as the angle between a principal vibration direction and that morphological feature, as shown in the following images of monoclinic tremolite and orthorhombic natrolite.

Tremolite is a monoclinic amphibole (Figure 17.61) that often exhibits (110), (100), and (010) crystal faces. When viewed down its c crystallo-

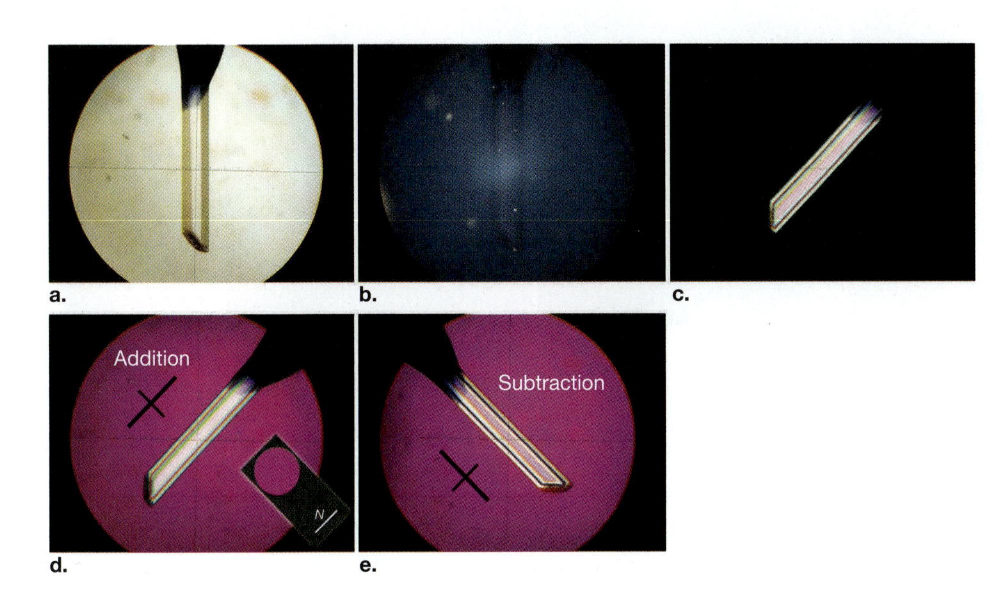

Figure 17.57. *Single crystal of natrolite mounted on the end of a needle with its long direction parallel to the needle, to allow measurement of the direction of elongation.* ***a.*** *Crystal in plane-polarized light.* ***b.*** *Crystal in cross-polarized light showing the crystal at extinction.* ***c.*** *The grain has now been rotated 45° clockwise.* ***d.*** *The quartz plate is inserted, and addition occurs. Thus* $N_{crystal}$ *corresponds to* N_{plate}*, and the mineral would be termed length slow.* ***e.*** *The crystal has been rotated 90° from Figure 17.57d, and subtraction occurs.*

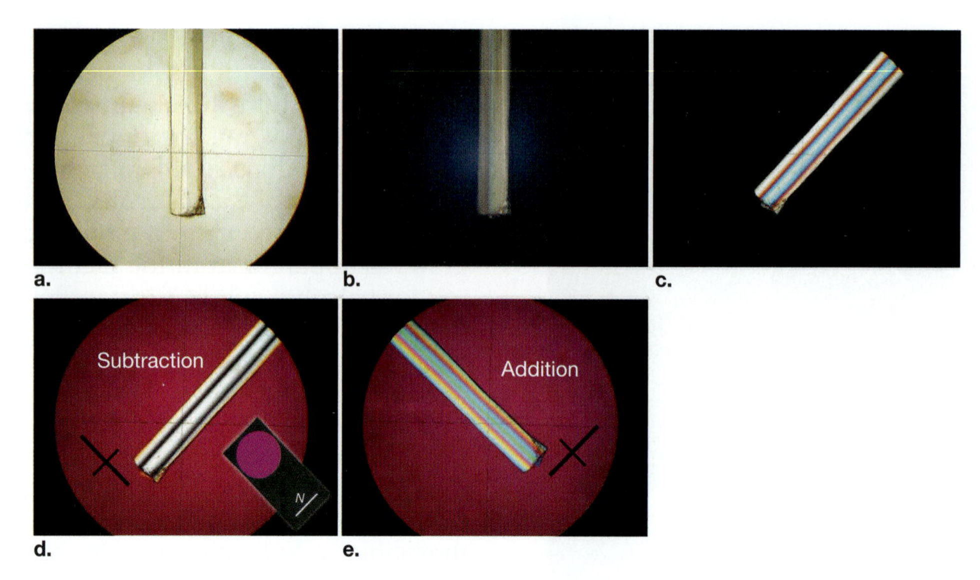

Figure 17.58. *Single crystal of scolecite mounted on the end of a needle with its long direction parallel to the needle.* ***a.*** *Crystal in plane-polarized light.* ***b.*** *Crystal in cross-polarized light at extinction.* ***c.*** *The grain rotated 45° clockwise.* ***d.*** *The quartz plate is inserted, and subtraction occurs. Thus $n_{crystal}$ corresponds to N_{plate}, and the mineral and the sample would be called length fast.* ***e.*** *The crystal rotated 90° from Figure 17.58e, so that addition occurs.*

graphic axis, traces of the (110) cleavage directions can be seen (Figure 17.62a). In cross-polarized light, the crystal is at extinction; any slight clockwise or counterclockwise rotation will move the crystal out of extinction, though the principal vibration directions bisect the (110) direction. This is called **symmetrical extinction**.

Another single crystal of tremolite is mounted with its *c* crystallographic axis parallel to the rotation axis of the spindle stage (Figure 17.63). In the upper row of photographs, the (100) plane is in the microscope stage, and the grain exhibits parallel extinction as seen in Figure 17.61. In the lower row of photos (Figure 17.63e–h), the grain

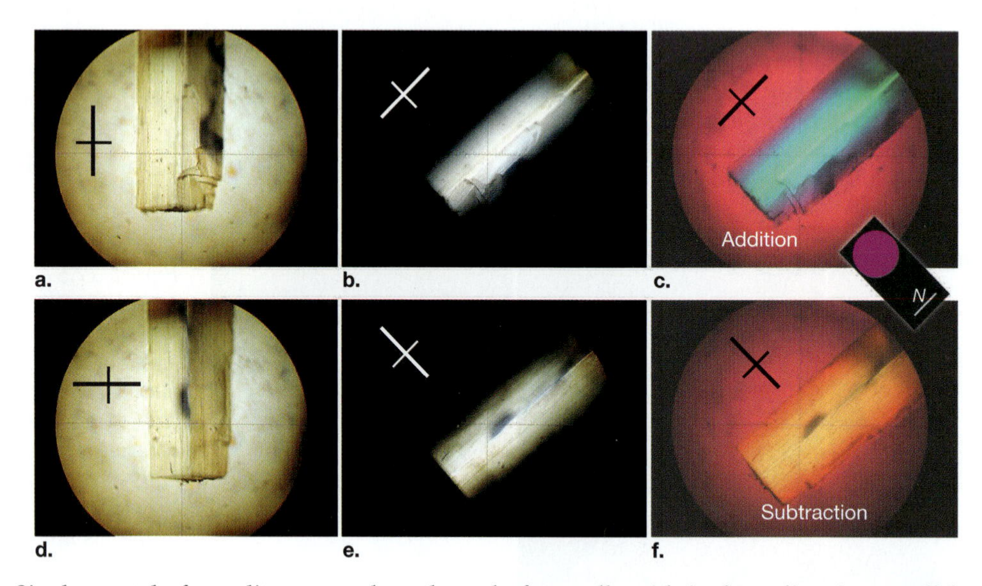

Figure 17.59. *Single crystal of mesolite mounted on the end of a needle with its long direction parallel to the needle.* ***a.*** *Crystal in plane-polarized light.* ***b.*** *Crystal in cross-polarized light rotated 45° clockwise from Figure 17.59a.* ***c.*** *The quartz plate has been inserted, and addition occurs.* ***d.*** *The crystal has been rotated 90° from Figure 17.59a around its long axis (i.e., S moved from 0° to 90°), and is viewed in plane-polarized light.* ***e.*** *The crystal in Figure 17.59d, rotated 45° clockwise, and viewed in cross-polarized light.* ***f.*** *The quartz plate has been inserted, and subtraction occurs. Thus this crystal appears to exhibit properties of both length fast and length slow behavior because the crystal is elongated parallel to the intermediate refractive index, β.*

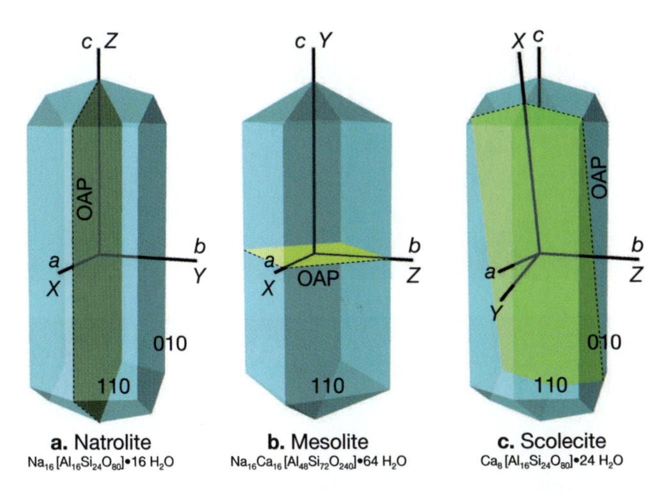

Figure 17.60. *Optical orientation of the three zeolite minerals in Figures 17.58–17.59.* ***a.*** *Natrolite is elongated parallel to the direction of its largest refractive index.* ***b.*** *Mesolite is elongated parallel to its intermediate refractive index, β, so it will appear length fast if the γ refractive index is in the plane of the microscope stage, and length slow if the α refractive index is in the plane of the microscope stage.* ***c.*** *Scolecite is elongated parallel to its smallest refractive index direction, so it always appears length fast.*

has been rotated 90° around the spindle so its (010) plane is in the plane of the microscope stage; it shows inclined extinction. Thus tremolite exhibits the three principal types of extinction angles in minerals: (1) symmetrical, when viewed down Z, (2) parallel when viewed on (100), and inclined when viewed on (010). Clearly, use of extinction angles to identify minerals requires knowledge of the orientation of the mineral.

Extinction angles can also be observed in crystals in grain mounts. If the minerals possess cleavage, they will come to rest on their cleavage faces, so they often will not produce centered interference figures, and only random extinction angles can be measured. Figure 17.64 shows grains of tremolite from the previous images with the grain in the center showing asymmetrical isochromes that are spaced farther apart on the left side of the grain than on the right. This indicates that the grain is thinner on the left side than on the right side, and it is resting on the (110) cleavage face. In this orientation, this grain shows random extinction.

One final example shows a thin section view of several grains of the same tremolite (Figure 17.65a–c). The grain in the center is viewed with its *c* axis tilted. In this orientation, traces of the (110) cleavage planes are still visible, which would aid in identification of this mineral. However, the mineral no longer shows symmetrical extinction because it does not produce a centered interference figure. In Figure 17.65d–f, a grain is shown with its Z axis near the plane of the microscope stage, again with a random extinction direction. Extinction angles could be measured in these latter grains (as in Figure

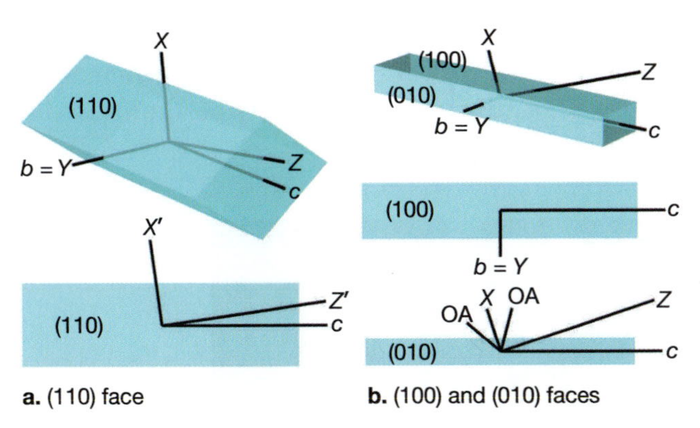

Figure 17.61. *The extinction angle of a mineral relates its principal vibration directions (i.e., X, Y, Z) to other linear features such as cleavage planes. Measurement of extinction angles requires that the grain of interest exhibit a centered interference figure. In other words, the grain must be viewed down a principal vibration direction, so that the other two principal vibration directions are in the plane of the microscope stage. Also, the optical orientation of the mineral grain must be known.* ***a.*** *The optical orientation of tremolite is shown in relationship to its (110) face.* ***b.*** *The optical orientations relative to the (100) and (010) faces.*

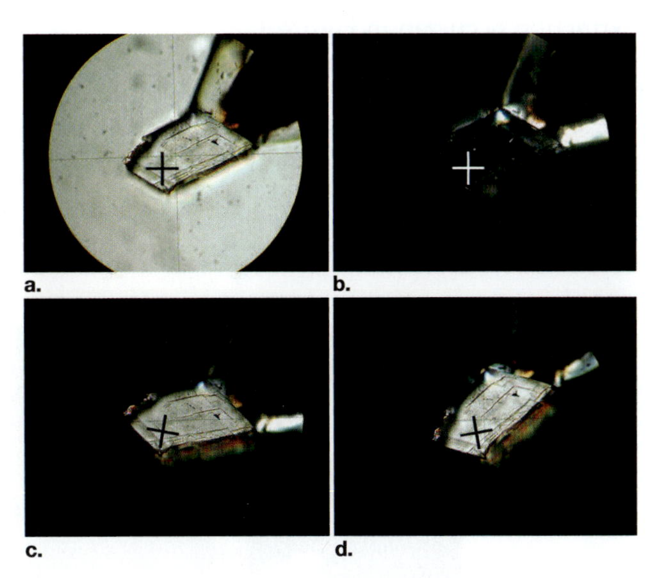

Figure 17.62. *Views down the Z direction of a tremolite grain, with traces of the (110) cleavage plane visible.* ***a.*** *Plane-polarized light with the vibration directions inscribed.* ***b.*** *The crystal in the same orientation as Figure 17.62a, in cross-polarized light at extinction.* ***c.*** *After ~10° clockwise rotation, the crystal is no longer at extinction.* ***d.*** *The crystal rotated ~10° counterclockwise, again no longer at extinction. This is an example of symmetrical extinction, where the vibration direction makes equal angles with the cleavage directions.*

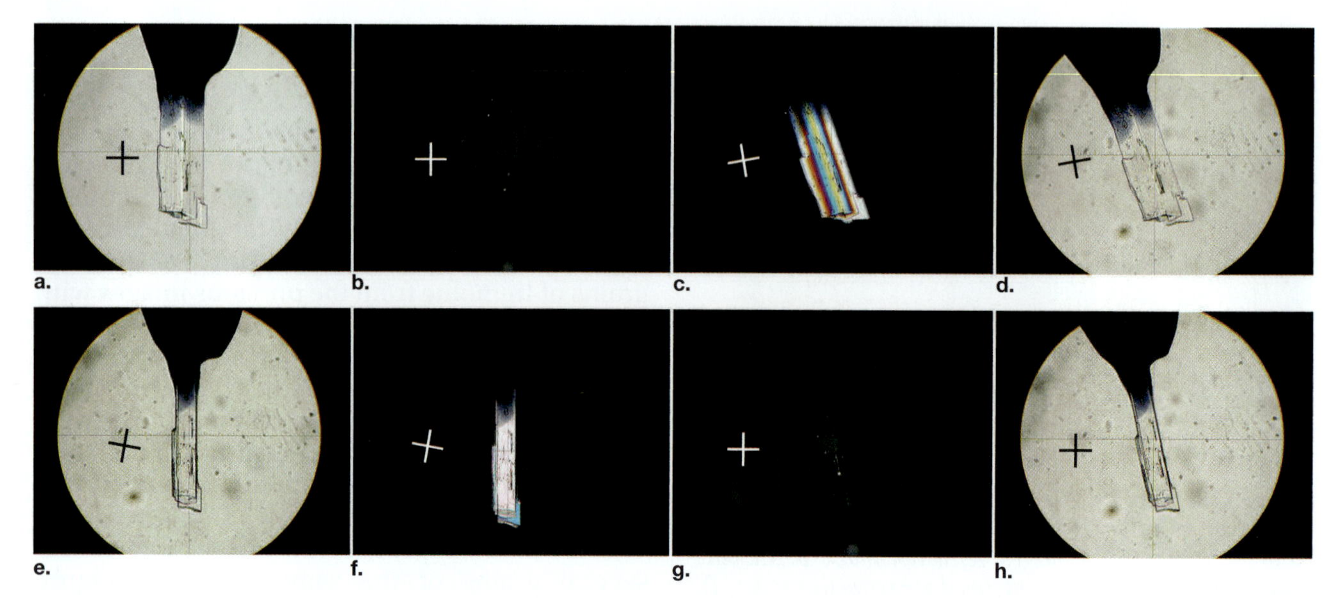

Figure 17.63. Tremolite single crystal mounted with its c axis parallel to the needle, allowing it to rotate around its long axis. In the top row (Figure 17.63a–d), the (100) plane is parallel to the microscope stage, whereas in Figure 17.63e–h (bottom row), the (010) plane is parallel to the microscope stage. ***a.*** Crystal in plane-polarized light. ***b.*** Same orientation as Figure 17.63a in cross-polarized light, showing parallel extinction. ***c.*** The crystal has been rotated ~10° counterclockwise so it shows retardation. ***d.*** The same crystal as in Figure 17.63c, viewed in plane-polarized light. ***e.*** The crystal has been rotated 90° about its long axis from that shown in Figure 17.63a, and is viewed in plane-polarized light. ***f.*** The same view as in Figure 17.63e, but in cross-polarized light. This crystal is not at extinction. ***g.*** The crystal is rotated to extinction. ***h.*** View in plane-polarized light, showing an angle of ~18° between the long axis of the crystal and its vibration direction. This is an example of inclined extinction. Tremolite can exhibit three different types of extinction in the same crystal as a function of orientation.

17.64), but they would change with sample orientation. The only true measurements of inclined extinction occurred in Figure 17.63e–h.

To show how an orthorhombic mineral appears to exhibit inclined extinction, natrolite is mounted on a spindle stage in two separate orientations (Figures 17.66 and 17.67). In the first, the *X* direction is parallel to the spindle (Figure 17.66); in the second, the crystal is mounted on its (110) cleavage face, which would bisect the *X* and *Y* vibration directions. The latter sample is in a more random orientation. In both sets of images, the *c* axis is first shown parallel to the stage of the microscope, and then progressively rotated 20° and 40° out of that plane. The vibration directions have been indicated on each photograph. In Figure 17.66, the grain remains at extinction as it is rotated, whereas in Figure 17.67, the grain exhibits increasing retarda-

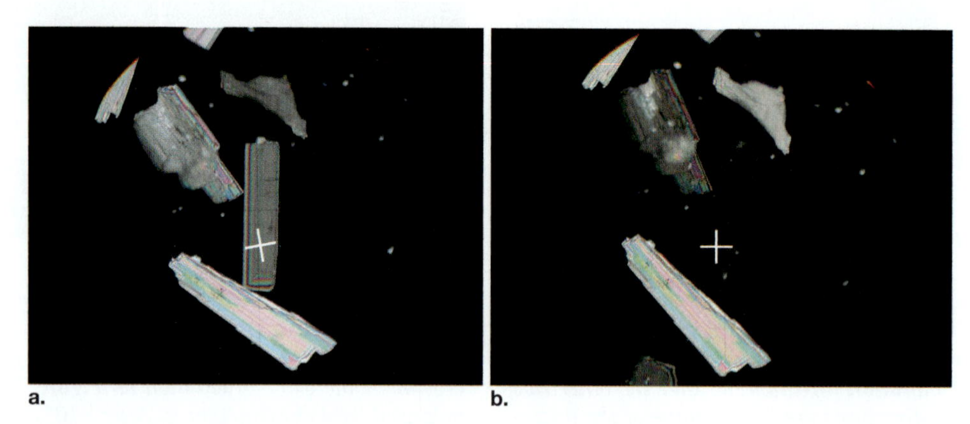

Figure 17.64. Immersion mount of the tremolite sample shown in the previous figure. ***a.*** The isochromes are not symmetrical, indicating that the crystal is thicker on the right side because it is resting on its (110) face. ***b.*** Here the crystal has been rotated to extinction. The extinction angle is slightly less (~10°) than the angle shown in Figure 17.63 for the case of inclined extinction, because this crystal is in a random orientation.

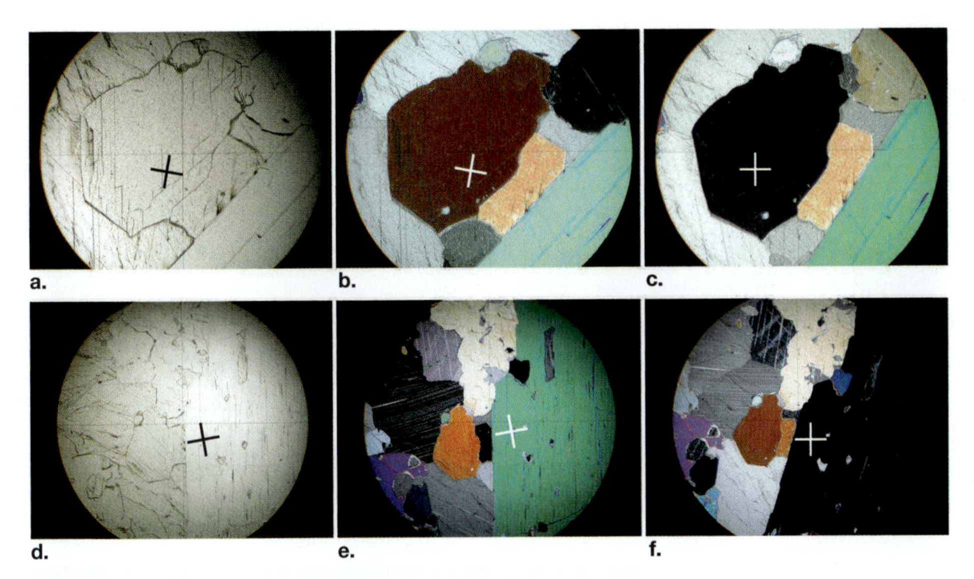

Figure 17.65. Thin section of the rock that contained the tremolite single crystals shown in the previous two figures. Figure 17.65a–c show views in which the c *axis is nearly perpendicular to the microscope stage, while in Figure 17.65d–f, the* c *axis is close to being parallel to the microscope stage.* ***a.*** *The vibration direction is inscribed in the single crystal, which is viewed in plane-polarized light.* ***b.*** *Sample is viewed in cross-polarized light and is not at extinction.* ***c.*** *The crystal rotated to extinction. It does not show symmetrical extinction because it is randomly oriented.* ***d.*** *Crystal viewed in plane-polarized light with the vibration directions noted.* ***e.*** *View in cross-polarized light, not at extinction.* ***f.*** *The crystal has been rotated to extinction showing at angle of ~15°, again not the true value of 18° because the crystal is in a random orientation.*

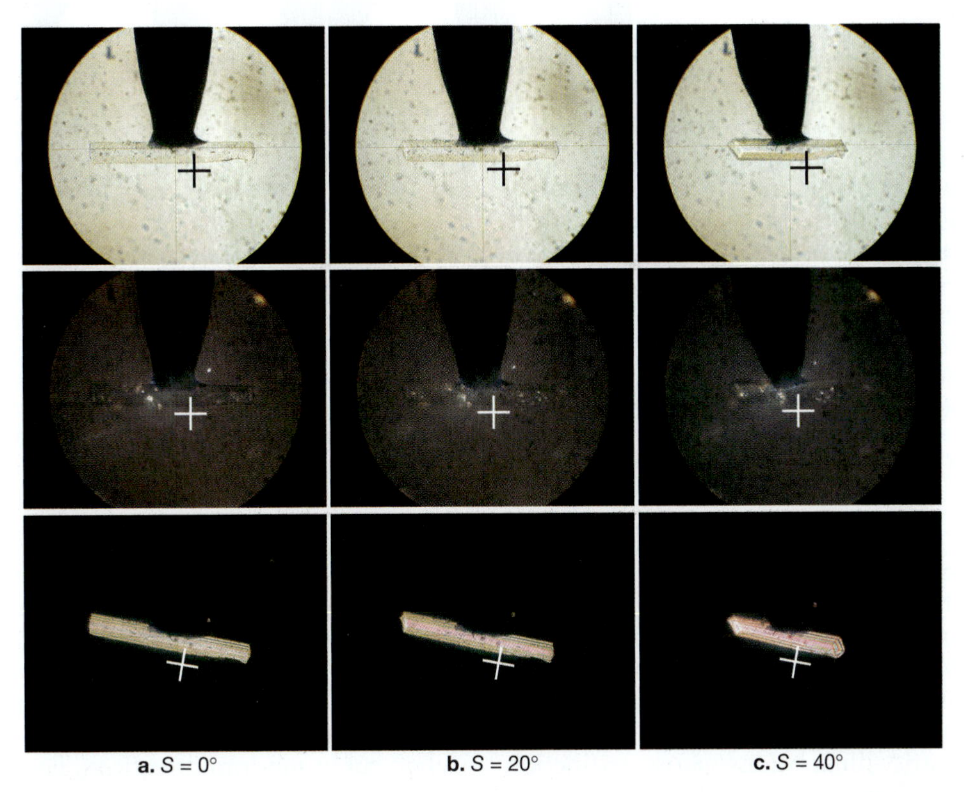

Figure 17.66. Single crystal of natrolite mounted with its long axis perpendicular and its b *crystallographic axis parallel to the needle, so the principal vibration direction can be seen. Here* α *and* γ *are in the plane of the microscope stage. The upper row of photos is in plane-polarized light; the middle row in cross-polarized light, and the bottom row in plane-polarized light with the microscope stage rotated ~10°.* ***a.*** *The* c *axis is horizontal, and the crystal exhibits parallel extinction.* ***b.*** *The* c *axis is rotated 20° from parallel (to the stage), but the crystal still exhibits parallel extinction.* ***c.*** *The crystal has been rotated 40° from its position in Figure 17.66a, and it still exhibits parallel extinction.*

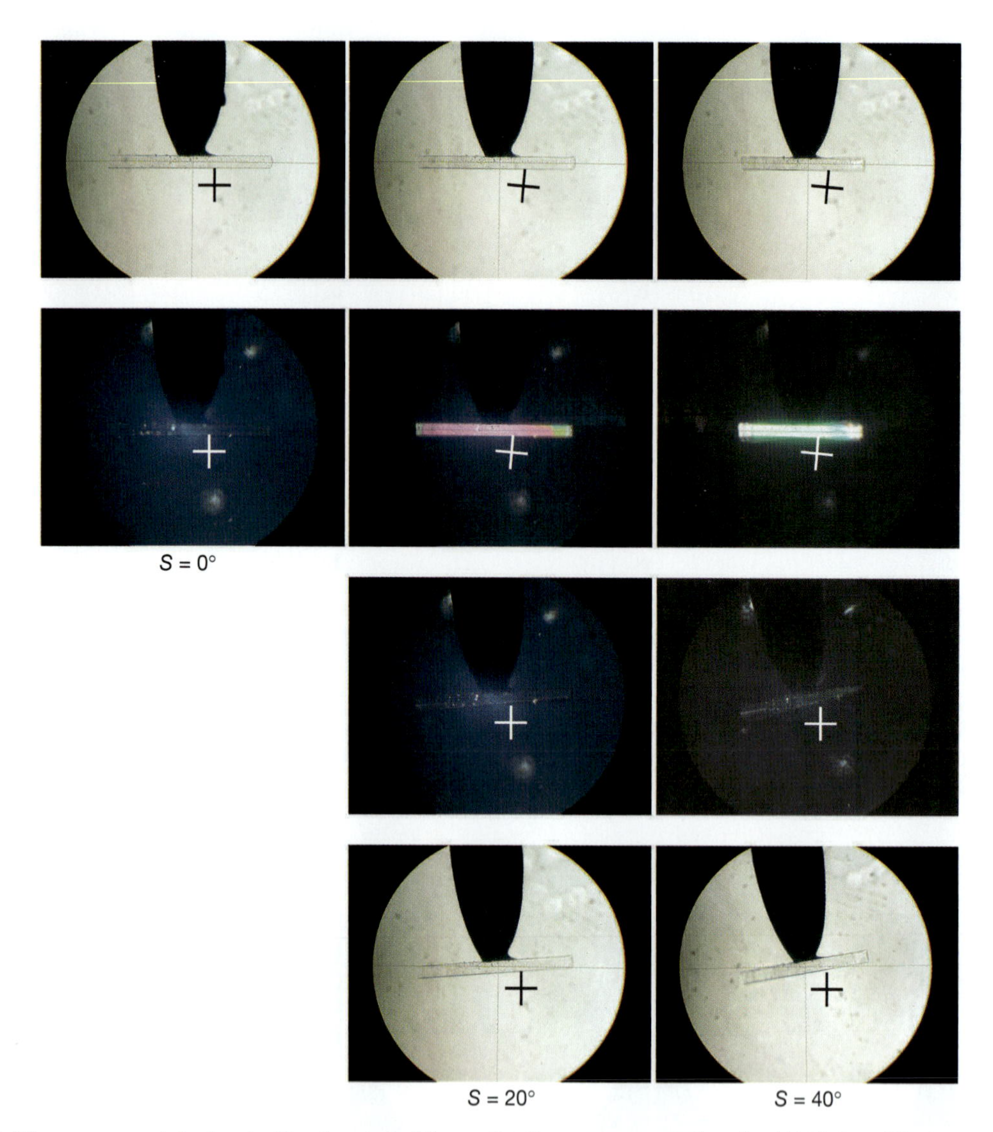

Figure 17.67. *The same crystal of natrolite shown in Figure 17.66, now mounted on its (110) face. The upper row of photos is in plane-polarized light; the middle row in cross-polarized light, the third row with each crystal rotated to extinction, and the bottom row viewed in plane-polarized light so as to show the extinction angle. For the* S = 0° *setting, the* c *axis is parallel to the microscope stage, and the crystal exhibits parallel extinction. For the* S = 20° *setting, the* c *axis has been rotated 20° out of the stage, and the crystal no longer exhibits parallel extinction, now showing ~5° extinction angle. For the* S = *40° case, the* c *axis has been rotated 40° out of the plane of the stage of the microscope, and the crystal now shows higher retardation and an extinction angle of ~10°. This illustration shows how an orthorhombic mineral will not exhibit parallel extinction when viewed at a random orientation.*

tion as it is rotated out of the plane of the microscope. In fact, when the grain is 40° out of the stage, you would see an apparent inclined extinction of 10°. Many orthorhombic pyroxenes have been misidentified as monoclinic because the grains were oriented like this.

Two schematic illustrations show extinction angles in orthorhombic and monoclinic amphiboles (Figures 17.68 and 17.69, modified from Nesse, 2004). A series of cuts through three-dimensional blocks shows the projections of the mutually-perpendicular vibration directions as well as the traces of the various cleavage planes. For the case of the orthorhombic amphibole, both parallel and symmetrical extinction are seen. The grains would exhibit parallel and symmetrical extinction only when they have centered interference figures. They would show random extinction angles and off-center interference figures when cut in random directions. For the monoclinic example, the extinction angles shown correspond to the examples previously seen for tremolite. In conclusion, Su and Bloss (1984) point out that angles greater than the *true* extinction angles can be observed in monoclinic pyroxenes and amphiboles based upon their optical orientations.

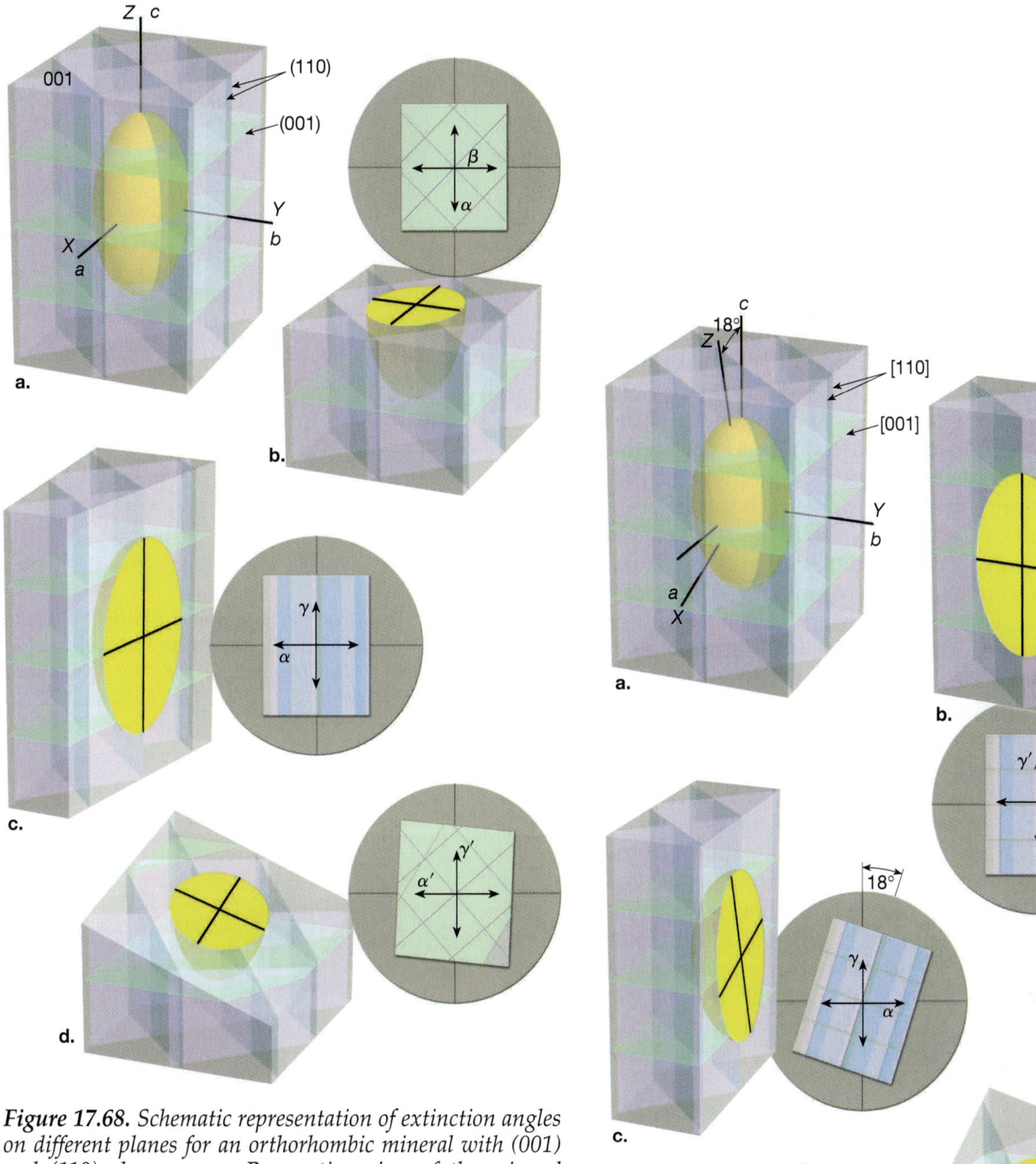

Figure 17.68. Schematic representation of extinction angles on different planes for an orthorhombic mineral with (001) and (110) cleavages. ***a.*** Perspective view of the mineral showing its cleavage, optical orientation (i.e., a = X, b = Y, c = Z), and biaxial indicatrix. Planes perpendicular to the X and Y directions would show parallel extinction. ***b.*** Cut perpendicular to Z showing how the mineral would exhibit symmetrical extinction in this orientation. ***c.*** Random cut showing "parallel" extinction. ***d.*** Random cut showing "inclined" extinction. The "" are used because extinction angles can only be determined when two principal vibrations are located in microscope stage.

Figure 17.69. Schematic representation of the extinction angles on different planes for a monoclinic mineral with (001) and (110) cleavages. ***a.*** Perspective view of the mineral showing its cleavage, optical orientation (b = Y and c, Z = 18°), and a biaxial indicatrix. ***b.*** Cut perpendicular to a showing "parallel" extinction. ***c.*** Cut perpendicular to Y showing extinction inclined at an 18° angle. ***d.*** Random cut, again showing "inclined," or more appropriately, random extinction.

So, just determining the maximum angle in a thin section may yield an incorrect result. Meaningful extinction angles are only determined when minerals are correctly oriented.

More ways to visualize an indicatrix. There are several geometric objects that can be helpful in learning about the indicatrix. A physical model can be made in the same way as a ball and stick crystal structure model (Figure 17.70) with three mutually–perpendicular directions X, Y, and Z that represent the α, β, and γ vibration directions. The optic axes have been added to these models. If a student is looking down the acute bisectrix direction of a positive mineral, the model can be held next to the microscope pointing down to show which other vibration directions would rest in the plane of the microscope stage. In other words, this model is a very useful teaching aid in the lab.

At the end of Chapter 16, we made several cuts through a lemon to show different sections of the indicatrix of a uniaxial positive mineral. The analogous exercise is to cut a potato, which has three mutually-perpendicular directions of different lengths, to illustrate the different cuts through a biaxial indicatrix. Figure 17.71 shows a simulated potato with cuts to match the views seen in Figure 17.27. You really can cut two circles out of this potato.

The last set of examples (Figure 17.72) shows a series of hollow glass vessels representing a uniaxial positive indicatrix, a uniaxial negative indicatrix, a biaxial positive indicatrix, and a biaxial negative indicatrix. Each one is half-filled with a colored liquid so that rotating it will make the water line intersect with the glass markings, showing a two-dimensional cut through the indicatrix. Figure 17.73 shows the uniaxial indicatrices in several different orientations, while Figure 17.74 shows five

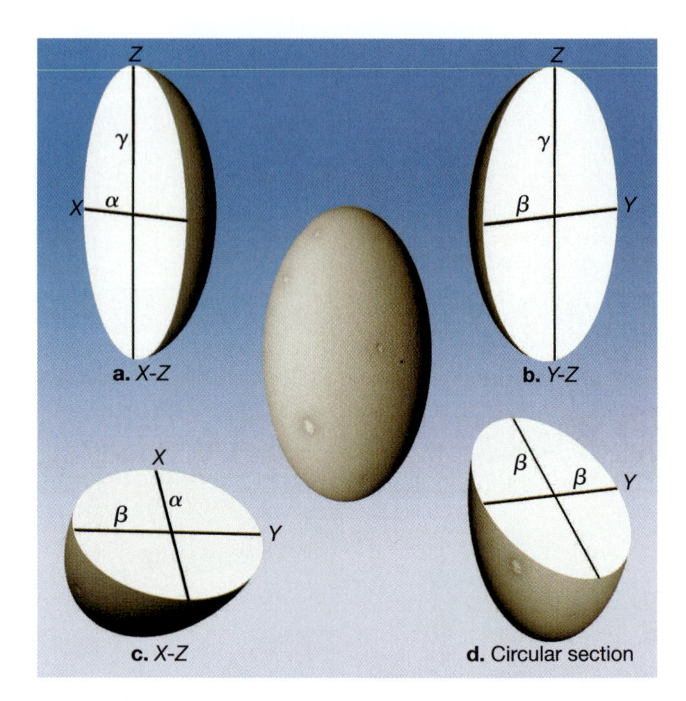

Figure 17.71. A 3-D representation of an animated Idaho potato, and four different 2-D cuts through it. (Thanks to Jane Selverstone for the inspiration for this figure and thanks to Idaho for potatoes.) ***a.*** *X-Z.* ***b.*** *Y-Z.* ***c.*** *X-Y.* ***d.*** *Circular section.*

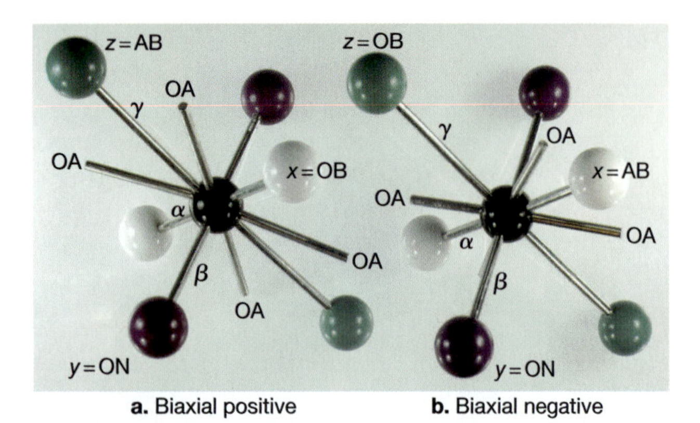

Figure 17.70. Photograph of two ball and stick models of a positive and negative biaxial indicatrix with the various parts labeled. These models show students which directions they are looking down, and help to visualize how the normal to it would be in the plane of the microscope stage. ***a.*** *Biaxial positive.* ***b.*** *Biaxial negative.*

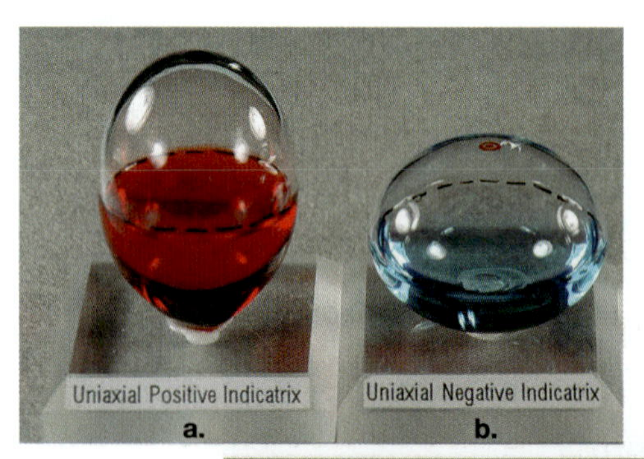

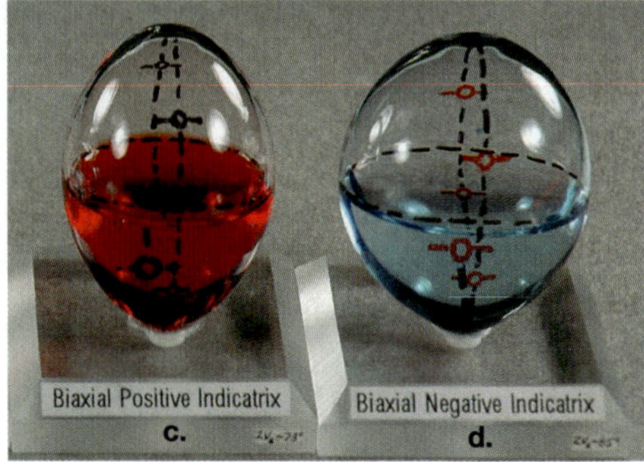

Figure 17.72. Four hollow glass spheres, each half-filled with a colored liquid, showing the size and shape of the 2-D plane (i.e., the plane of the indicatrix parallel to that of the microscope stage) and its normal (i.e., the light direction in the PLM) at different orientations. (We would like to thank Dan Kile for the loan of these indicatrix models.) ***a.*** *Uniaxial positive.* ***b.*** *Uniaxial negative.* ***c.*** *Biaxial positive.* ***d.*** *Biaxial negative.*

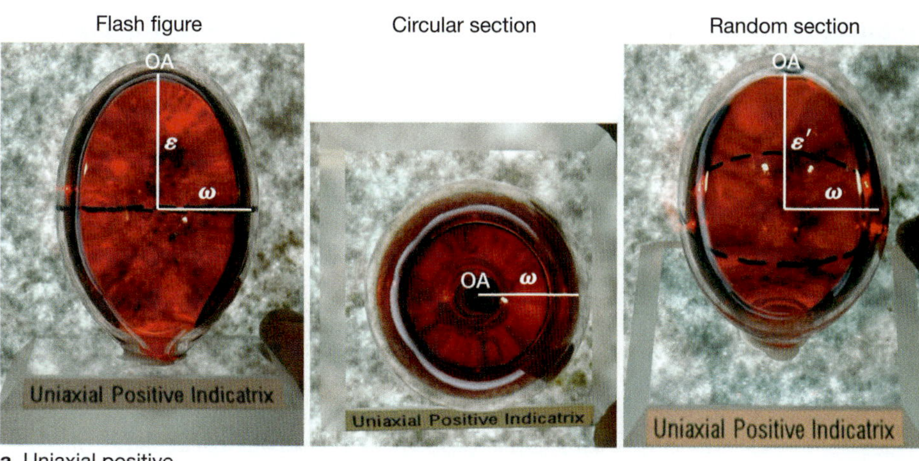

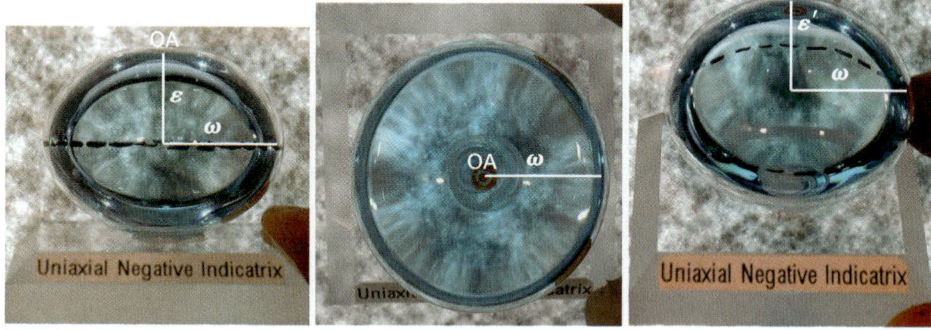

Figure 17.73. *Three different sections through the fluid-filled uniaxial indicatrices shown in Figure 17.72 with refractive index directions and optic axes (OA) labeled. The first column shows the orientation that would yield a flash figure, the second an optic axis figure, and the last a random figure.* ***a.*** *Positive indicatrix.* ***b.*** *Negative indicatrix.*

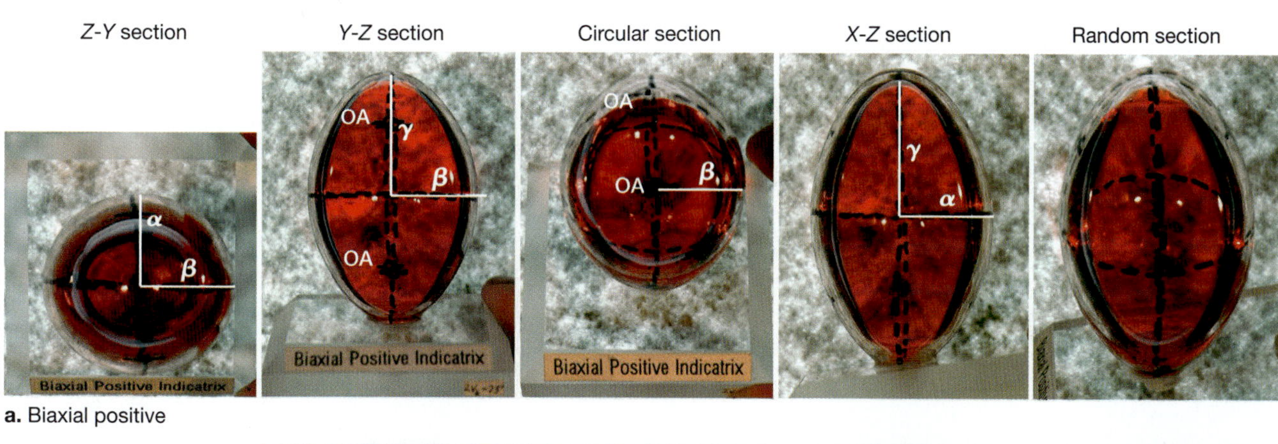

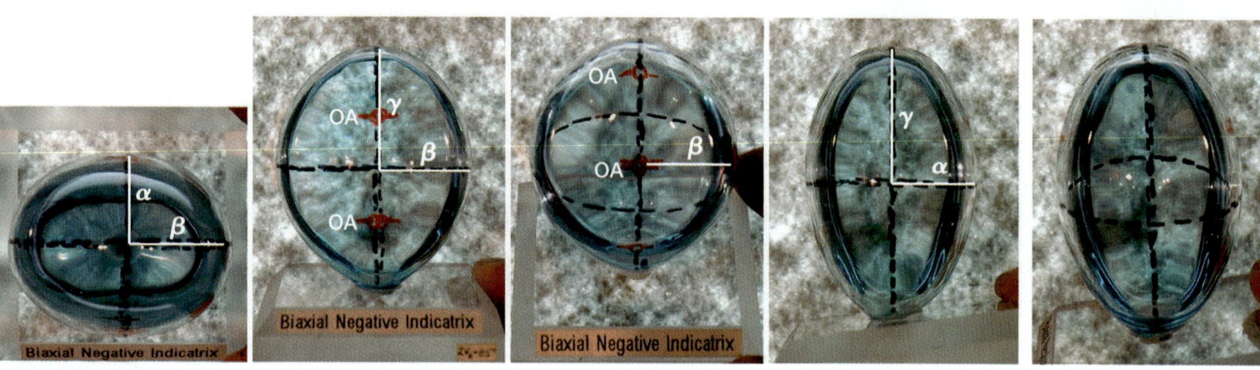

Figure 17.74. *Five different sections through the fluid-filled biaxial indicatrices shown in Figure 17.72 with refractive index directions and two optic axes (OA) labeled. The first column shows the X-Y section, the second the Y-Z section, the third the circular section, the fourth the X-Z section, and the last a random section.* ***a.*** *Positive indicatrix.* ***b.*** *Negative indicatrix.*

separate orientations for the biaxial indicatrix. The most interesting orientation for the latter is the one showing a circular section within a triaxial ellipsoid. Interpreting these two-dimensional images requires an understanding of the two-dimensional cuts that represent how the mineral rests on a microscope stage through the indicatrix of the mineral. In this chapter, we have tried to show these imaginary surfaces representing the speed of light in every way we could think of. Please be sure to supplement the images shown on these pages with the colorful animations on the MSA website, which will always be more informative.

References

Bloss, F.D. (1999) Optical Crystallography. Mineralogical Society of America, Washington D.C., 239 pp.

Brown, B.M. and Gunter, M.E. (2003) Morphological and optical characterization of amphiboles from Libby, Montana U.S.A. by spindle stage assisted polarized light microscopy. The Microscope, 51, 3, 121–140.

Gunter, M.E. (2003) A lucky break for polarization: The optical properties of calcite. ExtraLapis, Calcite, 4, 40–45.

Gunter, M.E. and Ribbe, P.H. (1993) Natrolite group zeolites: correlations of optical properties and crystal chemistry. Zeolites, 13, 435–440.

Kamb, W.B. (1958) Isogyres in interferences figures. American Mineralogist, 43, 1029–1067.

Kile, D.E. (2003) The Petrographic Microscope: Evolution of a mineralogical research instrument. Special Publication No. 1, The Mineralogical Record, Tucson, AZ, 96 pp.

Nesse, W.D. (2004) Introduction to Optical Mineralogy. Oxford University Press, 3rd Edition, 348 pp.

Su, S.C. and Bloss, F.D. (1984) Extinction angles for monoclinic amphiboles or pyroxenes: a cautionary note. American Mineralogist, 69, 399–403.

Chapter 18

Optical Crystal Chemistry

Because I had planned to pay tribute to the wonderful professors who taught me so much about the material in this chapter, I have looked forward to writing the personal statement for this chapter more than any other. However, I'm writing this in late April 2007, and it's a sad time for my alma mater, Virginia Tech. So there are going to be few words here, because words cannot express my feelings. For all of my career, I have been thankful for the education I received at Tech. Tech's geoscience department is one of the best in the U.S., specifically for mineralogy, and especially optical mineralogy. Shown here is a copy of the cover of The Microscope*; this was a special issue dedicated to my Ph.D. advisor, F.D. Bloss, on the occasion of his 70th birthday. The cover shows Don in his roles of research, writing, and lecturing. Also shown are his books and the spindle stage. Although I had originally wanted to dedicate this chapter to Don, it is now dedicated specifically to the memory of those who died at Tech, and in general to all the Hokies. Somewhat fittingly, it's worth pointing out that Tech's colors are orange and maroon as shown in the ribbon symbol memorializing the tragedy. So in this chapter, I'll refer to the orange-ish color Becke line, as "Virginia Tech" orange, another fitting tribute.*

M.E.G.

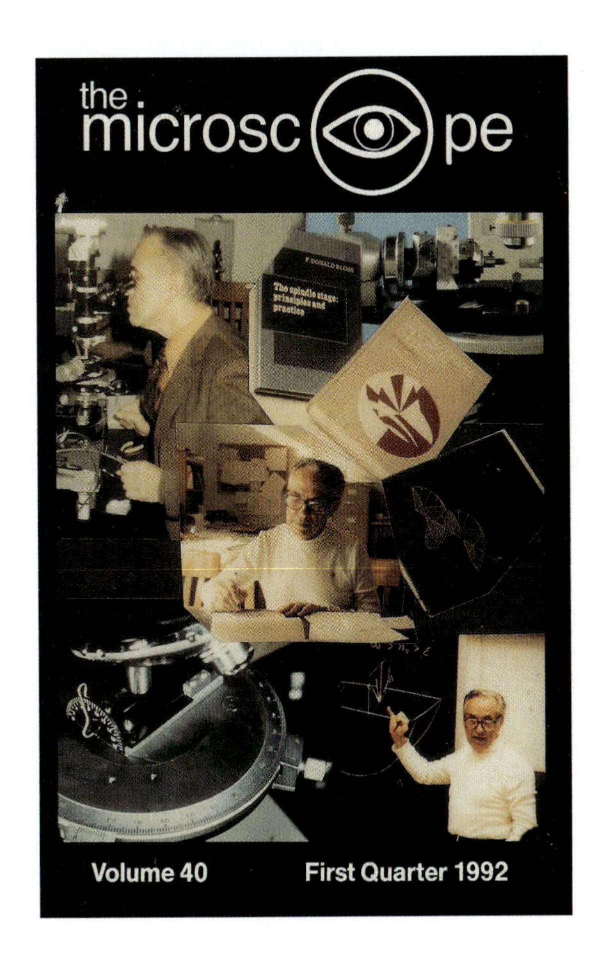

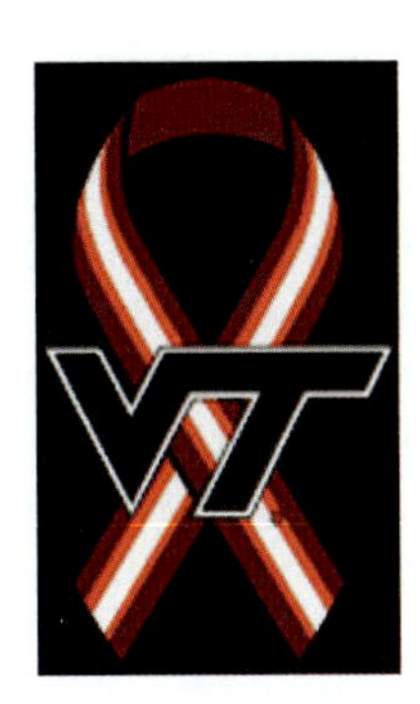

Introduction

This chapter deals with a subject we are terming **optical crystal chemistry**. We use this phrase to describe how the chemistry of a mineral affects its optical properties, especially its refractive index. Jaffe (1988) wrote an entire book on this subject, entitled *Crystal Chemistry and Refractivity*, which examined the relationships among refractive indices of minerals and their compositional variability. He also discussed theoretical and empirical ways to calculate refractive index. The related term **refractometry** is an expression often used for the measurements of refractive indices in minerals. We never directly measure the speed of light in a mineral, but we measure the effect of light passing through a mineral, especially how it refracts when it enters the mineral. This explains why the process is called refractometry.

This chapter will deal with two major topics: (1) how the refractive index of a mineral is measured, and (2) the relationship of refractive index to the crystal structures of minerals. Precise measurements of refractive index are difficult and time-consuming. Most scientists will never need to make these measurements themselves. We doubt that you will ever be called upon to produce charts of composition vs. refractive index like those shown later in this chapter. However, everyone who uses a polarizing light microscope (PLM) uses refractive index values to identify minerals.

Before we discuss how optical properties are interpreted, it is useful to give examples of how to measure them. We have already begun our discussion of the Becke line in Chapter 5 (Figure 5.10). Becke lines determine refractive index using successive closer comparisons between a grain and liquids with varying refractive indices. Historically, this has been viewed as one of the most tedious undertakings in an optical mineralogy course! This was the motivation for the unknown cartoonist who drew the "Becke Line Blues" cartoon (Figure 18.1) as a present for Don Bloss. In the column of figures to the right in Figure 18.1, $n_{grain} > n_{liquid}$ in the top image, $n_{grain} < n_{liquid}$ in the bottom image, and $n_{grain} = n_{liquid}$ in the middle image. These images provide a starting point for our discussion of advanced optical methods in this chapter.

Refractive Indices and Minerals

The refractive index of a mineral (the unit-less ratio of the speed of light in air divided by the speed of light in a material) is one of the most diagnostic of the many physical properties of minerals. It is particularly useful in identification because it directly relates to the crystal structure and composition of a mineral. We will begin this chapter with a qualitative discussion of refractive index, and proceed until we ultimately discuss current knowledge of the quantitative study of refractive index.

Photons of light enter a mineral and are slowed down (retarded) by interactions with electrons. In general, the higher the electron density, the greater the refractive index. A good analogy for this process is that of a man walking through an open field with his arms outstretched (Figure 18.2a). The direction he is walking would be the ray path direction for a ray of light; his arms represent the perpendicular vibration directions of the transverse light ray. Imagine that the man starts out walking at a speed of 3 mph. When he encounters a head-high overgrown field, he slows down to 2 mph because his arms keep hitting the brush. In this example, the "refractive index" of the overgrown field would be 1.5.

Bonding. Different types of chemical bonds affect the local electron density within a mineral. As an example, consider fluorite and diamond, which are both isometric minerals with simple crystal structures (Figure 14.3). Fluorite has one of the lowest refractive indices of any mineral (n = 1.434), while diamond has one of the highest (n = 2.418). Fluorite is held together by ionic bonds, in which the electrons are attracted to and therefore clumped around the F^{1-} anion, so relatively low electron densities are encountered by rays of light passing through the "space" between the Ca^{2+} and F^{1-} ions (Figure 18.2b). Diamond, on the other hand, is covalently bonded, so its electrons are distributed between the adjacent carbon atoms that share them (Figure 18.2c). The result is high electron density between the carbons, so light passing through is slowed considerably, resulting in a high refractive index.

Chemistry. As the number of electrons increases (and thus the likelihood of photons of light interacting with them), it is logical that the refractive index should also increase. The vast majority of the minerals we will study with a PLM exist as solid-solution series between end-members. As discussed in previous chapters, one of the most common substitutions is that of Fe^{2+} for Mg^{2+}. Because Fe^{2+} has more than twice as many electrons as Mg^{2+}, it is to be expected that a substitution of Fe^{2+} for Mg^{2+} would increase the refractive index of a mineral. Another common coupled substitution is that of Na^{1+} + Si^{4+} for Ca^{2+} + Al^{3+}. Because Ca^{2+} and Al^{3+} have about one-third more electrons than Na^{1+} and Si^{4+}, substitution of Ca^{2+}

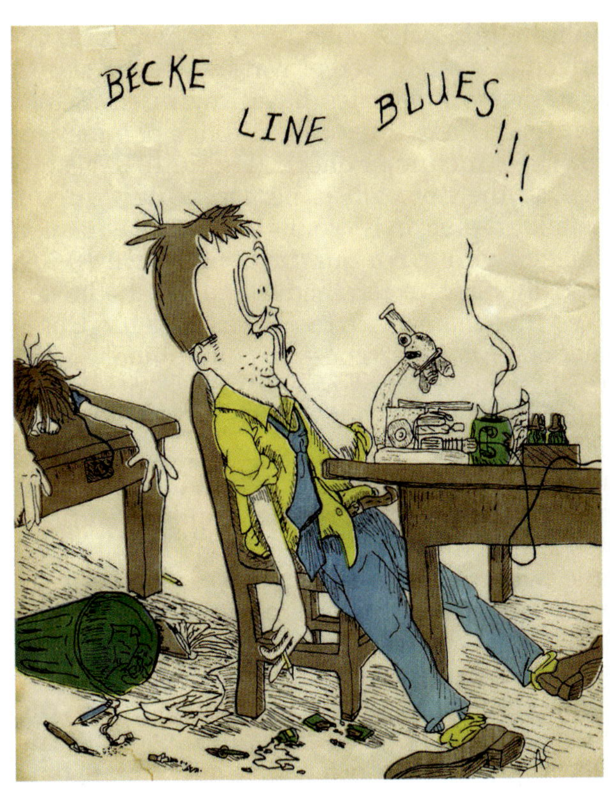

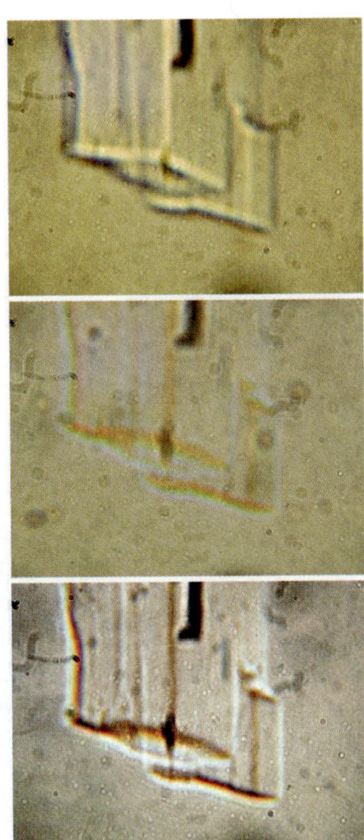

Figure 18.1. Cartoon given to Professor F.D. Bloss at Virginia Tech some time before 1979 and scanned for this book (thus the yellow faded paper background!). The image shows a frazzled student attempting to match the refractive index of a grain to that of a liquid. This is the Becke line method that is used to measure the refractive index of mineral grains. The three images to the right, from top to bottom, show a grain of tremolite in three separate refractive index liquids. In the top image, the liquid's refractive index is too low, in the bottom image, it is too high, and in the center, the refractive index is "just right." Two separate colors are seen in the latter: blue and orange. In fact the orange matches "Virginia Tech" orange.

and Al^{3+} should (and does!) increase refractive index (Figure 18.3).

Olivine group minerals commonly occur along the Mg^{2+}-Fe^{2+} solid-solution series between forsterite (Mg_2SiO_4) and fayalite ($Fe^{2+}{}_2SiO_4$). As explained above and shown in Figure 18.4a, all three refractive indices increase with fayalite content, though β increases at a slightly greater rate than α and γ. Density (ρ) is also plotted, and it also increases with fayalite content. The final parameter plotted on the graph is $2V_\gamma$, which is the $2V$ angle measured in the γ direction. Thus, for $2V_\gamma < 90°$, the mineral would be optically positive, while for $2V_\gamma > 90°$, the mineral would be optically negative. For forsterite, β is closer to α, and so the mineral is optically positive. For fayalite, β is closer to γ, so the mineral is optically negative. The diagram suggests that if you know how to measure the refractive indices, $2V_\gamma$, or density of an olivine grain, you can determine its approximate composition.

In the plagioclase feldspar series (Figure 18.4b), substitution of $Ca^{2+} + Al^{3+}$ (anorthite) for $Na^{1+} + Si^{4+}$ (albite) causes an increase in the refractive index. Also plotted on the graph are the ε and ω values for quartz. When the An content is <15%, $n_{plag} < n_{quartz}$, whereas when An > 45%, $n_{plag} > n_{quartz}$. These relationships turn out to be very useful for distinguishing among anorthite, albite, and quartz, especially in thin section.

If a mineral is well-characterized optically and its orientation is known, complex PLM images can sometimes be better understood. Figure 18.5 is a photomicrograph of a thin section of a synthetic olivine with a starting composition of ~Fo_{90}. Adjacent to the opaque magnetite crystals, the interference color is second-order yellow, while everywhere else it is second-order green. The increase in retardation from green to yellow is caused by diffusion of Fe from the magnetite crystals into the olivine. We can draw this conclusion because the sample was cut with α and β in the plane of the thin section. This is seen in Figure 18.4a, where $\beta - \alpha$ increases with Fe content. Chemical analysis by electron microprobe also showed higher Fe contents in the zones near the magnetite.

Structure. Most rock-forming minerals are optically anisotropic. In general, the amount of anisotropy (birefringence) can be related to the mineral structures. For example, the framework silicates such as quartz and feldspars typically exhibit low birefringence on the order of 0.01, so the speed of light is very similar in all directions. Thus, framework silicates consistently appear first-order gray in thin section. At the other end of the spectrum, one of the most anisotropic minerals is calcite (Figure 18.6). The CO_3 groups that form the basis of its structure are all in the same orientation, with the plane of the CO_3 groups

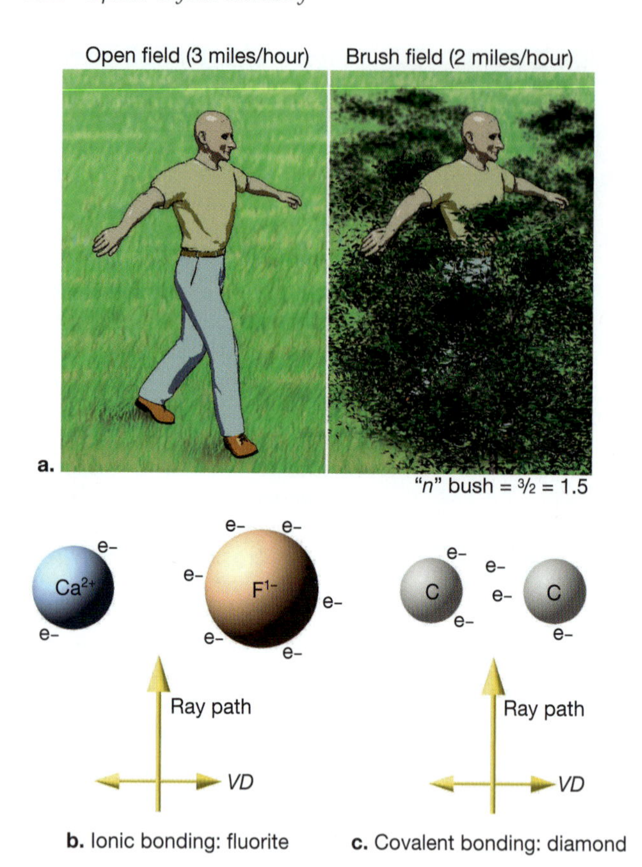

Figure 18.2. *Explanation of how the speed of light slows when it moves from air into a mineral.* ***a.*** *A man walks through a freshly-mown field at a speed of 3 mph with outstretched arms. When he gets to a densely-overgrown field, he can only walk at 2 mph because his arms keep hitting the brush. The latter field would have a "refractive index" of 1.5. Notice he is walking in one direction, but his arms are extended outward. His arms are slowing him down because they keep hitting the brush. This is similar to the way light interacts with a mineral—the vibration is perpendicular to the ray path, and in this case the arms are perpendicular to the direction the man is walking.* ***b.*** *Fluorite exhibits ionic bonding and has a low refractive index of 1.434. As light travels through the mineral, like the man walking through the brush, its vibration direction will interact with very few electrons because those electrons are attracted close to the* F^{1-} *ion.* ***c.*** *Diamond is an example of covalent bonding, where the electrons can be found distributed in the space between the C atoms. Diamond has one of the highest refractive indices (2.418) of any mineral, because photons of light can interact with the electrons filling the space.*

perpendicular to the *c* axis. When the ray path of light coincides with the *c* crystallographic axis, its associated vibration directions correspond to the bond directions in the CO_3 planar group and the ω direction. As a result, the refractive index in this direction is considerably higher than in the other directions. Therefore, calcite is uniaxial negative, and has a very high birefringence of ~0.172.

Silicate minerals contain SiO_4 polyhedra polymerized in many different ways. As noted above, the framework silicates have low birefringence because of their three-dimensional polymerization. The structure of quartz (Figure 18.7a) is oriented so the ray path is perpendicular to the *c* crystallographic axis, so its vibration direction coincides with ε. For quartz, $\varepsilon > \omega$ by only 0.009. When the silicate tetrahedra and octahedra are linked into chains with the same orientation, as is the case for pyroxenes and amphiboles, the refractive index is the greatest when light vibrates along the chains. This is shown for enstatite (Figure 18.7b), where the ray path is perpendicular to the *c* crystallographic axis, and the vibration direction thus corresponds to γ. In the layer silicates, electron density is greatest within the (001) plane, where it is approximately the same in all directions. In the structure of muscovite (Figure 18.7c), the ray path coincides with the *c* axis, and its vibration direction is along the *b* axis (γ).

Even subtle differences in optical properties can be related to the crystal structures. Dioctahedral muscovite and the trioctahedral mica biotite both have β and γ in the (001) plane *but* their relative orientations are flipped (Figure 18.7d,e). This results from two factors. In muscovite, the $(OH)^{1-}$ bond leans into the vacancy and there is a larger vector sum from the highly polarizable $(OH)^{1-}$ bond along the γ direction. In biotite, that $(OH)^{1-}$ bond is perpendicular to the sheets, influencing the α refractive index direction. The second factor is that the Fe-containing octahedra in biotite are more directly aligned along the γ direction. The retardation $\gamma - \beta$ is lower for biotite than for mus-

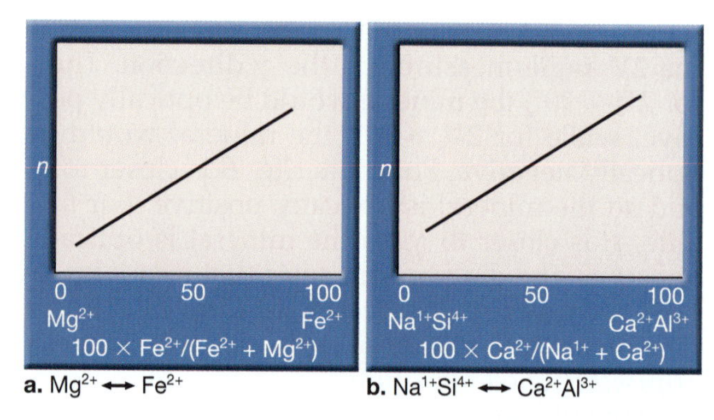

Figure 18.3. *Cation substitutions are common in silicate minerals.* Fe^{2+} *and* Mg^{2+} *commonly substitute for each other because they are of similar size and equal charge. Because* Fe^{2+} *has more than twice as many electrons as* Mg^{2+}*, substitution of* Fe^{2+} *for* Mg^{2+} *will cause the refractive index to increase. Another common substitution is that of* Na^{1+} *and* Si^{4+} *for* Ca^{2+} *and* Al^{3+}*. Because the latter contains half-again the electrons of the former, refractive index increases when* Ca^{2+} *and* Al^{3+} *are substituted.*

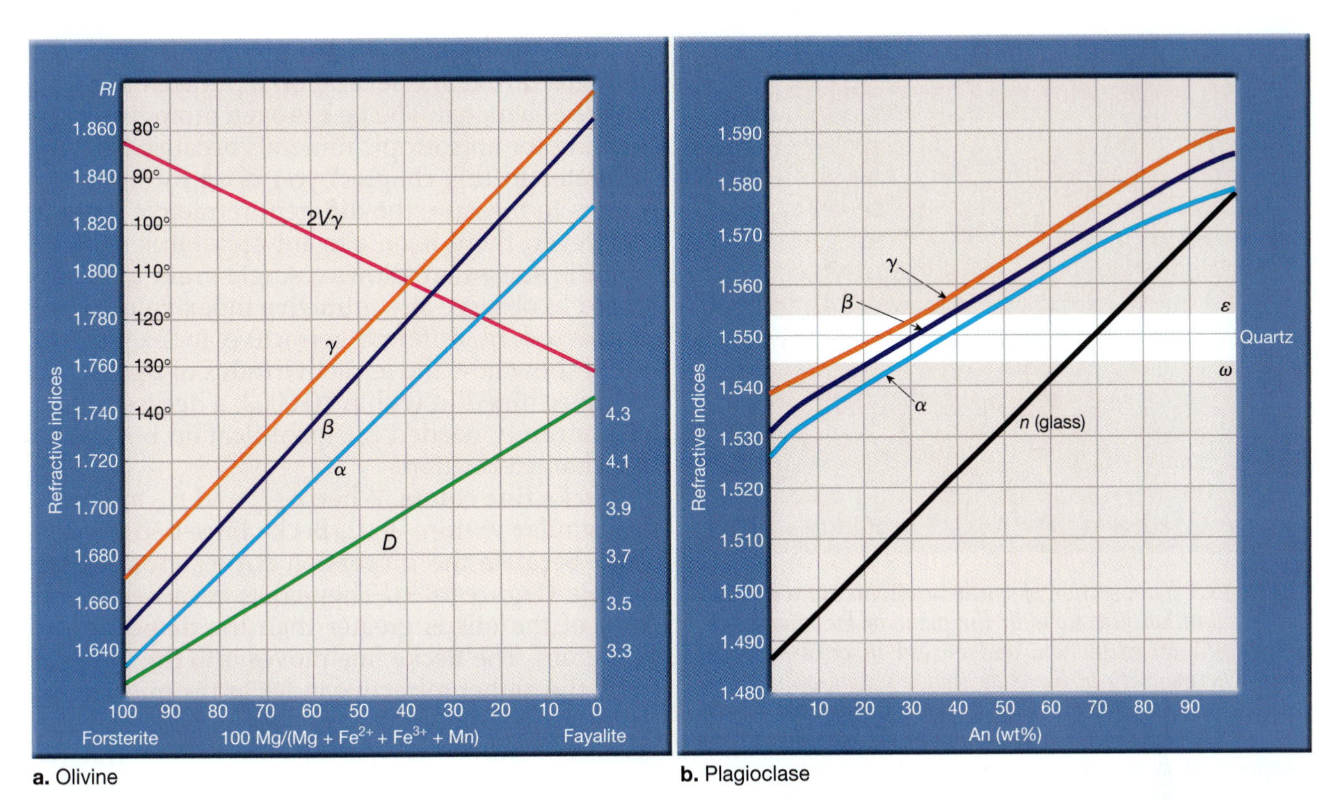

a. Olivine **b.** Plagioclase

Figure 18.4. *Changes in refractive index as a function of composition.* ***a.*** *The orthorhombic mineral group olivine is biaxial, and thus has three refractive indices. All of them increase with Fe content, though at different rates. Therefore, the 2V of the mineral also varies with composition.* ***b.*** *The plagioclase feldspars have three refractive indices that increase as Ca^{2+} and Al^{3+} substitute for Na^{1+} and Si^{4+}. The ε and ω values for quartz occur near the middle of the solid-solution series for the plagioclase feldspars. The refractive index of a glass made from feldspars of differing compositions, which has a significantly lower refractive index, is also shown for comparison.*

covite because the biotite structure is almost hexagonal, whereas muscovite is considerably more distorted. Biotite, like most trioctahedral micas, has $2V \approx 0°$, while muscovite and other dioctahedral micas can have significantly larger $2V$, up to as much as 60°. Thus, $2V$ is a useful diagnostic feature for distinguishing between from dioctahedral and trioctahedral micas.

Measuring Refractive Indices

In Chapters 5 and 17, we immersed minerals in liquids to aid in their study (i.e., to look at interference figures), and to approximate their refractive indices. This technique is called the immersion method, and it makes use of the previously-mentioned concept of Becke lines. This is by far the simplest, easiest, and most efficient technique for mineral identification, and it requires very little training (relative to other techniques like X-ray diffraction, the electron microprobe, etc.). Another instrument that we discussed for measurement of refractive index is a refractometer, but it requires ~1 cm^2 polished crystals. This makes it a powerful technique for gemologists, but far less useful for geologists.

Now you can apply what you have learned about the Becke line method (Figure 5.10) to compare the refractive index of some grains (n_{grain}, in Figure 18.8) to the refractive index of the liquid that surrounds them (n_{oil}, in Figure 18.8). Find and focus on some of the grains in an oil mount. If n_{grain} is similar to n_{oil}, the grains will exhibit low relief and be difficult to see. In such a case, reduce the light source by both turning down the bulb's intensity and closing the substage iris diaphragm. If the grains "stand out," they are exhibiting high relief, either positive ($n_{grain} > n_{oil}$) or negative ($n_{oil} > n_{grain}$). Once you've found a grain, compare n_{grain} to n_{oil} by using the Becke line method. First, reduce the light intensity, then slightly lower the stage to observe the Becke line; the bright Becke line will enter the material that has a higher refractive index.

If the Becke lines are not colored, then the refractive index of the oil is not very close to that of the grain. So, you will need to make another oil mount. It is helpful to record which oils you use and what you observe. For each new oil mount,

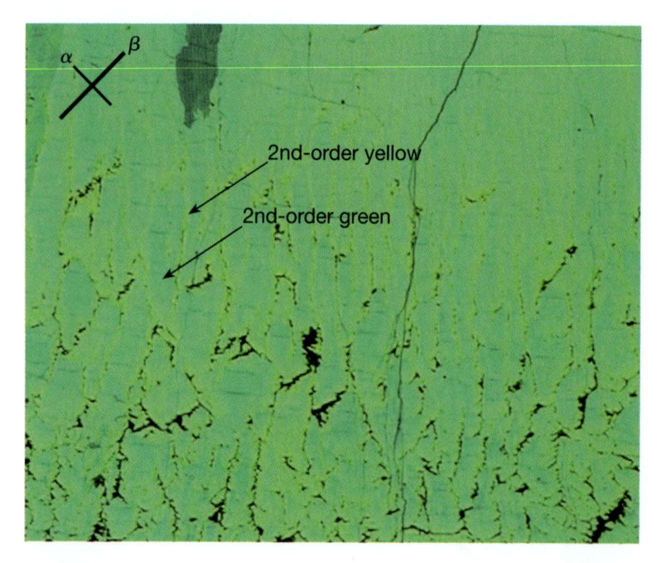

Figure 18.5. Thin section of synthetic olivine cut with its α and β vibration directions in the plane of the microscope, rotated 45° off extinction, and viewed in cross-polarized light. The majority of the slide shows second-order green retardation, but second-order yellow retardation occurs along zones adjacent to areas containing magnetite (the opaque black phase). The increase in retardation color from green to yellow reflects the higher Fe content of the olivine adjacent to the magnetite, as measured by SEM-EDS. This image shows the relationships that were expressed graphically in Figure 18.4a, (i.e., that the value of β – α, or birefringence, increases with Fe content).

bracket n_{grain} (e.g., if you start with a 1.55 oil and find $n_{grain} > n_{oil}$, then go to a 1.65 oil and find $n_{oil} > n_{grain}$, so that you make a mount approximately halfway between 1.55 and 1.65, which is 1.60). Continue bracketing n_{grain} by making new oil mounts until the Becke lines become colored. The value for n_{oil} that produces colored Becke lines will be close, usually within ±0.02 of n_{grain}, and may be good enough to identify the mineral.

If you need to determine n_{grain} within ±0.002, you must change n_{oil} until you observe an orange-ish ("Virginia Tech orange") line entering the grain and a blueish line entering the oil. If the Becke lines are colored but not quite blue and orange, you still need to slightly change the refractive index of the oil. As a general rule, if the Becke line entering the oil is dark blue, even bordering on black, then $n_{grain} > n_{oil}$. This is similar to having a dark Becke line enter the oil, meaning the grain has the greater refractive index, and n_{oil} needs to be increased. On the other hand, if the Becke line entering the grain is too red, bordering on dark, then $n_{oil} > n_{grain}$. This is similar to having a dark Becke line enter the grain, meaning the grain has the lower refractive index, and n_{oil} needs to be decreased.

This can be a laborious process, in that you may need to make several oil mounts before you arrive at a match. The task is even more time-consuming for anisotropic minerals because (as we'll see later in this chapter) you need to orient the grains to measure the different refractive indices. However, there is an elegant apparatus called a spindle stage that allows a single grain to be oriented to observe any refractive index value while immersed in different refractive index liquids. We'll show how the refractive index of a grain can be determined to within ±0.0002, a degree of precision rarely needed for identification but useful for characterization.

Becke line colors. When n_{grain} and n_{oil} around it are similar within ~0.02, Becke lines become colored because the dispersion curves cross in the visible (Figure 18.9). They cross because dispersion of the oils is greater than the dispersion of the grain. The Becke line moves into the material with the higher refractive index as the microscope stage is lowered. In the example in Figure 18.10a, the grain and the oil will have the same refractive index at 589 nm. When light has a wavelength of <589 nm, the Becke line would move into the oil.

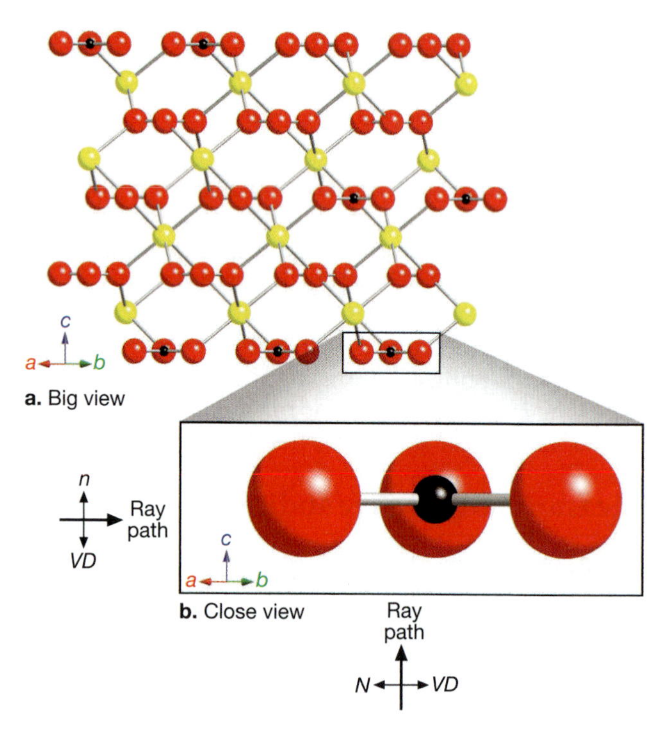

Figure 18.6. Views of the crystal structure of calcite, oriented with the plane of CO_3 clusters perpendicular to the ray path of the incident light, thus corresponding to the vibration direction of the polarized light. In this orientation, the vertical ray paths are slowed more than they would be if they were traveling parallel to the CO_3 plane, because of the high electronic density in the CO_3 plane. This is why calcite has a large birefringence, with ω >> ε.

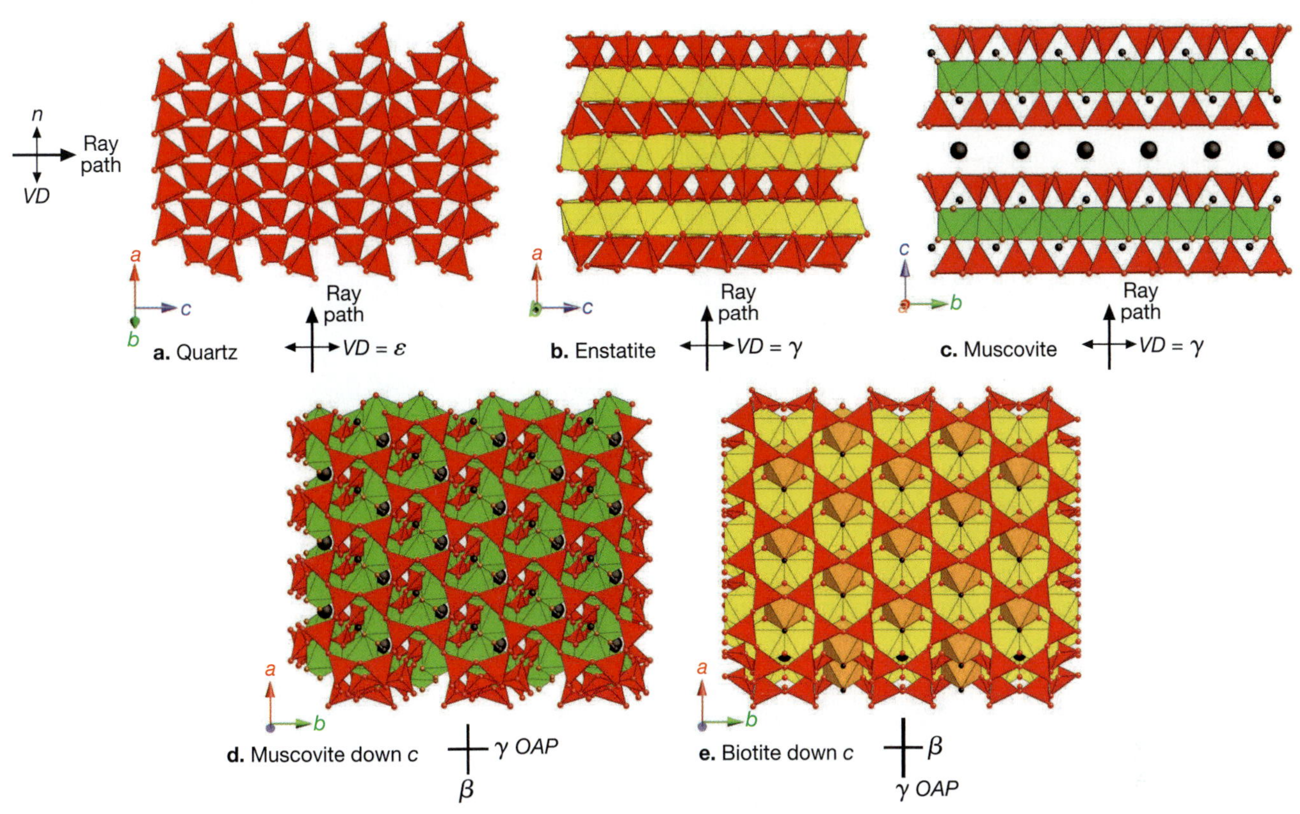

Figure 18.7. Crystal structure drawings for silicate minerals relating the orientational dependence of their refractive indices to their crystal structures. In Figure 18.7a–c, the light paths are vertical in the plane of the page, and the largest refractive index is horizontal. In Figure 18.7d and e, the light path is perpendicular to the page. ***a.*** *In the framework silicate quartz, the vibration direction coincides with ε, but because the structure is similar in all directions, quartz exhibits low birefringence.* ***b.*** *In the chain silicate enstatite, the largest refractive index direction occurs parallel to the chains, because this is the direction of greatest electron density. Its birefringence is thus greater than that of quartz.* ***c.*** *In the layer silicate muscovite, the largest refractive index direction is within the plane of the layers, and corresponds to the plane of greatest electron density. In Figure 18.7d and e, views perpendicular to the sheet show the dioctahedral mica muscovite and the trioctahedral mica biotite, respectively. The large refractive index directions β and γ both occur in this plane. Because β is similar to γ, both of these minerals will be optically negative, but the value of γ – β is smaller in biotite than in muscovite. Notice that the γ vibration direction switches between muscovite and biotite. In muscovite, the OH bonds have a larger projection along the γ direction, but in biotite the OH bonds are vertical along α. In biotite, the Fe-containing octahedra increase the electron density in the γ direction.*

When the wavelength is >589 nm, the Becke line will move into the grain. The visible spectrum is thus split in half. The color of each Becke line would be the sum of the portion of the spectrum it represents. For example, the color moving into the grain will be a mixture of red, orange, and some yellow, while the color moving into the oil will represent green, blue, and violet. The value of n at 589 nm will return again in this chapter, because for historical reasons it happens to be the reference wavelength for reporting refractive index (589 nm is the wavelength of Na in a sodium discharge bulb).

If the wavelength of match is at 475 nm (Figure 18.10b), the Becke lines will be a different color: light yellow moving into the grain, and dark blue going into the oil. If the wavelength of match is 650 nm, then the majority of the light will go into the oil, producing dark red into the grain, and light blue-green in the oil (Figure 18.10c).

To illustrate this discussion, a grain of quartz is shown immersed in a fluid whose refractive index matches in the visible (Figure 18.11), as evidenced by the colored Becke lines. The quartz grain is not at extinction, so the Becke lines are a combination of separate refractive indices in the mineral. Even for such a randomly-oriented grain, some insight into refractive index can be gained. To obtain the most meaningful data, a principal refractive index direction of the grain (i.e., in this case ε or ω) must be identified and made parallel to the lower polarizer. If a crystal is mounted on a spindle stage, it is relatively straightforward to orient it.

This is shown in Figure 18.12, where a quartz crystal has been rotated about the spindle and the

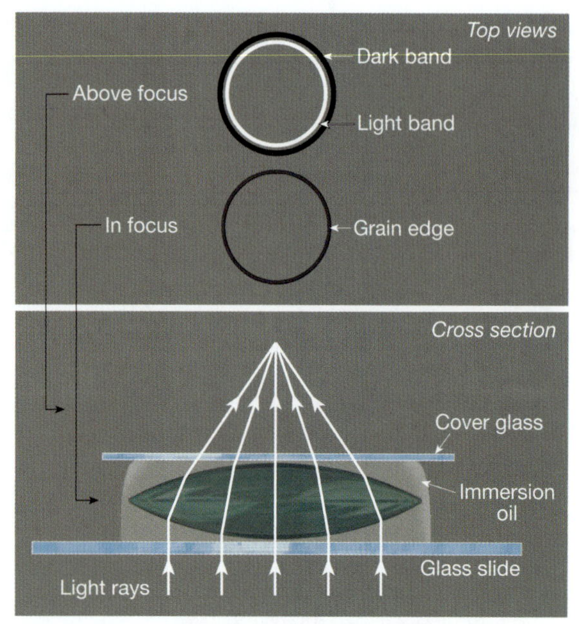

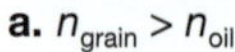

a. $n_{grain} > n_{oil}$

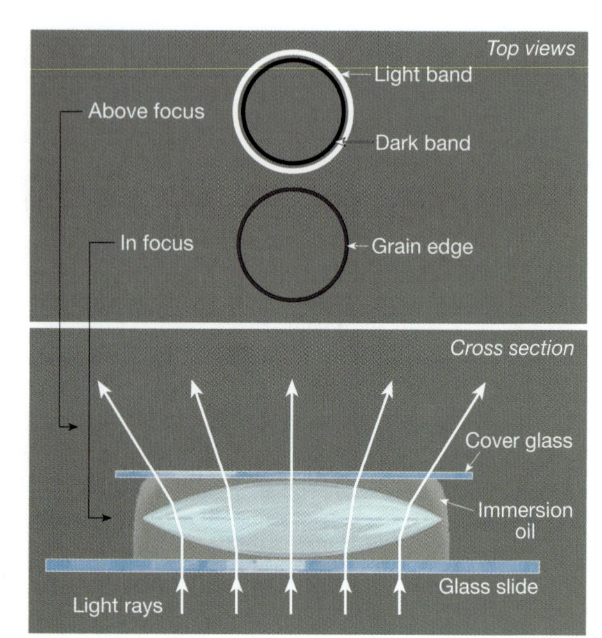

b. $n_{oil} > n_{grain}$

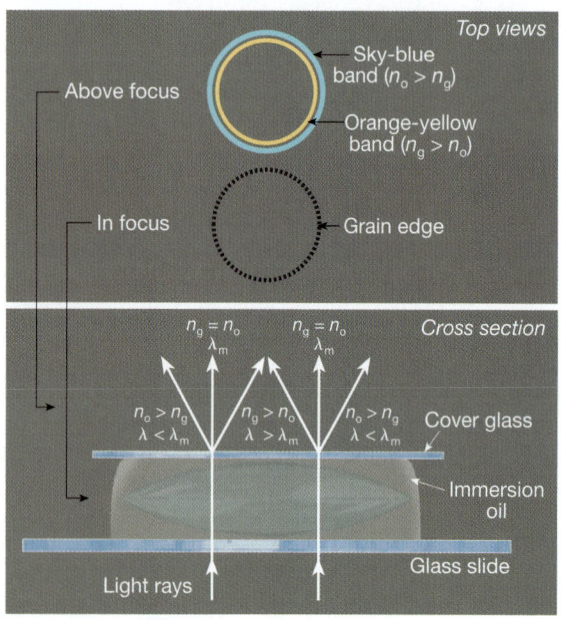

c. $n_{grain} = n_{oil}$ at 589 nm

Figure 18.8. *Formation of Becke lines. Light rays pass through a crystal fragment that is shaped like a converging lens. The crystal is placed in immersion oil between a glass slide and a cover slip. Two views of the crystal are shown, with the light rays in focus and slightly above focus. The latter is accomplished by lowering the microscope stage.* ***a.*** *Case for* $n_g > n_o$*: The light rays behave as if the grain is a converging lens. If the focus is raised, light is concentrated on the inside of the grain (light Becke line), and a dark band forms on the outside of the "grain" (dark Becke line). The positions of these light and dark lines of color indicate that* $n_g > n_o$*. The light Becke line moves into the material of higher refractive index as the stage is lowered.* ***b.*** *Case for* $n_o > n_g$*: The light rays behave as if the grain is a diverging lens. If the focus is raised, a dark band forms on the inside of the grain (dark Becke line), and a light band is on the outside of the grain (light Becke line). The positions of the light and dark bands of color indicate that* $n_o > n_g$*. The light Becke line moves into the material of higher refractive index as the stage is lowered.* ***c.*** *Case when* $n_g = n_o$ *at 589 nm: Because the refractive index of the oil and grain are equal at 589 nm, light of that wavelength passes straight through the grain without being refracted. For* $\lambda > \lambda_m$*, the light converges (as in Figure 18.8a), but for* $\lambda < \lambda_m$*, the light diverges (as in Figure 18.8b). This causes the Becke lines to become colored. The line on the inside of the grain is composed of wavelengths greater than 589 nm (which yield the Virginia Tech orange color), while the line on the outside of the grain is composed of wavelengths shorter than 589 nm (a blueish color). These two colors are shown in the center image of the grain in Figure 18.1.*

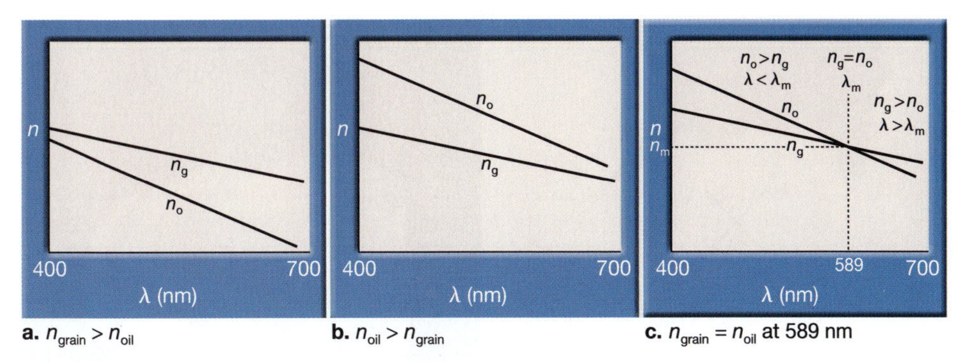

*Figure 18.9. Three figures corresponding to Figure 18.8a–c. These plots elaborate on Figure 16.6a, showing how refractive index decreases as wavelength increases. Two lines are now added, one for the grain and one for the oil. The slope of the oil line is greater than that of the grain. **a.** Case for $n_g > n_o$: At wavelengths of light in the visible region of the spectrum, the line for the grain plots above that for the oil. Thus, the refractive index for the grain exceeds that of the oil. **b.** Case for $n_o > n_g$: The line for the oil plots above that for the grain. The refractive index for the oil exceeds that of the grain. **c.** Case when $n_g = n_o$ at 589 nm: The two lines intersect in the visible at 589 nm. This intersection divides the visible spectrum into three sections: (1) at their intersection, light of this wavelength (λ_m) passes straight through the grain without being refracted; (2) for shorter wavelengths ($\lambda < \lambda_m$), the oil has a greater refractive index than the grain. Combining wavelengths from 400 to 589 nm would produce a sky-blue colored Becke line that enters the oil when the stage is lowered; and (3) for longer wavelengths ($\lambda > \lambda_m$), the grain has a greater refractive index than the oil. Combining wavelengths from 589 to 700 nm would produce an orange-colored Becke line that enters the grain.*

microscope stage, so its *ω* vibration direction is parallel to the lower polarizer. This results in subtle changes in the colors of the Becke lines relative to a randomly-oriented crystal. If the crystal is then illuminated by monochromatic light, the lighter-color Becke line will be visible at low wavelengths entering the oil because n_{oil} will be greater than n_{grain}. As the wavelength of light increases, n_{grain} will eventually be greater than n_{oil}, as exemplified by the lighter-color line entering the red-illuminated grain (Figure 18.12d). At the monochromatic wavelength where n_{grain} exactly equals n_{oil}, no refraction occurs and the grain disappears (Figure 18.12c). Observation of colored Becke lines in polychromatic light is only semi-quantitative. However, if a monochromator is available, a match can be precisely determined at a specific wavelength based on the disappearance of a grain. If the wavelength can be changed continuously, the light Becke line will move from the outside to the inside of the grain as it passes by the wavelength at which a match occurs. Seeing this phenomenon might help convince skeptics that the Becke line actually exists (see the Preface to Chapter 5).

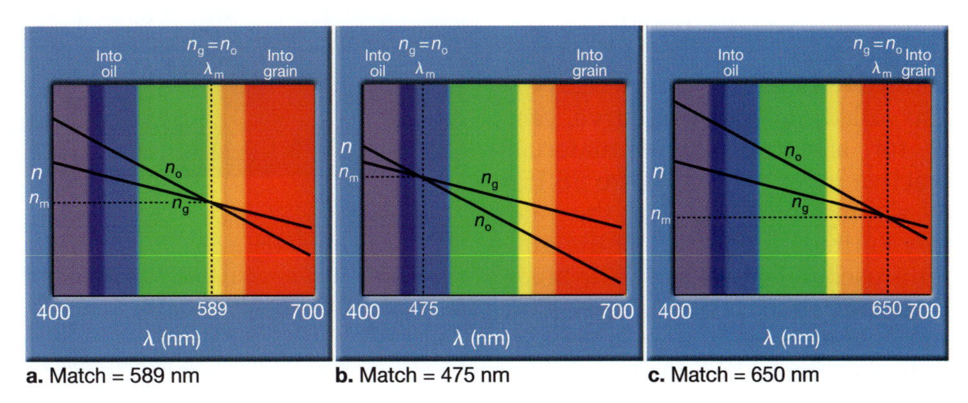

*Figure 18.10. Schematic showing dispersion curves for a mineral grain immersed in three different refractive index oils. **a.** The black lines representing dispersion for the grain and the oil interest at 589 nm. This is the chosen standard for matching the refractive index of the grain to the oil. The color produced in the Becke lines at this wavelength occurs because all of the wavelengths shorter than 589 nm move toward the oil as the stage is lowered, while those above 589 nm go into the grain. **b.** The refractive index of the oil is lower than that of the grain, so the lines match (intersect) at 475 nm, causing a dark blue line to enter the oil and a whitish yellow line to enter the grain. When the refractive index of the liquid is higher, as in Figure 18.10c, a match occurs at 650 nm and a light blue line enters the oil and a dark red line goes into the grain.*

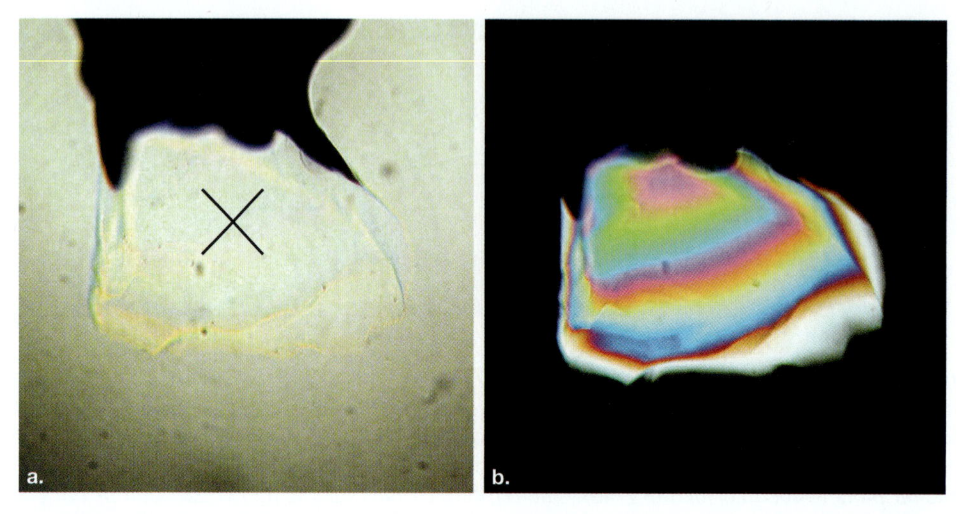

*Figure 18.11. The same single crystal of quartz that was shown in Figure 17.10a is here imaged in **a.** plane-polarized and **b.** cross-polarized light. Colored Becke lines are seen in the former image. In this oil, the refractive index curves of the oil and the grain intersect when viewed in visible light. However, the grain must be oriented properly (i.e., with a vibration direction parallel to the polarizer so the grain is at extinction) before precise refractive index results can be obtained.*

Another good example of Becke line behavior can be seen in the biaxial mineral tremolite. Each of three refractive indices is made parallel to the lower polarizer in Figure 18.13. The same refractive index oil is used in all three images, all use polychromatic light, and only the orientation of the crystal changes. The n_{oil} is greater than the n_{grain} for the α orientation, while at β, n_{oil} will be identical to n_{grain} at a long wavelength. The n_{oil} will also equal the n_{grain} at a shorter wavelength for γ. These determinations are based on the colors of the Becke lines. To quantify the wavelength at which a match occurs, monochromatic light can be used, and the wavelength where the grain disappears (or switches from the outside to the inside of the grain) can be noted (Figure 18.14).

A graph of refractive index vs. wavelength of illumination, also called a **dispersion curve**, can be constructed for any transparent material placed in a refractive index oil. In Figure 18.15, three negatively sloping lines represent the values of α, β, and γ for tremolite at each of the different wavelengths used in Figure 18.13. The bold line labeled n_0 is the dispersion curve for the oil in which the crystal is immersed. It is greater than and does not intersect the α line in the visible region, so no colored Becke lines form when the grain is oriented with α parallel to the lower polarizer. The dispersion curve of the oil intersects the lines corresponding to β and γ, producing different colors of Becke lines because the intersections occur at different wavelengths (475 vs. 600 nm).

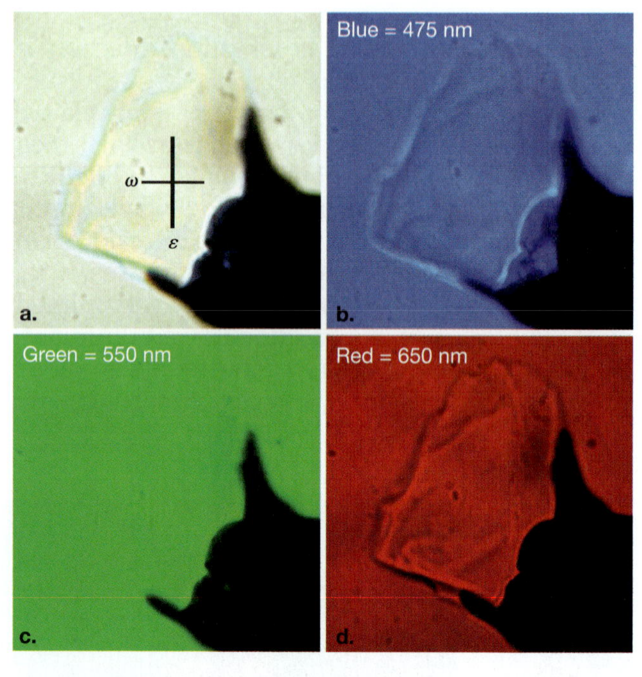

*Figure 18.12. **a.** The quartz grain from the preceding image is shown oriented (by a combination of rotation of the microscope stage and a rotation about the spindle stage) to bring its ω vibration direction parallel to the lower polarizer. Now its refractive index can be measured. In Figure 18.12 b–d, the crystal is illuminated in monochromatic light with three separate wavelengths. **b.** With blue light of 475 nm, the light Becke line is shown in the liquid, indicating that the $n_{oil} > n_{grain}$. **c.** The grain has completely disappeared because the wavelength of green light (550 nm) is exactly the same as that of the grain at this wavelength (i.e., $n_{grain} = n_{oil}$); thus, no refraction of the light occurs. **d.** Illuminated in red light of 650 nm, the light Becke line is now in the grain, indicating that $n_{grain} > n_{oil}$.*

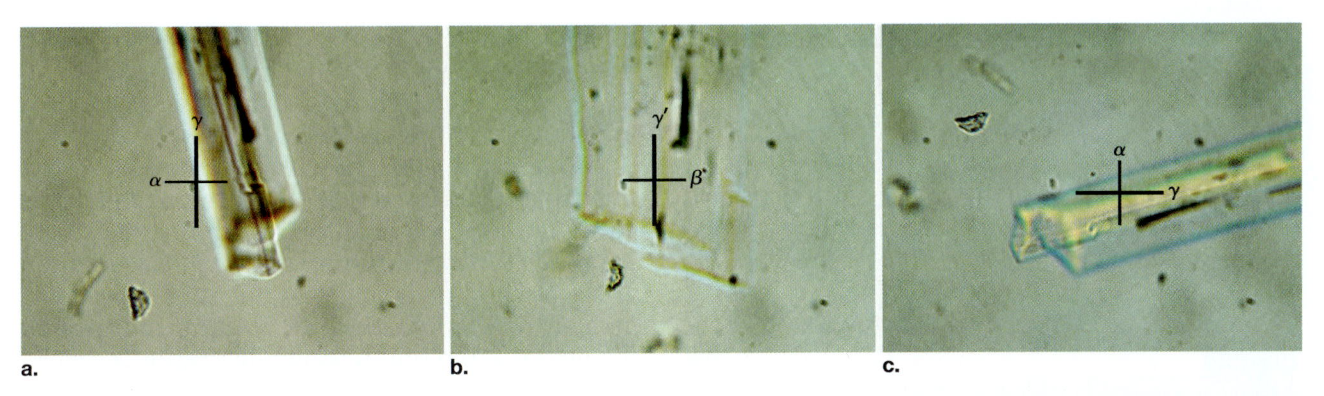

Figure 18.13. Three separate images of the same tremolite grain in the same oil, mounted with its c *crystallographic direction parallel to the spindle stage. In this orientation, the grain can be rotated on the spindle axis to bring all of its principal refractive index directions into the plane of the stage of the microscope, and then made parallel to the lower polarizer by rotation of the microscope stage.* ***a.*** *We are looking at the α vibration direction, and $n_{oil} > n_{grain}$.* ***b.*** *Where β is made parallel to the lower polarizer, the Becke lines become colored, indicating a near-match.* ***c.*** *γ is made parallel to the lower polarizer, and the Becke lines have a different color. (Thanks to Matt Sanchez for the idea of this figure.)*

The appearance of a grain in different refractive index liquids (over a range from higher to lower) will change as a function of the difference between n_{grain} and n_{oil} (Figure 18.16a,b), showing variations in relief as well as color. Figure 18.17 shows a series of matches between different oils and the grain at various wavelengths, with the resultant Becke line colors and the wavelengths that match each one. It can be rather difficult to distinguish the subtle colors of the Becke lines associated with the match in each case, especially at small increments of refractive index matches. The colors observed at 589 nm are particularly important because that is the standard wavelength for reporting refractive index. The resultant colors are (Virginia Tech) orange and blue. If a grain that matches the refractive index of its oil is viewed in polychromatic light, and in turn at

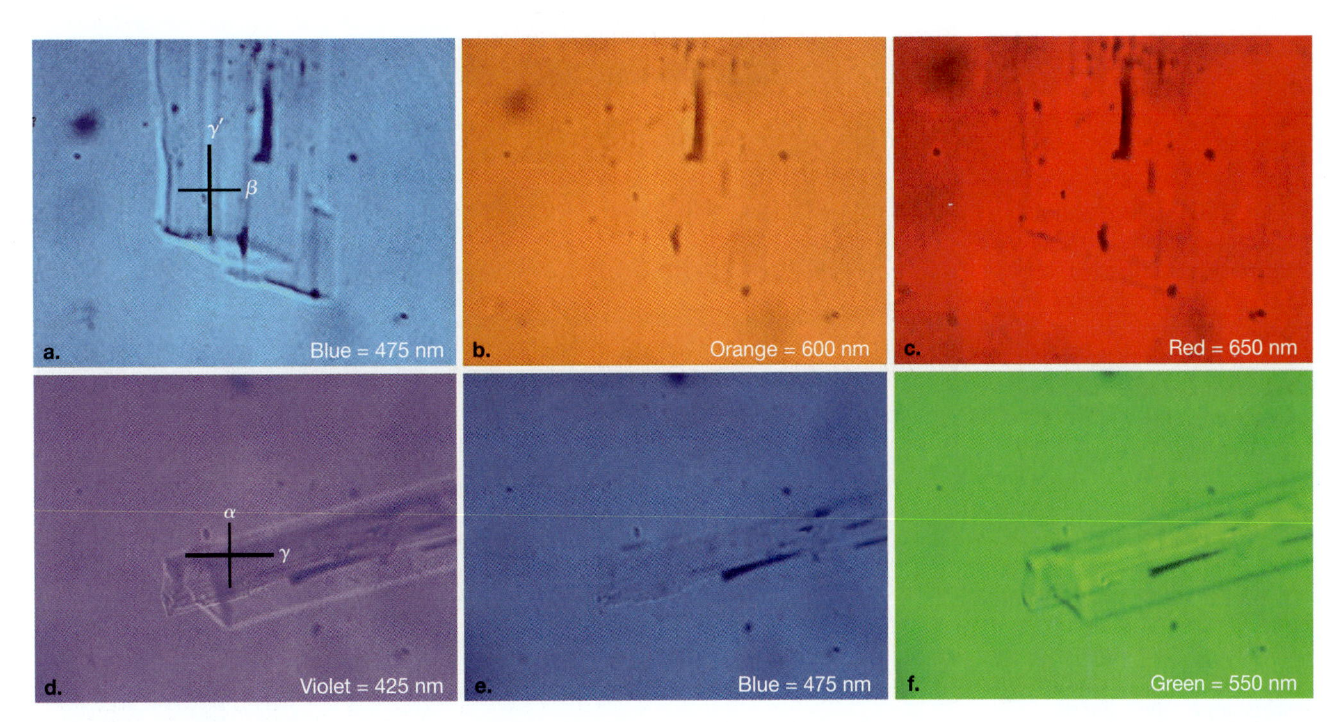

Figure 18.14. The same grain of tremolite shown in the previous figure is here imaged in monochromatic light, with Figure 18.14a–c showing the β refractive index direction, and Figure 18.14d–f showing the γ direction. For β, the match occurs at ~600 nm, and n_{oil} is too high at 475 nm and too low at 650 nm. For γ, the match occurs at 475 nm, with the oil too high at 425 nm and too low at 550 nm. Note again how the grains disappear at the specific wavelength of match and the light/dark Becke lines flip below and above the match.

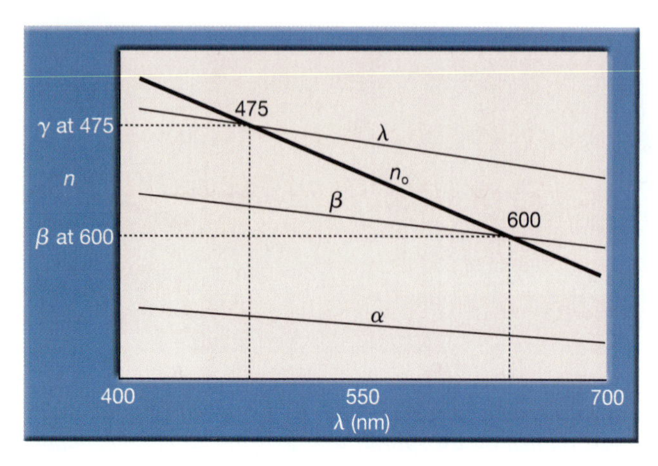

Figure 18.15. *Graphical representation of Figure 18.14. These dispersion curves show the relationship of the three refractive indices for tremolite, along with the oil used in Figures 18.13 and 18.14. Notice how γ matches the oil at 475 nm and β matches the oil at 600 nm. Thus, these Becke lines are colored while the one for α is not.*

three separate wavelengths of monochromatic light, it is apparent that $n_{grain} < n_{oil}$ in blue light, $n_{grain} > n_{oil}$ in red light, and matching ($n_{grain} = n_{oil}$) in yellow light (Figure 18.18).

To identify a material with an unknown refractive index, an immersion mount is usually prepared with a n_{oil} of ~1.6, a value in the middle of the range for rock-forming minerals. Observation of the Becke line is made to determine if n_{grain} is greater than or less than n_{oil}. To assist in choosing the next n_{oil}, the graph shown in Figure 18.19 can be used. For example, if a liquid has $n_{oil} = 1.6$, and a white Becke line moves into the grain, then n_{grain} must be greater than n_{oil}, and a refractive index oil ~0.03 higher would next be needed (sample #1 in Figure 18.19). An observation of the Becke line would then be made again, and potentially another oil chosen until a match is observed at 589 nm. This iterative (and often frustrating) process can require making several immersion mounts to

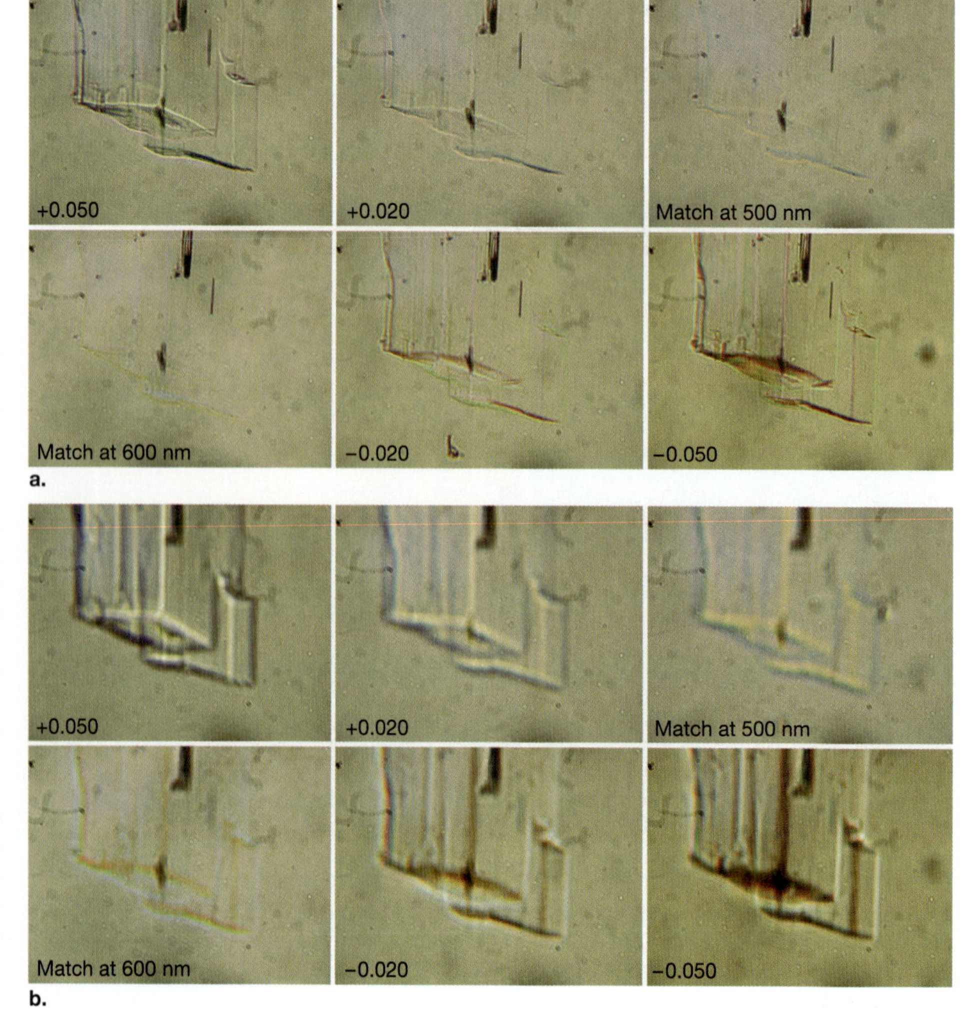

Figure 18.16. *The same tremolite grain shown previously, here with six different refractive index oils. Two of them produce colored Becke lines because n_{oil} and n_{grain} match in the visible; the other values listed are for $n_{grain} - n_{oil}$. For example, at +0.050, n_{grain} is 0.050 greater than that n_{oil}. The lighter-colored Becke line thus moves into the grain as the stage is lowered. These images are also useful to show how the relief of the grain varies as a function of the difference in refractive index between the grain and the liquid. The images in Figure 18.16a show the grain in focus, while those in Figure 18.16b show the Becke line after the stage has been slightly lowered.*

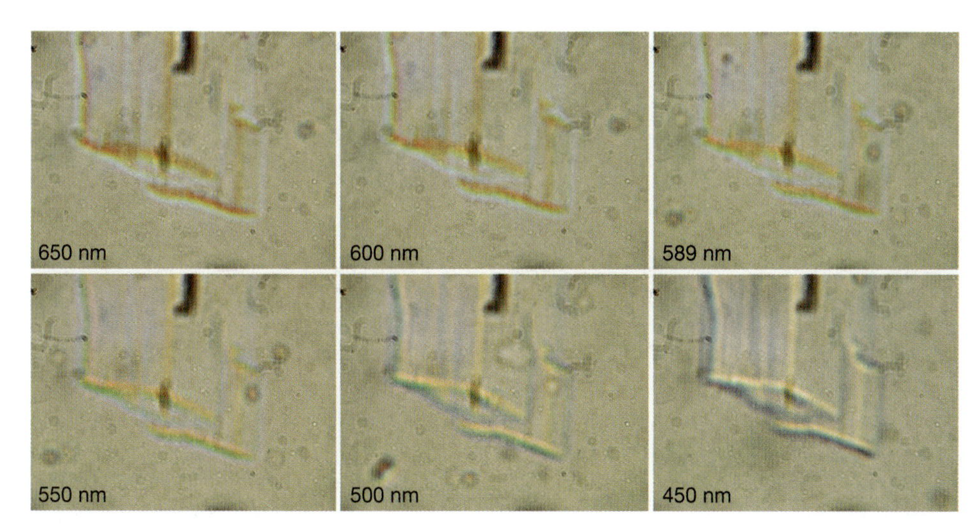

Figure 18.17. The same tremolite grain is again shown in six separate images where the match between the oil and the grain occurs at the indicated wavelength. The colors of the Becke lines differ in each image. The most important color to note is for the match at 589 nm, because that is the standard wavelength for measurement. A human eye would see these lines as "Virginia Tech" orange and light blue, with the orange light more intense and diagnostic.

arrive at the desired match. If the mineral is anisotropic, a grain with the same appropriate orientation must be found in every new mount. Use of a spindle stage greatly reduces the tedium of this process because the same grain can be used while the liquid is changed, and any grain can be oriented to determine any of its refractive indices as discussed later in this chapter.

Variation methods. The refractive indices of the liquids used in the immersion method change as a function of temperature (as well as wavelength), as do most materials. In general, as the temperature of any material increases, its refractive index will decrease because it becomes less dense as it expands. For the refractive index liquids, this change is ~0.0004 per °C as seen on the label of each refractive index liquid (Figure 18.20) along with its precise value of n (within 0.0002) at 25°C and 589 nm. If a match is obtained between an oil and a grain at any temperature other than 25°C, the refractive index of the liquid must be corrected based upon the temperature of the match and the gradient of change in n, as expressed here:

$$n_{oil} = n_{label} + (25 - T_0)dn/dt. \qquad \text{Equation 18.1}$$

Here n_{oil} is the temperature-corrected value, n_{label} is the value on the bottle, T_0 is the liquid's temperature, and dn/dt is listed on the side of the bottle (Figure 18.20).

The fact that n_{solid} typically changes more slowly with temperature than n_{oil} (by an order of magnitude) allows convenient measurements of n_{grain} to be made (Figure 18.20). The curves shown for n_{oil} are moved to lower values upon heating. In the example shown, the grain does not match n_{oil} before heating. As the liquid is heated, its refractive index decreases and a match occurs at 1. At point 2, n_{grain} matches n_{oil} at 589 nm, so when the temperature is increased, a correctly-colored Becke line in polychromatic light can be obtained (see points 3 and 4). This is called the **single-variation method** for refractive index determination (because a single variable, temperature, is varied!). If monochromatic light is available, this process can be done more precisely by matching n_{grain} at 589 nm with the monochromator.

The most precise method for determining refractive index of a transparent solid varies both the temperature of the liquid and the wavelength of the match simultaneously, in order to arrive at a series of match points (labeled 1–4 in Figure

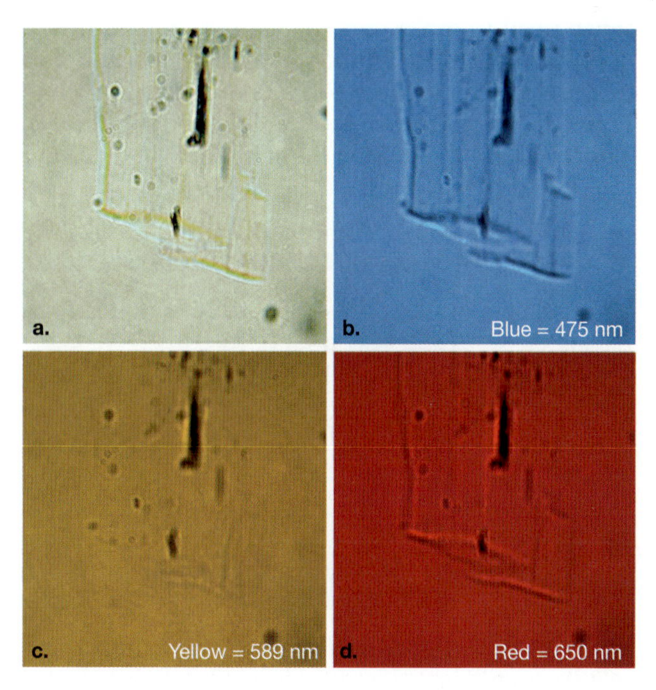

Figure 18.18. The same grain in the same oil as in Figure 18.17, shown in polychromatic light (a) and monochromatic light of three different wavelengths (b–d). At 589 nm, the grain is invisible.

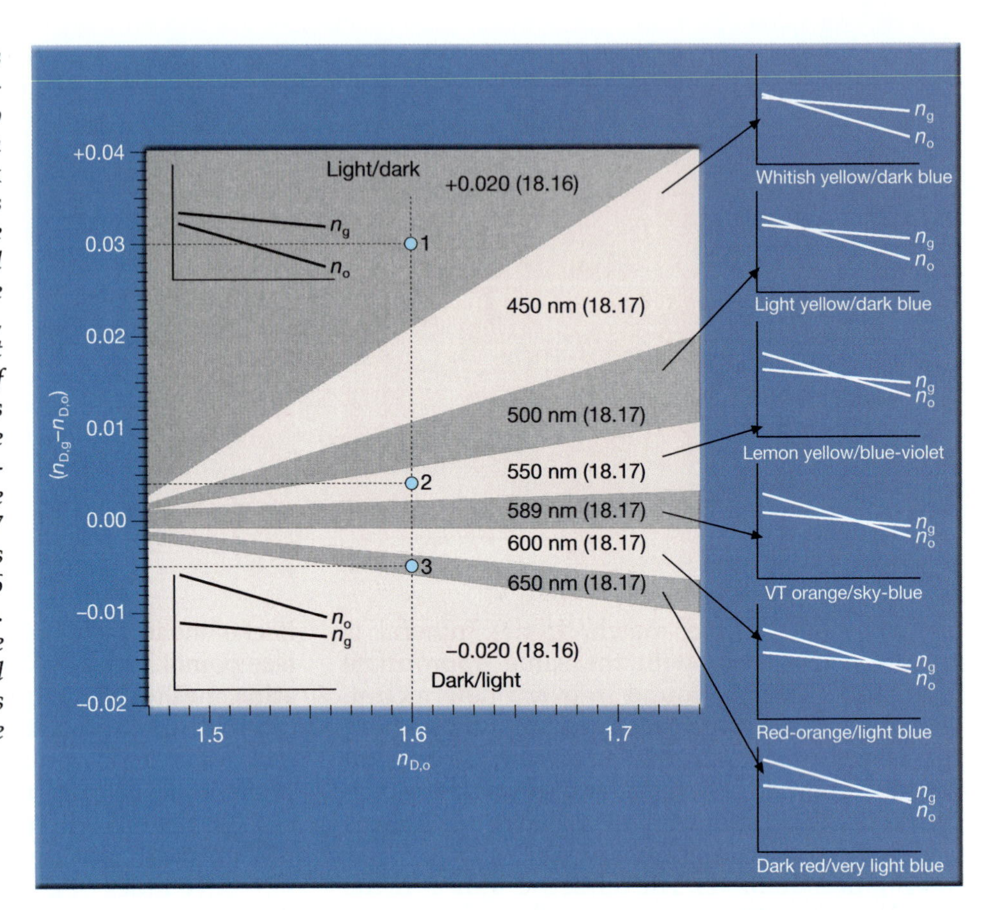

***Figure 18.19.** This graph is useful for selecting an appropriate refractive index oil to match the refractive index of an unknown mineral grain. The* x *axis is* n_{oil}, *and the* y *axis shows the difference in refractive index between the grain and the oil* ($n_{grain} - n_{oil}$). *Positive values occur when* $n_{grain} > n_{oil}$, *and negative values result when* $n_{grain} < n_{oil}$. *In each area of the graph, a small inset shows the dispersion curves for the grain and the oil and a reference to one of the tremolite grains shown in Figures 18.17 and 18.18. An example is shown for the case of* $n_{oil} = 1.6$ *and three separate grains (1–3). The colors of the Becke lines are also shown with the grain/oil relationships (i.e., what goes into the grain followed by the oil when the stage is lowered).*

18.20). This is called the **double-variation method**. These four points define a line that is the dispersion curve for that grain. The microscopic setup used to make such measurements is shown in Figure 18.21. For a more complete discussion of this method as well as some computer programs that can be used for it, see Bloss (1981, 1999), Gunter et al. (1989), and Su et al. (1987).

Grain orientations. Before the refractive index of an anisotropic material can be determined, the orientation of the grain must be known by use of interference figures or a spindle stage. An optic axis figure is perhaps the simplest case for which ω (for a uniaxial mineral) or β (for a biaxial mineral) can be measured in any orientation. To obtain the interference figures shown in Figures

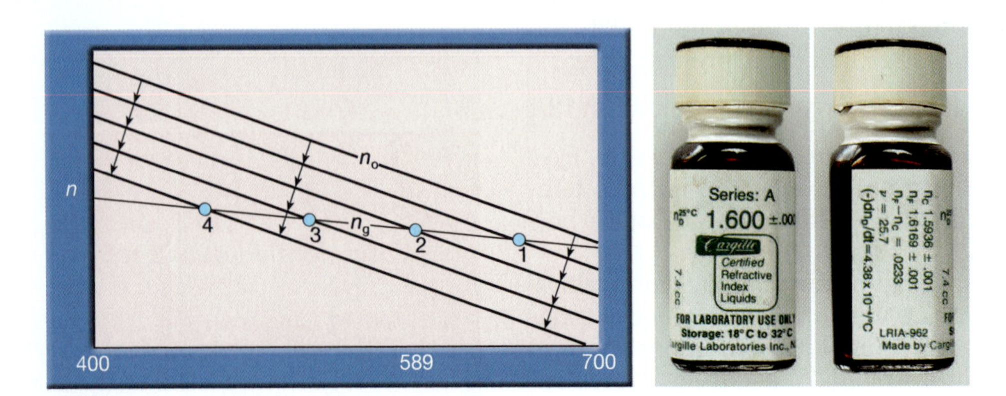

***Figure 18.20.** Precise determination of the refractive index of a grain requires use of the double variation method, in which both the temperature of the liquid and the wavelength of light are varied. In this chart, the line representing* n_{grain} *is intersected by a series of lines for* n_{oil}. *The highest* n_{oil} *does not match the grain in visible light, so a lower* n_{oil} *must be selected, or the oil must be heated to lower its refractive index until it matches* n_{grain} *at point 1. Continued heating of the oil will eventually cause it to match at points 2, 3, and 4. The photograph of a refractive index liquid shows its* n *value and* dn/dt *(i.e., how* n *changes with temperature).*

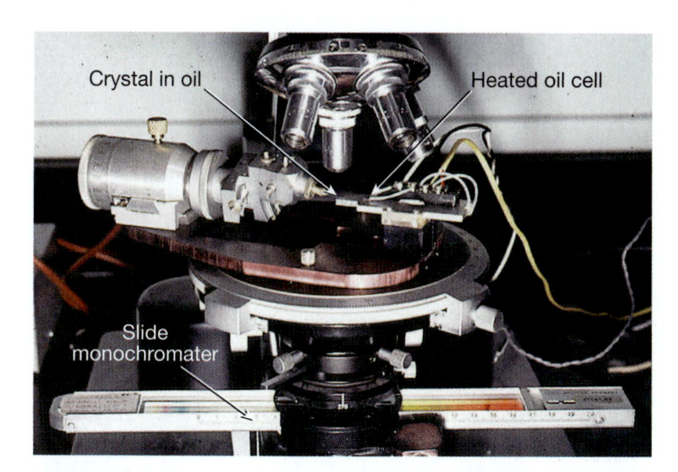

Figure 18.21. *The experimental setup used for the double-variation method, in which a crystal is immersed in an oil cell that can be heated. A thermocouple in the oil cell allows precise measurement of temperature. Monochromatic light is produced by a sliding monochromator below the microscope stage.*

18.22a and 18.23a, the largest grain with the smallest retardation in the view was selected (Figures 17.38 and 17.39). On the other hand, a small grain exhibiting high retardation might produce a flash figure for a uniaxial mineral and an optic normal for a biaxial figure (Figures 18.22b and 18.23b, respectively). For a uniaxial mineral, this is the only orientation in which the ε vibration direction can be measured. To distinguish ε from ω, the optic sign of the mineral must be known and an interference plate must be used. It may be difficult to find appropriate grains that display the necessary, nicely-centered interference figures! However, for a uniaxial interference figure, every grain will contain an ω vibration direction that can be easily oriented. If a grain is known to be uniaxial, ω can be oriented E-W by placing an isogyre vertically in the field of view (Figure 18.22c).

If an optic normal figure can be located in a biaxial mineral, the α and γ directions can be measured and discriminated by use of an accessory plate. Likewise, for a centered acute bisectrix or centered obtuse bisectrix figure, two of the principal vibration directions can be located and measured, as shown for an optically positive mineral in Figure 18.23c,d. It is often the case that none of the centered interference figures can be obtained. However, if a mineral is known to be biaxial and an isogyre can be brought parallel to the N-S cross-hairs (Figure 18.23e), then a principal vibration direction, either α or γ, will be E-W. An interference plate could then be used to distinguish between α and γ. Such figures occur when only one of the principal vibration directions is located in the same plane as the microscope stage. If there are no such principal vibration directions in the plane of the microscope stage, an off-centered interference figure will be seen (Figure 18.23f). Regardless, for any biaxial mineral the refractive indices will be between α and γ, and having some idea of those intermediate values might help identify a mineral.

In some of the previous figures (Figure 18.23b,c), a 2-fold axis emanates from the cross-hairs and vertical mirror planes correspond to the cross-hairs. A different situation is seen in Figure 18.23f, where the isogyre lacks symmetry. In fact, the point group symmetry of the biaxial indicatrix is $2/m\ 2/m\ 2/m$. Any views looking down the 2-fold axes will have two principal vibration directions in the plane of the microscope stage. Likewise, if any principal vibration direction is oriented E-W, a straight isogyre will be N-S (Figures 18.22c and 18.23e).

Table 18.1 lists some useful equations that relate the refractive indices of a mineral to the orientations of uniaxial and biaxial minerals. For biaxial minerals, another series of equations

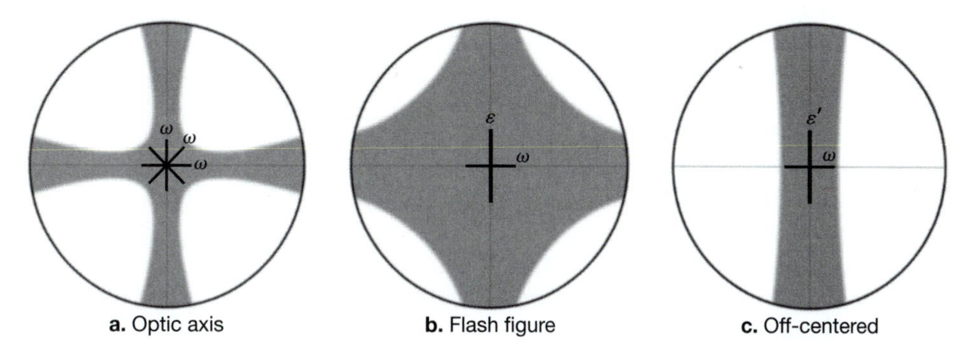

Figure 18.22. *Three possible interference figures can be used to determine the orientation of a uniaxial optical indicatrix.* ***a.*** *An optic axis figure, the ω vibration direction can be measured in any setting of the microscope stage.* ***b.*** *A flash figure, one vibration direction is ω and the other, ε. The ε value of the grain can only be measured with this type of interference figure.* ***c.*** *If the interference figure is off-center, when an isogyre is brought to N-S, ω is placed E-W. In all of these interference figures, ω is seen. So ω can be measured for any uniaxial mineral for any cut through the indicatrix.*

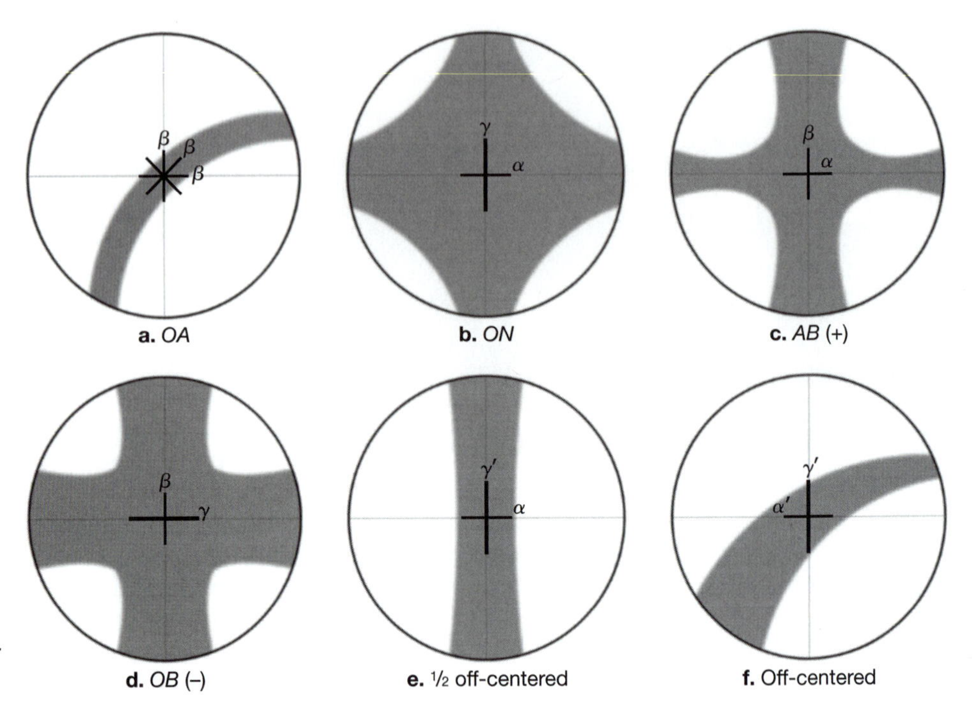

Figure 18.23. *Six separate schematic representations of biaxial interference figures.* ***a.*** *A centered optic axis figure, in which the β refractive index can be measured in any orientation of the microscope stage.* ***b.*** *An optic normal figure, in which both α and γ can be measured. To differentiate between the two, a quartz plate would be needed to distinguish α (n) from γ (N).* ***c.*** *An acute bisectrix for a biaxial positive mineral. To distinguish α from β, an accessory plate could be used, or the fact that β will be perpendicular to the thinner isogyre can be used.* ***d.*** *A centered obtuse bisectrix figure for a biaxial positive grain. Here β and γ could again be distinguished through use of an accessory plate.* ***e.*** *The case of an interference figure formed when only one principal vibration direction is in the plane of the stage of the microscope: in this case, α. Then α can be made parallel to the E-W polarizer by aligning the straight isogyre with the N-S direction.* ***f.*** *For a grain that is randomly oriented, no principal vibration direction can be measured, but insights into α' and γ' values can be gained as shown.*

relates $2V$ to α, β, and γ. Because these four variables are related, knowing any three of them will allow the fourth to be calculated using the equations in Table 18.1. These equations were derived in Gunter and Schares (1991), which also contains other more specialized equations.

Spindle Stage

The spindle stage has been in use since the early 1900s (Gunter, 2004; Gunter et al., 2004; and reference therein) as a means to orient single crystals on a polarizing light microscope, but it was not until the work of Wilcox (1959) in the late 1950s, followed by that of Bloss and coworkers in the late 20th century, that the spindle stage became a robust mineralogical research tool. The spindle stage is basically a one-axis rotation device (Figures 18.21 and 18.24). The preface of Bloss (1999), along with Gunter (2004) and Gunter et al. (2004) provide some historical perspectives on its use; Bloss (1981, 1999) and Gunter et al. (2004) give thorough discussions on the use of the spindle stage in research and teaching. The major contributions made by Bloss and coworkers over the past 35 years were the refinement of spindle stage hardware and the development of the computer program EXCALIBR (Bloss and Riess 1973), which solved extinction data sets (i.e., microscope stage settings that cause a crystal to go extinct as the spindle stage is incremented) to yield the orientation of the biaxial indicatrix and determined a precise value of $2V$. Much of the material presented in this section was extracted from Gunter et al. (2004, 2005) and the interested readers

Table 18.1. Relationships between optical values for uniaxial and biaxial minerals
$\varepsilon' = \dfrac{\varepsilon\omega}{\sqrt{\varepsilon^2 \sin^2\theta + \omega^2 \cos^2\theta}}$ where θ_x = angle between ε' and ε
$2V_x = \cos^{-1}\left(2\left[\dfrac{\alpha^2/\beta^2 - 1}{\alpha^2/\gamma^2 - 1}\right] - 1\right)$ or $\cos V_z = \dfrac{\alpha}{\beta}\sqrt{\dfrac{\gamma^2 - \beta^2}{\gamma^2 - \alpha^2}}$
$\gamma = \sqrt{\dfrac{\alpha^2}{\left[\left[\dfrac{\alpha^2/\beta^2 - 1}{0.5(\cos 2V_x + 1)}\right] + 1\right]}}$
$\beta = \sqrt{\dfrac{\alpha^2}{0.5[(\alpha^2/\gamma^2) - 1](\cos 2V_x + 1) + 1}}$
$\alpha = \sqrt{\dfrac{\beta^2 - \gamma^2 k}{1 - k}}$ where $k = \dfrac{(\beta^2/2)(\cos 2V_x + 1)}{\gamma^2}$
$\dfrac{1}{n} = \sqrt{\dfrac{\cos^2\theta_x}{\alpha^2} + \dfrac{\cos^2\theta_y}{\beta^2} + \dfrac{\cos^2\theta_z}{\gamma^2}}$ where n = a random direction, θ_x = angle between n and X, θ_y = angle between n and Y, θ_z = angle between n and Z

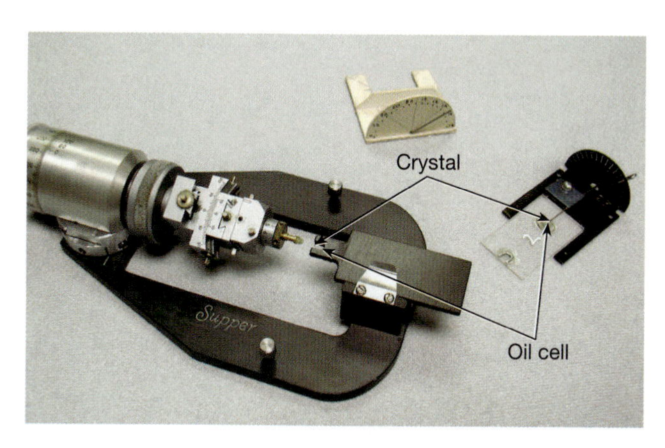

Figure 18.24. Photograph of three different types of spindle stages and their associated oil cells. The Supper model uses an X-ray goniometer head/brass pin combination to hold the crystal in place, while the detent and home-made models can use a needle of some type (e.g., sewing needle or straight pin) to mount the crystal. All three share the ability to rotate a grain about an axis parallel to the microscope stage.

should refer to those papers especially if they wish to make the most efficient use of the spindle stage, as well as Bloss (1981, 1999).

The spindle stage provides the ability to determine the orientation of a crystal's indicatrix so that its principal refractive indices (i.e., α, β, γ or ε, ω) can be measured by the immersion method (Figure 18.25). Measurement is performed after the spindle axis and microscope stage have been rotated to place a principal refractive index parallel to the plane of the microscope stage and parallel to the polarizer's privileged direction. Traditionally, the orientation of the crystal was found by observations of interference figures, in a manner similar to that used to find correctly oriented crystals in a grain mount (Figures 18.22 and 18.23). This technique became known as the **conoscopic method**. The **orthoscopic method** was soon developed. It required measurement of extinction positions at different spindle axis settings. These extinction positions could then be measured in either conoscopic or orthoscopic view and plotted using a stereonet. Graphical methods were used to locate the three (biaxial) or two (uniaxial) principal refractive index directions. The main advantage of the orthoscopic method, advanced during the mid-1960s, was that it could be used on grains too small to produce interference figures.

The 1973 version of EXCALIBR was improved upon and thoroughly discussed in Bloss (1981), further refined and made PC/Mac compatible in 1988 (Gunter et al., 1988), then completely rewritten and renamed EXCALIBR II in 1992 (Bartelmehs et al., 1992). EXCALIBR II has been further improved and the DOS interface has been replaced with a Windows Graphical User Interface (Gunter et al., 2004 and discussed herein). The most recent version of EXCALIBRW can be obtained for free by contacting Stanley Evans (find contact information at www.minsocam.org).

Collecting and plotting extinction data. The spindle stage uses a polar coordinate system represented by S and E angles. The S angle is read directly from the spindle stage dial (Figures 18.24 and 18.26) and E is the angle created between the spindle and any vector with its origin at the end of the spindle. Physically, S is the angle required to rotate a plane (or vector) to coincide (i.e., be made parallel) to the microscope stage, and E is the angle between a vibration direction (or any direction within the crystal) and the spindle axis (Figure 18.26). An appropriate rotation on S will bring a principal vibration direction into the plane of the microscope stage, and then the appropriate rotation of the microscope stage, the M_s angle, will orient that direction parallel to the

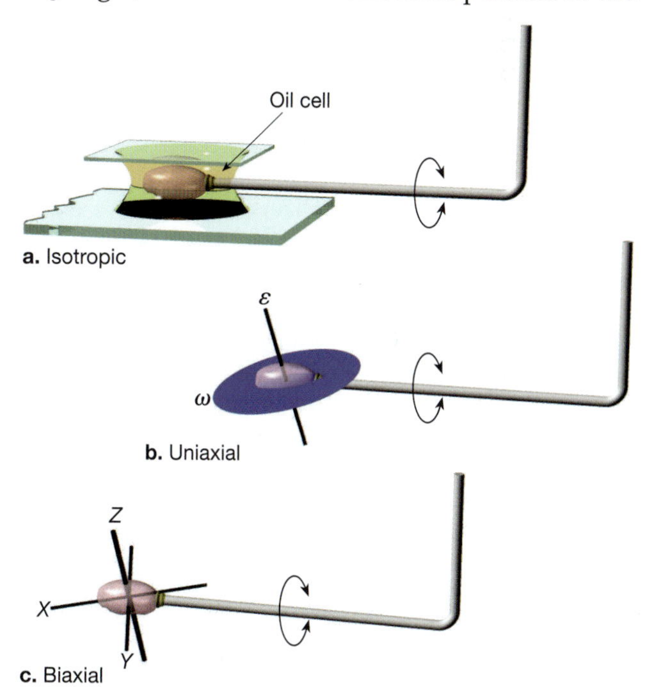

*Figure 18.25. The spindle stage can be used to orient a single crystal, and allows use of the same crystal as the liquid is changed. **a.** An isotropic crystal placed in the different refractive index oils. **b.**The skeleton of a uniaxial indicatrix is shown affixed to the spindle. In any orientation, ω can be measured. With the spindle, ε can be measured by rotating the crystal (around the spindle) into the plane of the microscope stage, which can then be rotated parallel to the lower polarizer so as to obtain a flash figure. **c.** The skeleton of a biaxial indicatrix is shown affixed to the spindle. With the spindle stage, any direction in the crystal (X, Y, or Z) can be rotated into the plane of the microscope stage, and in turn made parallel to the lower polarizer. While interference figures can be used for this, a simpler method exists.*

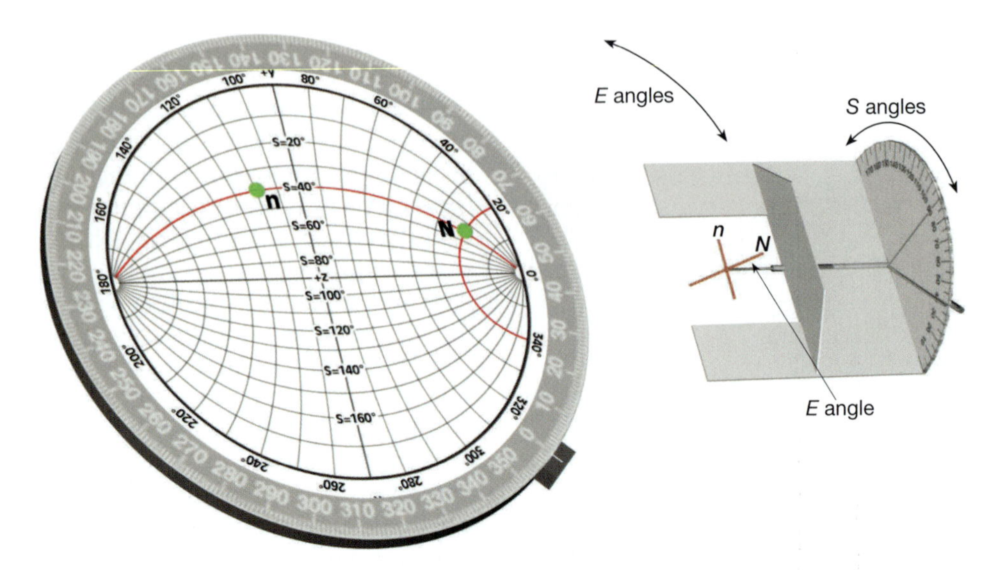

Figure 18.26. *Coordinate systems and stereographic plotting methods for the microscope stage/spindle stage combination. Two coordinate systems can be used: (1) A Cartesian system (x, y, z) or (2) a polar system (S, E). The right-handed Cartesian system is set up so the +x direction corresponds to the spindle axis, the +y direction plots in the plane of the stereographic projection at its top, and +z is perpendicular to the projection plotting in the middle. The* S*-angle is measured on the spindle stage, as shown on the right (the dial on the spindle stage is rotated inward). The* S*-angles are represented by great circles in the stereographic projection (the* S = 40° *plane is highlighted in the figure). The* S*-angle is the angle required to bring a plane (tilted at that* S *angle) into coincidence (i.e., parallel) with the microscope stage. The* E*-angle is measured on the microscope stage. The* E*-angle is represented by small circles on the stereographic projection (the* E = 20° *small circle is highlighted on the figure). The* E*-angle is the angle between the spindle axis and a vibration direction in the crystal.*

lower polarizer. An extinction data set for a crystal usually amounts to extinction positions (M_s values) observed at S settings of 0, 10, 20, …, 180° (i.e., extinctions positions are recorded at 10° increments of S). Even though EXCALIBR will solve extinction data sets to yield S and M_s values to orient a sample, it is worthwhile to understand how the data are processed by EXCALIBR. Also, the newest version of EXCALIBR outputs both numerical and graphical results. To understand the graphical results, which can be useful for refining crystal orientation, a basic understanding of stereographic projections is required.

Figure 18.26 illustrates both the polar (S, E) and Cartesian coordinate (x, y, z) systems used by EXCALIBR. The program performs all the calculations in the Cartesian system and outputs results in the polar system. The Cartesian system places $+x$ in the plane of the microscope stage and horizontal (i.e., E-W), $+y$ is also in the plane of the microscope stage, but vertical (i.e., N-S), and $+z$ is perpendicular to the stage, coinciding with the light path of the microscope. S angles represent rotations about the spindle axis and are read from the spindle stage's dial, and M_s values represent rotations of the microscope stage and are mathematically related to the E angles, as discussed below.

Before this relationship can be developed, the **reference azimuth**, M_r, must be defined. M_r is the microscope stage setting that places the spindle axis to coincide with the $+x$ direction with its tip pointing to the west. If $M_r = 0°$ (Figure 18.26), the relationship is simplified. However, the value of M_r depends on how the spindle stage was mounted on the microscope stage. A precision of 0.1 to 0.2° is required for accurate work, and it is very difficult to mount the spindle stage that precisely. Also, there might be mechanical limitations for mounting the spindle stage. For these reasons, M_r must be determined once the spindle stage is mounted on the microscope by use of extinction measurements using:

$$M_r = (M_{s0°} + M_{s180°a}) \ / \ 2. \qquad \text{Equation 18.2}$$

Here M_r is the reference azimuth, $M_{s0°}$ is an extinction position obtained at $S = 0°$, and $M_{s180°a}$ is an extinction position obtained at $S = 180°$ but in the opposite rotation direction from $M_{s0°}$. M_r is first estimated visually by reading the vernier scale on the microscope stage when the spindle is oriented parallel to the E-W microscope direction. Next, the spindle stage is set to the estimated $M_{r0°}$, then $M_{s0°}$ is determined. The microscope stage is reset to the estimated M_r, and $M_{s180°a}$ is found. The M_r value calculated in Equation 18.2 should be a refined value of the estimated M_r. Care must be taken that the refined M_r is not 90°, or some multiple of 90°, off from the estimated M_r, which is very possible. If this occurs, add or subtract 90° increments until the estimated and refined values agree.

Before equations can be given to relate M_s to E, the type of microscope stage being used must be considered. Some microscope stages are graduated to increase in a counterclockwise (ccw) direction (Figure 18.26). However, other stages increase in a clockwise (cw) fashion (Figure 18.A2c). There are two sets of equations that relate M_s to E based on stage types:

counterclockwise (ccw) stage:
$E = M_s - M_r$ Equation 18.3a

clockwise (cw) stage:
$E = M_r - M_s$. Equation 18.3b

Here E is the angle between an extinction direction and the spindle stage, M_s is the microscope stage reading causing extinction, and M_r is the reference azimuth. If M_r is 0°, then for the ccw stage $E = M_s$. However for the cw stage, when $M_r = 0°$, $E = -M_s$; thus a calculation is still required to convert M_s to E angles to produce stereographic plots.

The stereographic plotting technique for the spindle stage is similar to other plotting systems used in the geosciences, though the stereonet is oriented slightly differently and there is no need to rotate the overlying tracing paper (if plotted by hand) to plot the data. An equal angular stereonet (i.e., a Wulff net) is shown Figure 18.26 (modified from Gunter and Twamley (2001), in the appropriate orientation for plotting S and E values. Because the S angle is the number of degrees necessary to rotate a plane into coincidence with the microscope stage, $S = 0°$ is the starting position of the microscope stage; increasing S values are printed on the great circles of the stereonet. The E angles are plotted on the small circles of the stereonet and represent the angle between a vibration direction and the spindle axis. These are printed on the circumference of the stereonet. For example, two perpendicular vibration directions, N (the slow ray) and n (the fast ray), are shown on the tip of the spindle stage. The plane in which the vibration directions are located has been rotated 40° (i.e., they are in the $S = 40°$ plane), and they are placed in the plane of the page (i.e., microscope stage). N then makes an angle of 20° with the spindle axis, so $E = 20°$ for N; and n must make an angle of 110° (90° + 20°) with the spindle axis, so that $E = 110°$ for n.

To plot N on the stereonet, locate the $S = 40°$ great circle (shaded grey in Figure 18.26) and the $E = 20°$ small circle (shaded grey in Figure 18.26). N plots at their intersection, and n plots 90° from N on the $S = 40°$ plane at $E = 110°$. Next, the full extinction data set of S and E values can be plotted from $S = 0, 10, \ldots, 180°$, forming extinction curves that are used to graphically determine the location of the two optic axes and the three principal refractive indices of a biaxial crystal. Examples of plotting complete extinction data sets are given in the next section.

Computer solution of extinction data. EXCALIBR exploits a mathematical relationship between extinction data and the two optic axes of a biaxial crystal, using numerical methods to solve the Joel equation (Bloss, 1991). Basically, the program determines the orientation of the two optic axes cast into the spindle stage coordinate system discussed above. Given these two vectors (i.e., the two optic axes), the program calculates the orientations of: (1) the acute bisectrix (AB), the bisector of the acute angle formed by the two optic axes, (2) the obtuse bisectrix (OB), the bisector of the obtuse angle formed by the two optic axes, (3) the optic normal (ON), the normal (the cross product) of the two optic axes, and (4) $2V$, the angle (dot product) between the two optic axes. The program outputs the S, E, and M_s values needed to orient AB, OB, and ON in the plane of the stage of the microscope and parallel to the lower polarizer in order to check, or measure, the refractive index. The orientations of the lower polarizers (located below the microscope stage, and usually not removable) and the upper polarizers (located above the microscope stage, and always removable) are specific to different microscopes. Older microscopes tend to have N-S lower polarizers, while the newer scopes have E-W ones. Regardless, the orientation of the lower polarizer must be checked by using a Polaroid film of known vibration direction, or a strongly absorbing mineral in thin section (e.g., tourmaline or biotite). EXCALIBR lists M_s values for either a N-S or E-W setting, which differ by 90°.

The program also outputs $2V$, but lacks the ability to determine if the crystal is optically positive or negative. Optic sign can be determined by checking to see if AB or OB is N by setting the S and M_s values to orient AB parallel to the lower polarizer. Next, the microscope stage is rotated 45° to bring AB parallel to the N direction of an accessory plate. If addition occurs with insertion of the plate, AB is N (i.e., the γ refractive index of the crystal); if γ is AB, then by definition the crystal is optically positive. The crystal would be negative if AB were n (i.e., α).

Using EXCALIBR. Use of the EXCALIBR software, especially the new Windows-based version, may be the easiest aspect of spindle stage use. You launch it like any application, typically by clicking on its icon. Then select the "File" pull down menu and select "New *.dat." Figure 18.27 shows the input window (with a data set). There are input fields and radio buttons for the input of data and selection of criteria required by the program. The first field at the top is a title. Next there are radio buttons to select the microscope stage type, followed by a selection of mathematical models for either the uniaxial or biaxial case. Select the biaxial model unless you are sure the crystal is uniaxial. The next fields of the window deal with the light source. Often white light (i.e., polychromatic) is used for routine work. However, up to four wavelengths of monochromatic light can also be entered. When more than one wavelength of light is used, the program determines if movement of the optical directions occurred (i.e., determines the dispersion of the crystal). This may help identify the crystal system of a biaxial crystal. The program will calculate a refined reference azimuth, M_r, based on the input data, but an approximate reference azimuth should be input. This prevents the program from calculating an M_r value that may be 90° in error, or some multiple of 90°, as explained above. Finally, enter the M_s values in the window labeled "Extinction." The S values on the window to the left will start at "0" and increment by 10 after each M_s value is entered. The enter button to the window's right must be clicked to accept the entered M_s value. The S and M_s pairs will accumulate in the window below. If a mistake is made in entering an M_s value, double-click on that line in the lower window and edit it in the input window.

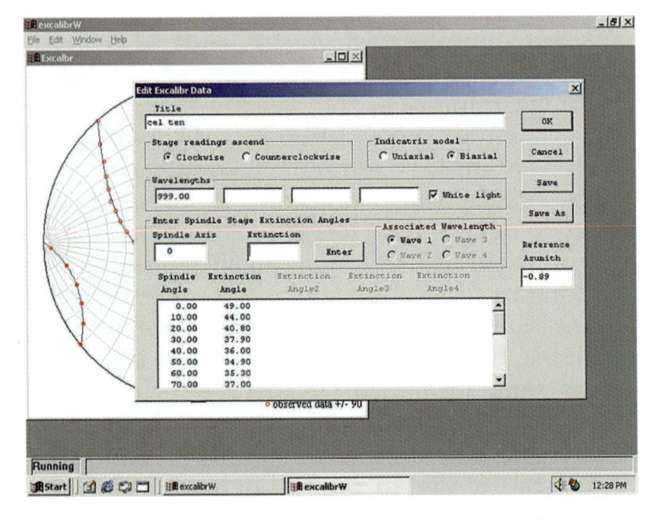

Figure 18.27. *The input window for EXCALIBR, with a portion of the graphical output window behind it. The input window's fields are labeled and, used in conjunction with the radio buttons, allow for input of the required data for EXCALIBR. Notice in this example that there is a different value of* M_s *(labeled "Extinction Angle") for each* S *setting (labeled "Spindle Angle").*

dow. Once all of the data are input, it is a good idea to click "Save" or "Save As" in the upper portion of the window. When "OK" is selected, the input window vanishes and a stereographic plot of the input data and its graphical solution is shown. To view the numerical results, go to the "Edit" pull-down menu and select "Output." If you then select "Data," the program will return you to the input window.

Numerical and graphical results from EXCALIBR are shown in Figure 18.28. At the top of the numerical results are the title and refined M_r value. In this case, the refined M_r value is –0.89°; this would correspond to a microscope stage setting of 359.11°. The next section gives a value for R^2, which is an indicator of the overall quality of the fit of the calculated data to the input data. The center portion of the output repeats the input S and M_s values followed by the E_s value obtained from Equation 18.3. Because E_s would be negative (a cw stage was used), 180° is added to the E_s values (i.e., the program uses the other end of the vector). In the next column are the calculated E_s values, "CAL(E_s)" determined after EXCALIBR solved the extinction data set, and the last column gives the observed E_s values minus the calculated ones, "E_s-CAL(E_s)." Large errors (i.e., greater than 2 to 3°) in this column might indicate a misread extinction position. The bottom portion of the figure provides the useful output. The calculated 2V and its estimated standard error (ese) are given along with the S, E_s, M_s (for both E-W and N-S lower polarizers) results and their estimated standard errors. The output in this data set shows that all of the optical directions are located to better than 1° and 2V is determined to within 0.6°. This type of precision would not be possible with graphical methods. However, the graphical results (Figure 18.28) do provide a nice way of visualizing the input data (shown as dots) and the output data, which are the calculated extinction curves and the locations of *OA*1, *OA*2, *AB*, *OB*, and *ON*. The graphical output is also useful to see if any M_s values were misread by observing any departures from the smooth curved pattern of the calculated extinctions curves. Misread extinction values would also be seen in the numerical analysis in the "E_s-CAL(E_s)" column.

Absorption and Pleochroism

Colors are produced by several methods: absorption, refraction, interference, retardation, and Becke lines. Causes of color in minerals were first discussed in the context of interactions of elec-

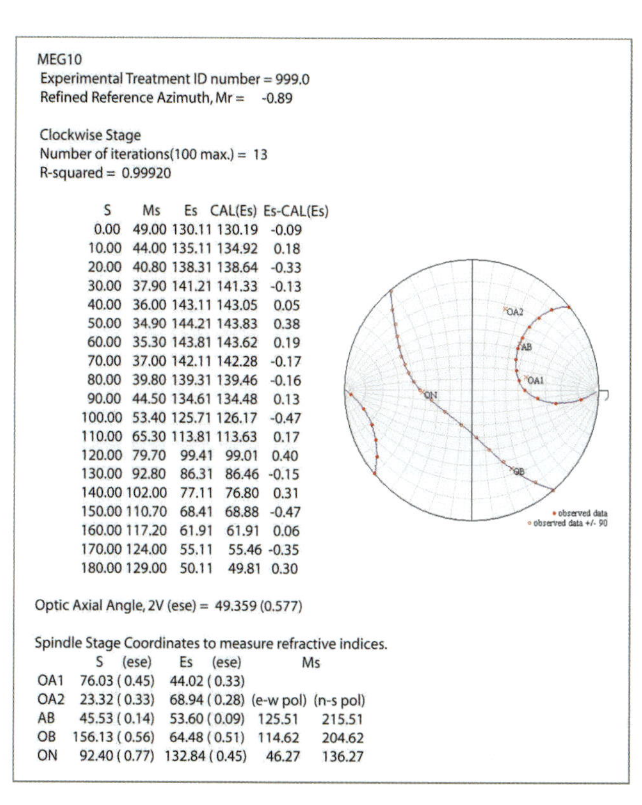
MEG10
Experimental Treatment ID number = 999.0
Refined Reference Azimuth, Mr = -0.89

Clockwise Stage
Number of iterations(100 max.) = 13
R-squared = 0.99920

S	Ms	Es	CAL(Es)	Es-CAL(Es)
0.00	49.00	130.11	130.19	-0.09
10.00	44.00	135.11	134.92	0.18
20.00	40.80	138.31	138.64	-0.33
30.00	37.90	141.21	141.33	-0.13
40.00	36.00	143.11	143.05	0.05
50.00	34.90	144.21	143.83	0.38
60.00	35.30	143.81	143.62	0.19
70.00	37.00	142.11	142.28	-0.17
80.00	39.80	139.31	139.46	-0.16
90.00	44.50	134.61	134.48	0.13
100.00	53.40	125.71	126.17	-0.47
110.00	65.30	113.81	113.63	0.17
120.00	79.70	99.41	99.01	0.40
130.00	92.80	86.31	86.46	-0.15
140.00	102.00	77.11	76.80	0.31
150.00	110.70	68.41	68.88	-0.47
160.00	117.20	61.91	61.91	0.06
170.00	124.00	55.11	55.46	-0.35
180.00	129.00	50.11	49.81	0.30

Optic Axial Angle, 2V (ese) = 49.359 (0.577)

Spindle Stage Coordinates to measure refractive indices.

	S (ese)	Es (ese)	Ms	
OA1	76.03 (0.45)	44.02 (0.33)		
OA2	23.32 (0.33)	68.94 (0.28)	(e-w pol)	(n-s pol)
AB	45.53 (0.14)	53.60 (0.09)	125.51	215.51
OB	156.13 (0.56)	64.48 (0.51)	114.62	204.62
ON	92.40 (0.77)	132.84 (0.45)	46.27	136.27

Figure 18.28. *An example of the numerical output from EXCALIBR based on the input data given in Figure 18.27. The program outputs the* M_s *and calculates* E_s *for each of the spindle stage settings in the first three columns. The fourth column, "CAL(*E_s*)," is the calculated value of* E_s *for each S determined after EXCALIBR has solved the extinction data to arrive at the crystal's orientation. The last column, "*E_s *− CAL(*E_s*)," is the difference between the observed and calculated values of* E_s*; outliers in this column might represent a misread* M_s *value. Next, the program gives 2V and is estimated standard error (ese). Finally, the program lists the S and* E_s *values for the five optical directions and the* M_s *values for the three principal refractive index directions. The S and* M_s *values can be used to orient each of the principal refractive index directions in the plane of the microscope stage parallel to the lower polar, in order to check or measure their respective refractive indices. On the right is a stereographic plot produced by EXCALIBR for the input given in 18.27. This provides a graphical representation of the data and shows input* E_s *values for each S setting, the calculated extinction curves (and how well they fit the observed data), and the locations of the five optical directions.*

trons with electromagnetic radiation in Chapter 7 (Table 7.8). In Chapter 16, we discussed the formation of a rainbow, which is caused by refraction of light into its components, and formation of an oil slick, which is caused by interference of light. Materials have different refractive indices as a function of wavelength, causing light to be refracted at different angles. Any transparent solid, whether a piece of glass or a raindrop, will refract light into a spectrum of colors.

We have also discussed retardation colors (at length), which are similar to interference colors seen in everyday life, such as those seen on a turkey's wing (Figure 16.1). They result from the interaction of two waves that are in phase, creating a resultant wave. If two waves are out of phase, they cancel each other. Because white light is polychromatic, and consists of many different wavelengths, different portions of the spectrum can be simultaneously in or out of phase, combining to make colors. In minerals, we used the term retardation colors to describe this phenomenon. Although lower retardation colors (first-order yellow, second-order blue, etc.) may resemble those of visible light, they are formed by a different mechanism.

Colored Becke lines are yet one more way for materials to produce color. They split the visible light spectrum into two parts, so the colors sum on either side of the matching refractive index to form colors such as the diagnostic orange vs. blue discussed above. Becke lines are observed in plane-polarized light, and retardation colors are typically only observed in cross-polarized light.

Finally, colors are also formed in plane-polarized light by differential absorption of light waves with visible and near-infrared energies. This concept is intimately related to both optical orientation and the chemical causes of color we discussed back in Chapter 7. Pleochroism is the property of anisotropic minerals that causes them to absorb specific wavelengths of light in different amounts depending on their crystallographic orientations. Pleochroism is particularly obvious when minerals are viewed in polarized light, although its effect is so strong that it can sometimes be seen in unpolarized light.

As discussed in Chapter 7, color most commonly arises from crystal field transitions between electron orbitals within individual cations of transition metals like Fe, Cu, Cr, Ti, etc. When excitations of electrons between *d* or *f* orbitals occur at wavelengths within the visible region, they will cause absorptions that can give rise to colors in minerals. If the coordination polyhedra are distorted, as is common in minerals, then the absorption of light will not be the same in all directions. Tourmaline (Figure 18.29) is the classic example of a mineral with differential absorption in different orientations, also called a **pleochroic mineral** (*pleo* = many and *chro* = color). This effect also occurs in the spectrum of an unusual blue elbaite (tourmaline) crystal that was acquired at two different orientations parallel and perpendicular to the *c* axis (Figure 7.10).

Kyanite, which occurs in bladed crystals that are elongated along the *c* axis, is blue because intense absorption occurs when light is polarized to vibrate parallel to the Z vibration direction. At that orientation (Figure 18.30), there is a very strong band at ~600 nm. It is the result of **intervalence charge transfers** (IVCT), which occur when a 3*d* (or higher) electron gains enough energy from light transmitted to the crystal that it can be transferred between cations in adjacent crystallographic sites. In general, IVCT bands are at least an order of magnitude more intense than those arising from crystal field bands. Recall that kyanite is composed of Al^{3+} octahedra that share edges to form chains (Figure 6.15). If Fe^{2+} and Ti^{4+} substitute for two adjacent Al^{3+} cations, they are close together in those edge-sharing chains, and can share an electron. If light is vibrating in a direction other than directly along the chains, then the amount of energy (the vector component) along the chains decreases, and the intensity of the blue color decreases. If the light is completely polarized and vibrating perpendicular to the length of the chains, then the electron is not excited between the two adjacent sites, and there is no blue color in that direction (in a perfect kyanite crystal).

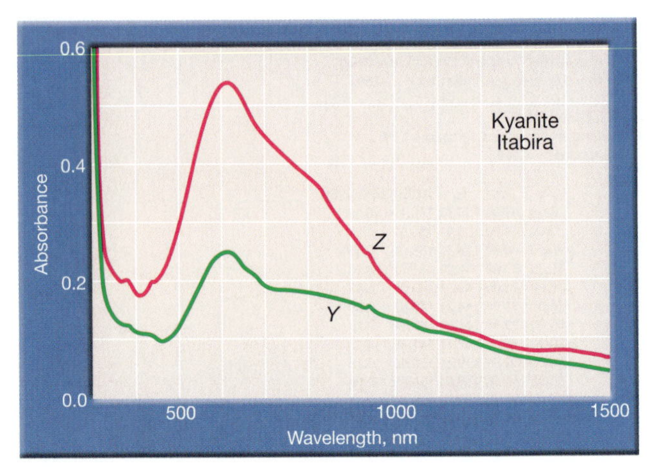

Figure 18.30. Visible region absorption spectrum of kyanite. The kyanite structure is composed of edge-sharing chains normally occupied by Al^{3+}. If Fe^{2+} and Ti^{4+} substitute for two adjacent Al^{3+} cations, then the two transition metals are close enough together to share an electron. As noted in Chapter 7, the colors that result from such intervalence charge transfers can cause intense spectral bands like the one seen at 620 nm. That band is the result of orienting the crystal so that light can vibrate along the chains (parallel to Z). If light is polarized and vibrating perpendicular to the length of the chains, then the electrons cannot be excited between the adjacent transition metals, and there is almost no blue color.

Pleochroism caused by both charge transfer and crystal field transitions is found in most minerals that can contain Fe or Ti. Figure 18.31 shows differential absorption and pleochroism in a biotite and an amphibole in two different orientations. In Figure 18.31a,b, some biotite flakes are resting on their (001) cleavage planes with their *c* axes perpendicular to the plane of the microscope stage. In this orientation, light is vibrating only in directions within the plane of the sheets, and so the color will stay the same upon rotation. The flake in the very center of the photograph is viewed edge-on looking down the layers, with its *c* axis parallel to the plane of the microscope stage, so that light can vibrate either parallel to Z (making the crystal black) or perpendicular to Z, where a lighter color is seen. Similar phenomena are seen in Figure 18.31c–f for hornblende; these views correspond to views of the crystal structures of biotite and hornblende in Figure 18.32.

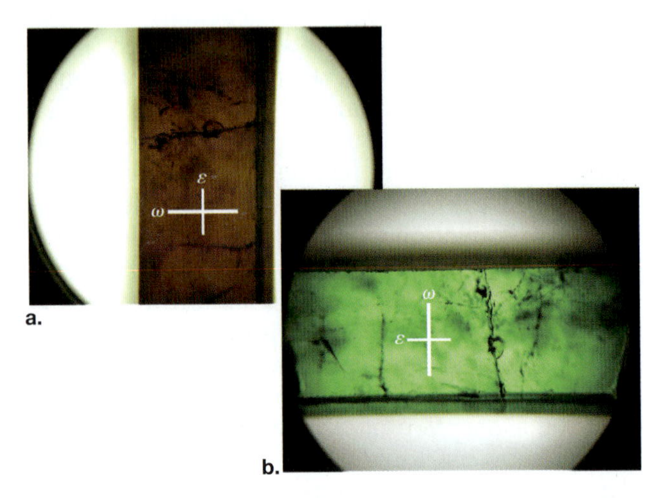

Figure 18.29. Two separate photomicrographs of the same tourmaline crystal showing differential absorption and pleochroism as a function of orientation. ***a.*** *The c axis is vertical, so ω is parallel to the lower polarizer. In this orientation, almost all of the incident light is absorbed and the crystal appears dark brown.* ***b.*** *The crystal is rotated 90° so the c axis is parallel to the lower polarizer. In this orientation the structure absorbs much less light and appears green. Field of view is 5 mm.*

Absorption spectra analogous to the views of biotite and amphibole (Figure 18.33a,b) demonstrate why the color differences occur. The color in micas is usually caused by two types of transitions: IVCT transitions at ~730 nm that cause absorptions between adjacent cations within the octahedral layer, and Fe-O crystal field transitions (~900 and 1175 nm) that arise from transitions within the Fe cations. Thus, the color of the grain shown in Figure 18.31a is dark from the IVCT transitions, while at the other orientation (Figure 18.31b), the light is no longer vibrating to excite electrons between adjacent sites within the plane of the octahedral sheets, and so only the weaker color from the crystal field transition is seen.

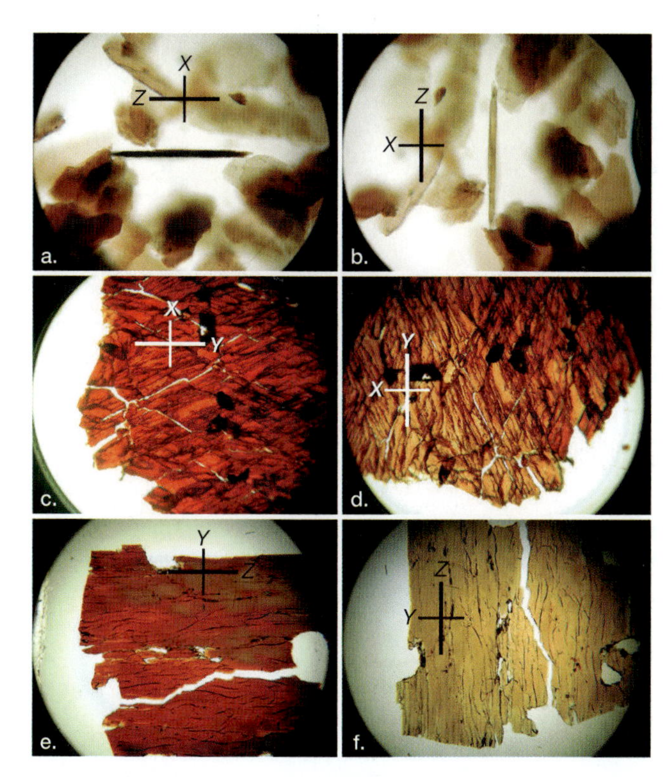

Figure 18.31. *Differential absorption and pleochroism in a biotite (Figure 18.31a,b) and an amphibole (Figure 18.31c–f) in two different orientations.* ***a.*** *and* ***b.*** *The majority of the biotite flakes are resting on their (001) surfaces (cleavage flakes), so their* c *axes are perpendicular to the plane of the microscope stage. In this orientation, their absorption (amount of light transmitted) does not change significantly with stage rotation. The flake in the very center of the photograph happens to be positioned edge-on, so we are viewing it with its* c *axis parallel to the plane of the microscope stage, coinciding with the* X *direction noted. When the light is vibrating parallel to* Z*, significant absorption occurs and the crystal appears to be dark brown or black. A 90° rotation results in a much lighter color (less absorption).* ***c.*** *and* ***d.*** *These views look down the* c *axis of hornblende. When the* Y *vibration direction is parallel to the lower polarizer, there is more absorption and a darker color than when* X *is in that orientation.* ***e.*** *and* ***f.*** *This amphibole crystal is orientated with its* c *axis in the plane of the microscope stage. Figure 18.31e shows the large vibration direction* Z *exhibiting a darker color and more absorption that the* Y *direction, which is shown in Figure 18.31f. Field of view is 5 mm.*

The same three bands, arising from the same phenomena, are seen in amphibole spectra (Figure 18.33b). Notice that absorption is most intense when light is constrained to vibrate parallel to the *Z* direction. The IVCT band is still visible, though less intense, at the *Y* orientation, and is absent at the *X* orientation. These relate orientations of grains in the photomicrographs in Figure 18.31c–f. The first pair of images looks down the *c* axis. When the *Y* vibration direction is parallel to the lower polarizer, there is more absorption and a darker color than when *X* is in that orientation. In Figure 18.31e–f, the crystal is oriented with its *c* axis in the plane of the microscope stage.

Some minerals have structures that allow the transition metals to be so very close together that electrons transfer not only between adjacent cations, but also circulate throughout the structure. This is called **electron delocalization**, and is characterized by electron hopping. Minerals in which electron delocalization occurs are generally opaque because so much energy is absorbed by the hopping electrons. For example, in magnetite, Fe^{2+} and Fe^{3+} cations in adjacent edge-sharing octahedral sites are delocalized in the three-dimensional chains along the [110] axes of the unit cell. Because magnetite is a isometric spinel, it is thus black in all directions. At temperatures above 119K, the sites are close enough together (~2.97 Å) for the electrons to delocalize; below that temperature, the electrons are too sluggish to move around, so the transfer of electrons drops dramatically. Though magnetite is still opaque, its electronic conductivity drops dramatically.

It is also worth noting that all wavelengths of energy are pleochroic to variable extents. For example, Figure 18.33c shows an X-ray absorption spectrum acquired in three different orientations of a muscovite crystal. All types of waves in the electromagnetic spectrum will show pleochroism when they interact with crystalline materials because the same basic rules of physics apply.

Relating Optic Properties to Crystal Chemistry

As noted at the outset, F.D. Bloss and co-workers have used the spindle stage to unravel several mineralogical problems over the past 35 years. These are reviewed in the preface to Bloss (1999) and by Gunter (2004) and Gunter et al. (2004). Many of these studies combine integrated crystal chemical and crystallographic analyses with optical data. Most importantly, they characterize the orientational dependence of optical properties. This is easily done with the spindle stage. It is possible to measure all properties (i.e., composition, structure, optics) on the same single crystal, thus avoiding variability that often occurs even for minerals from the same location. A few examples are given here to highlight the usefulness of the spindle stage.

Andalusite. Changes in the refractive indices of andalusite with substitution of Mn^{3+} and Fe^{3+}

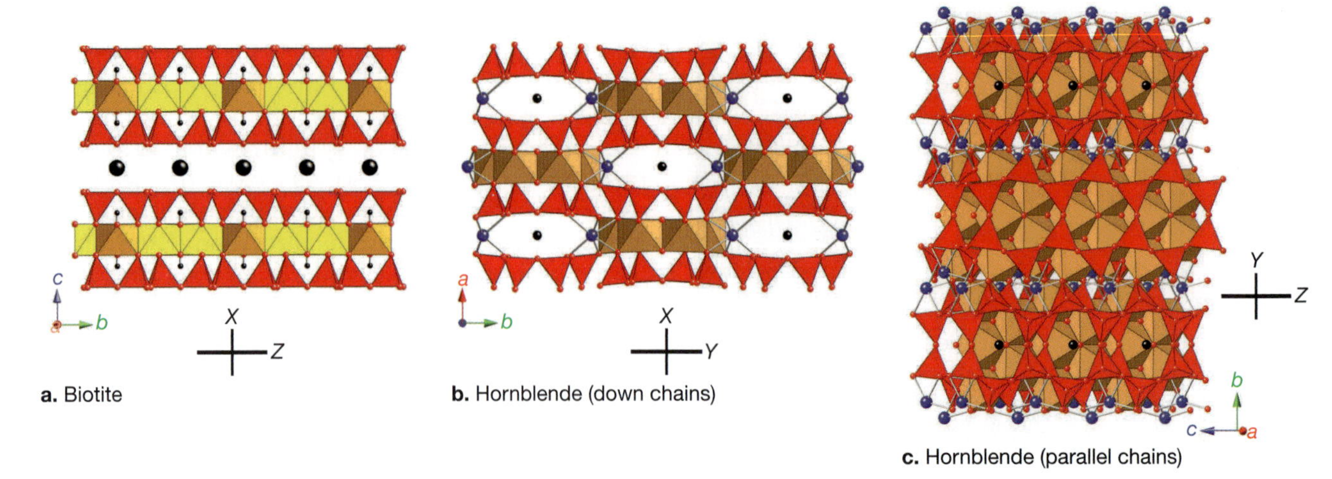

Figure 18.32. Structures of biotite and hornblende, to match the photos and orientation in Figure 18.30. ***a.*** *For the case of biotite, the large vibration direction Z corresponds to the sheet of octahedra.* ***b.*** *In hornblende, a similar phenomenon occurs for Y and* ***c.*** *Z.*

for Al^{3+} were studied by Gunter and Bloss (1982). They obtained a suite of samples with different chemistries and performed precise optical measurements of refractive indices with the aid of a spindle stage. Next, the optical orientation of each crystal was determined with X-ray diffraction (see Gunter and Twamley, 2001 for newer methods to determine optical orientations), and related to the structure of andalusite (Figure 18.34). Finally, each single crystal was removed from the glass fiber used to hold it in place on the goniometer head and prepared for chemical analysis with an electron microprobe (Figure 18.35). Two things are apparent from the graph: (1) the refractive indices increase with substitution of Mn^{3+} and Fe^{3+} for Al^{3+}, as would be predicted, and (2) the lines showing this increase cross around $Mn^{3+} + Fe^{3+} = 0.06$. Thus, andalusites with $Mn^{3+} + Fe^{3+} < 0.06$ would have their small refractive index α parallel to the *c* axis of the mineral (the elongated direction shown in the crystal drawing in Figure 18.34). For andalusites with $Mn^{3+} + Fe^{3+} > 0.08$, the large refractive index direction γ corresponds to the *c* axis. Thus, the mineral changes optical orientation. Note that there is a gap in composition in the region of $Mn^{3+} + Fe^{3+} = 0.06$. At the time of our work, we explained this by postulating that either: (1) those compositions do not occur in nature, or (2) petrologists had misidentified these samples in thin sections because they would appear optically isotropic. Within three years of our study, Grambling and Williams (1985) had found near-isotropic andalusites in metamorphic rocks in New Mexico (Figure 18.36). The detailed optical characterization of single crystals by a mineralogist allowed for the proper identification of a

LCD's

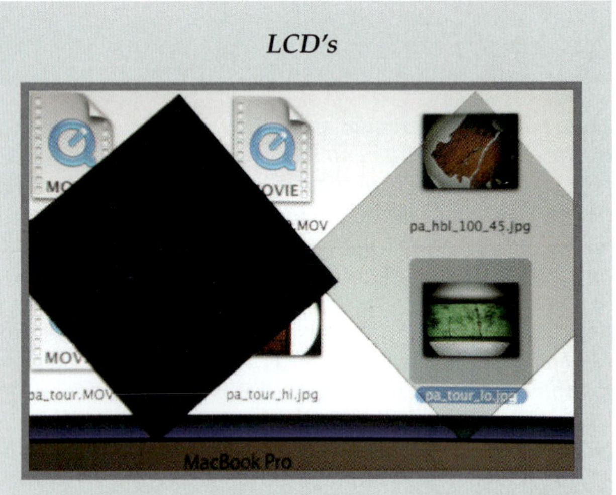

Liquid crystal displays (LCD's) are ubiquitous in everyday life. They work on two principles: polarization of light, and absorption of light. In general, the liquid crystals are randomly oriented between two polarizers with orthogonal privileged vibration directions; thus, no light is transmitted. When an electrical charge is placed on the crystals, they re-align themselves, rotating the plane of polarization so some light is transmitted. The light coming from all liquid crystal displays is polarized, as shown here. The polarizer on the left has its principal vibration direction perpendicular to the polarization direction of the liquid crystal display, so it is black (no light is transmitted). The polarizer on the right has its principal vibration direction parallel to the polarization direction of the liquid crystal display. The pixels representing red, green, and blue light can be charged separately, allowing colors to be seen. If you wear polarized sunglasses, you might sometimes notice that liquid crystal displays go black as you rotate them. Care is taken to be sure that the principal vibration direction of the polarized light in an LCD is usually at a 45° angle to horizontal.

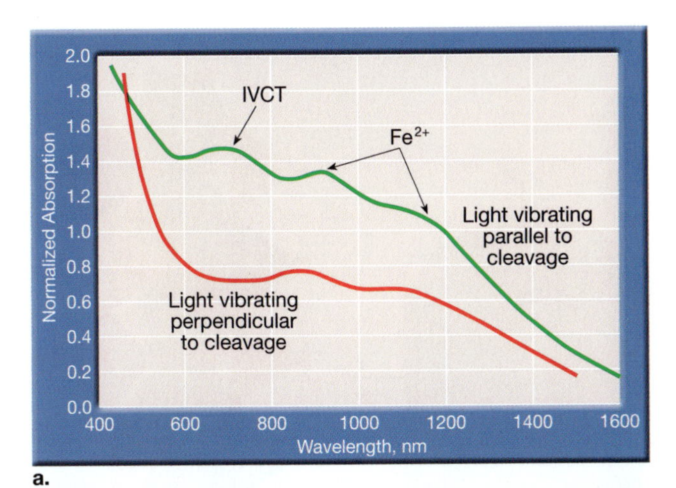

a.

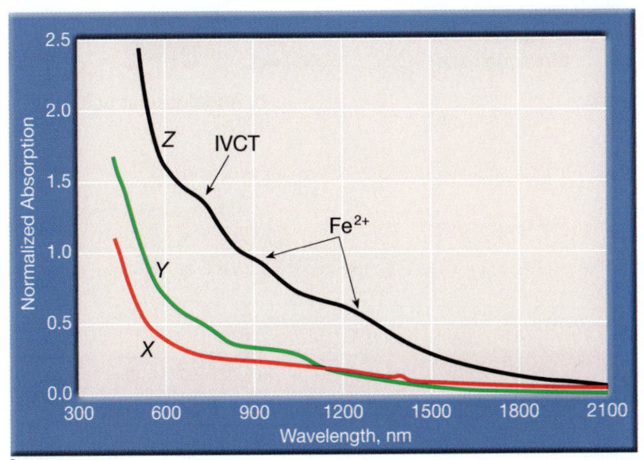

b.

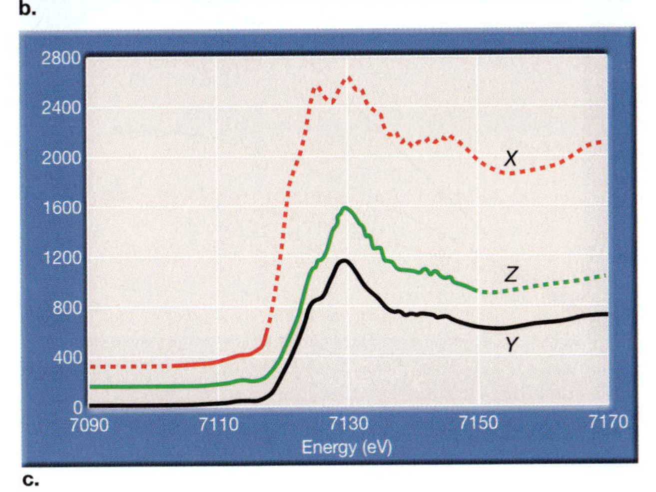

c.

Figure 18.33. Spectra of micas and amphiboles provide a graphical representation of the color variations that occur as a function of orientations. ***a.*** *Optical absorption spectra of biotite (unpublished data) and* ***b.*** *hornblende (from the Caltech Mineral Spectroscopy web site). The color in brown micas and amphiboles is usually caused by two types of transitions: intervalence charge transfer transitions at ~730 nm that cause absorptions between adjacent cations within the octahedral layer of micas and in the chains of the amphibole; and Fe-O crystal field transitions (~900 and 1175 nm) that arise from transitions within the Fe cations.* ***c.*** *X-ray near-edge absorption spectrum for Fe in muscovite. Note that the energy units on the X axis are now in eV; the range plotted represents ~0.002 nm. These data were acquired using a beam of polarized photons (at X-ray energies) at the National Synchrotron Light Source in Brookhaven, NY. There are distinct differences in the amount and character of the absorption at different orientations, especially in the X vibration direction. This figure is included here as a reminder that energy outside the visible wavelengths is also pleochroic. This muscovite, along with several others studied by Dyar et al. (2002), exhibits what we might call "X-ray pleochroism."*

mineral in thin section by a petrologist. Interestingly, note how the samples exhibit pleochroism even though they are optically isotropic.

Natrolite group. Zeolites are framework silicates and thus have low birefringence. Slight changes in optical properties (such as those caused by cation substitutions) can cause large changes in observable optical properties. Figure 18.37 shows a graph for the natrolite group of zeolites that is similar to the one for andalusite because the refractive index values cross. Gunter and Ribbe (1993) produced this graph based on literature data only, noting how the optical orientation of the natrolite group zeolites (i.e., natrolite, mesolite, and scolecite) changes (Figure 17.60). Shortly thereafter, Gunter et al. (1993) found a natrolite crystal tipped with "isotropic" mesolite (Figure 18.37b). Notice how mesolite appears isotropic simply because its birefringence is so low.

Pb Heulandite. In Chapters 14 and 15, the structure (Figure 14.17) and diffraction patterns (Figure 15.36) of natural and Pb-exchanged heulandite were shown. A single crystal of a natural heulandite exhibits very low retardation, as would be predicted for a zeolite, but the retardation greatly increases as Pb exchanges into the structure (Figure 18.38) (Gunter et al., 1994). However, if Pb-exchange is not complete, then the areas of exchange can be mapped out by using retardation, analogous to results of SEM back-scattered electron images as shown in Figure 9.16. The advantage of the optical method is that samples can be placed in a Pb-containing liquid and the rate of exchange observed in real time in the PLM. We also found that other elements caused similar changes in the retardation to those caused by Pb. By observing changes in retardation, we could measure exchange distances by placing a sample in different fluids and monitoring the distance of diffusion into the sample as a function time (Yang et al., 1997). This was a unique and simple way to calculate diffusion rates of cations into a zeolite, which are important for various industrial applications.

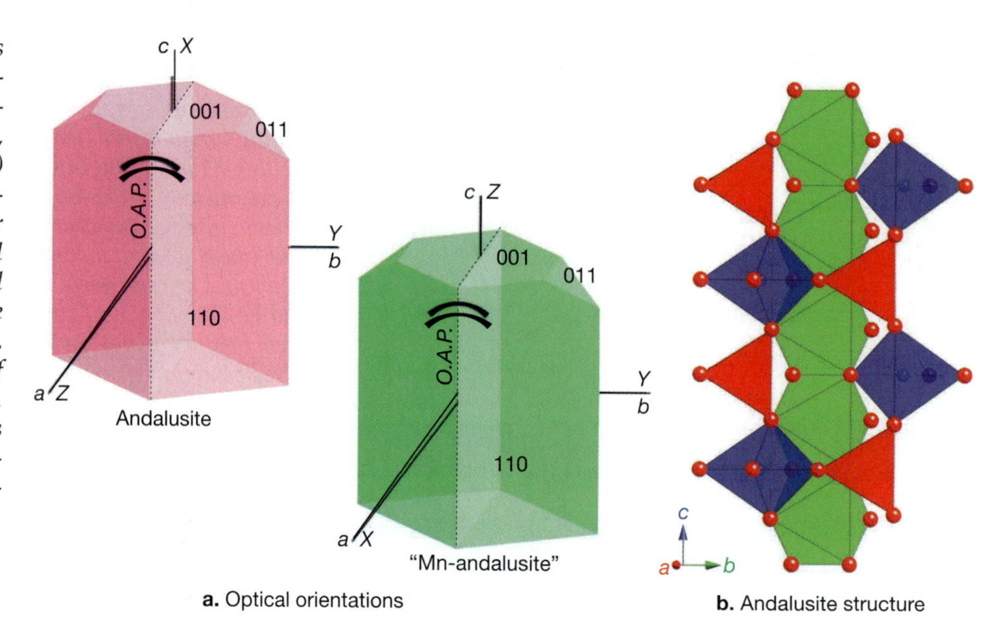

Figure 18.34. *a. Sketches showing the optical orientations for andalusite and "Mn-andalusite." For andalusite, the small refractive index (X) is parallel to the long dimension of the crystal, while for Mn-andalusite, the optical orientation has changed and Z (the largest refractive index) is parallel to c.* ***b.*** *Polyhedral representation of the structure of andalusite. The main structural unit is an edge-sharing chain of octahedral parallel to the c crystallographic axis.*

Calculating Refractive Indices

Often in the world of mineral sciences, we strive to be able to calculate a property of a material based on its crystal structure. This can be approached in two ways: from an empirical or predictive model that relates one set of parameters to another, or from a theoretical understanding of a system. In Chapter 15, the last section explained the process of theoretical calculation of X-ray diffraction patterns on the basis of the crystal structure; such calculations have only recently become computationally tractable. Attempts to predict the refractive indices of minerals date back to the 1800s before crystal structures of materials were known. Relationships to refractive index were based on what was known about mineral chemistry and density, both of which could be measured precisely at that time. Currently, attempts to predict refractive indices are being made using theoretical calculations of the interactions of light in minerals. These require an advanced background in theoretical physics and are beyond the scope of this introductory textbook. However, the interested student should refer to the following: Abbot (1993, 1994, 1996) and Teertstra (2004, 2005, 2006).

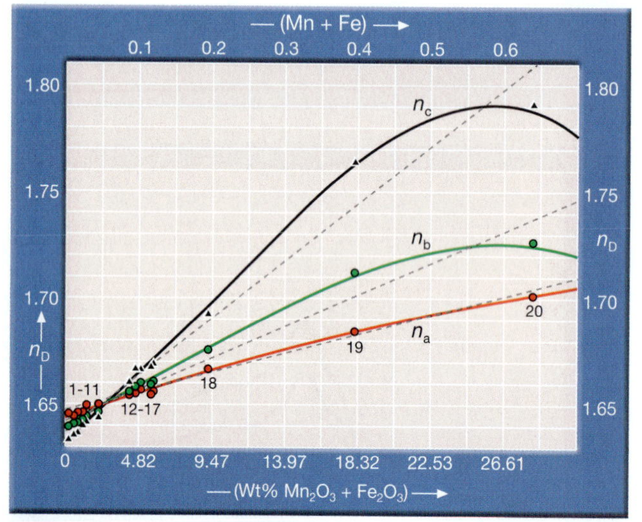

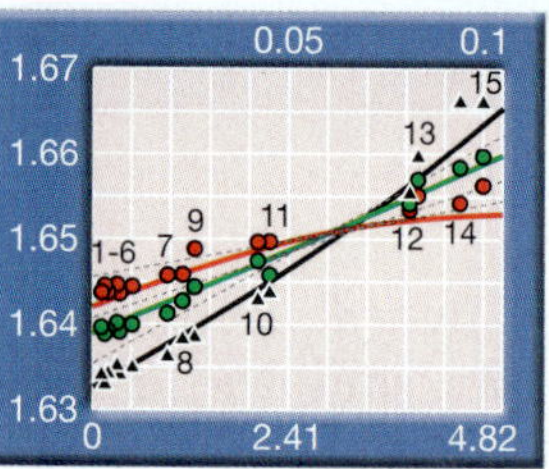

Figure 18.35. *Relationship between refractive indices and Mn + Fe content for twenty andalusite samples (modified from Gunter and Bloss, 1981). Because there is a change in the optical orientation as a function of Mn + Fe content, the refractive index curves are labeled* n_a, n_b, *and* n_c *to correspond to the refractive index directions in the a, b, and c crystallographic directions, respectively. Sample with compositions where the curves cross would appear isotropic.*

Relationship to Density and Bonding

In the silica polymorphs (Figure 22.5) and the olivine group minerals (Figure 18.4a), there is a

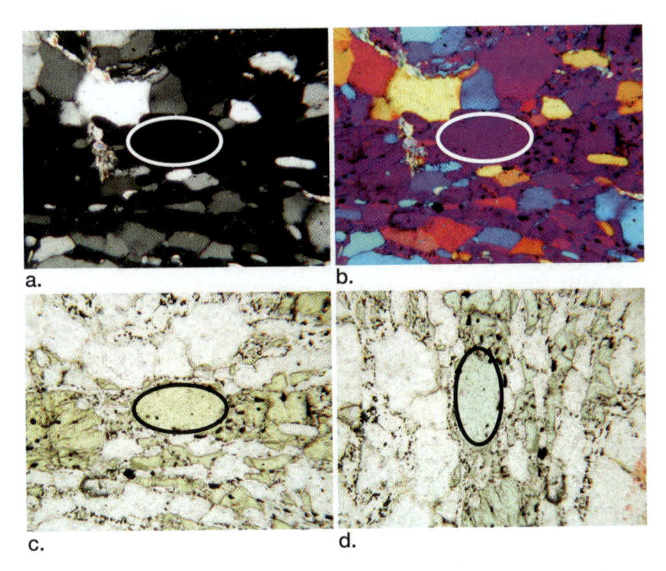

*Figure 18.36. Thin section showing isotropic andalusite (provided by Jeff Grambling). **a.** The circled crystal exhibits zero retardation, so it is optically isotropic even though it is an orthorhombic mineral. **b.** The quartz plate has been inserted, to better show the crystal. **c.** The same sample as Figure 18.36a and b, showing that the andalusite displays its typical light yellow-green color. **d.** The same area of the sample rotated 90°, showing its pleochroism. Although this mineral is optically isotropic, it still exhibits pleochroism.*

linear increase in refractive index as a function of increasing density. This observation was first made more than 150 years ago. High density implies that the elements contained in the material have a high atomic numbers, and/or the ions in the crystal structure are tightly-packed. In general, this trend holds true, but there are some exceptions (Figure 18.39a). The refractive index of NaF (the somewhat obscure mineral villiaumite) is significantly less than those of the other sodium halide compounds. Figure 18.2 showed the trend of decreasing refractive index with increasing ionic character of the bond. This trend is also seen in the sodium halide minerals (Figure 18.39b). Similar trends for refractive indices can also be seen for several minerals in which $(OH)^{1-}$ and F^{1-} can substitute for each other, such as topaz (Ribbe and Rosenberg, 1971) and amblygonite (Greiner and Bloss, 1987).

Gladstone-Dale Relationship

Gladstone-Dale factors were developed in the late 1800s to describe the relationships among

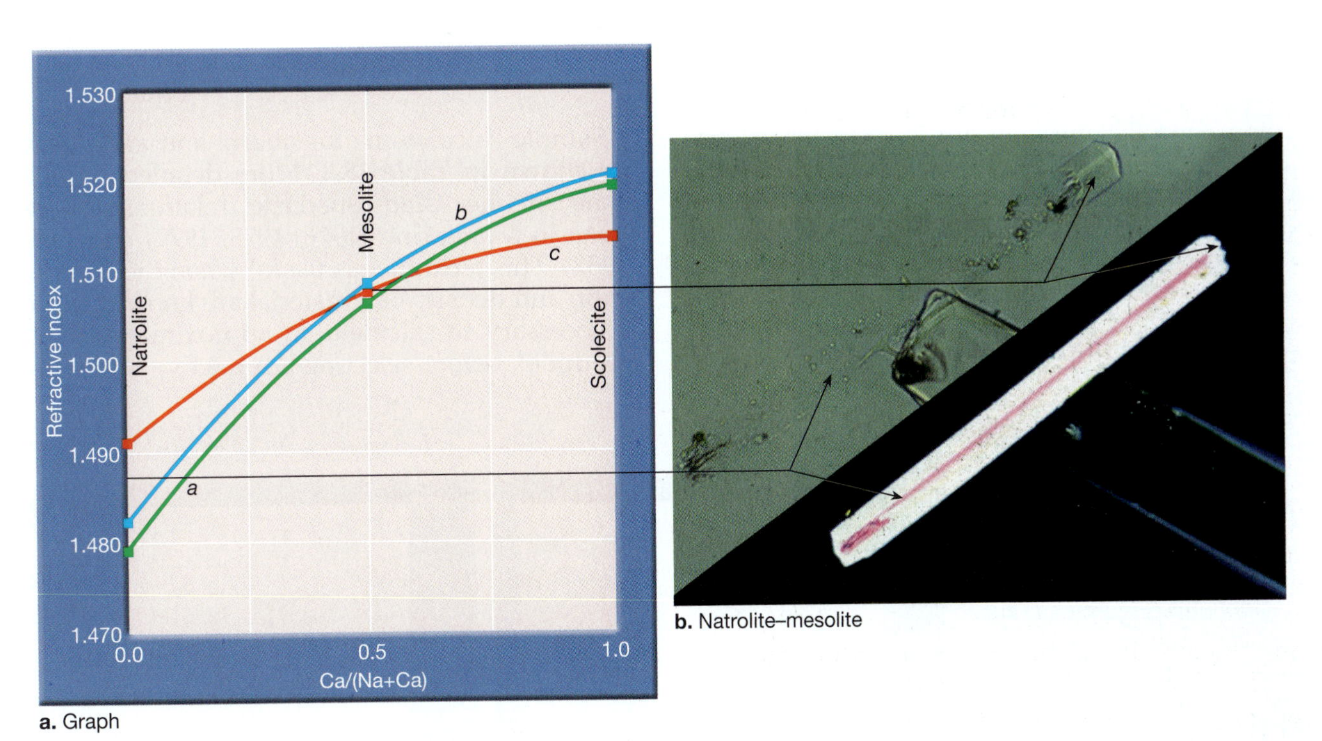

*Figure 18.37. Relationships of the refractive index among the fibrous zeolites: natrolite, mesolite, and scolecite (modified from Gunter and Ribbe, 1993). **a.** This graph was produced from data published in the literature. It is easy to distinguish between these three mineral species using sign of elongation; changes in refractive indices and birefringences also occur. **b.** Photomicrographs of a ~500 μm natrolite crystal capped with mesolite, taken in both plane-polarized and cross-polarized light (modified from Gunter et al., 1993). The crystal is mounted on a glass fiber and immersed in index-matching fluid for natrolite. In the plane-polarized photo, the mesolite tip has higher relief than the natrolite portion of the crystal. In the cross-polarized photo, the mesolite tip appears near isotropic because of its lower birefringence.*

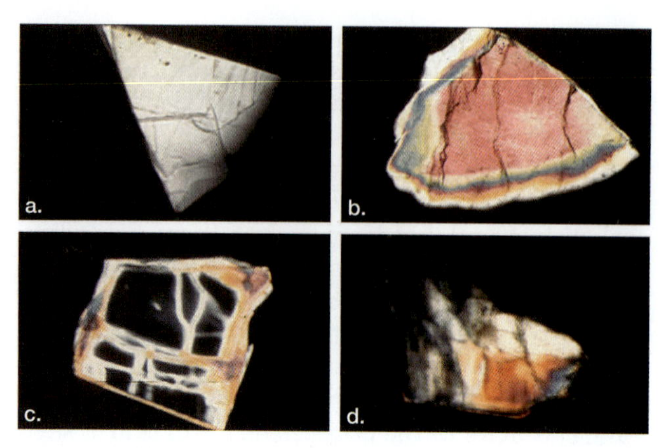

*Figure 18.38. A series of images of natural and Pb-exchanged heulandite zeolites showing changes in retardation a function of Pb exchange (modified from Gunter et al., 1994). **a.** A natural sample exhibiting the typical low retardation of a framework silicate. **b.** A fully Pb-exchanged sample showing a drastic increase in retardation. **c.** Pb diffusion along the edges and cracks in a sample, as indicated by increases in retardation. **d.** Increases in retardation can be used to monitor Pb exchange around the edges of a zeolite crystal; the greatest exchange is in the lower right-hand corner of the grain.*

refractive index, density, and weight fraction of the oxide components (i.e., composition) of minerals. They are empirical in much the same way as ionic radii. Ionic radii are determined by measuring bond distances of ions in different valence states and coordination polyhedra. Similarly, the Gladstone-Dale relationship was measured by determining Gladstone-Dale constants for the oxides, but taking into account how the values vary as a function of coordination polyhedra and valence state. Equation 18.4 shows these mathematical relationships:

$$\rho = \frac{n-1}{k} \text{ or } n = \rho k + 1 \qquad \text{Equation 18.4}$$

where ρ is the density, n is the refractive index of the mineral and k is the so-called Gladstone-Dale value. Some typical examples are given Table 18.2.

The density of a mineral can be calculated as shown below:

$$\rho = \frac{M_{uc}}{V_{uc} \times 10^{-24}\,(6.0228 \times 10^{23})}, \qquad \text{Equation 18.5}$$

where M_{uc} is the molecular weight of the contents of the unit cell and V_{uc} is the volume of the unit cell.

The average refractive indices for a material can then be described by:

$$n_{average} = \frac{\varepsilon + 2\omega}{3} \text{ and } n_{average} = \frac{\alpha + \beta + \gamma}{3}.$$

Equation 18.6

If a mineral is composed of a single oxide (e.g., quartz), then Equation 18.4 is used. If a mineral is composed of more than one oxide, then its Gladstone-Dale constant is the weighted sum of each constant times the weight fraction of the corresponding oxide, as in:

$$k_{\text{mineral}} = \text{sum}\,(k_{\text{ox1}} \times \text{wt.frac}_{\text{ox1}}) + (k_{\text{ox2}} \times \text{wt.frac}_{\text{ox2}}) + \ldots + (k_{\text{oxn}} \times \text{wt.frac}_{\text{oxn}})$$

Equation 18.7

Sample calculations for quartz and andalusite are shown in Table 18.3. More detailed discussions of the Gladstone-Dale relationship are given in Mandarino (1976, 1978, 1979) and Jaffe (1988). This technique is useful when the composition and density of a material are known, and it is necessary to calculate an approximate refractive index before selecting a refractive index liq-

*Figure 18.39. Refractive indices usually increase as density increases, as seen previously for olivine (Figure 18.4) and the silica polymorphs (Figure 22.5). **a.** However, this is not always the case, as demonstrated in the sodium halide materials. **b.** While density does play a role in affecting refractive index, so too does the type of bonding. In this case, there is a clear correlation between the percentage of ionic character of the bond and refractive index.*

a. Density vs. n for Na halides

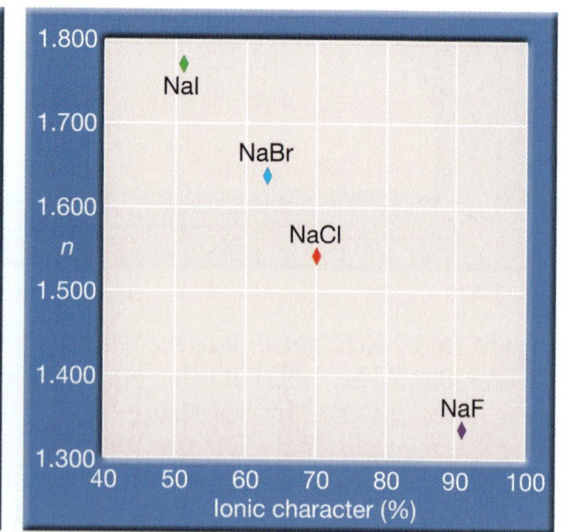

b. Ionic character vs. n for Na halides

uid. Calculation of the Gladstone-Dale constant is also required for new mineral descriptions as a cross-check on the chemistry, structure, and optical properties of a mineral (Mandarino 1981).

Conclusions

The preceding chapters have described many different uses for optical mineralogy. We'll return to them again the next chapter as well, because there is no other single technique that is as simple and efficient for mineral identification. These chapters illustrate that even a centuries-old instrument like the PLM can be a viable research instrument in the right hands; to learn more, consult Gunter (2003, 2004). Although some professors may doubt the importance of teaching of optical mineralogy in the undergraduate curricula, smart students will realize that these skills are extremely valuable in the private sector. At this writing, there is great unmet demand for trained microscopists! For those planning to pursue further work in geology, fundamental training in optical methods will prove useful in all fields that are concerned with microscale processes.

Appendix: How to Build a Spindle Stage and Oil Cell

The original prototype for the Wilcox spindle stage was built with a piece of used floor tile (Dan Kile, personal communication), proving that this apparatus can be fabricated from very simple materials. In the mid-1990s, Thomas Armbruster at Bern University in Switzerland fabricated a spindle stage out of poster board. His design was modified (Gunter, 1997) during a short course at the Teaching Mineralogy Workshop at Smith College and again during a workshop on the spindle stage at McCrone Research Institute (2004), where some modifications to the earlier design were made to allow students to build their own home-made spindle stages. The papers by Gunter (1997, 2004) were written for three reasons: (1) to show how simple the device can be, (2) because there is no current vendor for a simple, inexpensive spindle stage (the detent stage shown in Figure 18.24 is no longer available), and (3) because undergraduates build and use the spindle stage in mineralogy classes at the University of Idaho.

Table 18.2. Gladstone-Dale constants (modified from Jaffe 1988).

Z	Oxide	k (Jaffe)	CN
1	H_2O	0.340	NA
11	Na_2O	0.178	8
12	MgO	0.241	4
	MgO	0.200	6
13	Al_2O_3	0.218	4
	Al_2O_3	0.187	6
14	SiO_2	0.206	4
	SiO_2	0.187	6
19	K_2O	0.186	12
20	CaO	0.226	6
	CaO	0.215	8
26	Fe_2O_3	0.332	4
	Fe_2O_3	0.280	6
	FeO	0.185	6

Table 18.3. Example Gladstone-Date calculations

Quartz example:

density of quartz = [180.2529 / (113.01) (1 x 10^{-24}) (6.0228 x 10^{23})] = 2.65 g/cm^3

$n_{average}$ = [1.553 + 2(1.544)] / 3 = 1.547

n_{quartz} = (2.65) (0.206) + 1 = 1.546

Andalusite example:

$n_{average}$ = (1.6328 + 1.6386 + 1.6436) / 3 = 1.6383

$k_{andalusite}$ = (k_{SiO2} x wtfrac$_{SiO2}$) + (k_{Al2O3} x wtfrac$_{Al2O3}$) = (0.208 x 0.3708) + (0.176 x 0.6692) = 0.2002

$n_{andalusite}$ = (3.149)(0.2002) + 1 = 1.630

Figure 18.A1 shows how to cut the individual parts required for building a spindle stage. Gunter (1997) used poster board as a building material, mainly because it is easy for students to work with, but many other materials could be used (e.g., plastic, wood, metal, glass, etc.). The thickness of the material needs to be similar to that of the glass slide (Figure 18.A2a) that will be used to build the oil cell. If the thicknesses differ slightly, either the spindle stage or the glass slide can be shimmed by adding layers of Scotch tape underneath.

Building the base. Once a material has been selected, the first step is to cut out two 50 × 50 mm pieces. One of these will be the base of the spindle stage, and the other will form the protractor scale. For the base, scribe lines as shown in Figure 18.A1a to locate its center. Next, take a glass slide (Figure 18.A2a) and place it as shown in Figure 18.A1b. Use it to mark the sides and top of the base, then cut out this material to form the oil cell.

Building the dial. Make a 1:1 copy of the circle protractor scale in Figure 18.A1d (either with a copy machine or a scanner). The circle has a diameter of 50 mm. Glue it to a 50 × 50 mm square of the base material (e.g., poster board) as shown in Figure 18.A1e. Trim the edges and cut the circle in half to complete the spindle dial (Figure 18.A1f). If desired, you can glue another circle protractor onto the back of the 50 × 50 mm block before trimming and cutting it; that way you can read the dial from either side. If you mount a scale on the back, be sure to place it so the S angles correspond from front to back (see Figure 18.A2b).

Assembling the stage. Cut a 15 × 50 mm rectangle and trim the edges as shown by the scribe marks (Figure 18.A1g). Glue it vertically in the center of the base (Figures 18.A1a, A1c, and A2b). Then glue the protractor dial to the end of the base. Insert a hollow metal tube, which will serve two functions: one end serves as a sleeve to hold a needle with crystal attached, and the other end is a marker for reading the S angle. For example, a 20 gauge 2½" hypodermic needle can be used to make a hole in the center of the assembled stage, passing through the center of the protractor and

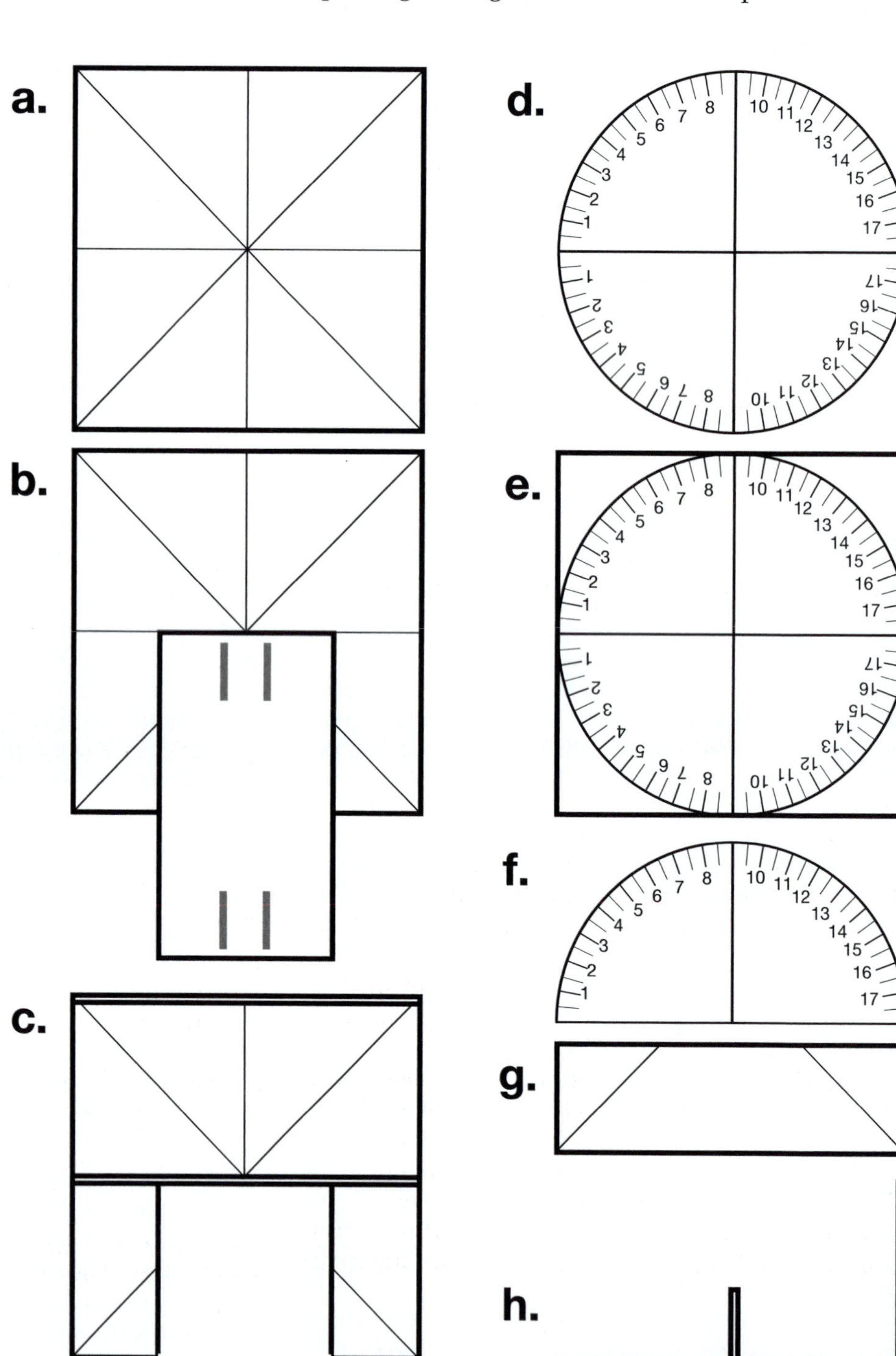

Figure 18.A1 Appendix. *"Blueprint" for each of the parts needed to build a UI spindle stage. (A detailed description is given in the text on how to assemble it, and a brief description of each part is given here.)* ***a.*** *The base of the stage is 50 × 50 mm. Scribed lines are used to locate its center.* ***b.*** *The base with an oil cell (see Figure 18.A2 for a photograph) placed over it. Lines are marked on the base (parallel to the oil cell) and cut out so the oil cell can be inserted.* ***c.*** *Sketch of the base with the oil cell dock area removed.* ***d.*** *A template for the S-angles suitable for reproduction (by photocopy or scanning).* ***e.*** *Another 50 × 50 mm square with the S-angle template attached.* ***f.*** *The template/square cut in half, trimmed to a semicircle, and now ready to mount onto the rear of the base.* ***g.*** *A 15 × 50 mm piece, with scribed edges to be removed. This piece is mounted in the middle of the stage.* ***h.*** *Side view of the base with the 15 × 50 mm piece mounted in the middle and the S-angle dial on the end. The image Figure 18.A1c is a top view of the assembled parts. Perspective photos are also in Figure 18.A2.*

the 15 × 50 mm mid-piece. Next, insert a piece of tubing approximately 2½″ long through this hole and allow it to extend a few millimeters into the cavity for the cell. Finally, put a 90° bend in the tubing where it passes through the back side of the protractor scale. If the tube is not long enough on the protractor scale, it can be extended by inserting a straight pin in its hollow end, as shown in Figure 18.A2b. If the pin fits too loosely, place a small bend in the end that is inserted into the tube.

Building the oil cell. A standard petrographic glass slide (or any glass slide) can be used for the oil cell. Cut two 8–10 mm pieces off the end of a large paper clip. Epoxy them to one end of glass slide (Figures 18.A1b and A2a). Make sure they form a cavity that is centered on the slide with an approximate width of 5–8 mm. Refractive index liquid can be placed between the paper clip pieces and a glass cover slide placed on top. If desired, another oil cell can be added to the other end of the slide.

Mounting the spindle stage on a microscope stage. Figure 18.A2c shows a completed home-made spindle stage and oil cell mounted on the stage of a polarizing light microscope. In this case, cellophane tape is used, but other more permanent methods could be used. The tape is first placed in front of the protractor dial of the spindle stage. Then the spindle stage, with an affixed crystal, is placed on the microscope stage and the crystal is positioned at the center of the cross-hairs. The tape is pushed down to mount the spindle stage to the microscope stage. Other pieces of tape can be placed on the ends of the spindle stage (next to the oil cell) to better secure it. It is also helpful to place the microscope stage to zero (i.e., obtain an M_r, or reference angle, near 0°), then orient the spindle stage so its axis of rotation is parallel to the E-W cross- hairs and the tip is pointed to the W, as shown in Figure 18.A2c. The spindle stage in Figure 18.A2c also shows how a straight pin has been bent into the form of a "U" so the *S* angles can be read on the back, as well as the front, of the protractor. If the "U"-shaped needle is removed, translation in the "*x*" direction will be possible, so the crystal can be better centered in the field of view.

The design presented above works well. It is probably the simplest possible design of a spindle stage, and as such might lack some useful features such as a translation in the *Y* direction, or detents for 10° increments of the *S* setting. We encourage you to experiment with this design and modify it in any way you see fit. Our main goal is to provide a starting point design for something that can easily be made in less than an hour and used to collect extinction data sets.

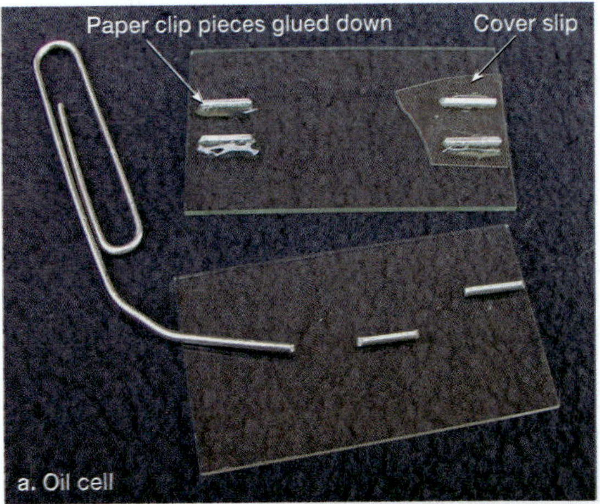

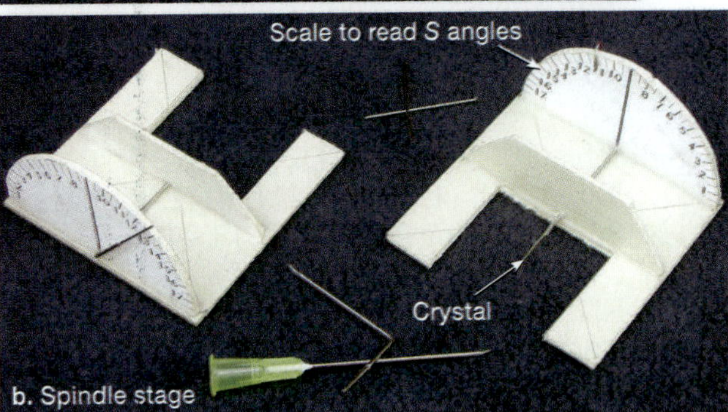

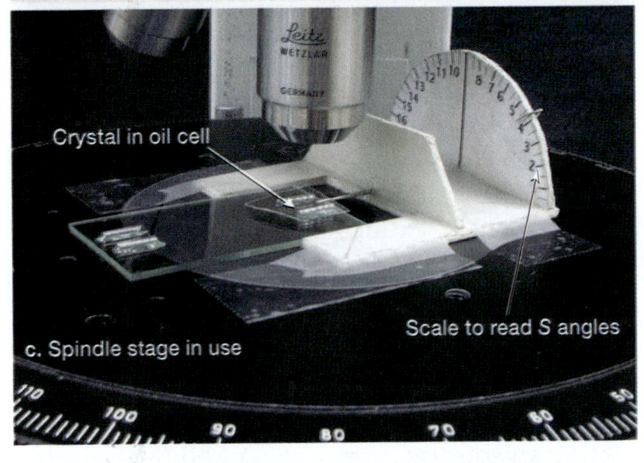

Figure 18.A2 Appendix. a. *To make an oil cell, cut two 8–10 mm pieces off a large paper clip and epoxy them 5–8 mm apart on a glass side, being careful to center the space between them. A small cover slip (in this case broken from a larger one) is shown placed on top of the oil cell. Lastly, a refractive index liquid can be placed in the void created by the cover slip and paper clip pieces.* ***b.*** *Front and back perspective view of two home-made spindle stages. In the foreground is a 2.5″ 20 gauge needle. The yellow tip can be pulled off the needle and the needle bent to make the spindle axis tube. In the background are two straight pins, which can be inserted into both ends of the spindle axis; the pin inserted in the S-angle end can, in turn, be bent over the dial so the S-angles can be read from both sides. The other pin can be used to mount single crystals and then slid into the spindle axis end that extends into the oil dock area.* ***c.*** *Close-up view of the home-made spindle stage/oil cell taped onto the stage PLM and ready for use.*

References

Abbott, R.N. Jr. (1993) Calculation of the orientation of the optical indicatrix in monoclinic and triclinic minerals: The point-dipole theory. American Mineralogist, 78, 952–956.

Abbott, R.N. Jr. (1994) Electronic polarizability of oxygen and various cations in selected triclinic minerals: Point-dipole theory. Canadian Mineralogist, 32, 87–92.

Abbott, R.N. Jr. (1996) Optical properties of C2/c pyroxenes: A point-dipole theoretical explanation. Canadian Mineralogist, 34, 595–603.

Bartelmehs, K.L, Bloss, F.D., Downs, R.T., and Birch, J.B. (1992) EXCALIBRII. Zeitschrift für Kristallographie 199, 185–196.

Bloss, F.D. (1981) The spindle stage: principles and practice. Cambridge University Press, Cambridge, England, 340 pp.

Bloss, F.D. (1999) Optical crystallography. Mineralogical Society of America, Washington, D.C., 239 pp.

Bloss, F.D. and Ries, D. (1973) Computer determination of 2V and indicatrix orientation from extinction data. American Mineralogist, 58, 1052–1061.

Dyar, M.D., Gunter, M.E., Delaney J.S., Lanzarotti, A., and Sutton, S.R. (2002a) Use of the spindle stage for orientation of single crystals for microXAS: Isotropy and anisotropy in Fe-XANES spectra. American Mineralogist, 87, 1500–1504.

Dyar, M.D., Gunter, M.E., Delaney, J.S., Lanzarotti, A., and Sutton, S.R. (2002b) Systematics in the structure and XANES spectra of pyroxenes, amphiboles, and micas. Canadian Mineralogist, 40, 1375–1393.

Gladstone, J.H. and Dale, T.P. (1864) Researches on the refraction, dispersion, and sensitiveness of liquids. Royal Society of London Philosophical Transactions, 153, 337.

Grambling, J.A. and Williams, M.L. (1985) The effects of Fe^{3+} and Mn^{3+} on aluminum silicate phase reactions in North-Central New Mexico, U.S.A. Journal of Petrology, 26, 324–354.

Greiner, D.J. and Bloss, F.D. (1987) Amblygonite-montebrasite optics: Response to (OH^-) orientation and rapid estimation of F from $2V$. American Mineralogist, 72, 617–624.

Gunter, M.E. (1997) Laboratory exercises and demonstrations with the spindle stage. In Teaching Mineralogy, editors J.B. Brady, D.W. Mogk, and D. Perkins III, 309–318.

Gunter, M.E. (2003) The future of polarized light microscopy: Bright, dim, or extinct? Journal of the Microscope Historical Society, 11, #1, 4–17 also published in The Microscope, 51, 3, 53–63.

Gunter, M.E. (2004) The polarizing light microscope: Should we teach the use of a 19th century instrument in the 21st century? Journal of Geoscience Education, 52, 34–44.

Gunter, M.E., Downs, R.T., Bartelmehs, K.L., Evans, S.H., Pommier, C.J.S., Grow, J.S., Sanchez, M.S., and Bloss, F.D. (2005) Optic properties of centimeter-sized crystals determined in air with the spindle stage using EXCALIBRW. American Mineralogist, 90, 1648–1654.

Gunter, M.E., Weaver, R., Bandli, B.R., Bloss, F.D., Evans, S.H., and Su, S.C. (2004) Results from a McCrone spindle stage short course, a new version of EXCALIBR, and how to build a spindle stage. The Microscope, 52, 1, 23–39.

Gunter, M.E. and Twamley, B. (2001) A new method to determine the optical orientation of biaxial minerals: A mathematical approach. Canadian Mineralogist, 39, 1701–1711.

Gunter, M.E., Armbruster, T., Kohler, T., and Knowles, C.R. (1994) Crystal structure and optical properties of Na- and Pb-exchanged heulandite group zeolites. American Mineralogist, 79, 675–682.

Gunter, M.E. and Ribbe, P.H. (1993) Natrolite group zeolites: correlations of optical properties and crystal chemistry. Zeolites, 13, 435–440.

Gunter, M.E., Knowles, C.R. and Schalck, D.K. (1993) Composite natrolite-mesolite crystals from the Columbia River Basalt Group, Clarkston, Washington. Canadian Mineralogist, 31, 467–470.

Gunter, M.E. and Schares, S.M. (1991) Computerized optical mineralogical calculations. Journal of Geological Education, 39, #4, 289–290.

Gunter, M.E., Bloss, F.D., and Su, S.C. (1989) Computer programs for the spindle stage and double-variation method. The Microscope, 37, 167–171.

Gunter, M.E., Bloss, F.D., and Su, S.C. (1988) EXCALIBR revisited. American Mineralogist, 73, 1481–1482.

Gunter, M.E. and Bloss, F.D. (1982) Andalusite-kanonaite series: Lattice and optical parameters. American Mineralogist, 67, 1218–1228.

Jaffe, H.W. (1988) Crystal Chemistry and Refractivity. Cambridge University Press, Cambridge, 335 pp.

Kile, D.E. (2003) The petrographic microscope: Evolution of a mineralogical research Instrument. Special Publication No.1, The Mineralogical Record, Tucson, Arizona, 96 pp.

Mandarino, J.A. (1976) The Gladstone-Dale relationship. Part I. Derivation of new constants. Canadian Mineralogist, 14, 498–502.

Mandarino, J.A. (1978) The Gladstone-Dale relationship. Part II. Trends among constants. Canadian Mineralogist, 16, 169–174.

Mandarino, J.A. (1979) The Gladstone-Dale relationship. Part III. Some general applications. Canadian Mineralogist, 17, 71–76.

Mandarino, J.A. (1981) The Gladstone-Dale relationship. Part IV. The compatibility concept and its application. Canadian Mineralogist, 19, 441–450.

Ribbe, P.H. and Rosenberg, P.E. (1971) Optical and X-ray determinative methods for fluorine in topaz. American Mineralogist 56, 1812–1821.

Su, S.C., Bloss, F.D., and Gunter, M.E. (1987) Procedures and computer programs to refine the double variation method. American Mineralogist, 72, 1011–1013.

Teertstra, D.K. (2004) Polarized light optics, the new physics of the photon. Euclid Geometrics Publishing, Ontario, Canada, 196 pp.

Teertstra, D.K. (2005) The optical analysis of minerals. Canadian Mineralogist, 543–552.

Teertstra, D.K. (2006) Index-of-refraction and unit-cell constraints on cation valence and order in garnet. Canadian Mineralogist, 341–346.

Yang, P., Armbruster, T., Stoltz, J., and Gunter, M.E. (1997) Na, K, Rb, and Cs exchange in heulandite single-crystals: Diffusion kinetics. American Mineralogist, 82, 517–525.

Chapter 19

Mineral Identification

In my own introductory geology class, we were handed a tray of minerals and instructed to "scratch this and sniff that" to arrive at a mineral name by process of elimination. Many of the minerals that were in those trays were ones that a typical geologist would never see again, especially in the field. This proved to be very confusing for me. When you first see minerals like galena, cinnabar, and sphalerite at the beginning of your career, you somehow get the idea that they are very important. The other troubling aspect of naming minerals in an introductory class is its dependence on the use of ambiguous physical properties such as color, streak, and cleavage. I also recall wondering which mineral in a hand sample I was supposed to identify! Even in mineralogy classes, we still seem "stuck" on using trays full of minerals the size of your fist, and using macroscopic physical properties to identify them.

The sad part is that this type of lab does not come even close to describing what mineralogists do today. One of my many mentors once told me that I should teach what mineralogists actually do! The truth is that we rarely encounter fist-size hand samples at random in the field. More often, our job is to identify tiny, nondescript minerals by using combinations of analytical techniques. Unlike the situations in many other disciplines in the geological or natural sciences when identification and classification can often be impossible, a mineralogist can always identify a mineral with the right tools—though the process may require anything from a few minutes to several hours, in rare cases resulting in the discovery of a new mineral.

Between us, Darby and I have been full-time mineralogists for more than 80 years, and we have been active contributors to the fields of mineral spectroscopy, mineral optics, planetary mineralogy, and minerals and health. But if you ever encounter one of us at a conference, please don't pull out a hand sample and ask for an identification. We both find it very difficult! Darby will run for the nearest spectrometer to identify your sample and determine its chemistry, and Mickey will head for a microscope, an X-ray diffractometer, and/or an SEM with an EDS to make a positive identification. One of my standard wisecracks when someone hands me a minerals is: "Do you want me to guess what it is or tell you? If you want me to tell you, it will take a little work, but we'll know for sure." The point is, most mineralogists today work with tools that are far more sophisticated than the physical property tests described in Chapter 2, and if the mineral identification really matters, they can determine it with the proper choice of analytical equipment.

As practicing mineralogists, members of the public often bring us minerals to identify, in the hopes that they will: (a) be worth something, and (b) be something that we might want to buy from them. Most of these samples are what people hope will turn out to be meteorites or precious gemstones. As a result of study with a few analytical methods, their samples typically turn out to be melted metal "meteor-wrongs" or large pieces of calcite or quartz. The good citizen often leaves very dissatisfied, saying "We'll have to go find someone who really knows what he's doing *to identify this…"*

M.E.G.

Introduction

With 18 chapters behind us now, we can integrate all the different characterization methods we have talked about and apply them to the identification of minerals. We will then tie the concept of identification to the idea of classification in Chapters 21–23. To aid in mineral identification, we developed a large, searchable mineral database now available as an app called Mineral Database; details on how to obtain are at www.minsocam.org/msa/DGTtxt/. This chapter integrates content from previous chapters with information on the app and found in other types of databases.

We will begin with museum-quality hand samples of minerals, and then consider the more common field occurrences, and the various types of equipment you might use to identify them all. We will discuss which analytical techniques are most useful in identification of common mineral groups, and the limitations of each. We will concentrate on the most common instruments you might have at your disposal for mineral *identification*: a polarizing light microscope (PLM), a powder X-ray diffractometer (XRD), and a scanning electron microscope (SEM) equipped with an energy-dispersive spectrometer (EDS). Most certainly, other instruments may be used to assist in the *characterization* of minerals. For example, the electron microprobe uses wavelength-dispersive spectrometry (WDS) to obtain precise chemical data. A transmission electron microscope (TEM) can collect diffraction and chemical data on a single particle, and a single-crystal X-ray diffractometer can determine a mineral's structure.

As we've said many times before, the most common mineral species in the Earth's crust is quartz, while the most common group is the feldspars (Figure 19.1). Many beautiful photographs are included in the database that accompanies this book; almost all of them were taken at the Harvard Mineralogical Museum (see the preface to Chapter 22). Many mineralogists would be able to identify these samples on the basis of their appearance, even from the photographs. If your main interest is in collecting minerals to exhibit, then these are the types of specimens you would become accustomed to identifying. For a geologist, the majority of mineral identifications will come from the field, not a museum. In our database, we show *hand sample* photographs of museum samples, but we realize that they hardly resemble samples you will see in the field (or in your mineralogy lab). On the other hand, we have made an effort to show *photomicrographs* of minerals in typical occurrences, so you will get a sense of their more common associations.

As you move from hand specimens in the lab (i.e., Geology 101) to the great outdoors, the same mineral can look very different. To this end, six different "views" of quartz are shown in Figure 19.2. Weathered quartz boulders are common in many places in granitic regions (Figure 19.2a), glacial outwash plains, and mountain streams. They occur because quartz is highly resistant to chemical and physical weathering. Large samples of feldspar, on the other hand, are extremely rare because they are so susceptible to weathering (see the Rock Cycle discussed in Chapter 20). Figure 19.2b,c shows a weathered granite in outcrop and a small piece of granite that is decomposing. Quartz and feldspars are the primary minerals in these rocks. We don't even have to look closely to know this, because those two minerals almost always occur together in granites. Based on Bowen's reaction series (Chapter 20), we can even predict that the plagioclase feldspar in these rocks would be Na-rich. The dark stains on the rock are secondary iron-bearing minerals; their precise identification would need to rely on some type of analytical method (most usefully, Mössbauer spectroscopy; see Chapter 9). Figure 19.2d,e shows unconsolidated sediments with grains of sand- and silt-size particles. Figure 19.2d shows sediments collected along the banks of the Salmon River in Idaho, and Figure 19.2e is soil along a dirt road in northern Idaho. Both have granitic sources, and are composed predominantly of quartz and feldspar. Grain mounts and a polarized light microscope would be useful to identify the mineralogical composition of larger than sand-sized particles, while powder X-ray diffraction would be most useful in studying the silt. The last figure (Figure 19.2f) shows dust on a dirty car. This dust is smaller yet (clay-size) and it was deposited on the car by raindrops in an semi-arid climate (northern Idaho). A combination of powder XRD and SEM analysis would be needed to characterize the mineralogy in the dust particles because of their small size, possibly assisted with a TEM.

What's in a Name?

Often, mineral *identification* is the goal of a mineralogy student (i.e., you are asked to give every sample a name). As we will see in this chapter, that task can be anywhere in the range from very simple to nearly impossible! To identify a mineral, we first inspect its physical properties, which are anything we can observe or measure, to give us hints about its crystal structure and composition. If a sample is large enough to observe in

a. Quartz

b. Orthoclase

Figure 19.1. ***a.*** *Images of quartz and* ***b.*** *orthoclase from the Harvard Mineralogical Museum, taken from our mineral database. At this point in the book, you can probably recognize these minerals by sight: they represent the most commonly-occurring mineral species (quartz) and mineral group (feldspar) in the Earth's crust.*

hand sample, the most useful physical properties will be those that can be seen with a naked eye: color, shape, and cleavage. These properties will sometimes be sufficient to identify a mineral (Figure 19.1), especially for hand samples. Increasingly smaller samples require use of other types of observations, such as PLM or SEM. As we proceed to using more complex analytical instruments, we could measure other properties such as refractive indices and chemical composition. Other methods, such as diffraction techniques, are used to explore crystal structures of minerals.

Closely related to identification (Figure 19.3) are two other concepts: mineral characterization and classification. *Characterization* is done using some method for describing a crystal's structure (most routinely, single-crystal X-ray diffraction) and some other method for determining its chemistry (electron microprobe equipped with WDS).

Figure 19.2. Typical "field occurrences" of quartz and feldspar—these are not as nice as the museum specimens! Because they are so common, quartz and feldspar occur naturally at many different "grain" sizes, though quartz predominates because it is less susceptible to weathering processes than feldspar. Here we show ***a.*** *a large quartz cobble,* ***b.*** *a weathered granite in outcrop, and* ***c.*** *a decomposing piece of granite.* ***d.*** *is at the same scale as Figure 19.2a–c, but shows individual sand grains, predominantly quartz, on a river beach (with the side of a bottle for scale). Figure 19.2e and f are examples of yet finer-grained materials in a soil, in both cases mostly quartz and feldspars.* ***e.*** *is dust along the side of dirt road and* ***f.*** *shows dried raindrops containing sediment from a dusty environment.*

By definition, every mineral must have a unique crystal structure and characteristic chemistry in order to be considered a mineral species (Chapter 2). However, mineral characterization is rarely undertaken for the purpose of classifying (i.e., naming) minerals alone. The majority of chemical analyses of minerals are made with a goal of characterizing cation quantities and distributions within individual mineral grains and among different phases in thin sections.

We have already discussed mineral classification, and will return again to this topic in great detail in Chapter 21. As with characterization, minerals are also classified using both composition and structure, but the end result is different. *Classification* is first based on composition, usually in terms of the main anion or anionic complex. For example, the sulfides all contain S^{2-} as their main anion, and the anionic complex $(SiO_4)^{4-}$ is diagnostic of all the silicate minerals. Classification also uses mineral structure. For example, in the silicates, we classify on the basis of the amount of polymerization (linkages) among the $(SiO_4)^{4-}$ tetrahedra. Thus proper classification depends on knowing enough of a mineral's characteristics to place it in its proper category. Less information is required to assign a group name (e.g., plagioclase) than is needed to determine a species name (e.g., albite).

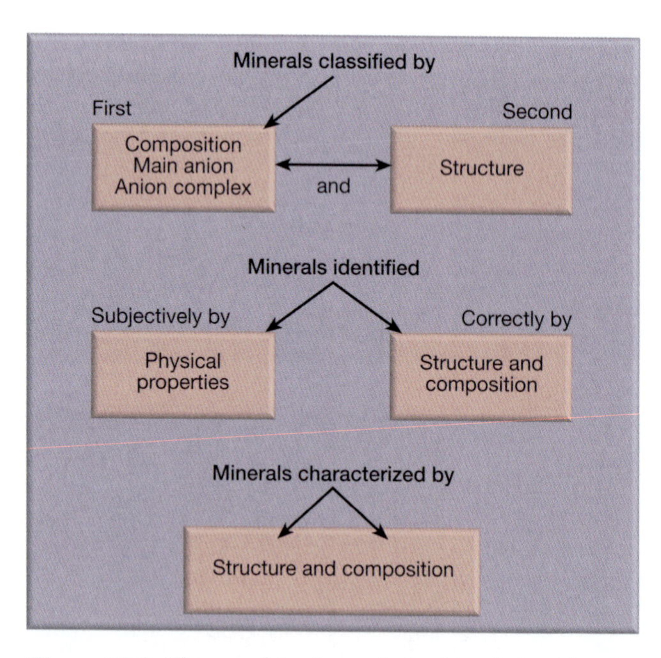

Figure 19.3. *This graphic shows that mineral classification is based first on composition and second on structure, while mineral characterization uses both composition and structure equally. Because the physical properties of a mineral are related to its composition and structure, we can use physical properties to aid in its identification. However, identification based on physical properties can be subjective, while minerals can always be correctly identified by determining composition and structure.*

The Process of Identification

In this book we deliberately do not provide a flow chart for mineral identification, as is typically done in a physical geology lab manual. These flow charts are nothing more than training tools to force the users to evaluate a series of criteria such as hardness, color, etc. that will point them to making a selection from a limited list of minerals. In the real life of a mineralogist, there is no flow chart, because a different process (method) for identification is needed in every situation, based upon the type and size of the sample, the tools available, and the level of identification needed (e.g., is it a pyroxene or an enstatite?).

In Figures 19.1 and 19.2, we saw the range from large hand samples to microscopic samples. Regardless of the size of the crystals, the first step in mineral identification is always a visual examination, followed by inspection at higher magnifications using a hand lens, a low-magnification binocular dissecting microscope, a polarizing light microscope, or maybe a scanning electron microscope. Often the distinctive shape of a mineral (e.g., a garnet dodecahedron) will exist over many different scales, so that magnification may help in identification.

For larger samples, we often resort to some of the same types of characterization used in introductory geology classes: hardness, acid test, streak, etc. However, some (most) samples are too small to apply such methods (Figure 19.4). They will require use of immersion mounts and a polarizing light microscope to diagnose their structures (e.g., optical class, as discussed in Chapter 17) and chemistries (based on refractive indices, Chapter 18). The PLM thus provides information about both structure and chemistry of a single particle, albeit indirectly. Direct measurements of a mineral's composition and/or structure could rely on, for example, an SEM equipped with an energy-dispersive X-ray spectrometer (Chapter 9) or an X-ray diffractometer (Chapter 15). This set of analytical instrumentation will arrive at a correct mineral *group* identification for almost any mineral imaginable. However, as we'll see later in this chapter, assigning mineral *species* names within a solid-solution series can be difficult, especially for silicate minerals, and may require more precise chemical data from an electron microprobe.

A summary of the different methods we use to identify minerals is presented graphically in

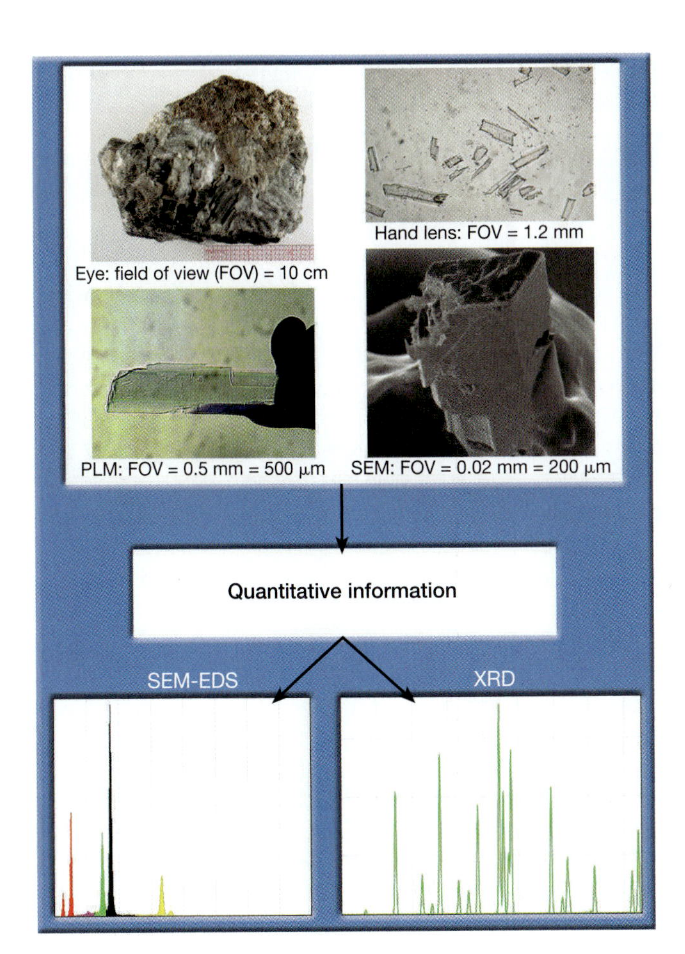

Figure 19.4. Different processes used to identify an unknown mineral. At hand sample scale, it sometimes can be easy to identify a mineral based on macroscopic properties. Confirming that identification may require use of other tools available to a mineralogist, such as magnification by a hand lens, microscope, or scanning electron microscope. In the PLM, we can indirectly measure composition and structure, whereas with an SEM we can obtain information about chemistry. When combined with diffraction data, these techniques allow positive identifications to be made.

Figure 19.5. The correct identification of a mineral would be at the very center of the diagram (the bull's eye!); distance from that center relates to how close you are to that correct identification. The biggest circle represents the physical properties of that mineral, and the inner circles indicate various types of information that could be collected to obtain a positive ID. For example, there are several minerals that would have the physical property of being gray with a black streak. If the composition of the mineral could be obtained, it would help discriminate among possible mineral species, and get you closer to a correct identification. Similarly, for clear minerals with low birefringence (i.e., almost all the framework silicates), compositional data would considerably narrow the search, but ultimately diffraction data would be needed to confirm a mineral identification because a synthetic material (e.g., a glass or a metal alloy) might be made of any composition.

Our Database

For this textbook, we have spent 15 years developing our own mineral database, with hand sample photos, photomicrographs of thin sections, physical properties, crystal structure animations, and other useful information (Figure 19.6). The majority of the photographs in Chapter 2 that displayed different physical properties were extracted from the database. In general, the database has two specific functions for you: (1) to aid in the identification of minerals by making it possible to use computer-based searches on various mineral properties, (2) to provide a means for you to interact with the crystal structures as well as to show EDS and XRD patterns for each mineral (Figure 19.7).

Because the mineral database was designed to be self-explanatory, we will discuss only a few examples of its use. Figure 19.8 shows a search for a mineral characterized on the basis of its hand sample characteristics: hardness, streak, color, luster, and cleavage. The illustrated search yielded two separate minerals: acanthite and galena. Inspection of other characteristics would then aid you in making a conclusive identification. For example, acanthite has a formula of Ag_2S, while galena is PbS, so a SEM-EDS analysis to look for the peaks caused by Ag vs. Pb would quickly permit an accurate identification.

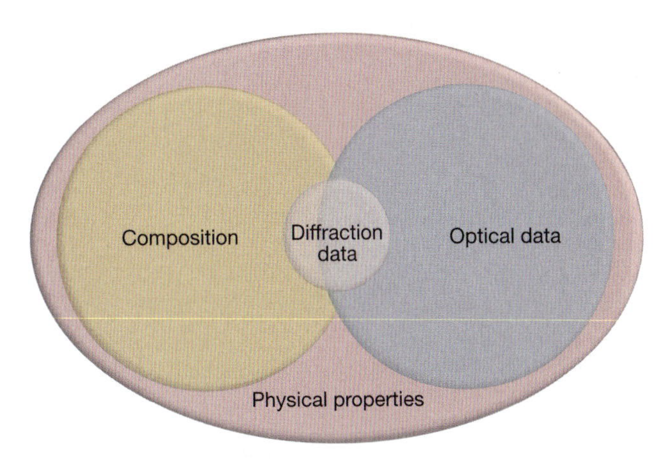

Figure 19.5. Schematic representation of the overlapping relationships among physical properties, composition, optical data, and diffraction data. Both structural and chemical data are usually needed to identify a mineral, while in other situations, a combination of diffraction data and optical observations might be sufficient. The biggest circle represents physical properties, because they provide the least confident identification.

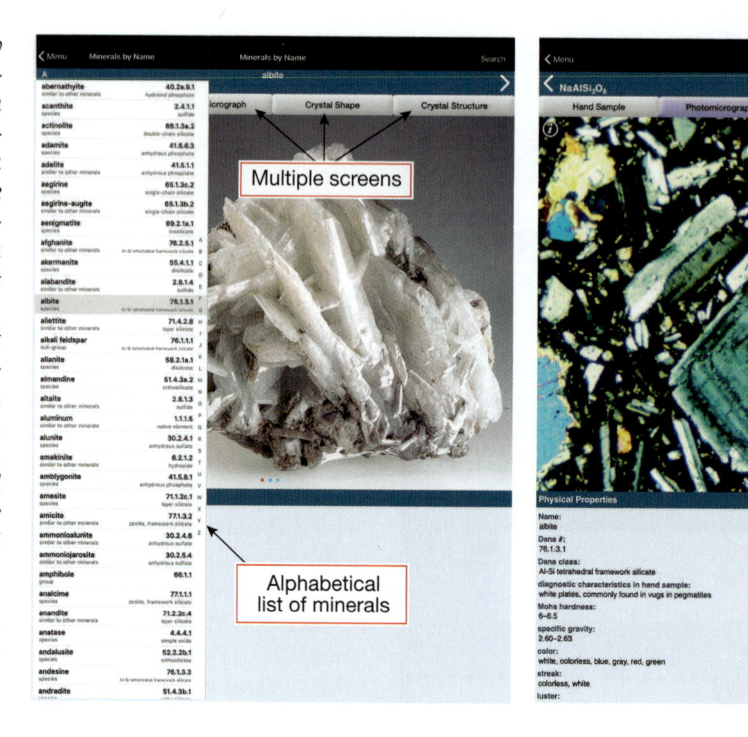

Figure 19.6. Sample view from our mineral database app shown on an iPad showing information on albite. The menu in the upper left can be used for several functions, but is here shown with an alphabetical list of included minerals. There are four different screens that can be selected for each mineral shown at the top: Hand Sample, Photomicrographs, Crystal Shape, and Crystal Structure. For each screen the Physical Properties, Optical Properties, and Classification and Occurrence are shown at the bottom and can be scrolled for details. At the top right of the screen is a button labeled "Search" that allow minerals to be found in the database based on of specific physical properties; this function and ohers can also be accessed from the Menu in the upper left. Note Photomicrographs can be viewed in both plane (PPL) and cross-polarized (XPL) light with the button in the upper right. Likewise there are two buttons on the Crystal Structure page, XRD (upper left) and EDS (upper right) that show a calculated powder X-ray diffraction and a simulated EDS spectra for each mineral.

Figure 19.9 shows a mineral that has been studied in thin section. Careful microscope work has determined that the phase is uniaxial positive with a low birefringence between 0.008 and 0.011. A search on these data yields five separate possibilities. Of them, quartz is the most common, and thus might be the most logical choice. To confirm this identification, the relief of the grain in thin section could be checked, or an immersion mount could be made to check the refractive index, because the other four minerals have significantly higher refractive indices than quartz. This process of mineral identification involves many different paths, and can make use of chemical, physical, or optical properties.

Other Databases

Our database app, discussed above, is a subset of ~10% of all the known minerals. However, we have weeded out the ones you are less likely to

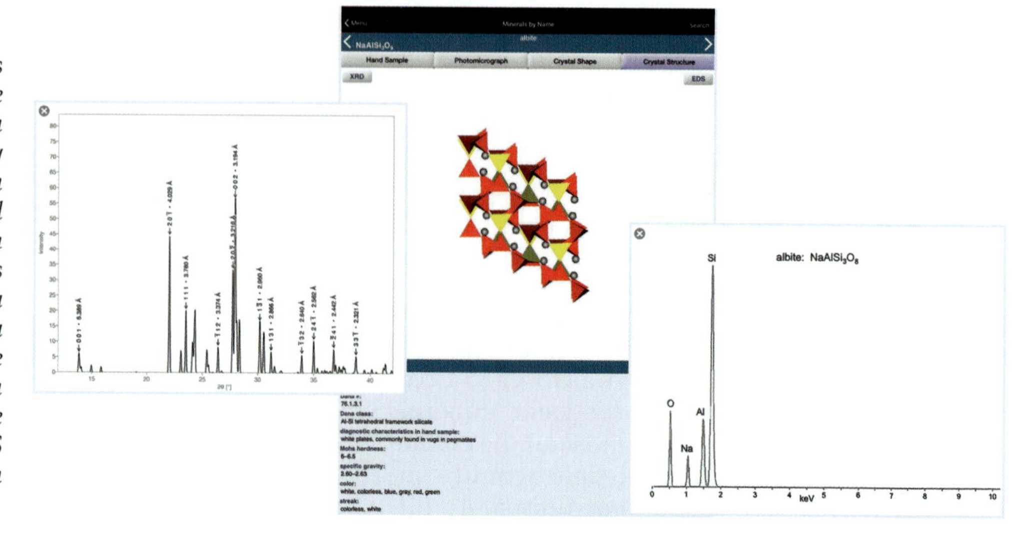

Figure 19.7. Shown here is the Crystal Structure page for albite. Note you can rotate the structure just by touching it. Also, shown (on the left) is a calculated X-ray powder diffraction pattern with the peaks labeled and (on the right) a simulated EDS spectra with the peaks labeled. The X-ray pattern can be seen by pressing XRD in the upper left, while the EDS is seen by pressing EDS in the upper right.

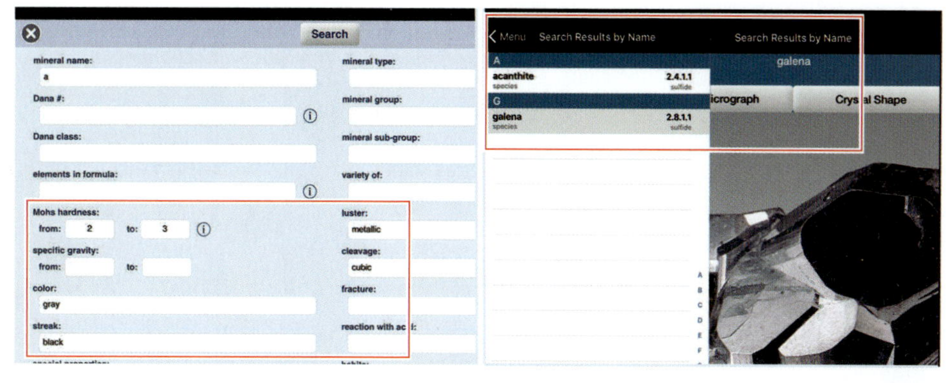

Figure 19.8. *Illustration of the Search capability in the database, using hand sample characteristics. The Search page is shown on the left with the results page on the right. In this example, hardness was determined to be between 2 and 3, the mineral was gray in appearance with a black streak, and it exhibited metallic luster with a cubic cleavage. The search resulted in the two possible minerals shown at right, acanthite and galena; their listings can then be inspected for further identifying characteristics. The more search fields you fill in, the better your chance of narrowing down the list of resultant possible minerals.*

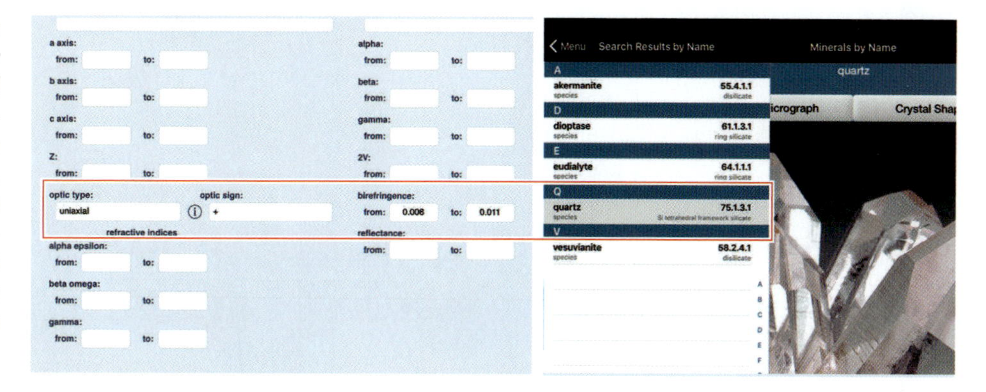

Figure 19.9. *Optical data obtained from a thin section (or grain mount) can also be used as search criteria in our mineral database. Here the unknown mineral is uniaxial positive, with birefringence in the range of 0.008–0.011. The search produces five possible candidates for identification, but quartz is the most likely because it is so common.*

encounter, to make it easier to find matches. We did not want to overwhelm you with too much information! Furthermore, the data tabulated in each listing are far from complete; for example, the data used to identify each mineral species on the basis of its powder or single-crystal diffraction pattern are not included to conserve space. However, such information is readily available to you from on-line sources or other well-developed databases. Please keep in mind that the goal of our database is to provide a foundation for you so you can move on to using other more specialized databases as needed.

Table 19.1 lists several other databases that are commonly used to identify minerals, with brief descriptions of each. When we took our mineralogy courses (back in the dark ages), we used printouts and other types of tables to search on mineral properties by hand. It was cumbersome, confusing, and not very definitive. With the advent of easily-accessible computing power, the majority of mineral searches are now computerized. This is in contrast to most introductory geology courses, where students are still asked to use tables in a book. These are too limiting and simplistic for real practical use beyond an introductory classroom!

Many attempts have been made over the years to devise ways to identify minerals based on optical properties, because they are relatively easy to obtain and quantify (i.e., the values of refractive index are quantities rather than descriptive terms). One such example that is truly useful is shown in Figure 19.10, which is a plot of refractive index vs. birefringence originally conceived by Tröger (1979). These data can often be estimated in thin sections and quantified by using immersion

Table 19.1.
Some example mineral databases and their websites

- ICSD (inorganic crystal structure database) www.fiz-karlsruhe.de/
- ICDD (International Centre for diffraction data) www.icdd.com/
- MinIdent (www.micronex.ca)
- Mindat.org (www.mindat.org/index.php)
- www.webmineral.com
- American Mineralogist Crystal Structure Database (www.minsocam.org)
- "Mineral Identification Key II" (www.minsocam.org/msa/collectors_corner/id/mineral_id_keyi1.htm)
- The Particle Atlas on CD (www.mcri.org)
- McCrone Atlas of Microscopic Particles on Web (www.mccroneatlas.com)
- Handbook of Mineralogy (www.handbookofmineralogy.org)

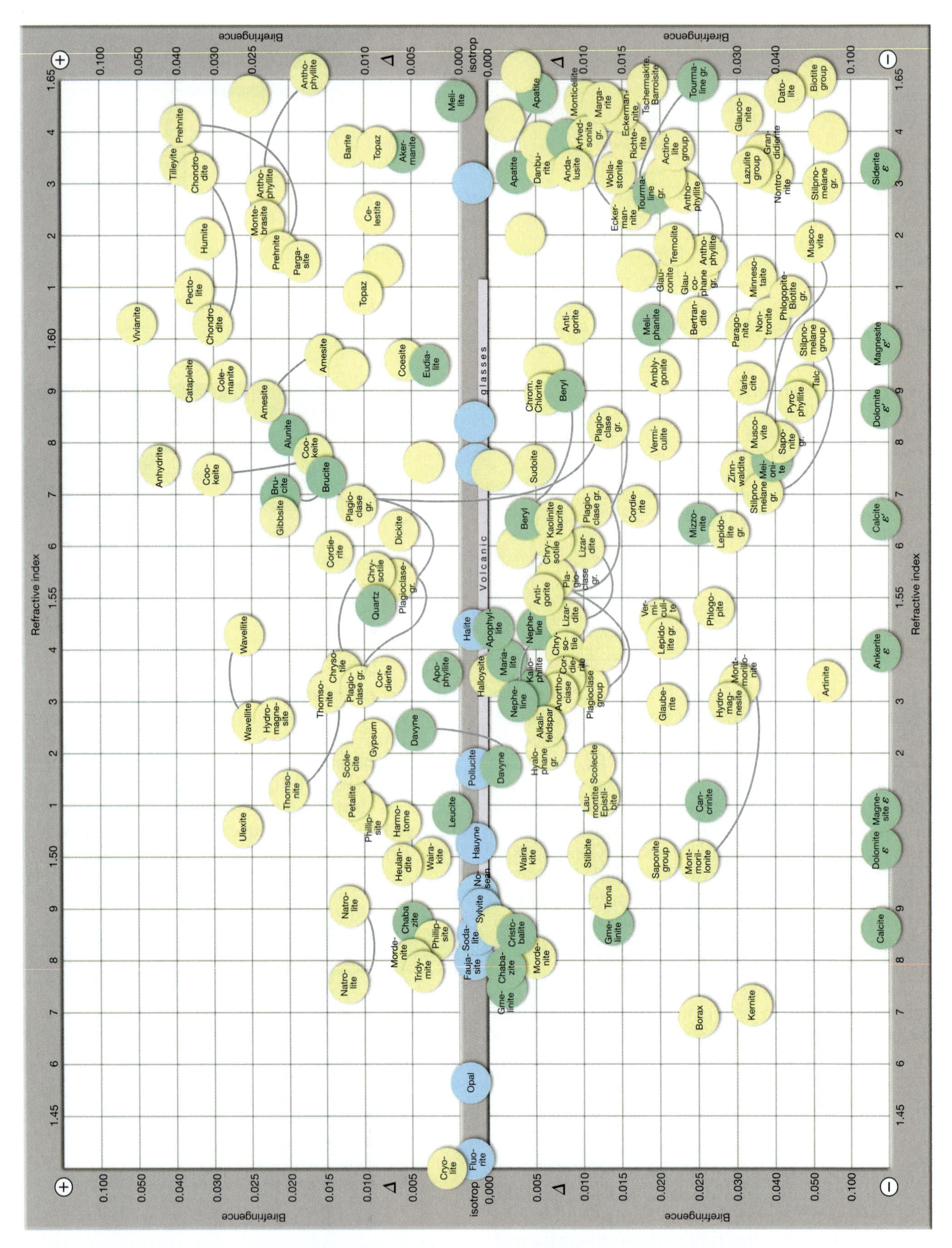

Figure 19.10. Graphical representation of the relationship between refractive index (x axis) vs. birefringence (y axis), with common rock-forming minerals plotted. At the center of the y axis (zero retardation), are the isotropic minerals. The upper half of the plot shows optically-positive minerals, and the bottom half shows optically-negative minerals. ***a.*** *Range of refractive index from 1.44–1.65.*

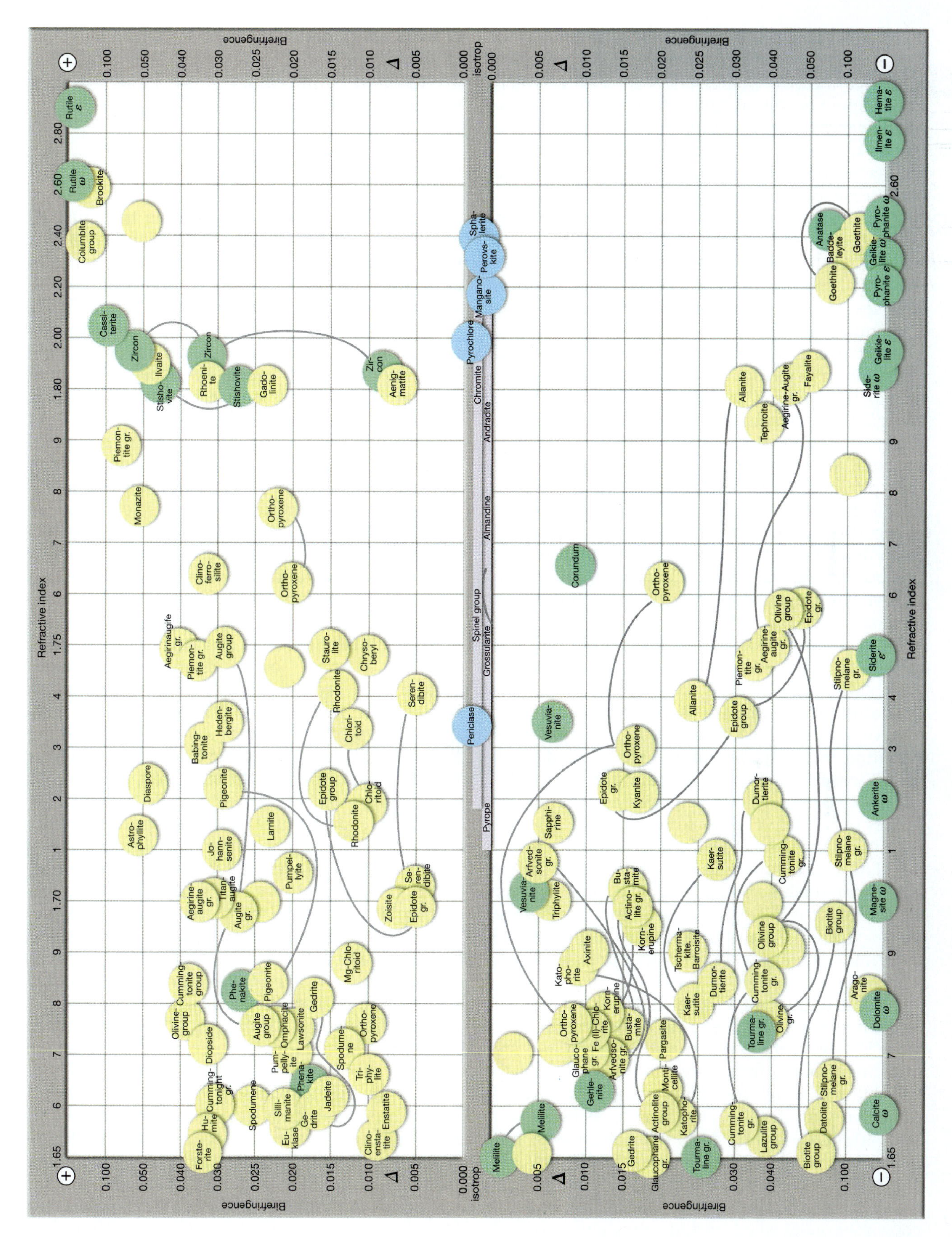

Figure 19.10. b. Range from 1.65 to 2.90. Isotropic minerals and glasses will plot along the y = 0 *axis. Glasses have a range of refractive index because they have different chemistries, as do the garnets. Many minerals, such as quartz, plot as single points, while others plot as ranges indicated by ellipses. (Figure adopted from Tröger, 1979.)*

methods. Many minerals, such as fluorite and halite, are easily distinguished on the basis of their refractive indices. Quartz and corundum, which have almost no cation substitutions and therefore have nearly constant compositions, plot as single points on the graph. Minerals that are part of solid-solution series present a greater challenge for precise identification based on optical properties. Figure 19.10 shows that mineral groups such as pyroxene and amphibole cover fairly large regions, with low refractive indices at their Mg-rich ends, and higher refractive indices for Fe-rich samples. However, the range of refractive indices for pyroxenes is higher than that for amphiboles. The same line of reasoning applies to K-Na-Ca substitution in feldspars; the plagioclase feldspars are all higher than the alkali feldspars. Finally, compare the feldspars as a group to the layer silicates. The framework silicates, with bonding similar in all directions, have low birefringence, and thus plot near the center of the chart. The micas, with significantly different bonding parallel and perpendicular to the sheets, have higher birefringence and are optically negative, so they plot at the bottom of the chart. Carbonates, with the largest birefringence, plot at the very bottom of the chart.

On a more detailed level, Figure 19.11 shows refractive index values for several amphiboles with similar chemistries, mainly the tremolite-actinolite-ferro-actinolite solid-solution series. The very upper portion of the figure shows the variation in refractive index for α, β, and γ, as the Fe content increases to the right across the solid-solution series. Also shown are individual ranges of refractive index for each of the mineral species, along with idealized end-member values. This chart shows graphically how difficult it is to distinguish between a high-Fe tremolite and a low-Fe actinolite! For these minerals, some method other than optics would be needed to determine chemistry and obtain a precise identification.

At the bottom of Figure 19.11, the range of refractive indices for the amphibole mineral richterite is shown; these values overlap with and are indistinguishable from those for tremolite. Such variability and overlap among these properties can be very frustrating! For instance, as F^{1-} substitutes for $(OH)^{1-}$, refractive index decreases. For richterite, the substitution of Na^{1+} and K^{1+} for Ca^{2+} also affects the refractive index. These are details that you need to be aware of, but at this stage in your mineralogical education, you should not feel burdened by them! This graph provides a detailed example for the amphiboles to illustrate the variability of physical properties that can occur in a solid-solution series, but we could easily have chosen a different mineral group. The point is that it will often be important to use more than one technique to identify a mineral.

Strategies

So far in this chapter, we've shown that mineral identification often begins with simple tasks if your samples are well-formed museum samples and rapidly becomes a more complicated process when sample size and quality decrease. Study with a polarized light microscope is the simplest method to use because it provides information on both structure and chemistry (albeit somewhat indirectly). It is also possible to use diffraction methods (most commonly, powder XRD) and compositional data (usually an SEM equipped with EDS) to identify minerals. Many geology departments will have one or both of these instruments, which are relatively simple to use in an automated fashion. So in this section, we will show examples of simulated powder X-ray diffraction patterns and their associated simulated and real EDS patterns (Figures 19.12–19.19), making the point that integration of these methods is often useful for mineral identification. We have selected a range of representative minerals, including feldspars, silica, layer silicates, amphiboles, and pyroxenes, to illustrate this process. The simulated EDS spectra were provided by Bryan Bandli, MVA Scientific Consultants, and the real data were obtained from Lowers and Meeker (2005) from particles of World Trade Center dust.

We have previously discussed powder X-ray diffraction at length in Chapter 15, and EDS data in Chapter 9. However, there are a few points worth reiterating here. The X-ray diffraction patterns plot 2θ values, which relate to the d-spacings of the minerals. In other words, the position of each peak (the x axis on these plots) relates to an interplanar spacing in the mineral. In the EDS spectra, the position of the peak on the x axis represents the energy associated with a transition in some individual atom in the mineral. Because we are only plotting $K\alpha$ transitions, higher keV values (to the right) represent higher atomic numbers. Thus, peak positions in both the XRD and EDS spectra are well-defined. In the XRD patterns, the height of each peak represents the intensity of the scattered electrons. This varies if preferred orientation of the mineral is present (see discussion at the end of Chapter 15 and in Figure 15.38). For certain minerals such as the layer silicates, the (001) diffraction lines are always the most intense. So, comparisons using

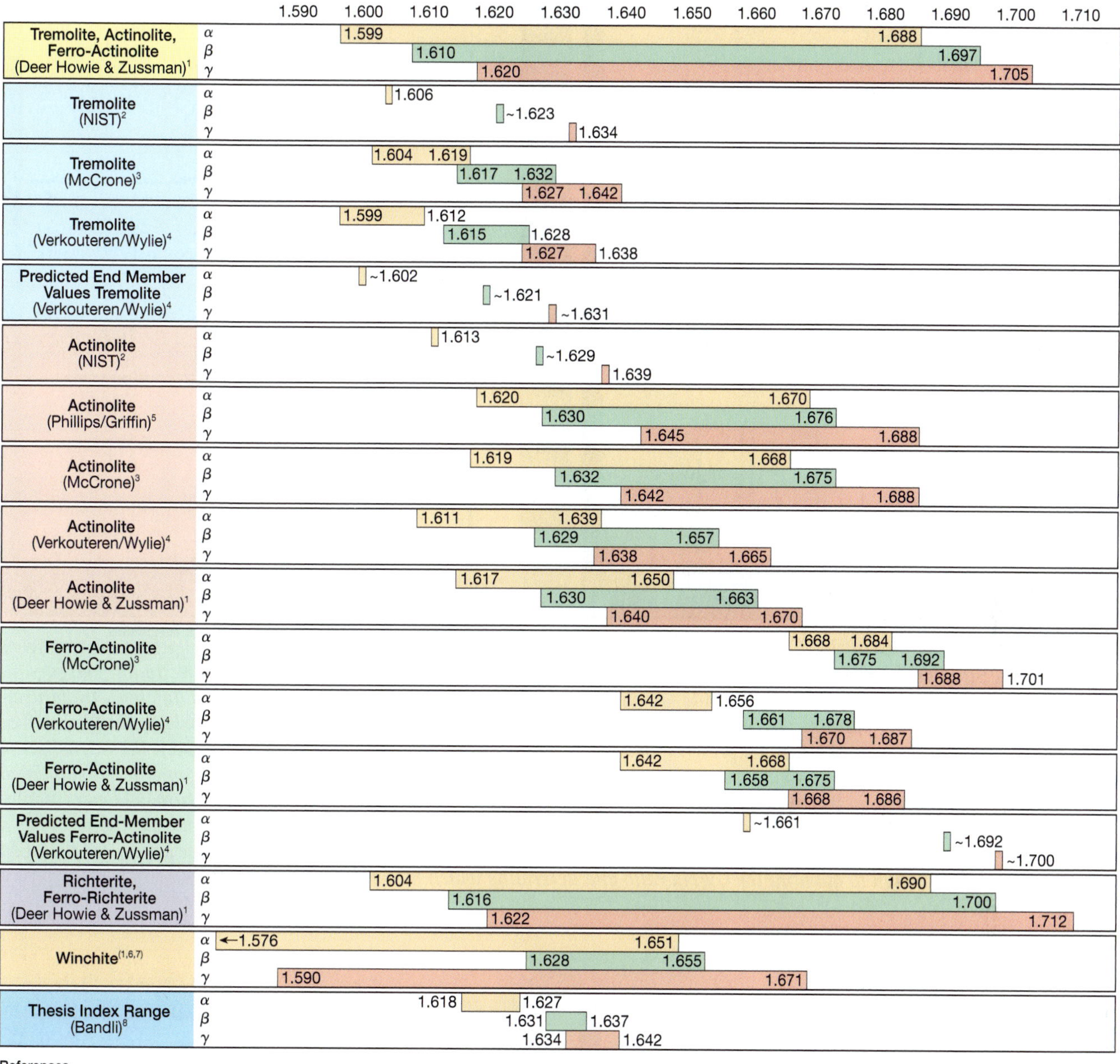

References

1. Deer, W.A.; Howie, R.A.; Zussman, J.; *Rock-Forming Minerals, Vol. 2B, Double-chain Silicates, 2nd Edition*; The Geological Society: London, United Kingdom, 1997.
2. NIST; Standard Reference Material 1867 Bulk Asbestos-Uncommon; U.S. Department of Commerce; National Institute of Standards and Technology; Gaithersburg, MD 20899.
3. McCrone, W.C.; *Asbestos Identification, 2nd Edition*, McCrone Research Institute, Chicago, Illinois, 1987.
4. Verkouteren, J.R.; Wylie, A.G.; "The tremolite-actinolite-ferro-actinolite series: Systematic relationships among cell parameters, composition, optical properties, and habit, and evidence of discontinuities; *American Mineralogist*, 2000, Vol. 85, 1239–1254.
5. Phillips, W.R.; Griffin, D.T.; *Optical Mineralogy*; The NonOpaque Minerals; WH Freeman and Company, 1981.
6. Wylie, A.G.; Verkouteren, J.R.; "Amphibole Asbestos from Libby, Montana: Aspects of Nomenclature"; *American Mineralogist* 2000, **85**, 1540–1542.
7. Nayak, V.K.; Leake, B.E.; "On Winchite' from the Original Locality at Kajlidongri, India"; *Mineralogical Magazine* 1975, 40, 395–399.
8. Bandli, B.R.; Characterization of Amphibole and Amphibole-Asbestos from the Former Vermiculite Mine at Libby Montana, University of Idaho, MS Thesis, 2002.

Figure 19.11. *Refractive index variations for different amphibole species from sources in the published literature, with yellow representing α, green β, and pink γ. The uppermost entry is for the tremolite-actinolite-ferro-actinolite series, followed by four separate ranges for tremolite, five for actinolite, four for ferro-actinolite, and one entry each for ferro-richterite, winchite, and richterite. Refractive indices of the different species often cover ranges as shown, and while they can often be used to identify a species, there are sometimes too many overlaps. For example, both high Fe tremolite and low Fe actinolite can have the same refractive index. Also, it would be difficult to distinguish winchite and richterite from actinolite and tremolite based on only these data. (Modified from Millette and Bandli, 2005.)*

peak intensities from powder XRD graphs should be used with caution. Other factors can also affect XRD patterns. If the sample is placed incorrectly in the diffractometer, such that the powder's surface is too high or too low, the entire diffraction pattern can shift to the right or left. Even though

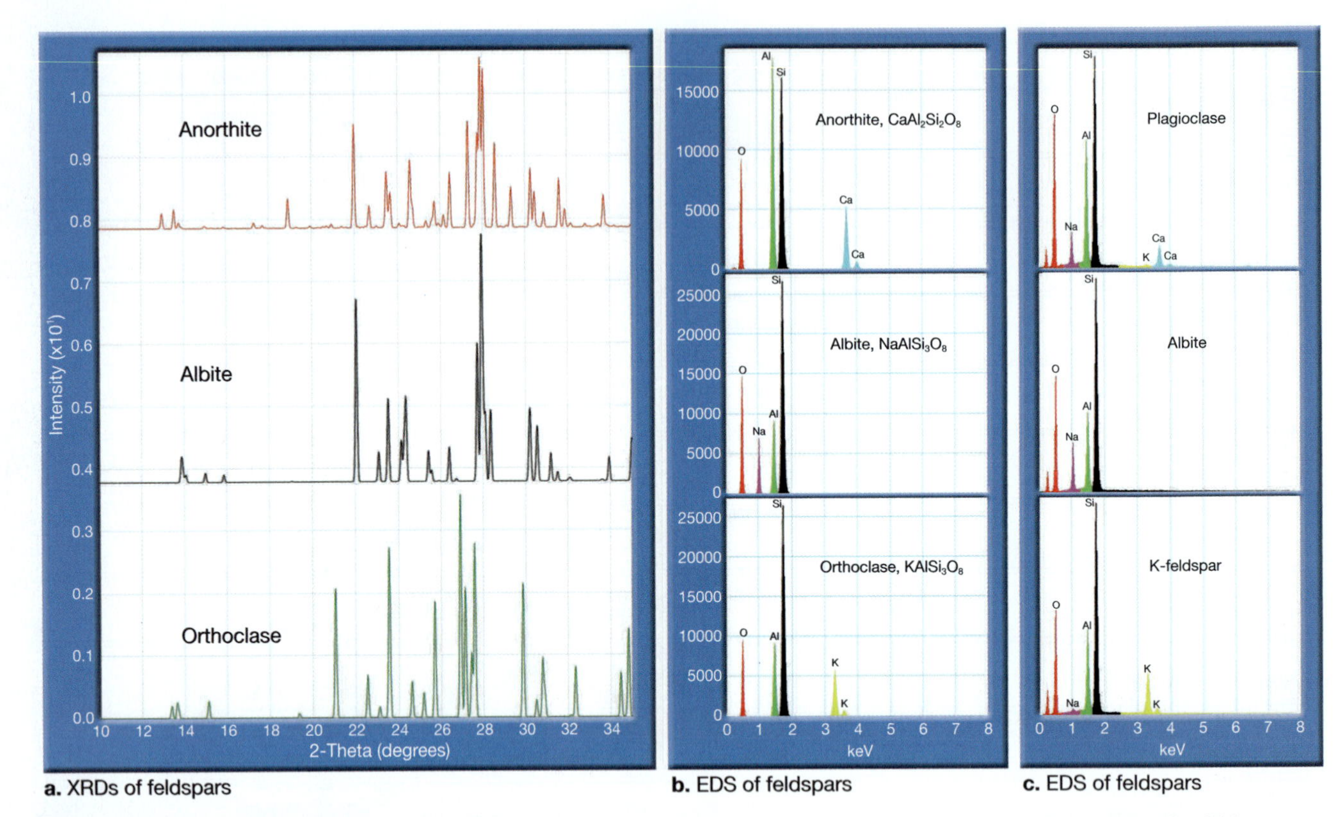

Figure 19.12. *Simulated powder X-ray diffraction patterns, along with simulated and observed EDS spectra for feldspars.* ***a.*** *XRD patterns of orthoclase, albite, and anorthite are all somewhat similar, though orthoclase is slightly different. In naturally-occurring samples, these patterns are even harder to tell apart.* ***b.*** *Energy-dispersive spectra (EDS) clearly distinguish orthoclase, albite, and anorthite based upon the presence of lines arising from K, Na, and Ca. The ratios between the intensities of the Si and Al peaks are similar for albite and orthoclase, but different in anorthite due to its higher Al content. All simulated EDS spectra were provided by Bryan Bandli, MVA Scientific Consultants.* ***c.*** *EDS spectra from three natural feldspar samples that are analogous to those shown in Figure 19.12b (modified from Lowers and Meeker, 2005). The K-feldspar (bottom) has a slight amount of Na. The middle sample is an albite. The top sample belongs somewhere in the plagioclase series, but the SEM data are not precise enough to quantify the Ca/Na ratio and determine which mineral species is present.*

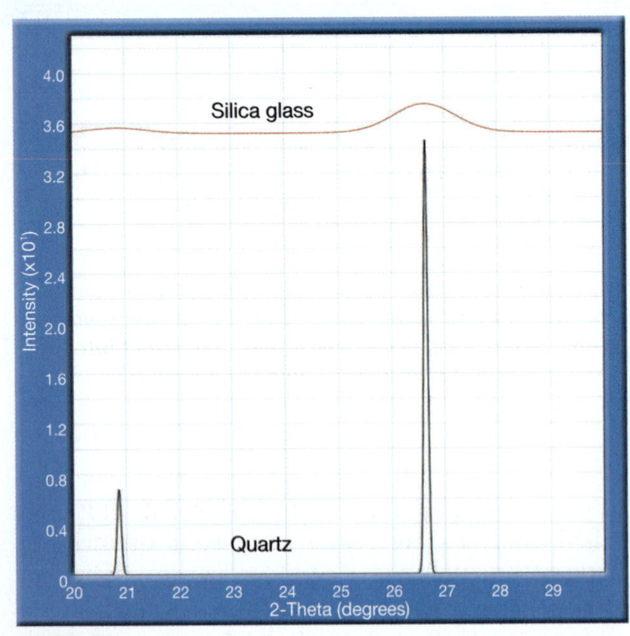

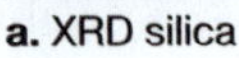

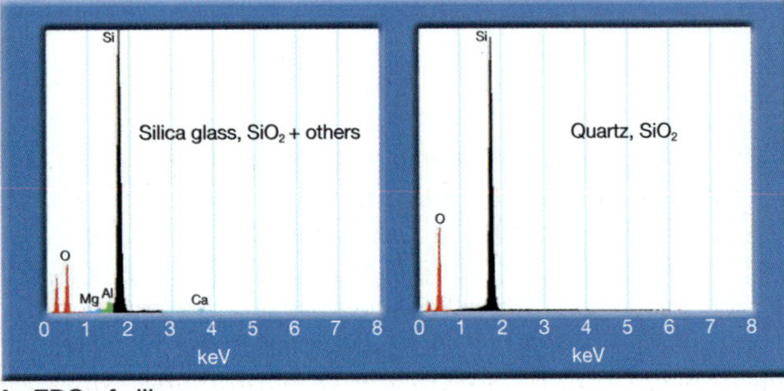

b. EDS of silica

Figure 19.13. *Simulated powder X-ray diffraction patterns and observed EDS spectra for* ***a.*** *crystalline (quartz) and* ***b.*** *amorphous (silica glass). The diffraction patterns show sharp peaks for quartz and broad features for glass. Both EDS spectra have Si and O lines. Quartz has only these two lines, but the glass shows small impurities such as Mg and Al. (EDS spectra modified from Lowers and Meeker, 2005.)*

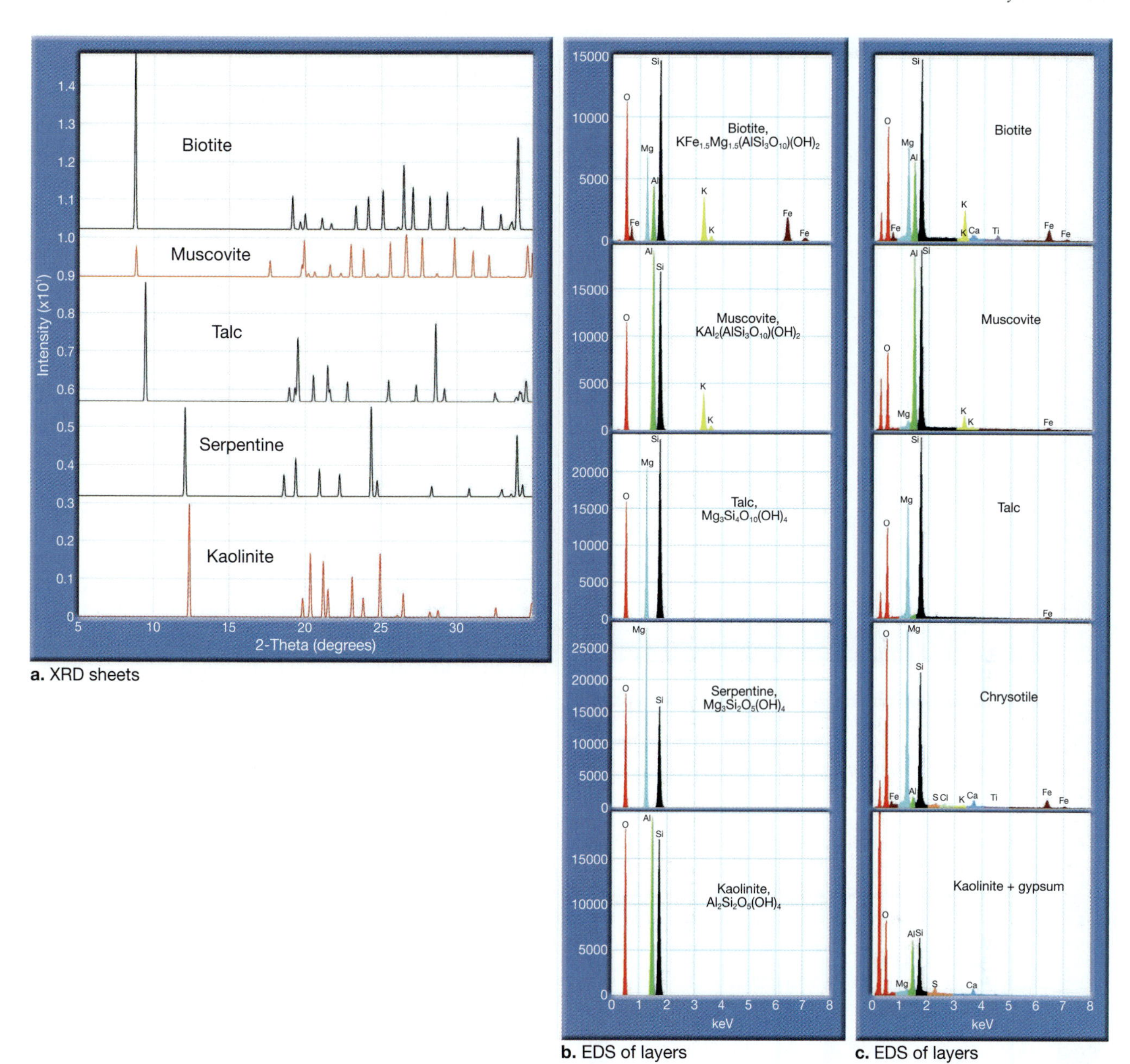

Figure 19.14. *Simulated powder X-ray diffraction patterns along with simulated and observed EDS spectra for five different layer silicates.* ***a.*** *The diffraction patterns of kaolinite and serpentine are similar because they are both 1:1 layer silicates, while the 2:1 micas muscovite and biotite are also similar. Talc, which is a 2:1 sheet silicate lacking an interlayer cation, is different from the other four. In real samples, the major peak occurring in all of these spectra is the one with the lowest 2θ value because of preferred orientation. Thus, it is difficult to distinguish kaolinite from serpentine or muscovite from biotite on the basis of XRD data.* ***b.*** *However, the simulated EDS data can be used to make these distinctions on the basis of the peak corresponding to Mg. The ratio of Mg/Si is slightly different between talc and serpentine, but this should probably not be used as the sole criterion to distinguish between them. Other characteristics, such as morphology, could be used to confirm identification.* ***c.*** *EDS spectra from natural layer silicates that are analogous to those shown in Figure 19.14b. Chrysotile, talc, and muscovite all contain small amounts of Fe, and kaolinite has what was assumed to be a small amount of intermixed gypsum (based on the presence of S in the EDS spectra). (EDS spectra modified from Lowers and Meeker, 2005.)*

biotite and muscovite have slightly different *d*-spacings, it can be difficult to distinguish between them because of this effect.

On the other hand, intensities of EDS peaks relate to the amount of each element present, though this relationship must also be used with caution. In general, the higher the EDS peak, the higher the abundance of that element. However, SEM analyses are really only semi-quantitative because many other factors must be taken into account, such as atomic number, absorption, and fluorescence phenomena (Chapter 9). If the grains

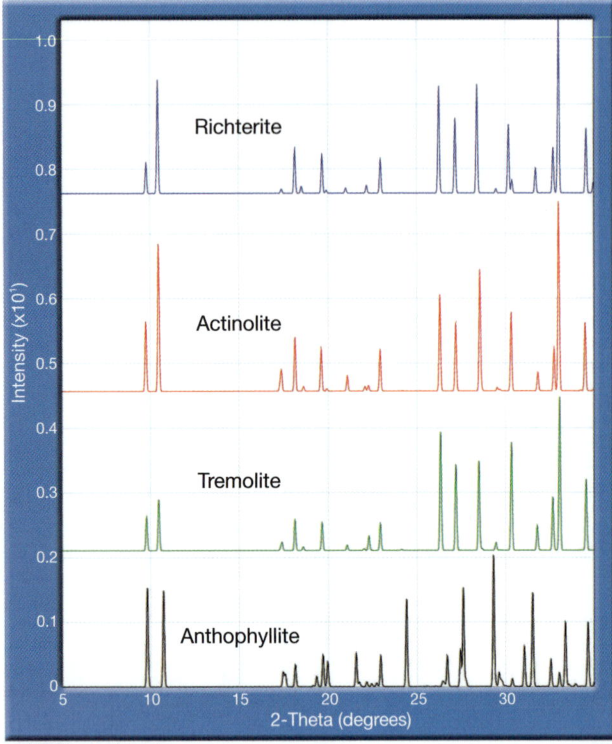

a. XRD amphiboles

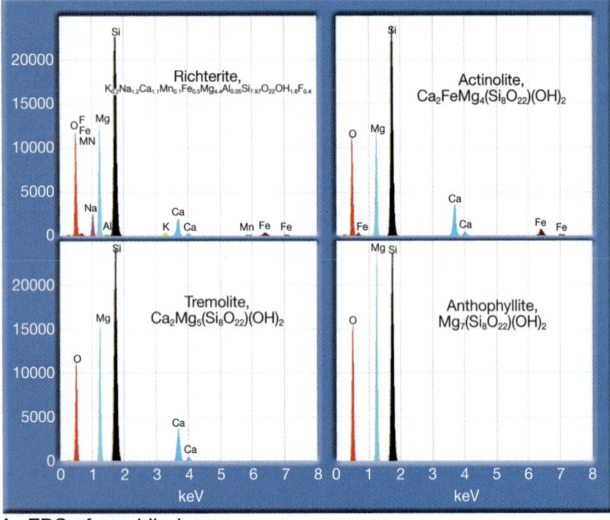

b. EDS of amphiboles

*Figure 19.15. Simulated powder X-ray diffraction patterns along with simulated EDS spectra for four amphiboles. **a.** Peak positions in all four diffraction patterns are similar, with only subtle variations in peak intensity. Anthophyllite is the most distinct pattern because it is orthorhombic. However, it would be very difficult to identify any of these amphiboles solely on the basis of XRD results. **b.** The simulated EDS spectra show a clear distinction between anthophyllite and the other three amphiboles because only Mg is present. Tremolite could also be distinguished from the other three amphiboles based on Ca and Mg content. However, the EDS spectra of an Fe-rich tremolite would appear similar to an Fe-poor actinolite. The richterite spectrum (based on data from Gunter et al., 2003) shows a more complex composition.*

are irregularly shaped or poorly polished, the peak heights will also be affected.

A few overarching trends are apparent from inspection of Figures 19.12–19.19. For minerals within solid-solution series (such as feldspar, amphiboles, and pyroxenes), it is very difficult to distinguish among the different species solely on the basis of a diffraction pattern. Luckily, the EDS spectra allow those distinctions to be made. XRD patterns work very well to distinguish between different mineral groups (Figure 19.18). EDS spectra cannot distinguish among different polymorphs with the same composition, so the pattern labeled "andalusite" in Figure 19.19a could easily be a kyanite, sillimanite, or synthetic glass with that composition. EDS can also be inconclusive when trying to distinguish between minerals of similar chemistry that belong to different mineral groups (Figure 19.19).

It is often useful to identify a mineral using more than one method. If you use a combination of optics, SEM, and diffraction, you can collect data on even an individual particle. Figure 19.20 (from Bandli and Gunter, 2001) shows a particle of an amphibole glued to the end of a needle on a spindle stage (Chapter 18). The sample is imaged in both a PLM and an SEM. An example of this idea is shown in Figure 19.21 (modified from Gunter et al., 2007). The purpose of this study was to identify acicular crystals that could be amphibole, serpentine, or pyroxene. In Figure 19.21a, the particle is shown in a matching refractive index oil (n_{oil} = 1.635), which identifies it to be amphibole (not serpentine or pyroxene). In Figure 19.21b, the same sample is shown in an SEM, with its EDS spectra given in Figure 19.21c. The chemical data confirms that the sample belongs to the tremolite-actinolite series and is not a member of the anthophyllite series. The EDS spectra does not allow a distinction to be made between tremolite and actinolite. Gunter et al. (2007) did do more precise chemical work to determine that this sample is indeed actinolite. Because there are well-defined chemical boundaries that separate the different species, precise chemical data (as obtained from polished samples and an electron microprobe with a standardized wavelength-dispersive system or standardized EDS data as obtained from a SEM or TEM) are needed to identify most mineral species in solid solutions.

Another example of mineral species identification of an amphibole is shown in Figure 19.22. In this case, 22 individual amphibole grains obtained from the same location were analyzed using a WDS system (Sanchez, 2007). The points are plotted along with their associated error bars based on repeated analyses on the same mineral grain for each one. Four different amphibole species are

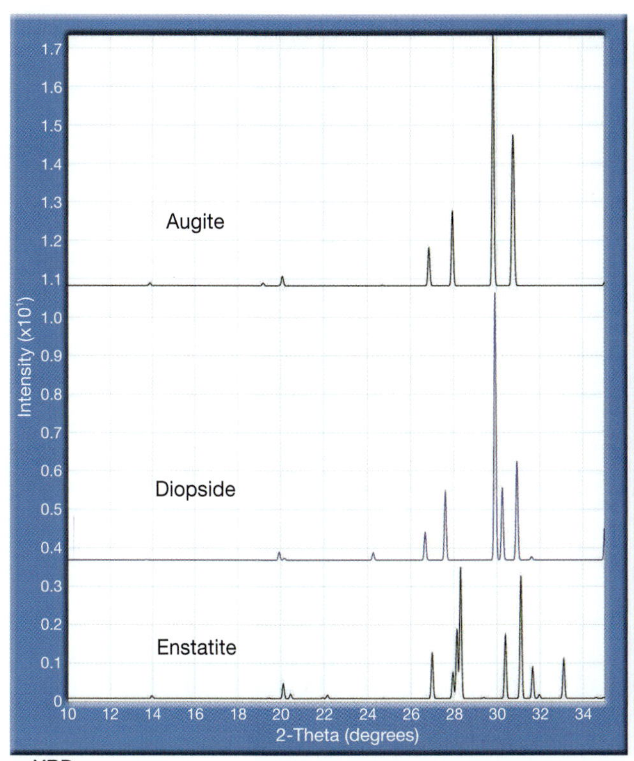

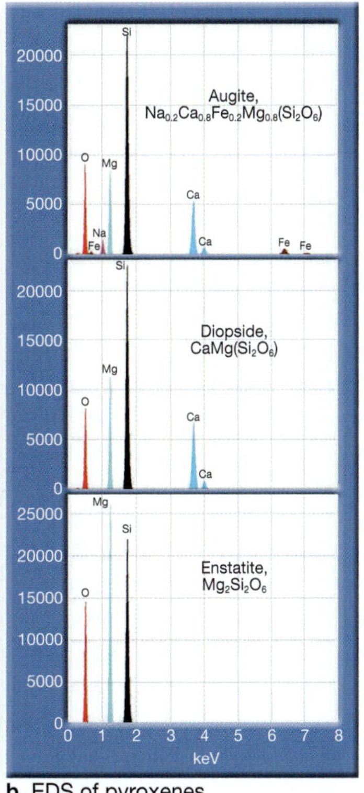

Figure 19.16. Simulated powder X-ray diffraction patterns along with simulated EDS spectra for three pyroxenes. ***a.*** *The enstatite (orthorhombic) diffraction pattern differs from those of the clinopyroxenes diopside and augite, but the two clinopyroxene XRD patterns are basically indistinguishable.* ***b.*** *The EDS spectra allow enstatite to be easily distinguished from the other two pyroxenes because the Ca peak is absent. However, it would still be somewhat difficult to distinguish between the clinopyroxenes based on the EDS spectra; thus, more quantitative chemical analysis would be required.*

defined here. Some of the points fall into the classification "boxes" for more than one mineral species. This is an extreme case that illustrates the difficulty of assigning a species name to a given sample. We mineralogists realize that there are limitations to any classification scheme. In most cases, the name of the mineral is far less important than its compositional and structural characteristics. However, we often use species names to aid in our discussions (and for labels in museums!)

New Minerals

As we've stated many times before, over 4,300 mineral species have been described. Every year, 50 new minerals without names are found. This brings us back to our earlier discussion of the importance of careful characterization of materials. When you undertake to describe a mineral, the first step is to fully analyze its chemistry, structure, and optical properties. With this information in hand, it is usually possible to match the sample to one of the 4,300+ known mineral species. Rarely, these characteristics will not match any existing mineral species, and the sample can be described and submitted for consideration to be a new mineral (see Table 2.1).

As an example of this process, consider the case of pertlikite, a new mineral recently approved by IMA Commission on New Minerals and Mineral Names (no. 2005–055). Its story began with two

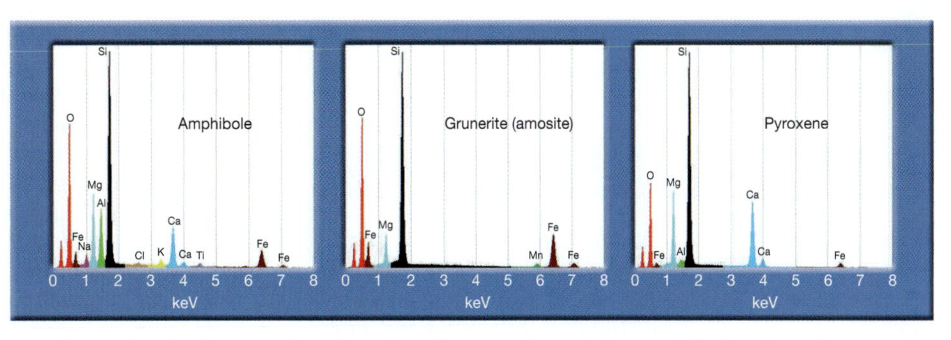

Figure 19.17. EDS spectra for three samples collected from the dust of the World Trade Center (modified from Lowers and Meeker, 2005). The pyroxene spectrum can be distinguished from those of the other two (amphiboles) because it contains too much Ca relative to Si. On the basis of this EDS spectrum and the morphology of the crystals, the middle sample was determined to be amosite, the asbestiform variety of the amphibole mineral grunerite, but no species name could be assigned to the upper pattern until more quantitative compositional data were obtained.

papers by Blaas (1881 and 1883), which gave the chemical composition and morphological aspects of a voltaite (Fe-rich hydrous sulfate) from a pyrite-bearing trachytic eruptive rock from Madeni Zakh, Iran that is enriched in Mg. The hand specimen was found by Andreas Ertl in the collection of the former Mineralogisches Museum der Universität Wien, now Institut für Mineralogie und Kristallographie, University of Vienna, Austria, which was acquired in the year 1886. A photo of a crystal of pertlikite is shown in Figure 19.23. Andreas named the new mineral pertlikite to appropriately honor Franz Pertlik, Professor of Mineralogy and Crystallography, University of Vienna, Austria, for his extensive work on the crystal chemistry of minerals. Pertlikite was approved by the IMA CNMMN and a report on its characterization was written by Ertl et al. (2008). Both of us were involved with this: Darby did Mössbauer analysis to determine its Fe^{2+} and Fe^{3+} contents, and Mickey did its optical characterization.

Along the way, there were some surprises. Pertlikite (an Mg-rich voltaite, with a composition of $K_2{}^{M2}(Fe^{2+},Mg)_2{}^{M3}(Mg,Fe^{3+})_4{}^{M1}Fe^{3+}{}_2Al$

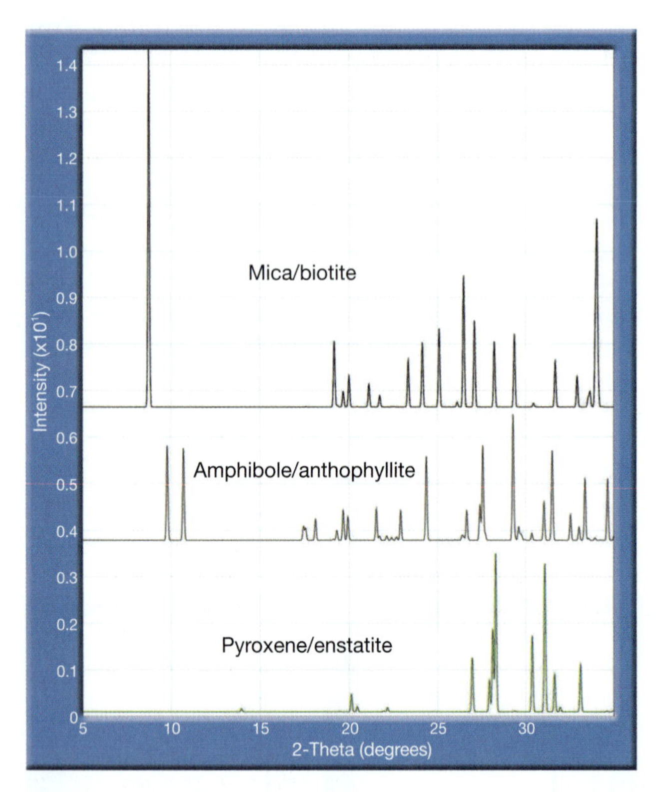

Figure 19.18. *The structural differences among pyroxenes (enstatite), amphiboles (anthophyllite), and micas (biotite) are shown in these three simulated diffraction patterns. Powder XRD can easily distinguish between these three mineral groups, but chemical data are needed to distinguish among members within each of these groups.*

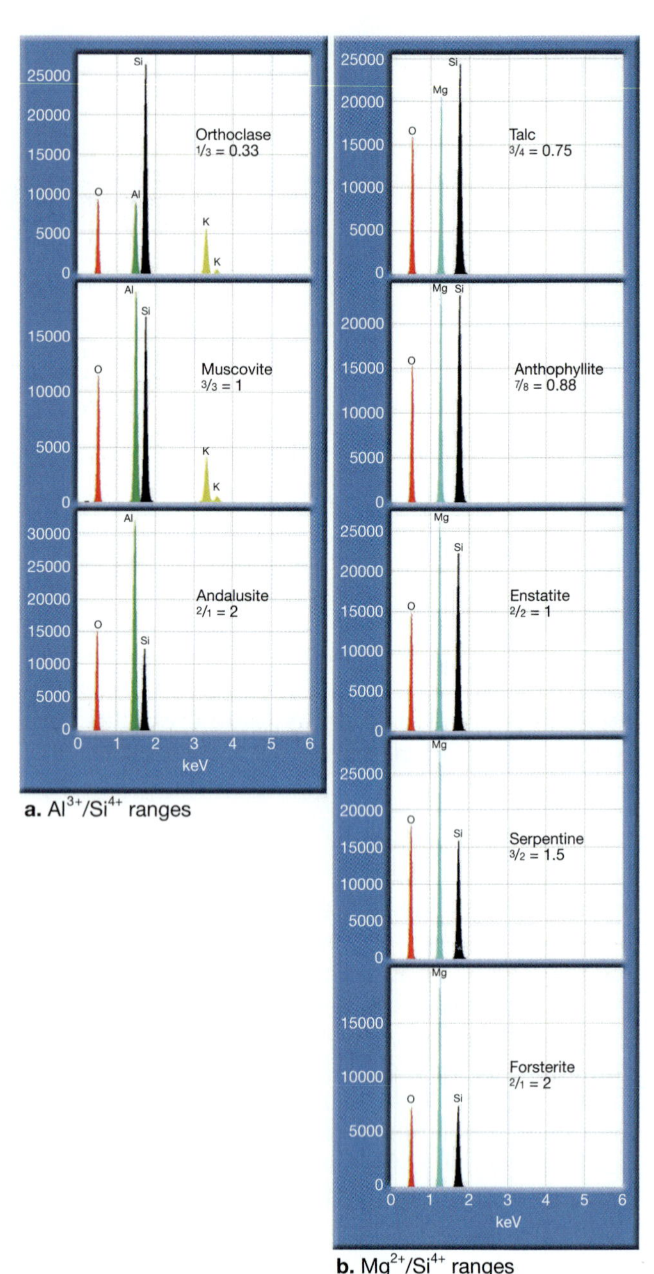

Figure 19.19. *This series of simulated EDS spectra shows the differences in ratios of Al^{3+}/Si^{4+} (Figure 19.19a) and Mg^{2+}/Si^{4+} (Figure 19.19b).* ***a.*** *Al^{3+}/Si^{4+} ratios can be used with care to help distinguish among different minerals such as the orthoclase, muscovite, and andalusite shown here. The spectrum of andalusite would also be identical to those for kyanite and sillimanite because they share the same chemistry, so another criterion such as XRD or refractive index would be needed to distinguish between the three polymorphs.* ***b.*** *These five different minerals containing only Mg, Si, and O (±H) can be distinguished with care on the basis of the ratio between Si and Mg peak intensities. Forsterite could be differentiated from enstatite, as could enstatite from talc, but it would be more difficult to distinguish enstatite from anthophyllite or talc from anthophyllite because of the variability in experimentally-collected data on natural samples.*

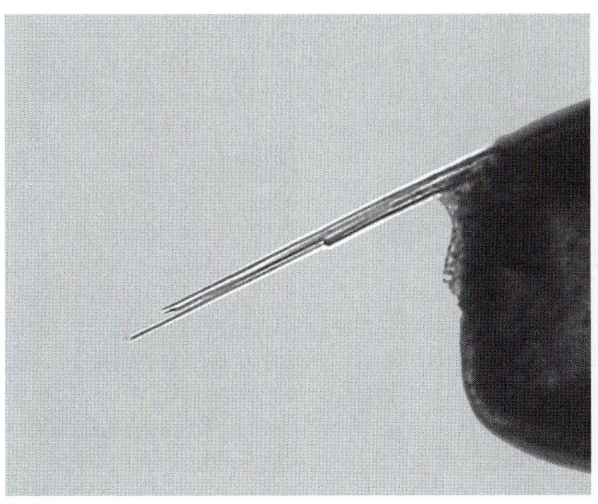

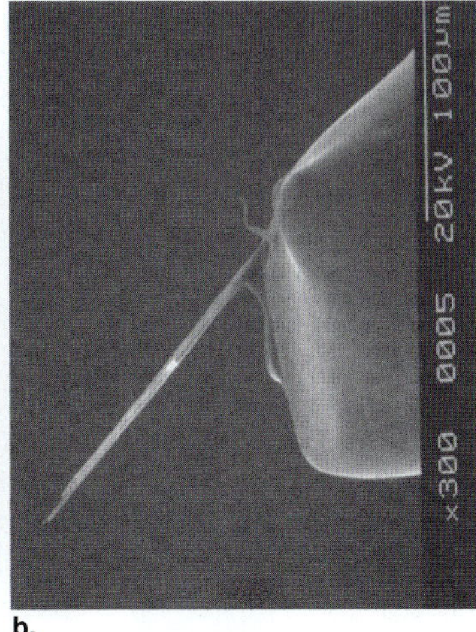

Figure 19.20. *An elongated amphibole particle mounted on the end of a glass fiber, viewed in* ***a.*** *a polarizing light microscope and* ***b.*** *by a scanning electron microscope. We could easily obtain XRD, optical, and chemical data on this crystal to aid in its identification. (Modified from Bandli and Gunter, 2001).*

$(SO_4)_{12}·18\ H_2O$) was expected to be very similar to voltaite, which has a formula of $K_2{}^{M2}(Fe^{2+}{}_5Fe^{3+})^{M1}$ $Fe^{3+}{}_2Al\ (SO_4)_{12}·18\ H_2O$. A single-crystal X-ray diffraction data set was collected on a crystal of pertlikite, and its parameters were refined in the same isometric space group as voltaite. However, that refinement suggested that further examination of the space group was necessary, and when a single crystal was observed in cross-polarized light, it was found to be optically anisotropic (Figure 19.23c). An extinction data set was then collected on this sample. Graphical output from the newest version of EXCALIBRW (see Chapter 18) showed pertlikite to be uniaxial because one of its extinction curves plotted as a great circle. The pole to this great circle is the optic axis (i.e., the ε vibration direction). From this, the optic sign was easily determined by use of an accessory plate, and pertlikite was shown to be optically (–). Clearly, this was not an isometric mineral!

Output from EXCALIBRW was also used to orient both ε and ω to determine their refractive indices. This method found $\varepsilon = 1.586(2)$ and $\omega = 1.590(2)$, so pertlikite exhibits very low birefringence. By using the cell parameters (Table 1, Ertl et al. 2008) and chemical composition (Table 6, Ertl et al. 2008), the density of pertlikite was calculated to be 2.56 g/cm^3. Given the density and the mean refractive index $[(\varepsilon + \omega^2)/\ 3 = 1.590]$, the Gladstone-Dale compatibility index was calculated to be –0.033. This index was used to determine

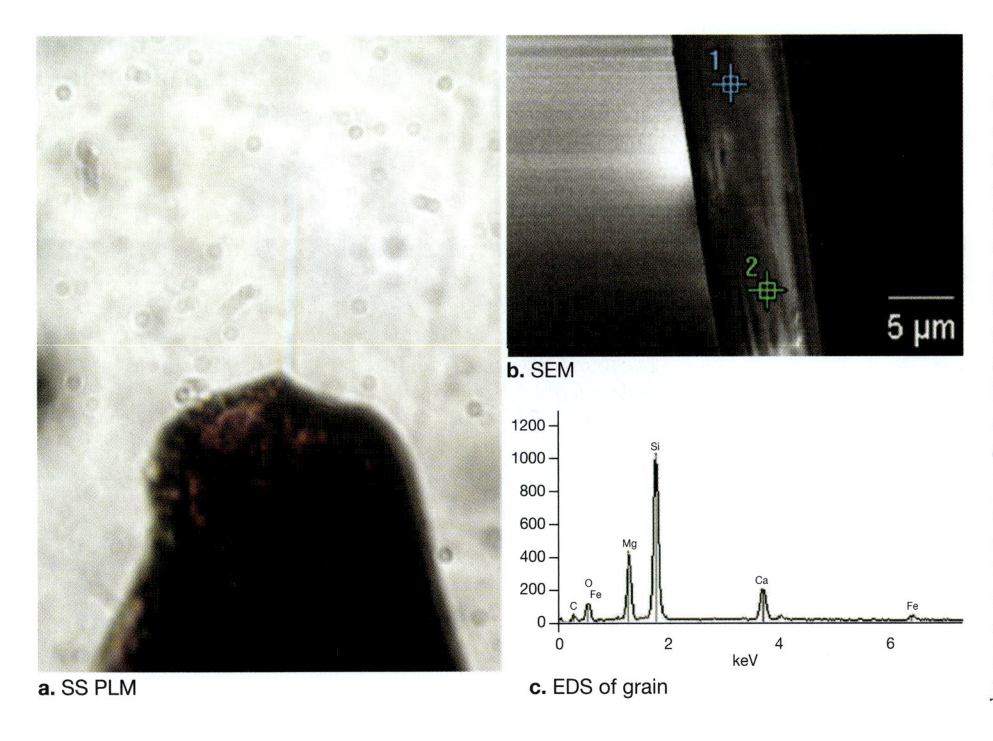

Figure 19.21. *Identification of a ~5 µm wide particle mounted on a fiber similar to the example shown in Figure 19.20. (Images modified from Gunter et al., 2007.)* ***a.*** *The particle is shown in a matching liquid used to determine its refractive indices.* ***b.*** *Higher magnification SEM image of the same particle, with analysis points for the SEM beam indicated by "1" and "2."* ***c.*** *EDS spectrum of the average of analyses from points 1 and 2. On this basis, the particle could be either a pyroxene or an amphibole. However, the refractive indices show that this particle is an amphibole. Choosing a species name between tremolite and actinolite would be difficult.*

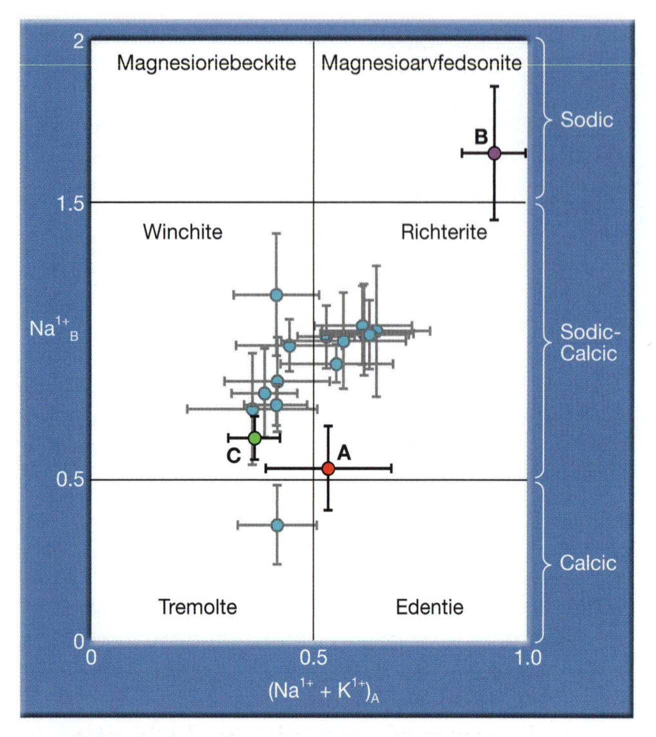

Figure 19.22. In order to actually assign a species name to an amphibole, both precise chemical data and site assignments of the various cations are needed. This plot of Na^{1+} in the B site vs. Na^{1+}+K^{1+} in the A site is used to discriminate among the sub-set of amphibole species for which Si > 7.85 pfu. Samples represented by data points on this plot all come from the same deposit, but they represent four different mineral species. The horizontal and vertical bars associated with each data point represent the analytical errors associated with that data point. Analysis A falls in at least three fields, B falls in two fields, and C falls in only one field. (Modified from Sanchez, 2007.)

the relationship between the measured refractive index and crystal structure.

This example for pertlikite is just an illustration of all the work and care that goes into naming a new mineral species. Armed with the information in this textbook, we are hopeful that you might be inspired to find and characterize a new mineral yourself. If you are, let us know and we'll be happy to help!

Conclusions

We hope that this chapter has clarified the process of mineral identification, and continued the task of integrating the many aspects of mineralogy discussed in this textbook. Our goal here has been to show that it is relatively straightforward to identify a mineral, at least by its group name, and sometimes by its species name, especially if you have access to some of the analytical equipment we've described throughout this book. It is important to realize that the characteristics of a mineral are more important than its species name (despite what your mineralogy professor may tell you). Identification is not really the ultimate objective: characterization is!

The same characterization tools used to identify minerals in geology are also used in other disciplines. For example, the image in Figure 19.24 shows a ~100 μm size particle with a composition similar to that of apatite. This mineral is resting in human lung tissue removed in an autopsy. The particle is so big that it could not have been inhaled, so it must have grown in place, in the lung, as do many other "stones" in the human body (e.g., bladder and kidney stones). As scientists, we want to characterize this particle in every possible way in order to better understand how it formed and its effect on the health of the lung (and the human!) in which it was found. This is a striking example of why characterization can be far more important than identification.

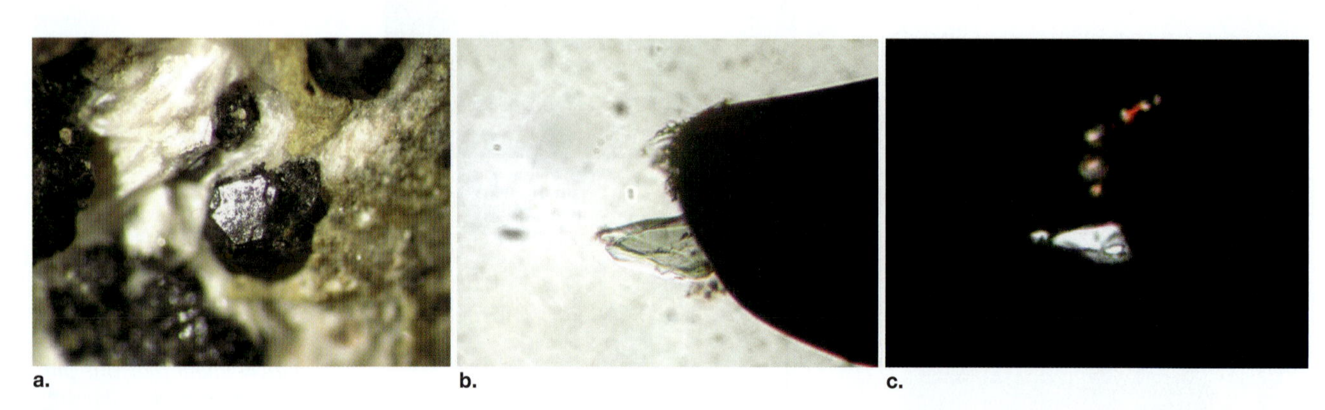

***Figure 19.23.** Photomicrograph of the new mineral pertlikite (Ertl et al., 2008), **a.** a single crystal about 2 mm wide, **b.** in plane-polarized light, and **c.** cross-polarized light; the latter image showing the crystal is non-isometric. Field of view in Figure 19.23b and c is about 300 μm.*

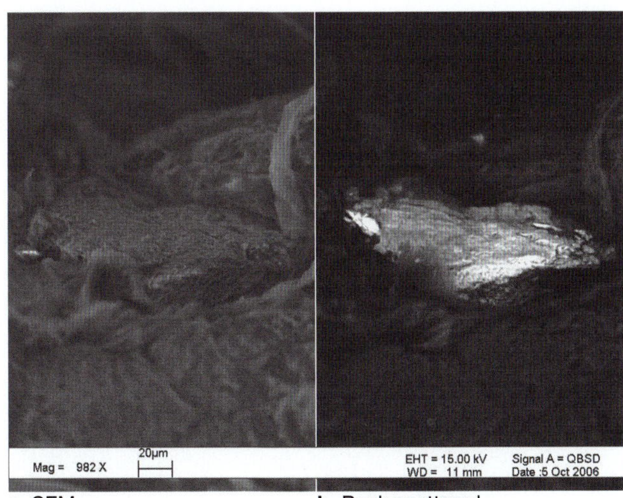

Figure 19.24. a. *A secondary electron image and* ***b.*** *a backscattered electron image of a Ca phosphate material formed within a human lung. In Figure 19.24a, it is difficult to distinguish the mineral from the surrounding lung tissue, but in Figure 19.24b, the particle is clearly visible. Diffraction data would be required to determine if the material is crystalline.*

In closing, we hope you now realize that in real life, there can be no simple flow chart for mineral identification. It has been our goal throughout this entire book to present mineralogy in a systematic manner, showing you the various relationships among composition, structure, and optical properties. Integration of these concepts will always make it easier to identify a mineral, but the various processes of characterization are far too complex to proceed in any linear fashion.

References

Bandli, B.R. and Gunter, M.E. (2001) Identification and characterization of mineral and asbestos particles using the spindle stage and the scanning electron microscope: The Libby, Montana, U.S.A. amphibole-asbestos as an example. The Microscope, 49, 191–199.

Blaas, J. (1881) XVIII. Petrographische Studien an jüngeren Eruptivgesteinen Persiens. Mineralogische und petrographische Mittheilungen, III, 1881, 457–503.

Blaas, J. (1883) Beiträge zur Kenntniss natürlicher wasserhaltiger Doppelsulfate. Sitzungsberichte der Mathematisch-naturwissenschaftlichen Klasse der kaiserlichen Akademie der Wissenschaften, 87, 141–163, Wien.

Ertl, A., Dyar, M.D., Hughes, J.M., Brandstätter, F., Gunter, M.E., Prem, M., and Peterson, R.C. (2008) Pertlikite, a new tetragonal Mg-rich member of the voltaite group from Madeni Zakh, Iran. Canadian Mineralogist, 46, 661–669.

Gunter, M.E., Dyar, M.D., Twamley, B., Foit, F.F., Jr., and Cornelius, S.B. (2003) Composition, $Fe^{3+}/\Sigma Fe$, and crystal structure of non-asbestiform and asbestiform amphiboles from Libby, Montana, U.S.A. American Mineralogist, 88, 1970–1978.

Gunter, M.E., Sanchez, M.S., and Williams, T.J. (2007) Characterization of chrysotile samples for the presence of amphiboles from the Carey Canadian deposit, southeastern Quebec, Canada. Canadian Mineralogist, 45, 263–280.

Lowers, H.A. and Meeker, G.P. (2005) Particle atlas of world trade center dust. USGS Open file report, 2005–1165.

Millette, J.R. and Bandli, B.R. (2005) Asbestos identification using available standard methods. The Microscope, 53, 179–185.

Sanchez, M. S. (2007) Characterization of amphiboles from Libby, Montana. Master Thesis, University of Idaho, 74 pp.

Tröger, W.E. (1979) Optical Determination of the Rock-Forming Minerals. Part 1 Determinative Tables. Translated by, H.U. Bambaur, F. Taborszky, and H.D. Trochim. E. Schweizerbrat'sche Verlagsbuchhandlung, Stuttgart, 188 pp.

Chapter 20

Environments of Mineral Formation

This morning when I got out of bed, it was –4°F outside (that's –20°C); New England was recovering from the throes of a winter storm, with over two feet of snow. Although it was a huge snowstorm, it was well-predicted, and the road crews had been out working for more than 24 hours. By the time I was driving to school, the sun was out and it had warmed up to 25°F (–4°C). I marveled at the barely-wet roads, which were rapidly drying to a salt-stained white. I found myself trying to remember exactly how salt makes the ice melt. How much does the pressure of a car's weight affect the melting point of water? How much salt do you have to add to bring those melting points down?

I knew from making ice cream at summer camp that salt is very effective in lowering the freezing point of water. The cream/sugar mixture has a lower freezing point than pure water, so the ice cream doesn't freeze if you surround it with ice or snow only. It turns out that salt is typical of many non-volatile molecules that dissolve (and don't evaporate). By adding the salt, you lower the temperature by removing the latent heat of crystallization. The Na^{1+} and Cl^{1-} ions also attract molecules from the liquid water, so there are fewer H_2O's available to make ice. Many other ionically-bonded materials behave in exactly the same way—but salt is cheap and easily accessible.

When I got to school, I pulled out my reference books; ice is, after all, a mineral, and its stability has been well-studied. It turns out that the answers I sought were fairly complicated. A smart highway engineer melts snow on the highway by taking advantage of a combination of effects that include:

- *warming of the pavement by the heat of the Sun as the snow melts—asphalt gets warmer than concrete because it absorbs more heat.*
- *melting of the snow by adding salt or other chemicals to lower the melting point of ice. NaCl only works if the temperature is higher than –6°F (–21°C); below that temperature, different (more expensive) solutes must be used, like $CaCl_2$. If you read on in this chapter, I will explain why this happens.*

So the science of clearing ice and snow from roads and highways depends on a knowledge of the relationship between temperature and composition. These variables, along with pressure, also constrain the minerals that occur in different geological environments. They make it possible to understand which minerals form under the conditions associated with igneous, metamorphic, and sedimentary rock types. You are already intuitively familiar with this concept: do you expect to find ice in Hawaii? (...only in the fancy drinks!)

A straightforward way to describe those conditions of mineral formation in rocks is through use of phase diagrams, which are graphical representations of the relationships among temperature, pressure, and composition. A petrologist uses phase diagrams to understand which minerals occur in association with others, and at what pressures and temperatures they might have formed. As mineralogists, we can also use phase diagrams to show how different species of the same mineral group are related, and why they might form in specific rock types. So in this chapter, we will describe the mineral associations that occur in each of the rock types. To explain them, we need to explore the underlying principles behind phase diagrams, and understand the relationships between minerals that coexist in rocks. From this, we can lay the foundations for understanding why different groups of minerals are commonly found in different rock types.

M.D.D.

Introduction

Any particular group of minerals that coexist stably reflects a specific range of pressure (*P*), temperature (*T*), and composition (*X*), representing a chemical equilibrium that was frozen into each rock when it was formed. This tells you a huge amount about the conditions under which the rocks formed, and can lend tremendous insights into the geological context of the rocks.

Our knowledge of the relationships among pressure, temperature, and composition started out with field observations. It has long been known that certain minerals occur together. Even 19th-century mineralogy books and articles describe the characteristic associations for various minerals. Quantitative understanding of the significance of mineral associations had to wait until the early 20th century, when it became technologically feasible to approximate geological conditions in the laboratory and to precisely quantify temperatures > 1150°C in furnaces (surprisingly, it was the latter capability that posed the biggest obstacle in this endeavor!). The history of the field of experimental petrology is described in a wonderful book by Davis Young (2003) called *Mind over Magma* that we recommend highly!

Today, mineral stabilities and relationships are studied using equipment that can reproduce conditions ranging from those on the Earth's surface to depths of roughly 650 km. Even higher *P-T* conditions can be reproduced in shock experiments that use guns to fire propellants at minerals, creating very high pressures, albeit very for less than one second!

Different types of apparatus are used for different *T-P* ranges. One-atmosphere experiments are commonly done in furnaces with tubes made from Al_2O_3, so they work well at $T < 1650$°C. Molybdenum disilicide ($MoSi_2$) furnaces usually operate from 900–1650°C, silicon carbide (SiC) furnaces can get up to 1500°C, and graphite furnaces can create temperatures up to 3000°C! For high pressure experiments, cold-seal pressure vessels are used to reproduce conditions in the Earth's crust (~10 kb), while conditions in the upper mantle are reproduced in piston-cylinder and multi-anvil devices that cover the range from about 30 kb with temperatures up to 2000°C.

Let's imagine a simplistic experiment involving the three crystal polymorphs of Al_2SiO_5: sillimanite (in which one of the two Al^{3+} cations is 4-coordinated), andalusite (one Al^{3+} 5-coordinated), and kyanite (both Al^{3+} cations are 6-coordinated) (see Figure 6.15). These minerals are predominantly found in metamorphic rocks at temperatures <800° and pressures <10 kb. To learn which minerals are stable under what conditions, we could grind up a mixture of equal parts of sillimanite, kyanite, and andalusite, and "cook" it at 3 kb and 750°C. After a month or so, we would remove our mixture from the furnace and check its mineralogy under a microscope. We would quickly find that the kyanite and andalusite crystals had begun to recrystallize to form sillimanite (if we ran the experiment long enough, the sample would be 100% sillimanite). This would tell us that sillimanite is the only stable phase at 3 kb and 750°C. We could then repeat the experiment many times, plotting our results on a graph of pressure vs. temperature (this experiment was indeed done in the 1960s and 1970s by several research groups). Eventually, we would create a diagram that would allow us to predict the mineralogy of Al_2SiO_5 under various conditions, as shown in Figure 20.1. The *x* axis is temperature, and the *y* axis is pressure. Each "field" on the diagram represents a region of stability for sillimanite, kyanite, or andalusite. The fields are separated by **univariant curves** where both phases coexist. The intersection of the three univariant curves is called an **invariant point**, and it represents the unique *P-T* conditions under which all three phases coexist stably.

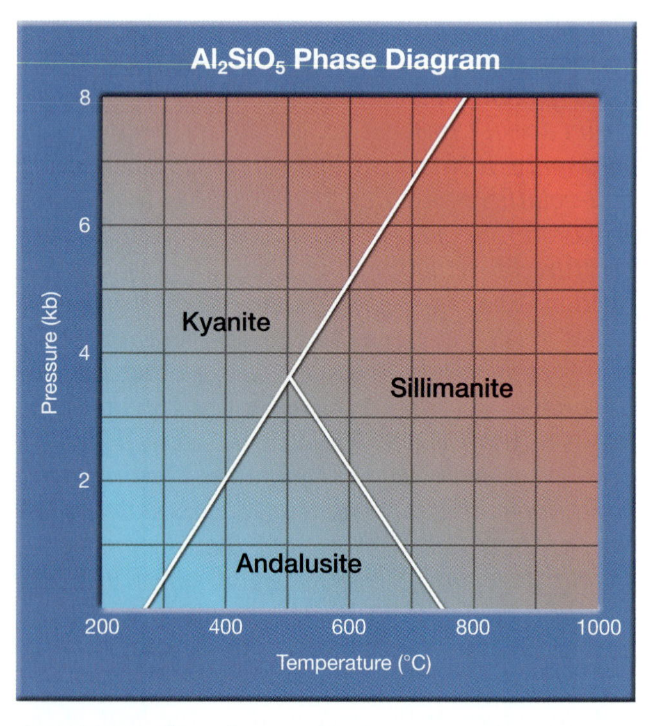

Figure 20.1. Phase diagram showing the equilibrium among the different polymorphs of Al_2SiO_5: sillimanite, andalusite, and kyanite. Notice that along any line separating two fields, two polymorphs would be stable and where all three lines intersect, all three polymorphs could coexist.

Underlying this discussion is the very important concept of **chemical equilibrium**. This occurs when the minerals that are present in a rock represent the lowest possible energy configuration for a specific composition under those particular conditions. In the example just given, kyanite and andalusite react to form sillimanite at 3 kb and 800°C because sillimanite is the most stable form of Al_2SiO_5.

The Al_2SiO_5 diagram can also be used to illustrate the underlying principle behind phase diagrams: the **phase rule**. The phase rule is part of the field of classical thermodynamics, and terms we are using here have formal definitions, as follows:

A **system** is an arbitrary part of the universe we are interested in: a magma chamber, a sedimentary bed, or a mineral. The environment is the rest of the universe surrounding our system. To illustrate this, let's return to the example of the salt-water system that began this chapter. If you put road salt and snow in a sealed thermos bottle, then you have an **isolated system**: neither mass nor energy can be exchanged between your system and the environment. If you take the top off, and cover it with plastic wrap so you let in heat and cold, then you'd have a **closed system**, which can exchange energy but not mass with the environment. Finally, if you take the plastic wrap off and add some fresh water, then you would have an **open system**. In geological situations, magma chambers can commonly be represented as closed systems, while magmas erupted into submarine environments might be open systems. For the aluminosilicate experiment we just discussed, we would have a closed system that contains kyanite, andalusite, and sillimanite.

A **component** (C) is a chemical constituent of our system; the number of components in the system has to be sufficient to completely define the compositional variability for our system. In the first example given above, there is a single component: Al_2SiO_5. You might think that the aluminosilicate minerals are composed of three chemical components: Si, Al, and O. However, for the phase rule, we choose the minimum number of components needed to describe the chemical variability of the components in our system. Thus, the aluminosilicates represent a single-component, or **unary**, system.

In the salt example, there are two components: water (H_2O) and road salt (NaCl)—a **binary system**.

A **phase** (P) is a mechanically-separable part of the system with distinct chemical and physical properties. Thus in the example at hand, if all the salt is dissolved in water, then there is only a single phase present: brine. If we've added so much salt that it doesn't all dissolve, we would have two components: salt + brine.

The phase rule, which was first proposed by Willard Gibbs in 1876, is used to describe a system that is in chemical equilibrium. The phase rule quantifies the difference between the number of thermodynamic variables in the system and the number of constraints on those variables. To state this simply for the geologic case, the thermodynamic variables are pressure, temperature, and the concentration of each of the components that are present in the rock. The constraints are the number of chemical reactions that can be written among the different components that are present. The relationship between them is the number of independent physical variables (also called **degrees of freedom**), or F. It tells you how many additional pieces of information are needed to completely define your system. This can be stated mathematically as:

$$F = C - P + 2.$$

To apply the phase rule, first ask yourself the following questions in reference to Figure 20.1:

1. If the only aluminosilicate in a rock is kyanite, what do you know about the specific pressure and temperature under which it formed?
2. If you have a rock containing both kyanite and sillimanite and you know the equilibration temperature, can you determine the pressure?
3. If your rock is an unusual sample containing sillimanite, andalusite, and kyanite, can you determine the temperature and pressure at which it equilibrated?

Each of these three scenarios can be understood using the phase rule. In 1, you may have concluded that the presence of only kyanite does not constrain either temperature or pressure. To think of this in terms of the phase rule, consider that there is only a single chemical component, Al_2SiO_5, so the value of $C = 1$. Similarly, there is only one phase: kyanite. So

$$F = 1 - 1 + 2 = 2.$$

Within limits, you can independently change both of the thermodynamic variables (here, pressure and temperature) and kyanite will be stable. So this system has two degrees of freedom (i.e., $F = 2$).

Now for scenario 2, there is again a single chemical component, Al_2SiO_5, so the value of C is again 1. There are now two phases, kyanite and sillimanite, so P is 2.

$$F = 1 - 2 + 2 = 1.$$

When $F = 1$, there is one degree of freedom, which means that there is only one independent variable. So, the temperature and pressure at which kyanite and sillimanite formed is restricted to be on the kyanite-sillimanite boundary curve. If you knew the pressure at which the rock formed, you would thus also know the temperature. When you know one, you know the other.

In the case where all three aluminosilicates coexist, we again have $C = 1$, but now there are three phases, so $P = 3$. So the phase rule variance is:

$$F = 1 - 3 + 2 = 0.$$

If a system has zero degrees of freedom (i.e., it's invariant), it's the same as saying that all the variables in the system are fixed. There is only one pressure and temperature where kyanite, sillimanite, and andalusite coexist in equilibrium. This invariant point is known as the aluminosilicate triple point.

There are many other examples of mineral polymorphs that can be conveniently described using phase diagrams. The unary $CaCO_3$ system has attracted a large number of experimental and thermodynamic studies and is shown in Figure 20.2. Once again, the x axis is temperature, and the y axis is pressure. One "field" on the diagram represents a region of stability for calcite and one for aragonite, separated by a univariant curve where they both coexist. The diagram shows that aragonite is only stable under high pressure conditions such as in high-pressure metamorphic rocks that form in subduction zones.

Binary Phase Diagrams

Let us see how the phase rule applies to binary (two-component) systems. Consider the example of the binary system $NaAlSi_2O_6$–SiO_2 that contains three minerals: albite, jadeite, and quartz. At high pressure, albite breaks down to form jadeite + quartz. All three minerals coexist stably along the albite = jadeite + quartz univariant ($C = 2$, $P = 3$, $F = 1$) line (Figure 20.3). Above the line, jadeite and quartz coexist stably in a divariant ($C = 2$, $P = 2$, $F = 2$) field. Below the univariant line, either albite + quartz or albite + jadeite will coexist stably, depending on the bulk composition.

From the preceding example, we have learned that binary systems have three variables: temperature (T), pressure (P), and composition. In Figure 20.3, we varied P and T and ignored composition. Now we will go on to some examples that hold P constant and allow T and composition to vary. As our initial example, we'll use the equilibria among melt, diopside, and anorthite, which are shown in Figure 20.4. [Petrologists often abbreviate mineral names, like An for anorthite and Di for diopside, when they use them in phase diagrams. You'll need to get used to seeing these abbreviations.] We have four different fields on our diagram: liquid, diopside + liquid, anorthite + liquid, and diopside + anorthite. The line dividing the liquid field from the crystal + liquid field is called the **liquidus**. It is the temperature at which the first crystals start to form if we are cooling the system. The line dividing the crystal + liquid fields from the diopside + anorthite field is called the **solidus**. Below that temperature (in the bottom field of the diagram), only those two solids are stable. The x axis shows the percentage of anorthite by weight (which also tells you how much diopside is present), and the y axis shows temperatures ranging from 1100–1650°C.

In this type of system, the location of your starting composition, X, determines which of your crystalline phases, diopside ($CaMgSi_2O_6$)—anorthite ($CaAl_2Si_2O_8$), will crystallize first. If we begin with liquid of composition X, it will cool until it reaches the liquidus. Within the two-phase field of anorthite + liquid, the liquid's composition can be determined by drawing a horizontal line at any temperature. The composition of the

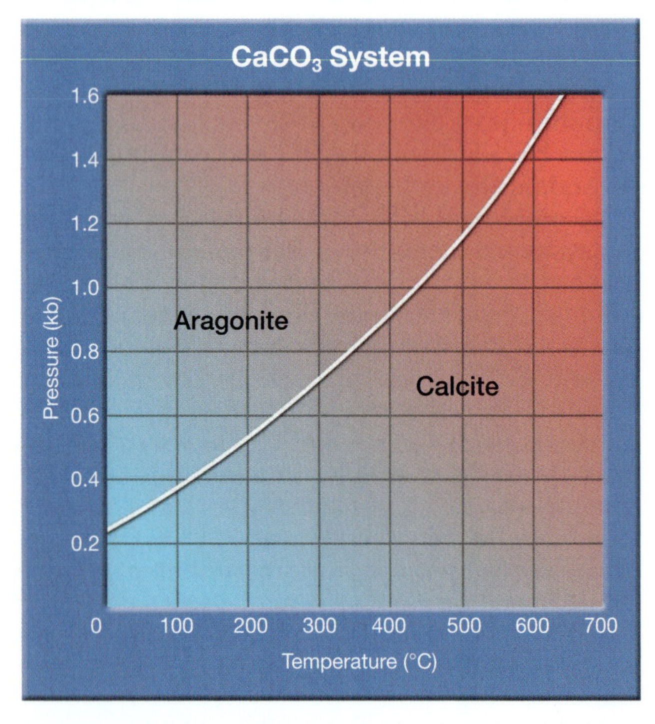

Figure 20.2. *Phase diagram of the $CaCO_3$ system, adapted from Redfern et al. (1989). Notice that with an increase in pressure, calcite could be converted to aragonite. Think about this the next time your professor hits the chalk board really hard with a chalk (which happens to be made from calcite).*

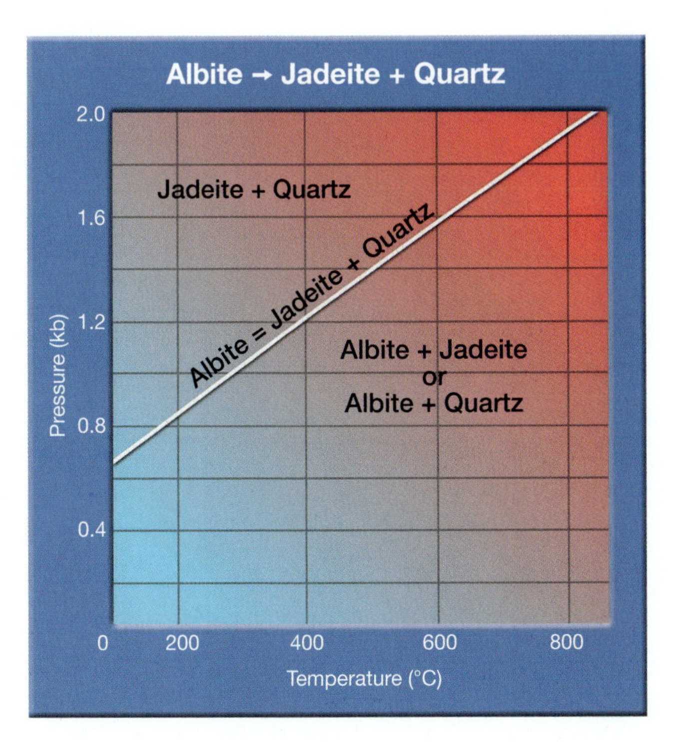

Figure 20.3. *Stability diagram for the reaction albite → jadeite + quartz. At high pressures, albite converts to jadeite and quartz. Therefore, feldspars are not found in high-pressure metamorphic rocks such as those of the blueschist facies.*

liquid will then be represented by the intersection of that horizontal line with the liquidus, while the composition of the solid phase, at the opposite end of the line, is 100% anorthite. So when the very first crystals form, they will be 100% anorthite, and the liquid will have a composition of X (An_{70}, or 70% anorthite).

As the system cools further, pure anorthite crystals will continue to crystallize, but the composition of the liquid will change continuously as the anorthite forms. At 1350°C, for example, the liquid will be An_{52} and the crystals will be An_{100}.

At 1274°C in Figure 20.4, we reach the intersection between the anorthite + liquid liquidus, the diopside + liquid liquidus, and the solidus line (below which everything is a solid). This is called the **eutectic** point. At this particular (and unique) position, the liquid is in equilibrium with both solids. The phase rule dictates that no change in temperature or composition of any of the three phases can occur until every little bit of liquid is completely crystallized. Below the eutectic point, the crystals are a mixture of diopside and anorthite, in a proportion fixed by the intersection of a vertical line from **X** and the *x* axis. Remember that compositions plotting close to the An end are An > Di, and compositions plotting close to the Di end are always Di > An.

We can also use this diagram to quantify the relative amounts of liquid and crystal at any given time, using a concept called the **lever rule**. Examine the horizontal lines drawn inside the anorthite + liquid stability field in Figure 20.4. In the top one (**B-C**), the liquid has just started to crystallize, so it is nearly all melt with just a few crystals. In the bottom line (**G-H**), the situation is reversed. What will the mixture look like in between, such as at 1350°C? We've already noted that the composition of the liquid at this point is An_{52}, and the crystals are pure anorthite. Draw a vertical line corresponding to the bulk composition (*X*, or 70% anorthite). Now, observe where that vertical line intersects the horizontal line **E-F**. If it crosses in the exact center, then the magma at that point in time will be 50% crystals and 50% liquid. If it crosses closer to point **E**, then the mixture will have more crystals than liquid, in an amount equal to however far along the **E-F** line you are. Similarly, if the bulk composition line crosses closer to **F**, then the magma will have crystals > liquid, again in an amount corresponding to where you are on line **E-F**. The fraction of liquid is the ratio of the lengths IF/EF.

One of the ways that phase diagrams are useful is in showing why groups of minerals do and don't occur together. To this end, consider Figure 20.5, which shows the change in SiO_2 content between forsterite (Mg_2SiO_4) and pure SiO_2 (quartz, tridymite, or cristobalite). This is another binary phase diagram, because the chemical components Mg_2SiO_4(Fo) and SiO_2 completely describe its chemical variability. Here we have several different stability fields, but interpreting the diagram is the same as for the previous example.

Figure 20.5 introduces one more concept that is sometimes encountered in phase diagrams: a reaction point, or **peritectic**. This is a point on a phase diagram where a reaction takes place between a previously-precipitated phase and the liquid, resulting in the formation of a new solid phase. There, the temperature has to stay the same until the reaction goes to completion, one of the reactants is exhausted, and the new phase has crystallized. In this example, the peritectic represents the temperature where forsterite reacts with the liquid to form enstatite. Follow the examples given in the figure caption and in the animation on the MSA website, and you'll observe that there is no starting bulk composition that can ultimately result in equilibrium between olivine and quartz. So, if you ever even think you see clear stuff mixed in with olivine in an igneous rock, don't even consider the possibility that it might be quartz. It's more likely to be a feldspar!

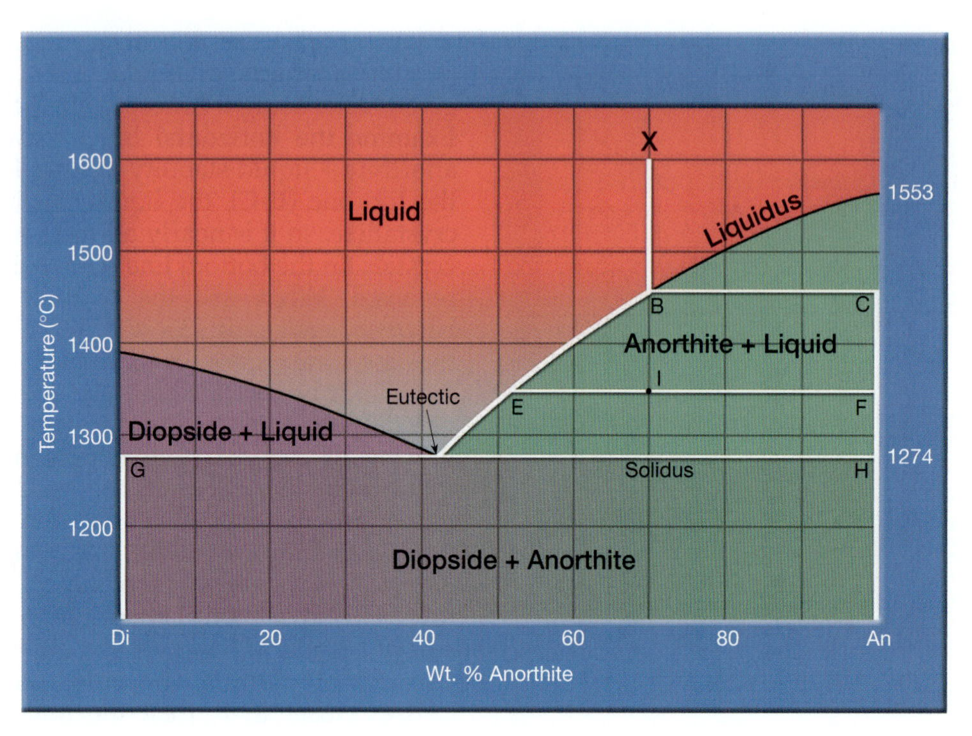

Figure 20.4. *The anorthite(An)-diopside(Di) system has two distinct components and is plotted here against temperature, with pressure held constant at 1 bar. The x axis represents composition. On the left side is 0% anorthite (pure diopside) and on the right side is pure anorthite. Liquid* ***X*** *can cool from high temperatures until it hits the liquidus, at which point anorthite will begin to crystallize. Continued crystallization of An will cause the liquid composition to move toward Di. At T = 1274°C, the liquid will reach the binary eutectic, where diopside and anorthite will both crystallize. The temperature cannot drop until all of the liquid has crystallized and is completely composed of the two solid phases. At temperatures below 1274°C, only crystals of An and Di are present. (Animation inspired by the work of Kenneth E. Windom, Iowa State University.)*

Think you've got these diagrams figured out? Let's move on to Figure 20.6. Unlike the preceding two figures, this phase diagram contains minerals with variable compositions. Feldspar here can be either orthoclase ($KAlSi_3O_8$) and albite ($NaAlSi_3O_8$). Suddenly, life is a lot more complicated!

The top half of this diagram will be explained shortly (Figure 20.8). At ~900°C, we have a single feldspar, and all solid solutions between $KAlSi_3O_8$ and $NaAlSi_3O_8$ are stable. But as the feldspar continues to cool, there comes a point at which there is no longer complete solid solution between orthoclase and albite. The difference in size between K^{1+} and Na^{1+} is so big and the structures are more discriminating at lower temperatures (for the different cation sizes) that the two cations cannot comfortably substitute for each other across the whole feldspar compositional range. So, beginning about 850°C, we see a hump labeled **solvus**, which is a line representing the limits of the solubility of the two phases. In this case, we see a phenomenon called exsolution, in which two phases *exsolve*, or recrystallize, from a single higher-temperature phase. Effectively, they *exit the solution* to form two separate phases.

We can best understand Figure 20.6 by working through two different starting magma compositions: **A** and **B**. Liquid **A** begins life with a composition of $K_{0.65}Na_{0.35}AlSi_3O_8$. As it cools, it encounters the orthoclase liquidus, whereupon it crystallizes K-rich feldspar in exactly the same fashion as in Figures 20.8 and 20.9. When it reaches point **D**, the last feldspar crystallizes with a composition of $K_{0.65}Na_{0.35}AlSi_3O_8$ crystallizes; this feldspar remains stable with further cooling until just below 800°C. At that point, the solid $K_{0.65}Na_{0.35}AlSi_3O_8$ crystals are no longer stable. It starts to exsolve into a slightly more K-rich orthoclase and a much more Na-rich albite. By the time the system cools to 700°C, it becomes a still-solid mixture of about 85% $K_{0.75}Na_{0.25}AlSi_3O_8$ crystals and 15% $K_{0.25}Na_{0.75}AlSi_3O_8$ crystals.

Over on the Na-rich part of the diagram, liquid **B** cools in an analogous fashion. It begins as a liquid with a composition of roughly $K_{0.15}Na_{0.85}AlSi_3O_8$ and crystallizes Na-rich feldspar until it reaches the solidus. At that point all the crystals have a composition of $K_{0.15}Na_{0.85}AlSi_3O_8$, and that composition remains stable until the system cools to 785°C and encounters the solvus. At that point, the albite starts to

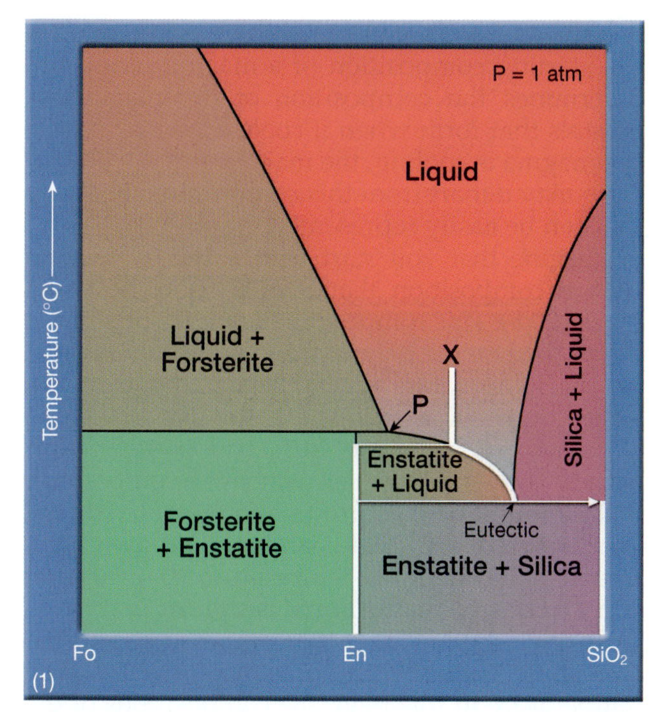

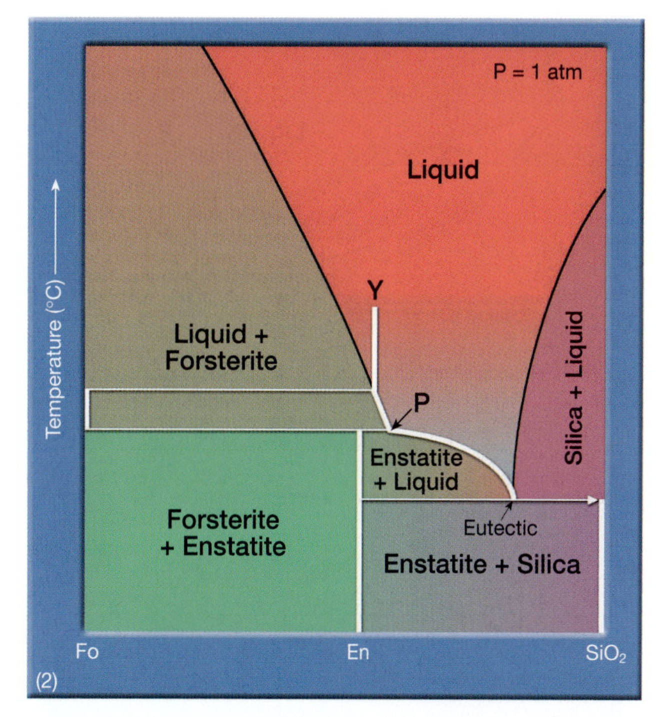

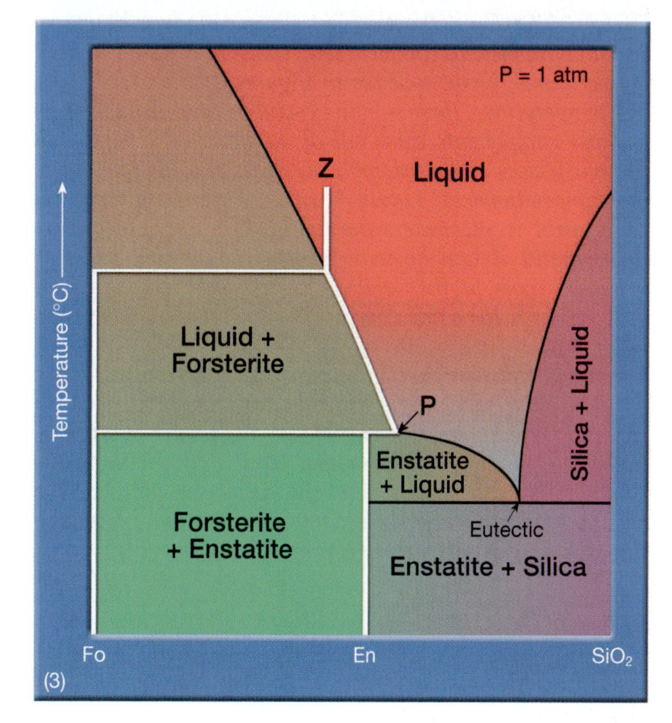

Figure 20.5. *The forsterite(Fo)-silica(SiO_2) phase diagram at 1 bar pressure. In this system, notice that forsterite and enstatite (En) are stable together, and enstatite and SiO_2 may coexist, but there are no possible ways to make forsterite and SiO_2 crystals in equilibrium. To demonstrate this to yourself, consider three examples: (1) Liquid* ***X*** *will cool from the liquid form until it encounters the liquidus and begins to grow En. The liquid will change composition and move toward the eutectic as En crystallizes. Once the eutectic is reached, the temperature will remain the same while enstatite and quartz crystallize. (2) Liquid* ***Y*** *contains slightly less SiO_2. It will first crystallize forsterite, while the liquid cools toward point* ***P****, which is a peritectic. At* ***P****, the forsterite reacts with the liquid to form enstatite; the reaction uses up all the forsterite. Upon further cooling, more enstatite will crystallize, and the remaining liquid will move toward the eutectic. Once there, enstatite and quartz will crystallize until all the liquid is consumed. The final result is a mixture of enstatite and silica. (3) Liquid* ***Z*** *represents a forsterite-richer liquid. When it cools, it will initially crystallize forsterite, causing the liquid composition to move toward the peritectic* ***P****. At* ***P****, forsterite will react with the remaining liquid to make enstatite, completely using up all the liquid. From that point, forsterite and enstatite will cool in equilibrium with each other. (Animation inspired by the work of Kenneth E. Windom, Iowa State University.)*

exsolve into a slightly more Na-rich albite and a much more K-rich orthoclase. Again, it may be helpful to follow through these examples using the explanation found in the caption for Figure 20.6 as well as on the MSA website.

Let's look at another example involving coexisting minerals in metamorphic rocks. Figure 20.7 shows the magnesite ($MgCO_3$)-calcite ($CaCO_3$) phase diagram. Although calcite may initially form the bulk of limestone mineralogy, in situations where Mg is present, it will rapidly dissolve or transform into calcite + dolomite ($CaMg(CO_3)_2$). This phase diagram shows that two separate **solvi** (that's the plural of solvus) exist in this system. At the temperatures shown, composition (i.e., where you are along the *x* axis in the diagram) determines whether you will get calcite + dolomite or magnesite + dolomite. Thus, you would never expect to see both magnesite and calcite in the same metamorphic rock!

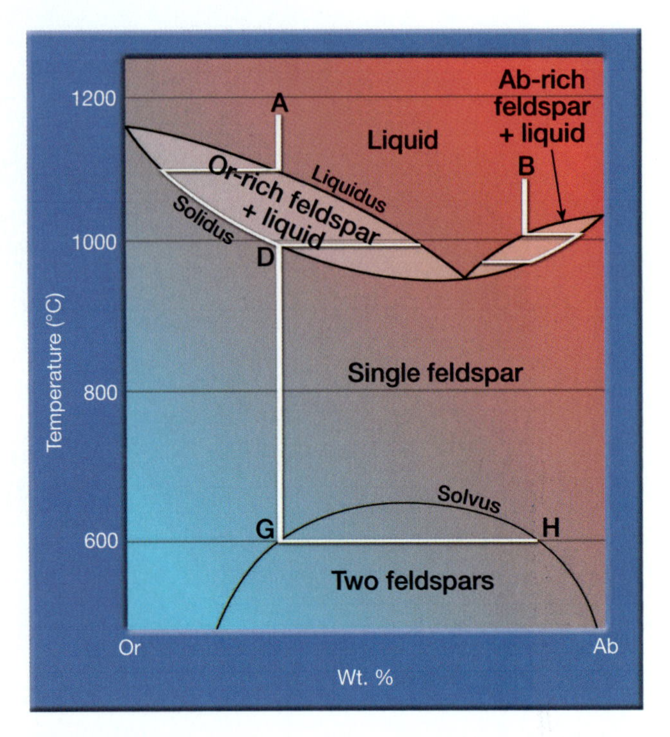

Figure 20.6. *The orthoclase(Or)-albite(Ab) phase diagram. This diagram shows complete solid solution behavior at high temperatures, as we will see in Figures 20.8 and 20.9. At low temperatures, there is limited solid solution, and the two feldspar components come out of the mixture to form two discrete phases. In this example, we show two different starting compositions. (1) Liquid **A** will cool from high temperatures until it encounters the liquidus around 1100°C, at which point it will begin to crystallize Or-rich feldspar. When it cools to the solidus, it will form crystals of a feldspar with the same composition as liquid **A**. However, upon further cooling, it reaches a solvus, which is a point below which the single crystalline phase is no longer stable. At that point, the single feldspar exsolves to an Or-rich feldspar (**G**) and an Ab-rich one (**H**). (2) Liquid **B** will initially crystallize Ab-rich feldspar until it cools below the solidus. When it encounters the solvus it will exsolve feldspars with the identical compositions to those found in liquid **A**, but in very different proportions.*

Finally, we conclude with two examples of binary phase diagrams that incorporate complete solid substitution. Arguably the most famous example of a binary phase diagram is the so-called "banana diagram," which shows the melting and crystallization of a binary mixture of $NaAlSi_3O_8$ (albite) and $CaAl_2Si_2O_8$ (anorthite). On this diagram (Figure 20.8), the x axis shows the solid solution between the two plagioclase endmembers. The y axis shows temperature. The diagram has three different fields. In the high-temperature field, only liquid is stable. The "banana" is outlined by two lines between which liquids and crystals coexist.

We can use the plagioclase phase diagram as an example to explain the normal crystallization behavior of a magma. As shown in Figure 20.8, the starting composition of a magma ultimately determines the composition of the plagioclase crystals that form when it cools. However, while the magma is cooling, the magma and the plagioclase may depart from this composition, in a way that can be easily represented graphically.

Imagine that you start with a 1600°C magma with a composition that is 61% anorthite (**A** in Figure 20.8). This composition lies in the one-phase liquid field on the diagram, and is entirely molten. As the temperature cools, the liquid retains the same composition until it hits the line labeled as "liquidus" (**B** in Figure 20.8), and crystallization begins. Those first crystals will have a composition that is roughly 87% anorthite, as represented by the intersection of the solidus and a line drawn horizontal to the liquidus temperature (**C** in Figure 20.8). At this point, the liquid is still An_{61}.

Further cooling puts the composition into the two-phase field of liquid plus plagioclase. At any temperature in this field, the compositions of the plagioclase and melt that coexist can be determined by drawing a horizontal line at that temperature. That line will intersect the boundaries of the two-phase field. Vertical lines drawn from the intersections of that horizontal line with the liqudus and the solidus give you the melt and plagioclase compositions, respectively.

As the system cools, the compositions that will be in equilibrium at any temperature follow the liquidus and solidus lines, which represent melts and crystals that are in equilibrium at each temperature, until about 1340°C. At that point, the placioclase will again have a composition of nearly 61% anorthite (**H** in Figure 20.8), while the very last little bit of liquid will be about An_{20} (**G** in Figure 20.8). Once that last bit cools, the entire system will be made up of plagioclase crystals with compositions of An_{61}.

Another useful phase relationship is found in the phase diagram for the olivine group minerals forsterite and fayalite; this equilibrium is very useful when trying to understand evolution of basaltic magmas (Figure 20.9). There are many mineral groups that display this type of phase diagram. An excellent, thorough discussion of phase equilibria like this one that are relevant to igneous rocks can be found in Morse (1994). Similarly, a thorough treatment of applications to metamorphic rocks is in Spear (1993).

Ternary Phase Diagrams

Of course, real geological situations sometimes involve more than two components. Thus, sys-

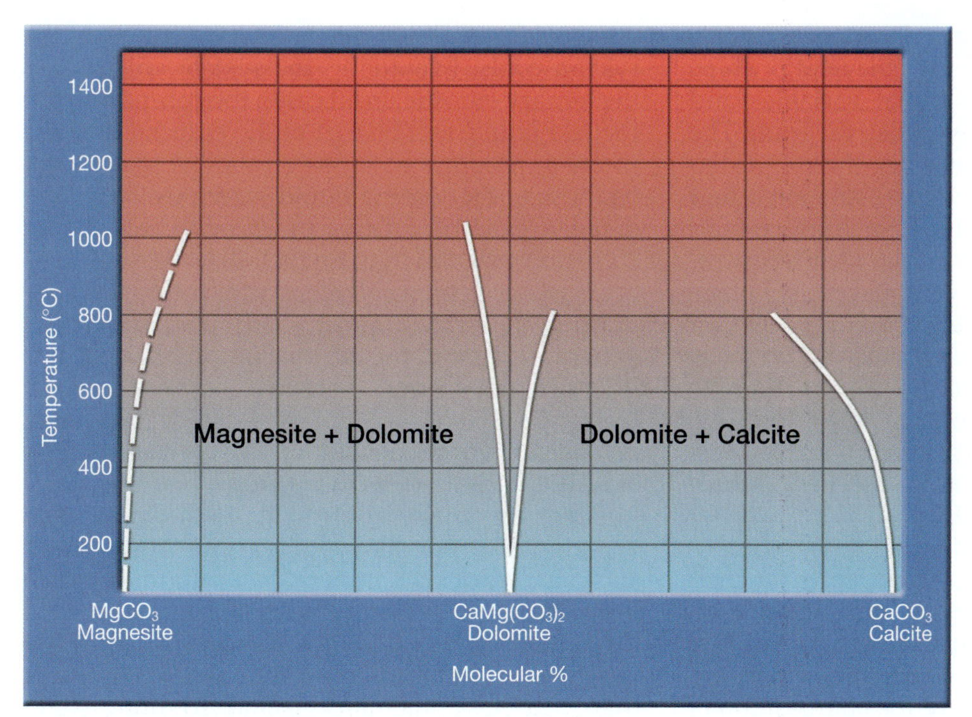

Figure 20.7. Very simplified magnesite-dolomite-calcite phase diagram. Normal $CaCO_3$-rich marbles are always a mixture of dolomite and calcite, whereas Mg-rich marbles that form from peridotite contain magnesite + dolomite.

tems with three, four, or many more components are common, and many ways of depicting them are available. One way is use a triangle, with individual chemical components at the corners. You will recall from our examples in Chapter 10 that these plots are called **ternary diagrams**. To uniquely specify a location on this plot, you need to specify two of the three components, and they all have to add up to 100%. To refresh your memory of this concept, you may want to refer back to Figure 10.2 and its explanation.

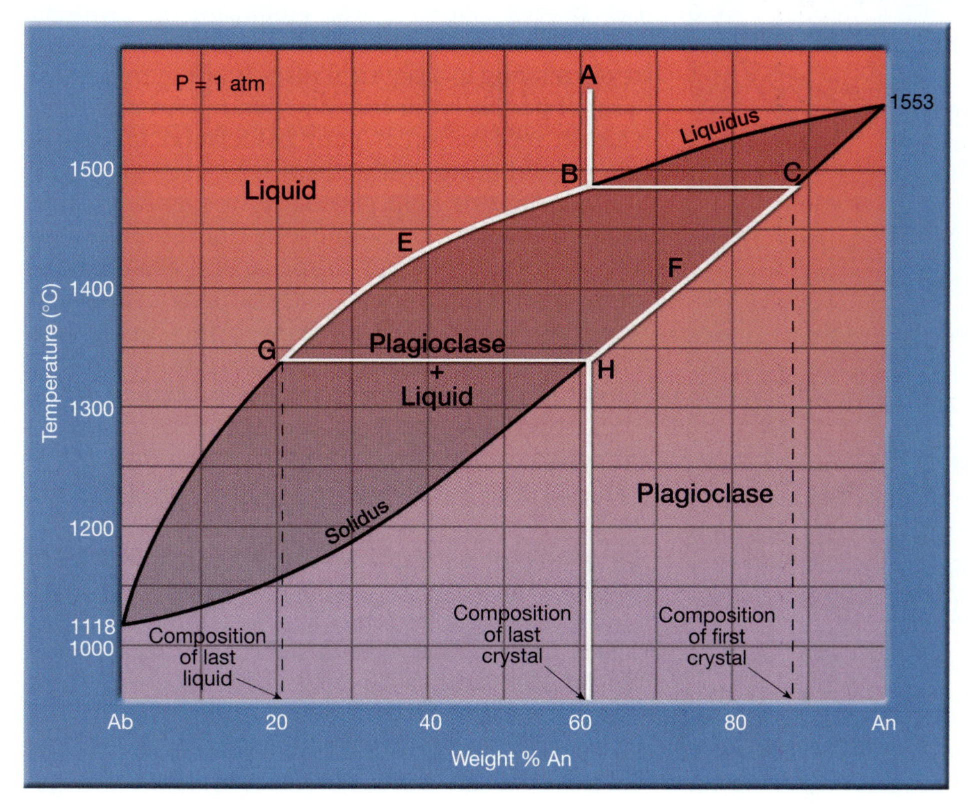

*Figure 20.8. Phase diagram of the binary system, albite (Ab)-anorthite (An) vs. temperature. If you begin with a magma of composition **A**, it will cool to the liquidus (An_{61}), where crystals with a composition of An_{87} begin to form. Those crystals have higher Ca/Na ratios than the liquid with which they coexist. So further precipitation of Ca-rich plagioclase causes the remaining liquid to grow more sodic (richer in Na). Both liquid and plagioclase become richer in Na with further cooling; the proportion of crystals increases as the proportion of liquid decreases. Finally, at a temperature of around 1340°C, all the liquid is crystallized, and An_{61} plagioclase will cool without further changes in composition. (Animation inspired by the work of Kenneth E. Windom, Iowa State University.)*

We can use a ternary diagram of composition as the basis for a plot that shows the relationships between our three compositional variables and temperature; this is a **ternary phase diagram**. The way this is done is to join together three *T-X* diagrams into a triangle (like wings), and then fold the *T-X* diagrams up, contour the surface, and make a 3-D diagram of a system with temperature on the *z* axis and composition in the *x-y* plane plotted as a triangle. This approach is shown in Figure 20.10, which shows the binary and ternary eutectics for the anorthite-titanite-wollastonite system. If a liquid's composition falls along the edges of this diagram, then it will follow the crystallization sequence indicated by that binary eutectic similar to Figure 20.4.

If the composition contains all three components, its crystallization sequence follows an analogous set of rules. When an arbitrary liquid composition falls within a one-phase field, it will first cool to the liquidus, then grow crystals of that single phase. As the crystals grow, the composition of the liquid will move away from the corner of the diagram represented by the mineral that is crystallizing. When it cools to the boundary curve between two stability regions, it will grow a mixture of crystals (again, the compositions are represented by corners of the diagram). The liquid's composition follows the boundary curve as the system cools. Eventually, the system will cool to the ternary eutectic. In all cases, the composition of the last liquid will be that of the eutectic, and the ultimate proportions of the crystalline phases will be those of the starting liquid's composition. For a detailed example of this progression, see Figure 20.10 and its animation on the MSA website.

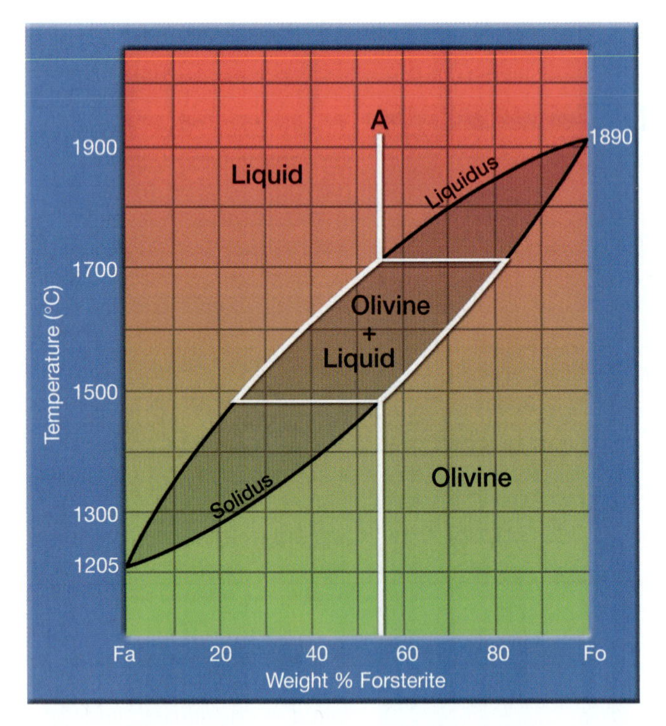

Figure 20.9. *Phase diagram of the binary system forsterite(Fo)-fayalite(Fa) vs. temperature. Cooling progresses in this system is exactly as shown in Figure 20.8. To test your understanding, walk through the cooling of liquid* ***A*** *shown here, and then check the MSA website to see if you are right.*

A more geologically-important ternary diagram is the system anorthite-diopside-forsterite, which is shown at 1 bar in Figure 20.11. In this diagram, the ternary eutectic is very low in forsterite, so it lies toward the left side. However, crystallization sequences in this system follow exactly the same rules just stated.

There are many useful phase diagrams that explain the mineralogy of various rock types; such studies of such mineral associations are really the focus of the discipline of petrology. So we'll leave further investigations of phase diagrams to your future courses, now that we're confident that you have the background to understand them in the proper mineralogical context.

Mineral Associations

Have you ever had the experience of meeting your gym teacher in the supermarket and not recognizing him because he's wearing regular clothes? Or encountering your best friend's girlfriend/boyfriend in the dentist's office and not being able to remember his/her name? These embarrassing incidents happen because we recognize people in certain contexts, relating to a physical place (your gym class) or a set of people (your friend's friends). An important clue in recognition is your association of people, places, and things with others.

Associations help geologists recognize minerals in the same way! Almost by definition, minerals are found together in rocks, and **associations** among different minerals are valuable in identifying them. One of the most important things you can learn in a mineralogy course (to aid you in mineral identification in rocks) is that different rock types are made up of distinctive groups of minerals. When someone hands you a rock, if you can identify one or two minerals in it, you ought to be able to predict what the other ones are. It's the equivalent of seeing Fred Astaire dancing in a movie and suspecting that the woman with him must be Ginger Rogers....

The phase diagrams that form the first half of this chapter are one way of graphically demonstrating relationships between minerals and magmas. When we discussed changes in pressure and temperature, we were really talking about changes

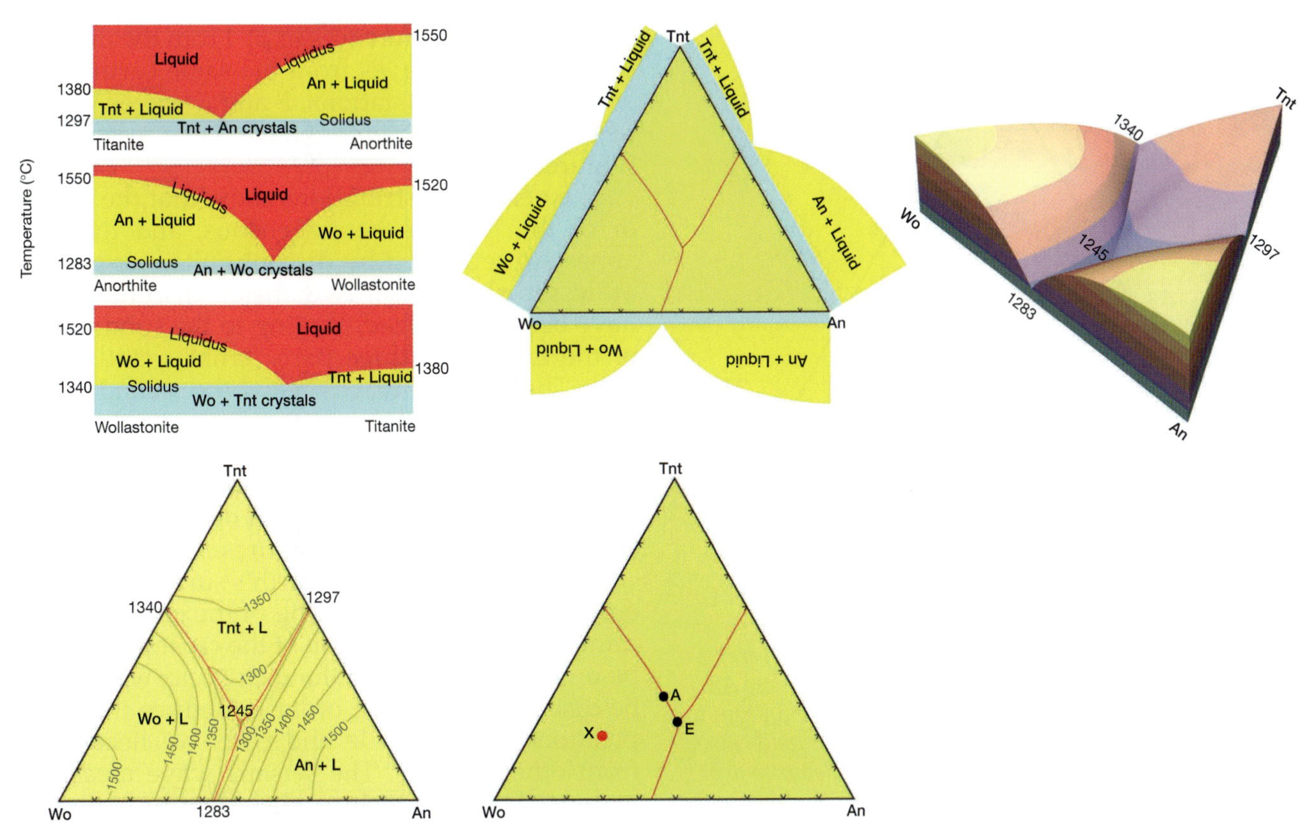

*Figure 20.10. The anorthite (An)-wollastonite (Wo)-titanite (Tnt) ternary phase diagram can be thought of as a triangular prism; each of the three sides is a 2-D phase diagram, and the top of the prism is contoured with lines representing the temperature of the liquidus. In this system, the sides of the prism show the three binary compositions: titanite-anorthite (Tnt-An), anorthite-wollastonite (An-Wo), and wollastonite-titanite (Wo-Tnt). From inspection, we see that the sides of these diagrams match up, so we can stand them up and make a 3-D diagram that allows us to represent compositions that contain all three components (Tnt, Wo, and An). When this diagram is viewed from the top, we see a triangle contoured with lines of temperature. To understand how this diagram works, let's follow through what happens to a liquid of composition **X** on the diagram (it's 60% Wo, 20% An, and 20% Tnt). When **X** cools to the liquidus, it begins to crystallize Wo crystals, so the composition of the liquid moves away from the Wo corner of the triangle (toward **A**). Eventually, the system cools so much that the liquid reaches the boundary curve (also called a cotectic) (at **A**) between Wo and Tnt. Both Wo and Tnt crystallize, and the composition of the liquid moves down the curve toward **E**. Wo and Tnt continue to crystallize until the liquid composition reaches the ternary eutectic, labeled **E**, at which point Wo + Tnt + An simultaneously crystallize. The relative proportions of those phases are constrained to be the same as the starting composition of the liquid (60% Wo, 20% An, and 20% Tnt).*

in geological conditions, usually as a function of processes that occur in the Earth's crust and mantle. To place these in context, we will review here the Earth's interior structure and the geological rock cycle, so we can talk about the minerals that are characteristic of each environment. Some minerals are stable over wide ranges of temperature and pressure, and thus occur in many different rock types (e.g., quartz), while others are stable only under very specific conditions (e.g., kaolinite) and occur in very specific mineral associations.

A Quick Journey to the Center of the Earth

In order to put mineralogy into a geological perspective, it's easiest to begin with a discussion of the Earth as a whole. You may already know that both pressure and temperature increase from the surface toward the center of the Earth. The change in pressure is caused by the same effect you experience when you swim to the bottom of a pool—at greater depth, there is more water pressing down on you, and so the pressure is greater. As you move toward the center of the Earth, the same thing occurs: the combination of the atmospheric pressure and the weight of soil and rocks pressing down causes pressure to increase. Temperature also increases as you move from the crust to the core; we know this because magmas that erupt from depth are hot! In fact, the geothermal gradient within the Earth can be predicted well, and is shown in Figure 20.12.

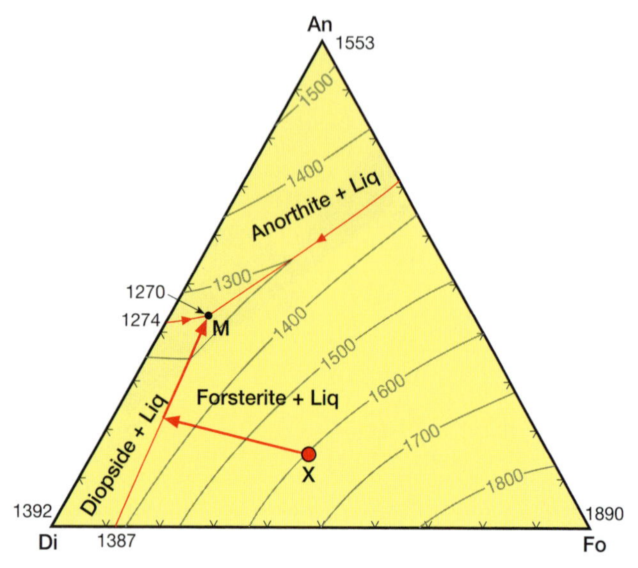

Figure 20.11. *Anorthite(An)-diopside(Di)-forsterite(Fo) phase diagram at 1 bar pressure. (Animation inspired by the work of Kenneth E. Windom, Iowa State University.) Consider a liquid with composition* ***X****, which is* Fo_{40}*,* An_{17}*, and* Di_{43}*. It will cool to the liquidus at 1600°C and begin to crystallize forsterite. The liquid composition will move directly away from the Fo corner as it cools down to about 1350°C. When the liquid composition encounters the cotectic phase boundary, diopside will also begin to form. Coprecipitation of forsterite and diopside causes the liquid composition to move down the cotectic curve toward the ternary eutectic at* ***M****, where anorthite begins to crystallize. When the system cools to 1270°C, the last bit of liquid will have composition* ***M****, and the solid phases will be anorthite, diopside, and forsterite in the proportions 17:43:40.*

To understand the composition and mineralogy of the Earth's interior structure, we must turn to largely indirect observations. Elementary geophysical measurements of the Earth's mass, volume, and moment of inertia indicate that the density (ρ) of the Earth's materials increases downward from the surface (where ρ averages about 2.5–3.0 g/cm^3) to the core, where densities near 15 g/cm^3. Perhaps the most graphic portrayal of the structure of the Earth's interior structure comes from the velocity distributions of earthquake waves, which travel at different speeds through different layers of the Earth (Figure 20.13). An earthquake generates waves of various kinds that travel through the body of the Earth to reach receivers (seismographs) dotted around the surface of the Earth. The two types of waves that give the most information about the Earth's interior are the faster, compressional *P* waves and slower, shear *S* waves. The velocities of *P* and *S* waves vary with the density and elasticity of the materials through which they pass, and refract and reflect at surfaces of velocity discontinuities (that represent differences in ρ). From the times at which *P* and *S* waves from the same earthquake arrive at different stations, *P* and *S* waves velocities at different depths have been calculated (Figure 20.13).

The velocity distributions show that the interior of the Earth is heterogeneous, and that the zones are separated by discontinuities of greater or lesser sharpness. Two major discontinuities have been recognized, leading to the subdivision of the Earth into three parts: crust, mantle, and core. The lack of *S* waves below the second major discontinuity suggests that the material of the underlying core behaves as a liquid. More recent, sophisticated refinements of the seismic data have led to further subdivision of the crust, mantle, and core, as shown in Figures 20.13 and 20.14.

Because we live on the Earth's surface, we have a pretty good understanding of the chemical and mineralogical composition of the crust, which can be understood through the rock cycle, which will be discussed below. In order to deduce the composition of the mantle and core, we need data from other sources. The seismic wave measurements tell us the density and elastic properties, so we know that the Earth's interior must contain denser minerals than the crust. These properties could be satisfied by a wide variety of substances with different compositions. So how do we best guess at the mineralogy involved?

Most of our information about planetary interiors comes from meteorites, which represent primitive parts of our solar system. Depending on their structure, mineralogy, and chemistry, three classes are distinguished: irons, stones, and stony-irons.

Irons are about 98% metal, which is usually mostly Fe with roughly 4–20% Ni. Their mineralogy is typically a mixture of two polymorphs of those elements: kamacite (α-Fe,Ni) and taenite (γ-Fe,Ni), with minor blebs of troilite (FeS). Iron meteorites probably come from the cores of large asteroids.

Stony meteorites probably come from the mantle and crust of asteroids (and in a few case, comets). Not surprisingly based on comparison with the Earth, there are several diverse types of stony meteorites, but they are predominantly composed of olivine and pyroxene, with lesser amounts of Fe-Ni metal and Fe sulfides.

Stony-iron meteorites consist of metal mixed with silicate minerals in roughly 50:50 proportions. **Pallasites** are made of olivine crystals surrounded by Fe-Ni metal, and probably formed at the boundary between the core and the mantle of a large asteroid. **Mesosiderites**, which are complex mixtures of metal grains, olivine, pyroxene,

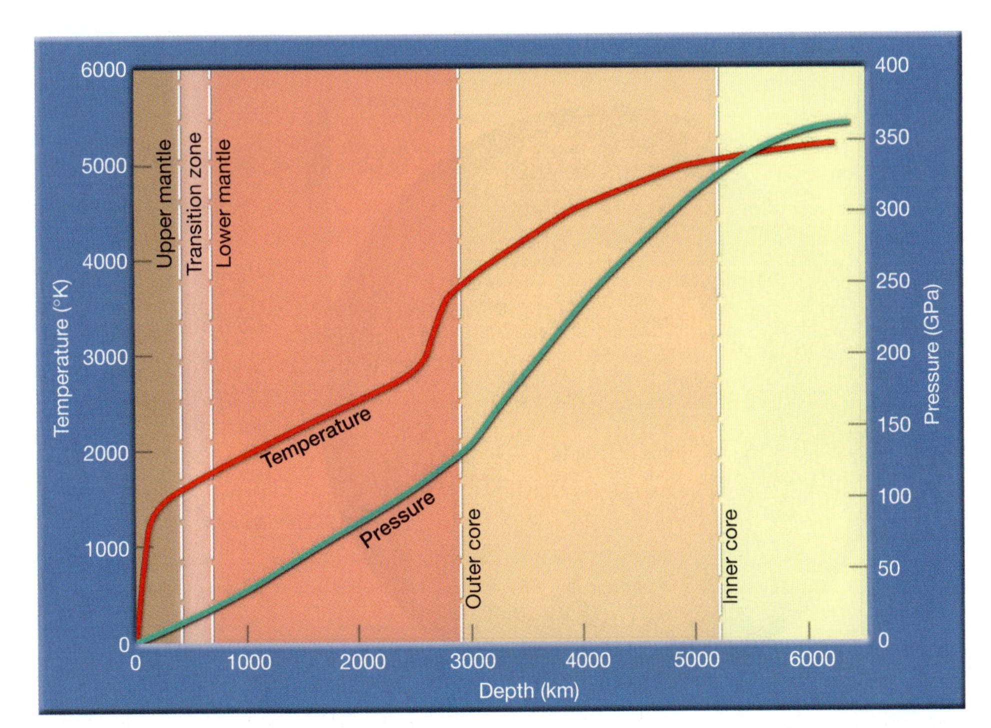

Figure 20.12. *Temperature and pressure changes as a function of depth in the Earth's interior.*

and plagioclase minerals, are believed to result from violent impacts between metal-rich and silicate-rich asteroids.

On the basis of evidence from stony meteorites, the Earth's mantle is believed to consist mainly of O, Si, Mg, and Fe, with smaller amounts of Ca and Al. The meteorite evidence is supplemented by mantle xenoliths (meaning "foreign rocks"), which are samples of mantle rocks that get caught up in mantle-derived magmas and brought rapidly to the Earth's surface. A third source of evidence for the mineralogy of the Earth's interior comes from experiments like the ones described at the start of this chapter, where rocks are squeezed and heated to approximate deep-Earth conditions. Finally, pieces of the mantle are sometimes faulted to the surface in mountain belts.

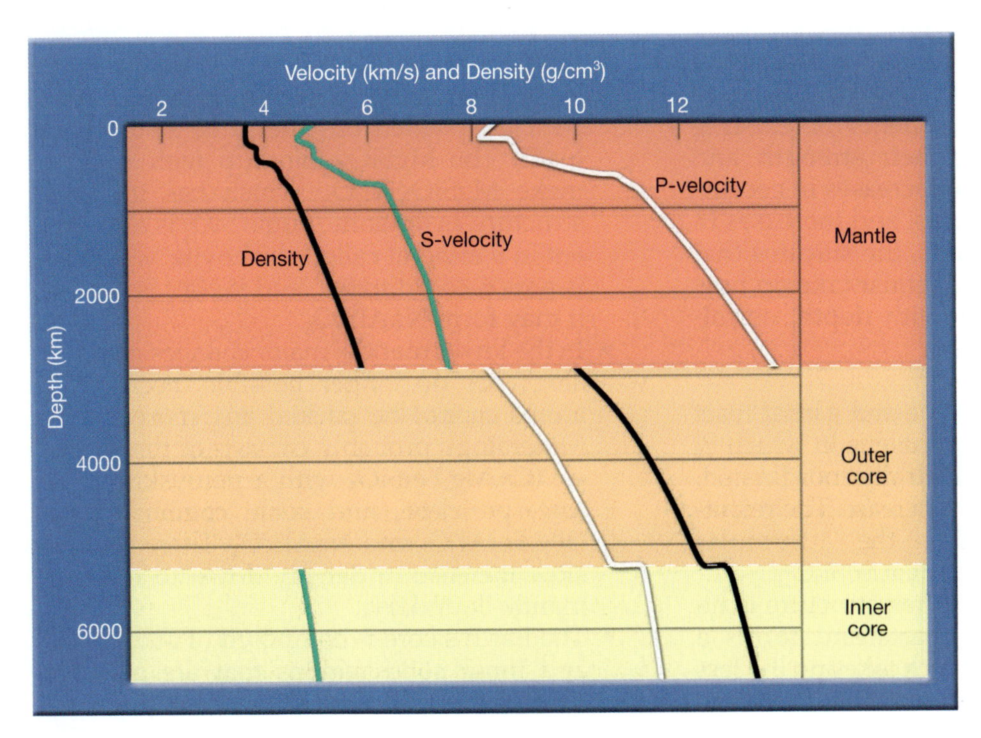

Figure 20.13. *Velocity of seismic S and P waves, as well as density, are plotted here against depth. Notice that the S waves cannot pass through liquids. They disappear in the outer core, inferring that the outer core is molten. Also, notice how density increases with depth.*

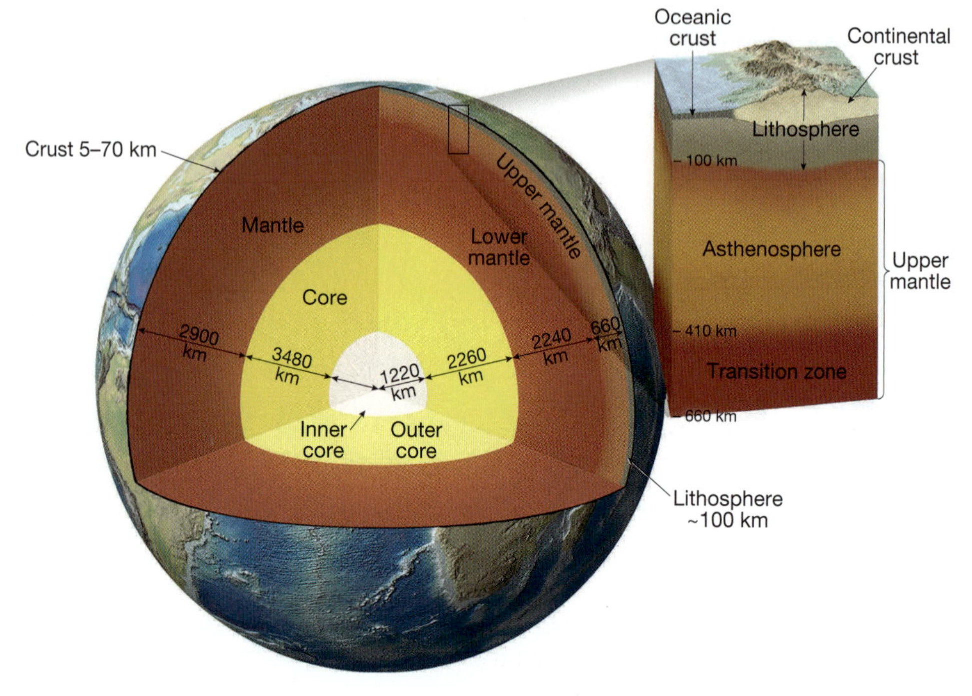

Figure 20.14. *Schematic diagram of the Earth's interior layers.*

From these lines of evidence, we know that the upper mantle's mineralogy is dominated by olivine and pyroxene, with lesser amounts of spinel, garnet, amphibole (usually kaersutite), and mica (usually phlogopite). As you move deeper, the pressure increases, and the minerals eventually become unstable. Why? You may recall from Chapter 8 that radius ratios are used to predict how oxygen atoms will pack into coordination polyhedra. However, those specific ratio ranges only apply at relatively low pressures—at high pressures, atoms contract in size, and the ranges of ratios for low pressure radii no longer apply. Larger atoms will contract more than smaller atoms. Pressure also acts to increase the covalent components of existing chemical bonds, changing their strength and character. All of this leads to increases in coordination numbers for the cations, and increases in density for the minerals. Thus, the silicates that are familiar to us in the crust must take on new forms as pressure increase with depth, as follows.

- At about 350 km, pyroxene and garnet react to form a new kind of pyroxene in which 1/4 of the Si atoms are octahedrally coordinated, leading to a 10% density increase. The resultant minerals have the formulas $Mg_3(Mg,Si)Si_3O_{12}$ and $Ca_3(Ca,Si)Si_3O_{12}$.
- At about 400 km, the olivine structure is no longer stable. Olivine reacts and forms a spinel-type structure, which takes up 8% less space; the composition remains the same (though we now call it β-Mg_2SiO_4). This causes a seismic discontinuity at about 400 km depth. This is generally defined as the boundary to the transition zone in the mantle, although some scientists use the velocity change at 350 km instead.
- At 500–550 km, $CaSiO_3$ in the garnet transforms to the same, very dense structure as the mineral perovskite ($CaTiO_3$), and β-Mg_2SiO_4 transforms to γ-Mg_2SiO_4, with a 2% density increase.
- At depths of 650–750 km, Mg_2SiO_4 is driven by high pressures to break down via the reaction $Mg_2SiO_4 \rightarrow MgO + MgSiO_3$. The MgO is probably in the form of the mineral periclase, which has the same structure as NaCl. The $MgSiO_3$ joins up with Al_2O_3 to make $MgSiO_3{\cdot}Al_2O_3$, which has the same structure as ilmenite. Some SiO_2 is also present in a mineral called shishovite, where Si^{4+} is now 6-coordinated, and if Na^{1+} is present, it may form $NaAlSiO_6$.
- In the lower mantle, major changes in mineralogy are no longer possible, because the atoms cannot be packed any more tightly! Mineralogy probably consists of three phases: $(Ca,Mg,Fe)SiO_3$ with a perovskite structure, periclase, and some combination of $(Mg,Fe)Al_2O_4$ and $NaAlSiO_4$. There is only a slow increase in density down to the core-mantle boundary.
- The Earth's core is composed of outer liquid and inner solid regions that are predominantly Fe with a little Ni and possibly S.

It is important to appreciate that changing conditions mean changes in the types of crystalline phases that are stable. So the mineralogy of the Earth's interior is largely constrained by changes in pressure (and to a lesser extent, temperature) with depth. We infer this by the velocity of seismic waves and our knowledge of what mineral phases are stable in different depths in the Earth.

The Rock Cycle, Mineralogy-Style

Although we cannot directly observe the mineralogy of the Earth's mantle and core, we can see what goes on right beneath our feet in the crust. There we can understand the different minerals and rocks in terms of the rock cycle. If you've ever taken an introductory geology class, you've probably become acquainted with geological processes on Earth (and other terrestrial planets!). There are three basic rock types: igneous, metamorphic, and sedimentary rocks. Igneous rocks crystallize from a magma, either after it flows onto the Earth's surface (extrusive rock), or by cooling beneath it (intrusive rocks). Once at the surface, igneous rocks may undergo weathering by mechanical processes (e.g., erosion by wind, water, wave, or glaciers) or chemical ones (in which rocks react chemically with water) to form sediment. When the resultant sediments are lithified (turned into rock), they become sedimentary rocks. If sedimentary or igneous rocks are subjected to intense heat (by proximity to magma, for example) or pressure (as might be caused by collisions between continents), the elements in the minerals may recombine in the solid state to form the third rock type: metamorphic rocks. Metamorphic rocks themselves can be weathered to form sedimentary rocks, and sediments or metamorphic rocks can be subducted and melted to form new igneous rocks (Figure 20.15). Some mineralogical examples serve to demonstrate how these processes might occur. Begin with a granite, which is an intrusive igneous rock that commonly contains feldspar, quartz, mica and/or amphibole in roughly that order of abundance. A typical feldspar might be the mineral species, orthoclase, which has the formula of $KAlSi_3O_8$. If the granite reacts with rainwater, it can chemically weather to produce the sedimentary mineral kaolinite ($Al_2Si_2O_5(OH)_4$) by this reaction:

$$2KAlSi_3O_8 + 5H_2O + 2H^{1+} \rightarrow Al_2Si_2O_5(OH)_4 + 4H_2SiO_3 + 2K^{1+}$$

orthoclase + water + H^{1+} ions → kaolinite + silicic acid + K^{1+} ions.

Kaolinite is only stable at relatively low temperatures. If you now metamorphose that rock by increasing the pressure and temperature, it will react to form pyrophyllite and then an aluminosilicate mineral, kyanite, which is characteristic of metamorphic rocks. For example:

$$Al_2Si_2O_5(OH)_4 + SiO_2 \leftrightarrow Al_2Si_4O_{10}(OH)_2 + H_2O$$

kaolinite + quartz ↔ pyrophyllite + water

and then

$$Al_2Si_4O_{10}(OH)_2 \leftrightarrow Al_2SiO_5 + 3SiO_2 + H_2O$$

pyrophyllite ↔ Ky + Qtz + fluid

If you melted this rock and added a little K^{1+}, you'd be back to an igneous magma again. Such reactions between and among mineral species are really the basis of the geologic rock cycle!

Each of these mineral reactions needs a driving mechanism to make it happen. In some cases, they occur because of a change in composition—the addition of water, for example. In other cases, the reactions involve changes in temperature and pressure that occur as a result of the geologic processes associated with plate tectonics: continental subduction, burial of sediments, or even unroofing due to erosion. As the conditions change, the stabilities of the minerals in each rock type also change. For this reason, certain *groups of minerals* are characteristic of particular rock types, as shown in Tables 20.1–20.3. In some cases, particular *mineral species* are even characteristic of particular rock types.

To see how this happens, let's begin with **igneous rocks**, whose minerals constitute the primary mineralogy in the Earth's crust. Most magmas originate from partial melting at depths of less than 250 km in the crust, and they want to rise toward the surface because, as liquids, they are less dense than the surrounding solid rocks. Whether the magma erupts on to the Earth's surface as an extrusive igneous rock, or stalls out and cools beneath the surface (intrusive igneous rock), the minerals that initially crystallize are made up of the same set of ions that predominate in the crust: O^{2-}, Si^{4+}, Al^{3+}, Ca^{2+}, Na^{1+}, K^{1+}, Mg^{2+}, and $Fe^{2+ \text{ or } 3+}$. The most common minerals to crystallize from magmas are feldspar, olivine, pyroxene, amphibole, mica, and quartz (Table 20.1). The crystal sizes and the relative proportions of these minerals are used to distinguish different igneous rock types.

The sequence by which these minerals crystallize is determined by chemical reactions that occur when a magma is allowed to cool undisturbed. Of the Fe-Mg-bearing minerals, olivine will generally crystallize first, followed by pyroxene, amphibole, and biotite. Meanwhile, of

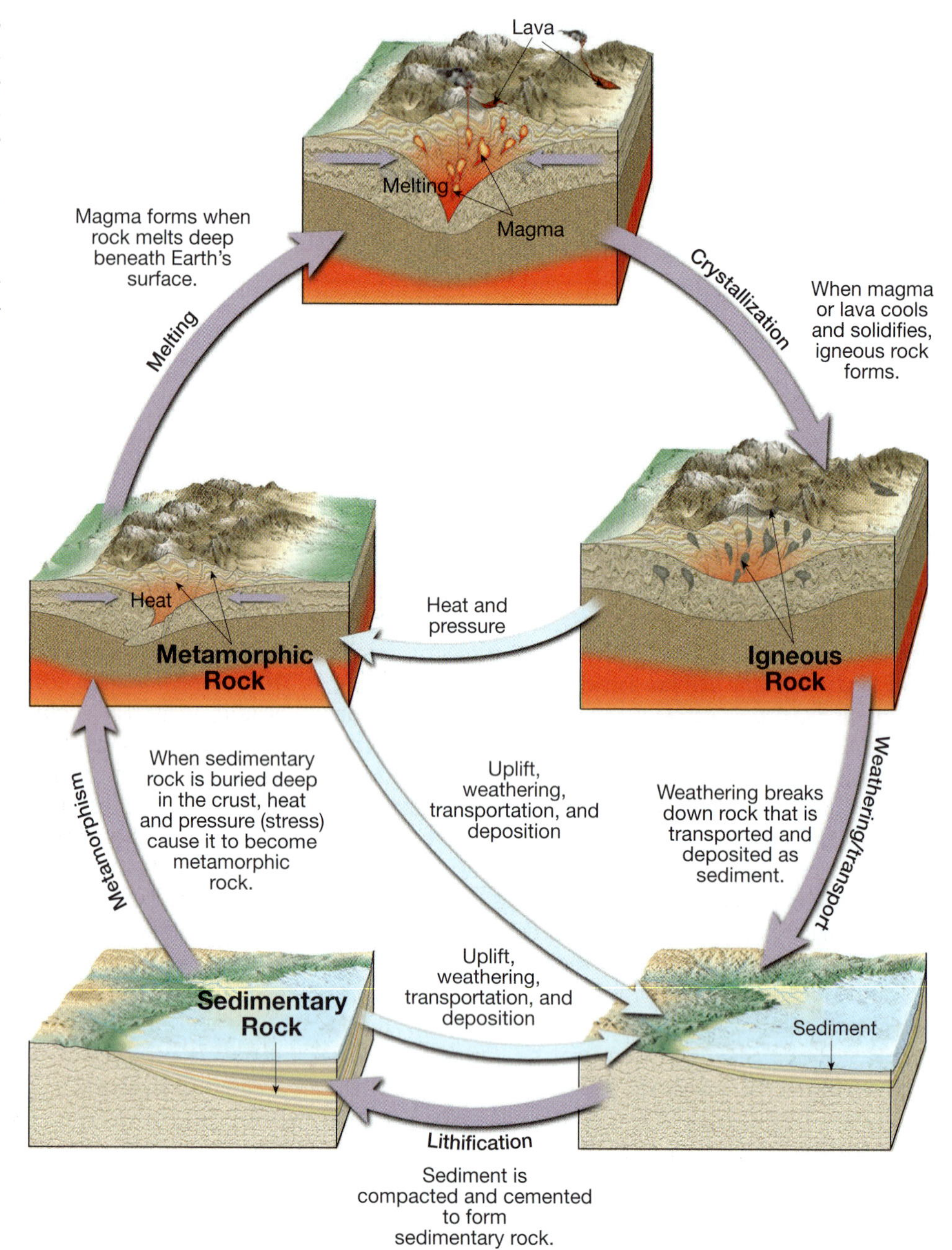
Lava
Melting
Magma
Magma forms when rock melts deep beneath Earth's surface.
Melting
Crystallization
When magma or lava cools and solidifies, igneous rock forms.
Heat
Metamorphic Rock
Heat and pressure
Igneous Rock
Metamorphism
When sedimentary rock is buried deep in the crust, heat and pressure (stress) cause it to become metamorphic rock.
Uplift, weathering, transportation, and deposition
Weathering breaks down rock that is transported and deposited as sediment.
Weathering/transport
Sedimentary Rock
Uplift, weathering, transportation, and deposition
Sediment
Lithification
Sediment is compacted and cemented to form sedimentary rock.

Figure 20.15. *The geologic rock cycle shows the interrelationships between the three basic rock types. From a mineralogist's point of view, these changes reflect changes in pressure and temperature, which in turn affect the stabilities of various minerals. For this reason, certain suites of minerals are diagnostic of particular rock types, as discussed in the text.*

the felsic (light-colored) minerals, Ca-rich plagioclase feldspar (anorthite) will crystallize first, and get gradually more Na-rich. Once these minerals have formed, the remaining magma will generally crystallize in the order K-feldspar first, then (rarely) muscovite, then quartz. This progression is called **Bowen's reaction series** (Figure 20.16).

The mineralogy of **sedimentary** rocks (Table 20.2) depends on the origins of the sediments that were lithified to make them. **Siliciclastic** sedimentary rocks are composed mostly of quartz and clay minerals like kaolinite because those minerals are what is left after the chemical reactions that break down the precursor rocks, as discussed with chemical formulas above. Detrital sediments are classified on the basis of the sizes of the particles. **Carbonate** sedimentary rocks are effectively chemical precipitates produced by organic and inorganic processes, and are by volume predominantly calcite ($CaCO_3$), which makes up the common rock type limestone. Most limestones form in marine environments from fragments of the shells and other hard parts of marine organisms (for example, coral). Other sedimentary minerals form from evaporation of shallow lakes, like· gypsum ($CaSO_4{\cdot}2H_2O$), halite ($NaCl$), and sylvite (KCl).

Composition	Rock names	Abundant minerals	Less abundant
Felsic (high in Si)	Granite, granodiorite, monzonite, rhyolite, dacite	Quartz/cristobalite/tridymite, microcline/orthoclase/sanidine, plagioclase, biotite, hornblende*	Apatite, zircon, magnetite, ilmenite, allanite, muscovite, epidote, titanite, rutile, fluorite, topaz, xenotime, monazite, spinel, garnet, fayalite, tourmaline, beryl, topaz, lepidolite, spodumene
Intermediate	Diorite, andesite	Plagioclase, hornblende, quartz, biotite, pigeonite, augite	Zircon, magnetite, ilmenite, epidote, apatite, allanite, titanite
Mafic	Basalt, gabbro	Forsterite, plagioclase, enstatite, hypersthene, augite, pigeonite	Magnetite, ilmenite, chromite, pseudobrookite, pyrrhotite
Ultramafic (low in Si)	Peridotite	Forsterite, enstatite-bronzite, diopside-augite, spinel, garnet	Kaersutite
Si-undersaturated**	Syenite, phonolite	Orthoclase, albite, nepheline, sodalite, leucite, aegirine, arfvedsonite, biotite, muscovite	Catapleiite, aenigmatite, monazite, apatite, eudialyte, magnetite, sodalite, nosean, hauyne, corundum

Table 20.1. Mineral Associations in Igneous Rocks

*Hornblende is a common name for calcic amphiboles in igneous and metamorphic rocks.
**Rocks with no quartz that may contain feldspathoid minerals, such as nepheline.

In **metamorphic** rocks, the mineralogy is more complicated because it depends on the starting composition, the minerals or elements present in the source rock (Table 20.3), and the intensity or grade of metamorphism (i.e., pressure and temperature). In some cases, the mineralogy does not change from metamorphism, though the fabric does: a metamorphosed limestone or dolomite ($CaCO_3$ or $MgCa(CO_3)_2$) can become a marble with the identical mineralogy. A metamorphosed quartz sandstone becomes a quartzite with the same mineralogy. In other cases, the starting rock may be a complex mixture of minerals. When you metamorphose a shale, which is a clay-rich sedimentary rock, the metamorphic rock that results will have variable mineralogy, depending on the pressure and temperature conditions. In these rocks, the elements recrystallize to form new (characteristic) minerals with discrete stability ranges. For example, low-temperature metamorphism might turn shale into a muscovite-rich rock with quartz and

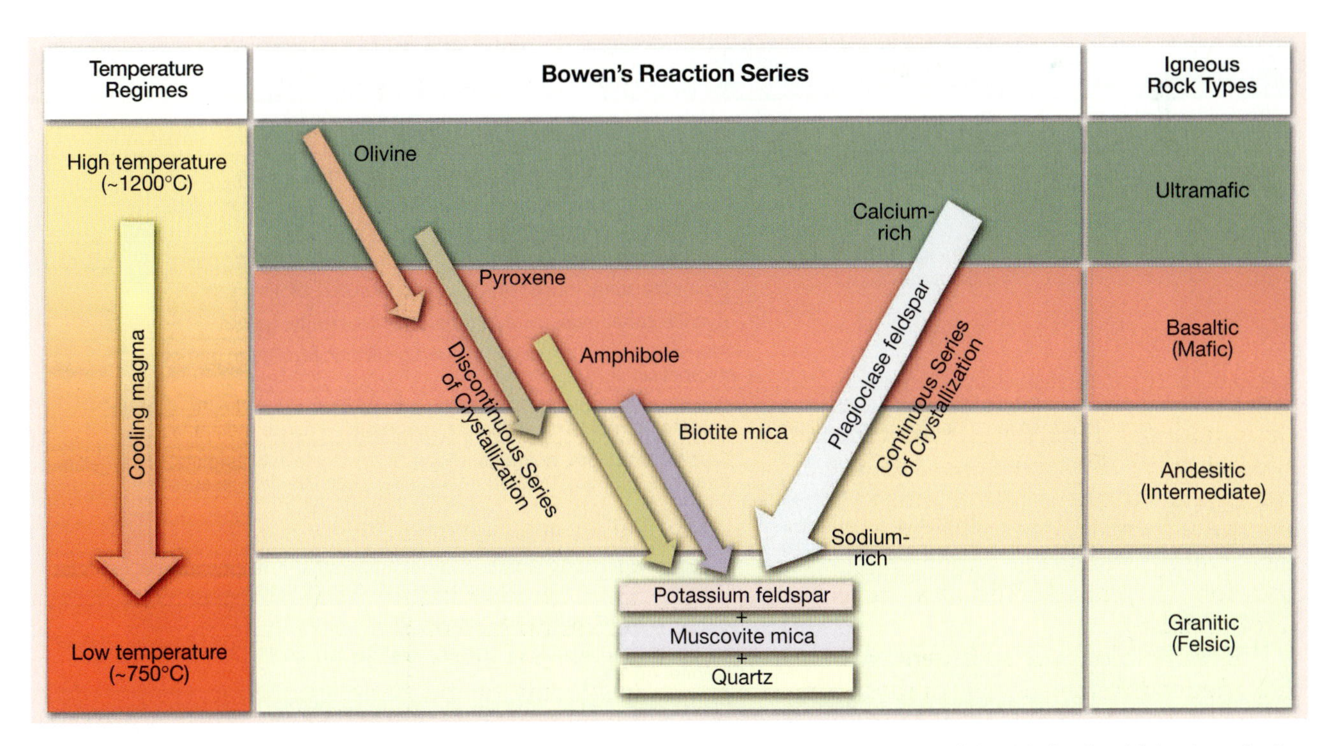

Figure 20.16. Bowen's reaction series shows the progression of Mg,Fe-rich (left side) and Si-rich (right side) minerals that will crystallize from a typical basaltic magma. The mineralogy of the resultant rock types is thus indicative of the temperatures at which they crystallize (see Table 20.1).

Table 20.2. Mineral Associations Sedimentary Minerals	
Quartz	Siliciclastic rocks*
Feldspars	
Clay minerals	
Calcite	Carbonate
Dolomite	
Gypsum	Evaporites
Anhydrite	
Halite	
Sylvite	
Opal	Chert
Chalcedony	
Hematite	Accessory minerals
Glauconite	

*Rocks composed of broken-up fragments of silicate minerals.

chlorite. With increasing temperature, garnet, biotite, staurolite, and sillimanite will form. In these rocks, only certain combinations of minerals can be found together.

Part of the study of petrology aims at understanding chemical reactions and depicting them using phase diagrams. The phase diagrams presented earlier in this chapter will prove useful in getting you started on this endeavor, but a full explanation will have to wait until you take petrology. For now, we have learned enough about mineral equilibria to allow us to explain the stability ranges for various minerals, as well as the typical associations that will be encountered in subsequent chapters. Now we can go on to present the nomenclature and classification systems that form the basis for the next chapters on systematic mineral identification of silicate and non-silicate minerals. In the end, we hope that these chapters will prepare you with the skills needed to recognize minerals both in isolation (on that mineralogy final exam!) and in rocks, which should also give you clues to their geological pasts.

Conclusion

At the beginning of this chapter, we promised to explain the phase relationships of water plus ice, the better to understand how highway engineers clear the roads! So the phase diagram for H_2O + NaCl is shown in Figure 20.17; by now, this should seem fairly simple! As we all know, plain water freezes at 0°C. As you add NaCl, the stability region for the liquid brine extends to lower temperatures. Practically, engineers rarely apply more than 10 wt% salt to a roadway, which limits the effective temperature for using NaCl to about –10°C. Even if you really dump the NaCl on a roadway, the lowest temperature at which liquid brine is stable is about –21°C. At really cold temperatures (below –6°F), even piles of salt on the roads won't melt the ice without help from some-

Table 20.3. Mineral Associations in Metamorphic Rocks

Rock Type	Source Rock	Mineralogy
Slate Phyllite Schist	Shale Siltstone	Chlorite zone: muscovite, quartz, chlorite, pyrophyllite
		Biotite zone: muscovite, quartz, biotite, phengite, pyrophyllite, chloritoid
		Garnet zone: muscovite, quartz, almandine, chloritoid
		Staurolite zone: muscovite, quartz, staurolite, biotite, garnet
		Kyanite zone: muscovite, quartz, kyanite, biotite, garnet
		Sillimanite zone: muscovite, quartz, sillimanite, biotite, garnet, microcline
Gneiss	Feldspathic sandstone Intermediate to felsic igneous rocks	Quartz, albite, microcline, biotite, orthopyroxene, cordierite, apatite, epidote, garnet, zircon, allanite, magnetite, ilmenite, hematite
Marble	Limestone, dolomite	Calcite, aragonite, dolomite, diopside, tremolite, wollastonite, grossularite, forsterite, phlogopite, epidote, brucite, periclase, talc, scapolite
Serpentinite, soapstone	Peridotite	Antigorite, lizardite, chrysotile, talc, anthophyllite, magnesite, brucite, chlorite, enstatite, forsterite
	Fe- and Mn-rich chert	Quartz, magnetite, hematite, garnet, hypersthene, stilpnomelane, epidote, pyroxenoids, Fe- and Mg-pyroxenes and amphiboles
Greenschist	Basalt, andesite, and sediments formed from volcanic rocks	Albite, epidote, actinolite, chlorite, biotite, ilmenite, titanite, quartz
Amphibolite		Hornblende, plagioclase, garnet, epidote/clinozoisite, biotite, ilmenite, titanite, quartz
Granofels		Plagioclase, garnet, ilmenite, titanite, rutile, quartz, hornblende, cordierite, clinopyroxene, orthopyroxene
Eclogite		Omphacite, garnet, quartz, rutile, muscovite, kyanite, zoisite/clinozoisite, glaucophane
Blueschist		Glaucophane, lawsonite, aragonite

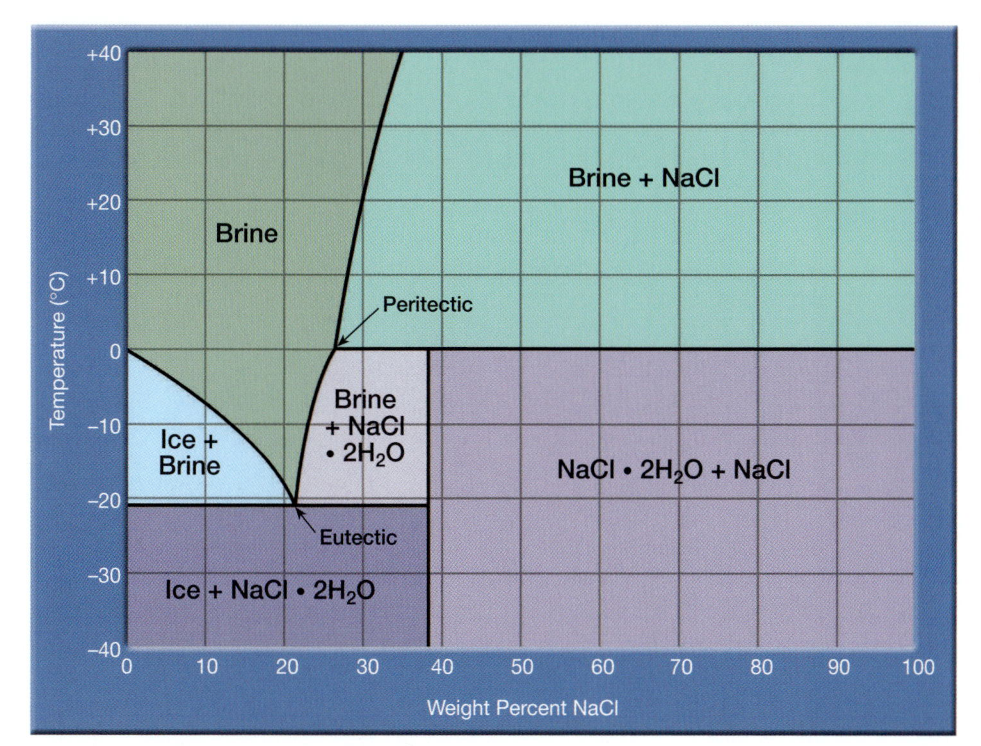

Figure 20.17. *Phase diagram for the binary H_2O-NaCl system. Notice that at high NaCl concentrations or very low temperatures, a new mineral species, hydrohalite (NaCl·$2H_2O$) is formed.*

thing else (like sunlight) to raise the temperature. This concept is very difficult to explain without phase diagrams! We'll find plenty more examples of their usefulness in the following chapters. And if you go on to take a class in petrology, understanding these diagrams will become a survival skill…

Acknowledgments. This chapter owes much to repeated lucid explanations of phase diagrams delivered to the author by Frank Spear, both in class and in writing, and to the helpful writings and advice of Tony Morse (Morse, 1994), Eric Essene, Stephen Nelson (www.tulane.edu/~sanelson/geol212), and Greg Finn. We thank Kenneth Windom for his inspirational animations of phase diagrams, and permission to adapt them. We acknowledge the author's lecture notes from courses with Roger Burns, which provide much of the context for the beginning of this chapter. We are indebted to Peter Crowley for his petrologic insights and patient explanations, which contributed hugely to this chapter, as well as for his careful revisions. Errors in content are, however, totally the responsibility of the authors!

References

Ehlers, E.G. (1972) The interpretation of geological phase diagrams. W.H. Freeman, 280 pp.

Morse, S.A. (1994) Basalts and Phase Diagrams: An Introduction to the Quantitative Use of Phase Diagrams in Igneous Petrology. Krieger Publishing Co., Malabar, FL, 493 pp.

Redfern, S.A.T., Salje, E. and Navrotsky, A. (1989) High-temperature enthalpy at the orientational order-disorder transition in calcite: implications for the calcite/aragonite phase equilibrium. Contributions to Mineralogy and Petrology 101, 479–484.

Spear, F.S. (1993) Metamorphic phase equilibria and pressure-temperature-time paths. Mineralogical Society of America, Washington, D.C., 799 pp.

Young, D.A. (2003) Mind over magma: the story of igneous petrology. Princeton University Press, Princeton, NJ, 686 pp.

Chapter 21

Nomenclature and Classification

Kingdom, phylum, class, order, family, genus, species... What's in a name? When I took 9th grade biology, I had the hardest time with the nomenclature for biological classification. I could not understand why there had to be so many, always in Latin, names for plants and animals. I found the whole system foreign and basically incomprehensible. No wonder I didn't do very well in the class (but then, I wasn't too keen on the dissections, either!).

In the world of minerals, you can probably imagine many different possible ways to classify. I often begin my mineralogy classes by giving each student a mineral and then asking the class to devise a classification system based on their observations of the entire suite of samples. I then leave the room. Eavesdropping from the hallway, I hear them arguing about the relative virtues of color, shape, cleavage, hardness as criteria for their new system. Eventually, they call me back into the room and unveil their new system, and nine times out of ten, it ends up being very close to the way mineralogists actually classify minerals. Why? Because, as we have endeavored to show you repeatedly in this textbook, macroscopic physical characteristics often reflect aspects of the structure, symmetry, and composition of minerals—and it is on those bases that we formally classify minerals.

Why are classification and nomenclature useful, anyway? First of all, they give us an intellectual framework within which to organize our knowledge of minerals. Second, they make our lives easier, because minerals in the same classes have similar structures and compositions—so when you learn one, you can more easily learn them all. Finally, they constitute an exact scientific vocabulary that allows comparison between mineral species at different localities. They make it possible to understand relationships between mineral species and everything else!

Here's one example. National attention was brought to the small town of Libby, Montana in November 1999 when a newspaper article in the Seattle Post-Intelligencer chronicled asbestos-related diseases found in former miners, with the asbestos contaminant being the amphibole mineral tremolite. Within days, the United States Environmental Protection Agency (EPA) arrived in Libby, and they have been there since. But it turns out that based on current regulatory definitions, the asbestos mineral species found at the mine was not asbestos. How can that be?

Government regulatory agencies currently regulate only five species of amphibole asbestos: riebeckite, grunerite, anthophyllite, actinolite, and tremolite. Much, if not all, of the amphibole at Libby is one of the non-regulated amphibole species winchite or richterite (Gunter et al., 2003). We mineralogists know that other amphibole species can, and do, exist in asbestos form (e.g., winchite-asbestos; Wylie and Huggins 1980). It would seem logical that all species of amphiboles that exhibit the asbestos properties should be regulated, and not just a select few. Ironically, medical review papers on the health effects of "tremolite" rely heavily on studies of the miners at Libby (Gunter et al., 2003), who probably inhaled winchite or richterite! There is no doubt that it is important for us as mineralogists and petrologists to set criteria, as was done for the amphiboles by Leake et al. (1997), for precise definition of mineral species names. We should insist that others in the sciences and regulatory fields use these names to avoid confusion.

So the point is, mineral names and nomenclature really are important! Please, read on...

M.D.D.

Introduction

In Chapter 2, we presented the basic taxonomy for classification of mineral species. As you will recall, the distinctions are made as follows:

- A **crystal** is defined as regions of matter composed of the same kind of atomic arrangement systematically repeated.
- A **mineral** is a homogeneous solid possessing a characteristic chemical composition (or a limited range of compositions) and a systematic three-dimensional order that is naturally-occurring either on Earth or in extraterrestrial bodies.
- A mineral **group** is a group of mineral species with the same type of crystal structure and variable compositions. Sometimes you will see the term **sub-group** used when many species exhibit the same chemistry and structure, but the chemistry covers only discrete ranges (Gaines et al., 1997).
- A mineral **series** involves two mineral species between which there is a complete range of naturally-occurring composition.
- A mineral **species** is a mineral distinguished from others by its well defined chemical and structural properties.
- A mineral **variety** is a specific type of a mineral species with some distinguishing characteristic such as color, habit, or other external physical characteristics.

So, for example, the mineral species muscovite is part of the mica group and the muscovite subgroup; other members of that subgroup include paragonite, glauconite, and celadonite, to name a few. Muscovite also has many variety names and old, now-discredited names as shown in Table 21.1. My favorite among these is *isinglass*, a term referring to micas that can be split into thin, transparent sheets. This characteristic allowed them to be used for windows and lampshades in the last two centuries, and more recently as insulators in electronics (this use has now been replaced by synthetics). Some of you may know the lyrics to a famous song from a Broadway musical that mentions isinglass curtains! Isinglass is still used in wood-stoves and lanterns due to its resistance to intense heat, and it is commercially available under that name. But strictly speaking, isinglass is not a mineral name—it is simply a variety name for the mineral species muscovite.

You may recall from Chapter 2 that mineral species names are regulated by the Commission on New Minerals, Nomenclature, and Classification (CNMNC). Of the more than 5500 approved mineral species, 45% are named in honor of a person (though only 2%, or about 86 species are named for women; see Table 21.2). Discovery locations lend their names to 23% of the species, 14% are named for their compositions, and 8% owe their names to distinctive physical properties (Blackburn and Dennen, 1997). Mineral species names are chosen by the person who first describes them (though frivolous names are not accepted); it is considered a great honor to have a mineral named after you!

Table 21.1. Some of the many variety names for the mineral species muscovite

adamsite	cat silver	glist
alourgite	chalcalaite	hammochrysos
ammochrysos	chalcaltocite	heshvitcite
ammocrisa	clay muscovite	higrophilita
ampohilogite	common mica	humbelite
anfilogita	czakaltait	isinglass
antonite	damourite	kalciumgumbelit
argent de chat	davreuxite	leucofilita
argile lithomarge	diszintribit	maria-glass
astrolite	epi-sericit	mirror stone
asztrolit	fahlunite	onkofillit
bildstein	falsa plata	schernikite
blister mica	fengit	sernikita
Ca-gumbelite	frauenglas	silkinit
cat gold	gabhardit	sternle

Historically, issues relating to mineral species names can sometimes be quite complicated. Such disputes are adjudicated by the CNMNC, the organization noted above. One infamous controversy arose over the mineral with the composition $CaTiSiO_5$. In 1795, a mineral with this formula was named **titanite** by M.H. Klaproth (Klaproth, 1795). A wet chemical analysis was presented, along with the first occurrence of the name *titanit* (later anglicized to titanite). The term titanite was an unfortunate choice because that is a chemical term that refers to an oxide of Ti^{3+}, not a silicate with Ti^{4+}. Meanwhile, in 1801, Haüy described the mineral sphene (Haüy, 1801), which was the commonly-used term for $CaTiSiO_5$ until 1980, when the IMA International Commission on New Minerals and Mineral Names had a vote on whether to choose titanite or sphene. Despite common usage of sphene and the misleading chemical terminology associated with this choice of name, titanite then became the approved species name on the basis of historical precedence. Thus, if you look in an older mineralogy book, you might find the name sphene in general usage, but since the publication of the IMA ruling in 1982 (Hey, 1982), titanite rules, and most mineralogists try to follow the recommendations of the governing body. Thus the oversight by the

CNMNC is very useful in keeping order among geologists. Can you imagine the big mess that mineral nomenclature would be if everyone chose to ignore the rules!

Another situation where group names and species names get mixed up occurs in field geology. Imagine that you are a metamorphic petrologist mapping in the backwoods of western Maine. As you walk a line perpendicular to a distant batholith, you notice that over a very short distance, the rocks change from containing mostly chlorite to containing mostly garnet, and you note this on your field map. Once you sample the rocks and return to the laboratory, you find out that the chlorite group mineral is clinochlore, and the garnet group mineral is almandine. In the field, you may have no way of knowing the mineral chemistry, so it's practical to use mineral group names in that context. But when you publish your analytical results, you need to use the mineral species names so that everyone will understand exactly what was there.

Of course, there are other ways of defining mineral names, including legal definitions (which are perhaps the worst!). This brings us back to the problem of asbestos identification.

Table 21.2.
Minerals named after women (Ciriotti, 2004a and b)

abswurmbachite	henmilite	nabokoite
allabogdanite	horváthite-(Y)	novgorodovaite
aurivilliusite	huangite	obertiite
benyacarite	jeanbandyite	olgite
bezsmertnovite	juanitaite	organovaite-Zn
bonshtedtite	kazakovite	palladium
bornemanite	kimrobinsonite	paganoite
borishanskite	kostylevite	perlialite
brassite	khristovite-(Ce)	petrovskaite
brodtkorbite	kogarkoite	rimkorolgite
bussenite	kraisslite	rosemaryite
caresite	krasnovite	sabinaite
charmarite	kupletskite	sakharovaite
christelite	kuzmenkoite-Mn	santabarbaraite
chursinite	labuntsovite-Mn	sazykinaite-(Y)
chvilevaite	laurite	schachnerite
clairite	liandratite	selenium
curite	lindbergite	shadlunite
dashkovaite	litvinskite	silvialite
donnayite-(Y)	lonsdaleite	sklodowskite
dugganite	malhmoodite	sofiite
effenbergerite	malinkoite	szenicsite
ekaterinite	marialite	tatyanaite
eveite	marsturite	thérèfsemagnanite
eylettersite	matulaite	vergasovaite
ferri-ottoliniite	mcnearite	volkovskite
gaidonnayite	mitryaevaite	weeksite
giniite	mozgovaite	winstanleyite
gutkovaite-Mn	mroseite	yakhontovite

The term "asbestos" should be used to describe any mineral with a particular set of properties. A mineralogist considers an amphibole to be asbestiform only if it exhibits the morphological characteristics of fibers that are easily separable. However, regulatory agencies base their classification on aspect ratio (i.e., the length of a particle divided by its width). If the aspect ratio is ≥ 3 and the particle is ≥ 5 μm in length, then it would be asbestos, regardless of the mineral's actual habit or size. Thus, many cleavage fragments are classified (incorrectly) as asbestos. Classifying asbestos based on aspect ratio has been, and will continue to be, hotly debated between mineralogists and the regulatory agencies.

To complicate the asbestos situation even further, amphibole nomenclature (i.e., the naming of species) is complex because of the variations in chemistry and the many substitutions that can occur. Leake et al. (1997 and 2003) revisited amphibole nomenclature. In their review, they divided amphiboles into five groups based on B-site occupancy, given the general formula $A\ B_2\ C_5\ T_8\ O_{22}\ (OH)_2$ as follows: (1) magnesium–iron–manganese–lithium group where $(Ca^{2+}+Na^{1+})_B < 1.0$ and $(Mg^{2+}+Fe^{2+}+Mn^{2+}+Li^{1+})_B \geq 1.0$, (2) calcic group where $(Ca^{2+}+Na^{1+})_B \geq 1.0$ and $Na^{1+}{}_B \leq 0.5$, (3) sodic–calcic group where $(Ca^{2+}+Na^{1+})_B \geq 1.0$ and $0.5 < Na^{1+}{}_B < 1.5$, (4) sodic group where $Na^{1+}{}_B \geq 1.5$, and (5) the Na^{1+}–Ca^{2+}–Mg^{2+}–Fe^{2+}–Mn^{2+}–Li^{1+} group, where $0.50 < (Mg^{2+}, Fe^{2+}, Mn^{2+}, Li^{1+})_B < 1.50$ and $0.50 \leq (Ca^{2+}, Na^{1+})_B \leq 1.50$ apfu. The 131 species names are then chosen in each of the five groups based on Si^{4+} content, $Mg^{2+}/(Mg^{2+}+Fe^{2+})$ ratio, and more detailed subdivisions of both the A- and B-site occupancy atom types and amounts. Interestingly, from a mineralogical perspective, asbestos-amphiboles occur in all four of the amphibole groups when the amphiboles crystallized with the asbestiform habit: anthophyllite and grunerite (the "amosite" asbestos variety name) belong to the magnesium–iron–manganese–lithium group, tremolite and actinolite belong to the calcic group, richterite and winchite to the sodic–calcic group, and riebeckite ("crocidolite" asbestos variety name) to the sodic group.

So the current situation is that regulatory agencies chose to regulate only five species of amphibole asbestos, given the proper morphology: riebeckite, grunerite, anthophyllite, actinolite, and tremolite. Sounds like legislation that was written without input from geologists, don't you think? Mineralogists have long known that tremolite and other amphibole species can, and do, have asbestiform habit. It would seem logical that all species of amphiboles should be regulated and not just a select few. Based on past and ongoing health studies of the Libby miners, the amphibole asbestos there has been very harmful, yet by current regulatory standards it would not be considered *asbestos*. This example well illustrates the importance of mineral nomenclature in something outside your classroom! Several other such stories are discussed in Chapter 24, including the regulations surrounding the SiO_2 polymorphs.

The CNMNC is also studying ways to create a better taxonomy for grouping mineral species names. Ongoing discussions suggest use of the following hierarchy:

- **Classes** of minerals defined on the basis of chemical composition (elements, oxides, etc.),
- **Sub-classes** of mineral species based on the main structural units (rings, chains, sheets, etc.),
- **Families** of mineral species sharing essential structural units, but with different overall structures,
- **Groups** of species with isostructural or closely related structures, compositions, and properties, with two or more minerals needed to define a group, and
- **Sub-groups** of species divided on the basis of chemistry (Na vs. Ca amphiboles) or symmetry (monoclinic vs. orthorhombic pyroxenes).

This taxonomy has not yet been formally approved, but it is interesting to consider and probably extremely useful.

Mineral Classification

Although the procedure for *naming* mineral *species* is completely consistent, thanks to the International Mineralogical Association (IMA), the *classification* of these minerals into groups is more troublesome. This is because the criteria for classification, which include chemistry and structure, vary according to the capabilities and needs of the user. The history of classification is well summarized by the IMA's Commission on Classification of Minerals (CCM, 2001), or CCM, which is the governing group for collection, documentation, and improvement of classification. We quote them here:

> "The ancient classification of minerals was mainly based on their practical uses, minerals being classified as gemstones, pigments, ores, etc., according to Theophrastos (372–287 B.C.) and to Plinius (77 A.D.). In the middle ages Geber (Jabir Ibn Hayyaan, 721–803) proposed a classification of minerals based on the external characters and on some physical properties such as fusibility, malleability, and fracture. This physical classification was developed by Avicenna (Ibn Sina, 980–1037), Agricola (1494–1555) and A. G. Werner (1749–1817), published by his student, L. A. Emmerling (1799). This system was substantially refined by F. Mohs (1773–1839) as *Natural-History System of Mineralogy* (Dresden, 1820), and used in the first editions of the *System of Mineralogy* by J. D. Dana (since 1837). With Werner the physical classification attained its maturity, and was generally adopted at the end of the 18th century. However, it became far too complicated. For instance, Werner mentioned 54 varieties for color. A. F. Cronstedt (1722–1765) seems to be the first to have outlined a classification whereby the chemical properties were taken first, followed by the physical properties. R. J. Hauy (1743–1822), in *Traité de Minéralogie* (1801), presented a mineral classification based on the nature of metals or, as he would say now, the nature of cations. In all fairness to some of these early workers, they didn't have the techniques to determine the composition of the minerals, so they based their classifications on such things as physical properties and uses.
>
> With the development of chemistry, the chemical properties became more and more important, and J. J. Berzelius (1779–1848) in 1819 proposed a chemical classification of minerals. He recognized that minerals with the same nonmetal (anion or anionic group) had similar chemical properties and resembled one another far more than minerals with a common metal. He considered minerals as salts of anions and anionic complexes: F^{1-}, Cl^{1-}, Br^{1-}, I^{1-}, O^{2-}, S^{2-}, Se^{2-}, Te^{2-}, NO^{3-}, CO_3^{2-}, BO_3^{3-}, SO_4^{2-}, PO_4^{3-}, SiO_4^{4-}, BO_4^{3-}, that is to say, as chlorides, sulphates, silicates, etc., and not as minerals of zinc, copper, etc. At this time Christian Samuel Weiss (1780–1856) introduced the seven crystal systems (1815) and Mitscherlich discovered isomorphy (1819) and polymorphy (1824)."

The stage was set for someone to devise a comprehensive system for organizing the rapidly-expanding number of known minerals.

The Dana System of Mineralogy

The first edition of the *System of Mineralogy* was published in 1837 by James Dwight Dana, a Yale professor who was one of the foremost geologists in the U.S. at the time. Dana got his education at Yale, studying under his future father-in-law Benjamin Silliman (after whom the mineral sillimanite was named). When he left Yale, he became an instructor for the U.S. Navy and traveled the Mediterranean, where he saw Vesuvius erupt (when he was in his 70s, he wrote a seminal text on volcanology). He returned to Yale to settle down in 1834, where his own childhood mineral collection, combined with Silliman's, inspired him to come up with a chemical and crystallographical classification system for minerals. He wrote the *System of Mineralogy* in only 10 months, and it was published when he was 24 years old. It is now in its 8th edition (Gaines et al., 1997). In his free time, he was also an expert on crustaceans and corals. An excellent summary of Dana's life can be found in Natland (2003).

Dana's original system predated accurate mineral chemistry, though his later editions (beginning with the 3rd and 4th, published in 1850 and 1854) eventually used chemistry as the main criterion for classification. This was helped along by the work of Gustav Rose (1798–1873), who proposed a chemical-morphological approach that included: I. elements, II. sulfides, III. halides, and IV. oxygen compounds, the latter being divided into simple and complex oxides such as carbonates, phosphates, silicates, borates, and sulfates. By the time Dana published his 5th edition of the *System of Mineralogy* in 1892, the chemical-morphological classification system was fully mature, and it became widely adopted. The approach was further supported by P. V. Groth (1843–1927) in his many editions of *Tabellarische Übersicht der Mineralien nach ihrer Kristallographisch-chemischen Beziehungen* (e.g., 1889).

The Dana classification system is still the most widely adopted (although its name may not be known to the folks who use it!), so we have chosen to organize the next chapters and our database largely within its framework. Each mineral species has a unique classification number as developed in earlier editions of the *System of Mineralogy*, and more recently by Abraham Rosenzweig and Eugene Foord as part of the 8th edition. As explained there (Gaines et al., 1997), each mineral species has a four-part number. The first number (Table 21.3) is its **class**, which is based on composition and dominant structural elements. The second number is the **type**, which specifies the ratio of cations to anions in the mineral. The third and fourth numbers group mineral species with similar structures within each type and class. Thus, each mineral species has a unique numerical value that is useful for its classification and for understanding its structure. Minerals with low Dana numbers generally have simpler structures than minerals with high ones, though that is not always the case.

All the Dana Classification numbers used in this text come from the 8th edition of the *System of Mineralogy* (Gaines et al., 1997), which has 1819 pages! However, this book is worth hefting because it also contains information on the derivation of each mineral name, its crystallographic parameters, physical and optical properties, occurrences and localities. Chapters 22 and 23, which follow this one, will be organized along the lines of Dana classifications, beginning with the highest Dana numbers and then working our way down. This order is based on our view that, in general, silicates are more important to the earth scientists than non-silicates. The Dana system is often the basis for the organization of mineral collections at museums and universities.

Strunz Classification

Another way of organizing mineral collections is to combine chemical and structural criteria. Again, the CCM (2001) summarizes the history of this process:

> "After 1913, when the first structures of minerals were determined, the structural criterium for classification was taken into account. The first classifications of this type, which take into consideration the distribution of bonds in a structure, are that of silicates proposed by Machatschki (1946), Náray-Szabó (1969), and developed by Bragg (1937). This chemical-plus-structural classification has been applied to many other branches of mineralogy such as fluoraluminates (Pabst, 1950), aluminosilicates (Liebau, 1956), silicates and other minerals with tetrahedral complexes (Zoltai, 1960), phosphates (Liebau, 1966; Corbridge, 1971), sulfosalts (Makovicky, 1981, 1993), and borates (Strunz, 1997). H. Strunz introduced a chemical-structural classification of the entire domain of minerals (*Mineralogische Tabellen*, 1941), followed by A. S. Povarennykh with a modified classification (1966 in Russian, 1972 in English)."

Some museums now use the Strunz chemical-structural classification system, which includes ten classes that are closely related to the Dana groups (Strunz, 2001): 1. elements, 2. sulfides and sulfosalts, 3. halides, 4. oxides and hydroxides, 5. carbonates and nitrates, 6. borates, 7. sulfates, chromates, molybdates, selenates, tellurates, and wolframates, 8. phosphates, arsenates, and vanadates. 9. silicates, and 10. organic compounds. The main subdivisions within each class are given in Table 21.4. Notice in the table that the symbols for the divisions are not all consecutive; unassigned letters have been built into the classification to allow for possible future discoveries of new minerals. A related classification system by Kostov (1975) emphasized not only structures, but also chemistry and geochemistry (paragenesis).

Table 21.3. Dana Classification for Minerals (Gaines et al., 1997)

Class Name	Class	Name	Examples
Native Elements and Alloys	1	Native Elements and Alloys	gold, silver, copper
Sulfides and Related Compounds	2	Sulfides, including Selenides and Tellurides	chalcocite, bornite, stibnite
	3	Sulfosalts	tennantite, proustite, energite
Oxides	4	Simple Oxides	ferrihydrite, corundum, ilmenite
	5	Oxides Containing Uranium, and Thorium	uraninite, curite
	6	Hydroxides and Oxides Containing Hydroxyl	goethite, lepidocrocite, brucite
	7	Multiple Oxides	gahnite, magnetite, chrysoberyl
	8	Multiple Oxides Containing Niobium, Tantalum, and Titanium	pyrochlore, betafite, columbite
Halogenides	9	Anhydrous and Hydrated Halides	sylvite, halite, fluorite
	10	Oxyhalides and Hydroxyhalides	atacamite, boleite
	11	Halide Complexes; Alumino-flourides	avogadrite, cryolite
	12	Compound Halides	creedite, arzrunite
Carbonates	13	Acid Carbonates	trona, nahcolite
	14	Anhydrous Carbonates	calcite, aragonite, dolomite
	15	Hydrated Carbonates	natron, ikaite, zellerite
	16	Carbonates Containing Hydroxyl or Halogen	parisite, azurite, malachite, rosasite
	17	Compound Carbonates	susannite, leadhillite
Nitrates	18	Nitrates	niter, nitratine
	19	Nitrates Containing Hydroxyl or Halogen	gerhardtite, sveite
	20	Compound Nitrates	darapskite
Iodates	21	Anhydrous and Hydrated Iodates	lautarite, bellingerite
	22	Iodates Containing Hydroxyl or Halogen	salesite, seeligerite
	23	Compound Iodates	dietzeite, fuenzalidaite
Borates	24	Anhydrous Borates	ludwigite, sinhalite
	25	Anhydrous Borates Containing Hydroxyl or Halogen	hambergite, fluoborite, boracite
	26	Hydrated Borates Containing Hydroxyl or Halogen	ulexite, colemanite
	27	Compound Borates	wiserite, iquiqueite
Sulfates	28	Anhydrous Acid and Sulfates	barite, anhydrite, glauberite
	29	Hydrated Acid and Sulfates	rhomboclase, gypsum, chalcanthite, melanterite
	30	Anhydrous Sulfates Containing Hydroxyl or Halogen	alunite, jarosite
	31	Hydrated Sulfates Containing Hydroxyl or Halogen	spangolite, amarantite, copiapite
	32	Compound Sulfates	burkeite, hanksite
Selenates and Tellurates; Selenites and Tellurites	33	Selenates and Tellurates	khinite, girdite, kuksite
	34	Selenites, Tellurites and Sulfites	scotlandite, smirnite
Chromates	35	Anhydrous Chromates	crocoite, chromatite
	36	Compound Chromates	iranite, hemihedrite
Phosphates, Arsenates, and Vanadates	37	Anhydrous Acid Phosphates, Arsenates, and Vanadates	monetite, weilite
	38	Anhydrous Normal Phosphates, Arsenates, and Vanadates	triphylite, lithiophilite, monazite
	39	Hydrated Acid Phosphates, Arsenates, and Vanadates	fluckite, brushite
	40	Hydrated Normal Phosphates, Arsenates, and Vanadates	autunite, torbernite, carnotite, vivianite
	41	Anhydrous Phosphates, Arsenates, and Vanadates containing Hydroxyl or Halogen	jarosewichite, adelite, descloizite, triplite, apatite
	42	Hydrated Phosphates, Arsenates, and Vanadates containing Hydroxyl or Halogen	eosphorite, turquoise, laueite
	43	Compound Phosphates, Arsenates, and Vanadates	beudantite, bradleyite
Antimonates, Antimonites, and Arsenites	44	Antimonates	stibiconite, brizziite
	45	Acid and Normal Antimonates and Arsenites	reinerite, trigonite, ludlockite

Table 21.3. (continued)

Class Name	Class	Name	Examples
Antimonates, Antimonites, and Arsenites (continued)	46	Basic or Halogen-Containing Antimonites, Arsenites	finnemanite, seelite
Vanadium Oxysalts	47	Vanadium Oxysalts	rossite, hummerite
Molybdates and Tungstates	48	Anhydrous Molybdates and Tungstates	ferberite, hübnerite, wulfenite
	49	Basic and Hydrated Molybdates and Tungstates	hydrotungstite, moluranite
Organic Compounds	50	Salts of Organic Acids	whewellite, dinite
Nesosilicates: Insular SiO_4	51	Insular SiO_4 Groups Only	phenakite, fayalite, almandine
	52	Insular SiO_4 Groups and O, OH, F, and H_2O	kyanite, staurolite, topaz, titanite
	53	Insular SiO_4 Groups and Other Anions or Complex Cations	uranophane, spurrite
	54	Borosilicates and Some Beryllosilicates	grandidierite, dumortierite, datolite
Sorosilicates: Isolated Tetrahedral Noncyclic Groups	55	Si_2O_7 Groups, Generally with No Additional Anions	barylite, melilite, gehlenite
	56	Si_2O_7 with Additional O, H, F and H_2O	bertrandite, hemimorphite, ferroaxinite, ilvaite
	57	Insular Si_3O_{10} and Larger Noncyclic Groups	aminoffite, kinoite, zunyite
	58	Insular, Mixed, Single, and Larger Tetrahedral Groups	kornerupine, epidote, allanite, zoisite
Cyclosilicates	59	Three-Membered Rings	benitoite, catapleiite
	60	Four-Membered Rings	joaquinite, papagoite
	61	Six-Membered Rings	beryl, cordierite, elbaite
	62	Eight-Membered Rings	muirite, megacyclite
	63	Condensed Rings	milarite, sugilite
	64	Rings with Other Anions and Insular Silicate Groups	eudialyte, traskite
Inosilicates: Two-Dimensionally Infinite Silicate Units	65	Single-Width Unbranched Chains W=1	enstatite, diopside, augite, spodumene, wollastonite
	66	Double-Width Unbranched Chains W=2	grunerite, gedrite, anthopyllite
	67	Unbranched Chains with W>2	jimthompsonite
	68	Structures with Chains of More Than One Width	chesterite, vinogradovite
	69	Chains with Side Branches or Loops	astrophyllite, aenigmatite
	70	Column or Tube Structures	litidionite, fenaksite
Phyllosilicates	71	Sheets of Six-Membered Rings	kaolinite, antigorite, biotite
	72	Two-Dimensional Infinite Sheets with Other Than Six-Membered Rings	prehnite, fluorapophyllite, petalite
	73	Condensed Tetrahedral Sheets	gyrolite, reyerite
	74	Modulated Layers	stilpnomelane, sepiolite
Tektosilicates	75	Si Tetrahedral Frameworks	quartz, coesite, opal
	76	Al-Si Frameworks	albite, anorthite, orthoclase
	77	Zeolites	analcime, natrolite, scolecite
Unclassified	78	Unclassified Silicates	apachite, fukalite, terskite

Structure Classification

Scientists who work with chemical compounds that do not fit the criteria for mineral species are effectively left out in the cold by the other classification systems. Instead, common structure types in minerals and related inorganic compounds are often grouped and studied on the basis of their structure types alone. This nomenclature is extensively used in the fields of material science, inorganic chemistry, and even the mineral group zeolites. It can also be useful in showing the relationships among mineral structures that are independent of the Dana system.

"The structural classification of minerals was first proposed by J. Lima de Faria in 1983. It corresponds to the application of the general structural classification of inorganic compounds (Lima de Faria and Figueiredo, 1978) to minerals, which are an integral part of them. The most general approach to the structural systematics is based on the analysis of the strength distribution in crystal structures and of the directional character of the bonds. There are atoms that are more tightly bounded, and these assemblages are called structural units. They are considered as the main basis for the structural classification of minerals. Thus there are five main categories of

Table 21.4. Strunz Mineralogical Classification (Strunz and Nickel, 2001)

Class	Division	Examples*
1. Elements	1.A: Metals and intermetallic alloys	copper, silver, gold
	1.B: Metallic carbides, silicides, nitrides, and phosphides	haxonite, roaldite, osbornite
	1.C: Metalloids and nonmetals	graphite, sulfur, diamond
	1.D: Nonmetallic carbides and nitrides	moissanite, sinoite, nierite
2. Sulfides and Sulfosalts	2.A: Metal/metalloid alloys	maldonite, orcelite, polarite
	2.B: Metal sulfides, M:S > 1:1 (mainly 2:1)	chalcocite, pentlandite, bornite
	2.C: Metal sulfides, M:S = 1:1 (and similar)	sphalerite, chalcopyrite, wurtzite
	2.D: Metal sulfides, M:S = 3:4 and 2:3	stibnite, linnaeite, edgarite
	2.E: Metal sulfides, M:S ≤ 1:2	pyrite, marcasite, arsenopyrite
	2.F: Sulfides of arsenic, alkalies, and sulfides with halide, oxide, hydroxide, H_2O	realgar, orpiment, tochlinite
	2.G: Sulfarsenates, sulfantimonites, sulfbismuthites	proustite, tennantite, simonite
	2.H: Sulfosalts of SnS archetypes	meneghinite, jamesonite, cylindrite
	2.J: Sulfosalts of PbS archetypes	pavonite, junoite, bursaite
	2.K: Sulfarsenates	fangite, billingsleyite
3. Halides	3.A: Simple halides, without H_2O	halite, fluorite, gananite
	3.B: Simple halides, with H_2O	hydrohalite, antarcticite
	3.C: Complex halides	cryolite, creedite, douglasite
	3.D: Oxyhalides, hydroxyhalides, and related double halides	atacamite, matlockite, kleinite
4. Oxides	4.A: Metal:oxygen = 2:1 and 1:1	ice, cuprite, periclase
	4.B: Metal:oxygen = 3:4 and similar	chrysoberyl, spinel, magnetite
	4.C: Metal:oxygen = 2:3, 3:5, and similar	corundum, ilmenite, hematite
	4.D: Metal:oxygen = 1:2 and similar	quartz, rutile, pyrochlore
	4.E: Metal:oxygen ≤ 1:2	tantite, molybdite
	4.F: Hydroxides (without V or U)	diaspore, brucite, gibbsite
	4.G: Uranyl hydroxides	schoepite, studtite, curite
	4.H: $V^{[5,6]}$ vanadates	uvanite, vanalite, duttonite
	4.J: Arsenites, antimonites, bismuthinites, sulfites, selenites, tellurites	nealite, marthozite, magnolite
	4.K: Iodates	lautarite, salesite, dietzeite
5. Carbonates + Nitrates	5.A: Carbonates without additional anions, without H_2O	calcite, aragonite, dolomite
	5.B: Carbonates with additional anions, without H_2O	azurite, malachite, parisite
	5.C: Carbonates without additional anions, with H_2O	barringtonite, tuliokite, ikaite
	5.D: Carbonates with additional anions, with H_2O	hydromagnesite, coalingite
	5.E: Uranyl carbonates	blatonite, voglite, sharpite
	5.N: Nitrates	niter, gerhardtite, sveite
6. Borates	6.A : Monoborates	sassolite, hambergite, ludwigite
	6.B : Diborates	wiserite, pinnoite, vimsite
	6.C : Triborates	inderite, hydroboracite, howlite
	6.D : Tetraborates	borcarite, kernite
	6.E : Pentabortes	ulexite, probertite, hilgardite
	6.F : Hexaborates	aksaite, fabianite, ginorite
	6.G : Heptaborates and other megaborates	boracite, rhodizite, metaborite
	6.H: Unclassified borates	chelkarite, satimolite, priceite
7. Sulfates	7.A: Sulfates (selenates, tellurates, chromates, molybdates, wolframates) without additional anions, without H_2O	thenardite, anhydrite, barite
	7.B: Sulfates (selenates, tellurates, chromates, molybdates, wolframates) with additional anions, without H_2O	caminite, alunite, linarite
	7.C: Sulfates (selenates, tellurates, chromates, molybdates, wolframates) without additional anions, with H_2O	bonattite, voltaite, alum, gypsum
	7.D: Sulfates (selenates, tellurates, chromates, molybdates, wolframates) with additional anions, with H_2O	amarantite, botryogen, copiapite

Table 21.4. (continued)

Class	Division	Examples*
7. Sulfates (continued)	7.E: Uranyl sulfates	uranopilite, johannite, zippeite
	7.F: Chromates	tarapacaite, crocoite, iranite
	7.G: Molybdates and wolframates	scheelite, paranite, lindgrenite
	7.H: Uranium and uranyl molybdates and wolframates	sedovite, cousinite, mourite
8. Phosphates	8.A: Phosphates, arsenates, vanadates without additional anions, without H_2O	monazite, berlinite, xenotime
	8.B: Phosphates, arsenates, vanadates with additional anions, without H_2O	amblygonite, triplite, fluorapatite
	8.C: Phosphates, arsenates, vanadates without additional anions, with H_2O	variscite, vivianite, wicksite
	8.D: Phosphates, arsenates, vanadates with additional anions, with H_2O	turquoise, foggite, wardite
	8.E: Uranyl phosphates and arsenates	autunite, walpurgite, upalite
	8.F: Polyphosphates, polyarsenates, [4]-polyvanadates	blossite, volborthite, alvanite
9. Silicates (Germanates)	9.A: Nesosilicates	forsterite, staurolite, chloritoid
	9.B: Sorosilicates	ilvaite, epidote, zoisite,
	9.C: Cyclosilicates	benitoite, beryl, cordierite
	9.D: Inosilicates	diopside, jadeite, anthophyllite
	9.E: Phyllosilicates	paragonite, biotite, glauconite
	9.F: Tektosilicates without zeolitic H_2O	nepheline, albite, scapolite
	9.G: Tektosilicates with zeolitic H_2O; zeolite family	natrolite, analcime, heulandite
	9.H: Unclassified silicates	ertixiite, apachite, thornasite
	9.J: Germanates	carboirite, bartelkeite
10. Organic Compounds	10.A: Salts of organic acids	acetamide, caoxite, mellite
	10.B: Hydrocarbons	hartite, dinite, karpatite
	10.C: Miscellaneous organic minerals	refikite, hoelite, urea, uricite

*Don't go looking for most of these minerals in the database that comes with this text! Many of these are pretty obscure, and would only act to confuse you in mineral searches, so they are not included. Many more comprehensive (and more confusing!) on-line databases, as well as any inclusive book on mineral species, will satisfy your curiosity as to the formulae, occurrences, and origins of some of these interesting mineral names.

structures: atomic or close-packed, group, chain, sheet, and framework according to their dimensionality. This approach to the analysis of the crystal structures was approved by IUCr Commission on Crystallographic Nomenclature (Lima de Faria et al., 1990). Hawthorne (1984, 1985) also proposed a structural classification of minerals based on the polymerization of coordination polyhedra. Lima de Faria, in 1994, applied the structural classification to the most common minerals (about 500 minerals organized in 230 structure types). Detailed structural classification of silicates was elaborated by Liebau (1983)." (CCM, 2001)

As described by Smith et al. (1998), a systematic structural approach is also used by the International Centre for Diffraction Data (ICDD). It uses twelve different letters of the alphabet to represent cations, anions, and molecules at various sites in the structure (T is for tetrahedral cations, G is for octahedral ones, etc.). More generally, the Smith et al. (1998) system defines:

A for any cation in any coordination
B for a second cation, in a different coordination
M for neutral molecular units
X for anions that are all the same element, and
Z for anions that are different elements.

These principles are described in detail in Smith et al. (1998).

An excellent succinct summary of the useful application of these ideas is given in Navrotsky (1994), as well as various material science texts including Allen and Thomas (1999) and Gersten and Smith (2001). We choose to discuss structure types in some detail here because they have such broad applicability in all fields dealing with crystal structures. The vocabulary of structure types presents an interesting complementary viewpoint to the Dana system, which we will discuss in the following two chapters. It allows us to draw parallels between minerals that have nothing in common geologically, but everything in common structurally. So, here we present an introduction to structure types, showing many common silicates in the context of this classification scheme. We will also revisit these minerals again in the subsequent chapters to show how they fit into the Dana scheme.

The basic idea behind structure types is that you can mix and match different cations and anions in similar structures. The various possible combinations are governed by the need for structures to be charge balanced and the anions and cations to fit together. A quick inspection of Table 3.4 shows that most of the commonly occurring *anions* have charges of –1 or –2, while *cations* are generally +1, +2, +3, or +4; thus, there are a limited number of possible combinations of ions with these charges (to maintain charge balance). Liu and Bassett (1986) give a comprehensive description of the phase relations of all these permutations, which are summarized in Table 21.5.

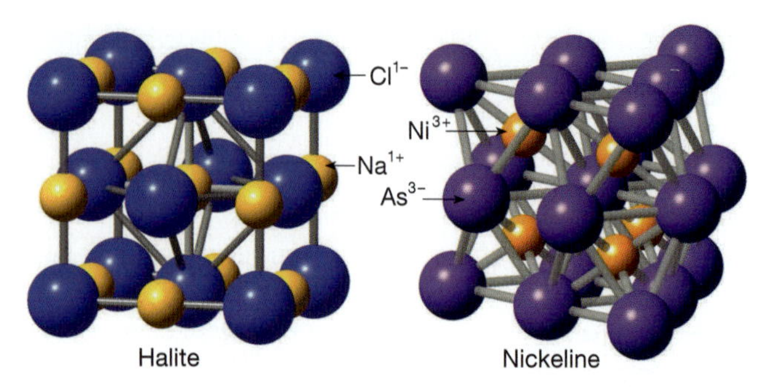

Figure 21.1. Cubic close packing (ccp) of Cl^{1-} with Na^{1+} in the octahedral interstices results in the halite (NaCl) structure (left), while hexagonal close packing (hcp) of As^{3-} with Ni^{3+} in the octahedral interstices results in the nickeline (NiAs) structure (right).

AX Compounds

The simplest compounds are composed of two ions with identical but opposite charge. These are binary compounds with a stoichiometry (i.e., chemical formula) of AX. We can predict their structures largely on the basis of radius ratio (Chapter 8), even though they may exhibit a combination of ionic and covalent bonding:

$Na^{1+}Cl^{1-}$	$R_{Na}/R_{Cl} = 0.56$	6-coordination
$Mg^{2+}O^{2-}$	$R_{Mg}/R_{O} = 0.51$	6-coordination
$Pb^{2+}S^{2-}$	$R_{Pb}/R_{S} = 0.65$	6-coordination
$Zn^{2+}S^{2-}$	$R_{Zn}/R_{S} = 0.40$	4-coordination
$Ni^{3+}As^{3-}$	$R_{Ni}/R_{As} = 0.31$	4-coordination

All of the AX compounds can be visualized in this context as cubic close-packed structures (ccp) of atoms (Chapter 8). The halite structure (Figure 21.1) results when the octahedral interstices of a cubic close-packed array of Cl^{1-} anions are filled by Na^{1+}; in this arrangement, both the Na^{1+} and the Cl^{1-} are in octahedral coordination. Many of the monoxides crystallize with the halite structure, including NiO (bunsenite), CoO, FeO (periclase), CdO (monteponite), CaO (lime), EuO, CuO (tenorite), AgO, SrO, and BaO.

If the octahedral interstices of a hexagonal close-packed (hcp) array are filled, then the resultant structure is that of nickeline (NiAs) (Figure 21.1). This structure is commonly found in sulfides and other chalcogenides such as millerite (NiS) and freeboldite (FeSe), perhaps because the edge-sharing octahedra are better suited to bonds with some covalent character.

A body-centered cubic lattice with alternating Cs^{1+} and Cl^{1-} at the lattice points forms the basis for the CsCl structure, which is found in metal alloys, and is the high-pressure form of CaO and several other oxides. This structure can be envisioned as two interpenetrating simple cubic arrays (i.e., an isometric primitive lattice) where the corner of a Cs^{1+} cell sits at the body center of the Cl^{1-} cube, and vice versa. The result is that each ion is 8-coordinated (Figure 21.2).

If the tetrahedral interstices of a cubic close-packed structure are filled, then the sphalerite (ZnS) structure results (Figure 8.6). As noted in Chapter 8, this can also be thought of as two interpenetrating lattices offset by half the body diagonal of the cube, but they are *face*-centered arrays. Each Zn^{2+} cation is in tetrahedral coordination with S^{2-}.

Finally, if the tetrahedral interstices of a hexagonal close-packed array are filled, the wurtzite (ZnS) structure results (Figure 8.6). As with sphalerite, the wurtzite structure can be visualized as two interpenetrating arrays of Zn^{2+} and S^{2-}, though in this mineral the arrays are hcp. Each atom is again in tetrahedral coordination with the four surrounding ions (Figure 8.6). ZnO (zincite) and BeO (bromellite) crystallize in the same structure as wurtzite, and can be visualized as two interpenetrating hcp arrays of Zn^{2+} or Be^{2+} and O^{2-}.

A_2X Compounds

Because most cations with a charge of 1+ are too large to form stable structures with anions (such as O^{2-}), there are not a large number of A_2X compounds known. Two of the A_2X compounds are important, however. We have previously encountered the H_2O (ice) structure (Figure 8.21), but we can now describe it in terms of packing of H_2O molecules. If the structure is viewed perpendicular to the *c* axis (Figure 21.3), H_2O molecules (here represented by H_4O tetrahedra because the positions of the two H's are really time-averaged over four positions) are seen to be joined laterally and stacked (parallel to *c*) in an ABABAB stacking sequence. This is very close to hexagonal close packing, though the ice structure is very open and has a low packing density of only about 34%—that is why ice floats!

Cuprite (Cu_2O) was one of the first mineral structures ever to have its structure determined by X-ray diffraction. To understand its structure in terms of packing (Figure 21.4), we need first to return to the sphalerite structure we studied earlier (Figure 8.6), which is also similar to the structure of diamond (Figure 8.6). Both of these are based on face-centered lattices. If we replace the Zn^{2+} cations with Si^{4+}, and the S^{2-} anions with O^{2-}, we would obtain a high temperature form of SiO_2, which is called cristobalite. Now, what would happen if we substitute Cu^{1+} cations for the Si^{4+}? Cu^{1+} is a much bigger cation than Si^{4+}, so a cristobalite-like framework of Cu^{1+} and O^{2-} ions must have a lot of unoccupied space. So there is room for a second, identical framework of Cu-O ions within the volume occupied by the original framework (Figure 21.4). Ag_2O also forms with a cuprite structure.

Chalcocite also has an A_2X structure, though it is based on a hexagonal close-packed arrangement of S^{2-} anions. The Cu^{1+} cations are in both 3- and 4-fold coordination with the S^{2-} anions (Figure 21.5).

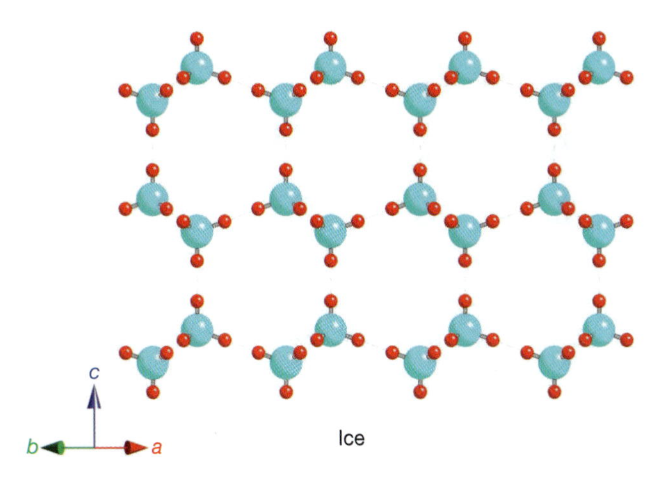

Figure 21.3. The ice structure is nearly a hexagonal close-packed arrangement of pairs of H_2O molecules. The H_2O's are shown here as H_4O tetrahedra because the 2 H's are really time-averaged over four sites. In the view above there are three close-packed layers of H_2O molecules stacked in an ABA sequence.

AX_2 Compounds

These structures are grouped based upon the anion present, which can be oxygen, fluorine, chlorine, bromine, iodine, sulfur, or a hydroxyl $(OH)^{1-}$. Of the many groups, five are important for mineral structure.

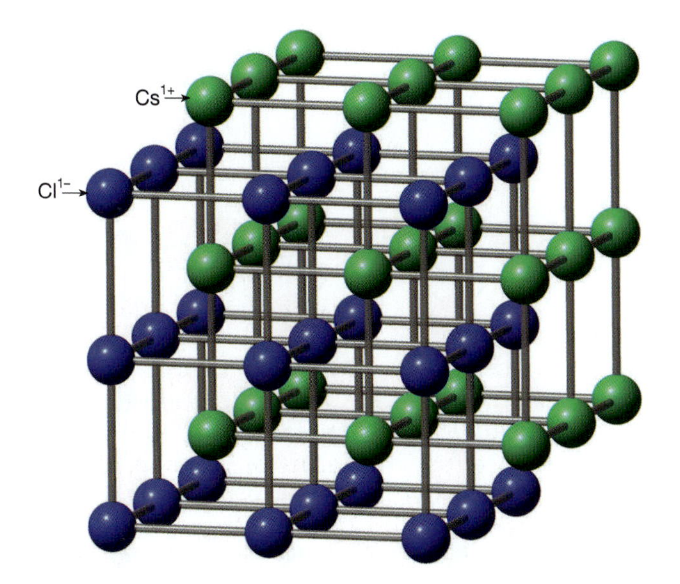

Figure 21.2. Because the sizes of Cs^{1+} and Cl^{1-} are so similar, the CsCl structure can be thought of as a body-centered cubic lattice with Cs^{1+} and Cl^{1-} as the motifs. Alternatively, it can be described as two interpenetrating simple cubic arrays, one of each type of atom at the corners of a cube.

1. The most important AB_2 structures are based upon linkages of SiO_4 tetrahedra into three-dimensional frameworks (for example, the framework silicates we have talked about since Chapter 1!). These also include the cristobalite structure just discussed, along with quartz, tridymite, coesite (all SiO_2), GeO_2 (argutite), and BeF_2.
2. The structure of the mineral rutile (TiO_2) is composed of hexagonally close-packed oxygen anions, with half of the octahedral interstices occupied by Ti^{4+} (Figure 21.6). Other rutile structure-type dioxides include cassiterite (SnO_2), plattnerite (PbO_2), pyrolusite (MnO_2), RuO_2, and paratellurite (TeO_2).
3. The Zr^{4+} cations in the baddeleyite (ZrO_2) structure are 7-coordinated with oxygen. There are two types of oxygen in the structure: one at the center of a three-Zr^{4+} triangle, and one in the center of a Zr^{4+} tetrahedron.
4. The fluorite (CaF_2) structure is based on a body-centered cubic lattice where the anions (F^{1-}) lie at the corners of a cube and the Ca^{2+} cation is at the center. As a result, the Ca^{2+} cation is 8-coordinated, and every F^{1-} anion has four Ca^{2+} neighbors. TbO_2, PrO_2, CeO_2, UO_2 (uraninite), and ThO_2 (thorianite) also crystallize with this structure.
5. Hydroxides like brucite ($Mg(OH)_2$), as well as oxyhydroxides like böhmite and diaspore (both $AlO(OH)$), can be considered as AX_2 structures because the $(OH)^{1-}$ hydroxyl essentially behaves like a single anion. Each $(OH)^{1-}$ is bonded to three Mg^{2+} cations, and each Mg^{2+} cation is in 6-coordination with $(OH)^{1-}$ groups, all of which

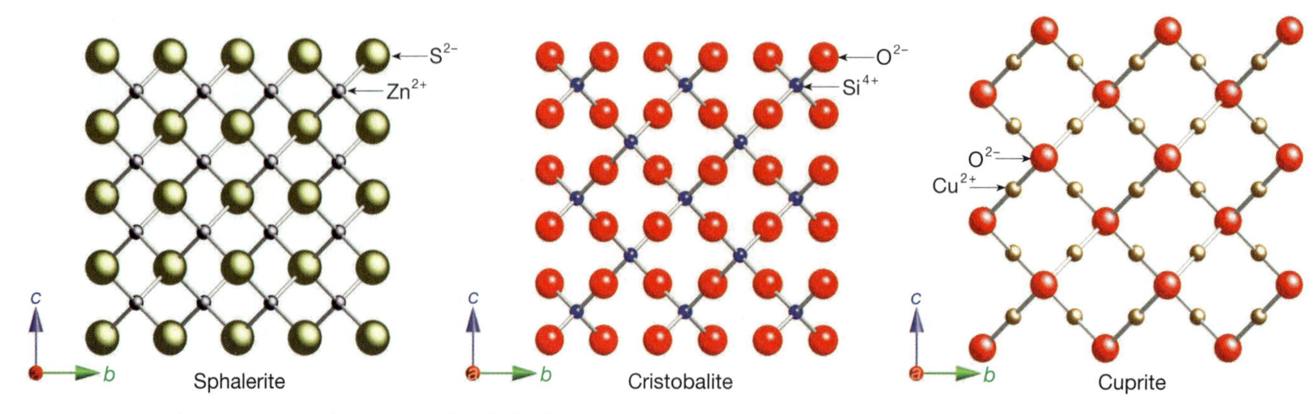

Figure 21.4. The structures of sphalerite (ZnS) (left), cristobalite (SiO_2) (middle), and cuprite (Cu_2O) (right) all appear related. The main differences is the ratio of anions to cations. These structures are originally based on the diamond structure with differing atoms occupying the C sites. The anions for all three are the larger spheres while the smaller spheres are the cations. It's ideal to look at the movies of each of these simultaneously on the MSA website and explore the similarities and differences of the structures.

have the hydrogen pointing in the direction of the next sheet. The Mg^{2+} octahedra all share edges to form a sheet, which is then completely charge-satisfied. As a result, adjacent octahedral sheets are only weakly bonded to each other.

AX_3 Compounds

The only common mineral to form with an AX_3 structure is skutterudite, $CoAs_3$, which is made up of corner-linked octahedral frameworks of Co surrounded by six As atoms (Figure 21.7). Each As atom is in tetrahedral coordination, with two short bonds to Co, and two short covalent bonds to other As atoms. The resultant structure is a distorted derivative of the perovskite structure (see ABX_3 compounds, below), which contains many large spaces (that can be thought of as "empty cages"). The skutterudite structure cannot be understood simply in terms of charge balance due to the significant covalent bonding that is present. However, this structure is important in the manufacture of thermoelectric devices, in which the holes in the structure are filled with poorly bonded rare earth elements such as lanthanum (La) or cerium (Ce). Vibrations of the loosely bound atoms scatter thermal waves as they propagate through the structure. The resultant materials have glass-like, low thermal conductivities.

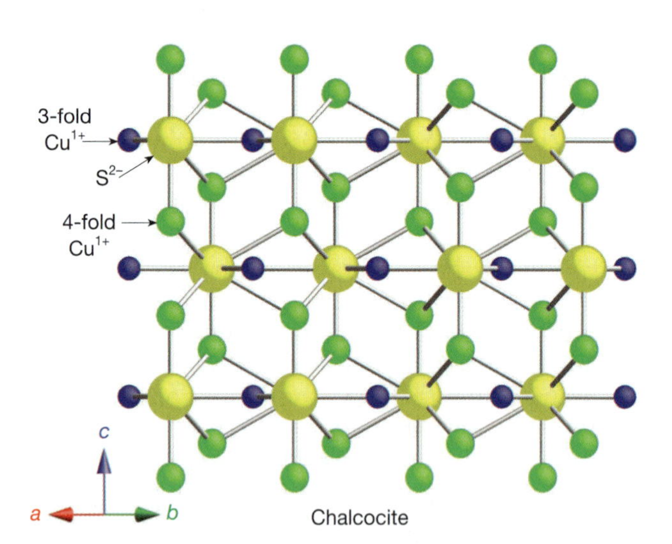

Figure 21.5. Chalcocite (Cu_2S) is based on a hexagonal (i.e., ABA) close packing of S^{2-} anions (large yellow spheres), with Cu^{1+} cations in 3-fold (small blue spheres) and 4-fold coordination (intermediate size green spheres).

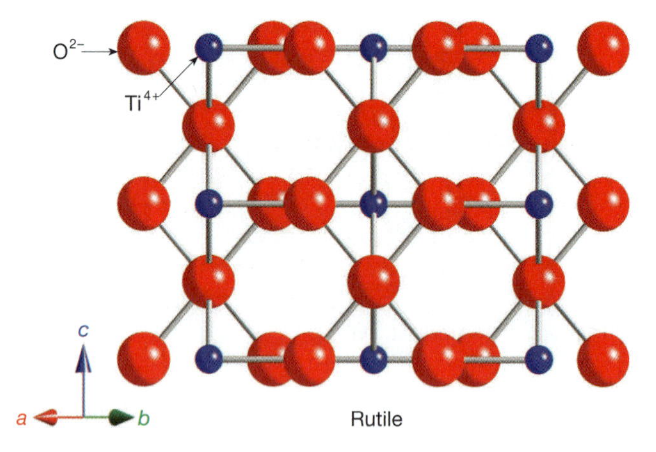

Figure 21.6. The rutile structure is a hexagonal close-packed array of oxygen anions (large red spheres) with half the octahedral holes occupied by Ti^{4+} (small blue spheres).

A_2X_3 Compounds

Several different oxides and sulfides take on this structural form, which is not surprising because it is conducive to combinations of two trivalent cations with three divalent anions. The oxide forms of arsenic and antimony have the same structures at low temperatures: As_2O_3 (arsenolite) and Sb_2O_3 (senarmontite), respectively. Arsenolite is industrially important as an agent for removing the color from glass, and both minerals are used to prepare numerous organic and inorganic derivatives. The structure of these minerals is based on the cubic close-packed structure of diamond (Figure 8.6). Four As^{3+} or Sb^{3+} cations occupy the corners of a tetrahedron, with each pair of As^{3+} or Sb^{3+} cations linked by a bridging oxygen (Figure 21.8). The centers of these tetrahedra occupy the lattice points on the face-centered lattice (Commission on Life Sciences, 1977). Bismite (Bi_2O_3) also belongs with this group of compounds, though its structure is more complicated, having three polymorphs, one in each of the isometric, tetragonal, and monoclinic systems.

Corundum (Al_2O_3) and hematite (Fe_2O_3) are geologically the most important of the A_2X_3 compounds, which also include Cr_2O_3 (eskolaite), Co_2O_3, Ni_2O_3, Rh_2O_3, and V_2O_3 (karelianite). These are all based on a hexagonal close-packed arrangement of oxygen anions, with cations occupying 2/3 of the octahedral interstices. So in any given row of the structure, there are two cations in adjacent sites, alternating with vacant sites (Figure 21.9). Adjacent rows are staggered. Each octahedron shares both edges and faces, but the vacancies on the octahedral holes are distributed so that each octahedron shares only one of its faces with a neighboring octahedron. This causes the face-sharing cations to shift away from each other, reducing the cation-cation repulsion. Ga_2O_3, In_2O_3, and Tl_2O_3 (avicennite) are also found with the corundum structure.

When the anion is sulfur instead of oxygen, several interesting sulfides result (Figure 21.10). In the orpiment (As_2S_3) structure, trigonal pyramids of AsS_3 are linked through sulfur anions to form corrugated layers that are perpendicular to the *b*-axis of the crystal. Stibnite (Sb_2S_3) and bismuthinite (Bi_2S_3) have related, somewhat more complex, structures.

ABX_3 Compounds

ABX_3 compounds form with a divalent and a quadravalent cation, usually in the company of oxygen as the anion. As a result of the difference in charge, the cation bonds more strongly with the anions than the A cation, and the A cations are larger then the B ones (recall the relationship between charge and cation size). Two groups of these compounds exist: those in which a BO_3 unit behaves like a single atom in the structure, and those that are close-packed arrays of oxygen anions with A and B cations in their interstices.

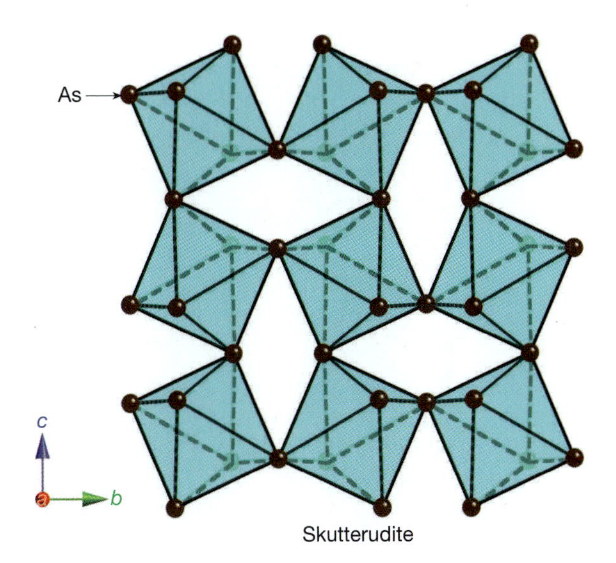

Figure 21.7. *The AX_3 crystal structure of skutterudite ($CoAs_3$) is composed of corner-sharing octahedra of As (small dark spheres on the vertices of the translucent octahedra) with the Co^{3+} cation 6-coordinated to the As.*

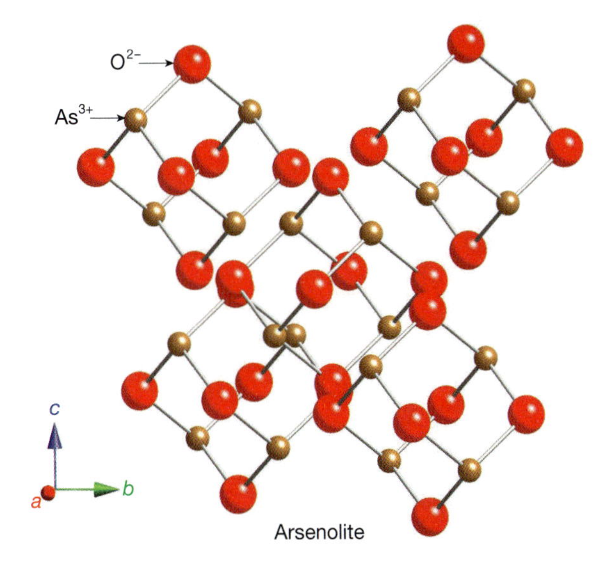

Figure 21.8. *Arsenolite, As_2O_3, is composed of four flattened trigonal pyramids of As_2O_3, with As^{3+} (small brown spheres) and O^{2-} (larger red spheres). These pyramids combine to form As_4O_6 molecules (the structure is tilted a bit so you can see one of the five, shown in the upper right corner) that share corners, and are held together by van der Waals bonding.*

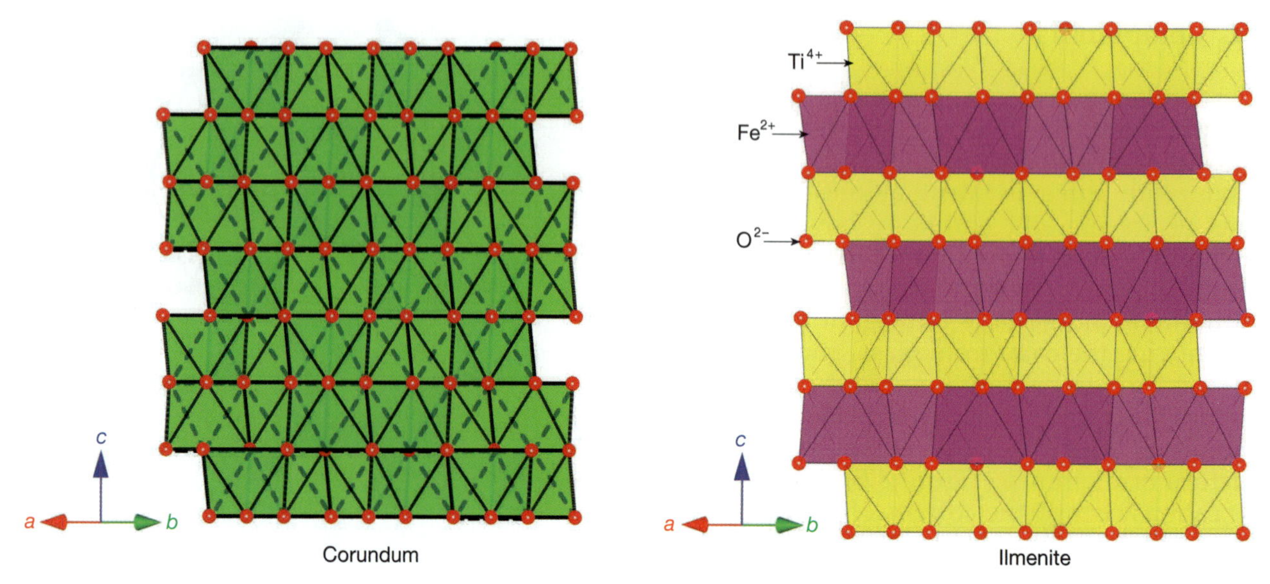

Figure 21.9. The corundum (Al_2O_3) structure (left) is a hexagonal close-packed array of oxygen anions in which 2/3 of the octahedral sites are filled by Al^{3+}. Ilmenite ($FeTiO_3$) (right) has the same structure, and Fe^{2+} occupies the magenta octahedra and Ti^{4+} cations the yellow ones. The Fe^{2+} and Ti^{4+} layers alternate. Note that we are viewing both structures parallel to the close-packed layers defined by the oxygens (small spheres at the apices of the octahedra).

The carbonate structure with its CO_3 units is a good example of the first group. If we treat the CO_3 as a single atom, then carbonates can be thought of as simple AX monoxides. Calcite ($CaCO_3$) can then be derived from the halite structure (Figure 21.11), while aragonite is derived from the nickeline structure, though with slightly lower symmetry. Halogenates ($LiIO_3$, $NaClO_3$), nitrates (e.g. nitratine, $NaNO_3$), and borates ($LaBO_3$, $NdBO_3$) all belong to this group.

As for the second group, the hexagonal close-packed arrangement of oxygen anions in the ilmenite structure is identical to that of corundum,

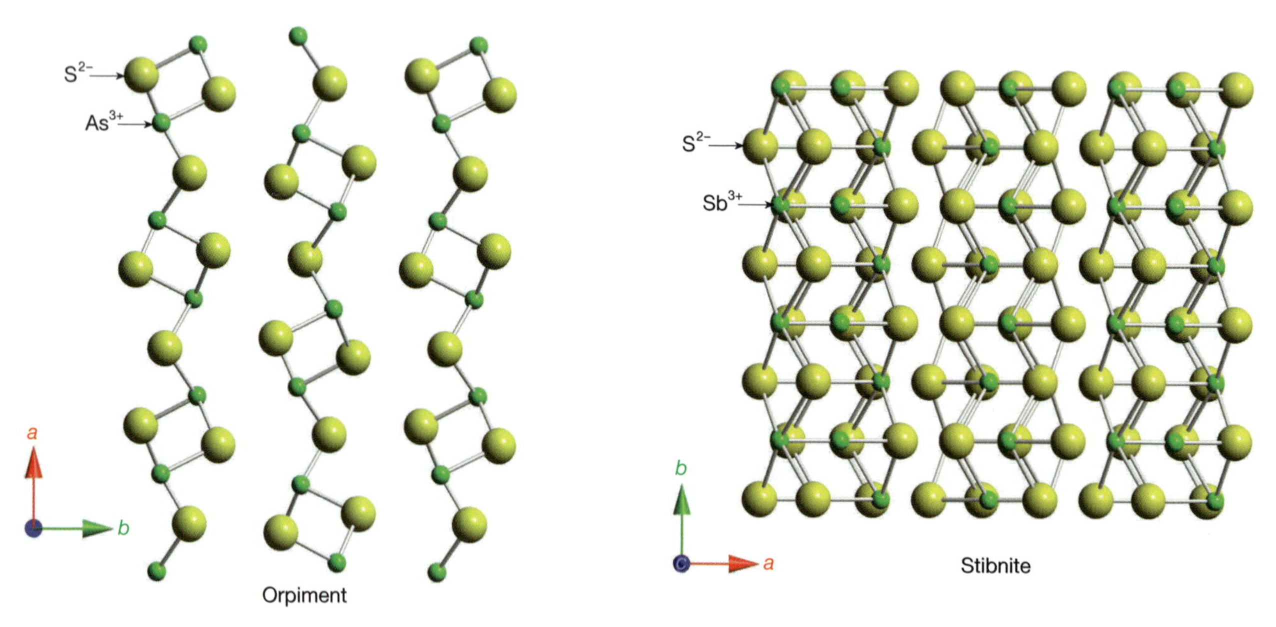

Figure 21.10. A_2X_3 compounds in which S^{2-} is the X anion can take on several different structures, including orpiment (left) and stibnite (right). In orpiment (As_2S_3), the As^{3+} cations (small green spheres) are bonded to three S^{2-} anions (larger yellow spheres) in a trigonal pyramids of AsS_3. These pyramids bond together in the a-c *plane. In stibnite (Sb_2S_3) the Sb^{3+} cations (small green spheres) bond with the S^{2-} anions (larger yellow spheres) in a more complex manner, but still forming layers like orpiment.*

with the metals (cations) in octahedral sites. However, the symmetry of ilmenite is lower than that of corundum because the Fe^{2+} and Ti^{4+} cations alternate in layers of the structure perpendicular to the *c* axis (Figure 21.9). The ilmenite group has three other members: ecandrewsite ((Zn,Fe^{2+},Mn^{2+})TiO_3), geikielite ($MgTiO_3$), and pyrophanite ($Mn^{2+}TiO_3$). The minerals brizziite ($NaSbO_3$) and melanostibite ($Mn^{2+}(Sb^{5+},Fe^{3+})O_3$) are also isostructural.

If the A cation is much larger than the B cation, then the perovskite ($CaTiO_3$) structure, which is important in the Earth's deep mantle, will form (Figure 21.12). B cations in octahedral coordination form a cubic array, and the sites between the octahedra are 12-coordinated, suitable for the large A cations. Distortions in the shape of the octahedra can change the symmetry from pseudocubic to orthorhombic or even monoclinic. Related minerals in this group include tausonite ($SrTiO_3$) and leushite ($NaNbO_3$).

Two other important rock-forming mineral groups also fall in this group of structures: garnet and pyroxene. In the numerous members of the garnet group (e.g., almandine, $Fe^{2+}_3Al_2Si_3O_{12}$), the first (A) cation is divalent, and occupies a distorted 8-coordinated site. The second cation (B) is trivalent and in octahedral coordination between the Si^{4+} tetrahedra. Alternatively, when the B cation is in tetrahedral coordination (i.e., when B is Si^{4+}), then the A cations form infinite chains with the resultant silicate tetrahedra, giving rise to the pyroxene structure, among others. Details about the garnet and pyroxene group minerals can be found in the next chapter.

ABX_4 Compounds

Numerous possible permutations on the ABX_4 compounds exist because the valence state (and size) of the A and B cations can be highly variable (Table 21.5). If the A cation is very large relative to B, the unusual $A^{1+}B^{7+}O_4$ composition of the perchlorates, periodates, and permanganates gives rise to compounds such as $NaClO_4$, $NaIO_4$, and $KMnO_4$, respectively.

More common are the many $A^{2+}B^{6+}O_4$ minerals (B is usually S^{6+}, but may also be Se^{6+}, Cr^{6+}, Mo^{6+}, or W^{6+}). Two different $A^{2+}B^{6+}O_4$ structures are relatively common; both are based on isolated tetrahedra held together in a framework by larger cations. The barite group of sulfates includes barite ($BaSO_4$), anglesite ($PbSO_4$), celestine ($SrSO_4$), and hashemite ($Ba(Cr,S)O_4$). To accommodate the disparity in cation sizes, the larger Ba^{2+} is in a 12-coordinated site (which shares three edges and four corners) and the smaller S^{6+} is in a SO_4 tetrahedron (Figure 21.13). A different way of accommodating this size difference is found in the scheelite ($CaWO_4$) structure, in which the A cation is 8-coordinated and the B cation is 4-coordinated. In this structure, the cations form a cubic close-packed cell with oxygen in the tetrahedral interstices (Figure 21.13). Powellite ($CaMO_4$), stolzite ($PbWO_4$), and wulfenite ($PbMoO_4$) are members of the scheelite group of minerals, and $NaIO_4$, KIO_4, $BaMoO_4$, and $BaWO_4$ also crystallize with this structure.

The scheelite structure is a distorted version of the $A^{4+}B^{4+}X_4$ structure found in zircon ($ZrSiO_4$) and its other group members hafnon ($HfSiO_4$), thorite ($(Th,U)SiO_4$), thorogummite ($Th(SiO_4)_{1-x}(OH)_{4x}$), and coffinite ($U(SiO_4)_{1-x}(OH)_{4x}$). Although the A and B cations have the same charge, the A cation in this structure (e.g., Zr^{4+}) is considerably larger (0.72 Å) than the B cation (e.g., Si^{4+} with an ionic radius of 0.26 Å). The resultant structural accommodation results in alternating chains of corner-sharing ZrO_8 triangular dodecahedra and SiO_4 tetrahedra (Figure 21.13). Other minerals with this arrangement include behierite ($TaBO_4$) and xenotime (YPO_4).

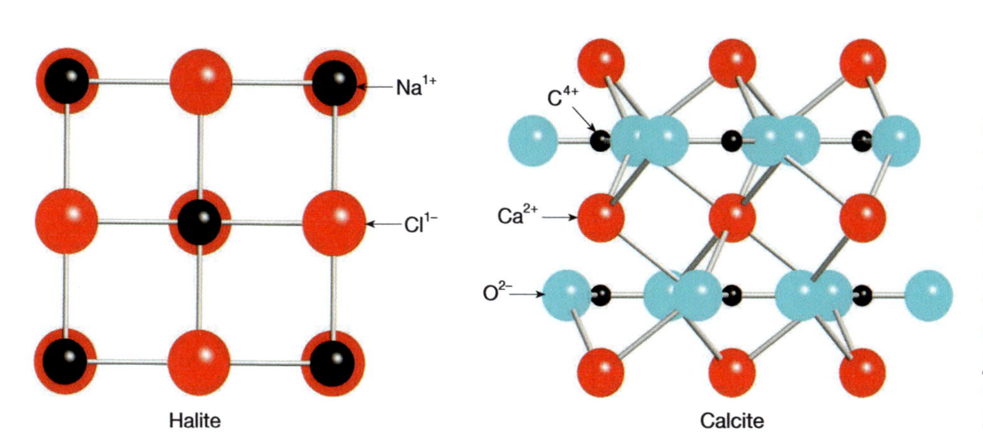

Figure 21.11. The structures of the familiar halite (left) and calcite (right) are related by replacing the Na^{1+} cations (black spheres) in halite with Ca^{2+} cations (large red spheres) and replacing the Cl^{1-} anions (large red spheres) in halite with CO_3 groups (a small black sphere of C^{4+} bonding to three blue spheres of O^{2-}).

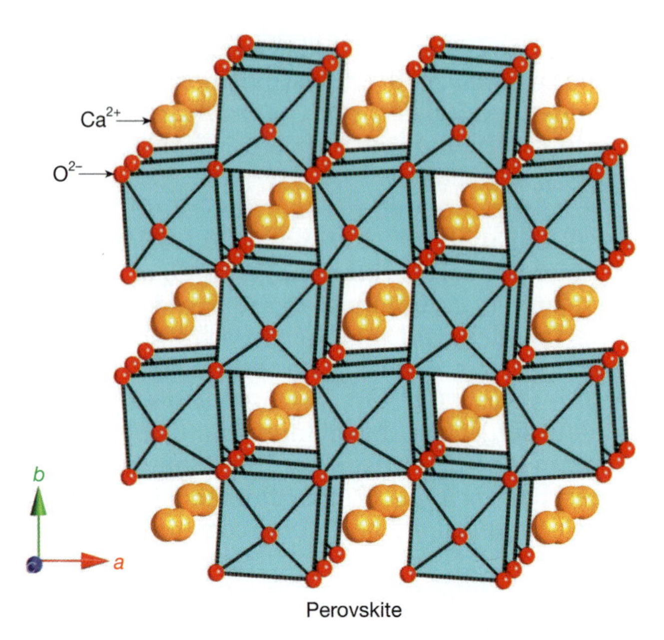

Figure 21.12. In perovskite ($CaTiO_3$), TiO_6 octahedra (shown as blue polyhedra with oxygens at their apices) share corners, forming large 12-coordinated sites between them that are occupied by Ca^{2+} (brown spheres).

When the sizes of the cations are more similar, as in the $A^{3+}B^{5+}O_4$ combination, then the resultant structures often closely resemble those of the dioxides (in fact, when the cations are disordered, the structures are *the same* as those of the dioxides). For example, berlinite ($AlPO_4$) has the same structure as quartz, with the *c* axis length doubled (Figure 21.14). Alasrite ($AlAsO_4$) and rodolicoite ($Fe^{3+}PO_4$) are also isostructural.

AB_2X_4 Compounds

The phenakite structure (Be_2SiO_4) and its relative willemite (Zn_2SiO_4) represent the AB_2X_4 compounds, in which both cations are small and 4-coordinated (Figure 21.15). Note here that the mineral formulas are written in a different order than the AB_2X_4 compound type. Alternatively, olivine group minerals such as fayalite (Fe_2SiO_4) and forsterite (Mg_2SiO_4) form when the cations are of disparate sizes, so that one occupies a tetrahedral site and the other an octahedral site. The olivine group minerals consist of chains of two types of octahedra linked by isolated SiO_4 tetrahedra. The resultant arrangement of oxygens is a distorted hexagonal close-packed array.

The denser spinel group structure is based on a distorted cubic close-packed array of oxygen atoms with tetrahedral occupancy by small cations and larger cations in only a single type of octahedron. Its unit cell contains 32 ccp oxygens, eight A cations, and 16 B cations. This group has 21 different mineral species, including chromite ($Fe^{2+}Cr_2O_4$), gahnite ($ZnAl_2O_4$), magnetite ($Fe^{2+}Fe^{3+}{}_2O_4$), and spinel ($MgAl_2O_4$). We will have much more to say about these minerals in Chapter 23!

Other Compounds

The structure classification system continues into far more complicated structures, as suggested in Table 21.5 and discussed in the Smith et al. (1998) paper. Consider the following structure type "for-

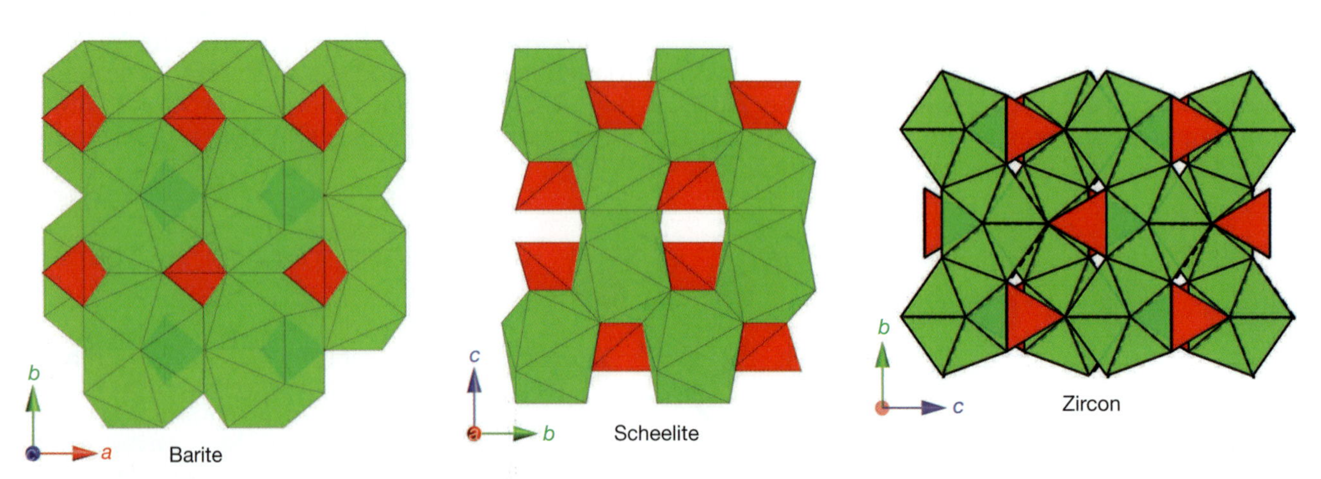

Figure 21.13. The structures of barite ($BaSO_4$) (left), scheelite ($CaWO_4$) (middle), and zircon ($ZrSiO_4$) (right) reflect three possible accommodations for when the two cations (i.e., A and B) in an ABX_4 compound are dramatically different in size. All three structures have two polyhedral types with the smaller cation (S^{6+}, W^{6+}, Si^{4+}) in tetrahedral coordination with four oxygens, and the larger cations in progressively smaller coordination polyhedral with decreasing size of the cation (i.e., Ba^{2+} = 12 CN, Ca^{2+} = 8 CN, Zr^{4+} = 8 CN). Tetrahedra are shown as solid red polyhedra and the larger polyhedra are shown as green and are transparent.

Table 21.5. Structural Classification of Some Minerals and Chemical Compounds

Structure	Charge on A & B	Charge on X	Examples: Formula	Examples: Mineral Species Name, if any
AX	2+	2–	ZnO	wurtzite
			NaCl	halite
			FeO	wüstite
A_2X	1+	2–	H_2O	ice
			Cu_2O	cuprite
			Cu_2S	chalcocite
AX_2	4+	2–	SiO_2	quartz, tridymite, cristobalite
			CaF_2	fluorite
			TiO_2	rutile
			FeS_2	pyrite
			ZrO_2	baddelyite
			$AuTe_2$	calaverite
			$Mg(OH)_2$	brucite
AX_3	3+	*	$CoAs_3$	skutterudite
A_2X_3	3+	2–	As_2O_3	arsenolite
			Sb_2O_3	senarmontite
			Al_2O_3	corundum
			As_2S_3	orpiment
			Bi_2S_3	bismuthinite
ABX_3	2+, 4+	2–	$CaCO_3$	calcite
			$FeTiO_3$	ilmenite
			$CaTiO_3$	perovskite
			$MgSiO_3$	clinoferrosilite
			$CaSiO_3$	wollastonite
ABX_4	1+, 7+	2–	$NaClO_4$	
	2+, 6+	2–	$BaSO_4$	barite
			$CaWO_4$	scheelite
	3+, 5+	2–	$AlPO_4$	berlinite
	4+, 4+	2–	$ZrSiO_4$	zircon
			$HfSiO_4$	hafnon
AB_2X_4	2+, 4+	2–	Be_2SiO_4	phenakite
			$Fe^{2+}{}_2SiO_4$	fayalite
			$Fe^{2+}Cr_2O_4$	chromite
Al_2SiO_5	3+, 4+	2–	Al_2SiO_5	kyanite, andalusite, sillimanite
$A_x...Si_2O_6$	1+, 3+, 4+	2–	$KAlSi_2O_6$	leucite
$A_x...Si_2O_7$	2+, 4+	2–	$Ca_2MgSi_2O_7$	akermannite
$A_x...Si_2O_8$	2+, 3+, 4+	2–	$CaAl_2Si_2O_8$	anorthite
$A_x...Si_2O_9$	1+, 4+	2–	$K_2Si_4O_9$	

*The skutterudite structure has significant covalent bonding, and therefore cannot be understood in terms of a simple ionic charge balance model.

mulas" for some common rock-forming mineral groups:

sphalerite $T_3(T'/L)X_4$
epidote $D_2G_3(SiO_4)[Si_2O_7]X_2$
mica $DG_{2,3}[T_4O_{10}]X_2$

The structural classification scheme is a lot more confusing for complicated compositions and for silicates in general. For these, the Dana classification is easier to understand, so we will leave further discussions of silicate and non-silicate crystal structures for the following chapters.

Silicates and the SiO_4^{4-} Tetrahedron

As we've discussed throughout the book, the silicates (i.e., minerals with $(SiO_4)^{4-}$ groups as their basic building blocks) are the most important groups of minerals for earth scientists because

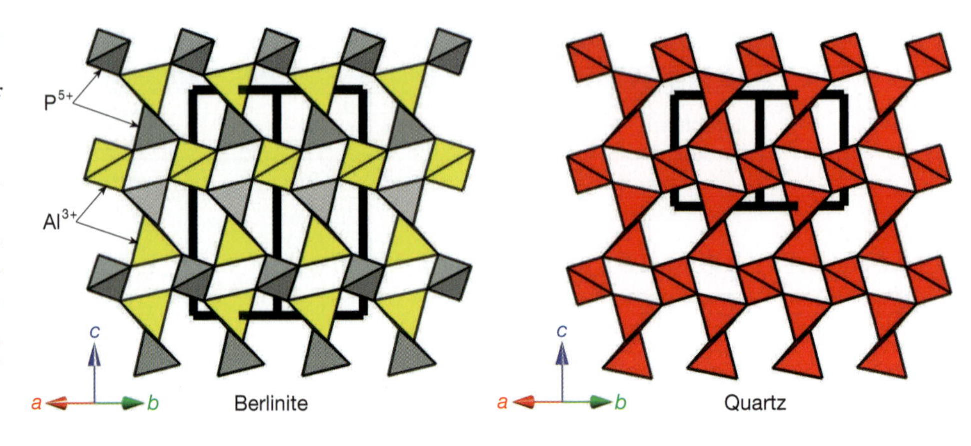

Figure 21.14. If the two cations in an ABX_4 structure are similar in size, as in the case of $A^{3+}B^{5+}O_4$, then the structures closely resemble those of quartz. Berlinite ($AlPO_4$) (left) has the same structure as quartz (right), though the length of the c axis is doubled (note the unit cells are outlined in each) because of the alternating PO_4 (dark gray) and AlO_4 (yellow) tetrahedra.

they comprise over 90% of the Earth's crust. Recall that we subdivide the silicates into framework, layer, ring, chains, di-, and orthosilciates based on the highest degree of polymerization of the silicate tetrahedra. However, the distinctions between these subdivisions can become blurry! For example, the silica polymorphs (except for stishovite) are clearly framework silicates because the Si^{4+} tetrahedra all share corners with other Si^{4+} tetrahedra. But how about the feldspars? They have anywhere from 1/4 to 1/2 of their tetrahedra occupied by Al^{3+}, and if we look at the SiO_4 *and* AlO_4 tetrahedra, it's true that they share all the corner oxygens. Thus we have now extended the polymerization of Si^{4+} tetrahedra to include Al^{3+} tetrahedra. (The following was discussed back in Chapter 6, but it is worth repeating here in this chapter on classification schemes.) Cordierite ($Mg_2Al_4Si_5O_{18}$) (Figure 6.10), is sometimes classed as a ring silicate, because not all of the Al^{3+} tetrahedra were considered in the polymerization classification. Because all the Al^{3+} are in tetrahedral coordination, the ratio of tetrahedral cations to O^{2-} is 1:2—the same as in a framework silicate. If we include all the Al^{3+} tetrahedra, then cordierite is, indeed a framework silicate. How about other atoms like Be that can occur in tetrahedral coordination? Look at the formula for beryl ($Al_2Be_3Si_6O_{18}$). Beryl (Figure 6.10) and cordierite are isostructural with the Mg^{2+} in cordierite replaced by Al^{3+} in beryl and the Al^{3+} in cordierite replaced by Be^{2+} in beryl. Thus, if Be^{2+} is included in the polymerization scheme, beryl also becomes a framework silicate. So these two "ring" silicates are really framework silicates if we include *all* the tetrahedral cations in the polymerization scheme.

One last example will illustrate the confusion in silicate classification. Sillimanite (Al_2SiO_5) is considered an orthosilicate in all classification schemes. However, if you go back and look closely at its structure (Figure 6.15c), you can see that it is really composed of four-membered rings of tetrahedra—two occupied by Si^{4+} and two by Al^{3+}. Thus sillimanite would really be a ring silicate! The moral of the story is that no classification system works perfectly for all minerals.

Concluding Comments on the Evolution of Mineral Classification Schemes

The preceding paragraph sounds a bit critical of our classification methods! That is not our intention. We merely wish to point out some of the shortcomings of any classification system that we, as humans, try to develop to "classify" the natural world. No doubt it is very frustrating as a student to encounter so many exceptions to the rules. But, it is important to understand the evo-

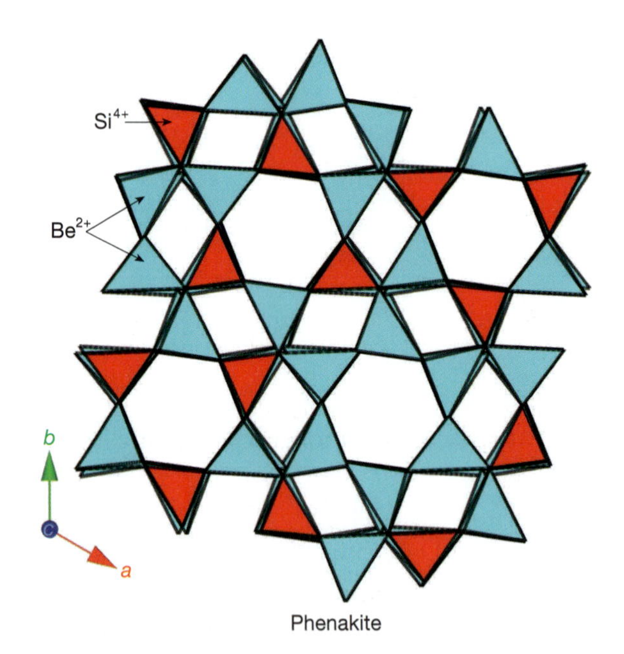

Figure 21.15. The phenakite (Be_2SiO_4) structure consists of SiO_4 (red) and BeO_4 (blue) tetrahedra that share corners forming a framework structure.

lution of the classification schemes along with our developing knowledge of minerals, starting first with uses, then physical properties, then moving to chemical compositions, and finally crystal structures. Dana's classification was developed before we really understood the structures of minerals. Amazingly, even now that much of our classification scheme is based on structures, his method still works pretty well. This really points out one of the themes of this book: the physical properties of mineral are directly related to their crystal structures. So these early scientists were truly skilled observers. Given some chemical data, they managed to devise classification methods that have pretty much stood the test of time.

References

Allen, S.M., and Thomas, E.L. (1999) The structure of materials. John Wiley and Sons, New York, 449 pp.

Berzelius, J.J. (1819) Nouveau systéme de minéralogie, traduit du suédois sous les yeux de lauteur, et publié par luiméme. Paris : Méquignon-Marvis, 314 pp.

Blackburn, W.H., and Dennen, W.H. (1997) Encyclopedia of Mineral Names. The Canadian Mineralogist Special Publication 1, 360 pp.

Bragg, W.L. (1937) Atomic structure of minerals. H. Milford, Oxford University Press, London, 292 pp.

Ciriotti, M.E. (2004a) Con più di una specie a testa ma...poche donne. Micro (letto e visto per voi), 2004, 3–6.

Ciriotti, M.E. (2004b) Con più di una specie a cranio ... ma poche donne (aggiornamento). MICRO (letto e visto per voi), 2004, 11.

Commission on Classification on Minerals, International Mineralogical Association (2001) Discussion paper on mineral groups. http://wwwobs.univ-bpclermont.fr/ima/w3/commissi/CCM2.html.

Commission on Life Sciences (1977) Arsenic: Medical and Biological Effects of Environmental Pollutants. National Academiy of Sciences, Washington, D.C., 4–15.

Corbridge, D.E.C. (1971) The structural chemistry of phosphates. Bulletin de la Societe Francaise de Mineralogie et de Cristallographie, 94, 271–299.

Gaines, R.V., Skinner, H.C.W., Foord, E.E., Mason, B., Rosenzweig, A., King, V.T., and Dowty, R. (1997) Dana's New Mineralogy. John Wiley and Sons, New York, 1819 pp.

Gersten, J.I., and Smith, F.W. (2001) The physics and chemistry of materials. John Wiley and Sons, New York, 856 pp.

Groth, P.H. (1889) *Tabellarische übersicht der mineralien nach ihren krystallographisch-chemischen beziehungen geordnet von P. Groth.* Braunschweig, F. Vieweg und sohn,, 167 pp.

Gunter, M.E. Dyar, M.D., Twamley, B., Foit, F.F. Jr., and Cornelius, S.B. (2003) Composition, $Fe^{3+}/\Sigma Fe$, and crystal structure of non-asbestiform and asbestiform amphiboles from Libby, Montana, U.S.A. American Mineralogist, 88, 1944–1952.

Haüy, R.J. (1801) Traité de Minéralogie. Chez Louis, Paris, 5 volumes, p. 114.

Hawthorne, F.C. (1984) Towards a structural classification of minerals. Acta Crystallographica, A40, C245–C245.

Hawthorne, F.C. (1985) Towards a structural classification of minerals—The $^{VI}M^{IV}T_2\varphi_n$ minerals minerals. American Mineralogist, 70, 455–473.

Hey, M.H. (1982) International Mineralogical Association: Commission on New Minerals and Mineral Names. Mineralogical Magazine, 46, 513–514.

Klaproth, M.H. (1795) Beiträge zur Chemischen Kenntniss der Mineralkörpers, I, p. 251.

Kostov, I. (1975) Crystal chemistry and classification of the silicate minerals. Geokhunuya, Mineralogiya I Petrologiya, 1, 5–41.

Leake, B.E., Woolley, A.R., Arps, C.E.S., Birch, W.D., Gilbert, M.C., Grice, J.D., Hawthorne, F.C., Kato, A., Kisch, H.J., Krivovichev, V.G., Linthout, K., Laird, J., Mandarino, J.A., Maresch, W.V., Nickel, E.H., Rock, N.M.S., Schumacher, J.C.,

Smith, D.C., Stephenson, N.C.N., Ungaretti, L., Whittaker, E.J.W., and Youzhi, G. (1997) Nomenclature of the amphiboles: Report of the subcommittee on amphiboles of the International Mineralogical Association, Commission on New Minerals and Mineral Names. Canadian Mineralogist, 35, 219–246.

Leake, B.E., Woolley, A.R., Birch, W.D., Burke, E.A.J., Ferraris, G., Grice, J.D., Hawthorne, F.C., Kisch, H.J., Krivovichev, V.G., Schumacher, J.C., Stephenson, N.C.N., and Whittaker, E.J.W. (2004) Nomenclature of ampiboles: Additions and revisions to the International Mineralogical Associations' amphibole nomenclature. American Mineralogist, 89, 883–887.

Liebau, F. (1956) Crystal structures of silicates with highly condensed anions. Zeitschrift für physikalische Chemie (Leipzig), 206, 73–92.

Liebau, F. (1966) The crystal chemistry of phosphates. Fortschritte der Mineralogie, 42, 266–302.

Liebau, F. (1983) Classification and mechanisms of phase transformations. Fortschritte der Mineralogie, 61, 29–84.

Lima de Faria J., and Figueiredo, M.O. (1978) General chart of inorganic structural units and building units. Garcia ed Orta, Serie de Geologia, 2, 1978, 69–72.

Lima de Faria J. (1994) Structural Classification of Minerals, An Introduction. Kluwer Acad., 346 pp.

Liu, L.-G., and Bassett, W.A. (1986) Elements, Oxides, and Silicates. Oxford University Press, 250 pp.

Machatschki, F. (1946) Grundlagen der allgemeinen Mineralogie und Kristallchemie, Wien, Springer. 209 pp.

Makovicky, E. (1981) The building principles and classification of bismuth-lead sulfosalts and related compounds. Fortschritte der Mineralogie, 59, 137–190.

Makovicky, E. (1993) Rod-based sulfosalt structures derived from the SnS and PbS archetypes. European Journal of Mineralogy, 5, 545–591.

Mitscherlich, E. (1818) Über die Krystallformen der Salze, in denen das Metall der Basis mit zwei Proportionen Sauerstoff verbunden ist. Abhandlungen der Königlichen Akademie der Wissenschaften zu Berlin.

Mitscherlich, E. (1825) Übersicht der Ausdehnung der krystallisirten Körper durch die Wärme. Abhandlungen der Königlichen Akademie der Wissenschaften zu Berlin.

Mohs, F. (1825) *Treatise on mineralogy; or, The natural history of the mineral kingdom. By Frederick Mohs ... Tr. from the German, with considerable additions, by William Haidinger.* Edinburgh, A. Constable and Co.

Náray-Szabó, I. (1969) Inorganic crystal chemistry [translation of Kristálykémia by P. Hervig and G. Zentai] Budapest, Akadémiai Kiadó, 479 pp.

Natland, J.H. (2003) James Dwight Dana (1813-1895) Mineralogist, zoologist, geologist, explorer. GSA Today, February, 20–21.

Navrotsky, A. (1994) Physics and Chemistry of Earth Materials. Cambridge University Press, New York, 417 pp.

Pabst, A. (1950) A structural classification of fluoaluminates. American Mineralogist, 35, 149–165.

Povarennykh, A.S. (1972) Crystal chemical classification of minerals, Translated from Russian by J. E. S. Bradley. Plenum Press, NY, 762 pp.

Povarennykh, A.S. (1966) Kristallokhimicheskaia klassifikatsiia mineralnykh vidov.

Smith, D.K., Roberts, A.C., Bayliss, P., and Liebau, F. (1998) A systematic approach to general and structure-type formulas for minerals and other inorganic phases. American Mineralogist, 83, 126–132.

Strunz, H. (1996) Chemical-structural mineral classification. Principles and summary of system. Neues Jahrbuch fur Mienralogie, Monatshefte, 10, 435–445.

Strunz, H. (1997) Classification of borate minerals. European Journal of Mineralogy, 9, 225–232.

Strunz, H. and Nickel, E. (2001) Strunz Mineralogical Tables, 9th Ed. Schweizerbart, Stuttgart, 870 pp.

Weiss, C.S. (1815) Übersichtliche Darstellung der verschiedenen natürlichen Abtheilungen der Kristallisationssysteme. Abhandlungen der Königlichen Akademie der Wissenschaften zu Berlin.

Wylie, A.G. and Huggins, C.W. (1980) Characteristics of a potassian winchite-asbestos from the Allamorre talc-district, Texas. Canadian Mineralogist, 18, 101–107.

Zoltai, T. (1960) Classification of silicates and other minerals with tetrahedral structures. American Mineralogist, 45, 960–973.

Chapter 22

Silicate Minerals

The next two chapters of this book provide a systematic examination of the structures and chemistry of major mineral groups. You might be wondering how we chose which minerals to include? This was largely my decision, I'm afraid. Here's how it came about…

Many years ago, I began the task of assembling the database that accompanies this book. The first job was to make a list of "important" minerals to include. I scoured my notes from classes and field camp, checked existing texts, and pondered recent articles in mineralogical publications. I scanned each and every page of the 8th edition of Dana's New Mineralogy*, and then made a list of every mineral that sounded familiar. Because I've worked on a number of really obscure minerals, this was a long list. Dennis Tasa and I then spent a memorable (grueling) week at the Harvard Mineralogical Museum in Cambridge, Massachusetts, with my friends (and curators) Carl Francis and Bill Metropolis. The Harvard collections are arranged in order by Dana classification, so it was a fairly simple matter to photograph the minerals we needed. Some of the species on my list were so obscure that Bill just laughed at me and crossed them off; he and Carl also noticed some important minerals that I had forgotten. We shot three exposures of each mineral (in order to obtain highest quality, technology-independent 35 mm slides!), and every day I took the subway to and from Kendall Square to get our film developed (some of you may know the famous song "Charlie and the MBTA"—that's just how I felt!). Dennis got muscle spasms from bending over the camera all day, and I was exhausted from shuttling samples and film back and forth and labeling and organizing everything. I think we wore out Carl and Bill, too!*

But the result was well worth it. The glorious photos you see in this text and in our database are almost all photos shot that week, and they represent our collective choices for what these minerals really look like. In a few cases, we fell for glamorous specimens (my favorites are the silver wires, the mica flowers, and some of the amazing zeolites), but mostly we tried to stick with samples that showed off the physical characteristics of the various species. We are grateful for the opportunity to share with you the broad representation, wealth of rare species, and high quality of display specimens at Harvard. If your travels ever take you to Cambridge, we highly recommend a visit to the mineral galleries at the Harvard Museum of Natural History. Many of the specimens on display there (including the one on the cover of this book!) will look like your old friends!

When it came time to write these two chapters, it quickly became apparent that we could not possibly cover everything that is in our own database. There were simply too many mineral species and too much information. So we have elected to keep these chapters simple. The following discussions cover: (1) why we think each mineral or mineral group is important, (2) chemistry and classification, (3) crystal structures, and (4) relationships among chemistry, structure, and physical properties of the minerals. Specifics of optical and physical characteristics, along with detailed information on occurrences (and a lot of other information) can be found in our mineral database app created specifically for our textbook (see MSA's website for details). We hope this format will allow you to focus on what (we think) is important. So these chapters will focus on integrating information rather than regurgitating it…

M.D.D.

Introduction

In the preceding chapter, we outlined the Dana and Strunz classification systems, which share (in the case of the silicates) a progression from simple structures with isolated SiO_4 tetrahedra to increasingly complex structures that share one, two, three, or all four corners of those SiO_4's. We choose to organize this chapter by thinking like geologists: the most commonly-occurring minerals in the Earth's crust are all framework silicates, and thus they deserve to go first. This is also the reason this chapter on silicates comes before the next chapter on non-silicates! So, it makes sense for us to organize our discussion of silicates in what is essentially the reverse of the standard classification schemes. We will begin with the highly polymerized (i.e., all corners of the commonly-SiO_4 tetrahedra are shared) minerals and work systematically backwards through the classification systems until we reach the orthosilicates (in which no tetrahedral corners are shared!). The next chapter will explore the non-silicates.

I visualize our methodology here in terms of Tinker Toys® and K'Nex™, building toys of which we have a billion in our house. My kids can quickly assemble a three-foot tall structure in which every hub (= atom) is connected to multiple other hubs by sticks (= bonds). Breaking this network up to put the toys away at bedtime is not always a simple task because the thing is so sturdy. Once you do get the structure to fall apart, it disconnects in pieces with different shapes, depending on the structural units that were used to assemble it. The more sticks (= bonds) that were used, the stronger the structure, and the harder it is to break up. I've been known to spend more than an hour trying to break up one of my kid's masterpieces so I can get the pieces back in the box. This process reminds me of the way we are approaching this chapter: start with the most highly-bonded, linked-up structure there is, and then break it apart and examine the resultant shapes.

Silica Polymorphs

The most abundant element in the Earth's crust is oxygen, which happens to be an anion. The next most abundant element is silicon, which happens to be a cation. Together, these two make SiO_2, which has the chemical name silica. It is not surprising that quartz is the most commonly-occurring mineral species in the Earth's crust.

Quartz also happens to be the most *stable* form of SiO_2 on the Earth's surface. However, there are several other different arrangements of pure SiO_2, including cristobalite, tridymite, coesite, moganite, and seifertite (Table 22.1). With the exception of moganite, which is metastable, each one is stable under a specific set of conditions (Figure 22.1). The ones shown in this phase diagram are all crystalline polymorphs (except the liquid). However, SiO_2 also occurs in a non-crystalline yet solid form, which is routinely called "amorphous."

Amorphous silica is not stable, and will alter to a crystalline form of silica given enough time. How much time does the alteration take? The oldest known opals found on the Earth are only 65 million years old. The most amorphous form of opal is called **opal-A** (where the A means amorphous). With time, opal-A transforms to **opal-CT** (cristobalite-tridymite), then to **opal-C**, then to a moganite/quartz mixture, and finally to the stable silica polymorph quartz (a gradational change as the scale of the crystallinity increases). As a side note, volcanic glass behaves in a similar way—over a period of time it will transform to crystalline materials.

Different arrangements of SiO_2 can be related to a wide range of pressure and temperature conditions found on the Earth, as you can see from Figure 22.1. Identifying the polymorphs can be extremely useful in unraveling geological problems. For example, every time a meteor hits the Earth, it exposes quartz grains in terrestrial rocks to transient but very high pressures, causing a conversion to coesite and/or stishovite. So, one way to find the sites of meteor impacts is to look for these high-pressure polymorphs. In fact, the Chixulub meteorite crater from the impact event that was at least partially the cause of the dinosaur's extinction was found by examining oil company drill cores from the Gulf of Mexico and

Table 22.1. Silica Polymorphs and Relatives*

Mineral	Formula
Cristobalite	SiO_2
Tridymite	SiO_2
Quartz	SiO_2
Coesite	SiO_2
Moganite	SiO_2
Seifertite	SiO_2
Stishovite**	SiO_2
Opal	$SiO_2 \cdot nH_2O$
Melanophlogite	$(2-x)(CH_4,N_2)\cdot(6-y)(N_2,CO_2)\cdot Si_{46}O_{92}$
Melanophlogite-β***	$(2-x)(CH_4,N_2)\cdot(6-y)(N_2,CO_2)\cdot Si_{46}O_{92}$
Silhydrite	$3SiO_2 \cdot H_2O$
Virgilite	$LiAlSi_2O_6$

*Unless otherwise noted, formulas in this chapter are taken from Mandarino and Back (2004) and are listed in order by Dana number.
**Structurally, stishovite actually belongs to the rutile group because its structure most solely resembles those other oxides, rather than a framework silicate, but it is composed of SiO_2.
***Species not yet approved by IMA.

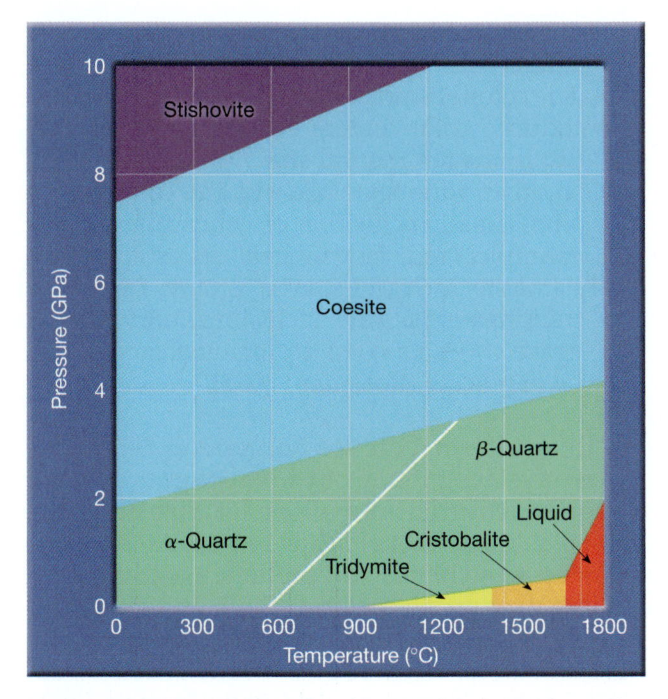

Figure 22.1. *The P-T phase diagram for the SiO_2 polymorphs. As the diagram shows, α-quartz is the stable form of SiO_2 on the Earth's surface. With increasing temperature α-quartz would convert to β-quartz, which in turn would convert to tridymite, then cristobalite, and finally melt to form a liquid. As pressure increases, α-quartz converts to coesite and then stishovite. Interestingly, in stishovite Si is 6- and not 4-coordinated as in the other silica polymorphs. Structures for these minerals are shown in our database app.*

discovering a coesite-rich layer, as well as trace element anomalies that originate from meteorites.

There are numerous industrial uses for quartz, including manufacturing of glass and computer chips. Girard-Perregaux developed the first mass-produced quartz watch in 1969. Since that time, quartz has revolutionized time-keeping, and is used in almost all watches. How does a quartz crystal keep time? If you squeeze a quartz crystal perpendicular to its *c* axis, a slight but measurable positive electrical charge occurs on one end, which results in a negative charge at the other end. Similarly, if you reverse the process and use a battery to apply the charge difference, the quartz crystal will flex and vibrate. Depending on the size, shape, and vibration frequency of the crystal, it may vibrate tens of thousands of times per second. By counting the vibrations, the watch can keep time with great precision. Quartz belongs to a space group that lacks a center of symmetry; if it had a center of symmetry, it could not exhibit this property.

Surprisingly, quartz dust is one of 95 compounds listed as a known human carcinogens by IARC (International Agency for Research on Cancer) along with cigarette smoke, arsenic compounds, and benzene. For what are still unknown reasons, quartz inhalation has been linked to tumors in the lung, causing cancer. Opal, on the other hand, does not appear to cause such an effect. So the difference between crystalline and non-crystalline has an effect on how the material interacts with living tissue. This is a fascinating field of research in dire need of mineralogical input, and will be discussed more in Chapter 24 (or see Gunter, 1999).

Classification and crystal structure. The silica polymorphs are somewhat unique as a group of minerals because they all have basically the same chemistry; only the structures are different. We first met the **quartz** structure in Chapter 1 of this book (Figure 1.9), and again in Chapters 6 (Figure 6.1) and 14 (Figure 14.11); it should rigorously be referred to as ***α*-quartz**. Its structure is a framework of corner-sharing SiO_4 tetrahedra that form spirals along [001]. The spirals are defined crystallographically as screw axes and can be either right-handed or left-handed. ***β*-quartz** is only stable at temperatures above 573°C at 1 bar.

If you compare the *α* and *β* forms of quartz (Figure 22.2), you'll see that the chains of Si-O bonds are somewhat kinked in *α*-quartz. In the higher temperature *β*-quartz, the ions move slightly farther away from one another, and the chains straighten out—thus giving *β*-quartz higher symmetry ($P6_222$) than *α*-quartz ($P3_221$). This illustration explains why *α*-quartz is sometimes called low quartz and *β*-quartz, high quartz—because of the differences in symmetry.

Quartz crystals typically are elongated when the mineral grows into an open space (e.g., as in a geode), and this growth is parallel to the screw axes. You already know that quartz does not have prominent cleavage because it usually breaks with conchoidal fracture. Thus a broken surface shows ripples that are evidence of shock waves emanating out from the point of breakage, which is sometimes called the **bulb of percussion**. Such "non-cleavage" is another result of the tight bonding between all the ions in quartz—and the lack of repetitive occurrence of weak bonding planes in the structure. However, quartz does indeed have a weak cleavage as pointed out in Bloss and Gibbs (1964).

Another result of the equivalent bonding in all directions is that physical properties that can vary as a function of direction (i.e., refractive index, hardness) are going to be similar in all orientations. One perfect example of this is the birefringence of quartz, which is very low, resulting in low retardation (usually first-order gray in thin section).

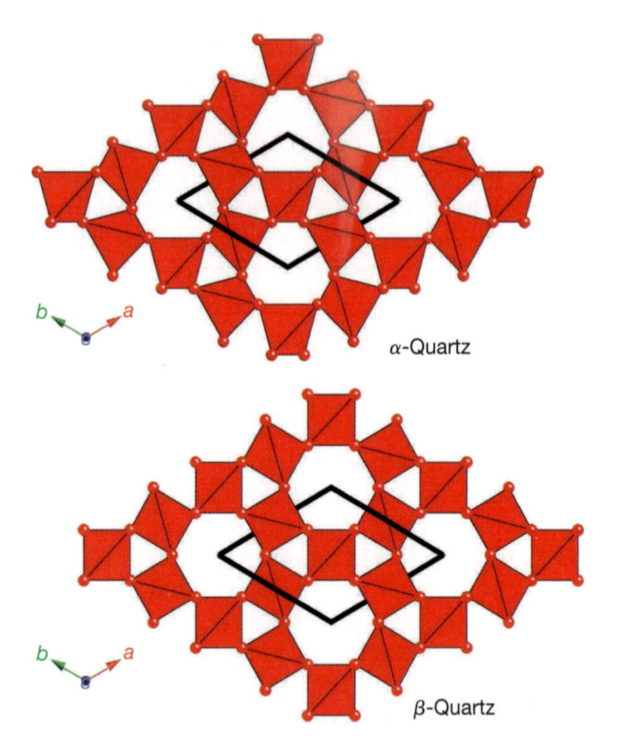

Figure 22.2. The crystal structures of α-quartz, or low quartz (top) and β-quartz, or high quartz (bottom). Both are polymorphs of SiO_2, though α-quartz is the lower temperature and lower symmetry form, and thus the most common in the Earth's crust. Both structures are projected onto the a-b *plane and the outlines of their unit cells are shown. Low quartz (with a 3-fold axis parallel to* c*) converts to high quartz (with a 6-fold axis parallel to* c*) at 574C°. The movie on the MSA website shows an animation of this conversion.*

The numerous varieties of quartz have colors that arise from some of the phenomena (Figure 22.3) that were discussed back in Chapter 7 (Table 7.8). These are nicely demonstrated in data from the Mineral Spectroscopy web site created by George Rossman at Caltech. In some cases, transition metals substitute for Si^{4+} or occupy defect (interstitial) sites between ions. For example, both amethyst and citrine are colored by small amounts of iron (approximately 40 parts per million).

- Amethyst color develops when Fe-containing quartz is exposed to ionizing radiation. In nature, gamma rays from the decay of ^{40}K are the most likely source of ionizing radiation. The model currently accepted is that radiation oxidizes Fe^{3+} to Fe^{4+}. There is still uncertainty about the location of the iron in the structure. Both interstitial sites are in the *c* axis channels, and the Si^{4+} tetrahedral sites have been proposed as the site of the amethyst center.
- Citrine's color is from Fe^{3+}. The properties of the Fe^{3+} spectra suggest that the Fe^{3+} ions are aggregated and hydrated in clusters of unknown size.
- The color of smoky quartz is due to radiation-induced color centers in which Al^{3+} replaces Si^{4+} in the tetrahedral site.
- In other varieties of quartz, the color comes from small inclusions of other phases. For example, red jasper gets its color from hematite inclusions, rose quartz from aluminum borosilicate inclusions, and chrysoprase owes its color to inclusions of a Ni^{2+} layer silicate (Rossman, 2005).

For more information, see Rossman (1994).

The crystal structure of **tridymite** goes through a sequence of transformations leading up to the most stable polymorph, *β*-tridymite. At temperatures up to ~1105°C, tridymite is usually monoclinic (Cc) or rarely triclinic ($F1$). Two orthorhombic phases form at slightly higher temperatures: $P2_12_12_1$ from 1105–1180°C, and $C222_1$ from 1180–1350°C. Hexagonal tridymite forms a $P6_322$ structure from 1350–1465°C, and a $P6_3/mmc$ structure from 1465–1470°C. The latter structure is the most common, and the simplest way to join tetrahedra so that all four of their oxygens are shared

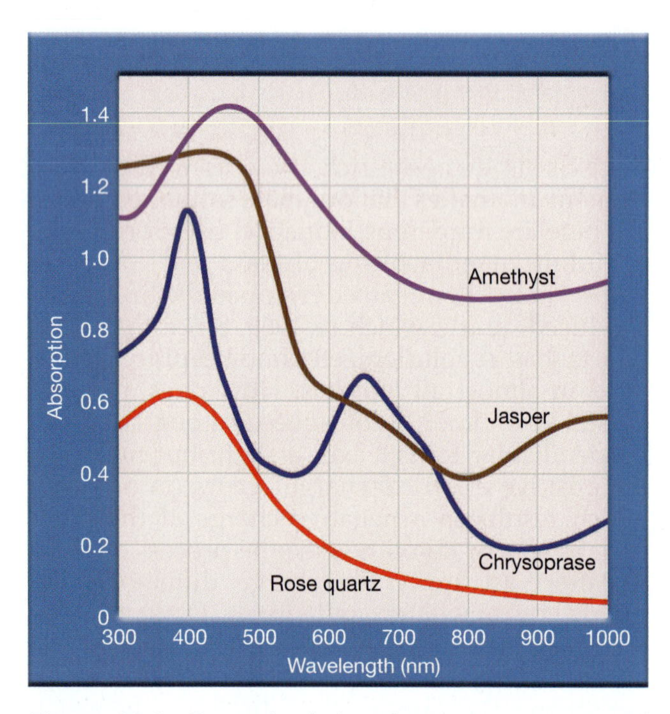

Figure 22.3. Transmitted (amethyst, rose quartz, and chrysoprase) and reflected light (jasper) spectra of quartz varieties. Intensities have been normalized to 1. Data from the Mineral Spectroscopy web site at Caltech. The purple color of amethyst develops in quartz with Fe impurities after exposure to radiation. The colors of jasper, chrysoprase, and rose quartz result from the presence of tiny inclusions in the crystals.

(i.e., to form framework silicate). In the $P6_3/mmc$ β-tridymite structure, rings of six SiO_4^{4-} tetrahedra create sheets in the *a-b* plane, with tetrahedra pointing alternately up and down to link adjacent sheets (Figure 22.4). For more details about these interesting relations, please consult the *Reviews in Mineralogy* volume on silica (Heaney et al., 1994).

It is believed that **α-tridymite** is always a pseudomorph after β-tridymite, and thus is found as pseudohexagonal crystals (although tridymite-like regions are also found in opal-CT). Furthermore, most specimens labeled tridymite are really quartz pseudomorphs, though sometimes true tridymite is preserved. True tridymite can be found at the Obsidian Cliffs in Yellowstone National Park, and on Mt. Lassen in Lassen Volcanic National Park in California.

Cristobalite also has two polymorphs: α-cristobalite (tetragonal, stable below 1268°C) and β-cristobalite ($P2_131$, stable from 1470–1728°C). In between those temperatures, it exists only in a metastable form. Cristobalite specimens are all polymorphs of β-cristobalite, and generally retain its diagnostic, cubic shape. The β-cristobalite structure is quite similar to that of diamond, in which every C atom is linked to four other C atoms (Figure 22.4). In the β-cristobalite structure, these C "tetrahedra" that are present in diamond are replaced by SiO_4 tetrahedra in a very similar arrangement! The resulting crystal structure is the least dense of any of the naturally-occurring silica polymorphs.

As shown in the phase diagram in Figure 22.1, **coesite** forms only under high temperature and pressure conditions. Thus coesite is relatively rare, except in specific types of geological environments. As noted above, coesite is commonly found in shocked materials (i.e., high pressure and low temperatures) at meteorite impact sites, such as those at Meteor Crater in Arizona. Some high-grade metamorphic rocks (e.g., eclogites in the western Alps) also contain coesite, as do diamond-bearing deposits like the Roberts Victor kimberlite in South Africa.

The coesite structure is composed of rings of four tetrahedra linked into a framework that nearly resembles the feldspar structure. Its close relative, **stishovite**, which only occurs at pressures >160 kb, has the same structure as rutile, with edge-sharing Si^{4+} chains linked by sharing corners (Figure 21.6). As we discussed in Chapter 8, the change of Si^{4+} coordination to a higher number, and thus, a larger site, reflects the changes in radius ratios of O^{2-} and Si^{4+} that occur at high pressures. The structure is quite dense, and as a result has a very high refractive index. We can similarly relate the densities of the remaining SiO_2 polymorphs to their structures, almost by simple observation of the structure drawings here. These densities are given for each of the polymorphs along the *x* axis in Figure 22.5, along with the refractive indices for each of the polymorphs. As density increases, so does refractive index! [This relationship holds true in many minerals, especially in a polymorphic series.]

At the other end of the spectrum, the structure of **opal** is quite loosely packed, imparting a low density and correspondingly low refractive index

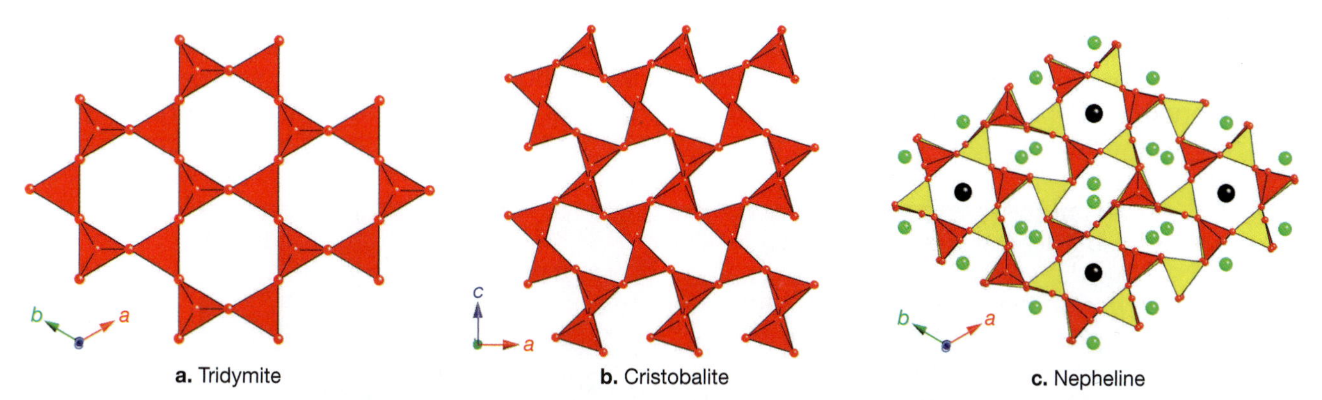

*Figure 22.4. The crystal structures of two silica polymorphs and the framework silicate nepheline, $KNa_3Al_4Si_4O_{16}$. **a.** Tridymite projected on its* a-b *plane (compare to quartz in Figure 22.2). Tridymite is the simplest of the silica polymorphs; it is composed of layers of linked 6-membered Si^{4+} tetrahedra, with half of the tetrahedra pointing up and the other half pointing down. These layers are then linked by apical oxygens that alternately point up and down (use CrystalViewer available from the MSA website to rotate the structure to see this). **b.** Cristobalite, like quartz, has a more complicated arrangement of corner-sharing Si^{4+} tetrahedra. Here it is projected on its* a-c *plane, which is analogous to the* a-b*-projection for quartz and tridymite. **c.** The structure of nepheline is similar to that of tridymite, except that half of the Si^{4+} cations have been replaced by Al^{3+}. Notice the distribution of the Si^{4+} tetrahedra (red) and those occupied by Al^{3+} (yellow). This substitution also causes a charge imbalance, so K^{1+} (large black spheres) and Na^{1+} (smaller green spheres) enter the structure for change balance. The rings of tetrahedra are distorted to accommodate these cations.*

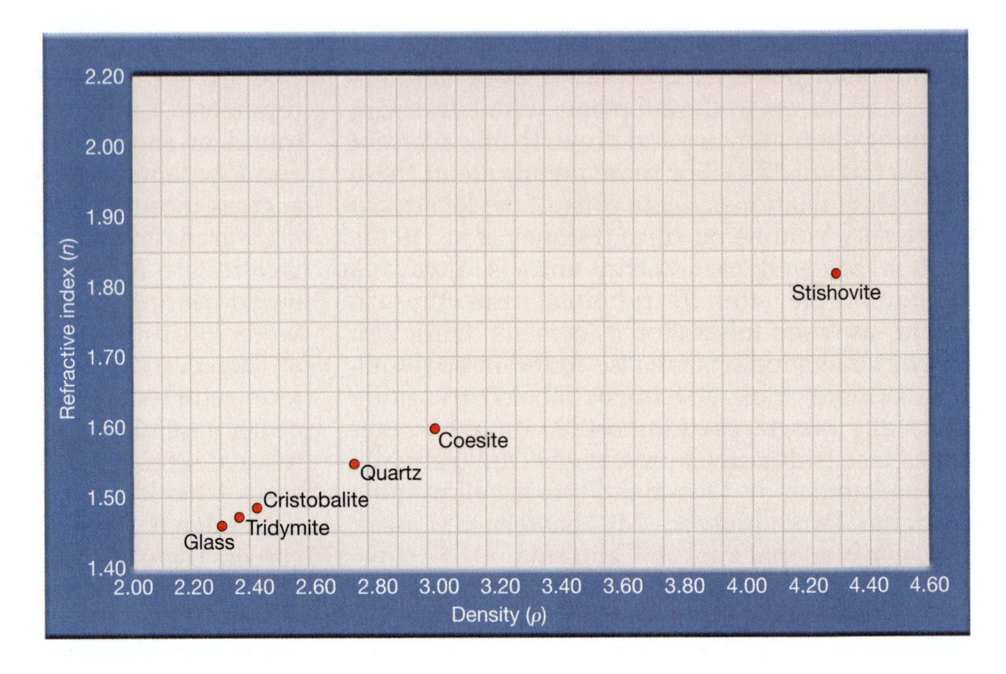

Figure 22.5. *Relationship between the average refractive index and density for the silica polymorphs (Bloss, 2000). The high temperature polymorphs have low densities, high pressure polymorphs have higher densities, and quartz (the form stable at surface conditions on Earth) plots between these two groups.*

(Graetsch, 1994). As noted above, opal has three different polymorphs, opal-A, opal-CT, and opal-C, which grade continuously into one another. Opal-A is considered amorphous because XRD patterns show no orderly internal structure. The "structure" consists of densely-packed spheres of SiO_2 and water that are 150–350 nm in diameter (Jones et al., 1964). In precious opal, the spheres are arranged into regular sheets (Figure 22.6); constructive interference of light causes reflections off the parallel planes that give rise to the characteristic play of colors. In common "potch" opal, the spheres are of varying sizes and their distribution is disorganized (Figure 22.6), so no reflections occur. Opal-CT consists of tiny balls (lepispheres) made of thin blades of disordered interlayers of cristobalite and tridymite (Guthrie et al., 1995; Graetsch and Topalovic, 1994; Elzea and Rice, 1996). According to Jones et al. (1963), the structure of opal-C is similarly based on that of low cristobalite.

The formula of opal ($SiO_2 \cdot nH_2O$) does not exactly make it clear where the water is located in the structure, and it turns out that this is a complex matter! For clarification, I asked George Rossman, who relayed a conversation he once had with Ralph Iler, a chemist at DuPont. Silicic acid, H_4SiO_4, polymerizes to form chains, then ribbons of a polysilicic acid. Think of a wide ribbon that has Si-O-Si links internally, but has H-O-Si....Si-O-H units on the surface. Ball-like structures appear when the number of units in a chain is about four; the ribbons will wrap into spheres where the spheres are mostly Si-O-Si internally, but with

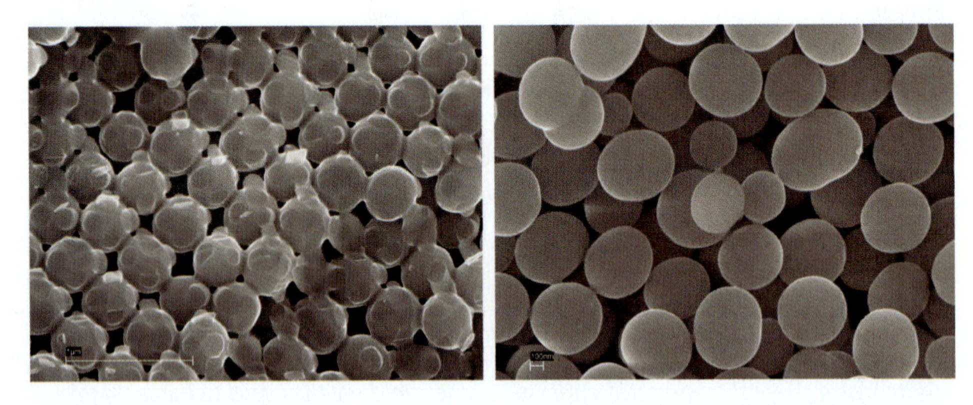

Figure 22.6. *SEM images of two different areas on an opal from altered volcanic ash beds in Lake Tecopa, 6 km northwest of Tecopa, Lake Tecopa, Inyo County, California. At left is a well-ordered region (ordered packing of the spheres), with the added twist of mini-spheres attached to the main spheres. At right is a more disorganized arrangement of spheres. Both images were taken at 100,000× magnification on a growth surface (not a broken surface). Photos courtesy of Ma Chi, Caltech.*

many Si-O-H units on the surface and probably some in the interior. Spheres settle to make rows. Water fills the gaps between the spheres. Adjacent spheres condense through ..Si-O-H + H-O-Si... reactions to form ...Si-O-Si... + H_2O units. In time, the cross-links grow and fuse the spheres together. There is still some water in the open spaces, and, probably, some bits of water in the spheres (as such reactions occur within the sphere) and there are also still some $(OH)^{1-}$ units in the spheres. The amount of water in the spaces ends up being far greater than most of the other water. If the opals are exposed to heat or very dry conditions, eventually even the water between the spheres will dry out, potentially causing the opals to "craze" or even lose their play of color.

Feldspars

Quartz may be the most abundant mineral *species*, but feldspars are the most abundant mineral *group*, composing approximately 50% of the Earth's crust. Although Si and O are the most abundant elements in the crust, the next most abundant is Al, so it is logical that many minerals containing Al, Si, and O would occur. In the feldspar structures, Al^{3+} substitutes for Si^{4+}, and the resultant charge imbalance requires the addition of some other cations that can't fit in tetrahedra. The structures must change to accommodate those larger cations in 7- to 9-coordination with O^{2-}. Given the formula $(AlSi_3O_8)^{1-}$ the structure requires a single monovalent cation, such as K^{1+} or Na^{1+}, to achieve charge balance, two other very common constitutes of the Earth's crust. Likewise, the formula $(Al_2Si_2O_8)^{2-}$ requires a single divalent cation, usually Ca^{2+}, for charge balance. So the three most common end-member formulas for feldspars are $KAlSi_3O_8$, $NaAlSi_3O_8$, and $CaAl_2Si_2O_8$.

The most common forms of these three feldspars, **orthoclase**, **albite**, and **anorthite**, have by far the greatest geological significance (even though there are nineteen feldspar species, see Table 22.2). Feldspars are especially useful in understanding cooling rates for igneous rocks. You may recall from earlier geology courses that rapid cooling generally results in fine-grained rocks with small crystal sizes, while slower cooling rates allow time for large crystals to grow. On the atomic scale, we see a similar phenomenon. In rapidly-cooled feldspars, Si^{4+} and Al^{3+} don't have time to find the most "comfortable" positions (from a thermodynamic perspective) in the musical chairs game of crystal growth. When the cooling rate is slower, Si^{4+} and Al^{3+} have time to find their most stable, lowest energy sites. The stability of the atoms in the site is based on finding the right position as a function of the slightly different charge and size of these two cations (Figure 22.7) in relationship to the entire feldspar structure.

For example, if a geologist finds a 100 μm grain of potassium feldspar in the sand on a beach, she *could* make careful, lab-based measurements to determine the arrangement of Al^{3+} and Si^{4+}. From this, she would learn whether the feldspar had a rapidly-quenched, volcanic origin (disordered array of Al^{3+} and Si^{4+}) or a more slowly-cooled, plutonic origin (orderly arrangement of Al^{3+} and Si^{4+}). The former of these would be the mineral species sanidine, while the latter would be microcline; they have the same formula, $KAlSi_3O_8$, but very different geological origins "locked" into their crystal structures. Of course, as in many geological processes, there is also an intermediate phase—a feldspar with that same formula but an intermediate amount of ordering; it is called orthoclase.

When you go the beach, the vast majority of the grains are actually quartz, not feldspar. However, there are about four times more feldspar minerals in the Earth's crust than there are quartz grains. Where does the feldspar go? The larger cations (with an increased number of weak bonds) in the structure of the feldspars (K^{1+}, Na^{1+}, Ca^{2+}) make them much more susceptible to chemical weathering. For instance, consider the feldspar $KAlSi_3O_8$. When it weathers, K^{1+} and some Si^{4+} and O^{2-} are removed from the structure, leaving Al^{3+}, Si^{4+}, and O^{2-} to combine with water to make the mineral kaolinite, $Al_2Si_2O_5(OH)_4$. Although this is a simplistic example, this process is responsible for the majority of clay mineral formation (i.e., feldspars weathering to form clays). So while we have quartz sand on the beach to play in, we have the weathering of feldspars to thank for the formation of clays and in turn, by the addition of organic components, the formation of soils in which we grow our food. This weathering process doesn't occur on the Earth's moon where the light-colored portions are made almost entirely of a feldspar with the composition, $CaAl_2Si_2O_8$; water to promote weathering is lacking there.

Feldspars are used in the manufacture of glasses and ceramics (including plumbing fixtures like your toilet bowl, as well as pottery). They are often mixed with pure SiO_2 because the Al^{3+} in feldspar makes glasses more chemically resistant (even though it lowers the hardness and durability somewhat). The addition of feldspar to a pure SiO_2 melt also lowers its melting point, making those glasses and ceramics cheaper and easier to

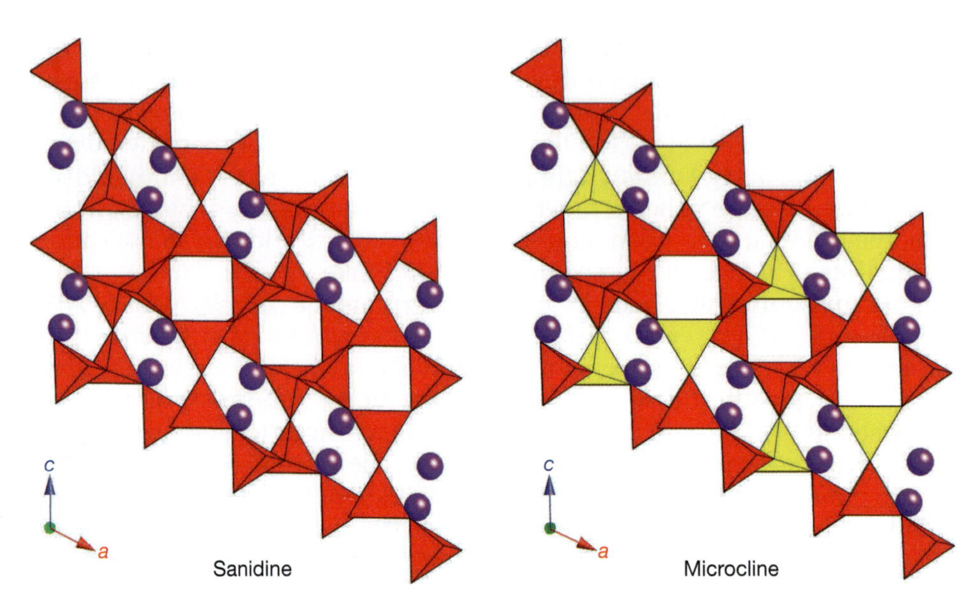

Figure 22.7. Crystal structures of two polymorphs of $KAlSi_3O_8$: the higher symmetry sanidine (left) and the lower symmetry microcline (right). The major difference in the two structures is that the Al^{3+} and Si^{4+} cations are disordered in sanidine (thus all the tetrahedra are represented as red, and each one would have a 25% chance of containing Al^{3+} and a 75% chance of containing Si^{4+}), while they are ordered in microcline (thus 1/4 of the tetrahedra contain Al^{3+} and are shown as yellow, while the Si^{4+} ones are red). The third polymorph, orthoclase, is intermediate in Al^{3+}/Si^{4+} ordering between sanidine and microcline. The K^{1+} cations (shown as spheres) fill voids and charge balance the mineral.

produce. Feldspars (and quartz) were once commonly used in toothpaste, until an optical microscopy study by R.C. Emmons was used to remind manufacturers that those minerals were harder than apatite, so that they abraded tooth enamel away! Feldspars have also been used in household cleaners for more than 100 years.

Classification and crystal structure. All feldspars are combinations of two basic formulas: $X^{1+}(AlSi_3O_8)$ and $X^{2+}(Al_2Si_2O_8)$, where the monovalent cations are usually Na^{1+} or K^{1+} (but sometimes, NH_4^+), and the other cations can be Ca^{2+}, Ba^{2+}, and/or Sr; Fe^{2+}, Fe^{3+}, and B^{3+} can sometimes substitute for the X^{2+} or Al^{3+}. The amount of Si^{4+} vs. Al^{3+} covaries with the Ca^{2+} vs. K^{1+} and Na^{1+} contents of the X site. The feldspar group is divided into subgroups on the basis of solid solutions that exist between compositional end members (Table 22.2). The most important of these subgroups are the **alkali feldspars**, which cover the range from $NaAlSi_3O_8$ (albite) to $KAlSi_3O_8$(orthoclase), and the **plagioclase feldspars**, between $NaAlSi_3O_8$ (albite) and $CaAl_2Si_2O_8$ (anorthite). The "alkali" and "plagioclase" subgroup names are useful in field identification because the species within each group can be nearly indistinguishable in hand samples.

Figure 22.8 shows a representation of the range of feldspar stabilities at room temperature. Below about 650°C there is no solid solution between albite and orthoclase, and homogeneous feldspars with compositions in the interior of the ternary phase diagram simply do not occur.

The symbols **Or** for orthoclase, **An** for anorthite, and **Ab** for albite are commonly used; subscripts following the symbols indicate the percentage of that component. For example, a plagioclase feldspar with a composition of slightly more Na than Ca might be given as $Ab_{60}An_{40}$, or even just Ab_{60} because the An_{40} part is implied. Names that are intermediate between albite and anorthite are not recognized as separate species. However, we include them in our table here (Table 22.2) because they are still commonly used in the literature, and you might need to know what they are. A complete presentation of feldspar nomenclature is given in the "bible" on feldspar minerals, Smith and Brown (1988).

The feldspars structures are all based upon a ring of four corner-sharing Si^{4+} and Al^{3+} tetrahedra. These link into chains with a double crankshaft (so-called because it resembles the crankshaft from an internal combustion engine) arrangement parallel to the *a* axis (Figure 22.9). The chains further link corners to make 6- and 8-membered rings (thus satisfying the "framework" silicate polymerization requirements); the large X cations occupy the latter. Compare this partial structure with the full structure of albite, which is shown is Figures 1.10 and 6.1.

The chains give rise to the dominant morphological properties of the feldspars. For example, the structure will break most easily parallel to them, resulting in (010) and (001) cleavages. Orthoclase derives its name from the fact that the angle between these two cleavages is 90°, while in microcline there is a small (i.e., "micro angle") deviation from 90°. Don't worry if you have a difficult time "seeing" these structures. Even though they are the most abundant mineral group on Earth, their structures are very complex!

As we noted above, ordering of Al^{3+} and Si^{4+} into particular arrangements among the tetrahedral sites causes changes in the crystal structure that create different minerals, particularly in the alkali feldspars. The ordering reflects the characteristic stabilities of each phase. Sanidine is stable above about 900°C, orthoclase occurs between 900° and 500°C, and below that temperature, microcline is usually stable. Which of these minerals occurs in your hand samples depends on how fast the rocks cooled (and also on the amount of H_2O present). At high temperatures, Al^{3+} and Si^{4+} are randomly distributed; thus sanidine has the highest symmetry of the three (monoclinic symmetry). As the cations make their way to specific sites and become more ordered, the symmetry is lowered: thus, orthoclase is in a different monoclinic space group and microcline is triclinic. The species names here are a bit confusing: for example, orthoclase has monoclinic crystal morphologies, but the name ("ortho" implies 90°) comes from the fact that it has 90° cleavages. Microcline is morphologically blocky and tends to occur in tablets or prisms, although it has triclinic symmetry.

Table 22.2. Feldspar group

Subgroup	Mineral	Formula
Alkali subgroup	orthoclase	$KAlSi_3O_8$
	sanidine	$(K,Na)AlSi_3O_8$
	hyalophane	$(K,Ba)Al(Si,Al)_3O_8$
	celsian	$BaAl_2Si_2O_8$
	microcline	$KAlSi_3O_8$
	anorthoclase	$(Na,K)AlSi_3O_8$
	buddingtonite	$(NH_4)AlSi_3O_8$
Plagioclase subgroup	anorthoclase	$(Na,K)AlSi_3O_8$
	albite	$NaAlSi_3O_8$
	anorthite	$CaAl_2Si_2O_8$
	dmisteinbergite	$CaAl_2Si_2O_8$
	svyatoslavite	$CaAl_2Si_2O_8$
	reedmergnerite	$NaBSi_3O_8$
Paracelsian subgroup	paracelsian	$BaAl_2Si_2O_8$
	slawsonite	$SrAl_2Si_2O_8$
Banalsite subgroup	banalsite	$BaNa_2Al_4Si_4O_{16}$
	stronalsite	$SrNa_2Al_4Si_4O_{16}$
	lisetite	$Na_2CaAl_4Si_4O_{16}$
	svyatoslavite	$CaAl_2Si_2O_8$
Plagioclase series names*	albite	An_{0-10}
	oligoclase	An_{10-30}
	andesine	An_{30-50}
	labradorite	An_{50-70}
	bytownite	An_{70-90}
	anorthite	An_{90-100}

*The non-endmembers here are no longer IMA-approved mineral species names, but because they persist in the literature, their compositional ranges are included here.

Because Ca^{2+}, Na^{1+}, and K^{1+} have varying sizes, charges, and atomic weights, the feldspar minerals have variations in density and refractive index (Figure 22.10) that correlate well to chemical differences. These variations can be useful in distinguishing feldspars from quartz, which always has the same refractive index because its composition does not vary. Both feldspar and quartz share low birefringence, which shows up as low retardation in thin sections, making these two mineral groups tricky to distinguish from each other unless twinning and/or exsolution are present. Fortunately, feldspar can usually be distinguished on the basis of appearance under cross-polarized light (Figure 22.11) due to the common presence of exsolution and/or twinning, which cause feldspars to show stripes or plaid patterns under cross-polarized light. These can have one or more causes.

Exsolution lamella form when feldspars that are stable at some high temperature cool to the point at which they are unstable, as would be the case for a feldspar with a composition of $Or_{50}Ab_{50}$. In our discussions of phase diagrams in Chapter 20, we highlighted the Or-Ab phase diagram (Figure 20.6) as an example of phases that are miscible at high temperatures, but must form discrete phases upon cooling. Exsolution can occur at such a fine scale that it can barely be detected, even by X-ray diffraction. Perhaps the most common exsolution texture is called **perthite** (exsolved Na-rich feldspar in a K-rich host) or **antiperthite** (K-rich feldspar in a Na-rich host) (Gaines et al., 1997). Order-disorder of Al^{3+} and Si^{4+}, caused by variations in cooling rate, can also cause exsolution to occur. Depending on its scale, the exsolution may be visible to the eye or only in thin section as strings, lamellae, blebs, or veins (Figure 22.11).

Twinning can occur while a crystal is growing or it can be induced by heat (e.g., metamorphism) or pressure (through deformation). For the feldspar minerals, many types of twinning have been distinguished on the basis of the orientation to the twin axis relative to the crystal structure.

- **Albite twinning**, with the twin axis perpendicular to the *b* axis, is found at some scale in all the plagioclase feldspars. Stacks of

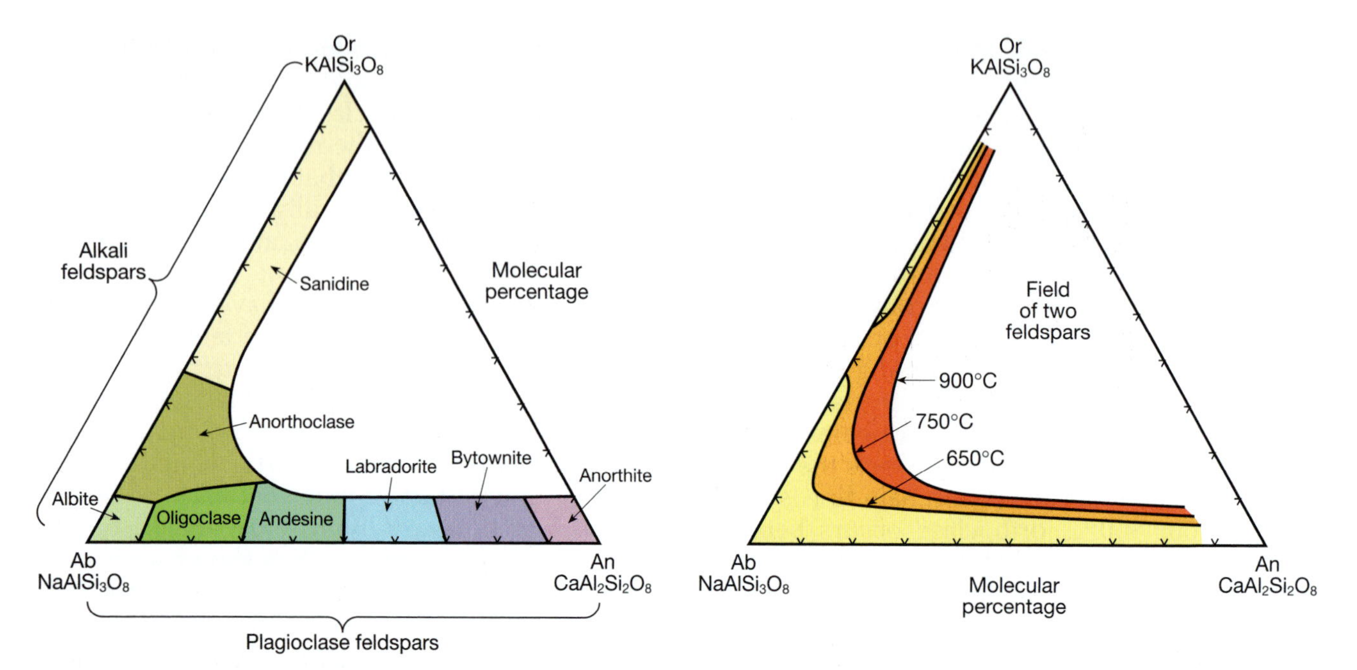

Figure 22.8. *Feldspar ternary plots showing relationships among mineral species in the compositional range between $KAlSi_3O_8$, $NaAlSi_3O_8$, and $CaAl_2Si_2O_8$ (left). The approved species names in these figures are orthoclase, sanidine, anorthoclase, albite, and anorthite. The IMA no longer endorses usage of the mineral names in the plagioclase series between albite and anorthite (oligoclase, andesine, etc.) but we show them here because they are still commonly used in the field (and in the older literature). Solid solutions among orthoclase, albite, and anorthite as a function of temperature at P_{H_2O} = 1 kb (right). Below about 650°C, albite and orthoclase have a large miscibility gap (which means that solid solutions aren't stable). Above 650°C, there is some solid solution. Cooling rate is the most important factor in determining if the high temperature mineral is quenched and preserved, or re-equilibrated into phases at equilibrium at lower temperatures. Figure based on work by Ribbe (1975).*

twinned layers (sub- to many-millimeters in thickness) produce tiny grooves on the surfaces of crystals and cleavage faces. The stacks of twins are called **multiple**, or **polysynthetic**, twinning, and appear as stripes under crossed polars in thin section. Albite twinning is a useful diagnostic tool for plagioclase identification (Figure 22.11).

- **Carlsbad twins**, where the twin axis is parallel to the *c* axis, are common in orthoclase (and occasionally in plagioclase as well). They can produce elongated crystals with a single composition boundary, or penetration twins with two crystals growing in opposite directions. Under crossed polars, the two parts will show different extinction angles.
- **Pericline twinning** occurs with the twin axis parallel to *b*, though its precise orientation depends on the variable geometry of the

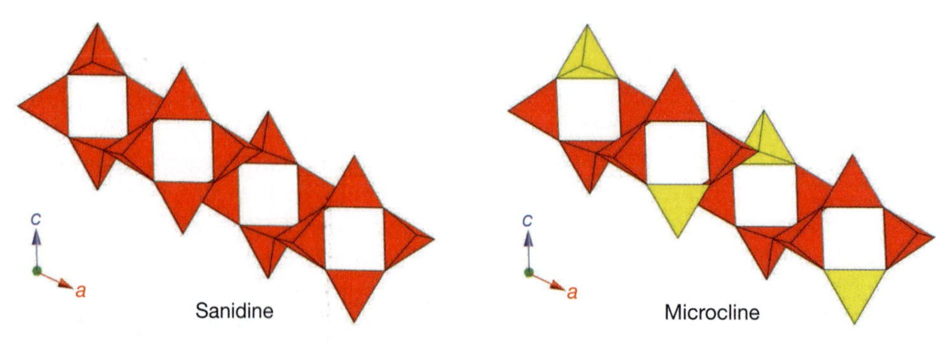

Figure 22.9. *The tetrahedral crankshaft chains that are the major building blocks for the feldspars. The chain on the left corresponds to the sanidine structure and the one on the right to the microcline structure (Figure 22.7). Note again how the tetrahedra are red for sanidine to illustrate disorder of the Al^{3+} and Si^{4+} cations, while for microcline 1/4 of the tetrahedra are yellow representing the location of the Al^{3+} cations.*

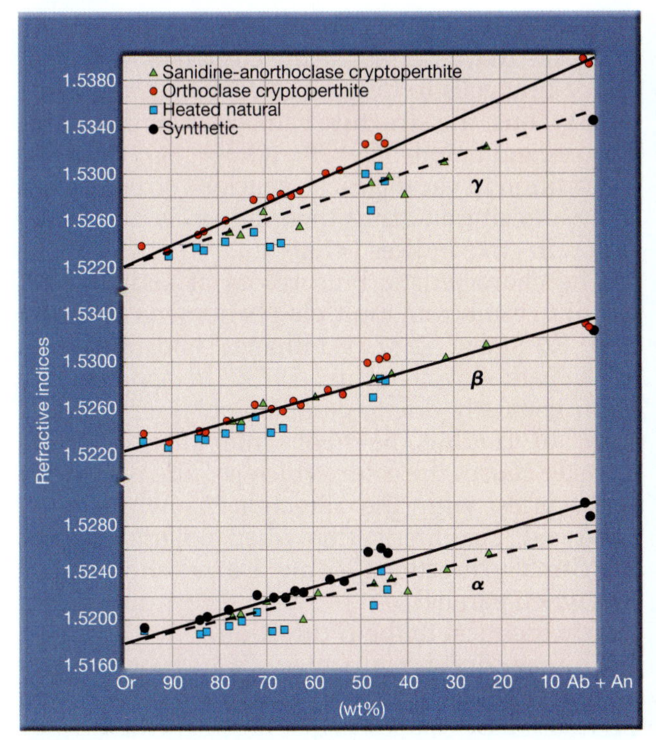

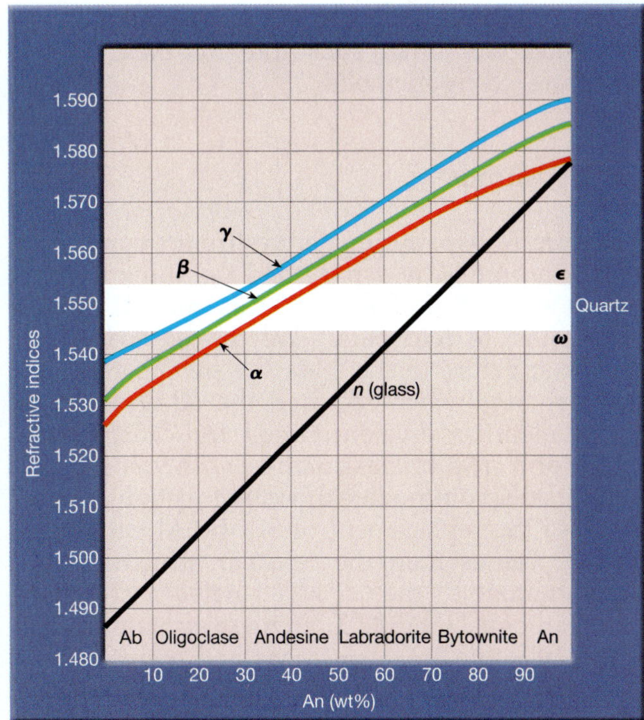

Figure 22.10. Refractive index of feldspar as a function of composition. Alkali feldspar diagram based on Figure 290 in Deer et al. (2001); plagioclase diagram after Figure 523.

unit cell. Frequently polysynthetic, these can resemble albite twins except that they become extinct at an inclination to the (001) or (010) edge. Pericline twinning in combination with albite twinning in two directions is common in microcline, where it is called **tartan** twinning (Figure 22.11), as well as in anorthoclase, where the twin lamella are even finer (thinner).

For more information on twinning, consult Deer et al. (2001) or Smith and Brown (1988).

In hand sample, feldspars can be difficult to distinguish from one another. As a general rule, alkali feldspars are often pink due to the presence of minute hematite inclusions, although microcline may be green. Plagioclase tends to be grayer, often as a result of magnetite or ilmenite inclusions. White feldspars can be distinguished on the basis of the presence (plagioclase) or absence (alkali feldspar) of twinning striations on cleavages. However, these distinctions are quite subtle, and geologic context aids in identifications. For example, recall from Chapter 20 that albite is often the last of the feldspar minerals to crystallize from a magma. Thus albite is often found in pegmatites, which are the coarse-grained, last remnants of a magma chamber to crystallize. In the alkali feldspars, as noted above, sanidine and anorthoclase are common in extrusive igneous rocks that cooled quickly. Orthoclase, on the other hand, is found in slowly-cooled intrusive rocks such as granites and syenites. In K-rich magmas, microcline may also be found in pegmatites, often as perthite. Positive identification of feldspar mineralogy often requires chemical analysis and

Figure 22.11. Photomicrographs of perthite texture and albite, Carlsbad, and pericline twins.

X-ray diffraction, but a thorough optical characterization will often substitute for these more elaborate, costly methods.

Feldspathoids

The term **feldspathoids** derives from a combination of a prefix, "feldspar," and the suffix (oids), which means "similar to." So this mineral group is "similar to feldspars." We moved from the quartz structure to the feldspars by substituting Al^{3+} for Si^{4+}, which required the addition of a monovalent or divalent cation for charge balance—and creation of structures with voids large enough to accommodate those large cations. If we increase the replacement of Si^{4+} by Al^{3+} to even greater values than the feldspar structure can accommodate, then we arrive at the feldspathoids (Table 22.3). In these structures, the resultant voids are even larger, and many possible cations (and even anionic clusters) can substitute into them. The feldspathoids form in environments that are Si-deficient, so there is not enough Si^{4+} to make the more typical feldspar minerals. For this reason, *feldspathoids do not coexist with quartz.*

The **nepheline** ($(Na,K)(AlSiO_4)$) structure can be thought of as a derivative of the β-tridymite structure in which half the Si^{4+} are replaced by Al^{3+} and the resultant missing charge is satisfied by Na^{1+} or K^{1+} (Figure 22.4). Nepheline is composed of rings of six tetrahedra that create sheets; Si^{4+} tetrahedra point in one direction along the *c* axis, and Al^{3+}-filled tetrahedra point in the opposite direction. Some rings are distorted to become nearly like ovals or triangles, thus lowering their symmetry relative to feldspars. The voids in the middle of the rings, of which there are two types (9- and 8-coordinated), are now occupied by large cations such as K^{1+}, Na^{1+}, or Ca^{2+}, although many of the rings are left vacant. The resultant structure, while strongly bonded, is less dense than that of quartz or a feldspar. Therefore its refractive index is also lower. However, unlike the situation with feldspar, cation substitutions in nepheline group minerals do little to affect its refractive index (Bannister and Hey, 1931).

Sodalite ($Na_8Al_6Si_6O_{24}Cl_2$) contains equal amounts of tetrahedral Al^{3+} and Si^{4+} in an isometric unit cell. Minerals in this group have relatively large unit cells composed of 24 corner-sharing tetrahedra alternately filled by Al^{3+} and Si^{4+}. Six rings of tetrahedra line up parallel to the *a* axis, while eight rings of six tetrahedra lie parallel to [111] creating 12-coordinated cuboctahedral cages. Stacking of the cages allows each 6-membered ring to be shared by two adjacent cages, creating channels that intersect at the corners and centers of the unit cells to form large cavities (Deer et al., 1963). Thus, the structure can accommodate large cations and anions like Cl^{1-}, $(OH)^{1-}$, SO_4^{2-}, and S^{2-}, in addition to vacancies. Its spacious structure has a low density, and thus these minerals have quite low refractive indices (<1.487).

The characteristic blue colors of sodalite and lazulite result from multiple phenomena (Mattson and Rossman, 1987; Amthauer and Rossman, 1984; Samoilovich et al., 1973). In sodalite, the color is mostly due to complex polymeric sulfide ions. In lazulite, Fe^{2+}–Fe^{3+} intervalence charge transfer causes the color, which is quite pleochroic and varies with the direction in which it is observed. Lazulite is blue when polarized along the β and γ directions (~*a* and *b* axes) and is colorless when parallel to α (~*c*-axis). The γ direction is nearly identical to the β direction.

The **cancrinite** group minerals are based on layers of rings of six tetrahedra, again alternating in occupancy by Si^{4+} and Al^{3+}. Between these rings are 11-coordinated sites that are occupied by H_2O and Na^{1+}. The 6-membered rings link sideways into 12-membered rings and stack along the *c* axis, forming channels that run through the structure. The channels can be occupied by CO_3^{2-} and/or Na^{1+} (Figure 22.12). The presence of CO_3^{2-} means that cancrinite will effervesce when hydrochloric acid is added; this reaction is quite diagnostic, because few other silicates contain CO_3^{2-}.

The **scapolite** series mineral end-members marialite and meionite are theoretical, which means that they do not exist in nature in pure form. For this reason, the series name scapolite is generally used. The end-members have formulas that can be derived from the plagioclase feldspars and have an analogous Na^{1+}-Ca^{2+} solid solution; in fact, scapolite is an alteration product of feldspar. If we write the formula of marialite as $3(Na(Al,Si)_4O_8)NaCl$, it nicely matches that of albite, $NaAlSi_3O_8$. Similarly, you can make meionite, $3(Ca(Al,Si)_4O_8)CaCO_3$, from anorthite by adding a little Na^{1+} (from dissolved NaCl) or $CaCO_3$ through metamorphism and changing the structure. Also as with the feldspars, there is a linear relationship between the proportion of Ca^{2+} and Na^{1+} in the composition and refractive index. This trend is nearly identical to that of the plagioclases shown in Figure 22.10, although scapolites are uniaxial. These minerals are composed of frameworks of Al^{3+} and Si^{4+} tetrahedra connected into rings with four or five members. The large cavities in the middle of the rings contain an anion such as Cl^{1-} or CO_3^{2-} that is surrounded by four cations like Na^{1+} or Ca^{2+} (Figure 22.12).

Table 22.3. Feldspathoid Minerals

Group	Mineral	Formula
Nepheline	Kalsilite	$KAlSiO_4$
	Nepheline	$(Na,K)AlSiO_4$
	Trikalsilite	$K_{0.67}Na_{0.33}AlSiO_4$
	Panunzite	$K_{0.7}Na_{0.3}AlSiO_4$
	Kaliophilite	$K(AlSiO_4)$
	Yoshiokaite	$(Ca,\square)(Al,Si)_2O_4$
Sodalite	Sodalite	$Na_8Al_6Si_6O_{24}Cl_2$
	Nosean	$Na_8(Al_6Si_6O_{24})(SO_4)\cdot H_2O$
	Haüyne	$Na_6Ca_2Al_6Si_6O_{24}(SO_4)_2$
	Lazurite	$(Na,Ca)_8Si_6Al_6O_{24}[(SO_4),S,Cl,(OH)]_2$
	Bicchulite	$Ca_2(Al_2SiO_6)(OH)_2$
	Kamaishilite	$Ca_2(Al_2SiO_6)(OH)_2$
	Tugtupite	$Na_4AlBeSi_4O_{12}Cl$
	Tsaregorodtsevite	$N(CH_3)_4[Si_2(Si_{0.5}Al_{0.5})O_6]_2$
Cancrinite	Afghanite	$[(Na,K)_{22}Ca_{10}][Si_{24}Al_{24}O_{96}](SO_4)_6Cl_6$
	Bystrite	$Ca(Na,K)_7Si_6Al_6O_{24}(S^{2-})_{1.5}\cdot H_2O$
	Cancrinite	$[(Ca,Na)_6(CO_3)_{1-1.7}][Na_2(H_2O)_2](Si_6Al_6O_{24})$
	Cancrisillite	$Na_7Al_5Si_7O_{24}(CO_3)\cdot 3H_2O$
	Davyne	$[(Na,K)_6(SO_4)_{0.5-1}Cl_{1-0}](Ca_2Cl_2)(Si_6Al_6O_{24})$
	Franzinite	$[(Na,K)_{30}Ca_{10}][Si_{30}Al_{30}O_{120}](SO_4)_{10}\cdot 2H_2O$
	Hydroycancrinite	$Na_8Al_6Si_6O_{24}(OH)_2\cdot 2H_2O$
	Liottite	$(Na,K)_{16}Ca_8Si_{18}Al_{18}O_{72}(SO_4)_5Cl_4$
	Microsommite	$[Na_4K_2(SO_4)](Ca_2Cl_2)Si_6Al_6O_{24}$
	Pitaglianoite	$K_2Na_6Si_6Al_6O_{24}(SO_4)\cdot 2H_2O$
	Quadridavyne	$[(Na,K)_6Cl_2](Ca_2Cl_2)(Si_6Al_6O_{24})$
	Sacrofanite	$(Na,K,Ca)_{112}(Si_{84}Al_{84}O_{336})[(SO_4)_{26}(Cl,F,H_2O)_{10}$
	Tounkite	$(Na,Ca,K)_8(Al_6Si_6O_{24})(SO_4)_2Cl\cdot H_2O$
	Vishnevite	$[(Na_6(SO_4)](Si_6Al_6O_{24})[Na_2(H_2O)_2]$
	Wenkite	$(Ba,K)_4(Ca,Na)_6(Si,Al)_{20}O_{41}(OH)_2(SO_4)_3\cdot H_2O$
	IMA2002-021	$(Na,K,Ca)_{48}Si_{36}Al_{36}O_{144}[(SO_4)8Cl_2]\cdot 3H_2O$
Scapolite	Marialite	$Na_4[Al_3Si_9O_{24}]Cl$
	Dipyre*	Me_{20}-Me_{50}
	Mizzonite*	Me_{50}-Me_{80}
	Meionite	$Ca_4Al_6Si_6O_{24}CO_3$

*These names have been discredited, but because they remain in popular use, they are included here.

Zeolites

Zeolites form at low pressures and temperatures in the presence of H_2O and possess channels and voids in their structures. There are two general places that zeolites form.

1. The most commonly-known occurrences to the geological community are large, often spectacularly beautiful single crystals of zeolite that form in vugs in basalts and other rock types. These crystals form as water that is rich in elements dissolved from basalt (Ca^{2+}, Al^{3+}, and Si^{4+}) percolates into open spaces in the host rocks and precipitates to form zeolite.
2. Hydrothermally-altered, very fine-grained zeolite deposits in sedimentary rocks are far more abundant. They typically form in arid regions where there are large expanses of volcanic ash. When it does rain on the ash, lakes form that lack external drainage. As they evaporate, the waters concentrate certain elements such as Na^{1+} and Ca^{2+}, and in turn, these saturated waters percolate into the silica-rich volcanic ash. The amorphous ash alters into various zeolite minerals. Rocks composed of zeolites were thought to be made of clay minerals until the late 1950s, when U.S.G.S. geologists carried samples from the deserts of Oregon back to their research labs and identified them using X-ray diffraction. These deposits are really zeolitic rocks that will often contain more than one zeolite plus various reaction products from the altered ash, which often include fine-grained opal.

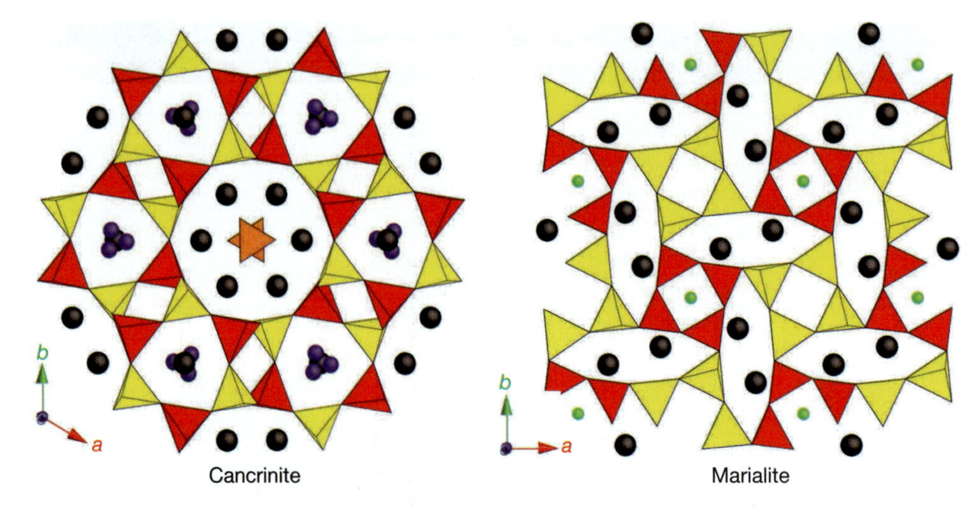

Figure 22.12. The crystal structures of two framework silicates that both contain atomic groups in their voids: cancrinite, $Na_6Al_6Si_6O_{24} \cdot CaCO_3 \cdot 2H_2O$ (left) and marialite (a scapolite group mineral), $Na_3Al_3Si_9O_{24} \cdot 3NaCl$ (right). First, examine these formulas and convince yourself that they are both framework silicates! In cancrinite, Al^{3+} occupies half of the tetrahedra (yellow). In marialite, the yellow tetrahedra are half occupied by Al^{3+} and half by Si^{4+} (i.e., there is a 50% chance of finding Al^{3+} in them). The center ring in cancrinite has CO_3 groups at its center surrounded by Na^{1+} cations, while in the smaller 6-membered rings, both Na^{1+} cations and H_2O occur. In marialite, the channels are filled with Na^{1+} (larger black spheres) and NaCl groups (denoted by smaller, green spheres).

Zeolites are used extensively in treatment of waste waters to remove unwanted cations (lead and radioactive cesium and strontium) or cationic complexes (such as ammonia, NH_4^{1+}). [Why would anyone want to remove ammonia from water? Think about extended periods of manned space travel—do you think they carry all their water with them in jugs?] Zeolites are also used as additives in cattle feed, as cat litter, and even in refrigeration units because they gain and lose heat upon hydration and dehydration.

Classification and crystal structure. The basic formula for the zeolite group minerals is $(Na,Ca,Ba)_{1-2}(Al,Si)_5O_{10} \cdot nH_2O$, very similar to the feldspars, with the addition of H_2O. According to the recommended nomenclature for zeolite minerals (Coombs et al., 1998), this basic formula gives rise to more than 80 distinct zeolite species (Table 22.4). The compositions are accommodated by open frameworks of AlO_4 and SiO_4 tetrahedra that share corners to form chains, sheets, and rings, creating what are called secondary building units (SBU)—however, all of them are framework silicates.

To try to make sense of the many zeolites, consider their basic building block, which is a framework with the general formula $[Al_{nx}Si_{n(4-x)}O_{n8}]^{nx-}$, where n is some multiple of this basic building unit needed to fill the unit cell and x is an integer. Different values of x and n will result in different zeolite species. For example, start with $[AlSi_3O_8]^{1-}$and let $n = 9$ and $x = 1$, creating the framework composition $[Al_9Si_{27}O_{72}]^{9-}$. This particular framework would need nine positive charges. One possible way to obtain this would be by adding Na^{1+}, K^{1+}, and Ca^{2+} to make $(Na,K)Ca_4[Al_9Si_{27}O_{72}]$. The channel cations bond into the frameworks but commonly have one or more sides "exposed" in the channel. These exposed sides lack local charge satisfaction. H_2O can enter the structure to form partial hydration spheres around the channel cations, with the negative dipole of the H_2O molecule pointing toward the exposed positive portion of the channel cation. To complete the above zeolite formula, we would just add H_2O's to fill the channels, resulting in $(Na,K)Ca_4[Al_9Si_{27}O_{72}]\cdot 24H_2O$. This is the general formula for the zeolite **heulandite** and its structure, SBU, and a mineral sample are shown in Figure 22.13.

Clinoptilolite has a similar framework of $[Al_6Si_{30}O_{72}]^{6-}$ but with less Al^{3+}, thereby requiring a lower channel cation charge (–6 vs –9). Thus, clinoptilolite's general formula can be written as $(Na,K)_6[Al_6Si_{30}O_{72}]\cdot 20H_2O$. Divalent cations like Ca^{2+} and Mg^{2+} will in general have larger hydration spheres than monovalent cations, like Na^{1+} and K^{1+}. Thus, zeolites with divalent channel cations will have more channel H_2O than those with monovalent cations. There is also more room in the channels for H_2O because one Ca^{2+} occupies less space than two Na^{1+}.

The tetrahedral framework of any zeolite is structurally and chemically much more rigid and stable than the channel cations and H_2O molecules. The "extraframework" cations in zeolites

Table 22.4. Zeolite group*		
	chiavennite	$CaMn[Be_2Si_5O_{13}(OH)_2]\cdot 2H_2O$
	leucite	$K[AlSi_2O_6]$
	ammonioleucite	$(NH_4)[AlSi_2O_6]$
	analcime	$Na[AlSi_2O_6]\cdot H_2O$
	pollucite	$(Cs,Na)[AlSi_2O_6]\cdot nH_2O$ ($Cs + n = 1$)
	wairakite	$Ca[Al_2Si_4O_{12}]\cdot 2H_2O$
	laumontite	$Ca_4[Al_8Si_{16}O_{48}]\cdot 18H_2O$
	hsianghualite	$Li_2Ca_3[Be_3Si_3O_{12}]F_2$
Chabazite series	chabazite series	$(Ca_{0.5},Na,K)_x[Al_xSi_{12-x}O_{24}]\cdot 12H_2O$, $x = 2.4–5.0$
	chabazite-Ca	$(Ca_{0.5},Na,K)_x[Al_xSi_{12-x}O_{24}]\cdot 12H_2O$, $x = 2.4–5.0$
	chabazite-Na	$(Na,K,Ca_{0.5})_x[Al_xSi_{12-x}O_{24}]\cdot 12H_2O$, $x = 2.5–4.8$
	chabazite-K	$(K,Na,Ca_{0.5})_x[Al_xSi_{12-x}O_{24}]\cdot 12H_2O$, $x = 3.0–4.5$
	chabazite-Sr	$(Sr_{0.5},Ca_{0.5},K)_4[Al_4Si_8O_{24}]\cdot 12H_2O$
	willhendersonite	$Ca_{3-x}K_x[Al_6Si_6O_{24}]\cdot 10H_2O$, $x = 0.0–2.0$
	offretite	$CaKMg[Al_5Si_{13}O_{36}]\cdot 16H_2O$
Erionite series	erionite series	$K_2(Na,Ca_{0.5})_8[Al_{10}Si_{26}O_{72}]\cdot 30H_2O$
	erionite-Ca	$K_2(Ca_{0.5},Na)_8[Al_{10}Si_{26}O_{72}]\cdot 30H_2O$
	erionite-K	$K_4(K,Na,Ca_{0.5})_7[Al_9Si_{27}O_{72}]\cdot 30H_2O$
	erionite-Na	$K_2(Na,Ca_{0.5})_7[Al_9Si_{27}O_{72}]\cdot 30H_2O$
Gmelinite series	gmelinite series	$(Na,Ca_{0.5},K)_8[Al_8Si_{16}O_{48}]\cdot 22H_2O$
	gmelinite-Ca	$(Ca_{0.5},Sr,K,Na)_8[Al_8Si_{16}O_{48}]\cdot 22H_2O$
	gmelinite-K	$(K,Ca_{0.5},Na)_8[Al_8Si_{16}O_{48}]\cdot 22H_2O$
	gmelinite-Na	$(Na,K,Ca_{0.5})_8[Al_8Si_{16}O_{48}]\cdot 22H_2O$
Faujasite series	faujasite series	$(Na,Ca_{0.5},Mg_{0.5},K)_x[Al_xSi_{12-x}O_{24}]\cdot 16H_2O$, $x = 3.2–3.8$
	faujasite-Ca	$(Ca_{0.5},Na,Mg_{0.5},K)_x[Al_xSi_{12-x}O_{24}]\cdot 16H_2O$, $x = 3.3–3.9$
	faujasite-Mg	$(Mg_{0.5},Ca_{0.5},Na,K)_{3.5}[Al_{3.5}Si_{8.5}O_{24}]\cdot 16H_2O$
	faujasite-Na	$(Na,Ca_{0.5},Mg_{0.5},K)_x[Al_xSi_{12-x}O_{24}]\cdot 16H_2O$, $x = 3.2–4.3$
Levyne series	levyne series	$(Ca_{0.5},Na,K)_6[Al_6Si_{12}O_{36}]\cdot 17H_2O$
	levyne-Ca	$(Ca_{0.5},Sr,K,Na)_6[Al_6Si_{12}O_{36}]\cdot 17H_2O$
	levyne-Na	$(Na,K,Ca_{0.5})_6[Al_6Si_{12}O_{36}]\cdot 17H_2O$
	tschörtnerite	$Ca_4(K,Ca,Sr,Ba)_3Cu_3(OH)_8[Al_{12}Si_{12}O_{48}]\cdot nH_2O$
	gismondine	$Ca_4[Al_8Si_8O_{32}]\cdot 18H_2O$
	amicite	$K_4Na_4[Al_8Si_8O_{32}]\cdot 10H_2O$
	garronite	$(Ca_{0.5},Na)_6[Al_6Si_{10}O_{32}]\cdot 14H_2O$
	gobbinsite	$Na_5[Al_5Si_{11}O_{32}]\cdot 12H_2O$
	harmotome	$(Ba_{0.5},Ca_{0.5},K,Na)(Al_5Si_{11}O_{32}]\cdot 12H_2O$
Phillipsite series	phillipsite series	$(K,Na,Ca_{0.5})_x[Al_xSi_{16-x}O_{32}]\cdot 12H_2O$, $x = 3.8–6.4$
	phillipsite-Ca	$(Ca_{0.5},K,Na)_x[Al_xSi_{16-x}O_{32}]\cdot 12H_2O$, $x = 4.1–6.8$
	phillipsite-K	$(K,Na,Ca_{0.5})_x[Al_xSi_{16-x}O_{32}]\cdot 12H_2O$, $x = 3.8–6.4$
	phillipsite-Na	$(Na,K,Ca_{0.5})_x[Al_xSi_{16-x}O_{32}]\cdot 12H_2O$, $x = 3.7–6.7$
	perlialite	$K_9Na(Ca,Sr)[Al_{12}Si_{24}O_{72}]\cdot 15H_2O$
Paulingite series	paulingite series	$(K,Ca_{0.5},Na,Ba_{0.5})_{10}[Al_{10}Si_{32}O_{84}]\cdot 27–44\ H_2O$
	paulingite-K	$(K,Ca_{0.5},Na)_{10}[Al_{10}Si_{32}O_{84}]\cdot 44\ H_2O$
	paulingite-Ca	$(Ca_{0.5},K,Na,Ba_{0.5})_{10}[Al_{10}Si_{32}O_{84}]\cdot 27–44\ H_2O$
	mazziite	$(Mg_{2.5}K_2Ca_{1.5})[Al_{10}Si_{26}O_{72}]\cdot 30H_2O$
	merlinoite	$(K,Ca_{0.5},Ba_{0.5},Na)_{10}[Al_{10}Si_{22}O_{64}]\cdot 22H_2O$
	montesommaite	$K_9[Al_9Si_{23}O_{64}]\cdot 10H_2O$
Heulandite series	heulandite series	$(Ca_{0.5},Sr_{0.5},Ba_{0.5},Mg_{0.5},Na,K)_9[Al_9Si_{27}O_{72}]\cdot 24H_2O$
	heulandite-Ca	$(Ca_{0.5},Sr_{0.5},Ba_{0.5},Mg_{0.5},Na,K)_9[Al_9Si_{27}O_{72}]\cdot 24H_2O$
	heulandite-K	$(K,Ca_{0.5},Sr_{0.5},Ba_{0.5},Mg_{0.5},Na)_9[Al_9Si_{27}O_{72}]\cdot 24H_2O$
	heulandite-Na	$(Na,Ca_{0.5},Sr_{0.5},Ba_{0.5},Mg_{0.5},K)_9[Al_9Si_{27}O_{72}]\cdot 24H_2O$
	heulandite-Sr	$(Sr_{0.5},Ca_{0.5},Ba_{0.5},Mg_{0.5},Na,K)_9[Al_9Si_{27}O_{72}]\cdot 24H_2O$

*Formulas in this table mostly follow those used by Coombs et al. (1997), with updates from Deer at al. (2006)

Table 22.4. (continued)*

Series	Mineral	Formula
Clinoptilolite series	clinoptilolite series	$(Na,K,Ca_{0.5},Mg_{0.5})_6[Al_6Si_{30}O_{72}]\cdot 20H_2O$
	clinoptilolite-Ca	$(Ca_{0.5},Na,K,Sr_{0.5},Ba_{0.5},Mg_{0.5})_6[Al_6Si_{30}O_{72}]\cdot 20H_2O$
	clinoptilolite-K	$(K,Na,Ca_{0.5},Sr_{0.5},Ba_{0.5},Mg_{0.5})_6[Al_6Si_{30}O_{72}]\cdot 20H_2O$
	clinoptilolite-Na	$(Na,K,Ca_{0.5},Sr_{0.5},Ba_{0.5},Mg_{0.5})_6[Al_6Si_{30}O_{72}]\cdot 20H_2O$
Stilbite series	stilbite series	$(Ca_{0.5},Na,K)_9[Al_9Si_{27}O_{72}]\cdot 30H_2O$
	stilbite-Ca	$(Ca_{0.5},Na,K)_9[Al_9Si_{27}O_{72}]\cdot 30H_2O$
	stilbite-Na	$(Na,K,Ca_{0.5})_9[Al_9Si_{27}O_{72}]\cdot 28H_2O$
	stellerite	$Ca_4[Al_8Si_{28}O_{72}]\cdot 28H_2O$
	barrerite	$Na_8[Al_8Si_{28}O_{72}]\cdot 26H_2O$
Natrolite series	natrolite	$Na_2[Al_2Si_3O_{10}]\cdot 2H_2O$
	mesolite	$Na_2Ca_2[Al_6Si_9O_{30}]\cdot 8H_2O$
	scolecite	$Ca[Al_2Si_3O_{10}]\cdot 3H_2O$
	edingtonite	$Ba[Al_2Si_3O_{10}]\cdot 4H_2O$
	gonnardite	$(Na,Ca_{0.5})_{8-10}[Al_{8+x}Si_{12-x}O_{40}]\cdot 12H_2O$, $x = 0–2$
	cowlesite	$Ca[Al_2Si_3O_{10}]\cdot 5.3H_2O$
Thomsonite series	thomsonite series	$(Ca,Sr)_2Na[Al_5Si_5O_{20}]\cdot 6–7H_2O$
	thomsonite-Ca	$Ca_2Na[Al_5Si_5O_{20}]\cdot 6H_2O$
	thomsonite-Sr	$(Sr,Ca)_2Na[Al_5Si_5O_{20}]\cdot 7H_2O$
	mordenite	$(Na_2,Ca,K_2)_4[Al_8Si_{40}O_{96}]\cdot 28H_2O$
	epistilbite	$(Ca,Na_2)_3[Al_6Si_{18}O_{48}]\cdot 16H_2O$
	maricopaite	$(Pb_7Ca_2)[Al_{12}Si_{36}(O,OH)_{100}]\cdot nH_2O$
Dachiardite series	dachiardite series	$(Ca_{0.5},Na,K)_{4-5}[Al_{4-5}Si_{20-19}O_{48}]\cdot 13H_2O$
	dachiardite-Ca	$(Ca_{0.5},Na,K)_5[Al_5Si_{19}O_{48}]\cdot 13H_2O$
	dachiardite-Na	$(Na,K,Ca_{0.5})_4[Al_4Si_{20}O_{48}]\cdot 13H_2O$
Ferrierite series	ferrierite series	$(K,Na,Mg_{0.5}Ca_{0.5})_6[Al_6Si_{30}O_{72}]\cdot 20H_2O$
	ferrierite-K	$(K,Na,Mg_{0.5}Ca_{0.5})_6[Al_6Si_{30}O_{72}]\cdot 20H_2O$
	ferrierite-Mg	$(Mg_{0.5},K,Na,Ca_{0.5})_6[Al_6Si_{30}O_{72}]\cdot 20H_2O$
	ferrierite-Na	$(Na,K,Mg_{0.5}Ca_{0.5})_6[Al_6Si_{30}O_{72}]\cdot 20H_2O$
	boggsite	$(Ca,Na_{0.5},K_{0.5})_9[Al_{18}Si_{78}O_{192}]\cdot 70H_2O$
	gottardiite	$(Na,K)_3Mg_3Ca_5[Al_{19}Si_{117}O_{272}]\cdot 93H_2O$
	terranovaoite	$NaCa[Al_3Si_{17}O_{40}]\cdot 13H_2O$
	mutinaite	$Na_3Ca_4Al_{11}Si_{85}O_{192}\cdot 60H_2O$
Brewsterite series	brewsterite series	$(Sr,Ba)_2[Al_4Si_{12}O_{32}]\cdot 10H_2O$
	brewsterite-Sr	$(Sr,Ba)_2[Al_4Si_{12}O_{32}]\cdot 10H_2O$
	brewsterite-Ba	$(Ba,Sr)_2[Al_4Si_{12}O_{32}]\cdot 10H_2O$
	yugawaralite	$Ca_2[Al_4Si_{12}O_{32}]\cdot 8H_2O$
	goosecreekite	$Ca[Al_2Si_6O_{16}]\cdot 5H_2O$
	roggianite	$Ca_2[Be(OH)_2Al_2Si_4O_{13}]\cdot <2.5H_2O$
	bellbergite	$(K,Ba,Sr)_2Sr_2Ca_2(Ca,Na)_4[Al_{18}Si_{18}O_{72}]\cdot 30H_2O$
	tschernichite	$(Ca,Mg,Na_{0.5})[Al_2Si_6O_{16}]\cdot 8H_2O$
	bikitaite	$Li[AlSi_2O_6]\cdot H_2O$
	parthéite	$Ca_2[Al_4Si_4O_{15}(OH)_2]\cdot 4H_2O$
	gaultite	$Na_4[Zn_2Si_7O_{18}]\cdot 5H_2O$
	kalborsite	$K_6[Al_4Si_6O_{20}]B(OH)_4Cl$
	lovdarite	$K_4Na_{12}[Be_8Si_{28}O_{72}]\cdot 18H_2O$
	pahasapaite	$(Ca_{5.5}Li_{3.6}K_{1.2}Na_{0.2}\square_{13.5})Li_8[Be_{24}P_{24}O_{96}]\cdot 38H_2O$
	weinebeneite	$Ca[Be_3(PO_4)_2(OH)_2]\cdot 4H_2O$

*Formulas in this table mostly follow those used by Coombs et al. (1997), with updates from Deer at al. (2006)

can usually be exchanged and dehydrated, which is why zeolites have found so many important industrial applications. Heulandite and clinoptilolite, which we just discussed, can be used as examples. The main difference in their formulas is the ratio of Si^{4+} to Al^{3+}. When Na^{1+}, Pb^{2+}, K^{1+}, Rb^{1+}, and Cs^{1+} are substituted into these structures, the framework remains essentially unchanged. Thus

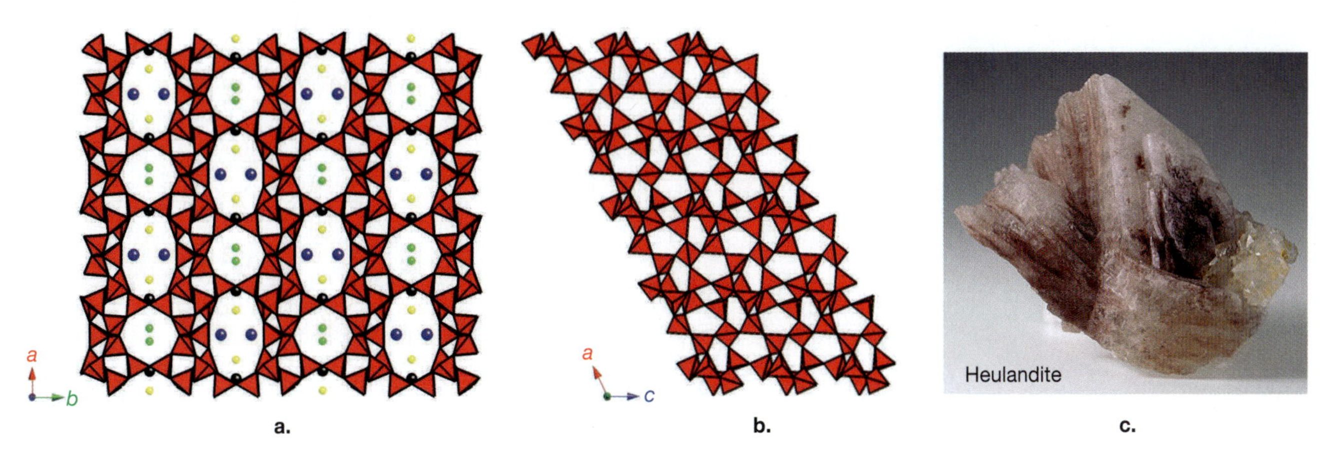

Figure 22.13. Crystal structure and photograph of the zeolite group mineral heulandite, (Na,K)$Ca_4[Al_9Si_{27}O_{72}]\cdot 24H_2O$, with its basic building block—a tetrahedral sheet—isolated. The tetrahedra are red. ***a.*** *The* a-b *projection of the structure shows two separate channels that house water molecules (large blue spheres),* K^{1+} *(smaller black spheres),* Ca^{2+} *(green spheres,) and* Na^{1+} *(yellow spheres).* **b.** *An* a-c *projection of an isolated "sheet-like" layer in the structure. Compare to the* a-b *projection, where these sheets are vertical and viewed on edge. The sheets are bonded together by tetrahedra that share an* O^{2-} *at the ends of the channels, as seen in the* a-b *projection, along the* b *direction.* ***c.*** *In this photograph of the mineral you can see some layering. These layers correspond to the tetrahedral layer shown in Figure 22.13b. Because of the weak bonding between these layers, heulandite breaks into sheets with perfect (010) cleavage.*

the channel occupants in those species can be altered both artificially and in nature.

However, other zeolites exist that have identical framework structures but do not exhibit a continuous solid-solution series with respect to the channel cations. An example of this is the series between natrolite $Na_{16}[Al_{16}Si_{24}O_{80}]\cdot 16H_2O$, mesolite $Na_{16}Ca_{16}[Al_{48}Si_{72}O_{240}]\cdot 64H_2O$, and scolecite $Ca_8[Al_{16}Si_{24}O_{80}]\cdot 24H_2O$ (Figure 22.14). From a first glance at the chemical formulas, it might appear that there is a solid solution, with two Na^{1+} cations substituting for one Ca^{2+} and one H_2O. However, the reality is that very little variation from these ideal formulas is found in nature or in samples synthesized in the lab (Ross et al., 1992). Not only are the formulas for these zeolites similar to those of the common feldspars, so too are their changes in refractive index as Ca^{2+} substitutes for Na^{1+} (Figure 22.15). Notice the refractive index curves cross; this feature is explained in detail in Chapter 18.

Chabazite (Figure 22.16) is the last example of the zeolites we'll discuss. Heulandite's SBU were sheets, natrolite's were chains, and chabazite is a bit more complicated—its SBU is a cage (Figure 22.16b). The cage is composed of six 8-membered tetrahedral rings. When the cages are bonded together, the structure appears blocky. So these three species are good examples of the variability in the structures and morphologies for the zeolites. Once again, their physical properties (in this case, refractive index and morphology) are directly related to their crystal structure.

How do the crystal structure and chemistry of zeolite minerals affect their optical and physical properties in general? The birefringence of zeolites is exactly analogous to those of quartz and feldspars. If you compare the structure of albite with that of natrolite, you would note that chemically, both contain the same elements, though natrolite has water in the channels. These channels are basically voids in the structure, so their presence reduces the overall density, and thus the refractive index, of the mineral (relative to feldspars). Zeolites typically occur with one of three morphologies: needles, plates, and blocky masses (Figures 22.13, 22.14, and 22.16). The channels control this morphology because the crystals break along those channels and not across the strongly-bonded tetrahedra.

For more information on the zeolite structures, please consult Deer et al. (2006) and Armbruster and Gunter (2001).

Layer Silicates (Phyllosilicates)

What kind of mineral structures result when Si^{4+} tetrahedra share only three corners (rather than four in framework silicates)? As you will recall, this creates a structure in which the tetrahedra form 6-membered rings that link up to form sheets (Figure 6.3). These are joined to sheets of edge-sharing octahedra in various combinations of stacking. The resultant layer structures have properties that make them useful in a variety of applications in industry: the layers are flexible, elastic,

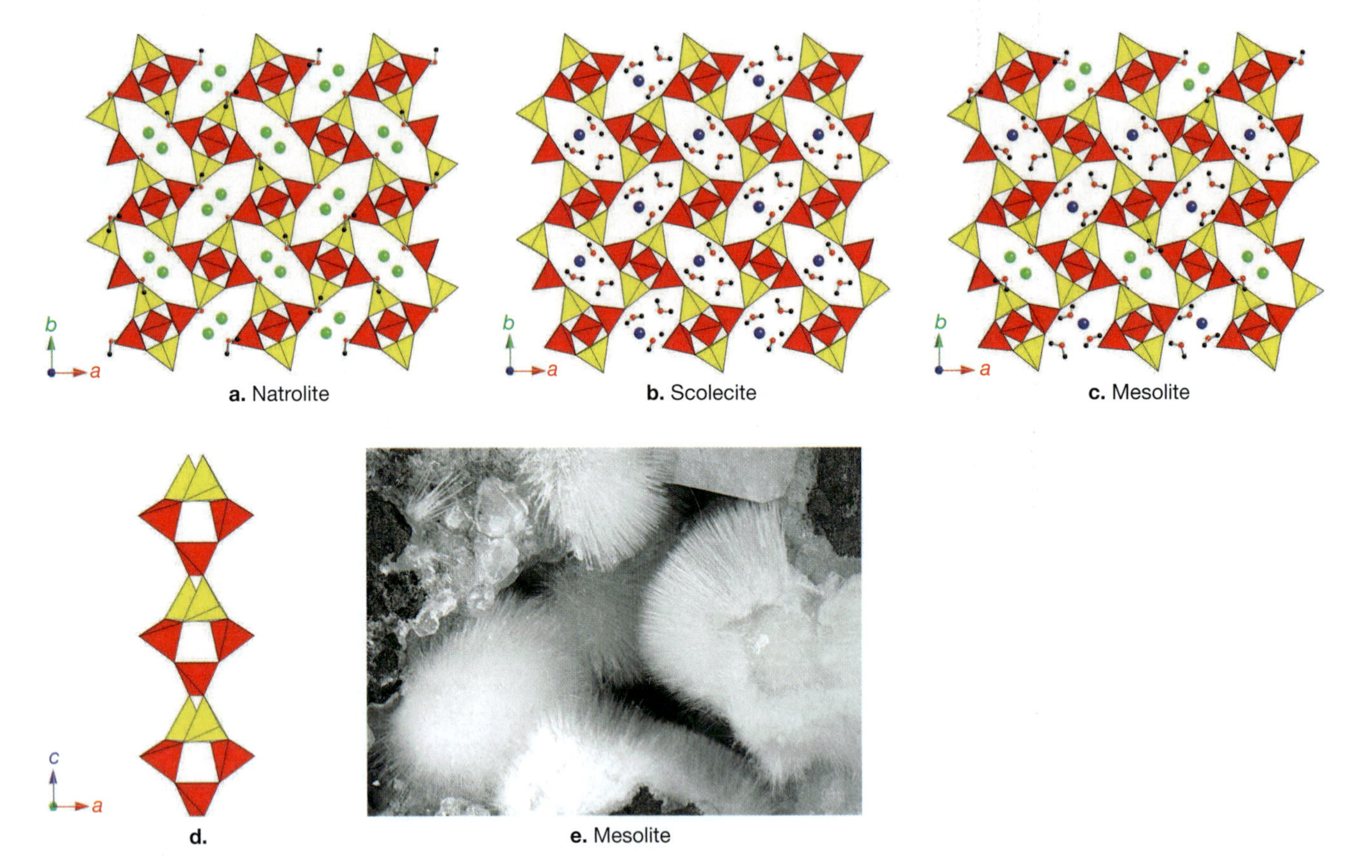

Figure 22.14. *Crystal structures, projected on the* a-b *plane, along with basic building blocks and photographs of three zeolites: natrolite,* $Na_{16}[Al_{16}Si_{24}O_{80}]\cdot16H_2O$, *scolecite,* $Ca_8[Al_{16}Si_{24}O_{80}]\cdot24H_2O$, *and mesolite,* $Na_{16}Ca_{16}[Al_{48}Si_{72}O_{240}]\cdot64H_2O$. *The* Si^{4+} *tetrahedra are red,* AlO_4 *tetrahedra are yellow,* Na^{1+} *cations are green,* Ca^{2+} *cations are blue, and* H_2O *molecules are represented by a very small red sphere* (O^{2-}) *bonded to two small black spheres (H's).* ***a.*** *In natrolite, each channel has two* Na^{1+} *cations and two water molecules.* ***b.*** *In scolecite, each channel has one* Ca^{2+} *and three water molecules. Here the channels are wider than in natrolite.* ***c.*** *Mesolite is sort of mixture of natrolite and scolecite. There is one "layer" of natrolite followed by two layers of scolecite, repeated in the* b *direction.* ***d.*** *The major building block for the zeolites is chains of tetrahedra that run parallel to* c*; here one such chain is isolated and projected on its* a-c *plane. The structures in Figure 22.14a–c are all viewed down these chains, and for each there are 13 such chains (with one in the very center of each structure), all cross-linked to form the structures.* ***e.*** *In this photograph of mesolite needles, the chains are elongated parallel to the tetrahedral chains shown in Figure 22.14d.*

transparent, good insulators, and can be split into extremely thin layers. Micas are thus used in the electronic industry as insulators: muscovite and phlogopite are used to line the gauge glasses of high-pressure steam boilers, in diaphragms for oxygen-breathing equipment, as optical filters, and as windows in stoves and kerosene heaters (Hedrick, 1999). Ground mica is also used to make joint compound for patching drywall, as a pigment extender in paint, and as a additive in rubber, paper (e.g., glossy magazine paper uses kaolinite), plastics, cosmetics (e.g., talcum powder) and drilling muds. You also see ground mica as a surface coating on rolled roofing and asphalt shingles!

On a more personal note, micas are added to cosmetics such as nail polish and lipstick to provide a high luster. In these applications, it is the highly reflective cleavage of the mica that gives luster to the products. The mica used in these applications is wet-ground and can be of different grain sizes depending on the effect desired (e.g., dry-ground mica is not so highly reflective). Coarser grains give a sparkly finish while finer-grained mica will provide a "smooth" and brilliant luster. Mica is also used for the same reasons in "metal-flake" paint for some automotive applications (André Lalonde, personal communication, 2005).

In a related but different application, titanium-coated micas are used in some fluorescent theatrical cosmetics. In these applications, the color effect does not come from the intrinsic color of the pigment, but from the interference of waves reflected by the surface of the mica support on one hand, and waves reflected by the coating surface. The resulting color depends on the thickness of the pigment layer (Bernard Grobéty, personal communication, 2005).

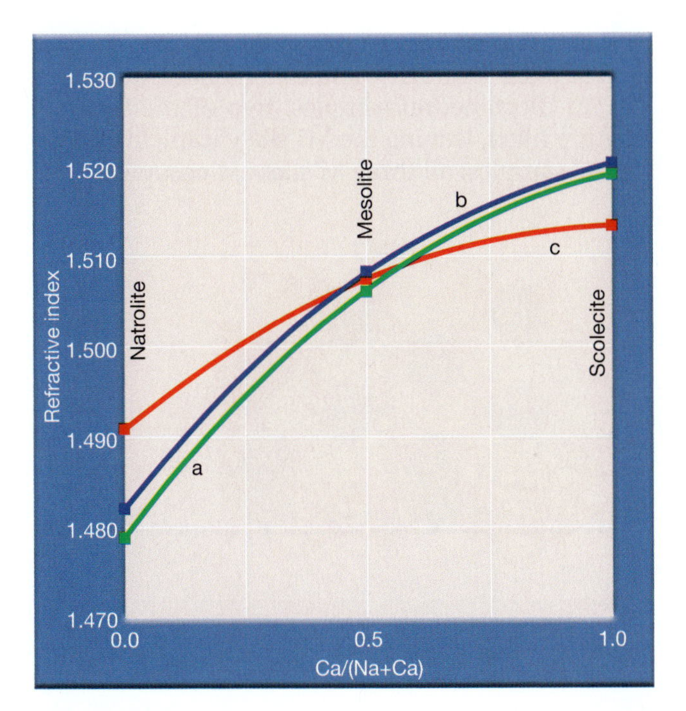

Figure 22.15. *Trends in the refractive indices of the three zeolites shown in Figure 22.14: natrolite, mesolite, and scolecite (modified from Gunter and Ribbe, 1993). Refractive indices increase as Ca^{2+} substitutes for Na^{1+}; this is the same trend that occurs for the feldspars. Also notice that the curves for the refractive index cross near mesolite; see Chapter 18 for a thorough of discussion of this.*

Other layer silicates with tremendous economic importance are the serpentine minerals lizardite, antigorite, and (especially) chrysotile, which form as fibers. Chrysotile is the most commonly-used industrial asbestos, and is also useful in insulation, fire retarding, and brake linings.

Crystal chemistry and structure. The basic building blocks of a layer silicate are Si^{4+} tetrahedra that link by sharing three corners (the basal oxygens) with neighboring tetrahedra; the fourth corner (apical oxygen) points in the same direction for all tetrahedra within a sheet (Figure 6.4). This gives a ratio of two tetrahedral cations to five (3 + 4 x 1/2) oxygens, or 2:5. So as we first noted in Chapter 6, you can always identify a layer silicate by its formula, which will have a 2:5 ratio of tetrahedral cations to oxygens (Table 6.1). In the case of an Si_2O_5 tetrahedral sheet, this gives a net charge of –2. The $(OH)^{1-}$ groups occur in the adjacent octahedral sheet and are thus not shared with the tetrahedral sheet.

To satisfy that –2 charge deficit, each of the layer silicates has at least one sheet of octahedra that bond to it, either in a 1:1 or 2:1 ratio of tetrahedral to octahedral sheets. Because the geometric arrangement of oxygens in a sheet of octahedra is *almost the same* as the geometrical arrangement of the *apical* oxygens in a sheet of tetrahedra, the sheets can stack together to form layered structures of many types.

The simplified general formula for layer silicates is thus:

$$IM_{2-3}\square_{1-0}T_4O_{10}A_2,$$

where I is the interlayer, which can be occupied by a cation such as K^{1+}, Na^{1+}, Ca^{2+}, Cs^{1+}, NH_4^{1+}, Rb^{1+}, Ba^{2+}, H_3O^+, etc. or sometimes be vacant; M is an octahedral cation such as Fe^{2+}, Fe^{3+}, Mg^{2+}, Li^{1+}, Mn^{2+}, Zn^{2+}, Al^{3+}, Cr^{3+}, or Ti^{4+}; $\square$ is a vacancy in an octahedral site; T is the tetrahedral cation, commonly Si^{4+}, Al^{3+}, Fe^{3+}, B^{3+}, and/or Be^{2+}; and A is F^{1-}, Cl^{1-}, $(OH)^{1-}$, O^{2-}, and/or S^{2-} in all cases except for chlorite (see below). Note that there are a total of three M sites, of

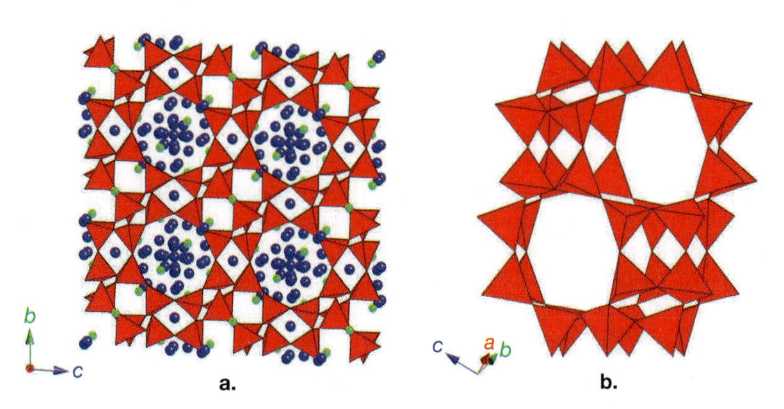

Figure 22.16. *Crystal structure and photograph of the zeolite group mineral chabazite, $Ca_2[Al_4Si_8O_{24}]\cdot 12H_2O$, with its basic building—a cage—isolated.* ***a.*** *A b-c projection of chabazite showing channels formed within the openings of 8-membered rings. Ca^{2+} cations (green spheres) and water molecules (blue spheres) fill the channels.* ***b.*** *The basic building block of chabazite is this rather complex looking linkage of tetrahedra. Six rings (three on top and three on bottom), each composed of eight tetrahedra, form cages in the chabazite. These cages are then bonded together to form the structure.* ***c.*** *A photograph of a crystal of chabazite. Its somewhat blocky external morphology can be correlated to the well-bonded net of the cages shown in Figure 22.16b.*

two types: one slightly larger M1 site, with $(OH)^{1-}$'s positioned on opposite (trans) sides of the octahedral (i.e., on the same edge), and two slightly smaller M2's, with (OH)'s on adjacent sides (*cis*) of the octahedra (Figure 22.17). Because of the difference in size, trivalent cations generally prefer to occupy the M2 sites, while divalent cations can enter either site. In **di**octahedral samples, **two** of the three M sites are filled, leaving the M1 site vacant. In **tri**octahedral samples, all **three** M sites are occupied.

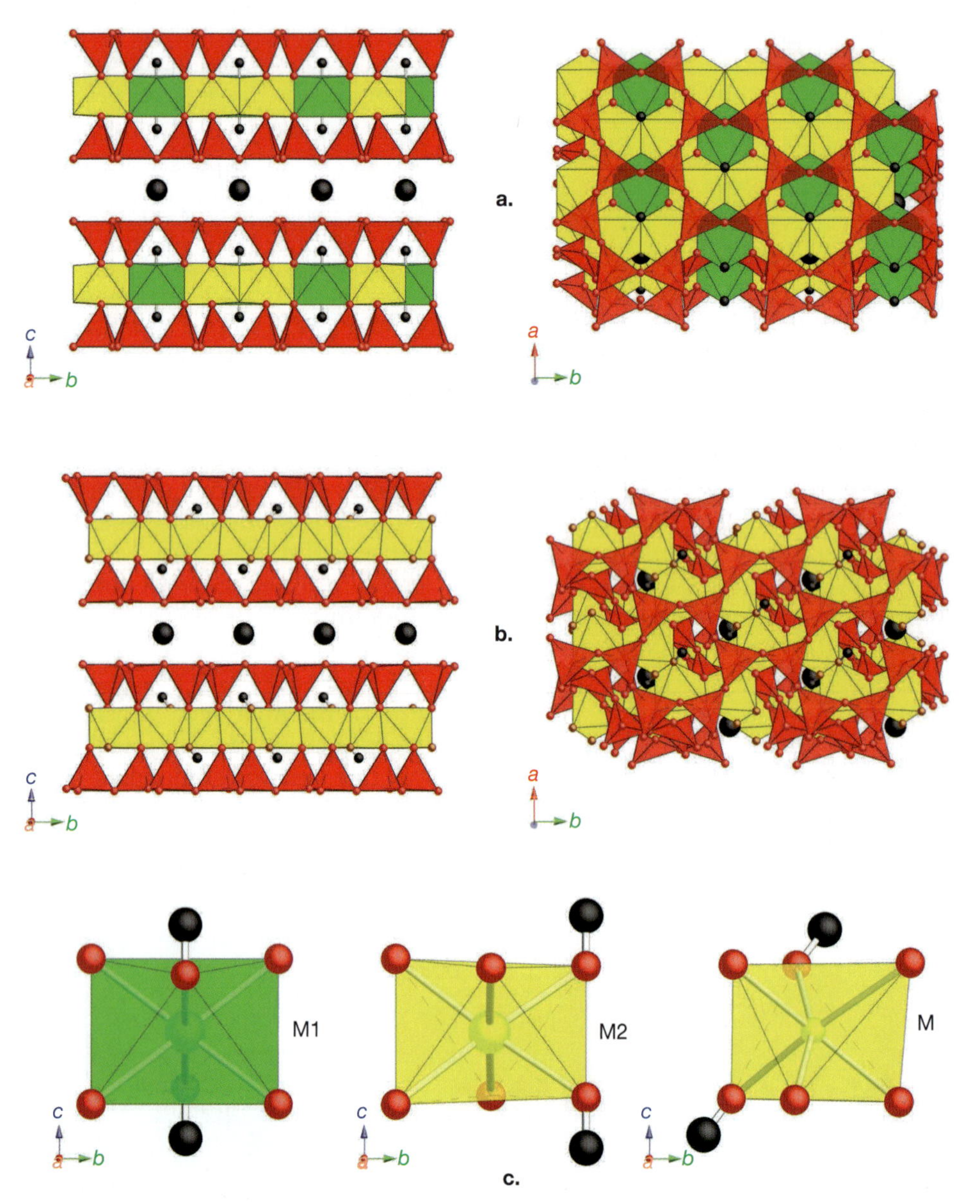

Figure 22.17. *The crystal structures of the trioctahedral mica phlogopite, $KMg_3(AlSi_3O_{10})(OH)_2$, and the dioctahedral mica muscovite, $KAl_2(AlSi_3O_{10})(OH)_2$, along with enlarged views of their octahedral sites. The Si^{4+} tetrahedra are transparent red, the M sites are green and yellow, K^{1+} is represented by the larger sphere, and H^{1+}, the smaller one.* ***a.*** *The* b-c *projection of phlogopite (left) and its* a-b *projection (right). All of the octahedral sites in phlogopite are filled with Mg^{2+}. There are two M sites; M1 is green and slightly larger than M2. The M1 sites have H^{1+} in a* trans *configuration (i.e., bonded to O^{2-}'s on opposite sides), while M2 sites have H^{1+} in a* cis *configuration (i.e., bonded to O^{2-}'s on the same side). The $(OH)^{1-}$ bond is perpendicular to the sheets.* **b.** *The* b-c *projection of muscovite (left) and its* a-b *projection muscovite (right). In this case, two-thirds of the octahedral sites are filled with Al^{3+}; notice the empty site in the* a-b *projection. There is only one M site (which is termed M2 in phlogopite) with H^{1+} in* cis *configuration. The $(OH)^{1-}$ bond is inclined relative to the sheets and projects into the void of the empty octahedral site.* **c.** *Expanded views of the M1 (left) and M2 (center) sites in phlogopite, along with the M site in muscovite (right). Note that M1 (for phlogopite) is larger than M2 and less distorted because the H's are in a* trans *(on opposite sides) configuration. For M2, notice how the O-O distance is decreased by $(OH)^{1-}$ bonds, thus distorting the site. The M site for muscovite appears similar to the M2 site in phlogopite, except that the $(OH)^{1-}$ bond has a different orientation due to the presence of the vacancy.*

The tetrahedral and octahedral sheets never fit perfectly together. In cases where the tetrahedral sheet is *larger than* the octahedra sheet, most of the misfit is accommodated by in-plane tetrahedral rotation, where adjacent tetrahedra rotate in opposite directions (Guggenheim, 1984). Second-order distortions can also be caused by vacancies in the octahedral sheet and shortening of the shared octahedral edges. When the tetrahedral sheets are *too small* for the octahedral sheets, modulations commonly occur—see the discussion of modulated structures, below.

Unlike other mineral classes that are grouped on the basis of composition alone, the layer silicates are classified on the basis of three main criteria: (1) the ratio of tetrahedral to octahedral sheets, (2) the charge required of the interlayer site on the basis of cation occupancies in the octahedral and tetrahedral sheets, and (3) the type of cation or molecule occupying the interlayer. A further distinction is made between di- and trioctahedral minerals on the basis of octahedral site occupancy, and species are distinguished by chemical composition. This classification scheme is presented in Table 22.5, and results in the mineral species and series listed in Tables 22.6–22.11. An excellent summary of the mica structures can be found in Fleet (2003) and in the *Reviews of Mineralogy* dedicated to this topic, which include volumes 13, 19, and 46.

Optical characteristics. As a result of their common structure, all of the layer silicates share optical characteristics. Recall that in the framework silicates, physical properties were similar in all three directions, echoing the crystal structure. In the layer silicates, we have gone from 3-D linkages of silicate tetrahedra to 2-D linkages (into sheets), so the physical properties should be vastly different *within* the layers vs. *perpendicular* to them. For example, the refractive indices are almost the same within the layers and significantly smaller perpendicular to them. They are larger within the layers because there is so much electron density in the layers; this retards the speed of light and makes the refractive index go up. As a group, these minerals are thus all biaxial negative, often with very small $2V$ because β and γ are nearly identical.

When observed with a polarizing light microscope, pleochroism of the layer silicates is very much a function of orientation. For example, if a mica flake happens to be resting on a cleavage plane (e.g., (001)), then rotating the stage will not result in significant changes in pleochroism. If the flake is on edge, light vibrating parallel to the cleavage planes is the most strongly absorbed, resulting in darker pleochroic colors; when rotated on the stage by 90°, pleochroism and absorption decrease significantly.

Of course, the most obvious manifestation of the crystal structure is the prominent cleavage between layers. The amount of electrostatic charge that holds the cleavage planes together varies in a systematic way. The 1:1 layer silicates such as kaolins and serpentines are held together by very weak hydrogen bonding as well as by bonds across the interlayer region. The layers of the 2:1 layer silicates (pyrophyllite and talc) are bonded by van der Waals bonds.

Table 22.5. Classification of Layer Silicates (after Bailey, 1991a)

Layer Type	*Layer Charge*	*Interlayer*	*Group*	*Sub-group*	*Examples of species*
1:1	0	none or H_2O only	kaolin-serpentine	serpentine	chrysotile
				kaolinite	kaolinite
2:1	<0.2	none	talc-pyrophyllite	talc	talc
				pyrophyllite	pyrophyllite
	0.2–0.6	hydrated exchangable cations	smectite	saponite	saponite
				montmorillonite	nontronite
	0.6–0.9	hydrated exchangable cations	vermiculite	trioctahedral vermiculite	vermiculite
				dioctahedral vermiculite	vermiculite
	0.6–1.0*	non-hydrated cations	true micas	trioctahedral micas	phlogopite
				dioctahedral micas	muscovite
	2.0	non-hydrated cations	brittle micas	trioctahedral brittle micas	clintonite
				dioctahedral brittle micas	margarite
	Variable	hydroxid sheet	chlorite	trioctahedral chlorites	clinochlore
				dioctahedral chlorites	donbasseite
				di-, trioctahedral chlorites	cookeite

*The charge on the layer for a true mica is 0.85 to 1.0 for dioctahedral micas. Trioctahedral micas may have a layer charge of near 0.6, but it is still an open question. There may be one exception, wonesite (layer charge of 0.5). Any layer charge of 0.6 to 0.85 represents an "interlayer-cation-deficient mica."

Table 22.6. Kaolin-serpentine group		
Kaolin sub-group dioctahedral	dickite	$Al_2Si_2O_5(OH)_4$
	halloysite (7 Å)	$Al_2Si_2O_5(OH)_4$
	halloysite (10 Å)	$Al_4Si_4O_{10}(OH)_8 \cdot 2H_2O$
	kaolinite	$Al_2Si_2O_5(OH)_4$
	nacrite	$Al_2Si_2O_5(OH)_4$
	odinite	$(Fe^{3+},Mg,Al,Fe^{2+},Ti,Mn)_5(Si,Al)_4O_{10}(OH)_8$
Serpentine-antigorite-related subgroup, trioctahedral	antigorite	$Mg_3Si_2O_5(OH)_4$
Serpentine-lizardite-related subgroup, trioctahedral	caryopilite*	$(Mn,Mg,Zn,Fe)_3(Si,As)O_5(OH,Cl)_4$
	lizardite	$Mg_3Si_2O_5(OH)_4$
	nepouite	$(Ni,Mg)_3Si_2O_5(OH)_4$
	greenalite*	$(Fe^{2+},Fe^{3+})_{2-3}Si_2O_5(OH)_4$
Serpentine-amesite-related subgroup, trioctahedral	amesite	$(Mg,Al)_3(Si,Al)O_5(OH)_4$
	berthierine	$(Fe^{2+},Fe^{3+},Al)_3(Si,Al)_2O_5(OH)_4$
	brindleyite	$Ni_2Al(AlSi)O_5(OH)_4$
	fraipontite	$(Zn,Cu,Al)_3(Si,Al)_2O_5(OH)_4$
	kellyite	$(Mn^{2+},Mg,Al)_3(Si,Al)_2O_5(OH)_4$
	manandonite	$Li_2Al_4(Si_2AlB)O_{10}(OH)_8$
	cronstedtite	$Fe^{2+}{}_2Fe^{3+}(SiFe^{3+})O_5(OH)_4$
Serpentine-chrysotile-related subgroup, trioctahedral	clinochrysotile	$Mg_3Si_2O_5(OH)_4$
	orthochrysotile	$Mg_3Si_2O_5(OH)_4$
	parachrysotile	$Mg_3Si_2O_5(OH)_4$
	pecoraite	$Ni_3Si_2O_5(OH)_4$

*These minerals are grouped here according to their Dana classification (Nickel et al., 1997); however, most workers consider them to be modulated 1:1 structures, quite unlike lizardite.

Within the 2:1 layer silicates, cation charges combine to give each layer a net **layer charge**, as given in column 2 of Table 22.5. Just as the substitution of Al^{3+} for the Si^{4+} cation in feldspars sets up a negative charge on the tetrahedra, so too does Al^{3+} in the tetrahedral sites of the layer silicates set up an extra negative charge on the structure. This net negative charge must be compensated either by cation substitutions to increase the charge in the octahedral sheet, or by interlayer cations that bond the layers together. For example, in muscovite, monovalent cations enter between the tetrahedral layers, while in the Al-richer **brittle micas**, the sheets are bonded together more strongly because there is more charge to be made up. Thus, the layers are more brittle because they break rather than cleaving nicely.

Polytypism in layer silicates. Layer silicates are commonly used as examples of the phenomenon of **polytypism**, in which layers of similar composition are stacked in various ways (see Polytypes box). In a simple sense, the hexagonal symmetry of these minerals permits three directions in which layers can be stacked, each of which is 120° apart. For example, if the stacking of the layers is always in the same direction, the result will be a *1M* structure with monoclinic symmetry, such as is found in phlogopite and the biotite series. If the stacking alternates between two vectors, each 120° apart, then the 2M monoclinic polytype results (as found in muscovite). Lepidolite has several polytypes, three of which are shown in Figure 22.18. More complicated polytypes arise because there are actually nine possible structural displacements in the sheet silicate structures. The result is an enormous number of polytypes of these structures. An excellent, thorough summary of mica polytypism can be found in Bailey (1988b), or for more advanced readers, Nespolo and Ďurovič (2002). Polytypes also occur in any mineral that has a layer repeat of some type in their structure; we'll see this shortly in the amphiboles and pyroxenes.

Classification. In keeping with the organization presented in Table 22.5, we begin with the minerals of the **kaolin-serpentine** group. These are all derived from the formula $A_{2-3}Si_2O_5(OH)_4$, where A, the octahedral cation, is Al^{3+}, Fe^{3+}, Mg^{2+}, Ti^{4+}, and/or Mn^{2+}. All of the kaolinite subgroup minerals are dioctahedral; in other words, there is a vacancy in one-third of the octahedral sites, though not always the same octahedral site. The location of the vacancy within the stacking of a layer depends on which of the species is being

Polymorphism

As we've discussed throughout this text, the ability of a single composition to occur with more than one crystal structure is called **polymorphism**. There are several different types:

1. **Displacive polymorphism** occurs when bonds are slightly kinked. A good example of this is the difference between α- and β-quartz, as shown in Figure 22.2. This type of polymorphism is commonly a function of temperature and/or pressure, and is usually reversible.
2. **Reconstructive polymorphism** requires structural rearrangement and breaking of bonds; such major changes require a lot of energy to affect and so may occur quite slowly, sometimes resulting in the existence of metastable phases. These changes are usually non-reversible. Good examples of these are diamond/graphite, calcite/aragonite, pyrite/marcasite, and quartz/coesite/stishovite.
3. **Order-disorder polymorphism** involves redistribution of cations or anions (usually the former) in a specific site. We discussed this process in the case of sanidine and microcline, in which Al^{3+} and Si^{4+} cations order or disorder among sites as a function of temperature (this also occurs between high and low albite, and in many metal alloys). In this process, the higher temperature polymorphs are usually the more disordered, giving them greater stability and higher symmetry.

Polytypes are a special kind of polymorphism with a strict definition (Guinier et al., 1984): "…an element or compound is polytypic if it occurs in several structural modifications, each of which can be regarded as built up by stacking layers of (nearly) identical structure and composition, and it the modifications differ only in their stacking sequence. Polytypism is a special case of polymorphism: the two-dimensional translations within the layers are essentially preserved." Polytypes commonly occur in layer silicates, as shown for lepidolite in Figure 22.18 and below.

Polytypes

1M 2O $2M_1$ $2M_2$ 3H 6H

Polymorph is the term used to refer to minerals with the same chemical composition but different crystal structures.

Polytypes are subset of polymorphs in which the structures are distinguished by differences in stacking of structural units. Polytypes are common in the sheet and chain silicates. To conceptualize polytypes, we'll use 2×4" blocks of wood cut into 12" lengths to represent structural units in minerals and stacked in different patterns. One method would be to stack them one on top of the other with each piece moved slightly to the right, as shown in 1M. Another option would be to translate the third board back directly over the top of the first board, as done in 2O. Here the unit cell is inscribed on the sides of the boards: the leading number represents the repeat distance (i.e., the thickness of the boards) and the letter denotes the crystal system (e.g., M = monoclinic and O = orthorhombic). The same structural units could form either a monoclinic cell or an orthorhombic cell with twice the vertical repeat.

For the next two examples, we begin with 2M, which has a 2 layer repeat and a monoclinic cell; for each, the second board is rotated 60° with respect to the first. The subscript denotes that the rotation was in a different direction: CW for 1 and CCW for 2. For 3H, each board is rotated 120° with respect to the one below; thus the fourth board is in the same orientation as the first, creating a hexagonal cell. For 6H, each board is rotated 60° with respect to the one below, so boards 1 and 7 are in the same orientation and this "array" also forms a hexagonal cell. See if you can predict the vertical repeat distances (i.e., the length of their "c" axis) for each of these.

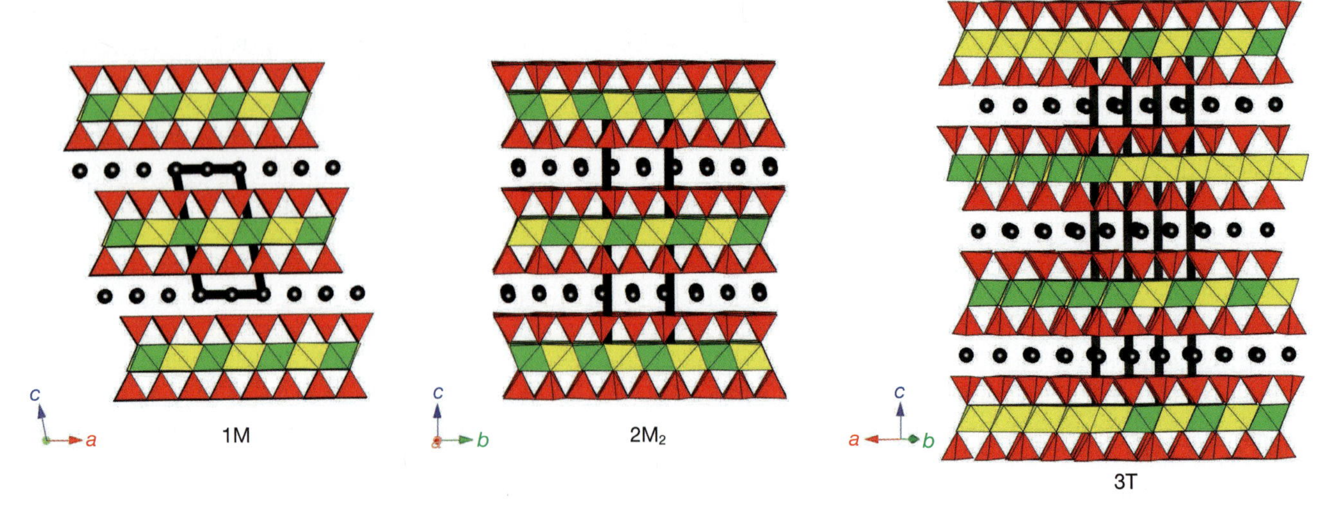

Figure 22.18. Three different polytypes of the mica mineral lepidolite, $K_2(Li,Al)_5(Al_2Si_6O_{10})(OH,F)_4$, projected in slightly different orientations and termed: 1M (left), $2M_2$ (center), 3T (right) based on the stacking arrangements of their layers. [Refer to the inset "Polytypes" to understand the how stacking effects the structures. The letter "T" in 3T refers to trigonal, which is a subset of the hexagonal crystal system and not commonly used.] The unit cells are also shown for each of the three polytypes, where the c axis increases in length from 10.1 to 20.2 to 29.8Å, corresponding to the one-layer, two-layer, and three-layer repeats of 1M, $2M_2$, and 3T. In the 1M polytype, all layers have the same orientation, while opposite layers have the same orientations for the $2M_2$ polytype. For 3T, the bottom (i.e., first) and top (i.e., fourth) layers are the same, while layers two and three are rotated 120° in the a-b plane relative to each other. So in the a-b plane for 1M, each layer is in the same orientation, for $2M_2$ each layer is rotated 180°, and for 3T each layer is rotated 120°. As a result of stacking, there would be a 1-fold axis in 1M, a 2-fold axis in $2M_2$, and a 3-fold axis in 3T perpendicular to the a-b plane.

studied. This structure has only a single tetrahedral sheet; its apical oxygens all point in the same direction and become the corners of octahedra in the single octahedral sheet. On the opposite side of the octahedral sheet from the tetrahedra are hydroxyl groups $(OH)^{1-}$, which point toward the basal oxygen anions of the adjacent layer and provide weak hydrogen bonding to hold the sheets together in layers. The stacking of the layers is variable depending on the polytype: they may be directly on top of each other, shifted by 1/3 of the *a* axis length, or shifted along the *b* axis (Figure 22.19). These minerals do not need interlayer cations because they are already electrostatically neutral. In these 1:1 silicate structure, several types of structural and chemical variations can occur, as nicely summarized by Veblen and Wylie (1993): (1) difference in the conformation of the layers (planar, curled, or corrugated); (2) variations, either ordered or disordered, in the way the layers are stacked together [polytypism], (3) variations in the occupancy of the M1 site (between di- and trioctahedral types), and (4) variations in the chemistry of the cations in the structures.

The **kaolinite** subgroup species are all dioctahedral, and include the industrially-important mineral species kaolinite, which is used to make bricks and ceramics as well as in fillers in plastics, paint, rubber, and paper. Kaolinite minerals usually form as weathering products of feldspars and muscovite, so they are extremely common in soils. Kaolinites tends to be quite Al-rich, showing only minor variations from the ideal $Al_2Si_2O_5(OH)_4$ composition. They form planar or slightly-curled crystals with at least twelve different polytypes based on simple stacking sequences; structural defects are also common in this group of minerals, which often show a high degree of disorder in XRD patterns (Bailey, 1988; Giese, 1988; Veblen and Wylie, 1993).

There are two forms of **halloysite**, a 7 Å and a 10 Å form (Table 22.5), and they differ by the amount of H_2O present in the interlayer. The 1:1 layer is generally believed to be kaolinite-like. Although several morphologies are known, the common variety is tubular, with layers rolled into spirals. Other forms are prisms and near-spheres.

The **serpentine** subgroup minerals are all trioctahedral. As discussed in depth in Wicks and O'Hanley (1988) and Guggenheim and Eggleton (1988), two factors are responsible for the curvature of these minerals. The first is the strength of the interlayer bonds. The second is the inherent misfit between the *b* axis length of an ideal, Mg-occupied trioctahedral sheet (≈9.43 Å) and the length of an ideal, pure-Si^{4+} tetrahedral sheet (≈9.10 Å). The presence of large divalent cations like Mg^{2+} and Fe^{2+} in the octahedra exacerbates the size difference between the octahedral sheets and the tetrahedral ones connected to them. In the case of lizardite, this misfit is relieved to some extent by substitution of Al^{3+} for Si^{4+} in the tetrahedral site. In antigorite and chrysotile, where there is significantly less Al^{3+} substitution, the tetrahedral-octahedral mismatch is more severe. The entire layer curves up and down in waves in antigorite. In chrysotile, the curvature is continuous, and results in cylinders with radii of roughly 90 Å (Whittaker, 1957) that may form fibers (see discussion of asbestiform minerals below in amphibole section and in Chapter 24).

Members of the **pyrophyllite-talc** group can be either dioctahedral (pyrophyllite, Figure 6.3, or ferripyrophyllite) or trioctahedral (talc, willemsite, minnesotaite, etc.—see Table 22.7). These structures also have a net neutral charge, but it is accomplished by stacking of two tetrahedral sheets pointing inward toward either side of an octahedral sheet, which is the 2:1 layer type. The 2:1 stacking, as compared with the 1:1 stacking of the serpentines, allows the tetrahedral sheets to balance each other out, so the layers lie flat. These minerals are used as fillers in paints, rubber, and cosmetics, including talcum powder.

The **smectite** and **vermiculite** group minerals are the so-called "swelling clays," in which the interlayer charge varies between 0 and 1 (Tables 22.5 and 22.8). In these minerals, some fraction of the interlayer sites is filled with cations (sometimes, even organic cations like tetramethylammonium), leaving the remainder vacant and available to be filled in with H_2O molecules. Vermiculite is considered a swelling clay of high layer charge, while smectite (see montmorillonite in Figure 6.3) is a swelling clay of low layer charge. The actual separation at a layer charge of –0.6 is not used precisely because it is very difficult to determine exactly what the layer charge is —the interlayer material is difficult to analyze. Therefore, a technique involving the reaction of a clay with (usually) glycerol ($C_3H_8O_3$) and the XRD layer-to-layer d(001) spacing is used. After Mg^{2+} exchange and glycerol solvation, the d-spacing must remain at 14.5 Å for vermiculite, and at about 17.7 Å for smectite. The swelling characteristics make the vermiculites and smectites reasonable groups in their own right.

Vermiculite has long been used for packing insulation and as a soil additive, but for these applications the layers are always "expanded" by heating above 870°C. The heat causes the H_2O in

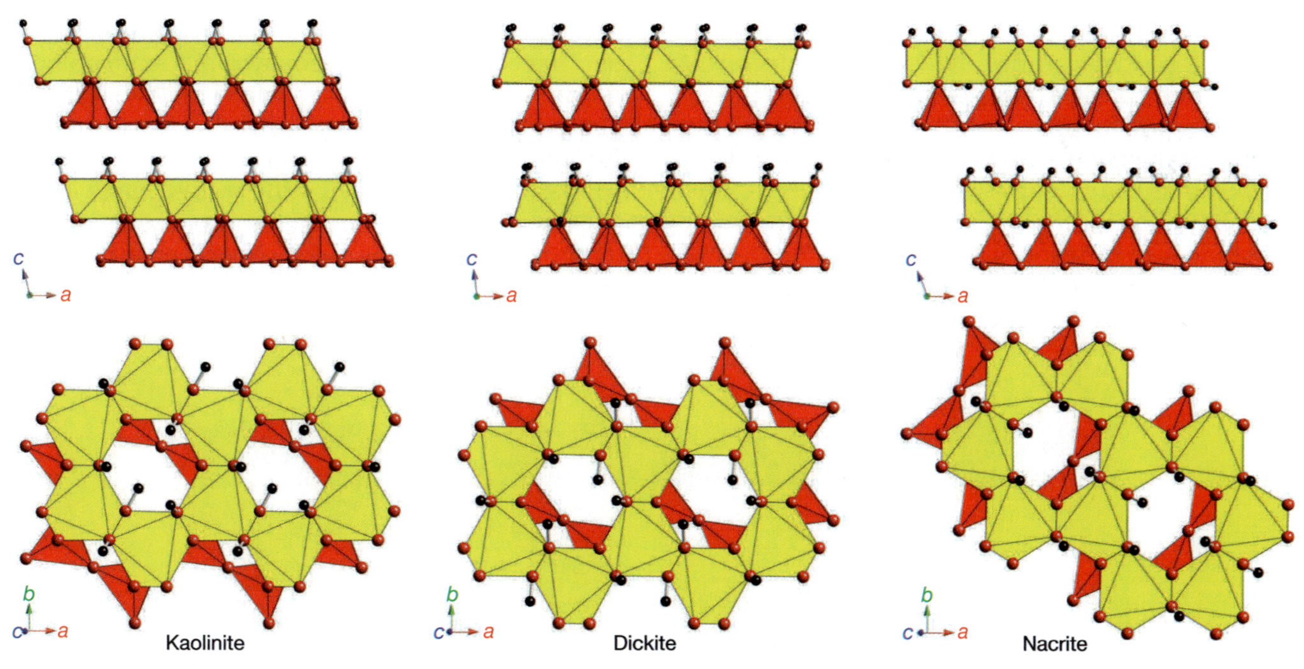

Figure 22.19. The structures of three different polytypes of $Al_2Si_2O_5(OH)_4$: kaolinite (left), dickite (center), and nacrite (right), projected onto their a-c *planes (top) with a close-up* a-b *projection of each below. These three minerals are distinguished on the basis of the stacking sequences of the tetrahedral and octahedral sheets. It can be tricky to see these differences in the structures. However, notice in the* a-c *projections that β differs for each. In the* a-b *projection, kaolinite and dickite at first look similar, but closer observation shows that the octahedral and tetrahedral sheets are offset in different directions—in a similar manner to the $2M_1$ and $2M_2$ polytypes discussed previously. Clearly the stacking in nacrite is in a different direction than in the other two, causing a 60° rotation of the structure. (The Si^{4+} tetrahedra are red, the Al^{3+} octahedra are yellow, and H^{1+} cations are represented by small black spheres.)*

Table 22.7. Talc-pyrophyllite group		
Dioctahedral	pyrophyllite	$Al_2Si_4O_{10}(OH)_2$
	ferripyrophyllite	$Fe^{3+}{}_2Si_4O_{10}(OH)_2$
Trioctahedral	talc	$Mg_3Si_4O_{10}(OH)_2$
	willemseite	$(Ni,Mg)_3Si_4O_{10}(OH)_2$
	minnesotaite*	$(Fe^{2+},Mg)_3Si_4O_{10}(OH)_2$
	brinrobertsite*	$(Na,K,Ca)_{0.35}(Al,Fe,Mg)_4(Si,Al)_8O_{20}(OH)_4\cdot 3.54(H_2O)$

*Minnesotaite is classified with this group in the Dana system, though it is generally considered to be a modulated layer silicate. Brinrobertsite is similarly misleadingly grouped in the Dana system with these minerals, but it is an interstratified clay containing layers of pyrophyllite alternating with smectite, and thus probably belongs on its own.

the structure to expand, and eventually be released as steam. The force of this reaction pushes the layers apart in a process called **exfoliation**. The resultant volume is 8–30 times bigger than the volume before heating! The expanded material resembles thread-like worms, giving us the name, which in Latin means "to breed worms." Vermiculite makes great thermal insulation, and also absorbs water, making it useful in helping soils maintain moisture.

The **true micas** are also 2:1 layer types. The charge on the interlayer ranges from 0.85–1, requiring an interlayer cation, and thus the formula becomes:

$$AM_{2-3}T_4O_{10}X_2,$$

where A is the 6- or 12-coordinated interlayer cation, which may be K^{1+}, Ca^{2+}, Na^{1+}, Ba^{2+}, $NH_4{}^+$, and/or a partial vacancy; M is a di- or trivalent cation in the octahedral site, including Mg^{2+}, Fe^{2+}, Al^{3+}, Fe^{3+}, Mn^{2+}, Mn^{3+}, Zn^{2+}, V^{3+}, and/or Li^+; T is a tetrahedral cation, usually Si^{4+} with Al^{3+}, Fe^{3+}, and/or B^{3+}; and X is the anion, generally $(OH)^{1-}$ with F^{1-}, Cl^{1-}, and/or S^{2-}. True micas can be either dioctahedral or trioctahedral (Table 22.9), and there are many polytypes (Figure 22.18). The mica nomenclature was revised in 1998 (Rieder et al.), and some of the traditional mineral names such as "biotite" and "illite" were discredited as species and renamed as series (Table 22.10). The series names are useful in situations where the precise chemistry is not known.

It would be difficult to overstate the importance of mica minerals in the geological record. Micas are significant rock-forming minerals in all three rock types. Granites typically contain muscovite, as do pegmatitic rocks. Intrusive igneous rocks with intermediate Fe contents, like granodiorites and syenites, contain biotite series micas because of the ample supply of Fe. Phlogopite (Figure 22.20) is found in ultramafic rocks like kimberlites. Biotite series minerals are common in regionally-metamorphosed rocks over a wide range of bulk compositions and metamorphic grades, and are also found in rocks that have experienced contact metamorphism. Sedimentary rocks often contain sericite, a fine-grained variety of muscovite that forms from alteration of feldspar.

Less common are the **brittle micas** (Table 22.9), which require an interlayer charge of about 2.0 due to cation substitutions. They contain no alkalis (K^+ or Na^+) because their charge is insufficient. Instead, the interlayer contains Ca^{2+} or Ba^{2+}. The

Table 22.8. Smectite group and vermiculite			
Smectite	Dioctahedral	beidellite	$(Na,Ca_{0.5})_{0.3}Al_2(Si,Al)_4O_{10}(OH)_2\cdot n(H_2O)$
		montmorillonite	$(Na,Ca)_{0.3}(Al,Mg)_2Si_4O_{10}(OH)_2\cdot n(H_2O)$
		nontronite	$Na_{0.3}Fe_2(Si,Al)_4O_{10}(OH)_2\cdot n(H_2O)$
		volkonskoite	$Ca_{0.3}(Cr^{3+},Mg,Fe^{3+})_2(Si,Al)_4O_{10}(OH)_2\cdot 4(H_2O)$
		swinefordite	$(Ca,Na)_{0.3}(Li,Mg)_2(Si,Al)_4O_{10}(OH,F)_2\cdot 2(H_2O)$
	Trioctahedral	saponite	$(Ca_{0.5},Na)_{0.3}(Mg,Fe^{2+})_3(Si,Al)_4O_{10}(OH)_2\cdot 4H_2O$
		sauconite	$Na_{0.3}Zn_3(Si,Al)_4O_{10}(OH)_2\cdot 4H_2O$
		hectorite	$Na_{0.3}(Mg,Li)_3Si_4O_{10}(F,OH)_2$
		stevensite	$(Ca_{0.5})_{0.3}Mg_3Si_4O_{10}(OH)_2$
		yakhontovite	$(Ca,Na)_{0.5}(Cu^{2+},Fe^{2+},Mg)_2Si_4O_{10}(OH)_2\cdot 3H_2O$
		zincsilite	$Zn_3Si_4O_{10}(OH)_2\cdot 4(H_2O)$
		IMA2002-025	$Ca_{0.3}(Fe,Mg,Fe)_3(Si,Al)_4O_{10}(OH)_2\cdot 4H_2O$
		vermiculite	$(Mg,Fe^{2+},Al)_3(Al,Si)_4O_{10}(OH)_2\cdot 4(H_2O)$

Table 22.9. Mica Group

Group	Sub-group	Mineral	Formula
True micas	Dioctahedral sub-group	muscovite	$KAl_2\square(AlSi_3)O_{10}(OH)_2$
		paragonite	$NaAl_2\square(AlSi_3)O_{10}(OH)_2$
		chernykite	$BaV_2\square(Al_2Si_2)O_{10}(OH)_2$
		roscoelite	$KV_2\square(AlSi_3)O_{10}(OH)_2$
		celadonite	$KFe^{3+}(Mg,Fe^{2+})\square Si_4O_{10}(OH)_2$
		ferroceladonite	$KFe^{3+}(Fe^{2+},Mg)\square Si_4O_{10}(OH)_2$
		ferro-aluminoceladonite	$K_2Fe^{2+}{}_2Al_2Si_8O_{20}(OH)_4$
		aluminoceladonite	$KAl(Mg,Fe^{2+})\square Si_4O_{10}(OH)_2$
		chromceladonite	$KCrMg\square Si_4O_{10}(OH)_2$
		tobelite	$(NH_4)Al_2\square(AlSi_3)O_{10}(OH)_2$
		nanpingite	$CsAl_2\square(AlSi_3)O_{10}(OH)_2$
		boromuscovite	$KAl_2\square(BSi_3)O_{10}(OH)_2$
		montdorite	$KFe^{2+}{}_{1.5}Mn^{2+}{}_{0.5}Mg_{0.5}\square_{0.5}(Si_4)O_{10}F_2$
		chromphyllite	$KCr_2\square(AlSi_3)O_{10}(OH)_2$
		shirokshinite	$K(Na,Mg_2)\square Si_4O_{10}F_2$
	Trioctahedral sub-group	phlogopite	$KMg_3(Si_3Al)O_{10}(F,OH)_2$
		tetra-ferriphlogopite	$KMg_3(Fe^{3+}Si_3)O_{10}(F,OH)_2$
		IMA2001-045	$KMn_3(AlSi_3)O_{10}(F,OH)_2$
		annite	$KFe^{2+}{}_3(AlSi_3)O_{10}(OH)_2$
		tetra-ferri-annite	$KFe^{2+}{}_3(Fe^{3+}Si_3)O_{10}(OH)_2$
		siderophyllite	$KFe^{2+}{}_2Al(Al_2Si_2)O_{10}(OH)_2$
		eastonite	$KMg_2Al(Al_2Si_2)O_{10}(OH)_2$
		hendricksite	$KZn_3(AlSi_3)O_{10}(OH)_2$
		polylithionite	$KLi_2AlSi_4O_{10}F_2$
		trilithionite	$KLi_{1.5}Al_{1.5}(AlSi_3)O_{10}F_2$
		norrishite	$KLiMn^{3+}{}_2(Si_4)O_{12}$
		masutomilite	$KLiAlMn^{2+}AlSi_3O_{10}F_2$
		aspidolite	$NaMg_3(AlSi_3)O_{10}(OH)_2$
		wonesite	$Na_{0.5}\square_{0.5}Mg_{2.5}Al_{0.5}(AlSi_3)O_{10}(OH)_2$
		preiswerkite	$NaMg_2Al(Al_2Si_2)O_{10}(OH)_2$
		ephesite	$NaLiAl_2(Al_2Si_2)O_{10}(OH)_2$
		fluorannite	$KFe^{2+}{}_3(AlSi_3)O_{10}F_2$
Brittle micas	Dioctahedral	margarite	$CaAl_2\square(Al_2Si_2)O_{10}(OH)_2$
		chernykhite	$BaV_2\square Al_2Si_2O_{10}(OH)_2$
	Triocahedral	clintonite	$CaMg_2Al(Al_3Si)O_{10}(OH)_2$
		bityite	$CaLiAl_2(BeAlSi_2)O_{10}(OH)_2$
		anandite	$BaFe^{2+}{}_3(Fe^{3+}Si_3)O_{10}S(OH)$
		kinoshitalite	$BaMg_3(Al_2Si_2)O_{10}(OH)_2$
		ferrokinoshitalite	$BaFe^{2+}{}_3(Si_2Al_2)O_{10}(OH)_2$
		ganterite	$[Ba_{0.5}(Na,K)_{0.5}]Al_2(Si_{2.5}Al_{1.5}O_{10})(OH)_2$
Interlayer -deficient	Trioctahedral	wonesite*	$Na_{0.5}\square_{0.5}Mg_{2.5}Al_{0.5}AlSi_3O_{10}(OH)_2$

*Not an end-member

layers are thus more strongly bonded than in the true micas, so the lamella are brittle rather than flexible, and harder than in micas (4–6 on the Mohs scale, vs. 2.5 for muscovite). Margarite, $CaAl_2\square(Al_2Si_2)O_{10}(OH)_2$, (Figure 22.20) may be thought of as a high Al, Ca muscovite. Clintonite, $CaMg_2Al(Al_3Si)O_{10}(OH)_2$), is effectively a Ca^{2+} version of phlogopite. These micas tend to be found in Ca-rich or Al-rich environments. Clintonite occurs in marbles and skarns that formed by metamorphism of dolomite ($(Ca,Mg)_2(CO_3)_2$), whereas margarite is stable with muscovite up to intermediate metamorphic grades.

The **chlorite** group minerals are composed of a talc-like layer alternating with a sheet of edge-sharing $M^{2+}(OH)_6$ octahedra (sometimes referred

Table 22.10. Series Names in Mica Nomenclature (after Rieder et al., 1998)	
Biotite	trioctahedral micas between, or close to, the annite-phlogopite and siderophyllite-eastonite join; dark micas without lithium
Glauconite	dioctahedral, interlayer-deficient micas with composition of $K_{0.8}R^{3+}_{1.33}R^{2+}_{0.67}\square Al_{0.13}(Si_{3.87})O_{10}(OH)_2$
Illite	dioctahedral, interlayer-deficient micas with composition of $K_{0.65}Al_2\square Al_{0.65}Si_{3.35}O_{10}(OH)_2$
Brammalite	dioctahedral, interlayer-deficient micas with composition of $Na_{0.65}Al_2\square Al_{0.65}Si_{3.35}O_{10}(OH)_2$
Lepidolite	trioctahedral mica on, or close to, the trilithionite-polylithionite join; light micas with substantial lithium
Phengite	potassic dioctahedral micas between, or close to, the joins muscovite-aluminoceladonite and muscovite-celadonite
Zinnwaldite	trioctahedral micas on, or close to, the siderophyllite-polylithionite join; dark micas containing lithium

to as a brucite-like sheet after that mineral species—see Chapter 23 and Figure 23.19). Thus, we *could* write the formula of a simplified chlorite as a combination of talc, $Mg_3Si_4O_{10}(OH)_2$, and brucite ($Mg(OH)_2$) to yield $Mg_3Si_4O_{10}(OH)_2 \cdot Mg_3(OH)_6$.

More conventionally, the formula for clinochlore (Figure 22.21) is written as $(Mg,Fe)_5Al(Si_3Al)O_{10}(OH)_8$. This formula lumps together the $(OH)_2$ from the talc layer and the $(OH)_6$ from the brucite layer as $(OH)_8$; it also sums the cations in both layers to a total of 6. Thus, a general formula for the chlorite group minerals is $M_{5-6}T_4O_{10}(OH)_8$, where M is a di- or trivalent cation in the octahedral site, including Mg^{2+}, Fe^{2+}, Al^{3+}, Fe^{3+}, Mn^{2+}, Mn^{3+}, Zn^{2+}, and/or Ni^{2+}; and T is a tetrahedral cation, usually three Si^{4+} with one Al^{3+} and/or Fe^{3+}, B^{3+}, Zn^{2+}, or Be^{2+} (Table 22.11). There is continuous solid solution between the Mg and Fe^{2+} end members. There are theoretically many polytypes of these compositions, but in reality, most chlorites are disordered and the number of actual polytypes is greatly limited. See Bailey (1991b) for details.

Chlorite group minerals can be either trioctahedral or dioctahedral, or some of both. Trioctahedral samples contain six octahedral cations, so the octahedral sites in both the "talc layer" and the "brucite layer" are completely filled. In dioctahedral chlorites (of which there is only one species, donbassite, $Al_2[Al_{2.33}](Si_3Al)O_{10}(OH)_8$), the 2:1 "talc" layer is dioctahedral, and the interlayer is 2/3 occupied as well. In tri,dioctahedral chlorites (only franklinfurnaceite, $Ca_2(Fe^{3+},Al)Mn^{3+}Mn^{2+}{}_3Zn_2Si_2O_{10}(OH)_8$), the 2:1 layer is trioctahedral and the interlayer is 2/3 occupied, while in di,trioctahedral chlorites, the 2:1 layer is 2/3 occupied and the interlayer is trioctahedral (Table 22.11).

The majority of rock-forming chlorites are trioctahedral and belong to the Mg-Fe series (e.g., Figure 6.3); dioctahedral, di-tri-, and tri-di-octahedral chlorites are rare. Chlorite commonly occurs in regionally-metamorphosed rocks of low to medium grades and in altered rocks, especially those around ore bodies. Its stability depends on the Mg/Fe ratio; Mg-rich chlorite is stable up to 700–750°C (in the absence of quartz), or ~600–

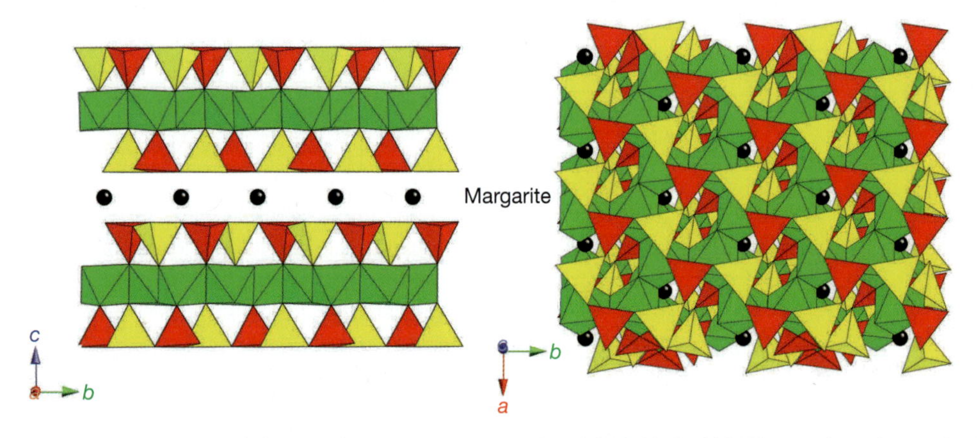

Figure 22.20. *The crystal structure of the brittle mica margarite, $CaAl_2(Al_2Si_2O_{10})(OH)_2$, projected onto its* b-c *plane (left) and* a-b *plane (right). At first glance this structure looks identical to muscovite (Figure 22.17); and it is very similar. Careful observation of the formula shows that an extra Al^{3+} has substituted for one of the Si^{4+} in the tetrahedral sheet. This is also seen in structure, with 50% of the tetrahedra yellow to represent Al^{3+} occupancy. In the micas we've seen so far, the Al^{3+} is randomly distributed in such a way that every tetrahedron has a 25% chance of containing Al^{3+}. Careful observation of the structure shows that it is a bit more distorted than muscovite. This results from the stronger bonding between the layers with the divalent interlayer cation, Ca^{2+} (as compared to the K^{1+} or Na^{1+} in true micas).*

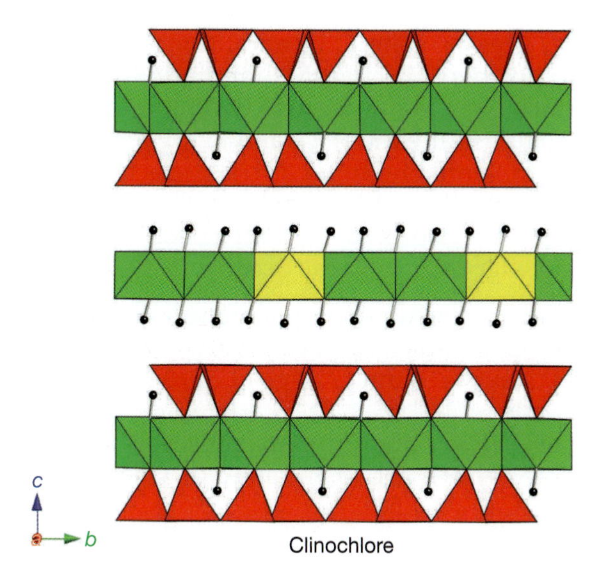

Figure 22.21. The crystal structure of clinochlore, $(Mg,Al)_6(Si,Al)_4O_{10}(OH)_8$. Based on the formula and the b-c *projection, it is apparent that clinochlore is a trioctahedral 2:1 layer silicate similar to phlogopite. What makes it different is that instead of inner cations between the t-o-t "sandwiches," there is an inner sheet with the ideal composition $(Mg,Al)(OH)_6$ (i.e., a sheet of Mg (green) and Al (yellow) octahedra formed with OH groups).*

650°C in the presence of quartz. The Fe-rich chlorites have lower thermal stabilities. Chlorite is rare in igneous rocks, occurring in fissures and vein deposits and as a replacement for biotite or hornblende during slow cooling. In sedimentary rocks, chlorite can form during diagenesis, and can be found in the clay fraction of sedimentary rocks. Chlorite can also be found as a minor component in soils (Bailey, 1991b).

Prehnite, $Ca_2Al_2Si_3O_{10}(OH)_2$, is a sheet silicate composed of 4-membered rings, unlike the preceding examples that were made from 6-membered rings. The tetrahedra in the rings share all of their oxygens, and are linked to other tetrahedra that share only two of their corners. Al^{3+} and Si^{4+} are disordered in the ring tetrahedra. Between the sheets are isolated $AlO_4(OH)_2$ octahedra and 7-coordinated Ca^{2+} anions. Prehnite is important because it is an index mineral for the prehnite-pumpellyite facies of metamorphic rocks, stable at relatively low temperatures of 200–300°C.

We should not leave the topic of layer silicates without mentioning a final group of minerals, the modulated layer silicates. These are minerals in which there is a periodic perturbation to the basic layer silicate structure (Guggenheim and Eggleton, 1991). Technically, some of the minerals we have already discussed are modulated silicates, including antigorite and minnesotaite. Most of these minerals are Al^{3+}-poor and contain large Mn^{2+} and Fe^{2+} cations, thereby having considerable misfit between the octahedral and tetrahedral sheets. They tend to have structures in which the lateral dimensions of the octahedral sheet are much larger than the lateral dimensions of the tetrahedral sheet. The resultant misfit is accommodated by severe readjustment of the tetrahedral sheets through a number of different mechanisms: tetrahedral inversions or tilting, tetrahedral omissions, tetrahedral additions, or combinations of the above (Guggenheim and Eggleton, 1991). Limited curving of the boundary between the tetrahedral and octahedral sheets can also relieve misfit. Tetrahedral inversions involve tetrahedra where all four corners are shared with other tetrahedra. These tetrahedra link the layers together by way of inversions across the interlayer space. The modulated phyllosilicates are still "layer-like" because there are continuous octahedral sheets, as in minnesotaite, or continuous tetrahedral sheets, as in palygorskite/sepiolite.

Stilpnomelane occurs in low-grade metapelite and metabasalts, commonly coexisting with chlorite. It has a variable formula of approximately $K(Fe^{2+},Mg,Fe^{3+})_8(Si,Al)_{12}(O,OH)_{27}$. Its structure

Table 22.11. Chlorite group

Dioctahedral	donbassite	$Al_2[Al_{2.33}](Si_3Al)O_{10}(OH)_8$
Di,trioctahedral	cookeite	$LiAl_4(Si_3Al)O_{10}(OH)_8$
	sudoite	$Mg_2Al_3(Si_3Al)O_{10}(OH)_8$
	borocookeite	$Li_{(1+3x)}Al_{(4-x)}(BSi_3)O_{10}(OH,F)_8$
Trioctahedral	clinochlore	$(Mg,Al)_6(Si,Al)_4O_{10}(OH)_8$
	nimite	$Ni_5Al(Si_3Al)O_{10}(OH)_8$
	baileychlore	$(Zn,Fe^{2+},Al,Mg)_6(Si,Al)_4O_{10}(OH)_8$
	chamosite	$(Fe,Al,Mg)_6(Si,Al)_4O_{10}(OH)_8$
	pennantite	$Mn_5Al(Si_3Al)O_{10}(OH)_8$
	orthochamosite	$(Fe,Al,Mg,Mn)_6(Si,Al)_4O_{10}(OH)_8$
Tri,dioctahedral	franklinfurnaceite	$Ca_2(Fe^{3+},Al)Mn^{3+}Mn^{2+}{}_3Zn_2Si_2O_{10}(OH)_8$

has a modulated layer composed of "islands" of 24 Si^{4+} tetrahedra with nearly coplanar bases. The islands are linked laterally to 6-membered rings of inverted tetrahedra. Those rings share corners with the tetrahedra in analogous rings in the next sheet of islands. The resultant structure is thus four tetrahedra thick (Gaines et al., 1997)!

Amphiboles

In micas, the sheets are continuous in 2-D. Once we begin to break up bonds between Si^{4+} tetrahedra, there are several possibilities. The one that results in the amphibole group structures involves breaking the sheets up into chains that are the width of a 6-membered tetrahedral ring, and thus resemble sets of double chains. The many members of the amphibole group are extremely common rock-forming minerals, and are found in igneous rocks from gabbro to granite. The Fe- and Mg-rich species (especially) are also found in their metamorphosed equivalents.

The fibrous amphiboles that make up the asbestiform species come from metamorphic occurrences, though they can also be found in hydrothermal deposits. Amphiboles are also found as minor components in coarse-grained sedimentary rocks such as sandstones. They comprise 5% of the Earth's crust, and constitute the dominant OH-bearing phase in the Earth's mantle (although, interestingly, the bulk of hydrogen in the mantle actually is dissolved in anhydrous phases like olivine and pyroxene, but that's another story...). Excellent overviews of this group can be found in volumes 9A, 9B, and 67 of the *Reviews in Mineralogy* series, as well as Deer et al. (1997) and Veblen and Wylie (1993).

Chemistry. The many amphibole species (Table 22.12) can encompass a huge range in chemistry—perhaps the largest of any mineral group except tourmaline! The complicated crystal chemistry in this group led Robinson et al. (1982) to describe amphibole as "a mineralogical garbage can" and a "mineralogical shark in a sea of unsuspecting elements." This may also help to explain the derivation of the term amphibole—which means "ambiguous."

The basic classification is given in Leake et al. (1997, 2004), and can be related to the standard formula, which is:

$$A_{0-1}B_2C_5T_8O_{22}(OH,F,Cl,O)_2$$

where A is N^{1+}, K^{1+}, and/or vacant; B is Na^{1+}, Li^{1+}, Ca^{2+}, Mn^{2+}, Fe^{2+}, Mg^{2+}, Zn^{2+}, Ni^{2+}, and/or Cu^{2+}; C is Mg^{2+}, Fe^{2+}, Mn^{2+}, Al^{3+}, Fe^{3+}, Mn^{3+}, Cr^{3+}, Zn^{2+}, and/or Ti^{4+}; and T is either Si^{4+}, Al^{3+}, or Ti^{4+}. The two B sites are called M4, and the C sites are composed of two M1, two M2, and one M3 site per formula unit. There are also two types of tetrahedral sites (four each of T1 and T2). Even with this formula, a mineral must also possess the double-chain structure to be an amphibole.

The amphiboles are further classified into five groups depending on the occupancy of the B sites:

1. The **magnesium-iron-manganese-lithium** subgroup includes species where $(Ca^{2+}+Na^{1+})_B < 1.00$ and the sum of Mg^{2+}, Fe^{2+}, Mn^{2+}, and Li^{1+} in B is ≥ 1.00.
2. The **calcic** subgroup includes species where $(Ca^{2+}+Na^{1+})_B \geq 1.00$ and $Na^{1+}{}_B < 0.50$, and usually, $Ca^{2+}{}_B > 1.50$.
3. The **sodic-calcic** subgroup includes amphiboles where $(Ca^{2+}+Na^{1+})_B \geq 1.00$ and $0.50 > Na^{1+}{}_B > 1.50$.
4. The **sodic** subgroup includes species with $Na^{1+}{}_B \geq 1.50$.
5. The **sodium-calcium-magnesium-iron-manganese-lithium (or "Group 5")** subgroup includes species where $(Mg^{2+}+Fe^{2+}+Mn^{2+}+Li^{1+})_B$ is between 0.5 and 1.50 and $(Ca^{2+}+Na^{1+})_B$ is also between 0.5 and 1.50.

Within these subgroups, species names are determined on the basis of cations. This nomenclature, while somewhat cumbersome due to the large number of possibilities within these boundaries, is at least capable of providing names for all new and future compositions within the amphibole group.

Many solid solutions can occur within the amphibole group minerals. The most important of these is $Mg \leftrightarrow Fe^{2+}$, but $Al^{3+} \leftrightarrow Si^{4+}$, $(Mg^{2+},Fe^{2+}) \leftrightarrow (Al^{3+},Fe^{3+})$, $Na^{1+} \leftrightarrow Ca^{2+}$ and $Na^{1+} \leftrightarrow K^{1+}$ are also common. Other coupled substitutions charge-balance these.

Structure. The basic building blocks of the amphibole structure are the double chains of tetrahedra, which have the formula Si_4O_{11} (Figure 6.4). These form "I-beams" that run lengthwise along the *c* axis, and stack in the *a* direction (for orthorhombic amphiboles) or the $a \sin \beta$ direction (for monoclinic amphiboles), with the A cations between them. Alternating stacks of I-beams are offset by half of a unit cell—this is like the polytypes of micas discussed above. Adjacent stacks are linked by the M4 cations (Figure 22.22), which occupy quite distorted polyhedra that are coordinated to either 6 or 8 nearest oxygens. The M1, M2, and M3 octahedral sites, which are all relatively symmetrical, lie within the I-beams. Each M1 octahedron has four O^{2-} and two $(OH)^{1-}$ in a *cis* arrangement; the M3 octahedra also have four O^{2-} and two $(OH)^{1-}$'s, but in a *trans* arrangement (Figure 22.22). The M2 sites are coordinated by six O^{2-}'s.

Table 22.12. Amphibole group* (adapted from Leake et al., 1997; 2004)

Subgroup	Species	Formula	Related Species	Formula
Mg-Fe-Mn-Li subgroup/anthophyllite subgroup	cummingtonite	$\square Mg_7Si_8O_{22}(OH)_2$	magnesiocummingtonite	$\square(Mg,Fe^{2+})_7Si_8O_{22}(OH)_2$
			manganocummingtonite (formerly tirodite)	$\square Mn_2Mg_5Si_8O_{22}(OH)_2$
	grunerite	$\square Fe^{2+}{}_7Si_8O_{22}(OH)_2$	permanganogrunerite	$\square Mn_4Fe^{2+}{}_3Si_8O_{22}(OH)_2$
			manganogrunerite (formerly dannemorite)	$\square Mn_2Fe^{2+}{}_5Si_8O_{22}(OH)_2$
	clino-holmquistite	$\square Li_2Mg_3Al_2Si_8O_{22}(OH)_2$	clinoferroholmquistite	$\square Li_2Fe^{2+}{}_3Al_2Si_8O_{22}(OH)_2$
			ferri-clinoholmquistite	$\square Li_2Mg_3Fe^{3+}{}_2Si_8O_{22}(OH)_2$
			ferri-clinoferroholmquistite	$\square Li_2Fe^{2+}{}_3Fe^{3+}{}_2Si_8O_{22}(OH)_2$
			sodic- ferri-clinoferroholmquistite	$(\square,Na)Li_2(Fe^{2+},Mg)_3Fe^{3+}{}_2Si_8O_{22}(OH)_2$
	ferripedrizite	$NaLi_2Mg_2Fe^{3+}{}_2LiSi_8O_{22}(OH)_2$	sodic-ferrodedrizite	$NaLi_2(LiFe^{2+}{}_2Fe^{3+}Al)Si_8O_{22}(OH)_2$
	anthophyllite	$\square Mg_7Si_8O_{22}(OH)_2$	magnesioanthophyllite†	$\square(Mg,Fe^{2+})_7Si_8O_{22}(OH)_2$
			ferro-anthophyllite	$\square Fe^{2+}{}_7Si_8O_{22}(OH)_2$
			sodicanthophyllite	$NaMg_7Si_7AlO_{22}(OH)_2$
			sodic-ferro-anthophyllite	$NaFe^{2+}{}_7Si_8O_{22}(OH)_2$
			proto-ferro-anthophyllite	$\square(Fe^{2+},Mn^{2+})_2(Fe^{2+},Mg)_5Si_8O_{22}(OH)_2$
			protomangano-ferro-anthophyllite	$\square(Mn,Fe^{2+})_2(Fe^{2+},Mg)_5Si_8O_{22}(OH)_2$
	gedrite	$\square Mg_5Al_2Si_6Al_2O_{22}(OH)_2$	magnesiogedrite†	$\square Mg_5Al_2Si_6Al_2O_{22}(OH)_2$
			ferrogedrite	$\square Fe^{2+}{}_5Al_2Si_6Al_2O_{22}(OH)_2$
			sodicgedrite	$NaMg_6AlSi_6Al_2O_{22}(OH)_2$
			sodic-ferrogedrite	$NaFe^{2+}{}_6AlSi_6Al_2O_{22}(OH)_2$
	holmquistite	$\square Li_2(Mg_3Al_2)Si_8O_{22}(OH)_2$	magnesioholmquistite	$\square Li_2(Mg,Fe^{2+})_3Al_2Si_6Al_2O_{22}(OH)_2$
			ferroholmquistite	$\square Li_2(Fe^{2+}{}_3Al_2)Si_8O_{22}(OH)_2$
	IMA2001-065	$\square(Mg,Fe^{2+})_7Si_8O_{22}(OH)_2$		
Calcic subgroup	tremolite	$\square Ca_2Mg_5Si_8O_{22}(OH)_2$		
	actinolite	$\square Ca_2(Mg,Fe^{2+})_5Si_8O_{22}(OH)_2$	ferro-actinolite	$\square Ca_2Fe^{2+}{}_5Si_8O_{22}(OH)_2$
	hornblende**	$Ca_2(Mg,Fe^{2+})_4(Al,Fe^{3+})Si_7AlO_{22}(OH)_2$	magnesiohornblende	$\square Ca_2Mg_4(Al,Fe^{3+})Si_7AlO_{22}(OH)_2$
			ferrohornblende	$\square Ca_2Fe^{2+}{}_4(Al,Fe^{3+})Si_7AlO_{22}(OH)_2$
	tschermakite	$\square Ca_2Mg_3AlFe^{3+}Si_6Al_2O_{22}(OH)_2$	alumino-tschermakite	$\square Ca_2Mg_3Al_2Si_6Al_2O_{22}(OH)_2$
			ferro-aluminotschermakite†	$\square Ca_2Mg_3AlFe^{3+}Si_6Al_2O_{22}(OH)_2$
			ferritschermakite	$\square Ca_2Mg_3Fe^{3+}{}_2Si_6Al_2O_{22}(OH)_2$
			ferro-ferritschermakite	$\square Ca_2Fe^{2+}{}_3Fe^{3+}{}_2Si_6Al_2O_{22}(OH)_2$
			alumino-ferrotschermakite	$\square Ca_2Fe^{2+}{}_3Al_2Si_6Al_2O_{22}(OH)_2$
			ferri-ferrotschermakite	$\square Ca_2Fe^{2+}{}_3Fe^{3+}{}_2Si_6Al_2O_{22}(OH)_2$
			ferrotschermakite	$\square Ca_2Fe^{2+}{}_3AlFe^{3+}Si_6Al_2O_{22}(OH)_2$
	edenite	$NaCa_2Mg_5Si_7AlO_{22}(OH)_2$	ferro-edenite	$NaCa_2Fe^{2+}{}_5Si_7AlO_{22}(OH)_2$
			fluoro-edenite	$NaCa_2Mg_5Si_7AlO_{22}F_2$
	pargasite	$NaCa_2(Mg_4Al)Si_6Al_2O_{22}(OH)_2$	ferropargasite	$NaCa_2(Fe^{2+}{}_4Al)Si_6Al_2O_{22}(OH)_2$
			potassicpargasite	$(K,Na)Ca_2(Mg,Fe^{2+},Al)_5(Si,Al)_8O_{22}(OH,F)_2$
			potassic-chloroparagasite	$(K,Na)Ca_2(Fe^{2+},Mg)_4Al(Si_6Al_2)O_{22}(Cl,OH)_2$
	hastingsite	$NaCa_2(Fe^{2+}{}_4Fe^{3+})Si_6Al_2O_{22}(OH)_2$	magnesiohastingsite	$NaCa_2(Mg_4Fe^{3+})Si_6Al_2O_{22}(OH)_2$
			potassic-chlorohastingsite	$(K,Na)Ca_2(Fe^{2+},Mg)_4Fe^{3+}Si_6Al_2O_{22}(Cl,OH)_2$
	sadanagaite	$NaCa_2Fe^{2+}{}_3(Fe^{3+},Al)_2Si_5Al_3O_{22}(OH)_2$	magnesiosadanagaite	$NaCa_2Mg_3(Fe^{3+},Al)_2Si_5Al_3O_{22}(OH)_2$
			potassic-magnesiosadanagaite	$NaCa_2Mg_3(Al,Fe^{3+})_2Si_5Al_3O_{22}(OH)_2$
			potassicsadanagaite	$(K,Na)Ca_2Fe^{2+}{}_3(Fe^{3+},Al)_2Si_5Al_3O_{22}(OH)_2$
			potassicferrisadanagaite	$(K,Na)Ca_2(Fe^{2+},Mg)_2(Fe^{3+},Al)_2Si_5Al_3O_{22}(OH,F,O)_2$
	kaersutite	$NaCa_2Mg_4TiSi_6Al_2O_{23}(OH)$	ferrokaersutite	$NaCa_2Fe^{2+}{}_4TiSi_6Al_2O_{23}(OH)$
	cannilloite	$CaCa_2Mg_4AlSi_5Al_3O_{22}(OH)_2$	fluorocannilloite	$CaCa_2(Mg_4Al)Si_5Al_3O_{22}F_2$

*Note: individual mineral species within subgroups are slightly rearranged relative to Dana order in this table, in order to clarify the presentation. □ indicates a vacant site.
**Hornblende is a general term for colored, Ca-rich amphibole, but it is not an approved species name
†Discredited by IMA, but still sometimes found in the literature.

Table 22.12. (continued)

	Species	Formula	Related Species	Formula
Sodic-calcic subgroup	winchite	$\square CaNaMg_4(Al,Fe^{3+})Si_8O_{22}(OH)_2$	alumino-winchite†	$\square(Na,Ca)Mg_3Fe^{2+}AlSi_8O_{22}(OH)_2$
			ferrowinchite	$\square CaNaFe^{2+}{}_4(Al,Fe^{3+})Si_8O_{22}(OH)_2$
			ferriwinchite	$\square NaCaMg_4Fe^{3+}Si_8O_{22}(OH)_2$
			ferroferriwinchite†	$\square CaNaFe^{2+}{}_3MgFe^{3+}Si_8O_{22}(OH)_2$
	barroisite	$\square CaNaMg_3AlFe^{3+}Si_7AlO_{22}(OH)_2$	aluminobarroisite	$\square CaNaMg_3Al_2Si_7AlO_{22}(OH)_2$
			ferrobarroisite	$\square CaNaFe^{2+}{}_3AlFe^{3+}Si_7AlO_{22}(OH)_2$
			alumino-ferrobarroisite	$\square CaNaFe^{2+}{}_3Al_2Si_7AlO_{22}(OH)_2$
			ferribarroisite	$\square CaNaMg_3Fe^{3+}{}_2Si_7AlO_{22}(OH)_2$
			ferroferribarroisite†	$\square CaNa(Fe^{2+},Mg)_3Fe^{3+}{}_2Si_7AlO_{22}(OH)_2$
			ferri-ferrobarroisite	$\square CaNaFe^{2+}{}_3Fe^{3+}{}_2Si_7AlO_{22}(OH)_2$
	richterite	$Na(CaNa)Mg_5Si_8O_{22}(OH)_2$	fluororichterite	$Na(CaNa)Mg_5Si_8O_{22}F_2$
			potassic-fluororichterite	$(K,Na)(Ca,Na)_2Mg_5Si_8O_{22}(F,OH)_2$
			potassicrichterite	$(K,Na)(Ca,Na)_2Mg_5Si_8O_{22}(OH)_2$
			ferrorichterite	$Na(CaNa)Fe^{2+}{}_5Si_8O_{22}(OH)_2$
	katophorite	$Na(CaNa)Fe^{2+}{}_4(Al,Fe^{3+})Si_7AlO_{22}(OH)_2$	magnesiokatophorite	$Na(CaNa)Mg_4(Al,Fe^{3+})Si_7AlO_{22}(OH)_2$
			magnesioferrikatophorite	$Na(CaNa)(Mg,Fe^{2+})_4Fe^{3+}Si_7AlO_{22}(OH)_2$
			ferrikatophorite	$Na(CaNa)(Fe^{2+},Mg)_4Fe^{3+}Si_7AlO_{22}(OH)_2$
	taramite	$Na(CaNa)Fe^{2+}{}_3AlFe^{3+}Si_6Al_2O_{22}(OH)_2$	magnesiotaramite	$Na(CaNa)Mg_3AlFe^{3+}Si_6Al_2O_{22}(OH)_2$
			magnesioferritaramite	$Na(CaNa)(Mg,Fe^{2+})_3Fe^{3+}{}_2Si_6Al_2O_{22}(OH)_2$
			ferritaramite	$Na(CaNa)Fe^{2+}{}_3Fe^{3+}{}_2Si_6Al_2O_{22}(OH)_2$
			aluminomagnesiotaramite	$Na(CaNa)Mg_3Al_2Si_6Al_2O_{22}(OH)_2$
			ferri-magnesiotaramite	$Na(CaNa)Mg_3Fe^{3+}{}_2Si_6Al_2O_{22}(OH)_2$
	IMA2002-051	$(Na,K)Ca_2Mg_3Al_2Si_5Al_3O_{22}(OH)_2$		
Sodic subgroup	glaucophane	$\square Na_2Mg_3Al_2Si_8O_{22}(OH)_2$	ferroglaucophane	$\square Na_2Fe^{2+}{}_3Al_2Si_8O_{22}(OH)_2$
	crossite†	$\square Na_2(Mg,Fe^{2+})_3(Al,Fe^{3+})_2Si_8O_{22}(OH)_2$		
	riebeckite	$\square Na_2Fe^{2+}{}_3Fe^{3+}{}_2Si_8O_{22}(OH)_2$	magnesioriebeckite	$\square Na_2Mg_3Fe^{3+}{}_2Si_8O_{22}(OH)_2$
	eckermannite	$NaNa_2Mg_4AlSi_8O_{22}(OH)_2$	ferro-eckermannite	$NaNa_2Fe^{2+}{}_4AlSi_8O_{22}(OH)_2$
	arfvedsonite	$NaNa_2Fe^{2+}{}_4Fe^{3+}Si_8O_{22}(OH)_2$	magnesio-arfvedsonite	$NaNa_2Mg_4Fe^{3+}Si_8O_{22}(OH)_2$
	kôzulite	$NaNa_2Mn_4(Fe^{3+},Al)Si_8O_{22}(OH)_2$		
	nyböite	$NaNa_2Mg_3Al_2Si_7O_{22}(OH)_2$	ferronyboite	
			ferric-ferronyboite	
	IMA2002-010	$NaNa_2Mg_3Al_2Si_7O_{22}(F,OH)_2$		
	leakeite	$NaNa_2Mg_2Fe^{3+}{}_2LiSi_8O_{22}(OH)_2$	ferroleakeite	$NaNa_2Fe^{2+}{}_3Fe^{3+}{}_2LiSi_8O_{22}(OH)_2$
			potassicleakeite	$KNa_2Mg_2Fe^{3+}{}_2LiSi_8O_{22}(OH)_2$
	kornite	$(Na,K)Na_2Mg_2Mn^{3+}{}_2LiSi_8O_{22}(OH)_2$		
	ungaretiite	$NaNa_2Mn_2Mn^{3+}{}_3Si_8O_{22}O_2$		
	fluoro-magnesio-arfvedsonite	$NaNa_2(Mg,Fe^{2+})_4Fe^{3+}Si_8O_{22}(F,OH)_2$		
	obertiite	$NaNa_2Mg_3Fe^{3+}Ti^{4+}Si_8O_{22}O_2$		
	IMA2001-066	$Li_2Fe^{2+}{}_3Fe^{3+}{}_2Si_8O_{22}(OH)_2$		
	IMA2001-067	$Li_2Mg_3Fe^{3+}{}_2Si_8O_{22}(OH)_2$		
	IMA2001-069	$Na(Na_{1.0\text{-}1.5}Li_{0.5\text{-}1.0}Mg_2LiFe^{3+}{}_2Si_8O_{22}(OH)_2$		
Na-Ca-Mg-Fe-Mn-Li subgroup	whittakerite	$Na(NaLi)(LiMg_2Fe^{3+}Al)Si_8O_{22}(OH)_2$	ferrowhittakerite	$Na(NaLi)(LiFe^{2+}{}_2Fe^{3+}Al)Si_8O_{22}(OH)_2$
	ottoliniite	$\square(Na,Li)(Mg_3Fe^{3+}Al)Si_8O_{22}(OH)_2$	ferro-ottoliniite	$\square(Na,Li)(Fe^{2+}{}_3Fe^{3+}Al)Si_8O_{22}(OH)_2$

*Note: individual mineral species within subgroups are slightly rearranged relative to Dana order in this table, in order to clarify the presentation. □ indicates a vacant site.
**Hornblende is a general term for colored, Ca-rich amphibole, but it is not an approved species name
†Discredited by IMA, but still sometimes found in the literature.

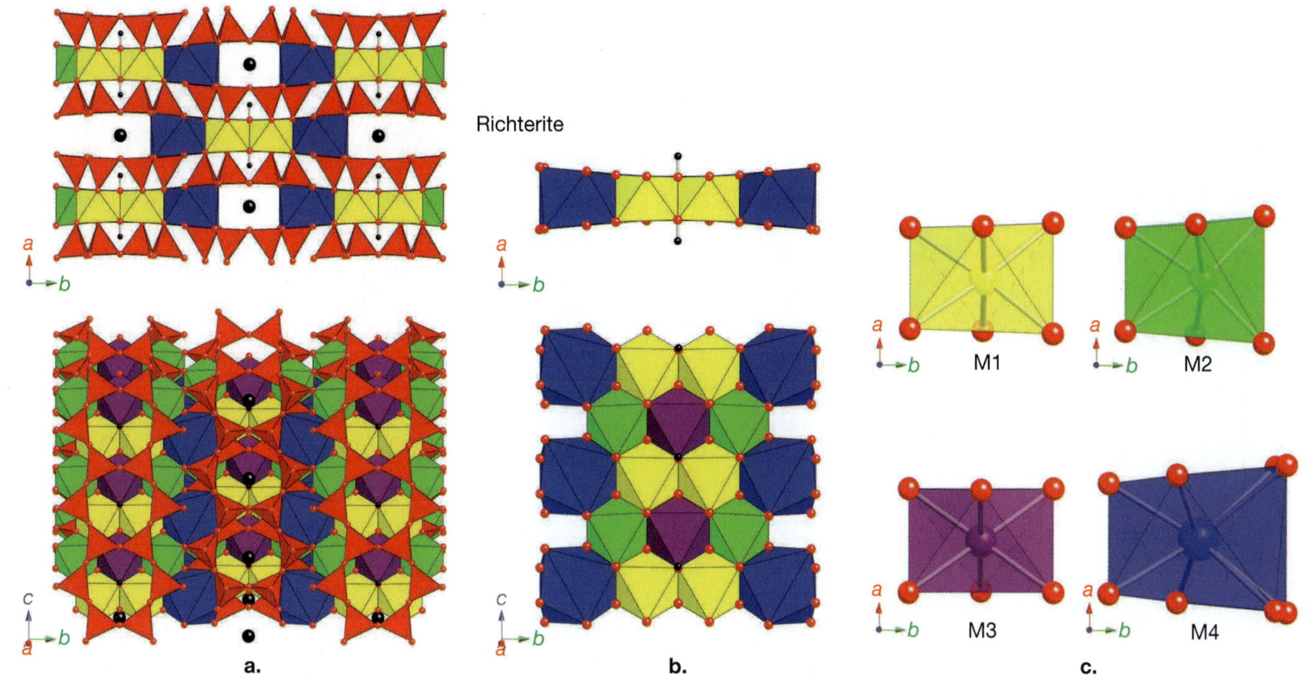

Figure 22.22. *The crystal structure of the amphibole mineral richterite shows the four different octahedral sites (M1, M2, M3, and M4) that compose the octahedral strip. The large black spheres represent the A site, which is not always occupied, and the small black spheres are* H^{1+}*, which bonds into the octahedral strip.* ***a.*** *In the* a-b *projection (left) three of the four M sites can be seen. The two yellow ones are the M1's and the large blue ones the M4's. On the opposite side of the strip, the green M2 can partially be seen. In the* a-c *projection (left), the M3 site is the purple octahedron in the strip's center. In this projection, the A site and H are located near the center of the 6-membered rings formed by the double chains.* ***b.*** *The octahedral strip is here isolated from the structure. On the top is the* a-b *plot that looks down the octahedral strip, again showing the two M1 sites in the middle with the two M4 sites on the outside and the* H^{1+} *bonded perpendicular to the sheets. The* b-c *projection on the bottom more clearly shows the types and locations of the four sites. At the bottom are the two M1 sites with two M4 sites on the outside. The next "strip" up shows the single M3 octahedral site (purple), surrounded by two M2 sites (green). In this projection, the* H^{1+} *bonds to the oxygen shared between the two M1's and one M3 site.* ***c.*** *Exploded views of the four octahedral sites. M1-M3 are similar in size and M4 is the largest; M3 is the least distorted. To further explore the characteristics of these polyhedra (i.e., bond lengths and bond angles), use the appropriate tools in CrystalViewer.*

The symmetry of each amphibole depends on the orientation of the I-beams within the structure, which is in turn a function of stacking patterns. Each octahedral layer introduces a stagger of *c*/3 between the tetrahedral layers on either side of it. If the stagger is always in the same (+) direction, the stacking can be called +++++, etc., and the amphibole will have a monoclinic unit cell with space group *C2/m*. If the stagger alternates in direction (symbolized by + – + – + – etc.), then the unit cell will be orthorhombic, with space group *Pnma*. A variation with a ++ – stacking sequence would also be orthorhombic, but would result in the *Pnmn* space group (this is only observed in synthetic samples). The formalism for this was delightfully described by J.B. Thompson (1981), as shown in Figure 22.23 and the "Swimming Octahedra" inset, where the ducks "swim" both in a "+" direction when they swim toward you, but in a "–" direction as they swim away from you.

As a general rule, the *c* axis is defined to be parallel to the silicate chains, while the *a* axis is in the stacking direction. The *b* axis is then perpendicular to *c* and is pointing across the chains. The choice of a unit cell is based on the stacking along the *a* direction and is somewhat complicated because the choice of the position of the *a* axis can lead to different types of lattices. For more information, see Thompson (1978) and Hawthorne (1981).

Optical properties. Because amphiboles are composed of infinitely long double chains of silica tetrahedra, their crystals in turn tend to be elongated in the direction of those chains. When the structure is projected down the chains, you can see that they have weak (110) planes along which the mineral would break—this is the prominent cleavage for amphiboles (Figure 22.24). Thus, the external morphology of the crystals can be directly linked to the structure.

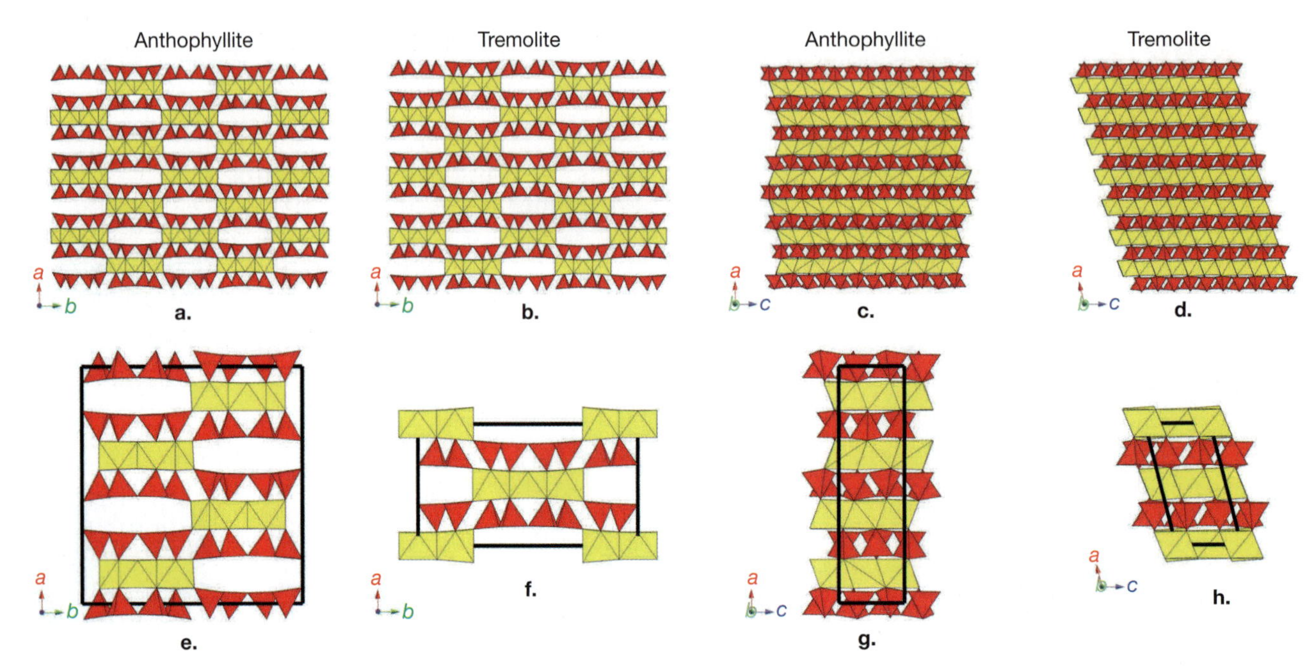

Figure 22.23. *Crystal structure drawings of the orthorhombic amphibole anthophyllite,* $Mg_7Si_8O_{22}(OH)_2$*, and the monoclinic amphibole tremolite,* $Ca_2Mg_5Si_8O_{22}(OH)_2$*, projected on their* a-b *and* a-c *planes, showing how the stacking orientations (see Swimming octahedra inset below) of the octahedra control the crystal systems.* Si^{4+} *tetrahedra are red and the* Mg^{2+} *octahedra (M1-M3) are yellow; M4 is excluded from these drawings.* ***a.*** *In this* a-b *projection of anthophyllite, the triangular faces of the octahedra point in the same direction in each horizontal layer, but alternate between them.* ***b.*** *In this* a-b *projection of tremolite, all the triangular faces of the octahedra point in the same direction.* ***c.*** *When anthophyllite is projected onto its* a-c *plane, the alternate stacking directions of the octahedral can be seen. These result in an orthorhombic unit cell (see Figure 22.23e and g).* ***d.*** *When tremolite is projected onto its* a-c *plane, it is apparent that all the octahedral layers are stacked in the same direction. These result in a monoclinic unit cell (see Figure 22.23f and h) with half the* a *repeat of anthophyllite.* ***e.*** *This view is an* a-b *projection of just the outlined unit cell of anthophyllite, showing the alternating stacking direction of the octahedra.* ***f.*** *This view is an* a-b *projection for the unit cell of tremolite; its* a *repeat is half that of anthophyllite and the octahedra all have the same orientation.* ***g.*** *In the* a-c *projection of the unit cell of anthophyllite,* a *and* c *are 90° and the octahedral stacking directions alternate.* ***h.*** *In the* a-c *projection of the unit cell of tremolite,* a *and* c *are not 90° and the octahedral stacking directions are all the same.*

Swimming Octahedra

An octahedron has eight triangular faces. If we place an octahedron on a flat surface, it will come to rest on one of those faces. To represent the orientation of the octahedron, we could add a swan's head and tail feathers as shown in the figure. We could then see if the swan is swimming toward us, away from us, or to the right or left. This "swan" story (borrowed from Thompson, 1981), is a great way to visualize the orientation of the octahedral layers that we discussed in Figures 22.23, 22.30, and 22.31. For tremolite in Figure 22.23b, all of the ducks are swimming away from us. For anthophyllite in Figure 22.23a, the bottom horizontal layer is toward us and the next layer up is swimming away; this stacking sequence continues along the a *axis. In the* a-c *projection of anthophyllite (Figure 22.23c), the bottom layer is "swimming" to the left and the layer above it to the right, while for the* a-c *projection of tremolite, all of the octahedra are "swimming" to the right.*

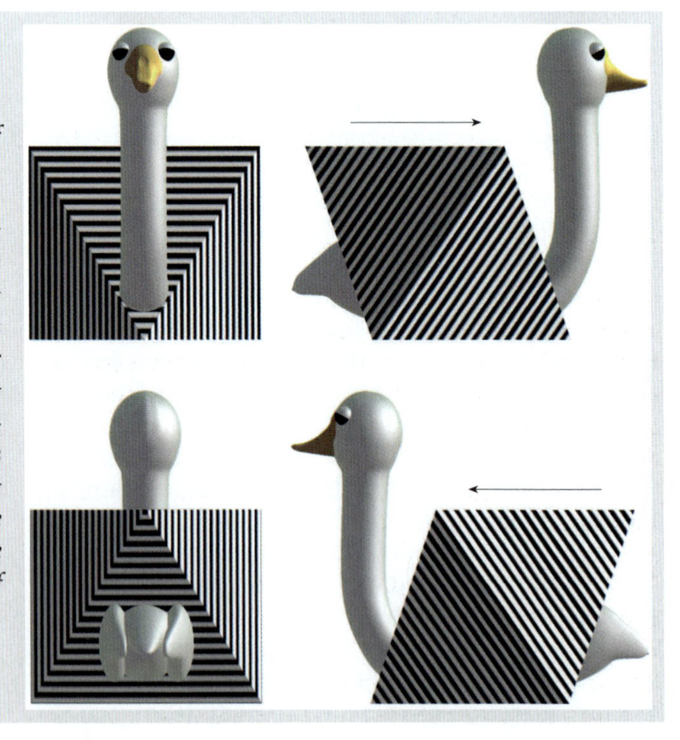

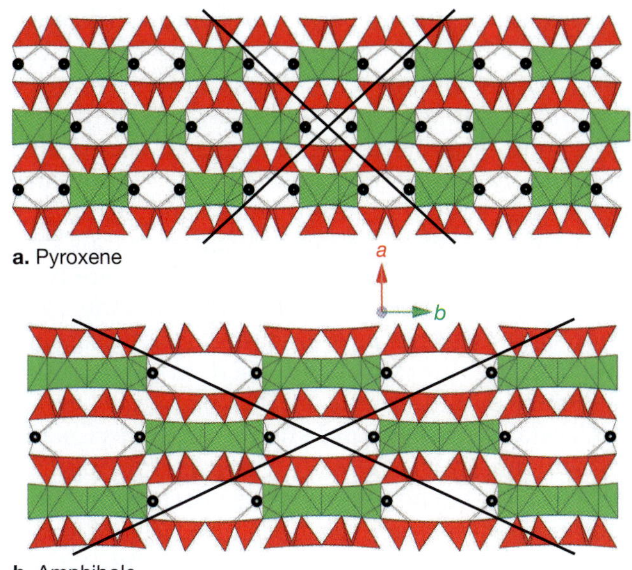

Figure 22.24. A projection of a pyroxene and an amphibole down the c axis (i.e., down the Si^{4+} chains). Both minerals are elongated parallel to these chains and have perfect cleavage imparted by the chains. In each case, planes of weakness go through the center of the cavities formed below the base of the chains. In pyroxenes this imparts a near 90° cleavage while in amphiboles it imparts a near 60°/120° cleavage.

As we noted above, there are several important solid solutions within the amphiboles, and their physical properties (e.g. refractive index, density, color) change systematically as a result. This is seen in Figure 22.25. In general, the amphiboles have their largest refractive indices parallel or near parallel to the double chains. For the pleochroic amphiboles, the absorption of light tends to be greatest parallel to the chains. When viewed perpendicular to the chains, they exhibit a similar phenomenon to absorption within the octahedral sheet of layer silicates. When polarized light vibrates parallel to the octahedral ribbons (in amphiboles the *b* direction), the most light is absorbed because the atomic density is the highest.

Asbestos. Because of their industrial and economic importance, the asbestos-forming amphiboles and their serpentine cousins deserve special mention here (see also *Reviews in Mineralogy* volumes 28 and 67). "Asbestos" is an industrial term describing a highly fibrous silicate mineral that readily separates into long, thin, strong fibers that have sufficient flexibility to be woven, are heat resistant and chemically inert, are electrical insulators, and are therefore suitable for uses where incombustible, nonconducting, or chemically resistant material is required (Gray et al., 1974). These applications number in the thousands and include construction materials (insulation, spackling compound, drywall, vinyl and roofing tiles, concrete pipes, textured paint, paper, cloth, brake pads, and filters).

Mineralogically, there are several types of minerals that can have asbestiform habits, which are distinguished by their fibrillar structure and the flexibility and strength of the fibers (Veblen and Wylie, 1993). **Fibrils** are single fibers with very thin widths. In amphibole asbestos, the fibrils form parallel to the double chains. In chrysotile, the fibrils are rolled-up silicate layers with cylindrical or circular cross-sections. The fibrils in all cases form in bundles with a common elongation axis, which is the *c* axis in amphiboles and the *a* axis in chrysotile (Veblen and Wylie, 1993). So asbestiform fibers can be either single fibrils or bundles of them, with widths ranging down to sub-micron levels. While the cleavage fragments of amphiboles expose the (110) surfaces, the smaller asbestiform fibers tend to expose the (100) surface (Brown and Gunter, 2003). The exact details of why the fibers form differently from longer crystals is still not well understood.

The amphibole group contains the largest number of asbestiform mineral species. As we write

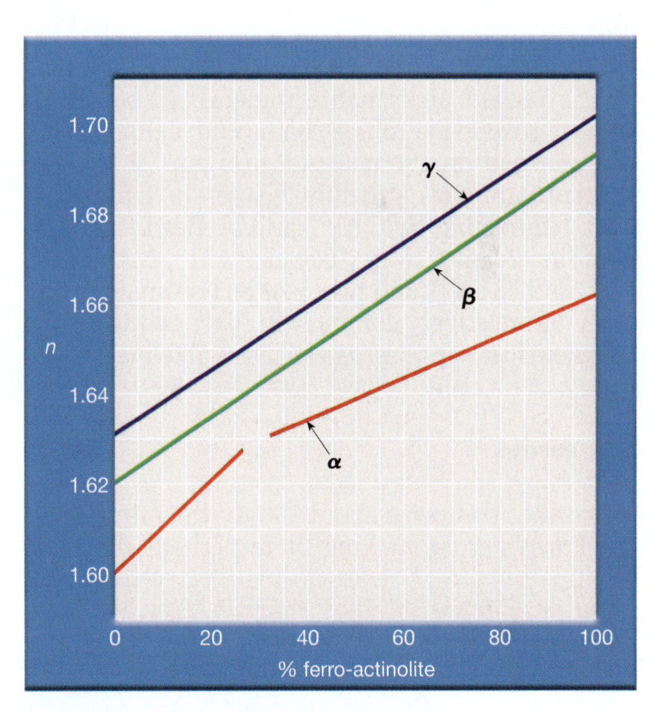

Figure 22.25. Variation of the principal refractive indices with composition for the tremolite-actinolite-ferroactinolite series (modified from Verkouteren and Wylie, 2000). Notice how all three refractive indices increase as Fe substitutes for Mg in the general formula for this group, $Ca_2(Mg,Fe)_5Si_8O_{22}(OH)_2$. Interestingly, in their careful study Verkouteren and Wylie found a discontinuity in the data for the a *axis; these sorts of breaks in variation of a physical parameter can often represent disruptions in a solid solution series.*

this, only five amphibole species are regulated in the U.S.: anthophyllite, tremolite, actinolite, grunerite, and riebeckite (along with chrysotile). A look at Table 22.12 makes it apparent that there are many other amphiboles with similar compositions that could possibly be (and probably are) asbestiform. Chrysotile is the most commonly used form of asbestos and its health effects appear to be far less severe than those of the asbestiform amphiboles (Gunter, 1994, and Gunter et al., 2007).

Chains with Side Branches or Loops

The 69th Dana class includes minerals in which the chains we saw in amphiboles are linked sideways by additional, different types of polyhedra. Two of these groups make good contrasting examples. In the **astrophyllite** group of minerals (Table 22.13), infinite $(Si_4O_{12})^{10-}$ chains that are cross-linked along the *b* axis by octahedra, forming sheets (Figure 22.26). Thus, these minerals have a formula of $A_3B_{6\text{-}7}C_2(Si_8O_{24})(O,OH)_7$. The A and B sites are similar to those in amphibole: A may contain Na^{1+}, K^{1+}, Cs^{1+}, Ca^{2+}, and H_3O^{+}, while B is occupied by Fe^{2+}, Mn^{2+}, and Mg^{2+}. The C sites in the cross-linking octahedra are quite small, and thus contain smaller, highly charged cations like Ti^{4+}, Zr^{4+}, and Nb^{4+}.

The related **aenigmatite** minerals (Table 22.14) have pyroxene-like single chains with extra tetrahedra attached on alternate sides (Figure 22.26). The resultant mineral structures have six tetrahedral sites (containing Si^{4+}, Al^{3+}, and/or Be^{2+}), seven 6-coordinated sites (Mg^{2+}, Fe^{2+}, Fe^{3+}, Ti^{4+}, Mn^{2+}, Sn^{4+}, Cr^{3+}, and/or Sb^{3+}), and two 7- or 8-coordinated sites. They are relatively common in Na^{1+}-rich intrusive rocks such as syenites, as well as in alkaline lavas.

Pyroxenes

Pyroxenes comprise about 10% of the minerals in the Earth's crust, making them perhaps the most important of the Fe and Mg-bearing silicates. They are very common in igneous rocks (thus, the origin of their names, from the Greek pyros for fire), and stable over a wide range of temperatures and pressures. For this reason, they also occur in meteorites and on the Earth's Moon and other terrestrial moons, planets, and asteroids. Because pyroxenes contain transition metals that have diagnostic reflectance spectra, they are relatively easy to detect from satellites. Spectral variations in pyroxenes as a function of chemistry are thus among the key tools used to understand planetary evolution (Figure 22.27).

Crystal chemistry and structure. The pyroxene group minerals share a general formula of $M2M1(Si_2O_6)$, where the M1 site is octahedral and the M2 site is a larger, quite distorted, 6- to 8-coordinated site (Figure 22.28). The M1 site is thus smaller, and can be occupied by Al^{3+}, Fe^{3+}, Cr^{3+}, and Ti^{4+}, while the M2 site may contain Ca^{2+}, Na^{1+}, Mg^{2+}, Fe^{2+}, Mn^{2+}, Ni, and/or Li^{1+}. Some substitution of Al^{3+} for Si^{4+} in the tetrahedral sites also occurs. The pyroxenes lack $(OH)^{1-}$ groups.

Perhaps the most common pyroxenes are those composed of Ca^{2+}, Fe^{2+}, and Mg^{2+}. Their compositions are commonly represented on a quadrilateral diagram as combinations of four components: $Mg_2Si_2O_6$, $CaMgSi_2O_6$, $CaFeSi_2O_6$, and $Fe_2Si_2O_6$ (Figure 22.29). Along the bottom of this figure is the solid solution between $MgSiO_3$ and $Fe^{2+}SiO_3$. These Ca-free minerals are usually orthorhombic, and are given the names enstatite (Figure 22.30) and ferrosilite; their monoclinic polytypes are called clinoenstatite (Figure 22.30) and clinoferrosilite ("clino-" is often added to an orthorhombic mineral's name if a monoclinic version occurs). If we move up the pyroxene quadrilateral (Figure 22.29) by adding Ca^{2+} to these formulas, the structures all become monoclinic, and are divided into Li^{1+}, Na^{1+}, Ca^{2+}-Na^{1+}, Ca^{2+}, Mn^{2+}-Mg^{2+}, and Mg^{2+}-Fe^{2+} subgroups. The resultant mineral species all have equivalents to the amphibole quadrilateral as shown in Figure 22.29. Note that there is complete solid solution between the

Table 22.13. Astrophyllite group	
Astrophyllite	$(K,Na)_3(Fe^{2+},Mn)_7Ti_2Si_8O_{24}(O,OH)_7$
Kupletskite	$(K,Na)_3(Mn,Fe^{2+})_7(Ti,Nb)_2Si_8O_{24}(O,OH)_7$
Cesium-kupletskite	$(Cs,K,Na)_3(Mn,Fe^{2+})_7(Ti,Nb)_2Si_8O_{24}(O,OH,F)_7$
Niobophyllite	$(K,Na)_3(Fe^{2+},Mn^{2+})_7(Nb,Ti)_2Si_8(O,OH,F)_7$
Zircophyllite	$(K,Na)_3(Fe^{2+},Mn^{2+})_7(Zr,Nb)_2Si_8O_{24}(O,OH,F)_7$
Hydroastrophyllite	$(H_3O^+,K,Ca)_3(Fe^{3+},Mn^{4+},\square)_7(Ti,Nb)_2(Si,\square)_8(O,OH,F)_{31}$
Magnesium astrophyllite	$Na_2K_2(Fe^{2+},Fe^{3+},Mn)_5Mg_2Ti_2Si_8O_{24}(O,OH,F)_7$
Niobokupletskite	$K_2Na(Mn,Zn,Fe^{2+})_7(Nb,Zr,Ti)_2Si_8O_{26}(OH)_4(O,F)$

Table 22.14. Aenigmatite and related species		
Aenigmatite subgroup	aenigmatite	$Na_2Fe^{2+}{}_5TiSi_6O_{20}$
	dorrite	$Ca_2Mg_2Fe^{3+}{}_4Al_4Si_2O_{20}$
	høgtuvaite	$(Ca,Na)_2(Fe^{2+},Fe^{3+},Ti,Mg,Mn,Sn)_6(Si,Be,Al)_6O_{20}$
	krinovite	$Na_2Mg_4Cr_2Si_6O_{20}$
	makarochkinite	$(Ca,Na)_2(Fe^{2+},Fe^{3+},Ti,Mg)_6(Si,Be,Al)_6O_{20}$
	rhönite	$Ca_2(Mg,Fe^{2+},Fe^{3+},Ti)_6(Si,Al)_6O_{20}$
	serendbite	$Ca_2(Mg,Al)_6(Si,Al,B)_6O_{20}$
	welshite	$Ca_2Sb^{5+}Mg_4Fe^{3+}Si_4Be_2O_{20}$
	wilkinsonite	$Na_2Fe^{2+}{}_4Fe^{3+}{}_2Si_6O_{20}$
Sapphirine subgroup	sapphirine	$(Mg,Al)_8(Al,Si)_6O_{20}$
	surinamite	$(Mg_3Al_3)(AlSi_3Be)O_{16}$
	khmaralite	$Mg_{5.5}Al_{14}Fe_2Si_5Be_{1.5}O_{40}$
Magbasite sub-group	magbasite	$KBa(Al,Sc)(Mg,Fe^{2+})_6Si_6O_{20}F_2$

Mg^{2+} and Fe^{2+} species, but not with respect to Ca^{2+}: there is a miscibility gap between pigeonite and augite. Note that when the Ca^{2+} content of the pyroxenes completely fills the M2 sites and Ca^{2+} makes up more than half of the cations, the pyroxene structure itself is no longer stable because the additional Ca^{2+} can't fit into the M1 site. In that situation, the wollastonite group minerals form (Table 22.16, and see below).

When Na^{1+} substitutes into pyroxene in the M2 site, the charge on the M1 site must increase to compensate; this is accomplished by inclusion of Al^{3+} or Fe^{3+}. Charge balance can also be maintained through the coupled Tschermak substitution, in which Al^{3+} replaces Si^{4+} and a trivalent cation replaces a divalent cation in M1. The Na^{1+}-rich pyroxenes can also be represented on a quadrilateral diagram, as seen in Figure 22.29.

Pyroxenes may adopt one of four different space groups (see the second column of Table 22.15, and Figure 22.31). The variations result from differences in the way the pyroxenes stack in the *a* direction in a similar manner to the space group variation based on stacking sequences in the *a* direction for the amphiboles. And as we'll see in this section, since the *a* direction in the chain silicates corresponds to the *c* direction in the layer silicates; then polytypes form in chain silicates based on stacking in *a* and in the layer silicates based on stacking in *c*.

Optical properties. As we noted in Chapter 6, the pyroxene structure is more tightly packed overall than the amphibole structure, so pyroxenes will generally have higher refractive indices.

In keeping with the same themes we saw in layer silicates and amphiboles, the pyroxenes also

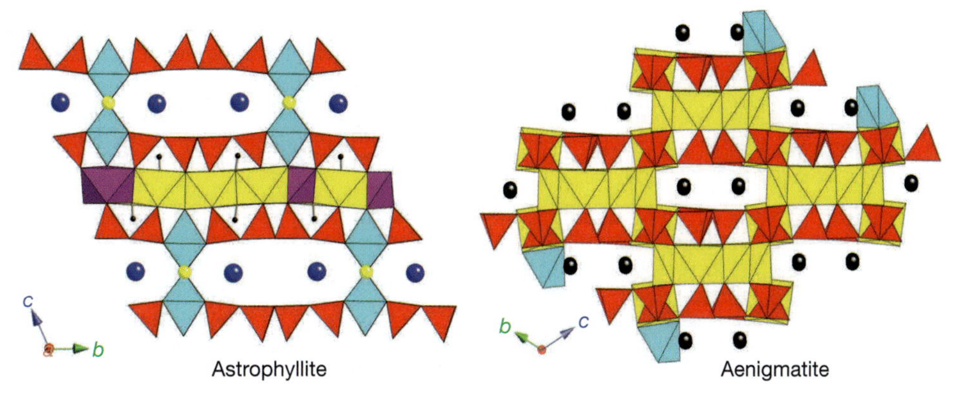

Figure 22.26. *The crystal structures of astrophyllite, $(K,Na)_3(Fe,Mn)_7Ti_2Si_8O_{24}(O,OH)_7$, (left) and aenigmatite, $Na_2Fe_5TiSi_6O_{20}$, (right). Both of these structures have characteristics between amphiboles and layer silicates, so they are projected here looking down the chains/sheets. Si^{4+} tetrahedra are red and Al^{3+} octahedra are the yellow polyhedra. Both minerals also contain Ti^{4+}, which is in octahedral coordination in astrophyllite and bonds the layers of tetrahedra together; in aenigmatite, Ti^{4+} forms distorted polyhedra bonded into tetrahedra. Both structures also have large cations in the voids that are similar to the A sites in amphiboles and the interlayer cations in micas.*

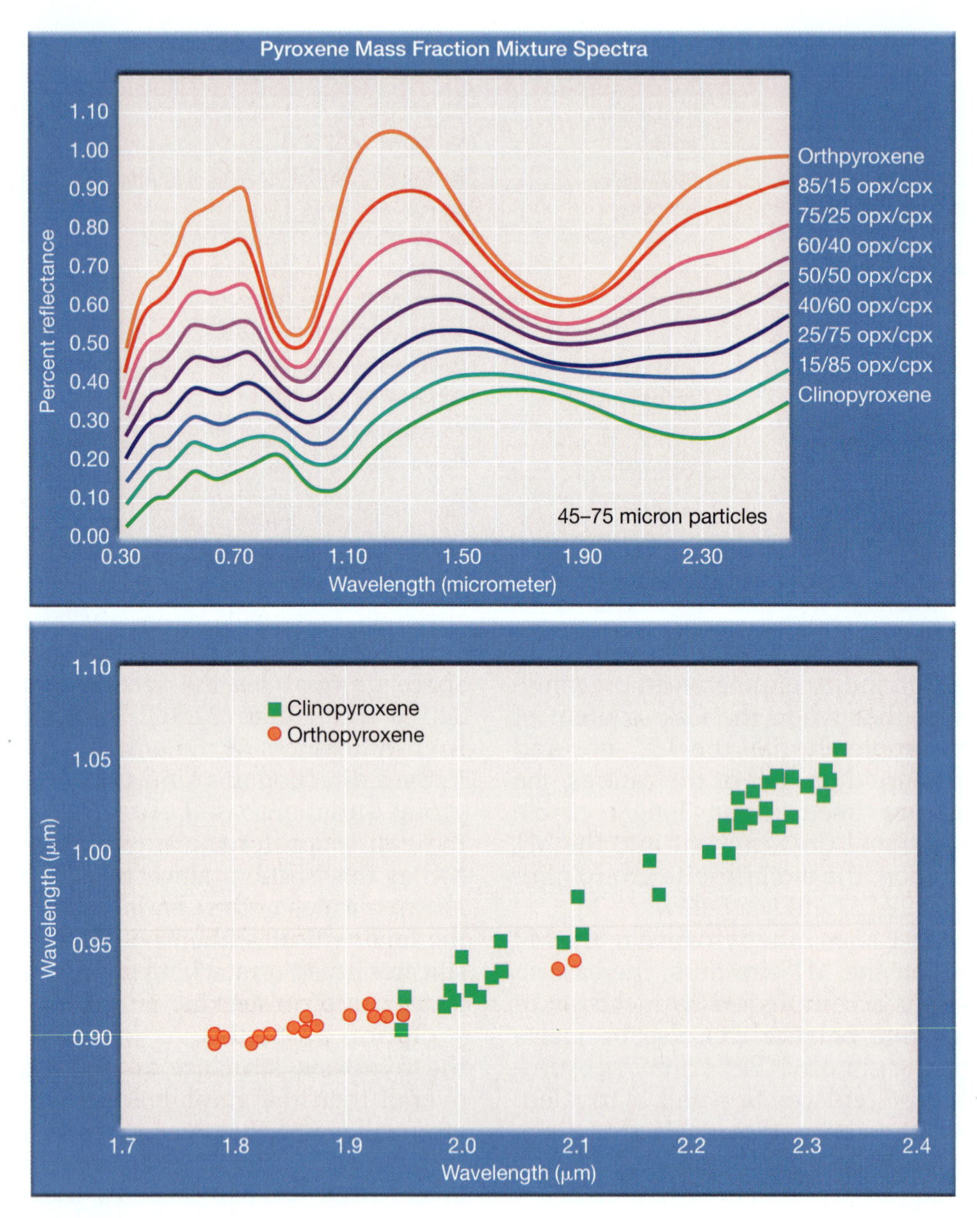

Figure 22.27. Reflectance spectra of pyroxenes with varying compositions as indicated, adapted from Adams (1974) and Sunshine et al. (1990)(top). The pyroxene structure has two types of octahedral sites for cations. The relatively undistorted M1 site is bounded entirely by non-bridging oxygens, and crystal field splitting results in a broad bands around 10,200 and 8,500 cm^{-1} (Rossman, 1980). Crystal field spectra of Fe^{2+} in the more distorted M2 site result in more intense bands at about ~5,000 cm^{-1} (2 μm) and ~10,000 cm^{-1} (1 μm). The energies of the M2 bands are particularly sensitive to Ca^{2+} substitution for Fe^{2+} or Mg^{2+} in the pyroxene because the substitution distorts the structure. Thus the positions of the two Fe^{2+} M2 bands can be used to determine the Ca^{2+} content of pyroxenes by remote observations using Earth-based telescopes and orbiters (bottom).

have physical properties that reflect the orientation of the single chains in their structures. For example, they form crystals that are elongated parallel to those chains. Their refractive indices are typically the largest parallel to those chains. Finally, they have cleavage planes that form parallel to the chains but in a 90° orientation rather than the 120° orientation seen in the amphiboles.

These optical properties showcase the interrelationships among layer silicates, amphiboles, and pyroxenes very well. For example, the directions along the length of the chains (*c* for pyroxene, *c* for amphibole, and *a* for layer silicates) are all similar for species with similar cations. Likewise, the *b* crystallographic axes (along the octahedral sheet) across the I-beam are the same for all groups, particularly in amphiboles and micas. Finally, light that is polarized along the stacking direction, which is the *a* crystallographic direction for pyroxene and amphiboles the *c* crystallographic direction in layer silicates, will be similar (Figure 22.32).

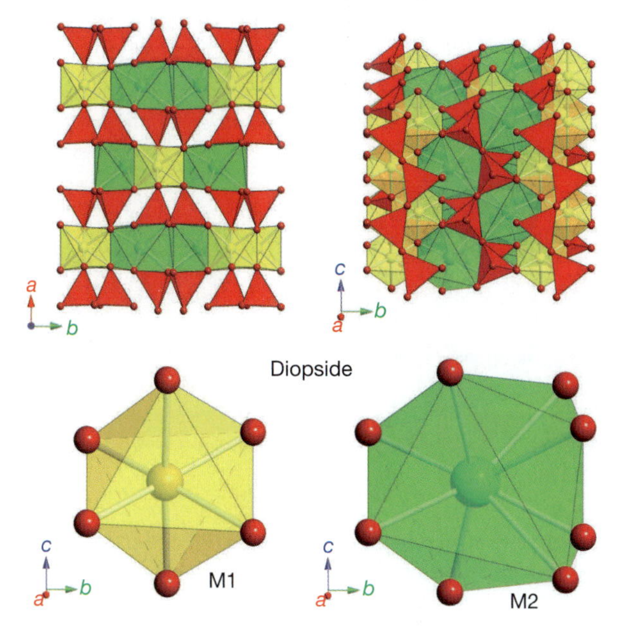

Figure 22.28. Crystal structure of the pyroxene mineral diopside, $CaMgSi_2O_6$, projected on the a-b *and* b-c *planes, along with its M1 and M2 sites isolated from the structure. The Si^{4+} tetrahedra are red while the M1 and M2 polyhedra are transparent yellow and green and show the central cations: M1 = Mg^{2+} and M2 = Ca^{2+} (note that $Ca^{2+} > Mg^{2+}$ in size). In the enlarged M1 and M2 views, it is apparent that M1 is smaller than M2; M2 is more distorted than M1, and the coordination number for M2 is 8 and for M1 is 6.*

The visible-wavelength region of pyroxenes is used in planetary science to determine pyroxene compositions from orbiting or passing satellites on the basis of the differences between the M1 and M2 coordination polyhedra. The M1 site is barely distorted; its geometry is that of a nearly perfect octahedron. Crystal field bands arising from Fe^{2+} in M1 sites are similar in nearly all pyroxene group minerals; they occur in pairs around 8550 cm^{-1} and 10,200–10,700 cm^{-1} (e.g., Adams 1975). These bands have been well-studied in pyroxenes such as $CaFe^{2+}Si_2O_6$ (hedenbergite), where Fe^{2+} occurs solely in the M1 site because the M2 site is filled by Ca^{2+}. The M1 bands are very weak because they arise from Fe^{2+} in such a symmetrical site (e.g., Straub et al. 1991).

The M2 site is far more distorted than M1 because the metal-to-oxygen distances around the M2 site vary considerably with composition. Because site distortion increases band intensity, pyroxene compositions in which Fe^{2+} substitutes for Ca^{2+} have bands that tend to dominate the spectra. Even when Fe^{2+} is concentrated in the M1 site, bands due to a small amount of Fe^{2+} in the M2 will dominate. The features are typically located at energies of 9600 cm^{-1} and 4400 cm^{-1} (1.9–2 μm) in Ca^{2+}-rich compositions (White and Keester, 1966; Burns and Huggins, 1973). When $^{[M2]}Fe^{2+} > {}^{[M2]}Ca^{2+}$, crystal field bands move to longer wavelengths. These spectral differences are readily apparent when looking at the pyroxene features found in the spectra of different types of meteorites and are used to infer the compositions of asteroids.

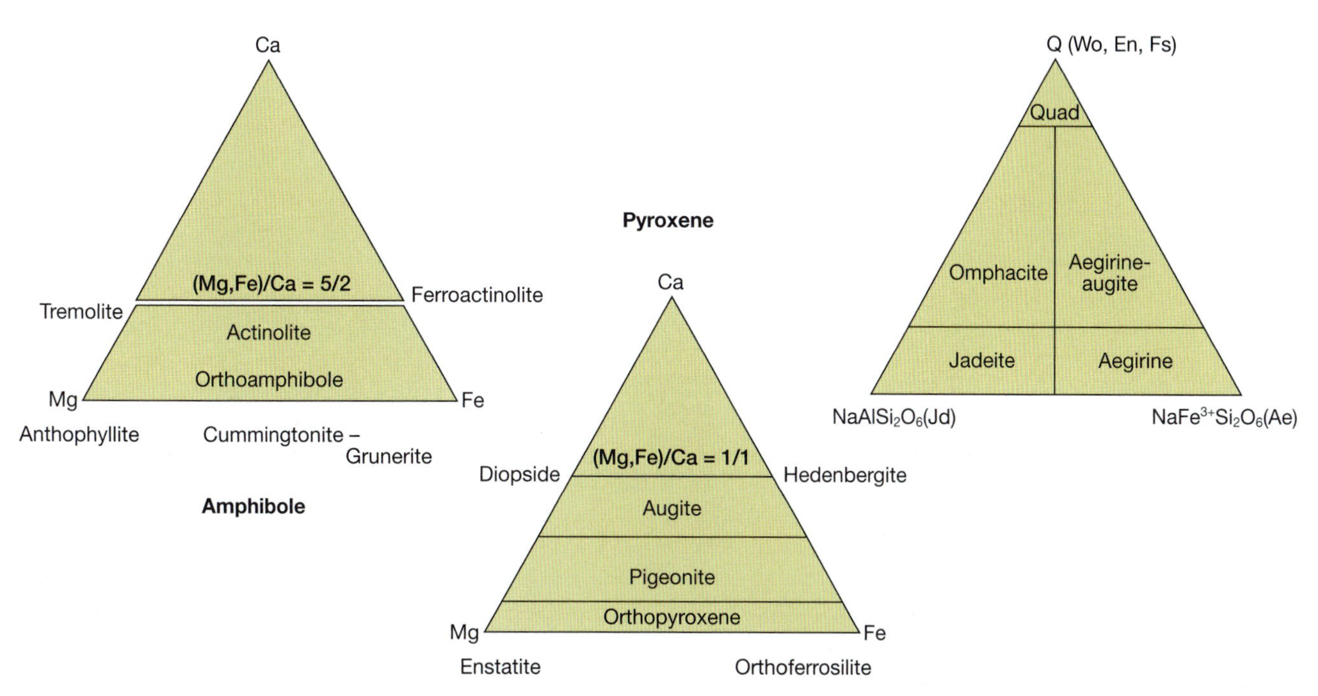

Figure 22.29. Amphibole (left) and Ca- (center) and Na- (right) pyroxene diagrams, side-by-side for clarity. Note that monoclinic analogues of the orthopyroxenes also exist, and these are aptly called clinoenstatite and clinoferrosilite.

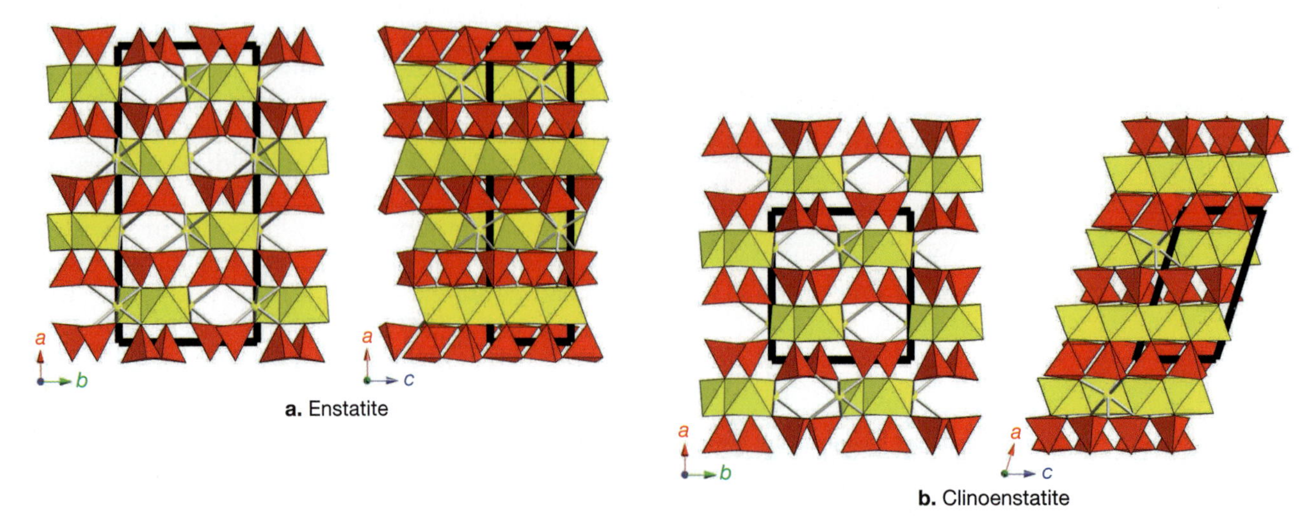

Figure 22.30. *Comparison of crystal structures of the orthorhombic pyroxene mineral enstatite, $Mg_2Si_2O_6$, and its monoclinic analogue, clinoenstatite. The unit cell is outlined in black, the Si^{4+} tetrahedra are red, the Mg^{2+} M1 octahedra are yellow, and the Mg^{2+} located in the M2 site is represented by small yellow spheres. Careful observation of the M1 site orientations is the key to distinguishing between the orthorhombic and monoclinic varieties.* ***a.*** *On the left, enstatite is projected on its* a-b *plane and on the right, its* a-c *plane; the unit cell is orthorhombic. In the* a-b *projection, the M1 octahedra face the same direction in the top and bottom layers, but face different directions in the two middle layers. This same feature can be seen in the* a-c *projection.* ***b.*** *On the left, clinoenstatite is projected on its* a-b *plane and on the right, its* a-c *plane; this unit cell is monoclinic (this can only be seen in the* a-c *projection). In both projections, the M1 octahedra all face the same direction. Notice that the* a *cell edge for the monoclinic cell is half that of the orthorhombic cell. If you look back at Figure 22.28, you can determine the crystal system for diopside based on the orientations of its octahedra.*

Table 22.15. Pyroxene group

	Space Group	*Species*	*Formula*
Mg-Fe	$P2_1/b2_1/c2_1/a$	enstatite	$Mg_2Si_2O_6$
	$P2_1/b2_1/c2_1/a$	ferrosilite	$(Fe^{2+})_2Si_2O_6$
	$P2_1/b2_1/c2_1/a$	hypersthene†	$(Mg,Fe^{2+})_2Si_2O_6$
	$P2_1/c$	clinoenstatite	$Mg_2Si_2O_6$
	$P2_1/c$	clinoferrosilite	$(Fe^{2+},Mg)_2Si_2O_6$
	$P2_1/c$	pigeonite	$(Mg,Fe^{2+},Ca)(Mg,Fe^{2+})Si_2O_6$
Mn-Mg	$P2_1/b2_1/c2_1/a$	donpeacorite	$(Mn^{2+},Mg)MgSi_2O_6$
	$P2_1/c$	kanoite	$(Mn^{2+},Mg)_2Si_2O_6$
Ca	$C2/c$	diopside	$CaMgSi_2O_6$
	$C2/c$	hedenbergite	$CaFe^{2+}Si_2O_6$
	$C2/c$	augite	$(Ca,Na)(Mg,Fe,Al,Ti)(Si,Al)_2O_6$
	$C2/c$	johannsenite	$CaMn^{2+}Si_2O_6$
	$C2/c$	petedunnite	$Ca(Zn,Mn^{2+},Fe^{2+},Mg)\ Si_2O_6$
	$C2/c$	esseneite	$CaFe^{3+}AlSiO_6$
Ca-Na	$C2/c$, $P2/n$	omphacite	$(Ca,Na)(Mg,Fe^{2+},Al)Si_2O_6$
	$C2/c$	aegirine-augite‡	$(Ca,Na)(R^{2+},Fe^{3+})Si_2O_6$
Na	$C2/c$	jadeite	$Na(Al,Fe^{3+})Si_2O_6$
	$C2/c$	aegirine	$NaFe^{3+}Si_2O_6$
	$C2/c$	kosmochlor	$NaCr^{3+}Si_2O_6$
	$C2/c$	jervisite	$(Na,Ca,Fe^{2+})(Sc,Mg,Fe^{2+})Si_2O_6$
	$C2/c$	namansilite	$NaMn^{3+}Si_2O_6$
	$C2/c$	natalyite	$Na(V^{3+},Cr^{3+})Si_2O_6$
Li	$C2/c$	spodumene	$LiAlSi_2O_6$

†Not an IMA-approved mineral species name, but commonly used to describe intermediate orthopyroxenes.
‡Also not an IMA-approved term, but frequently used in the literature.

Other Single-Chain Silicates

As we've already noted, the pyroxene structure simply cannot accommodate having more than half its octahedral cations as big as size of Ca^{2+}(i.e., substitution of Ca^{2+} into the M1 site cannot occur). However, more room can be made by making a fundamental change: instead of a straight repeat pattern, the chains are only straight for a certain number of tetrahedra as shown in Figure 22.33. This is the mineral group **wollastonite**, in which the chains are cross-linked by larger cations in two 6-coordinated sites (designated M1 and M2) and one 7-coordinated M3 site. Other types of repeats with multiplicities of 4, 5, 6, and 7 repeats are also possible. For example, the **rhodonite** group minerals contain Mn^{2+}, an element not easy to fit into the pyroxene structure, and have a 5-repeat pyroxenoid chain. **Pyroxmangite** has a 7-repeat chain. These configurations give rise to what are sometimes called the **pyroxenoids**—pyroxene-like structures with lower (triclinic) symmetry.

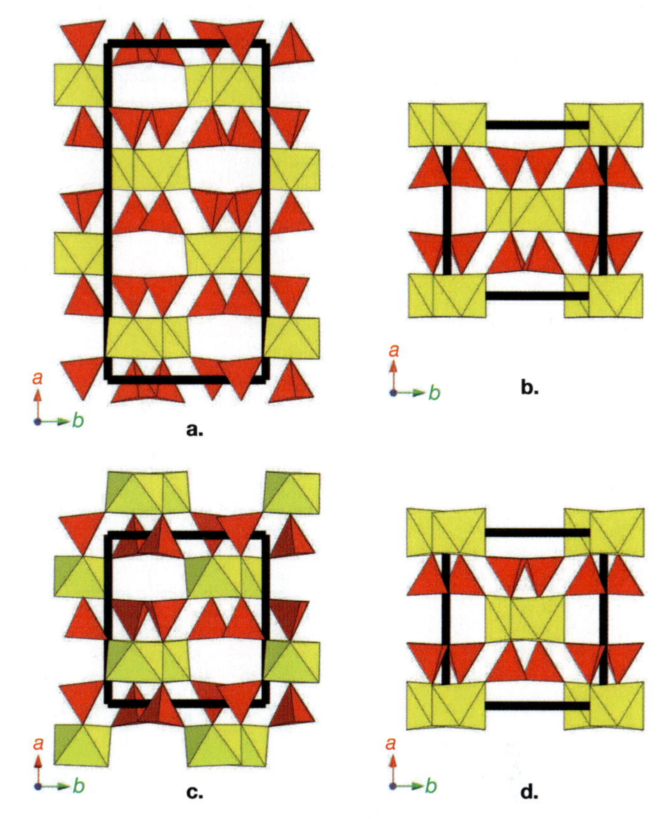

Figure 22.31. The pyroxenes have four space groups that are a function of the stacking sequences. They are best seen by looking at the a-b *projections of the M1 octahedra, observing which way they are "swimming" (refer to the Swimming Octahedra inset below Figure 22.23), and by examining the outlined unit cells.* ***a.*** *For the orthorhombic space* $P2_1/b2_1/c2_1/a$*, the lowest layer of octahedra are "swimming" away, while the next two are swimming toward you, with the fourth in the same orientation as the first. Thus there is a sequence of "2 away and 2 toward" swans, creating an orthorhombic cell.* ***b.*** *For the orthorhombic space* $P2_1/b2_1/c2_1/n$*, the first layer is swimming toward us, the next layer away, and the third has the same orientation as the first. Thus there is a "1 away and 1 toward" sequence, and the repeat is half the distance of* $P2_1/b2_1/c2_1/a$*.* ***c.*** *For the monoclinic space group* $P2_1/c$*, all of the octahedra are swimming away from us, making this a monoclinic unit cell.* ***d.*** *C2/c resembles* $P2_1/c$ *in that all of the octahedra have the same orientation, but this time they are swimming toward us. To distinguish* $P2_1/c$ *from C2/c, study the* a-b *projections available on the MSA website and note that the octahedral chains are offset in the +c direction for* $P2_1/c$ *and the –c direction for C2/c. This also results in a* β *angle of 72° for* $P2_1/c$ *and 106° for C2/c; these are approximate complements of each other.*

Ring Silicates

Continuing with our method of gradually depolymerizing silicate structures, we come next to the ring silicates, also known as cyclosilicates (Table 22.17). These result when you bend a chain of silicate tetrahedra into rings. You can create rings with 3, 4, 6, 8, 9, or even 12 members. We will focus here on only a few examples of these structures. For more information, please see Deer at al. (1986) and the *Reviews in Mineralogy*, vol. 33.

Table 22.16. Other Single-Chain Silicates

Group	Mineral	Formula
Wollastonite	wollastonite-1A	$CaSiO_3$
	wollastonite-2M	$CaSiO_3$
	wollastonite-3A-4A-5A-7A	$CaSiO_3$
	bustamite	$(Ca,Mn^{2+})_3Si_3O_9$
	ferro-bustamite	$Ca_{1.67}Fe^{2+}_{0.33}Si_2O_6$
	pectolite	$NaCa_2Si_3O_8(OH)$
	sérandite	$NaMn^{2+}_2Si_3O_8(OH)$
	cascandite	$Ca(Sc,Fe^{2+})Si_3O_8(OH)$
	denisovite	$(K,Na)Ca_2Si_3O_8(F,OH)$
Rhodonite	rhodonite	$CaMn^{2+}_4Si_5O_{15}$
	babingtonite*	$Ca_2(Fe^{2+},Mn)Fe^{3+}Si_5O_{14}(OH)$
	manganbabingtonite	$Ca_2(Mn, Fe^{2+})Fe^{3+}Si_5O_{14}(OH)$
	nambulite	$(Li,Na)Mn^{2+}_4Si_5O_{14}(OH)$
	natronambulite	$(Na,Li)Mn^{2+}_4Si_5O_{14}(OH)$
	marsturite	$NaCaMn^{2+}_3Si_5O_{14}(OH)$
	lithiomarsturite	$LiCa_2Mn^{2+}_2HSi_5O_{15}$
	scandiobabingtonite	$Ca_2(Fe^{2+},Mn)ScSi_5O_{14}(OH)$
Pyroxmangite	pyroxmangite	$Mn(Mn,Fe^{2+})_6Si_7O_{21}$
	pyroxferroite	$(Ca,Fe^{2+})(Fe^{2+},Mn)_6(Si_7O_{21})$

*Babingtonite is the state mineral of Massachusetts.

Table 22.17. Selected Ring Structures (exclusive of tourmaline)		
	brannockite	$KLi_3Sn_2Si_{12}O_{30}$
	chayesite	$KMg_2(Mg,Fe^{3+},Fe^{2+})_3Si_{12}O_{30}$
	darapiosite	$KNa_2(Mn,Zr,Y)_2(Li,Zn)_3Si_{12}O_{30}$
	eifelite	$KNa_2(MgNa)Mg_3Si_{12}O_{30}$
	merrihueite	$KNa(Fe^{2+},Mg)_5Si_{12}O_{30}$
	osumilite-(Fe)	$(K,Na)(Fe^{2+},Mg)_2(Al,Fe^{3+})_3(Si,Al)_{12}O_{30}$
	osumilite-(Mg)	$KMg_2Al_3(Si_{10}Al_2)O_{30}$
	poudretteite	$KNa_2B_3Si_{12}O_{30}$
	sugilite	$KNa_2(Fe^{2+},Mn^{2+},Al)_2Li_3Si_{12}O_{30}$
Milarite-osumilite	yagaiite	$(Na,K)_{1.5}Mg_2(Al,Mg)_3(Si,Al)_{12}O_{30}$
	dusmatovite	$K(K,Na,\square)(Mn^{2+},Y,Zr)_2(Zn,Li)_3Si_{12}O_{30}$
	milarite	$KCa_2AlBe_2Si_{12}O_{30}\cdot 0.5H_2O$
	sogdianite	$KNa(Zr,Fe^{3+},Ti,Al)_2Li_3Si_{12}O_{30}$
	roedderite	$KNa(Mg,Fe^{2+})_5Si_{12}O_{30}$
	berezanskite	$KLi_3Ti_2Si_{12}O_{30}$
	shibkovite	$K(Ca,Mn,Na)_2(K_{2-x}\square_x)_2Zn_3Si_{12}O_{30}$ where $x \sim 0.8$
	trattnerite	$(Fe^{2+},Mg)_2(Mg,Fe^{2+})_3Si_{12}O_{30}$
	almarudite	$(K,\square,Na)_2(Mn,Fe^{2+},Mg)_2(Be,Al)_3Si_{12}O_{30}$
	IMA-2003-045a	$K(Sc,Ca)_2Be_3Si_{12}O_{30}$
	eudialyte	$Na_{15}Ca_6(Fe^{2+},Mn^{2+})_3Zr_3(Si,Nb)(Si_{25}O_{73})(O,OH,H_2O)_3(Cl,OH)_2$
	alluaivite	$Na_{19}(Ca,Mn^{2+})_6(Ti,Nb)_3Si_{26}O_{74}Cl\cdot 2H_2O$
	kentbrooksite	$(Na,REE)_{15}(Ca,REE)_6Mn^{2+}{}_3Zr_3NbSi_{25}O_{74}F_2$
	khomyakovite	$Na_{12}Sr_3Ca_6Fe_3Zr_3W(Si_{25}O_{73})(O,OH,H_2O)_3(OH,Cl)_2$
	manganokhomyakovite	$Na_{12}Sr_3Ca_6Mn^{2+}{}_3Zr_3W(Si_{25}O_{73})(O,OH,H_2O)_3(OH,Cl)_2$
	oneillite	$Na_{15}Ca_3Mn_3Fe_3Zr_3Nb(Si_{25}O_{73})(O,OH,H_2O)_3(OH,Cl)_2$
	ferrokenbrooksite	$Na_{15}Ca_6(Fe,Mn)_3Zr_3Nb(Si_{25}O_{73})(O,OH,H_2O)_3(OH,F,Cl)_2$
	ikranite	$(Na,H_3O^+)_{15}(Ca,Mn,REE)_6Fe_2Zr_3(\square,Zr)(\square,Si)Si_{24}O_{66})(O,OH)_6\cdot 2\text{-}3H_2O$
	feklichevite	$Na_{11}Ca_9(Fe^{3+},Fe^{2+})_2Zr_3Nb(Si_{25}O_{73})(O,OH,Cl,H_2O)_5$
Eudialyte	taseqite	$Na_{12}Sr_3Ca_6Fe_3Zr_3Nb(Si_{25}O_{73})(O,OH,H_2O)_3Cl_2$
	carbokenbrooksite	$(Na,\square)_{12}(Na,Ce)_3Ca_6Mn_3Zr_3Nb(Si_{25}O_{73})(OH)_3CO_3\cdot H_2O$
	zirsilite-(Ce)	$(Na,\square)_{12}(Ce,Na)_3Ca_6Mn_3Zr_3Nb(Si_{25}O_{73})(OH)_3CO_3\cdot H_2O$
	labyrinthite	$(Na,K,Sr)_{35}Ca_{12}Fe_3Zr_6Ti(Si_{51}O_{144})(O,OH,H_2O)_9Cl_3$
	aqualite	$(H_3O^+)_8(Na,K,Sr)_5Ca_6Zr_3Si_{26}O_{66}(OH)_9Cl$
	raslakite	$Na_{15}Ca_3Fe_3(Na,Zr)_3Zr_3(Si,Nb)(Si_{25}O_{73})(OH,H_2O)_3(Cl,OH)$
	georgbarsanovite	$Na_{12}(Mn,Sr,REE)_3Ca_6Fe_3Zr_3NbSi_{25}O_{76}Cl_2\cdot H_2O$
	IMA-2004-026	$Na_{12}(Ce,REE,Sr)_3Ca_6Mn_3Zr_3WSi_{25}O_{73}(OH)_3CO_3\cdot H_2O$
	mogovidite	$Na_9(Ca,Na)_6Ca_6Fe_2Zr_3\square Si_{25}O_{72}(CO_3)(OH)$
	bazirite	$BaZrSi_3O_9$
Benitoite	benitoite	$BaTi\ Si_3O_9$
	pabstite	$Ba(Sn,Ti)\ Si_3O_9$
	wadeite	$K_2ZrSi_3O_9$
	beryl	$Be_3Al_2Si_6O_{18}$
	bazzite	$Be_3Sc_2Si_6O_{18}$
	indialite	$Mg_2Al_4Si_5O_{18}$
Beryl/Cordierite	stoppaniite	$Fe^{3+}{}_3(Mg,Fe^{2+})Na(Be_6Si_{12}O_{36})\cdot 2H_2O$
	pezzottaite*	$CsBe_2LiAl_2Si_6O_{18}$
	cordierite	$Mg_2Al_4Si_5O_{18}$
	sekaninaite	$(Fe^{2+},Mg)_2Al_4Si_5O_{18}$
	dioptase	$Cu_6Si_6O_{18}\cdot 6H_2O$

*Unlike the other beryl group members in which the three Be cations occupy three symmetrically-related positions, Be_2Li ordering in pezzottaite changes the symmetry to rhombohedral.

The **milarite** group of minerals contains double 6-membered rings of $Al_2Si_{10}O_{30}$ tetrahedra that are stacked in pairs along the *c* axis to create hexagonal structures, sharing apical oxygens. The double rings are then linked to each other laterally by Ca^{2+} (or Na^{1+}, Mg^{2+}, Fe^{2+}, Ti^{4+}, Zr^{4+}, Sn^{4+}) octahedra and

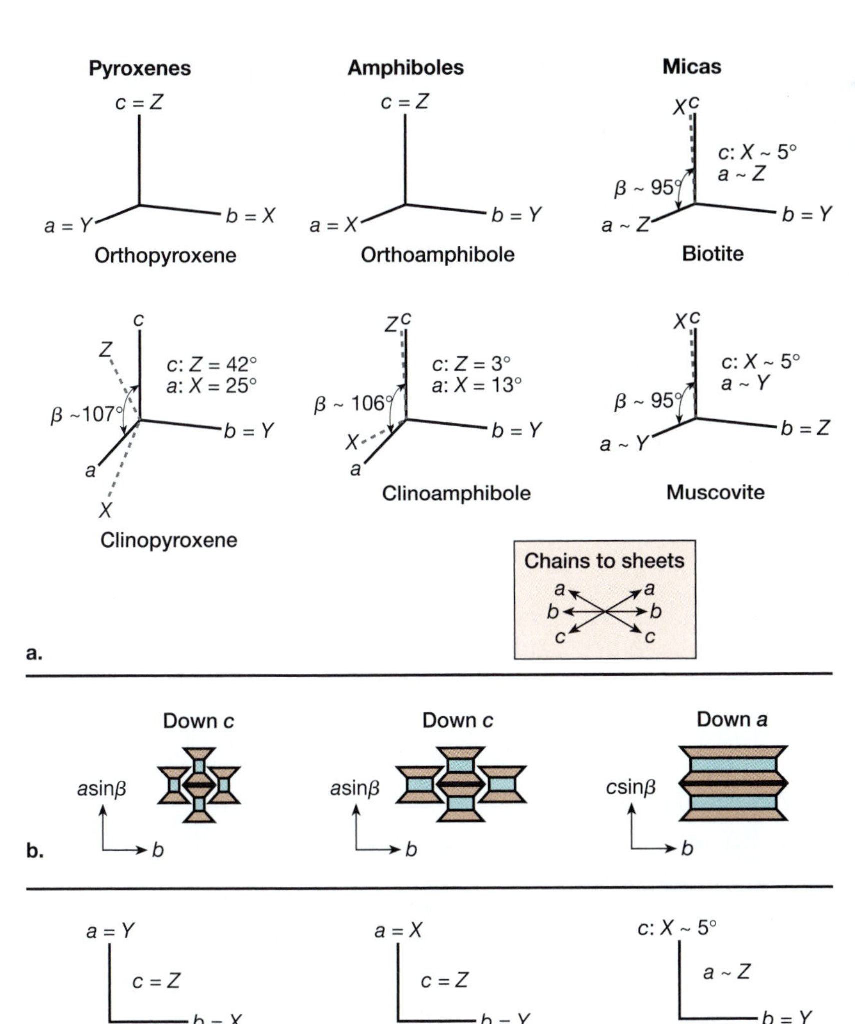

Figure 22.32. *Schematic representation of the optical orientations and their relationships to the main structural units in the biopyriboles (adapted from Dyar et al., 2002).* ***a.*** *Representative optical orientations for some minerals in the pyroxene, amphibole, and mica groups. This illustration is meant to show the similarities and differences between these closely-related structural groups. Note that the* a *and* c *crystallographic axes switch between the chain silicates and layer silicates, while* b *maintains the same orientation with regards to the structure. Solid lines represent the crystallographic axes (*a, b, c*) and dashed lines the optical directions (*X, Y, Z*).* ***b.*** *Schematic representation of the "I-beam" structures of pyroxenes, amphiboles, and layer silicates projected down the* c *axis for chain silicates and* a *axis for the layer silicates. Notice how the* b *axis is always across the octahedral strips.* ***c.*** *Two-dimensional projections down* c *for chain silicates and down* a *for the layer silicates. Again notice that* b *always corresponds to the octahedral strip, and for four of the six sketches* b = Y*, while* b = X *for orthopyroxene and* b = Z *for muscovite.*

(Be^{2+}, Al, Li^{1+}, Mg^{2+}, Fe^{2+}) tetrahedra. The stacked double-rings create channels that contain 12-coordinated cations like K^{1+}, Na^{1+}, and Mg^{2+} as well as H_2O and $(H_3O)^{1+}$. The structure resembles that of beryl (see below) except that it has double rather than single 6-membered rings (Figure 22.34a).

Beryl and cordierite group minerals along with dioptase all contain 6-membered rings of Si^{4+} tetrahedra with various permutations. As you'll recall from Chapter 6, **beryl** ($Al_2Be_3Si_6O_{18}$) is composed of six 6-membered tetrahedral rings that are linked together by two AlO_6 octahedra and three BeO_4 tetrahedra (Figure 6.10a). Beryl is especially valuable; its varieties result from minor substitutions of transition metals (Figure 22.35). **Cordierite** ($(Mg,Fe^{2+})_2(Al_2Si)Al_2Si_4O_{18}$) is similar to beryl except that the tetrahedra are linked by Mg^{2+} octahedra and $Al_2^{3+}Si^{4+}$ tetrahedra (Figure 6.10b). Cordierite is often hydrous, and you may sometimes see its formula written as $(Mg,Fe^{2+})_2(Al_2Si)Al_2Si_4O_{18} \cdot nH_2O$. In both beryl and cordierite, all four corners of the Si^{4+} tetrahedra are shared. The 6-membered rings in **dioptase** ($Cu_6Si_6O_{18}{\cdot}6H_2O$) are distorted to $\bar{3}$ symmetry, so that the rings are too narrow to allow H_2O molecules to be lost or cations exchanged, as was possible, for example, in zeolites. Cu^{2+} is in octahedral coordination with four O^{2-} and two H_2Os; these octahedra share edges in pairs and the pairs then link the Si rings together. The resultant structure (Figure 22.34b) also has 6-membered rings of H_2O.

Benitoite contains 3-membered rings of Si_3O_9, linked into a framework by octahedral containing

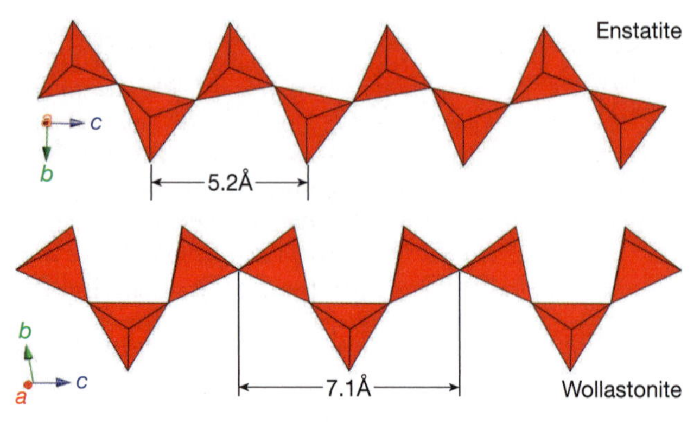

Figure 22.33. Comparison of the polymerized silicon tetrahedral chains in enstatite (top) and wollastonite (bottom). For enstatite (and all the pyroxenes), the repeat along c *is approximately 5.2Å, or two tetrahedra, while for wollastonite the repeat is approximately 7.3Å. In the pyroxene chain, the bases of the tetrahedra all lie in the same plane, while in wollastonite they are not coplanar. Notice that the pyroxene chain is not perfectly straight, but is somewhat kinked.*

Ti^{4+}, Sn^{4+}, and/or Zr^{4+}; Ba^{2+} polyhedra lie within the rings. This unusual mineral has so far been found in only two localities: San Benito, California and Ohmi, Japan. The structure is unusual because of the position of its mirror planes, and is the only mineral known to form crystals in the $\overline{6}m2$ point group (Figure 22.34c).

Eudialyte minerals contain both 3- (Si_3O_9) and 9-membered ($Si_9O_{24}(OH)_3$) rings of Si^{4+} tetrahedra in a complex structure. These Si^{4+} rings lie between 6-membered rings of unusual 4- or 5-coordinated Fe^{2+} and Mn^{2+} and edge-sharing Ca^{2+} octahedra to make double sheets. The sheets are then connected by Zr^{4+} octahedra, creating an open framework with large spaces in the centers of the 9-fold rings that can contain Na^{1+}, Ca^{2+}, or REE in 6-11-fold sites, F^{1-}, Cl^{1-}, O^{2-}, $(OH)^{1-}$, H_2O in various coordination, and Si^{4+}, Nb^{5+}, and/or W^{6+} in 4- or 6-coordinated sites (Figure 22.34d).

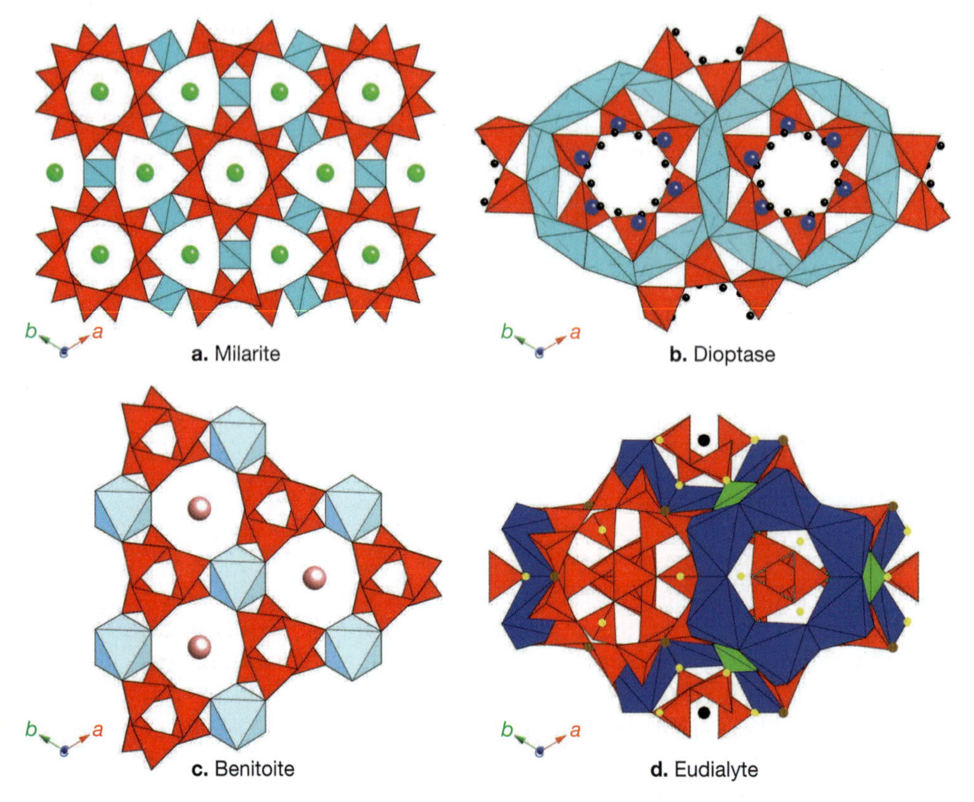

***Figure 22.34.** Milarite (**a**), dioptase (**b**), benitoite (**c**), and eudialyte (**d**) structures, all sharing different types of rings and projected on their* a-b *planes; all four are hexagonal. Si^{4+} tetrahedra are shown in red. You can use the formulas of dioptase, benitoite, and eudialyte to figure out the size of the rings, but the ring size based on milarite's formula is less obvious. **a.** Milarite, $K_2Ca_4Al_2Be_4Si_{24}O_{60}\cdot H_2O$, has a structure very similar to both beryl and cordierite, though it has double 6-membered rings (use CrystalViewer to rotate the structure to see this). Like beryl and cordierite, milarite is really a framework silicate because both Al^{3+} and Be^{2+} are in tetrahedral coordination (the teal tetrahedra), so all the tetrahedra in the structure share corners with each other. Also as in beryl and cordierite, other elements can fit into the center of the 6-membered rings, here those elements are Ca^{2+}, Na^{1+}, and H_2O. **b.** The main building unit of dioptase, $Cu_6Si_6O_{18}\cdot 6H_2O$, is 6-membered rings of silicate tetrahedral. Cu^{2+} cations in tetrahedral coordination (light blue, transparent tetrahedra) are bonded around them. All the corners of both the Si^{4+} and Cu^{2+} tetrahedra are shared; thus dioptase is also really a framework silicate. Notice that the H_2O molecules are H-bonded into the inside of the 6-membered rings. **c.** The main building unit of benitoite, $BaTiSi_3O_9$, is 3-membered Si^{4+} rings; these rings are in turn bonded to one other by Ti^{4+} octahedra (light blue). The Si^{4+} and Ti^{4+} polyhedra form 6-membered rings large enough to house big cations like Ba^{2+}. **d.** The formula and structure of eudialyte, $Na_{15}Ca_7Fe_3Zr_3Si(Si_3O_9)_2(Si_9O_{27})_2(OH)_2Cl_2$, are, well, complex! However, you can use your "mineral grammar" to dissect the formula; it has both 3- and 9-membered silicate tetrahedral rings. The structure drawing also shows a 6-membered ring of edge-sharing octahedra (blue) that contain Ca^{2+}. To gain an understanding of how complex this structure is, you might want to view it in different orientations using CrystalViewer. Also use "Atom Info" to see how the atoms bond.*

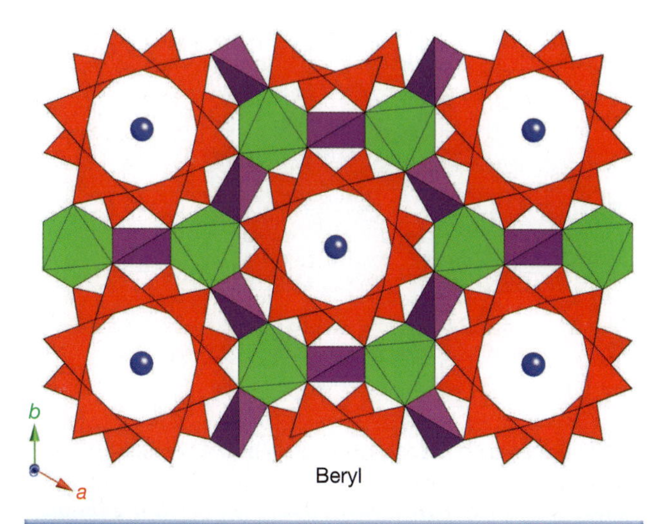

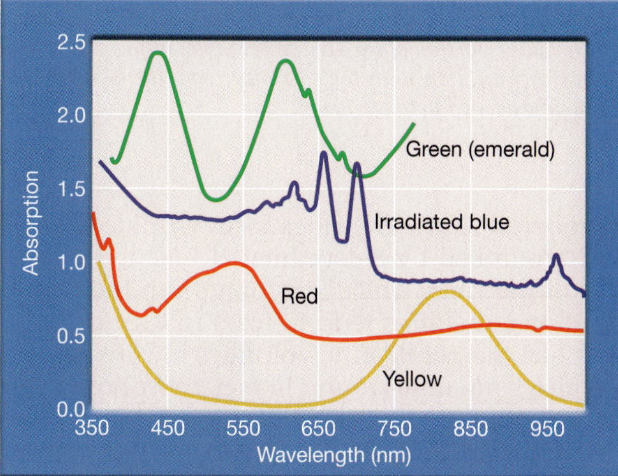

Figure 22.35. *Beryl, $Be_3Al_2Si_6O_{18}$, is sometimes considered a ring silicate with 6-membered Si^{4+} tetrahedra (red tetrahedra). It is really a framework silicate because the Be^{2+} is also in tetrahedral coordination (dark purple elongated tetrahedra) and bonds the Si^{4+} rings together so that all four O^{2-}'s are bonded to either Si^{4+} or Be^{2+}. The Al^{3+} cations are located in octahedra (green) that bond to the tetrahedra. Trace amounts of larger cations or water can occupy the centers of the rings (as shown by a blue sphere). Visible region spectra of beryl varieties are shown normalized to 1 mm thickness and then offset for clarity. Only the orientation with light vibrating perpendicular to the* c *axis is shown. Green (emerald) beryl can be caused by Cr^{3+} or V^{3+}, as seen in this sample from Miku, Zambia. Blue color can also be caused by radiation damage; shown here is a Brazilian sample colored by radiation—induced color centers involving ions such as carbonate in the* c *axis channels. The pale blue color of natural aquamarine arises from Fe^{2+}. The red color in the WahWah beryl shown here is caused by Mn^{3+}, while Fe^{3+} produces the golden yellow color of the heliodor variety (sample here is from unknown locality, Caltech collection). Data from Rossman (2005).*

Many of these ring structures should probably be called framework silicates because their structures are composed of Si^{4+} tetrahedra that share all corners. In beryl, the three Be^{2+} and six Si^{4+} cations in tetrahedral coordination result in a 1:2 ratio of tetrahedral cations to O^{2-}. In cordierite, there are four Al^{3+} and five Si^{4+} in tetrahedral coordination giving, again, a ratio of 1:2 cations:oxygens. These minerals even have open channels at the centers of their rings where larger cations or molecules can reside—just as in the zeolite group minerals (which are considered framework silicates).

Tourmaline group minerals are true ring silicates, also based upon the Si_6O_{18} ring (Figure 6.11). This structure has the general formula:

$$XY_3Z_6(BO_3)_3T_6O_{18}V_3W,$$

where X = Na^{1+}, Ca^{2+}, □, rarely K^{1+}; Y = Fe^{2+}, Mg^{2+}, Al^{3+}, Li, Fe^{3+}, Mn^{2+}, Mn^{3+}, Cr^{3+}, Ti^{4+},Cu, Zn, REE (trace); Z = Al^{3+}, Fe^{3+}, Fe^{2+}, Cr^{3+}, V^{3+}, Mn^{3+}, Ti^{4+}; T = Si^{4+}, Al^{3+}, and/or B; V = $(OH)^{1-}$, O^{2-}; and W = F^{1-}, $(OH)^{1-}$, and/or O^{2-}. The Si^{4+} tetrahedra in the 6-membered rings all have apical O's pointing toward [00$\bar{1}$], consistent with the absence of a center of symmetry. Between them are edge-sharing Y octahedra, with the BO_3 triangles between them. Z sites connect the silicate rings with the BO_3 triangles. The 9-coordinated X site is located in the hexagonal rings (Figure 22.36). There are also two different anionic sites, V and W. For more information on the tourmaline structure, see the chapters by Henry and Dutrow (1996) and Slack (1996) in the *Reviews in Mineralogy*, vol. 33.

The tourmaline group is subdivided into subgroups based on the occupancy of the X site, as shown in Figure 22.37 and Table 22.18: alkali, calcic, and vacant. Within each of these subgroups, a further subdivision can be made on the basis of the W site occupancy. This system results in the species names given in Table 22.18. Nomenclature of the tourmaline group minerals is currently undergoing revision, so keep an eye out for changes in this classification system that should be forthcoming.

Optically, tourmaline is highly pleochroic because the plane of the rings has a very different charge density than the direction perpendicular to it—and there are no cross-linking tetrahedra, as we found in beryl, to make the structure framework-like. The color of tourmaline also varies over a wide range as a function of substitutions by transition metal cations (Figure 22.38).

Disilicates

Disilicate structures result when two ("di-") SiO_4 tetrahedra pair up to form $(Si_2O_7)^{6-}$ units sometimes called "bow ties"; these minerals are also

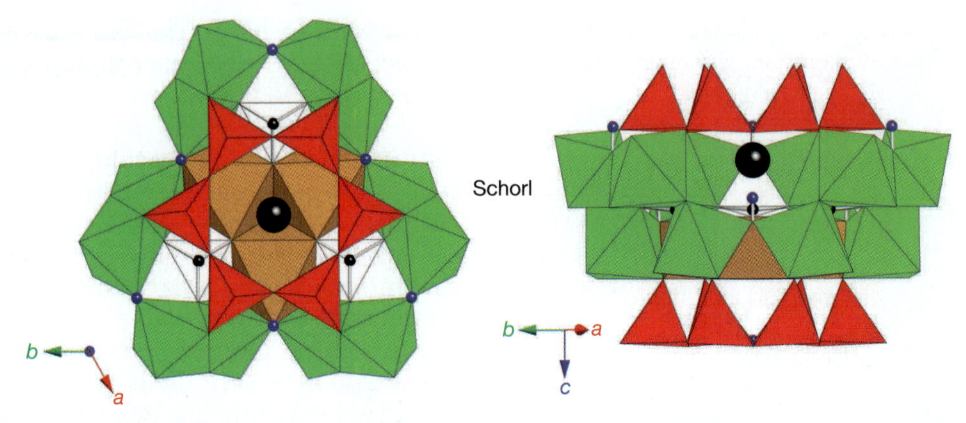

Figure 22.36. The tourmaline mineral group member schorl, $NaFe^{2+}_3Al_6Si_6O_{18}(BO_3)_3(OH)_4$, is shown here in two different projections: projected down c (left) and projected with c vertical (right). In the a-b projection, 6-membered Si^{4+} rings (red tetrahedra) are in the middle. In their center is the Na^{1+} cation (black spheres) and below it are three edge-sharing octahedra (brown) that house Fe^{2+}. Immediately on the outside of the 6-membered ring are three BO_3 groups represented by small black spheres, where the three O's are part of the Fe^{2+} and Al^{3+} octahedra (green) on the outer portion of the structure. Small blue spheres represent H^{1+} and have a single bond to an O^{2-} in an Al^{3+} octahedron. In the projection with the c axis vertical, the 6-membered rings are seen stacked on top of one another with the larger polyhedra sandwiched between them.

termed sorosilicates. There are four different Dana classes of these minerals Table 22.19, and many different structures that can form with this basic unit.

The various members of the **melilite** group have a general formula of $(Ca^{2+},Na^{1+}K^{1+})_2(Mg^{2+},Fe^{2+},Fe^{3+},Al^{3+})Si_2O_7$. Coupled substitutions among the end-members, especially $Mg^{2+} + Si^{4+} \Leftrightarrow 2Al^{3+}$, are so common that end-members are rarely found. The structures are all tetragonal. Pairs of Si_2O_7 tetrahedra in T2 sites share corners with $(Mg,Fe)O_4$ tetrahedra (T1 sites) to make 5-membered rings. The rings form wrinkled sheets and are linked by 8-coordinated Ca^{2+} in the centers of the rings. Unsurprisingly from their compositions, melilites tend to form in Ca-rich magmas like the peralkaline volcanics in Hawaii, and are also sometimes found in metamorphosed carbonate rocks.

Hemimorphite, $Zn_4Si_2O_7(OH)_2{\cdot}H_2O$, is named after its "hemimorphic" crystal habit (from the Greek *hemi*, half, plus *morphe*, form). It has tabular crystals that are pointed at only one end. Its structure is composed of double pairs of Si_2O_7

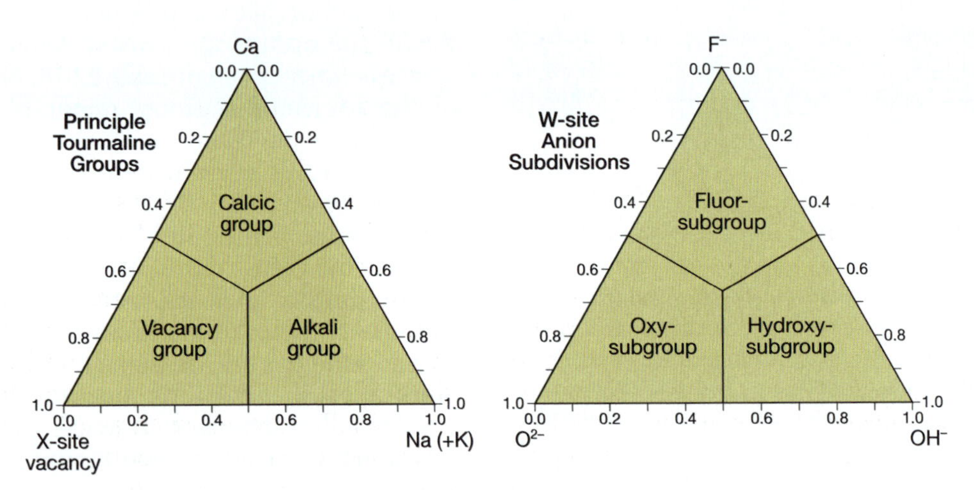

Figure 22.37. The tourmaline mineral species (with a general formula of $XY_3Z_6(BO_3)_3T_6O_{18}V_3W$) can be separated into principal groups based on the dominant occupancy of the X site, the alkali-, calcic- and vacant-tourmaline groups as shown here on the left. This subdivision is useful because the occupancy of the X site often reflects the rock type in which the tourmaline originally formed. On the right is another type of classification based on the occupancy of the W site. The dominant anion in the W site can be used to distinguish a secondary series of species with the hydroxy-, fluor- or oxy-prefix. Notice that the occurrence of O^{2-} rather than $(OH)^{1-}$ or F^{1-} in the W site requires that cations adjacent to those sites (Y and Z) provide substitutions necessary to maintain charge balance.

Subgroup	Species	Formula*	Related Species	Formula
Alkali subgroup	elbaite	$Na(Li_{1.5}Al_{1.5})Al_6(BO_3)_3Si_6O_{18}(OH)_4$	fluor-elbaite	$Na(Li_{1.5}Al_{1.5})Al_6(BO_3)_3Si_6O_{18}(OH)_3F$
			oxy-elbaite	$NaLiAl_2Al_6(BO_3)_3Si_6O_{18}(OH)_3O$
	schorl	$NaFe^{2+}{}_3Al_6(BO_3)_3Si_6O_{18}(OH)_4$	fluor-schorl	$NaFe^{2+}{}_3Al_6(BO_3)_3Si_6O_{18}(OH)_3F$
			oxy-schorl	$NaFe^{2+}{}_3Al_6(BO_3)_3Si_6O_{18}(OH)_3O$
	dravite	$NaMg_3Al_6(BO_3)_3Si_6O_{18}(OH)_4$	fluor-dravite	$NaMg_3Al_6(BO_3)_3Si_6O_{18}(OH)_3F$
			oxy-dravite	$NaMg_3Al_6(BO_3)_3Si_6O_{18}(OH)_3O$
	olenite	$NaAl_3Al_6(BO_3)_3Si_6O_{18}O_3OH$	fluor-olenite	$NaAl_3Al_6(BO_3)_3Si_6O_{18}O_3F$
	chromdravite	$NaMg_3Cr_6(BO_3)_3Si_6O_{18}(OH)_4$	fluor-chromdravite	$NaMg_3Cr_6(BO_3)_3Si_6O_{18}(OH)_3F$
			oxy-chromdravite	$NaMg_3Cr_6(BO_3)_3Si_6O_{18}(OH)_3O$
	buergerite	$NaFe^{3+}{}_3Al_6(BO_3)_3Si_6O_{18}O_3F$	hydroyxy-buergerite	$NaFe^{3+}{}_3Al_6(BO_3)_3Si_6O_{18}O_3(OH)$
	povondraite	$NaFe^{2+}{}_3Fe^{3+}{}_6(BO_3)_3Si_6O_{18}(OH)_3O$	Al-Cr-povondraite	$NaAl_3Mg_2Cr_4(BO_3)_3Si_6O_{18}(OH)_3O$
	vanadiumdravite	$NaMg_3V_6(BO_3)_3Si_6O_{18}(OH)_4$		
Liddicoatite subgroup	liddicoatite	$Ca(Li_2Al)Al_6(BO_3)_3Si_6O_{18}(OH)_3F$	hydroxy-liddicoatite	$CaLi_2AlAl_6(BO_3)_3Si_6O_{18}(OH)_4$
			oxy-liddicoatite	$Ca(Li_{1.5}Al_{1.5})Al_6(BO_3)_3Si_6O_{18}(OH)_3O$
	uvite	$CaMg_3(MgAl_5)(BO_3)_3Si_6O_{18}(OH)_3F$	hydroxy-uvite	$CaMg_3(MgAl_5)(BO_3)_3Si_6O_{18}(OH)_4$
			oxy-uvite	$CaMg_3(Mg_2Al_4)(BO_3)_3Si_6O_{18}(OH)_3O$
			ferri-uvite	$Ca(Mg,Fe^{3+}{}_2)(Mg_2Fe^{3+}{}_4)(BO_3)_3Si_6O_{18}(OH)_3O$
	hydroxy-feruvite	$CaFe^{2+}{}_3(MgAl_5)(BO_3)_3Si_6O_{18}(OH)_4$	fluor-feruvite	$CaFe^{2+}{}_3(MgAl_5)(BO_3)_3Si_6O_{18}(OH)_3F$
			oxy-feruvite	$Ca(Fe^{2+}Al_2)(MgAl_5)(BO_3)_3Si_6O_{18}(OH)_3O$
			ferri-feruvite	$Ca(Fe^{2+}Fe^{3+}{}_2)(Mg_2Fe^{3+}{}_4)(BO_3)_3Si_6O_{18}(OH)_3O$
X-site vacant subgroup	rossmanite	$(LiAl_2)Al_6(BO_3)_3Si_6O_{18}(OH)_4$	fluor-rossmanite	$(LiAl_2)Al_6(BO_3)_3Si_6O_{18}(OH)_3F$
			oxy-rossmanite	$(Li_{0.5}Al_{2.5})Al_6(BO_3)_3Si_6O_{18}(OH)_3O$
	foitite	$(Fe^{2+}{}_2Al)Al_6(BO_3)_3Si_6O_{18}(OH)_4$	fluor-foitite	$(Fe^{2+}{}_2Al)Al_6(BO_3)_3Si_6O_{18}(OH)_3F$
			oxy-foitite	$(Fe^{2+}Al_2)Al_6(BO_3)_3Si_6O_{18}(OH)_3O$
			oxy-ferri-foitite	$(Fe^{2+}Fe^{3+}{}_2)Fe^{3+}{}_6(BO_3)_3Si_6O_{18}(OH)_3O$
	magnesiofoitite	$(Mg_2Al)Al_6(BO_3)_3Si_6O_{18}(OH)_4$	fluor-magnesiofoitite	$(Mg_2Al)Al_6(BO_3)_3Si_6O_{18}(OH)_3F$
			oxy-magnesiofoitite	$(MgAl_2)Al_6(BO_3)_3Si_6O_{18}(OH)_3O$
			oxy-magnesio-ferri-foitite	$(MgFe^{3+}{}_2)Fe^{3+}{}_6(BO_3)_3Si_6O_{18}(OH)_3O$

Table 22.18. Tourmaline group

*Tourmaline formulas given here deviate slightly from those in Mandarino and Back (2004) but are consistent with formulas currently proposed to the IMA. See Hawthorne and Henry (1999) for more information.

tetrahedra linked with chains of four double $ZnO_3(OH)$ tetrahedra, creating a framework. Hemimorphite was once used as an ore of Zn^{2+}, which had the generic name of calamine; it often occurs with Zn^{2+} carbonate (smithsonite). The extracted zinc, in the form of oxide, is (still today) mixed with various iron oxides to make calamine lotion, which is used to sooth skin irritations. Hemimorphite commonly occurs with smithsonite and cerussite as a secondary alteration product in oxidized areas adjacent to sphalerite-containing zinc deposits.

Axinite group minerals are often found in low- to medium-grade regionally metamorphosed rocks. In this structure, pairs of Si_2O_7 tetrahedra share O^{2-} with pairs of BO_4 tetrahedra, creating 6-membered rings. Thus, this structure should really be considered a ring silicate. Additional Si_2O_7 tetrahedra share corners with the BO_4 tetrahedra to create planar clusters of $B_2Si_8O_{30}$. Another type of chain is formed by four small, regular $AlO_5(OH)$ octahedra (C) and M octahedra with compositions of $(Fe^{2+},Mg^{2+},Mn^{2+})O_6$. All the chains are connected by distorted $Ca^{2+}O_6$ and $CaO_5(OH)$ octahedra (A). The resultant formula is approximately $A_4M_2C_4B_2Si_8O_{30}(OH)_2$, but see Andreozzi et al. (2000) for a more precise description. The wedge-shaped "axe head" morphology gives the axinite group its name.

The **lawsonite-ilvaite** group minerals share structures with distorted Ca^{2+} (really 7-coordinated) octahedra bonded to double chains of edge-sharing $(Al,Fe^{2+},Fe^{3+})O_6$ octahedra along the *c* axis with smaller cations such as Al^{3+}, Fe^{2+}, and Fe^{3+}. The sorosilicate Si_2O_7 groups link to the sides of the octahedra. In ilvaite, the close proximity between Fe cations in adjacent, edge-sharing sites (Figure 22.39) allows electrons to hop between Fe^{2+} and Fe^{3+} in a process called **electron delocalization**. This gives rise to metallic properties like high electrical and thermal conductivity, and causes ilvaite to be opaque black in color. The

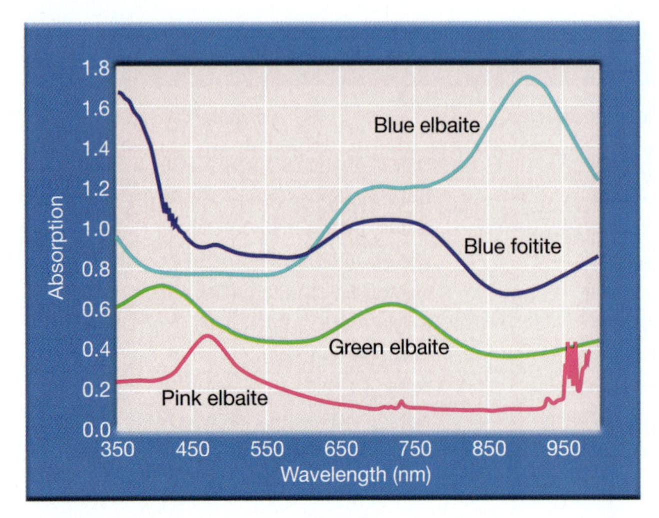

Figure 22.38. Color in tourmaline as a function of transition metal substitution, just as in the case of beryl! All spectra shown here are normalized to 100 μm thickness and then offset for clarity. Only the orientation where light is polarized perpendicular to c is shown. The blue foitite is from Schindler Mine, California; its color comes from Fe^{2+}. The green elbaite is from Afghanistan, and is colored by Fe^{2+} and Ti^{4+}-Fe^{2+} interactions. The blue elbaite from São José da Batalha, Paraíba, Brazil, is colored by Cu^{2+}; see Rossman et al. (1991). The pink crystal comes from the same locality in Brazil, but is colored by Mn^{3+} and Cu^{2+}. Data courtesy of M. Taran, Kiev.

same mechanism causes the color of magnetite and other Fe-rich silicates like riebeckite (amphibole group) and laihunite (olivine group).

Lawsonite only occurs in low temperature metamorphic rocks, where it is often found in associations with glauophane in schists. It is also an index mineral (typical of a specific grade of metamorphism) for rocks in the blueschist facies of metamorphism, such as those occurring along continent-ocean subduction zones. Lawsonite is indicative of pressures greater than 6 kb at roughly 200–400°C.

The **kornerupine group (kornerupine-prismatine)**, $(\square,Fe,Mg)(Al,Mg,Fe)_9(Si,Al,B)_5O_{21}(OH,F)$, has attracted attention for its occasional occurrence as a brown or green gemstone (the green color can be very close to that of emerald). Kornerupine also has a high refractive index, hardness of 6–7, and is yellow-green to red-brown pleochroic, making it a rare but very desirable gemstone. The structure consists of Mg^{2+} and Fe^{2+} octahedra in chains with four edge-sharing Al^{3+} octahedra parallel to the *b* axis. These chains share corners to make sheets parallel to the *a* axis. The sheets are linked by Si_2O_7 and $(Si,B,Al)_3O_{10}$ tetrahedral groups. Kornerupine and prismatine typically occur in Si-poor, Mg- and Al-rich gneisses in high-grade regionally metamorphosed Precambrian rocks, but are found in silica saturated and younger rocks as well.

Epidote group minerals are probably the most common of all the sorosilicates (c.f., *Reviews in Mineralogy* and *Geochemistry*, vol. 56), and are important rock-forming minerals because they are stable under a wide range of conditions. Many workers regard them as the low temperature, high-pressure equivalents of Ca-rich plagioclase minerals because of their petrologic usefulness. Epidotes are also among the most important silicates for understanding oxygen fugacity (the partial pressure of oxygen under which rocks crystallized) because they can contain significant amounts of Fe^{3+}. They are found in many rock types, ranging from mid-ocean ridge basalts to amphibolites to granites to Ca-rich pelitic rocks.

The many members of the epidote group share a formula that can be expressed in terms of the different types of available coordination polyhedra, as:

$$A1A2M1M2M3(SiO_4)(Si_2O_7)(O,OH)(O,F),$$

where A1 is Ca^{2+}; A2 is Ca^{2+}, REE, Pb^{4+}, or Sr^{2+}; M1 is Al^{3+}, Fe^{3+}, Mn^{2+}, or Mg^{2+}; M2 is Al^{3+}; and M3 is Al^{3+}, Fe^{3+}, Mn^{3+}, or Cr^{3+}.

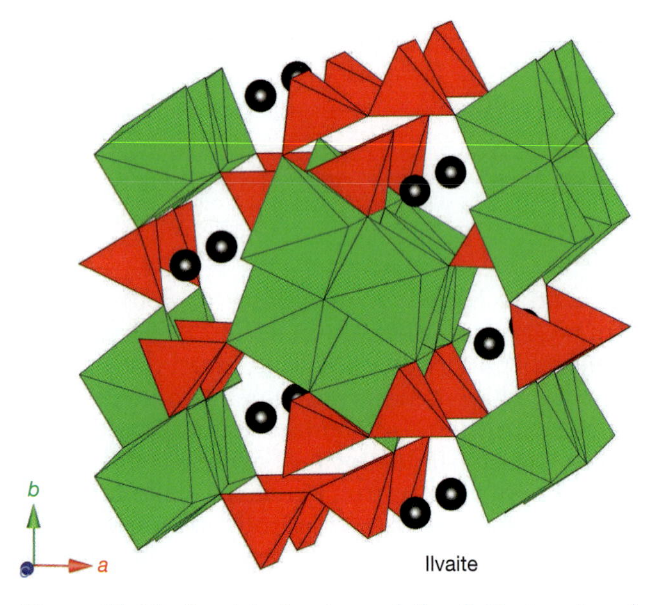

Figure 22.39. Crystal structure of the disilicate mineral ilvaite, $CaFe^{2+}Fe^{3+}{}_2O(Si_2O_7)(OH)$. The projection is tilted slightly out of the a-b plane to show the main structural features of this mineral: edge-sharing octahedral chains of Fe^{2+} and Fe^{3+} (green polyhedra). The other Fe^{2+} octahedra bond into the sides of these chains by sharing edges. With the disilicate groups (corner sharing pairs of red Si^{4+} tetrahedra), the latter Fe^{2+} octahedra cross-link the main edge-sharing chains. This mineral is very dark black because Fe^{2+} and Fe^{3+} in adjacent sites can share electrons. The black spheres represent Ca^{2+} cations occurring between the voids in the structures and further serve to bond the structure together.

There are both monoclinic and orthorhombic epidote minerals. In monoclinic species such as the species **epidote**, the structure contains alternating edge-sharing octahedral chains of M1 octahedra with M3 octahedra attached at the edges. The M2 octahedra are also in edge-sharing chains, in this case connected to SiO_4 tetrahedra on alternating sides (Figure 22.40). Si_2O_7 groups cross-link the chains (Figure 6.12), and the large A1 and A2 sites between them have 9- and 10-coordination, respectively (Franz and Liebscher, 2004).

In the orthorhombic species **zoisite**, there is only a single type of octahedral chain because the M1 and M2 octahedra are now identical. Thus, the chains of M1 and M2 octahedra extend parallel to the *b* axis, with M3 octahedra on one side only (Figure 22.40). The chains are cross-linked by SiO_4 tetrahedra and Si_2O_7 groups, with the A1 and A2 positions between them.

The edge-sharing chains of octahedra control the external morphology of this group in the same way that tetrahedral chains control the morphology of the chain silicates. Bonding *within* the sheets of cross-linked chains is very strong, but the sheets are held together only by the bonds to Ca^{2+} cations, so epidote group minerals have only one direction of perfect cleavage.

The color of epidote is variable depending on the transition metal cations occupying the M sites. The color of clinozoisite and epidote results from Fe^{3+} content, and commonly varies from amber to yellow-green to brown. Piemontite contains Mn^{3+} and is red. Allanite is colored by Fe^{3+} and radiation damage and possibly by rare earth elements. Zoisite is best known as the pleochroic gem variety tanzanite, which owes its purple-blue color to vanadium (Rossman, 2005 and Liebscher, 2004). Use of tanzanite as a gemstone is limited only by its moderate hardness (6.5–7 on the Mohs scale) and the fact that its only cleavage is along the direction of strongest pleochroism (making it difficult to facet to show the deepest colors). Great demand for gem-quality tanzanite has caused its only known locality (at Merelani, 10 miles south of the Kilimanjaro International Airport in Tanzania) to be nearly mined out, so it is anticipated that tanzanite will become increasingly valuable. Most samples are heat-treated to transform them from the natural green color to the purple-blue prized in jewelry.

Minerals of the **pumpellyite group** are common in low-grade metamorphic rocks. Their structures are related to those of the epidote group. They contain edge-sharing octahedra in chains extending along the *b* axis. These chains are connected to each other by sharing corners with SiO_4 tetrahedra and Si_2O_7 groups and by sharing edges with 7-coordinated Ca^{2+} cations. Some chains contain solely Al^{3+}, while others contain Al^{3+} in combination with Fe^{3+}, Mg^{2+}, etc. The pumpellyite formula can be expressed as $W_2XY_2(SiO_4)(Si_2O_7)(OH)_2{\cdot}H_2O$, where W is Ca^{2+}, K^{1+}, or Na^{1+}, X is Mg^{2+}, Mn^{2+}, Fe^{2+}, Al^{3+}, or Cr^{3+}; and Y is Al^{3+}, Fe^{3+}, Cr^{3+}, Mn^{3+}, or Ti^{4+} (Table 22.19).

Pumpellyite's composition is linked to its paragenesis. Fe-rich species (such as pumpellyite-

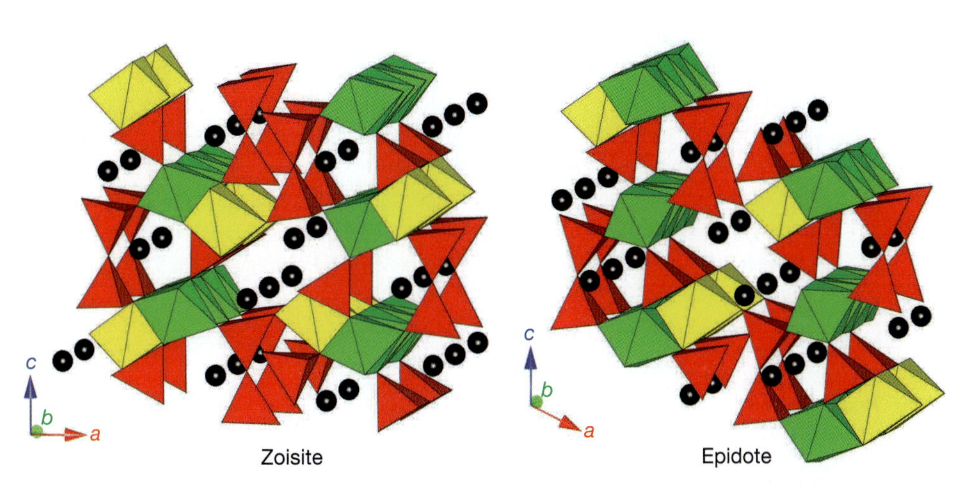

Figure 22.40. *The crystal structure of zoisite, $Ca_2Al_3O(SiO_4)(Si_2O_7)(OH)$, (left) and epidote, $Ca_2(Fe^{3+},Al)_3O(SiO_4)(Si_2O_7)(OH)$, (right). Both are disilicates, as can be seen by the presence of Si_2O_7 groups (i.e., the two red Si^{4+} tetrahedra that share one oxygen). However both also contain isolated Si^{4+} tetrahedra. The major building block for each structure is composed of edge-sharing chains of Al^{3+} octahedra that are parallel to* b*; these cause both minerals to be elongated along* b*. The major differences between zoisite and epidote occur as Fe^{3+} substitutes for Al^{3+}, causing the orthorhombic symmetry of zoisite to become monoclinic in zoisite. Both Fe^{3+} and Al^{3+} occur in the octahedra shown in the drawings; notice how the* a *and* c *directions are defined slightly differently in each mineral. The large black spheres represent Ca^{2+}, which fits into the larger voids in the structure. Other elements, especially the rare earth elements, can substitute for Ca^{2+}.*

Table 22.19. Disilicates

Group	Space group	Mineral	Formula
Melilite		akermanite	$Ca_2MgSi_2O_7$
		gehlenite	$Ca_2Al(AlSi)O_7$
		melilite	$(Ca,Na)_2(Al,Mg)(Si,Al)_2O_7$
		okayamalite	$Ca_2B_2SiO_7$
		hemimorphite	$Zn_4Si_2O_7(OH)_2\cdot(H_2O)$
Axinite		ferroaxinite	$Ca_2Fe^{2+}Al_2BO(OH)(Si_2O_7)_2$
		magnesio-axinite	$Ca_2MgAl_2BO(OH)(Si_2O_7)_2$
		manganoaxinite	$Ca_2MnAl_2BO(OH)(Si_2O_7)_2$
		tinzenite	$CaMn^{2+}{}_2Al_2BO(OH)(Si_2O_7)_2$
Lawsonite-ilvaite		lawsonite	$CaAl_2Si_2O_7(OH)_2\cdot H_2O$
		hennomartinite	$SrMn_2Si_2O_7(OH)_2\cdot H_2O$
		ilvaite	$CaFe^{3+}Fe^{2+}{}_2OSi_2O_7(OH)$
		noelbensonite	$BaMn_2Si_2O_7(OH)_2\cdot H_2O$
		itoigawaite	$SrAl_2Si_2O_7(OH)_2\cdot H_2O$
		IMA2002-016	$CaFe^{2+}Fe^{3+}(Mn,Fe^{2+})(Si_2O_7)O(OH)$
		kornerupine	$(\square,Mg,Fe^{2+})(Al,Mg,Fe^{3+})_9(Si,Al,B)_5(O,OH,F)_{22}$
Epidote	$P2_1/m$	allanite-(Ce)	$(Ca,Ce,La)_2(Al,Fe^{2+},Fe^{3+})_3(SiO_4)_3(OH)$
	$P2_1/m$	allanite-(La)	$(La,Ce,Ca)_2(Al,Fe^{2+},Fe^{3+})_3(SiO_4)_3(OH)$
	$P2_1/m$	allanite-(Y)	$(Y,Ce,Ca)_2(Al,Fe^{3+})_3(SiO_4)_3(OH)$
	$P2_1/m$	clinozoisite	$Ca_2Al_3(SiO_4)_3(OH)$
	$P2_1/m$	dissakisite-(Ce)	$Ca(Ce,La)MgAl_2(SiO_4)_3(OH)$
	$P2_1/m$	dollaseiite-(Ce)	$CaCeMg_2AlSi_3O_{11}(F,OH)_2$
	$P2_1/m$	epidote	$Ca_2Al_2(Fe^{3+},Al)(SiO_4)_3(OH)$
	$P2_1/m$	hancockite	$(Pb,Ca,Sr)_2(Al,Fe^{3+})_3(SiO_4)_3(OH)$
	$P2_1/m$	khristovite-(Ce)	$(Ca,REE)REE(Mg,Fe^{2+})AlMn^{2+}Si_3O_{11}(OH)(F,O)$
	$P2_1/m$	mukhinite	$Ca_2Al_2V^{3+}(SiO_4)_3(OH)$
	$P2_1/m$	piemontite	$Ca_2(Al,Mn^{3+},Fe^{3+})_3(SiO_4)_3(OH)$
	$P2_1/m$	strontiopiemontite	$CaSr(Al,Mn^{3+},Fe^{3+})_3Si_3O_{11}O(OH)$
	$P2_1/m$	androsite-(La)	$(Mn,Ca)(La,Ce,Ca,Nd)AlMn^{3+}Mn^{2+}(SiO_4)(Si_2O_7)O(OH)$
	$P2_1/m$	ferriallanite-(Ce)	$CaCeFe^{3+}AlFe^{2+}(SiO_4)(Si_2O_7)O(OH)$
	$P2_1/m$	tweeddillite	$CaSr(Mn^{3+},Fe^{3+})_2Al(SiO_4)_3(OH)$
	$P2_1/m$	gatelite-(Ce)	$(Ca,Ce,La,Na)_4[Al_2(Al,Mg)(Mg,Fe,Al)]_4(SiO_4)_3(Si_2O_7)(O,F,OH)_3$
	$P2_1/m$	niigataite	$CaSrAl_3(Si_2O_7)(SiO_4)O(OH)$
	$P2_1/m$	vastmanlandite-(Ce)	$Ce_3CaMg_2Al_2Si_5O_{19}(OH)_2F$
	$P2_1/m$	IMA2002-049	$(Mn,Ca)(Ce,La,Ca,Nd)AlMn^{3+}Mn^{2+}(SiO_4)(Si_2O_7)O(OH)$
	$P2_1/m$	dissakisite-(La)	$(Ca,Fe^{2+},Th)(La,Ce,Nd,Ca)(Al,Cr,Ti)_2(Mg,Fe,Al)(SiO_4)_3(OH,F)$ with La>Ce
	$P2_1/m$	IMA2004-015	$(Mn,Ca)(REE)VAlMn(SiO_4)(Si_2O_7)O(OH)$
	$P2_1/n2_1/m/2_1/a$	zoisite	$Ca_2Al_3Si_3O_{12}(OH)$
Pumpellyite		julgoldite-(Fe^{2+})	$Ca_2(Fe^{2+},Fe^{3+})(Fe^{3+},Al)_2Si_3(O,OH)_{14}$
		julgoldite-(Fe^{3+})	$Ca_2Fe^{3+}(Fe^{3+},Al)_2(SiO_4)(Si_2O_7)(O,OH)_2\cdot H_2O$
		okhotskite-(Mg)	$Ca_2(Mg,Mn^{2+})(Mn^{3+},Al,Fe^{3+})Si_3O_{10}(OH)_4$
		okhotskite-(Mn^{2+})	$Ca_2(Mn^{2+},Mg)(Mn^{3+},Al,Fe^{3+})Si_3O_{10}(OH)_4$
		pumpellyite-(Fe^{2+})	$Ca_2(Fe^{2+},Fe^{3+})(Al,Fe^{3+})_2Si_3(O,OH)_{14}$
		pumpellyite-(Fe^{3+})	$Ca_2(Fe^{3+},Mg)(Al,Fe^{3+})_2Si_3(O,OH)_{14}$
		pumpellyite-(Mg)	$Ca_2(Mg,Al)Al_2Si_3(O,OH)_{14}$
		pumpellyite-(Mn^{2+})	$Ca_2(Mn^{2+},Mg)(Al,Mn^{3+})_2Si_3(O,OH)_{14}$
		shuiskite	$Ca_2(Mg,Al)(Cr,Al)_2(Si,Al)_3(O,OH)_{14}$
Vesuvianite		vesuvianite	$Ca_{19}(Al,Mg,Fe)_{13}Si_{18}O_{68}(O,OH,F)_{10}$
		wiluite	$Ca_{19}(Al,Mg,Fe,Ti)_{13}(B,Al,\square)_5Si_{18}O_{68}(O,OH)_{10}$
		fluorvesuvianite	$Ca_{19}(Al,Mg)_{13}Al_4(SiO_4)_{10}(Si_2O_7)_4(F,OH)_{10}$
		manganvesuvianite	$Ca_{19}Mn(Al,Mn,Fe)_{10}(Mg,Mn)_2Si_{18}O_{69}(O,OH)_9$

(Fe^{2+})) are very common in low-pressure rocks such as mid-ocean ridge basalts that have been metamorphosed. The Mg-rich species (e.g., pumpellyite-(Mg)) are more common in blueschist facies metamorphic rocks, often occurring with lawsonite. Under these conditions, the pumpellyite group includes index minerals characteristic of temperatures from 250–350°C and pressures from 2–7 kb.

Minerals of the **vesuvianite** group, including vesuvianite and wiluite, are also closely related to epidote and pumpellyite. The general formula is $X_{19}Y_{13}Z_{18}T_{0\text{-}5}O_{68}W_{10}$, where X is Ca^{2+}, Na^{1+}, Pb^{2+}, Sb^{3+}; Y can be Al^{3+}, Mg^{2+}, Fe^{3+}, Fe^{2+}, Ti^{4+}, Mn^{2+}, Cu^{2+}, or Zn^{2+}; Z is Si^{4+}; T is B^{3+}; and W is $(OH)^{1-}$, F^{1-}, or O^{2-} (Groat et al., 1992). The primary substitutions in the structure occur at the Y site, where the dominant exchange can be written as $(Mg + Fe^{2+} + Mn^{2+}) + Ti^{4+} \leftrightarrow 2(Al^{3+} + Fe^{3+})$. Wiluite contains up to 4 wt% B_2O_3; the B^{3+} is incorporated primarily by the coupled substitution $B^{3+} + Mg^{2+} \leftrightarrow 2H^{1+} + Al^{3+}$.

Groat et al. (1992) also reported that some vesuvianite analyses gave sums of Y cations significantly in excess of 13, the number of sites available to accommodate the cations in the current model of the vesuvianite structure. Such crystals show an additional site, T(1), that is tetrahedrally coordinated by four oxygen anions. This site is occupied by Al^{3+} or Fe^{3+}, which replaces two H^{1+} at adjacent H(1) positions. Local bond-valence arguments show that the substitution is accompanied by the incorporation of a vacancy at an adjacent X(3) site.

Vesuvianite is commonly found in calc-silicate rocks and skarn deposits, which result from metamorphism and metasomatism of carbonate rocks in both contact aureoles and on a regional scale. It is commonly associated with Ca- and Mn-rich garnets (andradite, grossular, spessartine) and with pyroxenoids like wollastonite and rhodonite.

Orthosilicates

The orthosilicates (also called nesosilicates or island silicates) represent the far opposite end of the polymerization spectrum from framework silicates, with structures composed of isolated $(SiO_4)^{4-}$ tetrahedra interspersed with other cations for charge balance. In this group of minerals (Table 22.20), unlike those we have previously studied, substitutions of larger cations (such as Al^{3+}, Fe^{3+}, and B^{3+}) for Si^{4+} almost never occur. Because the SiO_4 unit has a net charge of –4, typically two cations are needed for charge balance, though many variations are possible (such as a single Zr^{4+} cation).

Phenakite group minerals have close-packed structures of oxygen anions with Be^{2+}, Zn^{2+}, Li^{1+}, Al^{3+}, and/or Si^{4+} in the tetrahedral interstices. This results in a framework of corner-sharing BeO_4 and SiO_4 tetrahedra in which each oxygen is shared between two Be^{2+} and one Si^{4+} tetrahedra. The structure has 4- and 6-membered channels parallel to *c*. Phenakite, Be_2SiO_4, occurs in granitic pegmatites that have abundant Be^{2+}, and also in schists derived from them. The Zn-polymorph, willemite (Zn_2SiO_4) is found oxidized zones around zinc deposits, while the third species, eucryptite ($LiAlSiO_4$) is found in pegmatites. In eucryptite, Li^{1+} may occur in the Si^{4+} site, while Al^{3+} and Si^{4+} are randomly distributed in the Be^{2+} site.

Olivine is a particularly important mineral group because of its dominance in the Earth's upper mantle, where the Mg-Fe^{2+} compositions control many of the geophysical properties of the interior, such as heat transfer, electrical conductivity, and seismic boundaries. Recall from Chapter 20 that forsterite (Mg_2SiO_4) transforms from an olivine structure to a spinel structure in the upper mantle, and eventually to an ilmenite structure at even greater temperatures and pressures. Mg-olivine has a ubiquitous presence in silica-poor extrusive igneous rocks as well as metamorphic rocks from granulite facies terranes; the Fe-rich olivines are less common, but may be found in Fe-rich basalts, alkali granites (rarely), and metamorphosed banded iron formations. The phase relations of olivine are of great interest as we discussed in Chapter 20 (Figure 20.9). Olivine even forms beach sands on the celebrated green sand beaches on the southern tip of the Big Island in Hawaii.

Olivine has a general formula of $M2M1SiO_4$, where both the M1 and M2 sites are 6-coordinated, with M2 slightly larger than M1. Because Si^{4+} is nearly always stoichiometric, compositional variations occur in the two octahedral sites. There is continuous solid solution between the Fe_2SiO_4 (fayalite) and Mg_2SiO_4 (forsterite) end members, as we showed in Chapter 10 (see Figure 10.2), and between fayalite and tephroite (Mn_2SiO_4), which is found in association with Mn-rich sediments and ore deposits.

There is also a rare Fe^{3+} form of olivine, laihunite, that is occurs in oxidized olivines such as those from metasomatized mantle xenoliths. It has an approximate formula of $^{[M1]}[Fe^{2+}_{0.5}\square_{0.5}]$ $^{[M2]}Fe^{3+}SiO_4$. The substitution of Fe^{3+} for Fe^{2+} is charge-balanced by ordered vacancies in half of the M2 sites. A Ca^{2+}-rich olivine, monticellite

Table 22.20. Orthosilicates

Group	Series	Mineral	Formula
Phenakite		phenakite	Be_2SiO_4
		willemite	Zn_2SiO_4
		eucryptite	$LiAlSiO_4$
Olivine		fayalite	Fe_2SiO_4
		forsterite	Mg_2SiO_4
		liebenbergite	$(Ni,Mg)_2SiO_4$
		tephroite	Mn_2SiO_4
		laihunite	$Fe^{2+}Fe^{3+}{}_2(SiO_4)_2$
		monticellite	$CaMgSiO_4$
		kirschsteinite	$CaFe^{2+}SiO_4$
		glaucochroite	$CaMn^{2+}SiO_4$
Garnet	Pyralspite series	pyrope	$Mg_3Al_2(SiO_4)_3$
		almandine	$Fe^{2+}{}_3Al_2(SiO_4)_3$
		spessartine	$Mn^{2+}{}_3Al_2(SiO_4)_3$
		knorringite	$Mg_3Cr_2(SiO_4)_3$
		majorite	$Mg_3Fe^{3+}{}_2(SiO_4)_3$
		calderite	$Mn^{2+}{}_3Fe^{3+}{}_2(SiO_4)_3$
	Ugrandite series	andradite	$Ca_3Fe^{3+}{}_2(SiO_4)_3$
		grossular	$Ca_3Al_2(SiO_4)_3$
		uvarovite	$Ca_3Cr_2(SiO_4)_3$
		goldmanite	$Ca_3V^{3+}{}_2(SiO_4)_3$
		yamatoite	$(Mn^{2+},Ca)_3(V^{3+},Al)_2(SiO_4)_3$
	Schorlomite-kimzeyite series	schorlomite	$Ca_3(Ti^{4+},Fe^{3+})_2[(Si,Fe^{3+})O_4]_3$
		kimzeyite	$Ca_3(Zr,Ti)_2(Si,Al,Fe^{3+}O_4)_3$
		morimotoite	$Ca_3TiFe^{2+}(SiO_4)_3$
	Hydrogarnet series	hibschite	$Ca_3Al_2(SiO_4)_{3-x}(OH)_{4x}$, where x = 0.2–1.5
		katoite	$Ca_3Al_2(SiO_4)_{3-x}(OH)_{4x}$, where x = 1.5–3
Zircon		zircon	$ZrSiO_4$
		hafnon	$HfSiO_4$
		thorite	$(Th,U)SiO_4$
		coffinite	$U(SiO_4)_{1-x}(OH)_{4x}$
		thorogummite	$Th(SiO_4)_{1-x}(OH)_{4x}$

($CaMgSiO_4$), is found in metamorphosed limestones in contact with intrusive rocks, and also in some kimberlites and other magmas from deep source regions. Its composition tends to stay strictly that of the end member, with little substitution; with Ca^{2+} occurring in the slightly larger M2 sites. Finally, many synthetic members of the olivine group have been synthesized, though they are not technically "minerals:" Ni_2SiO_4, $CaFeSiO_4$, Co_2SiO_4, Ge_2SiO_4, $LiScSiO_4$, Ca_2SiO_4 etc.

The olivine structure (Figure 22.41) is nearly hexagonally close-packed, with half of its octahedral interstices occupied by either Mg^{2+} or Fe^{2+}, and 1/8 of the tetrahedral sites occupied by Si^{4+}. The $(Fe,Mg)O_6$ octahedra share edges to create chains along the *c* axis; the chains have two alternating, distinct types of octahedra. The M1 site is slightly smaller and flattened along its three-fold axis. The M2 site is slightly larger, and is much more distorted; it generally contains the larger cations, such as Ca^{2+} in monticellite (Figure 10.7). Divalent cations in olivine may be slightly ordered between the M1 and M2 sites. For example, Ni^{2+}, Co^{2+}, and Fe^{2+} appear to have a preference for the M1 site over M2 in Mg-bearing olivines, resulting from its higher crystal field splitting in the smaller site (see discussion of octahedral site preference in Chapter 23). Note that other minerals, including chrysoberyl (also see Chapter 23) and members of the triphylite ($LiFePO_4$)-lithiphilite ($LiMnPO_4$) series also have the olivine structure.

The physical properties of olivine change systematically as a function of composition. The size of the unit cell varies in a nearly linear relationship with the size of the cations, especially the length of the *b* axis. Refractive index, density, and $2V$ can be related to $Mg^{2+}/(Mg^{2+}+Mn^{2+}+Fe^{2+})$ (Figure 22.42). Olivine is usually colorless at the

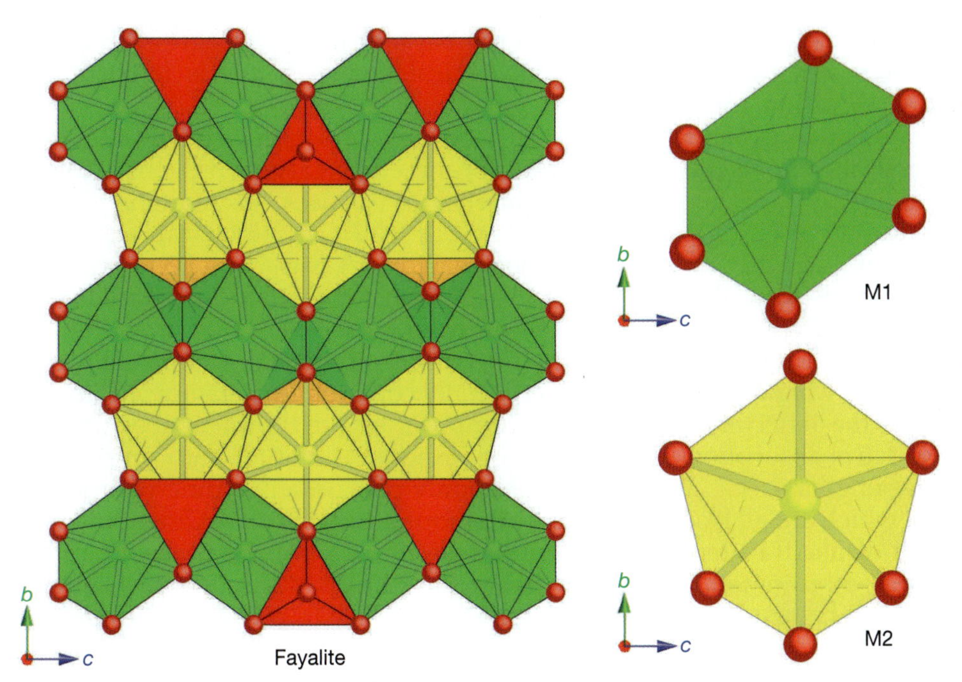

Figure 22.41. The crystal structure of an olivine group mineral—in this case, fayalite, Fe_2SiO_4. The isolated tetrahedra are represented by the red polyhedra, while the M1 and M2 Fe^{2+} sites are shown as green and yellow, respectively, distorted octahedra. The M1 and M2 octahedra form edge-sharing chains parallel to c. *In the M1 and M2 sites, the central Fe^{2+} cation and its bonds to O^{2-} are shown. On the right of the figure, the M1 and M2 octahedra have been isolated from the structure to show their different sizes and bond lengths. Using CrystalViewer you can measure the different bond lengths yourself.*

forsterite end member and increasingly yellow-green with Fe^{2+} substitution. Fe^{3+}-bearing olivines tend to be tan or brown in color, depending on the number of defects in the olivine. Even the cleavages tend vary: the {010} and {100} cleavages tend to be more distinct in Fe-rich olivines than in Mg-rich samples.

The visible and infrared regions of olivine spectra are used as important diagnostic tools in exploration of terrestrial bodies (moons, planets, asteroids) in our solar system, and provide us with yet another demonstration of the relationships between chemistry and physical properties. The M1 site is a tetragonally-elongated octahedron (Burns, 1993). The resultant site distortion creates a split in the energies of the two e_g orbitals, giving rise to two distinctive absorptions in the visible region with energies of roughly 11,000 cm^{-1} and 8000 cm^{-1} (0.9 and 1.25 μm). The M2 site is slightly larger and is generally more distorted than the M1 site, but the distortion results in the two e_g orbitals being similar in energy. Thus, although there are again two possible transitions to e_g orbitals, their energies are so similar (~8830 and ~9270 cm^{-1}) (Burns, 1985) that only a single M2 absorption feature is usually observed in the

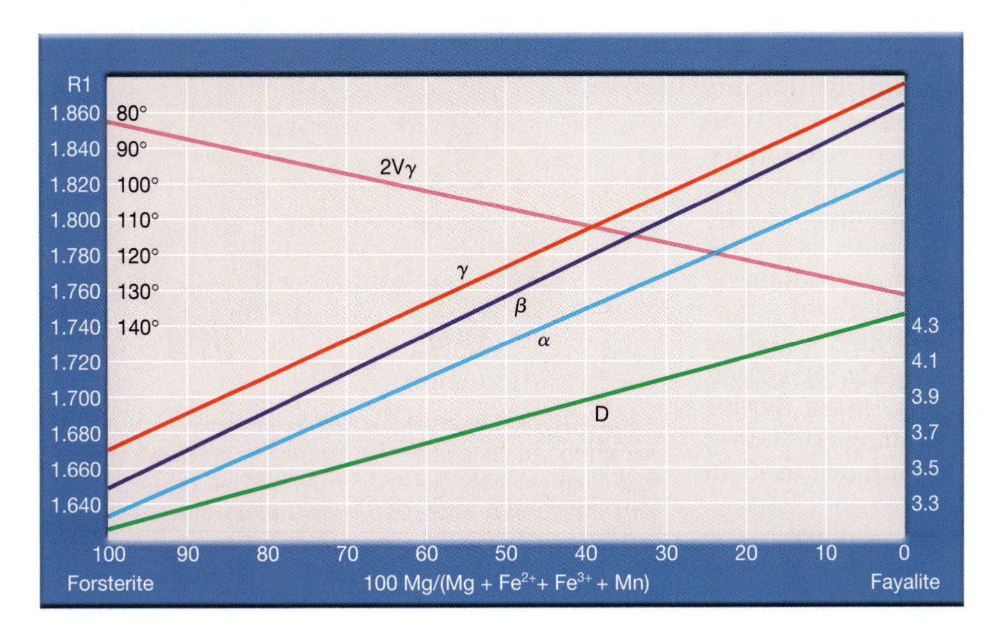

Figure 22.42. Variation in refractive index, density, and 2V as a function Fe content, adapted from Deer et al (1982). As Fe^{2+} replaces Mg^{2+}, the density and refractive indices both increase because Fe^{2+} is heavier than Mg^{2+} (thus causing an increase in density) and Fe^{2+} has more electrons than Mg^{2+} (thus causing the light to be retarded, or slowed down). Notice that 2V also decreases with Fe^{2+} content. 2V is mathematically related to the three principal refractive indices; because these change at different rates, 2V is not constant as Fe^{2+} substitutes for Mg^{2+}.

visible region around 9000 cm^{-1} (~1.1 μm). However, band *energy* does change with composition. The substitution of a larger cation for a smaller one (e.g., Fe^{2+} for Mg^{2+}) tends to expand the crystal structure of a mineral and move the wavelength energy of band minima to longer wavelengths. For olivine, substitution of Fe^{2+} for Mg^{2+} increases the wavelength of all three olivine bands (e.g., Sunshine et al. 1990). Ni^{2+} can also substitute for the Mg^{2+} cation in olivine and change the spectral characteristics (King and Ridley 1987). Ni^{2+} has an absorption band centered near 0.7 μm, which causes the reflectance maximum between the UV and 1 μm absorption features to move to shorter wavelengths.

In the **garnet** group minerals, which are all isometric, there are corner-sharing Y octahedra alternating with SiO_4 tetrahedra, with 8-coordinated X sites between them (Figure 22.43). The resultant formula is $^{[8]}X_3{}^{[6]}Y_2(ZO_4)_3$. The three different types of polyhedra are strongly linked together by edge-sharing of the polyhedra: the 8-fold X site shares ten of its edges, the Y octahedra share six edges, and the tetrahedra share two opposite edges. Every oxygen is bonded to one Z cation (typically Si^{4+}, but rarely Al^{3+} or Fe^{3+}) in tetrahedral coordination, one 6-coordinated Y cation (Al^{3+}, Cr^{3+}, Fe^{3+}, Mn^{3+}, Si^{4+}, Ti^{4+}, V^{3+}, Zr^{4+}), and two 8-coordinated X cations (Fe^{2+}, Mg^{2+}, Mn^{2+}, Ni^{2+}, Zn^{2+}). The sharing and strong bonding make for a structure with a Mohs hardness of 6.5 to 7.5 that allows garnets to be used as abrasives. The close-packed structure with edge sharing polyhedra results in a high refractive index.

Garnets are classified according to their chemistry and the cation distribution between the X and Y sites. The compositions between **pyr**ope, $Mg_3Al_2Si_3O_{12}$, **al**mandine, $Fe^{2+}{}_3Al_2Si_3O_{12}$, and **sp**essartine, $Mn^{2+}{}_3Al_2Si_3O_{12}$, all of which have Al in the Y site, are called the pyralspite series. The ugrandite series minerals, which include **uv**arovite, $Ca_3Cr^{3+}{}_2Si_3O_{12}$, **gr**ossular, $Ca_3Al_2Si_3O_{12}$, and **andr**adite, $Ca_3Fe^{3+}{}_2Si_3O_{12}$, have Ca in the X site. Within each of these series there is continuous substitution, but less variation occurs between the two series. End-member compositions are rare in nature. Hydrogarnets containing $(OH)^{1-}$ or H_2O have been recognized with a formula of $X_3Y_2(ZO_4)_{3-m}(OH)_{4m}$. A high pressure phase with a composition of $Mg_3(Mg,Si)Si_3O_{12}$ is important in the transition zone in the Earth's interior. Synthetic garnets containing Al^{3+}, Ga^{3+}, Ge^{4+}, Y^{3+}, and Fe^{3+} in the Si^{4+} site are used in lasers.

The length of the *a* axis and the volume of the garnet unit cell can be directly related to the size of the cations. Refractive indices can also be predicted on the basis of multiple regression equations of chemical analyses, *a* axis length, and density.

As would be expected from a chemically complex group of minerals, garnets owe their colors to a variety of causes (Rossman, 2005). Absorption by Cr^{3+}, V^{3+}, Fe^{3+} (6-coordinated site), Fe^{2+} (8-coordinated site), Mn^{2+}, and Fe^{2+}-Ti^{4+} intervalence charge transfer can all be found in the spectra of garnets (Figure 22.44).

Garnets occur in both crustal and mantle rocks, but are especially common in metamorphic rocks. Different compositions are typical of specific conditions. For example, biotite and amphibole schists commonly contain almandine with lesser amounts of pyrope and grossular. Almandine and spessartine predominate in pegmatites, granites, and Si-rich rocks that have undergone contact metamorphism. Grossular prevails in eclogites, while pyrope is the most common garnet in kimberlites and peridotites. Andradite is relatively rare, but is found in Ca-rich thermal or contact-metamorphosed rocks, especially in skarn deposits.

In **zircon** group minerals, 8-coordinated Zr^{4+} occupies edge-sharing, triangular dodecahedra that create a framework by sharing edges and corners with SiO_4 tetrahedra (Figure 22.45). The Zr^{4+} dodecahedra share edges with each other to form chains, with each ZrO_8 polyhedra sharing edges with four other ZrO_8 polyhedra (Finch and

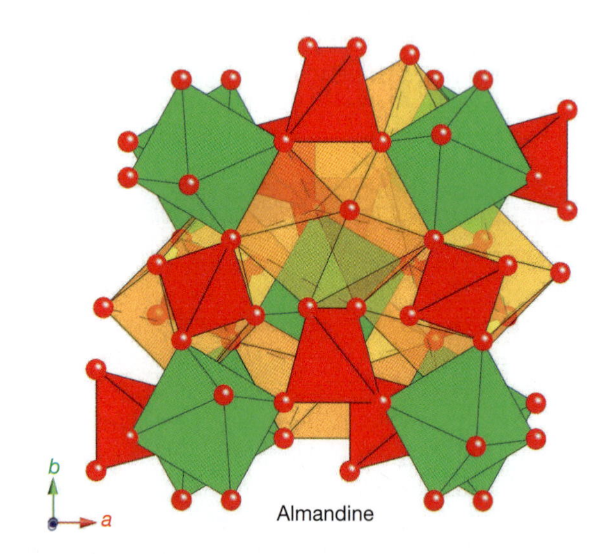

Figure 22.43. *The crystal structure of members of the garnet group. In almandine, $Fe_3Al_2(SiO_4)_3$, the small red polyhedra represent the isolated Si^{4+} tetrahedra, the large transparent polyhedra house Fe^{2+} bonded to eight oxygens (small red spheres), and Al^{3+} is 6-coordinated to O^{2-}'s in regular octahedra. Many of the polyhedra share edges, making garnet very dense.*

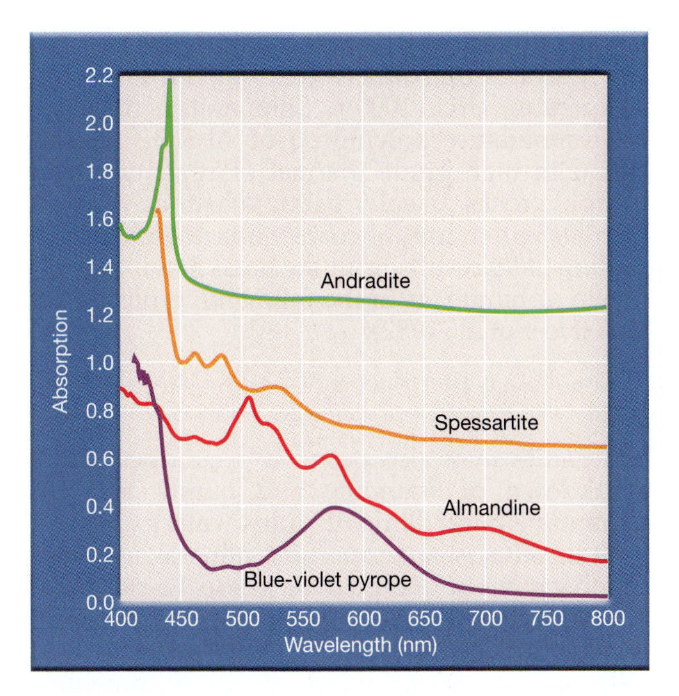

Figure 22.44. Rainbow of colors displayed in naturally-occurring garnets. All spectra are normalized to 100 μm thickness and then offset for clarity. Almandine from Umba River Valley, Tanzania, is colored red by Fe^{2+}. Spessartite, here from Rutherford Mine, Virginia, is colored orange by Mn^{2+}. Andradite from Val Malenco, Italy, is greenish-yellow due to Fe^{3+}. Pyrope from Tunduru, Tanzania shows a color change from pale purplish-blue to light purple depending upon the illumination. Data from the Mineral Spectroscopy web site at Caltech (Rossman, 2005).

Hanchar, 2003). The resultant structure has chains running both parallel and perpendicular to the *c* axis (Gaines et al., 1997).

The zircon minerals are extremely common, making them useful in many types of geological studies. They occur as accessory minerals in granites, rhyolites, and tuffs, and are common in sediments. Zircon is perhaps the most commonly occurring accessory mineral that can act as a host for radioactive elements that are used in geochronology. Its 8-coordinated sites can accommodate many trace elements such as U^{6+}, Lu^{3+}, Hf^{4+}, Sm^{3+}, Nd^{3+}, etc. (Bowring and Schmitz, 2003; Kinny and Maas, 2003); Pb^{2+} is sometimes present as a decay product. Furthermore, the zircon chronometer appears to be very difficult to reset, so that old rocks may even survive subsequent metamorphism without losing their isotopic compositions. Melt inclusions in zircon are used to understand trace element and isotopic compositions of parent magmas (Thomas et al., 2003). Zircon is also used in thermometry (Hanchar and Watson, 2003) and oxygen isotope studies (Valley, 2003). As a host for radioactive elements, zircon is also a natural laboratory for understanding the long-term effects of radiation damage, and it has been proposed as a suitable structure for storage of nuclear waste (Ewing et al., 2003). These and other aspects of zircon mineralogy are discussed in *Reviews in Mineralogy and Geochemistry*, vol. 53.

Euclase group minerals such as the species euclase, $BeAlSiO_4(OH)$, are nesosilicates that incorporate (OH) in their structures. The result is zig-zagging chains of edge-sharing $AlO_5(OH)$ octahedra, linked by SiO_4 to form sheets (Strunz and Nickel, 2001). The sheets are held together by corner-sharing tetrahedra of $BeO_3(OH)$ that lie along the *c* axis. Euclase crystals are thus commonly prismatic and elongated along [001]. Euclase is found in low temperature hydrothermal vein deposits, pegmatites, and greisens.

As first introduced in Chapter 1 (Figure 1.7), there are three **aluminosilicate** minerals with exactly the same composition, but different structures. All of them have isolated SiO_4 tetrahedra and one Al^{3+} cation in an octahedral site. They differ because of the coordination environment of the second Al^{3+} cation, which can be 6-, 5-, or 4-coordinated:

kyanite: $^{[6]}Al^{3+}\ ^{[6]}Al^{3+}\ ^{[4]}SiO_5$
andalusite: $^{[6]}Al^{3+}\ ^{[5]}Al^{3+}\ ^{[4]}SiO_5$
sillimanite: $^{[6]}Al^{3+}\ ^{[4]}Al^{3+}\ ^{[4]}SiO_5$.

All of these polymorphs of Al_2SiO_5 are based on edge-sharing octahedra parallel to the *c* axis that are occupied by Al^{3+}. Their crystals are usually elongated along *c* as a result, just as in epidote. The linkages between the octahedral chains distinguish the different species. Each species is unique to a specific pressure and temperature range, as discussed in Chapter 6 (Figure 6.16).

Kyanite has chains of Al^{3+} octahedra linked together by sharing edges with other Al^{3+} octahedra and corners with SiO_4 tetrahedra (Figure 6.15a). This arrangement has the closest packing of oxygen anions of the three aluminosilicates, and results in the highest density and refractive index. Because the chains are oriented in the [001] direction, kyanite tends to grow as elongated prisms parallel to [001].

A minor amount of Fe^{2+} and Ti^{4+} substitution for Al^{3+} is known in kyanite; $(OH)^{1-}$ for O^{2-} exchange also occurs, though these are not coupled substitutions. The colors are characteristic of these substitutions: the blue color of metamorphic kyanite comes from the Fe^{2+}-Ti^{4+} intervalence interaction, while yellow-green comes from Fe^{3+}. Sky blue kyanite found in diamond pipes contains Cr^{3+} (Rossman, 2005). The same chromophores are also responsible for the colors in the other aluminosilicates.

Kyanite is usually found in rocks that have experienced regional metamorphism. It is also the only aluminosilicate to be found in eclogites, which are the result of medium temperature and high pressure metamorphism of rocks Fe-Mg-rich rocks (usually basalt) in subduction zone settings.

In **andalusite** (Figure 6.15b), half of the Al^{3+} cations occupy 5-coordinated sites that alternate with Si^{4+} tetrahedra to form double chains adjacent to the chains of octahedral Al^{3+} along the *c* direction. Significant substitution of Mn^{3+} and Fe^{3+} for Al^{3+} can occur in the 6-coordinated site. Because the change in refractive index varies with this substitution, andalusite of certain compositions can appear isotropic, even while maintaining orthorhombic symmetry (Gunter and Bloss, 1982). Also, substitution of Fe^{3+} and Mn^{3+} strongly affect the triple point in pressure-temperature space for the aluminosilicates. Most andalusites also contain minor amounts of Cr^{3+}, Ti^{4+}, and V^{3+}.

Andalusite tends to occur in low- and medium-grade metamorphic rocks that have experienced contact metamorphism, and it is often found in association with cordierite. Andalusite has also been reported in granites and granitic pegmatites, and is found as a detrital mineral in sandstones and other sediments that originate from the breakdown of andalusite-bearing metamorphic rocks.

The **sillimanite** structure has half of its Al^{3+} in tetrahedral sites that form 4-membered rings with Si^{4+} tetrahedra (Figure 6.15c)—really making it a ring silicate! From a different angle, it can be seen that sillimanite has two adjacent chains that contribute to its prismatic habit: one of Al^{3+} octahedra, and one of alternating Si^{4+} and Al^{3+} tetrahedra. Fe^{3+} infrequently substitutes for Al^{3+} in both the 4-coordinated and 6-coordinated sites (less so than in andalusite), and some limited $(OH)^{1-}$ substitution for O^{2-} may occur. The resultant double chain is analogous to the double-chain structure of amphiboles. The length of the chains and the unit cell dimensions vary linearly with temperature up to 1000°C. Sillimanite is the highest temperature polymorph of Al_2SiO_5, so it is found in high grade thermal or regional metamorphic rocks, usually pelitic schist, in which it coexists with micas, or coarser quartz-rich gneisses. It usually results from the breakdown of muscovite or biotite, as in the following typical reaction (Deer et al., 1982):

$$KAl_2AlSi_3O_{10}(OH)_2 + SiO_2 \longleftrightarrow Al_2SiO_5 + KAlSi_3O_8 + H_2O$$
$$\text{muscovite + quartz} \longleftrightarrow \text{sillimanite + orthoclase + water.}$$

Sillimanite may also occur as a product of the breakdown of staurolite and quartz to form almandine plus sillimanite (plus water).

Staurolite is an important mineral in metamophic rocks of intermediate grade. Fe-rich compositions are stable at temperatures between roughly 550–700°C at pressures greater than 0.15 GPa, and at lower temperatures at pressures down to 0.9 GPa. Mg-rich staurolites have a larger stability range, and can be found between 700–1000°C at pressures above 1.2 GPa (Richardson, 1966; Ganguly, 1972; Schreyer, 1988).

The complicated staurolite structure took many years to resolve before work by Holdaway et al. (1986) finally pinned it down. The structure is generally described as "slabs" of (Al_2SiO_5) alternating with layers of $^{[VI]}Al_{0.7}{}^{[IV]}Fe_2O_2(OH)_2$ in the (001) plane along the *b* axis (Figure 22.46). The slabs contain octahedral Al^{3+} with minor Fe^{2+} and Mg^{2+} substitution, and the tetrahedral sites are occupied, as expected, by Si^{4+} and Al^{3+}. In the other layer, edge-sharing octahedra form chains along [001]; the octahedra also share faces with FeO_4 tetrahedra. In total, the unit cell contains seven octahedral sites (mostly filled by Al^{3+}) and two tetrahedral sites, one containing Si^{4+} and the other occupied by Fe^{2+}, Mg^{2+}, Zn^{2+}, and Li^{1+}. The basic staurolite formula can thus be indicated as

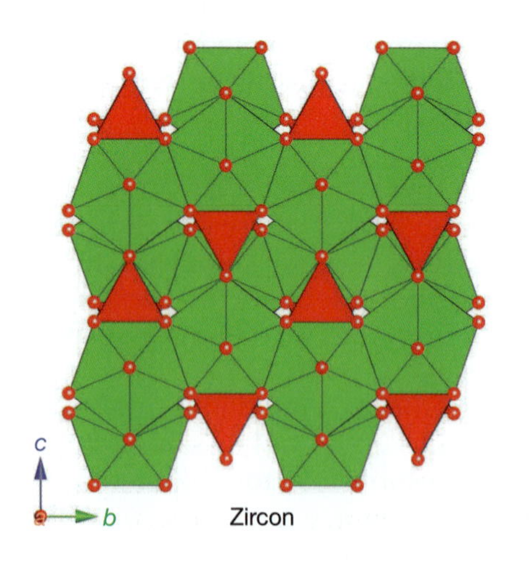

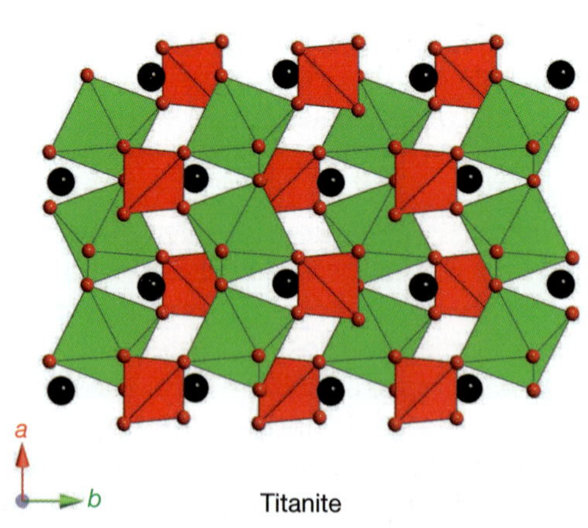

Figure 22.45. *The structures of zircon, $ZrSiO_4$, (left) and titanite, $CaTiSiO_5$, (right) both contain isolated Si^{4+} tetrahedra (red). In zircon, Zr^{4+} is located in 7-coordinated, edge-sharing polyhedra (green). In titanite, Ti^{4+} is 6-coordinated into octahedra that corner share O^{2-} anions with each other and the Si^{4+} tetrahedra. Ca^{2+} (black spheres) fills the voids and bonds the entire structure together.*

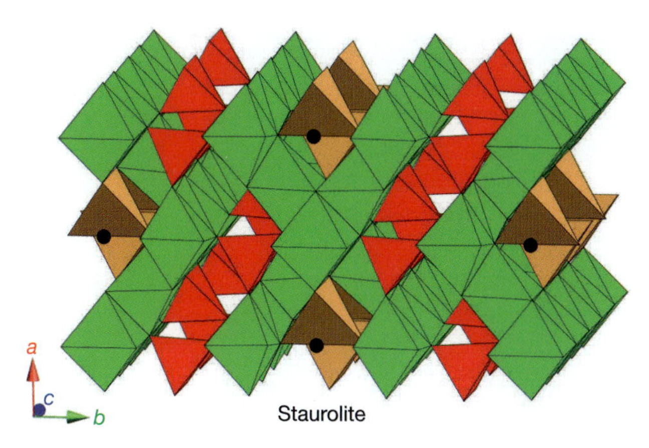

Figure 22.46. The structure of staurolite, $Fe_2Al_9(Al,Si)_4O_{22}(OH)_2$, projected slightly tilted from its (001) plane. The main structure units are edge-sharing chains of Al^{3+} octahedra (green) that are cross-linked by the isolated Si^{4+} tetrahedra (red), Fe^{2+} tetrahedra (brown), and larger cations (black) spheres.

roughly $({}^{[T2]}Fe^{2+}_{3-4}{}^{[M4]}Fe_{0-0.5}{}^{[M3]}Al_2\square_2)^{[M1,M2]}Al_{16}{}^{[T1]}Si_8O_{48}H_{2-4}$, with the site names [given in brackets] preceding the cations (Comodi et al., 2002). However, staurolite stoichiometry is highly variable.

In the **topaz** structure, pairs of edge-sharing AlO_4F_2 octahedra share corners with SiO_4 tetrahedra to form chains parallel to *c* (Strunz and Nickel, 2001). The chains link up by sharing corners. Limited substitution of $(OH)^{1-}$ for F^{1-} may occur, and there is a linear relationship between the length of the *b* axis and the OH/(OH+F) ratio. Topaz has a prismatic habit that reflects the orientation of the chains in the structure, and its perfect (010) cleavage is the result of breaking Al^{3+}-O^{2-} and Al^{3+}-F^{1-} bonds preferentially over Si^{4+}-O^{2-} bonds.

Topaz is typically found in granites and granitic pegmatites, often associated with tourmaline. It is sometimes suitable for use as gemstones. Topaz is naturally blue, pink, and various shades of brown or amber. The pink color comes from Cr^{3+} and the amber and brown colors come from radiation damage. For use as a gemstone, natural topaz is commonly irradiated and heat-treated to induce the desirable blue color. A gemstone known as Mystic topaz is also commercially available; it is colored by a coating applied to the bottom portion of the cut stone.

Humite group minerals, rich in $(OH)^{1-}$ and F^{1-} compared with their close relatives olivines, are nearly always found in association with impure, metamorphosed and metasomatized limestones and dolomites, usually adjacent to felsic plutonic rocks, or in association with ore deposits. Their structures are based on hexagonal close packing of O^{2-}, $(OH)^{1-}$, and F^{1-}, with half of the octahedral interstices occupied by divalent cations. Si^{4+} cations occupy only some of the tetrahedral sites, avoiding those that are coordinated with either $(OH)^{1-}$ or F^{1-}. The similarity in the olivine and humite structures explains the resemblance of their formulas as well as their *a* and *b* cell parameters and cell volumes. The structure can also be viewed as being composed of seven edge-sharing $(Mg,Fe,Mn)O_6$ octahedra that form zig-zag chains along the *a* axis. Si^{4+} tetrahedra lie within the "V" formed by the zigzags. This structure can be either monoclinic or orthorhombic.

The formula of humite is often expressed as stacks of alternating layers of M_2SiO_4 olivine and $M(OH,F)_2$. One way of writing the formula is $n[M_2SiO_4]\cdot[M_{1-x}Ti_x(OH,F)_{2-2x})O_{2-x}$, where M is Mg^{2+}, Fe^{2+}, Mn^{2+}, Ca^{2+}, and Zn^{2+}, x is always less than 1, and $n = 1$ for norbergite, 2, for chondrodite, 3, for humite, and 4 for clinohumite. We give the simpler, identical, yet less revealing formulas in Table 22.21!

The structure of **chloritoid** consists of two types of edge-sharing octahedral layers perpendicular to the *c* axis that give this mineral mica-like cleavage along layers. One layer contains Fe^{2+} and Al^{3+} in a 2:1 ratio, with a formula of roughly $Fe^{2+}{}_2AlO_2(OH)_4]^{1-}$; Mg^{2+} and Mn^{2+} substitute for the Fe^{2+} here. The alternate layer, with a formula of $[Al_3O_8]^{7-}$, is 3/4 occupied by octahedral Al^{3+}. Isolated SiO_4 tetrahedra bond the two layers together. The stacking of these layers allows multiple polytypes, as in the case of micas. Chloritoid is a key mineral in Al-rich metamorphic rocks at low grades (chlorite zone). Fe-rich compositions are stable between roughly 250–590°C. Mg-chloritoid is stable up to about 730°C at pressures from roughly 17–50 kb; its lower temperature limit has not yet been determined.

The **titanite** structure is composed of kinked chains of corner-sharing Ti^{4+} octahedra cross-linked by SiO_4 tetrahedra; the structure cleaves parallel to these chains (Figure 22.45). Each of the Si^{4+} tetrahedra shares O^{2-} with four separate TiO_6 groups in three separate chains. The resultant framework encloses 7-coordinated polyhedra occupied by Ca^{2+}, or rarely by Sr^{2+}, Ba^{2+}, Mn^{2+}, Na^{1+}, and rare earth elements such as Ce^{3+}, Nd^{3+}, or Y^{3+}. Such highly-charged substitutions for Ca^{2+} are usually charge-balanced by substitution of Al^{3+} or Fe^{3+} for Ti^{4+}. The presence of the radioactive cations sometimes causes damage to the crystal structure, resulting in metamict titanite. **Metamictization** is the process by which radiation gradually breaks down the crystal lattice of a mineral structure. $(OH)^{1-}$ and F^{1-} are also common in titanite, with Cl^{1-} rarely observed.

Titanite group minerals have a large stability field, and are therefore found in a large range of igneous and metamorphic rock types, usually as an accessory mineral. Titanite is rare in sedimen-

Table 22.21. Other Orthosilicates

Group	Subgroup	Mineral	Formula
Euclase		euclase	$BeAlSiO_4(OH)$
		clinohedrite	$CaZnSiO_4\cdot(H_2O)$
		hodgkinsonite	$MnZn_2SiO_4(OH)_2$
		gerstmannite	$(Mg,Mn)_2ZnSiO_4(OH)_2$
Andalusite	*Sillimanite subgroup*	sillimanite	Al_2SiO_5
		mullite	$Al_{4+2x}Si_{2-2x}O_{10-x}$
	Andalusite subgroup	andalusite	Al_2SiO_5
		kanonaite	$(Mn^{3+},Al)AlSiO_5$
		yoderite	$Mg_2(Al,Fe^{3+})_6Si_4O_{18}(OH)_2$
	Kyanite subgroup	kyanite	Al_2SiO_5
Staurolite		staurolite	$(Fe,Mg,Zn)_{3-4}(Al,Fe)_{18}(Si,Al)_8O_{48}H_{2-4}$
		magnesiostaurolite	$\square_4Mg_4Al_{16}(Al,\square)_2Si_8O_{40}O_6(OH)_2$
		zincostaurolite	$\square_4Zn_4Al_{16}(Al,\square)_2Si_8O_{40}O_6(OH)_2$
		topaz	$Al_2SiO_4(F,OH)_2$
Humite	*Humite subgroup*	chondrodite	$(Mg,Fe)_5(SiO_4)_2(F,OH)_2$
		clinohumite	$(Mg,Fe)_9(SiO_4)_4(F,OH)_2$
		humite	$Mg_7(SiO_4)_3(F,OH)_2$
		hydroxylclinohumite	$(Mg,Fe)_9(SiO_4)_4(F,OH)_2$
		norbergite	$Mg_3(SiO_4)(F,OH)_2$
	Manganhumite subgroup	alleghanyite	$Mn_5(SiO_4)_2(OH,F)_2$
		manganhumite	$Mn_7(SiO_4)_3(OH)_2$
		sonolite	$Mn_9(SiO_4)_4(OH)_2$
	Meucophoenicite subgroup	jerrygibbsite	$Mn_9(SiO_4)_4(OH)_2$
		leucophoenicite	$Mn_7(SiO_4)_3(OH)_2$
		leucophoenicite	$Mn_5(SiO_4)_2(OH)_2$
	Reinhardbarunsite subgroup	reinhardbarunsite	$Ca_5(SiO_4)_2(OH,F)_2$
Chloritoid		chloritoid	$(Fe,Mg,Mn)_2Al_4Si_2O_{10}(OH)_4$
		magnesiochloritoid	$(Mg,Fe^{2+},Mn^{2+})_2Al_4Si_2O_{10}(OH)_4$
		ottrelite	$(Mn,Fe,Mg)_2Al_4Si_2O_{10}(OH)_4$
		carboirite-VIII	$Fe^{2+}Al_2GeO_5(OH)_2$
Titanite		titanite	$CaTiSiO_5$
		malayaite	$CaSnSiO_5$
		vanadomalayaite	$CaVOSiO_4$

tary rocks, however, because of its softness, which is probably the result of the relatively open structure and the corner- rather than edge-sharing polyhedra. Like monazite (see Chapter 23), titanite can be a useful geochronometer in U- and Th-bearing rocks that start out with negligible Pb^{2+} (so that any Pb^{2+} that is present can be considered to be due to the breakdown of U and Th). Titanite can survive temperatures of up to ~600–700°C before diffusive loss (or gain) of Pb becomes significant, making ages difficult to interpret.

Borosilicates and Beryllosilicates

The ring silicate tourmaline may be the most common of the B-rich silicates, but there are several other silicates with structures based upon BO_3 triangles, as well as several based on BO_4 and BeO_4 tetrahedra. These minerals have received increasing attention recently as the role of boron in transport processes (e.g., magma ascent) has been recognized. Exhaustive coverage of these minerals is given in *Reviews in Mineralogy*, volumes 33 and 50, particularly in the chapters by Grew (1996 and 2002).

Members of the **datolite-gadolinite group** (monoclinic) have a general formula of $A_2BC_2(TO_4)_2X_2$, where A is a REE, Ca^{2+}, or Bi^{3+}; B is Fe^{2+}, Fe^{3+}, Mn^{2+}, Ca^{2+}, or a vacancy; B is Be^{2+} or B^{3+}; T is Si^{4+}, P^{5+}, or As^{5+}; and X is O^{2-}, $(OH)^{1-}$, or F^{1-} (Table 22.22). These minerals have sheet-like structures that lie in the *b-c* plane; the sheets are made from alternating, corner-sharing $(B,Be)(O,OH)_4$ and SiO_4 tetrahedra that form 4- and 8-membered rings (Figure 22.47). The 8-coordinated A cations connect the sheets with strong bonds, so these minerals lack prominent cleavage. Datolite occurs predominantly as a replace-

Table 22.22. Borosilicates			
Datolite-gadolinite	*Datolite series*	datolite	$CaBSiO_4(OH)$
		hingganite-(Ce)	$(Ce,Y)_2(\square,Fe^{2+})Be_2Si_2O_8(OH)_2$
		hingganite-(Y)	$(Y,Ce)_2(\square,Fe^{2+})Be_2Si_2O_8(OH)_2$
		hingganite-(Yb)	$(Yb,Y)_2Be_2Si_2O_8(OH)_2$
		calcybeborosilite-(Y)	$(REE,Ca)_2(\square)(B,Be)_2Si_2O_8(OH,O)_2$
	Homolite series	bakerite	$Ca_4B_5Si_3O_{15}(OH)_5{\cdot}2H_2O$
		gadolinite-(Ce)	$(Ce,La,Nd,Y)_2Fe^{2+}Be_2Si_2O_{10}$
		gadolinite-(Y)	$Y_2Fe^{2+}Be_2Si_2O_{10}$
		homilite	$Ca_2FeB_2Si_2O_{10}$
		minasgeraisite-(Y)	$(Y,Yb,Bi)_2CaBe_2Si_2O_{10}$
		grandidierite	$MgAl_3(BO_3)(SiO_4)O_2$
Dumortierite		dumortierite	$Al_7(BO_3)(SiO_4)_3O_3$
		magnesiodumortierite	$(Mg,Ti,\square)_{<1}(Al,Mg)_2Al_4Si_3O_{18-y}(OH)_yB$ where y=2-3

ment of calcite and cal-silicates in metamorphic deposits such as skarns.

Several of the Al-borosilicates are closely related to sillimanite and andalusite. **Grandidierite,** $(Mg,Fe)Al_3BSiO_9$, its Fe^{2+}-dominant analog ominelite, $(Fe,Mg)Al_3BSiO_9$, werdingite, $Al_8(Mg,Fe)_2Al_4(Al,Fe)_2Si_4(B,Al)_4O_{37}$, and **boralsilite**, $Al_{16}B_6Si_2O_{37}$, all also have structures based on chains of edge-sharing Al^{3+} octahedra along the direction of the *c* axis. Differences between these minerals result from the ways the cross-linkages are accomplished. For example, in boralsilite the additional Al^{3+} cations are 5-coordinated (as in andalusite) but they occupy shifted trigonal-bipyramical coordination, nearly creating distorted tetrahedra (Peacor et al., 1999). The Si^{4+} tetrahedra link up to form disilicate groups (as in the sorosilicates), and B^{3+} is found in both tetrahedral and trigonal planar coordination. In grandidierite and ominelite, the chains are cross-linked by trigonal planar B^{3+}, 4-coordinated Si^{4+}, pairs of distorted 5-coordinated Mg^{2+} or Fe^{2+} and Al^{3+} polyhedra (Stephenson and Moore, 1968; Hiroi et al., 2001). No other silicate mineral contains significant amounts of Fe^{2+} in five-fold coordination. Grandidierite and ominelite occur in pegmatites, granites, granulite-facies metapelites and migmatites, pelitic hornfels, and calc- silicate rocks (Grew, 1996).

In **dumortierite,** $(Al,\square)Al_6(BO_3)(SiO_4)_3(O,OH)_3$, there are three different octahedral chains parallel to the *c* axis: face-sharing AlO_6 octahedra, double chains of face-sharing $(Al,\square)_4O_{12}$ octahedra, and a strip of $(Al)_2(Al)_2O_{12}$ linked by isolated Si^{4+} tetrahedra and B^{3+} triangles. There is limited substitution of Ti^{4+}, Fe^{2+}, Mg^{2+}, Sb, and As for Al^{3+}, P for Si^{4+}, and $(OH)^{1-}$ for O^{2-}. Dumortierite is found in volcanic rocks, in quartzofeldspathic vein rocks and pegmatites associated with metamorphic and migmatitic complexes, in Al-rich rocks dominated by quartz, muscovite, and/or Al_2SiO_5 minerals,

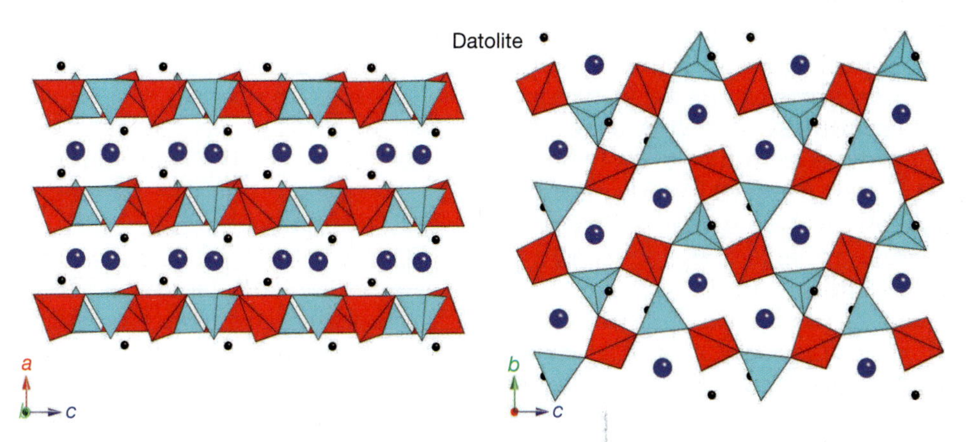

Figure 22.47. The datolite, $CaBSiO_4$, structure is composed of sheets of polymerized Si^{4+} (red) and B^{3+} (light blue) tetrahedra. **a.** *The datolite structure projected parallel to the sheets and* **b.** *the structure projected perpendicular to the sheets. The sheets are bonded to each other by Ca^{2+} and H^{1+} ions. Even though this structure is classified as an orthosilicate, it really should be considered a layer silicate because each tetrahedron shares three of its four O's with adjoining tetrahedra.*

and in moderately Al-rich metamorphic rocks that may be enriched in boron (Grew, 1996).

Concluding Remarks

The silicates summarized in this chapter do not in any way presume to represent all of the variations that are possible with a SiO_4 tetrahedron as a structural unit. However, we hope that they give you a good foundation for understanding the crystal structures of the common rock-forming silicates that you are likely to encounter. In all cases, a great deal more information is given in our mineral database or the references at the end of this chapter, and the many reference books that are probably available to you in your library.

Acknowledgments. We thank George Rossman for his assistance with mineral spectra and general advice about mineral structures. We found the on-line postings of mineralogy lecture outlines by Eric Essene to be especially helpful in organizing and placing various minerals in appropriate geological contexts. We are also grateful to Ma Chi at Caltech for permission to use the images of opal, to Michail Taran for use of his tourmaline spectra, and to Bernard Grobéty and André Lalonde for illuminating comments about uses of mica. Steve Guggenheim, Ed Grew, Peter Heaney, Paul Ribbe, and Dave Jenkins all contributed thoughtful reviews of various sections of this chapter, which greatly improved it.

References

Adams, J.B. (1974) Visible and Near-Infrared diffuse reflectance spectra of pyroxenes as applied to remote sensing of solid objects in the solar system. Journal of Geophysical Research, 79, 4829–4836.

Adams, J.B. (1975) Interpretations of visible and near-infrared diffuse reflectance spectra of pyroxenes and other rock-forming minerals. In Infrared and Raman Spectroscopy of Lunar and Terrestrial Minerals, C. Karr, Jr., ed., Academic Press, NY, pp. 91–116.

Amthauer, G., Rossman, G.R. (1984) Mixed valence of iron in minerals with cation clusters. Physics and Chemistry of Minerals, 11, 37–51.

Andreozzi, G.B., Ottolini, L., Lucchesi, S., Graziani, G., and Russo, U. (2000) Crystal chemistry of the axinite-group minerals: A multi-analytical approach. American Mineralogist, 85, 698–706.

Armbruster, T. and Gunter, M.E. (2001) Crystal structures of natural zeolites. In Reviews in Mineralogy and Geochemistry, Natural Zeolites: Occurrence, Properties, Applications, 45, 1–68.

Bailey, S.W. (1988a) Introduction. Reviews in Mineralogy, 19, 1–8.

Bailey, S.W. (1988b) Polytypism of 1:1 layer silicates, Reviews in Mineralogy, 19, 9–27.

Bailey, S.W. (1988c) Chlorites: Structures and crystal chemistry. Reviews in Mineralogy, 19, 347–403.

Bannister, F.A., and Hey, M.H. (1931) A chemical, optical, and X-ray study of nepheline and kaliophilite. Mineralogical Magazine, 22, 569–581.

Bloss, F.D. and Gibbs, G.V. (1963) Cleavage in quartz. American Mineralogist, 48, 821–838.

Bowring, S.A., and Schmitz, M.D. High-precision U-Pb zircon geochronology and the stratigraphic record. Reviews in Mineralogy and Geochemistry, 53, 305–325.

Brown, B.M. and Gunter, M.E. (2003) Morphological and optical characterization of amphiboles from Libby, Montana U.S.A. by spindle stage assisted polarized light microscopy. The Microscope, 51, 3, 121–140.

Burns, R.G. (1993) Mineralogical applications of crystal field theory. Cambridge University Press, 551 pp.

Burns, R.G. (1985) Thermodynamic data from crystal field spectra. Reviews in Mineralogy, 14, 277–316.

Burns, R.G., and Huggins, F.E. (1973) Visible-region absorption spectra of a Ti^{3+} fassaite from the Allende meteorite: A discussion. American Mineralogist, 58, 955–961.

Cameron, M., and Papike, J.J. (1980) Crystal chemistry of silicate pyroxenes. Reviews in Mineralogy, 7, 5–116.

Churg, A. (1993) Asbestos lung burden and disease patterns in man. Reviews in Mineralogy, 28, 409–426.

Comodi, P., Montagnoli, M., Zanazzi, P.F., and Ballaran, T.B. (2002) Isothermal compression of staurolite: A single crystal study. American Mineralogist, 87, 1164–1171.

Coombs, D.S., Alberti, A., Armbruster, T., Artili, G., Colella, C. Galli, E., Grice, J.D., Liebau, F., Mandarino, J.A., Minato, H., Nickel, E.H., Passaglia, E., Peacor, D.R., Quartieri, S., Rinaldi, R., Ross, M., Sheppard, R.A., Tillmanns, E., and Vezzalini, G. (1998) Recommended nomenclature for zeolite minerals: Report of the subcommittee on zeolites of the International Mineralogical Association, Commission on New Minerals and Mineral Names. American Mineralogist Special Feature, 28 pp.

Deer, W.A., Howie, R.A., and Zussman, J. (1982) Orthosilicates. Vol. 1A, 2nd ed. Longman, London and New York, 932 pp.

Deer, W.A., Howie, R.A., and Zussman, J. (1986) Disilicates and Ring Silicates. Vol. 1B, 2nd ed. Longman, London and New York, 630 pp.

Deer, W.A., Howie, R.A., and Zussman, J. (1997) Double-chain silicates. Vol. 2B, 2nd ed. The Geological Society, London, 784 pp.

Deer, W.A., Howie, R.A., and Zussman, J. (1998) Single-chain silicates. Vol. 2A, 2nd ed. John Wiley and Sons, New York, 680 pp.

Deer, W.A., Howie, R.A., and Zussman, J. (2001) Framework silicates: Feldspars. Vol. 4A, 2nd ed. The Geological Society, London, 992 pp.

Deer, W.A., Howie, R.A., and Zussman, J. (2006) Framework silicates: Silica Minerals, Feldspathoids, and the Zeolites. Vol. 4B, 2nd ed. The Geological Society, London, 982 pp.

Elzea, J.M., and Rice, S.B. (1996) TEM and X-ray diffraction evidence for cristobalite and tridymite stacking sequences in opal. Clays and Clay Minerals, 44, 492–500.

Ewing, R.C., Meldrum, A., Wang, L., Weber, W.J., and Corrales, R. (2003) Radiation effects in zircon. Reviews in Mineralogy and Geochemistry, 53, 387–425.

Finch, R.J., and Hanchar, J.M. (2003 Structure and chemistry of zircon and zircon-group minerals. Reviews in Mineralogy and Geochemistry, 53, 1–26.

Fleet, M.E. (2003) Micas. Rock-Forming Minerals, vol. 3A. The Geological Society, London, 980 pp.

Franz, G., and Liebscher, A. (2004) Physical and chemical properties of the epidote minerals—an introduction. Reviews in Mineralogy and Geochemistry, 56, 1–82.

Gaines, R.V., Skinner, H.C.W., Foord, E.E., Mason, B., Rosenzweig, A., King, V.T., and Dowty, R. (1997) Dana's New Mineralogy. John Wiley and Sons, New York, 1819 pp.

Ganguly, J. (1972) Staurolite stability and related parageneses: Theory, experiments, and applications. Journal of Petrology, 13, 335–365.

Giese, R.F.Jr. (1988) Kaolin minerals: Structures and stabilities. Reviews in Mineralogy, 19, 29–66.

Graetsch, H. (1994) Structural characteristics of opaline and microcrystalline silica minerals. Reviews in Mineralogy, 29, 209–232.

Graetsch, H., Gies, H. and Topalovic, I. (1994) NMR, XRD, and IR study on microcrystalline opals. Physics and Chemistry of Minerals, 21, 166–175.

Grew, E.S. (1996) Borosilicates (exclusive of tourmaline) and boron in rock-forming minerals in metamorphic environments. Reviews in Mineralogy, 33, 387–501.

Grew, E.S. (2002) Mineralogy, petrology and geochemistry of beryllium: An introduction and list of beryllium minerals, 1–50.

Groat, L.A., Hawthorne, F.C., and Ercit, T.S. (1992) The chemistry of vesuvianite. Canadian Mineralogist, 30, 19–48.

Guggenheim, S. and Eggleton, R.A. (1988) Crystal chemistry, classification, and identification of modulated layer silicates. Reviews in Mineralogy, 19, 675–725.

Guinier, A. (1984) Nomenclature of polytype structures. Acta Crystallographica, Section A: Foundations of Crystallography, 40, 399–404.

Gunter, M.E. (1994) Asbestos as a metaphor for teaching risk perception. Journal of Geological Education, 42, 17–24.

Gunter, M.E. (1999) Quartz—the most abundant mineral species in the earth's crust and a human carcinogen? Journal of Geoscience Education, 341–349.

Gunter, M.E. and Bloss, F.D. (1982) Andalusite-kanonaite series: Lattice and optical parameters. American Mineralogist, 67, 1218–1228.

Gunter, M.E. and Ribbe, P.H. (1993) Natrolite group zeolites: correlations of optical properties and crystal chemistry. Zeolites, 13, 435–440.

Gunter, M.E., Belluso, E., and Mottana, A. (2007) Amphiboles: Environmental and health concerns. In Amphiboles: Crystal chemistry, occurrences, and health concerns. Reviews in Mineralogy and Geochemistry, 67, 453–516.

Guthrie, G.D.Jr., Bish, D.L., and Reynolds, R.C.Jr. (1995) Modeling the X-ray diffraction pattern of opal-CT. American Mineralogist, 80, 869–872.

Hanchar, J.M., and Watson, E.B. (2003). Zircon saturation thermometry. Reviews in Mineralogy and Geochemistry, 53, 89–111.

Hawthorne, F.C. (1981) Crystal chemistry of the amphiboles. Reviews in Mineralogy, 9A, 1–102.

Hawthorne, F., and Henry, D. (1999) Classification of the minerals of the tourmaline group. European Journal of Mineralogy, 11, 201–215.

Heaney, P.J., Prewitt, C.T., and Gibbs, G.V., eds. (1994) Silica: Physical Behavior, Geochemistry, and Materials Applications. Reviews in Mineralogy, 29, Mineralogical Society of America, 606 pp.

Hedrick, J.B. (1999) Mica. U.S. Geological Survey Minerals Yearbook, 1999, 51.1–51.11.

Henry, D.J., and Dutrow, B.L. (1996) Metamorphic tourmaline and its petrologic applications. Reviews in Mineralogy, 33, 503–558.

Hiroi, Y., Grew, E.S., Motoyoshi, Y., Peacor, D.R., Rouse, R.C., Matsubara, S., Yokoyama, K., Miyawaki, R., McGee, J.J., Su, S.-C., Hokada, T., Furukawa, N., and Shibasaki, H. (2001) Ominelite, $(Fe,Mg)Al_3BSiO_9$ (Fe^{2+}analog of grandidierite), a new mineral from porphyritic granite in Japan. American Mienralogist, 87, 160–170.

Holdaway, M.J., Dutrow, B.L., and Shore, P. (1986) A model for the crystal chemistry of staurolite, American Mineralogist, 71, 1142–1159.

Jones, J.B., Segnit, E.R., and Nickson, N.M. (1963) Differential thermal and X-ray structure of opal. Nature, 198, 1191.

Jones, J.B., Sanders, J.V., and Segnit, E.R. (1964) Structure of opal. Nature, 204, 990–991.

King, T.V.V. and Ridley, W.I. (1987) Relation of the spectroscopic reflectance of olivine to mineral chemistry and some remote sensing implications. Journal of Geophysical Research, 92, 11457–11469.

Kinny, P.D., and Maas, R. (2003) Lu-Hf and Sm-Nd isotope systematics in zircon. Reviews in Mineralogy and Geochemistry, 53, 327–341.

Leake, B.E., Woolley, A.R., Arps, C.E.S., Birch, W.D., Gilbert, M.C., Grice, J.D., Hawthorne, F.C., Kato, A., Kisch, H.J., Krivovichev, V.G., Linthout, K., Laird, J., Mandarino, J.A., Maresch, W.V., Nickel, E.H., Rock, N.M.S., Schumacher, J.C., Smith, D.C., Stephenson, N.C/N., Ungaretti, L., Whittaker, E.J.W., and Youzhi, G. (1997) Nomenclature of amphiboles: Report of the Subcommittee on Amphiboles of the International Mineralogical Association, Commission on New Minerals and Mineral Names. The Canadian Mineralogist, 35, 219–246.

Leake, B.E., Woolley, A.R., Birch, W.D., Burke, E.A.J., Ferraris, G., Grice, J.D., Hawthorne, F.C., Kisch, H.J., Krivovichev, V.G., Schumacher, J.C., Stephenson, N.C.N., and Whittaker, E.J.W. (2004) Nomenclature of amphiboles: Additions and revisions to the International Mineralogical Associations' amphibole nomenclature. American Mineralogist, 89, 883–887.

Liebscher, A. (2004) Spectroscopy of epidote minerals. Reviews in Mineralogy and Geochemistry, 56, 125–170.

Mandarino, J.A., and Back, M.E. (2004) Fleischer's Glossary of Mineral Species 2004. The Mineralogical Record, Tucson, 309 pp.

Mattson, S.M., Rossman, G.R. (1987) Identifying characteristics of charge transfer transitions in minerals. Physics and Chemistry of Minerals, 14, 94–99.

Morimoto, N., Fabries, J., Ferguson, A.K., Ginzburg, I.V., Ross, M., Seifert, F.A., and Zussman, J. (1989) Nomenclature of pyroxenes. Canadian Mineralogist, 27, 143–156.

Nespolo, M. and Durovič, S. (2002) Crystallographic basis of polytypism and twinning in micas. Reviews in Mineralogy, 46, 155–279.

Peacor, D.R., Rouse, R.C., and Grew, E.S. (1999) Crystal sturcutre of boralsilite and its relation to a family of boroaluminosilicates, sillimanite, and andalusite. American Mineralogist, 84, 1152–1161.

Richardson, S.W. (1966) The stability of Fe-staurolite + quartz. Carnegie Institute of Washington Yearbook, 66, 397–398.

Rieder, M., Cavazzini, G., D'Yakonov, Y.S., Frank-Kamenetskii, V.A., Gottardi, G., Guggenheim, S., Koval, P.V., Müller, G., Neiva, A.M.R., Radoslovich, E.W., Robert, J.-L., Sassi, F.P., Takeda, H., Weiss, Z., and Wones, D.R. (1998) Nomenclature of the micas. The Canadian Mineralogist, 36, 905–912.

Robinson, P.R., Spear, F.S., Schumacher, J.C., Laird, J., Klein, C., Evans, B.W., and Doolan, B.L. (1982) Phase relations of metamorphic amphiboles: Natural occurrence and theory. Reviews in Mineralogy, 9B, 1–227.

Ross, M., Flohr, M.J.K., and Ross, D.R. (1992) Crystalline solution series and order-disorder within the natrolite mineral group. American Mineralogist, 77, 685–703.

Rossman, G.R. (1994) Colored varieties of the silica minerals. Reviews in Mineralogy, 29, 433–468.

Rossman, G.R., Fritsch, E., Shigley, J.E. (1991) Origin of color in cuprian elbaite from São José da Batalha, Paraíba, Brazil. American Mineralogist, 76, 1479–1484.

Rossman, G.R. (2005) Mineral Spectroscopy web site, California Institute of Technology, http://minerals.gps.caltech.edu/

Samoilovich MI, Novozhilov AI, Radyanskii VM, Davydchenko AG, Smirnova CA (1973) On the nature of the blue color in lazurite. Izvestiya Akademii Nauk SSSR. Seriya Geologicheskaya 1937–7, 95–102 (in Russian).

Schreyer, W. (1988) Experimental studies on metamorphism of crustal rocks under mantle pressures. Mineralogical Magazine, 52, 1–26.

Slack, J.F. (1996) Tourmaline associations with hydrothermal ore deposits. Reviews in Mineralogy, 33, 559–644.

Smith, J.V., and Brown, W.L. (1988) Feldspar Minerals: Crystal Structures, Physical, Chemical, and Microtextural Properties, 2nd ed. Springer, 828 pp.

Stephenson, D.A., and Moore, P.B. (1968) The crystal structure of grandidierite, $(Mg,Fe)Al_3SiBO_9$. Acta Crystallographica, Section B: Structural Science, 24, 1518–1522.

Straub, D.W., Burns, R.G., and Pratt, S.F. (1991) Spectral signatures of oxidized pyroxenes: Implications to remote-sensing of terrestrial planets. Journal of Geophysical research, 96, 18819–18830.

Strunz, H. and Nickel, E. (2001) Strunz Mineralogical Tables, 9th Ed. Schweizerbart, Stuttgart, 870 pp.

Sunshine, J.M., Pieters, C.M., and Pratt, S.F. (1990) Deconvolution of mineral absorption bands; an improved approach. Journal of Geophysical Research, 95, 6955–6966.

Thomas, J.B., Bodnar, R.J., Shimizu, N., and Chesner, C.A. (2003) Melt inclusions in zircon. Reviews in Mineralogy and Geochemistry, 53, 63–87.

Thompson, J.B. Jr. (1978) Biopyriboles and polysomatic series. American Mineralogist, 63, 239–249.

Thompson, J.B. Jr. (1981) An introduction to the mineralogy and petrology of the biopyriboles. Reviews in Mineralogy, 9A, 239–249.

Valley, J.W. (2003) Oxygen isotopes in zircon. Reviews in Mineralogy and Geochemistry, 53, 343–385.

Veblen, D.R., and Wylie, A.G. (1993) Mineralogy of amphiboles and 1:1 layer silicates. Reviews in Mineralogy, 28, 61–137.

Verkouteren, J.R. and Wylie, A.G. (2000) The tremolite-actinolite-ferro-actinolite series: Syste-matic relationships among cell parameters, composition, optical properties, and habit, and evidence of discontinuities. American Mineralogist, 85, 1239–1254.

White, W.B., and Keester, K.L. (1966) Optical absorption spectra of iron in the rock-forming silicates. American Mineralogist, 51, 774–791.

Whittaker, E.J.W. (1957) The structure of chrysotile. 5. Diffuse reflections and fibre texture. Acta Crystallographica, 10, 149–156.

Wicks, F.J., and O'Hanley, D.S. (1988) Serpentine minerals : structure and petrology. Reviews in Mineralogy, 19, 91–167.

Chapter 23

Non-Silicate Minerals

Non-silicate minerals can all be considered "minor constituents" when it comes to the Earth's crust, but of course, they loom large in terms of economic importance. Gold, for example, is the 73rd most abundant element in the Earth's crust, and there are only 0.005 g of gold per metric ton of rock in the Earth's crust, about 0.0000005% (that's 5 parts per billion!) if it was evenly distributed (Weiner, 2003). Only geological processes can concentrate gold and bring it into the range of >0.1 g/metric ton to make it economical enough to be mined. Interestingly, one of the largest "deposits" of gold on our planet is seawater, which has for millennia been concentrating gold from run-off of continental erosion. Although gold is present in only small absolute percentages (0.005–0.01 parts per billion), the sheer volume of seawater means that the oceans may hold some 25 billion ounces of gold (roughly 9 trillion dollars at this writing!) (Burk, 1989). Unfortunately, there is no known method by which the gold can be economically extracted from seawater.

More tractable resources to exploit are deposits that contain gold metal as placers or veins, and formations that contain minerals with stoichiometric Au, or even Au substituted into other minerals. There are at this writing 27 mineral species, in addition to the pure metal, with Au in their formulas. Au can also substitute into many minerals such as platinum, tellurium, and even arsenopyrite, pyrite, and chalcopyrite. These variations are made possible by the fact that Au^0 *is similar in size to other pure metals on the periodic chart, while* Au^{3+} *has similar charge and ionic radius (0.68–0.84 Å) to* Sb^{3+} *(0.76 Å) and* Bi^{3+} *(1.03 Å).*

Gold-bearing minerals are thus typical of many non-silicates because they give us a chance to explore variations in crystal structures that result from replacement of ions by other ions. As we learned earlier in this book, cations and anions combine to form polyhedra in ways that can be well-predicted by the ratios of their ionic radii. Non-silicates provide some permutations on possible arrangements because the cations and anions are variable in size, though some of them mimic arrangements found in silicates. For example, minerals in the ***kieserite*** *group of hydrous sulfates (e.g.,* $Mg[SO_4]\cdot H_2O$*) are isostructural with those in the* ***titanite*** *group of nesosilicates (let's write it as* $CaTiO[SiO_4]$*). So* $(SO_4)^{4-}$ *tetrahedra can play the role of* $(SiO_4)^{2-}$ *tetrahedra to create many interesting analogs to more-familiar silicates, with the added twist that* S^{2-} *bonds are predominantly covalent. Many* PO_4*,* AsO_4*, and* VO_4 *minerals are also isostructural with various silicates. Other interesting structures can be found in minerals composed of all the same ions with metallic bonding (as we saw in the various packing schemes for pure metals shown in Chapter 21) or in minerals with anions that differ considerably from the ionic radius of* O^{2-} *(1.40 Å), such as* Cl^{1-} *(1.81 Å),* Br^{1-} *(1.96 Å), and* I^{1-} *(2.20 Å).*

So even if you don't plan to become an economic geologist, the non-silicates can teach you some valuable lessons about the breadth and variety in crystal structures of all minerals, using the knowledge you've (hopefully) gained studying the silicates. In this chapter, we have included examples of nearly all the non-silicate classes so as to illustrate the array of possibilities in naturally-occurring crystal structures. Enjoy!

M.D.D.

Introduction

In the previous chapter on silicates, our emphasis was on the interrelationships among physical and optical properties and crystal chemistry of minerals, though we also stressed the broader importance of each mineral group. In this chapter, our focus will shift to a predominantly structural approach, for several reasons.

First, the majority of these minerals are extremely rare, so it is quite unlikely that you will encounter them in places other than museums. (Unless, of course, you make a career out of economic geology!) For this reason, it seems silly to belabor the physical or optical characteristics of each of these minerals—you can always find detailed information on our mineral database app or other databases. On the other hand, we will be sure to discuss the significance of each mineral (whether as an interesting hand sample or structure, or as an economically- or petrologically-important phase), so you'll know why we chose to include it here. Additional information on many of these minerals can be found in the excellent summary volumes by Chang et al. (1996) and Chang (2002). Many of the *Reviews in Mineralogy and Geochemistry* series are also dedicated to these mineral groups; you can check your library or the Mineralogical Society of America's web site for those volumes.

Second, many of these phases are opaque under polarized light, so it will be very difficult to distinguish among them using the transmitted light microscope you are likely to have in your hands. Training your eye to recognize the subtle differences in reflectivity between opaque phases is an art, and a critical skill for economic geologists in particular. However, this is such a specialized skill that we choose to leave the explanation of reflected light microscopy to others (we particularly recommend the classic texts by Stanton, 1972 and Craig and Vaughan, 1994).

Finally, as noted above, the non-silicates are perhaps most useful as a playing field for learning about the various ways in which cations and anions can combine to form mineral structures. These minerals provide superb demonstrations of many of the principles we have continually reinforced in this book, such as:

- Radius ratios govern the sizes of anion polyhedra and the resultant site occupancies of variably-sized cations.
- Elements with similar charge and size will commonly substitute for each other, creating new minerals.
- Mineral structures are commonly composed of linked tetrahedra that share 0–4 of their corners. In the non-silicates, the tetrahedra might be SO_4, PO_4, AsO_4, and VO_4, etc.—but the structures will still look familiar!

A Word About Nomenclature and the Organization of this Chapter

We explained in Chapter 21 that there are several different classification systems used by mineralogists: most notably, the Dana and Strunz schemes. New mineral species that are approved by the international commission are usually assigned classification numbers. However (and unfortunately for us), mineral *group* names are not strictly regulated, and the predominant classification schemes do not always agree on groupings. As a result, different books and references will include slightly different mineral species in different groups. These discrepancies have made it a challenge to develop a consensus for the tables in this textbook. Furthermore, for the sake of brevity, the tables in this chapter *do not represent* all the species in each mineral class, but only a small sub-set of species.

So in writing this chapter, we will follow the non-silicates from simple structures to more complex ones, roughly in keeping with the order of both the Dana and Strunz systems. The tables in this chapter are grouped by mineral class, generally using the Dana terminology, and they bring together minerals with similar structures and chemistries. For the purpose of thinking about chemical substitutions, it is quite useful to see all the related species together, *but please don't memorize them all*. Your instructor will stress the ones that he or she thinks are important. The tables of formulae should really be viewed as lessons in how nature makes systematic atomic, cation and anion substitutions...

Close Packing, Revisited

Many of the non-silicates are built from close-packed structures, which we first described in Chapter 8. These are the most efficient ways for densely packing atoms that are all the same size, in a way that minimizes the amount of pore space between atoms. *Many non-silicates have perfect close-packed structures*, so understanding of this concept makes it easier to describe them. We should note that some workers also use this methodology to describe close packing of oxygens in *silicate* structures, but the complexity of the deviations from ideal close packing makes this notation cumbersome at the introductory level.

Recall from the example presented in Figure 8.3 that structures can be built from layers of equal-sized atoms. In any one layer there are six atoms arranged in a hexagonal shape with a void in the middle large enough to accommodate a seventh atom of equal size. Also, recall that there were different ways these layers of atoms could be stacked one on top of the other to form different structures. When minerals have the same layers stacked in a different way, we say these materials are polytypes of each other.

To explore this in more detail, consider first a general hexagonal cell, which can be represented as a rhombus with 120° and 60° angles. At each of the four corners of the rhombus we will place an atom labeled A. If we think of this as a unit cell, then we would have balls at (0,0), repeated by symmetry to lie at all four corners of the rhombus. You can imagine that this simple hexagonal lattice could be extended infinitely in two dimensions to form a closely-packed sheet. (We urge you experiment with this problem by gluing together some gumdrops with frosting and pursuing this analogy with your own hands!)

What happens when we stack another, identical sheet above the first one? Returning to our simple rhombus, we note that there are always three possible locations for further stacking:

1. Stack the balls in the new sheet directly above the balls in the existing one. Thus, the new balls would lie above (0,0) in our new cell. It turns out this isn't a very efficient way of stacking.
2. Place the balls in the new sheet *in between* the corners of the bottom sheet. One way of doing this is to position them directly above (1/3,1/3) in the rhombus (the "B" location; see Figure 23.1).
3. An alternative stacking pattern could be created by positioning our new sheet directly about (2/3,2/3) in the bottom sheet (the "C" location).

Different types of close packing use different repetitions of this patterns. Hexagonal close packing (hcp) repeats stacks the sheets together by placing atoms alternating between the A (0,0,0) and B (1/3,1/3,1/2) positions (Figure 23.1), resulting in repetitions of ABABAB etc. Cubic close packing (ccp) repeats atoms by positioning them above the A (0,0,0), B (1/3,1/3,1/3), and C (2/3,2/3,2/3) positions, so you have repetitions of ABCABC etc. Sometimes you can even have complicated stacking patterns, like those for lanthanum (ABACABAC...) and samarium (ABAB CBCAC...)

You may wonder why stacking sequence 3 is called "cubic" when it is based on a repetition of hexagonal unit cells? In ccp structures, the plane in which the starting sheets of close-packed atoms lies is actually perpendicular to the 3-fold axis of a cube. The ABC stacking thus occurs along the body diagonal of a cube, as is shown in Figure 23.2. For this reason, ccp structures are generally in the isometric system.

Many possible crystal structures can be assembled from close-packed structures by placing cations in the holes between the large, close-packed anions. Close inspection of this structure will show you that it contains two types of "holes"—those surrounded by four anions, which form sites for 4-coordinated cations, and those bounded by six anions, forming 6-coordinated sites. It is helpful to think about where these interstices are located—if you have made a gumdrop model, you can start slicing it up until you find the holes! Perhaps more easily, the locations of those holes can be mathematically deter-

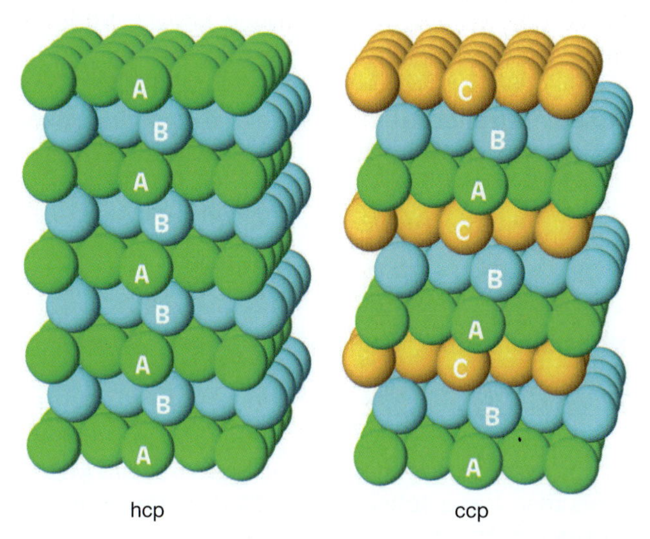

Figure 23.1. *Two types of close packing of spheres begin with a general hexagonal cell, expanded in two directions to create the layer of balls as shown here with the label A. If you repeat this layer of spheres with the balls in the new sheet lying directly above the balls below, there will be a lot of space between the balls—this would be an inefficient packing scheme! If the balls in the second layer are placed* above the holes *in the bottom layer (i.e., above the location that is 1/3 in the* a *direction and 1/3 in the* b *direction), the balls will be packed with maximum efficiency (see layer B). It is equally possible to stack the second layer above the (2/3,2/3) location in the original layer, creating the C layer in the drawing. Different types of close packing use the A, B, and C layers to create structures. Hexagonal close packing (hcp) repeats the layers by alternating atoms between the A (0,0,0) and B (1/3,1/3,1/2) positions resulting in repetitions of ABABAB etc. Cubic close packing (ccp) repeats atoms by positioning them above the A (0,0,0), B (1/3,1/3,1/3), and C (2/3,2/3,2/3) positions., so you have repetitions of ABCABC etc. Other, more complicated stacking sequences are also possible!*

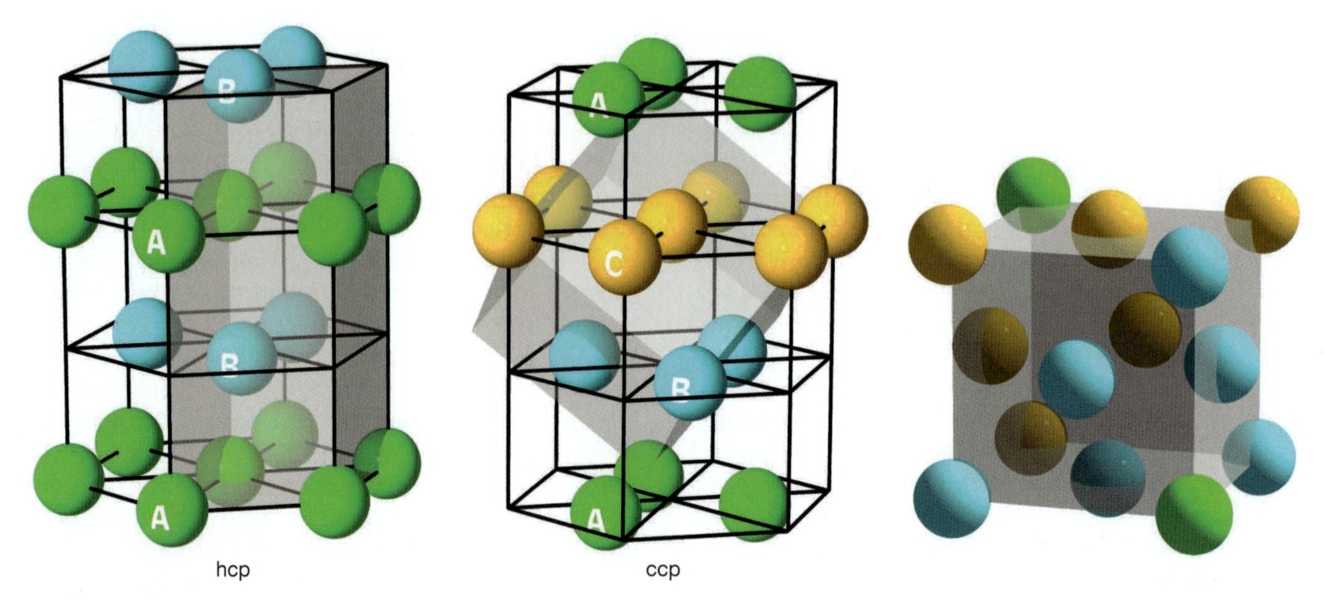

Figure 23.2. *Comparison of hexagonal close packing (left) and cubic close packing (center). Hcp stacking repeats only two layers, offsetting each layer by one-third of the long diagonal in the base of the cell. Note that the atoms in each "A" layer line up directly with all the other "A" atoms. Ccp packing (center) arises from stacking of three different A, B, and C layers, but it results in a shape that is cubic, as shown when the structure is rotated (right).*

mined. Tetrahedral "interstices" in a *cubic* close-packed structure lie at:

$$(1/4,1/4,1/4)\ (1/4,3/4,1/4)$$
$$(3/4,1/4,1/4)\ (3/4,3/4,1/4)$$
$$(1/4,1/4,3/4)\ (1/4,3/4,3/4)$$
$$(3/4,1/4,3/4)\ (3/4,3/4,3/4)$$

while octahedral "holes" occur at:

$$(1/2,0,0)\ (0,1/2,0)$$
$$(0,0,1/2)\ (1/2,1/2,1/2).$$

In a similar way, hexagonal close-packed structures also have spaces between the close-packed atoms, with 4-coordinated holes at (0,0,3/8), (0,0,5/8), (1/3,2/3,1/8) and (1/3,2/3,7/8). The 6-coordinated holes are at (2/3,1/3,1/4) and (2/3,1/3,3/4).

It is rare that cations will occupy all of the tetrahedral *and* octahedral interstices in any given structure; that proximity would bring positively-charged cations into uncomfortable proximity. Instead, some sites will be occupied and others will be left empty (vacant). The sulfide minerals will provide us with particularly good examples of these variations. The distribution of cations vs. vacancies also varies with the mineral. Sometimes, the cations and vacancies will be **ordered**, which means they have a systematic, organized repetition of occupancy. For example, the octahedral sites in a ccp array might be occupied at the (0,0,0) and (1/2,1/2,1/2) locations and vacant everywhere else, or you might always have Fe at (1/2,0,0) and (0,1/2,0) and Zn at (0,0,0) and (1/2,1/2,1/2). In other cases, the cations and vacancies will be randomly distributed; this is called **disorder**. Typically, disordered polymorphs will occur at high temperatures, while the ordered ones are usually (but not always!) stable at room temperatures.

Finally, it is important to mention again the two related lattice types that can be used to describe simple mineral structures. **Face-centered** lattices have motifs at all corners and at the center of each face. **Body-centered** lattices have motifs at each corner of the unit cell, as well as a single motif in the very center. These two lattice types form the bases for different types of close packing.

Because the majority of minerals are built around anions (like O^{2-}) that are so much bigger than commonly-occurring cations (like S^{6+} and Si^{4+}), we will find that an understanding of these structure types will come in very handy! Please take the time to study and visualize the different types of close-packed arrays to pave the way for the study of non-silicate and other structures.

Native Elements

This group includes mineral species that occur in "pure" forms in deposits; examples are given in Table 23.1. "Pure" here implies that an element is not bonded to any other elements. From an economic standpoint, pure minerals are the best because they do not need to be processed (i.e., no energy needs to be used to separate the elements of interest from minerals containing unwanted elements). The U.S. and other governments monitor mining and quarrying trends of these minerals very carefully. Some excellent statistics and information on the worldwide supplies, demand, and flow of minerals and materials essential to the U.S. economy can be found on the website of the U.S. Geological Survey (USGS) under "Minerals Information." Many of the USGS publications make for fascinating reading! For example, only ten states accounted for ~63% of the tonnage of non-fuel minerals mined in the U.S. in 2003—you can check the website to learn which ones these were! The economic importance of these operations is also made quite clear: in 2003, 4.1 billion metric tons of ore were mined or quarried, along with 1.4 billion metric tons of mine and waste were produced. Of these, 62% were for the production of minerals used in industry and 38% were for metals.

Another excellent source of information on the structures of these minerals is the website of the American Society for Metals in Metals Park, Ohio. In particular, their *Metals Handbook*, which is available in two desk editions as well as twenty-one packed volumes, summarizes all the essential information you might want to know about metals, including their crystal structures. Your library

Table 23.1. Native elements

Group	Mineral	Formula
Gold	gold	Au
	silver	Ag
	copper	Cu
	lead	Pb
	aluminum	Al
Iron-nickel	iron	Fe
	kamacite	α-(Fe,Ni)
	taenite	γ-(Fe,Ni)
	tetrataenite	FeNi
	awaruite	Ni_2Fe-Ni_3Fe
	nickel	Ni
	wairauite	CoF
Platinum	platinum	Pt
	iridium	(Ir,Os,Ru)
	rhodium	(Rh,Pt)
	palladium	Pd
Arsenic	arsenic	As
	antimony	Sb
	stibarsen	SbAs
	bismuth	Bi
	stistaite	SnSb
	sulfur	S
	diamond	C
	graphite	C
	lonsdaleite	C
	chaoite	C
	silicon	Si

may have these books in its collections, or you may be lucky enough to have on-line access to the handbooks. These references constitute the most important source of information on all aspects of the industries that depend on metals.

Minerals in the gold group are shown in Table 23.1. The term derives its name from the fact that all of these minerals have the same crystal structure as gold. Because they contain different atoms, the size of the unit cell and other physical properties may vary. They all share the same simple cubic close-packed structure, with horizontal layers of ABCABC close-packed atoms perpendicular to the [111] direction; the resultant space group is $F\ 4/m\ \bar{3}\ 2/m$. The repetitive ABC arrangement results in a pattern of cubes with atoms in the center of all four faces (Figure 23.2). Using such x, y, z orthogonal coordinates, we would describe the locations of the atoms as (0,0,0), (½,½,0), (0,½,½,), (½,0,½) etc.

Mineral species in these groups are held together by metallic bonding. This means that each individual atoms has effectively "lost" its outermost, valence electrons to the constantly-moving sea of electrons that circulates between the nuclei. These pure metals are the type examples of metallic bonding, which gives rise to a set of diagnostic physical properties, including, most prominently, metallic luster.

Metals are shiny because the range of possible transitions between the electrons and the possible, unoccupied valence shells is large, and so the range of energies (light) that can be absorbed covers the whole visible region. The energy put into the crystal can be either absorbed, reflected, transmitted, fluoresced, or converted to thermal energy. When that energy is released, in the case of metals, it comes back out as scattered light. If the scattering is intense and unidirectional (mirror surface), it looks like a metallic reflection.

The lifetime of the absorbed electrons must be so short that there is not time to convert the excited state into thermal energy. If the excited state is so long-lived that a lot of conversion occurs, visible photons change to infrared photons, the crystal heats up a tiny bit, and the surface appears comparatively dark. Which is to say, not much of the light will be scattered; it will all be absorbed. Even if a lot of the energy is re-emitted as scattered light, the appearance is not metallic unless the surface is smooth, so all the light reflects back in nearly a single direction, like a mirrored surface. Finely-powdered metals are dark black; smooth metals are silver (or gold, or whatever).

High electrical and thermal conductivity result because photons or delocalized electrons pass efficiently from one atom to another as the result of close packing, which also makes these minerals very dense. The malleable and ductile tenacity of metals occurs because the metallic bonds are equal in all directions, but rather weak. Each atom also possesses several unoccupied valence orbitals that make it quite easy for electrons to move. This allows atoms to be "repositioned" and slide around when the structure is stressed by pulling or pushing.

The **gold group** minerals include gold, silver, copper, lead, and aluminum. With the exception of aluminum, which was not discovered as a mineral until 1978 (!), these have been known and used since antiquity. The differences between their structures can best be understood by thinking about their sizes. For each species, the size of the ions can be defined by looking at the distance between metals in the pure substance—for example, half the distance between two adjacent Au atoms in pure gold. The resultant radii are Au (1.442 Å), Ag (1.444 Å), Cu (1.278 Å), Pb (1.750 Å), and Al (1.432 Å). Note how the size varies as a function of each element's location on the periodic chart. With this information and the knowledge that all these minerals have the same structure, we also can predict the variations in the sizes of their unit cell volumes: a_{Au} = 4.0786 Å, a_{Ag} = 4.0862 Å, a_{Cu} = 3.615 Å, a_{Pb} = 4.9505 Å, and a_{Al} = 4.040 Å because there are two metal-metal bonds coinciding with the face diagonal of the cell, $a\sqrt{2}$.

The size of the element also affects substitutions that can occur between elements. For example, the sizes above explain why there is complete solid substitution between Ag and Au, but only rare substitution between Ag or Au and Cu. In fact, native gold of the kind that was found by miners in the California Gold Rush is rarely pure; it almost always occurs with some substitution of Ag. Metallic copper, on the other hand, is almost always found in nearly pure form with only trace amounts of substitution (usually by As), although the ionic form of copper, Cu^{2+}, does occur in other minerals as we'll see later in this chapter.

Uses of the precious metals in the gold group are well known, but the more abundant Cu, Pb, and Al metals are cornerstones of many industrial uses from wires to baseball bats. Aluminum bats are banned from professional baseball because they can flex and spring back elastically (thanks to their metallic bonding). This quality allows more energy to be transferred from the bat to the baseball (interestingly, the addition of Sc to the Al further increases the strength of the bat).

Iron-nickel group mineral species are nearly all combinations of Fe and/or Ni (Table 23.1); these are close neighbors in the same row of the periodic chart, with only Co between them. The two elements have similar atomic radii (Fe^0 is 1.241 Å, Ni^0 is 1.246 Å) so that they freely substitute for each other at high temperatures (above ~900°C). Below that temperature, the size mismatch causes separate structures to form. The resultant exsolution is similar to that seen in silicates such as feldspars!

Each of these mineral species has a different crystal structure. Pure metallic iron (sometimes called α-Fe) has an atomic arrangement based upon a body-centered cubic lattice type (bcc). The resultant cubic structure has atoms at the corners (0,0,0) and in the center (½,½,½) of a cube (Figure 23.3). Kamacite (Figure 23.3), which is nearly pure Fe with a bit of Ni, also has the bcc structure. Taenite (Figure 23.3), alternatively, is based on a face-centered cubic lattice type.

Iron meteorites, which are composed of Fe-Ni mixtures that originate in the cores of differentiated planets, contain the minerals kamacite (α-(Fe,Ni)) and taenite (γ-(Fe,Ni)). The difference between the two lies in the amount of Ni substitution each structure can accommodate. Kamacite is usually less than 5–7% Ni , while taenite has higher Ni from 7–37%. As meteorites slowly cool, the exsolution pattern between kamacite and taenite forms an intriguing pattern of interlocking triangular and square crystals known as a Widmanstätten pattern. The larger the crystals, the slower the cooling rate! These patterns are so diagnostic of meteoritic formation that they can be used to trace the origins of meteorites in ancient objects. Many different primitive cultures made ceremonial objects for royalty, such as sword blades, out of meteoritic material.

Platinum group minerals (Table 23.1) have ccp structures that are all isostructural with species in the gold group. These minerals contain the elements Pt, Ir, Os, Ru, Rh, and Pd, which all freely substitute in complete solid solutions. Platinum itself rarely occurs in pure form but is usually mixed with the elements listed above, as well as with Fe. You may be familiar with its use as a durable white metal in jewelry, but it also has many industrial applications. It is used in thermocouples, as a chemical catalyst, in cigarette lighters, and dental fillings.

Arsenic group minerals are considered semimetals because they have not quite metallic properties and are not malleable (Table 23.1). They share cubic close-packed arrangement of atoms arranged in a rhombohedral lattice (Figure 23.4). Each atom is closely bonded to three neighbors, and less closely bonded to three other neighbors. The inequity in bond length, which results from variations in the strength of the connecting covalent bonds, causes the sheets to pucker, thus deviating from simple close packing. The longer dis-

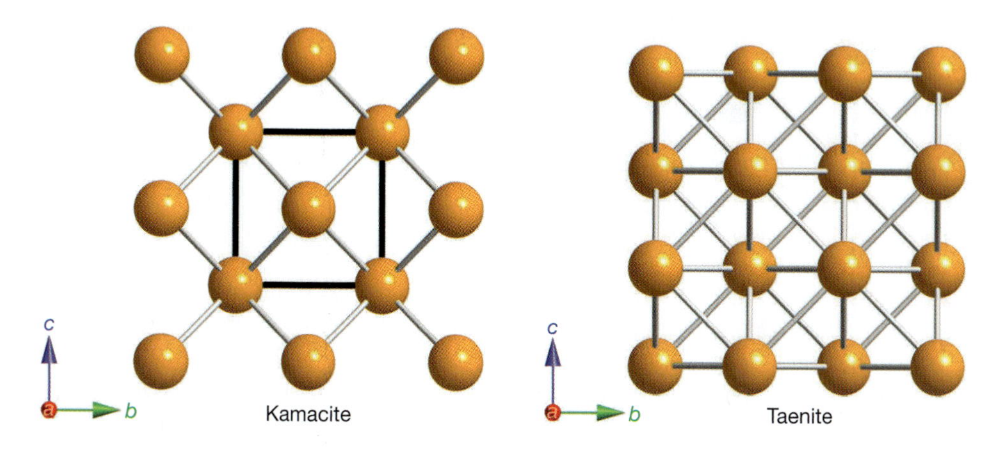

Figure 23.3. The three most common forms of Fe metal include pure iron metal (α-Fe) and kamacite (left), which have similar, body-centered cubic lattice types, and taenite (right), which has a face-centered lattice type because of an ordered substitution of Ni for Fe.

tances between the puckered sheets result in weaker bonds *between* the sheets than *within* the sheets, so these minerals have perfect basal cleavages.

The remaining minerals in Table 23.1 are defined individually and don't belong to the above groups. **Diamond** is formed from pure carbon at high pressures. The great hardness of diamonds (number 10 on the Mohs scale!), is the result of strong covalent bonding among carbon atoms in its close-packed structure. As we noted in Chapter 8, each carbon atom at a corner of the cube is linked with the three nearest face-centered atoms, as well as a fourth carbon atom that lies at the center of that tetrahedron. The short bond lengths in these tetrahedra result in a very strong framework (Figure 23.5).

Diamond is isostructural with **silicon**, which was not even known to be a naturally-occurring mineral until it was described and approved in 1983. It is found (rarely!) in volcanic rocks and as inclusion in xenoliths that come from the mantle. Synthetic silicon has tremendous importance in the manufacture of semiconductors. Its value lies in the fact that silicon crystals, like those of diamond, are bound together very tightly by covalent bonding. The four outer valence electrons are completely filled by electrons. Semiconducting devices are made by introducing very tiny amounts of impurity to ultra-pure silicon crystals. In some cases, this dopant has one or more valence electron than silicon, such as P or As. This introduces a fifth electron to the valence band, and the addition of a small amount of energy to the crystal will easily allow that electron to move to the higher, unoccupied orbitals that make up the conduction band (n-type silicon). Alternatively, silicon can be doped with an element with only three valence electrons, such as B, Ga, or Al. The resultant loss of an electron from the valence band creates what is called an electron hole (p-type silicon). Silicon wafers of these two types of silicon are commonly stacked so that when current is applied, the electrons and holes move through their respective crystals, trying to get together. When current is removed, they return to the energy states where they started, releasing the extra energy in the form of light to create a diode.

Graphite is a lower pressure polymorph of diamond. In this case, the C bonds link up groups of four C atoms, which then join up to form 6-membered rings (Figure 23.6). Recall from Chapter 8 that this results from the overlap of p orbitals in this highly-covalent structure. The three orbitals form coplanar lobes 120° apart, leaving the 4th electron in the p orbital lying out of the plane. This p orbital overlaps sideways to form what are called π-bonds, so that the 4th electron is detached from its atom. The charge associated

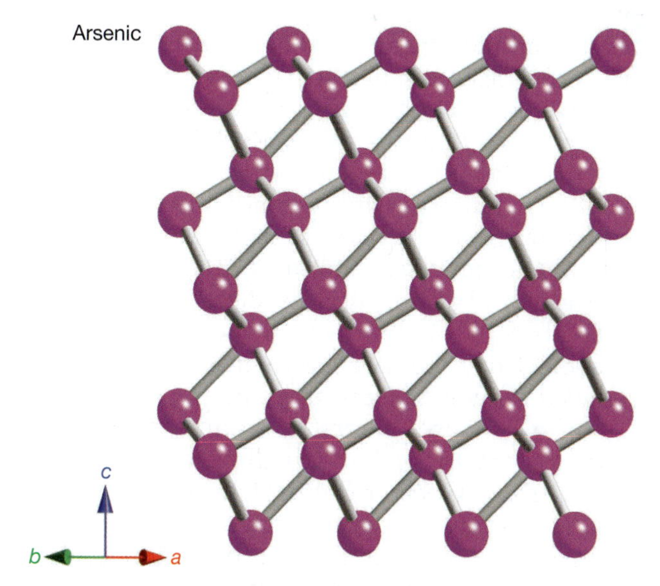

Figure 23.4. Arsenic has a nearly cubic close-packed atomic structure. However, variations in bond length and degree of covalency cause some layers in the structure to be compressed, while others are lengthened. The structure is thus composed of alternating thin and thick layers of As atoms. To better see this, use CrystalViewer and link to this file found on the MSA website.

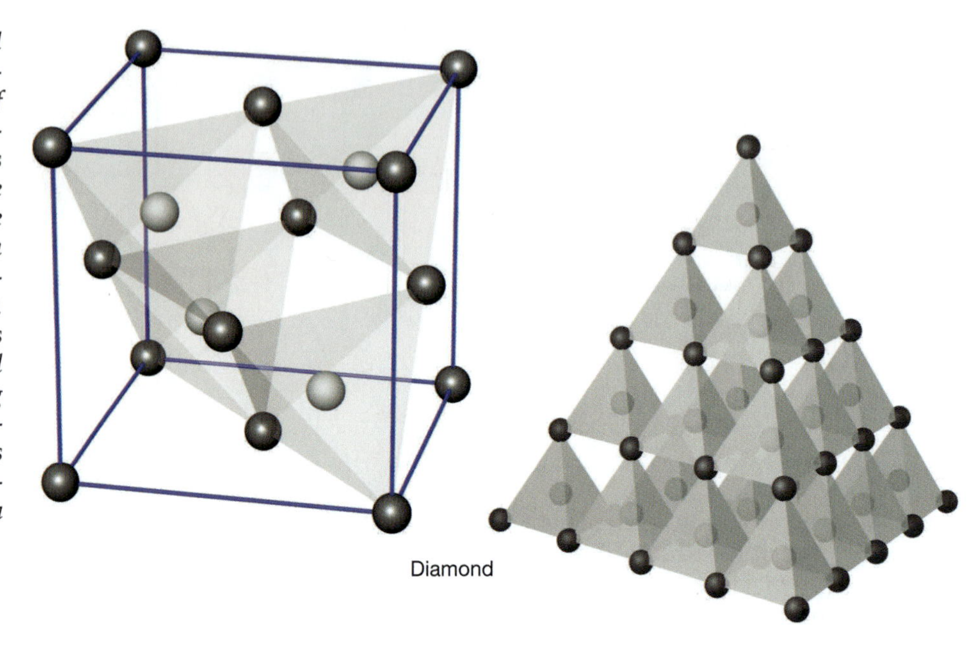

Figure 23.5. The diamond structure is composed of a face-centered cubic arrangement of carbon atoms. Tetrahedra formed by the atoms in the corners and centers of each face of the cube can be joined up to make tetrahedra, as shown here. Each tetrahedron then has an additional carbon atom at its center. The angle of the bonds is 109.28°. This structure is held together by covalent bonding that is very strong in all directions. Thus, diamond earns its place as one of the hardest mineral species known, with a value of 10 on the Mohs scale.

with this 4th electron forms a weak van der Waals bond. Thus, the carbon atoms have a strong, covalently-bonded hexagonal sheet structure, but the sheets are held to other sheets by only the weak forces of van der Waals bonds, allowing graphite to cleave easily along those bonds.

Elemental **sulfur** comes mostly from refining of crude oil, where it is present as part of complex organic compounds, and from purification of natural gas, which contains hydrogen sulfide (H_2S). Sulfur is also mined in underground deposits where it is commonly associated with salt domes. A less common occurrence is in hot springs and modern volcanic regions; if you've ever visited Yellowstone National Park, you have smelled the sulfur in the deposits there. Note that the diagnostic "rotten eggs" smell of sulfur is actually the result of interaction with water in air, which produces H_2S gas—it is this gas that smells, rather than the sulfur itself. Sulfur is in great demand for industry because it is used to make sulfuric acid (H_2SO_4), which is in turn used to make batteries, fertilizer, rubber, insecticides, and gunpowder, among many others.

Sulfur usually occurs at room temperature with an orthorhombic structure consisting of kinked rings of eight S atoms held together by covalent bonding to form what are effectively S_8 molecules. Within the rings, each S atom is closely bonded to two other S atoms. Van der Waals bonding holds the S_8 molecules together (Figure 23.7), and the unit cell contains 16 rings. Although the orthorhombic structure is most common, there are also monoclinic (formed above 95.4°C) and amorphous forms. Rhombohedral S_6 can also be synthesized, as can S_7, S_{18}, and even S_{20} (see citations in Wuensch, 1982).

Sulfides and Related Structures

The element sulfur has the ability to form mineral structures while acting as a pure element, S^0, (as just discussed above), as an anion (S^{2-}, where it has an "ionic" radius of 1.84 Å) or as a cation

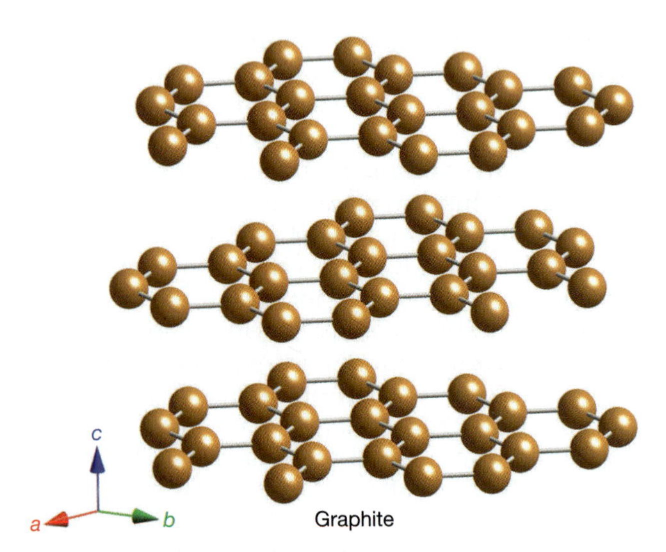

Figure 23.6. The structure of graphite is also composed of covalently-bonded carbon atoms. In this case, sets of C bonds create coplanar triangles that share corners to build sheets. The sheets of rings are very tightly bound by overlapping p orbitals. Only weak van der Waals bonds hold the sheets together; these are easily broken, allowing graphite to cleave easily.

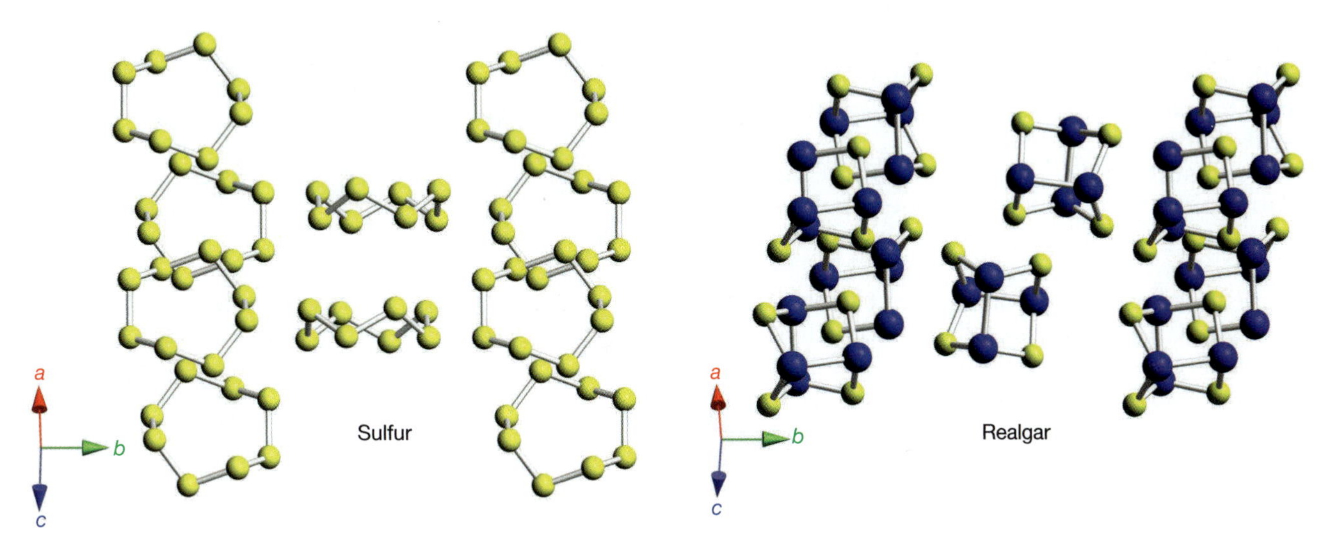

Figure 23.7. *The structure of elemental sulfur (left) is composed of molecules of eight S atoms bound together in rings resembling crowns. Within the rings, each S atom is closely bonded to two other S atoms, but the bonds between the molecules are weak van der Waals bonds. This structure has a huge unit cell, with 16 S_8 rings. Realgar (right, As_4S_4) has a very similar structure, with As substituting for every other S atom in the rings.*

(S^{4+} or S^{6+}, with ionic radii of 0.37 Å and 0.12–0.29 Å, respectively). This "adaptability" arises from the atomic configuration of S, which is $1s^22s^22p^63s^23p^4$. With this arrangement, ionic sulfur can occur by several means. If it gains two electrons, its outer *p* orbitals will be filled; this is the S^{2-} anion, which forms the structures in the sulfide mineral group. The S^{2-} is a lot smaller than O^{2-} (~1.35 Å), and its electronegativity (2.5) is smaller than that of O^{2-} (3.5), facilitating covalent and metallic bonding. So crystal structures composed of sulfur arrays look quite different from those formed from oxygen arrays. We got our first glimpse of some of these minerals in Chapter 8 when we reviewed the nomenclature for structure types used by materials scientists.

On the other hand, elemental sulfur can lose electrons to empty out its outer shell, resulting in the S^{4+} (if it loses its outer four *p* electrons) or S^{6+} valence states (loss of outer *p* and *s* electrons). Minerals with S^{4+} are rigorously called **sulfites**; the S^{4+} occupies a triangular pyramid in coordination with oxygen. Minerals with $(S^{6+}O_4)^{2-}$ are **sulfates**. The tiny S^{6+} is found in tetrahedral coordination and forms SiO_4-like structures; these will be discussed later in this chapter when we get to the sulfates. Interestingly, the charge on the S cation depends mainly of the formation conditions (i.e., oxidizing vs. reducing) environments. Thus we can often interpret formation conditions depending on the speciation of S; this also holds true for all elements (e.g., Fe, Se, etc.)

Minerals in the sulfide groups, despite containing S^{2-}, are largely bonded by covalent or metallic bonding. Thus the number of atoms in a formula is not constrained by charge balance to simple ratios, and many sulfides can have variable compositions. Substitutions in small amounts are also common for these minerals. In some cases, it is speculated that such "impurities" are needed to stabilize the sulfide structures.

Many sulfides are based upon hcp or ccp structures. These close-packed arrays can then be modified by various means, including substitutions of cations or anions, orderly omission of atoms (resulting in vacancies), the addition of atoms to formerly empty sites (for example, the tetrahedral or octahedral interstices discussed above in the ideal closest-packed layers), or distortion of the entire array. Many of the sulfide compositions also result in **polymorphism**, in which two minerals with distinct crystal structures can occur with the same chemical composition. These permutations are apparently facilitated by the combinations of covalent, metallic, and ionic bonding that occur with these structures, as well as formation conditions.

The following discussion of sulfide minerals will be organized using the Strutz system. We will first discuss the **simple sulfides**, which are divided up and organized on the basis of the ratio of metal (Fe^{2+}, Cu^{2+}, etc.) to sulfur: M:S = 2:1, 1:1, 3:4 and 2:3, 1:2, and higher. These will be followed by the **sulfosalts**, in which the fundamental structural unit is a trigonal or square pyramid. In these minerals, a metal cation such as As^{3+}, Sb^{3+}, or Bi^{3+} is found at the tip of the pyramid, while the base of the pyramid consists of three (or sometimes four) sulfur neighbors. For more information, we also highly recommend Volumes 1 and 65 of the *Reviews in Mineralogy and Geochemistry* series.

Metal:sulfur>1:1, mainly 2:1. Chalcocite group minerals form when copper (Cu^{1+}) and S^{2-} combine in various ratios with the approximate formula $Cu_{2-x}S$, as shown in Table 23.2. The **chalcocite** species, Cu_2S, and the closely-related djurleite ($Cu_{1.97}S$), are based on hcp arrays of S^{2-} anions (Figure 21.5). One third of the Cu^{1+} cations occupy 3-fold (triangular) interstices in the ccp array of S^{2-} anions; these form sheets like those observed in covellite. The remaining Cu^{1+} cations occupy sites between the sheets. As the amount of Cu^{1+} in the interstices of the close-packed array decreases, the array changes. Digenite ($Cu_{1.80}S$) has two polymorphs, both based on ccp S^{2-} arrays with Cu^{1+} in 6-, 4-, and 3-coordinated interstices. It usually contains about 1% substitution by Fe^{2+}, which may actually be needed for the mineral to form. These structures are quite similar to those of pyrrhotite (Fe_7S_8), which contains Fe^{2+} instead of Cu^{1+}, and nickeline (NiAs) (Figure 21.1).

Bornite (Cu_5FeS_4) has an fcp structure of S^{2-} with Cu^{1+} and Fe^{2+} cations in 4-fold interstices; it has an orthorhombic unit cell. The structure of the room temperature polymorph can be visualized as two types of ccp cubes that alternate along each of the three crystallographic axes; each cube has eight tetrahedral interstices. One sub-cell has an ordered arrangement of four Cu^{1+} and four vacancies (the four vacancies cluster together to form a little tetrahedron), while the other sub-cell has all eight tetrahedral sites filled by either Cu^{1+} or Fe^{2+} in at least a partially-ordered arrangement (Figure 23.8).

Table 23.2. Sulfides with metal:sulfur >1:1, mainly 2:1

Group	Mineral	Formula	
Chalcocite	chalcocite	Cu_2S	
Chalcocite	djurleite	$Cu_{31}S_{16}$	($Cu_{1.97}S$)
Chalcocite	digenite	Cu_9S_5	($Cu_{1.80}S$)
Chalcocite	roxbyite	Cu_9S_5	($Cu_{1.78}S$)
Chalcocite	anilite	Cu_7S_4	($Cu_{1.75}S$)
Chalcocite	geerite	Cu_8S_5	($Cu_{1.60}S$)
Chalcocite	spionkopite	$Cu_{39}S_{28}$	($Cu_{1.40}S$)
	bornite	Cu_5FeS_4	
	acanthite-argentite	Ag_2S	
Pentlandite	pentlandite	$(Fe,Ni)_9S_8$	
Pentlandite	argentopentlandite	$Ag(Fe,Ni)_8S_8$	
Pentlandite	cobalt pentlandite	$(Co,Fe,Ni)_9S_8$	
Pentlandite	shadlunite	$(Pb,Cd)(Fe,Cu)_8S_8$	
Pentlandite	manganese-shadlunite	$(Mn,Pb,Cd)(Cu,Fe)_8S_8$	
Pentlandite	geffroyite	$(Cu,Fe,Ag)_9(Se,S)_8$	

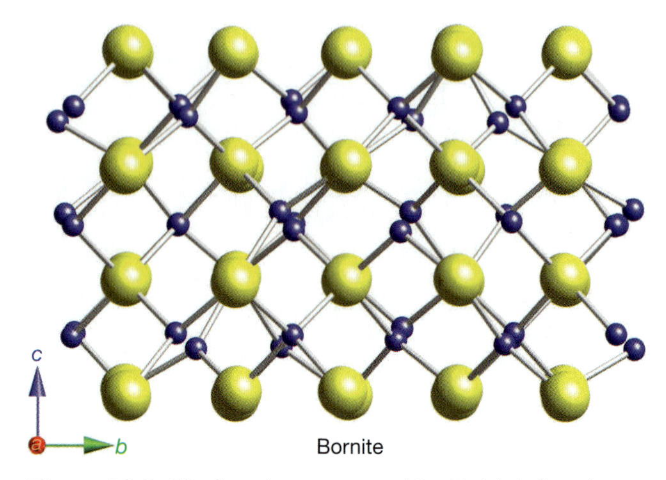

Figure 23.8. The bornite structure (Cu_5FeS_4) is based on a cubic close-packed arrangement of S^{2-} anions (large yellow spheres), with Cu^{1+} and Fe^{2+} (small blue spheres, which are disordered) in the tetrahedral interstices.

Bornite is commonly found in copper deposits, and has a diagnostic brownish-purple surface when fresh. However, it tarnishes quickly upon exposure to air, taking on a characteristic, iridescent purple color that makes it a favorite among collectors. This "peacock" iridescence arises because the air causes oxidation of the surface, and the bornite reacts to form copper oxides, hydroxides, and sulfates in a thin surface layer (~100 nm thick!; see Buckley et al., 1984). The layer is roughly the same thickness as the wavelength of light in the visible region. Thus, interactions of light waves between the bornite surface and the oxides/hydroxides cause iridescence when light waves pass through the surface layers at various angles, resulting in reflection of different colors. In the case of bornite, these are commonly purple to blue. This same interference of light effect causes the color in peacock feathers; thus the derivation of the term as applied to this mineral!

The combination of Ag^{1+} and S^{2-} in exactly a 2:1 ratio gives rise to the mineral species **acanthite** (Ag_2S) below 173.5°C and **argentite** (β-Ag_2S) above that temperature. The use of a Greek letter in front of a chemical formula as a prefix refers to different phases (i.e., crystal structures) with the same formulas (i.e., these minerals are polymorphs of each other, and more often in mineralogy have different names). Both structures are bcc arrays of 4-coordinated S^{2-} anions with AgS_3 triangles linked up to form planes. The planes are linked together by Ag^{1+} cations that share S^{2-} anions in adjoining sheets. You have probably seen acanthite in your own house or on jewelry, as it is the mineral that makes up tarnish on sterling silver. Acanthite, which is mined predominantly in Mexico, is one of the most important ore sources of silver; it also occurs in gold deposits.

The **pentlandite** group consists of six minerals (Table 23.2) with the same structure and the general formula of AB_8X_8. The A element can be Ag^{1+}, Cd^{2+}, Mn^{2+}, Fe^{2+}, Ni^{2+}, Co^{2+}, or Pb^{2+}; the B can be Fe^{2+}, Ni^{2+}, Co^{2+}, or Cu^{2+}; and the X is either S^{2-} or Se^{2-}. All these minerals have the same ccp arrangement of S^{2-} (or Se^{2-}) anions. Within this framework, eight edge-sharing BS_4 tetrahedra form clusters; these are bound into a framework by linking A-occupied octahedra (Figure 23.9). This structure closely corresponds to those of chalcocite (similar except for its hcp arrangement) and sphalerite (see below).

The most common of these is **pentlandite,** $(Fe^{2+}, Ni^{2+})_9S_8$. It is the most abundant nickel mineral and occurs worldwide. The extracted Ni^{2+} is used to make Ni-steel, cast iron, and alloys with copper and silver. It is sometimes added to coins and stainless steel for hardening, and is used for armor plating of various types. Ni^{2+} adds a green color to glass.

Metal:sulfur=1:1 and smaller. Covellite (CuS) and its near-twin **klockmannite** (CuSe) represent the group of minerals based on a 1:1 ratio between metals and other "anions" such as S^{2-}, Se^{2-}, and Te^{2-}. Despite their simple formulas, these structures are quite complex. Covellite is composed of sheets of CuS_4 tetrahedra that share corners, linking up their bases to form continuous layers. The layers of CuS_4 tetrahedra join up by sharing apices to form double sheets, and another Cu^{2+} cations occupies the trigonal interstices among the apices. (This is analogous to the structure of many sheet silicates, which contain rings of SiO_4 tetrahedra that share corners.) These doubled sheets alternate with sheets containing Cu^{2+} bonded to three S^{2-} anions in a flat triangle (Figure 23.10), much like what we see in carbonates. The doubled tetrahedral and triangular layers are held together by S-S bonds. The sheet-like structure is responsible for the fact that covellite has perfect basal cleavage and can be easily split into sheets (like a mica). Iridescence is sometimes observed on the sheets, and is caused by oxidation of the Cu^{2+}, in a manner similar to that of bornite. Covellite is used as a copper ore, but is also highly prized by collectors due to its attractive color and luster.

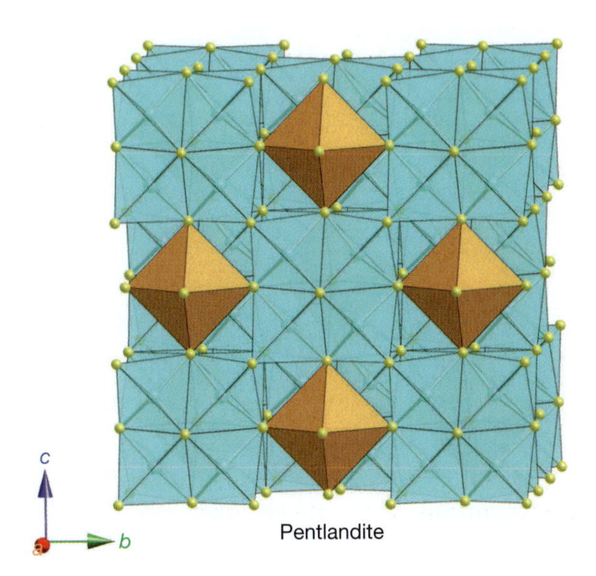

Figure 23.9. Crystal structure of pentlandite (which can be written $(Fe^{2+},Ni^{2+})_9S_8$ contains two types of sites: $(Ni,Fe)S_4$ tetrahedra (blue) in edge-sharing clusters at the corners and face centers of the drawing, and FeS_6 octahedra (brown, midway along the edges) that bond the clusters together into a framework. Each of the clusters shares S^{2-} anions with twelve neighboring cations. The S^{2-} anions form a cubic close-packed array with the Fe^{2+} and Ni^{2+} in the interstices.

As we noted in Chapter 21, the **sphalerite,** (Zn,Fe)S, structure is a ccp array of S^{2-} with half the tetrahedral interstices filled by Zn^{2+} (Figure 21.4). Thus each Zn^{2+} cation is in tetrahedral coordination with S^{2-}. If this structure looks familiar, it is because it is the same structure as diamond. These structures may also be visualized as having two interpenetrating fcc lattices, each displaced by 1/4 along the body diagonal of a cube. As noted in Table 23.3, sphalerite leads a group consisting of minerals with the AB formula, where A = Zn^{2+}, Fe^{2+}, Hg^{2+}, and Cd^{2+}, and B = S^{2-}, Se^{2-}, or Te^{2-}. Sphalerite is the most important ore of zinc, which is used mostly as a coating on Fe and steel to fight corrosion; zinc is also an important ingredient in curing rubber. It also a component of medicines, batteries, and toys and was used in the original color TV sets!

The sphalerite structure lends itself well to imitation, and has many derivatives. The chalcopyrite ($CuFeS_2$) structure is effectively a doubled version of the sphalerite structure (Figure 23.11), though it contains an ordered arrangement of Cu^{2+} and Fe^{2+} instead of Zn^{2+}. This gives it a tetragonal unit cell. Despite the fact that it contains relatively low amounts of copper relative to minerals like chalcocite, chalcopyrite is extremely common, making it the leading source of copper. Copper has many important uses including wires, coins, agriculture, and water purification. **Stannite** (Cu_2FeSnS_4) group minerals are isostructural with chalcopyrite, but the cations have three ordered locations: A = Cu^{2+}, Ag^{1+}, B = Fe^{2+}, Cd^{2+}, Cu^{2+}, Zn^{2+}, and Hg^{2+}; and C = Sn^{4+}, Ge^{4+}, and In^{3+} (Table 23.3).

Wurtzite, ZnS (Figure 8.6), is like the sphalerite structure except that the S^{2-} anions are arranged in a hexagonal close-packed configuration. The Zn^{2+}

cations fill half the tetrahedral interstices. The S^{2-} anions are sometimes replaced by Se^{2-}. As with sphalerite, this structure can be thought of as two interpenetrating arrays of hcp Zn^{2+} and S^{2-}. Note that the tetrahedra in the wurtzite structure all point *up*, so this mineral has hexagonal pyramidal ($P6_3mc$) symmetry rather than hexagonal prismatic. As a result, wurtzite crystals may form hemimorphic crystal shapes such as six-sided pyramids.

The many isostructural minerals in the **nickeline group** are based on hexagonal close-packed arrays of As (or S^{2-}, Se^{2-}, or Te^{2-}) with cations (Co^{2+}, Fe^{2+}, Ni^{2+}, Cu^{2+}, Pd^{2+}, Pt^{2+}) in all of the octahedral interstices. The octahedral sites link up along the *c* axis by sharing faces in a structure that would not be possible for ionic materials because the cations would be too close together. Instead, metal-metal bonding makes this structure stable. The **nickeline** (NiAs) structure is shown in Figure 21.1.

Many other minerals and materials can be derived from the nickeline structure by removing the octahedral cations in a systematic way. **Pyrrhotite** ($Fe_{1-x}S$, where x = 0.1–0.2) is an interesting derivative of the nickeline structure (Figure 8.11). It is basically a nickeline structure that is "missing" some of its Fe^{2+} cations, introducing vacancies into the octahedral sites of the hcp array of S^{2-}. Depending on temperature, the vacancies may be either in an orderly or disordered arrangement. **Troilite** (FeS) can also be derived from the nickeline structure by incremental displacement of the Fe^{2+} cations perpendicular to the *c* axis and the S^{2-} anions parallel to *c* (Figure 8.11).

Millerite (NiS) has a rhombohedral structure in which square pyramids of S^{2-} anions containing Ni^{2+} share edges to form chains. Ni^{2+} is thus in rare 5-coordination with S^{2-}. The chains are then

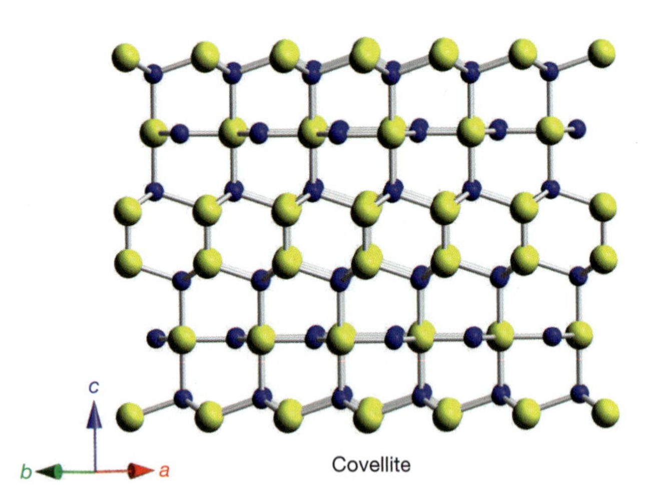

Figure 23.10. *Covellite (CuS) is composed of layers of bonded S-S (large yellow spheres) and layers of Cu^{2+} (small blue spheres) bonded into S^{2-}.*

Table 23.3. Sulfides with metal:sulfur = 1:1 and smaller

Group	Mineral	Formula
	covellite	CuS
Sphalerite	sphalerite	ZnS
	stilleite	ZnSe
	metacinnabar	HgS
	tiemannite	HgSe
	coloradoite	HgTe
	hawleyite	CdS
Chalcopyrite	chalcopyrite	$CuFeS_2$
	eskebornite	$CuFeSe_2$
	gallite	$CuGaS_2$
	roquesite	$CuInS_2$
	lenaite	$AgFeS_2$
	laforetite	$AgInS_2$
Stannite	stannite	Cu_2FeSnS_4
	cernyite	Cu_2CdSnS_4
	briartite	Cu_2ZnGeS_4
	kuramite	Cu_2CuSnS_4
	sakuraiite	$(Cu,Zn,In,Fe,Sn)_4S_4$
	hocartite	Ag_2FeSnS_4
	pirquitasite	Ag_2ZnSnS_4
	velikite	Cu_2HgSnS_4
	kesterite	$Cu_2(Zn,Fe)SnS_4$
	ferrokesterite	$Cu_2(Fe,Zn)SnS_4$
	barquillite	Cu_2CdGeS_4
Wurtzite	wurtzite	(Zn,Fe)S
	greenockite	CdS
	cadmoselite	CdSe
	rambergite	MnS
Nickeline	nickeline	NiAs
	breithauptite	NiSb
	sederholmite	β-NiSe
	hexatestibiopanickelite	Ni(Te,Sb)
	sudburyite	PdSb
	kotulskite	Pd(Te,Bi)
	sobolevskite	PdBi
	stumpflite	PtSb
	langisite	CoAs
	freboldite	CoSe
	sorosite	Cu(Sn,Sb)
	pyrrhotite	$Fe_{1-x}S$, where x = 0.1–0.2
	troilite	FeS
	millerite	NiS
Galena	galena	PbS
	clausthalite	PbSe
	altaite	PbTe
	alabandite	$Mn^{2+}S$
	oldhamite	(Ca,Mg)S
	niningerite	$(Mg,Fe^{2+},Mn)S$
	borovskite	Pd_3SbTe_4
	crerarite	$(Pt,Pb)Bi_3(S,Se)_{4-x}$, where x~0.7
	keilite	(Fe,Mg)S
	cinnabar	HgS

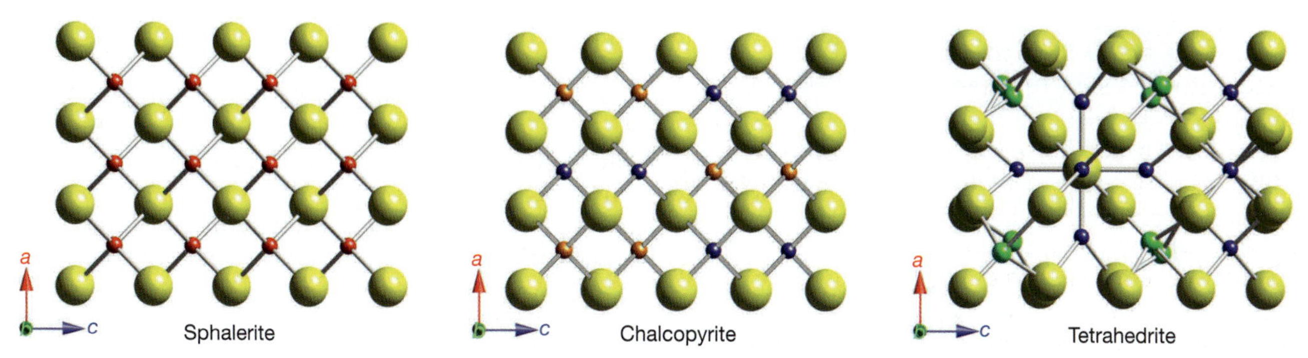

Figure 23.11. *The crystal structures of sphalerite (left, ZnS), chalcopyrite (middle, $CuFeS_2$), and tetrahedrite (left, $Cu_{12}Sb_4S_{13}$) after Wuensch (1992). Sphalerite, which has the same structure as diamond, contains ZnS_4 tetrahedra (Zn^{2+} is shown as small red spheres and S^{2-} as large yellow spheres) and is based on filling half the tetrahedral interstices of a cubic close-packed S^{2-} framework. The chalcopyrite structure is similar to the sphalerite structure because of an ordered substitution of Cu^{2+} (small blue spheres) and Fe^{2+} (small orange spheres) for the Zn^{2+}. The tetrahedrite structure is similar to chalcopyrite, though with Sb^{3+} (small green spheres) replacing Fe^{2+}; however, some of the S^{2-} locations become vacant and new S^{2-} anions are introduced elsewhere in the structure, which makes tetrahedrite more complex than chalcopyrite.*

linked together by sharing edges and corners. Again, such a cozy arrangement of Ni^{2+} cations would not be possible without the metallic bonding in this mineral.

Minerals in the **galena** group (PbS) all have the halite structure we have (too?) often discussed in this text. The Pb^{2+} and S^{2-} ions alternate in a face-centered cubic arrangement. Galena (Figure 8.8) often contains as much as 1% Ag^{2+} substituting for lead; this concentration is high enough to make galena the major ore of silver as well as lead. Although uses of lead in things like paint and water pipes have all but ended with the discovery of its toxicity, lead is still very much in demand for use in lead-acid storage batteries (e.g., those eco-friendly hybrid cars). Lead is also used in ammunition and to make shielding from X-ray and other types of radiation.

Cinnabar (HgS) is a slight distortion of the halite/galena structure. The Hg^{2+} cations are still in 6-coordination with S^{2-}, but those octahedra are distorted by *sp* bonding of Hg^{2+}. The resultant structure resembles that of elemental sulfur: helical chains of Hg^{2+} alternating with S^{2-} spiral up the unit cell in the *c* direction (Figure 23.12). Cinnabar is the main ore of mercury, which is used in barometers, pigments, dental fillings, mercury vapor lamps, photography, and many other applications. In some cases, liquid Hg actually occurs on the surface of this mineral. M.E.G. once panned Hg, when looking for gold, in the Snake River in Idaho. Interestingly, there is an old mining town in Idaho with the name of Cinnabar that was upstream! So not all Hg occurring is anthropogenic.

Metal:sulfur =3:4 and 2:3. Stibnite (Sb_2S_3) is the most important species in this group (Table 23.4), in part because many other mineral structures are derived from it (e.g., bismuthinite). Its structure contains two types of Sb^{3+} sites. One Sb^{3+} cation forms a trigonal pyramid with S^{2-}. The second Sb^{3+} occupies a distorted square pyramid (5-coordination), which shares an apical edge with another square pyramid to link up into a double chain. The 3-coordinated Sb's share corners at the ends of these double chains (Figure 21.10). Stibnite is the main ore of the element antimony, which is used in lead alloys in batteries, as a coating for metals and plaster casts, in solders, semiconductors, and ceramic pigments. **Stibiconite** (an antimonate class mineral, $Sb^{3+}Sb^{5+}{}_2O_6(OH)$) commonly forms pseudomorphs after stibnite, through oxidation and hydration.

Metal:sulfur <=1:2. Minerals in the **kostovite-sylvanite group**, which include **sylvanite**

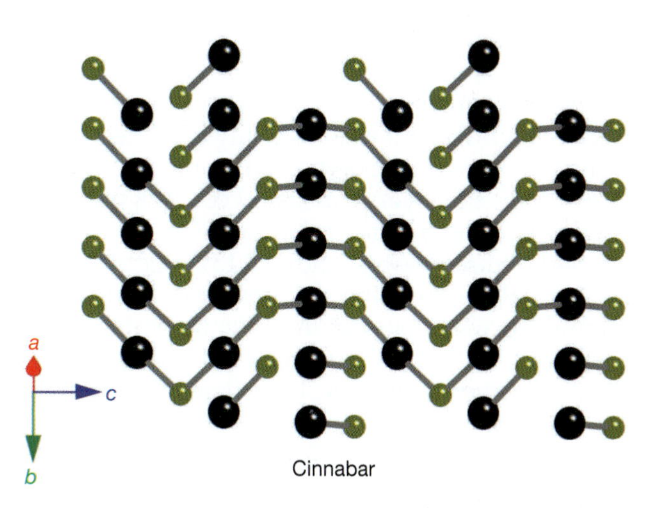

Figure 23.12. *Cinnabar (HgS) is composed of Hg^{2+} (large black spheres) bonded to S^{2-} (small greenish spheres) in a one-to-one ratio.*

Table 23.4.
Sulfides with metal:sulfur = 2:3

Group	Mineral	Formula
Stibnite	stibnite	Sb_2S_3
	antimonselite	Sb_2Se_3
	bismuthinite	Bi_2S_3
	guanajuatite	Bi_2Se_3

($AuAgTe_4$) and **calaverite** ($AuTe_2$) (Table 23.5), are the most important of the rare mineral ores of gold. Gold vastly prefers to occur in the elemental state or as a substituting element in other metals such as Ag^{1+} or Cu^{1+}. However, when Au^{1+} does occur in a non-metal, it is usually found in one of the members of this group. Note that all of these minerals include the element Te (tellurium), for which Au has an apparent affinity.

Sylvanite is the more common of these ores. Its monoclinic structure belies its {010} cleavage, which is the result of continuous, brucite-like octahedra sheets of edge-sharing $AgTe_6$ and $AuTe_6$ octahedra. The sheets are held together by Te-Te bonds. Calaverite has a similar structure with only $AuTe_6$ octahedra, but the octahedra are slightly offset.

Molybdenite (MoS_2) is the primary ore of molybdenum, and the mineral is primarily used as a lubricant. It is often mixed with oil and greases, allowing engines to continue running even when oil runs low (a useful characteristic for applications involving aircraft engines, for example!) MoS_2 almost always contains the element rhenium (Re) as an impurity, though it can sometimes reach levels of 1–2%. Re^{4+} substitutes for the Mo^{4+}, and it is the primary ore for that element as well. Re is used with platinum as a catalyst for production of high octane gasoline, and is also used in high-temperature alloys for aircraft engines. Re also makes an excellent radiation shield that is much lighter than lead.

The structure of molybdenite explains its usefulness as a lubricant: it cleaves easily into sheets. The common hexagonal polytype (i.e., one of the different ways the layers can stack) consists of sheets of edge-sharing trigonal prisms of MoS_6; each S^{2-} is bonded to three Mo^{4+} (Figure 23.13). This structure is unusual because the 6-coordinated polyhedra around the Mo^{4+} cations are prismatic rather than octahedral in shape. Only weak S-S bonds link the sheets together, facilitating easy cleaving and making the mineral quite soft (1–1.5 on the Mohs scale). Molybdenite is isostructural with tungstenite (WS_2), which has similar properties. These structures strongly resemble that of gibbsite.

There are 19 species in the **pyrite group**, all of which are based on cubic or nearly-cubic structures. The basic AX_2 composition can contain A cations of Au^{1+}, Co^{2+}, Cu^{2+}, Fe^{2+}, Mn^{2+}, Ni^{2+}, Os^{4+}, Pd^{2+}, Pt^{4+}, or Ru, while X can be As^{2-}, Bi^{2-}, S^{2-}, Sb^{2-}, Se^{2-}, or Te^{2-}. The **pyrite** (FeS_2) structure that we saw in Chapter 8 (Figure 8.8), is derived from the NaCl structure. It is based on a face-centered cubic lattice type. Fe^{2+} cations with S^{2-} anions are covalently-bonded in tight pairs (like little dumbbells) along each of the four body diagonals of the cube. The distance between the S^{2-} anions is only 2.14 Å. Each dumbbell has its own characteristic orientation, which still preserves the overall cubic symmetry. As in the NaCl structure, each Fe^{2+} cation has six S_2 neighbors, and each S_2 pair is coordinated to six Fe^{2+} cations. Each individual S^{2-} anion occupies a distorted tetrahedron composed of three Fe^{2+} and one S^{2-} ion.

Pyrite is the most commonly-occurring sulfide mineral, and is found in igneous, sedimentary, and metamorphic rock types. As a resource, it is a primary source of sulfur, which is used to make sulfuric acid, a critically-important industrial raw material. Consumption of sulfuric acid has been regarded as one of the best indexes of a nation's industrial development. As a measure of its importance, more sulfuric acid is produced in the United States every year than any other industrial chemical (USGS website).

Marcasite group minerals also have the general AX_2 formulas, but their structures are orthorhombic. In these minerals, A = Fe^{2+}, Co^{2+}, Ni^{2+}, Ru^{3+}, and Os^{4+}, and X = S^{2-}, Se^{2-}, Te^{2-}, As^{2-}, and Sb^{2-}. The **marcasite** (FeS_2) structure is based on an orthorhombic body-centered lattice type. S^{2-} octahedra surround each Fe^{2+}. This results in the creation of FeS_6 octahedra that share edges along the (001) direction. The S_2 pairs are bonded like the dumbbells in the pyrite structure, though they lie in the *a-b* plane in marcasite (Figure 23.14).

Another derivative of the pyrite structure is the mineral **cobaltite** (CoAsS), in which the S-S dumbbells are replaced by As-S dumbbells, and the symmetry is the same as for pyrite, $2/m\bar{3}$. Cobaltite is the most significant ore of Co, which has a number of industrial and other uses. An alloy of Co, Cr, and W (tungsten) called stellite is used to make high speed cutting tools and dyes. Another alloy, Alnico, combines the properties of Al, Ni, and Co to make magnets. Other alloys of Co are used in jet engines, gas turbines, and various types of steel. Isotopes of Co are also important sources of gamma rays: ^{60}Co is used in radiation treatments of cancer patients, and ^{57}Co is used in Mössbauer and other types of spectroscopy (Chapter 9). If you've ever taken a painting class, you know that cobalt is used as a blue,

Table 23.5. Selected sulfides with metal: sulfur <=1:2 and sulfides with non-metals

Group	Mineral	Formula
Kostovite-sylvanite	kostovite	$CuAuTe_4$
Kostovite-sylvanite	sylvanite	$AuAgTe_4$
Kostovite-sylvanite	calaverite	$AuTe_2$
Kostovite-sylvanite	krennerite	$(Au,Ag)Te_2$
	molybdenite	MoS_2
Pyrite	pyrite	FeS_2
Pyrite	vaesite	NiS_2
Pyrite	cattierite	CoS_2
Pyrite	penroseite	$(Ni,Co,Cu)Se_2$
Pyrite	trogtalite	$CoSe_2$
Pyrite	villamaninite	CuS_2
Pyrite	fukuchilite	$(Cu,Fe)S_2$
Pyrite	krutaite	$CuSe_2$
Pyrite	hauerite	MnS_2
Pyrite	laurite	RuS_2
Pyrite	aurostilbite	$AuSb_2$
Pyrite	krutovite	$NiAs_2$
Pyrite	sperrylite	$PtAs_2$
Pyrite	geversite	$Pt(Sb,Bi)_2$
Pyrite	insizwaite	$Pt(Bi,Sb)_2$
Pyrite	erlichmanite	OsS_2
Pyrite	dzharkenite	$FeSe_2$
Pyrite	gaotaiite	Ir_3Te_8
Pyrite	mayaingite	$IrBiTe$
Marcasite	marcasite	FeS_2
Marcasite	ferroselite	$FeSe_2$
Marcasite	frohbergite	$FeTe_2$
Marcasite	hastite	$CoSe_2$
Marcasite	mattagamite	$CoTe_2$
Marcasite	kullerudite	$NiSe_2$
Marcasite	omeiite	$(Os,Ru)As_2$
Marcasite	anduoite	$(Ru,Os)As_2$
Marcasite	löllingite	$FeAs_2$
Marcasite	seinajokite	$FeSb_2$
Marcasite	safflorite	$CoAs_2$
Marcasite	rammelsbergite	$NiAs_2$
Marcasite	nisbite	$NiSb_2$
Cobaltite	cobaltite	$CoAsS$
Cobaltite	gersdorffite	$NiAsS$
Cobaltite	ullmannite	$NiSbS$
Cobaltite	willyamite	$(Co,Ni)SbS$
Cobaltite	tolovkite	$IrSbS$
Cobaltite	platarsite	$(Pt,Rh,Ru)AsS$
Cobaltite	irarsite	$(Ir,Ru,Rh,Pt)AsS$
Cobaltite	hollingworthite	$(Rh,Pt,Pd)AsS$
Cobaltite	jolliffeite	$NiAsSe$
Cobaltite	padmaite	$PdBiSe$
Cobaltite	michenerite	$PdBiTe$
Cobaltite	maslovite	$PtBiTe$
Cobaltite	testibiopalladite	$Pd(Sb,Bi)Te$
Arsenopyrite	arsenopyrite	$FeAsS$
Arsenopyrite	gudmundite	$FeSbS$
Arsenopyrite	osarsite	$(Os,Ru)AsS$
Arsenopyrite	ruarsite	$RuAsS$
Arsenopyrite	iridarsenite	$(Ir,Ru)As_2$
Arsenopyrite	clinosafflorite	$(Co,Fe,Ni)As_2$
	realgar	As_4S_4
	orpiment	As_2S_3

yellow, and green pigment and dye. Cobalt is also used to make the vitamin known as B_{12}, which is the chemical cyanocobalamin, which then breaks down in your body to form 5-deoxyadenosyl and methylcobalamin.

The arsenopyrite structure, shared by all six members of the **arsenopyrite group**, is a monoclinic (though pseudo-orthorhombic) derivative of the marcasite structure. In **arsenopyrite** (FeAsS), half of the S^{2-} atoms in FeS_2 are replaced by As^{3+}. In the resultant structure, each Fe^{2+} cation is surrounded by a distorted octahedron in which three adjoining corners are As^{3+} and three are S^{2-} (Figure 23.15). Again as with marcasite, the structure can also be visualized as having chains of these $FeAs_3S_3$ octahedra that run parallel to [001].

Arsenopyrite is the most important ore of arsenic, though is usually mined as a byproduct of gold, silver, lead, and copper production. Arsenic was once used in insecticides and rat poisons, though it has subsequently been replaced for those purposes because of its extreme toxicity. Now arsenic is used as a hardener of lead, and as a minor component to raise the melting point of copper alloys.

Sulfides of arsenic. Realgar (As_4S_4) and **orpiment** (As_2S_3) almost always occur together in nature, though their structures are quite different. Realgar can be derived from the S_8 elemental sulfur structure by substituting As for every other S in the 8-membered rings; this substitution lowers the symmetry of realgar to monoclinic (Figure 23.7). Each As^{3+} cation is bonded to one As^{3+} and two S^{2-} anions, while each S^{2-} is bonded to two As^{3+}. The resultant molecule is often described as

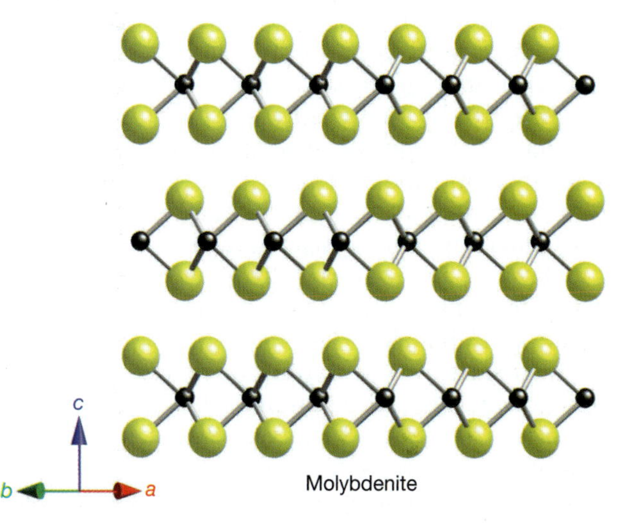

Figure 23.13. *Molybdenite (MoS_2) is composed of layers of 6-coordinated Mo^{4+} (small black spheres) bonded to S^{2-} anions (large yellow spheres). The layers are in turn held together by van der Waals bonding between the S^{2-} anions.*

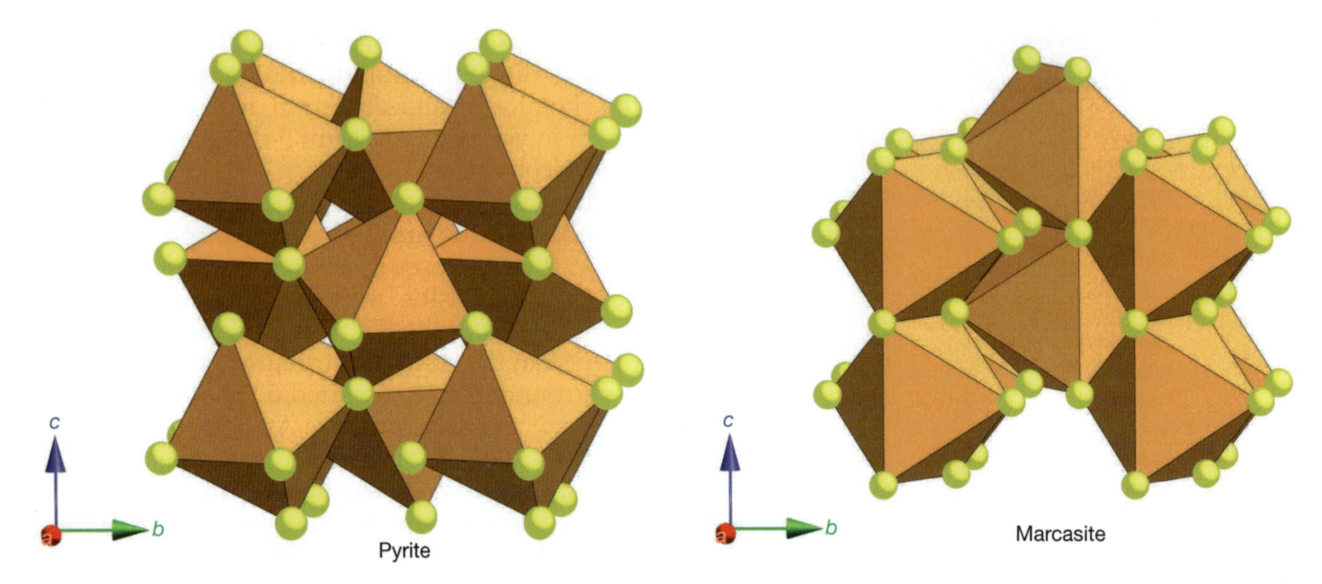

Figure 23.14. The crystal structures of pyrite (left, FeS_2) and marcasite (right, FeS_2). These two minerals are polymorphs, though both are composed of FeS_6 octahedra. In pyrite, all the octahedra link by corner-sharing, while in marcasite there are edge-sharing chains of octahedra that are laterally linked by corner-sharing.

"cage-like" (Wuensch, 1992) or "cradle-like" (Strunz and Nickel, 2001). The molecules are held together by van der Waals bonds, making realgar quite soft (1.5–2 on the Mohs scale). Orpiment has a structure that is related to that of stibnite. Trigonal pyramids of AsS_3 share corners, creating rings of six sulfurs that extend in two dimensions to form sheets in the *a-c* plane (Figure 21.10). This gives orpiment perfect {010} cleavage and the same hardness as realgar.

Orpiment was (and still is) used as a pigment by many Middle East and Asian cultures, including Egypt in the 16th century B.C., and Mesopotamia in the 8th century B.C. Realgar is unstable and will decompose to form pararealgar; for this reason, samples should be kept in dark closed containers. Realgar was used as a pigment in antiquity by many different cultures and was prized for its bright red color. Many of those paintings and manuscripts have now deteriorated to pararealgar, and the change from red to yellow makes it hard to visualize how those works must once have looked. In the early 20th century, another (albeit toxic!) use of realgar was as a blush and lipstick.

Sulfosalts. As we mentioned above, the sulfosalts (Table 23.6) share structures based on trigonal or square pyramids. Most are very rare, but a few deserve mention as examples of this group. The key to understanding sulfosalts is to notice that, in general, they have three atoms in their formulas. In keeping with the grammar of mineral formulas, the anion is the last element listed and the cations are first.

Proustite (Ag_3AsS_3) and **pyrargyrite** (Ag_3SbS_3) are both important ores of silver, and good examples of sulfosalt structures. Both are isometric. Their isomorphous structures contain sheets of AsS_3 or SbS_3 linked by spiralling trigonal chains of Ag^{1+} in 2-coordination with S^{2-}. All the pyramids point in the same direction (Figure 23.16).

The tetrahedrite-tennantite group minerals share a general formula of $A_{12}B_4X_{13}$, where A = Cu^{2+}, Ag^{1+}, Fe^{2+}, Hg^{2+}, and Zn^{2+}, B = As^{3+}, Sb^{3+}, and Te^{4+}, and X = S^{2-} and Se^{2-}. Their structures are all based on sphalerite, with one-fourth of the

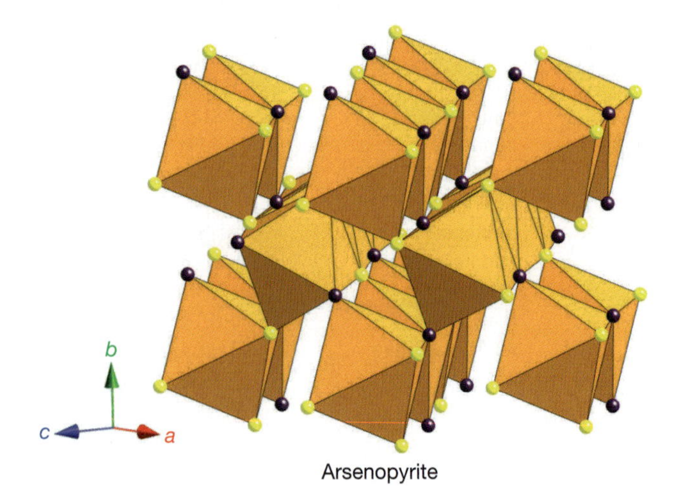

Figure 23.15. Arsenopyrite (FeAsS) is composed of edge-sharing chains of As^{3+} octahedra with three of the ligands being S^{2-} (small yellow spheres) and the other 3 being Fe^{2+} (small purple spheres). The chains are in turn linked by sharing corners.

Table 23.6. Sulfosalts		
	proustite	Ag_3AsS_3
	pyargyrite	Ag_3SbS_3
Tetrahedrite	tetrahedrite	$Cu_6Cu_4(Fe,Zn)_2(Sb,As)_4S_{13}$
Tetrahedrite	tennantite	$Cu_6Cu_4(Fe,Zn)_2(As,Sb)_4S_{13}$
Tetrahedrite	frebergite	$(Ag,Cu,Fe)_{12}(Sb,As)_4S_{13}$
Tetrahedrite	hakite	$Cu_6Cu_4Hg_2Sb_4(Se,S)_{13}$
Tetrahedrite	giraudite	$Cu_6Cu_4Zn_2(As,Sb)_4(Se,S)_{13}$
Tetrahedrite	goldfieldite	$Cu_6Cu_4Te_2(Sb,As)_4S_{13}$
Tetrahedrite	argentotennantite	$Ag_6Cu_4(Zn,Fe)_2(As,Sb)_4S_{13}$
	cylindrite	$(Pb,Sn^{2+})_8Sb_4Fe_2Sn^{4+}{}_5S_{27}$
	enargite	Cu_3AsS_4

Cu^{2+} being replaced by Sb^{3+}, As^{3+}, or Te^{4+}. Thus, the structure of **tetrahedrite**, $(Cu,Fe)_{12}Sb_4S_{13}$, is made from a cubic close-packed array of S^{2-} anions with Cu^{2+} and Fe^{2+} in 6-coordination and Sb^{3+} in tetrahedral interstices. The trigonal pyramids formed by the tetrahedra are composed of a Sb^{3+} linked to three S^{2-} anions. Tetrahedrite is also an important Ag-containing mineral and the main mineral in the largest silver-mining district in the world—the so-called "Silver Valley" in northern Idaho.

Cylindrite ($Pb_3Sn_4FeSb_2S_{14}$) is an unusual sulfosalt worth mentioning because it is one of the few minerals to crystallize with circular habit. The structure is composed of two types of sheets: one with 4-coordinated sites and one with 6-coordinated sites. The misfit between the sites is believed to cause its curvature, much like the mismatch between the tetrahedral and octahedral sheets in the fibrous form of serpentine—the most common of the asbestos minerals (i.e., chrysotile). This structure causes black crosses to be seen in reflected cross-polarized light.

The **enargite** (Cu_3AsS_4) structure is derived from that of wurtzite by replacing a quarter of the Zn^{2+} with As^{3+} and three-quarters with Cu^{2+}. Thus, it has a ccp array of S^{2-} anions with the Cu^{2+} and As^{3+} in tetrahedra interstices. It is sometimes mined as an ore of Cu^{2+}.

Oxides

Oxide minerals are grouped into classes on the basis of the number of different types of sites they possess. Thus simple oxides are generally those with only a single cation site; they tend to have formulas like XO_2, XO, or X_2O_3. Multiple oxides usually contain two or more different sites, usually one with 6-coordination and one with 4-coordination; their compositions are usually expressed as XY_2O_4.

Simple oxides. You might think that simple oxides would have simple structures, but there is a wide variety among minerals in this class (Table 23.7). They all have O^{2-} as the main anion. They also have at least one predictable characteristic—these phases all have primarily ionic bonding, so coordination polyhedra obey radius ratio rules. The sizes and charges of the cations govern the size and complexity of the structures. For a more detailed overview of these minerals, we recommend volume 3 of *Reviews in Mineralogy* (Rumble,

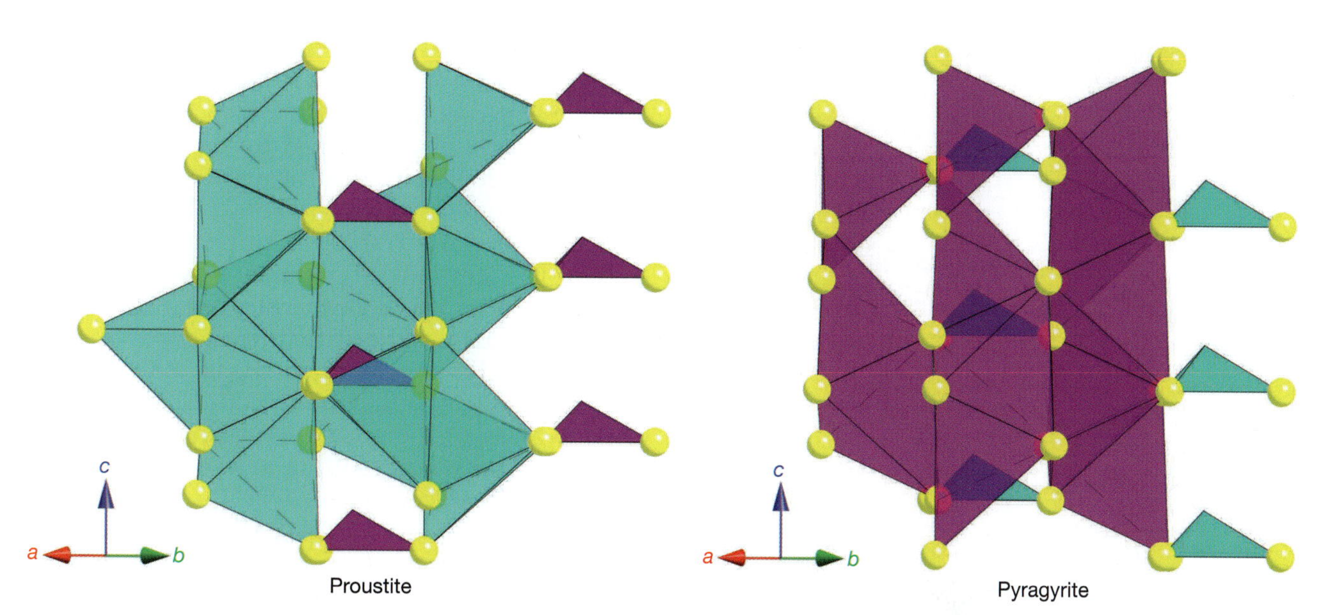

Figure 23.16. Proustite (left, Ag_3AsS_3) and pyrargyrite (right, Ag_3SbS_3) are both composed of distorted Ag^{1+} polyhedra defined by S^{2-} ligands and flattened tetrahedra with three corners defined by S^{2-} and the fourth occupied by As^{3+} in proustite and Sb^{3+} in pyrargyrite. The As^{3+} in proustite and the Sb^{3+} in pyrargyrite are in tetrahedral coordination.

Table 23.7. Simple and Multiple Oxides

Group	Subgroup	Mineral	Formula
		cuprite	$Cu^{1+}{}_2O$
		ice	H_2O
Periclase		periclase	MgO
		bunsenite	NiO
		manganosite	$Mn^{2+}O$
		monteponite	CdO
		lime	CaO
		wüstite	$Fe^{2+}O$
		hongquiite	TiO
Corundum-hematite		corundum	Al_2O_3
		hematite	α-Fe_2O_3
		eskolaite	Cr_2O_3
		karelianite	V_2O_3
Rutile		rutile	TiO_2
		ilmenorutile	$(Ti,Nb,Fe^{3+})O_2$
		struverite	$(Ti,Ta,Fe^{3+})O_2$
		pyrolusite	$Mn^{4+}O_2$
		cassiterite	SnO_2
		plattnerite	PbO_2
		argutite	GeO_2
		stishovite	SiO_2
		brookite	TiO_2
		anatase	TiO_2
Ilmenite		ilmenite	$Fe^{2+}TiO_3$
		geikielite	$MgTiO_3$
		pyrophanite	$Mn^{2+}TiO_3$
		ecandrewsite	$ZnTiO_3$
		melanostibite	$Mn(Sb^{5+},Fe^{3+})O_3$
		brizziite-III	$NaSb^{5+}O_3$
		akimotoite	$MgSiO_3$
Spinel	*Al subgroup*	spinel	$Mg[Al]_2O_4$
		galaxite	$(Mn^{2+},Fe^{2+},Mg)[Al,Fe^{3+}]_2O_4$
		hercynite	$Fe^{2+}[Al]_2O_4$
		gahnite	$Zn[Al]_2O_4$
	Fe subgroup	magnesioferrite	$Fe^{3+}[MgFe^{3+}]O_4$
		jacobsite	$Fe^{3+}[(Mn^{2+},Fe^{2+},Mg)Fe^{3+}]O_4$
		magnetite	$Fe^{3+}[Fe^{2+}Fe^{3+}]O_4$
		franklinite	$(Zn,Mn^{2+},Fe^{2+})[Fe^{3+},Mn^{3+}]_2O_4$
		trevorite	$Fe^{3+}[NiFe^{3+}]O_4$
		cuprospinel	$Fe^{3+}[(Cu^{2+},Mg)Fe^{3+}]O_4$
		brunogeierite	$Ge^{2+}[Fe^{3+}]_2O_4$
	Cr subgroup	magnesiochromite	$Mg[Cr^{3+}{}_2]O_4$
		manganochromite	$(Mn^{2+},Fe^{2+})[Cr^{3+},V^{3+}]_2O_4$
		chromite	$Fe^{2+}[Cr^{3+}]_2O_4$
		nichromite	$(Ni,Co,Fe^{2+})[Cr^{3+},Fe^{3+},Al]_2O_4$
		cochromite	$(Co,Ni,Fe^{2+})[Cr^{3+},Al]_2O_4$
		zincochromite	$Zn[Cr^{3+}]_2O_4$
	V subgroup	vuorelainenite	$(Mn,Fe^{2+})[V^{3+},Cr^{3+}]_2O_4$
		coulsonite	$Fe^{2+}[V^{3+}]_2O_4$
		magnesiocoulsonite	$Mg[V^{3+}]_2O_4$
	Ti subgroup	qandilite	$(Mg,Fe^{2+})_2(Ti,Fe^{3+},Al)O_4$
		ulvöspinel	$Fe^{2+}{}_2TiO_4$
		Chrysoberyl	$BeAl_2O_4$
		Maghemite	γ-$Fe^{3+}{}_2O_3$
		Arsenolite	As_2O_3
Perovskite		Perovskite	$CaTiO_3$
		Latrappite	$(Ca,Na)(Nb,Ti,Fe)O_3$
		loparite-(Ce)	$(Ce,Na,Ca)(Ti,Nb)O_3$
		lueshite	$NaNbO_3$
		tausonite	$SrTiO_3$
		isolueshite	$(Na,La,Ca)(Nb,Ti)O_3$

*Note that some of the formulas given here deviate in format from the "standard" formulations of Mandarino and Back (2004), in order to present the formulas in a comprehensible way that relates them to the crystal structure. See text for more details.

1976), as well as the chapter by Smyth et al. (2000) in vol. 40 of that same series.

In the **cuprite** (Cu_2O) structure (Figure 21.4), Cu^{1+} has an ionic radius of at least 0.77 Å, and is found in linear 2-coordination with O. Each oxygen is bonded to a surrounding tetrahedron of Cu^{1+} cations in a cubic unit cell. This structure can be envisioned as a body-centered lattice type with close packing of O^{2-}. If you divide that cubic structure into eighths, there will be a Cu^{1+} cation at the center of every other sub-cube.

The structure of **ice** (H_2O) is quite different from that of cuprite because the H "cation" (really just a proton) is so tiny relative to the O^{2-} anion (Figure 8.21 and 21.3). In ordinary ice, H_2O molecules with H-O-H angles of 104.5° form tetrahedra with four adjoining H_2O molecules (very much like Si in tridymite). Hydrogen bonding plays an important role in bonding the H_2O molecules to one other. Because of the large interstitial spaces that result, ice has a low density (which is why it floats). There are numerous ice polymorphs (at least nine), each designated with a roman numeral. Ordinary ice, the polymorph we find on Earth, is "ice I."

The **periclase group** includes minerals with divalent cations, the majority of which crystallize with a face-centered cubic lattice type like the halite structure. Both the cation and the anion are in 6-coordination (just as in halite). Of these mineral species, **periclase** (MgO) (Figure 8.13) is unusual because of its great stability when subjected to high temperatures and pressures. As a result, it is likely that periclase is an important phase in the Earth's interior, even under the high pressure conditions in the lower mantle.

Wüstite (FeO) is never completely stoichiometric (i.e., a 1:1 ratio of Fe:O), but is nearly so at pressures exceeding 10 GPa. Below that pressure, vacancies (usually indicated by the symbol "□") are needed to stabilize the structure, and a nominal composition of $Fe^{2+}{}_{1-3x}Fe^{3+}{}_{2x}\square_xO^{2-}$, where x ranges from $0.04<x<0.12$, is found (Smyth et al., 2000). The amount of vacancy has a dramatic effect on its physical properties because it introduces defects into the structure, and Fe^{2+} may be oxidized to Fe^{3+} to maintain charge balance. The Fe^{3+} can occupy either the 6-coordinated site in the fcp array or the normally vacant 4-coordinated interstitial sites, resulting in a complex formula that can be expressed as $^{VI}[Fe^{2+}{}_{1-3x}Fe^{3+}{}_{2x-t}\square_{x+t}]^{IV}Fe^{3+}{}_tO^{2-}$ (Smyth et al., 2000). Wüstite has been well-studied because of its potential significance at high pressures and temperatures in the Earth's mantle.

Corundum group minerals are composed of a slightly-distorted, hexagonal close-packed array of O^{2-} anions in which 2/3 of the 6-coordinated sites are occupied by trivalent A cations, which can be Al^{3+}, Fe^{3+}, Cr^{3+}, or V^{3+} (Figure 21.9). The resultant structure is comprised of gibbsite-like sheets of rings of AO_6 octahedra in the (001) plane. The sheets link into a framework by sharing faces and corners of octahedra (Strunz and Nickel, 2001), giving the structure a hardness second only to diamond on the Mohs scale.

Corundum (Al_2O_3) has economic significance for many reasons. Because of its hardness, it is extensively used as an industrial abrasive. Historically, the brown form of corundum, which has the variety name emery, was mined for this purpose. Emery thus lends its name to the boards used as nail files, and is used in sandpaper. Corundum is fairly easy to create in the laboratory, so most modern commercial applications use synthetics. Purer varieties of corundum, usually in the form of sapphire rods and plates, are also used as gemstones and in the semiconductor industry as lasers (thanks to their fluorescence).

One interesting use of ruby lasers deserves mention here. The Apollo 11, 14, and 15 astronauts placed reflectors on the lunar surface; the Soviets also put reflectors on their Lunikhod roving vehicles. In several different experiments, a ruby laser pulse was sent by the MacDonald Observatory in Texas to the lunar surface, and the length of time it took for the reflection to be detected back on Earth was measured. From these experiments, we now know the distance from the Moon to Mount Livermore (near Ft. Davis, Texas) within about 15 cm! Interestingly, these experiments have also been used to demonstrate the "reality" of the Apollo landings to skeptics, because the return of the laser pulses cannot be explained by anything except human-made reflectors.

The gem varieties of corundum have been given a host of colloquial variety names. Strictly speaking, emery is brown corundum, ruby is red, padparadschah is orange, and sapphire is used for all the other colors, which include blue, green, purple, etc. The colors arise from very minor amounts of impurity (<1% of the Al^{3+} replaced by other cations) because the Al_2O_3 structure apparently does not tolerate substitutions. However, as noted in Chapter 7, these trace substitutions can cause intense colors. Ruby is red because of its Cr^{3+} content. Yellow sapphire owes its color to Fe^{3+}. Blue sapphire derives its color from Fe^{2+}-Ti^{4+} and Fe^{2+}-Fe^{3+} intervalence charge transfer. Green sapphires contain a mixture of the blue and green colors. Corundum varieties sold as gemstones are nearly always heated-treated to intensify (or in some cases, lighten) their colors; the heat treatment is done at the mines when they remove the stones in a highly-accepted practice. Spectra of these minerals and more information on color can be found on the Mineral Spectroscopy server at Caltech.

The other common member of the corundum group is **hematite** ($Fe^{3+}{}_2O_3$), which is usually found in sedimentary or metamorphosed sedimentary rocks. It is a common mineral in soils all over the world, because it results from the breakdown of Fe-rich minerals in many different rock types. When present in equilibrium in metamorphic or igneous rocks, it is indicative of highly oxidizing conditions. Some hematite deposits reach tremendous thicknesses, such as those in Brazil, and thus they make convenient sources of iron ore for the iron and steel industries. The red color of hematite is very intense because the $Fe^{3+}O_6$ octahedra are somewhat asymmetric, intensifying the bands in the hematite spectrum. Magnetic coupling between Fe^{3+} cations further intensifies the red color. Thus hematite is used as a red pigment in many applications (including rouge).

Species in the **rutile group** have the general formula AO_2, as discussed in Chapter 21, and the cation can be Ti^{4+}, Mn^{2+}, Sn^{4+}, Pb^{4+}, Ge^{4+}, Nb^{4+}, Fe^{2+}, Fe^{3+}, or even Sb^{5+}. Due to the different sizes and charges of these disparate cations, these structures generally do not tolerate cation substitutions, and thus are typically quite pure. The only exception to this rule is found in the species **rutile** (TiO_2), where Fe^{2+}, Fe^{3+}, Nb^{4+}, and Ta^{5+} commonly substitute (Figure 21.6). Although ionic bonding predominates in the other members of the rutile group, there is evidence to suggest that covalent bonding may be significant in rutile. The result is that the size of the unit cell does not appear to vary at high temperatures and pressures (Hazen and Finger, 1981).

Most of the AO_2 oxides can occur in many different structure types. Most preserve the octahedral coordination of the cation, but vary in the number of edges that are shared. For example, rutile, **brookite**, and **anatase** are all polymorphs of TiO_2, but their octahedra share two, three, and four edges, respectively. As the number of shared edges goes up, so does the density. The denser structures are correlated with increased pressure of formation. The most common structure (as in rutile) is tetragonal, and composed of edge-sharing octahedra that link up to form chains in the *c* direction (Figure 21.6). The corners of the chains join to create a framework. In the resultant configuration, each oxygen anion is surrounded by three Ti^{4+} cations in a nearly planar arrangement.

Several species in the rutile group have special importance. Rutile is a major ore of titanium, and is used as a white pigment in plastics, paper, and paint because its high refractive index causes it to reflect a greater portion of the incident light. **Pyrolusite** (MnO_2) is an important source of manganese, which is a key ingredient in the manufacture of steel. Manganese is also used in alloys with aluminum, and as an oxide in dry cell batteries. The ore is also used in livestock feed, fertilizers, and as a pigment for coloring bricks! Finally, **cassiterite** (SnO_2) is similarly important for tin production. Tin is mostly used in coatings or in alloys with other metals to make solders, glasses, and various chemicals.

Stishovite is a high pressure form of SiO_2 that forms when meteorites impact quartz-rich rocks. It is included here to compare its structure with that of the other minerals in this group, but it is really a silicate mineral because it forms from polymerization of the anionic complex $(SiO_6)^{12-}$.

The **ilmenite group** minerals ($ATiO_3$, where A is Fe^{2+}, Mg^{2+}, Mn^{2+}, Zn^{2+}, and Sb^{5+}) are isostructural with corundum, though with the lower symmetry that arises from alternating the A cations with Ti^{4+} in the unit cell (Figure 21.9). This gives these minerals hexagonal unit cells. There are two different octahedral sites (the A location and the Ti^{4+} location) for the cations, and the structure is dense because the octahedra share faces. **Ilmenite** ($Fe^{2+}Ti^{4+}O_3$) is extremely common in both igneous and metamorphic rocks, where its presence often serves as a signature of high activity of TiO_2 saturation and/or relatively reducing conditions. Ilmenite typically contains some $Fe_2^{3+}O_3$ in solid solution and in favorable conditions can help indicate both the temperature and redox conditions under which it formed (see magnetite, below).

Multiple and related oxides. The large **spinel group** of multiple oxides comprises 22 minerals with the formula of $X^{2+}Y^{3+}{}_2O_4$, where X = Mg^{2+}, Mn^{2+}, Fe^{2+}, Zn^{2+}, Co^{2+}, Cu^{2+}, and/or Ge^{2+}, while Y = Al^{3+}, Fe^{3+}, Cr^{3+} and/or V^{3+}. The A cations have divalent charge and the B cations are trivalent. Each unit cell contains a (nearly perfect) close-packed array of 32 oxygen anions with 64 tetrahedral and 32 octahedral sites between them. As we have discussed before, it is unlikely that all the tetrahedral and octahedral sites would be occupied. This would require face-sharing between adjacent tetrahedra and octahedra that would be unlikely in an ionically-bonded crystal.

The spinel formula suggests that there should be three cations for every four oxygens. Thus, the unit cell must contain 3/4 of 32, or 24, cations. Many possible combinations of site occupancy present themselves for these 24 cations! An optimal configuration is for eight of the tetrahedral (A) sites to be filled, along with 16 of the octahedral (B) sites. In the spinel group, there are thus two possibilities.

1. The mineral species **spinel**, which has a formula of $MgAl_2O_4$, was the first spinel structure to be described (Barth and Posnjak, 1932), and thus was given the name "normal." In **normal spinels**, the X (divalent) cations fill 8 A tetrahedral sites, and the Y (trivalent) cations occupy 16 B octahedral sites. Thus a "normal" distribution of cations in a spinel would be $X[Y^{3+}{}_2]O_4$. The tetrahedral site is listed first, and the octahedral site occupancy is given in brackets; this is reversed from the typical way formulas are written (i.e., going from the largest to smallest from left to right, respectively). Formulas in Table 23.7 are given using this nomenclature so you can figure out which ones are normal and which are inverse.
2. In **inverse spinels**, the Y (trivalent) cations occupy the tetrahedral site, and the octahedral sites contains a mixture of Y and X (divalent) cations, with a formula of Y^{3+} $[XY^{3+}]O_4$. Although these are more common in nature, they were described after the spinel species, and thus were given the name "inverse" to indicate that they were the opposite of what had already been described.

In other words, if the trivalent cations are all together on the 6-coordinated site, the occupancy is normal, and if they are split between the two types of sites, the spinel is said to be inverse. When the cations are disordered between the two sites, they are neither inverse nor spinel, but "random." This nomenclature is important because mixing of cations with different charges on the 6-coordinated sites is very different (from a thermodynamic standpoint) than segregarion of charges on distinct sites. Thus, cation distribution in spinels has profound consequences for their relative stabilities in different rock types. In practice, natural samples generally have a distribution of cations that lies between normal and inverse!

You cannot predict on the basis of a formula alone if a spinel will be normal or inverse. In fact, the patterns don't follow any obvious trends, and were considered to be anomalous until crystal field theory came along. As we discussed in Chapter 7, you can calculate a crystal field stabilization energy (CFSE) for any transition metal cation by weighting the contributions of electrons in various orbitals (see Table 7.9). Recall that CFSE can be calculated for a cation in various types of coordination polyhedra, and a high value of CFSE means that a cation will be energetically preferred in a coordination polyhedron with that coordination type. For example, in Table

7.10 we noted that Cr^{3+} has one of the highest values for CFSE of any transition metal in high-spin octahedral coordination, which means that Cr^{3+} *really prefers* to be in octahedral coordination. You can do the same set of CFSE calculations for tetrahedral coordination, and then compare the results, as shown in Table 23.8 for oxides.

The difference between the octahedral and tetrahedral crystal field splitting energies for various cations is termed the **octahedral site preference energy** (OSPE) and is a measure of a cation's preference for octahedral sites relative to tetrahedral ones. We see here again how Cr^{3+} much prefers octahedral over tetrahedral coordination. Thus, in a spinel containing Cr^{3+} we would expect to find the Cr^{3+} in the octahedral site, forming a normal spinel. Similarly, both Ni^{2+} and Cu^{2+} have relatively large values of OSPE; spinels with divalent cations in octahedral coordination are inverse. Cations like Fe^{3+} and Ti^{4+}, which have zero OSPE, will have no preference for either tetrahedral or octahedral coordination, and thus will form both normal and inverse spinels depending on the site preferences of the other coexisting cations. The spinel structures predicted by crystal field theory match the experimentally-measured ones extremely well (see Table 6.2 in Burns, 1993). In Table 23.7 of this text, we have written the formulas to show the typical distribution of cations in each mineral species. For more information on this topic, consult the book by Burns (1993).

The subgroups in the spinel classification (Table 23.7) are defined on the basis of the cation in the B (octahedral) position. Their cation site occupancies are easily understood in terms of the preceding discussion. For example, all the minerals in the Cr^{3+} subgroup are normal spinels because Cr^{3+} has such a high preference for octahedral coordination. On the other hand, those in the Fe^{3+} subgroup are a mixture of inverse and normal spinels depending on the OSPE of the other cation in each one.

Table 23.8. Octahedral Site Preference Energies (OSPE) for Transition Metals in Oxide Structures (after Burns, 1993, Table 6.3)

Number of 3d electrons	Cations	Octahedral CFSE (kJ/mole)	Tetrahedral CFSE (kJ/mole)	OSPE (kJ/mole)
0	Ca^{2+}, Sc^{3+}, Ti^{4+}	0.0	0.0	0.0
1	Ti^{3+}	-87.4	-58.6	-28.8
2	V^{3+}	-160.2	-106.7	-53.5
3	Cr^{3+}	-224.7	-66.9	-157.8
4	Cr^{2+}	-100.4	-29.3	-71.1
4	Mn^{3+}	-135.6	-40.2	-95.4
5	Mn^{2+}, Fe^{3+}	0.0	0.0	0.0
6	Fe^{2+}	-49.8	-33.1	-16.7
6	Co^{3+}	-188.3	-108.8	-79.5
7	Co^{2+}	-92.9	-61.9	-31.0
8	Ni^{2+}	-122.2	-36.0	-89.2
9	Cu^{2+}	-90.4	-26.8	-63.7
10	Zn^{2+}, Ga^{3+}, Ge^{4+}	0.0	0.0	0.0

The spinel structure itself has great significance for studies of the Earth's interior. As you will recall from Chapter 20, the upper mantle's mineralogy is dominated by olivine, orthopyroxene, clinopyroxene, and spinel or garnet, with small amounts of amphibole (usually kaersutite) and mica (usually phlogopite). Deeper in the mantle, the pressure increases, and the minerals eventually become unstable. At about 400 km, the olivine structure is no longer stable. It changes spontaneously from the olivine structure to a spinel-type structure, which takes up 8% less space; the composition remains the same (though we now call it β-Mg_2SiO_4). This causes a seismic discontinuity at about 300–400 km depth that is the boundary to the transition zone.

Magnetite has particular importance because of its magnetic properties (see Types of Magnetism Box). You can probably guess how this mineral earned its name. Because of its ubiquitous presence in the geologic record, magnetite and other oxides have been useful in preserving a geological record of the changes in the direction of Earth's magnetic field. This is the field of paleomagnetism, which measures magnetic behavior in oxide minerals belonging to two solid solutions: ulvöspinel-magnetite and ilmenite-hematite, as shown in Figure 23.17. The bottom of the ternary represents the solid solution between the Fe^{2+} oxide, wüstite (FeO), and its more oxidized equivalent, hematite ($Fe^{3+}_2O_3$). As you go to the top of the diagram, Ti^{4+} substitutes for Fe^{3+}, creating titanomagnetite and hemoilmenite. Magnetite is magnetic because it is a ferrimagnet (see Types of Magnetism Box). Thus, as a magnetite crystal grows, it will record the direction of the magnetic field at that time. If the magma happens to be Ti-rich, the effect of the unpaired Ti^{4+} electrons is to reduce the strength of the magnetic field to zero; the pure ulvöspinel end member is antiferromagnetic (again, see Types of Magnetism Box). A related phenomenon occurs between hematite, which is antiferromagnetic, and ilmenite. It has only recently been recognized that fine-scale intergrowths of hematite and ilmenite can produce ferromagnetic interactions at the atom scale due to coupling in the layers between them (Robinson et al., 2002). Thus magnetite and hemo-ilmenite/ilmeno-hematites contribute to retaining the magnetic signatures of planets.

Many organisms (including humans) are known or suspected to have magnetite crystals in their brains that apparently trigger impulses in

Types of Magnetism

Why are some minerals magnetic, and how are some minerals able to record the direction of the Earth's magnetic field? Really, all minerals are magnetic, but we routinely think of magnetism as the property of being attracted to a magnet. Magnetism arises fundamentally from the spin and orbital motions of electrons, though spin is most important in common Fe and Ti-rich materials. There are five types of magnetism and all materials will exhibit one of these.

1. **Diamagnetism** is found in all matter, such as helium, water, and copper. It occurs in materials in which all outer orbital shells are filled and there are no unpaired electrons. In this situation, electrons with opposite spins and equal energy pair up to cancel each other out. Although these materials generally have no magnetic moments, they can acquire weak magnetization in the opposite direction to an applied magnetic field because of weak currents created by changes in the orbital motions.

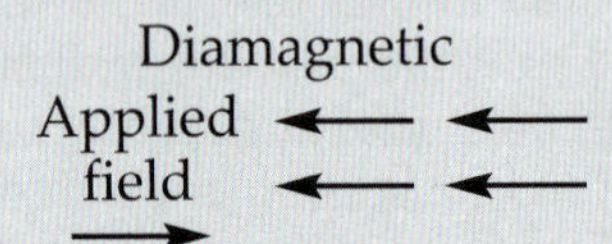

2. **Paramagnetism** results when some of the atoms in a material have unpaired electrons in partially-filled orbitals. If the electrons do not interact, there is no magnetization. But when a magnetic field is applied, the atomic motions may align with it. If you place an ion with unpaired outer electrons in a crystal structure, it will normally be paramagnetic. Thus, most minerals are paramagnetic at room temperature, including such examples as pyrite, siderite, and phlogopite.

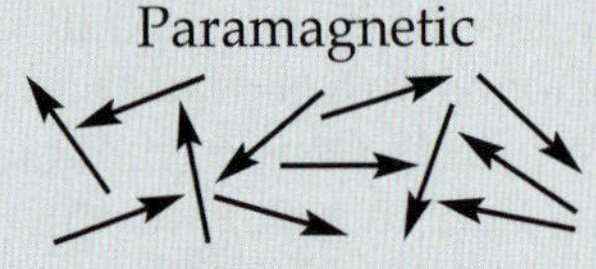

3. If the atoms containing unpaired outer electrons are close together within a crystal structure, their spins can interact, and electrons in adjacent sites will acquire parallel alignments. These materials will take on their own, strong magnetization, and are considered to be **ferromagnetic**. Examples of this include Fe, Ni, Co, and many of the related alloys. Note that when ferromagnetic materials are cooled, they eventually reach a point at which the electronic forces between adjacent electrons can no longer interact, and effectively "quenched." This occurs at a (somewhat) diagnostic temperature for each material, and is called the **Curie temperature**.

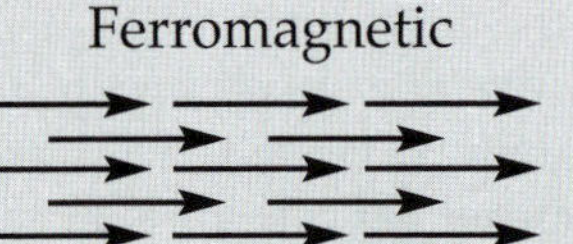

4. **Ferrimagnetism** is a more complex form of ferromagnetism in which two sets of magnetic fields are created within the same crystal structure, usually separated by oxygen anions. One set of atoms spontaneously aligns in one direction, and a second set, equally spontaneously, pairs off in the opposite direction. However, the magnetic moments of the two opposing fields are not equal, and thus these materials have a strong net magnetic field above the Curie temperature. Magnetite, pyrrhotite, and feroxyhyte are all ferrimagnets.

Ferrimagnetic

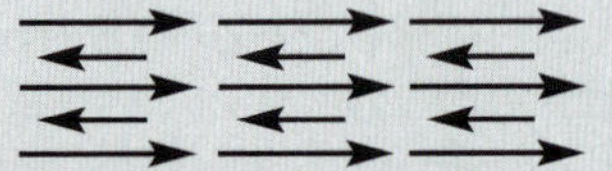

5. **Antiferromagnetic** materials also contain two set of atoms with exactly equal and opposite magnetic moments. In this case, the moments cancel out perfectly, and the material will have no magnetic moment. Many minerals such as ilmenite, goethite, and troilite are all antiferromagnets at room temperature. When they are cooled to low temperatures, however, thermal vibrations cause them to lose their antiferromagnetic structure, and they become paramagnetic; the temperature at which this occurs is called the **Néel point**.

Antiferromagnetic

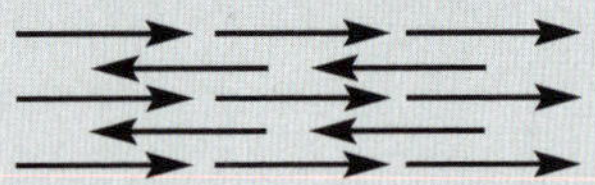

surrounding nerves in response to changes in magnetic field. This capability is probably what allows many migratory animals to navigate. Magnetite crystals have been found in the brains of homing pigeons, for example, as well as many fish species.

Finally, the spinel species **magnetite** has petrologic importance because it is found in a huge range of different rock types. The equilibrium between quartz, olivine, and magnetite is used as a reference to specify a commonly-occurring oxygen pressure typical of rocks formed at or near the Earth's surface. In addition, magnetites in many igneous and metamorphic rocks are titaniferous (containing ulvöspinel, Fe_2TiO_4, in solid solution). Titaniferous magnetites coexisting with ilmenite solid solution are sensitive to both temperature and redox conditions (oxygen fugacity), and in favorable conditions the compositions of both phases can indicate the temperature and oxygen fugacity at which they formed.

There are several minerals closely related to those in the spinel group. **Chrysoberyl** ($BeAl_2O_4$) has the same structure as olivine, with Al^{3+} in the octahedral (Fe,Mg) sites and Be^{2+} in the Si^{4+} tetrahedral sites. It is similar to the spinel structure as well, though the smaller Be^{2+} in the tetrahedral site makes the structure orthorhombic rather than isometric. Other simple oxides like **maghemite** ($(Fe^{3+},\square)_3O_4$) have a spinel structure with both Fe^{3+} and vacancies in both the 4- and 6-coordinated sites. **Arsenolite**, As_2O_3 (Figure 21.8), is made from four pyramids with O^{2-} at the base and As^{3+} at the tip; these form As_4O_6 molecules linked by weak van der Waals bonds.

The **perovskite** group minerals (Table 23.7; note that these are, strictly speaking, "simple" oxides) have a nearly-cubic structure in which Ca^{2+} cations replace O^{2-} in a cubic close-packed

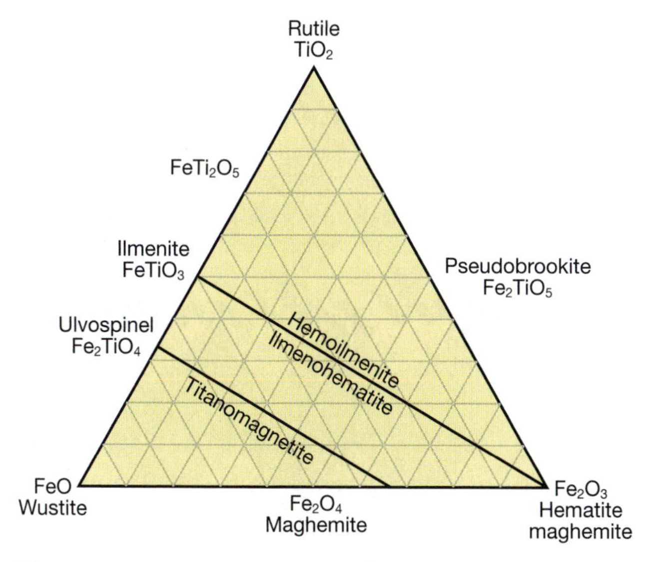

Figure 23.17. Ternary diagram for the iron oxides showing the interrelationships among Fe^{2+}, Fe^{3+}, and Ti^{4+}. Substitution along the hemoilmenite and titanomagnetite lines toward Ti causes a decrease in the magnetic remanence of the phases.

arrangement; thus, the Ca^{2+} cations are in 12-coordination with O^{2-}. Ti^{4+} and other small cations occupy some of the octahedral interstices in coordination with oxygen anions; these octahedra share corners to create a framework (Figure 21.12). The resulting minerals have a general formula of ABO_3, where A = Ca^{2+}, Na^{1+}, Ce^{3+}, and/or Sr^{2+}, and B is either Ni^{2+} or Nb^{3+}. The perovskite structure is important within the transition zone in the Earth's mantle, where Ca garnet transforms to a perovskite structure, resulting in a density increase.

One last oxide that has great political and economic importance is the oxide **uraninite** (UO_2), which is the major ore of uranium. Uraninite crystallizes with a body-centered cubic lattice type; O^{2-} anions lie at the corners of the cube and U^{1+} is in the center in 8-coordination. Uraninite was first mined in the early 20th century for its minor radium contents (radium was used in making luminescent dials for watches, and other unhealthy applications). Of course, uranium itself became important during the Manhattan Project in World War II for the development of the atomic bomb and after for both weapons and generation of electricity. Extensive mining operations were then undertaken in the U.S., Canada, and the Belgian Congo. We now know that Australia has the world's largest reserve of uranium, though strict laws regulate its export only to uses involving the production of electricity (e.g., in nuclear power plants). Uranium is also exploited by geologists because the long half-life of the ^{238}U isotope (4.51×10^9 years) makes it well-suited for radiometric dating of rocks.

Hydroxides and oxides containing hydroxyl. The addition of H^{1+} to oxide structures, usually incorporated in the form of hydroxyl $(OH)^{1-}$, changes the possible structures only slightly because the H^{1+} is so small. These minerals contain structural $(OH)^{1-}$ and not structural H_2O. There is often confusion over these two terms because the $(OH)^{1-}$ content of a mineral may be expressed as the weight percent H_2O. However, OH-bearing minerals *do not contain molecular water* so they are not considered hydrated minerals.

Perhaps the most important of these phases are the three mineral species that make up the valuable aluminum ore, bauxite: **diaspore**, **gibbsite**, and **böhmite**. The rock type "bauxite" occurs in nature as a heterogeneous mixture of these three phases, and forms in area of very high weathering rates—like the tropics. It is primarily mined to make the aluminum metal used in many materials in our everyday lives: kitchen appliances, cars, ships, airplanes, glass, aluminum foil, and even electrical transmission lines. Aluminum is preferred for the latter because it is light in weight, even though it has only about 60% of the electrical conductivity of copper.

The **diaspore group** contains five isostructural mineral species with the formula AO(OH), where A = Al^{3+}, Fe^{3+}, Mn^{3+}, V^{3+}, and/or Cr^{3+}. In the species **diaspore**, (α-Al(O,OH)), $Al(O,OH)_6$ octahedra each share three edges to create double chains running parallel to [001]; the chains share corners of the octahedra, and are also held together by weak H bonds (Figure 23.18).

Goethite (α-FeO(OH)) is isostructural with diaspore, and is a very common alteration product of rocks that have been oxidized and/or weathered in the presence of water. Again, the mineral does not contain water, but water interacts with it and disassociates into $(OH)^{1-}$ groups. In the older literature, goethite and mixtures of it with other hydrous Fe oxides, are sometimes called limonite. Recall that the Greek letters used as a prefix in the above chemical formulas are just symbols that indicate which polymorphs are represented by each of these minerals.

Böhmite (γ-AlO(OH)) and its isostructural Fe-bearing twin, **lepidocrocite** (γ-FeO(OH)), are also composed of $Al/Fe(O,OH)_6$ octahedra, though in these minerals they form sheets by sharing six of their twelve edges (Figure 23.18). As a result, the $O^{2-}/(OH)^{1-}$ anions lie in a cubic close-packed arrangement. Again, weak hydrogen bonds hold the sheets together.

In **brucite group** minerals (Table 23.9), all the anions are hexagonal close-packed hydroxyls with cations (generally Mg^{2+}, but with smaller amounts of Fe^{2+}, Mn^{2+}, Ca^{2+}, and/or Ni^{2+}) in all the octahedral interstices (Figure 23.19). Layers of edge-sharing octahedra create sheets. Each hydroxyl is linked to three Mg^{2+} in its own layer, and weakly bonded to three $(OH)^{1-}$ in the adjacent layer. Thus the layers are held together not only by van der Waals bonds, but also by hydrogen bonding. This structure unit is important because it appears in a number of silicates, notably those in the chlorite group. **Gibbsite** (γ-$Al(OH)_3$) is closely related to **brucite** ($Mg(OH)_2$) except that the higher charge of the Al^{3+} cation means that only 2/3 of the octahedral sites in the hcp array are filled. The gibbsite structure has lower symmetry than brucite because the $Al(OH)_6$ octahedra are slightly distorted (by rotating into the void created by 1/3 of the octahedral sites being empty), and this causes the stacking of layers to be slightly offset (Figure 23.19). This structure should look familiar, because we have already seen it as part of the "sandwich" in the structure of chlorite.

Multiple oxides with Nb, Ta, and Ti. The **pyrochlore** group includes 23 mineral species in four subgroups that are subdivided as shown in

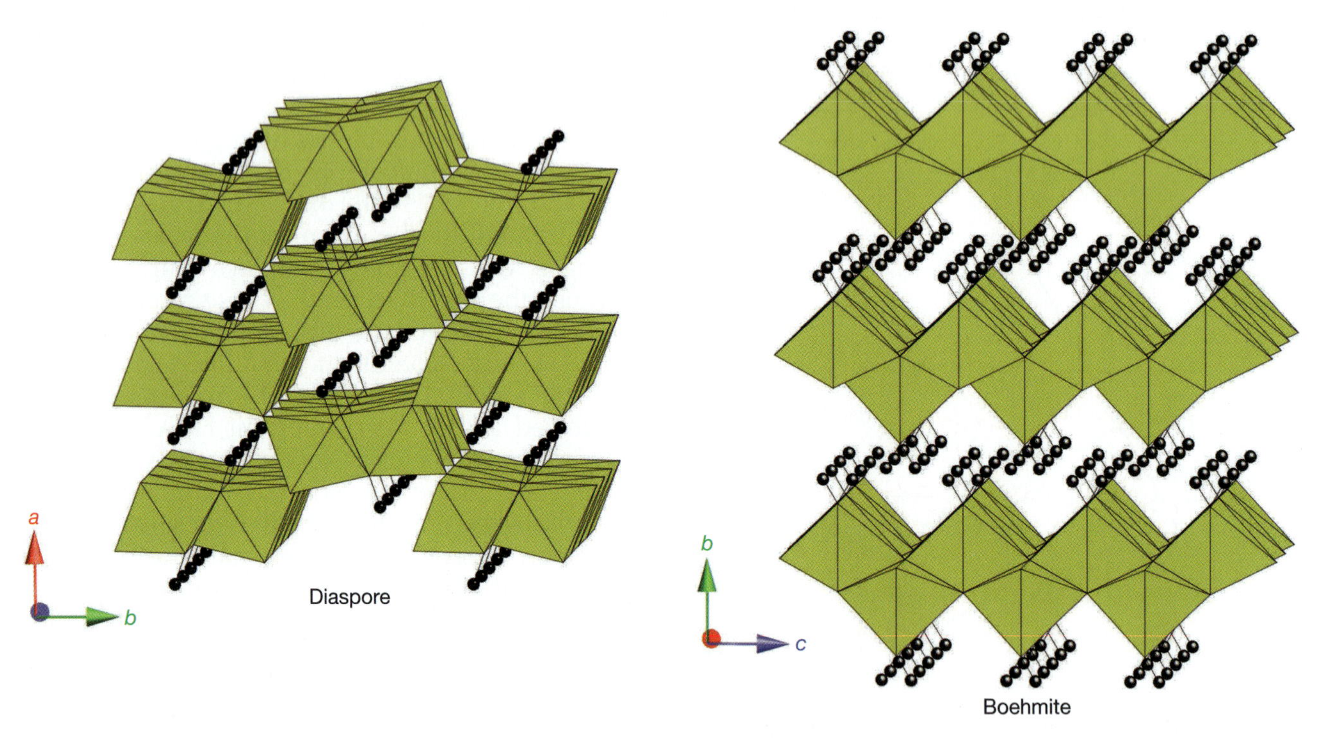

Figure 23.18. *The crystal structures of the polymorphs of AlO(OH): diaspore (left) and böhmite (right). Notice that both of these minerals contain* Al^{3+} *in 6-coordination with differing amounts of* O^{2-} *and* $(OH)^{1-}$*. The OH's are represented by* H^{1+} *ions bonding off the* O^{2-} *anions that define the corners of the octahedra. The major difference in the structures is the number of edges shared by the* Al^{3+} *octahedra—three for diaspore and six for böhmite.*

Table 23.9. Hydroxides		
Diaspore	diaspore	α-AlO(OH)
	goethite	α-$Fe^{3+}O(OH)$
	groutite	α-MnO(OH)
	montroseite	$(V^{3+},Fe^{2+},V^{4+})O(OH)$
	bracewellite	$Cr^{3+}O(OH)$
	tsumgallite	GaO(OH)
	böhmite	AlO(OH)
	lepidocrocite	γ-$Fe^{3+}O(OH)$
	gibbsite	γ-$Al(OH)_3$
Brucite	brucite	$Mg(OH)_2$
	amakinite	$(Fe^{2+},Mg)(OH)_2$
	pyrochroite	$Mn^{2+}(OH)_2$
	portlandite	$Ca(OH)_2$
	theophrastite	$(Ni,Mg)(OH)_2$

Table 23.10. They share a basic formula of $A_2^{3+}B_2^{4+}O_7$. The pyrochlore structure is a derivative of the fluorite structure type (ZrO_2, or Zr_4O_8) in which half the B^{4+} cations are replaced by A^{3+}, and 1/8 of the O^{2-} are omitted for charge balance. There are two types of cations: large A cations in 8-coordinated sites (Ca^{2+}, K^{1+}, Ba^{2+}, Y^{3+}, Ce^{3+}, Pb^{2+}, U^{4+}, Sr^{2+}, Cs^{1+}, Na^{1+}, Sb^{3+}, Bi^{3+}, and/or Th^{4+}) and smaller B cations in the distorted octahedra sites (Nb^{5+}, Ta^{5+}, Ti^{4+}, Sn^{4+}, Fe^{3+}, and W^{6+}). The large range of possible cation substitutions allows the physical properties to vary considerably. Thus, they are used in electronic materials, electrodes, electrolytes, and in structures that can store radioactive waste. The complex pyrochlore structure can accommodate large cations as well as the $(OH)^{1-}$ and F^{1-} (Figure 23.20). Thus, many radioactive cations are found in these minerals, and their hand samples must be handled carefully.

The **columbite** group has only six known mineral species, with the general formula of AB_2O_4; A is Fe^{2+}, Mn^{2+}, and/or Mg^{2+}, while B is Nb^{5+} or Ta^{5+}. These are used as ores of niobium, which is used in metals alloys to make them stronger, in jewelry, and (increasingly) in superconductors. Pyrochlore is also an ore of niobium.

Anhydrous and hydrated oxides. After all the discussions involving **halite** (NaCl) in this textbook, we will assume that its structure needs no further explanation (Figures 14.1 and 14.9). There are actually five mineral species in the **halite group** (Table 23.11), all of which are isostructural. The real significance of halite relates to its use as table salt. It is mined from many different types of deposits in sedimentary rocks, where is occurs as the result of evaporation of brines. It is mined in two ways. The majority of salt comes from solution mining, in which fresh water is injected into a well, forcing salty water to rise to the surface, where it can be dried to crystallize salt. Alternatively, huge salt deposits such as those located a quarter of a mile beneath the city of Detroit were, until recently, mined like coal. They supplied road salt to Michigan and many other

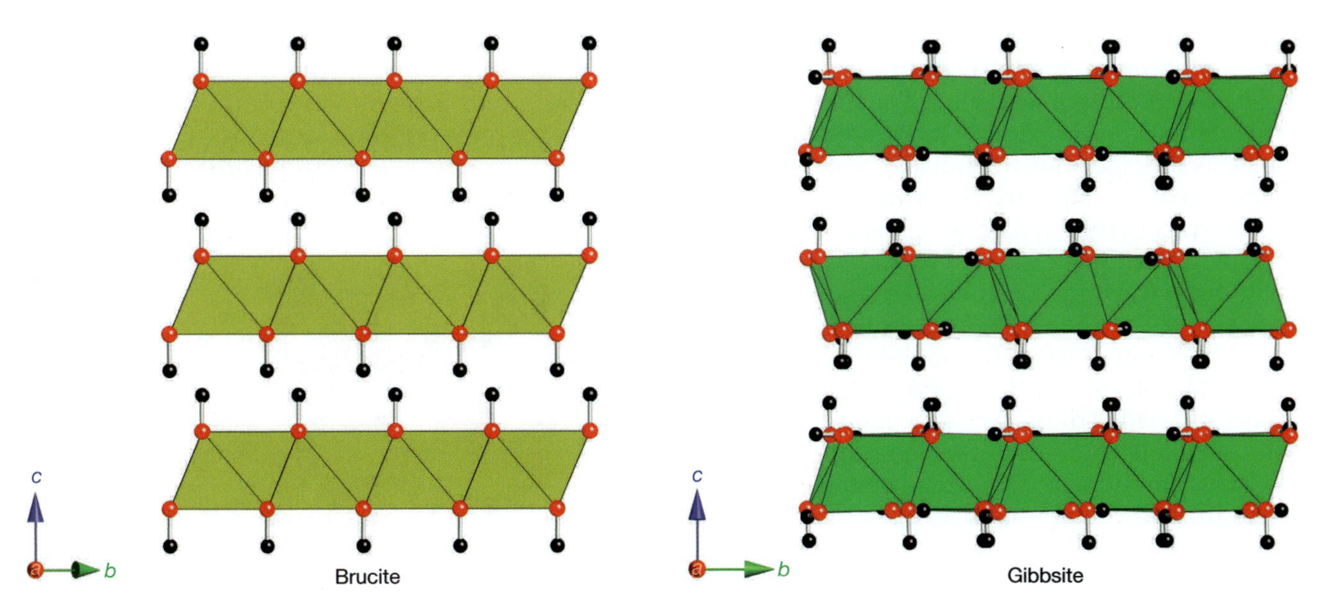

Figure 23.19. The crystal structures of brucite (left, $Mg(OH)_2$) and gibbsite (right, $Al(OH)_3$), are both formed of layers of octahedra with OH groups forming the six bonding ligands. In this projection, all of the O-H bonds are perpendicular to the octahedral sheets in brucite, but some of them are parallel to the sheets in gibbsite. It might be helpful to use CrystalViewer and link to these files on the MSA website then rotate them about the c-axis. Think about the dioctahedral and trioctahedral micas to guess at the major difference between these two structures.

Table 23.10. Additional Multiple Oxides			
Pyrochlore	*Pyrochlore subgroup Nb>Ta;(Nb+Ta)>2(Ti)*	pyrochlore	$(Ca,Na)_2Nb_2O_6F$
		kalipyrochlore	$(K,Sr)_{2-x}Nb_2O_6[O,(OH)]\cdot nH_2O$
		bariopyrochlore	$BaNb_2(O,OH)_7$
		yttropyrochlore-(Y)	$YNb_2O_6(OH)$
		ceriopyrochlore-(Ce)	$(Ce,Ca,Y)_2(Nb,Ta)_2O_6(OH,F)$
		plumbopyrochlore	$(Pb,Y,U,Ca)_2Nb_2O_6(OH)$
		uranpyrochlore	$Ca_{0.5}U_{0.5}Nb_2O_6(OH)$
		bismutopyrochlore	$(Bi,U,Ca,Pb)_{1+x}(Nb,Ta)_2O_6(OH)\cdot n(H_2O)$
	Microlite subgroup Ta>Nb;(Ta+Nb)>2(Ti)	microlite	$NaCaTa_2O_6(OH)$
		bariomicrolite	$BaTa_2(O,OH)_7$
		plumbomicrolite	$PbTa_2O_6(OH)$
		uranmicrolite	$U_{0.5}Ca_{0.5}Ta_2O_6(OH)$
		bismutomicrolite	$BiTa_2O_6(OH)$
		stannomicrolite	$Sn^{2+}Ta_2[O,(OH)]_7$
		stibiomicrolite	$SbTa_2O_6(OH)$
		IMA1998-018	$(Na,Ca,Bi)_2Ta_2O_6F$
	Betafite subgroup 2(Ti)>(Ta+Nb)	betafite	$U^{4+}{}_2(Ti,Nb)_2O_6(OH)$
		yttrobetafite-(Y)	$Y_2(Ti,Nb)_2O_6(OH)$
		plumbobetafite	$Pb_2(Ti,Nb)_2O_6(OH)$
		stibiobetafite	$CaSb^{3+}(Ti,Nb)_2O_6(O,OH)$
		calciobetafite	$Ca_2(Nb,Ti)_2(O_5OH)(OH)$
	Cesstibtantite subgroup	cesstibtantite	$(Sb^{3+}{}_{0.5}Na_{0.5})Ta_2O_6[Cs_{0.5}(OH)_{0.5}]$
		natrobistantite	$(Na,Cs)Bi(Ta,Nb,Sb)_4O_{12}$
Columbite		ferrotantalite	$Fe^{2+}Ta_2O_6$
		ferrocolumbite	$Fe^{2+}Nb_2O_6$
		manganotantalite	$Mn^{2+}Ta_2O_6$
		manganocolumbite	$(Mn^{2+},Fe^{2+})(Nb,Ta)_2O_6$
		magnesiocolumbite	$(Mg,Fe,Mn)(Nb,Ta)_2O_6$
		magnesiotantalite	$(Mg,Fe)(Ta,Nb)_2O_6$

midwestern states. There are over 100 miles of rock salt tunnels under Detroit (unfortunately, these are closed to the public)! There are also salt deposits in New Mexico, which have been proposed as storage sites for nuclear waste because of their geologic stability. Some of these deposits also contain **sylvite** (KCl), which is often used as a rather bitter tasting salt substitute.

Fluorite, CaF_2, has been known since antiquity, and was one of the earliest mineral structures to be studied with X-rays. Its rainbow of naturally-occurring colors makes fluorite a favorite with collectors. The colors arise from a variety of defects called **color centers**. These are places in the structure where a defect occurs, with the result that a F^{1-} anion is missing. When an electron moves in to occupy that F^{1-} vacancy for charge compensation, it can absorb light by being excited into higher energy states.

Many types of these color centers are known. **F centers** are named after the characteristic purple color of common fluorite. Work by George Rossman (see the Mineral Spectroscopy website at Caltech) suggests that green fluorite is colored by trapped electrons in the presence of yttrium (Y^{3+}) ions, and proximal to cerium (Ce^{3+}) ions, or by the presence of trace amounts of Sm^{2+}. Pink results from trapped electrons adjacent to Y^{3+}. A

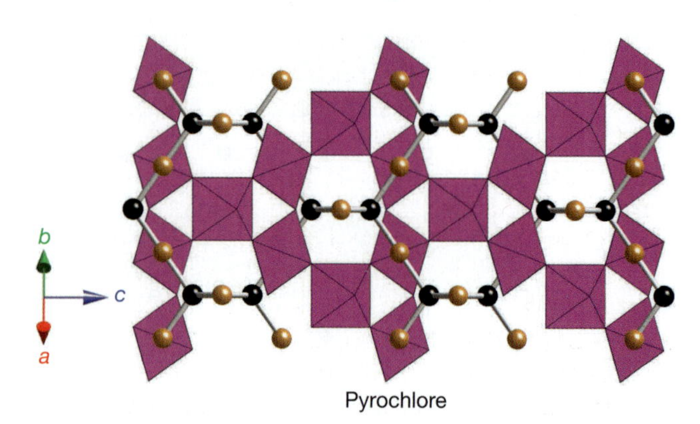

Figure 23.20. The crystal structure of pyrochlore. Channels are formed by corner-sharing Nb-octahedra. Ca^{2+} (brown spheres) also bond to form a network with F^{1-} (black spheres) anions that fill the channels.

Table 23.11. Halite group	
halite	NaCl
sylvite	KCl
villiaumite	NaF
carobbiite	KF
griceite	LiF

nice review of color center theory is given in Schulman and Compton (1962).

Fluorite is also the major "ore" of fluorine, which is used as a flux in metals to help remove impurities. Fluorite crystals are used in optics to make lenses because fluorite lenses can transmit light better at certain wavelengths than glass lenses. Flourine is also an essential ingredient of many glasses and enamels.

As we have noted previously, the fluorite structure consists of Ca^{2+} cations in a cubic face-centered arrangement. If that arrangement is divided into eight sub-cubes, F^{1-} will be found in the center of each one. Thus, Ca^{2+} is in 8-coordination with F^{1-}, and every F^{1-} is in 4-coordination with Ca^{2+} (Figure 14.2 and 14.3). Many other minerals, as we have already noticed, take on this important structure type.

There are some known hydrated oxides, but they are extremely rare. The list includes such unusual minerals as antarcticite ($CaCl_2 \cdot 6H_2O$), eriochalcite ($CuCl_2 \cdot 2H_2O$), bischofite ($MgCl_2 \cdot 6H_2O$), and hydromolysite ($FeCl_3 \cdot 6H_2O$).

Carbonates

The carbonate $(CO_3)^{2-}$ "ion" is composed of three O^{2-} anions in a triangular arrangement with a tiny C^{4+} cation in the middle. The distance between the C^{4+} and O^{2-} ions (about 1.28Å) and the O-C-O angle of 120° are nearly constant in all minerals within this group. The bonds between C^{4+} and O^{2-} are strong and covalent (perhaps 4× stronger than the Ca-O bonds), making the $(CO_3)^{2-}$ unit fairly rigid, similar to the $(SiO_4)^{4-}$ anionic complex that form the silicates The difference in bond strength also controls the cleavage directions of minerals in this class, because they will tend to cleave along ionic bonds, which in general are parallel to the planes of the $(CO_3)^{2-}$ groups.

Different arrangements of the $(CO_3)^{2-}$ triangles give rise to different minerals within this class, as well as many other minerals including such silicates as scapolite. All of these minerals will react and effervesce in the presence of hydrochloric acid (HCl in water); for this reason, many field geologists carry acid bottles for carbonate identification. This reaction occurs because the carbonates react with the H^{1+} and Cl^{1-} in the acid to form CO_2 gas, which bubbles. The reaction can be described as $CaCO_3 + H^{1+} + Cl^{1-} \Rightarrow Ca^{2+} + CO_2(gas) + (OH)^{1-} + Cl^{1-}$. In fact, CO_2 fluctuation in the atmosphere is related to limestone formation and weathering; thus, these minerals play role in climate change issues. Carbonate deposits effectively "store" CO_3 in the oceans, buffering the CO_2 content of the atmosphere and helping keep global warming in abeyance.

You will notice that incorporation of carbonate into mineral structures adds an additional level of complexity, so the remaining structures in this chapter are more difficult to describe in words. We thus particularly urge you to study not only the illustrations here, but also the animations on the MSA website. In addition, we highly recommend the *Reviews in Mineralogy* volume (11) edited by R.J. Reeder (1983) for detailed information about all these structures.

Acid carbonates. Minerals in the acid carbonate class (a somewhat antiquated term) contain the radical group HCO_3^{1-}, which is more properly called bicarbonate. These minerals usually form in the presence of carbonic acid (H_2CO_3), which is the weak acid that forms when CO_2 is dissolved in water. When one proton is removed from H_2CO_3, acid or bicarbonates form; further deprotonation results in formation of carbonate!

One of the most economically important carbonates is the acid carbonate mineral **trona** ($Na_3H(CO_3)_2 \cdot 2H_2O$, a hydrated sodium bicarbonate. The structure of trona consists of one NaO_6 octahedron, two NaO_5H_2O trigonal prisms, two $(CO_3)^{2-}$ flat pyramids, and one H^{1+}, as described by Struntz and Nickel (2001) (Figure 23.21). It has many uses. In soaps, detergents, and various cleansers, trona emulsifies oil stains, helps remove dirt, and softens the water (some people add trona/baking soda to their wash loads to help their detergents do a better job!). Trona is used to make baking soda and baking powder, cattle feed, swimming pool products, medicines, paper, textiles, and toothpaste.

Anhydrous carbonates. The **calcite group** includes eight members with the general formula ACO_3 (Table 23.12). Although the cation varies, they all share the characteristic of very high optical birefringence that results largely from the presence of carbonate ions in parallel planes. There is limited solid solution among the A cations, which include Ca^{2+}, Mg^{2+}, Mn^{2+}, Fe^{2+}, Co^{2+}, Ni^{2+}, Zn^{2+}, and Co^{2+}.

The structures are composed of planes of $(CO_3)^{2-}$ that lie perpendicular to the *c* axis. The

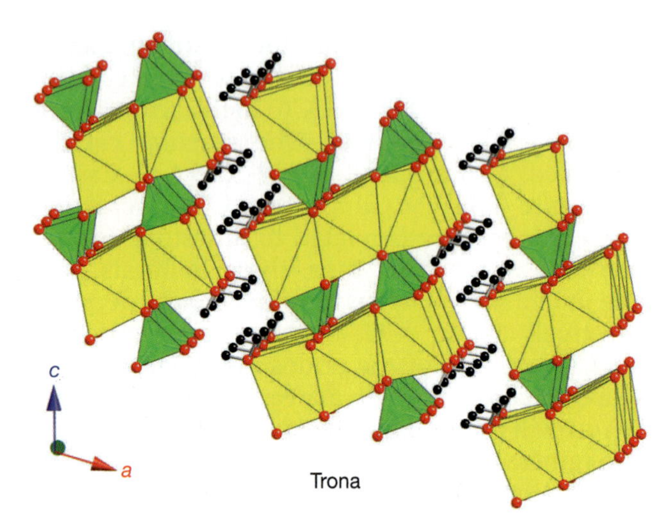

Figure 23.21. *The crystal structure of trona ($Na_3H(CO_3)_2 \cdot 2H_2O$), a hydrated carbonate. ($(CO_3)^{2-}$ groups are represented by the small green triangles, while Na^{1+} is housed in the yellow shaded polyhedra. The two H^{1+} ions in each of the H_2O molecules are represented by small black spheres and the H-bonds that they form hold the structure together.*

planes are connected by Ca^{2+} in octahedral sites. All the planes are stacked along the *c* axis, and they alternate with $(CO_3)^{2-}$ triangles pointing in opposite directions (Figure 23.22). The resultant structure is *nearly* hexagonal close packing of oxygens, with Ca^{2+} in some of the octahedral interstices. Carbon atoms are then distributed to avoid being stacked over either Ca^{2+} or another C^{4+} from the adjacent layer. Six of these layers make up the unit cell. Its size is fairly well related to the size of the A cation in the structure, although octahedral distortion prevents the relationship from being perfect. There is only limited solid substitution of the A cations for one another.

Calcite ($CaCO_3$) is by far the most common carbonate mineral, and is the primary mineral in both limestone and its metamorphic equivalent, marble. In addition to being mined as building and ornamental stone, crushed calcite has long been used as a source of lime (CaO) to make cement. The production of cement is major contributor to CO_2 in the atmosphere, because CaO is produced by heating $CaCO_3$ and driving off CO_2 as a gas, leaving the CaO. Calcite is also used in steel manufacture and in many chemical and optical applications. The white paint on your ceiling may well contain calcite as a whitener! Finally, we owe our first understanding of the polarization of light to calcite (Chapter 16).

Aragonite is the second common polymorph of $CaCO_3$ (a third rare polymorph of $CaCO_3$, vaterite, also exists, and it has hexagonal symmetry). As in calcite, there are alternating sheets of $(CO_3)^{2-}$ pointing in opposite directions (Figure 23.22). Not all the ions lie perfectly in the plane; the C^{4+} is raised slightly from the oxygen triangle (~0.026 Å), and the O^{2-} triangle itself is tilted 2.5° from the plane. The cations are also slightly out of the plane, about ±0.05 Å. Thus the slightly-puckered carbonate and cation layers are stacked perpendicular to the *c* axis in an ABAB arrangement, with each Ca^{2+} cation coordinated by nine oxygen corners of the $(CO_3)^{2-}$ groups. As a result, aragonite and other members of this group have different (orthorhombic) symmetry than those in the calcite group. Aragonite is the high pressure polymorph of calcite. If you crush calcite in a mortar and pestle you can convert some of it to aragonite. You can observe this either by using a PLM to note the different optical classes of the two or by study with powder XRD.

Variations in stacking patterns give rise to the commonly-observed twinning in this group; sometimes, the AB stacking becomes ABC, causing 120° bends in the crystals. These twins are common to all species in this group (Table 23.12).

Recall from Chapter 20 that the relationship between aragonite and calcite depends on temperature and pressure (Figure 20.2), with aragonite as the higher pressure phase. Aragonite is preserved in some metamorphic rocks from the blueschist facies, which form at high pressure and low temperature in subduction zone settings. Carlson and Rosenfeld (1981) showed that the post-metamorphism cooling path must be below 180°C when it crosses back into the stability field of calcite in nature; dry conditions are also required in order for metamorphic aragonite to be preserved. Aragonite readily grows metastably at

Table 23.12. Anhydrous carbonates

Group	Mineral	Formula
Calcite	calcite	$CaCO_3$
	magnesite	$MgCO_3$
	siderite	$Fe^{2+}CO_3$
	rhodochrosite	$Mn^{2+}CO_3$
	spaherocobaltite	$CoCO_3$
	smithsonite	$ZnCO_3$
	otavite	$CdCO_3$
	gaspeite	$(Ni,Mg,Ca,Fe^{2+})CO_3$
Aragonite	aragonite	$CaCO_3$
	witherite	$BaCO_3$
	strontianite	$SrCO_3$
	cerussite	$PbCO_3$
Dolomite-ankerite	dolomite	$CaMg(CO_3)_2$
	ankerite	$Ca(Fe^{2+},Mg,Mn)(CO_3)_2$
	kutnohorite	$Ca(Mn^{2+},Mg,Fe)(CO_3)_2$
	minrecordite	$CaZn(CO_3)_2$

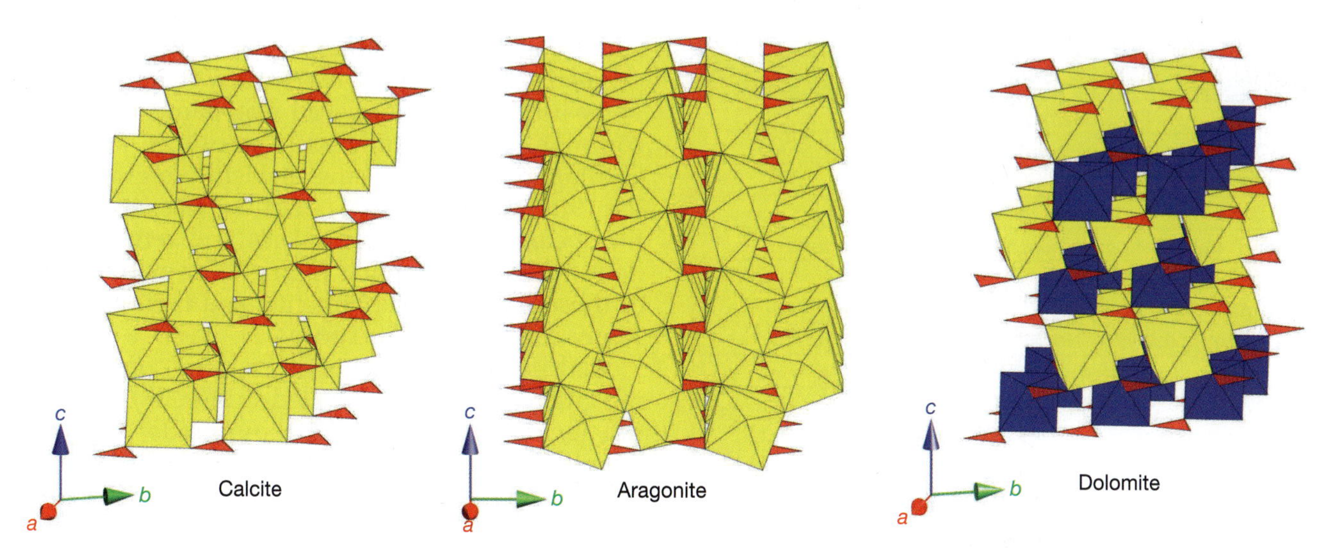

Figure 23.22. The crystal structures of three common carbonates: calcite (left, $CaCO_3$), aragonite (middle, $CaCO_3$), and dolomite (right, $CaMg(CO_3)_2$). In all the structures, the small red triangles represent the $(CO_3)^{2-}$ groups, and these triangles are perpendicular to the c axis. Ca^{2+} is located in the yellow shaded octahedron. Notice how the structure of aragonite appears more "dense" than that of calcite—in fact, it is more dense and is the high pressure polymorph of $CaCO_3$. In dolomite, the blue octahedra house Mg^{2+} cations; notice how the layers of Ca^{2+} and Mg^{2+} octahedra alternate in dolomite.

1 atm and room temperature in magnesian solutions (Bischoff and Fyfe, 1968). It may also form under conditions of extreme supersaturation in caves and some coral reefs. The formation of aragonite in shells indicates the capability of some marine species to control the growth of aragonite over calcite despite its greater solubility in seawater. It may be favored in evolutionary terms because its lower symmetry leads to formation of strong elongate fibers that are difficult to cleave, unlike calcite. Vaterite is a rare polymorph of $CaCO_3$, which has been shown to be metastable at all pressures and temperatures (Wold et al., 2000). It has been reported in cold British lakes and in a kidney stone from a dog. The phase relationships between calcite and aragonite were discussed in Chapter 20 (Figure 20.2).

The members of the **dolomite** group are considered double carbonates based on their formula, $CaB(CO_3)_2$, in which B can be Mg^{2+}, Fe^{2+}, Mn^{2+}, or Zn^{2+}. These minerals nearly have a calcite structure except that the cations are ordered among the octahedral sites, slightly lowering the space group symmetry (Figure 23.22). The Ca^{2+}-occupied octahedral layers alternate with layers filled by the B cations along the *c* axis. The B cations are all uniformly smaller than Ca^{2+} and the resultant bond lengths with oxygen are shorter; this causes distortion in the octahedra and a slight amount of rotation of the $(CO_3)^{2-}$ groups (about 6–7°). The net effect is a reduction in symmetry. This is complicated in natural samples by the fact that Ca^{2+} frequently substitutes for the B cations, and the B cations may sometimes occupy the Ca^{2+} site.

Dolomite is also extremely common in metamorphic and sedimentary rock types. **Dolomitization**, or replacement of calcite and aragonite by dolomite, can result from hydrothermal metamorphism and the addition of Mg^{2+} cations from seawater. Because the volume of dolomite (323.85 Å^3) is only 88% of that of calcite (367.78 Å^3), this process causes an increase in the amount of pore space in a rock, thus making dolomite an important reservoir for petroleum and natural gas. A solid solution between calcite and dolomite does exist at high temperatures (Figure 20.7), but the two phases exsolve below temperatures of about 1000°C.

Carbonates containing hydroxyl or halogen. The many members of the bästnasite/synchysite/parasite group are valued as commercial sources for rare earth elements (REE), especially cerium and lanthanum. Minerals in this group combine three components, REE(OH,F), $(CO_3)^{2-}$, and $CaCO_3$, in varying ratios as described by Gaines et al. (1997):

bastnäsites are 1:1:0
synchysites are 2:1:1
parasites are 2:2:1, and
röntgenite (Table 23.13).

These minerals have layered structures that may display numerous polytypes and are often intergrown with one another. Again, all of them have high birefringence because of the parallel layers of the CO_3 groups.

Table 23.13. Bästnasite/synchysite/parasite group

Mineral	Formula
Bästnasite subgroup	
bastnäsite-(Ce)	$(Ce,La)(CO_3)F$
bastnäsite-(La)	$(La,Ce)(CO_3)F$
bastnäsite-(Y)	$(Y,Ce)(CO_3)F$
Bastnäsite subgroup, hydroxylbastnäsite series	
hydroxylbastnäsite-(Ce)	$(Ce,La)(CO_3)(OH)$
hydroxylbastnäsite-(La)	$(La,Nd)(CO_3)(OH)$
hydroxylbastnäsite-(Nd)	$(Nd,La)(CO_3)(OH)$
Synchysite subgroup	
synchysite-(Ce)	$Ca(Ce,La)(CO_3)_2F$
synchysite-(Y)	$CaY(CO_3)_2F$
synchysite-(Nd)	$Ca(Nd,La)(CO_3)_2F$
Unnamed subgroup	
huanghoite-(Ce)	$Ba(Ce,La, Nd)(CO_3)_2F$
IMA2004-019	$Ba(Ce,REE)(CO_3)_2F$
Parisite subgroup	
parisite-(Ce)	$Ca(Ce,La)_3(CO_3)_2F$
parisite-(Nd)	$Ca(Nd,Ce,La)_3(CO_3)_2F$
Unnamed subgroups	
röntgenite-(Ce)	$Ca_2(Ce,La)_3(CO_3)_5F_3$
cordylite-(Ce)	$NaBaCe_2(CO_3)_4F$
lukechangite-(Ce)	$Na_3Ce_2(CO_3)_4F$
zhonghuacerite-(Ce)	$Ba_2Ce(CO_3)_3F$
kukharenkoite-(Ce)	$Ba_2Ce(CO_3)_3F$
kukharenkoite-(La)	$Ba_2(La,Ce)(CO_3)_3F$
cebaite-(ce)	$Ba_3Ce_2(CO_3)_5F_2$
cebaite-(Nd)	$Ba_3(Nd,Ce)_2(CO_3)_5F_2$
horvathite-(Y)	$NaY(CO_3)F_2$
decrespignyite-(Y)	$(Y,REE)_4Cu(CO_3)_4Cl(OH)_5 \cdot H_2O$

The **bastnäsite** subgroup minerals contain layers of $REE(F_3O_6)$ polyhedra that alternate with layers of $(CO_3)^{2-}$ groups parallel to the *c* axis. Large 9-fold coordination polyhedra are needed to accommodate the large rare earth elements. In the hydroxylbastnäsite series, the structure is the same except that $(OH)^{1-}$ ions replace F^{1-}. **Synchysite** minerals have stacks of two $REE(F_3O_6)$ layers, a $CaCO_3$ layer, and a CaO_8 layer perpendicular to *c*. The stacks are held together by sharing of O^{2-} in $(CO_3)^{2-}$ groups. Parisite and röntgenite are characterized by ordering of Ca^{2+} and REE cations into separate layers.

These minerals, especially bastnäsites, are mined for their cerium and lanthanum contents; these are used in glass polishing, television screens, light bulbs in carbon lighting applications, and flints for lighters ("mischmetal"). Cerium metal is also used as an additive on the walls of self-cleaning ovens!

Azurite ($Cu_3(CO_3)_2(OH)_2$) and **malachite** ($Cu_2(CO_3)(OH)_2$) are closely-related copper carbonates that are highly prized for their attractive blue and green colors (Table 23.14). They were used as pigments in antiquity. Early Egyptian civilizations used malachite as an eye paint, and it is common in cave and tomb paintings. Chinese and Japanese began using malachite pigments in the 9th century A.D. Azurite was used by Europeans in the middle Ages and renaissance. Around 1800, synthetic pigments replaced both of these minerals in pigments, and they are now rarely used for that purpose.

Azurite is composed of zigzagging, edge-sharing chains of highly distorted $Cu(O,OH)_6$ octahedra (actually, nearly square pyramids) that are linked by square planes of $Cu(O,OH)_4$. The framework is held together by $(CO_3)^{2-}$ (Figure 23.23). Malachite's structure is quite different, although its chemistry is similar. In malachite, all the Cu^{2+} occupy distorted octahedra that alternately share edges and corners to form chains. $(CO_3)^{2-}$ links the chains into sheets, and weak H-bonding holds the sheets together (Figure 23.23).

Both azurite and malachite are colored by Cu^{2+}, but the difference in their Cu^{2+} coordination polyhedra gives rise to different colors. In azurite, Cu^{2+} is 4-coordinated while Cu^{2+} in malachite is 6-coordinated. The resultant crystal field splitting (see Chapter 7) is quite different, such that azurite has absorption bands at 620 and 770 nm, while malachite has bands at about 495, 550, 607, and 840 nm (Burns, 1993).

Of the two minerals, azurite is not particularly stable in the presence of water or humid air, breaking down to malachite via the reaction:

$$2\,[Cu_3(CO_3)_2(OH)_2] + H_2O => 3\,[Cu_2(CO_3)(OH)_2] + CO_2.$$

Fortunately for artists, this reaction occurs relatively slowly. However, many of the old paintings that used azurite for beautiful blue pigment have now been altered to shades of green by this reaction!

Hydrozincite ($Zn_5(CO_3)_2(OH)_6$) and **aurichalcite** ($(Zn,Cu)_5(CO_3)_2(OH)_6$) have structures closely related to those of malachite and azurite, in which

Table 23.14. Copper Carbonates

Group	Mineral	Formula
	azurite	$Cu^{2+}{}_3(CO_3)_2(OH)_2$
Rosasite	rosasite	$(Cu^{2+},Zn)_2(CO_3)(OH)_2$
Rosasite	glaukosphaerite	$(Cu^{2+},Ni)_2(CO_3)(OH)_2$
Rosasite	kolwezite	$(Cu^{2+},Co)_2(CO_3)(OH)_2$
Rosasite	zincrosasite	$(Zn,Cu^{2+})_2(CO_3)(OH)_2$
Rosasite	mcguinnessite	$(Mg,Cu^{2+})_2(CO_3)(OH)_2$
Malachite	malachite	$(Cu^{2+}{}_2(CO_3)(OH)_2$
Malachite	nullaginite	$Ni_2(CO_3)(OH)_2$
Malachite	pokrovskite	$Mg_2(CO_3)(OH)_2 \cdot 0.5H_2O$

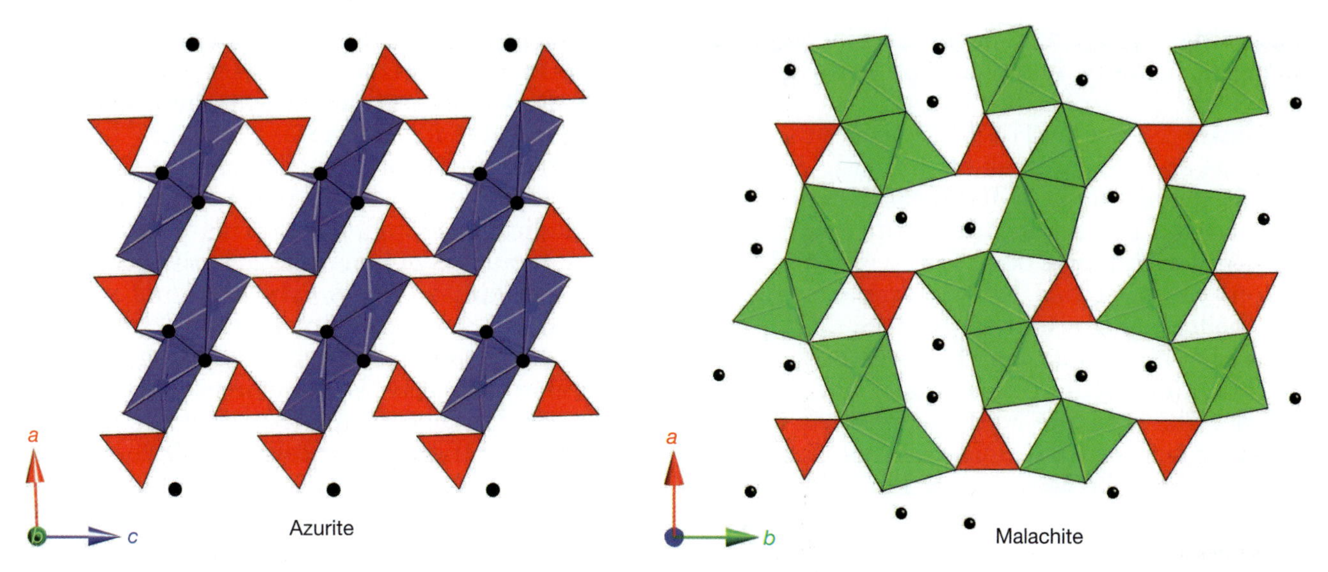

Figure 23.23. *The crystal structures of two hydroxylated copper-carbonate minerals: azurite (left, $Cu_3(CO_3)_2(OH)_2$) and malachite (right, $Cu_2(CO_3)(OH)_2$). For both minerals, the C^{4+} is in three-coordination with O^{2-} and represented by small red triangles. Cu^{2+} is in square planar coordination (blue) with O^{2-} in azurite and octahedral coordination (green) in malachite. Interestingly, this difference in coordination polyhedra causes the differences in color between these two minerals: azurite is blue and malachite is green. $(OH)^{1-}$ ions are represented in each structure by small black spheres.*

Zn^{2+} replaces Cu^{2+}. They contain octahedral layers of $Zn(OH)_4O_2$ linked into a framework by two $Zn(OH)_3O$ and two $(CO_3)^{2-}$ groups (Struntz and Nickel, 2001). Hydrozincite is colorless because Zn^{2+} has a full complement of 10 3*d* electrons, so crystal field transitions do not occur. The substitution of Cu^{2+} into aurichalcite gives it a green-blue appearance.

Borates

At this writing, there are more than 96 B-bearing mineral species recognized, and a *Reviews in Mineralogy and Geochemistry* volume (33) (Grew and and Anovitz, 2002) is dedicated to boron. Four of the borates have important uses discussed below (Table 23.15). Boron is interesting because its size is slightly larger than C^{4+} and smaller then Si^{4+}. Thus, B^{3+} occurs in borates in both 3- and 4-coordination with O^{2-}. Boron is an important industrial mineral. Boron is useful in glass and porcelain production because it is a small and highly-charged cation, and thus bonds to form strong networks with O^{2-} anions, even in disordered glass structures. This characteristic enables boron-enriched materials to better resist thermal expansion, and it increases their resistance to shock. For this reason, borosilicate glasses like Pyrex® are often used in laboratories and kitchens because they can tolerate the thermal shock involved in going directly from the freezer to the oven.

Boron is also used in refining gold and silver. Borates are used as gasoline additives and in automotive hydraulic brake fluids. ^{10}B is used in shielding at nuclear reactors because it can absorb neutrons without emitting secondary radiation. Many other types of materials incorporate boron: detergents, cosmetics, pesticides, medicines, etc. Interestingly, the world's largest reserves of boron-rich minerals are located in the U.S. (this is one of the few elements for which this is the case!) For more information, please check out the interesting chapter by Hawthorne et al. in Grew and Anovitz (2002).

Boracite ($Mg_3B_7O_{13}Cl$) has an unusual open framework of $(BO_4)^{5-}$ tetrahedra. In some cases, three $(BO_4)^{5-}$ tetrahedra share the same corner oxygen anion (Figure 23.24)—an arrangement that is not possible with $(SiO_4)^{2-}$ tetrahedra because of bond strength. Within this open framework are Mg^{2+} cations in 5-coordination with four O^{2-} and one Cl^{1-}. Each Cl^{1-} is in 6-coordination

Table 23.15. Anhydrous and Hydrous Borates

Boracite	boracite	$Mg_3B_7O_{13}Cl$
	ericaite	$(Fe^{2+},Mg,Mn)_3B_7O_{13}Cl$
	chambersite	$Mn_3B_7O_{13}Cl$
	congolite	$(Fe^{2+},Mg,Mn)_3B_7O_{13}Cl$
	trembathite	$(Mg,Fe^{2+})_3B_7O_{13}Cl$
	borax	$Na_2B_4O_5(OH)_4 \cdot 8H_2O$
	kernite	$Na_2B_4O_6(OH)_2 \cdot 3H_2O$
	ulexite	$NaCaB_5O_6(OH)_6 \cdot 5H_2O$

with Mg^{2+}. Boracite is found in evaporite deposits, usually in association with gypsum, anhydrite, and/or halite.

Borax ($Na_2B_4O_5(OH)_4{\cdot}8H_2O$) contains B^{3+} in both 3- and 4-coordination. Its basic structural unit is an anionic $B_4O_5(OH)^{-2}_4$ complex of two triangular and two tetrahedral groups in which one corner of each unit is occupied by $(OH)^{1-}$. The O^{2-} anions are all shared by two tetrahedra or by a tetrahedron and a triangle (Gaines et al., 1997). Between these units are chains of edge-sharing $Na(H_2O)_6$ octahedra that link by sharing a single $(OH)^{1-}$. The structure holds together via weak H bonding between the units, making borax very soft (2–2.5 on the Mohs scale). Borax forms from precipitation of water in saline lake environments (playas).

Kernite ($Na_2B_4O_6(OH)_2{\cdot}3H_2O$) is composed of chains of corner-sharing $(BO_4)^{5-}$ tetrahedra that link alternately with B(OH) units to make BO_2OH triangles; the complex thus created has the formula $B_4O_6(OH)^{2-}{}_2$. Na^{1+} octahedra with five O^{2-} and one H_2O connect those chains along [100]. It is thought to form in metamorphic environments from the breakdown of borax.

Ulexite ($NaCaB_5O_6(OH)_6{\cdot}5H_2O$) has Ca^{2+} in 8-coordination with three O^{2-}, three $(OH)^{1-}$, and two H_2O's; these polyhedra are linked in chains parallel to [001]. Na^{1+} cations occupy octahedra of two $(OH)^{1-}$ and four H_2O's; these create chains by sharing OH-OH and H_2O- H_2O edges. $(BO_3)^{3-}$ ions link the two types of chains. Given this chain-like structure, it is not surprising that ulexite is a fibrous mineral that can form small, fluffy puffballs as well as bundles of parallel fibers. In the latter habit, it is known as "TV stone" because it has the property of projecting an image through an aggregate of crystals if polished faces are cut perpendicular to the fibers. This optical effect arises from reflections along (010) twinning planes within the fibers, which are not actually tubular, but serrated from twinning. The interfaces between the twins act as reflecting surfaces. A similar property is sometimes observed in trona and halotrichite (Baur et al., 1957). Interestingly, fiber optic cables operate by this same principle. If materials with varying refractive indices are introduced, then reflection at their interfaces causes effective transmission of light.

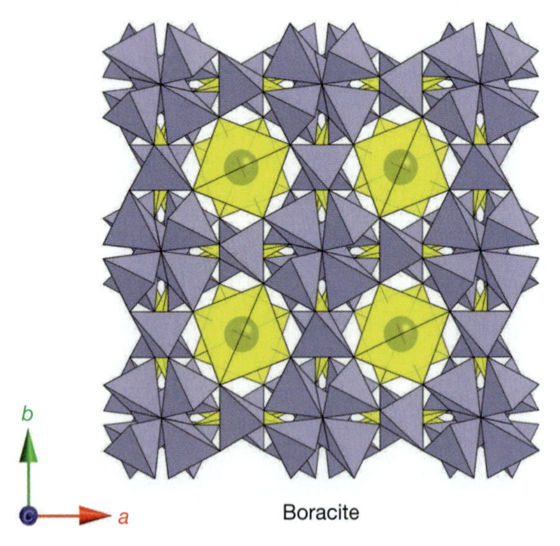

Figure 23.24. The crystal structure of boracite ($Mg_3B_7O_{13}Cl$), an example of the borate group minerals that are formed of $(BO_4)^{5-}$ tetrahedra (purple) and light yellow translucent octahedra containing Mg^{2+}. Notice that the Cl^{1-} anion (represented as a sphere), which is visible by looking through the translucent octahedral, resides in cavities in the structure.

Sulfates

As we've mentioned earlier in this chapter, S^{6+} is a small, highly-charged cation that prefers to take on tetrahedral coordination with oxygen. As a result, it forms SiO_4-like structures based upon $(SO_4)^{2-}$ tetrahedra, which are collectively called sulfates. These occur in anhydrous and hydrous species, just as in silicates, and make up at least 370 different species. These are thoroughly discussed in the chapter by Hawthorne et al. (and other chapters as well) in yet one more *RIMG* volume (volume 40, Alpers et al., 2000).

Within the sulfate tetrahedra, the bond distances between the S^{6+} and the O^{2-} range from 1.430–1.501 Å. The S-O bond lengths are correlated with:

$$\frac{1}{\cos(O-S-O)},$$

where O-S-O is the average value of the three O-S-O bond angles (Hawthorne et al., 2000). Only in rare instances does anything actually substitute for the S^{6+} in the tetrahedra (e.g., vergasovaite and hashemite).

Anhydrous sulfates. Barite group minerals all possess structures with 12-coordinated sites for large cations (Pb^{2+}, Ba^{2+}, Sr^{2+}, K^{1+}, Cs^{1+}) and tetrahedra for Ba^{2+} and other small cations (Cr^{3+}, Se^{4+}, etc.) (Table 23.16). In **barite** ($BaSO_4$), the Ba^{2+} is in 10-coordination in a polyhedron that shares three edges and four corners with $(SO_4)^{2-}$ tetrahedra (Struntz and Nickel, 2001). The structure forms sheets parallel to (001) that give barite its perfect cleavage (Figure 23.25). This structure grows in tabular crystals, and the radiating forms of the tablets closely resemble flowers, which are commonly called "barite roses."

Table 23.16. Anhydrous Acids and Sulfates		
Barite	avogradrite	$(K,Cs)BF_4$
	barite	$BaSO_4$
	celestine	$SrSO_4$
	anglesite	$PbSO_4$
	olsacherite	$Pb_2[SO_4](SeO_4)_2$
	hashemite	$Ba(Cr,S)O_4$
	anhydrite	$CaSO_4$
	glauberite	$Na_2Ca(SO_4)_2$

Barium is quite dense (4.50 g/cc), which makes it useful as a weighting agent in oil and gas well-drilling fluids (its number one use), as well as in cements, rubber, and foam. Barite is also added to glasses to increase shine and brilliance. Barite has the property of being able to block X-rays and γ-rays, which makes it useful not only in shielding applications, but in medical testing. If you ever have had a gastrointestinal "barium enema," you have seen barite in action, quite literally.

The **anhydrite** ($CaSO_4$) structure contrasts with that of barite in that the large cation, Ca^{2+}, occupies an 8-coordinated trigonal-dodecahedron. These dodecahedra alternate with $(SO_4)^{2-}$ tetrahedra by sharing edges to make chains in all three directions; the resultant structure (Figure 23.26) is nearly identical to that of zircon (Figure 22.45). Anhydrite is an important mineral in marine evaporates, where it may form by dehydration of gypsum. It may also precipitate directly from evaporating sea water. **Glauberite** ($Na_2Ca(SO_4)_2$) often coexists with anhydrite in evaporite deposits. It is highly soluble in water, and thus is frequently pseudomorphed by such minerals as calcite, quartz, or gypsum. The glauberite casts are usually recognized by their monoclinic, inclined wedge or tabular dipyramidal crystals. Glauberite's claim to fame is as a source of Glauber's salt, $Na_2SO_4{\cdot}10H_2O$, which is used in stomach-calming medications as well as paper and glass production.

Hydrated sulfates. By far the most common of the hydrous sulfate minerals is **gypsum** ($CaSO_4{\cdot}2H_2O$), which has basically the same structure as anhydrite, with the addition of H_2O molecules parallel to the (010) direction. There are five isostructural gypsum group minerals with Ca^{2+}, Nd^{3+}, Y^{3+}, and Er^{3+} as the large cations and $(SO_4)^{2-}$, $(HPO_4)^{2-}$, $(HAsO_4)^{4-}$, and $(PO_4)^{3-}$ as the structural unit. In gypsum ($CaSO_4{\cdot}2H_2O$), polyhedra of $CaO_6(H_2O)_2$ share edges with $(SO_4)^{2-}$ tetrahedra to make chains along [001]. The chains form double layers by further edge-sharing of polyhedra and tetrahedra (Strunz and Nickel, 2001). The double layers are linked by H bonding, which makes gypsum quite easy to cleave and also very soft (2 on the Mohs scale).

Gypsum is found in marine evaporite deposits, usually in the presence of halite, calcite, and other minerals. It too sometimes forms rosettes when combined with sand grains; these are often called "desert roses." The primary use of gypsum is as an insulator in plaster, drywall, and sheet-rock—you probably have gypsum in the walls of your house! Sometimes this building material is called plaster-of-Paris, because it was used widely in that city years ago. Gypsum serves its purpose well, for two reasons:

1. It has very low thermal conductivity, and therefore will help keep heat escaping from your house. Thermal conductivity, which is a reflection of bond strength, is usually directly correlated with density (e.g., diamond is an excellent conductor of heat). Because gypsum has a low density (~2.308 g/cc), it also has a low conductivity.
2. Gypsum has the ability to maintain a low temperature when exposed to heat because it can effectively lose heat by dehydration:

$$2(CaSO_4{\cdot}2H_2O) + \text{heat} \rightarrow 2(CaSO_4{\cdot}{}^1\!/\!_2H_2O) + 3H_2O\ (\text{steam}).$$

As gypsum loses its water, it will eventually heat up until it becomes completely dehydrated. Until it reaches that point, however, it acts like a cool insulator for walls, preventing wood studs from burning and potentially preventing building structures from collapsing.

The partially-dehydrated form of gypsum, $CaSO_4{\cdot}{}^1\!/\!_2H_2O$, is the mineral species bassanite.

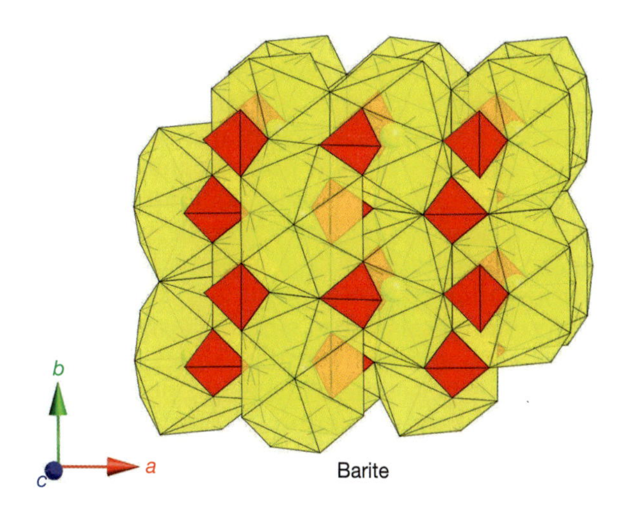

Figure 23.25. The structure of barite ($BaSO_4$). Ba^{2+} occupies the large yellow translucent polyhedra while $(SO_4)^{2-}$ is represented by red tetrahedra. This structure is quite similar to that of anhydrite (Figure 23.26).

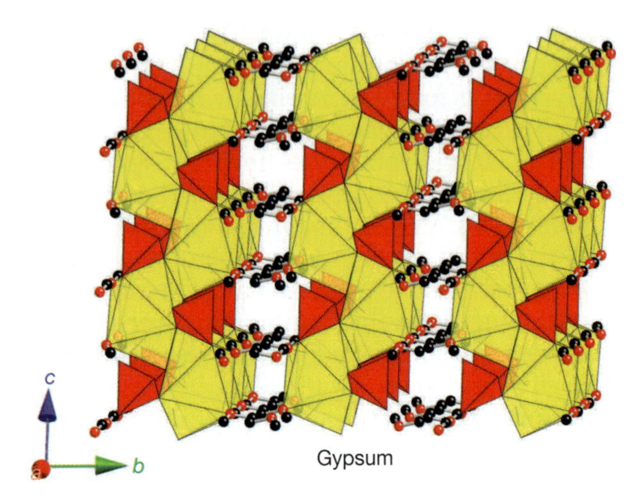

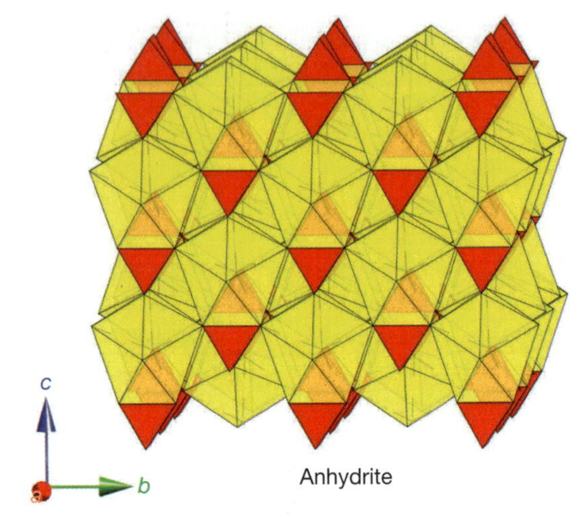

Figure 23.26. *The structures of gypsum (left, $CaSO_4 \cdot 2H_2O$) and anhydrite (right, $CaSO_4$); both are sulfates; the former hydrated and the later dehydrated. As sulfates, they contain $(SO_4)^{2-}$ tetrahedra (red). They also both contain Ca^{2+} in the yellow shaded, translucent polyhedra. The major difference between the two is that gypsum contains layers of H_2O molecules parallel to the (010) plane. H-bonding weakly bonds these layers in gypsum.*

When it is combined with H_2O (i.e., when the reaction above goes in the opposite direction), it quickly recrystallizes to the gypsum structure. For this reason, it is useful in making blackboard chalk (which is extruded and then cut into short lengths) and different types of molds. If you have ever broken an arm or leg, you probably had a cast made from plaster of paris/gypsum. Gypsum is also used as a source for calcium in vitamins, and is added to various food products (such as Wonder® Bread) for that purpose.

A majority of the other hydrous sulfates form as the result of weathering and oxidation of sulfide minerals such as pyrite, chalcopyrite, pyrrhotite, and marcasite. These interactions, which are accelerated when sulfide deposits are mined and thus exposed to air and water (not to mention catalyzing microbes), produce waters that are rich in acid sulfates. Such mine drainages pose threats to fish and other aquatic species when they drain into streams and lakes (see Bigham and Nordstrom, 2000 for the whole story!). If acid sulfate waters are allowed to precipitate, they will form a variety of different hydrous sulfates. Notice in Table 23.17 that many of these minerals have similar formulas with only slight-varying hydration states. Compare, for example, kieserite, starkeyite, pentahydrite, hexahydrite, and pickeringite! The assemblages of minerals that crystallize from acid mine drainages depend on complex equilibria among the Fe_2O_3, $(SO_4)^{2-}$, and H_2O contents of the deposit. Changing conditions (i.e., changes in any of these variables) will cause changes in mineralogy!

Anhydrous sulfates with OH, F, and/or Cl. Sulfate minerals that contain either hydroxyl $(OH)^{1-}$ or halogens $(F, Cl)^{1-}$ are considered to be "anhydrous" despite the fact that they may contain H^{1+}, because $(OH)^{1-}$ (the hydroxide anion) differs structurally from molecular H_2O. Of the minerals in this class, those in the alunite group are by far the most abundant. The 16 alunite minerals are all isostructural. The formula unit consists of one monovalent cation (K^{1+}, Na^{1+}, Tl^{1+}, H_3O^{1+}, NH_4^{1+} Ag^{1+}, Pb^{1+}, etc.), three trivalent cations (Al^{3+}, Fe^{3+}, or Cu^{3+}), and a tetrahedral cation, usually S^{2-}, as well as some $(OH)^{1-}$ (Table 23.18 and Figure 23.27). The trivalent cation is in octahedral coordination with four $(OH)^{1-}$ and two O^{2-} from $(SO_4)^{2-}$ ions. These octahedra share corners and make a pattern of 3- and 6-membered rings (Strunz and Nickel, 2001). The monovalent cation is bonded to six O's and six OH's, creating a trigonal structure that closely resembles turquoise.

Hydrated sulfates with OH, F, and/or Cl. Minerals that include both H_2O and either $(OH)^{1-}$ or a halogen element (F^{1-} or Cl^{1-}) fall in this class (Table 23.19); in this case it is the H_2O that truly makes these mineral hydrated. The most common of these is probably copiapite ($Fe^{2+}Fe^{3+}_4(SO_4)_6(OH)_2{\cdot}20H_2O$), which forms as a secondary mineral at coal mines. Species in this group have a simple structure composed of layers of $Fe^{2+}(H_2O)_6$ octahedra alternating with layers of octahedral Fe^{3+} and SO_4 (Struntz and Nickel, 2001). H bonding holds the layers together and makes these minerals quite soft (H = 2.5–3).

Table 23.17. Hydrated Sulfates

Group	Mineral	Formula
Gypsum	gypsum	$CaSO_4 \cdot 2H_2O$
	brushite	$CaHPO_4 \cdot 2H_2O$
	pharmacolite	$CaHAsO_4 \cdot 2H_2O$
	churchite-(Nd)	$NdPO_4 \cdot 2H_2O$
	churchite-(Y)	$YPO_4 \cdot 2H_2O$
Kieserite	kieserite	$MgSO_4 \cdot H_2O$
	szomolnokite	$FeSO_4 \cdot H_2O$
	szmikite	$MnSO_4 \cdot H_2O$
	poitevinite	$(Cu,Fe)SO_4 \cdot H_2O$
	gunningite	$ZnSO_4 \cdot H_2O$
	dwornikite	$NiSO_4 \cdot H_2O$
	cobaltkiesierite	$CoSO_4 \cdot H_2O$
Rozenite	rozenite	$FeSO_4 \cdot 4(H_2O)$
	starkeyite	$MgSO_4 \cdot 4(H_2O)$
	ilesite	$MnSO_4 \cdot 4(H_2O)$
	aplowite	$CoSO_4 \cdot 4(H_2O)$
	boyleite	$ZnSO_4 \cdot 4(H_2O)$
	IMA2002-034	$CdSO_4 \cdot 4(H_2O)$
Chalcanthite	chalcanthite	$Cu^{2+}SO_4 \cdot 5(H_2O)$
	siderotil	$FeSO_4 \cdot 5(H_2O)$
	pentahydrite	$MgSO_4 \cdot 5(H_2O)$
	jokokuite	$Mn^{2+}SO_4 \cdot 5(H_2O)$
Hexahydrite	hexahydrite	$MgSO_4 \cdot 6(H_2O)$
	bianchite	$ZnSO_4 \cdot 6(H_2O)$
	ferrohexahydrite	$FeSOv \cdot 6(H_2O)$
	nickelhexahydrite	$NiSO_4 \cdot 6(H_2O)$
	moorhouseite	$CoSO_4 \cdot 6(H_2O)$
	chvaleticeite	$MnSO_4 \cdot 6(H_2O)$
Melanterite	melanterite	$FeSO_4 \cdot 7(H_2O)$
	boothite	$CuSO_4 \cdot 7(H_2O)$
	zinc-melanterite	$(Zn,Cu,Fe)SO_4 \cdot 7(H_2O)$
	bieberite	$CoSO_4 \cdot 7(H_2O)$
	mallardite	$MnSO_4 \cdot 7(H_2O)$
	IMA2003-040	$(Mg,Cu)SO_4 \cdot 7(H_2O)$
Halotrichite	pickeringite	$MgAl_2(SO_4)_4 \cdot 22(H_2O)$
	halotrichite	$Fe^{2+}Al_2(SO_4)_4 \cdot 22(H_2O)$
	apjohnite	$Mn^{2+}Al_2(SO_4)_4 \cdot 22(H_2O)$
	dietrichite	$ZnAl_2(SO_4)_4 \cdot 22(H_2O)$
	bilinite	$Fe^{2+}Fe^{3+}{}_2(SO_4)_4 \cdot 22(H_2O)$
	redingtonite	$Fe^{2+}(Cr,Al)_2(SO_4)_4 \cdot 22(H_2O)$
	wupatkiite	$(Co,Mg,Ni)Al_2(SO_4)_4 \cdot 22(H_2O)$

Phosphates

Phosphate, $(PO_4)^{3-}$, is an important component of fertilizer and animal feed supplements, lending significance to the many minerals in this class. Phosphorus (P) is also used in food products as a preservative, stabilizer, and thickener, and phosphoric acid is used in many carbonated drinks.

The phosphate, arsenate, and vanadate minerals are grouped (Tables 23.20–23.22) by composition and designated on the basis of the cation that occupies the tetrahedral sites. Thus, normal phosphates contain P^{5+} in 4-coordination with O^{2-} $(PO_4)^{3-}$, hydrated normal phosphates have $(PO_4)^{3-}$ as well as H_2O, anhydrous phosphates contain $(PO_4)^{3-}$ and $(OH)^{1-}$, while hydrated phosphates contain $(PO_4)^{3-}$, $(OH)^{1-}$, and H_2O. Parallel nomenclature is used for the arsenates and vanadates.

Anhydrous normal physophates, arsenates, and vanadates. Both the monazite group and the xenotime group in this class are important sinks for rare earth elements (REE), for which they are used as ores (especially thorium, cerium, and lanthanum). Because monazite is relatively common, it is also extremely useful for geochronology in igneous and metamorphic rocks. Monazite can contain relatively large amounts of U and Th that decay to various isotopes of lead. If an assumption is made that all the Pb in the sample is the result of the decay from U and Th parents, and Pb, U, and Th can all be measured with high precision (as is possible with modern electron microprobes), then the following equation can be used to solve for age:

$$Pb = \frac{Th}{232}[\exp(\lambda^{232}\tau)-1]208 + \frac{U(0.9928)}{238.04}[\exp(\lambda^{238}\tau)-1]206 + \frac{U(0.0072)}{238.04}[\exp(\lambda^{235}\tau)-1]207$$

where λ^{232}, λ^{235}, and λ^{238} are the decay constants for ^{232}Th, ^{235}U, and ^{238}U, respectively, and τ is the time in years (e.g., Montel et al., 1996). For more information on this technique, see the special April 2005 issue of the *American Mineralogist.*

All these species have structures that can accommodate the large REE cations in different ways. The **monazite group** minerals have ATO_4 structures, where A is Ce^{3+}, La^{3+}, Nd^{3+}, Th^{4+}, Ca^{2+},

Table 23.18. Alunite group

Group	Mineral	Formula
Alunite	alunite	$KAl_3(SO_4)_2(OH)_6$
	nantroalunite	$NaAl_3(SO_4)_2(OH)_6$
	schlossmacherite	$(H_3O,Ca)Al_3[(SO_4,AsO_4)_2(OH)_6$
	osarizawaite	$Pb(Al,Cu)_3(SO_4)_2(OH)_6$
	minamiite	$(Na,Ca,\square)_2Al_6(SO_4)_4(OH)_{12}$
	ammonioalunite	$(NH_4)Al_3(SO_4)_2(OH)_6$
	walthierite	$Ba_{0.5}\square_{0.5}Al_3(SO_4)_2(OH)_6$
	huangite	$Ca_{0.5}\square_{0.5}Al_3(SO_4)_2(OH)_6$
Jarosite	jarosite	$KFe^{3+}{}_3(SO_4)_2(OH)_6$
	natrojarosite	$NaFe^{3+}{}_3(SO_4)_2(OH)_6$
	hydronium jarosite	$(H_3O)Fe^{3+}{}_3(SO_4)_2(OH)_6$
	ammoniojarosite	$(NH_4)Fe^{3+}{}_3(SO_4)_2(OH)_6$
	argentojarosite	$AgFe^{3+}{}_3(SO_4)_2(OH)_6$
	plumbojarosite	$PbFe^{3+}{}_3(SO_4)_2(OH)_6$
	beaverite	$Pb(Fe,Cu^{2+})_3(SO_4)_2(OH,H_2O)_6$
	dorallcharite	$(Tl,K)Fe^{3+}{}_3(SO_4)_2(OH)_6$

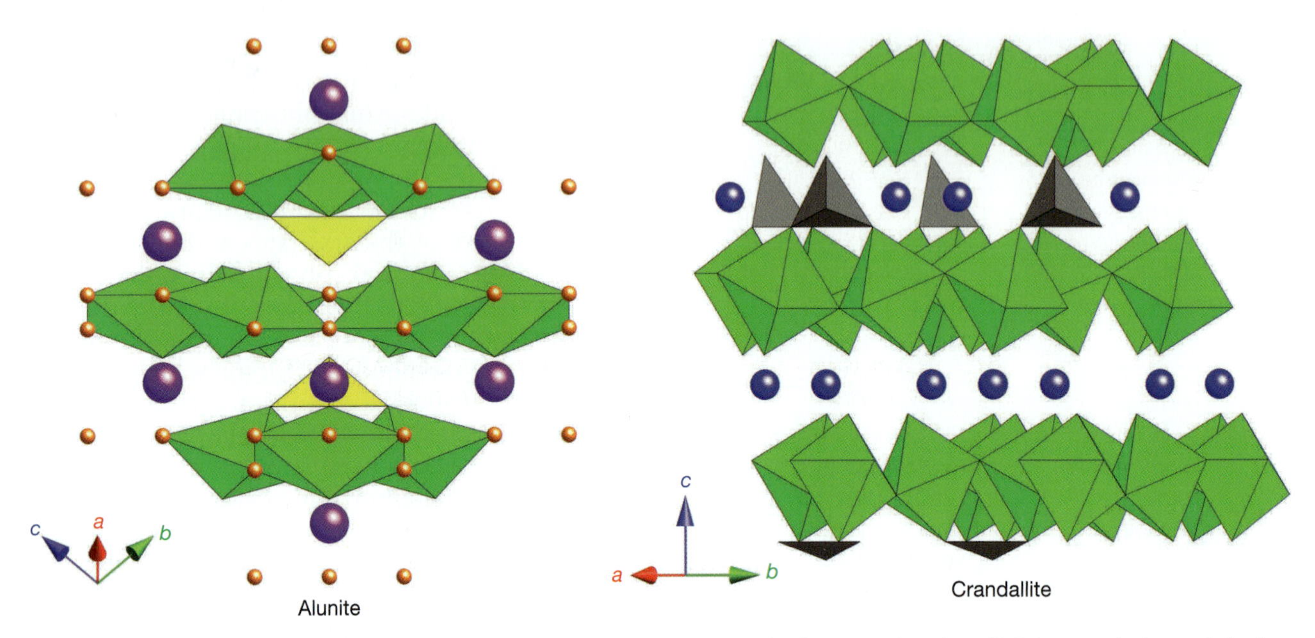

***Figure 23.27.** Alunite (left) and crandallite (right) are similar structures; the former is based on $(SO_4)^{2-}$ tetrahedra and the later on $(PO_4)^{3-}$ tetrahedra. Both also have Al^{3+} in octahedral coordination (green) (with the octahedra flattened somewhat in alunite), while alunite has K^{1+} (shown as a sphere) in a larger site and crandallite has Ca^{2+} (shown as large sphere). $(OH)^{1-}$ in alunite is shown as small spheres.*

Bi^{3+} and/or REEs and T is tetrahedral P or As. The structure is a three-dimensional framework of $(PO_4)^{3-}$ groups with the large cation in 9-coordination. In the **xenotime** group, the REE polyhedra are linked together by TO_4 tetrahedra. A chain of alternating $REEO_8$ and $(PO_4)^{3-}$ polyhedra runs parallel to the *c* axis, while a chain of $REEO_8$ polyhedra alone runs parallel to *a*. All polyhedra share edges.

Hydrated normal physophates, arsenates, and vanadates. The **roselite** group of phosphates (e.g., $Ca_2(Co,Mg)(AsO_4)_2{\cdot}2H_2O$) is notable for its striking red color, deeper than a rhodochrosite red. This is one of the very few naturally-occurring minerals in which the red color arises from the presence of the transition metal Co^{2+} in octahedral coordination. This may seem counter-intuitive to those who are familiar with the painter's pigment called "cobalt blue"—but the blue color only arises when Co is in *tetrahedral* coordination! Again this is similar to the different colors of the copper minerals azurite and malachite in that the color changes as a function of the coordination polyhedra of the color-causing transition metal. The Co^{2+} *octahedra* in roselite group minerals share corners with $(AsO_4)^{5-}$ tetrahedra to form chains connected by Ca^{2+} polyhedra into 6-membered rings.

The **autunite** group minerals are all good sources of uranium, which occupies large sites with the U^{6+} cation between two oxygens in a $(UO_2)^{2+}$ dumbbell. These are linked with tetrahedral groups that contain P, As, or V, thus forming layers stacked perpendicular to *c*. Thus, the U^{6+} cations occupy tetragonal dipyramids. Between the sheets is a layer of H_2O and cations (Ca^{2+}, Ce^{3+}, Ba^{2+}, K^{1+}, NH_4^{1+}, Sr^{2+}, Pb^{2+}, Mg^{2+}, Na^{1+}, Co^{2+}, Zn^{2+}, or H_3O^{+}). There are two subgroups: autunite and subautunite, which have slightly different crystal structures. Those in the autunite group are usually fully hydrated, while meta-autunites contain less water (Table 23.20).

The **autunite group** minerals are all a bright yellow-green color that is usually fluorescent (Figure 23.28). The color arises from very complex crystal field effects that are typical of minerals with partial occupancy of 5*f* orbitals—for example, 23 electronic bands have been recognized in $U^{4+}GeO_4$, which contains U^{4+} in a tetragonal structure (Gajek et al., 1993). In addition to the resultant yellow-green absorption colors, fluorescence occurs when energy (in this case, light or sunlight) causes an electron to be excited out of a ground state, low energy orbital into one with higher energy. When the electron decays back to its lowest energy state, it gives off the excess energy in the form of fluorescent light.

Carnotite group minerals are also yellow-green, though they contain a different type of U^{6+} coordination polyhedra: U_2O_5 pentagonal dipyramids! These share edges with pairs of VO_4 square pyramids to form sheets parallel to (001) that are linked into frameworks by the large cations (K^{1+}, Cs^{1+}, and H_3O^{1+}).

Table 23.19. Hydrated Sulfates with OH, F, and/or Cl

Botryogen	botryogen	$MgFe^{3+}(SO_4)_2(OH)\cdot 7H_2O$
	zincobotryogen	$(Zn,Mg,Mn^{2+})Fe^{3+}(SO_4)_2(OH)\cdot 7H_2O$
	xitieshanite	$Fe^{3+}(SO_4)Cl\cdot 6H_2O$
Ettringite	ettringite	$Ca_6Al_2(SO_4)_3(OH)_{12}\cdot 26(H_2O)$
	bentorite	$Ca_6(Cr,Al)_2(SO_4)_3(OH)_{12}\cdot 26(H_2O)$
	jouravskite	$Ca_6Mn^{4+}{}_2(SO_4,CO_3)_4(OH)_{12}\cdot 26(H_2O)$
	carraraite	$Ca_3Ge(OH)_6(SO_4)(CO_3)\cdot 12(H_2O)$
	charlesite	$Ca_6(Al,Si)_2(SO_4)_2B(OH)_4(OH,O)_{12}\cdot 26(H_2O)$
	sturmanite	$Ca_6(Fe^{3+},Al,Mn^{2+})_2(SO_4)_2B(OH)_4(OH)_{12}\cdot 25(H_2O)$
	buryatite	$Ca_3(Si,Fe^{3+},Al)(SO_4)[B(OH)_4](OH)_5O\cdot 12(H_2O)$
Copiapite	copiapite	$Fe^{2+}Fe^{3+}{}_4(SO_4)_6(OH)_2\cdot 20H_2O$
	magnesiocopiapite	$MgFe^{3+}{}_4(SO_4)_6(OH)_2\cdot 20H_2O$
	cuprocopiapite	$Cu^{2+}Fe^{3+}{}_4(SO_4)_6(OH)_2\cdot 20H_2O$
	ferricopiapite	$(Fe^{3+}{}_{2/3}\square_{1/3})Fe^{3+}{}_4(SO_4)_6(OH)_2\cdot 20H_2O$
	calciocopiapite	$CaFe^{3+}{}_4(SO_4)_6(OH)_2\cdot 19H_2O$
	zincocopiapite	$ZnFe^{3+}{}_4(SO_4)_6(OH)_2\cdot 20H_2O$
	aluminocopiapite	$(Al_{2/3}\square_{1/3})Fe^{3+}{}_4(SO_4)_6(OH)_2\cdot 20H_2O$

Vivianite group minerals are classic examples of another color phenomenon we discussed in Chapter 7: intervalence charge transfer (see Table 7.8 and related discussion). This occurs when valence electrons jump *between atoms*. In vivianite ($Fe^{2+}{}_3(PO_4)_2\cdot 8H_2O$), there are $(PO_4)^{3-}$ tetrahedra and two types of octahedra (Figure 23.29). One isolated Fe^{2+} octahedron has four corners with H_2O and two corners with O's that are part of $(PO_4)^{3-}$ groups. A second type of octahedron coordinates with two H_2O and four $(PO_4)^{3-}$ groups, forming a double octahedron through sharing a common O-O edge (Gaines et al., 1997). In the latter, the adjacent Fe^{2+} cations are separated by a distance of only 2.85 Å along the *b* axis (Burns, 1993). If the mineral is oxidized, some of the Fe^{2+} will lose an electron to form Fe^{3+}. When light is polarized along *b*, the adjacent Fe^{2+} and Fe^{3+} cations share their electrons, giving rise to an intense absorption band at 630 nm (Amthauer and Rossman, 1984). This, in addition to the typical Fe^{2+} crystal field bands at 800 and 1200 nm, gives rise to the intense blue color of vivianite!

The variscite group minerals have a three-dimensional framework composed of distorted octahedra occupied by trivalent cations (Al^{3+}, Fe^{3+}, Sc^{3+}, etc,) and $(PO_4)^{3-}$ tetrahedra. The octahedra have four corners bonded with $(PO_4)^{3-}$ groups and two corners bonded with H_2O's. There is complete solid solution between **variscite** ($AlPO_4\cdot 2H_2O$) and **strengite** ($Fe^{3+}PO_4\cdot 2H_2O$), and most natural samples contain both Al^{3+} and Fe^{3+}. Fe^{3+} in the distorted octahedra gives variscite a blue-green color that is sometimes confused with turquoise, though variscite is usually greener.

Anhydrous phosphates. The members of the **adelite-descloizite group** (Table 23.21) all have the general formula $ABXO_4(OH)$, where A is a large cation like Ca^{2+} or Pb^{2+} in 8-coordination, B is Co^{2+}, Cu^{2+}, Fe^{2+}, Mg^{2+}, Mn^{2+}, Zn^{2+}, and/or Ni^{2+} in a BO_6 site, and X is a tetrahedral cation such as As^{3+} or V^{3+}. The octahedra form chains by sharing of edges, with cross-linking by the XO_4 tetrahedra. This creates an open, mesh-like framework with plenty of room for the big A cations to occupy highly distorted polyhedra.

The **apatite group** members are arguably the most important of all phosphates, with the species **fluorapatite** ($Ca_5(PO_4)_3F$), **chlorapatite** ($Ca_5(PO_4)_3Cl$), and **hydroxylapatite** ($Ca_5(PO_4)_3(OH)$) being the most common. In these minerals, the largest polyhedra are 9-coordinated, and they

Table 23.20. Anhydrous Normal Phosphates and Arsenates

Monazite	monazite-(Ce)	$(Ce,La,Nd)PO_4$
	monazite-(La)	$(La,Ce,Nd)PO_4$
	cheralite-(Ce)	$(Ce,Ca,Th)(P,Si)O_4$
	brabantite	$Ca_{0.5}Th_{0.5}(PO_4)_2$
	monazite-(Nd)	$(Nd,Ce,La)(P,Si)O_4$
	gasparite-(Ce)	$CeAsO_4$
	monazite-(Sm)	$(Sm,Gd,Ce)PO_4$
Xenotime	xenotime-(Y)	YPO_4
	chernovite-(Y)	$YAsO_4$
	wakefieldite-(Y)	YVO_4
	wakefieldite-(Ce)	$(Ce^{3+},Pb^{2+},Pb^{4+})VO_4$
	pretulite	$ScPO_4$
	xenotime-(Yb)	$YbOP_4$

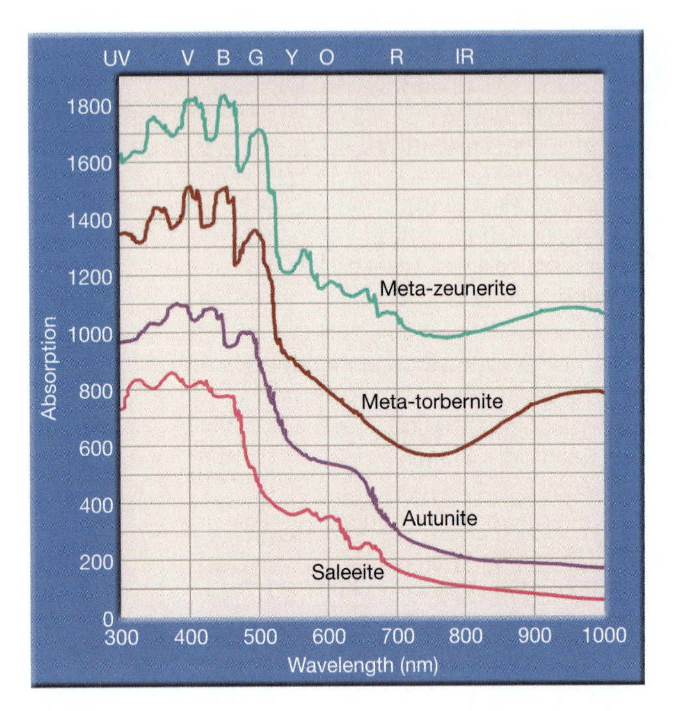

Figure 23.28. Optical absorption spectrum of autunite group minerals, adapted from Platonov (1976). Note the many crystal field bands in the ultraviolet-violet region of the spectrum, arising from complicated crystal field transitions of U^{6+}.

share faces to create chains parallel to the *c* axis. The chains further share edges and corners with $(PO_4)^{3-}$ tetrahedra to make a hexagonal array (Figure 23.30). Between these arrays are channels that can be occupied by $(OH)^{1-}$, Cl^{1-} or F^{1-}. As is evident from Table 23.21, there are many possible substitutions into this structure, and samples with true end member composition are quite rare. Half the elements in the periodic table have been shown to occur in the apatites (Gaines et al., 1997)! Apatite is sometimes used as a gemstone (Figure 7.12).

Apatite group minerals are the most important crystalline phases in the teeth and bones of vertebrate animals (including humans, of course). However the inorganic apatite mineral is intergrown with collagen to make bone. You might recall that the first figure in this book showed a schematic representation of this intergrowth. In many ways we could view our bones as "organic rocks" because they, like rocks, contain crystalline compounds (in this case, apatite and collagen) that are cemented together.

Apatite group minerals are used in the remediation of soils and sediments that have been contaminated by heavy metals leaching out of mining operations, agriculture, industrial manufacturing, and landfills. The goal in the remediation of soils that may contain elements such as Pb, Cd, Cu, U, Pu, and Zn is stabilization or reduction of the solubility of the metals in the soil matrix. Thus, the soils are mixed with apatite group minerals to chemically bind heavy metals into low solubility phosphate minerals. Mixing of heavy metal-contaminated soils with apatite group minerals can be achieved by excavation and *ex situ* mixing in mechanical pugmills, as well as *in situ* mixing techniques such as large diameter auger mixing. The process is known as **Phosphate-Induced Metals Stabilization (PIMS)**. Its development resulted from paleochemical oceanographic studies, in the 1970s and 1980s, of phosphatic sedimentary materials from the Cambrian period (570 my ago) to the present. They showed that apatite parts of marine animals, and even abiotic phosphorite deposits, developed identical trace metal signatures to those of the seawater

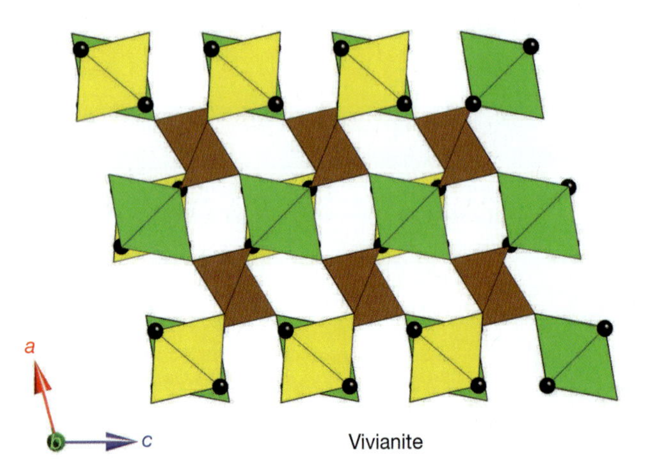

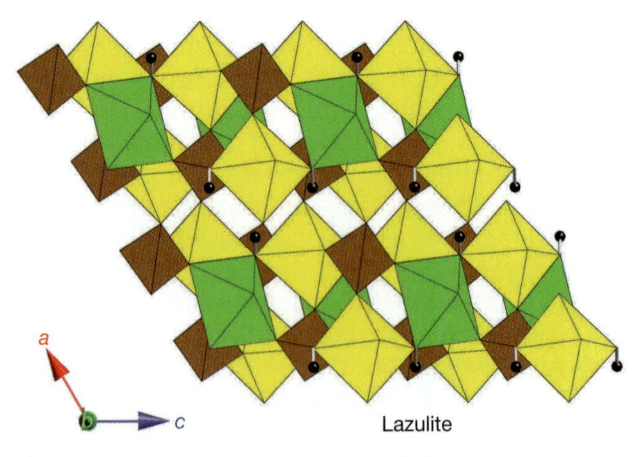

Figure 23.29. Vivanite and lazulite are two intense blue minerals colored by intervalence charge transfer (IVCT). In vivianite, there are two edge-sharing octahedra (yellow and green shaded in image above) and the charge transfer occurs between them. In lazulite, the IVCT occurs between Fe^{2+} and Fe^{3+} cations that occupy adjacent, face-sharing octahedra (yellow and green shaded in the image above). In both structures the $(PO_4)^{3-}$ tetrahedra are brown.

Table 23.21. Hydrated Normal Phosphates and Arsenates

Group	Subgroup	Mineral	Formula
Roselite		roselite	$Ca_2(Co^{2+},Mg)(AsO_4)_2·2H_2O$
		brandtite	$Ca_2(Mn^{2+},Mg)(AsO_4)_2·2H_2O$
		zincroselite	$Ca_2Zn(AsO_4)_2·2H_2O$
		wendwilsonite	$Ca_2(Mg,Co)(AsO_4)_2·2H_2O$
		manganoloharmeyerite	$Ca_2(Mn,Mg)(AsO_4)_2·2H_2O$
Autunite	Autunite	autunite	$Ca(UO_2)_2(PO_4)_2·10–12H_2O$
		uranospinite	$Ca(UO_2)_2(AsO_4)_2·10H_2O$
		uranocircite	$Ba(UO_2)_2(PO_4)_2·12H_2O$
		heinrichite	$Ba(UO_2)_2(AsO_4)_2·10–12H_2O$
		sodium autunite	$Na_2(UO_2)_2(PO_4)_2·8H_2O$
		novacekite	$Mg(UO_2)_2(AsO_4)_2·12H_2O$
		saleeite	$Mg(UO_2)_2(PO_4)_2·10H_2O$
		seelite	$Mg(UO_2)_2(As^{3+}O_3)_{1.4}(As^{5+}O_4)_{0.6}·7H_2O$
		torbernite	$Cu^{2+}(UO_2)_2(PO_4)_2 8–12H_2O$
		zeunerite	$Cu^{2+}(UO_2)_2(AsO_4)_2·10–16H_2O$
		kahlerite	$Fe^{2+}(UO_2)_2(AsO_4)_2·10–12H_2O$
		uranospathite	$HAl(UO_2)_2(PO_4)_4·40H_2O$
		arsenuranuospathite	$HAl(UO_2)_4(AsO_4)_4·40H_2O$
		sabugalite	$H_{0.5}Al_{0.5}(UO_2)_2(PO_4)_2·8H_2O$
		hallimondite	$Pb_2(UO_2)(AsO_4)_2$
	Meta-autunite	meta-autunite	$Ca(UO_2)_2(PO_4)_2·2–6H_2O$
		meta-uranospite	$Ca(UO_2)_2(AsO_4)_2·6H_2O$
		meta-uranocircite	$Ba(UO_2)_2(PO_4)_2·8H_2O$
		meta-heinrichite	$Ba(UO_2)_2(AsO_4)_2·8H_2O$
		sodium uranospite	$(Na_2Ca)(UO_2)_2(AsO_4)_2·5H_2O$
		uramphite	$(NH_4)_2(UO_2)_2(PO_4)_2·6H_2O$
		meta-ankoleite	$K_2(UO_2)_2(PO_4)_2·6H_2O$
		abernathyite	$K_2(UO_2)_2(AsO_4)_2·8H_2O$
		meta-novacekite	$Mg(UO_2)_2(AsO_4)_2·4–8H_2O$
		meta-torbernite	$Cu^{2+}(UO_2)_2(PO_4)_2·8H_2O$
		meta-zeunerite	$Cu^{2+}(UO_2)_2(AsO_4)_2·8H_2O$
		meta-kahlerite	$Fe^{2+}(UO_2)_2(AsO_4)_2·8H_2O$
		basseite	$Fe^{2+}(UO_2)_2(PO_4)_2·8H_2O$
		lehnerite	$Mn^{2+}(UO_2)_2(PO_4)_2·8H_2O$
		meta-kirscheimerite	$Co(UO_2)_2(AsO_4)_2·8H_2O$
		meta-lodevite	$Zn(UO_2)_2(AsO_4)_2·10H_2O$
		chernikovite	$H_3O_2(UO_2)_2(PO_4)_2·6H_2O$
		trögerite	$(UO_2)_3(AsO_4)_2·12H_2O$
Carnotite		carnotite	$K_2(UO_2)_2(VO_4)_2·3H_2O$
		margaritasite	$(Cs,K,H_3O^+)_2(UO_2)_2V_2O_8·H_2O$
Vivianite		vivianite	$Fe^{2+}{}_3(PO_4)_2·8H_2O$
		baricite	$(Mg,Fe^{2+})_3(PO_4)_2·8H_2O$
		erythrite	$Co_3(AsO_4)_2·8H_2O$
		annabergite	$Ni_3(AsO_4)_2·8H_2O$
		kottigite	$Zn_3(AsO_4)_2·8H_2O$
		parasymplesite	$Fe^{2+}{}_3(AsO_4)_2·8H_2O$
		hornesite	$Mg_3(AsO_4)_2·8H_2O$
		arupite	$Ni_3(PO_4)_2·8H_2O$
		IMA2004-02	$Co_3(PO_4)_2·8H_2O$
Variscite		variscite	$AlPO_4·2H_2O$
		strengite	$Fe^{3+}PO_4·2H_2O$
		scorodite	$Fe^{3+}AsO_4·2H_2O$
		mansfieldite	$AlAsO_4·2H_2O$
		yanomamite	$InAsO_4·2H_2O$

Table 23.21. (continued) Anhydrous Phosphates, Arsenates, and Vanadates

Group	Subgroup	Mineral	Formula
Adelite-descloizite	Adelite subgroup	adelite	$CaMg(AsO_4)(OH)$
		conichalcite	$CaCu^{2+}(AsO_4)(OH)$
		austinite	$CaZn(AsO_4)(OH)$
		duftite-β	$PbCu(AsO_4)(OH)$
		gabrielsonite	$PbFe^{2+}(AsO_4)(OH)$
		tangeite	$CaCu(VO_4)(OH)$
		nickelaustinite	$CaNi(AsO_4)(OH)$
		cobaltaustinite	$CaCo(AsO_4)(OH)$
		arsendescloizite	$PbZn(AsO_4)(OH)$
		gottlobite	$CaMg(VO_4,AsO_4)(OH)$
	Descloizite subgroup	descloizite	$PbZn(VO_4)(OH)$
		mottramite	$PbCu(VO_4)(OH)$
		pyrobelonite	$PbMn^{2+}(VO_4)(OH)$
		cechite	$Pb(Fe^{2+},Mn)(VO_4)(OH)$
		duftite-α	$PbCu(AsO_4)(OH)$
Apatite	Ca-phosphate subgroup	fluorapatite	$Ca_5(PO_4)_3F$
		chlorapatite	$Ca_5(PO_4)_3Cl$
		hydroxylapatite	$Ca_5(PO_4)_3(OH)$
		carbonate-fluorapatite	$Ca_5(PO_4,CO_3)_3(F,OH)$
		carbonate-hydroxylapatite	$Ca_5(PO_4,CO_3)_3(OH)$
		fluorcaphite	$Ca(Sr,Na,Ca)(Ca,Sr,Ce)_3(PO_4)_3F$
		deloneite-(Ce)	$NaCa_2SrCe(PO_4)_3F$
	Ca-arsenate subgroup	svabite	$Ca_5(AsO_4)_3F$
		turneaureite	$Ca_5[(As,P)O_4]_3Cl$
		johnbaumite	$Ca_5(AsO_4)_3(OH)$
		fermorite	$(Ca,Sr)_5(AsO_4,PO_4)_3(OH)$
		hedyphane	$Ca_2Pb_3(AsO_4)_3Cl$
	Sr-containing subgroup	belovite-(Ce)	$Sr_3Na(Ce,La)(PO_4)_3(F,OH)$
		belovite-(La)	$Sr_3Na(La,Ce)(PO_4)_3(F,OH)$
		strontium-apatite	$(Sr,Ca)_5(PO_4)_3(F,OH)$
	Pb-containing subgroup	pyromorphite	$Pb_5(PO_4)_3Cl$
		mimetite	$Pb_5(AsO_4)_3Cl$
		vanadinite	$Pb_5(VO_4)_3Cl$
		clinomimetite	$Pb_5(AsO_4)_3Cl$
		IMA2004-006	$(Ca,Na)_5[(P,S)O_4]_3(OH,Cl)$
	Ba-containing subgroup	morelandite	$(Ba,Ca,Pb)_5(AsO_4,PO_4)_3Cl$
		alforsite	$Ba_5(PO_4)_3Cl$
		kuannersutite-(Ce)	$Ba_6Na_2REE_2(PO_4)_6(F,Cl)$
Lazulite		lazulite	$MgAl_2(PO_4)_2(OH)_2$
		scorzalite	$(Fe^{2+},Mg)Al_2(PO_4)_2(OH)_2$
		hentschelite	$Cu^{2+}Fe^{3+}{}_2(PO_4)_2(OH)_2$
		barbosalite	$Fe^{2+}Fe^{3+}{}_2(PO_4)_2(OH)_2$

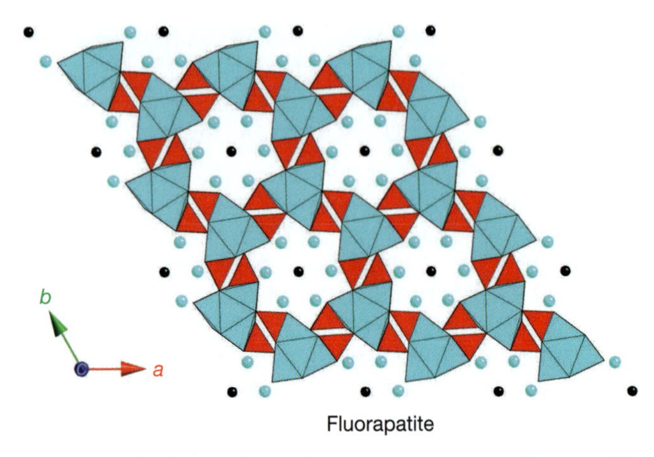

Figure. 23.30. Crystal structure of fluorapatite ($Ca_5(PO_4)_3F$), viewed down the c-axis onto the (001) plane, showing the hexagonal nature of the channels parallel to c formed by rings of $(PO_4)^{3-}$ tetrahedra (red) and Ca^{2+} octahedra (light blue). F^{1-} anions (small black spheres) occur in the center of the channels and are bonded (bonds not shown) to 9-coordinated Ca^{2+} (light blue spheres) at the edges of the channels.

where they formed, often enriching the concentrations of metals by several orders of magnitude (Wright et al., 1987).

Apatites are a major source of phosphate used in the manufacture of fertilizers. They begin with the apatite ore and add H_2SO_4, which dissolves the apatite and forms phosphoric acid (H_3PO_4), which is used as a liquid fertilizer. The remaining Ca^{2+} (from the apatite) and the $(SO_4)^{2-}$ from the sulfuric acid combine, and with a little bit of water, to make gypsum, which is a byproduct of the process.

The apatite group minerals have channels in their structures (Figure 23.30) that provide sites for a wide variety of different sized cations, also occurring in different valance states. A very small amount of apatite (~1 % by weight) is capable of removing metals from a much larger mass of soil. Once metals have been incorporated into the apatite structure, they tend to stay there: apatite is stable over a range of temperatures (up to 1000 °C), pHs (2–12), and in the presence of water and other liquids. Apatite minerals may also act as buffers to neutralize acidic soils.

Lazulite ($MgAl_2(PO_4)_2(OH)_2$) is another intense blue mineral whose color arises from intervalence charge transfer (IVCT) between Fe^{2+} and Fe^{3+}, which substitute for Mg^{2+}. The IVCT occurs between Fe cations that occupy adjacent, face-sharing octahedra. Lazulite is only blue when light is polarized along the β and γ directions (~*a* and *b* axes) and is colorless when parallel to α(~*c*-axis). Again, the crystal structure makes the charge transfer possible. Three identical face-sharing clusters of $(Cu,Fe^{2+},Mg)O_6$ lie between two corner-sharing $(Al,Fe^{3+})O_6$ octahedra (Figure 23.29) that lie within the *a-c* plane. The $(PO_4)^{3-}$ tetrahedra help form a three-dimensional framework.

Hydrated phosphates. The **crandallite group** (Table 23.22) consists of minerals that are closely related to alunite: compare alunite, $KAl_3(SO_4)_2(OH)_6$, with **crandallite**, $CaAl_3(PO_4)_2(OH)_5{\cdot}H_2O$ (Figure 23.27). Both structures are composed of corner-sharing octahedra that make 3- and 6-membered rings and form sheets parallel to the *c* axis. The sheets are linked by large cations such as Ca^{2+}, Ba^{2+}, Sr^{2+}, and Pb^{2+}. In turn, these structures are related to that of **turquoise** ($CuAl_6(PO_4)_4(OH)_8{\cdot}4H_2O$), which also contains two different types of octahedra: those containing larger cations like Cu^{2+}, Fe^{2+}, and Zn^{2+}, and those containing smaller trivalent cations like Al^{3+}, Fe^{3+}, and Cr^{3+}. Alternate pairs of edge-sharing Al^{3+} or $Fe^{3+}O_6$ octahedra share corners with $(PO_4)^{3-}$ tetrahedra. The divalent cations in distorted $(OH)_4(H_2O)_2$ octahedra join two or four chains together. Cu^{2+} in those distorted octahedra gives turquoise its blue color.

Anhydrous Tungstates

Wolframite is the name given to the solid solution series between the minerals **hübnerite** ($MnWO_4$) and **ferberite** ($FeWO_4$); a species with Zn^{2+} (sanmartinite, $(Zn,Fe)WO_4$) also exists. These minerals comprise the major ore of the element tungsten, which is used to make tungsten carbide, a wear-resistant material used in the construction and metals industries. Tungsten has an extremely high melting temperature, so it can be drawn into thin metal wires that are useful in such applications as light bulb and vacuum tube filaments. The high

Table 23.22. Hydrated Phosphates

Group	Mineral	Formula
Crandallite	crandallite	$CaAl_3(PO_4)_2(OH,H_2O)_6$
	gorceixite	$BaAl_3(PO_4)_2(OH,H_2O)_6$
	goyazite	$SrAl_3(PO_4)_2(OH,H_2O)_6$
	plumbogummite	$PbAl_3(PO_4)_2(OH,H_2O)_6$
	kintoreite	$PbFe^{3+}{}_3(PO_4)_2(OH)_5{\cdot}H_2O$
	benauite	$HSrFe^{3+}{}_3(PO_4)_2(OH)_6$
	springcreekite	$BaV^{3+}{}_3(PO_4)_2(OH,H_2O)_6$
Turquoise	turquoise	$Cu^{2+}Al_6(PO_4)_4(OH)_8{\cdot}4H_2O$
	coeruleolactite	$(Ca,Cu^{2+})Al_6(PO_4)_4(OH)_8{\cdot}4\text{-}5H_2O$
	faustite	$(Zn,Cu^{2+})Al_6(PO_4)_4(OH)_8{\cdot}4H_2O$
	chalcosiderite	$Cu^{2+}Fe^{3+}{}_6(PO_4)_4(OH)_8{\cdot}4H_2O$
	aheylite	$Fe^{2+}Al_6(PO_4)_4(OH)_8{\cdot}4H_2O$
	planerite	$\square Al_6(PO_4)_2(PO_3OH)_2(OH)_8{\cdot}4H_2O$

temperature stability also makes tungsten useful as a target for X-ray production, as heating elements in furnaces, and for parts of spacecraft and missiles that need to withstand high temperatures. The other important source of tungsten, especially in the U.S., is **scheelite**, $CaWO_4$. Despite their similar formulas, these minerals have dissimilar crystal structures. In the wolframite minerals, there are two types of edge-sharing octahedral chains: those filled by Mn^{2+} or Fe^{2+}, and those filled by W^{6+}. In scheelite, Ca^{2+} cations alternate with $(WO_4)^{2-}$ tetrahedra along the [110] direction, forming a tetrahedral framework.

Salts of Organic Acids

Whewellite, $CaC_2O_4{\cdot}H_2O$, is one of the very few "organic" minerals. It is recognized as a mineral species because it is found as a primary deposit in hydrothermal ore veins. It is also found in coal seams and sedimentary deposits associated with organic debris. The structure is composed of edge-sharing Ca^{2+} cations in 8-coordination that form layers connected by oxalic groups $(C_2O_4)^{2-}$ and H_2O molecules.

Concluding Thoughts

Although the non-silicates do not comprise a significant percentage of rock-forming minerals, their importance lies in their usefulness as ores of important elements. It is also interesting to compare some of the non-silicates to the silicates. For example, gypsum and apatite both contain tetrahedra, the former of $(SO_4)^{2-}$ and the later of $(PO_4)^{3-}$; this in many ways makes them similar to the silicates. So a detailed understanding of the silicates helps to understand the non-silicates. Throughout the book we have tended to stress the silicates over the non-silicates, mainly because the silicates comprise the vast majority of the Earth's crust, as stated above, and are the minerals you will typically encounter. However the non-silicates play many important roles along with being ore minerals and if you continue on in mineralogy you will have many encounters with them.

Acknowledgments. We thank George Rossman for his patient clarifications of issues relating to interactions of light with mineral structures, Donald Lindsley for his illuminating petrologic discussions, Eric Essene for his thoughts on carbonates, and Chris Voci for his insights into uses of apatite, all of which greatly improved this chapter. The discussion of sulfides here owes much to papers by Wuensch (1972, 1982) and my lecture notes from his courses at M.I.T. Descriptions of crystal structures are largely adapted from Struntz and Nickel (2001) and Gaines et al. (1997).

References

Alpers, C.N., Jambor, J.L., and Nordstrom, D.K. (2000) Sulfate Minerals—Crystallography, Geochemistry, and Environmental Significance. Reviews in Mineralogy and Geochemistry, 40, Mineralogical Society of America, 608 pp.

Amthauer, G. and Rossman, G.R. (1984) Mixed valence of iron in minerals with cation clusters. Physics and Chemistry of Minerals, 11, 37–51.

Barth, T.F.W., and Posnjak, E. (1932) Spinel structures with and without variate atom equipoints. Zeitschrift fur Kristallographie, 82, 325–341.

Baur, G.S., Larsen, W.N., and Sand, L.B. (1957) Image projection by fibrous minerals. American Mineralogist, 42, 697–699.

Bigham, J.M. and Nordstrom, D.K. (2000) Iron and aluminum hydroxysulfates from acid sulfate waters. Reviews in Mineralogy and Geochemistry, 40, 351–404.

Bischoff, J.L. and Fyfe, W.S. (1968) Temperature controls on aragonite-calcite transformation in aqueous solution. *Am. Mineral.* 54, 149–155.

Buckley, A.N., Hamilton, I.C. and Woods, R. (1984) Investigation of the surface oxidation of bornite by linear potential sweep voltammetry and x-ray photoelectron-spectroscopy. Journal of Applied Electrochemistry, 14(1), 63–74.

Burk, M. (1989) Gold, silver, and uranium from the seas and oceans. ARDOR Publishing, Los Angeles, CA, 252 pp.

Burns, R.G. (1993) Mineralogical applications of crystal field theory. Cambridge University Press, 551 pp.

Carlson, W.D. and Rosenfeld, J. (1981) Optical determination of topotactic aragonite-calcite growth kinetics: metamorphic implications. *J. Geol.* 89, 615–638.

Chang, L.L.Y. (2002) Industrial Mineralogy: Materials, Processes, and Uses. Prentice-Hall, Inc., 472 pp.

Chang, L.L.Y., Howie, R.A., and Zussman, J. (1996) Sulphates, carbonates, phosphates, and halides. Rock-Forming Minerals, vol. 5B: Non-silicates. Longman Group, London, 392 pp.

Craig, J.R., and Vaughan, D.J. (1994) Ore Microscopy and Ore Petrography. Wiley, New York, 406 pp.

Gaines, R.V., Skinner, H.C.W., Foord, E.E., Mason, B., Rosenzweig, A., King, V.T., and Dowty, R. (1997) Dana's New Mineralogy. John Wiley and Sons, New York, 1819 pp.

Gajek, Z., Krupa, J.C., Zolnierek, Z., Antic-Fidancev, E., and Lemaitre-Balise, M. (1993) Interpretation of the optical absorption spectrum of uranium germanate. Journal of Physics: Condensed Matter, 5, 9223–9234.

Grew, E.S., and Anovitz, L.M., eds. (2002) Boron Mineralogy and Geochemistry. Reviews in Mineralogy, 33, 2nd ed., Mineralogical Society of America, 864 pp.

Hawthorne, F.C., Krivovichev, S.V., and Burns, P.C. (2000) The crystal chemistry of sulfate minerals. Reviews in Mineralogy, 40, 1–112.

Hawthorne, F.C., Burns, P.C., and Grice, J.D. (2002) The crystal chemistry of boron. Reviews in Mineralogy, 33, 2nd ed., Mineralogical Society of America, 263–298.

Montel, J.-M., Foret, S., Veschambre, M., Nicolet, C. and Provost, A. (1996) Electron microprobe dating of monazite. Chemical Geology, 131, 37–53.

Platonov A.N. (1976) Origin of Color in Minerals. Moscow, Nedra (in Russian).

Reeder, R.J., ed. (1983) Carbonates: Mineralogy and Chemistry. Reviews in Mineralogy, 11, Mineralogical Society of America, 394 pp.

Ribbe, P.H., ed. (1982) Sulfide Minerals. Reviews in Mineralogy, 1, 2nd ed., Mineralogical Society of America, 384 pp.

Robinson, P., Harrison, R.J., McEnroe, S.A. and Hargraves, R. (2002) Lamellar magnetism in the hematite-ilmenite series as an explanation for strong remnant magnetization. Nature, 418, 517–520, 2002.

Rumble, D. III., ed. (1976) Oxide Minerals. Reviews in Mineralogy, 3, Mineralogical Society of America, 502 pp.

Schulman, J. H. and W. D. Compton (1962) *Color Centers in Solids.* Pergamon Press, New York, 368 pp.

Schwertmann, U. (1988) Goethite and hematite formation in the presence of clay minerals and gibbsite at 25 degrees C. Soil Science Society of America Journal, 52, 288–291.

Smyth, J.R., Jacobsen, S.D., and Hazen, R.M. (2000) Comparative crystal chemistry of dense oxide minerals. Reviews in Mineralogy, 41, 157–185.

Stanton, R.L. (1972) Ore Petrology, McGraw-Hill, 713 pp.

Strunz, H. and Nickel, E. (2001) Strunz Mineralogical Tables, 9th Ed. Schweizerbart, Stuttgart, 870 pp.

USGS Mineral Information web site, http://minerals.usgs.gov/minerals/pubs/commodity/sulfur/

Weiner, K.-L. (2003) What is gold ? *extraLapis English*, 5, 4–10.

Wolf, G., Konigsberger, E., Schmidt, H.G., Konigsberger, L.C. and Gamsjager. H. (2000) Thermodynamic aspects of the vaterite-calcite phase transition. *J. Thermal An. Calorim.* 60, 463–472.

Wright, J., Schrader, H., and Holser, W.T. (1987) Paleoredox variations in ancient oceans recorded by rare earth elements in fossil apatite. Geochimica et Cosmochica Acta, 51, 631–644.

Wuensch, B.J. (1982) Sulfide crystal chemistry. in Sulfide Mienralogy, P.H. Ribbe, Ed., Reviews in Mineralogy, 1, W-21–W-44.

Wuensch, B.J. (1972) The crystal chemistry of sulfur. Chapter 16A. In Handbook of Geochemistry, vol. II/3. Springer-Verlag, Berlin.

Chapter 24

Mineralogy Outside of Geology

I now spend a considerable percentage of my time in rooms full of people discussing minerals. When I gaze around those rooms, I realize that not only am I the only mineralogist, but I may be the only person in the room who's had more than one course in geology (or who could correctly explain the difference between quartz and silica). The other people in that room with me, although well-educated, are often medical researchers: concrete, insulation, paint, ceramics and industry reps; or staff from the regulatory sector: EPA, NIOSH, and OSHA. Many of these folks feel comfortable discussing minerals, but they often prickle when I use such words as "mesothelioma" or "standard mortality ratio" in a sentence. They all feel comfortable discussing minerals with only a limited background in the discipline—something they would never do when discussing their own areas! This is distressing because the field of mineralogy can be every bit as complex as these other disciplines. Moreover, it has enormous potential for contributing to our understanding of many health-related problems. As a mature discipline, we have pretty good understanding of minerals; however, our understanding often becomes simplified by others without the deep knowledge in our field. For example, there was panic in the Northwest when Mt. St. Helens erupted in 1980 and the media reported the ash contained over 50% silica—which would cause an outbreak of silicosis. Yes, the ash did contain over 50% silica, but not in the form of quartz, but as an oxide component of the ash.

I have argued for years that mineralogy is the most important sub-discipline in geology. Why else is it often the first "real" geology course a student takes? In days of old, mineralogy was a discrete discipline unto itself, and it still should be today. However, especially in the U.S., mineralogy is taught in geology departments, most frequently as a course that should more appropriately be titled "Introduction to Petrology." The main reason the courses are taught this way is that mineralogy is fundamental to an understanding of the processes that form rocks, but this approach diminishes the importance of mineralogy as a discipline, and fails to represent the countless other fields in our modern society that use minerals. The most obvious of some of these, related somewhat closely to geology, are the extractive industries producing such items as gold and copper for the electronics industry. When I tell folks that one of my areas of interest is medical mineralogy, they often question the existence of a connection between these two areas. It is apparent that few geologists consider the subject of the formation and interaction of minerals in the human body to be closely related to their discipline, and this is quite unfortunate. We hope this book has shown you some of the many ways that minerals can be significant, not only to geologists, but to all of society. If the related disciplines such as biomineralogy, optics, medical mineralogy, forensic mineralogy, mining engineering interest you, we hope you'll be motivated to pursue them, for they have great importance to us all.

M.E.G.

Introduction

In this last brief chapter of the book, we want to highlight areas outside the field of geology that use mineralogy, so we've chosen the title of "Mineralogy Outside of Geology." Other titles that we pondered for this chapter included "Mineralogy without Rocks," "Industrial Mineralogy," "Environmental Mineralogy," and even "Medical Mineralogy." We will see how all of these areas are linked together by one common denominator: minerals.

In science, we often discuss disciplines as being "vertical" or "horizontal." In that vocabulary, mineralogy can be considered a vertical discipline, in that it contains in-depth knowledge within a specific field. However, mineralogy is also a "horizontal" discipline because it crosses many other disciplines: industry, medicine, forensics, economic resources, public policy, etc. Herein we only want to briefly introduce you to these areas and provide a few examples of each. Entire books and research fields are dedicated to them, and these will be referenced here to provide you with sources for future exploration.

Figure 24.1 shows five separate photographs of seemingly-unrelated items. We assure you that they all share a common thread of logic within the field of mineralogy, as we will show in the next few pages.

Industry

Use of minerals by humans dates to prehistory. Today, minerals are used or related to every single product in modern society. Many of these are obvious: copper in electrical wiring, diamonds used in engagement rings, and talc used in talcum powder. Many (perhaps most) mineral uses are not so obvious. For example, Figure 24.1a shows a photograph of what looks like black sand. This material happens to be the major ore for the production of phosphorus. Phosphorus is contained in this carbon-rich sedimentary rock as microcrystalline apatite (see Chapter 23 and Kohn et al., 2003). Phosphorus is extracted from this material using extreme heat, which produces a phosphorus gas. Fertilizer, in the form of phosphoric acid, is made by mixing this ore with sulfuric acid, which forms phosphoric acid (a key ingredient in fertilizer) and gypsum as a by-product. Basic mineralogy and simple chemistry are at work to produce this much-needed material as they are for all the mineral-based products we use daily.

There are countless other examples of mineral resources that are mined and then processed by various chemical methods, all requiring basic knowledge of mineralogy outside of the field of geology. In this industry, mining engineers find the minerals and figure out how to get them out of the ground. Chemical engineers work on extraction and the chemistry needed to turn them into commodities. Kesler (1994) and Chang (2001) give good reviews on mineral mining and uses.

The tooth shown in Figure 24.1b has an outer surface of enamel that is partially composed of the mineral apatite just discussed above. In fact, all the bones in the human body are composed of a very poorly-crystalline form of apatite. People in the medical community make a great effort to understand how bones form, break, and heal (as well as how teeth decay). To date, the mineralogical community has not devoted enough energy to these subjects. This is changing fast. For example, see the following recent books: Sahai and Schoonen (2006), Selinus (2005), Skinner and Berger (2003). If mineralogists spent more time thinking about these issues, we might be able to contribute to solving common human maladies such as osteoporosis or kidney stones. We could do this by simply applying our knowledge of the interaction of fluids with minerals gained from working with complex natural systems. This might very well improve our understanding of processes that occur within the human body. For example, Wood et al. (2006) explored the reactions of minerals in the human lung—basically treating the human lung as a reaction chamber.

Many of you reading this book have never had the experience of having a cavity in a tooth. This is because you belong to a generation that has benefited from the addition of fluorine to municipal water supplies. The actual role of fluorine in preventing tooth decay is still the subject of debate. It might enter the enamel structure of the tooth, preventing decay, or it might prevent certain types of bacteria that attack tooth enamel from living in the mouth. Regardless, there is a good liquid/solid interaction between the fluorine and the tooth enamel.

Notice that the tooth in Figure 24.1b has a silver-colored filling. Although your dentist may tell you that this is a "silver" filling, it is actually made from a mixture of five separate metals: silver, tin, copper, zinc, and mercury. Typically, metals are mixed to make alloys, each designed with a specific set of properties. Brass and steel are also examples of such alloys. All of these materials have as their starting points as metals (minerals) that were obtained from the Earth. Even mercury occurs naturally and is considered a mineral in its liquid form. There are locations on Earth where mercury is so abundant that it is visible in rocks and can be panned (with a gold pan) from streams. Mercury is an element that is much-feared as a toxic contributor to the environment. However, the mercury in dental fillings poses little health risk because it is chemically combined with the other metals, making it non-bioavailable. The field of combining metals to make alloys is called metallurgical engineering, and it is closely related to mineralogy and chemistry. Biomineralogy is, of course, intimately related to the medical fields. For a good review of this field, see Dove et al. (2005).

Figure 24.1c shows a piece of synthetic ruby machined into the shape of a cylinder. Even though this material is really the mineral species corundum, the processes used in its manufacture come from the field of ceramics, which along with metallurgical engineering is a subset of the broad discipline of materials science. This piece of ruby will be used as a bearing in small instruments. Your grandfather may have owned a 12-jewel wristwatch; those "jewels" were probably pieces of ruby with different shapes and sizes. Of course, synthetic forms of corundum also have a market in the jewelry industry.

Another major industry that uses minerals as a source of raw materials is the field of optics. Lenses with different refractive indices (based upon composition) can be directly related to various minerals. Many lasers also use minerals to produce high-intensity light. The YAG (yttrium-aluminum-garnet) laser is used in laser surgery on eyes, cosmetic surgery, dentistry, manufacturing of steel and other alloys, and many other applications. Other, more specialized lasers use anisotropic materials in their lens systems. If these materials have appropriate birefringence and dispersion, they produce an effect called frequency doubling, which allows an input wavelength of 1000 nm to come out of a material with a wavelength of 500 nm and thus higher energy. Such materials are used in defense systems and experimental nuclear reactors.

Many minerals are used to extract unwanted elements from water. For example, zeolites are used in such varied applications as removing radioactive elements from waste water in the nuclear industry, and ammonium from the recycled "water" from astronaut urine. Another interesting use of zeolites exploits the fact that they contain loosely-bonded water molecules in their structures. As the H_2O molecules enter or leave the structures, heating and cooling occurs, making them useful in solid-state refrigerants in arid climates. For a review of some of the uses of zeolites, see Bish and Ming (2001). An interesting specific application of zeolites occurs in the paper recycling in Japan. Most papers have clay added as a filler and binder to produce a glossy surface. In the recycling process, these inorganic materials are concentrated in the paper-sludge ash (an incombustible part that remains after burning off the organics). These materials are composed of oxides somewhat similar in compo-

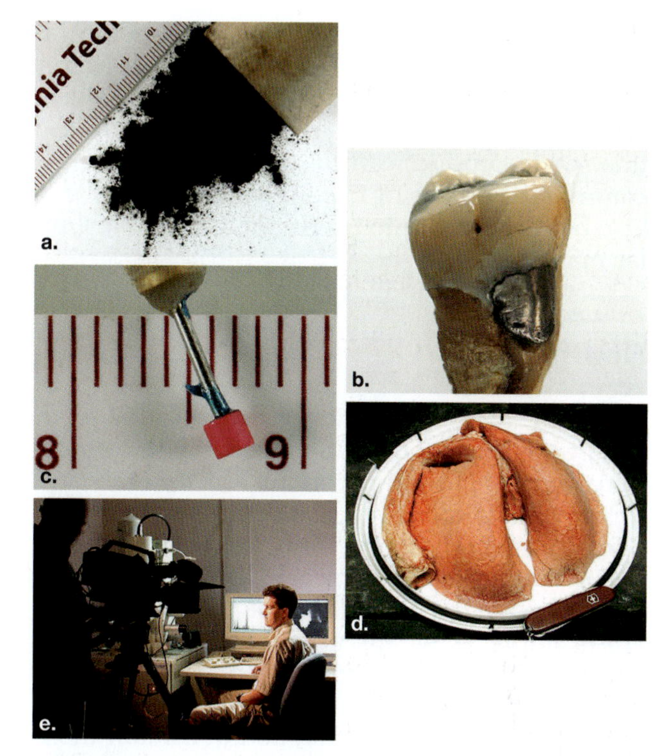

Figure 24.1. *We end the book with only one figure in this penultimate chapter, showing five initially unrelated images: some recognizable, others totally obscure.* ***a.*** *This is a phosphate ore from southeastern Idaho. This image of apatite ore is shown to represent industrial minerals.* ***b.*** *The outer portion of the tooth enamel is also composed of the mineral apatite, used here to represent the field of medical mineralogy. On the bottom part of the tooth is some rough material called calculus, which is at the boundary between organic and inorganic materials (biomineralogy). The tooth has a filling made of a metal alloy (metallurgical engineering) containing, among other things, the element (and mineral) mercury. This reminds us of the potential health effects relating minerals to humans.* ***c.*** *A synthetic ruby (also valuable in the optics industry for lasers) has been fabricated in the shape of a cylinder for use as a bearing in small machines (industrial engineering).* ***d.*** *The lungs of a sheep are shown, along with a knife for scale. The lungs contain minerals that either form there or were inhaled. Proper identification of these origins requires use of the previously-discussed methods.* ***e.*** *One of the methods shown is an electron microscope (forensic mineralogy). In this case, a fragment of metal from a crime scene is being investigated under the additional scrutiny of a television news crew from the CBS show* 48 Hours. *(Photo courtesy of Bill Turner, MVA Scientific Consultants.)*

sition to silica-rich volcanic ash. As with natural volcanic ashes, reaction with alkali-rich water will convert them to zeolites, which then have commercial applications (Wajima et al., 2007). So this waste product can be turned into a usable commodity. Husks of rice are silica-rich and can also be converted into useful materials in a similar manner.

Health

Currently, there is an emerging field called medical geology, as well as a subfield called medical mineralogy. (To learn more, see Sahai and Schoonen, 2006; Selinus, 2005; and Skinner and Berger, 2003). These disciplines explore both the positive and negative impacts of geological and mineralogical environments on human beings. For example, as we discussed in Chapter 3, deficiencies in iodine and selenium affect 20% of the global population. The problems occur when crops are raised in soils that do not contain those elements. Curing the problem requires only the addition of the missing elements to some product in the food chain, such as the addition of iodine to salt (which began in the 1950s in the U.S.).

One of the negative aspects of mineral interactions with humans is that of dust inhalation. The mineralogical community has been involved with this issue for several decades as shown by a sampling of reviews: Guthrie and Mossman (1994), Gunter (1994), Skinner et al. (1988), and Ross (1981). In the U.S. in the 1990s, the asbestos issue had somewhat waned until events surrounding the now-closed vermiculite mine near Libby, Montana emerged in 1999. The issue was that the vermiculite ore there contains trace amounts of asbestiform amphiboles. Occupational exposure to asbestos (a class of fibrous minerals including the serpentines and some amphiboles) can cause asbestosis, mesothelioma, and lung cancer. At this writing, one of the most controversial parts of this discipline deals with non-occupational (i.e., non-anthropogenic) exposure to amphiboles. To learn more about this issue, see Bandli and Gunter (2006) and Gunter et al. (2007).

One of the more intriguing results of dust inhalation is the finding that quartz, which is the most abundant mineral species in the Earth's crust, has been deemed a human carcinogen. Many geologists and mineralogists find this hard to believe. Data do show that high-level occupational exposure to quartz dust creates a non-cancerous disease called silicosis. Research on this topic is discussed in Norton and Gunter (1999), while material that can be used in teaching the issue of quartz exposure can be found in Gunter (1999). (See also Taunton and Gunter, 2007, for an example of how to conduct a geologically-based epidemiology study.)

Another interesting application of basic mineralogical methods was the study by Pasteris et al. (1999) of the mineral particles in breast explants of women who were having silicon implants removed. They correctly identified the minerals that had formed, though these minerals had been misidentified by pathologists (who probably never took a mineralogy course!).

Forensics

Another field in which mineralogy can play a major role is that of forensics. For instance, during World War II, bombs carried across the Pacific Ocean by unmanned hydrogen balloons were falling on the northwestern part of the United States. Analysis of the ballast (sand) used in the balloons showed a distinctive composition that was traced to a single location in Japan. Knowing the location allowed Allied Forces to bring the balloon bombings to a halt.

Over the past few years, the field of forensics has been glamorized by the popular media. Figure 24.1e shows a photograph of an SEM operator (on the right) and a television camera filming his actions on the left. In the microscope is a piece of metal from shavings found at a crime scene. The SEM results were used to convict a murderer by matching the chemical composition of the unusual alloy found in his workplace with shavings found among the victim's possessions. It is worth noting that the microscopist, Bryan Bandli, has a master's degree in mineralogy. For a review of forensics in geology, see Murray (2004).

Conclusions

In this last brief chapter, we have provided you with a glimpse of the many other applications of mineralogy outside of the field of geology. Although many of you may have entered the field of geology with the idea that you would spend your life outdoors looking at rocks, there are many, many more jobs in the analytical fields outside of geology that require a basic knowledge of mineralogy. If you can manage to earn a "B" grade in this class, you'll probably understand this discipline better than many of the professionals who use it. We've shown you here only the "tip of the iceberg" of areas outside classical geology that require knowledge of minerals (yes, ice is a mineral!). We encourage you to think "outside the rock box" and put the skills you have learned in this class to use in any application that appeals to you!

The End
4:56 p.m. EDT, Wednesday, May 17, 2007
Pelham, Massachusetts and Moscow, Idaho

References

Bandli, B.R. and Gunter, M.E. (2006) A review of scientific literature examining the mining history, geology, mineralogy, and amphibole asbestos health effects of the Rainy Creek Igneous Complex, Libby, Montana USA. Inhalation Toxicology, 18, 949–962.

Bish, D.L. and Ming, D.W. (2001) Natural Zeolites: Occurrence, Properties, Applications. Reviews in Mineralogy and Geochemistry, 45, Mineralogical Society of America, 654 pp.

Chang, L.L.Y. (2002) Industrial Mineralogy: Materials, Processes, and Uses. Prentice-Hall, Inc., Upper Saddle River, New Jersey, 472 pp.

Dove, P.M., De Yoreo, J., and Weiner, S. (2005) Biomineralization. Reviews in Mineralogy and Geochemistry, 54, Mineralogical Society of America, 331 pp.

Gunter, M.E. (1999) Quartz - the most abundant mineral species in the earth's crust and a human carcinogen? Journal of Geoscience Education, 341–349.

Gunter, M.E. (1994) Asbestos as a metaphor for teaching risk perception. Journal of Geological Education, 42, 17–24.

Gunter, M.E., Belluso, E., and Mottana, A. (2007) Amphiboles: Environmental and health concerns. In Amphiboles: Crystal Chemistry, Occurrences, and Health Concerns, Reviews in Mineralogy and Geochemistry, 67, 453–516.

Guthrie, G.D., Jr. and Mossman, B.T. (1994) Health Effects of Mineral Dusts. Reviews in Mineralogy, 28, Mineralogical Society of America, 584 pp.

Kesler, S.E. (1994) Mineral Resources, Economics, and the Environment. Macmillian College Publishing Company, Inc., New York, 391 pp.

Kohn, M.L., Rakovan, J., and Hughes, J.M. (2003) Phosphates: Geochemical, Geobiological, and Materials Importance. Reviews in Mineralogy and Geochemistry, 48, Mineralogical Society of America, 742 pp.

Murray, R.C. (2004) Evidence from the Earth: Forensic Geology and Criminal Investigation. Mountain Press Publishing Company, Missoula, Montana, 226 pp.

Norton, M.R. and Gunter, M.E. (1999) Relationships between respiratory diseases and quartz-rich dust in Idaho. American Mineralogist, 84, 1009–1019.

Pasteris, J.D., Wopenka, B., Freeman, J.J., Young, V.L., and Brandon, H.J. (1999) Medical mineralogy as a new challenge to the geologist: Silicates in human mammary tissue? American Mineralogist, 84, 997–1008.

Ross, M. (1981) The geological occurrences and health hazards of amphibole and serpentine asbestos. In Reviews in Mineralogy, 9A, Mineralogical Society of America, 279–324.

Sahai, N. and Schoonen, A.A. (2006) Medical Mineralogy and Geochemistry. Reviews in Mineralogy and Geochemisty, 64, Mineralogical Society of America, 332 pp.

Selinus, O. (2005) Essentials of Medical Geology: Impacts of the Natural Environment on Public Health. Academic Press, 832 pp.

Skinner, H.C.W. and Berger, A.R. (2003) Geology and Health: Closing the Gap. Oxford University Press, USA, 192 pp.

Skinner, H.C.W., Ross, M., and Frondel, C. (1988) Asbestos and other Fibrous Materials: Mineralogy, Crystal Chemistry, and Health. Oxford University Press, USA, 222 pp.

Taunton, A.E. and Gunter, M.E. (2007) Introducing medical geology to undergraduates as a critical thinking and risk assessment tool. Journal of Geoscience Education, 55, 169–180.

Wajima, T., Kuzawa, K., Ito, K., Tamada, O., Gunter, M.E., and Rakovan, J. (2007) Material conversion from paper sludge ash in NaOH, KOH, and LiOH solutions. American Mineralogist, 92, 1105–1111.

Wood, S.A., Taunton, A.E., Normand, C., and Gunter, M.E. (2006) Mineral-fluid interaction in the lungs: Insights from reaction-path modeling. Inhalation Toxicology, 18, 975–984.

Index